EIGHTH EDITION

The LEGAL ENVIRONMENT of BUSINESS
A CRITICAL THINKING APPROACH

NANCY K. KUBASEK

BARTLEY A. BRENNAN

M. NEIL BROWNE

PEARSON

Boston   Columbus   Indianapolis   New York   San Francisco
Amsterdam   Cape Town   Dubai   London   Madrid   Milan   Munich   Paris   Montreal   Toronto
Delhi   Mexico City   São Paulo   Sydney   Hong Kong   Seoul   Singapore   Taipei   Tokyo

Vice President, Business Publishing: Donna Battista
Editor-in-Chief: Stephanie Wall
Senior Sponsoring Editor: Neeraj Bhalla
Editorial Assistant: Eric Santucci
Vice President, Product Marketing: Maggie Moylan
Director of Marketing, Digital Services and Products: Jeanette Koskinas
Executive Field Marketing Manager: Adam Goldstein
Field Marketing Manager: Lenny Ann Raper
Product Marketing Assistant: Jessica Quazza
Team Lead, Program Management: Ashley Santora
Team Lead, Project Management: Jeff Holcomb
Project Manager: Ilene Kahn
Operations Specialist: Carol Melville
Creative Director: Blair Brown
Art Director: Janet Slowik
Vice President, Director of Digital Strategy and Assessment: Paul Gentile
Manager of Learning Applications: Paul DeLuca
Full-Service Project Management and Composition: Cenveo® Publisher Services
Interior Designer: Cenveo Publisher Services
Cover Designer: Cenveo Publisher Services
Cover Photo: Maksym Yemelyanov/Fotolia
Printer/Binder: RR Donnelley/Willard
Cover Printer: Phoenix Color/Hagerstown

Copyright © 2017, 2015, 2012 by Pearson Education, Inc. or its affiliates. All Rights Reserved. Manufactured in the United States of America. This publication is protected by copyright, and permission should be obtained from the publisher prior to any prohibited reproduction, storage in a retrieval system, or transmission in any form or by any means, electronic, mechanical, photocopying, recording, or otherwise. For information regarding permissions, request forms, and the appropriate contacts within the Pearson Education Global Rights and Permissions department, please visit www.pearsoned.com/permissions/.

Acknowledgments of third-party content appear on the appropriate page within the text.

PEARSON, ALWAYS LEARNING, are exclusive trademarks owned by Pearson Education, Inc. or its affiliates in the U.S. and/or other countries.

Unless otherwise indicated herein, any third-party trademarks, logos, or icons that may appear in this work are the property of their respective owners, and any references to third-party trademarks, logos, icons, or other trade dress are for demonstrative or descriptive purposes only. Such references are not intended to imply any sponsorship, endorsement, authorization, or promotion of Pearson's products by the owners of such marks, or any relationship between the owner and Pearson Education, Inc., or its affiliates, authors, licensees, or distributors.

Library of Congress Cataloging-in-Publication Data
Kubasek, Nancy K., author.
  The legal environment of business : a critical thinking approach / Nancy K. Kubasek,
Bartley A. Brennan, M. Neil Browne. — Eight edition.
    pages cm
  ISBN 978-0-13-407403-0 (alk. paper)
  1. Industrial laws and legislation—United States. 2. Trade regulation—United States.
3. Commercial law—United States. 4. Critical thinking. I. Brennan, Bartley A., author.
II. Browne, M. Neil, author. III. Title.
KF1600.K83 2015a
346.7307—dc23
                                                                    2015036209

10 9 8 7 6 5 4 3 2 1
ISBN 10:   0-13-407403-3
ISBN 13: 978-0-13-407403-0

To the numerous students who appreciate the importance of developing their critical thinking skills for their personal growth and development.

NANCY K. KUBASEK AND M. NEIL BROWNE

To Sandra for everything.

BARTLEY A. BRENNAN

Brief Contents

PART ONE

Introduction to the Law and the Legal Environment of Business   1

1. Critical Thinking and Legal Reasoning   2
2. Introduction to Law and the Legal Environment of Business   17
3. The American Legal System   35
4. Alternative Tools of Dispute Resolution   76
5. Constitutional Principles   105
6. White-Collar Crime and the Business Community   145
7. Ethics, Social Responsibility, and the Business Manager   188
8. The International Legal Environment of Business   213

PART TWO

Private Law and the Legal Environment of Business   249

9. The Law of Contracts and Sales—I   250
10. The Law of Contracts and Sales—II   279
11. The Law of Torts   299
12. Product and Service Liability Law   332
13. Law of Property: Real and Personal   360
14. Intellectual Property   385
15. Agency Law   408
16. Law and Business Associations—I   432
17. Law and Business Associations—II   453

PART THREE

Public Law and the Legal Environment of Business   481

18. The Law of Administrative Agencies   482
19. The Employment Relationship and Immigration Laws   503
20. Laws Governing Labor–Management Relations   538
21. Employment Discrimination   569
22. Environmental Law   617
23. Rules Governing the Issuance and Trading of Securities   647
24. Antitrust Laws   699
25. Laws of Debtor–Creditor Relations and Consumer Protection   741

APPENDIX A   The Constitution of the United States   788
GLOSSARY   794
INDEX   808

Contents

PREFACE xvii
ACKNOWLEDGMENTS xxi
ABOUT THE AUTHORS xxiii

PART ONE
Introduction to the Law and the Legal Environment of Business 1

1 Critical Thinking and Legal Reasoning 2
The Importance of Critical Thinking 2
A Critical Thinking Model 4
United States of America v. Martha Stewart and Peter Bacanovic 5
The Critical Thinking Steps 7
   Facts 7
   Issue 7
   Reasons and Conclusion 7
   Rules of Law 8
   Ambiguity 8
   Ethical Norms 9
   Analogies 10
   Missing Information 10
Using Critical Thinking to Make Legal Reasoning Come Alive 11
   Legal Reasoning 11
Applying the Critical Thinking Approach 15
Assignment on the Internet 16
On the Internet 16
For Future Reading 16

2 Introduction to Law and the Legal Environment of Business 17
Definition of the Legal Environment of Business 18
Definition of Law and Jurisprudence 19
   Natural Law School 20
   Positivist School 21
   Sociological School 21
   American Realist School 22
   Critical Legal Studies School 22
   Feminist School 22
   Law and Economics School 23
Sources of Law 23
   The Legislature as a Source of Statutory Law 23
   The Judicial Branch as a Source of Case Law 25
   The Executive Branch as a Source of Law 27
   Administrative Agencies as a Source of Law 27
Classifications of Law 27
   Criminal Law and Civil Law 28
   Public and Private Law 28
   Substantive and Procedural Law 29
   Cyberlaw 29
Global Dimensions of the Legal Environment of Business 30
Summary 30
Review Questions 30
Review Problems 31
Case Problems 31
Thinking Critically about Relevant Legal Issues 33
Assignment on the Internet 33
On the Internet 34
For Future Reading 34

3 The American Legal System 35
Jurisdiction 35
   Original versus Appellate Jurisdiction 35
   Jurisdiction over Persons and Property 36
World-Wide Volkswagen Corp. v. Woodson, District Judge of Cook County 37
   Subject Matter Jurisdiction 40
Hertz Corporation v. Friend 41
Venue 44
The Structure of the Court System 46
   The Federal Court System 46
   State Court Systems 46
The Actors in the Legal System and Their Relationship to the Business Community 49
   The Attorney 49
   The Jury 52

The Adversary Process 52
   Criticisms of the Adversary System 53
Steps in Civil Litigation and the Role of Businesspersons 53
   The Pretrial Stage 53
   The Trial 59
J.E.B. v. Alabama, ex rel. T.B. 60
   Appellate Procedure 64
   Class Actions 66
Wal-Mart Stores Inc. v. Dukes 67
Global Dimensions of the American Legal System 69
Summary 71
Review Questions 71
Review Problems 71
Case Problems 72
Thinking Critically about Relevant Legal Issues 73
Assignment on the Internet 74
On the Internet 74
For Future Reading 75

4 Alternative Tools of Dispute Resolution 76

Negotiation and Settlement 78
Mediation 78
   Selection of a Mediator 78
   Common Uses of Mediation 79
   Advantages of Mediation 79
   Criticisms of Mediation 80
Arbitration 80
Hall Street Associates, L.L.C. v. Mattel, Inc. 82
   Methods of Securing Arbitration 83
Ignazio v. Clear Channel Broadcasting, Inc. et al. 84
American Express Co. vs. Italian Colors Restaurant 89
   Selection of an Arbitrator 91
   Common Uses of Arbitration 92
   Problems with Arbitration 92
Minitrials 93
Early Neutral Case Evaluation 94
Private Trials 95
Summary Jury Trials 95
Court-Annexed Alternative Dispute Resolution 95
   Use of Court-Annexed ADR in the State and Federal Systems 95
   Differences between Court-Annexed and Voluntary ADR 97

The Future of Alternative Dispute Resolution 98
Global Dimensions of Alternative Dispute Resolution 98
Mitsubishi Motors Corp. v. Soler Chrysler-Plymouth 99
Summary 100
Review Questions 100
Review Problems 100
Case Problems 101
Thinking Critically about Relevant Legal Issues 102
Assignment on the Internet 103
On the Internet 103
For Future Reading 104

5 Constitutional Principles 105

The Constitution 105
Federalism 105
   Supremacy Clause 106
   Federal Preemption 106
Separation of Powers 107
William Jefferson Clinton v. Paula Corbin Jones 107
The Impact of the Commerce Clause on Business 110
   The Commerce Clause as a Source of Federal Authority 110
Gonzales v. Raich 112
   The Commerce Clause as a Restriction on State Authority 116
Nat'l Ass'n of Optometrists & Opticians v. Brown 116
The Taxing and Spending Powers of the Federal Government 119
   Taxation of the Internet? 120
The Impact of the Amendments on Business 121
   The First Amendment 121
Central Hudson Gas & Electric Corp. v. Public Service Commission of New York 123
   The Fourth Amendment 128
Florida v. Jardines 129
   The Fifth Amendment 132
United States v. Windsor 134
   The Fourteenth Amendment 139
Summary 140
Review Questions 140
Review Problems 140

Case Problems 141
Thinking Critically about Relevant Legal Issues 143
Assignment on the Internet 143
On the Internet 144
For Future Reading 144

6 White-Collar Crime and the Business Community 145

Crime and Criminal Procedure 148
    Crime 148
    Criminal Procedure 149
Distinguishing Features of White-Collar Crime 154
    The Corporation as a Criminal 154
    Arguments in Support of Corporate Liability 155
    Arguments in Opposition to Corporate Liability 156
    Imposition of Liability on Corporate Executives 157
United States v. Park 157
    Imposition of Liability on Lower-Level Corporate Criminals 160
    Factors Encouraging the Commission of White-Collar Crime 160
    Sentencing of White-Collar Criminals 162
Common White-Collar Crimes 164
Sekhar v. United States 164
    Bribery 165
    Violations of Federal Regulations 166
    Criminal Fraud 167
United States v. Gray 167
    Larceny 169
    Embezzlement 170
    Computer Crimes 170
Prevention of White-Collar Crime 173
Federal Laws Used in the Fight against White-Collar Crime 175
    The Racketeer Influenced and Corrupt Organizations Act (RICO) 175
    False Claims Act 177
    Sarbanes-Oxley Act 178
    Whistleblower Protection Act 179
State Laws Used in the Fight against White-Collar Crime 180
Global Dimensions of White-Collar Crime 181
Summary 182
Review Questions 182
Review Problems 182
Case Problems 183
Thinking Critically about Relevant Legal Issues 186
Assignment on the Internet 186
On the Internet 187
For Future Reading 187

7 Ethics, Social Responsibility, and the Business Manager 188

Definition of Business Ethics and the Social Responsibility of Business 190
    Business Ethics 190
    The Social Responsibility of Business 191
In re Exxon Valdez 191
Theories of Ethical Thought 193
    Consequential Theories 193
    Deontological Theories 194
    Humanist Theories 195
Codes of Ethics 195
    Individual Codes of Ethics 195
    Corporate Codes of Ethics 197
    Industry Codes of Ethics 198
    Professional Codes of Ethics 198
Schools of Social Responsibility 201
    Profit-Oriented School 201
    Managerial School 204
    Institutional School 204
Cooper Industries v. Leatherman Tool Group, Inc. 205
    Professional Obligation School 205
    Regulation School 206
Johnson Construction Co. v. Shaffer 206
Global Dimensions of Ethics and Social Responsibility 207
    Code of Conduct for Transnational Corporations 207
Summary 207
Review Questions 208
Review Problems 208
Case Problems 209
Thinking Critically about Relevant Legal Issues 210
Assignment on the Internet 211
On the Internet 211
For Future Reading 211

viii CONTENTS

8 The International Legal Environment of Business 213

Dimensions of the International Environment of Business 214
   Political Dimensions 214
   Economic Dimensions 215
   Cultural Dimensions 215
   Corruption and Trade 216
United States v. Kay 216
   Legal Dimensions 219
   Selected National Legal Systems 221
Crosby v. National Foreign Trade Council 222
Methods of Engaging in International Business 223
   Trade 223
   International Licensing and Franchising 224
Russian Entertainment Wholesale, Inc. v. Close-Up International, Inc. 225
   Foreign Direct Investment 226
In re Union Carbide Corp. Gas Plant Disaster v. Union Carbide Corp. 227
Risks of Engaging in International Business 229
   Expropriation of Private Property 229
   Sovereign Immunity Doctrine 230
Keller v. Central Bank of Nigeria 230
Act-of-State Doctrine 231
Linde v. Arab Bank, PLC 231
   Export and Import Controls 232
Legal and Economic Integration as a Means of Encouraging International Business Activity 233
   The World Trade Organization 233
   The European Union 236
   North American Free Trade Agreement 240
Global Dispute Resolution 242
   Arbitration 243
   Litigation 243
   Globalization: Hurts or Helps 243
Summary 244
Review Questions 244
Review Problems 245
Case Problems 245
Thinking Critically about Relevant Legal Issues 246
Assignment on the Internet 246
On the Internet 247
For Future Reading 247

PART TWO

Private Law and the Legal Environment of Business 249

9 The Law of Contracts and Sales—I 250

Definition, Sources, and Classifications of Contract Law 251
   Definition 251
   Sources of Contract Law 251
Paramount Contracting Co. v. DPS Industries, Inc. 252
   Classifications of Contracts 253
Pan Handle Realty, LLC v. Olins 254
Audito v. City of Providence 255
Elements of a Legal Contract 257
   Legal Offer 257
Beer v. Chase 258
   Legal Acceptance 260
The Private Movie Company, Inc. v. Pamela Lee Anderson et al. 260
   Consideration 263
   Genuine Assent 264
Stambovsky v. Ackley and Ellis Realty 265
   Competent Parties 267
   Legal Object 269
Brown & Brown, Inc. v. Johnson 270
Contracts That Must Be in Writing 271
   Contracts for the Sale of an Interest in Land 271
   Contracts to Pay the Debts of Another 271
   Contracts Not Performable in One Year 271
   Sale of Goods of $500 or More 272
Iacono v. Lyons 272
   Nonbusiness Contracts 273
Parol Evidence Rule 273
Third-Party Beneficiary Contracts and Assignment of Rights 274
   Types of Third-Party Beneficiary Contracts 274
   Assignment of Rights 274
Summary 275
Review Questions 276
Review Problems 276
Case Problems 277
Thinking Critically about Relevant Legal Issues 278
On the Internet 278
For Future Reading 278

10 The Law of Contracts and Sales—II 279

Methods of Discharging a Contract 280
    Discharge by Performance 280
Kohel v. Bergen Auto Enterprises, L.L.C. 280
Performance to Satisfaction of Another 281
    Material Breach of Contract 281
    Discharge by Mutual Agreement 282
    Discharge by Conditions Precedent and Subsequent 282
Architectural Systems, Inc. v. Gilbane Building Co. 283
    Discharge by Impossibility of Performance 283
    Discharge by Commercial Impracticability 284
    Contracts with the Government and the Sovereign Acts Doctrine 284
Facto v. Pantagis 285
Remedies for a Breach of Contract 286
    Monetary Damages ("Legal" Remedies) 286
Hallmark Cards, Inc. v. Murley 287
Arrowhead School District No. 75, Park County, Montana v. James A. Klyap, Jr. 288
    Equitable Remedies 289
    Remedies for Breach of a Sales Contract (Goods) 290
Fitl v. Strek 290
E-Contracts 291
    E-Signatures 292
    The Uniform Computer Information Transaction Act 292
Global Dimensions of Contract and Sales Law 293
Summary 294
Review Questions 295
Review Problems 295
Case Problems 296
Thinking Critically about Relevant Legal Issues 297
Assignment on the Internet 297
On the Internet 297
For Future Reading 298

11 The Law of Torts 299

The Goals of Tort Law 299
Damages Available in Tort Cases 300
    Compensatory Damages 300
    Nominal Damages 301
    Punitive Damages 301
Young v. Becker & Poliakoff 303
Classifications of Torts 307
Intentional Torts 307
    Intentional Torts against Persons 307
Nemet Chevrolet, Ltd. v. Consumeraffairs.com, Inc. 310
    Intentional Torts against Property 317
    Intentional Torts against Economic Interests 318
Negligent Torts 319
    Elements of Negligence 319
    Defenses to Negligence 322
Coomer v. Kansas City Royals 323
Venkateswarlu Thota and North Texas Cardiology Center v. Margaret Young 324
Strict Liability Torts 326
Global Dimensions of Tort Law 326
Summary 327
Review Questions 327
Review Problems 328
Case Problems 328
Thinking Critically about Relevant Legal Issues 330
Assignment on the Internet 331
On the Internet 331
For Future Reading 331

12 Product and Service Liability Law 332

Theories of Recovery in Product Liability Cases 333
    Negligence 333
Mutual Pharmaceutical Company, Inc. v. Bartlett 340
    Strict Liability in Contract for Breach of Warranty 341
Williams v. Braum Ice Cream Store, Inc. 343
    Strict Liability in Tort 346
Welge v. Planters Lifesavers Co. 347
    Liability to Bystanders 351
Market Share Liability 351
Service Liability 352
    Accountants' Liability 353
Global Dimensions of Product Liability Law 353
Summary 355
Review Questions 355
Review Problems 356
Case Problems 356

Thinking Critically about Relevant Legal Issues 357
Assignment on the Internet 358
On the Internet 358
For Future Reading 359

13 Law of Property: Real and Personal 360

Real Property 361
   Definition of Real Property 362
   Extent of Ownership 362
Interests in Real Property 362
   Fee Simple Absolute 362
   Conditional Estate 362
   Life Estate 363
   Future Interest 363
   Leasehold Estates 364
   Easements 364
   License 364
   Co-Ownership 364
   Condominiums and Cooperatives 365
Plum Creek C.A. v. Oleg Borman 365
Voluntary Transfer of Real Property 368
   Execution 368
   Delivery 370
   Acceptance 370
   Recording 370
Involuntary Transfer of Real Property 370
   Adverse Possession 370
   Condemnation 371
Susette Kelo et al. Petitioners, v. City of New London, Connecticut et al. 372
Restrictions on Land Use 374
   Restrictive Covenants 374
   Zoning 374
Emine Bayram v. City of Binghamton and City of Binghamton Zoning Board of Appeals 375
   Other Statutory Restrictions on Land Use 377
Personal Property 378
   Voluntary Transfer of Personal Property 378
   Involuntary Transfers of Personal Property 378
   Bailments 379
Global Dimensions of Property Law 379
Summary 380
Review Questions 380
Review Problems 381
Case Problems 381
Thinking Critically about Relevant Legal Issues 382
Assignment on the Internet 383
On the Internet 383
For Future Reading 383

14 Intellectual Property 385

Introduction to Intellectual Property 385
Trademarks 385
Toys "R" Us, Inc. v. Canarsie Kiddie Shop, Inc. 386
   Trade Dress 389
   Federal Trademark Dilution Act of 1995 391
Victor Moseley and Cathy Moseley et al., dba Victor's Little Secret v. V Secret Catalogue, Inc. et al. 391
Patents 393
Bilski v. Kappos 393
Trade Secrets 396
   Economic Espionage Act 397
Copyrights 397
   Fair Use Doctrine 398
   Copyrights in the Digital Age 398
American Broadcasting Company, Inc. et. al. v. Aereo, Inc. 399
RealNetworks, Inc. v. DVD Control Copy Association, Inc. et al. 402
Global Dimensions of Intellectual Property Law 402
Summary 403
Review Questions 404
Review Problems 404
Case Problems 404
Thinking Critically about Relevant Legal Issues 406
Assignment on the Internet 407
On the Internet 407
For Future Reading 407

15 Agency Law 408

Definition and Types of Agency Relationships 409
   Definition of Agency 409
   Types of Agency Relationships 409
Coker v. Pershad 411
Creation of an Agency Relationship 412
   Expressed Agency or Agency by Agreement 413
   Agency by Implied Authority 413

Penthouse International v. Barnes 414
   Agency through Ratification by Principal 415
   Agency by Estoppel or Apparent Authority 415
Motorsport Marketing, Inc. v. Wiedmaier, Inc. 416
Duties of Agents and Principals 416
   Principal's Duties to Agent 416
   Agent's Duties to Principal 417
Cousins v. Realty Ventures, Inc. 417
Gossels v. Fleet National Bank 419
Principal's and Agent's Liability to Third Parties 420
   Contractual Liability 420
McBride v. Taxman Corp. 421
   E-Commerce: Intelligent Agents 421
   Liability of Disclosed, Partially Disclosed and Undisclosed Principals 422
   Liability of Undisclosed Principal 422
   Tort Liability 422
   Tort Liability and Negligence 423
Auer v. Paliath 423
   Criminal Liability 424
Termination of the Principal–Agent Relationship 424
   Termination by Agreement 424
   Termination by Operation of Law 425
Gaddy v. Douglass 425
Global Dimensions of Agency Law 426
   Japan 427
   European Union 427
   U.S. Agents Abroad 427
Summary 428
Review Questions 428
Review Problems 428
Case Problems 429
Thinking Critically about Relevant Legal Issues 430
Assignment on the Internet 431
On the Internet 431
For Future Reading 431

16 Law and Business Associations—I 432

Factors Influencing a Business Manager's Choice of Organizational Form 434
Some Common Forms of Business Organization in the United States 434
   Sole Proprietorships 434

Quality Car & Truck Leasing, Inc. v. Sark 435
   General Partnerships 436
In re KeyTronics 438
Enea v. Superior Court of Monterey County 439
   Limited Partnerships and Limited Liability Limited Partnerships 443
Specialized Forms of Business Associations 444
   Joint Stock Company 444
   Syndicate 445
   Joint Venture 445
   Franchising 445
Holiday Inn Franchising, Inc. v. Hotel Associates, Inc. 446
Global Dimensions of Business Associations 447
   Outsourcing 448
Summary 449
Review Questions 449
Review Problems 449
Case Problems 450
Thinking Critically about Relevant Legal Issues 451
Assignment on the Internet 451
On the Internet 452
For Future Reading 452

17 Law and Business Associations—II 453

The Corporation 453
Classification of Corporations 454
   Closely Held Corporation 454
   Publicly Held Corporation 455
   Multinational or Transnational Corporation 455
   Subchapter S Corporation 455
   ROBS Corporation 455
   Professional Corporation 455
   Nonprofit Corporation 456
Creation of Corporations 456
Brennan's Inc. v. Colbert 457
Financing of Corporations 458
   Debt Financing 458
   Equity Financing 459
   Consideration 461
Operation of Corporations 462
   The Role of the Shareholders 462
   The Role of the Board of Directors 464

In re Abbott Laboratories Derivative Shareholders Litigation 465
   The Role of the Officers and Managers 466
   Fiduciary Obligations of Directors, Officers, and Managers 466
Beam v. Stewart 467
Smith v. Van Gorkom 469
Limited Liability Companies 473
   The Uniform Limited Liability Act 473
   LLC Characteristics 473
   Creating a Limited Liability Company 473
   Duration of the LLC 474
   Financing of the LLC 474
   Control Considerations 474
   Tax Ramifications 474
Gatz Properties, LLC v. Auriga Capital Corporation 474
Global Dimensions of Corporations: A "Big Fat Greek" Bailout II and III 476
Summary 476
Review Questions 477
Review Problems 477
Case Problems 478
Thinking Critically about Relevant Legal Issues 478
Assignment on the Internet 479
On the Internet 479
For Future Reading 480

PART THREE

Public Law and the Legal Environment of Business 481

18 The Law of Administrative Agencies 482

Introduction to Administrative Law and Administrative Agencies 483
   Administrative Law 483
   Administrative Agencies 484
Creation of Administrative Agencies 485
Functions of Administrative Agencies 486
   Rulemaking 486
City of Arlington v. Federal Communications Commission 488
   Adjudication 489
Fox Television Stations, Inc. v. Federal Communications Commission 493
   Administrative Activities 494

Limitations on Administrative Agencies' Powers 494
   Statutory Limitations 494
   Institutional Limitations 494
State and Local Administrative Agencies 496
Vonage Holdings Corp. v. Minnesota Public Utilities Commission 497
Global Dimensions of Administrative Agencies 498
Summary 498
Review Questions 498
Review Problems 498
Case Problems 499
Thinking Critically about Relevant Legal Issues 501
Assignment on the Internet 501
On the Internet 501
For Future Reading 502

19 The Employment Relationship and Immigration Laws 503

Wage and Hour Laws 504
Unemployment Compensation 506
Cassandra Jenkins v. American Express Financial Corp. 509
Consolidated Omnibus Budget Reconciliation Act of 1985 510
Workers' Compensation Laws 510
   Coverage 510
   Recoverable Benefits 512
   The Claims Process 12
   Benefits of the Workers' Compensation System 512
The Family and Medical Leave Act of 1993 513
   Major Provisions 513
Jeffrey Bonkowski v. Oberg Industries, Inc. 514
   Remedies for Violations of the FMLA 516
   The Future of the FMLA 517
The Occupational Safety and Health Act of 1970 518
   Occupational Safety and Health Administration 518
   Occupational Safety and Health Review Commission 521
   National Institute for Occupational Safety and Health 521
   Implementation of the OSH Act 523

Employee Privacy Rights 523
    Electronic Monitoring and Communication 523
Brian Pietrylo and Doreen Marino, Plaintiffs v. Hillstone Restaurant Group d/b/a Houston's, Defendant 524
    Drug Testing 527
    Other Testing 528
Immigration Law 529
    Immigration Reform and Control Act of 1986 529
    Authorized Noncitizen Workers 530
Global Dimensions of the Employment Relationship 532
Summary 533
Review Questions 533
Review Problems 533
Case Problems 534
Thinking Critically about Relevant Legal Issues 535
Assignment on the Internet 536
On the Internet 536
For Future Reading 537

20 Laws Governing Labor–Management Relations 538

Structure of the Primary U.S. Labor Legislation and the Mechanisms for Its Enforcement 540
    The Wagner Act of 1935 540
    The Taft–Hartley Act of 1947 541
    The Landrum–Griffin Act of 1959 542
    The National Labor Relations Board 542
Labor Organizing 548
    Board Rules 548
    Unfair Labor Practices by Employers 549
Gaetano & Associates, Inc. v. National Labor Relations Board 553
    Unfair Labor Practices by Employees 554
Laborers' International Union of North America, Local 872, AFL–CIO, and Stephanie Shelby 554
    Organizing the Appropriate Unit 556
Specialty Healthcare and Rehabilitation Center of Mobile and United Steelworkers, District 9, Petitioner 557
The Collective Bargaining Process 559
    Subjects of Bargaining 560
Strikes, Boycotts, and Picketing 560
    Strikes 561
    Boycotts 563
    Picketing 563
Global Dimensions of Labor–Management Relations 564
Summary 564
Review Questions 564
Review Problems 565
Case Problems 565
Thinking Critically about Relevant Legal Issues 567
Assignment on the Internet 568
On the Internet 568
For Future Reading 568

21 Employment Discrimination 569

The Employment-at-Will Doctrine 570
Constitutional Provisions 572
The Civil Rights Acts of 1866 and 1871 572
    Applicability of the Acts 572
The Equal Pay Act of 1963 573
    Equal Work 574
    Defenses 574
    Remedies 575
The Civil Rights Act of 1964, as Amended (Title VII), and the Civil Rights Act of 1991 575
    Applicability of the Act 575
    Proof in Employment Discrimination Cases 576
Teresa Harris v. Forklift Systems, Inc. 580
Vance v. Ball State University 583
    Retaliation 583
Reya C. Boyer-Liberto v. Fontainbleu Corporation 584
    Statutory Defenses 586
    Protected Classes 588
    Enforcement Procedures 592
    Remedies 595
    Lilly Ledbetter Fair Pay Act of 2009 595
The Age Discrimination in Employment Act of 1967 596
    Applicability of the Statute 597
    Proving Age Discrimination 597
Jones v. National American University 598
    Statutory Defenses 599
    Enforcement Procedures 600
    Remedies under ADEA 601
The Rehabilitation Act of 1973 601
The Americans with Disabilities Act of 1991 602
    Covered Individuals 602

McMillan v. City of New York 604
    Enforcement Procedures 606
    Remedies 606
Affirmative Action 606
Global Dimensions of Employment Discrimination Legislation 610
Summary 611
Review Questions 612
Review Problems 612
Case Problems 613
Thinking Critically about Relevant Legal Issues 615
Assignment on the Internet 615
On the Internet 616
For Future Reading 616

22 Environmental Law 617

Alternative Approaches to Environmental Protection 618
    Tort Law 618
Boomer et al. v. Atlantic Cement Co. 618
    Government Subsidies Approach 620
    Emission Charges Approach 620
    Marketable Discharge Permits Approach 620
    Direct Regulation Approach 621
The Environmental Protection Agency 621
The National Environmental Policy Act of 1970 623
    Threshold Considerations 623
    Content of the EIS 623
Brodsky v. United States Nuclear Regulatory Commission 624
    Effectiveness of the EIS Process 625
Regulating Water Quality 625
    The Federal Water Pollution Control Act 625
Los Angeles County Flood Control District v. Natural Resources Defense Council, Inc. et al. 626
    The Safe Drinking Water Act 627
Regulating Air Quality 627
    The National Ambient Air Quality Standards 628
    New Source Review 629
    The Acid Rain Control Program 629
    Climate Change 630
Regulating Hazardous Waste and Toxic Substances 631
    The Resource Conservation and Recovery Act of 1976 633
    The Comprehensive Environmental Response, Compensation, and Liability Act of 1980, as Amended by the Superfund Amendment and Reauthorization Act of 1986 635
    The Toxic Substances Control Act of 1979 636
    The Federal Insecticide, Fungicide, and Rodenticide Act of 1972 637
The Pollution Prevention Act of 1990 638
    Business Aspects of Voluntary Pollution Prevention 639
    Sustainable Development 639
Global Dimensions of Environmental Regulation 639
    The Need for International Cooperation 639
    The Transnational Nature of Pollution 639
    The Global Commons 641
    Primary Responses of the United States 641
Summary 642
Review Questions 643
Review Problems 643
Case Problems 643
Thinking Critically about Relevant Legal Issues 645
Assignment on the Internet 645
On the Internet 646
For Future Reading 646

23 Rules Governing the Issuance and Trading of Securities 647

Introduction to the Regulation of Securities 648
    Summary of Federal Securities Legislation 649
    The Securities and Exchange Commission 652
Dodd-Frank Wall Street Reform and Consumer Protection Act of 2010 655
    Oversight of Financial Problems by Regulatory Agencies 656
    Risk Taking by Large Banks and Nonbanks 656
    Executive Compensation 656
    Too Big to Fail 657
    Credit Rating Agencies 657
    Derivatives 657
    Consumer Protection 658
    Exemptions 658

Regulation of the Regulators by a Court of Law 659
The Sarbanes-Oxley Act of 2002 659
　Corporate Accountability 660
　New Accounting Regulations 660
　Criminal Penalties 660
The Securities Act of 1933 660
　Definition of a Security 661
Securities and Exchange Commission v. Edwards 662
　Registration of Securities under the 1933 Act 662
　Securities and Transactions Exempt from Registration under the 1933 Act 666
　Resale Restrictions 668
　Liability, Remedies, and Defenses under the 1933 Securities Act 669
Litwin v. Blackstone Group, LP 670
The Securities Exchange Act of 1934 673
　Registration of Securities Issuers, Brokers, and Dealers 673
　Disclosure: Compensation 675
　Securities Markets 676
　Proxy Solicitations 677
　Tender Offers and Takeover Bids 678
　Remedies and Defensive Strategies 679
Barbara Schreiber v. Burlington Northern, Inc. 680
　Securities Fraud 682
Siracusano v. Matrixx Initiatives, Inc. 683
Securities and Exchange Commission v. Texas Gulf Sulphur Co. 687
　Liability and Remedies under the 1934 Exchange Act 689
The Wharf (Holdings) Limited v. United International Holdings, Inc. 690
　Short-Swing Profits 691
State Securities Laws 692
E-Commerce, Online Securities Disclosure, and Fraud Regulation 692
　Marketplace of Securities 692
　E-Commerce and Fraud in the Marketplace 692
Global Dimensions of Rules Governing the Issuance and Trading of Securities 693
　Legislation Prohibiting Bribery and Money Laundering Overseas 693
　Legislation Governing Foreign Securities Sold in the United States 694
　Regulations and Offshore Transactions 695

Summary 695
Review Questions 695
Review Problems 696
Case Problems 697
Thinking Critically about Relevant Legal Issues 697
Assignment on the Internet 698
On the Internet 698
For Future Reading 698

24 Antitrust Laws 699

Introduction to Antitrust Law 700
　A Definition of Antitrust 700
　Law and Economics: Setting and Enforcing Antitrust Policy 700
American Needle, Inc. v. National Football League 702
　Goals of the Antitrust Statutes 703
Enforcement of and Exemptions from the Antitrust Laws 705
　Enforcement 705
　Exemptions 707
The Sherman Act of 1890 708
　Section 1: Combinations and Restraints of Trade 708
Williamson Oil Co. v. Philip Morris, USA 710
Leegin Creative Leather Products, Inc. v. PSKS, Inc., dba Kay's Kloset, Kay's Shoes 714
Continental TV, Inc. v. GTE Sylvania 716
　Section 2: Monopolies 718
E.I. DuPont de Nemours and Co. v. Kolon Industries 719
Newcal Industries, Inc. v. Ikon Office Solutions 721
United States v. Microsoft Corporation 723
The Clayton Act of 1914 725
　Section 2: Price Discrimination 725
　Section 3: Tying Arrangements and Exclusive-Dealing Contracts 727
　Section 7: Mergers and Acquisitions 727
　Section 8: Interlocking Directorates 732
Other Antitrust Statutes 732
　Federal Trade Commission Act of 1914 732
California Dental Association v. Federal Trade Commission 732
　Bank Merger Act 733

Global Dimensions of Antitrust Statutes 734
   Transnational Reach of U.S. Antitrust Legislation 734
   Carrier Corp. v. Outokumpu Oyj 735
   Global Dimensions of U.S. Antitrust Laws 735
   Enforcement 736
Summary 738
Review Questions 738
Review Problems 738
Case Problems 739
Thinking Critically about Relevant Legal Issues 740
Assignment on the Internet 740
On the Internet 740
For Future Reading 740

25 Laws of Debtor–Creditor Relations and Consumer Protection 741

Debtor–Creditor Relations 742
   Rights of and Remedies for Creditors 743
   Rights and Remedies for Debtors 744
The Federal Bankruptcy Code and the Incorporation of the Bankruptcy Abuse Prevention and Consumer Protection Act of 2005 745
   History and Background 745
   Bankruptcy Management and Proceedings 746
   Chapter 7 749
In re Savage v. United States Bankruptcy 750
   Chapter 13 752
   Chapter 11 753
RadLAX Gateway Hotel, LLC v. Amalgamated Bank 754
   Chapter 12 756
   The New Bankruptcy Law—2011 757
The Evolution of Consumer Law 758
   Economics 758
Federal Regulation of Business Trade Practices and Consumer–Business Relationships 759
   The Federal Trade Commission: Functions, Structure, and Enforcement Powers 759
   Deceptive and Unfair Advertising 760
Federal Trade Commission v. Verity International, Ltd. 761
Federal Trade Commission v. QT, Inc. 763

Paduano v. American Honda Motor Co. 765
   Consumer Legislation 765
Federal Laws Regulating Consumer Credit and Business Debt-Collection Practices 769
   Truth-in-Lending Act 769
Household Credit Services, Inc. v. Pfenning 771
   Credit Card Accountability, Responsibility and Disclosure Act of 2009 772
   The Electronic Fund Transfer Act 773
   A Plastic Society 773
   The Fair Credit Reporting Act 774
Safeco Insurance Co. v. Burr 775
   Identity Theft and Credit Ratings 776
   Equal Credit Opportunity Act 777
   The Fair Credit Billing Act 778
   The Fair Debt Collection Practices Act 778
Jerman v. Carlisle, McNellie, Rini, Kramer & Ulrich, LPA 779
Dodd-Frank Act and Consumer Protection 781
   Credit and Debit Cards 781
   Consumer Loans 781
   Credit Scores 781
   Residential Mortgages 782
State Consumer Legislation 782
   Uniform Consumer Credit Code 782
   Unfair and Deceptive Practices Statutes 782
   Arbitration of Disputes 783
Global Dimensions of Consumer Protection Laws 783
Summary 783
Review Questions 784
Review Problems 784
Case Problems 785
Thinking Critically about Relevant Legal Issues 786
Assignment on the Internet 787
On the Internet 787
For Future Reading 787

APPENDIX A THE CONSTITUTION OF THE UNITED STATES 788

GLOSSARY 794

INDEX 808

Preface

***The Legal Environment of Business: A Critical Thinking Approach*, 8th Edition**, is exactly what its name implies: a comprehensive textbook that not only helps students develop a thorough understanding of the legal environment of business but also enhances their ability to engage in critical thinking and ethical analysis. Students thus develop the knowledge and skills necessary to survive in an increasingly competitive global environment.

The initial motivation for this book was the authors' perceptions that there was no legal environment book available that explicitly and adequately facilitated the development of students' critical thinking skills. Nor was there a book that really integrated ethical analysis throughout the text.

Some people may argue that the traditional method of case analysis allows students to develop their critical thinking skills. The problem with this approach, however, is that it focuses only on the analytical skills, while ignoring the evaluative component that is the essence of critical thinking; it also lacks an ethics component. To engage in critical thinking necessarily includes consideration of the impact of values on the outcome being considered.

The use of cases in the legal environment of a business classroom, however, can provide an excellent opportunity for the development of students' critical thinking abilities when the traditional case method is modified to emphasize development of these critical thinking skills. In addition, as students enhance their critical thinking skills, their understanding of the substance of the law also improves.

The following components of *The Legal Environment of Business: A Critical Thinking Approach* ensure that the textbook's goal of developing critically thinking students who understand the important concepts of business law and the legal environment of business is attained.

- **An explicit critical thinking model, set forth in the first chapter, developed by the author of the best-selling critical thinking textbook.** An eight-step model has as its base the traditional method of case analysis, but adds crucial critical thinking questions that also incorporate ethical analysis. The steps are clearly explained, and students are encouraged to apply the steps to every case in the text.
- **Additional critical thinking and ethical analysis questions incorporated at the beginning of each chapter and after selected cases.** These additional questions help to reinforce the skills emphasized in the model.
- **"Thinking Critically about Relevant Legal Issues" essays at the end of each chapter, which give students additional opportunities to develop their critical thinking skills.** These essays, found at the end of each chapter, allow students to extend their use of their newly developed critical thinking skills beyond cases to the kinds of arguments they will encounter in their daily lives.

Other Points of Distinction

- **Explicit links connecting the law to other disciplines.** This text is the only legal environment book to respond to the call for more integration among courses in colleges of business. "Linking Law and Business" boxes explicitly state how the law in an area directly affects or is affected by a concept in one of the core areas of business, such as accounting, management, and marketing. These boxes appear in every chapter.

- **A balanced mix of classic and current cases.** This book contains many of the most significant classic and contemporary cases, including key U.S. Supreme Court decisions handed down as recently as 2015. Whenever possible, cases were chosen that not only demonstrate important concepts but also contain fact situations that interest students.
- **Applying the Law to the Facts.** This pedagogical feature appears in every chapter beginning with Chapter 4. It provides periodic hypothetical situations to which students apply legal concepts they just learned. This feature allows students to continually check their understanding of new legal concepts as they read the material.
- **Emphasis on the global environment.** Many students will be working in countries other than the United States, and U.S. companies will have many dealings with foreign companies. Thus, an understanding of the global environment is essential for today's business student. This text emphasizes the importance of the global environment by using both the stand-alone and infusion approaches. Chapter 9 focuses explicitly on the global environment of business, and global considerations are integrated into every chapter with the global dimensions sections and the "Comparative Law Corner," which allows students to see how U.S. law compares to that of other nations around the world. The feature can also sensitize students to the idea that if something is not working well in the United States, it might make sense to see how other countries address similar issues. Examples include:
  - Eminent domain in Germany
  - The judicial system in Germany
  - Corporate speech in Canada
  - Unions in Sweden
  - Pollution controls in Japan
- *For Future Reading* **feature.** It is important that students become lifelong learners and that they continue learning about the law. But how do they know where to go? This feature, found at the end of each chapter, provides a short list of books and articles related to the material in each chapter that interested students may read to learn more about the new areas of law they just discovered.

New to This Edition

- **America Invents Act**—Discussion of the America Invents Act, which had a significant impact on patent law, has been added to Chapter 12.
- **Immigration law**—A new section about immigration law has been added to Chapter 19.
- **New procedures in Representation Cases**—New procedures in Representation Cases that went into effect in 2015 have been added to Chapter 19.
- **Updated cases**—Cases in this edition have been significantly updated. The classic cases from the previous edition have been retained, as have those that students find especially interesting or that do an exceptional job of illustrating an important point of law. All of the other cases have been replaced by more current cases that will be of greater interest to students and that capture the most current changes in the law. A few examples of new cases include:
  - *American Express Co. v. Italian Colors Restaurant* (United States Supreme Court, 2013) (Chapter 4)
  - *Plum Creek C.A. v. Oleg Borman* (Illinois State Court of Appeals, 2013) (Chapter 8)

- *Mutual Pharmaceutical Company, Inc. v. Bartlett* (United States Supreme Court, 2013) (Chapter 12)
- *American Broadcasting Company, Inc. et al. v. Aereo, Inc.* (United States Supreme Court, 2014) (Chapter 14)
- *Quality Car & Truck Leasing, Inc. v. Sark* (Ohio Court of Appeals, 2013) (Chapter 16)
- *Reya C. Boyer-Liberto v. Fontainebleau Corporation* (Fourth Circuit Court of Appeals, 2015) (Chapter 21)
- *Los Angeles County Flood Control District v. Natural Resources Defense Council, Inc. et al.* (United States Supreme Court, 2013) (Chapter 22)

- **Case problems**—Approximately one-third of the case problems from the seventh edition have been replaced with more current case problems.
- **Revised "For Future Reading" sections**—Suggested readings at the end of each chapter have been updated to emphasize more current legal issues.

For Instructors

At the Instructor Resource Center, www.pearsonhighered.com/irc, instructors can easily register to gain access to a variety of instructor resources available with this text in downloadable format.

If assistance is needed, our dedicated technical support team is ready to help with the media supplements that accompany this text. Visit http://247.pearsoned.com for answers to frequently asked questions and toll-free user support phone numbers.

The following supplements are available with this text:

- Instructor's Resource Manual
- Test Bank
- TestGen® Computerized Test Bank
- PowerPoint Presentation

Acknowledgments

The authors would like to acknowledge, with thanks, the following reviewers of the current and past editions of this text:

Robert Aalberts, University of Nevada, Las Vegas
Victor Alicea, Normandale Community College
Carlos Alsua, University of Alaska, Anchorage
S. Catherine Anderson, Queens University of Charlotte
Teddy Jack Armstrong, Carl Albert State College
Janie Blankenship, Del Mar College
William Bockanic, John Carroll University
Heidi Bulich, College of Business, Michigan State University
Kimble Byrd, Rowan University
Greg Cermignano, Widener University
Glenn Chappell, Coker College
William Christian, College of Santa Fe
Linda Christiansen, Indiana University Southeast
Patrick Cihon, Whitman School of Management, Syracuse University
Michael Costello, University of Missouri–St. Louis
Robert Cox, Salt Lake Community College
Jamey Darnell, Barton College
Regina Davenport, Pearl River Community College
Kevin Derr, Pennsylvania College of Technology
Julia Derrick, Brevard Community College
David F. Dieteman, Penn State Erie, The Behrend College
Joseph Dworak, San Jose State University
Bruce Elder, University of Nebraska, Kearney
Gail Evans, University of Houston, Downtown
David Forsyth, ASU Polytechnic
Lucky Franks, Bellevue University
Samuel B. Garber, DePaul University
Rosario Girasa, Pace University
Van Graham, Gardner-Webb University
John Gray, Loyola College in Maryland
David Griffis, University of San Francisco
Jason Harris, Augustana College
Norman Hawker, Haworth College of Business, Western Michigan University
Richard Hunter, Seton Hall University
Marilyn Johnson, Mississippi Delta Community College
Nancy Johnson, Mt. San Jacinto Community College
Catherine Jones-Rikkers, Grand Valley State University
James Kelley, Notre Dame de Namur University
Lara Kessler, Grand Valley State University
Ernest King, University of Southern Mississippi
Audrey Wolfson Latourette, Richard Stockton College of New Jersey
Larry Laurent, McCoy College of Business, Texas State University
Marty Ludlum, Oklahoma City Community College
Leslie S. Lukasik, Skagit Valley College
Vicki Luoma, Minnesota State University
Daniel Lykins, Oregon State University
Bryan Jon Maciewski, Fond du Lac Tribal and Community College
Maurice McCann, Southern Illinois University
George McNary, College of Business Administration, Creighton University
Don Miller-Kermani, Brevard Community College

David Missirian, Bentley College
Odell Moon, Victor Valley College
Henry Moore, University of Pittsburgh, Greensburg
Mark Muhich, Mesabi Range Community & Technical College
Kimber Palmer, Texas A&M International University
Steve Palmer, Eastern New Mexico University
Darka Powers, Northeastern Illinois University
Charles Radeline, St. Petersburg College
Linda Reid, University of Wisconsin–Whitewater
Bruce Rockwood, College of Business, Bloomsburg University
Robert Rowlands, Harrisburg Area Community College
Kenneth J. Sanney, Central Michigan University
Ira Selkowitz, University of Colorado at Denver and Health Sciences Center
Mary Sessom, Cuyamaca College
James Smith, Bellevue University
Craig Stilwell, Michigan State University
Pamela Stokes, Texas A&M–Corpus Christi
Keith Swim, Jr., Mays Business School, Texas A&M University
Harold Tepool, Vincennes University
Daphyne Saunders Thomas, James Madison University
David Torres, Angelo State University
Kyle Usre, Whitworth College
Deborah Walsh, Middlesex Community College
Joe Walsh, Lees-McRae College
Dalph Watson, Madonna University
Mary Ellen Wells, Alvernia College
John Whitehead, Kilgore College
John Williams, Northwestern State University
Levon Wilson, Georgia Southern University
Rob Wilson, Whitworth College
Andrew Yee, University of San Francisco

The authors would also like to acknowledge Tami Thomas and Meghan Moore for their assistance in typing the manuscript.

About the Authors

NANCY KUBASEK is a Professor of Legal Studies at Bowling Green State University, where she teaches the Legal Environment of Business and Environmental Law courses. For eight years, she team-taught a freshman honors seminar on critical thinking and values analysis. She has published another undergraduate textbook with Pearson Education, *Environmental Law* (8th ed., 2012) and more than 75 articles. Professor Kubasek's articles have appeared in such journals as the *American Business Law Journal*, the *Journal of Legal Studies Education*, the *Harvard Women's Law Journal*, the *Georgetown Journal of Legal Ethics*, and the *Harvard Journal on Legislation*. She received her J.D. from the University of Toledo College of Law and her B.A. from Bowling Green State University.

Active in her professional associations, Professor Kubasek has served as president of the TriState Regional Academy of Legal Studies in Business and president of the national professional association of undergraduate professors of law, the Academy of Legal Studies in Business (ALSB). Committed to helping students become excited about legal research, she organized the first Undergraduate Student Paper Competition of the ALSB's Annual Meeting, an event that now provides an annual opportunity for students to present their original legal research at a national convention. She has also published several articles with students and has received her university's highest award for faculty–student research. She states:

> The most important thing that a teacher can do is to help his or her students develop the skills and attitudes necessary to become lifelong learners. Professors should help their students learn the types of questions to ask to analyze complex legal issues and to develop a set of criteria to apply when evaluating reasons. If we are successful, students will leave our legal environment of business classroom with a basic understanding of important legal concepts, a set of evaluative criteria to apply when evaluating arguments that includes an ethical component, and a desire to continue learning.
>
> To attain these goals, the classroom must be an interactive one, where students learn to ask important questions, define contexts, generate sound reasons, point out the flaws in erroneous reasoning, recognize alternative perspectives, and consider the impacts that their decisions (both now and in the future) have on the broader community beyond themselves.

BARTLEY A. BRENNAN is an Emeritus Professor of Legal Studies at Bowling Green State University. He is a graduate of the School of Foreign Service, Georgetown University (B.S. International Economics); the College of Law, State University of New York at Buffalo (J.D.); and Memphis State University (M.A. Economics). He was a volunteer in the U.S. Peace Corps, was employed by the Office of Opinions and Review of the Federal Communications Commission, and worked in the general counsel's office of a private international corporation. Professor Brennan has received appointments as a visiting associate professor at the Wharton School, University of Pennsylvania, and as a Research Fellow at the Ethics Resource Center, Washington, DC. He is the author of articles dealing with the Foreign Corrupt Practices Act of 1977, as amended; the business judgment rule; law and economics; and business ethics. He has published numerous articles in such journals as the *American Business Law Journal*, *University of North Carolina Journal of International Law*, and the *Notre Dame University Journal of Legislation*. He is a coauthor of *Modern Business Law* (3rd ed.). He has testified on amending the Foreign Corrupt Practices Act before the Subcommittee on International Economics and Finance of the House Commerce, Energy, and Telecommunications Committee.

ABOUT THE AUTHORS

M. NEIL BROWNE is a Distinguished Teaching Professor of Economics; director of IMPACT, an Honors Residential Learning Community Centered around the Principles of Intellectual Discovery and Moral Commitment; and coach of the Mock Trial Team at Bowling Green State University. He received a J.D. from the University of Toledo and a Ph.D. from the University of Texas. He is the coauthor of seven books and more than a hundred research articles in professional journals. One of his books, *Asking the Right Questions: A Guide to Critical Thinking* (6th ed.), is a leading text in the field of critical thinking. His most recent book, *Striving for Excellence in College: Tips for Active Learning*, provides learners with practical ideas for expanding the power and effectiveness of their thinking. Professor Browne has been asked by dozens of colleges and universities to aid their faculty in developing critical thinking skills on their campuses. He also serves on the editorial board of the *Korean Journal of Critical Thinking*. In 1989, he was a silver medalist for the Council for the Advancement and Support of Education's National Professor of the Year award. Also, in 1989, he was named the Ohio Professor of the Year. He has won numerous teaching awards on both local and national levels. He states:

> When students come into contact with conflicting claims, they can react in several fashions; my task is to enable them to evaluate these persuasive attempts. I try to provide them with a broad range of criteria and attitudes that reasonable people tend to use as they think their way through a conversation. In addition, I urge them to use productive questions as a stimulus to deep discussion, a looking below the surface of an argument for the assumptions underlying the visible component of the reasoning. The eventual objectives are to enable them to be highly selective in their choice of beliefs and to provide them with the greater sense of meaning that stems from knowing that they have used their own minds to separate sense from relative nonsense.

PART ONE

Introduction to the Law and the Legal Environment of Business

Part One introduces the concept of critical thinking. A business manager needs to learn and practice asking questions that ensure the selection of the best advice and subsequent strategy. In addition, we provide an overview of how the American legal system works. This overview requires us to understand alternative philosophies of law, alternative approaches to ethics, the constitutional foundations of our legal system, and alternative methods of resolving disputes. Part One concludes with a discussion of white-collar crime, a major problem in the legal environment of business.

CHAPTER ONE

Critical Thinking and Legal Reasoning

- THE IMPORTANCE OF CRITICAL THINKING
- A CRITICAL THINKING MODEL
- THE CRITICAL THINKING STEPS
- USING CRITICAL THINKING TO MAKE LEGAL REASONING COME ALIVE
- APPLYING THE CRITICAL THINKING APPROACH

The Importance of Critical Thinking

critical thinking skills The ability to understand the structure of an argument and apply a set of evaluative criteria to assess its merits.

Success in business requires the development of **critical thinking skills**. Business leaders regularly list these skills as the first set of competencies needed in business. A simple Google search for "critical thinking in business" produces more than 80 million suggested URLs.

Critical thinking refers to the ability to understand what someone is saying and then to ask specific questions enabling you to evaluate the quality of the reasoning offered to support whatever advice someone has given you. Because firms are under increasing competitive pressure, business and industry need managers with advanced thinking skills.[1] Highlighting this need, a report by the U.S. secretary of education states that because "one of the major goals of business education is preparing students for the workforce, students and their professors must respond to this need for enhancing critical thinking skills."[2]

Calls for improvements in critical thinking skills also come from those concerned about business leadership. David A. Garvin of the Harvard Business School argues that there is a general feeling in the business community that business leaders need to sharpen their critical thinking skills.[3] As a future business manager, you will experience many leadership dilemmas: All such questions require legal analysis and business leadership, guided by critical thinking.

A business leader must listen to many sources of information and many advisors. They are not all going to give advice that leads in a single direction. Critical thinking skills enable you to weigh the relative worth of alternative courses of action. For example, there will always be reasons you should encourage the growth of your labor force, but there will also be reasons you should not. You do not want these options to paralyze you, nor do you want to latch onto one approach for insubstantial reasons and then pay the price later.

---

[1] C. Sormunen and M. Chalupa, "Critical Thinking Skills Research: Developing Evaluation Techniques," *Journal of Education for Business* 69: 172 (1994).
[2] *Id.*
[3] John Baldon, "How Leaders Should Think Critically," *HBR Blog Network*, January 20, 2010.

Courtesy Holly Barnes

The message is clear: Success in business today requires critical thinking skills, and there is no better context in which to develop them than in the study of business law. Critical thinking skills learned in the Legal Environment of Business course will be easily transferred to your eventual role as a manager, entrepreneur, or other business professional. The law develops through argument among various parties. Critical thinking facilitates the development of more effective law.

Legal reasoning is like other reasoning in some ways and different in others. When people, including lawyers and judges, reason, they do so for a purpose. Some problem or dilemma bothers them. The stimulus that gets them thinking is the *issue*. It is stated as a question because it is a call for action. It requires them to do something, to think about answers.

For instance, in our Legal Environment of Business course, we are interested in such issues as:

1. Under the National Labor Relations Act, when are union organizers permitted to enter an employer's property?
2. When do petroleum firms have liability for the environmental and economic effects of oil spills?
3. Must a business fulfill a contract when the contract is made with an unlicensed contractor in a state requiring that all contractors be licensed?

Such questions have several possible answers. Which one should you choose? Critical thinking and ethical reasoning moves us toward better choices, more thoughtful decisions reflecting knowledge of specific skills for weighing and selecting productive approaches. Issues or business dilemmas require answers. Business leaders often do not have the luxury of waiting around until perfect information floats by. They have to respond effectively or risk business failure. Some answers could get you into trouble; others could advance your purpose. Each answer is called a *conclusion*. A **conclusion** is a position or stance on an issue, the takeaway that the person giving you advice wants you to believe.

conclusion A position or stance on an issue; the goal toward which reasoning pushes us.

Business firms encounter legal conclusions in the form of laws or court decisions and in the advice they receive from people with formal legal training. As businesses learn about and react to decisions or conclusions made by courts, they have two primary methods of response:

1. Memorize the conclusions or rules of law as a guide for future business decisions.
2. Make judgments about the quality of the conclusions. When legal rules fail to reflect understanding of the practicalities of doing business, business leaders play an important civic role in trying to modify those laws.

This book encourages you to do both. What is unique about this text is its practical approach to evaluating legal reasoning. This approach is based on using critical thinking skills to understand and evaluate the law as it affects business.

There are many forms of critical thinking, but they all share one characteristic: They focus on the quality of someone's reasoning. Critical thinking is active; it challenges each of us to form judgments about the quality of the link between someone's reasons and conclusions. In particular, we will be focusing on the link between a court's reasons and its conclusions.

A Critical Thinking Model

You will learn critical thinking by practicing it. This text will tutor you, but your efforts are the key to your skill as a critical thinker. Because people often learn best by example, we will introduce you to critical thinking by demonstrating it in a model that you can easily follow.

> **EXHIBIT 1-1**
>
> **THE EIGHT STEPS TO LEGAL REASONING**
>
> 8. Is there relevant missing information?
> 7. How appropriate are the legal analogies?
> 6. What ethical norms are fundamental to the court's reasoning?
> 5. Does the legal argument contain significant ambiguity?
> 4. What are the relevant rules of law?
> 3. What are the reasons and conclusion?
> 2. What is the issue?
> 1. What are the facts?

We now turn to a sample of critical thinking in practice. The eight critical thinking questions listed in Exhibit 1-1 and applied in the sample case that follows illustrate the approach you should use when reading cases to develop your critical thinking abilities.

As a citizen, entrepreneur, or manager, you will encounter cases like the one that follows. How would you respond? What do you think about the quality of Judge Cedarbaum's reasoning?

CASE 1-1

United States of America v. Martha Stewart and Peter Bacanovic
United States District Court for the Southern District of New York, 2004
U.S. Dist. LEXIS 12538

Defendants Martha Stewart and Peter Bacanovic were both convicted of conspiracy, making false statements, and obstruction of an agency proceeding, following Stewart's sale of 3,928 shares of ImClone stock on December 27, 2001. Stewart sold all of her ImClone stock after Bacanovic, Stewart's stockbroker at Merrill Lynch, informed Stewart that the CEO of ImClone, Samuel Waksal, was trying to sell his company stock. On December 28, 2001, ImClone announced that the Food and Drug Administration (FDA) had not approved the company's cancer-fighting drug Erbitux. Thereafter, the Securities and Exchange Commission (SEC) and the United States Attorney's Office for the Southern District of New York began investigations into the trading of ImClone stock, including investigations of Stewart and Bacanovic.

Following Stewart's and Bacanovic's criminal convictions, the defendants filed a motion for a new trial, alleging that expert witness Lawrence F. Stewart, director of the Forensic Services Division of the United States Secret Service, had committed perjury in his testimony on behalf of the prosecution. As the "national expert for ink analysis," Lawrence Stewart testified about the reliability of defendant Bacanovic's personal documents that contained information about Martha Stewart's investments in ImClone.

Judge Cedarbaum

Rule 33 provides: "Upon the defendant's motion, the court may vacate any judgment and grant a new trial if the interest of justice so requires."* However, "in the interest of according finality to a jury's verdict, a motion for a new trial based on previously-undiscovered evidence is ordinarily 'not favored and should be granted only with great caution.'" In most situations, therefore, "relief is justified under Rule 33 only if the newly-discovered evidence could not have been discovered, exercising due diligence, before or during trial, and that evidence 'is so material and non-cumulative that its admission would probably lead to an acquittal.'"

But the mere fact that a witness committed perjury is insufficient, standing alone, to warrant relief under Rule 33. "Whether the introduction of perjured testimony requires a new trial initially depends on the extent to which the prosecution was aware of the alleged perjury. To prevent prosecutorial misconduct, a conviction obtained when the prosecution's case includes testimony that was known or should have been known to be perjured must be reversed if there is any reasonable likelihood that the perjured testimony influenced the jury." When the Government is unaware of the perjury at the time of trial, "a new trial is warranted only if the

---
*Excerpt from United States District Court-Southern District of New York.

testimony was material and 'the court [is left] with a firm belief that but for the perjured testimony, the defendant would most likely not have been convicted.'"

Defendants have failed to demonstrate that the prosecution knew or should have known of Lawrence's perjury. However, even under the stricter prejudice standard applicable when the Government is aware of a witness's perjury, defendants' motions fail. There is no reasonable likelihood that knowledge by the jury that Lawrence lied about his participation in the ink tests and whether he was aware of a book proposal could have affected the verdict.

The verdict, the nature of Lawrence's perjury, and the corroboration that Lawrence's substantive testimony received from the defense's expert demonstrate that Lawrence's misrepresentations could have had no effect on defendants' convictions.

First, the jury found that the Government did not satisfy its burden of proof on the charges to which Lawrence's testimony was relevant. Defendants do not dispute that Bacanovic was acquitted of the charge of making and using a false document, and that none of the false statement and perjury specifications concerning the existence of the $60 agreement were found by the jury to have been proved *beyond a reasonable doubt*. . . . In other words, the jury convicted defendants of lies that had nothing to do with the $60 agreement. The outcome would have been no different had Lawrence's entire testimony been rejected by the jury, or had Lawrence not testified at all.

Defendants argue that acquittal on some charges does not establish that the jury completely disregarded Lawrence's testimony. They contend that the $60 agreement constituted Stewart and Bacanovic's core defense and that the "@60" notation was evidence which supported that defense.

This argument is wholly speculative and logically flawed. The existence of the $60 agreement would not have exonerated defendants. It would not have been inconsistent for the jury to find that defendants did make the $60 agreement, but that the agreement was not the reason for the sale. In addition to the substantial basis for concluding that the jury's decision could not have been affected by the revelation of Lawrence's misrepresentations, ample evidence unrelated to the $60 agreement or to Lawrence's testimony supports defendants' convictions.

The testimony of Faneuil, Perret, and Pasternak supports the jury's determinations that Stewart lied when she told investigators that she did not recall being informed of Waksal's trading on December 27. . . .

Finally, Faneuil's testimony supports the jury's determination that Stewart lied when she claimed not to have spoken with Bacanovic about the Government investigation into ImClone trading or Stewart's ImClone trade (Specifications Six and Seven of Count Three). Faneuil stated that Bacanovic repeatedly told him in January 2002 and afterward that Bacanovic had spoken to Stewart and that everyone was "on the same page."

Motion for a new trial *denied*.

Before we apply critical thinking to this case, notice that the law is a place where people actively disagree. They are fighting over responsibilities, rights, and fairness. Business law provides a scenario in which parties can peacefully settle the disputes they will inevitably have. Law is and always has been an alternative to war and physical fights. It is a human invention that should make us proud that we can do better than use physical force to settle disagreements.

Now let's get to work, learning how to use the law optimally in business.

First, review the eight steps of a critical thinking approach to legal reasoning in Exhibit 1-1. Throughout the book, we will call these the *critical thinking questions*. They are questions we are asking of those who have particular legal conclusions. Notice the primary importance of the first four steps; their purpose is to discover the vital elements in the case and the reasoning behind the decision. Failure to consider these four foundational steps might result in our reacting too quickly to what a court or legislature has said. The rule here is: We should never evaluate until we first understand the argument being made.

The answers to these four questions enable us to understand how the court's argument fits together and to make intelligent use of legal decisions. These answers are the necessary first stage of a critical thinking approach to legal analysis. The final four questions are the critical thinking component of legal reasoning. They are questions that permit us to evaluate the reasoning and to form our reaction to what the court decided.

You will develop your own workable strategies for legal reasoning, but we urge you to start by following our structure. Every time you read a case, ask yourself these eight questions. The remainder of this section will demonstrate the use of each of the eight steps. Notice that the order makes sense. The first four follow the path that best allows you to discover the basis of a particular legal decision; the next four assist you in deciding what you think about the worth of that decision.

The Critical Thinking Steps
FACTS

First we look for the most basic building blocks in a legal decision or argument. These facts provide the context in which the legal issue is to be resolved. Alter those facts, and the legal conclusion might be very different. Certain events occurred; certain actions were or were not taken; particular persons behaved or failed to behave in specific ways. We always wonder, what happened in this case? Let's now turn our attention to the *Stewart* case:

1. Martha Stewart sold 3,928 shares of her ImClone stock on December 27, 2001.
2. On December 28, 2001, ImClone announced the FDA's rejection of its new cancer-fighting drug, which caused the company's stock to lose value.
3. Stewart and Bacanovic were convicted of conspiracy, making false statements, and obstruction of an agency proceeding.
4. Expert witness Lawrence Stewart was accused of perjuring himself in the testimony he gave prior to the defendants' conviction.
5. According to a federal rule and case law, perjury of a witness could constitute grounds for a new trial.

ISSUE

In almost any legal conflict, finding and expressing the issue is an important step in forming a reaction. So important is the definition of the issue, that many times the lawyers in a legal suit spend considerable effort trying to get the judge or jury to see the issue a particular way so that they have a better chance of winning the case. The issue is the question that caused the lawyers and their clients to enter the legal system. Usually, there are several reasonable perspectives concerning the correct way to word the issue in dispute.

1. In what instances may a court grant a new trial?
2. Does perjury of a witness mean that defendants should have a new trial?
3. Do the regulations in Rule 33 and relevant case law permit the defendants to have a new trial?

Do not let the possibility of multiple useful ways to word the issue cause you any confusion. The issue is certainly not just anything that we say it is. If we claim something is an issue, our suggestion must fulfill the definition of an issue in this particular factual situation.

REASONS AND CONCLUSION

Judge Cedarbaum held that the defendants should not have a new trial. This finding by Judge Cedarbaum is her conclusion; it serves as her answer to the legal issue. Why did she answer this way? Here we are calling for the **reasons**, explanations or justifications provided as support for a conclusion.

reason An explanation or justification provided as support for a conclusion.

1. Under Rule 33 and relevant case law, perjury is not sufficient to justify a new trial, unless (a) the government knew about the perjury or (b) the perjured testimony was so material that without it the verdict would probably have resulted in acquittal of the defendants.

2. The defendants did not demonstrate that the government knew or should have known about the perjured testimony.
3. The jury would still have convicted the defendants apart from Lawrence's testimony.
4. Defense experts agreed with Lawrence on the "most critical aspects of his scientific analysis."

Let's not pass too quickly over this very important critical thinking step. When we ask *why* of any opinion, we are showing our respect for reasons as the proper basis for any assertion. The judge did not rely on astrology or palm readers to guide her. Instead, she relied on our special ability to identify and sort through reasons and evidence.

We want a world rich with opinions so we can have a broad field of choice. We should, however, agree with only those legal opinions that have convincing reasons supporting the conclusion. Thus, asking *why* is our way of saying, "I want to believe you, but you have an obligation to help me by sharing the reasons for your conclusion."

RULES OF LAW

Judges cannot offer just any reasoning they please. They must always look back over their shoulders at the laws and previous court decisions that together provide a foundation for current and future decisions. They must follow precedents, the decisions in past cases with similar facts.

This particular case is an attempt to match the words of the Federal Rules of Criminal Procedure, specifically Rule 33, and its regulations with the facts in this particular case. The court also references case law. What makes legal reasoning so complex is that statutes and findings are never crystal clear. Judges and businesspeople have room for interpretive flexibility in their reasoning.

AMBIGUITY

The starting point for thinking about this important critical thinking standard is recognizing that a word does not have just one meaning. Thus, when I say a particular word to you, there is no reason I should presume that the meaning I had in mind is transferred into your mind in exactly the same form as it left my mouth. As soon as we realize the flexibility of words, a huge responsibility falls onto our shoulders. We have to seek clarity in what people say to us, or we risk reacting to what they said in a manner they never intended to encourage. Exploring the meaning of what people say is only fair to them.

ambiguous Possessing two or more possible interpretations.

The court's reasoning must rest on its implied assumptions about the meaning of several ambiguous words or phrases. (An **ambiguous** word is one capable of having more than one meaning in the context of these facts.) For instance, Judge Cedarbaum stated that Rule 33 permits the court to grant a new trial if the "interest of justice so requires." But what is the "interest of justice"?

Does the interest of justice entail strict conformity to legal precedents? Or could the court's reliance on certain precedents result in some form of injustice in the *Stewart* case? If we adopt the former definition, we would be more inclined to conclude that the judge's denying the defendants' motion for a new trial was consistent with the "interest of justice." However, if the legislators who created Rule 33 intended a definition of "justice" that placed a stronger emphasis on judicial fairness, for example, perhaps we would be less supportive of Judge Cedarbaum's decision.

Another illustration of important ambiguity in the decision is the court's use of the term *reasonable likelihood,* referring to the probability that Lawrence's alleged perjury could not have affected the jury's verdict—but what degree of probability is a "reasonable likelihood"? Does this level of probability suggest that knowledge of Lawrence's testimony could have affected the jury's verdict? If we interpret "reasonable likelihood" as still including the possibility that knowledge of Lawrence's perjury could have affected the jury, we might reach a conclusion that differs from the court's decision. If we assume a definition of "reasonable likelihood" similar to "beyond a reasonable doubt," however, we would be more inclined to agree with the judge's decision. Hence, until we know what "reasonable likelihood" means, we cannot fairly decide whether the judge made the appropriate decision.

ETHICAL NORMS

The primary ethical norms that influence judges' decisions are justice, stability, freedom, and efficiency. Notice that each of these words is an abstraction, something we cannot touch, smell, hear, or see. As important as these ethical norms are, they are simply an invitation to a conversation—a conversation focusing on the meaning being used in this particular instance. Judge Cedarbaum expresses herself as a defender of stability or order. (Here is a good place to turn to Exhibit 1-2 to check alternative definitions of *stability*.) She is unwilling to grant a new

EXHIBIT 1-2

CLARIFYING THE PRIMARY ETHICAL NORMS

A judge's allegiance to a particular ethical norm focuses our attention on a specific category of desired conduct. We have, or think we have, an understanding of what is meant by freedom and other ethical norms.

But do we? Ethical norms are, without exception, complex and subject to multiple interpretations. Consequently, to identify the importance of one of the ethical norms in a piece of legal reasoning, we must look at the context to figure out which form of the ethical norm is being used. The types of conduct called for by the term *freedom* not only differ depending on the form of freedom being assumed, but at times they can contradict each other.

As a future business manager, your task is to be aware that there are alternative forms of each ethical norm. Then a natural next step is to search for the form used by the legal reasoning so you can understand and later evaluate that reasoning.

The following alternative forms of the four primary ethical norms can aid you in that search.

Ethical Norms	*Forms*
1. Freedom	To act without restriction from rules imposed by others
	To possess the capacity or resources to act as one wishes
2. Security	To provide the order in business relationships that permits predictable plans to be effective
	To be safe from those wishing to interfere with your property rights
	To achieve the psychological condition of self-confidence such that risks are welcomed
3. Justice	To receive the product of your labor
	To provide resources in proportion to need
	To treat all humans identically, regardless of class, race, gender, age, and so on
	To possess anything that someone else was willing to grant you
4. Efficiency	To maximize the amount of wealth in our society
	To get the most from a particular input
	To minimize costs

trial simply on the fact that one of the witnesses allegedly committed perjury. Instead of granting the defendants' motion, Judge Cedarbaum elevates the "interest of according finality to a jury's verdict," even if the prosecution knew or should have known about the alleged perjury. Citing previous case law, she is able to grant new trials only in rare instances.

ANALOGIES

Ordinarily, our examination of legal analogies will require us to compare legal precedents cited by the parties with the facts of the case we are examining. Those precedents are the analogies on which legal decision making depends. In this case, Judge Cedarbaum relies on several legal precedents as analogies for her ruling.

In *United States v. Wallach*, the Second Circuit held that even if the prosecution knew of a witness's perjury, the court should not grant a new trial when other "independent" evidence is sufficient to convict a defendant. The worth of this analogy depends on a greater understanding of independent evidence. In other words, what constitutes independent evidence? And is the strength of independent evidence in the *Stewart* case comparable to the independent evidence in *Wallach*? Or are there significant differences between the two cases such that the court's reliance on *Wallach* is unwarranted in this case?

To feel comfortable with the analogy, we would need to be persuaded that the independent evidence in the *Stewart* case is basically similar to the independent evidence in the precedent, *United States v. Wallach*. Law is an interpretive practice. Each of us brings a different set of experiences, aspirations, and perspectives to our interpretations. We move forward in the midst of our differences by assembling reasonable arguments for why our understanding of the analogy makes sense.

MISSING INFORMATION

When any of us makes a decision, we always do so with less information than we would love to have. In the search for relevant missing information, it is important not to say just anything that comes to mind. For example, where did the defendants eat Thanksgiving dinner? Anyone hearing that question would understandably wonder why it was asked. Ask only questions that would be helpful in understanding the reasoning in this particular case.

To focus on only relevant missing information, we should include an explanation of why we want it with any request for additional information. We have listed a few examples here for the *Stewart* case. You can probably identify others.

1. How well informed is Judge Cedarbaum with respect to the deliberations of the jury? If her understanding of the jurors' preverdict discussions is very limited, the defendants' request for a new trial might be more convincing, because Judge Cedarbaum repeatedly contends that jurors' knowledge of Lawrence's alleged perjury would not have affected the jurors' decision.

2. Congress, as it does with any legislation, discussed the Rules of Criminal Procedure before passing them. Does that discussion contain any clues as to congressional intent with respect to the various conditions required for a defendant to receive a new trial? The answer would conceivably clarify the manner in which the court should apply Rule 33.

3. Are there examples of cases in which courts have examined fact patterns similar to those in the *Stewart* case but reached different conclusions about a new trial? The answer to this question would provide greater clarity about the appropriateness of using certain case precedents, thereby corroborating or undermining Judge Cedarbaum's decision.

Many other critical thinking skills could be applied to this and other cases. In this book, we focus on the ones especially valuable for legal reasoning. Consistently applying this critical thinking approach will enable you to understand the reasoning in legal cases and increase your awareness of alternative approaches our laws could take to many legal environment of business problems. The remaining portion of this chapter examines each of the critical thinking questions in greater depth to help you better understand the function of each.

Using Critical Thinking to Make Legal Reasoning Come Alive

Our response to an issue is a *conclusion*. It is what we want others to believe about the issue. For example, a court might conclude that an employee, allegedly fired for her political views, was actually a victim of employment discrimination and is entitled to a damage award. Conclusions are reached by following a path produced by reasoning. Hence, examining reasoning is especially important when we are trying to understand and evaluate a conclusion.

There are many paths by which we may reach conclusions. For instance, I might settle all issues in my life by listening to voices in the night, asking my uncle, studying astrological signs, or just playing hunches. Each method could produce conclusions. Each could yield results.

But our intellectual and legal tradition demands a different type of support for conclusions. In this tradition, the basis for our conclusions is supposed to consist of reasons. When someone has no apparent reasons, or the reasons don't match the conclusion, we feel entitled to say, "But that makes no sense." We aren't impressed by claims that we should accept someone's conclusion "just because."

This requirement that we all provide reasons for our conclusions is what we mean, in large part, when we say we are going to think. We will ponder what the reasons and conclusion are and whether they fit together logically. This intense study of how a certain conclusion follows from a particular set of reasons occupies much of the time needed for careful decision making.

Persons trained to reason about court cases have uncommon appreciation for the unique facts that provoked a legal action. Those facts, and no others, provide the context for our reasoning. If an issue arises because environmentalists want to prevent an interstate highway from extending through a wilderness area, we want to know right away: *What are the facts?* Tell us more about this wilderness area. What procedures were followed before approving the route for this highway? What evidence was presented to project the possible harm to the ecological system?

Legal reasoning encourages unusual and necessary respect for the particular factual situation that stimulated disagreement between parties. These *fact patterns,* as we call them, bring the issue to our attention and limit the extent to which the court's conclusion can be applied to other situations. Small wonder that the first step in legal reasoning is to ask and answer the question: What are the facts?

LEGAL REASONING

Step 1: What Are the Facts? The call for the facts is not a request for all facts, but only those that have a bearing on the dispute at hand. The precise nature of the dispute tells us whether a certain fact is pertinent. In some cases, the plaintiff's age may be a key point; in another, it may be irrelevant.

Only after we have familiarized ourselves with the relevant legal facts do we begin the familiar pattern of reasoning that thoughtful people use. We then ask and answer the following question: What is the issue?

Step 2: What Is the Issue? The issue is the question that the court is being asked to answer. For example, courts face groups of facts relevant to issues such as the following:

1. Does Title VII apply to sexual harassment situations when the accused and the alleged victim are members of the same sex?
2. Does a particular merger between two companies violate the Sherman Act?
3. When does a governmental regulation require compensation to the property owner affected by the regulation?

As we pointed out earlier, the way we express the issue guides the legal reasoning in the case. Hence, forming an issue in a very broad or an extremely narrow manner has implications for the scope of the effect stemming from the eventual decision. You can appreciate now why parties to a dispute work very hard to get the court to see the issue in a particular way.

You will read many legal decisions in this book. No element of your analysis of those cases is more important than careful consideration of the issue. The key to issue spotting is asking yourself: What question do the parties want the court to answer? The next logical step in legal analysis is to ask: What are the reasons and conclusion?

Step 3: What Are the Reasons and Conclusion? The issue is the stimulus for thought. The facts and the issue in a particular case get us to start thinking critically about legal reasoning, but the conclusion and the reasons for that conclusion put flesh on the bones of the court's reaction to the legal issue. They tell us how the court has responded to the issue.

To find the conclusion, use the issue as a helper. Ask yourself: How did the court react to the issue? The answer is the conclusion. The reasons for that conclusion provide the answer to the question: Why did the court prefer this response to the issue rather than any alternative? One part of the answer to that question is the answer to another question: What are the relevant rules of law?

Step 4: What Are the Relevant Rules of Law? The fourth step in legal reasoning reveals another difference from general nonlegal reasoning. The issue arises in a context of existing legal rules. We do not treat each legal dispute as if it were the first such dispute in human history. A court has already responded to situations much like the ones now before it. The historical record of pertinent judicial decisions provides a rich source of reasons on which to base the conclusions of courts.

These prior decisions, or *legal precedents,* provide legal rules to which those in a legal dispute must defer. Thus, the fourth step in legal reasoning requires a focus on those rules. These legal rules are what the parties to a dispute must use as the framework for their legal claims. How those rules and the reasoning and conclusions built on them are expressed, however, is not always crystal clear. Hence, another question—one that starts the critical thinking evaluation of the conclusion—is: Does the legal argument contain significant ambiguity?

Step 5: Does the Legal Argument Contain Significant Ambiguity? Legal arguments are expressed in words, and words rarely have the clarity we presume. Whenever we are tempted to think that our words speak for themselves, we should remind ourselves of Emerson's observation that "to be understood is a rare luxury." Because legal reasoning is couched in words, it possesses elasticity. It can be stretched and reduced to fit the purpose of the attorney or judge.

As an illustration, a rule of law may contain the phrase *public safety*. At first glance, as with any term, some interpretation arises in our minds; however, as we continue to consider the extent and limits of public safety, we realize it is not so clear. To be more certain about the meaning, we must study the intent of the person making the legal argument. Just how safe must the public be before an action provides sufficient threat to public safety to justify public intervention?

As a strategy for critical thinking, the request for clarification is a form of evaluation. The point of the question is that we cannot agree with a person's reasoning until we have determined what we are being asked to embrace.

What we are being asked to embrace usually involves an ethical component. Therefore, an important question to ask is: What ethical norms are fundamental to the court's reasoning?

Step 6: What Ethical Norms Are Fundamental to the Court's Reasoning? The legal environment of business is established and modified according to ethical norms. A **norm** is a standard of conduct, a set of expectations that we bring to social encounters. For example, one norm we collectively understand and obey is that our departures are ordinarily punctuated by "good-bye." We may presume rudeness or preoccupation on the part of someone who leaves our presence without bidding us some form of farewell.

Ethical norms are special because they are steps toward achieving what we consider good or virtuous. Goodness and virtue are universally preferred to their opposites, but the preference has little meaning until we look more deeply into the meaning of these noble aims.

Conversations about ethics compare the relative merit of human behavior guided by one ethical norm or another. Ethical norms represent the abstractions we hold out to others as the most fundamental standards defining our self-worth and value to the community. For example, any of us would be proud to know that others see us as meeting the ethical norms we know as honesty, dependability, and compassion. Ethical norms are the standards of conduct we most want to see observed by our children and our neighbors.

The legal environment of business receives ethical guidance from many norms. Certain norms, however, play a particularly large role in legal reasoning. Consequently, we highlight what we will refer to as the four **primary ethical norms**: freedom, stability, justice, and efficiency. (See Exhibit 1-2 for clarification of these norms.) As you examine the cases in this text, you may identify other ethical norms that influence judicial opinions. To discover the relevant ethical norm, we must infer it from the court's reasoning. Courts often do not announce their preferred pattern of ethical norms, but the norms are there anyway, having their way with the legal reasoning. As critical thinkers, we want to use the ethical norms, once we find them, as a basis for evaluating the reasoning.

We do so by thinking about the business effects from relying on a different ethical norm. Would a different ethical norm lead the law in a direction that would be more consistent with the goals of our community?

Another element used in arriving at legal conclusions is the device of reasoning by analogy. Part of the critical thinking process in the evaluation of a legal conclusion is another question: How appropriate are the legal analogies?

Step 7: How Appropriate Are the Legal Analogies? A major difference between legal reasoning and other forms of analysis is the heavy reliance on analogies. Our legal system places great emphasis on the law, as it has evolved from previous decisions. This evolutionary process is our heritage, the collective judgments of our historical mothers and fathers. We give them and their intellects our respect by using legal precedents as the major support structure for

norm An expected standard of conduct.

ethical norms Standards of conduct that we consider good or virtuous.

primary ethical norms The four norms that provide the major ethical direction for the laws governing business behavior: freedom, stability, justice, and efficiency.

judicial decisions. By doing so, we do not have to approach each fact pattern with entirely new eyes; instead, we are guided by similar experiences that our predecessors have already resolved.

The use of precedent to reach legal conclusions is so common that legal reasoning can be characterized as little but analogical reasoning. An **analogy** is a verbal device for transferring meaning from something we understand quite well to something we have just discovered and have, as yet, not understood satisfactorily. What we already understand in the case of legal reasoning is the *precedent*; what we hope to understand better is the current legal dispute. We call on precedent for enlightenment.

To visualize the choice of legal analogy, imagine that we are trying to decide whether a waitress or waiter can be required to smile for hours as a condition of employment. (What is artificial about such an illustration, as we hope you already recognize, is the absence of a more complete factual picture to provide context.) The employer in question asks the legal staff to find appropriate legal precedents. They discover the following list of prior decisions:

1. Professional cheerleaders can be required to smile within reason, if that activity is clearly specified at the time of employment.
2. Employees who interact regularly with customers can be required as a condition of employment to wear clothing consistent with practice in the trade.
3. Employers may not require employees to lift boxes over 120 pounds without the aid of a mechanical device, under the guidelines of the Employee Health Act.

Notice that each precedent has similarities to, but also major differences from, the situation of the waiter or waitress. To mention only a few:

- Is a smile more natural to what we can expect from a cheerleader than from a waiter or waitress?
- Were the restaurant employees told in advance that smiling is an integral part of the job?
- Is a smile more personal than clothing? Are smiles private, as opposed to clothing, which is more external to our identity?
- Is a plastered-on smile, held in place for hours, a serious risk to mental health?
- Is a potential risk from smiling as real a danger as the one resulting from physically hoisting huge objects?

The actual selection of precedent and, consequently, the search for appropriate analogies are channeled by the theory of logic that we find most revealing in this case. For example, if you see the requirement to smile as an invasion of privacy, you will likely see the second precedent as especially appropriate. Both the precedent and the case in question involve employment situations with close customer contact.

The differences, however, could be significant enough to reject that analogy. Do you see your clothing as part of your essence, in the same fashion as you surely see the facial form you decide to show at any given moment? Furthermore, the second precedent contains the phrase *consistent with practice in the trade*. Would not a simple field trip to restaurants demonstrate that a broad smile is a pleasant exception?

As you practice looking for similarities and differences in legal precedents and the legal problem you are studying, you will experience some of the fun and frustration of legal reasoning within a business context. The excitement comes when you stumble on just the perfect, matching fact pattern; then, after taking a closer look, you are brought back to earth by those annoying analogical differences that your experience warns you are always present.

---

analogy A comparison based on the assumption that if two things are alike in some respect, they must be alike in other respects.

Ambiguity, ethical norms, and legal analogies are all areas in which legal arguments may be deficient; but even if you are satisfied that all those considerations meet your standards, there is a final question that must not be overlooked in your critical analysis of a conclusion: Is there relevant missing information?

Step 8: Is There Relevant Missing Information? When we ask about the facts of a case, we mean the information presented in the legal proceedings. We are, however, all quite aware that the stated facts are just a subset of the complete factual picture responsible for the dispute. We know we could use more facts than we have, but at some point we have to stop gathering information and settle the dispute.

You might not be convinced that the facts we know about a situation are inevitably incomplete; however, consider how we acquire facts. If we gather them ourselves, we run into the limits of our own experience and perceptions. We often see what we want to see, and we consequently select certain facts to file in our consciousness. Other facts may be highly relevant, but we ignore them. We can neither see nor process all the facts.

Our other major source of information is other people. We implicitly trust their intentions, abilities, and perspective when we take the facts they give us and make them our own. No one, however, gives us a complete version of the facts. For several reasons, we can be sure that the facts shared with us are only partial.

Armed with your awareness of the incompleteness of facts, what can you do as a future businessperson or employee to effectively resolve disagreements and apply legal precedents?

You can seek a more complete portrayal of the facts. Keep asking for detail and context to aid your thinking. For example, once you learn that a statute requires a firm to use the standard of conduct in the industry, you should not be satisfied with the following fact:

> *On 14 occasions, our firm attempted to contact other firms to determine the industry standards. We have bent over backwards to comply with the ethical norms of our direct competitors.*

Instead, you will persist in asking probing questions designed to generate a more revealing pattern of facts. Among the pieces of missing information you might ask for would be the extent and content of actual conversations about industry standards, as well as some convincing evidence that "direct" competitors are an adequate voice, representing "the industry."

Applying the Critical Thinking Approach

Now that you have an understanding of the critical thinking approach, you are ready to begin your study of the legal environment of business. Remember to apply each of the questions to the cases as you read them. As an incentive to do the work associated with careful thinking, imagine what it would be like to NOT apply critical thinking in your business careers. You would receive advice, and you would always believe it, as long as the person speaking seemed nice and authoritative. As soon as someone told you what "the law" is, you would, like a sponge, simply proceed to do business as if that single statement about the content of the law is the one and only possible understanding of the law. You would be the mental puppet of the last clever person with whom you spoke. You will agree that this portrait of a businessperson who does not use critical thinking is a recipe for disaster.

After you become proficient at asking these questions of every case you read, you may find that you start asking these evaluative questions in other contexts. For example, you might find that, when you read an editorial in the

Wall Street Journal, you start asking whether the writer has used ambiguous terms that affect the quality of the reasoning, or you start noticing when important relevant information is missing. Once you reach this point, you are well on your way to becoming a critical thinker whose thinking skills will be extremely helpful in the legal environment of business.

ASSIGNMENT ON THE INTERNET

You have now been introduced to the critical thinking steps that create a working strategy to evaluate legal reasoning. In the same manner that you evaluated *United States of America v. Martha Stewart and Peter Bacanovic*, practice evaluating the legal reasoning in a case of your choosing.

Go to **www.law.cornell.edu** for current legal issues and cases. Find a case of interest to you, and evaluate the reasoning using the critical thinking steps outlined in this chapter. *Burwell v. Hobby Lobby*, for example, would be a fun and important business law case to look at. The following websites on critical thinking may assist you in evaluating legal reasoning.

ON THE INTERNET

Use this site for practice, identifying reasons and conclusions in arguments.
www.austhink.org/critical This site is a compendium of useful critical thinking websites.
pegasus.cc.ucf.edu/~janzb/reasoning Both sites contain numerous links for those wishing additional reading and practice with the critical thinking skills learned in this chapter.

FOR FUTURE READING

Browne, M. Neil, and Stuart Keeley. *Asking the Right Questions: A Guide to Critical Thinking* (11th ed.). Upper Saddle River, NJ: Prentice Hall, 2014.

Damer, T. Edward. *Attacking Faulty Reasoning: Practical Guide to Fallacy-Free Arguments* (7th ed.). Belmont, CA: Wadsworth, 2012.

LeGault, Michael R. *Think: Why Crucial Decisions Can't Be Made in the Blink of an Eye*. New York: Threshold Editions, 2006.

CHAPTER TWO

Introduction to Law and the Legal Environment of Business

- **DEFINITION OF THE LEGAL ENVIRONMENT OF BUSINESS**
- **DEFINITION OF LAW AND JURISPRUDENCE**
- **SOURCES OF LAW**
- **CLASSIFICATIONS OF LAW**
- **GLOBAL DIMENSIONS OF THE LEGAL ENVIRONMENT OF BUSINESS**

This book is about the legal environment in which the business community operates today. Although we concentrate on law and the legal variables that help shape business decisions, we have not overlooked the ethical, political, and economic questions that often arise in business decision making. In this chapter, we are especially concerned with legal variables in the context of critical thinking, as outlined in Chapter 1. In addition, we examine the international dimensions of several areas of law. In an age of sophisticated telecommunication systems, computer networking, and wrist watch software it would be naïve for our readers to believe that, as citizens of a prosperous, powerful nation situated between two oceans, they can afford to ignore the rest of the world. Just as foreign multinational companies must interact with U.S. companies and government agencies, so must U.S. entities interact with regional and international trade groups and agencies of foreign governments.

The United States, Canada, and Mexico created the North American Free Trade Agreement (NAFTA) to lower trade barriers among themselves. In the Asian-Pacific Economic Cooperation (APEC) forum, the United States and 22 Pacific Rim nations are discussing easing barriers to trade and investments among themselves and creating a Pacific free trade zone extending from Chile to China. The European Union has added new member nations, bringing its total to 28 as of mid-2015. The World Trade Organization continues to lower trade barriers among the 144 nations that have joined it. No nation is an island unto itself today, and economic globalization is accelerating in the twenty-first century. (See Chapter 8 for a discussion of the global legal environment of business.)

CRITICAL THINKING ABOUT THE LAW

This chapter serves as an introduction to the legal and ethical components of the business environment. You will learn about different schools of jurisprudence and about sources and classifications of law. In addition, this chapter offers the opportunity to practice the critical thinking skills you learned in Chapter 1. The following critical thinking questions will help you better understand the introductory topics discussed in this chapter.

1. Why should we be concerned with the ethical components of the legal environment of business? Why shouldn't we just learn the relevant laws regarding businesses?

   Clue: Which critical thinking questions address the ethical components of the legal environment of business?

2. As you will soon discover, judges and lawyers often subscribe to a particular school of legal thought. Judges and lawyers, however, will probably not explicitly tell us which school of thought they prefer. Why do you think this knowledge might be beneficial when critically evaluating a judge's reasoning?

   Clue: Think about why we look for missing information. Furthermore, why do we want to identify the ethical norms fundamental to a court's reasoning?

3. You tell your landlord that your front door lock is broken, but he does not repair the lock. A week later, you are robbed. You decide to sue the landlord, and you begin to search for an attorney. As a legal studies student, you ask the potential lawyers what school of jurisprudence they prefer. Although you find a lawyer who prefers the same school of jurisprudence you prefer, your decision is not final. What else might you want to ask the lawyer?

   Clue: Think about the other factors that might affect a lawyer's performance.

Definition of the Legal Environment of Business

The *legal environment of business* is defined in various ways. For our purposes, the study of the legal environment includes:

- The study of the legal reasoning, critical thinking skills, ethical norms, and schools of ethical thought that interact with the law.
- The study of the legal process and our present legal system, as well as alternative dispute resolution systems such as private courts, mediation, arbitration, and negotiation.
- The study of the administrative law process and the role of businesspeople in that process.
- The study of selected areas of public and private law, such as securities regulation, antitrust, labor, product liability, contracts, and consumer and environmental law. In each of these areas, we emphasize the processes by which business managers relate to individuals and government regulators.
- The examination of the international dimensions of the legal environment of law.

Our study of the legal environment of business is characterized by five features:

1. *Critical thinking skills.*
2. *Legal literacy.* A survey by the Hearst Corporation found that 50 percent of Americans believe that it is up to the criminally accused to prove their innocence, despite our common-law heritage that a person is presumed innocent until proven guilty. Of those responding to the survey, 49.9 percent had served on a jury, and 31 percent were college graduates.
3. *An understanding that the law is dynamic, not static.* The chapters on discrimination law, securities regulation, antitrust law, and labor law in particular have had to be constantly updated during the writing of this book, because federal regulatory agencies issue new regulations, rules, and guidelines almost daily.

> **EXHIBIT 2-1**
>
> **TOP 10 REASONS FOR STUDYING THE LEGAL ENVIRONMENT OF BUSINESS**
>
> 1. Becoming aware of the rules of doing business.
> 2. Familiarizing yourself with the legal limits on business freedom.
> 3. Forming an alertness to potential misconduct of competitors.
> 4. Appreciating the limits of entrepreneurship.
> 5. Being able to communicate with your lawyer.
> 6. Making you a more fully informed citizen.
> 7. Developing an employment-related skill.
> 8. Exploring the fascinating complexity of business decisions.
> 9. Providing a heightened awareness of business ethics.
> 10. Opening your eyes to the excitement of the law and business.

4. *Real-world problems.* You will be confronted with real (not theoretical) legal and ethical problems. As the great American jurist Oliver Wendell Holmes once pointed out, the law is grounded in "experience." In reading the cases excerpted in this book, you will see how business leaders and others either were ignorant of the legal and ethical variables they faced or failed to consider them when making important decisions.

5. *Interdisciplinary nature.* Into our discussions of the legal environment of business we interweave materials from other disciplines that you either are studying now or have studied in the past, especially economics, management, finance, marketing, and ethics. You may be surprised to learn how often officers of the court (judges and attorneys) are obliged to consider material from several disciplines in making decisions. Your own knowledge of these other disciplines will be extremely helpful in understanding the content of this book.

The connections to other areas of business are so significant that we have chosen to highlight many of them in subsequent chapters of this book. As you are reading, you will encounter boxes entitled "Linking Law and Business." These boxes contain material from other business disciplines that are related to the business law material you are studying. By highlighting these connections, we hope to provide greater cohesiveness to your education as a future business manager. As listed in Exhibit 2-1, there are a number of benefits to be gained by studying the legal environment of business.

Definition of Law and Jurisprudence

Jurisprudence is the science or philosophy of law, or law in its most generalized form. *Law* itself has been defined in different ways by scholarly thinkers. Some idea of the range of definitions can be gained from the following quote from a distinguished legal philosopher:

jurisprudence The science or philosophy of law; law in its most generalized form.

> *We have been told by Plato that law is a form of social control; by Aristotle that it is a rule of conduct, a contract, an ideal of reason; by Cicero that it is the agreement of reason and nature, the distinction between the just and the unjust; by Aquinas that it is an ordinance of reason for the common good; by Bacon that certainty is the prime necessity of law; by Hobbes that law is the command of the sovereign; by Hegel that it is an unfolding or realizing of the idea of right.[1]**

The various ideas of law expressed in this passage represent different schools of jurisprudence. To give you some sense of the diversity of meaning

---

[1] See H. Cairns, *Legal Philosophy from Plato to Hegel* (Baltimore: Johns Hopkins University Press, 1949).

*Excerpt from "Legal Philosophy from Plato to Hegel" by Huntington Cairns. Published by John Hopkins University Press, © 1949.

EXHIBIT 2-2

SCHOOLS OF JURISPRUDENCE

School	Characteristics
Natural law school	Source of law is absolute (nature, God, or reason)
Positivist school	Source of law is the sovereign
Sociological school	Source of law is contemporary community opinion and customs
American realist school	Source of law is actors in the legal system and scientific analysis of their actions
Critical legal studies school	Source of law is a cluster of legal and nonlegal beliefs that must be critiqued to bring about social and political change
Feminist school	Jurisprudence reflects a male-dominated executive, legislative, and judicial system in which women's perspectives are ignored and women are victimized
Law and economics school	Classical economic theory and empirical methods are applied to all areas of law in order to arrive at decisions

the term *law* has, we will examine seven accepted schools of legal thought: (1) natural law, (2) positivist, (3) sociological, (4) American realist, (5) critical legal studies, (6) feminist, and (7) law and economics. Exhibit 2-2 summarizes the outstanding characteristics of each of these schools of jurisprudence.

NATURAL LAW SCHOOL

For adherents of the natural law school, which has existed since 300 B.C., law consists of the following concepts: (1) There exist certain legal values or value judgments (e.g., a presumption of innocence until guilt is proved); (2) these values or value judgments are unchanging because their source is absolute (e.g., nature, God, or reason); (3) these values or value judgments can be determined by human reason; and (4) once determined, they supersede any form of human law. Perhaps the most memorable statement of the natural law school of thought in this century was made by Martin Luther King, Jr., in his famous letter from a Birmingham, Alabama, city jail.

In that letter he explained to a group of ministers why he had violated human laws that discriminated against his people. He explained that not all laws were the same, and that some laws were consistent with God's law, and those laws were just and should be obeyed. But the laws that were inconsistent with God's law were unjust and should not be obeyed. In particular, laws that degrade the human personality are inconsistent with God's law and therefore not just. He cited segregation laws as an example of laws that harm the human spirit and therefore are unjust, which is why he urged disobedience to those laws.

Now, what is the difference between the two? How does one determine when a law is just or unjust? A just law is a manmade code that squares with the moral law or the law of God. An unjust law is a code that is out of harmony with the moral law. To put it in the terms of the writings of Saint Thomas Aquinas, an unjust law is a human law that is not rooted in eternal and natural law. Any law that uplifts human personality is just. Any law that degrades human personality is unjust. All segregation statutes are unjust because segregation distorts the soul and damages the personality.

Let us turn to a more concrete example of just and unjust law. An unjust law is a code that a majority inflicts on a minority but that is not binding on the majority itself. In contrast, a just law is a code that a majority compels a minority to follow that it is willing to follow itself. This is sameness made legal.

Let me give another explanation. An unjust law is a code inflicted upon a minority that that minority had no part in enacting or creating because they did not have the unhampered right to vote.[2]

Adherents of other schools of legal thought view King's general definition of law as overly subjective. For example, they ask, "Who is to determine whether a manmade law is unjust because it is 'out of harmony with the moral law'?" Or: "Whose moral precepts or values are to be included in the 'moral law'?" The United States is a country of differing cultures, races, ethnic groups, and religions, each of which may hold or reflect unique moral values.

POSITIVIST SCHOOL

Early in the 1800s, followers of positivism developed a school of thought in opposition to the natural law school. Its chief tenets are (1) law is the expression of the will of the legislator or sovereign, which must be followed; (2) morals are separate from law and should not be considered in making legal decisions (thus, judges should not take into consideration extralegal factors such as contemporary community values in determining what constitutes a violation of law); and (3) law is a closed logical system in which correct legal decisions are reached solely by logic and the use of precedents (previous cases decided by the courts).

Disciples of the positivist school would argue that when the Congress of the United States has not acted on a matter, the U.S. Supreme Court has no power to act on that matter. They would argue, for example, that morality has no part in determining whether discrimination exists when a business pays workers differently on the basis of their sex, race, religion, or ethnic origin. Only civil rights legislation passed by Congress, and previous cases interpreting that legislation, should be considered. Laws of other nations should not be considered when U.S. courts must make decisions, as Justice Scalia of the U.S. Supreme Court has argued.

Positivism has been criticized by adherents of other schools of thought as too narrow and literal minded. Critics argue that the refusal to consider social, ethical, and other factors makes for a static jurisprudence that ill serves society.

SOCIOLOGICAL SCHOOL

Followers of the sociological school propose three steps in determining law:

1. A legislator or a judge should make an inventory of community interests.
2. Judges and legislators should use this inventory to familiarize themselves with the community's standards and mores.
3. The judge or legislator should rule or legislate in conformity with those standards and mores.

For those associated with this school of legal thought, human behavior or contemporary community values are the most important factors in determining the direction the law should take. This philosophy is in sharp contrast to that of the positivist school, which relies on case precedents and statutory law. Adherents of the sociological school seek to change the law by surveying human behavior and determining present community standards. For example, after

---

[2] See M. L. King, Jr., *Letters from a Birmingham Jail* (April 16, 1963), reprinted in M. McGuaigan, *Jurisprudence* (New York: Free Press, 1979), p. 63.

a famous U.S. Supreme Court decision stating that material could be judged "obscene" on the basis of "contemporary community standards,"[3] a mayor of a large city immediately went out and polled his community on what books and movies they thought were obscene. (He failed to get a consensus.)

Critics of the sociological school argue that this school would make the law too unpredictable for both individuals and businesses. They note that contemporary community standards change over time and, thus, the law itself would be changing all the time and the effects could harm the community. For example, if a state or a local legislature offered a corporation certain tax breaks as an incentive to move to a community and then revoked those tax breaks a few years later because community opinion on such matters had changed, other corporations would be reluctant to locate in that community.

AMERICAN REALIST SCHOOL

The American realist school, though close to the sociological school in its emphasis on people, focuses on the actors in the judicial system instead of on the larger community to determine the meaning of law. This school sees law as a part of society and a means of enforcing political and social values. In the landmark book *The Bramble Bush*, Karl Llewellyn wrote: "This doing of something about disputes, this doing it reasonably, is the business of the law. And the people who have the doing of it are in charge, whether they be judges, or clerks, or jailers, or lawyers, they are officials of the law. What these officials do about disputes is, to my mind, the law itself."[4]* For Llewellyn and other American realists, anyone who wants to know about law should study the judicial process and the actors in that process. This means regular attendance at courthouses and jails, as well as scientific study of the problems associated with the legal process (e.g., plea bargaining in the courtroom).

Positivists argue that if the American realist definition of law were accepted, there would be a dangerous unpredictability to the law and legal decisions.

CRITICAL LEGAL STUDIES SCHOOL

As a contemporary extension of American legal realism, critical legal studies seek to connect what happens in the legal system to the political–economic context within which it operates. Adherents of critical legal jurisprudence believe that law reflects a cluster of beliefs that convince human beings that the hierarchical relations under which they live and work are natural and must be accommodated. According to this school, this cluster of beliefs has been constructed by elitists to rationalize their dominant power. Using economics, mass communications, religion, and, most of all, law, members of society's elite have constructed an interlocking system of beliefs that reinforces established wealth and privilege. Only by critiquing these belief structures, critical legal theorists believe, will people be able to break out of a hierarchical system and bring about democratic social and political change.

Traditional critics argue that the critical legal theorists have not developed concrete strategies to bring about the social and political changes they desire. Essentially, they have constructed only a negative position.

FEMINIST SCHOOL

There is a range of views as to what constitutes feminist jurisprudence. Most adherents of this school, believing that significant rights have been denied to women, advocate lobbying legislatures and litigating in courts for changes in

---

[3] *Roth v. United States*, 354 U.S. 476, 479 (1957).

[4] K. Llewellyn, *The Bramble Bush* (Oceana Publications, 1950), p. 12.

*Excerpt from "The Bramble Bush: On Our Law and Its Study" by Karl Nickerson Llewellyn. Published by Oxford University Press, © 1951.

laws to accommodate women's views. They argue that our traditional common law reflects a male emphasis on individual rights, which at times is at odds with women's views that the law should be more reflective of a "culture of caring." To other adherents of this school of jurisprudence, the law is a means of male oppression. For example, some feminists have argued that the First Amendment, forbidding Congress from making any laws abridging the freedom of speech, was authored by men and is presently interpreted by male-dominated U.S. courts to allow pornographers to make large profits by exploiting and degrading women.

Traditional critics of feminist jurisprudence argue that it is too narrow in scope and that it fails to account for changes taking place in U.S. society, such as the increasing number of women students in professional and graduate schools and their movement into higher-ranking positions in both the public and private sectors.

LAW AND ECONOMICS SCHOOL

The law and economics school of jurisprudence started to evolve in the 1950s, but has been applied with some rigor only for the past 30 years. It advocates using classical economic theory and empirical methods of economics to explain and predict judges' decisions in such areas as torts, contracts, property law, criminal law, administrative law, and law enforcement. The proponents of the law and economics school argue that most court decisions, and the legal doctrines on which they depend, are best understood as efforts to promote an efficient allocation of resources in society.

Critics of the school of law and economics argue that there are many schools of economic thought, and thus no single body of principles governs economics. For example, neo-Keynesians and classical market theorists have very different views of the proper role of the state in the allocation of resources. A related criticism is that this school takes a politically conservative approach to the legal solution of economic or political problems. Liberals and others argue that it is a captive of conservative thinkers.

Sources of Law

The founders of this country created in the U.S. Constitution three direct sources of law and one indirect source (see Appendix A). The legislative branch (Article I) is the maker or creator of laws; the executive branch (Article II) is the enforcer of laws; and the judicial branch (Article III) is the interpreter of laws. Each branch represents a separate source of law while performing its functions (Table 2-1). The fourth (indirect) source of law is administrative agencies, which will be briefly discussed in this chapter and examined in detail in Chapter 18.

THE LEGISLATURE AS A SOURCE OF STATUTORY LAW

Article I, Section 1, of the U.S. Constitution states, "All legislative Powers herein granted shall be vested in a Congress of the United States which shall consist of a House and Senate." It is important to understand the process by which a law (called a *statute*) is made by the Congress, because this process and its results have an impact on such diverse groups as consumers, businesspeople, taxpayers, and unions. It should be emphasized that at every stage of the process, each of the groups potentially affected seeks to influence the proposed piece of legislation through lobbying. The federal legislative process described here (Exhibit 2-3) is similar in most respects to the processes used by state legislatures, though state constitutions may prescribe some differences.

TABLE 2-1 WHERE TO FIND THE LAW

Levels of Government	Legislative Law	Executive Orders	Common Law/Judicial Interpretations	Administrative Regulations
Federal	• United States Code (U.S.C.) • United States Code Annotated (U.S.C.A.) • United States Statutes at Large (Stat.)	• Title 3 of the Code of Federal Regulations • Codification of presidential proclamations and Executive Orders	• United States Reports (U.S.) • Supreme Court Reporter (S. Ct.) • Federal Reporter (F., F.2d, F.3d) • Federal Supplement (F. Supp., F. Supp. 2d) • Federal agency reports (titled by agency; e.g., F.C.C. Reports) • Regional reporters • State reporters	• Code of Federal Regulations (C.F.R.) • Federal Register (Fed. Reg.)
State	• State code or state statutes (e.g., Ohio Revised Code Annotated, Baldwin's)	• Executive Orders of governors and proclamations		• State administrative code or state administrative regulations
Local	• Municipal ordinances		• Varies; often difficult to find. Many municipalities do not publish case decisions, but do preserve them on microfilm. Interested parties usually must contact the clerk's office at the local courthouse.	• Municipality administrative regulations

Note: Databases (e.g., Westlaw and LexisNexis) online assist in finding all sources of law listed here.

EXHIBIT 2-3
HOW A BILL BECOMES A LAW

This graphic sets out steps in the legislative process outlined in the text. Although this route is simpler, it should be noted that there are other, more complex ways for a bill to become a law. Bills are subject to amendments and changes as part of the process shown here.

When such changes are made, a compromise version of the original bill is sent back to both the House and Senate for a vote. If a compromise bill is approved, the compromise version is sent to the president of the United States for signature, or it becomes law in 10 days without the president's signature. The president may veto the bill, which may then become law only if two-thirds of the House and Senate approve it following the veto.

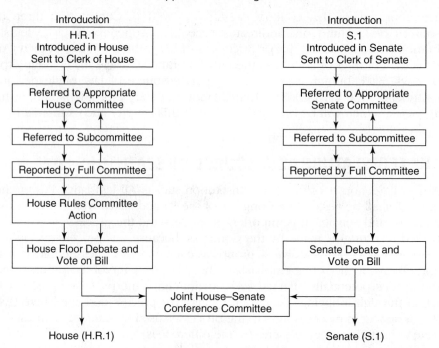

Steps in the Legislative Process

STEP 1 A bill is introduced into the U.S. House of Representatives or Senate by a single member or by several members. It is generally referred to the committee of the House or Senate that has jurisdiction over the subject matter of the bill. (In most cases, a bill is simultaneously introduced into the Senate and House. Within each body, committees may vie with each other for jurisdictional priority.)

STEP 2 Let's briefly follow through the House of Representatives a bill proposing to deregulate the trucking industry by doing away with the rate-making power of the Interstate Commerce Commission (ICC). This bill would be referred to the House Committee on Energy and Commerce, which, in turn, would refer it to the appropriate subcommittee.

STEP 3 The House subcommittee holds hearings on the bill, listening to testimony from all concerned parties and establishing a hearing record.

STEP 4 After hearings, the bill is "marked up" (drafted in precise form) and then referred to the subcommittee for a vote.

STEP 5 If the vote is affirmative, the subcommittee forwards the bill to the full House Energy and Commerce committee, which either accepts the subcommittee's recommendation, puts a hold on the bill, or rejects it. If the House committee votes to accept the bill, it reports the bill to the full House of Representatives for a vote by all members.

STEP 6 If the bill is passed by the House of Representatives and a similar bill is passed by the Senate, the bills go to a Senate–House Conference Committee to reconcile any differences in content. After compromise and reconciliation of the two bills, a single bill is reported to the full House and Senate for a vote.

STEP 7 If there is a final affirmative vote by both houses of Congress, the bill is forwarded to the president, who may sign it into law or veto it. When the president signs the bill into law, it becomes known as a *statute*, meaning it is written down and codified in the United States Code. In the event of a presidential veto, a two-thirds vote of the Senate and House membership is required to override the veto. If the president takes no action within 10 days of receiving the bill from Congress, the bill automatically becomes law without the president's signature.

The single exception to this procedure occurs when Congress adjourns before the 10-day period has elapsed: In that case, the bill would not become law. It is said to have been "pocket-vetoed" by the president: The president "stuck the bill in a pocket" and vetoed it by doing nothing. With either type of veto, the bill is dead and can be revived only by being reintroduced in the next session of Congress, in which case the procedure begins all over again.

THE JUDICIAL BRANCH AS A SOURCE OF CASE LAW

The federal courts and most state courts make up the judicial branch of government. They are charged by their respective constitutions with interpreting the constitution and statutory law on a case-by-case basis. Most case interpretations are reported in large volumes called *reporters*. These constitute a compilation of our federal and state case law.

When two parties disagree about the meaning of a statute, they bring their case to court for the court to interpret. For example, when the bill to deregulate the trucking industry and take away the rate-making function of the ICC was signed by the president and became law, two parties could have disagreed about its meaning and asked the federal courts to interpret it. If the law had been challenged, the court would first have looked at the law's legislative history to determine the intent of the legislature. This history can be found in the hearings held by the subcommittees and committees previously referred to, as well as any debates on the Senate and House floors. Hearings are published in the *U.S. Code Congressional News and Administrative Reports,* which may be ordered from the Government Printing Office or found in most university libraries in the government documents section. Debates on a bill are published in the daily *Congressional Record,* which may also be found in most university libraries.

The U.S. Supreme Court and most state supreme courts have the power of judicial review—that is, the power to determine whether a statute is constitutional. Although this power was not expressly provided for in the U.S. Constitution, the Supreme Court established it for the judiciary in the landmark case *Marbury v. Madison*[5] (see Chapter 3 for a discussion of this case). The right of judicial review gives the U.S. Supreme Court the ultimate power to check the excesses of either the legislative or the executive branch.

Furthermore, this decision establishes case law precedents, which are followed by all federal and state courts. Thus, through its case-by-case interpretation of the Constitution and statutes, the U.S. Supreme Court establishes a line of authoritative cases on a particular subject that has to be followed by the lower courts, both federal and state. Similarly, state supreme courts establish precedents that must be followed by lower courts in their particular state systems.

Case Law Precedents and the Internet. Decisions by state and federal appellate courts were (and to some extent still are) printed in volumes that were placed on (law) library shelves. Today, decisions are not always published as in the past; many are *unpublished* at the appellate level (e.g., only about 10 percent of California's appellants' decisions are published). Many decisions are *posted* (published) to online databases (e.g., Westlaw or LexisNexis).

This has led to a debate as to whether these posted decisions should be given the same precedential value as published opinions in traditional volumes. In 2006, the U.S. Supreme Court announced that it would allow lawyers, judges, and other officers of the courts to cite (refer to) unpublished opinions of *federal* courts. See Rule 32.1 of the Federal Rules of Civil Procedure. However, this rule does not specify the weight that federal courts must give to unpublished opinions. Over time (often a lengthy time), the weight to be given such opinions and their value as precedents become evident.

Restatements of the Law. Scholars writing in various areas of the law—including torts, contracts, agency, property, security, and conflicts of laws—have published summaries of the case law generally followed by the 50 states. The American Law Institute published these scholarly compilations. The Restatements are secondary sources, which in and of themselves may not have the force of law, but are often still relied upon by judges in making decisions. Throughout this text, you may see references to the Restatements (e.g., the *Restatement (Third) of Contracts*). Over a number of years, the areas referred to here have been updated to the second or third edition as the case law has evolved.

---

[5] U.S. (1 Branch) 137 (1803).

THE EXECUTIVE BRANCH AS A SOURCE OF LAW

The executive branch is composed of the president, the president's staff, and the cabinet, which is made up of the heads of each of the executive departments (e.g., the secretary of state, the secretary of labor, the secretary of defense, and the secretary of the treasury) and the counselor to the president. The Executive Office is composed of various offices, such as the Office of Management and Budget (OMB) and the Office of Personnel Management (OPM). The executive branch is a source of law in two ways.

Treaty Making. The president has the power, subject to the advice and consent of the Senate, to make treaties. These treaties become the law of the land, on the basis of the Supremacy Clause of the U.S. Constitution (Article VI), and supersede any state law. When President Carter entered into a treaty returning the Panama Canal Zone to the nation of Panama under certain conditions, it became the law of the land, and the treaty provisions superseded any federal or state laws inconsistent with the treaty.

Executive Orders. Throughout history, the president has made laws by issuing executive orders. For example, as we shall see in Chapter 18, President Reagan, by virtue of an executive order, ruled that all executive federal agencies must do a cost-benefit analysis before setting forth a proposed regulation for comment by interested parties. President Truman, by executive order, directed the secretary of commerce to seize all the nation's steel mills to prevent a strike in this essential industry during the Korean Conflict. President Johnson issued Executive Order No. 11246 requiring government contractors to set out an affirmative action plan for hiring and promoting minorities and women. (This executive order is discussed in Chapter 21.)

The executive order as a source of law is also used by state governors to deal with emergencies and budget functions. Often, a governor will call out the National Guard or, in some states, implement particular aspects of the budget by executive order. For example, a governor may order a freeze on the hiring of employees in the state university system or order an across-the-board cut in budgets in all state departments.

ADMINISTRATIVE AGENCIES AS A SOURCE OF LAW

Less well known as a source of law are the federal regulatory agencies, among which are the Securities and Exchange Commission (SEC), the Federal Trade Commission (FTC), the Equal Employment Opportunity Commission (EEOC), and the Occupational Safety and Health Administration (OSHA). Congress has delegated to these agencies the authority to make rules governing the conduct of business and labor in certain areas. This authority was delegated because it was thought to be in the public interest, convenience, and necessity. Because each of the agencies must notify the public of proposed rulemaking and set out a cost-benefit analysis, all proposed and final rules can be found in the *Federal Register*.

Administrative agencies constitute what many have called a fourth branch of government. They exist at the state and local levels as well. (See Chapter 18 on administrative law.)

Classifications of Law

Besides **statutory law** made by the legislative branch and **case law** resulting from judicial interpretation of constitutions and statutes, there are several other classifications of law that are necessary to an understanding of the legal environment of business.

statutory law Law made by the legislative branch of government.

case law Law resulting from judicial interpretations of constitutions and statutes.

CRIMINAL LAW AND CIVIL LAW

criminal law Composed of federal and state statutes prohibiting wrongful conduct ranging from murder to fraud.

Criminal law comprises those federal and state statutes that prohibit wrongful conduct such as arson, rape, murder, extortion, forgery, and fraud. The purposes of criminal law are punitive (punishing offenders by imprisonment or fines), rehabilitative (rehabilitating offenders), and restitutive (making restitution to victims). The plaintiff in a criminal case is the United States, State X, County X, or City X, representing society and the victim against the defendant, who is most likely to be an individual but may also be a corporation, partnership, or single proprietorship. The plaintiff must prove beyond a reasonable doubt that the defendant committed a crime.

Crimes are generally divided into felonies and misdemeanors. In most states, felonies are serious crimes (e.g., rape, arson, and criminal fraud) that are punishable by incarceration in a state penitentiary. Misdemeanors are less serious crimes (e.g., driving while intoxicated) that are usually punishable by shorter periods of imprisonment in a county or city jail or by fines. An act that is a misdemeanor in one state could be a felony in another state. White-collar felonies and misdemeanors are discussed in Chapter 6.

civil law Law governing litigation between two private parties.

Civil law comprises federal and state statutes governing litigation between two private parties. Neither the state nor the federal government is represented in *most* civil cases (exceptions will be pointed out in future chapters). Rather than prosecutors, there are plaintiffs, who are usually individuals or businesses suing other individuals or businesses (the defendants) to obtain compensation for an alleged breach of a private duty. For example, A, a retailer, enters into a contract with B, a manufacturer, who agrees to supply A with all the bicycles of a certain brand that the retailer can sell. A advertises and sales exceed all expectations. B refuses to ship any more bicycles, and A's customers sue him for reneging on the rain check he gave them. In turn, A sues B for breach of contract. A must show by a preponderance of evidence (a lower standard of proof than the "beyond a reasonable doubt" standard that prevails in criminal cases) that B is liable (legally obligated) to fulfill the contract. Note that A is not seeking to put B in prison or to fine B. A is seeking only to be compensated for his advertising costs, his lost sales, and what it may cost him in lawyers' fees, court costs, and damages to settle with his customers (Table 2-2).

PUBLIC AND PRIVATE LAW

public law Law dealing with the relationship of government to individual citizens.

Public law deals with the relationship of government to individual citizens. Constitutional law, criminal law, and administrative law fit this classification. Constitutional law (discussed in Chapter 5) comprises the basic principles and laws of

TABLE 2-2

COMPARISON OF CIVIL AND CRIMINAL LAW

	Civil Law	*Criminal Law*
Parties	Individual or corporate plaintiff (in most cases) versus individual or corporate defendant (in most cases)	County, city, state, or federal prosecutor versus individual or corporate defendant (in most cases)
Purpose	Compensation	Punishment
	Deference-deterrence	Deference-deterrence
		Rehabilitation
Burden of proof and sanctions	Preponderance of evidence	Beyond a reasonable doubt
	Monetary damages	Imprisonment
	Equitable terms	Fines

the nation as set forth in the U.S. Constitution. It determines the powers and obligations of the government and guarantees certain rights to citizens. Examples of questions that fall under constitutional law: Does an individual citizen have a Sixth Amendment right to counsel when stopped by a police officer, taken into custody, and interrogated? Is it cruel and unusual punishment under the Eighth Amendment to electrocute a person when that person has been found guilty of certain crimes, such as first-degree murder or killing a police officer in the line of duty? We have already touched on criminal law (which is discussed more fully in Chapter 6).

Administrative law (examined in Chapter 18) covers the process by which individuals or businesses can redress grievances against regulatory agencies such as the FTC and the SEC. It prevents the agencies from acting in an arbitrary or capricious manner and from extending their power beyond the scope that Congress has given them. For example, when the Federal Communications Commission (FCC) ruled that cable television corporations had to set aside so many channels for access by any public group that requested time, the courts reversed this FCC rule, deciding that it was beyond the agency's authority and in violation of a provision of the Federal Communications Act. Administrative law also covers the process whereby government agencies represent individuals or classes of individuals against business entities—for example, when the EEOC represents individuals alleging discrimination in pay under the provisions of the Civil Rights Act of 1964.

Private law is generally concerned with the enforcement of private duties between individuals, between an individual and a business, or between two businesses. Contracts, torts, and property law fall under this classification. Note that the government is not a concerned party in *most* private law cases.

private law Law dealing with the enforcement of private duties.

SUBSTANTIVE AND PROCEDURAL LAW

Substantive Law. Substantive law creates and regulates legal rights. For example, the rules of contract law (set out for your study in Chapters 9 and 10) determine whether an agreement between two parties is binding and, thus, an enforceable contract.

Procedural Law. Procedural law sets forth the rules for enforcing substantive rights in a court of law. In effect, procedural law defines the manner by which one obtains a remedy in a court of law. For example, when there is a possible breach of contract, the plaintiff will have to file a complaint indicating the basis for the suit, and the defendant will set forth an answer responding to the complaint, indicating why the defendant should not have to compensate the plaintiff.

CYBERLAW

Over the past 20 years, the use of the Internet to carry out commercial transactions has brought about a body of law that is largely traditional in the above categories, but often unique to cyberspace communication. Cyberlaw is not really a new type of law, but rather traditional categories (e.g., private law—contracts and torts) applied to a relatively new form of communication (online). Many chapters in this text include discussions of cyberlaw. In Chapters 9 and 10 on the law of contracts and sales, we have dedicated parts of each chapter to online applications. In Chapters 6 and 11 we see further examples of the application of traditional law to online situations involving white-collar crime and torts, respectively. Those prepared to enter business today need to know traditional laws and their application when dealing with cyberlaw issues.

Global Dimensions of the Legal Environment of Business

At the beginning of this chapter, we stated that managers need to be aware of the impact of international variables on their business. As of 2011, approximately 40 percent of all jobs in the United States depended on exports, and in the view of many experts, that percentage will soon rise to 50 percent. Additionally, many jobs are being outsourced to other countries by American corporations for cost purposes. Trade treaties will make the international dimensions of the legal environment of business increasingly important to U.S. firms. Throughout this book, therefore, we discuss the international dimensions of product liability, tort, contracts, labor, securities, and antitrust law, as well as ethics whenever appropriate. For example, current U.S. securities laws include the Foreign Corrupt Practices Act of 1977 (FCPA), as amended in 1988 and 1998. If the laws of Country X do not forbid bribery in order to obtain a $10 million contract to build an oil pipeline, should U.S. companies be constrained by the FCPA prohibitions against such bribery? Ethical and cultural relativists would say no: "When in Rome, do as the Romans do." Normative ethical theorists, such as rule utilitarians, would say yes, arguing that rules agreed upon by the world community, or a preponderance of its members, cannot be compromised by a particular situation. They would point out that both the United Nations Multinational Code and the laws of most of the UN member states prohibit bribery.

SUMMARY

The study of the legal environment of business includes the study of legal reasoning, critical thinking skills, and ethical norms; the legal and administrative law processes; selected areas of public and private law; and relevant international law. Jurisprudence is the science or philosophy of law, or law in its most generalized form. The major schools of jurisprudence are natural law, positivism, sociological, American realism, critical legal studies, feminism, and law and economics.

The three direct sources of law are the legislative (statutory), judicial (case law), and executive (executive orders) branches of government. Administrative agencies, which promulgate regulations and rules, constitute the fourth (indirect) source of law. The international dimensions of law include legal, financial, economic, and ethical variables that have an impact on business decision making.

REVIEW QUESTIONS

2-1 Contrast the positivist school's definition of law with that of the feminist school.

2-2 Explain how the critical legal studies and the feminist schools of jurisprudence are similar.

2-3 Describe how the judicial branch of government is a source of law.

2-4 What is the difference between statutory law and case law? Explain.

2-5 If the president vetoes a bill passed by Congress, is there any way the bill can become law? Explain.

2-6 Distinguish between the pairs of terms in each of these three classifications of law:
   a. public law, private law
   b. civil law, criminal law
   c. felonies, misdemeanors

REVIEW PROBLEMS

2-7 Three men are trapped in a cave with no hope of rescue and no food. They roll dice to determine who will be killed and eaten by the others so that some may survive. The two survivors are unexpectedly rescued 10 days later and tried for murder. Judge A finds them guilty, saying that the unjustifiable killing of another is against the homicide laws of State X. He bases his decision solely on statutory law and case precedents interpreting the law. To which school of legal thought does Judge A belong? Explain.

2-8 Basing his decision on the same set of facts as given in Problem 2-7, Judge B rules that the survivors are not guilty because they were cut off from all civilized life, and in such a situation, the laws of nature apply, not manmade laws. To which school of legal thought does Judge B belong? Explain.

2-9 Basing her decision on the same set of facts as given in Problem 2-7, Judge C rules that the two survivors are not guilty because, according to a scientific survey of the community by a professional polling organization, the public believes that the survivors' actions were defensible. To which school of legal thought does Judge C belong? Explain.

2-10 Suppose that the California legislature passes a law that severely restricts carbon dioxide emissions from automobiles in that state. A group of automobile manufacturers files suit against the state of California to prevent the enforcement of the law. The automakers claim that a federal law already sets fuel economy standards nationwide and that fuel economy standards are essentially the same as carbon dioxide emission standards. According to the automobile manufacturers, it is unfair to allow California to impose more stringent regulations than those set by the federal law. State tort or case law? Which school of thought?

2-11 Madison and his adult son lived in a house owned by Madison. At the request of the son, Marshall painted the house. Madison did not authorize the work, but he knew that it was being done and raised no objection. However, Madison refused to pay Marshall, arguing that he had not contracted to have the house painted.

Marshall asked his attorney if Madison was legally liable to pay him. The attorney told Marshall that, in their state, several appellate court opinions had established that when a homeowner allows work to be done on his home by a person who would ordinarily expect to be paid, a duty to pay exists. The attorney stated that, on the basis of these precedents, it was advisable for Marshall to bring a suit to collect the reasonable value of the work he had done. Explain what the attorney meant by precedent and why the fact that precedent existed was significant.

2-12 Smith was involved in litigation in California. She lost her case in the trial court. She appealed to the California appellate court, arguing that the trial court judge had incorrectly excluded certain evidence. To support her argument, she cited rulings by the Supreme Court of North Dakota and the Supreme Court of Ohio. Both the North Dakota and Ohio cases involved facts that were similar to those in Smith's case. Does the California court have to follow the decisions from North Dakota and Ohio? Support your answer.

CASE PROBLEMS

2-13 In 2006, Myspace, an Internet website for young adult social networking, was sued by the Universal Music Group, Inc., for copyright infringement. Basically, Myspace was allowing users to upload musical content such as videos and songs to their profiles. Myspace was somewhat protected under a federal law that makes such practices illegal only if the copyright holder actually complains and the content is not subsequently removed from the site. Universal Music Group brought suit in an effort to change that law and claimed that such practices were still copyright infringement. Were the practices of Myspace copyright infringement? How would a judge's legal philosophy affect how he or she would rule in this case? *UMG Recordings Inc. v. Myspace, Inc.,* 526 F. Supp. 2d 1046; U.S. Dist. LEXIS 91179 (C.D. Cal. 2007).

2-14 Three same-sex couples who are residents of Vermont have lived together in committed relationships for a period. Two of the couples have raised children together. All three couples applied for marriage licenses and were refused a license on the grounds that they were ineligible under the state marriage laws. Plaintiffs sought a declaratory judgment that the refusal to issue them a license violated the marriage statutes and the Vermont constitution. They argued that it violated the Common Benefits Clause of the Vermont constitution, which provides "[t]hat government is, or ought to be, instituted for the common benefit, protection, and security of the people, nation, or community, and not for the particular emolument or advantage of a single person, family, or set of persons, who are part of that community. . . ." They argued that in not having access to a civil marriage license, they are denied many legal benefits and protections, including coverage under a spouse's medical, life, and disability insurance; hospital visitation and other medical decision-making privileges; and spousal support. Argue whether Vermont's marriage license law violates the same-sex couples' rights under the Vermont constitution. Which school of jurisprudence would you apply in your reasoning? *Baker v. State of Vermont*, 170 Vt. 194; 744 A.2d 864; 1999 Vt. LEXIS 406.

2-15 Various merchants entered into an agreement with American Express that required all disputes to be resolved by arbitration. The agreement also specifically provided that no claims could be arbitrated on a class-action basis. Despite the class-action waiver, the merchants instituted a class-action lawsuit against American Express, alleging violations of certain federal antitrust laws (the Sherman Act and the Clayton Act). They claimed that American Express had used its "monopoly power in the market for charge cards [cards requiring payment in full at the end of each billing cycle] to force merchants to accept credit cards [cards requiring payment of only a portion of the balance each billing cycle, with interest charges added] at rates approximately 30% higher than the fees for competing credit cards." (This is termed a *tying arrangement*.) American Express responded by moving to compel individual arbitration under the provisions of the Federal Arbitration Act (FAA). The merchants sought to invalidate the ban on class arbitration by introducing evidence that (1) the cost of proving the antitrust claims via expert testimony would range from "at least" several hundred thousand dollars to more than one million dollars while (2) the maximum recovery for an individual plaintiff would be only about $12,500 (or about $38,500 if the plaintiffs were awarded treble damages, which they sought). Who won? Explain. *American Express v. Colons Italian Restaurant* 133 S. Ct. 2304 (2013).

2-16 The United States age discrimination law protects individuals beginning at the age of 40. However, this law has specific requirements regarding whether an employer or the employee determines whether age-related factors were involved with a termination or layoff. Prior to this case, the employer who let go an individual 40 years of age or more had the burden of saying the separation was based on a reasonable factor unrelated to age. However, during this case in 2008, 28 individuals at or above the age of 40 were laid off, and in turn sued the employer for age discrimination. These individuals argued that instead of the employer having the burden, and thus being able to argue that there was a reasonable factor unrelated to age, the individuals being let go should have the burden. This way, the individuals would then be able to argue in court that there was no "reasonable" factor other than age involved in the individuals' terminations from their jobs. In this case, did the Supreme Court agree with the individuals? Currently, who has the burden in such cases, the employer or the employees? *Meacham v. Knolls Atomic Power Laboratory*, 552 U.S. 1306, 128 S. Ct. 1764, 170 L. Ed. 2d 558; 2008 U.S. LEXIS 3090; 76 U.S.L.W. 3554 (2008).

2-17 A&M Records, plaintiffs, are in the business of the commercial recording, distribution, and sale of copyrighted musical compositions and sound recordings. It filed suit against Napster Inc. (Napster) as a contributory and vicarious copyright infringer. Napster operates an online service for "peer-to-peer file sharing" (www.Napster.com) so that users can, free of charge, download recordings via the Internet through a process known as "ripping," which is the downloading of digital MP3 files. *MP3* is the abbreviated term for audio recordings in a digital format known as MPEG-3. Napster's online service provides a search vehicle for files stored on others' computers and permits the downloading of the recordings from the hard drives of other Napster users. Napster provides technical support as well as a chat room for users to exchange information. The result is that users, who register and have a password through Napster, could download single songs and complete CDs or albums via the peer-to-peer file sharing. The district court granted a preliminary injunction to the plaintiffs, enjoining Napster from "engaging in, or facilitating others in copying, downloading, uploading, transmitting, or distributing plaintiffs' copyrighted musical compositions and sound recordings, protected by either federal or state law, without express permission of the rights owner." Who won? *A&M Records v. Napster*, 239 F.3d 1004 (9th Cir. 2001).

2-18 In 2007, two fair housing groups in California sued a website that helps people find roommates, called Roommate.com. The housing groups alleged that the website was in clear violation of the Fair Housing Act. Apparently, the website violated the act because the website allowed users to discriminate among people and who individuals were able to live with based on religion, sexual preference, gender, and so on. However, the website argued that its practices were protected under the Communication Decency Act. In other words, the website stated that the users were freely describing their own wants, and not the views or opinions of the website itself. However, the court took note that the website provided menus for users to choose preferences among religions, race, gender, and the like. How did the court decide in this case? Explain how the legal philosophy one adheres to would make one more or less supportive of the court's ruling. *Fair Housing Council of San Fernando Valley v. Roommate.com LLC,* 521 F.3d 1157; 2008 U.S. App. LEXIS 7066 (9th Cir. 2008).

THINKING CRITICALLY ABOUT RELEVANT LEGAL ISSUES

Although there is debate over the various schools of jurisprudence, not all options are of equal merit. In that the law is meant to offer protection and to guide society, not all philosophies can best achieve this desired outcome. To have a just legal system, laws must be based on absolute principles that provide clarity in the prescribed rules to follow, as well as justice in the result of following the laws. Therefore, the natural law school best provides for the maintenance of law, order, and justice in society.

One advantage of the natural law school is the acknowledgment of the black-and-white nature of legal issues. When someone commits a crime and harms another, one party is wrong and the other is harmed because of the wrong. Certain actions, such as murder, are simply wrong acts that are never permissible. The natural law school of jurisprudence readily recognizes moral absolutes and seeks to create a legal system around these absolutes, ultimately strengthening the resulting laws. Good and evil exist, and natural law sides with the good against the evil.

The basis in moral absolutes grounds natural law in the pursuit of the right and the good. These moral absolutes exist and are available to those who study and think about what is right and just. People, by considering these moral absolutes, can come upon the naturally right code of conduct, and make laws to ensure that people will live up to this naturally right code of conduct. No other school of jurisprudence adequately tells people the proper way to conduct their lives. After all, the role of the law is to maintain peace and justice in society by creating the laws that best channel people toward following right actions and avoiding wrong actions.

Only an application of natural law jurisprudence can guide society for the good of all.

In addition to prescribing proper conduct for citizens, law grounded in moral absolutes can avoid subjective approaches to laws. A quick review of almost any legal issue will demonstrate that judges and lawyers do not always agree upon what a law means. When laws are firmly grounded in moral absolutes, however, the subjective element of the law is removed. No longer would judges need to ponder over what a law means and when it applies. Instead, judges would have to look at the law and determine the relevant moral truth the law upholds or on which the law is based. By enforcing the moral absolutes underlying the law, judges would no longer apply their subjective beliefs to laws, and instead would create a more consistent and predictable legal system.

1. How would you frame the issue and conclusion of this essay?

2. What is the primary ethical norm underlying the author's argument?

3. Does the argument contain significant ambiguity in the reasoning?

    Clue: Which word or phrases could have multiple meanings, where changing the meaning used either strengthens or weakens the argument?

4. Write an essay that someone who holds an opinion opposite to that of the essay author might write.

    Clue: How might reasonable people disagree with the author's conclusion? See Exhibit 1-2 in Chapter 1.

ASSIGNMENT ON THE INTERNET

This chapter introduces you to seven different schools of jurisprudence, each with distinct elements. Yet, the various schools also share a number of similarities that often blur the lines separating one from the other. Using the Internet, research at least two of the schools of jurisprudence discussed here to go beyond the information provided in this chapter. Then apply the critical thinking skills highlighted in Chapter 1 to compare the two schools you researched. How are they similar? How are they different?

For example, if you wanted to compare the critical legal studies school to the feminist school, you could begin by visiting this page on critical legal theory: www.law.cornell.edu/, then search for the topic(s) you are interested in.

ON THE INTERNET

www.seanet.com/~rod/ Learn about the jurisprudence of access to justice from this page. Once at this page, click on "Books and Articles," then go to "The Jurisprudence of Access to Justice: From *Magna Carta* to *Romer v. Evans* via *Marbury v. Madison*" by Leonard W. Schroeter.

www.iep.utm.edu The various components of feminist jurisprudence, discussed in greater detail, can be found at this site, along with reading recommendations for further study. Search "feminist jurisprudence" in the search bar, then go to the first link, titled "Feminist Jurisprudence."

www.fact-index.com/ Seven schools of legal thought or jurisprudence are discussed in this chapter, but other theories of jurisprudence exist. This site provides an overview of virtue jurisprudence. Search "virtue," then go to the second link, entitled "virtue jurisprudence."

www.archives.gov/ This is the website of the *Federal Register,* which allows you to search and read executive orders. At the bottom of the page, find "Publications," then click on "Federal Register."

FOR FUTURE READING

Elias, Stephen, and Susan Levinkind. *Legal Research: How to Find & Understand the Law.* Berkeley, CA: Nolo Press, 2007.

Fuller, Lon. "The Case of the Speluncean Explorers." *Harvard Law Review* 62, no. 4 (1949): 616-45.

Learned, E. P., K. R. Christensen, and W. D. Guth. *Business: The Uniform Commercial Code.* St. Paul, MN: West Publishing Company, 2004. This book is significant to an understanding of Chapters 9 and 10 of this text.

Murphy, Jeffrie, and Jules Coleman. *Philosophy of Law: An Introduction to Jurisprudence.* Boulder, CO: Westview Press, 1990.

Suber, Peter. *The Case of the Speluncean Explorers: Nine New Opinions.* 1998. Reprint, New York: Routledge, 2002.

CHAPTER THREE

The American Legal System

- JURISDICTION
- VENUE
- THE STRUCTURE OF THE COURT SYSTEM
- THE ACTORS IN THE LEGAL SYSTEM AND THEIR RELATIONSHIP TO THE BUSINESS COMMUNITY
- THE ADVERSARY PROCESS
- STEPS IN CIVIL LITIGATION AND THE ROLE OF BUSINESSPERSONS
- GLOBAL DIMENSIONS OF THE AMERICAN LEGAL SYSTEM

We are all subject to both state and federal laws. Under our dual court system, all lawsuits must be brought in either the federal or the state court system. In some cases, an action may be brought in either. Thus, it is important that those in the business community understand how the decisions are made as to which court system can resolve their grievances. This chapter first considers the principles that determine which court system has the power to hear various types of cases and then examines in greater detail the structure of the two basic divisions of our dual court system. Next, it focuses on the primary actors who play major roles in our litigation process. Finally, it examines the philosophy behind our American legal system and traces the procedures that must be followed when using one of our courts.

Jurisdiction

The concept of jurisdiction is exceedingly simple, yet at the same time exceedingly complex. At its simplest level, **jurisdiction** is the power of the courts to hear a case and render a decision that is binding on the parties. Jurisdiction is complex, however, because there are several types of jurisdiction that a court must have if it is to hear a case.

jurisdiction The power of a court to hear a case and render a binding decision.

ORIGINAL VERSUS APPELLATE JURISDICTION

Perhaps the simplest type of jurisdiction to understand is the distinction between original and appellate jurisdiction, which refers to the role the court plays in the judicial hierarchy. A court of **original jurisdiction**, usually referred to as a *trial court*, has the power to initially hear and decide a case. It is in the court of original jurisdiction that a case originates; hence its name.

A court with **appellate jurisdiction** has the power to review a previously made decision to determine whether the trial court erred in making its initial decision.

original jurisdiction The power to initially hear and decide (try) a case.

appellate jurisdiction The power to review a decision previously made by a trial court.

35

CRITICAL THINKING ABOUT THE LAW

Our American legal system can seem confusing at first. Using your critical thinking skills to answer the following questions as you read this chapter will help you understand how our legal system operates.

1. Critical thinkers recognize that ambiguous words—words that have multiple possible meanings—can cause confusion. Sam boldly asserts that the court of common pleas has jurisdiction over *Jones v. Smith*, while Clara asserts equally strongly that the court of common pleas does not have jurisdiction over the case. Explain the ambiguity that allows these two apparently contradictory statements to both be true.

    Clue: Is it possible for a court to have one type of jurisdiction and not another?

2. Our legal system contains numerous procedural requirements. Which of the primary values is furthered by these requirements?

    Clue: Review the four primary values described in Chapter 1.

3. Many say that the adversary system is consistent with the American culture. What value that is furthered by our adversary system is important to our culture?

    Clue: Can you go beyond the four primary values described in Chapter 1 and think of any other important values?

JURISDICTION OVER PERSONS AND PROPERTY

in personam jurisdiction Jurisdiction over the person; the power of a court to render a decision that affects the legal rights of a specific person.

plaintiff Party on whose behalf the complaint is filed.

defendant Party against whom an action is being brought.

complaint The initial pleading in a case that states the names of the parties to the action, the basis for the court's subject matter jurisdiction, the facts on which the party's claim is based, and the relief that the party is seeking.

summons Order by a court to appear before it at a certain time and place.

service Providing the defendant with a summons and a copy of the complaint.

Before the court can render a decision affecting a person, the court must have **in personam jurisdiction** (jurisdiction over the person). In personam jurisdiction is the power to render a decision affecting the specific persons before the court. When a person files a lawsuit, that person, called the **plaintiff**, gives the court in personam jurisdiction over him or her. By filing a case, the plaintiff is asking the court to make a ruling affecting his or her rights. The court must acquire jurisdiction over the party being sued, the **defendant**, by serving him or her with a copy of the plaintiff's complaint and a summons. The **complaint**, discussed in more detail later in this chapter, is a detailed statement of the basis for the plaintiff's lawsuit and the relief being sought. The **summons** is an order of the court notifying the defendant of the pending case and telling him or her how and when to respond to the complaint.

Personal **service**, whereby a sheriff or other person appointed by the court hands the summons and complaint to the defendant, has been the traditional method of service. Today, other types of service are more common. Residential service may be used, whereby the summons and complaint are left by the representative of the court with a responsible adult at the home of the defendant. Certified mail or, in some cases, ordinary mail are also used to serve defendants. Once the defendant has been properly served, the court has in personam jurisdiction over him or her and may render a decision affecting his or her legal rights, regardless of whether the defendant responds to the complaint.

When one thinks about how the rules of service would apply to a suit against a corporation, the question arises: How do you serve a corporation? The legal system has solved this question. Most states require that corporations appoint an agent for service when they are incorporated. This agent is a person who has been given the legal authority to receive service for the corporation. Once the agent has been served, the corporation is served. In most states, service on the president of the corporation also constitutes service on the corporation.

A court's power is generally limited to the borders of the state in which it is located. So, traditionally, a defendant had to be served within the state in which the court was located in order for the court to acquire jurisdiction over the person of the defendant. This restriction imposed severe hardships when a defendant who lived in one state entered another state and injured the plaintiff. If the

TECHNOLOGY AND THE LAW

Electronic Service of Process?

While the federal courts have updated their practices significantly to make use of new technologies by doing such things as having websites for federal courts, allowing federal filing, and allowing Internet streaming of court coverage, there is one surprising gap in the federal system's use of technology: there is no provision in the Federal Rules of Evidence for regular use of electronic means for service of process on defendants.

In 2012, in the case of *Chase Bank USA v. Fortunato*,[1] a district court judge refused the application of Chase Bank to authorize either email or Facebook message as an alternative means of service on a third party defendant who had used false addresses to fraudulently secure multiple credit cards and whom the bank had been unable to locate physically. In denying the request, the court said that the bank had had not given the court "a degree of certainty" about the defendant's alleged Facebook profile and the email address attached to that profile that would ensure that the defendant would receive and read the notice. Oddly, the court did allow notice by publication "in two newspapers, at least one in the English language, designated in the order as most likely to give notice to the person to be served, for a specified time, at least once in each of four successive weeks."

The federal courts however, on occasion, have allowed electronic service as an alternative in international cases under Fed. R. Civ. p. 4(f)(3), which governs service internationally. For example, in 2001, the court allowed service of Osama Bin Laden and the Taliban via television (including the Turkish version of CNN International and BBC World) for deaths resulting from the 9/11 attacks.

There have also been a limited number of states that have authorized electronic service under limited circumstances. For example, a Minnesota state court in May 2011 allowed service of process via email, Facebook, MySpace, or any other social networking site in a divorce proceeding, when the wife had not seen her husband in years, and believed he had returned to Africa's Ivory Coast.

South Carolina permits electronic service of process only on corporations and partnerships (but not individuals), and all registered corporations and partnerships are required to register an email address with the Secretary of State (SCRCP 4(d)(3) 4(d)(8) and 5(b)(1)). To use electronic service of process in South Carolina under this program, however, a party cannot just send an email to a potential defendant; it must utilize a "certifying" authority such as the U.S. Postal Service's "Electronic Post Mark" or "EPM."

---

[1]Not Reported in F.Supp.2d, 2012 WL 2086950 (S.D.N.Y.).

---

defendant never again entered the plaintiff's state, the plaintiff could bring an action against the defendant only in the state in which the defendant lived. Obviously, this restriction would prevent many legitimate actions from being filed.

To alleviate this problem, most states enacted **long-arm statutes**, which enable the court to serve a defendant outside the state as long as the defendant has engaged in certain acts within the state. Those acts vary from state to state, but most statutes include such acts as committing a tort within the state or doing business within the state. The following case demonstrates the application of such a statute.

long-arm statute A statute authorizing a state court to obtain jurisdiction over an out-of-state defendant when that party has sufficient minimum contacts with the state.

CASE 3-1

World-Wide Volkswagen Corp. v. Woodson, District Judge of Cook County
Supreme Court of the United States
444 U.S. 286 (1980)

Mr. and Mrs. Robinson, the plaintiffs in the original case, filed a product liability action against defendant World-Wide Volkswagen in a state court in Oklahoma to collect compensation for damages they incurred as a result of an accident involving an automobile they had purchased in New York. The defendants in that case, the retailer and the wholesaler of the car, were both New York corporations.

Defendants claimed that the Oklahoma court could not exercise jurisdiction over them because they were nonresidents and they lacked sufficient "minimum contacts" with the state to be subject to its in personam jurisdiction.

The trial court rejected defendant petitioner's claims. The Oklahoma Supreme Court likewise rejected their claims, and so they petitioned the U.S. Supreme Court. Note that the case that went to the Supreme Court is against the trial court, because the issue on appeal is whether the trial court acted properly in asserting jurisdiction.

Justice White

The issue before us is whether, consistently with the Due Process Clause of the Fourteenth Amendment, an Oklahoma court may exercise in personam jurisdiction over a nonresident automobile retailer and its wholesale distributor in a products liability action, when the defendants' only connection with Oklahoma is the fact that an automobile sold in New York to New York residents became involved in an accident in Oklahoma.

As has long been settled, and as we reaffirm today, a state court may exercise personal jurisdiction over a nonresident defendant only so long as there exist "minimum contacts" between the defendant and the forum State. The concept of minimum contacts, in turn, can be seen to perform two related, but distinguishable, functions. It protects the defendant against the burdens of litigating in a distant or inconvenient forum. And it acts to ensure that the States, through their courts, do not reach out beyond the limits imposed on them by their status as coequal sovereigns in a federal system.

The protection against inconvenient litigation is typically described in terms of "reasonableness" or "fairness." We have said that the defendant's contacts with the forum State must be such that maintenance of the suit "does not offend 'traditional notions of fair play and substantial justice'.?"

The limits imposed on state jurisdiction by the Due Process Clause, in its role as a guarantor against inconvenient litigation, have been substantially relaxed over the years. This trend is largely attributable to a fundamental transformation in the American economy.

Today many commercial transactions touch two or more States and may involve parties separated by the full continent. With this increasing nationalization of commerce has come a great increase in the amount of business conducted by mail across state lines. At the same time modern transportation and communication have made it much less burdensome for a party sued to defend himself in a State where he engages in economic activity.

Nevertheless, we have never accepted the proposition that state lines are irrelevant for jurisdictional purposes, nor could we, and remain faithful to the principles of interstate federalism embodied in the Constitution.

Applying these principles to the case at hand, we find in the record before us a total absence of those affiliating circumstances that are a necessary predicate to any exercise of state court jurisdiction. Petitioners carry on no activity whatsoever in Oklahoma. They close no sales and perform no services there. They avail themselves of none of the privileges and benefits of Oklahoma law. They solicit no business there either through salespersons or through advertising reasonably calculated to reach the State; nor does the record show that they regularly sell cars at wholesale or retail to Oklahoma customers or residents or that they indirectly, through others, serve or seek to serve the Oklahoma market. In short, respondents seek to base jurisdiction on one isolated occurrence and whatever inferences can be drawn therefrom: The fortuitous circumstance that a single Audi automobile sold in New York to New York residents happened to suffer an accident while passing through Oklahoma.

It is argued, however, that because an automobile is mobile by its very design and purpose it was "foreseeable" that the Robinsons' Audi would cause injury in Oklahoma. Yet "foreseeability" alone has never been a sufficient benchmark for personal jurisdiction under the Due Process Clause.

If foreseeability were the criterion, a local California tire retailer could be forced to defend in Pennsylvania when a blowout occurs there, a Wisconsin seller of a defective automobile jack could be hauled before a distant court for damage caused in New Jersey, or a Florida soft-drink concessionaire could be summoned to Alaska to account for injuries happening there.

This is not to say, of course, that foreseeability is wholly irrelevant. But the foreseeability that is critical to due process analysis is not the mere likelihood that a product will find its way into the forum State. Rather, it is that the defendant's conduct and connection with the forum State are such that he should reasonably anticipate being hauled into court there. When a corporation "purposefully avails itself of the privilege of conducting activities within the forum State," it has clear notice that it is subject to suit there and can act to alleviate the risk of burdensome litigation by procuring insurance, passing the expected costs on to customers, or, if the risks are too great, severing its connection with the State. Hence, if the sale of a product of a manufacturer or distributor such as Audi or Volkswagen is not simply an isolated occurrence but arises from the efforts of the manufacturer or distributor to serve directly or indirectly the market for its product in other States, it is not unreasonable to subject it to suit in one of those States if its allegedly defective merchandise has there been the source of injury to its owner or to others.

But there is no such or similar basis for Oklahoma jurisdiction over World-Wide or Seaway in this case. Seaway's sales are made in Massena, New York. World-Wide's market, although substantially larger, is limited to dealers in New York, New Jersey, and Connecticut. There is no evidence of record that any automobiles distributed by World-Wide are sold to retail customers outside this tristate area. It is foreseeable that the purchasers of automobiles sold by World-Wide and Seaway may take them to Oklahoma. But the mere "unilateral activity of those who claim some relationship with a nonresident defendant cannot satisfy the requirement of contact with the forum State."*

Reversed in favor of World-Wide Volkswagen Corporation

Contrast the facts in the foregoing case with those in the 2010 case of *Southern Prestige Industries, Inc. v. Independence Plating Co.*,[2] in which the court came to a contrary decision. In this case, the plaintiff, a North Carolina company, filed an action for breach of contract in a North Carolina court against the defendant, a company whose only offices and personnel were located in New Jersey. Both parties admitted that they had an ongoing business relationship that included 32 purchase orders over an eight-month period. In accordance with these orders, the plaintiff would ship parts to the defendant, who would oxidize them and ship them back. The defendant filed a motion to dismiss on grounds that there were insufficient minimum contacts to satisfy due process. In upholding the trial court's denial of the defendant's motion, the appellate court said that in order to establish the minimum contacts between the nonresident defendant and the forum state, such that the maintenance of the suit does not offend "traditional notions of fair play and substantial justice," "the defendant must have purposefully availed itself of the privilege of conducting activities within the forum state and invoked the benefits and protections of the laws of North Carolina. The relationship between the defendant and the forum state must be such that the defendant should reasonably anticipate being hauled into a North Carolina court."[3]

To determine whether those minimum contacts existed, the court looked at "(1) the quantity of the contacts, (2) the nature and quality of the contacts, (3) the source and connection of the cause of action to the contacts, (4) the interest of the forum state, and (5) the convenience to the parties."[4]† Using those factors, the court determined that the defendant had indeed availed itself of the benefits of doing business in North Carolina and should have anticipated being hauled into a North Carolina court.

APPLYING THE LAW TO THE FACTS...

Now that you have seen cases that have come to different conclusions, how would you apply the law to the following set of facts? A California resident files suit against Harrah's and several other Nevada casino operators in a California state court for unfair competition, breach of contract, and false advertising. The defendants advertised their casinos extensively in California, had an interactive website and also offered promotional activities for California residents. Harrah's also had a subsidiary marketing corporation that maintained offices in California to "assist customers from California who contacted them" and "attempted to attract a limited number of high-end gaming patrons to Harrah's properties." The district court initially dismissed the case due to lack of personal jurisdiction. How would you apply the law to the facts and decide the case if you were on the court of appeals? (*Snowney v. Harrah's Entertainment, Inc.*, 11 Cal. Rptr. 3d 35 [Cal. Ct. App. 2004].)

---

[2]690 S.E.2d 768 (2010).
[3]*Id.*
[4]*Id.*

*World-Wide Volkswagen Corp. v. Woodson, District Judge of Cook County. Supreme Court of the United States 444 U.S. 286 (1980).

†Southern Prestige Industries, Inc. v. Independence Plating Corporation, Court of Appeals of North Carolina.

In Rem Jurisdiction. If a defendant has property within a state, the plaintiff may seek to bring the action directly against the property rather than against the owner. For example, if a Michigan defendant owned land in Idaho on which taxes had not been paid for 10 years, the state could bring an action to recover those taxes. The Idaho court would have **in rem jurisdiction** over the property and, in an in rem proceeding, could order the property sold to pay the taxes. Such proceedings are often used when the owner of the property cannot be located for personal service.

in rem jurisdiction The power of a court to render a decision that affects property directly rather than the owner of the property.

SUBJECT MATTER JURISDICTION

One of the most important types of jurisdiction is **subject matter jurisdiction**, the power of the court to hear certain kinds of cases. Subject matter jurisdiction is extremely important because if a judge renders a decision in a case over which the court does not have subject matter jurisdiction, the decision is void or meaningless. The parties cannot give the court subject matter jurisdiction. It is granted by law, as described in the subsequent sections.

subject matter jurisdiction The power of a court to render a decision in a particular type of case.

At the beginning of this chapter, you learned that the United States has a dual court system, comprised of both a state and a federal system. The choice of the system in which to file a case is not purely a matter of deciding which forum is most convenient or which judge would be most sympathetic. Subject matter jurisdiction determines which court may hear the case. When you think about the concept of subject matter jurisdiction, it is easiest to think of it in two steps. First, within which court system does the case fall? Once you know which court system has jurisdiction over the case, you then need to ask whether there is a special court within that system that hears that specific type of case. When asking which court system has subject matter jurisdiction, there are three possible answers: state jurisdiction, exclusive federal jurisdiction, or concurrent federal jurisdiction (Exhibit 3-1).

state court jurisdiction Applies to cases that may be heard only in the state court system.

State Jurisdiction. The state court system has subject matter jurisdiction over all cases not within the exclusive jurisdiction of the federal court system. Only a very limited number of cases fall within the exclusive jurisdiction of the federal courts. Consequently, almost all cases fall within the **state court jurisdiction**.

EXHIBIT 3-1

SUBJECT MATTER JURISDICTION

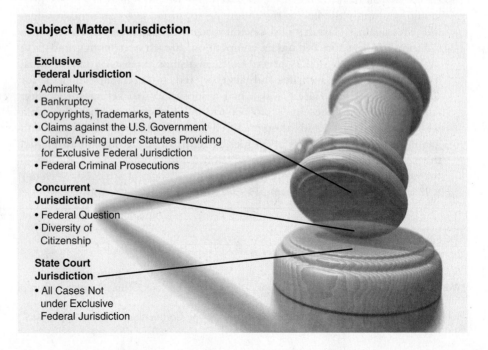

Subject Matter Jurisdiction

Exclusive Federal Jurisdiction
- Admiralty
- Bankruptcy
- Copyrights, Trademarks, Patents
- Claims against the U.S. Government
- Claims Arising under Statutes Providing for Exclusive Federal Jurisdiction
- Federal Criminal Prosecutions

Concurrent Jurisdiction
- Federal Question
- Diversity of Citizenship

State Court Jurisdiction
- All Cases Not under Exclusive Federal Jurisdiction

Suits for breach of contract, product liability actions, and divorces are just a few of the types of cases falling within the state court system's jurisdiction. However, it is important to remember that when we say that the state court system has jurisdiction, it is not just any state's court system, but only the court system of the state whose law will be used to decide the case. So if there is a breach of contract case arising out of a contract made in Ohio, Ohio's law would govern the contract, so the state court system of Ohio would have jurisdiction over the case.

Exclusive Federal Jurisdiction. A few types of cases may be heard only in the federal courts. Such cases are within the exclusive jurisdiction of the federal court system. If these cases were tried in a state court, any decision rendered by the judge would be void. Cases that fall within the exclusive jurisdiction of the federal courts include such matters as admiralty, bankruptcy, federal criminal prosecutions, claims against the United States, and claims arising under those federal statutes that include a provision for **exclusive federal jurisdiction**. Many of these latter cases are of particular concern to businesspeople. For example, one statute that gives exclusive jurisdiction to the federal court system is the National Environmental Policy Act, discussed in Chapter 22. Cases brought under this act *must* be filed in a federal district court.

exclusive federal jurisdiction Applies to cases that may be heard only in the federal court system.

Concurrent Federal Jurisdiction. Many cases may be heard in either a federal or a state court. These cases are said to fall within the federal court's **concurrent jurisdiction**, meaning that both court systems have jurisdiction, so the plaintiff may file in the trial court of either system. There are two types of such cases. The first are federal question cases. If a case requires an interpretation of the U.S. Constitution, a federal statute, or a federal treaty, it is said to involve a federal question and may be heard in either state or federal court. Many people make the mistake of thinking that when a person believes his or her rights under the federal Constitution have been violated, the case must go to the federal courts. They are wrong. Such a case involves a federal question and is, therefore, within the concurrent jurisdiction of both court systems.

concurrent jurisdiction Applies to cases that may be heard in either the federal or the state court system.

The second means by which a case may fall within the federal court's concurrent jurisdiction is through diversity of citizenship. If the opponents in a case are from different states, there is said to be *diversity of citizenship*. The diversity must be complete. If any two parties on opposing sides reside in the same state, diversity is lost. For example, if the plaintiff is an Ohio resident and one of the defendants lives in Michigan and the other in Indiana, diversity exists. If, however, an Ohio plaintiff is bringing an action against a Michigan defendant and an Ohio defendant, there is not complete diversity and therefore no concurrent federal jurisdiction. When the basis for federal jurisdiction is diversity of citizenship, an amount in excess of $75,000 must be in controversy.

When one of the parties to a case is a corporation, there is sometimes a question as to which state constitutes the corporation's residence. In Case 3-2, the U.S. Supreme Court explains how to solve that issue.

CASE 3-2

Hertz Corporation v. Friend
United States Supreme Court
130 S. Ct. 1181 (2010)

Plaintiffs, California citizens, sued Hertz Corporation for state law violations in a California State Court. Defendant Hertz filed a motion to remove the case to federal court on diversity of citizenship grounds, claiming that it was not a resident of California, the residence of all the plaintiffs. Plaintiffs argued that Hertz was a California citizen, like themselves, and that, hence, diversity jurisdiction was lacking under

§ 1332(c)(1), which provides that "a corporation shall be deemed to be a citizen of any State by which it has been incorporated and of the State where it has its principal place of business."

To show that its "principal place of business" was in New Jersey, not California, Hertz submitted a declaration stating, among other things, that it operated facilities in 44 States, that California accounted for only a portion of its business activity, that its leadership is at its corporate headquarters in New Jersey, and that its core executive and administrative functions are primarily carried out there. The District Court concluded that it lacked diversity jurisdiction because Hertz was a California citizen under Ninth Circuit precedent, which asks, *inter alia*, whether the amount of the corporation's business activity is "significantly larger" or "substantially predominates" in one State. Finding that California was Hertz's "principal place of business" under that test because a plurality of the relevant business activity occurred there, the District Court remanded the case to the state court. The Ninth Circuit affirmed. Hertz appealed to the United States Supreme Court.

Justice Breyer

The federal diversity jurisdiction statute provides that "a corporation shall be deemed to be a citizen of any State by which it has been incorporated *and of the State where it has its principal place of business*." We seek here to resolve different interpretations that the Circuits have given this phrase. In doing so, we place primary weight upon the need for judicial administration of a jurisdictional statute to remain as simple as possible. And we conclude that the phrase "principal place of business" refers to the place where the corporation's high level officers direct, control, and coordinate the corporation's activities. Lower federal courts have often metaphorically called that place the corporation's "nerve center."

We begin our "principal place of business" discussion with a brief review of relevant history. The Constitution provides that the "judicial Power shall extend" to "Controversies . . . between Citizens of different States." . . . In 1928 this Court made clear that the "state of incorporation" rule was virtually absolute. It held that a corporation closely identified with State A could proceed in a federal court located in that State as long as the corporation had filed its incorporation papers in State B, perhaps a State where the corporation did no business at all. . . . Subsequently, many in Congress and those who testified before it pointed out that this interpretation was at odds with diversity jurisdiction's basic rationale, namely, opening the federal courts' doors to those who might otherwise suffer from local prejudice against out-of-state parties. . . . [i]n 1958, Congress both codified the courts' traditional place of incorporation test and also enacted into law a slightly modified version of the Conference Committee's proposed "principal place of business" language. A corporation was to "be deemed a citizen of any State by which it has been incorporated and of the State where it has its principal place of business."

The phrase "principal place of business" has proved more difficult to apply than its originators likely expected. . . . In an effort to find a single, more uniform interpretation of the statutory phrase, we have reviewed the Courts of Appeals' divergent and increasingly complex interpretations. Having done so, we now return to, and expand, Judge Weinfeld's approach, as applied in the Seventh Circuit. . . . We conclude that "principal place of business" is best read as referring to the place where a corporation's officers direct, control, and coordinate the corporation's activities. It is the place that Courts of Appeals have called the corporation's "nerve center." And in practice it should normally be the place where the corporation maintains its headquarters—provided that the headquarters is the actual center of direction, control, and coordination, *i.e.*, the "nerve center," and not simply an office where the corporation holds its board meetings (e.g., attended by directors and officers who have traveled there for the occasion).

Three sets of considerations, taken together, convince us that this approach, while imperfect, is superior to other possibilities. First, the statute's language supports the approach. The statute's text deems a corporation a citizen of the "State where it has its principal place of business." The word "place" is in the singular, not the plural. The word "principal" requires us to pick out the "main, prominent" or "leading" place. And the fact that the word "place" follows the words "State where" means that the "place" is a place *within* a State. It is not the State itself.

A corporation's "nerve center," usually its main headquarters, is a single place. The public often (though not always) considers it the corporation's main place of business. And it is a place within a State. By contrast, the application of a more general business activities test has led some courts, as in the present case, to look, not at a particular place within a State, but incorrectly at the State itself, measuring the total amount of business activities that the corporation conducts there and determining whether they are "significantly larger" than in the next-ranking State. This approach invites greater litigation and can lead to strange results, as the Ninth Circuit has since recognized.

Second, administrative simplicity is a major virtue in a jurisdictional statute. . . . Simple jurisdictional rules also promote greater predictability. Predictability is valuable to corporations making business and investment decisions. . . .

A "nerve center" approach, which ordinarily equates that "center" with a corporation's headquarters, is simple to apply *comparatively speaking*. The metaphor of a corporate "brain," while not precise, suggests a single location. By contrast, a corporation's general business activities more often lack a single principal place where they take place. That is to say, the corporation may have several plants, many sales locations, and employees located in many different places. If so, it will not be as easy to determine which of these different business locales is the "principal" or most important "place."

Third, the statute's legislative history, for those who accept it, offers a simplicity-related interpretive benchmark. . . .

We recognize that there may be no perfect test that satisfies all administrative and purposive criteria. We recognize as well that, under the "nerve center" test we adopt today, there will be hard cases. . . . Our approach provides a sensible test that is relatively easier to apply, not a test that will, in all instances, automatically generate a result.

Petitioner's unchallenged declaration suggests that Hertz's center of direction, control, and coordination, its "nerve center," and its corporate headquarters are one and the same, and they are located in New Jersey, not in California. Because respondents should have a fair opportunity to litigate their case in light of our holding, however, we vacate the Ninth Circuit's judgment and remand the case for further proceedings consistent with this opinion.*

Judgment *reversed*, in favor of Hertz Corporation

When a case falls within the federal court's concurrent jurisdiction because of either a federal question or diversity of citizenship, the suit may be filed in either state or federal court. If the case is filed in state court, the defendant has a right of removal, which means that he or she may have the case transferred to federal court. All the defendant has to do is file a motion with the court asking to exercise his or her right of removal. Thereupon, the case must be transferred to federal court; the judge has no discretion but must comply with the request.

CRITICAL THINKING ABOUT THE LAW

Justice Breyer is focusing on fairness to the parties in a case where state courts may provide a jury panel that is more homogeneous and perhaps rural than would give the litigants a fair hearing for their arguments. But notice that his reasoning also pays attention to other ethical norms that guide courts when they make legal decisions.

1. What legal norms in addition to fairness does Justice Breyer reference explicitly or implicitly?

    Clue: Return to the discussion of legal norms in Chapter 1. Then go through each of Justice Breyer's arguments for the nerve center test, asking yourself: does this reasoning point to any of the four primary legal norms that shape judicial decision making.

2. Justice Breyer admits that there will probably be some tough cases in the future that will push the boundaries of the nerve center test. Try to create a set of facts that would result in highlighting the ambiguity in the nerve center test.

    Clue: Make a list of all the factors that lead to the conclusion that a particular state is the nerve center of a business. Now design your set of facts around a situation where some factors are present in one state, while others are present in another state.

3. Justice Breyer can form his reasoning based only on the information he has, not the information he might wish to have but lacks. Justice Breyer spells out his reasoning in an orderly fashion. Choose one of his reasons. What facts, were they true, would damage at least one of Justice Breyer's reasons for the nerve center test.

    Clue: Look at his discussion of potential jury bias. What fact would make this reasoning invalid in the Hertz decision?

The right of removal arises only when the case is filed in state court; there is no right of removal to state court. As a result, whenever a case is under concurrent jurisdiction, if either party wants the case heard in federal court, it will be heard there.

Why should both parties have the right to have such a case heard in federal court? In certain cases, a party may fear local prejudice in a state court. Juries for a state court are generally drawn from the county in which the court is located. The juries for federal district courts are drawn from the entire district, which encompasses many counties. Juries in state court are, therefore, usually more homogeneous than those in a district court. One problem that

---

*Hertz Corporation v. Friend, United States Supreme Court 130 S. Ct. 1181 (2010).

this homogeneity may present to the out-of-state corporate defendant occurs when the county in which the court is located is predominantly rural. If the case involves an injury to a member of this rural community, the defendant may feel that the rural jurors would be more sympathetic to the local injured party, whereas jurors drawn from a broader area, including cities, may be more likely to view the victim less sympathetically. City residents are also more likely to work for a corporation, and thus may not regard corporations as unfavorably as might rural residents.

Some people also believe that federal judges are better qualified to hear cases that involve a federal question, because they have more experience in resolving questions that require an interpretation of federal statutes. Finally, if a party anticipates that it may be necessary to appeal the case to the U.S. Supreme Court, bringing the case first in a federal district court may save one step in the appeals process.

Venue

venue Where a case is brought (usually the county of the trial court); prescribed by state statute.

Subject matter jurisdiction should not be confused with venue. Once it is determined which court system has the power to hear the case, **venue** determines which of the many trial courts in that system is appropriate. Venue, clearly prescribed by statute in each state, is a matter of geographic location. It is usually based on the residence of the defendant, the location of the property in dispute, or the location in which the incident out of which the dispute arose occurred. When there are multiple defendants who reside in various geographic locations, the party filing the lawsuit may usually choose from among the various locales.

If the location of the court in which the case is filed presents a hardship or inconvenience to one of the parties, that person may request that the case be moved under the doctrine of *forum non conveniens*, which simply means that the location of the trial court is inconvenient. The judge in the case will consider the party's request and decide whether to grant the party's request. Unlike the right of removal, the request for change of venue is granted at the judge's discretion. There will usually be a hearing on the issue of whether the judge should grant the motion, because the plaintiff generally files the case in a particular court for a reason and will, therefore, be opposed to the defendant's motion.

TECHNOLOGY AND THE LEGAL ENVIRONMENT

The Internet and In Personam Jurisdiction

Is the sponsor of a website that can be visited from every state subject to in personam jurisdiction in every state? As long as the sponsor is not conducting any business or trying to reach customers in a state, many courts have held that mere access to the website is not sufficient to grant in personam jurisdiction.

One case that illustrates this point involved two organizations that both used the name Carefirst. Carefirst of Maryland, a nonprofit insurance company, accused Carefirst Pregnancy Center (CPC), a Chicago-based nonprofit organization, of trademark infringement.[a] Carefirst of Maryland operated a website from which the company promoted its products to consumers who are located primarily in the mid-Atlantic region, with the majority of its consumers living in Maryland. CPC also operated a website, which was accessible anywhere in the world, for the purpose of promoting its services for women with pregnancy-related crises and to generate donations for the organization. CPC's operations were confined almost entirely to the state of Illinois.

Since CPC began using the name Carefirst, the Chicago-based organization received only one

---

[a]*Carefirst of Maryland, Inc. v. Carefirst Pregnancy Centers, Inc.*, 334 F.3d 390 (4th Cir. 2003).

donation from a Maryland resident via the company's website. From 1991 to 2001, CPC claimed that only 0.0174 percent of its donations came from Maryland residents. The only means through which CPC has contact with Maryland residents is CPC's website. Therefore, a district court in Maryland and the appellate court both dismissed the case for lack of personal jurisdiction, concluding that even though CPC's website could be contacted from anywhere, its purpose was to provide information about the organization and solicit donations primarily from Illinois residents. Although the court noted that the donations received from Maryland residents were negligible, the court also held that CPC made no effort to target Maryland donors. Furthermore, the court observed that CPC had no agents, employees, or offices located in Maryland. Hence, there was not sufficient contact with Maryland to support personal jurisdiction.

If the potential defendant, however, is actively trying to do business in other states via a website, the outcome of a case may be different. For example, in *Gator.com Corp. v. L.L. Bean, Inc.*,[b] the Ninth Circuit Court of Appeals held that L.L. Bean was subject to in personam jurisdiction in California. Gator.com, a company that develops software for consumers who make online purchases, also created pop-up coupons that would appear on L.L. Bean's website for L.L. Bean's competitors, such as Eddie Bauer. In response to its receiving a cease-and-desist letter from L.L. Bean, Gator.com sought a declaratory judgment that its actions were not illegal according to state and federal laws. L.L. Bean filed a motion to dismiss, after which a district court in California ruled that the court did not have in personam jurisdiction. The Ninth Circuit reversed on appeal, noting that 6 percent of L.L. Bean's $1 billion in annual sales is attributable to California customers. The court also observed that L.L. Bean "targets" California consumers with its direct email solicitations, and by maintaining a highly interactive website, from which numerous California customers make online purchases and interact with L.L. Bean sales representatives. The Ninth Circuit found these email solicitations and website services to California consumers to be sufficient minimum contacts for in personam jurisdiction.

[b]341 F.3d 1072 (9th Cir. 2002).

One example of a case in which a party sought to have the trial location changed due to *forum non conveniens* is *Ex parte Kia Motors America, Inc.*[5] In this case, four people were riding in Florida in a 1998 Kia Sephia that was involved in a high-speed car accident; the car was forced from the road, caught fire, and burned. Three of the four passengers did not survive. The families of the deceased sued Kia for product liability and negligence, among other claims, in the Alabama courts. Kia filed a motion for *forum non conveniens* to have the case moved to Florida, stating that the car was purchased in Florida, the deceased were residents of Florida, the claims were to be tried according to Florida law, and 25 of the witnesses were also Florida residents. The Supreme Court of Alabama ruled that the motion for *forum non conveniens* was appropriate and the case should be moved to Florida.

APPLYING THE LAW TO THE FACTS...

Before we leave this section of the chapter, ask yourself which court (or courts) would have subject matter jurisdiction and venue in the following situation. Jenny is a resident of Wayne County, Michigan. She is driving through Lucas County, Ohio, when she is hit by another driver who ran a red light. The driver who hit her is a resident of Wood County, Ohio. How would your answer change if the driver who hit her was driving a truck owned by his employer, who is located in Wayne County, Michigan?

[5]881 So. 2d 396 (Ala. 2003).

The Structure of the Court System

As noted previously, our system has two parallel court structures, one federal system and one state system. Because of subject matter jurisdiction limitations, one often does not have a choice as to the system in which to file the case. Once a case is filed in a system, it will stay within that system, except for appeals to the U.S. Supreme Court. The following sections set forth the structure of the two systems. As you will see, they are indeed very similar. Their relationship is illustrated in Exhibit 3-2.

THE FEDERAL COURT SYSTEM

Federal Trial Courts. As you already know, trial courts are the courts of original jurisdiction. In the federal court system, the trial courts are the U.S. district courts. The United States is divided into 96 districts, and each district has at least one trial court of general jurisdiction. *General jurisdiction* means that the court has the power to hear cases involving a wide variety of subject matter and that it is not limited in the types of remedies it can grant. All cases to be heard in the federal system are filed in these courts, except those cases for which Congress has established special trial courts of limited jurisdiction.

Trial courts of limited jurisdiction in the federal system are limited in the type of cases they have the power to hear. Special federal trial courts of limited jurisdiction have been established for bankruptcy cases; claims against the U.S. government; and copyright, patent, and trademark cases. In an extremely limited number of cases, the U.S. Supreme Court also functions as a trial court of limited jurisdiction. Such cases include controversies between two or more states and suits against foreign ambassadors.

Intermediate Courts of Appeal. The second level of courts in the federal system is made up of the U.S. circuit courts of appeal. The United States is divided into 12 geographic areas, including the District of Columbia, each of which has a circuit court of appeals. Exhibit 3-2 illustrates this division. There is also a federal circuit court of appeals and a U.S. Veterans Court of Appeals. Each circuit court of appeals hears appeals from all of the district courts located within its geographic area. These courts also hear appeals from administrative agencies located within their respective circuits. In some cases, appeals from administrative agencies are heard by the Federal Circuit Court of Appeals. The Veterans Court of Appeals hears appeals of benefits decisions made by the Veterans Administration.

Court of Last Resort. The U.S. Supreme Court is the final appellate court in the federal system. In a limited number of instances, discussed in the last section of this chapter, the U.S. Supreme Court also hears cases from the court of last resort in a state system. As previously noted, the U.S. Supreme Court also functions as a trial court in a limited number of cases. The federal court system is illustrated in Exhibit 3-3.

STATE COURT SYSTEMS

There is no uniform state court structure because each state has devised its own court system. Most states, however, follow a general structure similar to that of the federal court system.

State Trial Courts. In state court systems, most cases are originally filed in the trial court of general jurisdiction. As in the federal system, state trial courts of general jurisdiction are those that have the power to hear all the cases that would be tried in the state court system, except those cases for which special

EXHIBIT 3-2
GEOGRAPHIC BOUNDARIES OF UNITED STATES COURTS OF APPEALS AND UNITED STATES DISTRICT COURTS

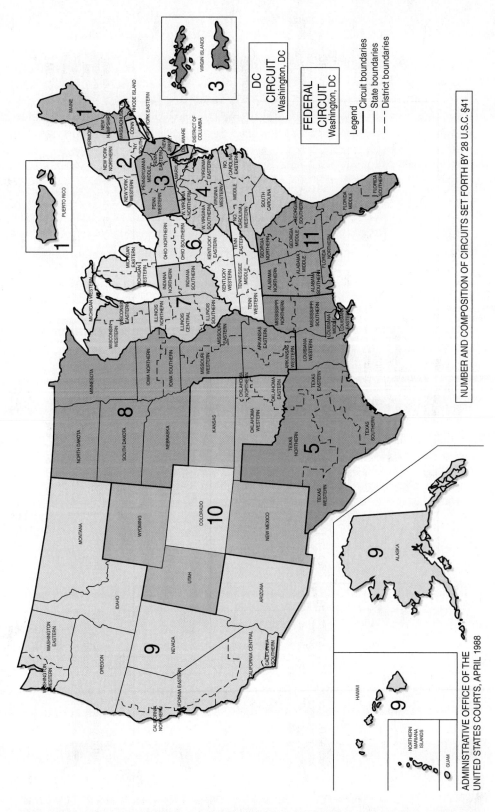

trial courts of limited jurisdiction have been established. These trial courts of general jurisdiction are distributed throughout each state, usually by county. The names of these courts vary from state to state but are usually called courts of common pleas or county courts. New York uniquely calls its trial courts of general jurisdiction supreme courts. In some states, these courts may have specialized divisions, such as domestic relations or probate.

EXHIBIT 3-3

THE STRUCTURE OF THE COURT SYSTEM

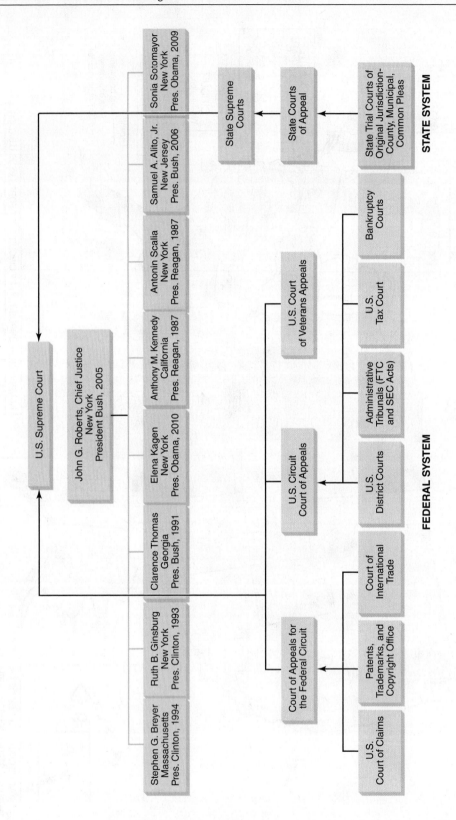

Most states also have trial courts of limited jurisdiction. These courts are usually limited in the remedies they may grant. Some may not issue injunctions or orders for specific performance. A common court of limited jurisdiction in most states is the small claims court, which may not grant damage awards in excess of specified amounts. Some courts of limited jurisdiction are limited to certain types of cases, such as traffic cases. Some criminal courts of limited jurisdiction may be limited to hearing misdemeanors. It is difficult to generalize about these courts because they vary so much from state to state. The main distinction between trial courts of general and limited jurisdiction, however, is that the former hear almost all types of cases that are filed in the state system and are unlimited in the remedies they can provide, whereas the latter hear only a particular type of case or may award only limited remedies.

Intermediate Courts of Appeal. Intermediate courts of appeal, analogous to the federal circuit courts of appeal, exist in approximately half the states. These courts usually have broad jurisdiction, hearing appeals from courts of general and limited jurisdictions, as well as from state administrative agencies. The names of these courts also vary by state. They may be called courts of appeal or superior courts.

Courts of Last Resort. In almost all cases filed in the state court system, the last appeal is to the state court of last resort. This court is frequently called the supreme court. In some states, it is known as the court of appeals. In approximately half of the states, it is the second court to which an appeal can be made; in the remaining states, it is the only appellate court.

The Actors in the Legal System and Their Relationship to the Business Community

THE ATTORNEY

An understanding of the structure of the legal system would be incomplete without an awareness of the primary actors within the system. The party with whom businesspersons usually have the most frequent contact is the attorney. Although the exact qualifications for being an attorney vary from state to state, most require that an attorney have a law degree, have passed the state's bar examination, and be of high moral character. Attorneys are the legal representatives of the parties before the court. Some corporations have full-time attorneys, referred to as *in-house counsel*. Other corporations send all their legal work to an outside law firm. Many larger businesses have in-house counsel and also use outside counsel when a problem arises that requires a specialist.

Attorney–Client Privilege. The attorney can provide effective representation only when he or she knows all the pertinent facts of the case. The businessperson who withholds information from his or her attorney may cause irreparable harm if the hidden facts are revealed by the opposing side in court. To encourage client honesty, the **attorney–client privilege** was established. This privilege provides that information furnished in confidence to an attorney, in conjunction with a legal matter, may not be revealed by that attorney without permission from the client. There is, however, an important exception to this rule. If the lawyer knows that the client is about to commit a crime, the lawyer may reveal confidential information in order to prevent the commission of that crime. Revealing such information, however, is not required of the attorney; it is simply allowed. This protection also extends to the attorney's work product under what is known as the **work-product doctrine**. Work product includes both formal and informal documents prepared by the attorney in conjunction with a client's case.

One of the problems arising out of use of the attorney–client privilege in the corporate setting is the definition of the client. The client is the corporation, but the communication sought to be protected is that between the attorney and upper-, middle-, or lower-level employees of the corporation. In

attorney–client privilege Provides that information furnished by a client to an attorney in confidence, in conjunction with a legal matter, may not be revealed by the attorney without the client's permission.

work-product doctrine Provides that formal and informal documents prepared by an attorney in conjunction with a client's case are privileged and may not be revealed by the attorney without the client's permission.

such cases, the corporate attorney usually tries to rely on the work-product doctrine to protect the information that he or she has gathered from employees, especially when such information is in the form of written communications. Such an approach has generally been successful, but the courts have not yet precisely defined the parameters of the attorney–client privilege and the work-product doctrine as they apply to the corporate setting.

Additional Functions of the Attorney. Attorneys are probably best known for representing clients in litigation, but they also provide many other services for business clients. Attorneys represent their clients not only in courtroom litigation but also before administrative boards. Attorneys may be called on to represent their corporate clients in negotiations with labor unions or with other firms.

Corporate attorneys also serve as advisors or counselors, reviewing proposed corporate activities and advising management of legal problems that may arise as a result of such activities. In-house counsel familiar with the various activities of the firm are often in the best position to fulfill this role. Thus, businesspersons should attempt to establish a good working relationship with in-house counsel, using them as a resource whenever legal issues arise. Managers should not assume that they know all the legal ramifications of all the business activities in which they engage. Most in-house counsel would prefer to be consulted before an activity is undertaken rather than after it results in a legal problem.

Finally, the attorney may serve as a draftsperson, drawing up contracts, deeds, the corporate charter, securities registration statements, and all other legal documents needed by the corporation. Thus, it is clear that the attorney is one actor in the American legal system who is of special importance to the business manager.

The Judge. The role of the judge is especially important in our legal system. The judge's function depends on whether he or she is a trial or appellate court judge. A trial court judge presides over the trial, making sure the case is heard with reasonable speed; rules on all motions made in the case; and decides all questions of law. One of the most crucial functions of the trial court judge is ruling on whether evidence is admissible. A judge's failure or refusal to admit certain items into evidence may determine the outcome of a case, and a judge's ruling on any particular piece of evidence may subsequently become the basis for the appeal of an unfavorable decision. If the parties waive their rights to a jury trial, or if they are not entitled to a jury, the judge also decides the facts in the case and renders a decision accordingly. A single judge presides over each case.

Appellate judges serve on panels. They review lower-court cases to determine whether errors of law were committed by the lower courts. Their review consists primarily of reading the transcript of the trial, reading written arguments by counsel for both parties, and sometimes hearing oral arguments from both parties' attorneys.

State court judges are usually elected, although some are appointed, whereas federal court judges are appointed by the president with the advice and consent of the Senate. This appointment process is a good example of how the legislative and executive branches serve as checks on each other. The president has the greatest role because he makes the nomination, but he cannot choose just anyone. The president will usually select a list of potential nominees who will then be rated by a committee of the American Bar Association. The bar association looks at the nominees' legal experience and reads their written opinions and published articles in an attempt to ensure that only the most qualified candidates will be named to the federal bench. The

Senate also scrutinizes the list and gives the president some idea, in advance, of whether the various potential nominees have a high likelihood of being confirmed.

Once the president actually makes a nomination, the Senate judiciary subcommittee holds formal hearings on the nominee's fitness for office. After the hearings, the full Senate votes on the nomination. Although the president generally tries to nominate someone with an ideological background similar to his own, if the Senate is dominated by the opposite political party, a nominee who has too strong an ideology is not likely to be confirmed. In recent years, the appointment process has become familiar to most Americans as the hearings on Supreme Court nominees have been televised. Federal court judges serve for life, whereas state court judges generally serve definite terms, the length of which varies from state to state.

There is a lot of debate over whether judges should be appointed for life or elected for specific terms. The rationale behind appointment for life is that it takes the politics out of the judicial process. A judge will be selected based on his or her credentials as opposed to the quality of his or her campaign skills or the size of the campaign budget. Once in office, the judge is free to make honest decisions without having to worry about the impact of any decision on reelection.

Of course, this independence is just what makes some people prefer elected judges. They point out that the members of every other branch of government are elected and are, therefore, forced to represent the will of the people; they argue that judges should represent the people no less than members of the other branches.

The Power of Judicial Review. One very important power that judges have is the power of *judicial review*, that is, the power to determine whether a law passed by the legislature violates the Constitution. Any law that violates the Constitution must be struck down as null and void. The justices of the U.S. Supreme Court are the final arbiters of the constitutionality of our statutory laws.

This power of judicial review was not explicitly stated in the Constitution. Rather, it was established in the classic 1803 case of *Marbury v. Madison*,[6] wherein the Supreme Court stated:

> *It is emphatically the province and duty of the judicial department to say what the law is. Those who apply the rule to particular cases must of necessity expound and interpret that rule. If two laws conflict with each other, the courts must decide which of these conflicting rules governs the case. This is of the very essence of judicial duty.**

When individual justices exercise this power of judicial review, they do so with different philosophies and attitudes. Their philosophies can have a powerful effect on how they make decisions. One distinction frequently made with respect to judicial philosophies is the difference between a judge who believes in judicial activism and one who believes in judicial restraint.

A judge who believes in **judicial restraint** believes that the three branches are coequal, and the judiciary should refrain from determining the constitutionality of an act of Congress unless absolutely necessary, to keep from interfering in the congressional sphere of power. These judges tend to believe that social, economic, and political change should result from the political process, not from judicial action. They consequently give great deference to actions of the state and federal legislatures.

Those who believe in judicial restraint are much less likely to overturn an existing precedent. They tend to focus much more on the facts than

judicial restraint A judicial philosophy that says courts should refrain from determining the constitutionality of a legislative act unless absolutely necessary and that social, political, and economic change should be products of the political process.

---
[6]5 U.S. 137 (1803).

*Marbury v. Madison, 5 U.S. (1 Cranch) 137; 2 L. Ed. 60 (1803).

on questioning whether the law should be changed. They tend to uphold lower-court decisions unless those decisions are clearly wrong on the facts.

In contrast, **judicial activists** tend to see a need for the courts to take an active role in encouraging political, economic, and social change, because the political process is often too slow to bring about necessary changes. They believe that constitutional issues must be decided within the context of today's society and that the framers meant for the Constitution to be an evolving document.

Judicial activists are much less wedded to precedent and are more result oriented. They are much more likely to listen to arguments about what result is good for society. Activist judges have been responsible for many social changes, especially in the civil rights area.

judicial activism A judicial philosophy that says the courts should take an active role in encouraging political, economic, and social change.

THE JURY

The jury is the means by which citizens participate in our judicial system. It had its roots in ancient Greek civilization, and it is often seen as the hallmark of democracy. A *jury* is a group of individuals, selected randomly from the geographic area in which the court is located, who will determine questions of fact. There are two types of juries: petit and grand.

Petit Juries. Businesspersons are primarily concerned with **petit juries**. These juries serve as the finders of fact for trial courts. Though originally composed of 12 members, most juries in civil cases are allowed to have fewer members in many jurisdictions. Traditionally, jury decisions had to be unanimous. Today, however, more than half the jurisdictions no longer require unanimity in civil cases. This change in the jury system has been made primarily to speed up trial procedures.

petit jury A jury of 12 citizens impaneled to decide on the facts at issue in a criminal case and to pronounce the defendant guilty or not guilty.

An important decision to be made by any corporate client and his or her attorney is whether to have a jury. In any civil action in which the plaintiff is seeking a remedy at law (money damages), a jury may hear the case. If both parties to the case agree, however, the jury may be waived, and a judge decides the facts of the case. There is no hard-and-fast rule about when a businessperson should opt for a jury, but a few factors should frequently be considered. One is the technical nature of the case. If the matter is highly technical, a judge may be able to decide the case more fairly, especially if that judge has expertise in the area in dispute. Another factor is the emotional appeal of the case. If the opponent's arguments in the case may have strong emotional appeal, a judge may render a fairer decision.

Grand Juries. Grand juries are used only in criminal matters. The Fifth Amendment requires that all federal prosecutions for "infamous" crimes (including all federal offenses that carry a term of imprisonment in excess of one year) be commenced with an **indictment** (a formal accusation of the commission of a crime, which must be made before a defendant can be tried for the crime) by a **grand jury**. This jury hears evidence presented by the prosecutor and determines whether there is enough evidence to justify charging a defendant. The prudent business manager who carefully heeds the advice of an attorney should not be faced with a potential indictment by a grand jury. Increasingly, however, corporate managers are facing criminal charges for actions taken to benefit their corporate employers. Such cases are discussed in Chapter 6.

indictment A formal written accusation in a felony case.

grand jury A group of 12 to 23 citizens convened in private to decide whether enough evidence exists to try the defendant for a felony.

adversarial system System of litigation in which the judge hears evidence and arguments presented by both sides in a case and then makes an objective decision based on the facts and the law as presented by each side.

The Adversary Process

Our system of litigation is accurately described as an adversary system. In an **adversarial system**, a neutral factfinder, such as a judge or jury, hears evidence and arguments presented by both sides and then makes an objective decision based on the facts and the law as presented by the proponents of each side. Strict rules govern the types of evidence that the factfinder may consider.

Theoretically, the adversary system is the best way to bring out the truth, because each side will aggressively seek all the evidence that supports its position. Each side attempts to make the strongest possible argument for its position.

CRITICISMS OF THE ADVERSARY SYSTEM

Many people criticize this system. They argue that because each side is searching only for evidence that supports its position, a proponent who discovers evidence helpful to the other side will not bring such evidence to the attention of the court. This tendency to ignore contrary evidence prevents a fair decision—one based on all the available evidence—from being rendered.

Another argument of the critics is that the adversary process is extremely time-consuming and costly. Two groups of "investigators" are seeking the same evidence. Thus, there is a duplication of effort that lengthens the process and unnecessarily increases the cost.

Others argue that the adversary system, as it functions in this country, is unfair. Each party in the adversarial process is represented by an attorney. Having the most skillful attorney is a tremendous advantage. The wealthier a party is, the better the attorney she or he can afford to hire; hence, the system unjustifiably favors the wealthy.

Law professor Marc Galanter has written an interesting critique of our adversary system that has generated a lot of discussion.[7] He argues that, given the structure of our system, certain parties tend to have a distinct advantage. Galanter divides litigants into two groups: the repeat players (RPs), those who are engaged in similar litigations over time; and the one-shotters (OSs), those who have only occasional recourse to the courts. RPs would typically be large corporations, financial institutions, landlords, developers, government agencies, and prosecutors. Typical OSs would be debtors, employees with grievances against their employers, tenants, and victims of accidents.

According to Galanter, the RPs have a distinct advantage over the OSs in litigation. Because of their experience, RPs are better prepared for trial; they know what kinds of records to keep and how to structure transactions so that they will have an advantage in court. RPs will have developed expertise in the area and will have access to specialists. They will have low "start-up costs" for a case because they have been through it before. RPs will have developed helpful informal relationships with those at the courthouse. RPs know the odds of success better because of their experience and can use that knowledge to calculate whether to settle. Finally, RPs can litigate for favorable rulings or for an immediate outcome.

Thus, in a typical case involving an RP and an OS, the RP has a distinct advantage. Some people believe this advantage is significant enough to prevent our current system from dispensing justice in these cases.

Steps in Civil Litigation and the Role of Businesspersons

THE PRETRIAL STAGE

Every lawsuit is the result of a dispute. Business disputes may result from a breach of contract, the protested firing of an employee, or the injury of a consumer who uses the corporation's product. This section focuses on dispute

---
[7]Marc Galanter, "Why the Haves Come Out Ahead: Speculation on the Limits of Legal Change," *J. L. & Soc. Rev.* 9: 96 (1974).

rules of civil procedure The rules governing proceedings in a civil case; federal rules of procedure apply in all federal courts, and state rules apply in state courts.

resolution in this country under the adversary system. It examines the procedure used in a civil case, the stages of which are outlined in Exhibit 3-4. The rules that govern such proceedings are called the **rules of civil procedure**. There are federal rules of civil procedure, which apply in all federal courts, as well as state rules, which apply in the state courts. Most of the state rules are based on the federal rules.

Informal Negotiations. For the businessperson involved in a dispute, the first step is probably to discuss the dispute directly with the other party. When it appears that the parties are not going to be able to resolve the problem

EXHIBIT 3-4

ANATOMY OF A CIVIL LAWSUIT

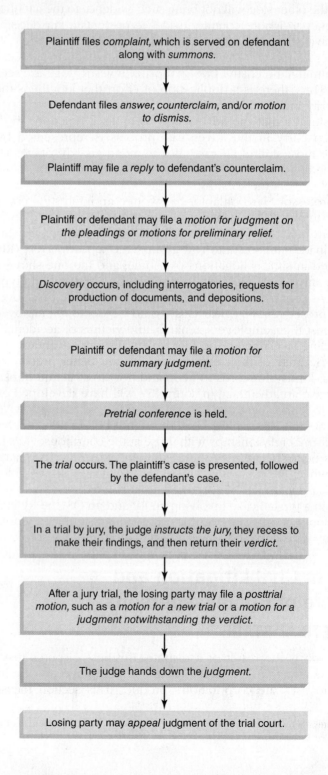

Plaintiff files *complaint*, which is served on defendant along with *summons*.

Defendant files *answer, counterclaim,* and/or *motion to dismiss*.

Plaintiff may file a *reply* to defendant's counterclaim.

Plaintiff or defendant may file a *motion for judgment on the pleadings* or *motions for preliminary relief*.

Discovery occurs, including interrogatories, requests for production of documents, and depositions.

Plaintiff or defendant may file a *motion for summary judgment*.

Pretrial conference is held.

The *trial* occurs. The plaintiff's case is presented, followed by the defendant's case.

In a trial by jury, the judge *instructs the jury,* they recess to make their findings, and then return their *verdict*.

After a jury trial, the losing party may file a *posttrial motion*, such as a *motion for a new trial* or a *motion for a judgment notwithstanding the verdict*.

The judge hands down the *judgment*.

Losing party may *appeal* judgment of the trial court.

themselves, the businessperson will then discuss the dispute with an attorney. It is important that the attorney be given all relevant information, even if it does not make the businessperson look good. The more relevant facts the attorney has, the better the attorney's advice will be. Together, the attorney and the client may be able to resolve the dispute informally with the other party.

Initiation of a Legal Action. Once a party decides that an informal resolution is not possible, the parties enter what is often called the pleading stage of the lawsuit. **Pleadings** are papers filed by a party in court and then served on the opponent. The basic pleadings are the complaint, the answer, the counterclaim, and the motion to dismiss. Exhibit 3-5 provides an illustration of a typical complaint. The attorney of the businessperson who feels that he or she has been wronged initiates a lawsuit by filing a complaint in the appropriate court. A *complaint* is a document that states the names of the parties to the action, the basis for the court's subject matter jurisdiction, the facts on which the party's claim is based, and the relief that the party is seeking. Remember that the party

pleadings Papers filed by a party in court and then served on the opponent in a civil lawsuit.

EXHIBIT 3-5

COMPLAINT

THE COURT OF COMMON PLEAS
OF LUCAS COUNTY, OHIO

Pam Streets, Plaintiff v. Daniel Lane, Defendant
COMPLAINT FOR NEGLIGENCE
Case No. _____

Now comes the plaintiff, Pam Streets, and, for her complaint, alleges as follows:

1. Plaintiff, Pam Streets, is a citizen of Lucas County, in the state of Ohio, and Defendant, Daniel Lane, is a citizen of Lucas County in the state of Ohio.
2. On December 1, 1987, the Plaintiff was lawfully driving her automobile south on Main Street in Toledo, Ohio.
3. At approximately 4:00 p.m., on December 1, 1987, the Defendant negligently ran a red light on Starr Avenue, and as a result crashed into Plaintiff's car.
4. As a result of the collision, the Plaintiff suffered lacerations to the face and a broken leg, incurring $10,000 in medical expenses.
5. As a result of the above described collision, her car was damaged in the amount of $12,000.
6. As a result of the foregoing injuries, the Plaintiff was required to miss eight weeks of work, resulting in a loss of wages of $2,400.

WHEREFORE, Plaintiff demands judgment in the amount of $24,400, plus costs of this action.

Sam Snead
Attorney for Plaintiff
124 East Broadway
Toledo, OH 43605

JURY DEMAND
Plaintiff demands a trial by jury in this matter.

Sam Snead
Attorney for Plaintiff

on whose behalf the complaint is filed is the plaintiff, and the defendant is the party against whom the action is being brought.

In determining the appropriate court in which to file the complaint, the attorney must determine which court has subject matter jurisdiction over the case. Once that determination has been made, the attorney must ascertain the proper venue for the case. The means used by the attorney to determine subject matter jurisdiction and venue were discussed earlier in this chapter.

Service of Process. Once the complaint is filed, the court serves a copy of the complaint and a summons on the defendant. The reader should remember that service is the procedure used by the court to ensure that the defendant actually receives a copy of the summons and the complaint. Service of process gives the court in personam jurisdiction over the defendant and provides the notice of the charges required by the defendant's due process rights.

Defendant's Response. Once the defendant has been properly served, he or she files an answer and possibly a counterclaim. The answer is a response to the allegations in the plaintiff's complaint. The answer must admit, deny, or state that the defendant has no knowledge about the truth of each of the plaintiff's allegations. The answer may also contain affirmative defenses, which consist of facts that were not stated in the complaint that would provide justification for the defendant's actions and a legally sound reason to deny relief to the plaintiff. These defenses must be stated in the answer. If they are not raised in the answer, the court may choose not to allow these defenses to be raised later. The defendant is required to plead his or her affirmative defenses in the answer in order to give the plaintiff notice of all the issues that will be raised at the trial.

As an illustration, two affirmative defenses to a breach-of-contract action might be that the plaintiff procured the defendant's signature on the contract through fraud and that the contract was illegal because its enforcement would result in a violation of the antitrust laws. As another example, suppose that a manufacturer is being sued because the plaintiff was injured by the manufacturer's negligently produced defective product. The defendant manufacturer might raise the affirmative defense of contributory negligence, arguing that the plaintiff's injury would not have occurred if the plaintiff had not also been negligent. Notice the use of an affirmative defense in the sample answer in Exhibit 3-6. It is important that the businessperson who is being sued immediately try to think of any potential affirmative defenses that might excuse his or her actions.

Upon receiving the complaint, a defendant may believe that even if all of the plaintiff's factual allegations were true, the plaintiff would still not be entitled to a favorable judgment. In that situation, the defendant may file a **motion to dismiss**. No factual issues are being debated, so the judge accepts the facts as stated by the plaintiff and makes a ruling on the legal questions in the case. Judges are generally not receptive to such motions, granting them only when it appears beyond doubt that the plaintiff can prove no set of facts, in support of his or her claim, that would entitle him or her to relief.

If the defendant believes that he or she has a cause of action against the plaintiff, this will be included as a **counterclaim**. The form of a counterclaim is just like that of a complaint. The defendant states the facts supporting his or her claim and asks for the relief to which he or she feels entitled. Exhibit 3-6 also contains a counterclaim.

If the defendant files a counterclaim, the plaintiff generally files a reply. A *reply* is simply an answer to a counterclaim. In the reply, the plaintiff admits, denies, or states that he or she is without knowledge of the truth of the facts asserted by the defendant in the counterclaim. Any affirmative defenses that are appropriate must be raised in the reply.

motion to dismiss Defendant's application to the court to put the case out of judicial consideration because even if the plaintiff's factual allegations are true, the plaintiff is not entitled to relief.

counterclaim Defendant's statement of facts showing cause for action against the plaintiff and a request for appropriate relief.

EXHIBIT 3-6

AFFIRMATIVE DEFENSES AND COUNTERCLAIM

THE COURT OF COMMON PLEAS
OF LUCAS COUNTY, OHIO

Pam Streets, Plaintiff v. Daniel Lane, Defendant
ANSWER AND COUNTERCLAIM
Case No. _____

Now comes the Defendant, Daniel Lane, and answers the complaint of Plaintiff herein as follows:

FIRST DEFENSE
1. Admits the allegations in paragraphs 1 and 2.
2. Denies the allegation in paragraph 3.
3. Is without knowledge as to the truth or falsity of the allegations contained in paragraphs 4, 5, and 6.

SECOND DEFENSE
4. If the court believes the allegations contained in paragraph 3, which the Defendant expressly denies, Plaintiff should still be denied recovery because she was negligently driving in excess of the speed limit and without her glasses, both of which contributed to the cause of the accident.

COUNTERCLAIM
5. Defendant lawfully drove his automobile in an eastbound direction on Starr Avenue on December 1, 1987.
6. At approximately 4:00 p.m., on December 1, 1987, Plaintiff negligently drove her automobile at an excessive speed through a red light on Main Street where said street crosses Starr Avenue, colliding into Defendant's automobile.
7. As a result of the collision, Defendant suffered bruises and a concussion, resulting in $5,000 in medical bills.
8. Defendant further suffered $6,000 in property damage to his automobile.

WHEREFORE, Defendant prays for a judgment dismissing the Plaintiff's complaint and granting the Defendant a judgment against Plaintiff in the amount of $11,000 plus costs of this action.

Shelly Shaker
Attorney for Defendant
216 Nevada
Toledo, OH 43605

Pretrial Motions. The early pleadings just described serve to establish the legal and factual issues of the case. Once these issues have been established, either the plaintiff or the defendant may file a motion designed to bring the case to an early conclusion or to gain some advantage for the party filing the motion. A *motion* is simply a party's request for the court to do something. A party may request, or move, that the court do almost anything pertaining to the case, such as a motion for some form of temporary relief until a decision has been rendered. For example, if a suit is brought over the right to a piece of property, the court may grant a motion prohibiting the current possessor of that property from selling it. A party may file a motion to proceed in forma pauperis, which is a motion to proceed without payment of fees if the party feels it has good reasons for why the court should allow the case to proceed even if it does not have the money to pay the fees up front (the assumption being that after successful suit the party will be able to pay the fees). As noted earlier, a defendant may believe that even if everything the plaintiffs plead in their complaint were true, there would still be no legitimate basis for a lawsuit. In such a situation, the defendant may file a motion to dismiss.

Many of the really frivolous lawsuits that are filed are dismissed in response to pretrial motion. For example, in 2007, Jehovah J. God, Jesus J. Christ, the Jehovah Witness Foundation Inc., and William E. Moore filed a $9 million lawsuit against the University of Arizona, arguing that the university was using God's autobiography, the Bible, without paying him royalties. In dismissing the case upon hearing of the motion to proceed in forma pauperis, the court said that the allegations were "both fanciful and factually frivolous," a finding that is made when the facts alleged by the plaintiff rise to the level of the irrational or wholly incredible, whether or not there are judicially recognized facts available to contradict them.[8]

When a party files any motion with the court, a copy is always sent to the opposing attorney. That attorney may respond to the motion, usually by requesting that the judge deny the motion. In many cases, the judge will simply rule on the motion, either granting or denying it. In some cases, the judge may hold a hearing at which the two sides orally present arguments.

Discovery. Once the initial pleadings and motions have been filed, the parties gather information from each other through **discovery**. As a result of discovery, each party should have knowledge of most of the facts in the case. This process is supposed to prevent surprises from occurring in the courtroom.

At this stage, the businessperson is frequently asked by his or her attorney to respond to the opponent's discovery requests. There are a number of tools of discovery. One of the most common is interrogatories, which are a series of written questions that are sent to the opposing party, who must truthfully answer them under oath. The interrogatories are frequently accompanied by a request to admit certain facts. The attorney and the client work together to answer these interrogatories and requests for admission of facts.

Another discovery tool is the request to produce documents or other items. Unless the information requested is privileged or is irrelevant to the case, it must be produced. Photographs, contracts, written estimates, and forms that must be filed with governmental agencies are among the items that may be requested. One party may also request that the other party submit to a mental or physical examination. This motion will be approved only when the party's mental or physical health is at issue in the case.

Finally, testimony before trial may be obtained by taking a **deposition**. At a deposition, a witness is examined under oath by attorneys. A court reporter (stenographer) records every word spoken by the attorneys and witnesses. The testimony is usually transcribed so that both parties have a written copy. If a businessperson is to be deposed in a case, it is very important that he or she and the attorney talk extensively about what kinds of questions may come up at the deposition and how such questions are to be answered. The party who requested the deposition is not only seeking information, but is also laying the groundwork for identifying any inconsistencies that may arise between a person's testimony at the deposition and in court. If such inconsistencies exist, they will be brought to the attention of the factfinder and may result in a loss of credibility for the courtroom testimony.

Depositions may also be used when a potential witness is old or ill and may die before the trial. They are useful if witnesses may be moving or for some other reason may not be available at the time of the trial.

discovery The pretrial gathering of information from each other by the parties.

deposition Pretrial testimony by witnesses who are examined under oath.

---

[8]*God, et al. v. Arizona State University*, U.S. Dist. LEXIS 38679 (2007).

Parties must comply with discovery requests, or the court may order that the facts sought to be discovered be deemed admitted. Thus, it is important that the businessperson involved in litigation produce for the attorney all requested discovery material. An attorney who feels that certain material should not be discovered makes arguments about its lack of relevance to the case, but if the court disagrees, the information must be supplied.

Pretrial Conference. If the judge finds that questions of fact do exist, he or she usually holds a pretrial conference. This is an informal meeting of the judge with the lawyers representing the parties. At this meeting, they try to narrow the legal and factual issues and work out a settlement if possible. When a lawsuit begins, there are many conflicting assertions as to what events actually led up to the lawsuit. Questions about what actually happened are referred to as *questions of fact*. Many times, as a result of discovery, parties come to agree on most of the facts. Remaining factual disputes may often be resolved at the conference. Then the only questions left are how to apply the law to the facts and what damages, if any, to award.

By the time of the pretrial conference, the businessperson should have determined the limits on any settlement to which he or she is willing to agree and should have communicated those limits to his or her attorney, who may be able to reach a settlement at the conference. Judges frequently try very hard to help the parties reach agreement before trial. If no settlement can be reached, the attorneys and the judge discuss the administrative details of the trial, its length, the witnesses, and any pretrial stipulations of fact or law to which the parties can agree.

THE TRIAL

Once the pretrial stage has concluded, the next step is the trial. As stated previously, if the plaintiff is seeking a legal remedy (money damages), he or she is usually entitled to a jury trial. The judge is the factfinder when an equitable remedy (an injunction or other court order) is being sought or the parties have waived their right to a jury. For example, when a plaintiff in a product liability action requests a judgment for $10,000 in medical expenses, he or she would be seeking a legal remedy and would be entitled to a jury trial. However, a plaintiff seeking an injunction, under the antitrust laws, to prohibit two defendant corporations from merging would be requesting an equitable remedy and thus would not be entitled to a jury. It is important for the business manager to determine at the outset whether a jury is desirable, because a jury must be demanded in the complaint.

The stages of the trial are (1) jury selection, (2) the opening statements, (3) the plaintiff's case, (4) the defendant's case, (5) the conference on jury instructions, (6) closing arguments, and (7) posttrial motions.

Jury Selection. An important part of a jury trial is the selection of the jury. A panel of potential jurors is selected randomly from a list of citizens. In the federal court system, voter lists are used. In a process known as **voir dire**, the judge or the attorneys, or both, question potential jurors to determine whether they could render an unbiased opinion in the case.

When a juror's response to a question causes an attorney to believe that this potential juror cannot be unbiased, the attorney will ask that the potential juror be removed "for cause." For example, in an accident case, a potential juror might reveal that he had been in a similar accident, or the potential juror may have filed a similar lawsuit against one of the defendant's competitors five years ago. Attorneys are given an unlimited number of challenges for cause. In most states, each attorney is allowed to reject a minimal number of potential jurors without giving a reason. These rejections are called *peremptory challenges*.

voir dire Process whereby the judge and/or the attorneys question potential jurors to determine whether they will be able to render an unbiased opinion in the case.

The legitimate rationale for "peremptories" is that they recognize and accommodate (to a certain extent) a lawyer's "gut reaction" to a potential juror who does not say anything that technically reveals a bias. Nevertheless, there has been some abuse of peremptories in the past. One potential source of abuse was to use peremptories to discriminate against certain classes of people, such as by race or gender.

In 1986, in the case of *Batson v. Kentucky*,[9] the U.S. Supreme Court ruled that prosecutors could not use race-based peremptory challenges in criminal cases. Subsequently, the Supreme Court extended the ban to the use of race-based challenges by either party in civil cases. Several unsuccessful attempts were made to extend the prohibition to challenges based on gender. Finally, in 1994, the Court in the following case extended the equal protection guarantee to cover gender.

CASE 3-3

J.E.B. v. Alabama, ex rel. T.B.
Supreme Court of the United States
511 U.S. 127 (1994)

On behalf of T.B., the unwed mother of a minor child, the State of Alabama filed a complaint for paternity and child support against J.E.B. A panel of 12 males and 24 females was called by the court as potential jurors. After the court removed three individuals for cause, only 10 males remained. The state used its peremptory challenges to remove nine male jurors and J.E.B. removed the tenth, resulting in an all-female jury. The court rejected J.E.B.'s objection to the gender-based challenges, and the jury found J.E.B. to be the father.

J.E.B. appealed to the court of appeals, who affirmed the trial court's decision that the Equal Protection Clause does not prohibit gender-based challenges. The Alabama Supreme Court denied certiorari, and J.E.B. then appealed to the U.S. Supreme Court.

Justice Blackmun

Today we reaffirm what should be axiomatic: Intentional discrimination on the basis of gender by state actors violates the Equal Protection Clause, particularly where, as here, the discrimination serves to ratify and perpetuate invidious, archaic, and overbroad stereotypes about the relative abilities of men and women.

Discrimination on the basis of gender in the exercise of peremptory challenges is a relatively recent phenomenon. Gender-based peremptory strikes were hardly practicable for most of our country's existence, since, until the 20th century, women were completely excluded from jury service.

Many States continued to exclude women from jury service well into the present century, despite the fact that women attained suffrage upon ratification of the Nineteenth Amendment in 1920.

Despite the heightened scrutiny afforded distinctions based on gender, respondent argues that gender discrimination in the selection of the petit jury should be permitted, though discrimination on the basis of race is not. Respondent suggests that "gender discrimination in this country . . . has never reached the level of discrimination" against African-Americans, and therefore gender discrimination, unlike racial discrimination, is tolerable in the courtroom.

While the prejudicial attitudes toward women in this country have not been identical to those held toward racial minorities, the similarities between the experiences of racial minorities and women, in some contexts, "overpower those differences." Certainly, with respect to jury service, African-Americans and women share a history of total exclusion.

Discrimination in jury selection, whether based on race or on gender, causes harm to the litigants, the community, and the individual jurors who are wrongfully excluded from participation in the judicial process. The litigants are harmed by the risk that the prejudice which motivated the discriminatory selection of the jury will infect the entire proceedings. The community is harmed by the State's participation in the perpetuation of invidious group stereotypes and the inevitable loss of confidence in our judicial system that state-sanctioned discrimination in the courtroom engenders.

When state actors exercise peremptory challenges in reliance on gender stereotypes, they ratify and reinforce prejudicial views of the relative abilities of men and women. Because these stereotypes have wreaked injustice in so many other spheres of our country's public life, active discrimination by litigants on the basis of gender

---

[9]476 U.S. 79 (1986).

during jury selection "invites cynicism respecting the jury's neutrality and its obligation to adhere to the law."

In recent cases we have emphasized that individual jurors themselves have a right to nondiscriminatory jury selection procedures.

As with race-based *Batson* claims, a party alleging gender discrimination must make a prima facie showing of intentional discrimination before the party exercising the challenge is required to explain the basis for the strike. When an explanation is required, it need not rise to the level of a "for cause" challenge; rather, it merely must be based on a juror characteristic other than gender, and the proffered explanation may not be pretextual.

Equal opportunity to participate in the fair administration of justice is fundamental to our democratic system. It reaffirms the promise of equality under the law—that all citizens, regardless of race, ethnicity, or gender, have the chance to take part directly in our democracy. When persons are excluded from participation in our democratic processes solely because of race or gender, this promise of equality dims, and the integrity of our judicial system is jeopardized.

In view of these concerns, the Equal Protection Clause prohibits discrimination in jury selection on the basis of gender, or on the assumption that an individual will be biased in a particular case for no reason other than the fact that the person happens to be a woman or happens to be a man. As with race, the "core guarantee of equal protection, ensuring citizens that their State will not discriminate . . ., would be meaningless were we to approve the exclusion of jurors on the basis of such assumptions, which arise solely from the jurors' [gender]."*

Reversed and remanded in favor of Defendant, J.E.B.

Justice Scalia, Dissenting

Today's opinion is an inspiring demonstration of how thoroughly up-to-date and right-thinking we Justices are in matters pertaining to the sexes, and how sternly we disapprove the male chauvinist attitudes of our predecessors. The price to be paid for this display—a modest price, surely—is that most of the opinion is quite irrelevant to the case at hand. The hasty reader will be surprised to learn, for example, that this lawsuit involves a complaint about the use of peremptory challenges to exclude men from a petit jury. To be sure, petitioner, a man, used all but one of his peremptory strikes to remove women from the jury (he used his last challenge to strike the sole remaining male from the pool), but the validity of his strikes is not before us. Nonetheless, the Court treats itself to an extended discussion of the historic exclusion of women not only from jury service, but also from service at the bar (which is rather like jury service, in that it involves going to the courthouse a lot). All this, as I say, is irrelevant since the case involves state action that allegedly discriminates against men.

The Court also spends time establishing that the use of sex as a proxy for particular views or sympathies is unwise and perhaps irrational. The opinion stresses the lack of statistical evidence to support the widely held belief that, at least in certain types of cases, a juror's sex has some statistically significant predictive value as to how the juror will behave. This assertion seems to place the Court in opposition to its earlier Sixth Amendment "fair cross-section" cases. ("Controlled studies . . . have concluded that women bring to juries their own perspectives and values that influence both jury deliberation and result.")

Of course the relationship of sex to partiality would have been relevant if the Court had demanded in this case what it ordinarily demands: that the complaining party have suffered some injury. Leaving aside for the moment the reality that the defendant himself had the opportunity to strike women from the jury, the defendant would have some cause to complain about the prosecutor's striking male jurors if male jurors tend to be more favorable towards defendants in paternity suits. But if men and women jurors are (as the Court thinks) fungible, then the only arguable injury from the prosecutor's "impermissible" use of male sex as the basis for his peremptories is injury to the stricken juror, not to the defendant. Indeed, far from having suffered harm, petitioner, a state actor under precedents, has himself actually inflicted harm on female jurors. The Court today presumably supplies petitioner with a cause of action by applying the uniquely expansive third-party standing analysis of according petitioner a remedy because of the wrong done to male jurors. Insofar as petitioner is concerned, this is a case of harmless error if there ever was one; a retrial will do nothing but divert the State's judicial and prosecutorial resources, allowing either petitioner or some other malefactor to go free.

The core of the Court's reasoning is that peremptory challenges on the basis of any group characteristic subject to heightened scrutiny are inconsistent with the guarantee of the Equal Protection Clause. That conclusion can be reached only by focusing unrealistically upon individual exercises of the peremptory challenge, and ignoring the totality of the practice. Since all groups are subject to the peremptory challenge (and will be made the object of it, depending upon the nature of the particular case), it is hard to see how any group is denied equal protection.

Even if the line of our later cases guaranteed by today's decision limits the theoretically boundless *Batson* principle to race, sex, and perhaps other classifications subject to heightened scrutiny, much damage has been done. It has been done, first and foremost, to the peremptory challenge system, which loses its whole character when (in order to defend against "impermissible stereotyping" claims) "reasons" for strikes must be given. The right of peremptory challenge "is," as Blackstone says, "an arbitrary and capricious right; and it must be exercised with full freedom, or it fails of its full purpose."

And damage has been done, secondarily, to the entire justice system, which will bear the burden of the

---

*J.E.B. v. Alabama, ex rel. T.B, Supreme Court of the United States 511 U.S. 127 (1994).

expanded quest for "reasoned peremptories" that the Court demands. The extension of *Batson* to sex, and almost certainly beyond, will provide the basis for extensive collateral litigation. . . . Another consequence, as I have mentioned, is a lengthening of the voir dire process that already burdens trial courts.

The irrationality of today's strike-by-strike approach to equal protection is evident from the consequences of extending it to its logical conclusion. If a fair and impartial trial is a prosecutor's only legitimate goal; if adversarial trial stratagems must be tested against that goal in abstraction from their role within the system as a whole; and if, so tested, sex-based stratagems do not survive heightened scrutiny—then the prosecutor presumably violates the Constitution when he selects a male or female police officer to testify because he believes one or the other sex might be more convincing in the context of the particular case, or because he believes one or the other might be more appealing to a predominantly male or female jury. A decision to stress one line of argument or present certain witnesses before a mostly female jury—for example, to stress that the defendant victimized women—becomes, under the Court's reasoning, intentional discrimination by a state actor on the basis of gender.*

I dissent.

CRITICAL THINKING ABOUT THE LAW

The reasoning in Case 3-3 is played out with *Batson v. Kentucky* standing tall and visible in the background. The legal system reinforces our ethical preference for order. The resulting dependability of our legal rules serves as a guide for business decisions, facilitating the many transactions required by modern business.

Nevertheless, the courts recognize that rules must evolve as our social needs and understandings change. Hence, the courts must struggle with achieving a balance between order and flexibility. *J.E.B.* provides an opportunity to use our critical thinking skills to see this tension in action.

1. What facts in our society have become more visible such that Justice Blackmun feels it appropriate to expand the application of *Batson*?

    Clue: What about our history makes Blackmun's reasoning less likely to have been the basis for a Supreme Court decision in 1950?

2. Justice Blackmun disagrees with the respondent concerning the comparative "level of discrimination" experienced by nonwhites and women. Legal reasoning frequently contains phrases like level of discrimination that require some numerical determination—but recognize that clear numbers measuring such a level are hard to come by. As critical thinkers, you can often see soft spots in reasoning by asking, "Now, how are they measuring that concept?" Could you help Justice Blackmun measure "level of discrimination" by suggesting what data might be useful for this determination?

    Clue: Start with the number of people potentially affected, the probability that they would be affected, and the extent of the harm.

3. Justice Scalia does not categorically disagree with extension of *Batson*. What facts would have had to be different for Scalia to have concurred with the majority?

    Clue: Find the section in his dissent in which he explains the inadequacies in the majority's reasoning.

The voir dire process has changed significantly over the years, and to many lawyers, a successful voir dire is the essential element in winning a case. Jury selection today has become a "science," and in most cases involving large potential judgments, at least one side, and often both, use a professional jury selection service. An example of one such service is Litigation Sciences, a firm established in 1979. By 1989, 10 years later, the firm claimed to have handled more than 900 cases, with a win figure of 90 to 95 percent. It employed a full-time staff of more than 100, and the average cost of its services was approximately $200,000, although some cases ran into the millions.[10] Its clients include both major law firms and corporations.

---

[10]Maureen E. Lane, "Twelve Carefully Selected Not So Angry Men: Are Jury Consultants Destroying the American Legal System?" *Suffolk U. L. Rev.* 32: 463 (1999).

*J.E.B. v. Alabama, ex rel. T.B, Supreme Court of the United States 511 U.S. 127 (1994).

Some of the services include identifying demographic data to help lawyers build a profile of the ideal juror, helping design questions for the lawyers to ask during voir dire, and providing such post–voir dire services as mock juries and shadow juries.

A **mock jury** is a body of individuals whose demographic makeup matches that of the actual jury. The lawyers practice their case before the mock jury to find out how receptive the "jurors" are to their arguments and how the mock jurors relate to the witnesses. Lawyers can gain valuable information about what they need to change before actually presenting the case. Depending on how much money a client has, lawyers may go through multiple "trials" before a mock jury.

A **shadow jury** again matches the demographics of the real jury, but the shadow jury actually sits in the courtroom during the trial. They "deliberate" at the end of each day, so the lawyer has an ongoing idea of how the case is faring. The shadow jury's deliberations may let a lawyer know when damage has been done to the case that should be repaired. After the trial is finished, the shadow jury deliberates for a predetermined, brief period. Their "verdict" then helps the lawyer decide whether to try to settle the case before the jury comes back with a verdict. (Remember, the parties can agree to settle at any time until the judge hands down the final decision in the case.)

You can see from this brief discussion how valuable a jury selection service can be. You can also see why many argue that such services should not be allowed. After all, they give a tremendous advantage to the client with more money to spend on the trial.

Opening Statements. Once a jury has been impaneled, or selected, the case begins with the opening statements. Each party's attorney explains to the judge and the jury what facts he or she intends to prove, the legal conclusions to which these facts will lead, and how the case should be decided.

Plaintiff's Case. The plaintiff then presents his or her case, which consists of examining witnesses and presenting evidence. The procedure for each witness is the same. First, the plaintiff's attorney questions the witness in what is called *direct examination*. The plaintiff's lawyer asks questions designed to elicit from the witnesses facts that support the plaintiff's case. The opposing counsel may then cross-examine the witness, but may ask only questions pertaining to the witness's direct examination. The purpose of cross-examination is often to "poke holes" in the witness's testimony or to reduce the credibility of the witness. The plaintiff's attorney then has the opportunity for **redirect examination** to repair any damage done by the cross-examination. The opposing counsel then has a last opportunity to cross-examine the witness to address facts brought out in redirect examination. This procedure is followed for each of the plaintiff's witnesses.

Immediately following the plaintiff's case, the defendant may make a motion for a directed verdict. In making such a motion, the defendant is stating to the court that even if all the plaintiff's factual allegations are true, the plaintiff has not proved his or her case. For example, as will be discussed in Chapter 11, to prove a case of negligence, the plaintiff must prove that the defendant breached his or her duty to the plaintiff, causing compensable injury. If the plaintiff offers no evidence of any compensable injury, then there can be no judgment for the plaintiff. In such a case, a motion for a directed verdict would be granted, and the case would be dismissed. Such motions are rarely granted, because the plaintiff usually introduces some evidence of every element necessary to establish the existence of his or her case.

A motion for a directed verdict also may be made by either party after the presentation of the defendant's case. The party filing the motion (the moving party) is saying that even if the judge looks at all the evidence in the light most

mock jury Group of individuals, demographically matched to the actual jurors in a case, in front of whom lawyers practice their arguments before presenting the case to the actual jury.

shadow jury Group of individuals, demographically matched to the actual jurors in a case, that sits in the courtroom during a trial and then "deliberates" at the end of each day so that lawyers have continuous feedback as to how their case is going.

redirect examination Questioning by the directing attorney following cross-examination. The scope of the questions during redirect is limited to questions asked in the cross-examination.

favorable to the other party, it is overwhelmingly clear that the only decision the jury could come to is that the moving party is entitled to judgment in his or her favor.

Defendant's Case. If the defendant's motion for a directed verdict is denied, the trial proceeds with the defendant's case in chief. The defendant's witnesses are questioned in the same manner as were the plaintiff's, except that it is the defendant's attorney who does the direct and redirect examinations, and the plaintiff's attorney is entitled to cross-examine the witnesses.

Conference on Jury Instructions. If the case is being heard by a jury, the attorneys and the judge then retire for a conference on jury instructions. *Jury instructions* are the court's explanation to the jury of what legal decision they must make if they find certain facts to be true. Each attorney presents to the judge the set of jury instructions he or she feels will enable the jury to accurately apply the law to the facts. Obviously, each attorney tries to state the law in the manner most favorable to his or her client. The judge confers with the attorneys regarding their proposed instructions and then draws up the instructions for the jury.

Closing Arguments. The attorneys' last contact with the jury then follows, as they present their closing arguments. The party who has the burden of proof, the plaintiff, presents the first closing argument; the defendant's closing argument follows. Finally, the plaintiff is entitled to a **rebuttal**. The judge then reads the instructions to the jury, and the jurors retire to the jury room to deliberate. When they reach a decision, the jurors return to the courtroom, where their verdict is read.

rebuttal A brief additional argument by the plaintiff to address any important matters brought out in the defendant's closing argument.

Posttrial Motions. The party who loses has a number of options. A motion for a judgment notwithstanding the verdict may be made. This motion is a request for the judge to enter a judgment contrary to that handed down by the jury on the ground that, as a matter of law, the decision could only have been different from that reached by the jury. For example, if a plaintiff requests damages of $500 but introduces evidence of only $100 in damages, the jurors cannot award the plaintiff the $400 for unsubstantiated damages. If they do so, the defendant would file a motion for a judgment notwithstanding the verdict. Alternatively, the dissatisfied party may file a motion for a new trial, on the ground that the verdict is clearly against the weight of the evidence. If neither of these motions is granted and the judge enters a judgment in accordance with the verdict, the losing party may appeal the decision.

APPELLATE PROCEDURE

As explained earlier, the court to which the case is appealed depends on the court in which the case was originally heard. If a case was heard in a federal district court, it is appealed to the U.S. Circuit Court of Appeals for the geographic region in which the district court is located. If heard in a state trial court, the case is appealed to that state's intermediate appellate court or, if none exists, to the state's final appellate court.

APPLYING THE LAW TO THE FACTS...

Let's say that Annie's case is heard in a federal district court in Connecticut. During the case Annie is in the process of moving to Oregon. After the judge dismisses her case, she seeks to appeal the case in the 9th Circuit Court of Appeals, because she now lives in the 9th Circuit. Is Annie allowed to do this? Does the location of the court to which she can appeal change because of her move?

To appeal a case, the losing party must allege that a prejudicial error of law occurred during the trial. A *prejudicial error* is one that is so substantial that it could have affected the outcome of the case. For example, the judge may have ruled as admissible in court certain evidence that had a major impact on the decision, when that evidence was legally inadmissible. Or the party may argue that the instructions the judge read to the jury were inaccurate and resulted in a misapplication of the law to the facts.

When a case is appealed, there is not a new trial. The attorney for the appealing party (the *appellant*) and the attorney for the party who won in the lower court (the *appellee*) file *briefs*, or written arguments, with the court of appeals. They also generally present oral arguments before the appeals court. The court considers these arguments, reviews the record of the case, and renders a decision. The decisions of the appellate court can take a number of forms. The court may accept the decision of the lower court and **affirm** that decision. Alternatively, the appellate court may conclude that the lower court was correct in its decision, except for granting an inappropriate remedy, and so it will **modify** the remedy. If the appellate court decides that the lower court was incorrect in its decision, that decision will be **reversed**. Finally, if the appeals court feels that an error was committed, but it does not know how that error would have affected the outcome of the case, it will **remand** the case to the lower court for a new trial.

Although the appeals procedure may sound relatively simple compared with the initial trial procedure, appeals require a great deal of work on the part of the attorneys. They are consequently expensive. Thus, when deciding whether to appeal, the businessperson must consider how much money he or she wishes to spend. If a judgment is rendered against a businessperson, it may be less expensive to pay the judgment than to appeal.

Another factor to consider when one is deciding whether to appeal is the precedential value of the case. The case may involve an important issue of law that a party hopes may be decided in her or his favor by an appeals court. If she or he anticipates similar suits arising in the future, it may be important to get a favorable ruling, and if the case appears to be strong, an appeal may be desirable.

Appellate courts, unlike trial courts, are usually composed of a bench of at least three judges. There are no juries. The decision of the appeals court is determined by the majority of the judges. One of the judges who votes with the majority records the court's decision and their reasons in what is called the *majority opinion*. These have precedential value; they are used by judges to make future decisions and by attorneys in advising their clients as to the appropriate course of behavior in similar situations. If any of the judges in a case agrees with the ultimate decision of the majority but for different reasons, he or she may write a *concurring opinion*, stating how this conclusion was reached. Finally, the judge or judges disagreeing with the majority may write *dissenting opinions*, giving their reasons for reaching a contrary conclusion. Dissenting opinions may be cited in briefs by attorneys arguing that the law should be changed. Dissents may also be cited by an appellate judge who decides to change the law.

For most cases, only one appeal is possible. In some states in which there is both an intermediate and a superior court of appeals, a losing party may appeal from the intermediate appellate court to the state supreme court. In a limited number of cases, a losing party may be able to appeal from a state supreme court or a federal circuit court of appeals to the U.S. Supreme Court.

Appeal to the U.S. Supreme Court. Every year, thousands of individuals attempt to have their appeals heard by the U.S. Supreme Court. The Court hears, however, only about 80 cases every year on average. When a party wishes to

affirm Term for an appellate court's decision to uphold the decision of a lower court in a case that has been appealed.

modify Term for an appellate court's decision that, although the lower court's decision was correct, it granted an inappropriate remedy that should be changed.

reverse Term for an appellate court's decision that the lower court's decision was incorrect and cannot be allowed to stand.

remand Term for an appellate court's decision that an error was committed that may have affected the outcome of the case and that the case should therefore be returned to the lower court.

have its case heard by the highest court in the nation, it files a petition with the Court, asking it to issue a *writ of certiorari*, which is an order to the lower court to send the record of the case to the Supreme Court.

As you may guess from the number of cases heard by the Supreme Court, very few writs are issued. The Justices review the petitions they receive and will issue a writ only when at least four Justices vote to hear the case. The Court is most likely to issue a writ when (1) the case presents a substantial federal question that has not yet been addressed by the Supreme Court; (2) the case involves a matter that has produced conflicting decisions from the various circuit courts of appeal and is, therefore, in need of resolution; (3) a state court of last resort holds that a federal law is invalid or upholds a state law that has been challenged as violating federal law; or (4) a federal court has ruled that an act of Congress is unconstitutional.

It is often difficult to predict whether the Court will hear a case. In the first instance in the preceding list, for example, a *federal question* is simply an issue arising under the federal Constitution, treaties, or statutes. Substantiality is more difficult to define. If the decision would affect a large number of people or is likely to arise again if not decided, it may be considered substantial. Sometimes, however, a case may in fact involve a very important federal question of statutory interpretation, yet the Supreme Court may believe that the problem was unclear drafting by Congress, and so it may choose not to hear the case in anticipation of an amendment of the federal statute whose interpretation is at issue. If the Supreme Court refuses to hear a case, this refusal has no precedential effect.

CLASS ACTIONS

In discussing the stages of civil litigation, we have been talking as if there were only one plaintiff and one defendant, but multiple parties may join as plaintiffs and multiple parties may be named or joined as defendants. For example, if a person gets injured using a defective product, he or she would probably sue both the manufacturer and the retailer.

There is a special kind of case, however, in which the plaintiff is not a single party, or even a few parties, but rather a large group of individuals who may not even know each other but who all share a common complaint against the defendant. This kind of case is referred to as a *class action*. For example, all of the shareholders of a corporation may want to sue a member of the board of directors. One of the most common class actions involves product liability cases, situations in which numerous people injured by the same product join together to sue the manufacturer of that product. Other kinds of cases that may give rise to class action suits include discrimination claims and antitrust claims. Sometimes, people come together to bring a class action because their individual claims are so small that separate litigation really is not feasible; when all the claims are combined, though, the amount is large enough that it will be profitable for a lawyer to take the case.

Some see class actions as efficient because, instead of all the individuals filing and trying individual cases based on the same issue, all of the claims can be dealt with in one action. This efficiency seems even more significant when complex issues are involved and the costs of trial preparation are high. When a class action is brought, the case is usually filed in the name of one or two of the parties and all others who are similarly situated. The named plaintiffs in the case have to pay all the court costs, including the costs of finding the names and addresses of everyone in the class and notifying them.

The first step in a class action suit, which differentiates it from other suits, is certification of the class. The court will review the claims to ensure that all the named plaintiffs indeed share a common interest that can adequately be raised by the named plaintiffs. Defendants will often challenge the certification of the class, knowing that if the class does not get certified, the named plaintiffs may not have the resources to bring their cases as separate, individual actions.

The following Supreme Court case illustrates how difficult it sometimes is to meet the standard for class certification.

CASE 3-4

Wal-Mart Stores Inc. v. Dukes
United States Supreme Court
131 S. Ct. 254 (2011)

Three current or former Wal-Mart employees represented 1.5 million claimants who alleged that the company discriminated against them on the basis of sex. They claimed that local managers' discretion over pay and promotions is exercised disproportionately in favor of men, leading to an unlawful disparate impact on female employees, and they sought injunctive and declaratory relief, punitive damages, and back pay.

The District Certified the class, and a divided appellate court upheld the certification. The majority concluded that the respondents' evidence of commonality was sufficient to "raise the common question of whether Wal-Mart's female employees nationwide were subjected to a single set of corporate policies (not merely a number of independent discriminatory acts) that may have worked to unlawfully discriminate against them in violation of Title VII." Wal-Mart appealed the decision.

Justice Scalia

Importantly for our purposes, respondents claim that the discrimination to which they have been subjected is common to all Wal-Mart's female employees. The basic theory of their case is that a strong and uniform "corporate culture" permits bias against women to infect, perhaps subconsciously, the discretionary decision-making of each one of Wal-Mart's thousands of managers—thereby making every woman at the company the victim of one common discriminatory practice. Respondents therefore wish to litigate the Title VII claims of all female employees at Wal-Mart's stores in a nationwide class action.

. . . Under Rule 23(a), the party seeking certification must demonstrate, first, that: "(1) the class is so numerous that joinder of all members is impracticable, (2) there are questions of law or fact common to the class, (3) the claims or defenses of the representative parties are typical of the claims or defenses of the class, and (4) the representative parties will fairly and adequately protect the interests of the class."

Second, the proposed class must satisfy at least one of the three requirements listed in Rule 23(b). Respondents rely on Rule 23(b)(2), which applies when "the party opposing the class has acted or refused to act on grounds that apply generally to the class, so that final injunctive relief or corresponding declaratory relief is appropriate respecting the class as a whole.

. . . [R]espondents moved . . . to certify a plaintiff class consisting of [a]ll women employed at any Wal-Mart domestic retail store at any time since December 26, 1998, who have been or may be subjected to Wal-Mart's challenged pay and management track promotions policies and practices. As evidence that there were indeed "questions of law or fact common to" all the women of Wal-Mart, . . . respondents relied chiefly on three forms of proof: statistical evidence about pay and promotion disparities between men and women at the company, anecdotal reports of discrimination from about 120 of Wal-Mart's female employees, and the testimony of a sociologist, Dr. William Bielby, who conducted a "social framework analysis" of Wal-Mart's "culture" and personnel practices, and concluded that the company was "vulnerable" to gender discrimination.

The crux of this case is commonality—the rule requiring a plaintiff to show that "there are questions of law or fact common to the class." That language is easy to misread, since [a]ny competently crafted class complaint literally raises common questions. For example: Do our managers have discretion over pay? . . . Reciting these questions is not sufficient to obtain class certification. Commonality requires the plaintiff to demonstrate that the class members have suffered the same injury . . . This does not mean merely that they have all suffered a violation of the same provision of law . . . Their claims must depend upon a common contention—for example, the assertion of discriminatory bias on the part of the same supervisor. That common contention, moreover, must be of such a nature that it is capable of classwide resolution—which means that determination of its truth or falsity will resolve an issue that is central to the validity of each one of the claims in one stroke.

"What matters to class certification . . . is not the raising of common 'questions'—even in droves—but, rather the capacity of a classwide proceeding to generate common answers apt to drive the resolution of the litigation. Dissimilarities within the proposed class are what have the potential to impede the generation of common answers."

. . . Here respondents wish to sue about literally millions of employment decisions at once. Without some glue holding the alleged reasons for all those decisions together, it will be impossible to say that examination of all the class members' claims for relief will produce a common answer to the crucial question why was I disfavored.

Falcon suggested two ways in which that conceptual gap might be bridged. First, if the employer "used a biased testing procedure to evaluate both applicants for employment and incumbent employees, a class action on behalf of every applicant or employee who might have been prejudiced by the test clearly would satisfy the commonality and typicality requirements. . . ." Second, "[s]ignificant proof that an employer operated under a general policy of discrimination conceivably could justify a class of both applicants and employees if the discrimination manifested itself in hiring and promotion practices in the same general fashion, such as through entirely subjective decision-making processes."

We think that statement precisely describes respondents' burden in this case. The first manner of bridging the gap obviously has no application here; Wal-Mart has no testing procedure or other companywide evaluation method that can be charged with bias. The whole point of permitting discretionary decision-making is to avoid evaluating employees under a common standard. The second manner of bridging the gap requires "significant proof" that Wal-Mart "operated under a general policy of discrimination." That is entirely absent here. Wal-Mart's announced policy forbids sex discrimination . . . and as the District Court recognized the company imposes penalties for denials of equal employment opportunity. The only evidence of a "general policy of discrimination" respondents produced was the testimony of Dr. William Bielby. Relying on "social framework" analysis, Bielby testified that Wal-Mart has a "strong corporate culture," that makes it "'vulnerable'" to "gender bias." He could not, however, "determine with any specificity how regularly stereotypes play a meaningful role in employment decisions at Wal-Mart. . . . [W]e can safely disregard what he has to say. It is worlds away from "significant proof" that Wal-Mart "operated under a general policy of discrimination."

The only corporate policy that the plaintiffs' evidence convincingly establishes is Wal-Mart's "policy" of allowing discretion by local supervisors over employment matters. On its face, of course, that is just the opposite of a uniform employment practice that would provide the commonality needed for a class action. . . . It is also a very common and presumptively reasonable way of doing business—one that we have said should itself raise no inference of discriminatory conduct.

. . . Respondents have not identified a common mode of exercising discretion that pervades the entire company. . . . In a company of Wal-Mart's size and geographical scope, it is quite unbelievable that all managers would exercise their discretion in a common way without some common direction. Respondents attempt to make that showing by means of statistical and anecdotal evidence, but their evidence falls well short. The statistical evidence consists primarily of regression analyses performed by Dr. Richard Drogin. . . . Drogin concluded that "there are statistically significant disparities between men and women at Wal-Mart . . . [and] these disparities . . . can be explained only by gender discrimination". . . . Bendick compared workforce data from Wal-Mart and competitive retailers and concluded that Wal-Mart "promotes a lower percentage of women than its competitors." Even if they are taken at face value, these studies are insufficient to establish that respondents' theory can be proved on a classwide basis.

In *Falcon*, we held that one named plaintiff's experience of discrimination was insufficient to infer that "discriminatory treatment is typical of [the employer's employment] practices." . . . A similar failure of inference arises here. . . . A regional pay disparity, for example, may be attributable to only a small set of Wal-Mart stores, and cannot by itself establish the uniform, store-by-store disparity upon which the plaintiffs' theory of commonality depends.

There is another, more fundamental, respect in which respondents' statistical proof fails. . . . Other than the bare existence of delegated discretion, respondents have identified no "specific employment practice"—much less one that ties all their 1.5 million claims together. Merely showing that Wal-Mart's policy of discretion has produced an overall sex-based disparity does not suffice. Respondents' anecdotal evidence suffers from the same defects, and in addition is too weak to raise any inference that all the individual, discretionary personnel decisions are discriminatory. . . .

In sum, we agree with Chief Judge Kozinski that the members of the class: "held a multitude of different jobs, at different levels of Wal-Mart's hierarchy, for variable lengths of time, in 3,400 stores, sprinkled across 50 states, with a kaleidoscope of supervisors (male and female), subject to a variety of regional policies that all differed. . . . Some thrived while others did poorly. They have little in common but their sex and this lawsuit."

Rule 23(b)(2) applies only when a single injunction or declaratory judgment would provide relief to each member of the class. It does not authorize class certification when each individual class member would be entitled to a different injunction or declaratory judgment against the defendant.*

Reversed, in favor of Defendant Wal-Mart

Once the class has been certified, the parties often enter into settlement negotiations. The court will approve a classwide settlement only if it is fair and equitable and benefits the entire class, not just the named plaintiffs and their lawyers. Once a settlement has been approved by the court, it legally satisfies the claims of all the class members.

---

*Wal-Mart Stores Inc. v. Dukes, United States Supreme Court, 131 S. Ct. 254 (2011).

In an effort to reform class action lawsuits, Congress passed, and President Bush signed, the Class Action Fairness Act of 2005. The Class Action Fairness Act had several intended goals. First, it sought to limit the enormous legal fees attorneys representing plaintiff classes frequently receive as part of their service. Second, the act was written to allow the defendants in class action lawsuits to have greater access to federal courts. Finally, the act was intended to protect the interests of the individual class members and guarantee them equitable compensation.

As part of the new regulations on attorney fees, attorneys now receive compensation based on the actual amount class members claim as compensation, rather than on the gross sum awarded to the plaintiff class. In addition, the court must first approve the award the plaintiff class members are to receive to ensure that the award does not violate the defendants' due process rights. If the attorney's fees are not to be determined by the amount of the award, the attorney's compensation is to be limited to actual time spent working on the case.

Moving beyond attorney's fees, the Class Action Fairness Act changes the requirements for diversity of citizenship in class actions, as well as the general requirements for federal jurisdiction. The act provides federal courts with jurisdiction over any class action where there is partial diversity of citizenship; that is, if any of the class members is a citizen of a different state than any of the defendants, the federal courts have jurisdiction. The federal courts also have jurisdiction whenever any plaintiff or defendant is a foreign state or the citizen of a foreign state. There must also be at least $5 million in controversy. These new requirements under the act allow more class actions to fall within federal jurisdiction and make it much harder for plaintiffs to pursue their actions in state courts. Business defendants are typically happy with this last change, as they tend to face smaller awards in federal courts than in state courts.

Despite the greater access to federal courts, the act also allows federal courts some discretion in accepting jurisdiction. A district court can choose not to accept jurisdiction over a class action if between one-third and two-thirds of the plaintiff class members and the primary defendants are citizens of the same state. There are also two situations in which the federal court has no choice regarding jurisdiction. When fewer than one-third of the plaintiff class members are residents of the same state as the primary defendants, then the class action will be subjected to federal jurisdiction. If more than two-thirds of the plaintiff class members are residents of the same state as the primary defendants, however, there will be no federal jurisdiction in the case.

As mentioned previously, the act contains a number of provisions that are intended to protect the interests of the individual plaintiff class members. For example, the act forbids any settlement that would result in a net loss to the class members. That is, class members cannot accept a settlement that does not cover their damages, unless the court decides that other, nonmonetary benefits are valuable enough to outweigh the net monetary loss. In addition, class members cannot be awarded different monetary amounts based upon their geographical location. All class members, barring differences in their actual damages, are awarded the same monetary amount.

Global Dimensions of the American Legal System

This chapter has focused on the American legal system. With the growth of multinational corporations and trade among nations, Americans will likely increasingly become involved in disputes in foreign nations, and foreigners will likely increasingly become involved in disputes with Americans and American corporations.

When parties make international agreements, they can incorporate as a term of the agreement their choice of which nation's court will hear any disputes arising under the agreement. Because of differences between the U.S.

litigation system and others, it is important to compare the procedures in each country before choosing a forum. For example, in Japan, there is no procedure comparable to discovery, so parties go to trial not knowing what evidence the other side has.

LINKING LAW AND BUSINESS

Management

In your management class, you may have learned about a concept known as cost-benefit analysis. This idea is defined as the process by which managers weigh the benefits or revenues of a particular activity in comparison to the costs of performing the action. Usually, managers will decide to pursue an action if the benefits outweigh the costs.

Managers or other decision makers can effectively come to a conclusion as to which alternative to pursue only after the options have been evaluated. Management texts often state that there are three basic steps decision makers should follow in this evaluation: (1) Estimate, as accurately as possible, the potential effects of each of the possible actions. (2) Assign probabilities to each of the expected effects of each decision if the idea were implemented. (3) Compare the possible effects of each alternative decision and the probabilities of each. Meanwhile, consideration should be given to organizational objectives. After taking these three steps, managers will have a better understanding about the benefits and risks of alternative decisions. Therefore, it is hoped that they will be able to understand which choice will be most advantageous to the organization.

A cost-benefit analysis can also be done when a businessperson is faced with the decision of whether to appeal a court decision. The businessperson should examine the costs of the appellate procedures, the probability of the outcome in the appeals court, and the time involved with the appeals process. Thereafter, the businessperson can choose the course that will potentially be most beneficial.

COMPARATIVE LAW CORNER

The Judicial System in Germany

The United States and Germany are major trading partners and have many similarities, but the German judicial system is very different from the American judicial system. German law is based on a civil-law tradition rather than a common-law tradition. The United States has a common-law system, which relies on precedents set by previous cases to rule on current cases. German judges make decisions based on the country's extensive civil codes, rather than previous decisions. German judges are not elected as many American judges are; instead, most are appointed for life, after a probationary period.

The judicial system is a federal system, as in the United States, but German courts are separated by field. The ordinary courts hear most criminal and civil cases, each specialty court (for labor, patents, social, administrative, and fiscal issues) hears cases related to its individual area, and constitutional courts hear cases involving constitutional issues. The courts all have local, land (state), and federal levels. The highest court in Germany is the Bundesverfassungsgericht, the Federal Constitutional Court, which deals only with constitutional issues, unlike the U.S. Supreme Court. Each specialty court has its own highest court of appeals, such as the Federal Court of Germany for the ordinary courts.

Germany does not have any jury trials; all cases are heard by a judge or a panel of judges. Also, the judges are the primary questioners of witnesses. Lawyers can question witnesses after the judges have finished. This legal factfinding method differs from the American method of examination and cross-examination.

With the increase in trade, many foreigners now purchase American goods. Because of differences between court systems, many citizens of foreign countries who have allegedly been injured by U.S. corporations prefer to sue in the United States. In Japan, for example, there are no contingency fees, and an injured plaintiff must pay his or her lawyer's fees up front, at a cost of 8 percent of the proposed recovery plus nonrefundable court costs. Also, in Japan, there are no class actions.

SUMMARY

Our American legal system is really composed of two systems: a federal system and a state system. When one has a legal dispute, subject matter jurisdiction determines which court system will hear the case. Almost all cases fall within the state court's jurisdiction. Only the limited number of cases within the exclusive jurisdiction of the federal courts do not. A case may be heard in either court when there is concurrent jurisdiction. Concurrent jurisdiction exists when (1) the case involves a federal question, or (2) there is diversity of citizenship between the plaintiff and the defendant. Besides having subject matter jurisdiction, a court must also have in personam jurisdiction and proper venue to hear a case.

Cases are filed in courts of original jurisdiction. In the state system, these courts are usually called the courts of common pleas or county courts. In the federal system, the courts of original jurisdiction are called the district courts. In the state system, state courts of appeals and state supreme courts have appellate jurisdiction. Depending on the state, there may be either one or two levels of appeal. In the federal system, cases are appealed to the circuit court of appeals and then to the U.S. Supreme Court.

Cases are guided through the courts by attorneys. Juries act as finders of fact in trials. Judges resolve questions of law and, in bench trials, also serve as finders of fact.

There are four basic stages in a lawsuit. (1) In the pretrial stage, there are (a) informal negotiations, (b) pleadings, (c) pretrial motions, (d) discovery, and (e) a pretrial conference. (2) Next comes the trial, with (a) jury selection, (b) opening statements, (c) the plaintiff's case, (d) the defendant's case, (e) jury instructions, and (f) closing arguments. (3) Third are the posttrial motions, which may include a motion for a judgment notwithstanding the verdict or a motion for a new trial. (4) The final stage is the appellate stage, during which the party who lost at the trial appeals the case.

REVIEW QUESTIONS

3-1 Identify the different types of jurisdiction and explain why each is important.

3-2 Explain the two situations that cause the state and federal courts to have concurrent jurisdiction.

3-3 What is venue?

3-4 What is the relationship between federal district courts and courts of common pleas?

3-5 What is the attorney–client privilege and what is the rationale for its existence?

3-6 Explain the importance of the work-product doctrine.

REVIEW PROBLEMS

3-7 Jacobson, a Michigan resident, sues Hasbro Corporation for negligence after one of Hasbro's truck drivers fell asleep and ran his semi off the road and into Jacobson's house, causing structural damage of approximately $80,000. Hasbro has small plants in Michigan, Ohio, and Indiana. The company is incorporated in Illinois and has its central offices there. Jacobson files his case in the state court in Michigan. Hasbro files a motion for removal, which Jacobson contests, arguing that the case does not fall within the concurrent jurisdiction of the federal courts. Should the case be transferred? Why or why not?

3-8 Bill, a white male, is charged with spousal abuse. His attorney uses his peremptory challenges to remove all white females from the jury. The prosecution objects. Was there any impropriety in the jury selection process?

3-9 Marx Corporation is incorporated in the state of Delaware, but all of the firm's business is conducted within the state of New York. Sanders, a Delaware resident, is injured by one of Marx Corporation's products and subsequently files suit against Marx Corporation in Delaware state court. Marx files a motion to dismiss the case on the ground that

Delaware cannot assert jurisdiction over the corporation because it does not conduct business in Delaware; it is incorporated there only because it gained certain legal advantages from incorporating in that state. Explain why the Delaware state court system does or does not have jurisdiction over this case.

3-10 Carson is a resident of Clark County, Nevada. He sues Stevens, a resident of Washoe County, Nevada, for injuries he received in an accident that took place in Washoe County. Carson files the case in Clark County. Can Stevens get the case moved to Washoe County? How would he try to do so?

3-11 Attorney Fox represented Davis in a number of drunk driving cases. Davis shows up at Fox's office to discuss having the attorney draw up a will for him. The attorney recognizes that Davis is clearly intoxicated. The attorney offers to pay for a cab to take Davis home, but he refuses the offer. Fox's secretary suggests that he call the state highway patrol. If Fox calls the highway patrol, is he violating the attorney–client privilege? Why or why not?

3-12 Watson brought a negligence case against the PrescripShun Drug Store to recover damages for injuries he received from falling on the wet floor of the store. He believed that the store was negligent for marking the floor with only a small sign that said "Slippery When Wet." The trial court refused to let Watson introduce evidence that after his fall the store started marking wet floors with cones and a large sign saying, "Caution—Floor Is Wet and Slippery." Watson lost in the trial court and lost his appeals to the state appeals court and state supreme court. Will he be able to appeal to the U.S. Supreme Court? Why or why not?

CASE PROBLEMS

3-13 In 2007, an Alabama court ordered a father to pay $600 for monthly postminority educational support. Ten months later, in December of that year, the mother brought the father to court because the father was not paying the educational support. He also had not shown up for the court hearing, and was therefore in contempt of court. However, the father argued that under Ala. R. Civ. p. 60(b), he did not have to make the payments because his attorney did not tell him about the court hearing. The rule he cited could have protected him because the rule provides relief for a defendant based on "mistake, inadvertence, surprise, or excusable neglect." Unfortunately for the father, the rule also states that an appeal must be made four months after the original order; the father in this case made the appeal 10 months later. Thus, due to the missed deadline, the court would have no jurisdiction over the case. Nevertheless, the appellant court then considered taking the father's appeal under a subsection of the 60(b) rule, 60(b)(6). This section states that an order may be undone for "any . . . reason justifying relief." This subsection also contains a time limit more ambiguous than four months, stating that "[the appeal] only has to be brought within a 'reasonable time.'" Was the father's appeal successful? Did the court determine whether it had jurisdiction over the case? *Noll v. Noll*, 2010 Ala. Civ. App. LEXIS 32.

3-14 The three Missouri plaintiffs owned Whispering Lane Kennel, a Missouri business that breeds and shows AKC dogs, especially the Chinese Crested breed. Defendants show and sell Chinese Crested dogs in competition with plaintiffs. Defendants ran a website, www.stop-whisperingland.com, that posted negative information about the plaintiffs' kennel that was viewed by people around the country, including at least 25 Missouri residents who were in the dog breeding business. The owners of Whispering Lane Kennel sued out-of-state defendants for libel in state district court in Missouri. The court held that minimum contacts had not been established, so Missouri courts did not have jurisdiction over the defendants. How do you think the court ruled on the plaintiffs' appeal? Why? *Baldwin v. Fisher-Smith*, S.W.3d (2010 WL 2662977, Ct. App., Mo., 2010).

3-15 A debtor who had defaulted on a student loan asked a bankruptcy court to decide that the student-loan principal was fully paid. The court made a determination about the student-loan principal, determining that the debtor still owed what was due on the student loan. However, the court also decided that postpetition interest was to be awarded, and collection costs were to be denied. Subsequently, the Educational Credit Management Corporation appealed this decision, claiming that the court lacked jurisdiction over postpetition interest and collection costs because the court was a bankruptcy court. In other words, postpetition interest and collection costs did not fall under United States Code Title 11 (over which the bankruptcy court had jurisdiction) or 28 U.S.C. § 1334(a) or (b) (over which the bankruptcy court also had jurisdiction). Furthermore, the Credit Management Corporation argued that a bankruptcy proceeding is completely unrelated to collection costs and postpetition interest. Do you think the Credit Management Corporation's appeal was successful? Was the decision of the first court reversed? *Educational Credit Management Corp. v.*

Kirkland, 600 F.3d 310; 2010 U.S. App. LEXIS 5212 (4th Cir. 2010).

3-16 Plaintiff resident of New Jersey was a collector of classic automobiles. Specifically, the plaintiff owned a rare Alfa Romeo 6C 3000 Competizione automobile. The defendant, a company that auctions automobiles, makes its business by sending catalogs to New Jersey residents. The plaintiff alleged that the auction company made a defamatory statement about the plaintiff's rare automobile in three different places: the company's website; the company's catalog of listed cars for auction; and at the company's auction in California. The plaintiff claimed that at the same time he was trying to sell his automobile, the company's catalog that included the defamatory statement was published. The plaintiff filed a claim for defamation and trade libel against the auction company. The defendant motioned to dismiss the claim due to lack of personal jurisdiction, reasoning that its business activities did not satisfy the minimum contacts requirement. How do you think the court ruled? Why? *Wolfe v. Gooding & Co.* 2015 U.S. Dist. LEXIS 26165.

3-17 In 2004, EJS Properties LLC sued the city of Toledo and Bob McCloskey, alleging the city violated constitutional rights to property, liberty, and due process by denying a rezoning application from EJS in 2002. The company said that then-councilman McCloskey demanded a $100,000 payment to approve the deal and the rest of the council followed his lead in eventually denying the rezoning request. Eventually the case made it to the U.S. 6th Circuit Court of Appeals. However, the judge who had previously heard the case was still making decisions on some small parts of the case. The Appellate court grappled with whether it had jurisdiction over the case yet with the prior court still acting on it. How did this case turn out? *EJS PROPERTIES, LLC v. City of Toledo*, 689 F.3d 535 - Court of Appeals, 6th Circuit, 2012.

3-18 A debtor, P.K., filed a motion for change of venue, seeking leave to file within the Northern District of Ohio, Western Division (Toledo), asserting that this venue was proper. The debtor reasoned that her place of residence, Lambertville, Michigan, was closer in proximity to Toledo, Ohio than Detroit, Michigan. The debtor also reasoned that she had a principal place of business in Ohio because she was employed at the Toledo Northwest Ohio Food Bank. Last, the debtor argued that her income was below the poverty level, and litigating the case in Michigan versus Ohio would "place undue hardship on the debtor." Do you think the court granted the debtor's motion for change of venue? Why or why not? *In re: Golembiewski* 2015 Bankr. LEXIS 695.

3-19 Plaintiff Advance Industrial Coating, LLC brought a suit against defendant Westfield Insurance Company based on diversity jurisdiction. The plaintiff alleged that the defendant was an Ohio corporation "authorized to issue surety bonds in the State of Florida." The court reasoned that the principal place of business for a corporation is determined by "the nerve center test," and that a company is a "citizen of every state in which one of its members is located." Do you think that the court found that it had federal jurisdiction to facilitate the case, based on diversity jurisdiction? *Advance Indust. Coating v. V.* 2015 U.S. Dist. LEXIS 27655.

3-20 Vernon Minton created a software and subsequently leased it to the National Association of Securities Dealers, Inc. ("NASD") and NASDAQ Stock Market, Inc. ("NASDAQ"). Later, he patented the software and then filed a lawsuit for patent infringement against the two entities. His case was dismissed by a trial court due to the "on-sale bar" rule. Minton then sued his lawyer for legal malpractice saying that his lawyer should have argued on another basis. The attorney argued that the case was not important enough to be heard in federal court, yet federal courts handle patent litigation. What was the final decision in this case? *Gunn v. Minton*, 105 U.S.P.Q.2d 1665 (U.S. 2013) [2013 BL 43481].

THINKING CRITICALLY ABOUT RELEVANT LEGAL ISSUES

The Election of Judges

Despite a lack of consensus among legal commentators, there is a clear answer to the debate about whether judges should be elected or appointed. America is a democracy; it is only right that the American people should get to elect all judges. Like the legislative and executive branches of government, the judicial branch functions to provide services to the American people. Those very people deserve to elect their judges just as they elect members of Congress and the president. One of the keys to democracy is having a responsible and responsive government. Judges will be neither responsible nor responsive unless they must face the American people periodically and ask for their votes.

Part of holding judges accountable for their legal decisions is not only choosing to elect or reelect them, but also having the ability to remove a judge who is not performing his duty adequately.

The American people should have the option of removing judges who fail to uphold the standards and morals of the community. Accordingly, allowing voters to have recall elections would further prevent judges from engaging in undue judicial activism. Judges who are appointed for life terms are beholden to no one; that life tenure challenges the very essence of democracy.

Sometimes a judge's ideology or judicial philosophy will change over time; this change affects the decisions the judge will make. When a judge changes her decision-making process, she is no longer staying true to why she was elected in the first place. One way to fix unexpected changes is to have periodic elections for judges. These elections will help to keep judges consistent in their rulings and interpretations, while also preventing surprises for the American people who elected the judges in the first place.

By having elections, out-of-touch judges who do not reflect the current social climate can be removed in favor of judges who are in touch with the American people. This last point has the added benefit of possibly bringing younger people to the bench, thus opening up the possibility for a wider group of Americans to shape the law.

1. How would you frame the issue and conclusion of this essay?
2. What ethical norms does the author primarily rely on in arguing for why judges should be elected?
3. Part of being a critical thinker is avoiding the temptation to dichotomize (look at everything as an either-or situation) and to look for other reasonable alternatives. Does the author engage in any dichotomous thinking? If so, what are other reasonable alternatives?

   Clue: What either-or situations does the author create, and are there third and fourth possibilities?

4. Write an essay that someone who holds an opinion opposite to that of the essay author might write.

   Clue: What other ethical norms could influence an opinion on this issue?

ASSIGNMENT ON THE INTERNET

As you learned in this chapter, the question of jurisdiction determines whether a court has the power to render a meaningful decision. The growth of Internet commerce, however, brings additional jurisdictional questions and concerns that have yet to be resolved. Use the website **corporate.findlaw.com**. Search "standards for internet jurisdiction," then scroll down to the section titled "Other Legal Research." Click on the first link, titled "Standards for Internet Jurisdiction." Read this article in order to familiarize yourself with the interactive and passive-use distinctions made in cases of Internet jurisdiction.

Next, apply this distinction to the case of *Barton Southern Co., Inc. v. Manhole Barrier Systems, Inc. and JFC Co.*, 318 F. Supp. 2d 1174 (N.D. Ga. 2004), which can be found using **findlaw.com** or through the LexisNexis database. Does the interactive/passive distinction help resolve the issue of jurisdiction in this case? Why or why not?

ON THE INTERNET

www.allbusiness.com/legal/961805-1.html This website contains an article discussing the difficulties of and possible ways in which businesses can seek to obtain in personam jurisdiction over foreign "cybersquatters." *Cybersquatters* are people in other countries who have bought up popular domain names and are using them for their own personal gain and to the detriment of the owner of the products with the purchased domain names.

writ.news.findlaw.com/amar/20040220.html In his column, Vikram David Amar discusses some problems with juries as well as possible changes to the jury selection process that might improve juries.

www.litigationresources.com Litigation Resources is a company that offers business and legal professionals with many tools and services they might need to ease the litigation process.

www.martindale.com This address is for the site of the Martindale-Hubble Law Directory, which provides information about lawyers and law firms. Typical entries include area of special expertise, address, and telephone number.

www.lexisnexis.com This site will lead you to a number of sites that contain the Federal Rules of Civil Procedure in a variety of formats. Search "federal rules of civil procedures."
http://www.justlawlinks.com/STATUTE/statstat.htm Go to this page to find a link to your state's constitution, statutes, and rules of civil procedure.
www.supremecourt.gov This site provides a wealth of information about the U.S. Supreme Court. Click on "About the Court" to learn more.

FOR FUTURE READING

Anenson, T. Leigh. "Creating Conflict of Interest: Litigation as Interference with the Attorney-Client Relationship." *American Business Law Journal* 43 (2006): 173.

Babcock, Chip, and Luke Gilman. "Symposium: Social Media: Use of Social Media in Voir Dire." *The Advocate* 60 (2012): 44.

Bonfield, Lloyd. *American Law and the American Legal System in a Nutshell*. St. Paul, MN: West, 2006.

Henderson, Jr., James A. "A Process Perspective on Judicial Review: The Rights of Party Litigants to Meaningful Participation." *Michigan State Law Review* (2014): 979.

Magier, David S. "Tick, Tock, Time Is Running Out to Nab Cybersquatters: The Dwindling Utility of the Anticybersquatting Consumer Protection Act."

IDEA: The Intellectual Property Law Review 46 (2006): 415.

McTarnaghan, Christian. "Equity Run Amuck: The Necessary Reevaluation of the Preliminary Injunction Standard to Reflect Modern Day Legal Realities—A Comparison of the Massachusetts and Delaware Noncompete Agreement Preliminary Injunction Standard." *Suffolk University Law Review* 47 (2014): 871.

Stoltz, Brian W. "Rethinking the Peremptory Challenge: Letting Lawyers Enforce the Principles of *Batson*." *Texas Law Review* 85 (2007): 1031.

Walkowiak, Vincent S., ed. *Attorney-Client Privilege in Civil Litigation: Protecting and Defending Confidentiality* (3rd ed.). New York: American Bar Association, 2005.

CHAPTER FOUR

Alternative Tools of Dispute Resolution

- NEGOTIATION AND SETTLEMENT
- MEDIATION
- ARBITRATION
- MINITRIALS
- EARLY NEUTRAL CASE EVALUATION
- PRIVATE TRIALS
- SUMMARY JURY TRIALS
- COURT-ANNEXED ALTERNATIVE DISPUTE RESOLUTION
- THE FUTURE OF ALTERNATIVE DISPUTE RESOLUTION
- GLOBAL DIMENSIONS OF ALTERNATIVE DISPUTE RESOLUTION

Pacific Gas and Electric Company (PG&E) was faced with six disputes stemming from the crash of a helicopter that hit one of its electrical lines. Officials at PG&E knew that they were facing the expensive and time-consuming litigation process described in Chapter 3. They were aware that a trial would take about two years to wind its way through the court system, with litigation costs of $300,000 if the matter was settled before the trial and double that amount if the case was tried through to a verdict. Instead, the case was resolved within 10 months, and legal fees and administrative costs were kept to around $20,000.[1]

Like a growing number of would-be litigants who see the trial process as unwieldy, time-consuming, and expensive, PG&E decided to consider resolving its dispute outside the court, through what is often referred to as **alternative dispute resolution (ADR)**. Many judges now encourage the increased use of these alternatives to litigation, which include (1) negotiation and settlement, (2) mediation, (3) arbitration, (4) minitrials, (5) early neutral case evaluation, (6) private trials, (7) summary jury trials, and (8) court-annexed alternative dispute resolution. The ADR method that PG&E successfully opted for was mediation.

These options are becoming so common that many law and some business schools now offer courses covering alternative dispute resolution methods. In the business world, 600 of the nation's largest corporations have signed a

alternative dispute resolution (ADR) Resolving legal disputes through methods other than litigation, such as negotiation and settlement, arbitration, mediation, private trials, minitrials, summary jury trials, and early neutral case evaluation.

---

[1] E. J. Pollock, "Mediation Firms Alter the Legal Landscape," *Wall Street Journal* 1 (Mar. 22, 1993).

pledge not to sue another corporation before first trying to resolve the conflict out of court.[2] Nearly every state court has some ADR program in place.

Almost all of these alternative methods share certain advantages over litigation. First, they are generally less expensive. For example, PG&E shifted toward an aggressive litigation-alternative approach in 1988, and found that by 1993, its legal department's operation costs, including legal fees, had fallen by 9 percent. The amount it paid out in judgments and settlements during that time fell by 25 percent.[3] While researchers would agree that ADR saves money, the estimates of cost savings vary depending on a number of factors such as the exact type of case, the form of ADR used, and local conditions. Estimates of the reduction of costs of using ADR rather than litigation range from 2 percent to 50 percent of the cost of litigation.[4] Second, most ADR methods are more convenient for the participants: They are less time-consuming, and the formal-hearing times and places can be set to accommodate the parties. Admittedly, it is difficult to estimate the time required by a given ADR case. Studies of private commercial arbitration cases, however, have shown that the average time lag from filing date to the decision date is 145 days. Third, the persons presiding over the resolution process can be chosen by the parties, and, in many cases, they are chosen because they are more familiar with the subject matter of the dispute than a randomly assigned judge would be.

These alternatives may also prevent adverse publicity, which could be ruinous to a business. Imagine how much better Denny's image might be, for example, without the adverse publicity it received as a result of litigation over charges of discrimination! Similarly, these alternatives preserve confidentiality, which may be extremely important when a company's trade secrets are involved. A lot of information that a firm might wish to keep from its competitors could come out in a lawsuit but would not be subject to public disclosure if the matter was settled through an ADR method. Another related potential advantage might be that the immediate case could be resolved through ADR without a precedent being set. Finally, the less adversarial ADR methods might also help to preserve the relationship between the parties, who often desire to continue doing business with one another. In the following sections, we will examine the most important ADR methods in greater detail. From these discussions, future businesspersons should gain some awareness of typical situations in which each of these alternatives may be preferable to litigation.

CRITICAL THINKING ABOUT THE LAW

Why would one want to use an alternative method for dispute resolution? Alternative dispute resolution seems to be advantageous compared with the traditional method of litigation for several reasons. For example, ADR methods are less time-consuming than litigation. The following questions will help you learn to think critically about alternative dispute resolution.

1. What reasons are offered to suggest that ADR is advantageous compared with litigation?

    Clue: Reread the introductory section.

2. What primary ethical norms underlie the reasons given to suggest that ADR is advantageous?

    Clue: Examine the reasons given in response to Question 1. Consider the list of ethical norms given in Chapter 1. For example, ADR methods are less time-consuming. Which ethical norm is being upheld with this reason?

---

[2]M. Chambers, "Sua Sponte," *National Law Journal* 21 (Sept. 27, 1993).
[3]Pollock, *supra* note 1.
[4]I. Love, "Settling Out of Court, How Effective Is Alternative Dispute Resolution," Viewpoint, Public Policy for the Private Sector, Note Number 329 (October 2011). Available at http://www.worldbank.org/fpd/publicpolicyjournal.

Negotiation and Settlement

The oldest, and perhaps the simplest, alternative to litigation is negotiation and settlement. **Negotiation and settlement** is a process by which the parties to a dispute come together informally, either with or without their lawyers, and attempt to resolve their dispute. No independent or neutral third party is involved.

To successfully negotiate a settlement, each party must, in most cases, give up something in exchange for getting something from the other side. Almost all lawyers will attempt to negotiate a settlement before taking a case to trial or before adopting a more formal type of dispute resolution method. Attempts at negotiation and settlement are so common that we often do not even consider negotiation as an alternative to litigation.

negotiation and settlement An alternative dispute resolution method in which the disputant parties come together informally to try to resolve their differences.

Mediation

A key advantage to the negotiation and settlement method is that it is not as adversarial as litigation; many would like to resolve their disputes through negotiation and settlement to help preserve their relationship. If this informal process does not work, however, the least adversarial of the formal methods of dispute resolution is mediation.

Mediation is a process in which the two disputants select a party, usually one with expertise in the disputed area, to help them reconcile their differences. It is sometimes characterized as a creative and collaborative process involving joint efforts of the mediator and the disputants. Mediation primarily differs from litigation and arbitration (which we discuss next) in that the mediator makes no final decision. She or he is simply a facilitator of communication between the disputing parties.

mediation An alternative dispute resolution method in which the disputant parties select a neutral party to help them reconcile their differences by facilitating communication and suggesting ways to solve their problems.

Even though different mediators use different techniques, their overall goals are the same. They try to get the disputants to listen to each other's concerns and understand each other's arguments with the hope of eventually getting the two parties communicating. Once the parties are talking, the mediator attempts to help them decide how to solve their problem. The agreed-upon resolution should be "fair, equitable," and "based upon sufficient information."

Although there is no guarantee that a decision will be reached through mediation, when a decision is reached the parties generally enter into a contract that embodies the terms of their settlement. Skilled mediators will attempt to help the parties to draft agreements that reflect the principles underlying mediation; that is, the agreement should not attempt to assess blame either implicitly or explicitly, but should reflect mutual problem solving and consensual agreement. If one party does not live up to the terms of the settlement, he or she can then be sued for breach of contract. Often, however, the parties are more likely to live up to the terms of the agreement than they are to obey a court order, because they were the ones who reached the agreement. The agreement is their own idea of how to resolve the dispute, not some outsider's solution.

If the mediation is not successful and the parties subsequently take their dispute to arbitration or litigation, nothing that was said during mediation can be used by either party at a later proceeding. Whatever transpires during the mediation is confidential.

SELECTION OF A MEDIATOR

Mediators are available from several sources. In any community, you will be able to find organizations offering mediation services by looking in the Yellow Pages or online. Often the local bar association or Better Business Bureau offers mediation services. A governmental organization, the Federal Mediation and Conciliation Service (FMCS), also provides mediation services, as do national private services such as JAMS Resolution Services, which has locations across the country and can be found on the Internet. A person in need of mediation services can also find a mediator by going to a site such as Mediate.com, which contains a mediator directory for mediators in the United States and abroad. The state in which one resides may also have a professional organization that can provide a list of recognized mediators, such as California's Academy of Professional Neutrals.

One factor to consider when selecting a mediator is what type of background he or she has. Some companies use only former judges, whereas others use judges, lawyers, and nonlawyers as mediators. There is, not surprisingly, a lot of contention over who makes the best mediator. Many people prefer judges because of their legal knowledge, but others say that judges are too ready to make a decision because they spend most of their careers making clear-cut decisions rather than seeking to find compromises.

COMMON USES OF MEDIATION

Perhaps the best-known use of mediation is in collective bargaining disputes. Under the National Labor Relations Act, before engaging in an economic strike to achieve better wages, hours, or working conditions under a new collective bargaining agreement, a union must first contact the FMCS and attempt to mediate contract demands.

Mediation is also used increasingly to resolve insurance claim disputes and commercial contract problems. More and more frequently, a mediation clause is being included in commercial contracts in conjunction with a standard arbitration clause. Under such clauses, only if mediation is unsuccessful do the parties then submit the dispute to arbitration. Going through both of these processes is still generally less time-consuming and expensive than litigation, especially when one considers the drawn-out discovery process.

Some argue that mediation should play a greater role in employment disputes. Often such disputes arise out of miscommunication, and mediation helps to open the lines of communication. A wide variety of creative remedies might be applicable to the employment situation. In fact, JAMS/Endispute offers three basic employment dispute options: standard mediation, streamlined mediation, and a mediation-arbitration combination.

A growing area for mediation is in the resolution of environmental disputes. Many advocate the use of mediation for environmental matters because the traditional dispute resolution methods are designed to handle a problem between two parties, whereas environmental disputes often involve multiple parties. Mediation can easily accommodate multiple parties. Environmental matters are often likely to involve parties who will have to deal with one another in the future, so preservation of their relationship is of utmost importance. For example, you might have a dispute over a new development in which a local citizen group, a developer, the municipality, and an environmental interest group all have concerns. Mediation could theoretically resolve their problems in a way that would prevent greater problems in the future. Mediation also offers the potential for creative solutions, which are often needed for environmental disputes.

ADVANTAGES OF MEDIATION

The primary advantage of mediation is that, because of its nonadversarial nature, it tends to preserve the relationship between the parties to a greater extent than would a trial or any of the other alternatives to litigation. Because the parties are talking to each other, rather than talking about each other and trying to make the other "side" look bad, there is less of the bitterness that often results from a trial. Parties will almost always come away from a trial with increased hostility toward the other side, whereas even if the parties cannot resolve their differences completely, they almost always leave mediation with a little better understanding of each other's positions. Thus, mediation is increasingly being used in cases in which the parties will have an ongoing relationship once the immediate dispute is settled.

Another important advantage to mediation is its ability to find and implement creative solutions. Because the parties are not searching for a decision in their favor or an award of money damages, they are more open to finding some sort of creative solution that may allow both parties to receive some benefit from the situation. Finally, as with other ADR methods, mediation is usually less expensive and less time-consuming than litigation. It is difficult to measure precisely how successful mediation is, however, because settlements are not made public.

CRITICISMS OF MEDIATION

The process of mediation has its critics. They argue that the informal nature of the process represses and denies certain irreconcilable structural conflicts, such as the inherent strife between labor and management. They also argue that this informal process tends to create the impression of equality between the disputants when no such equality exists. The resultant compromise between unequals is an unequal compromise, but it is clothed in the appearance of equal influence.

The mediation process can also arguably be abused. A party who believes that he or she will ultimately lose a dispute may enter the mediation process in bad faith, dragging the process out as long as possible.

APPLYING THE LAW TO THE FACTS...

Jim and Jack are the owners of two different companies. They become involved in a dispute and are not sure whether to go to court. Let's assume the dispute is confusing, and both men would like to further understand each other's position. They also would really like to preserve their relationship with each other and be able to do business amicably in the future. What form of dispute resolution should the men engage in?

Arbitration

arbitration A dispute resolution method whereby the disputing parties submit their disagreement to a mutually agreed-upon neutral decision maker or one provided for by statute. The decision of the arbitrator is binding on both parties.

One of the most well-known alternatives to litigation, **arbitration**, is the resolution of a dispute by a neutral third party outside the judicial setting. According to a 2010 survey by the American Bar Association Litigation Section Task Force on Alternative Dispute Resolution, 78 percent of surveyed lawyers said arbitration was more efficient than litigation, while 56 percent said it was more cost effective.[5] The arbitration hearing is similar to a trial, but there is no prehearing discovery phase, although in some cases an arbitrator may allow limited amounts of discovery. In addition, the stringent rules of evidence applicable in a trial are generally relaxed. Each side presents its witnesses and evidence and has the opportunity to cross-examine its opponent's witnesses. The arbitrator frequently takes a much more active role in questioning the witness than would a judge. If the arbitrator wants more information, he or she generally does not hesitate to ask for that information from witnesses.

---

[5]Jen Woods, "Arbitration Gains Popularity," General Counsel Consulting. Accessed November 23, 2010 at http://www.gcconsulting.com/articles/120084/73/Arbitration-Gains-Popularity.

Ordinarily, no official record of the arbitration hearing is made. Rather, the arbitrator and each of the parties usually take their own notes of what happens. The parties and the arbitrator, however, may agree to have a stenographer record the proceedings at the expense of the parties. Although attorneys may represent parties in arbitration, legal counsel is not required; individuals may represent themselves or have nonlawyers represent them. For example, in a labor dispute (discussed in Chapter 20) the union may be represented by one of its officers. In addition to the oral presentation, in some cases the arbitrator may request written arguments from the parties. These documents are called *arbitration briefs*.

APPLYING THE LAW TO THE FACTS...

Susan wants her dispute resolved quickly by a third party in a courtroom-type setting, but she would prefer more relaxed rules of evidence, no official record of the hearing, and no requirement for representation by legal counsel. What kind of ADR might Susan prefer?

The arbitrator usually provides a decision for the parties within 30 days of the hearing. In many states, this deadline is mandated by statute. The arbitrator's decision is called the **award**, even if no monetary compensation is ordered. The award does not have to state any findings of fact or conclusions of law, nor must the arbitrator cite any precedent for the decision or give any reasons. He or she must only resolve the dispute. However, if the arbitrator hopes to continue to be selected by parties and function in that role, he or she should provide reasons for the decision.

award The arbitrator's decision.

For a number of reasons, the arbitrator's decision is much more likely to be a compromise than a decision handed down by a court. First, the arbitrator is not as constrained by precedent as are judges. An arbitrator is interested in resolving a factual dispute, not in establishing or strictly applying a rule of law. Second, the arbitrator may be more interested than a judge in preserving an ongoing relationship with the parties. A compromise is much more likely to achieve this result than is a clear win-or-lose decision. Finally, because an arbitrator frequently decides cases in a particular area, he or she wants to maintain a reputation of being fair to both sides. For example, a person who focuses on labor arbitrations would not want to gain a reputation as being prolabor or promanagement.

The decision rendered by the arbitrator is legally binding. In some cases, such as labor cases, a decision may be appealed to the district court, but there are few such appeals. Of the more than 25,000 labor cases decided by arbitrators each year, fewer than 200 are challenged.

TECHNOLOGY AND THE LEGAL ENVIRONMENT

Resolving Disputes Online

Today, dispute resolution can be as convenient as sitting at your desktop and going online, thanks to a burgeoning number of online dispute resolution service providers.

Many online dispute resolution providers specialize in resolving particular kinds of disputes. For example, Mediations Arbitration Services (MARS) offers consumers and merchants an opportunity for the online resolution of disputes that resulted from online transactions. This service can be found at **www.resolvemydispute.com**. Another online dispute resolution service is PeopleClaim, available at **www.peopleclaim.com**.

The policy of deferring to the arbitrator's decision applies to disputes in areas other than labor contracts. Unless there is a clear showing that the arbitrator's decision is contrary to law or there was some defect in the arbitration process, the decision will be upheld. Section 10 of the Federal Arbitration Act, the federal law enacted to encourage the use of arbitration, sets forth the four grounds on which the arbitrator's award may be set aside:

1. The award was the result of corruption, fraud, or other "undue means."
2. The arbitrator exhibited bias or corruption.
3. The arbitrator refused to postpone the hearing despite sufficient cause, refused to hear evidence pertinent and material to the dispute, or otherwise acted to substantially prejudice the rights of one of the parties.
4. The arbitrator exceeded his or her authority or failed to use such authority to make a mutual, final, and definite award.

The U.S. Supreme Court reiterated and reinforced its desire to limit the ability of the judiciary to overturn arbitrators' decisions in the following case.

CASE 4-1

Hall Street Associates, L.L.C. v. Mattel, Inc.
Supreme Court of the United States
552 U.S. 576 (2008)

Mattel was leasing a property to Hall Street Associates. Hall Street sued Mattel on the ground that Mattel illegally terminated the lease. Hall Street additionally demanded compensation for what it said were necessary "environmental cleanup costs." Although the court decided in favor of Mattel on the termination issue, the parties made an agreement to arbitrate the environmental cleanup issue. Both parties determined that the court could modify the arbitration award as it saw fit. In other words, the court would be granted expanded judicial review of the arbitration award. After the arbitrator decided in favor of Hall Street, the court dismissed the arbitration award on the ground that the decision was based on errors of law. Subsequently, the arbitrator instead decided in favor of Mattel. This time, the court let the second decision of the arbitrator stand. However, a second court reversed the decision, stating that the scope of judicial review permitted by the Federal Arbitration Act (FAA) did not encompass the actions of the earlier court. In other words, the earlier court should not have had the power to dismiss or modify the arbitration award. The case was appealed to the Supreme Court.

Justice Souter

". . . Congress enacted the FAA to replace judicial indisposition to arbitration with a national policy favoring [it] and plac[ing] arbitration agreements on equal footing with all other contracts." As for jurisdiction over controversies touching arbitration, the Act does nothing, being "something of an anomaly in the field of federal-court jurisdiction" in bestowing no federal jurisdiction but rather requiring an independent jurisdictional basis (providing for action by a federal district court "which, save for such [arbitration] agreement, would have jurisdiction under title 28"). But in cases falling within a court's jurisdiction, the Act makes contracts to arbitrate "valid, irrevocable, and enforceable," so long as their subject involves "commerce." And this is so whether an agreement has a broad reach or goes just to one dispute, and whether enforcement be sought in state court or federal. . . .

The Act also supplies mechanisms for enforcing arbitration awards: a judicial decree confirming an award, an order vacating it, or an order modifying or correcting it. An application for any of these orders will get streamlined treatment as a motion, obviating the separate contract action that would usually be necessary to enforce or tinker with an arbitral award in court. Under the terms of § 9, a court "must" confirm an arbitration award "unless" it is vacated, modified, or corrected "as prescribed" in §§ 10 and 11. Section 10 lists grounds for vacating an award, while § 11 names those for modifying or correcting one. . . .

The Courts of Appeals have split over the exclusiveness of these statutory grounds when parties take the FAA shortcut to confirm, vacate, or modify an award, with some saying the recitations are exclusive, and others regarding them as mere threshold provisions open to expansion by agreement. As mentioned already, when this litigation started, the Ninth Circuit was on the threshold side of the split, from which it later departed en banc in favor of the exclusivity view, which it followed in this case. We now hold that §§ 10 and 11 respectively provide the FAA's exclusive grounds for expedited vacatur and modification. . . .

[I]t makes more sense to see the three provisions, §§ 9–11, as substantiating a national policy favoring arbitration with just the limited review needed to maintain arbitration's essential virtue of resolving disputes straight-away. Any other reading opens the door to the full-bore legal and evidentiary appeals that can "rende[r] informal arbitration merely a prelude to a more cumbersome and time-consuming judicial review process," and bring arbitration theory to grief in postarbitration process. . . .

We do not know who, if anyone, is right, and so cannot say whether the exclusivity reading of the statute is more of a threat to the popularity of arbitrators or to that of courts. But whatever the consequences of our holding, the statutory text gives us no business to expand the statutory grounds. . . .

In holding that §§ 10 and 11 provide exclusive regimes for the review provided by the statute, we do not purport to say that they exclude more searching review based on authority outside the statute as well. The FAA is not the only way into court for parties wanting review of arbitration awards: they may contemplate enforcement under state statutory or common law, for example, where judicial review of different scope is arguable. But here we speak only to the scope of the expeditious judicial review under §§ 9, 10, and 11, deciding nothing about other possible avenues for judicial enforcement of arbitration awards. . . .

Although we agree with the Ninth Circuit that the FAA confines its expedited judicial review to the grounds listed in 9 U.S.C. §§ 10 and 11, we vacate the judgment and remand the case for proceedings consistent with this opinion. It is so ordered.*

Judgment *vacated* and case remanded.

Case Questions

1. What does it mean for a court to be able to vacate or modify an arbitration award, rather than the two parties being limited to an arbitrator's decision?

   Clue: What would be the effect on cooperation of the parties in an arbitration case if they both knew that courts are eager to vacate or modify an arbitration decision?

2. Why would the FAA want to limit a court's power over an arbitration award?

   Clue: Look at the various advantages of arbitration when compared to court resolution of disputes.

METHODS OF SECURING ARBITRATION

Arbitration may be secured voluntarily, or it may be imposed on the parties. The first voluntary means of securing arbitration is by including a **binding arbitration clause** in a contract. Such a clause provides that all or certain disputes arising under the contract are to be settled by arbitration. The clause should also include the means by which the arbitrator is to be selected. More than 95 percent of the collective bargaining agreements in force today have some provision for arbitration.[6] Exhibit 4-1 contains a sample binding arbitration clause that could be included in almost any business contract.

If their contract contains no binding arbitration clause, the parties may secure arbitration by entering into a **submission agreement**. This contract can be entered into at any time. It is a written contract stating that the parties wish to settle their dispute by arbitration. Usually, it specifies the following conditions:

- How the arbitrator will be selected
- The nature of the dispute
- Any constraints on the arbitrator's authority to remedy the dispute
- The place where the arbitration will take place
- A time by which the arbitration must be scheduled

binding arbitration clause
A provision in a contract mandating that all disputes arising under the contract be settled by arbitration.

submission agreement
Separate agreement providing that a specific dispute be resolved through arbitration.

EXHIBIT 4-1

SAMPLE BINDING ARBITRATION CLAUSE

> Any controversy, dispute, or claim of whatever nature arising out of, in connection with, or in relation to the interpretation, performance, or breach of this agreement, including any claim based on contract, tort, or statute, shall be resolved, at the request of any party to this agreement, by final and binding arbitration conducted at a location determined by the arbitrator in (City), (State), administered by and in accordance with the existing rules of Practice and Procedure of Judical Arbitration & Mediation Services, Inc. (J.A.M.S.), and judgment upon any award rendered by the arbitrator may be entered by any state of federal court having jurisdiction thereof.

---

[6]M. Jacobs, "Required Job Arbitration Stirs Critics," *Wall Street Journal* B3 (June 22, 1993).
*Hall Street Associates, L.L.C. v. Mattel, Inc., Supreme Court of the United States 552 U.S. 576 (2008).

Usually, both parties declare their intent to be bound by the arbitrator's award.

If the parties have entered into a submission agreement or have included a binding arbitration clause in their contract, they will be required to resolve their disputes through arbitration. Both federal and state courts must defer to arbitration if the contract in dispute contains a binding arbitration clause.

The drafters of either a submission agreement or a binding arbitration clause must be precise because courts will enforce the agreements as written. Moreover, parties who decide to specify that a certain state's laws govern an agreement should be familiar with all of the laws that might be applicable, including the state laws governing the arbitration procedures themselves.

The courts will, however, make sure that the parties intended to be bound before they find that a binding arbitration agreement exists. For example, in 2013, a judge refused to enforce a mandatory arbitration clause in an employee handbook because the employer retained the ability to alter the terms of the handbook. According to the judge, "Defendants thus had the power to require plaintiff to arbitrate the covered dispute, while simultaneously reserving the right to modify the agreement. Such an agreement is not enforceable."[7]

Drafters must also be sure that the binding arbitration clause is fair to both sides. Courts are willing to scrutinize binding arbitration clauses and submission agreements to make sure that they are not so extraordinarily one-sided as to make them unconscionable, and therefore unenforceable. In examining claims of unconscionability, however, the concept of severability may come into play. If only one part of the arbitration agreement is unenforceable, the court might decide to sever the problematic provision to allow the rest of the arbitration agreement to stand. The following case shows what the Supreme Court of Ohio did when faced with an arbitration agreement that had a clause incompatible with state law.

CASE 4-2

Ignazio v. Clear Channel Broadcasting, Inc. et al.
Supreme Court of Ohio
113 Ohio St. 3d 276 (2007)

When Clear Channel became Diane Ignazio's employer, they required her to enter into an arbitration agreement with Clear Channel. Both parties waived their right to sue for any claim covered by the agreement and agreed to enter into final and binding arbitration. Section 10B, however, allowed either party to bring an action in court to set aside the award "where the standard of review will be the same as that applied by an appellate court reviewing a decision of a trial court sitting without a jury."*

Ignazio was terminated on October 7, 2003, and she filed an action against Clear Channel asserting claims of age and sex discrimination, retaliation, and wrongful termination of employment in violation of public policy.

Clear Channel filed a motion to stay the action pending arbitration, as Ignazio's claims fell under the arbitration agreement. The trial court granted the motion to stay. Ignazio appealed.

The court of appeals found that Section 10B violated Ohio arbitration laws (Ohio Rev. Code ch. 2711) by allowing too great a standard of judicial review, rendering any arbitration award not final and binding. An agreement that did not provide for a final and binding decision could not be classified as an arbitration agreement. The appeals court reversed the judgment of the trial court and remanded to allow Ignazio's lawsuit to proceed. Clear Channel appealed to the Ohio Supreme Court.

---

[7]*Domenichetti v. The Salter School, LLC et al.*, 2013 WL 1748402, 20 Wage & Hour Cas.2d (BNA) 1047, D. Mass., April 19, 2013 (NO. CIV. A. 12-11311-FDS).

*"Ignazio v. Clear Channel Broadcasting, Inc. et al., Supreme Court of Ohio 113 Ohio St. 3d 276 (2007)." http://www.supremecourt.ohio.gov/rod/docs/pdf/0/2007/2007-Ohio-1947.pdf.

Justice Stratton

The issue before us is whether a clause in an arbitration agreement that provides for greater judicial review of an award than is permitted under R.C. Chapter 2711 renders the entire agreement unenforceable, or whether the offensive clause may be severed and the remainder of the agreement enforced.

. . . [F]or purposes of this appeal, the parties do not dispute that the second sentence in Section 10B provides for a standard of review that is not permitted under Ohio's arbitration laws. We must decide whether the sentence in Section 10B is fundamental to the overall meaning of the agreement, or whether it may be severed so that the remainder of the agreement may be given effect. Whether a part of a contract may be severed from the remainder "depends generally upon the intention of the parties, and this must be ascertained by the ordinary rules of construction."

The arbitration agreement in this case expressly provides, "Should any provision of this Agreement be found to be unenforceable, such portion will be severed from the Agreement and the remaining portions shall remain in full force and effect." We presume the intent of the parties from the language employed in the contract. Clearly, the parties contemplated and provided for severing a provision that was unenforceable and giving effect to the remaining provisions.

Severing only the second sentence of Section 10B will not modify or alter the remainder of the provision for enforcing an arbitration award. The agreement still requires the parties to arbitrate disputes. Severing does not modify or change the terms of the agreement for demanding and conducting the arbitration process. The agreement continues to provide a means of enforcement in that "[e]ither party may bring an action in any court of competent jurisdiction to compel arbitration under this Agreement and to enforce an arbitration award." R.C. 2711.09, 2711.10, and 2711.11 permit a party to challenge the arbitration agreement in certain circumstances. If the phrase is severed, the only difference will be that a party may not seek to have a court review the award using the same standard of review as an appellate court. Therefore, this single phrase in one sentence of a multipage agreement does not alter the fundamental nature of the agreement.

Furthermore, severing the offending provision and enforcing the remainder of the agreement is consistent with this state's strong public policy in favor of arbitration. The law favors and encourages arbitration as a means of resolving disputes.

Therefore, we hold that the offending provision in this agreement does not fundamentally alter the otherwise valid and enforceable provisions of the agreement. Based on the express severability clause and Ohio's strong public policy in favor of arbitration of disputes, the second sentence of Section 10B is severed from the agreement, and the remaining provisions of the agreement are to be given full force and effect. Following arbitration, the parties may seek to confirm, vacate, or modify the award in court or otherwise challenge the process as provided by law.*

Judgment *reversed* in favor of Appellants, Clear Channel.

CRITICAL THINKING ABOUT THE LAW

Courts tend to hold parties to the bargain they make in a contract. In the contract in this particular case, specific language supports the idea that under certain conditions, a clause can be severed from the contract without affecting the enforceability of the rest of the contract.

1. **What are the reasons for this decision? How does the attitude of the courts toward arbitration serve as a reason in this instance?**

    Clue: Ask yourself, "What grounds did the court provide to substantiate its conclusion?"

2. **Why do the parties care whether the clause is severed? What ethical norms support the court's reasoning?**

    Clue: What happens if the court rules that the clause cannot be severed from the contract?

Although in the foregoing case the court decided to sever the section of the agreement that was incompatible with state law, other courts have found arbitration agreements with objectionable sections to be entirely unenforceable. For example, the Fourth Circuit Court of Appeals refused to enforce an agreement between Hooters Restaurant and one of its waitresses to settle all employment disputes through arbitration, because the court found that the rules for arbitration under the agreement were "so one-sided that their only possible purpose is to undermine the neutrality of the proceeding."[8] Some of

---

[8]*Hooters of America, Inc. v. Phillips*, 173 F.3d 933 (4th Cir. 1999).
*Ignazio v. Clear Channel Broadcasting, Inc. et al., Supreme Court of Ohio 113 Ohio St. 3d 276 (2007). http://www.supremecourt.ohio.gov/rod/docs/pdf/0/2007/2007-Ohio-1947.pdf.

the more objectionable rules incorporated in the agreement included the following:

- The employee had to provide notice of the specifics of the claim, but Hooters did not need to file any type of response to these specifics or notify the claimant of what kinds of defenses the company planned to raise.
- Only the employee had to provide a list of all fact witnesses and a brief summary of the facts known to each.
- Although the employee and Hooters could each choose an arbitrator from a list and those arbitrators would then select a third to create the arbitration panel that would hear the dispute, Hooters alone selected the arbitrators on the list.
- Only Hooters had the right to widen the scope of arbitration to include any matter, whereas the employee was limited to those raised in her or his notice.
- Only Hooters had the right to record the arbitration. Only Hooters had the right to sue to vacate or modify an arbitration award because the arbitration panel exceeded its authority. Only Hooters could cancel the agreement to arbitrate or change the arbitration rules.[9]*

If you think about these conditions, each of them individually seems inherently unfair. When you consider their impact as a group, it is easy to see why the court refused to uphold the agreement.

As increasing numbers of employers began incorporating mandatory, binding arbitration clauses in their employment contracts, a question arose as to whether certain statutory rights, such as the right to not be discriminated against on the basis of age, sex, race, religion, color, or national origin, were so important that one could not contract away one's ability to protect those rights through litigation. The U.S. Supreme Court addressed that issue in the case of *Gilmer v. Interstate/Johnson Lane Corp.*[10] Gilmer was a 62-year-old securities broker who sued his employer under the Age Discrimination in Employment Act (ADEA) when he was terminated. The employer argued that Gilmer had to submit his age discrimination claim to arbitration because as part of his application to be a securities representative with the New York Stock Exchange, he had signed an agreement to arbitrate "any claim, dispute, or controversy" arising between himself and Interstate.

The high court upheld the binding arbitration clause, noting that the employee was not agreeing to give up any statutory rights; instead, the employee was simply agreeing to resolve disputes involving those rights through arbitration. The court had no problem with such a waiver of the employee's right to sue to enforce his statutory rights, as long as the employee understood that the binding arbitration clause also encompassed statutory rights.

The initial impact of this ruling was to increase the use of mandatory arbitration agreements in employment contracts in all industries. In fact, about 100 large companies soon moved to follow the securities industry's policy. Some courts and the Equal Employment Opportunity Commission (EEOC), however, became increasingly concerned about whether arbitration agreements that had to be accepted as a condition of employment were really voluntary. On July 10, 1997, the EEOC issued a policy statement saying that the mandatory "arbitration of discrimination claims as a condition of employment are [sic] contrary to the fundamental principles of employment discrimination laws." The EEOC chairman stated that the agency strongly supported agreements to arbitrate once a dispute had arisen, but did not believe that a prospective employee

---

[9]*Ibid.*

[10]500 U.S. 20 (1991).

*Paraphrase from "Arbitration and Employment Disputes: Drafting to Maximize Employer Protection," by Nancy K. Kubasek.

could make a voluntary agreement to arbitrate future disputes as part of a "take it or leave it" contract of employment. In 2003, however, in *EEOC v. Luce, Forward, Hamilton & Scripps*, the Ninth Circuit[11] cleared the way for the inclusion of mandatory arbitration clauses in employment agreements. The court held that federal civil rights claims can be the subject of mandatory arbitration agreements, and an employer can require all employees to sign such an agreement as a condition of employment.

Thus, an employer need not provide its employees with a choice in statutory discrimination disputes. However, the Supreme Court recently upheld the EEOC's right to seek remedies in discrimination cases, even when a binding arbitration agreement exists to settle all employment-related disputes. In *EEOC v. Waffle House, Inc.*,[12] an employee brought suit against his employer, claiming that he was unlawfully discharged after he had a seizure at work. With regard to the defendant's allegations that this discharge violated the Americans with Disabilities Act, the Supreme Court ruled that the EEOC could seek statutory remedies, including "victim-specific" relief, back pay, compensatory damages, and punitive damages. In the year after this case was handed down, the EEOC settled a case by agreeing to allow an employer to require employees to sign a mandatory arbitration agreement as long as the employees were notified of their right to still file a charge with the EEOC that the agency might wish to pursue.

In 1997, the National Association of Securities Dealers changed its policy to allow employees to choose between entering into private arbitration agreements with their employer or reserving the right to file a claim in state or federal court. Nevertheless, the trend in many industries is toward an increasing use of binding arbitration clauses in employment agreements. After all, the EEOC files suit in less than 1 percent of the charges it receives each year. Hence, even if an employee wants to litigate for statutory discrimination, the likelihood that the EEOC will represent him or her is very small. To avoid any chance of an arbitration clause being challenged by an employee wishing to bring a lawsuit under a civil rights statute, the binding arbitration clause should be drafted in such a way as to make it clear that the employee knew he or she was agreeing to arbitrate all disputes arising out of the employment situation, including statutory rights.

If the binding arbitration clause is contained in an agreement between the employer and the union, and individuals under the agreement are giving up their rights to sue for statutory violations, the agreement to arbitrate statutory disputes must be even more explicit; it must be clear and unmistakable. In *Wright v. Universal Maritime Service*,[13] the U.S. Supreme Court said that "[the] right to a federal judicial forum is of sufficient importance to be protected against a less-than-explicit union waiver in a CBA [collective bargaining agreement]. The CBA in this case does not meet that standard. Its arbitration clause is very general, providing for arbitration of '[m]atters under dispute'—which could be understood to mean matters in dispute under the contract. And the remainder of the contract contains no explicit incorporation of statutory antidiscrimination requirements."* Thus, an employer wishing to avoid litigation of allegations of civil rights violations must be even more careful in crafting the binding arbitration clause when it is a union that is waiving its members' rights.

Binding arbitration clauses, however, are included not only in employment agreements, but also sometimes in consumer and business-to-business

---

[11]345 F.3d 742 (9th Cir. 2003).
[12]534 U.S. 279 (2002).
[13]525 U.S. 70 (1998).
*Wright v. Universal Maritime Service, 525 U.S. 70 (1998). https://supreme.justia.com/cases/federal/us/525/70/.

contracts. For instance, in *Eagle v. Fred Martin Motor Co.*,[14] plaintiff Lisa Eagle sued Fred Martin Motors, alleging that the defendant violated the Consumer Sales Practices Act, after the plaintiff's new car repeatedly stalled and experienced other mechanical problems that the dealer failed to fix. After Eagle filed suit, Fred Martin Motors moved to compel arbitration in accordance with a binding arbitration clause that was included in the sales agreement. The common pleas court granted the defendant's motion, and Eagle appealed, claiming that the arbitration agreement was unconscionable.

The appellate court reversed the common pleas court's decision. Explaining that consumer transactions should be subject to closer scrutiny, the judge claimed that the agreement was substantively unconscionable because the agreement included excessive filing fees, especially for the plaintiff, who earned only about $14,000 per year. The judge stated that the binding arbitration clause was procedurally unconscionable because the preprinted purchase contract was an adhesion contract, in the sense that the plaintiff had no actual choice about the terms of arbitration. Furthermore, the judge cited several additional factors relating to procedural unconscionability, including allegations that the dealer's sales representative hurried through the purchasing papers with the plaintiff, the plaintiff was not made aware of the binding arbitration clause during the purchase (nor did the plaintiff know what arbitration meant), and the arbitration clause did not mention the excessive costs. Considering the plaintiff's limited educational and economic background, the judge agreed that these two factors were also relevant to his conclusion that the arbitration clause was procedurally unconscionable. Hence, the court ruled that the binding arbitration clause was substantively and procedurally unconscionable and therefore, unenforceable. See Exhibit 4-2 for a list of points to consider to ensure that any mandatory arbitration clause you draft will not be struck down as unconscionable.

The U.S. Supreme Court reinforced its commitment to upholding binding arbitration clauses in 2011. In *AT&T Mobility LLC v. Concepcion et ux*,[15] the high court upheld a clause requiring all disputes under the contract to be settled through arbitration while also requiring all arbitration claims to be brought in the parties' individual capacity and not as a plaintiff or class member in any class action, thereby effectively doing away with class action arbitration in consumer cases. This case was seen by many commentators as reflecting a tendency of the Roberts Court to favor the interests of businesses over those of consumers.

Two years later, in the following case, the high court once again seems to be putting the interests of corporations above those of consumers.

EXHIBIT 4-2

DRAFTER'S CHECKLIST

Checklist for Ensuring That a Mandatory Arbitration Clause Is Not Unconscionable

- The contract is not a contract of adhesion, that is, one where the consumer could not meaningfully negotiate the inclusion and terms of the mandatory arbitration clause.
- The mandatory arbitration clause and its provisions are not buried in fine print.
- The provisions of the clause do not unduly favor the interests of the drafter of the contract.
- The provisions of the clause do not impose significant costs on the consumer.
- The provisions of the clause do not include a confidentiality requirement (which may make it difficult for consumers to share information).
- The provisions do not limit the ability of the consumer to obtain punitive damages.

---

[14]157 Ohio App.3d 150 (2004).
[15]United States Supreme Court 131 S. Ct. 1740 (2011).

CASE 4-3

American Express Co. vs. Italian Colors Restaurant
United States Supreme Court
133 S. Ct. 2304 (2013)

American Express entered into an agreement with merchants who accept their credit cards that all disputes were to be settled through arbitration. Despite this, several merchants filed a class action suit on the grounds that American Express violated § 1 of the Sherman Act, and in turn sought damages under § 4 of the Clayton Act. American Express then attempted to seek individual arbitration by way of the Federal Arbitration Act (FAA). The merchants claimed that the costs of the analysis necessary to prove the antitrust claims would far exceed the maximum recovery for the plaintiff. While the District Court granted the motion, therefore dismissing the lawsuits, the Second Circuit reversed the decision. The reversal was based on the idea that due to the prohibitive costs that the respondents would incur, the class action waiver was not an enforceable one, and therefore arbitration could not occur. By writ of certiorari, the Supreme Court heard the case in 2013.

Justice Scalia

We consider whether a contractual waiver of class arbitration is enforceable under the Federal Arbitration Act when the plaintiff's cost of individually arbitrating a federal statutory claim exceeds the potential recovery.

. . . Congress enacted the FAA in response to widespread judicial hostility to arbitration. . . . As relevant here, the Act provides:

> A written provision in any maritime transaction or contract evidencing a transaction involving commerce to settle by arbitration a controversy thereafter arising out of such contract or transaction . . . shall be valid, irrevocable, and enforceable, save upon such grounds as exist at law or in equity for the revocation of any contract.

This text reflects the overarching principle that arbitration is a matter of contract . . . And consistent with that text, courts must "rigorously enforce" arbitration agreements according to their terms, . . . including terms that "specify with whom [the parties] choose to arbitrate their disputes," Stolt-Nielsen, supra, at 683, and "the rules under which that arbitration will be conducted."

Congress has taken some measures to facilitate the litigation of antitrust claims. . . . Congress has told us that it is willing to go, in certain respects, beyond the normal limits of law in advancing its goals of deterring and remedying unlawful trade practice. But to say that Congress must have intended whatever departures from those normal limits advance antitrust goals is simply irrational. "[N]o legislation pursues its purposes at all costs."

The antitrust laws do not "evinc[e] an intention to preclude a waiver" of class-action procedure. . . . The Sherman and Clayton Acts make no mention of class actions. In fact, they were enacted decades before the advent of Federal Rule of Civil Procedure 23, which was "designed to allow an exception to the usual rule that litigation is conducted by and on behalf of the individual named parties only." . . . The parties here agreed to arbitrate pursuant to that "usual rule," and it would be remarkable for a court to erase that expectation.

. . . Congressional approval of Rule 23 establishes an entitlement to class proceedings for the vindication of statutory rights. . . . It imposes stringent requirements for certification that in practice exclude most claims. And we have specifically rejected the assertion that one of those requirements (the class-notice requirement) must be dispensed with because the "prohibitively high cost" of compliance would "frustrate [plaintiff's] attempt to vindicate the policies underlying the antitrust" laws.

Our finding of no "contrary congressional command" does not end the case. Respondents invoke a judge-made exception to the FAA which, they say, serves to harmonize competing federal policies by allowing courts to invalidate agreements that prevent the "effective vindication" of a federal statutory right. Enforcing the waiver of class arbitration bars effective vindication, respondents contend, because they have no economic incentive to pursue their antitrust claims individually in arbitration.

The "effective vindication" exception to which respondents allude originated as dictum in Mitsubishi Motors, where we expressed a willingness to invalidate, on "public policy" grounds, arbitration agreements that "operat[e] . . . as a prospective waiver of a party's right to pursue statutory remedies." . . . Dismissing concerns that the arbitral forum was inadequate, we said that "so long as the prospective litigant effectively may vindicate its statutory cause of action in the arbitral forum, the statute will continue to serve both its remedial and deterrent function." Subsequent cases have similarly asserted the existence of an "effective vindication" exception, . . . but have similarly declined to apply it to invalidate the arbitration agreement at issue.

And we do so again here. As we have described, the exception finds its origin in the desire to prevent "prospective waiver of a party's right to pursue statutory remedies," . . . That would . . . cover filing and administrative fees attached to arbitration that are so high as to make access to the forum impracticable. . . . ("It may well be that the existence of large arbitration costs could preclude a litigant . . . from effectively vindicating her federal statutory rights.") But the fact that it is not worth the expense involved in proving a statutory remedy does not constitute the elimination of the right to pursue that remedy. [The class-action waiver] no more eliminates those parties' right to pursue their statutory remedy than did federal law before its adoption of the class action for legal relief in 1938. . . .

A pair of our cases brings home the point. In *Gilmer*, we had no qualms in enforcing a class waiver in an arbitration agreement even though the federal statute at issue, the Age Discrimination in Employment Act, expressly permitted collective actions. We said that statutory permission did "'not mean that individual attempts at conciliation were intended to be barred.'" . . . And in *Vimar Seguros y Reaseguros*, . . . we held that requiring arbitration in a foreign country was compatible with the federal Carriage of Goods by Sea Act. That legislation prohibited any agreement "'relieving'" or "'lessening'" the liability of a carrier for damaged goods, which is close to codification of an "effective vindication" exception. The Court rejected the argument that the "inconvenience and costs of proceeding" abroad "lessen[ed]" the defendants' liability, stating that "[i]t would be unwieldy and unsupported by the terms or policy of the statute to require courts to proceed case by case to tally the costs and burdens to particular plaintiffs in light of their means, the size of their claims, and the relative burden on the carrier." . . . Such a "tally[ing] [of] the costs and burdens" is precisely what the dissent would impose upon federal courts here.

Truth to tell, our decision in *AT&T Mobility* all but resolves this case. There we invalidated a law conditioning enforcement of arbitration on the availability of class procedure because that law "interfere[d] with fundamental attributes of arbitration.". . . We specifically rejected the argument that class arbitration was necessary to prosecute claims "that might otherwise slip through the legal system."

The regime established by the Court of Appeals' decision would require—before a plaintiff can be held to contractually agreed bilateral arbitration—that a federal court determine (and the parties litigate) the legal requirements for success on the merits claim-by-claim and theory-by-theory, the evidence necessary to meet those requirements, the cost of developing that evidence, and the damages that would be recovered in the event of success. Such a preliminary litigating hurdle would undoubtedly destroy the prospect of speedy resolution that arbitration in general and bilateral arbitration in particular was meant to secure. The FAA does not sanction such a judicially created superstructure.*

Reversed in favor of American Express.

CRITICAL THINKING ABOUT THE LAW

A judge, Justice Scalia in this instance, does not have the luxury of knowing all relevant information before ruling. Some information that would be helpful in deciding this case is not in the record. Judges can respond to only the information presented to them. Suppose Justice Scalia were to have a magic wand that would produce information that would prove useful in making a decision that reflected a more complete version of the relevant facts.

1. What information would he have liked to have had so that he could be even more certain that he is making the correct decision? For example, his decision for the Court does indicate that there are circumstances where a contract specifying required arbitration would be invalidated because more important public policy issues required the opportunity to offer class action remedies. What possible facts would have triggered a contrary decision for Justice Scalia?

    Clue: Think of facts that would have moved Justice Scalia in one direction or another, but because they were not present, they had no apparent role on Scalia's decision.

2. Justice Scalia comes very close at times in his decision to saying that he is finding in favor of American Express because it is right to do so. In other words, he is not in those instances providing reasoning so much as he is "begging the question." To beg the question is to say that something is correct because it is correct. In other words, one is repeating the conclusion rather than providing a basis for the conclusion. See whether you can find places in his decision where Scalia is repeating the conclusion and pretending that in doing so he has provided a reason.

    Clue: Whenever Justice Scalia gives a reason, check to see whether that "reason" is simply another way to say his conclusion.

3. What ethical norms does Justice Scalia's ruling lean on for its ethical anchor?

    Clue: Think about who gains and who loses from a decision such as this one. Which stakeholders in the decision are happy that the decision went as it did? Look at a list of ethical norms. Go through them one by one, asking yourself: "Does this decision assist our country in moving toward or away from this individual ethical norm?"

---

*American Express Co. vs. Italian Colors Restaurant United States Supreme Court 133 S. Ct. 2304 (2013).

In addition to the two voluntary means of securing arbitration, state law may mandate the process for certain types of conflicts. For instance, in some states, public employees must submit collective bargaining disputes to binding arbitration. In other states, disputes involving less than a certain amount of money automatically go to arbitration.

SELECTION OF AN ARBITRATOR

Once the decision to arbitrate has been made, an arbitrator must be selected. Arbitrators are generally lawyers, professors, or other professionals. They are frequently selected on the basis of their special expertise in some area. If the parties have not agreed on an arbitrator before a dispute, they generally use one of two sources for selecting one: the Federal Mediation and Conciliation Services, a government agency; or the American Arbitration Association (AAA), a private, nonprofit organization founded in 1926, whose stated purpose is "to foster the study of arbitration in all of its aspects, to perfect its techniques under arbitration law, and to advance generally the science of arbitration for the prompt and economical settlement of disputes." Its slogan is "Speed, Economy, and Justice." It currently employs more than 8,000 arbitrators and mediators worldwide, of whom 1,100-plus are bilingual or multilingual and collectively speak more than 40 languages. The parties may also turn to one of the private arbitration services.

When the disputants contact one of these agencies, they receive a list of potential arbitrators along with a biographical sketch of each. The parties then jointly select an arbitrator from the list. Once the arbitrator has been selected, the parties and the arbitrator agree on the time, date, and location of the arbitration. They also agree on the substantive and procedural rules to be followed in the arbitration. Regardless of whether the arbitrator is selected through the AAA, the FMCS, or some private association, he or she will be subject to the Arbitrator's Code of Ethics (Exhibit 4-3). Organized by canons of ethics,

EXHIBIT 4-3

THE ARBITRATOR'S CODE OF ETHICS

CANON 1
- An arbitrator will uphold the integrity and fairness of the arbitration process.

CANON 2
- If the arbitrator has an interest or relationship that is likely to affect his or her impartiality or that might create an appearance of partiality or bias, it must be disclosed.

CANON 3
- An arbitrator, in communicating with the parties, should avoid impropriety or the appearance of it.

CANON 4
- The arbitrator should conduct the proceedings fairly and diligently.

CANON 5
- The arbitrator should make decisions in a just, independent, and deliberate manner.

CANON 6
- The arbitrator should be faithful to the relationship of trust and confidentiality inherent in that office.

CANON 7
- In a case where there is a board of arbitrators, each party may select an arbitrator. That arbitrator must ensure that he or she follows all the ethical considerations in this type of situation.

the code is designed to ensure that arbitrators perform their duties diligently and in good faith, thereby maintaining public confidence in the arbitration process. Exhibit 4-3 sets out the seven canons of the Arbitrator's Code of Ethics.

COMMON USES OF ARBITRATION

As noted earlier, arbitration is frequently used to resolve grievances under collective bargaining agreements. It is so frequently used in these cases that it is sometimes identified primarily as a means for resolving labor disputes. Furthermore, it is not used only in traditional labor disputes; we are increasingly seeing it used in disputes over athletes' salaries. For example, salary arbitration made headlines when the Washington Nationals baseball player Brian Bruney went to arbitration in an attempt to increase his salary to $1.85 million for 2010, but the Washington Nationals argued for a salary of $1.5 million.[16]

However, arbitration is also used to resolve a much broader range of issues: insurance and uninsured motorist claims, construction disputes, securities, real estate, intellectual property, and other commercial disputes. For example, many consumer complaints are now being handled through arbitration sponsored by Better Business Bureaus (BBBs) in more than 100 cities. Under the BBB's National Consumer Arbitration Program, which has been in effect since 1972, the consumer first files a complaint with the BBB, whose representative attempts to negotiate a solution between the consumer and the business. If negotiation fails, the consumer and the business representative sign a submission agreement specifying the disputed issues. They then select an arbitrator from a panel of five trained volunteers from the local community. After the hearing (at which the parties are usually not represented by counsel), the arbitrator has 10 days to render a decision.

Another area that is seeing an increasing use of arbitration is technology disputes. In 2003, the AAA handled technology-related claims that totaled more than $1 billion. The average claim was $1.2 million, and the average counterclaim $2.3 million.

The National Consumer Arbitration Program is limited in that arbitrators cannot award damages beyond the value of the product in question. They cannot, for example, award punitive damages or damages for personal injuries suffered as a result of a defective product. Despite these limitations, this program provides an excellent forum for disputes that often would not be resolved in court, because the cost of litigation would probably exceed or come close to the amount of the damage award. The use of arbitration in consumer cases is discussed further in Chapter 25.

As mentioned in the previous section, another area in which arbitration is increasingly being used is in employment disputes. Other areas in which arbitration is prevalent include malpractice cases, environmental disputes, community disputes and elections, and commercial contract conflicts. It is also being increasingly used to handle insurance liability claims arising from accidents.

PROBLEMS WITH ARBITRATION

Despite the growing use of arbitration, the process does not escape criticism, especially from the securities industry. From 1991, when the U.S. Supreme Court affirmed the authority of the securities industry to do so, until 1997, securities firms required that employees who execute, buy, or sell orders at brokerages or investment banks take all their employment disputes—including allegations of race, sex, and age discrimination—to arbitration instead of to court.

---

[16]http://voices.washingtonpost.com/nationaljournal/2010/02/bruney_loses_arbitration_case.html.

According to a study by the General Accounting Office (GAO), these disputes are most likely to be filed by women and minorities. They are also the cases most likely to be heard by white males over the age of 60. Only 11 percent of arbitrators are female, and fewer than 1 percent are Asian or black. Arbitrators in the securities industry come from the New York Stock Exchange and the National Association of Securities Dealers. Most are retired or semiretired executives or professionals, so they frequently lack expertise in discrimination matters.

Although not criticizing the fairness of the outcome in any particular case, the GAO did recommend that the industry appoint as arbitrators in discrimination cases people who had some knowledge about discrimination. They also recommended that the industry track the outcome of discrimination cases and establish criteria to keep individuals with records of regulatory violations or other disciplinary actions from becoming arbitrators. Finally, the report expressed concern over the lack of regularized Securities and Exchange Commission (SEC) oversight of the arbitration process in general, including the thousands of broker–customer disputes arbitrated each year. The GAO asked the SEC to establish a regular cycle of reasonably frequent reviews of all securities arbitrations, which the SEC has agreed to do.

The trend toward greater use of compulsory arbitration in employment and termination cases concerns some lawyers and legal scholars, because they believe that such a requirement erodes workers' rights. Thus, an employee who is subject to mandatory arbitration gives up the right to a public trial, the ability to get an injunction to stop unlawful practices, and the right to bring a class action suit.

Another potential problem is the arbitrator's background. For example, although about half of the AAA's arbitrators are lawyers, often an arbitrator is merely someone with expertise in the area in which the dispute arose. He or she may be solving a legal dispute without any real understanding of the applicable law.

A related problem is that if more employers (and other institutions) turn to mandatory arbitration, arbitration may start to become more and more like litigation. Those forced to give up their day in court may start pressing for their "due process" rights in arbitration. They may also argue that arbitrators should be allowed to grant the same remedies as courts. Ultimately, arbitration proceedings could be burdened with the same kinds of formalities that plague litigation—that is, the same disadvantages that gave rise to ADR in the first place.

Some also question the absence of written opinions. Without such opinions, legal precedents cannot develop to reflect changing circumstances. Finally, there is a question of whether the public interest is harmed by allowing an industry to use arbitration to "hide" its disputes from the public. If, for example, a bank required all billing disputes to be handled by arbitration, the public would never know if that bank was continually making overbilling errors. If such cases went to court, the public would be informed about the bank practices and inefficiencies. In addition, of course, firms that are secure in the knowledge that their operations will not be publicized are more likely to be lax than are firms that know they are subject to public scrutiny.

Minitrials

A relatively new means of resolving commercial disputes, probably first used in the United States in 1977 to resolve a dispute between TRW and Telecredit, is the minitrial. A **minitrial** is presided over by a neutral adviser, but the settlement authority resides with senior executives of the disputing corporations. Lawyers for each side make a presentation of the strengths and weaknesses of their respective positions. The adviser may be asked to give his or her opinion as to

minitrial An alternative dispute resolution method in which lawyers for each side present the case for their side at a proceeding refereed by a neutral adviser, but settlement authority usually resides with senior executives of the disputing corporations.

LINKING LAW AND BUSINESS

Management

Your management professor may have taught you about the concept of perceptions, the psychological process of selecting stimuli, gathering data from observation, and interpreting the organized information. From the perspective of an employee, one of the most important perceptions about the workplace is *procedural justice*, which is the apparent fairness of processes used to determine outcomes. Examples of these processes include performance appraisals, interviews, payment systems, methods of making decisions, and ADR. Procedural justice evaluates the fairness of these methods, with an employee's expectation that the managers will make fair decisions as leaders in these processes. Despite the difficulty of determining the consequences of unfair behavior, some research suggests that a perception of unfairness often leads to an increased amount of job absenteeism and turnover, less production efficiency, lower employee morale, and a lack of employee self-confidence. If there is some truth in this research, it is in the best interest of managers to continually examine the attitudes of employees as a means of understanding the perceived fairness of the organization from their perspective. Thereafter, measures can be taken to remedy unfair processes, which changes may result in greater organizational productivity. Additionally, a greater perception of fairness from the employees' point of view may act as a preventative measure for disputes within the organization. As a result, managers can focus more of their time and finances on attaining organizational objectives.

---

what the result would be if the case went to trial. The corporate executives then meet, without their attorneys, to discuss settlement options. Once they reach a settlement, they can make it binding by entering into a contract that encompasses the terms of the settlement.

The use of a minitrial before resorting to arbitration or litigation may be provided for in a contractual clause. One modification of the minitrial is to give the neutral adviser the authority to settle the case if the corporate executives cannot agree on a means of resolving the dispute within a given period of time.

The minitrial is seen by some as more desirable than arbitration when the disputes involve complex matters that may be better understood by parties directly involved in the contract than by an outside arbitrator. The process also requires intensive, direct communication between the disputants, which may help them better understand the other party's position and thus may ultimately help their relationship. Minitrials, in some instances, also have the advantage of being less costly than arbitration. For example, the Army Corps of Engineers settled several cases through the use of minitrials, including one trial that involved a $105 million claim that was lowered to a $7 million settlement.[17] Another Army Corp of Engineers case was settled in two and a half days, reducing a complex $515,213 claim to $155,000.[18]

early neutral case evaluation When parties explain their respective positions to a neutral third party who then evaluates the strengths and weaknesses of the cases. This evaluation then guides the parties in reaching a settlement.

Early Neutral Case Evaluation

Early neutral case evaluation is similar to a minitrial. In **early neutral case evaluation**, parties select a neutral third party and explain their respective positions to this neutral party, who then evaluates the strengths and weaknesses of their cases. This evaluation guides the parties in reaching a settlement.

---

[17] Eldon H. Crowell and Charles Pou, Jr., "Appealing Government Contract Decisions: Reducing the Cost and Delay of Procurement Litigation with Alternative Dispute Resolution Techniques," *Md. L. Rev.* 49: 183 (1990).

[18] *Id.*

Private Trials

In several states, legislation now permits **private trials** in which cases are tried by a referee selected and paid by the disputants. These referees are empowered by statute to enter legally binding judgments. Referees usually need not have any special training, but they are frequently retired judges—hence the disparaging reference to this ADR method as "rent-a-judge." The time and place of the trial are set by the parties at their convenience. Cases may be tried in private, a provision that ensures confidentiality.

On hearing the case, the referee states findings of facts and conclusions of law in a report with the trial court. The referee's final judgment is entered with the clerk when the report is filed. Any party who is dissatisfied with the decision can move for a new trial before the trial court judge. If this motion is denied, the party can appeal the final judgment.

Until recently, private trials did not involve juries. Private jury trials, however, are now offered by a small number of private firms, most notably JAMS/Endispute. Jurors in these cases tend to be slightly better educated than the typical jury, and many will have served on several previous private juries. Whether having "semiprofessional" jurors serving on many cases perverts the idea of a "jury of one's peers" is currently being debated.

Like other forms of ADR, private trials are subject to criticisms. Many people, for example, are concerned that the use of private trials may lead to the development of a two-tier system of justice. Disputants with sufficient resources will be able to channel their disputes through an efficient, private system; everyone else will have to resolve their complaints through a slower, less-efficient public system. Similarly, critics suggest that as wealthier individuals and corporations opt for the private courts, they will be less willing to channel their tax dollars into the public system. With less funding, the public system will then become even less effective. Finally, there is the same question that was raised with respect to arbitration: whether the public interest is harmed by allowing an industry to use private trials to "hide" its disputes from the public.

private trial An alternative dispute resolution method in which cases are tried, usually in private, by a referee who is selected by the disputants and empowered by statute to enter a binding judgment.

Summary Jury Trials

Originating in a federal district court in Cleveland, Ohio, in 1983, as a way to clean up an overcrowded docket, **summary jury trials** are today used in many state and federal courts across the nation for that purpose. The primary advantage of a summary jury trial is that it lasts only one day. The judge first instructs a jury on the law. Each side then has a limited amount of time to make an opening statement and present a summary of the evidence it would have presented at a regular trial.

The jury, which is usually composed of no more than six people, then retires to reach a verdict. The verdict, however, is only advisory, although jurors are usually not aware that their decision does not have a binding effect.

Immediately after receiving the verdict, the parties retire to a settlement conference. Roughly 95 percent of all cases settle at this point. If the parties do not settle, the case is then set for trial. If a case goes to trial, nothing from the summary trial is admissible as evidence.

summary jury trial An alternative dispute resolution method that consists of an abbreviated trial, a nonbinding jury verdict, and a settlement conference.

Court-Annexed Alternative Dispute Resolution

USE OF COURT-ANNEXED ADR IN THE STATE AND FEDERAL SYSTEMS

In an attempt to relieve the overburdened court systems and in partial recognition of the success of voluntary ADR, many state and federal jurisdictions are mandating that disputants go through some formal ADR process before certain types of cases may be brought to trial. It is difficult to generalize about the use of ADR in the court system, however, because practices vary from jurisdiction to jurisdiction, and even from court to court, as courts experiment with a broad range of programs and approaches. Some courts mandate ADR; others make it voluntary. Some refer almost all civil cases to ADR; others target certain cases by subject matter or by amount of controversy.

In the federal system, for example, in 1984, some federal district courts began adopting programs for mandatory, nonbinding arbitration of disputes involving amounts of less than $100,000. The subsequent passage of the Alternative Dispute Resolution Act of 1998 required all federal district courts to have an ADR program along with a set of rules regarding this program. Under these mandatory programs, fewer than 10 percent of the cases referred to arbitration end up going to trial. When Congress enacted the ADR Act of 1998, Congress did not create any programs for appellate courts, because all federal circuit courts had already implemented ADR programs. The one exception is the Federal Circuit, which hears cases of more specific and complex subject matter and, therefore, only suggests that parties discuss settlements before a trial. Nevertheless, ADR programs appear to be successful at the appellate level. For example, the Court of Appeals in the Ninth Circuit, like most courts at the appellate level, offers opposing parties an opportunity to reach a settlement during the appeals process. Relying primarily on mediation, the Ninth Circuit settled 739 cases in 2000, returning only 98 for additional litigation.

Today, all states have some form of court connected ADR, and judges may mandate that parties use some form of ADR. Mediation is the form of ADR most commonly used, in part because of the informality of proceedings, the greater emphasis on cooperation over competition, and the larger amount of control that lawyers and parties have over their cases.

The aggressiveness with which a state encourages the use of ADR methods varies a great deal. For instance, Florida has one of the most comprehensive systems of court-connected ADR. In 1987, legislation granted civil trial judges the statutory authority to refer cases to mediation or arbitration, subject to rules and procedures established by the Supreme Court of Florida. Since that time, the legislature has continued to improve the regulation of court-connected ADR, setting procedural rules and certification qualifications, ethical standards, grievance procedures, training standards, and continuing education requirements for mediators. In 2004, additional legislation mandated that ADR services were available in all counties in the state. As of August of 2012, there were 6,360 state certified mediators in the state, in some states, such as Missouri, however, a completely voluntary approach is used. Courts are simply required to provide all parties with a notice of the availability of ADR services and names and addresses of persons or agencies able to provide such services. Although such a casual approach is criticized on the ground that few people will voluntarily use such services, defenders point out that if people choose ADR, their commitment to the process will be stronger than if they are forced into it, and therefore it is more likely to work. Many states have adopted programs that require parties to take certain types of cases either to mandatory nonbinding arbitration or to mediation. In fact, at least 28 states require arbitration or mediation for certain kinds of cases or for cases under a particular monetary amount. Only when the parties cannot reach an agreement through mediation or when one party disagrees with the arbitrator's decision will the court hear the case.

New Jersey is one example of a state that requires nonbinding arbitration. For two decades, New Jersey courts have required all civil suits involving vehicular accidents to be sent to arbitration. In 2000, the requirements were extended to include most claims involving personal injury, contracts, and commercial disputes. Consequently, New Jersey experienced a 75 percent increase in the number of arbitration claims filed each year. In Minnesota, while ADR is not mandated, parties are required to discuss the use of ADR and address this issue in the informational statement filed with the court. If the parties are unable to make a decision on the use of an ADR process or a neutral, the court may order the parties to any number of ADR alternatives. This does not mean parties are required to settle their differences through ADR. They are required, however, to at least discuss their differences with the neutral and attempt to resolve their differences prior to a trial. Similarly, in the Western District of Missouri, approximately one-third of litigants must select a form of ADR to use. If the litigants cannot decide, an administrator of the court will meet with the parties and their lawyers and make a decision for them. To find out more about the use of court-annexed ADR in any state, go to the resource center of Court ADR Across the U.S. at http://courtadr.org/court-adr-across-the-us/state.php?state=449.

Some state systems also use ADR at the appellate level. South Carolina was the first state to do so. Under its system, arbitration at the appellate level is optional, but once arbitration has been selected, the decision of the arbitrator is binding.

DIFFERENCES BETWEEN COURT-ANNEXED AND VOLUNTARY ADR

Arbitration. Probably the major difference between court-annexed arbitration and voluntary arbitration is that, in most cases, the court-mandated arbitration is not binding on the parties. If either party objects to the outcome, the case will go to court for a full trial. As you remember from earlier sections of this chapter, the outcome of voluntary arbitration is binding. The right of a dissatisfied party to reject a court-mandated arbitration decision is really necessary, however, to preserve the disputants' due process rights. The government cannot take away a person's right to his or her day in court in the interests of streamlining the justice system.

Of course, a person who chooses to go forward with a trial after rejecting an arbitrator's decision may still be penalized to some extent in a number of states. In some systems, a party that rejects an arbitrator's decision and does not receive a more favorable decision from the trial court may be forced to pay the opposing counsel's court fees. In other systems, that party may be required to pay the costs of the arbitration or other court fees.

Another difference between court-mandated and voluntary arbitration lies in the rules of evidence, which are generally more relaxed in arbitration than in litigation. A few states do treat evidence in court-mandated arbitration in this same relaxed fashion, either allowing almost any evidence to be admitted regardless of whether it would be admissible in court, or allowing the arbitrator to decide what evidence is admissible. In most states, however, the same rules of evidence apply to court-annexed arbitration as apply to trials.

Mediation. The primary difference between court-mandated mediation and voluntary mediation is in the attitudes of the parties toward the process. In voluntary mediation, the parties are likely to enter the mediation process with a desire to work out an agreement; in court-ordered mediation, they are much more likely to view mediation as simply a hurdle to go through before the trial. Although some have tried to challenge the power of the courts to mandate some form of ADR, the courts have generally upheld this power. For example, in 2002, the First Circuit Court of Appeals, in the case of *In re Atlanta Pipe Corp.*, held that the district court had inherent power to require the parties to participate in nonbinding mediation and to share the costs.[19]

---

[19] 304 F.3d 135 (1st Cir. 2002).

The Future of Alternative Dispute Resolution

You are already familiar with some of the problems associated with each of the ADR methods. You should also be aware of some of the concerns raised about the overall increase in the use of ADR.

First, some legal scholars are concerned about whether a dispute resolution firm can be truly unbiased when one of the parties to the dispute is a major client of the dispute resolution provider. For example, if a major insurance company includes a binding arbitration clause in all its contracts and specifies that JAMS/Endispute will provide the arbitrator, it may be tempting for JAMS/Endispute to favor the insurance company to try to ensure that it will continue to benefit from the firm's business in the future. The more intense the competition among ADR providers, the more tempted the providers may be to favor large firms.

Another issue raised by some critics is whether it is fair for consumers to be coerced into an ADR forum and thereby forced to give up their right to a trial, especially when (1) they may be much more likely to get a higher award from a jury than from an arbitrator; and (2) they may find themselves having "agreed" to arbitration not really voluntarily but because of a clause stuck in a purchase agreement that they failed to read. Despite these concerns, however, interest in ADR continues to grow, and there is no reason to think this growth will end in the near future. During 2000, the AAA handled nearly 200,000 cases through ADR, more than double the number of cases handled in 1998.

Global Dimensions of Alternative Dispute Resolution

Internationally, alternative dispute resolution methods are highly favored. Seventy-three countries currently belong to the United Nations Convention on the Recognition and Enforcement of Foreign Arbitral Awards, commonly referred

---

COMPARATIVE LAW CORNER

Alternate Dispute Resolution in India

India has supported alternate dispute resolution for decades. One of the major ADR mechanisms is the *Lok Adalat* or "people's court." The Lok Adalat system originated in the 1980s and was designed to give poor people a quick way to settle disputes. Originally, the ruling was not legally binding and the parties first had to file the case in a court before turning to the Lok Adalat. Then, once a settlement had been reached, the court ruled the way the Lok Adalat had ruled. The court step has now been eliminated and the ruling of a Lok Adalat is legally binding, as long as the parties both agreed to go to the Lok Adalat—similar to arbitration in the United States.

The Lok Adalat system has a lot of support in India. It takes a large burden off the court system, and it gives people too poor to afford legal fees a chance to recover damages. There are no fees for the Lok Adalat and if a court fee has already been paid, it will be refunded if the case is settled through the Lok Adalat. The process is less formal than a court proceeding and the rules of procedure and evidence are not strictly applied.

International organizations, such as the World Bank, have offered some criticisms of the Lok Adalat system, however. Because the system is village-level, traditional prejudices against women and other underprivileged groups sometimes influence the rulings. Also, there has been evidence that some parties are abusing the Lok Adalat system, using it to delay resolution to get the injured party to abandon its claim.

The Lok Adalat system is an interesting innovation in ADR mechanisms, and it does have many benefits, but it also has drawbacks and can be misused, just like ADR mechanisms in the United States.

to as the New York Convention. The primary function of this treaty is to ensure that an arbitration award made in any of the signatory countries is enforceable in the losing party's country. Defenses to enforcement that are allowed under the treaty include one of the parties to the contract lacked the legal capacity to enter into a contract, the losing party did not receive proper notice of the arbitration, or the arbitrator was acting outside the scope of his or her authority in making the award.

Organizations exist to provide alternative dispute resolution services for firms of different nations. The most commonly known to U.S. businesspersons is probably the AAA. Others include the United Nations Commission of International Trade Law, the London Court of International Arbitration, the Euro-Arab Chamber of Commerce, and the International Chamber of Commerce.

U.S. policy favors arbitration of international disputes. The following case demonstrates this policy.

CASE 4-4

Mitsubishi Motors Corp. v. Soler Chrysler-Plymouth
United States Supreme Court
473 U.S. 614 (1985)

Plaintiff Mitsubishi, a Japanese corporation, and Chrysler International, a Swiss corporation, formed a joint venture, Mitsubishi Motors, to distribute worldwide motor vehicles manufactured in the United States and bearing Mitsubishi and Chrysler trademarks. Defendant Soler Chrysler-Plymouth, a dealership incorporated in Puerto Rico, entered into a distributorship agreement with Mitsubishi that included a binding arbitration clause. When Soler began having difficulty selling the requisite number of cars, it first asked Mitsubishi to delay shipment of several orders and then subsequently refused to accept liability for its failure to sell vehicles under the contract.

Plaintiff Mitsubishi filed an action to compel arbitration. The district court ordered arbitration of all claims, including the defendants' allegations of antitrust violations. The court of appeals reversed in favor of the defendant. The plaintiff, Mitsubishi, appealed to the U.S. Supreme Court.

Justice Blackmun

We granted certiorari primarily to consider whether an American court could enforce an agreement to resolve antitrust claims by arbitration when that agreement arises from an international transaction. Soler reasons that because it falls within a class [for] whose benefit the federal and local antitrust laws were passed, the clause cannot be read to contemplate arbitration of these statutory claims.

We do not agree, for we find no warrant in the Arbitration Act for implying in every contract a presumption against arbitration of statutory claims. The "liberal federal policy favoring arbitration agreements," manifested by the act as a whole, is at bottom a policy guaranteeing the enforcement of private contractual arrangements: the act simply "creates a body of federal substantive law establishing and regulating the duty to honor an agreement to arbitrate."

There is no reason to depart from these guidelines where a party bound by an arbitration agreement raises claims founded on statutory rights. Of course, courts should remain attuned to well-supported claims that the agreement to arbitrate resulted from the sort of fraud or overwhelming economic power that would provide grounds "for the revocation of any contract." But, absent such compelling considerations, the act itself provides no basis for disfavoring agreements to arbitrate statutory claims.

By agreeing to arbitrate a statutory claim, a party does not forgo the substantive rights afforded by the statute, it only submits to their resolution in an arbitral, rather than a judicial, forum. It trades the procedures and opportunity for review of the courtroom for the simplicity, informality, and expedition of arbitration.

We now turn to consider whether Soler's antitrust claims are nonarbitrable even though it agreed to arbitrate them. . . . [W]e conclude that concerns of international comity, respect for the capacities of foreign and transnational tribunals, and sensitivity to the need of the international commercial system for predictability in the resolution of disputes require that we enforce the parties' agreement, even assuming that a contrary result would be forthcoming in a domestic context.

There is no reason to assume at the outset of the dispute that international arbitration will not provide an adequate mechanism. To be sure, the international arbitral tribunal owes no prior allegiance to the legal norms of particular states; hence, it has no direct obligation to vindicate their statutory dictates. The tribunal, however, is bound to effectuate the intentions of the parties. Where the parties have agreed that the arbitral body is to decide a defined set of claims that includes, as in these cases, those arising from the application of American antitrust law, the tribunal therefore should be bound to decide that dispute in accord with the national law giving rise to the claim.

As international trade has expanded in recent decades, so too has the use of international arbitration to resolve disputes arising in the course of that trade. The controversies that international arbitral institutions are called upon to resolve have increased in diversity as well as in complexity. Yet the potential of these tribunals for efficient disposition of legal disagreements arising from commercial relations has not yet been tested. If they are to take a central place in the international legal order, national courts will need to "shake off the old judicial hostility to arbitration," and also their customary and understandable unwillingness to cede jurisdiction of a claim arising under domestic law to a foreign or transnational tribunal. To this extent, at least, it will be necessary for national courts to subordinate domestic notions of arbitrability to the international policy favoring commercial arbitration.

Accordingly, we "require this representative of the American business community to honor its bargain," . . . by holding this agreement to arbitrate "enforce[able] in accord with the explicit provisions of the Arbitration Act."*

[As to the issue of arbitrability] *Reversed* and *remanded* in favor of the Plaintiff, Mitsubishi.

SUMMARY

As the burden on our court system increases, many disputants are turning to alternative ways of resolving disputes. These alternatives include (1) negotiation and settlement, (2) arbitration, (3) mediation, (4) minitrials, (5) early neutral case evaluation, (6) private trials, and (7) summary jury trials.

Which, if any, alternative is best for an individual depends on the situation. The primary benefits that come from these alternatives, in varying degrees, include (1) less publicity, (2) less time, (3) less expense, (4) more convenient proceedings, and (5) a better chance to have a reasonable relationship with the other disputant in the future.

Alternative dispute resolution also has its critics. Some of their concerns are that compulsory ADR may force parties to give up their "day in court," that a party who consistently procures ADR services from one firm will end up with a "neutral" that is biased toward that party, and that we will end up with a two-tier system of justice—an efficient private system and an overburdened public one.

REVIEW QUESTIONS

4-1 Explain why the use of ADR is increasing.

4-2 When will a judge overturn an arbitrator's decision?

4-3 Explain how to secure arbitration as a means of resolving a dispute.

4-4 Why are some people opposed to the growing use of arbitration?

4-5 What are the basic obligations of an arbitrator, according to the Arbitrator's Code of Ethics?

4-6 Identify the factors that would lead a disputant to favor mediation as a dispute resolution method.

REVIEW PROBLEMS

4-7 McGraw and Duffy have a contract that includes a binding arbitration clause. The clause provides for arbitration of any dispute arising out of the contract. The clause also provides that the arbitrator is authorized to award damages of up to $100,000 in any dispute arising out of the contract. McGraw allegedly breaches the contract. Duffy seeks arbitration, and the arbitrator, given the willfulness of the breach and the magnitude of its consequences, awards Duffy $150,000 in damages. Would a court uphold the award? Why or why not?

4-8 Sam and Mary enter into a contract that does not include a provision for arbitration. Mary wants to arbitrate the dispute, but Sam believes that arbitration is not possible because there is no binding

---

*Mitsubishi Motors Corp. v. Soler Chrysler-Plymouth, United States Supreme Court 473 U.S. 614 (1985).

arbitration clause in the contract. How can the parties secure arbitration?

4-9 Howard and Hannah decide to resolve a contract dispute through arbitration. They select their arbitrator through a private service. The arbitrator returns a significant award for Howard. The weekend after receiving a notice of the award, Hannah finds out from one of Howard's coworkers that, although the arbitrator and Howard acted as if they did not know each other, they had actually been college roommates. Does Hannah have any basis for getting the award set aside?

4-10 Eloise is hired as a pharmacist and signs an employment agreement that includes a provision stating that she will submit any employment disputes to arbitration. After being on the job for three years, she is denied a promotion that she feels she deserved. She files a sex discrimination charge with the EEOC and a lawsuit in the federal district court. The employer files a motion to compel arbitration. Will Eloise be forced to arbitrate her claims? Why or why not?

4-11 Boxley Corporation and Eberly Corporation have a contract dispute before an arbitrator. Eberly wants to present evidence of prior contract disputes the two have had, but the arbitrator refuses to receive that evidence, saying it is not relevant to the alleged breach of contract at issue before him. When Boxley receives an award from the arbitrator that entitles it to significant damages from Eberly, the latter appeals the arbitrator's decision on the ground that the excluded evidence would have changed the outcome. Why is Eberly likely or unlikely to have the award overturned?

4-12 Marshall files a complaint against S.A. & E., a brokerage firm registered with the Securities and Exchange Commission. The complaint alleges a violation of the Securities Exchange Act by engaging in fraudulent excessive trading. S.A. & E. files a motion to dismiss the case because Marshall had signed a customer agreement that included in its terms a promise to submit all disputes arising regarding their accounts to arbitration. Marshall argues that, owing to the egregious nature of S.A. & E.'s conduct, Marshall should be entitled to his day in court and the binding arbitration clause should be nullified. Why will the court grant or deny S.A. & E's motion?

CASE PROBLEMS

4-13 In 2011, the U.S. Supreme Court heard a case involving employees who worked for a company who had to sign an employment contract waiving their right to seek an administrative Berman hearing before the Labor Commissioner. The California Supreme Court ruled that employees have a statutory right to defend their right to the use of a Berman hearing. In other words, the court stipulated that the right is an unwaivable one, and such waivers may not be presented in employment contracts. The case then moved up to the Supreme Court. How do you think the Supreme Court decided? *Sonic-Calabasas v. Moreno*, 80 U.S.L.W. 3260 (U.S. Oct. 31, 2011).

4-14 E.Z. was a resident of the Fredericksburg Care Company, L.P., a nursing home. E.Z. died, and her beneficiaries sued the nursing home for negligent care and wrongful death. The nursing home moved to compel arbitration based on an arbitration clause contained in the agreement that E.Z. signed before her admission to the nursing home. It was undisputed that this agreement signed by E.Z. did not comply with the requirement that an agreement to arbitrate a health care liability claim "must contain a written notice in bold-type, ten-point font that conspicuously warns the patient of several important rights." However, the nursing home still motioned to compel arbitration, asserting that federal law should determine the enforceability for the agreement because "the underlying patient-provider transaction involved interstate commerce, which made the FAA applicable to the pre-admission nursing home agreement." The nursing home therefore argued that the FAA prevented the arbitration agreement from being invalid. The trial court denied the defendant's motion to compel, and the defendant nursing home appealed. How do you think the court of appeals ruled? Why? *Fredericksburg Care Co., L.P. v. Perez* (2015 Tex. LEXIS 221).

4-15 In 1999, American Express (AE) began requiring customers to sign a waiver for their right to sue the company through a class action lawsuit. However in 2003, a group of AE customers sued the company on an antitrust charge where AE was inflating credit card fees. AE moved to have the case moved to arbitration, but the plaintiffs argued that AE's antitrust laws went against New York antitrust law since individual plaintiffs would likely not have the resources to pursue a claim as an individual. How do you think this case was decided? *American Express Co. et al. v. Italian Colors Restaurant et al.*, U.S. Supreme Court, No. 12-133 (2012).

4-16 In 2009, a hairstylist and makeup artist named Rita Ragone brought suit against her employer for sexual harassment. She worked for Atlantic Video, ESPN. Her contract stated that all sexual harassment claims would be subject to arbitration and not a court trial. In fact, Ragone was hired on the condition that she sign the contract containing the arbitration provision. The provision specifically said that her claims would be submitted to one arbitrator, and that the arbitrator's decision would be binding.

Ragone wanted to dismiss the arbitration provision. However, the court could waive the provision only if it was unconscionable. Ragone argued that three clauses were unconscionable: Specifically, she must file the claim within 90 days instead of the usual 300 days, the losing party would pay all attorneys fees, and there was no right to a jury trial. Do you think these three clauses are unconscionable? How do you think the court decided? *Ragone v. Atlantic Video*, 2010 U.S. App. LEXIS 3018 (2d Cir. Feb. 17, 2010).

4-17 Plaintiff N.V. filed suit against her former employer, Autozone, Inc. The plaintiff alleged sex and disability discrimination. It was undisputed in court that the defendant company's Occupational Injury Benefit Plan contained an enforceable arbitration provision. The parties disagreed about whether this arbitration provision applied to all of the plaintiff's claims. The defendant company argued that the plaintiff's claims of "sexual harassment, retaliation, and hostile work environment" were not covered by this arbitration provision, reasoning that this provision applied only to claims "made in connection with a job-related injury." Do you think that the defendant substantially invoked the judicial process "to the detriment or prejudice of the other party," such that it waived its right to arbitration? Why or why not? *Valdez v. Autozone Inc.* (2015 U.S. Dist. LEXIS 18608).

4-18 Ally Cat LLC bought a condo from Chauvin, the seller of the unit. A Home Owners Limited Warranty document, containing an arbitration provision, was transferred to Ally Cat. However, the document did not specifically state that the signer of the document had to comply with the terms of the document. The document alternatively meant that the signer (or purchaser of the condo) was simply acknowledging that the document and condo had been received. In other words, under Kentucky law, the document did not qualify as a binding arbitration agreement. If the document did qualify under Kentucky law, Ally Cat would be forced to resolve disputes through arbitration and Kentucky courts would have jurisdiction to enforce the arbitration. Therefore, when Ally Cat attempted to bring Chauvin to court for damages, how do you think the court resolved the dispute? Did the parties have to settle through arbitration? *Ally Cat LLC et al. v. Chauvin et al.*, 274 S.W.3d 451; 2009 Ky. LEXIS 10 (Ky. 2009).

4-19 Plaintiff health care workers sought damages, statutory penalties, and relief from their former employer for alleged wage law violations. The plaintiffs sought a declaration that the arbitration clauses in their employment contracts were "unconscionable and unenforceable." The plaintiffs reasoned that other sections of the employment contract "permit the employer to seek limited judicial relief without affording the employees that same option." The defendant employer moved to compel arbitration. The trial court denied the defendant's motion to compel arbitration, holding that the arbitration agreement was unconscionable. The defendant employer appealed. On appeal, the court considered that the state of Washington has a strong public policy favoring arbitration. How do you think the court of appeals ruled? *Romney v. Franciscan Med. Grp.* (2015 Wash. App. LEXIS 322).

4-20 Kinecta's employment agreement contained a clause banning employees from filing class action lawsuits against the company. When employees attempted to sue the company, Kinecta compelled arbitration. However, the employees fought back saying the class action ban and arbitration clauses were illegal and unfair. The court decided to allow the company to compel arbitration in this case. However, the court also allowed the class action allegations to apply to the case. Kinecta appealed. What was the final result of this case? *Kinecta Alternative Financial Solutions, Inc. v. Superior Court (Malone)*, Cal. App. 4th (2012).

THINKING CRITICALLY ABOUT RELEVANT LEGAL ISSUES

Arbitration in the Malpractice Arena

Recently, a group of New Jersey gynecologists have come under fire for requiring their patients to sign a binding arbitration agreement before treating them. The agreement waived the patients' rights to a jury trial and limited the amount of damages they could be awarded. The use of binding arbitration agreements has grown in recent years, and they are desperately needed in the field of medicine.

Malpractice insurance for doctors has become unreasonably expensive because of frivolous lawsuits that award plaintiffs huge damages. These lawsuits are especially a problem in the fields of obstetrics and gynecology. Consequently, many doctors avoid obstetrics and gynecology, and those doctors who remain in this field are forced to charge more for their care to pay for their insurance. Obstetrics and gynecology are crucial areas in medicine and the shortage of obstetricians and gynecologists poses a major threat to the health of American women.

The arbitration agreement is beneficial for both doctor and patient. Arbitration agreements limit the award a patient can receive, which helps to keep insurance rates down and lowers the cost of care for all patients. Arbitration is faster and less expensive for plaintiffs than a lawsuit. Arbitration also allows patients to recover damages much more quickly than a lawsuit, and although the damages are limited, arbitration avoids the many appeals that larger punitive damages generate.

Some have questioned the legality of these doctor–patient arbitration agreements. Binding arbitration agreements between doctor and patient are no different from those between employer and employee, which the Supreme Court ruled are completely legal in *Gilmer v. Interstate/Johnson Lane Corp*. The court ruled that Gilmer's age discrimination suit was subject to compulsory arbitration because of the employment agreement he signed and that the employment agreement requiring arbitration was valid. Therefore, a binding arbitration agreement between doctor and patient is valid.

Just as an employee does not have to accept employment if she does not like the offered contract, a patient does not have to be treated by that doctor if she does not like the doctor's terms. There is, however, no reason a patient should have a problem with an arbitration agreement. It helps lower costs for both doctors and patients and speeds up the process for everyone involved. Not only should doctor–patient arbitration agreements be upheld as legal, they should be encouraged.

1. How would you frame the issue and conclusion of this essay?

2. What ethical norms would cause the author of this essay to take this position?

   Clue: Look at the ethical norms discussed in Chapter 1.

3. What legal analogy is used to support the author's argument? How appropriate is that legal analogy?

4. Write an essay that someone who holds an opinion opposite to that of the essay author might write.

   Clue: What other ethical norms could influence an opinion on this issue?

ASSIGNMENT ON THE INTERNET

After reading this chapter, you should have an understanding of the advantages and disadvantages of ADR. Using the Internet, find a private mediation and arbitration firm, such as National Arbitration and Mediation. Visit its website (www.namadr.com) and create a list of questions you would want to ask before hiring its services by applying the information in this chapter and your critical thinking skills.

What kind of problems might be associated with this company's process? Where do its mediators or arbitrators come from? What qualities would you look for in choosing a mediator or arbitrator? What are the advantages of using this company? What are the disadvantages?

Now that you have evaluated one ADR firm, make a comparison to another firm that you found on the Internet (e.g., www.resolvemydispute.com). What distinguishes one firm from the other? Why would you choose one and not the other? What factors affect your answer? When would you forego ADR and use the court system?

ON THE INTERNET

www.adrr.com This ADR site contains substantial online materials for ADR and particularly for mediation.
www.odr.info This site has information on online dispute resolution.
www.calbar.ca.gov This website has recent ADR cases from California.

www.adr.org This is the home page of the American Arbitration Association.

www.bbb.org To access the Better Business Bureau's dispute resolution page from the BBB home page, scroll to the bottom of the page and click "Council of Better Business Bureaus." From here, highlight "Programs and Services" and click on "Dispute Resolution."

www.law.cornell.edu The Federal Arbitration Act can be found at this website. Scroll to the bottom of the page and click on "Legal Information Institute." Once you have followed the link, click on "Federal law" under the heading "Legal Resources." Under the heading "Legislative Branch," click on "U.S. Code." In the box named "Title," type in "9." This will take you to the resources for the Federal Arbitration Act.

FOR FUTURE READING

Karcher, Molly, Alexandra Klaus, Ryan J. Nichols, and Stanley A. Prenger. "State Legislative Update." *University of Missouri Journal of Dispute Resolution* 375 (2013).

Lavi, Dafna. "Divorce Involving Domestic Violence: Is Med-Arb Likely to Be the Solution?" *Pepperdine Dispute Resolution Law Journal* 14 (2014): 91.

Levenson, Wendy D. "Let's Make a Deal: Negotiating Resolution of Intellectual Property Disputes through Mandatory Mediation at the Federal Circuit." *John Marshall Law School Review of Intellectual Property Law* 6 (2007): 365.

Molinoff, Morgan. "The Age of (Guilt or) Innocence: Using ADR to Reform New York's Juvenile Justice System in the Wake of *Miller v. Alabama*." *Cardozo Journal of Conflict Resolution* 15 (2013): 297.

Moore, Christopher W. *The Mediation Process: Practical Strategies for Resolving Conflict* (3rd ed.). San Francisco: Jossey-Bass, 2003.

Nolan-Haley, Jacqueline M. *Alternative Dispute Resolution in a Nutshell* (2nd ed.). St. Paul: West Group, 2001.

Schwartz, Amy. "'Drive-Thru' Arbitration in the Digital Age: Empowering Consumers through Binding ODR." *Baylor Law Review* 63, no. 9 (2010): 178.

CHAPTER FIVE

Constitutional Principles

- THE CONSTITUTION
- FEDERALISM
- SEPARATION OF POWERS
- THE IMPACT OF THE COMMERCE CLAUSE ON BUSINESS
- THE TAXING AND SPENDING POWERS OF THE FEDERAL GOVERNMENT
- THE IMPACT OF THE AMENDMENTS ON BUSINESS

Fiercely independent, highly individualistic, and very proud of their country would be a good characterization of Americans. Many say there is no place they would rather live than the United States. Much of their pride stems from a belief that they have a strong Constitution, which secures for all individuals their most fundamental rights. Most people, however, are not aware of precisely what their constitutional rights are or of how to go about enforcing those rights. This chapter provides the future business manager with basic knowledge of the constitutional framework of our country, as well as an overview of the significant impact of some of the constitutional provisions on the legal environment of business.

The Constitution

The Constitution provides the legal framework for our nation. The articles of the Constitution set out the basic structure of our government and the respective roles of the state and federal governments. The Amendments to the Constitution, especially the first 10, were primarily designed to establish and protect individual rights.

Federalism

Underlying the system of government established by the Constitution is the principle of **federalism**, which means that the authority to govern is divided between two sovereigns or supreme lawmakers. In the United States, these two sovereigns are the state and federal governments. Federalism allocates the power to control local matters to local governments. This allocation is embodied in the U.S. Constitution. Under the Constitution, all powers that are neither given exclusively to the federal government nor taken from the states are reserved to the states. The federal government has only those powers granted to it in the Constitution. Therefore, whenever federal legislation that affects business is passed, the question of the source of authority for that regulation always arises. The Commerce Clause is the predominant source of authority for the federal regulation of business, as we will see later.

federalism A system of government in which power is divided between a central authority and constituent political units.

CRITICAL THINKING ABOUT THE LAW

The Constitution secures numerous rights for U.S. citizens. If we did not have these rights, our lives would be very different. Furthermore, businesses would be forced to alter their practices because they would not enjoy the various constitutional protections. As you will soon learn, various components of the Constitution, such as the Commerce Clause and the Bill of Rights, offer guidance and protection for businesses. The following questions will help sharpen your critical thinking about the effects of the Constitution on business.

1. One of the basic elements in the Constitution is the separation of powers in the government. What ethical norm would guide the framers' thinking in creating a system with a separation of powers and a system of checks and balances?

    Clue: Consider what might happen if one branch of government became too strong.

2. If the framers of the Constitution wanted to offer the protection of unrestricted speech to citizens and businesses, what ethical norm would they view as most important?

    Clue: Return to the list of ethical norms in Chapter 1. Which ethical norm might the framers view as least important in protecting unrestricted speech?

3. Why should you, as a future business manager, be knowledgeable about the basic protections offered by the Constitution?

    Clue: If you were ignorant of the constitutional protections, how might your business suffer?

In some areas, the state and federal governments have concurrent authority; that is, both governments have the power to regulate the matter in question. This situation arises when authority to regulate in an area has been expressly given to the federal government by the Constitution. In such cases, a state may regulate in the area as long as its regulation does not conflict with any federal regulation of the same subject matter. A conflict arises when a regulated party cannot comply with both the state and the federal laws at the same time. When the state law is more restrictive, such that compliance with the state law is automatically compliance with the federal law, the state law will usually be valid. For example, as discussed in Chapter 22, in many areas of environmental regulation, states may impose much more stringent pollution-control standards than those imposed by federal law.

SUPREMACY CLAUSE

Supremacy Clause Provides that the U.S. Constitution and all laws and treaties of the United States constitute the supreme law of the land; found in Article VI.

federal supremacy Principle declaring that any state or local law that directly conflicts with the federal Constitution, laws, or treaties is void.

federal preemption Constitutional doctrine stating that in an area in which federal regulation is pervasive, state legislation cannot stand.

The outcome of conflicts between state and federal laws is dictated by the **Supremacy Clause**. This clause, found in Article VI of the Constitution, provides that the Constitution, laws, and treaties of the United States constitute the supreme law of the land, "any Thing in the Constitution or Laws of any State to the Contrary notwithstanding." This principle is known as the principle of **federal supremacy**: Any state or local law that directly conflicts with the federal Constitution, laws, or treaties is void. Federal laws include rules promulgated by federal administrative agencies. Exhibit 5-1 illustrates the application of the Supremacy Clause to state regulation.

FEDERAL PREEMPTION

The Supremacy Clause is also the basis for the doctrine of **federal preemption**. This doctrine is used to strike down a state law that, although it does not directly conflict with a federal law, attempts to regulate an area in which federal legislation is so pervasive that it is evident that the U.S. Congress wanted only federal regulation in that general area. It is often said in these cases that federal law

EXHIBIT 5-1

APPLICATION OF THE SUPREMACY CLAUSE TO STATE REGULATION

- Does the state law directly conflict with a federal law?
  - If yes, the state law is invalid under the Supremacy Clause.
  - If no, does the state law attempt to regulate in an area where federal law is so pervasive that it appears that Congress intended only federal regulation in that area?
    - If yes, the state law is invalid under the doctrine of federal preemption.
    - If no, the state law is valid.

"preempts the field." Cases of federal preemption are especially likely to arise in matters pertaining to interstate commerce, such as when a local regulation imposes a substantial burden on the flow of interstate commerce through a particular state. This situation is discussed in some detail in the section on the Commerce Clause.

Separation of Powers

The U.S. Constitution, in its first three articles, establishes three independent branches of the federal government, each with its own predominant and independent power. These three are the legislative, executive, and judicial branches. Each branch was made independent of the others and was given a separate sphere of power to prevent any one source from obtaining too much power and consequently dominating the government.

The doctrine of **separation of powers** calls for Congress, the legislative branch, to enact legislation and appropriate funds. The president is commander-in-chief of the armed forces and is also charged with ensuring that the laws are faithfully executed. The judicial branch is charged with interpreting the laws in the course of applying them to particular disputes. No member of one branch owes his or her tenure in that position to a member of any other branch; no branch can encroach on the power of another. This system is often referred to as being a system of *checks and balances*; that is, the powers given to each branch operate to keep the other branches from being able to seize enough power to dominate the government. Exhibit 5-2 provides a portrait of this system.

Despite this delicate system of checks and balances, on numerous occasions a question has arisen as to whether one branch was attempting to encroach on the domain of another. This situation arose in an unusual context: a sexual harassment charge against President Bill Clinton.

separation of powers Constitutional doctrine whereby the legislative branch enacts laws and appropriates funds, the executive branch sees that the laws are faithfully executed, and the judicial branch interprets the laws.

CASE 5-1

William Jefferson Clinton v. Paula Corbin Jones
Supreme Court of the United States
520 U.S. 681 (1997)

Plaintiff Paula Jones filed a civil action against defendant (sitting) President Bill Clinton, alleging that he made "abhorrent" sexual advances. She sought $75,000 in actual damages and $100,000 in punitive damages.

Defendant Clinton sought to dismiss the claim on the ground of presidential immunity, or, alternatively, to delay the proceedings until his term of office had expired.

The district court denied the motion to dismiss and ordered discovery to proceed, but it also ordered that the trial be stayed until the end of Clinton's term. The court of appeals affirmed the denial of the motion to dismiss and reversed the stay of the trial. President Clinton appealed to the U.S. Supreme Court.

Justice Stevens

Petitioner's principal submission—that "in all but the most exceptional cases," the Constitution affords the President temporary immunity from civil damages litigation arising out of events that occurred before he took office—cannot be sustained on the basis of precedent.

Only three sitting presidents have been defendants in civil litigation involving their actions prior to taking office. Complaints against Theodore Roosevelt and Harry Truman had been dismissed before they took office; the dismissals were affirmed after their respective inaugurations. Two companion cases arising out of an automobile accident were filed against John F. Kennedy in 1960 during the Presidential campaign. After taking office, he unsuccessfully argued that his status as Commander in Chief gave him a right to a stay. The motion for a stay was denied by the District Court, and the matter was settled out of court. Thus, none of those cases sheds any light on the constitutional issue before us.

The principal rationale for affording certain public servants immunity from suits for money damages arising out of their official acts is inapplicable to unofficial conduct. In cases involving prosecutors, legislators, and judges we have repeatedly explained that the immunity serves the public interest in enabling such officials to perform their designated functions effectively without fear that a particular decision may give rise to personal liability.

That rationale provided the principal basis for our holding that a former president of the United States was "entitled to absolute immunity from damages liability predicated on his official acts." Our central concern was to avoid rendering the President "unduly cautious in the discharge of his official duties."

This reasoning provides no support for an immunity for unofficial conduct. . . . "[T]he sphere of protected action must be related closely to the immunity's justifying purposes." But we have never suggested that the President, or any other official, has an immunity that extends beyond the scope of any action taken in an official capacity.

Moreover, when defining the scope of an immunity for acts clearly taken within an official capacity, we have applied a functional approach. "Frequently our decisions have held that an official's absolute immunity should extend only to acts in performance of particular functions of his office." Petitioner's strongest argument supporting his immunity claim is based on the text and structure of the Constitution. The President argues for a postponement of the judicial proceedings that will determine whether he violated any law. His argument is grounded in the character of the office that was created by Article II of the Constitution and relies on separation-of-powers principles.

As a starting premise, petitioner contends that he occupies a unique office with powers and responsibilities so vast and important that the public interest demands that he devote his undivided time and attention to his public duties. He submits that—given the nature of the office—the doctrine of separation of powers places limits on the authority of the Federal Judiciary to interfere with the Executive Branch that would be transgressed by allowing this action to proceed.

We have no dispute with the initial premise of the argument. We have long recognized the "unique position in the constitutional scheme" that this office occupies.

It does not follow, however, that separation-of-powers principles would be violated by allowing this action to proceed. The doctrine of separation of powers is concerned with the allocation of official power among the three coequal branches of our Government. The Framers "built into the tripartite Federal Government . . . a self-executing safeguard against the encroachment or aggrandizement of one branch at the expense of the other." Thus, for example, the Congress may not exercise the judicial power to revise final judgments, or the executive power to manage an airport.

. . . [I]n this case there is no suggestion that the Federal Judiciary is being asked to perform any function that might in some way be described as "executive." Respondent is merely asking the courts to exercise their core Article III jurisdiction to decide cases and controversies. Whatever the outcome of this case, there is no possibility that the decision will curtail the scope of the official powers of the Executive Branch. The litigation of questions that relate entirely to the unofficial conduct of the individual who happens to be the President poses no perceptible risk of misallocation of either judicial power or executive power.

Rather than arguing that the decision of the case will produce either an aggrandizement of judicial power or a narrowing of executive power, petitioner contends that—as a by-product of an otherwise traditional exercise of judicial power—burdens will be placed on the President that will hamper the performance of his official duties. We have recognized that "[e]ven when a branch does not arrogate power to itself . . . the separation-of-powers doctrine requires that a branch not impair another in the performance of its constitutional duties." As a factual matter, petitioner contends that this particular case—as well as the potential additional litigation that an affirmance of the Court of Appeals judgment might spawn—may impose an unacceptable burden on the President's time and energy and thereby impair the effective performance of his office.

Petitioner's predictive judgment finds little support in either history or the relatively narrow compass of the issues raised in this particular case. If the past is any indicator, it seems unlikely that a deluge of such litigation will ever engulf the presidency. As for the case at hand, if properly managed by the District Court, it appears to us highly unlikely to occupy any substantial amount of petitioner's time.

Of greater significance, petitioner errs by presuming that interactions between the Judicial Branch and the Executive, even quite burdensome interactions, necessarily rise to the level of constitutionally forbidden impairment of the Executive's ability to perform its constitutionally mandated functions. Separation of powers does not mean that the branches "ought to have no partial agency in, or no control over the acts of each other." The fact that a federal court's exercise of its traditional Article III jurisdiction may significantly burden the time and attention of the Chief Executive is not sufficient to establish a violation of the Constitution. Two long-settled propositions . . . support that conclusion.

First, we have long held that when the President takes official action, the Court has the authority to determine whether he has acted within the law. Perhaps the most dramatic example of such a case is our holding that President Truman exceeded his constitutional authority when he issued an order directing the Secretary of Commerce to take possession of and operate most of the Nation's steel mills, in order to avert a national catastrophe.[1]

Second, it is also settled that the President is subject to judicial process in appropriate circumstances. We . . . held that President Nixon was obligated to comply with a subpoena commanding him to produce certain tape recordings of his conversations with his aides. As we explained, "neither the doctrine of separation of powers, nor the need for confidentiality of high-level communications, without more, can sustain an absolute, unqualified presidential privilege of immunity from judicial process under all circumstances."

Sitting Presidents have responded to court orders to provide testimony and other information with sufficient frequency that such interactions between the Judicial and Executive Branches can scarcely be thought a novelty. President Ford complied with an order to give a deposition in a criminal trial, and President Clinton has twice given videotaped testimony in criminal proceedings.

"[I]t is settled law that the separation-of-powers doctrine does not bar every exercise of jurisdiction over the President of the United States." If the Judiciary may severely burden the Executive Branch by reviewing the legality of the President's official conduct, and if it may direct appropriate process to the President himself, it must follow that the federal courts have power to determine the legality of his unofficial conduct. The burden on the President's time and energy that is a mere by-product of such review surely cannot be considered as onerous as the direct burden imposed by judicial review and the occasional invalidation of his official actions. We therefore hold that the doctrine of separation of powers does not require federal courts to stay all private actions against the President until he leaves office.*

Reversed in part. *Affirmed* in part in favor of Respondent, Jones.

COMMENT: After this case was sent back for trial on the merits, the case was ultimately dismissed on April 1, 1998, on a motion for summary judgment on the ground that the plaintiff's allegations, even if true, failed to state a claim of criminal sexual assault or sexual harassment. It is ironic that despite the high court's claim that the case would be "highly unlikely to occupy any substantial amount of the petitioner's time," matters arising out of this case managed to occupy so much of the president's time and become such a focus of a media frenzy that many people were calling for the media to reduce coverage of the issues so the president could do his job.[2]

LINKING LAW AND BUSINESS

Finance

The principle behind the separation of powers in government is also modeled in another realm of business. In your accounting class, you learned that internal controls are the policies and procedures used to create a greater assurance that the objectives of an organization will be met. One feature of internal controls is the separation of duties. This feature calls for the functions of authorization, recording, and custody to be exercised by different individuals. The likelihood of illegal acts by employees is reduced when the responsibility of completing a task is dependent on more than one person. If there are three people responsible for carrying out a particular task, then each person acts as a deterrent to the other two in regard to the possibility of embezzlement by one or more employees. Therefore, the chance of dishonest behavior is minimized when employees act as a check on the other employees involved in striving to meet organizational objectives.

---

[1]*Youngstown Sheet & Tube Co. v. Sawyer*, 343 U.S. 579 (1952).

[2]*Jones v. Clinton and Danny Ferguson*, 12 F. Supp. 2d 931 (E.D. Ark. 1998).

*William Jefferson Clinton v. Paula Corbin Jones, Supreme Court of the United States 520 U.S. 681 (1997). https://www.law.cornell.edu/supct/html/95-1853.ZO.html.

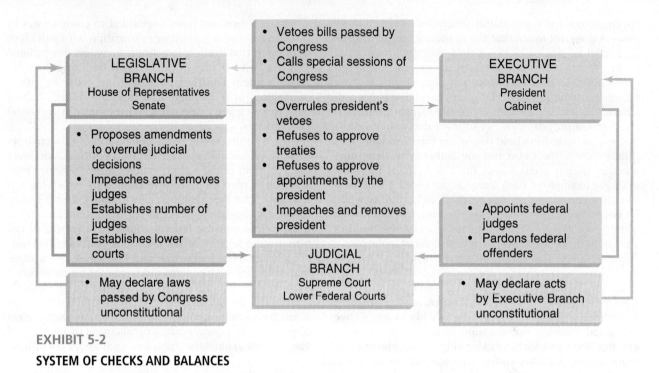

EXHIBIT 5-2

SYSTEM OF CHECKS AND BALANCES

Cases like *Jones v. Clinton* are not common. The reason is not that each branch generally operates carefully within its own sphere of power. Rather, the explanation lies in the fact that because it is difficult to determine where one branch's authority ends and another's begins, each branch rarely challenges the power of its competing branches. The powers of each branch were established so that, although the branches are separate and independent, each branch still influences the actions of the others and there is still a substantial amount of interaction among them. You can review this system by examining Exhibit 5-2.

The Impact of the Commerce Clause on Business

THE COMMERCE CLAUSE AS A SOURCE OF FEDERAL AUTHORITY

The primary powers of Congress are listed in Article I of the Constitution. It is important to remember that Congress has only limited legislative power. Congress possesses only that legislative power granted to it by the Constitution. Thus, all acts of Congress not specifically authorized by the Constitution or necessary to accomplish an authorized end are invalid.

Commerce Clause Empowers Congress to regulate commerce with foreign nations, with Indian tribes, and among the states; found in Article I.

The **Commerce Clause** provides the basis for most of the federal regulation of business today. This clause empowers the legislature to "regulate Commerce with foreign Nations, and among the several States, and with the Indian Tribes." Early in our history, the Supreme Court was committed to a laissez-faire ideology, an ideology that was grounded in individualism. A narrow interpretation of the Commerce Clause means that only a limited amount of trade or exchange can be regulated by Congress.

Under the Court's initial narrow interpretation, the Commerce Clause was interpreted to apply only to the transportation of goods. Manufacturing of goods, even of goods that were going to be sold in another state, was not considered to have a direct effect on interstate commerce and, thus, was not subject to federal regulation. Businesses conducted solely in one state were similarly excluded from the authority of Congress. Under this restrictive interpretation, numerous federal regulations, such as laws attempting to regulate the use of child labor in manufacturing plants,[3] were struck down.

During the 1930s, the Supreme Court's interpretation of the Commerce Clause was broadened to allow a greater scope for federal regulations. In *NLRB v. Jones & Laughlin Steel Corp.*,[4] for example, the Court said that:

> *Although activities may be intrastate in character when separately considered, if they have such a close and substantial relationship to interstate commerce that their control is essential or appropriate to protect that commerce from burdens or obstructions, Congress cannot be denied the power to exercise that control.**

Over the next several decades, the Supreme Court continued to expand Congress's power under the Commerce Clause to regulate intrastate activities that affect interstate commerce. For example, in *Perez v. United States*,[5] loan-sharking, conducted on a local basis, was deemed to affect interstate commerce because of its connection to organized crime on a national scale, as the funds from loan-sharking help to pay for crime across the United States. According to the subsequent ruling in *International House of Pancakes v. Theodore*,[6] a locally owned and operated franchise located near two interstates and three hotels qualifies the restaurant as related to interstate commerce and thus subject to the Americans with Disabilities Act. This expansive view continued with *United States v. Lake*,[7] in which the court ruled that a locally operated coal mine that sells its coal locally and buys its supplies locally can still be subjected to federal regulations (Federal Mine Safety and Health Act) because the local activities of all coal mines help to influence the interstate market for coal.

As these cases illustrate, during most of the twentieth century, almost any activity, even if purely intrastate, could be regulated by the federal government if it substantially affected interstate commerce. The effect may be direct or indirect, as the U.S. Supreme Court demonstrated in the classic 1942 case of *Wickard v. Filburn*,[8] when it upheld federal regulation of the production of wheat on a farm in Ohio that produced only 239 bushels of wheat solely for consumption on the farm. The Court's rationale was that even though one wheat farmer's activities might not matter, the combination of a lot of small farmers' activities could have a substantial impact on the national wheat market. This broad interpretation of the Commerce Clause has made possible much of the legislation covered in other sections of this book.

Since the mid-1990s, however, the Supreme Court has appeared to be scrutinizing congressional attempts to regulate based on the Commerce Clause a little more closely. In the 1995 case of *United States v. Lopez*,[9] the U.S. Supreme Court found that Congress had exceeded its authority under the Commerce Clause when it passed the Gun-Free School Zone Act, a law that banned the possession of guns within 1,000 feet of any school. The Court found the statute to be unconstitutional because Congress was attempting to regulate

---

[3]*Hammer v. Dagenhart*, 247 U.S. 251 (1918).
[4]301 U.S. 1 (1937).
[5]402 U.S. 146 (1971).
[6]844 F. Supp. 574 (S.D. Cal. 1993).
[7]985 F.2d 265 (6th Cir. 1995).
[8]317 U.S. 111 (1942).
[9]514 U.S. 549 (1995).
*NLRB v. Jones & Laughlin Steel Corp, 301 U.S. 1 (1937). https://supreme.justia.com/cases/federal/us/301/1/case.html.

in an area that had "nothing to do with commerce, or any sort of economic enterprise." At first, commentators did not see this case, decided in a 5–4 vote, as a major shift in the Supreme Court's Commerce Clause interpretation. As the Court's ruling in *Brzonkala v. Morrison*[10] demonstrates, however, *Lopez* may indeed have indicated that the courts are going to look more closely at congressional attempts to regulate interstate commerce, an action that seems consistent with the high court's increasing tendency to support greater power for states in conflicts between the state and federal governments. In *Morrison*, the Court ruled that the Violence Against Women Act, which Congress justified through its Commerce Clause power, was unconstitutional. Despite the rulings in *Lopez* and *Morrison*, however, the following case, *Gonzales v. Raich*, shows that the Court may not be categorically opposed to an expansion of congressional power through the Commerce Clause.

CASE 5-2

Gonzales v. Raich
Supreme Court of the United States
545 U.S. 1 (2005)

In 1996, California voters passed the Compassionate Use Act of 1996, which allowed seriously ill residents of the state to have access to marijuana for medical purposes. Angel Raich and Diane Monson are California residents who were using medical marijuana pursuant to their doctors' recommendations for their serious medical conditions.

County deputy sheriffs and federal Drug Enforcement Administration (DEA) agents investigated Raich's and Monson's use of medical marijuana. Although Raich and Monson were found to be in compliance with the state law, the federal agents seized and destroyed their cannabis plants.

Raich and Monson brought suit against the attorney general of the United States and the head of the DEA, seeking injunctive and declaratory relief prohibiting the enforcement of the federal Controlled Substances Act (CSA) to the extent it prevents them from possessing, obtaining, or manufacturing cannabis for their personal medical use. The district court denied the respondents' motion for a preliminary injunction. A divided panel of the court of appeals for the Ninth Circuit reversed and ordered the district court to enter a preliminary injunction. The United States appealed.

Justice Stevens

Respondents in this case do not dispute that passage of the CSA, as part of the Comprehensive Drug Abuse Prevention and Control Act, was well within Congress' commerce power. Nor do they contend that any provision or section of the CSA amounts to an unconstitutional exercise of congressional authority. Rather, respondents' challenge is actually quite limited; they argue that the CSA's categorical prohibition of the manufacture and possession of marijuana as applied to the intrastate manufacture and possession of marijuana for medical purposes pursuant to California law exceeds Congress' authority under the Commerce Clause.

[There are] three general categories of regulation in which Congress is authorized to engage under its commerce power. First, Congress can regulate the channels of interstate commerce. Second, Congress has authority to regulate and protect the instrumentalities of interstate commerce and persons or things in interstate commerce. Third, Congress has the power to regulate activities that substantially affect interstate commerce. Only the third category is implicated in the case at hand.

Our case law firmly establishes Congress' power to regulate purely local activities that are part of an economic "class of activities" that have a substantial effect on interstate commerce. As we stated in *Wickard v. Filburn*, 317 U.S. 111 (1942), "even if appellee's activity be local and though it may not be regarded as commerce, it may still, whatever its nature, be reached by Congress if it exerts a substantial economic effect on interstate commerce." We have never required Congress to legislate with scientific exactitude. When Congress decides that the "total incidence" of a practice poses a threat to a national market, it may regulate the entire class.

Our decision in *Wickard* is of particular relevance. In *Wickard*, we upheld the application of regulations promulgated under the Agricultural Adjustment Act of

---
[10] 529 U.S. 598 (2000).

1938, which were designed to control the volume of wheat moving in interstate and foreign commerce in order to avoid surpluses and consequent abnormally low prices. The regulations established an allotment of 11.1 acres for Filburn's 1941 wheat crop, but he sowed 23 acres, intending to use the excess by consuming it on his own farm. Filburn argued that even though we had sustained Congress' power to regulate the production of goods for commerce, that power did not authorize "federal regulation [of] production not intended in any part for commerce but wholly for consumption on the farm." Justice Jackson's opinion for a unanimous Court rejected this submission. He wrote:

> *The effect of the statute before us is to restrict the amount which may be produced for market and the extent as well to which one may forestall resort to the market by producing to meet his own needs. That appellee's own contribution to the demand for wheat may be trivial by itself is not enough to remove him from the scope of federal regulation where, as here, his contribution, taken together with that of many others similarly situated, is far from trivial.*

Wickard thus establishes that Congress can regulate purely intrastate activity that is not itself "commercial," in that it is not produced for sale, if it concludes that failure to regulate that class of activity would undercut the regulation of the interstate market in that commodity.

The similarities between this case and *Wickard* are striking. Like the farmer in *Wickard*, respondents are cultivating, for home consumption, a fungible commodity for which there is an established, albeit illegal, interstate market. Just as the Agricultural Adjustment Act was designed "to control the volume [of wheat] moving in interstate and foreign commerce in order to avoid surpluses . . ." and consequently control the market price, a primary purpose of the CSA is to control the supply and demand of controlled substances in both lawful and unlawful drug markets. In *Wickard*, we had no difficulty concluding that Congress had a rational basis for believing that, when viewed in the aggregate, leaving home-consumed wheat outside the regulatory scheme would have a substantial influence on price and market conditions. Here too, Congress had a rational basis for concluding that leaving home-consumed marijuana outside federal control would similarly affect price and market conditions.

More concretely, one concern prompting inclusion of wheat grown for home consumption in the 1938 Act was that rising market prices could draw such wheat into the interstate market, resulting in lower market prices. The parallel concern making it appropriate to include marijuana grown for home consumption in the CSA is the likelihood that the high demand in the interstate market will draw such marijuana into that market. While the diversion of homegrown wheat tended to frustrate the federal interest in stabilizing prices by regulating the volume of commercial transactions in the interstate market, the diversion of homegrown marijuana tends to frustrate the federal interest in eliminating commercial transactions in the interstate market in their entirety. In both cases, the regulation is squarely within Congress' commerce power because production of the commodity meant for home consumption, be it wheat or marijuana, has a substantial effect on supply and demand in the national market for that commodity.

Nonetheless, respondents suggest that *Wickard* differs from this case in three respects: (1) the Agricultural Adjustment Act, unlike the CSA, exempted small farming operations; (2) *Wickard* involved a "quintessential economic activity"—a commercial farm—whereas respondents do not sell marijuana; and (3) the *Wickard* record made it clear that the aggregate production of wheat for use on farms had a significant impact on market prices. Those differences, though factually accurate, do not diminish the precedential force of this Court's reasoning.

The fact that Wickard's own impact on the market was "trivial by itself" was not a sufficient reason for removing him from the scope of federal regulation. That the Secretary of Agriculture elected to exempt even smaller farms from regulation does not speak to his power to regulate all those whose aggregated production was significant, nor did that fact play any role in the Court's analysis. Moreover, even though Wickard was indeed a commercial farmer, the activity he was engaged in—the cultivation of wheat for home consumption—was not treated by the Court as part of his commercial farming operation.

In assessing the scope of Congress' authority under the Commerce Clause, we stress that the task before us is a modest one. We need not determine whether respondents' activities, taken in the aggregate, substantially affect interstate commerce in fact, but only whether a "rational basis" exists for so concluding. Given the enforcement difficulties that attend distinguishing between marijuana cultivated locally and marijuana grown elsewhere, and concerns about diversion into illicit channels, we have no difficulty concluding that Congress had a rational basis for believing that failure to regulate the intrastate manufacture and possession of marijuana would leave a gaping hole in the CSA. Thus, as in *Wickard*, when it enacted comprehensive legislation to regulate the interstate market in a fungible commodity, Congress was acting well within its authority to "make all Laws which shall be necessary and proper" to "regulate Commerce . . . among the several States." That the regulation ensnares some purely intrastate activity is of no moment. As we have done many times before, we refuse to excise individual components of that larger scheme.*

Reversed and *remanded* in favor of Attorney General Gonzalez.

---

*Gonzales v. Raich, Supreme Court of the United States 545 U.S. 1 (2005). https://www.law.cornell.edu/supct/html/03-1454.ZO.html.

CRITICAL THINKING ABOUT THE LAW

Analogies are a standard method for creating a link between the case at hand and legal precedent.

Wickard v. Filburn is a long-established precedent. The court's reasoning in Case 5-2 is that the use of medical marijuana by the plaintiffs is sufficiently similar to the facts in *Wickard* to rely on this precedent.

1. What are the similarities between the case at hand and *Wickard*?

   Clue: Try to make a large list of similarities. Later, after you have made a large list, think about the logic the analogy is trying to support. Eliminate those similarities that do not assist that logic because they are not relevant to an assessment of the quality of the analogy.

2. Are there significant differences that the Court ignores or downplays?

   Clue: First think about the purpose this analogy is serving. Then think about the differences in the facts for this case and the facts for *Wickard*.

Although the current Supreme Court seems to prefer greater regulatory power for states, *Gonzales v. Raich* and another recent case stand as examples in which the Supreme Court upheld congressional acts on the basis of the Commerce Clause. In *Pierce County v. Guillen*,[11] the Supreme Court held that the Hazard Elimination Program was a valid exercise of congressional authority under the Commerce Clause. This program provided funding to state and local governments to improve conditions of some of their most unsafe roads. To receive federal funding, however, state and local governments were required to regularly acquire information about potential road hazards. The state and local governments were reluctant to avail themselves of the program for fear that the information they acquired to receive funding would be used against them in lawsuits based on negligence. To alleviate these fears, Congress amended the program, allowing state and local governments to conduct engineering surveys without publicly disseminating the acquired information, even for discovery purposes in trials.

Following his spouse's death in an automobile accident, Ignacio Guillen sued Pierce County and sought information related to previous accidents at the intersection where his wife died. The county argued that such information was protected under the provisions of the Hazard Elimination Program. Reversing the appellate court's holding that Congress exceeded its powers when amending the act, the Supreme Court concluded that the amended act was valid under the Commerce Clause. The Supreme Court reasoned that Congress had a significant interest in assisting local and state governments in improving safety in the channels of interstate commerce, the interstate highways. The Court validated Congress's belief that state and local governments would be more likely to collect relevant and accurate information about potential road hazards if those governments would not be required to provide such information in discovery. Hence, the Supreme Court held that the amended act of Congress was valid on the basis of the Commerce Clause.

Despite the Court's ruling in *Pierce County v. Guillen*, many Supreme Court commentators had thought that the Court's turn toward a more restrictive interpretation of the Commerce Clause would lead the Court to rule that Congress cannot justify regulating states' decisions regarding medical marijuana through the Commerce Clause. Instead, Justice Stevens distinguished *Gonzales v. Raich* from *United States v. Lopez* and *Brzonkala v. Morrison*, explaining that the federal regulations at issue in *Lopez* and *Morrison* were not related to economic activity, even understood broadly, and thus in both cases Congress had overstepped its bounds. Raich's activities, however, did involve economic activity, even if it is the economic activity of an illegal, controlled substance. Exhibit 5-3

---

[11] 537 U.S. 129 (2003).

Case	Summary	Significance of the Case
Reno v. Condon, 528 U.S. 141 (2000)	Supreme Court upheld the Driver Privacy Protection Act (DPPA), which prevented intentional disclosure of personal information obtained by the Department of Motor Vehicles (DMV) about another person.	The personal information of drivers, and the subsequent sale of said information, was deemed to be a valid part of interstate commerce and thus within Congress's regulatory reach.
United States v. Morrison, 529 U.S. 598 (2000)	Congressional law that added a federal civil penalty to gender-motivated crimes was ruled to be unconstitutional.	Under its Commerce Clause power, Congress cannot regulate gender-motivated violence that is in no way an economic activity.
Solid Waste Agency of Northern Cook County v. United States Army Corps of Engineers, 531 U.S. 159 (2001)	An abandoned sand-and-gravel pit that had subsequently been filled with water forming an intrastate body of water was deemed out of the regulatory reach of Congress and the Army Corps of Engineers, which sought to regulate the water under the Clean Water Act through Congress's Commerce Clause power.	The Supreme Court effectively limited Congress's power to regulate intrastate waterways through the elimination of the Migratory Bird Rule provision of the Clean Water Act.
Pierce County v. Guillen, 537 U.S. 129 (2003)	The federal Hazard Elimination Program, which requires states to collect information regarding dangerous roadways, contains a provision that prevents the data from being used as evidence in state or federal courts. The Supreme Court ruled that the protection does not violate Congress's Commerce Clause powers.	Although not expanding Congress's Commerce Clause powers per se, the Court's ruling did not limit the power Congress has over the regulation of roadways and the dissemination of information regarding the safety of roadways.
Gonzales v. Raich, 545 U.S. 1 (2005)	The California medicinal marijuana laws were ruled to violate the Constitution's Commerce Clause, as Congress has the power to regulate illegal drugs.	Congress's Commerce Clause power includes the ability to regulate illegal substances, whereas states do not have a reciprocal ability to do so.
Granholm v. Heald, 544 U.S. 460 (2005)	New York and Michigan state laws that allow in-state wineries to ship wine directly to consumers, but that prohibit out-of-state wineries from engaging in the same practice, were ruled to unconstitutionally violate the Commerce Clause by discriminating against interstate commerce.	The Twenty-first Amendment, which ended prohibition, does not allow states to interfere with interstate commerce associated with the sale of wine.
American Trucking Association, Inc. v. Michigan Public Service Commission, 545 U.S. 429 (2005)	The Supreme Court ruled that a flat $100 fee that Michigan imposed on all in-state truck deliveries did not discriminate against interstate commerce and thus did not violate the Commerce Clause.	State regulations that apply to in-state and out-of-state groups equally do not violate the Commerce Clause.
Rapanos v. United States, 126 S. Ct. 2208 (2006)	The expanded use of the term *navigable waters* by the Army Corps of Engineers in order to regulate an increased number of waterways violates Congress's Commerce Clause powers.	The Supreme Court greatly restricted Congress's ability to regulate waterways under the Clean Water Act, thus diminishing Congress's Commerce Clause authority.
United Haulers Association, Inc. v. Oneida-Herkimer Solid Waste Management Authority, 127 S. Ct. 1786 (2007)	State flow control ordinances regulating solid waste disposal that favor an in-state public facility were upheld, as they do not violate the Commerce Clause.	When state laws validly favor a public facility, and consequently treat in-state and out-of-state businesses the same, there is no unconstitutional interference with interstate commerce.

EXHIBIT 5-3

MODERN SUPREME COURT INTERPRETATIONS OF THE COMMERCE CLAUSE

offers a summary of a number of Commerce Clause cases the Supreme Court has decided since *Lopez*. One inference a person could draw after examining the cases in Exhibit 5-3 is that the Court is willing to limit congressional power but not necessarily in every instance where such restriction is possible.

THE COMMERCE CLAUSE AS A RESTRICTION ON STATE AUTHORITY

Because the Commerce Clause grants authority to regulate commerce to the federal government, a conflict arises over the extent to which granting such authority to the federal government restricts the states' authority to regulate commerce. The courts have attempted to resolve the conflict over the impact of the Commerce Clause on state regulation by distinguishing between regulations of commerce and regulations under the state police power. **Police power** means the residual powers retained by the state to enact legislation to safeguard the health and welfare of its citizenry. When the courts perceived state laws to be attempts to regulate interstate commerce, these laws would be struck down; however, when the courts found state laws to be based on the exercise of the state police power, the laws were upheld.

police power The states' retained authority to pass laws to protect the health, safety, and welfare of the community.

Since the mid-1930s, whenever states have enacted legislation that affects interstate commerce, the courts have applied a two-pronged test. First, they ask: Is the regulation rationally related to a legitimate state end? If it is, then they ask: Is the regulatory burden imposed on interstate commerce outweighed by the state interest in enforcing the legislation? If it is, the state's regulation is upheld. Case 5-3 is an example of a state statute that has been upheld by considering this two-pronged test.

CASE 5-3

Nat'l Ass'n of Optometrists & Opticians v. Brown
United States Court of Appeals for the Ninth District
682 F.3d 1144 (2012)

California had a regulation that prohibited licensed opticians from offering prescription eyewear in the same city or location where professional eye examinations are provided. The National Association of Optometrists and Opticians, LensCrafters, Inc., and EyeCare Centers of America, Inc. challenged this California statute, stating that it places a burden on interstate commerce that "excessively outweighs the local benefits of the law." In this case, the district court granted the State's motion for summary judgment. The plaintiff companies appealed.

Judge Hug

Plaintiffs challenge these laws to the extent they prohibit opticians and optical companies from offering prescription eyewear at the same location in which eye examinations are provided and from advertising that eyewear and eye examinations are available in the same location. The district court denied Plaintiffs' motion for summary judgment and granted the State's motion for summary judgment. The court effectively concluded that, based on the facts and the law, there were no genuine issues of material fact. Plaintiffs argued that the challenged laws impermissibly burdened interstate commerce because: (1) the challenged laws preclude an interstate company from offering one-stop shopping, which is the dominant form of eyewear retailing; and (2) interstate firms would incur a great financial loss as a result of the challenged laws. The district court concluded that it need not consider the evidence supporting these theories because both theories failed as a matter of law. In reaching this conclusion, the court reasoned that, because there was no cognizable burden on interstate commerce, it need not attempt to balance the "non-burden" against the putative local interests under the test derived from Plaintiffs timely appealed, and that appeal is now before us.

Modern dormant Commerce Clause jurisprudence primarily "is driven by concern about economic protectionism—that is, regulatory measures designed to benefit in-state economic interests by burdening out-of-state competitors." "The principal objects of dormant Commerce Clause scrutiny are statutes that discriminate against interstate commerce." "The central rationale for the rule against discrimination is to prohibit state or municipal laws whose object is local economic protectionism," because these are the "laws that would excite those jealousies and retaliatory measures the Constitution was designed to prevent."

Although dormant Commerce Clause jurisprudence protects against burdens on interstate commerce, it also respects federalism by protecting local autonomy. Thus, the Supreme Court has recognized that "under our constitutional scheme the States retain broad power to legislate protection for their citizens in matters of local concern such as public health" and has held that "not every exercise of local power is invalid merely because it affects in some way the flow of commerce between the States."

In a long line of dormant Commerce Clause cases, the Supreme Court has sought to reconcile these competing interests of local autonomy and burdens on interstate commerce. In one of those cases, *Pike v. Bruce Church, Inc.*, the Supreme Court set forth the following summary of dormant Commerce Clause law, stating:

> *Although the criteria for determining the validity of state statutes affecting interstate commerce have been variously stated, the general rule that emerges can be phrased as follows: Where the statute regulates even-handedly to effectuate a legitimate local public interest, and its effects on interstate commerce are only incidental, it will be upheld unless the burden imposed on such commerce is clearly excessive in relation to the putative local benefits. If a legitimate local purpose is found, then the question becomes one of degree. And the extent of the burden that will be tolerated will of course depend on the nature of the local interest involved, and on whether it could be promoted as well with a lesser impact on interstate activities.*

Unfortunately, the Pike test has not turned out to be easy to apply. As the Supreme Court has acknowledged, there is "no clear line" in Supreme Court cases between cases involving discrimination and cases subject to Pike's "clearly excessive" burden test. Justice Scalia has candidly observed that "once one gets beyond facial discrimination our negative-Commerce-Clause jurisprudence becomes (and long has been) a quag-mire." According to the Supreme Court, only a small number of its cases invalidating laws under the dormant Commerce Clause have involved laws that were "genuinely nondiscriminatory, in the sense that they did not impose disparate treatment on similarly situated in-state and out-of-state interests." The threshold issue in this appeal is whether Plaintiffs have produced sufficient evidence that the challenged laws, though non-discriminatory, impose a significant burden on interstate commerce. As discussed below, we hold that the Plaintiffs have not produced such evidence.

We conclude that Supreme Court precedent establishes that there is not a significant burden on interstate commerce merely because a non-discriminatory regulation precludes a preferred, more profitable method of operating in a retail market. Where such a regulation does not regulate activities that inherently require a uniform system of regulation and does not otherwise impair the free flow of materials and products across state borders, there is not a significant burden on interstate commerce.

We find no support in the law for Plaintiffs' proposition that there is a significant burden on interstate commerce whenever, as a result of nondiscriminatory retailer regulations, there is an incidental shift in sales and profits to in-state entities from retailers that operate in-state but are owned by companies incorporated out-of-state. In light of this law, it is apparent that, in the case before us, there is no material issue of fact regarding whether the challenged laws place a significant burden on interstate commerce. Plaintiffs have not produced evidence that the challenged laws interfere with the flow of eyewear into California; any optician, optometrist, or ophthalmologist remains free to import eyewear originating anywhere into California and sell it there. In addition, we are not concerned here with activities that require a uniform system of regulation. Thus, Plaintiffs have failed to raise a material issue of fact concerning whether there is a significant burden on interstate commerce. Relying on Pike, Plaintiffs argue that, in determining whether a regulation violates the dormant Commerce Clause, courts are required to examine the actual benefits of nondiscriminatory regulations. However, [HN20] Pike discusses whether the burden on interstate commerce is "clearly excessive in relation to the putative local benefits." See Pike, 397 U.S. at 142. It does not mention actual benefits as part of the test for determining when a regulation violates the dormant Commerce Clause.

Even if Pike's "clearly excessive" burden test were concerned with weighing actual benefits rather than "putative benefits," we need not examine the benefits of the challenged laws because, as discussed above, the challenged laws do not impose a significant burden on interstate commerce. If a regulation merely has an effect on interstate commerce, but does not impose a significant burden on interstate commerce, it follows that there cannot be a burden on interstate commerce that is "clearly excessive in relation to the putative local benefits" under Pike. Accordingly, where, as here, there is no discrimination and there is no significant burden on interstate commerce, we need not examine the actual or putative benefits of the challenged statutes. For the foregoing reasons, the district court's order granting the State's motion for summary judgment and denying Plaintiffs' motion for summary judgment is affirmed.*

Affirmed.

Another example of a state statute that has been upheld is Chicago's ban on the use of spray paint in the city. Paint retailers challenged the statute, arguing that it could have caused $55 million in lost sales over the next six years for spray paint retailers. The U.S. Court of Appeals eventually found the law to be constitutional. The state had a legitimate interest in trying to clean up graffiti,

---

*Nat'l Ass'n of Optometrists & Opticians v. Brown, United States Court of Appeals for the Ninth District 682 F.3d 1144 (2012). http://cdn.ca9.uscourts.gov/datastore/opinions/2012/06/13/10-16233.pdf.

and it did not "discriminate against interstate commerce" nor violate the Commerce Clause. The appeals court reversed the previous ruling and allowed Chicago's enactment of this ordinance to remain intact.[12] Despite the rulings in the *United Haulers* and *Chicago* cases, it is not necessarily easy to craft a state statute that affects interstate commerce that will be upheld. Frequently, the courts will find that state legislation that impinges upon interstate commerce in some way is an unconstitutional interference with interstate commerce.

The most recent illustration of the United States Supreme Court's addressing a state's attempt to evade the power of the dormant commerce clause occurred in 2015 in the case of *Comptroller of the Treasury of Maryland v. Wynne et ux.*[13] Maryland's personal income tax on state residents consisted of a "state" tax and a "county" tax. Maryland residents who earned money in other states received a "state" tax credit for the taxes they paid to the state where the income was earned, but not a "county" tax credit, resulting in double taxation of out-of-state income.

The law was challenged by the Wynnes, who claimed a tax credit for the taxes paid on their out-of-state earnings. The state comptroller denied their claim of a credit against their "county" tax and assessed them a deficiency. The appeals board of the comptroller's office affirmed the assessment, as did the Maryland Tax Court, but the Court of Appeals for Howard County reversed on grounds that Maryland's tax system violated the Constitution. The court applied the four-part test previously used by the Supreme Court in examing state tax systems, a test that asks whether a "tax is applied to an activity with a substantial nexus with the taxing State, is fairly apportioned, does not discriminate against interstate commerce, and is fairly related to the services provided by the State." The court found that the law failed the fair apportionment test because if every state had a law like Maryland's, interstate commerce would be taxed at a higher rate than intrastate commerce and create a risk for multiple taxation. It also failed the nondiscrimination part because by denying residents tax credit on interstate commerce it made them pay a higher tax rate on income earned on interstate commerce.

Noting that a state "may not tax a transaction or incident more heavily when it crosses state lines than when it occurs entirely within the State," the Supreme Court affirmed the decision of the appellate court. In striking down Maryland's law, the court reiterated the point that when a tax system has the potential to result in the discriminatory double taxation of income earned out of state, it creates a powerful incentive to engage in intrastate rather than interstate economic activity, and thereby places an undue burden on interstate commerce in violation of the dormant Commerce Clause.

Take a look at the cases described in Table 5-1. They further illustrate how the courts attempt to determine when a state statute that affects interstate commerce will be upheld.

APPLYING THE LAW TO THE FACTS...

The Alabama legislature proposed an amendment to the state constitution that would require certain public works projects, such as highway construction and public transportation, to not be funded unless the company receiving the funds "is an Alabama based company or corporation employing only Alabama residents." What is the potential constitutional problem with this proposed amendment?

---

[12]*National Paint & Coatings Association et al. v. City of Chicago*, 45 F.3d 1124 (7th Cir. 1995).
[13]135 S. Ct. 1787 (2015).

TABLE 5-1 DORMANT COMMERCE CLAUSES CASES

Case	Facts	Holding
Black Star Farms v. Oliver, 600 F.3d 1225 (9th Cir. 2010)	The state of Arizona developed a scheme of regulating wine sales whereby suppliers sell wine to wholesalers, who then sell the wine to retailers, who sell wine to the public. Exceptions to this distribution scheme were made for (1) small wineries that produce no more than 20,000 gallons of wine annually, who are allowed to sell directly to the public, and (2) direct shipments to consumers who visit a winery and request, in person, that purchased winery be shipped to their address. A winery that produces 35,000 bottles a year challenged the state scheme as unconstitutionally interfering with interstate commerce, thus violating the dormant Commerce Clause.	Because the law applied equally to wineries in all states, including Arizona, it did not unfairly burden interstate commerce. The appellate court thus upheld the District Court's grant of a motion for summary judgment to the state of Arizona and denying the winery's motion for summary judgment.
Family Winemakers of California, et al. v. Jenkins, 592 F.3d 1 (2010)	Massachusetts passed a statute allowing only "small" wineries, defined as those producing 30,000 gallons or less of grape wine a year, to obtain a "small winery shipping license" that allows them to sell their wines in Massachusetts in three ways: by shipping directly to consumers, through wholesaler distribution, and through retail distribution. All of Massachusetts's wineries, and some out-of-state wineries, are "small" wineries. "Large" wineries, those producing more than 30,000 gallons per year, and which are all located out of state, must choose between relying upon wholesalers to distribute their wines in-state or applying for a "large winery shipping license" to sell directly to Massachusetts consumers. They cannot, by law, use both methods to sell their wines in Massachusetts, and they cannot sell wines directly to retailers under either option.	The District Court granted the plaintiffs request for injunctive relief. The Court of Appeals upheld the order, finding that the statute afforded significant competitive advantages to "small" wineries for no nondiscriminatory purpose.
Florida Transportation Services, Inc. v. Miami-Dade County, 703 F.3d 1230 (11th Cir. 2012)	If a licensed stevedore wanted to operate at the Port of Miami, a Florida statute required that person to also have a permit issued by the Director of the Port of Miami. In practice, the Port Director automatically renewed permits of existing holders and repeatedly denied permits to new applicants. Florida Transportation Services brought suit against the County, alleging that the County's stevedore permit regulation, as applied by the Port Director, violated the dormant Commerce Clause by placing an undue burden on interstate commerce.	On appeal, the court affirmed the judgment of the district court in favor of the plaintiffs. Because the ordinance effectively shut out new entrants, even if they could have provided better service, better equipment, or lower prices than incumbent stevedores, the ordinance of the County did place an unconstitutional burden on interstate commerce.
Rousso v. State, 170 Wn.2d 70 (2010)	Rousso brought suit against a Washington statute that criminalized the knowing transmission and reception of gambling information by means such as the Internet. The plaintiff claimed that this regulation violated the dormant Commerce Clause. The plaintiff asserted that this state regulation, which effectively bans Internet gambling, excessively burdens interstate commerce.	The court ruled that the interests of Washington were best served by banning Internet gambling, and that the burden on interstate commerce was not clearly excessive in light of the state's interests. The Washington statute did not violate the dormant Commerce Clause.

The Taxing and Spending Powers of the Federal Government

Article I, Section 8, of the Constitution gives the federal government the "Power to lay and collect Taxes, Duties, Imports and Excises." The taxes laid by Congress, however, must be uniform across the states. In other words, the U.S. government cannot impose higher taxes on residents of one state than another.

Although the collection of taxes is essential for the generation of revenue needed to provide essential government services, taxes can be used to serve additional functions. For example, the government may wish to encourage the development of certain industries and discourage the development of others, so it may provide tax credits for firms entering the favored industries. As long as the "motive of Congress and the effect of its legislative action are to secure revenue for the benefit of the general government,"[14] the tax will be upheld as constitutional. The fact that it also has what might be described as a regulatory impact will not affect the validity of the tax.

While we may all think we know what a tax is, sometimes a tax is not easy to recognize. For example, in 2012, many commentators were surprised when the United States Supreme Court determined that the penalty imposed by the Affordable Care Act (ACA) on those who did not purchase health insurance was a tax.[15] Chief Justice Roberts explained that even though the imposition of the tax payment was designed to encourage certain behavior, taxes may legitimately be used by the federal government to encourage behavior. Nothing in the ACA made the failure to purchase insurance a violation of any law; it simply imposed the payment of a tax, collected by the IRS like all other taxes, on those who opted to not purchase health care insurance. The fact that the ACA referred to the tax as a penalty did not change the fundamental nature of the payment as a tax.

This finding that the payment was a tax was an important decision because it preserved the constitutionality of the ACA, which most commentators thought would have been upheld not as a tax, but rather as an exercise of the federal government's authority to regulate interstate commerce under the Commerce Clause. In fact, the high court rejected the argument that the ACA constituted an exercise of congressional authority under the Commerce Clause.

Article I, Section 8, also gives Congress its spending power by authorizing it to "pay the Debts and provide for the common Defence and general Welfare of the United States." Just as Congress can indirectly use its power to tax to achieve certain social welfare objectives, it can do the same with its spending power. For example, the U.S. Supreme Court in 1987 upheld the right of Congress to condition the states' receipt of federal highway funds on their passing state legislation making 21 the legal drinking age.

TAXATION OF THE INTERNET?

The rapid rise in Internet commerce has many states wondering how they will collect their fair share of sales taxes. According to the U.S. Department of Commerce, Internet retail sales have continued to increase. U.S. retail e-commerce sales reached almost $142 billion in 2008, up from a revised $137 billion in 2007—an annual gain of 3.3 percent. From 2002 to 2008, retail e-sales increased at an average annual growth rate of 21.0 percent, compared with 4.0 percent for total retail sales. Although Internet sales constituted only 3.7 percent of overall retail sales in the United States during 2009, advocates of taxes on Internet sales insist that states are losing a considerable amount of revenue each year.[16]

Currently, states are only allowed to require a business to submit sales tax payments if the business has a store or distribution center in the state. Otherwise, states are prohibited from collecting sales taxes, although residents are supposed to report the taxes on personal income tax returns. In addition to the e-commerce business, increased access to the Internet has some clamoring for a use tax on Internet access, in addition to a sales tax on Internet purchases.

In 1998, Congress approved the Internet Tax Freedom Act, which established a moratorium on Internet taxes until November 2001. The 1998 bill

---

[14] *J. W. Hampton Co. v. United States*, 276 U.S. 394 (1928).

[15] *National Federation of Independent Business v. Sebelius*, 132 S. Ct. 2566 (2012).

[16] U.S. Census Bureau, E-Stats, May 27, 2010; access to these statistics is available at http://www.census.gov/econ/estats/2008/2008reportfinal.pdf.

provided a grandfather clause that allowed several states to continue levying taxes on Internet access if those taxes were established before the Internet Tax Freedom Act was passed. In November 2001, Congress extended the moratorium for two more years to allow for more discussion and research on the effects of the ban on state governments.

In September 2003, the House of Representatives passed the Internet Tax Nondiscrimination Act (H.R. 49), a bill designed to replace the Internet Tax Freedom Act that would have expired in November 2003 with a permanent ban on taxes on Internet access and a permanent extension of the moratorium on multiple and discriminatory taxes on electronic commerce. On April 29, 2004, the Senate passed a different version of the Internet Tax Nondiscrimination Act (S. 150) that extended the moratorium on Internet taxes until November 2007. The Senate bill was a compromise between supporters of a permanent Internet tax ban and a group of senators who questioned how a permanent ban would affect state and local budgets.

The final version of the legislation, signed into law by President George W. Bush on December 3, 2004, had two different grandfather exemptions. States that taxed Internet service before October 1, 1998, were allowed to continue their taxes until November 1, 2007, whereas states that taxed Internet service before October 1, 2003, were allowed to continue their taxes until November 1, 2005. The law banned all other states from imposing Internet taxes from November 1, 2003 to November 1, 2007.

In 2007, Congress and the president extended the act until November 1, 2014, with the Internet Tax Nondiscrimination Act.

The Impact of the Amendments on Business

The first 10 amendments to the U.S. Constitution, known as the Bill of Rights, have a substantial impact on governmental regulation of the legal environment of business. These amendments prohibit the federal government from infringing on certain freedoms that are guaranteed to individuals living in our society. The **Fourteenth Amendment** extends most of the provisions in the Bill of Rights to the behavior of states, prohibiting their interference in the exercise of those rights. Many of the first 10 amendments have also been held to apply to corporations because corporations are treated, in most cases, as "artificial persons." The activities protected by the Bill of Rights and the Fourteenth Amendment are not only those that occur in one's private life, but also those that take place in a commercial setting. Several of these amendments have a significant impact on the regulatory environment of business, and they are discussed in the remainder of this chapter.

Fourteenth Amendment Applies the entire Bill of Rights, excepting parts of the Fifth Amendment, to the states.

THE FIRST AMENDMENT

The **First Amendment** guarantees freedom of speech and of the press. It also prohibits abridgment of the right to assemble peacefully and to petition for redress of grievances. Finally, it prohibits the government from aiding the establishment of a religion and from interfering with the free exercise of religion.

Although we say these rights are guaranteed, they obviously cannot be absolute. Most people would agree that a person does not have the right to yell "Fire!" in a crowded theater. Nor does one's right of free speech extend to making false statements about another that would be injurious to that person's reputation. Because of the difficulty in determining the boundaries of individual rights, a large number of First Amendment cases have been decided by the courts.

Not surprisingly, student speech has given rise to a number of free speech cases. For example, in *Tinker v. Des Moines Independent School District*,[17] the

First Amendment Guarantees freedom of speech, press, and religion and the right to peacefully assemble and to petition the government for redress of grievances.

[17]393 U.S. 503 (1969).

Court ruled that a school policy that prohibited students from wearing antiwar armbands was unconstitutional because the students' message was political and was not disruptive to normal school activities. In *Tinker*, the Court famously stated that students do not "shed their constitutional rights . . . at the schoolhouse gate."[18] Subsequently, the Court ruled in *Bethel School District No. 403 v. Fraser*[19] that speech that would otherwise be protected can be restricted within the school context. Fraser gave a speech at a school assembly that contained a graphic and extended sexual metaphor. The Court held that, although the speech would have been protected if given in the public forum, the fact that the speech was delivered at school allowed for administrators to censor the speech and restrict Fraser's right to give the speech. Most recently, relying upon their rulings in *Tinker* and *Fraser*, the Court ruled in *Morse v. Frederick*[20] that student speech advocating drug use during a school function can constitutionally be restricted. *Morse v. Frederick* is the much-discussed "Bong Hits 4 Jesus" case. The Court determined that the "Bong Hits 4 Jesus" banner clearly advocated drug use, and because the poster was displayed at a school function, the students responsible could be punished. Furthermore, the banner did not portray a political or religious message and thus was not protected speech.

Attempts to regulate new technologies also raise First Amendment issues. For example, Congress passed the Communications Decency Act of 1996 (CDA) to protect minors from harmful material on the Internet; however, the U.S. Supreme Court found that provisions of the CDA that criminalized and prohibited the "knowing" transmission of "obscene or indecent" messages to any recipient under age 18 by means of telecommunications devices or through the use of interactive computer services were content-based blanket restrictions on freedom of speech. Because these provisions of the statute were too vague and overly broad, repressing speech that adults have the right to make, these provisions were found to be unconstitutional.[21]

After the Supreme Court held that the CDA was unconstitutional, Congress responded by passing the Child Online Protection Act (COPA), which imposed a $50,000 fine and six months of imprisonment on individuals who posted material for commercial purposes that was harmful to minors. Websites that required individuals to submit a credit card number or some other form of age verification, however, were not in violation of the act. Nevertheless, the Supreme Court ruled that this act was also unconstitutional, as the provisions of the act likely violated the First Amendment.[22] The Court reasoned that COPA was not narrowly tailored to meet a compelling governmental interest, and the regulations were not the least restrictive methods of regulating in this area, as filtering programs could more easily restrict minors' access to obscene material than could criminal penalties.

Congress also passed the Child Internet Protection Act (CIPA), requiring libraries to implement filtering software to prevent minors from accessing pornography or other obscene and potentially harmful material. Libraries that did not comply with the provisions of CIPA would not receive federal funding for Internet access. In *United States v. American Library Association*,[23] numerous libraries and website publishers brought suit, claiming that the CIPA was unconstitutional. Reversing the district court's decision that the act was unconstitutional because it violated the First Amendment, the Supreme Court ruled in

---

[18]*Id.* at 506.
[19]478 U.S. 675 (1986).
[20]127 S. Ct. 2618 (2007).
[21]*Reno v. American Civil Liberties Union*, 521 U.S. 844 (1997).
[22]*Ashcroft v. ACLU*, 124 S. Ct. 2783 (2004).
[23]539 U.S. 194 (2003).

a split decision that the act was constitutional. Although six justices ruled that the act was not unconstitutional, there was greater disagreement about the Court's opinion. The majority reasoned that the act did not violate an individual's First Amendment rights, as libraries are afforded broad discretion about the kinds of materials they may include in their collections. In other words, a library is not a public forum in the traditional sense.

An interesting issue that has arisen on many campuses is whether so-called *hate speech*—derogatory speech directed at members of another group, such as another race—is unprotected speech that can be banned. Thus far, hate-speech codes on campuses that were challenged as unconstitutional have been struck down by state courts or federal appeals courts, although the issue has not yet reached the Supreme Court. Hate speech is a serious issue that affects more than 1 million students every year, prompting 60 percent of universities to ban verbal abuse and verbal harassment and 28 percent of universities to ban advocacy of an offensive viewpoint.[24] Because universities are often viewed as breeding grounds for ideas and citizen development, courts have not looked favorably on limits to speech on campuses.

The international community has been quicker than the United States to call hate speech unprotected, with a declaration from the United Nations and laws in several countries passed years ago.[25] In October 22, 2009, however, the House and Senate passed the federal Matthew Shepard and James Byrd, Jr. Hate Crime Prevention Statute, and on October 28, 2009, President Obama signed the legislation.[26] Under the law, a hate crime is defined as a crime of violence that is motivated by hatred of the group to which the victim belongs. Protected groups are those based on race, color, religion, national origin, gender, disability, sexual orientation, and gender identity.

Corporate Commercial Speech. Numerous cases have arisen over the extent to which First Amendment guarantees are applicable to corporate commercial speech. The doctrine currently used to analyze commercial speech is discussed in the following case.

CASE 5-4

Central Hudson Gas & Electric Corp. v. Public Service Commission of New York
Supreme Court of the United States
447 U.S. 557 (1980)

Plaintiff Central Hudson Gas and Electric Corporation filed an action against Public Service Commission of New York to challenge the constitutionality of a regulation that completely banned promotional advertising by the utility but permitted "informational" ads—those designed to encourage shifting consumption from peak to nonpeak times. The regulation was upheld by the trial court. On appeal by the utility, the New York Court of Appeals sustained the regulation, concluding that governmental interests outweighed the limited constitutional value of the commercial speech at issue. The utility appealed.

Justice Powell
The Commission's order [enforcing the regulation's advertising ban] restricts only commercial speech, that is, expression related solely to the economic interests of the speaker and its audience. The First Amendment, as applied to the States through the Fourteenth

---

[24]Timothy C. Shiell, *Campus Hate Speech on Trial 2*, 49 (Lawrence: University Press of Kansas, 1998).

[25]*Id.* at 32.

[26]Matthew Shepard and James Byrd, Jr. Hate Crimes Prevention Act. Accessed December 7, 2010 at http://www.hrc.org/laws_and_elections/5660.htm.

Amendment, protects commercial speech from unwarranted governmental regulation. Commercial expression not only serves the economic interest of the speaker, but also assists consumers and furthers the societal interest in the fullest possible dissemination of information. In applying the First Amendment to this area, we have rejected the "highly paternalistic" view that government has complete power to suppress or regulate commercial speech. Even when advertising communicates only an incomplete version of the relevant facts, the First Amendment presumes that some accurate information is better than no information at all. Nevertheless, our decisions have recognized "the 'common sense' distinction between speech proposing a commercial transaction, which occurs in an area traditionally subject to government regulation, and other varieties of speech."

The Constitution therefore accords a lesser protection to commercial speech than to other constitutionally guaranteed expression. The protection available for particular commercial expression turns on the nature both of the expression and of the governmental interests served by its regulation. Two features of commercial speech permit regulation of its content. First, commercial speakers have extensive knowledge of both the market and their products. Thus, they are well situated to evaluate the accuracy of their messages and the lawfulness of the underlying activity. In addition, commercial speech, the offspring of economic self-interest, is a hardy breed of expression that is not "particularly susceptible to being crushed by overbroad regulation."

If the communication is neither misleading nor related to unlawful activity, the government's power is more circumscribed. The State must assert a substantial interest to be achieved by restrictions on commercial speech. Moreover, the regulatory technique must be in proportion to that interest. The limitation on expression must be designed carefully to achieve the State's goal. Compliance with this requirement may be measured by two criteria. First, the restriction must directly advance the state interest involved; the regulation may not be sustained if it provides only ineffective or remote support for the government's purpose. Second, if the governmental interest could be served as well by a more limited restriction on commercial speech, the excessive restrictions cannot survive.

The second criterion recognizes that the First Amendment mandates that speech restrictions be "narrowly drawn." The regulatory technique may extend only as far as the interest it serves. The State cannot regulate speech that poses no danger to the asserted state interest, nor can it completely suppress information when narrower restrictions on expression would serve its interest as well. In commercial speech cases, then, a four-part analysis has developed. At the outset, we must determine whether the expression is protected by the First Amendment. For commercial speech to come within that provision, it at least must concern lawful activity and not be misleading. Next, we ask whether the asserted governmental interest is substantial. If both inquiries yield positive answers, we must determine whether the regulation directly advances the governmental interest asserted and whether it is not more extensive than is necessary to serve that interest.

The Commission does not claim that the expression at issue is inaccurate or relates to unlawful activity. Yet the New York Court of Appeals questioned whether Central Hudson's advertising is protected commercial speech. Because appellant holds a monopoly over the sale of electricity in its service area, the state court suggested that the Commission's order restricts no commercial speech of any worth.

In the absence of factors that would distort the decision to advertise, we may assume that the willingness of a business to promote its products justifies belief that consumers are interested in the advertising. Since no such extraordinary conditions have been identified in this case, appellant's monopoly position does not alter the First Amendment's protection for its commercial speech.

The Commission offers two state interests as justifications for the ban on promotional advertising. The first concerns energy conservation. Any increase in demand for electricity—during peak or off-peak periods—means greater consumption of energy. The Commission argues that the State's interest in conserving energy is sufficient to support suppression of advertising designed to increase consumption of electricity. In view of our country's dependence on energy resources beyond our control, no one can doubt the importance of energy conservation. Plainly, therefore, the state interest asserted is substantial.

We come finally to the critical inquiry in this case: whether the Commission's complete suppression of speech ordinarily protected by the First Amendment is no more extensive than necessary to further the State's interest in energy conservation. The Commission's order reaches all promotional advertising, regardless of the impact of the touted service on overall energy use. But the energy conservation rationale, as important as it is, cannot justify suppressing information about electric devices or services that would cause no net increase in total energy use. In addition, no showing has been made that a more limited restriction on the content of promotional advertising would not serve adequately the State's interests.

Appellant insists that but for the ban, it would advertise products and services that use energy efficiently. These include the "heat pump," which both parties acknowledge to be a major improvement in electric heating, and the use of electric heat as a "backup" to solar and other heat sources. Although the Commission has questioned the efficiency of electric heating before this Court, neither the Commission's Policy Statement nor its order denying rehearing made findings on this issue. The Commission's order prevents appellant from promoting electric services that would reduce energy use by diverting demand from less-efficient sources, or that would consume roughly the same amount of energy as do alternative sources. In neither situation would the utility's advertising endanger conservation or mislead the public. To the extent that the Commission's order suppresses speech that in no way impairs the State's interest in energy conservation, the Commission's order violates the First and Fourteenth Amendments and must be invalidated.

The Commission also has not demonstrated that its interest in conservation cannot be protected adequately by more limited regulation of appellant's commercial expression. To further its policy of conservation, the Commission could attempt to restrict the format and content of Central Hudson's advertising. It might, for example, require that the advertisements include information about the relative efficiency and expense of the offered service, both under current conditions and for the foreseeable future.*

Reversed in favor of Plaintiff, Central Hudson.

CRITICAL THINKING ABOUT THE LAW

In Case 5-4, the Court had to balance government interests in energy efficiency, as well as fair and efficient pricing, with the conflicting constitutional value of Central Hudson's right to free commercial speech. Having affirmed the validity of the government's substantial interests in regulating the utility company, the Court sought to determine whether these interests could have been sufficiently served with more limited restrictions. Because this determination is of central importance to the Court's reversal of the earlier court's judgment, it will be the focus of the questions that follow.

1. What primary ethical norm is implicit in the legal requirement that regulations on commercial speech be of the most limited nature possible in carrying out the desired end of advancing the state's substantial interest?

   Clue: Review the four primary ethical norms. You want to focus not on the government regulation but on the rationale for limits on that regulation.

2. What information missing from the Court's opinion must you, as a critical thinker, know before being entirely satisfied with the decision?

   Clue: You want to focus on the issue about which the Public Service Commission and Central Hudson have conflicting viewpoints. What information would you want to know before accepting the soundness of the Court's judgment in resolving this conflict?

The test set forth in *Central Hudson* was reaffirmed by the U.S. Supreme Court in two decisions handed down in the summer of 1995, when it applied the test in *Rubin v. Coors Brewing Co.*[27] and *Florida Bar v. Went for It and John T. Blakely*.[28] In the first case, Coors challenged a regulation of the Federal Alcohol Administration Act that prohibited beer producers from disclosing the beer's alcohol content. The Court found that the government's interest in suppressing "strength wars" among beer producers was "substantial" under the *Central Hudson* test, but held that the ban failed to meet the asserted government interest and be no more extensive than necessary to serve that interest.

A restriction that passed the *Central Hudson* test was the Florida ethics rule upheld in the *Went for It* case. The rule requires lawyers to wait 30 days before sending targeted direct-mail solicitation letters to victims of accidents or disasters. The high court found a substantial interest both in protecting the privacy and tranquility of victims and their loved ones against invasive and unsolicited contact by lawyers and in preventing the erosion of confidence in the profession that such repeated invasions have caused. The bar association had established, by unrebutted survey data, that Floridians considered immediate postaccident direct-mail solicitation to be an invasion of victims' privacy that reflects poorly on lawyers. The Court also found that the ban's scope was reasonably well tailored to meet the stated objectives. It was limited in duration, and there were other ways for injured Floridians to learn about the availability of legal services during the time period set by the ban. Thus, the ban was

---

[27]514 U.S. 476, 115 S. Ct. 1585 (1995).
[28]515 U.S. 618, 115 S. Ct. 2371 (1995).

*Central Hudson Gas & Electric Corp. v. Public Service Commission of New York, Supreme Court of the United States 447 U.S. 557 (1980).

upheld as directly advancing the asserted legitimate interest in a manner no more extensive than necessary to serve that interest.

The U.S. Supreme Court once again reaffirmed and applied the four-part test of *Central Hudson* in the case of *Lorillard Tobacco Co. et al. v. Thomas F. Reilly*,[29] a case challenging Massachusetts's comprehensive set of regulations regarding cigarette, cigar, and smokeless tobacco advertising and distribution. The high court found that Massachusetts's outdoor-advertising regulations that prohibited smokeless tobacco or cigar advertising within 1,000 feet of a school or playground violated the First Amendment because they failed the fourth part of the *Central Hudson* test. The Court reasoned as follows:

> *Their broad sweep indicates that the Attorney General did not "carefully calculat[e] the costs and benefits associated with the burden on speech imposed." The record indicates that the regulations prohibit advertising in a substantial portion of Massachusetts' major metropolitan areas; in some areas, they would constitute nearly a complete ban on the communication of truthful information. This substantial geographical reach is compounded by other factors. "Outdoor" advertising includes not only advertising located outside an establishment, but also advertising inside a store if visible from outside. Moreover, the regulations restrict advertisements of any size, and the term advertisement also includes oral statements. The uniformly broad sweep of the geographical limitation and the range of communications restricted demonstrate a lack of tailoring. The governmental interest in preventing underage tobacco use is substantial, and even compelling, but it is no less true that the sale and use of tobacco products by adults is a legal activity. A speech regulation cannot unduly impinge on the speaker's ability to propose a commercial transaction and the adult listener's opportunity to obtain information about products. The Attorney General has failed to show that the regulations at issue are not more extensive than necessary.*[30]*

In that same case, the high court also struck down regulations prohibiting placement of indoor, point-of-sale advertising of smokeless tobacco and cigars lower than five feet from the floor of a retail establishment located within 1,000 feet of a school or playground because they failed both the third and fourth steps of the *Central Hudson* analysis. The Court found that the five-foot rule did not seem to advance the goals of preventing minors from using tobacco products and curbing demand for that activity by limiting youth exposure to advertising, because not all children are less than five feet tall, and those who are can look up and take in their surroundings. In the case, the Court overruled the circuit court's finding that the regulations met the four-part test of *Central Hudson*, so it clearly is not always easy to know how the test is going to be applied.[31]

Corporate Political Speech. Not all corporate speech is considered commercial speech. Sometimes, for example, corporations might spend funds to support political candidates or referenda. At one time, states restricted the total amount of money corporations could spend on advertising because of a fear that, with their huge assets, corporations' speech on behalf of a particular candidate or issue would drown out other voices. In the 1978 case of *First National Bank of Boston v. Bellotti*,[32] however, the U.S. Supreme Court struck down a state law that prohibited certain corporations from making contributions or expenditures influencing voters on any issues that would not materially affect the corporate assets or business. Stating that "the concept that the government may restrict speech of some elements of our society in order to enhance the relative voice

---

[29]533 U.S. 525 (2001).

[30]*Id.*

[31]*Id.*

[32]435 U.S. 765 (1978).

**Lorillard Tobacco Co. et al., v. Thomas F. Reilly*, 533 U.S. 525 (2001).

of others is wholly foreign to the First Amendment," the high court ruled that corporate political speech should be protected to the same extent as the ordinary citizen's political speech.

The ability of the government to regulate corporate political speech was further restricted in 2010 by the landmark decision in *Citizens United v. Federal Election Commission*.[33] By a 5–4 vote, the majority ruled that the corporate funding of independent political broadcasts in candidate elections cannot be limited under the First Amendment, thus finding political spending to be a form of protected speech. The decision struck down a provision of a 2002 campaign financing law that prohibited all corporations, both for-profit and not-for-profit, and unions from broadcasting "electioneering communications," which were defined as "a broadcast, cable, or satellite communication that mentioned a candidate within 60 days of a general election or 30 days of a primary." Arguably, this decision paved the way for the 2010 midterm election setting a record for midterm election spending, with an estimated $3.8 billion being spent.[34]

Our discussion of the First Amendment has focused on its effect on people's ability to speak, but just as the First Amendment may also prevent the government from prohibiting speech, the Amendment may also prevent the government from compelling individuals to express certain views or from compelling certain individuals to pay subsidies for speech to which they object. This protection is illustrated by the 2001 case of *United States Department of Agriculture v. United Foods, Inc.*,[35] in which the Court struck down a governmental assessment against mushroom growers that was used primarily for generic advertising to promote mushroom sales. The mushroom producer who challenged the assessment wanted to be able to promote his mushrooms as different from other producers' mushrooms and, hence, did not want to be forced to help fund generic advertising that promoted the idea that all mushrooms were good. The Supreme Court agreed, stating that "First Amendment values are at serious risk if the government can compel a citizen or group of citizens to subsidize speech on the side that it favors."[36]

COMPARATIVE LAW CORNER

Corporate Speech in Canada

As the limits on political speech made by corporations in the United States have been expanded, Canada has been tightening its regulations. The recent Federal Accountability Act in Canada bans contributions by corporations, trade unions, and associations. This act signals a change in policy by eliminating the exception that allowed annual contributions of $1,000 to various political entities.

The law also tightens the limits on individual contributions, but not to the extent of a total ban. This reveals a difference in the treatment of corporations and people in Canada, despite the fact that corporations are considered juristic persons under Canadian law, just as they are in the United States. Although there are differences in the restrictions on individual and corporate speech in the United States, the recent trend in Supreme Court rulings has been to eliminate the differences between individual and corporate speech. Canada's conservative government seems to be taking the opposite approach and restricting political speech for corporations.

---

[33] 130 S. Ct. 876 (2010).

[34] US Mid-Term Elections Most Expensive: 3.98 Billion Dollars (Extra), *M&C News*. Accessed December 7, 2010 at http://www.monstersandcritics.com/news/usa/news/article_1596039.php/US-mid-term-elections-most-expensive-3-98-billion-dollars-Extra.

[35] 533 U.S. 405, 121 S. Ct. 2334 (2001).

[36] *Id.* at 411.

THE FOURTH AMENDMENT

Fourth Amendment Protects the right of individuals to be secure in their persons, homes, and personal property by prohibiting the government from conducting unreasonable searches of individuals and seizing their property.

The **Fourth Amendment** protects the right of individuals to be secure in their persons, their homes, and their personal property. It prohibits the government from conducting unreasonable searches of individuals and seizing their property to use as evidence against them. If such an unreasonable search and seizure occurs, the evidence obtained from it cannot be used in a trial.

An *unreasonable* search and seizure is basically one conducted without the government official's having first obtained a warrant from the court. The warrant must specify the items sought, and this requirement is strictly enforced, as the Supreme Court case from 2004 illustrates. In *Groh v. Ramirez*,[37] the high court ruled that a search warrant was invalid on its face when it utterly failed to describe the persons or things to be seized, despite the fact that the requisite particularized description was contained in the search warrant application. The residential search that was conducted pursuant to this facially invalid warrant was therefore unreasonable, despite the fact that the officers conducting the search exercised restraint in limiting the scope of the search to materials listed in the application.[38]

Government officials are able to obtain such a warrant only when they can show probable cause to believe that the search will turn up the specified evidence of criminal activity. Supreme Court decisions, however, have narrowed the protection of the Fourth Amendment by providing for circumstances in which no search warrant is needed. For example, warrantless searches of automobiles, under certain circumstances, are allowed. Moreover, the Supreme Court even held in one case that an out-of-car arrest and subsequent warrantless search of the individual's car did not violate the individual's Fourth Amendment rights.[39] The Supreme Court decided in another case that police highway checkpoints, where police officers questioned drivers on a particular highway for information about a recent incident, did not violate drivers' Fourth Amendment rights, even though the police arrested one of the passing drivers for drunk driving.[40]

Improvements in technology have also caused problems in application of the Fourth Amendment, because it is now simpler to eavesdrop on people and to engage in other covert activities. One such case was decided by the U.S. Supreme Court in mid-2001.[41] In that case, the police had information from informants that led them to believe that Danny Kyllo was growing marijuana in his home. Kyllo also had unusually high electricity bills (common when one is using heat lamps to grow the plants indoors). The police used a thermal imager, an instrument that can detect unusually high levels of heat emissions and translate them into an image, to provide them with the evidence necessary to get a warrant to physically search his house. The question the Court had to address was whether the use of the thermal imaging instrument on the property constituted a "search." Or, if we think of the case the way judges do, by comparing it to past precedents, it is a question of whether thermal imaging is more like going through someone's garbage or more like using a high-powered telescope to look through someone's window. If the former situation is more analogous, then the behavior does not constitute a search; however, if the case is more analogous to the latter scenario, then using thermal imaging on a home is a search that requires a warrant. The Ninth Circuit, in a case of first impression (the first time an issue is ruled on by the court),

---

[37] 450 U.S. 551, 124 S. Ct. 1284 (2004).

[38] *Id.*

[39] *Thornton v. United States*, 124 S. Ct. 2127 (2004).

[40] *Illinois v. Lidster*, 124 S. Ct. 885 (2004).

[41] *Kyllo v. United States*, 533 U.S. 27, 121 S. Ct. 2038 (2001).

found that using thermal imaging was not a search that was prohibited by the Fourth Amendment in the absence of a warrant.

The U.S. Supreme Court, however, in a 5–4 decision, ruled that police use of a thermal imaging device to detect patterns of heat coming from a private home is a search that requires a warrant. The Court said further that the warrant requirement would apply not only to the relatively crude device at issue, but also to any "more sophisticated systems" in use or in development that let the police gain knowledge that in the past would have been impossible without a physical entry into the home. In explaining the decision, Justice Scalia wrote that in the home, "all details are intimate details, because the entire area is held safe from prying government eyes." He went on to add that the Court's precedents "draw a firm line at the entrance to one's house."[42]

Although many were happy with the Supreme Court's decision in this case, some were quick to point out that this case is not necessarily the final word when it comes to the use of technology. They noted that Scalia seemed to rely heavily on the fact that the thermal imaging was used to see inside a home. It is, therefore, not clear whether thermal imaging of some other locale would be upheld.

As the U.S. Supreme Court continued to refine the definition of unreasonable searches in 2013, canine searches were the focus of two important cases. In the first case, *Florida v. Harris*,[43] the court found that that the police had reasonable cause to search the defendant's truck based on the reaction of a dog trained in drug detection during a routine traffic stop. When the policeman pulled Harris over for a routine traffic stop, he saw an open beer can in the truck and noted nervousness of the driver, so he asked permission to search the truck. When it was refused, he had the dog do a sniff test and the dog alerted at the driver's side handle, leading the officer to conclude that he had reasonable cause for a search. The U.S. Supreme Court agreed, noting that, "All we have required is the kind of 'fair probability' on which 'reasonable and prudent [people,]' not legal technicians, act." The court went on to say that as long as the State has produced proof from controlled settings that a dog performs reliably in detecting drugs, and the defendant has not contested that showing, then there is probable cause for the search.

The second case, which follows, led to a contrary outcome. As you read the following case, think about the similarities and differences between the two to help you better understand how the court analyzes Fourth Amendment challenges.

CASE 5-5

Florida v. Jardines
United States Supreme Court
133 S. Ct. 1409 (2013)

A detective received an unverified tip that marijuana was being grown in Jardines' home. A month later, the Police Department and the Drug Enforcement Administration sent a joint surveillance team to Jardines' home. The detective approached Jardines' home accompanied by a trained canine handler who had just arrived at the scene with his drug-sniffing dog. The dog was trained to detect the scent of marijuana, cocaine, heroin, and several other drugs, indicating the presence of any of these substances through particular behavior recognizable by his handler.

As the dog approached Jardines' front porch, he apparently sensed one of the odors he had been trained to detect, and began energetically exploring the area for the strongest point source of that odor. After sniffing the base of the front door, the dog sat, which is the

---
[42]*Id.*
[43]568 U.S. 133 S. Ct. 1050 (2013).

trained behavior upon discovering the odor's strongest point, which the handler described as a positive indicator of narcotics.

On the basis of what he had learned from the dog, the detective applied for and received a warrant to search the home. A later search revealed marijuana plants, and Jardine was charged with trafficking in cannabis.

At trial, Jardines moved to suppress the marijuana plants on the ground that the canine investigation was an unreasonable search. The trial court granted the motion, and the Florida Third District Court of Appeal reversed. On a petition for discretionary review, the Florida Supreme Court quashed the decision of the Third District Court of Appeal and approved the trial court's decision to suppress, holding (as relevant here) that the use of the trained narcotics dog to investigate Jardines' home was a Fourth Amendment search unsupported by probable cause, rendering invalid the warrant based upon information gathered in that search. The police department appealed.

Justice Scalia

We granted certiorari, limited to the question of whether the officers' behavior was a search within the meaning of the Fourth Amendment.

The Fourth Amendment provides in relevant part that the "right of the people to be secure in their persons, houses, papers, and effects, against unreasonable searches and seizures, shall not be violated." The Amendment establishes a simple baseline, one that for much of our history formed the exclusive basis for its protections: When "the Government obtains information by physically intruding" on persons, houses, papers, or effects, "a 'search' within the original meaning of the Fourth Amendment" has "undoubtedly occurred" . . .

That principle renders this case a straightforward one. The officers were gathering information in an area belonging to Jardines and immediately surrounding his house—in the curtilage of the house, which we have held enjoys protection as part of the home itself. And they gathered that information by physically entering and occupying the area to engage in conduct not explicitly or implicitly permitted by the homeowner.

The Fourth Amendment "indicates with some precision the places and things encompassed by its protections": persons, houses, papers, and effects . . . The Fourth Amendment does not, therefore, prevent all investigations conducted on private property; for example, an officer may gather information in what we have called "open fields"—even if those fields are privately owned—because such fields are not enumerated in the Amendment's text . . .

But when it comes to the Fourth Amendment, the home is first among equals. At the Amendment's "very core" stands "the right of a man to retreat into his own home and there be free from unreasonable governmental intrusion." . . . This right would be of little practical value if the State's agents could stand in a home's porch or side garden and trawl for evidence with impunity; the right to retreat would be significantly diminished if the police could enter a man's property to observe his repose from just outside the front window.

We therefore regard the area "immediately surrounding and associated with the home"—what our cases call the curtilage—as "part of the home itself for Fourth Amendment purposes" . . . This area around the home is "intimately linked to the home, both physically and psychologically," and is where "privacy expectations are most heightened" . . .

Here there is no doubt that the officers entered [the cutilege]: The front porch is the classic exemplar of an area adjacent to the home and "to which the activity of home life extends."

Since the officers' investigation took place in a constitutionally protected area, we turn to the question of whether it was accomplished through an unlicensed physical intrusion . . . As it is undisputed that the detectives had all four of their feet and all four of their companion's firmly planted on the constitutionally protected extension of Jardines' home, the only question is whether he had given his leave (even implicitly) for them to do so. He had not.

"A license may be implied from the habits of the country," . . . This implicit license typically permits the visitor to approach the home by the front path, knock promptly, wait briefly to be received, and then leave. Complying with the terms of that traditional invitation does not require fine-grained legal knowledge; it is generally managed without incident by the Nation's Girl Scouts and trick-or-treaters. Thus, a police officer not armed with a warrant may approach a home and knock, precisely because that is "no more than any private citizen might do."

But introducing a trained police dog to explore the area around the home in hopes of discovering incriminating evidence is something else. There is no customary invitation to do *that*. An invitation to engage in canine forensic investigation assuredly does not inhere in the very act of hanging a knocker. To find a visitor knocking on the door is routine (even if sometimes unwelcome); to spot that same visitor exploring the front path with a metal detector, or marching his bloodhound into the garden before saying hello and asking permission, would inspire most of us to—well, call the police. The scope of a license—express or implied—is limited not only to a particular area but also to a specific purpose. Consent at a traffic stop to an officer's checking out an anonymous tip that there is a body in the trunk does not permit the officer to rummage through the trunk for narcotics. Here, the background social norms that invite a visitor to the front door do not invite him there to conduct a search . . . That the officers learned what they learned only by physically intruding on Jardines' property to gather evidence is enough to establish that a search occurred . . .

The government's use of trained police dogs to investigate the home and its immediate surroundings is a "search" within the meaning of the Fourth Amendment.

Judgment of the Florida Supreme Court *affirmed*, in favor of the Defendant.

During the second term of 2013, the high court handed down another important Fourth Amendment case, but that decision was not quite as clear as the holding in *Jardines*, and some lawyers thought it created as many questions as it answered. In *Maryland v. King*,[44] the high court upheld a Maryland law allowing the warrantless collection of DNA samples from suspects accused of certain serious crimes. Writing the majority opinion in the 5–4 decision, Justice Kennedy said that the collection of DNA via a cheek swab was clearly a search, but the government's substantial interest in using DNA evidence to identify arrestees and connect them to other crimes outweighs suspects' privacy interests given the nonintrusive nature of the search. Kennedy observed that DNA testing was analogous to fingerprinting except more accurate. More than two dozen other states also have laws allowing the warrantless collection of DNA, some of which are broader than the Maryland statute, so some of those may still be challenged by defense lawyers.*

Another important Fourth Amendment case was handed down in 2014. In *Riley v. California*,[45] the high court ruled that the police generally may not, without a warrant, search digital information on a cell phone seized from an individual who has been arrested. While the officers may examine the phones' physical aspects to ensure that the phones would not be used as weapons, digital data stored on the phones could not itself be used as a weapon to harm the arresting officers or to effectuate the defendants' escape, so police have no justification for examining it. Further, the potential for destruction of evidence by remote wiping or data encryption was not shown to be prevalent and could be countered by disabling the phones. Moreover, the immense storage capacity of modern cell phones implicated privacy concerns with regard to the extent of information which could be accessed.

The Fourth Amendment protects corporations as well as individuals. This protection is generally applicable, as noted earlier, in criminal cases. Fourth Amendment issues also arise, however, when government regulations authorize, or even require, warrantless searches by administrative agencies.

Although administrative searches are presumed to require a search warrant, an exception has been carved out. If an industry has been subject to pervasive regulation, a warrantless search is considered reasonable under the Fourth Amendment. In such industries, warrantless searches are required to ensure that regulations are being upheld, and a warrantless search would not be unreasonable because the owner has a reduced expectation of privacy. When a warrantless search is challenged, and the state argues that the "pervasively regulated" exception should apply, before the court will find that the search was reasonable, the agency will have to demonstrate that:

1. there is "a substantial government interest that informs the regulatory scheme pursuant to which the inspection is made"[46]

2. the warrantless inspections must be "necessary to further the regulatory scheme"[47]

3. "the statute's inspection program, in terms of the certainty and regularity of its application, must provide a constitutionally adequate substitute for a warrant";[48] that is, it advises the business owner that "the search is being made pursuant to the law and has a properly defined scope," and limits the discretion of the inspecting officers.[49]

---

[44] 133 S. Ct. 1958 (2013).
[45] 134 S. Ct. 2473 (2014).
[46] *New York v. Burger*, 482 U.S. 691, 107 S. Ct. 2636 (1987).
[47] *Id.*
[48] *Id.*
[49] *Id.*

*Florida v. Jardines, United States Supreme Court 133 S. Ct. 1409, 2013.

As the Fourth Amendment is currently interpreted, a warrantless search authorized by the Gun Control Act or the Federal Mine Safety and Health Act would be legal. A warrantless search under the Occupational Safety and Health Act, however, would violate the Fourth Amendment because there is no history of pervasive legislation on working conditions before passage of that act.

APPLYING THE LAW TO THE FACTS . . .

One of a store's employees asked the mall security guards to stop a man that she thought had shoplifted some merchandise. They stopped him, handcuffed him, and searched his pockets while waiting for police to arrive. In his pocket they found a small pill bottle that contained bags of cocaine. They gave the bottles to the police when the police arrived. In his subsequent trial, the suspect argued that that the bottles of cocaine should not be used as evidence against him because the mall security guards searched him in violation of his Fourth Amendment rights. Do you think he has a good argument? Why or why not?

Fifth Amendment Protects individuals against self-incrimination and double jeopardy and guarantees them the right to trial by jury; protects both individuals and businesses through the Due Process Clause and the Takings Clause.

Due Process Clause Provides that no one can be deprived of life, liberty, or property without "due process of law"; found in the Fifth Amendment.

procedural due process Procedural steps to which individuals are entitled before losing their life, liberty, or property.

THE FIFTH AMENDMENT

The **Fifth Amendment** provides many significant protections to individuals. For instance, it protects against self-incrimination and double jeopardy (that is, being tried twice for the same crime). Of more importance to businesspersons, however, is the **Due Process Clause** of the Fifth Amendment. This provision provides that one cannot be deprived of life, liberty, or property without due process of law.

There are two types of due process: procedural and substantive. Originally, due process was interpreted only procedurally. **Procedural due process** requires that a criminal whose life, liberty, or property would be taken by a

TECHNOLOGY AND THE LEGAL ENVIRONMENT

In 2010, in the case of *United States v. Warshak*,[50] the Sixth Circuit Court of Appeals became the first appellate court to rule that the people who use email have a reasonable expectation of privacy in its contents, and that the government, therefore, must obtain a warrant under probable cause to have an ISP turn over the communications to it. The court analyzed the expectation of privacy regarding the various forms of communication, finding that email is the direct descendant of telephone and letters, because email has replaced those forms of communication. The same privacy considerations, therefore, apply to email and trigger the Fourth Amendment warrant requirement. Consequently, the court held that a subscriber enjoys a reasonable expectation of privacy in the contents of emails "that are stored with, or sent or received through, a commercial ISP." However, despite ruling that the Stored Communications Act under which the warrantless search was unconstitutional, and finding that because they did not obtain a warrant, the government agents violated the Fourth Amendment when they obtained the contents of Warshak's emails, the court in Warshak still allowed the government to use the email evidence it had acquired because the agents had relied in good faith on the Stored Communications Act, under which the ex parte order to the company to turn over the emails had been issued.

---

[50]*United States v. Warshak*, 631 F.3d 266 (6th Cir. 2010).

conviction be given a fair trial; that is, he or she is entitled to notice of the alleged criminal action and the opportunity to confront his or her accusers before an impartial tribunal. The application of procedural due process soon spread beyond criminal matters, especially after passage of the Fourteenth Amendment, discussed in the next section, which made the requirement of due process applicable to state governments.

Today, the Due Process Clause has been applied to such diverse situations as the termination of welfare benefits,[51] the discharge of a public employee from his or her job, and the suspension of a student from school. It should be noted, however, that the types of takings to which the Due Process Clause applies are not being continually increased. In fact, after a broad expansion of the takings to which this clause applied, the courts began restricting the application of this clause during the 1970s and have continued to do so since then. The courts restrict application of the clause by narrowing the interpretation of *property* and *liberty*. This narrowing is especially common in interpreting the Due Process Clause as it applies to state governments under the Fourteenth Amendment.

What procedural safeguards does procedural due process require? The question is not easily answered. The procedures that the government must follow when there may be a taking of an individual's life, liberty, or property vary according to the nature of the taking. In general, as the magnitude of the potential deprivation increases, the extent of the procedures required also increases.

The second type of due process is **substantive due process**. The concept of substantive due process refers to the basic fairness of laws that may deprive an individual of his or her liberty or property. In other words, when a law is passed that will restrict individuals' liberty or their use of their property, the government must have a proper purpose for the restriction, or it violates substantive due process.

substantive due process Requirement that laws depriving individuals of liberty or property be fair.

During the late nineteenth and early twentieth centuries, this concept was referred to as *economic substantive due process* and was used to strike down a number of pieces of social legislation, including laws that established minimum wages and hours. Business managers successfully argued that such laws interfered with the liberty of employer and employee to enter into whatever type of employment contract they might choose. Analogous arguments were used to defeat many laws that would allegedly have helped the less fortunate at the expense of business interests. Economic substantive due process flourished only until the late 1930s. Today, many pieces of social legislation are in force that would have been held unconstitutional under the old concept of economic substantive due process.

The concept of substantive due process is not dead. Its use today, however, protects not economic interests, but personal rights, such as the still-evolving right to privacy. The right to privacy is a liberty now deemed to be protected under the Constitution. In order for a law restricting one's right to privacy to conform to substantive due process, the restriction in question must bear a substantial relationship to a compelling governmental purpose.

And most recently, in the following case, we saw the Due Process Clause used to strike down the Defense of Marriage Act (DOMA).

---

[51]*Goldberg v. Kelly*, 90 U.S. 101 (1970). In this case, the Supreme Court stated that the termination of a welfare recipient's welfare benefits by a state agency without affording him or her opportunity for an evidentiary hearing before termination violates the recipient's procedural due process rights.

CASE 5-6

United States v. Windsor
United States Supreme Court
133 S. Ct. 2675 (2013)

Edith Windsor and Thea Spyer were legally married in Canada, and moved to New York, which recognizes same sex marriage. When Spyer died in 2009, she left her entire estate to Windsor, who sought to claim the federal estate tax exemption for surviving spouses, but was barred from doing so by § 3 of the federal Defense of Marriage Act (DOMA), which amended the federal law that provides rules of construction for over 1,000 federal laws and all federal regulations—to define "marriage" and "spouse" as excluding same-sex partners. Windsor paid $363,053 in estate taxes and applied for a refund. When the Internal Revenue Service refused to give her the refund, Windsor filed suit, arguing that DOMA violates the principles of equal protection incorporated in the Fifth Amendment.

The District Court found that DOMA unconstitutionally violated principles of due process and equal protection enshrined in the Constitution. The Second Circuit Court of Appeals upheld the decision, saying the courts should apply strict scrutiny to classifications based on sexual orientation. (In an unrelated case, the United States Court of Appeals for the First Circuit also found § 3 of DOMA unconstitutional.) The United States then appealed the case to the U.S. Supreme Court.

Justice Kennedy

III

. . . [S]ome States concluded that same-sex marriage ought to be given recognition and validity in the law for those same-sex couples who wish to define themselves by their commitment to each other. The limitation of lawful marriage to heterosexual couples, which for centuries had been deemed both necessary and fundamental, came to be seen in New York and certain other States as an unjust exclusion.

Against this background of lawful same-sex marriage in some States, the design, purpose, and effect of DOMA should be considered as the beginning point in deciding whether it is valid under the Constitution. By history and tradition the definition and regulation of marriage, . . . has been treated as being within the authority and realm of the separate States. Yet it is further established that Congress, in enacting discrete statutes, can make determinations that bear on marital rights and privileges. . . . Congress has the power both to ensure efficiency in the administration of its programs and to choose what larger goals and policies to pursue.

Though these discrete examples establish the constitutionality of limited federal laws that regulate the meaning of marriage in order to further federal policy, DOMA has a far greater reach; for it enacts a directive applicable to over 1,000 federal statutes and the whole realm of federal regulations. And its operation is directed to a class of persons that the laws of New York, and of 11 other States, have sought to protect. . . .

In order to assess the validity of that intervention it is necessary to discuss the extent of the state power and authority over marriage as a matter of history and tradition. . . . State laws defining and regulating marriage, of course, must respect the constitutional rights of persons, . . . but, subject to those guarantees, "regulation of domestic relations" is "an area that has long been regarded as a virtually exclusive province of the States."

. . . DOMA rejects the long-established precept that the incidents, benefits, and obligations of marriage are uniform for all married couples within each State, though they may vary, subject to constitutional guarantees, from one State to the next. . . . Here the State's decision to give this class of persons the right to marry conferred upon them a dignity and status of immense import. When the State used its historic and essential authority to define the marital relation in this way, its role and its power in making the decision enhanced the recognition, dignity, and protection of the class in their own community. . . .

The Federal Government uses this state-defined class for the opposite purpose—to impose restrictions and disabilities. That result requires this Court now to address whether the resulting injury and indignity is a deprivation of an essential part of the liberty protected by the Fifth Amendment. What the State of New York treats as alike the federal law deems unlike by a law designed to injure the same class the State seeks to protect.

In acting first to recognize and then to allow same-sex marriages, New York was responding "to the initiative of those who [sought] a voice in shaping the destiny of their own times." . . . These actions were without doubt a proper exercise of its sovereign authority within our federal system, all in the way that the Framers of the Constitution intended. The dynamics of state government in the federal system are to allow the formation of consensus respecting the way the members of a discrete community treat each other in their daily contact and constant interaction with each other.

The States' interest in defining and regulating the marital relation, subject to constitutional guarantees, stems from the understanding that marriage is more than a routine classification for purposes of certain statutory benefits. Private, consensual sexual intimacy between two adult persons of the same sex may not be punished by the State, and it can form "but one element in a personal bond that is more enduring." By its recognition of the validity of same-sex marriages performed in other jurisdictions and then by authorizing same-sex unions and same-sex marriages, New York sought to give further protection and dignity to that bond. For same-sex couples who wished to be married, the State acted to give their lawful conduct a lawful status. This status is a far-reaching legal acknowledgment of the intimate relationship between two people, a relationship deemed by the State worthy of dignity in the community equal with all other marriages. It reflects both the community's considered perspective on the historical roots of the institution of marriage and its evolving understanding of the meaning of equality.

IV

DOMA seeks to injure the very class New York seeks to protect. By doing so it violates basic due process and equal protection principles applicable to the Federal Government. . . . The Constitution's guarantee of equality "must at the very least mean that a bare congressional desire to harm a politically unpopular group cannot" justify disparate treatment of that group. . . . In determining whether a law is motived by an improper animus or purpose, "'[d]iscriminations of an unusual character'" especially require careful consideration. . . . DOMA cannot survive under these principles. The responsibility of the States for the regulation of domestic relations is an important indicator of the substantial societal impact the State's classifications have in the daily lives and customs of its people. DOMA's unusual deviation from the usual tradition of recognizing and accepting state definitions of marriage here operates to deprive same-sex couples of the benefits and responsibilities that come with the federal recognition of their marriages. This is strong evidence of a law having the purpose and effect of disapproval of that class. The avowed purpose and practical effect of the law here in question are to impose a disadvantage, a separate status, and so a stigma upon all who enter into same-sex marriages made lawful by the unquestioned authority of the States.

The history of DOMA's enactment and its own text demonstrate that interference with the equal dignity of same-sex marriages, a dignity conferred by the States in the exercise of their sovereign power, was more than an incidental effect of the federal statute. It was its essence. . . .

As the title and dynamics of the bill indicate, its purpose is to discourage enactment of state same-sex marriage laws and to restrict the freedom and choice of couples married under those laws if they are enacted. The congressional goal was "to put a thumb on the scales and influence a state's decision as to how to shape its own marriage laws." . . . The Act's demonstrated purpose is to ensure that if any State decides to recognize same-sex marriages, those unions will be treated as second-class marriages for purposes of federal law. This raises a most serious question under the Constitution's Fifth Amendment.

DOMA's operation in practice confirms this purpose. When New York adopted a law to permit same-sex marriage, it sought to eliminate inequality; but DOMA frustrates that objective through a system-wide enactment with no identified connection to any particular area of federal law. DOMA writes inequality into the entire United States Code. The particular case at hand concerns the estate tax, but DOMA is more than a simple determination of what should or should not be allowed as an estate tax refund. Among the over 1,000 statutes and numerous federal regulations that DOMA controls are laws pertaining to Social Security, housing, taxes, criminal sanctions, copyright, and veterans' benefits.

DOMA's principal effect is to identify a subset of state-sanctioned marriages and make them unequal. The principal purpose is to impose inequality, not for other reasons like governmental efficiency. Responsibilities, as well as rights, enhance the dignity and integrity of the person. And DOMA contrives to deprive some couples married under the laws of their State, but not other couples, of both rights and responsibilities. By creating two contradictory marriage regimes within the same State, DOMA forces same-sex couples to live as married for the purpose of state law but unmarried for the purpose of federal law, thus diminishing the stability and predictability of basic personal relations the State has found it proper to acknowledge and protect. By this dynamic DOMA undermines both the public and private significance of state-sanctioned same-sex marriages; for it tells those couples, and all the world, that their otherwise valid marriages are unworthy of federal recognition. This places same-sex couples in an unstable position of being in a second-tier marriage. The differentiation demeans the couple, whose moral and sexual choices the Constitution protects, and whose relationship the State has sought to dignify. And it humiliates tens of thousands of children now being raised by same-sex couples. The law in question makes it even more difficult for the children to understand the integrity and closeness of their own family and its concord with other families in their community and in their daily lives.

Under DOMA, same-sex married couples have their lives burdened, by reason of government decree, in visible and public ways. By its great reach, DOMA touches many aspects of married and family life, from the mundane to the profound. It prevents same-sex married couples from obtaining government healthcare benefits they would otherwise receive. It deprives them of the Bankruptcy Code's special protections for domestic-support obligations. It forces them to follow a complicated procedure to file their state and federal taxes jointly. It prohibits them from being buried together in veterans' cemeteries.

The power the Constitution grants it also restrains. And though Congress has great authority to design laws to fit its own conception of sound national policy, it cannot deny the liberty protected by the Due Process Clause of the Fifth Amendment.

What has been explained to this point should more than suffice to establish that the principal purpose and the necessary effect of this law are to demean those persons who are in a lawful same-sex marriage. This requires the Court to hold, as it now does, that DOMA is unconstitutional as a deprivation of the liberty of the person protected by the Fifth Amendment of the Constitution.

The liberty protected by the Fifth Amendment's Due Process Clause contains within it the prohibition against denying to any person the equal protection of the laws. . . . While the Fifth Amendment itself withdraws from Government the power to degrade or demean in the way this law does, the equal protection guarantee of the Fourteenth Amendment makes that Fifth Amendment right all the more specific and all the better understood and preserved.

The class to which DOMA directs its restrictions and restraints are those persons who are joined in same-sex marriages made lawful by the State. DOMA singles out a class of persons deemed by a State entitled to recognition and protection to enhance their own liberty. It imposes a disability on the class by refusing to acknowledge a status the State finds to be dignified and proper. DOMA instructs all federal officials, and indeed all persons with whom same-sex couples interact, including their own children, that their marriage is less worthy than the marriages of others. The federal statute is invalid, for no legitimate purpose overcomes the purpose and effect to disparage and to injure those whom the State, by its marriage laws, sought to protect in personhood and dignity. By seeking to displace this protection and treating those persons as living in marriages less respected than others, the federal statute is in violation of the Fifth Amendment.*

Affirmed, in favor of Respondents.

The complete ramifications of this holding were not immediately clear at the time of the decision. Twelve states and the District of Columbia recognized same-sex marriage at that time, and employers at that time were scrambling to find out what their obligations would be under the new law. Over 1,000 federal regulations are affected by the ruling, many of those directly or indirectly affecting employers. Employers in states recognizing same-sex marriage were being forced to look carefully for policy changes they would need to make in three areas: health benefits, retirement plans, and family and medical leave policies. For example, the Family Medical Leave Act (discussed in greater detail in Chapter 19), now covers same-sex marriage partners, whereas before it did not. There are also benefits from this ruling for many employers.

In 2015, however, the impact of this decision became even more extensive, as the recognition of same-sex marriages spread from those 12 states and the District of Columbia to every state in the Union, as the result of the ruling by the U.S. Supreme Court in the case of *Obergefell et al. v. Hodges et al.*[52] In that case, the high court held that the Fourteenth Amendment requires a State to license a marriage between two people of the same sex and to recognize a marriage between two people of the same sex when their marriage was lawfully licensed and performed out-of-State.[53]

Takings Clause Provides that if the government takes private property for public use, it must pay the owner just compensation; found in the Fifth Amendment.

The Fifth Amendment further provides that if the government takes private property for public use, it must pay the owner just compensation. This provision is referred to as the **Takings Clause**. Unlike the protection against self-incrimination, which does not apply to corporations, both the Due Process Clause and the provision for just compensation are applicable to corporations. This provision for just compensation has been the basis of considerable litigation. One significant issue that has arisen is the question of what constitutes a "public use" for which the government can take property. This issue is discussed in greater detail in Chapter 13.

A second issue under this takings provision is the question of when a government regulation can become so onerous as to constitute a taking for which just compensation is required. These takings, which do not involve a physical taking of the property, are called *regulatory takings*. Environmental regulations, because they often have an impact on the way landowners may use their property, have been increasingly challenged as unconstitutional regulatory takings.

Perhaps one of the most important takings cases was *Lucas v. South Carolina Coastal Commission*,[54] which was decided in 1992. The case arose out of a dispute between a beachfront property owner and the state of South Carolina,

---

[52]135 S. Ct. 2584 (2015).

[53]*Ibid*.

[54]505 U.S. 1003 (1992).

*United States v. Windsor, Supreme Court of the United States, 133.S. Ct. 2675.

after the state passed a regulation prohibiting permanent construction on any eroding beach. Lucas had bought two beachfront lots for $975,000 a few years before the law was passed, and planned to build a couple of condominiums on the land. He challenged the law as constituting an unlawful taking because it prohibited him from building the condominiums or really doing anything with the property. The state court agreed with Lucas that the regulation denied him the economic value of his land and thus constituted an unconstitutional taking without compensation. It ordered the state to pay him $1.2 million in compensation. The South Carolina Supreme Court, however, citing U.S. Supreme Court precedents, overturned the lower court's decision.

Lucas appealed the decision to the U.S. Supreme Court, which reversed the state supreme court ruling in a 6–3 decision. The Court held that a state regulation that deprives a private property owner of all economically beneficial uses of property, except those that would not have been permitted under background principles of state property and nuisance law, constitutes a taking of private property for which the Fifth Amendment requires compensation. The highest court in the land said that the state court had erred in applying the principle that the Takings Clause does not require compensation when the regulation at issue is designed to prevent "harmful or noxious uses" of property.

One of the factors that many commentators believed was critical to the Court's ruling in *Lucas* was the fact that the law that led to his inability to develop his land had been enacted after Lucas had acquired his property. In the 2001 case of *Palazzolo v. Rhode Island*,[55] however, the Supreme Court held by a 5–4 vote that someone who bought property after restrictions on development were in place could still challenge the restrictions as an unconstitutional "taking" of private property.

Many advocates of private property rights now believe that the Takings Clause has taken on new importance because of cases such as *Lucas* and *Whitney Benefits, Inc. v. United States*,[56] wherein a federal court found that the federal Surface Mining and Reclamation Act constituted a taking with respect to one mining company whose land became completely useless as a result of the act. Whether the "Property Firsters," as they call themselves, will be successful in the future remains to be seen, but they have clearly brought back attention to an argument against regulation that had been fairly dormant for the past 50 years. Moreover, they are using their arguments primarily to challenge a broad range of environmental laws involving matters from forcing cleanups of hazardous waste sites to restricting grazing and rationing water, as well as land use planning statutes.

For example, in *Dolan v. Tigard*,[57] the owner of a store in the city's business district sued when her receipt of a permit to double the size of her store and pave its gravel parking lot was made contingent on the condition that she dedicate a sixth of her land to the city. She was to make part of the land, which was in a floodplain, a public recreational greenway and part of a bike trail that could help reduce the increased congestion in the area that might result from the expansion of her store. The U.S. Supreme Court found that there was no evidence of a reasonable relationship between the floodplain easement required of Dolan and the impact of the new building. They held that the city had the right to take the easement for the greenway, but it had to provide just compensation for the regulatory taking.

Using the Fifth Amendment to bring individual challenges to land use regulations and zoning laws is a very time-consuming process, so many property

---

[55] 533 U.S. 606 (2001).
[56] 926 F.2d 1169 (Fed. Cir. 1991).
[57] 512 U.S. 374, 114 S. Ct. 2309 (1994).

rights organizations have instead focused on trying to pass state laws that would make it easier for property owners to get compensation when their property values fall because of new regulations. In 2004, such groups achieved their greatest success with the passage of Ballot Measure 37[58] in the state of Oregon. The measure, which was initially struck down by the state court, but was subsequently upheld by the Oregon State Supreme Court, provides that any property owners who can prove that environmental or zoning laws have hurt their investments can force the government to compensate them for their losses or get an exemption from the rules. Other states that have laws providing compensation for aggrieved property owners are Florida, Texas, Louisiana, and Mississippi, but these laws provide compensation only after a particular loss threshold has been reached, usually a 25 percent reduction in the property's value. Those laws also allow compensation only for losses caused by new land use laws. Whether the passage of the legislation in Oregon indicates the beginning of a new wave of legislation remains to be seen.

Another factor that many who studied the *Lucas* case saw as important was the fact that the regulation really deprived its owner of any possible economically viable use for the land. Challenges based on the idea of a regulatory taking have not been quite as successful when there is just arguably a diminishment of the value of the property, as property rights advocates have found when attempting to use the law to challenge smoking bans as constituting a regulatory taking in violation of the Fifth Amendment. For example, operators of bars and restaurants in Toledo, Ohio were unsuccessful in challenging a citywide smoking ban as a violation of the Fifth Amendment, as applied to the states by the Fourteenth Amendment. They unsuccessfully argued that the statute denied them any "economically viable" use of their land because they would have to either spend huge amounts of money to establish smoking lounges in their establishments or lose all of their customers who smoked.

A final problem that often causes confusion among businesspersons is the question of the extent of Fifth Amendment protections for corporations. The Fifth Amendment protection against self-incrimination has not been held to apply to corporations. Some decisions, however, have raised questions about this long-standing interpretation. In *United States v. Doe*,[59] the U.S. Supreme Court determined that even though the contents of documents may not be protected under the Fifth Amendment, a sole proprietor should have the right to show that the act of producing the documents would entail testimonial self-incrimination as to admissions that the records existed. Therefore, the sole proprietor could not be compelled to produce the sole proprietorship's records.

In the subsequent case of *Braswell v. United States*,[60] however, the Court clearly distinguished between the role of a custodian of corporate records and a sole proprietor. In *Braswell*, the defendant operated his business as a corporation, with himself as the sole shareholder. When a grand jury issued a subpoena requiring him to produce corporate books and records, Braswell argued that to do so would violate his Fifth Amendment privilege against self-incrimination. The U.S. Supreme Court denied Braswell's claim and said that, clearly, subpoenaed business records are not privileged, and, because Braswell was a custodian for the records, his act of producing the records would be in a representative capacity, not a personal one, so the records must be produced. The Court stated that, had the business been a sole proprietorship, Braswell would have had the opportunity to show that the act of production would have been self-incriminating. Because his business was a corporation, he was acting as a representative of a corporation, and regardless of how small the corporation,

---

[58]Subsequently codified as Oregon Revised Statutes (ORS) 195.305.
[59]465 U.S. 605 (1984).
[60]487 U.S. 99 (1988).

he could not claim a privilege. In 2003, the Court of Appeals for the Eighth Circuit applied the Supreme Court's decision in the *Braswell* case, as the appellate court ordered a woman to produce corporate documents even though the corporation's charter had been revoked.[61] The appellate court reasoned that the subpoena of documents from an inactive corporation does not constitute a violation of the corporate custodian's Fifth Amendment rights.

THE FOURTEENTH AMENDMENT

The Fourteenth Amendment is important because it applies the Due Process Clause to the state governments. It has been interpreted to apply almost the entire Bill of Rights to the states, with the exceptions of the Fifth Amendment right to indictment by a grand jury for certain types of crimes and the right to trial by jury.

The Fourteenth Amendment is also important because it contains the Equal Protection Clause, which prevents the states from denying "the equal protection of the laws" to any citizen. This clause, discussed in more detail in Chapter 21, has been a useful tool for people attempting to reduce discrimination in this country.

The most recent significant use of the Fourteenth Amendment to fight discrimination was the previously mentioned case of *Obergefell et al. v. Hodges et al.*[62] In that case, the court demonstrated that the right to marry is a fundamental right inherent in the liberty of the person, and under the Due Process and Equal Protection Clauses of the Fourteenth Amendment couples of the same sex may not be deprived of that right and that liberty.[63]

Standard of Review. The Equal Protection Clause prohibits "invidious" discrimination, that is, discrimination not based on a sufficient justification. To determine whether a specific classification system being used by the government has sufficient justification, the Supreme Court has established standards of review based upon the nature of the classification. The standards are: strict scrutiny, intermediate scrutiny, and the rational basis test.

Strict scrutiny is used when a government activity classifies people based on their belonging to a suspect class (race, color, national origin). For the court to uphold such a classification, the government must have a compelling reason for the classification and the regulation must be narrowly drawn so that it goes no further than necessary to meet the compelling government interest. This is the same level of scrutiny applied when the government is attempting to deprive a person of their fundamental rights, as in the *Obergefell* case.

Intermediate scrutiny is applied when the classification is based on a protected class other than race, color, or national origin, such as sex, age, or religion. In such a case the government must have an important reason for treating those in the classification differently, and the regulation must be "reasonably related" to furthering that reason.

If the classification does not involve a suspect classification, then the court will simply apply a rational basis test, meaning that there has to be some legitimate reason why the government would treat members of that class differently. The regulation at issue must simply be reasonably related to furthering that reasonable government interest. For example, in 2008, the Iowa Supreme Court ruled that a law that taxed apartment buildings at a higher commercial rate rather than at the lower residential rate charged to owners of owner-occupied condominiums did not violate the Equal Protection Clause.[64]

---

[61]*In re Grand Jury Subpoena*, 75 Fed. Appx. 562 (2003).
[62]*Supra*, note 53.
[63]*Ibid.*
[64]*State v. DeAngelo*, 2009 WL 291169 (N.J. 2009).

SUMMARY

The framework of our nation is embodied in the U.S. Constitution, which established a system of government based on the concept of federalism. Under this system, the power to regulate local matters is given to the states; the federal government is granted limited powers to regulate activities that substantially affect interstate commerce. All powers not specifically given to the federal government are reserved to the states.

The Commerce Clause is the primary source of the federal government's authority to regulate business. The same clause restricts states from passing regulations that would interfere with interstate commerce. The state and federal governments are limited in their regulations by the amendments to the Constitution, especially the Bill of Rights. The First Amendment, for example, protects our individual right to free expression; commercial speech is also entitled to a significant amount of protection in this area.

Other important amendments for the businessperson are the Fourth Amendment, which protects one's right to be free from unwarranted searches and seizures, and the Fifth Amendment, which establishes one's right to due process. A final amendment that has a significant impact on the legal environment is the Fourteenth Amendment, which applies most of the Bill of Rights to the states and also contains the Equal Protection Clause.

REVIEW QUESTIONS

5-1 Explain the relationship between the Supremacy Clause and the doctrine of federal preemption.

5-2 How does the Commerce Clause affect federal regulation of business activities?

5-3 What is police power?

5-4 How does the Commerce Clause affect state regulation of business?

5-5 Explain why you believe that the courts have found each of the following to either constitute or not constitute a regulatory taking: (a) a city ordinance reducing the size limit for freestanding signs within the city limits, forcing the plaintiff to replace his sign; (b) a city ordinance prohibiting billboards along roads in residential areas of the city; (c) the refusal of the Army Corps of Engineers to grant Florida Rock Company a permit to allow it to mine limestone that lay beneath a tract of wetlands.

5-6 How does the First Amendment's protection of private speech differ from its protection of commercial speech?

REVIEW PROBLEMS

5-7 Voters in the state of California decide that the tobacco industry is having too great an impact on the outcome of local referenda limiting smoking in public places. To curb the influence of that powerful lobby, so that the fate of the legislation will more clearly reflect the will of "the people," they pass a law prohibiting firms in the tobacco industry and tobacco industry associations from (a) purchasing advertising related to smoking-related referenda and (b) making cash contributions to organizations involved in campaigns related to antismoking legislation. Several tobacco firms challenge the law. What is the constitutional basis for their challenge? Why will they be likely to succeed or fail?

5-8 Ms. Crabtree is given a one-year, nontenured contract to teach English at Haddock State University, a public institution. After her contract year ends, she is not offered a contract for the next year and is not given any explanation as to why she is not being rehired. She sues the school, arguing that her right to procedural due process has been violated. Is she correct?

5-9 The state of Ohio decides that Ohio's landfills are becoming too full at too rapid a pace, so it passes a law banning the import of waste generated out of state. Several landfill operators have contracts with out-of-state generators, so they sue the state to have the statute declared void. What arguments will each side make in this case? What is the most likely decision? Why?

5-10 Chen opens a small business. As business thrives, he decides to incorporate. He becomes the corporation's president and also its sole shareholder. Chen comes under investigation for tax fraud and is subpoenaed to produce the corporate tax records for the previous three years. He challenges the subpoena on the ground that it violates his Fifth Amendment right not to incriminate himself. Must he comply with the subpoena?

5-11 Congress passed the Americans with Disabilities Act requiring, among other provisions, that places of public accommodation remove architectural barriers to access where such removal is "readily achievable." If such removal is not "readily achievable," the firm must make its goods or services available through alternative methods if such methods are readily achievable. To comply with the law, Ricardo's Restaurant will have to construct a ramp at the entrance to the restaurant and two ramps within the dining area, rearrange some of the tables and chairs, and remodel the restrooms so that they will accommodate wheelchairs. Because of the expenditures Ricardo will have to make to comply with the law, he challenges the law as being a taking of private property without just compensation in violation of the Fifth Amendment. Does the law violate the Fifth Amendment?

5-12 Plaintiffs owned a piece of lakeside property. The land was subsequently rezoned to prohibit high-density apartment complexes and restrict the use of the property to single-family dwellings. On what basis would the plaintiffs claim that their constitutional rights had been violated by the zoning change? Would they be correct?

CASE PROBLEMS

5-13 Twins Steven and Sean Wilson were students at Lee's Summit High School in Missouri in 2011. In their time away from school, the two brothers decided to create a blog site, NorthPress, in which they could talk about the goings-on at their high school. However, the two brothers posted comments on their blog site that were highly racist against African Americans. They also made sexually explicit comments about several girls from their high school. The twins' blog did not have a password-protected login, so anyone could have seen the posts or added new posts, which a third student did. Only a few days after the Wilsons had put up their blog, teachers reported difficulty controlling their classes, in addition to observing the adverse effects of the blog posts on one of the girls who was mentioned in them. Subsequently the high school received calls from concerned parents and a visit from a local television station. The Wilsons were at first suspended for 10 days. After further investigation, the boys' suspension was extended to 180 days. The Wilsons sued the school in federal court, claiming that their suspensions violated the boys' First Amendment rights. How should the court have ruled in this case? Under what conditions would the Wilsons' injunction have been justified? *S. J. W. v. Lee's Summit R-7 Sch. Dist.,* 696 F.3d 771 (8th Cir. Mo. 2012).

5-14 Plaintiffs James Brooks and Donald Hamlette were employees of the Virginia Department of Corrections (VDOC). Brooks was a senior corrections officer, supervised by Hamlette. Hamlette was African American and a minister. Both Brooks and Hamlette were to report to the defendants, Howard Arthur, the Superintendent, and Major Randal Mitchell, the Assistant Superintendent. One of the plaintiff officers alleged that the defendants publicly embarrassed him and discriminated against him, while the other plaintiff officer filed an Equal Employment Opportunity (EEO) complaint, alleging that the defendants discriminated against him based on race and religion. The defendants then subsequently terminated the plaintiff officers for disciplinary violations. The plaintiff officers sued the defendants, alleging First Amendment retaliation. The plaintiff officers appealed. On appeal, the court concluded that the officers' First Amendment retaliation claims failed because the officers' speech pertained to personal grievances and complaints about employment rather than "broad matters of policy meriting the protection of the First Amendment." How do you think the court should have ruled? Why? What legal precedents discussed in this case were likely to be relevant to this opinion? *Brooks v. Arthur,* 685 F.3d 367 (4th Cir. 2012).

5-15 John Doe was convicted of child exploitation in Marion County, Indiana, and once released was required to register as a sex offender with the state of Indiana. Under the state law, as a registered sex offender, he was not allowed to use certain websites or programs. He sued the prosecutor of Marion County in federal court, alleging that the prohibitions under this law violated his First Amendment rights, and sought an injunction prohibiting enforcement of the law. The District Court found in favor of the defendant, stating that the statute was narrowly tailored to serve a significant state interest. Doe appealed. How do you think the Circuit Court of Appeals ruled on this case? Why? *Doe v. Prosecutor, Marion County, Indiana*, 705 F.3d 694 (2014).

5-16 The state of California passed a statute making it illegal to sell or rent violent video games to minors. Prohibited games included those in which players could kill, maim, dismember, or sexually assault an image of a human being in a manner that a reasonable person would find "appeals to a deviant or morbid interest of minors" that is "patently offensive to prevailing standards in the community as to what is suitable to minors for minors" and "causes the game, as a whole, to lack serious literary, artistic, political, or scientific value for minors." The penalty for violating the act is a civil fine of up to $1,000. An association of video game and software firms challenged the law as violative of the First Amendment rights. The Federal District Court concluded that the Act violated the First Amendment and permanently enjoined its enforcement. The Ninth Circuit affirmed. How do you think the U.S Supreme Court ruled on the final appeal and why? *Brown, Governor of California v. Entertainment Merchants Association*, 131 S. Ct. 2729 (2011).

5-17 Aurora, Colorado, passed an ordinance banning pit bulls and other selected breeds of dogs from being inside the city limits. The American Canine Foundation, an organization aimed at improving the canine industry, sued to have the ordinance struck down as unconstitutional. The Foundation argued that the ordinance infringed on their constitutional rights, in violation of the Fifth and Fourteenth Amendments. Specifically, they argued the law resulted in an uncompensated taking and violated the Due Process and Equal Protection Clauses. Explain why the ordinance is or is not constitutional. *American Canine Foundation v. City of Aurora, Colorado*, 618 F. Supp. 2d 1271; 2009 WL 1370893 (D. Colo. 2009).

5-18 A defendant, Antoine Jones, was under suspicion of trafficking narcotics. To investigate the defendant, plaintiff government agents installed a GPS tracking device on the undercarriage of a vehicle registered to the defendant's wife while it was parked in a public parking lot. For the next 28 days the government used this device to track the vehicle's movements. In court, the defendant sought a motion to suppress evidence obtained through the GPS device, alleging that it violated the Fourth Amendment. The court found that admission of the evidence obtained by warrantless use of the GPS device by the government agents violated the Fourth Amendment, and constituted an unreasonable search. Do you agree with the court's decision on appeal? Why or why not? Based on your reading in this chapter, what evidence would the court need to determine whether a violation of the Fourth Amendment existed? *United States v. Jones*, 132 S. Ct. 945 (2012).

5-19 David Milam was a member of the Delta Tau Delta fraternity at the University of Kentucky, and leased a room at the fraternity house. Upon receiving a tip that Milam was selling marijuana at the fraternity house, University of Kentucky detective McBride and two other university detectives went to the fraternity house to perform a "knock and talk." They accidentally went to the back door, thinking it was the front, and knocked several times but no one responded. After talking among themselves, they decided the fraternity building was more like an apartment complex than a private dwelling, so they entered through the slightly ajar door.

Once inside, they identified themselves as police officers and asked where Milam was. It is unclear whether they were told where his room was, but they followed another resident up to the second floor where the private rooms were. The resident pointed to Milam's room, from which detectives smelled marijuana. When detectives knocked on the door, Milam opened it, and the police saw a jar of marijuana in full view.

Milam then provided consent for the detectives to search his bedroom. During the search, they discovered marijuana, $1,700, Adderall pills, drug paraphernalia, and a fake driver's license. Milam was charged with one count of trafficking in a controlled substance within 1,000 yards of a school and other drug-related charges. He moved to suppress the evidence discovered in his bedroom, arguing that the detectives unlawfully entered and searched the house in violation of the Fourth Amendment. The trial court denied the motion to suppress. Milam subsequently pled guilty to the trafficking charge. The State Court of Appeals affirmed the trial court's denial of Appellant's motion to suppress. Milam appealed again. How do you think the State Supreme Court ruled on his appeal? *David Milam v. Commonwealth of Kentucky*, Kentucky State Supreme Court (2015). Case available at http://cases.justia.com/kentucky/supreme-court/2015-2013-sc-000681-dg.pdf?ts=1431612086.

5-20 In 1996, a practicing lawyer from Colorado was disbarred. In 2011, he brought a $5 million suit against the U.S. government, claiming that the revocation of his license to practice law constituted a governmental taking, and that he was entitled to just compensation. The plaintiff argued that under the Tucker Act, the revocation amounted to the taking of public property. In the end, the court threw out the case because of the six-year statute of limitations. However, the court had an alternative explanation it could have used, had the case been within the six-year limit. According to the court, the Fifth Amendment does not mandate the payment of money as compensation. Had the case not been barred because of time, could the plaintiff have sought compensation other than monetary payment? Why or why not? *Smith v. United States,* 2012 U.S. Claims (Fed. Cl. May 30, 2012).

THINKING CRITICALLY ABOUT RELEVANT LEGAL ISSUES

Protection of Corporate Political Speech

Over the past decade, a number of groups and individuals have argued against many of the legal protections that corporations receive. These critics argue that problems such as those associated with Enron, Tyco, and WorldCom would not have been as severe if corporations did not have legal protection for their political speech, which limits their accountability to the people. What these critics miss, however, is that going back as far as 1889, in *Minneapolis & St. Louis Railroad Co. v. Beckwith,* 129 U.S. 26, the Supreme Court has found corporations to be persons. This finding of personhood means that corporations are properly granted the same protections for their political speech as are enjoyed by other individuals.

The First Amendment grants individuals the right to free speech. When the right to free speech is combined with the Fourteenth Amendment's Equal Protection Clause, corporations are also provided protection for their political speech. The framers of our Constitution intended for all parties to have the right to free speech, and the critics who want to strip corporations of this right are acting in an unconstitutional and unpatriotic manner. After all, one of the principles underlying the American Revolution was the belief in "no taxation without representation," and corporations are taxpayers.

Another reason corporations deserve protections for their political speech is that, although corporations are treated as artificial persons, they are not capable of voting. Instead, the only means corporations have available to influence laws and the political culture is through their political speech. To limit corporations' political speech would be similar to barring corporations from having any fair say in the political process. Individuals are allowed to engage in political speech and vote; if corporations are not given a similar right, they are at a disadvantage in putting forward their point of view. The fair thing to do in this situation is to continue to grant corporations protections for their political speech. Corporations may not be able to directly have equal say and equal representation in government, but continuing to grant them their political speech protections is an important step toward equality in the political realm.

Finally, allowing corporations to engage in political speech is not the same as saying that the general population will agree with the corporations' political speech. Rather, corporate political speech allows corporations to express their opinions openly and then allows the people to agree or disagree with these opinions. Accordingly, there is no threat to the political speech of others to have corporations have greater access to corporate political speech. Also, in *First National Bank of Boston v. Bellotti,* 435 U.S. 765 (1978), the Supreme Court held that corporate political speech cannot be limited to try to grant others more speech. Hence, any such attempts should be prohibited.

1. How would you frame the issue and conclusion of this essay?
2. What are the relevant rules of law used to justify the argument in this essay?
3. Does the argument contain significant ambiguity in the reasoning?
   Clue: What word(s) or phrase(s) could have multiple meanings that would change the meaning of the reason entirely?
4. Write an essay that someone who holds an opinion opposite to that of the essay author might write.
   Clue: What other ethical norms could influence an opinion about this issue?

ASSIGNMENT ON THE INTERNET

This chapter introduces you to the many constitutional principles that govern business activities. One such principle is free speech and the extent to which it applies to commercial speech. For example, should advertisements for pharmaceuticals be subject to Federal Drug Administration regulations and limitations?

Read the summary from a Supreme Court decision about the advertising of pharmaceuticals from the pharmaceutical industry, found at **www.oyez.org**

(type in "advertisement of pharmaceutical" in the search bar and click on the first link, titled "Bolger vs. Youngs Drug Products Corp.). Can you determine from the summary whether the *Central Hudson* test was applied to the case? If so, how?

ON THE INTERNET

www.gpo.gov GPO is a website run by the Government Printing Office, and is intended to make available a large amount of government information to the general public. The link is to a document that has information summarizing important constitutional court cases. Do a search for "important constitutional cases" to find this link.

http://firstamendmentcoalition.org This is the Web page for the First Amendment Coalition, a nonprofit organization devoted to defending the First Amendment and seeking more transparency in government.

http://www.streetlaw.org/en/landmark/home Landmark Supreme Court cases interpreting the Commerce Clause can be found at this site. Pay close attention to how Congress has used the Commerce Clause to police and regulate businesses. On the right-hand side of the page, under the heading "The Cases," click on "Gibbons vs. Ogden (1824)."

www.umkc.edu This website links to information regarding the impact of the Commerce Clause on states' abilities to regulate commerce. In addition, the site contains useful links to information regarding important Commerce Clause cases. Search for "exploring constitutional law" in the search bar at the top of the page.

FOR FUTURE READING

Huberfeld, Nicole. "Medicaid Matters: Where There is a Right, There Must Be a Remedy (Even in Medicaid)." *Kentucky Law Journal* 102 (2013): 327.

Johnson, Asmara Tekle. "Privatizing Eminent Domain: The Delegation of a Very Public Power to Private, Non-Profit and Charitable Corporations." *American University Law Review* 56 (2007): 455.

Jones, Jackson. "The Fourth Amendment and Search Warrant Presentment: Is a Man's House Always His Castle?" *American Journal of Trial Advocacy* 35 (2012): 525.

Marcus, Nancy C. "Deeply Rooted Principles of Equal Liberty, Not 'Argle Bargle': The Inevitability of Marriage Equality after Windsor." *Tulane Journal of Law and Sexuality* 23 (2014): 17.

Ross II, Bertrand L. "Embracing Administrative Constitutionalism." *Boston University Law Review* 95 (2015): 519.

Sprague, Robert. "Business Blogs and Commercial Speech: A New Analytical Framework for the 21st Century." *American Business Law Journal* 44 (2007): 127.

Walker Wilson, Molly J. "Behavioral Decision Theory and Implications for the Supreme Court's Campaign Finance Jurisprudence." *Cardozo Law Review* 31 (2010): 2365.

Weeden, L. Darnell. "The Commerce Clause Implications of the Individual Mandate under the Patient Protection and Affordable Care Act." *Journal of Law and Health* 26 (2013): 29.

CHAPTER SIX

White-Collar Crime and the Business Community

- CRIME AND CRIMINAL PROCEDURE
- DISTINGUISHING FEATURES OF WHITE-COLLAR CRIME
- COMMON WHITE-COLLAR CRIMES
- PREVENTION OF WHITE-COLLAR CRIME
- FEDERAL LAWS USED IN THE FIGHT AGAINST WHITE-COLLAR CRIME
- STATE LAWS USED IN THE FIGHT AGAINST WHITE-COLLAR CRIME
- GLOBAL DIMENSIONS OF WHITE-COLLAR CRIME

Amidst the turmoil and fallout of the Enron scandal that led to the company's declaration of bankruptcy, a number of former Enron officials faced charges for various offenses. One such official was former CEO Jeffrey Skilling, who was ultimately found guilty of 19 fraud related charges, including conspiracy, insider trading, securities fraud, and making false statements to auditors. As punishment for his misdeeds, the 52-year-old Skilling was sentenced in 2006 to 24 years and 4 months in a federal prison. In addition, he was fined $45 million, which was to be put into a fund to benefit those who had been harmed by Enron's collapse. While serving his sentence in 2010, he won a minor victory when the U.S. Supreme Court found that instructions to the jury with respect to one of the charges were inaccurate, and threw out the conviction on that charge. The case was then sent back to the trial court judge to determine whether the inaccurate instructions regarding the one charge tainted the convictions on the other charges. In 2013, the case was finally resolved as he was resentenced to 14 years in a federal prison as part of a court ordered reduction and a separate plea agreement with the prosecution. Unfortunately, this story is just one of many recent large and complex white-collar crime scandals. During 2009, Internet crime resulted in losses in the United States of $559.7 million, more than two times as much as in 2008.[1] At the end of 2008, the FBI was investigating 545 corporate fraud cases each of which involved investor losses that exceeded $1 billion.[2] The Coalition

---

[1] Internet Crime Complaint Center, *IC3 Annual Internet Crime Report 2009*; retrieved May 10, 2010, from National White Collar Crime Center, http://www.nw3c.org/research/site_files.cfm?fileid=d1991bea-8a22-4e54-82f5-678d4d83581a&mode=r.

[2] Federal Bureau of Investigation, *Financial Crimes Report to the Public Fiscal Year 2008*; retrieved May 10, 2010, from http://www.fbi.gov/publications/financial/fcs_report2008/financial_crime_2008.htm#health.

Against Insurance Fraud reports that insurance fraud costs Americans more than $80 billion per year.[3]

White-collar crimes—crimes committed in a commercial context—occur every day. Collectively, these crimes often result in millions of dollars of damages. In recent years, as corporate crimes such as the ones detailed in Exhibit 6-1

Allen Stanford. Sentence: 110 years
Allen Stanford, 63, was a Texan financier accused of running a $7 billion Ponzi scheme. He had investors invest billions of dollars into his bank, and then spent the money on private jets, yachts, and acres of undeveloped Antiguan land among other expenditures. In December 2008, Stanford International Bank had $88 million in cash, but it fudged its numbers to say it had $1 billion in assets. In the same month it finally owed investors $7 billion when they tried to pull out their money, and the bank had no money to cover the costs.

In 2012, a jury found Stanford guilty of conspiracy, along with 12 other criminal charges including obstruction. He was found innocent of one wire fraud charge. Stanford was sentenced to 110 years in federal prison.

Bernard Madoff, Businessman. Sentence: 150 years
Madoff, 72, directed one of the largest Ponzi schemes in U.S. history. Madoff, in his role as CEO of Bernard L. Madoff Investment Securities LLC, stole from his clients in a $65 billion Ponzi scheme. Despite a continuing decline in the economy, Madoff continued to assure his clients that his numbers (investment returns) would continue rising. As the economy continued to decline, Madoff's increases became suspicious and clients began to contact him to get their money back. When the requests for returned funds reached $7 billion, Madoff met with his sons and told them that his business was fraudulent. The sons turned Madoff in to the authorities.

In 2009, Madoff pleaded guilty to, among other things, securities fraud, wire fraud, money laundering, making false filings with the SEC, and making false statements. He was sentenced to the maximum 150 years in prison for his offenses. His projected release date is November 14, 2159. Since Madoff's plea, David Friehling from his accounting department has pled guilty to securities fraud, investment advisor fraud, and making false filings with the SEC. Additionally, Frank DiPascali has pled guilty to securities fraud, investment advisor fraud, mail fraud, wire fraud, income tax evasion, international money laundering, falsifying books and records, and more.

Joseph Nacchio. Sentence: 6 years
Joseph Nacchio, 60, was the chief financial officer and chairman of the board for Qwest Communications International. Qwest is a telecommunications provider in the western United States. When the economy began to decline, Nacchio continued to assure Wall Street that the company would continue making large returns even though he knew that such returns would not occur. Based on inside information, Nacchio sold $52 million of Qwest stock just before the prices fell.

In 2007, Nacchio was convicted on 19 counts of insider trading and sentenced to 6 years in federal prison. Additionally, Nacchio was ordered to pay a $19 million fine and restitution of the $52 million he had made as a result of illegal stock transactions. Although his conviction was overturned in 2008 because of improperly excluded expert testimony, the conviction was reinstated in 2009 when he finally began serving his six-year term.

Jamie Olis, Vice President of Finance. Sentence: 24 years
Jamie Olis, 38, was vice president of finance and senior director of tax planning at Dynergy, a natural gas energy company. Olis attempted to conceal more than $300 million in company debt from public investors. When the attempted concealment was discovered, millions of investor dollars were lost, including a $105 million loss suffered by 13,000 participants in the California Retirement Plan.

In 2004, Olis was sentenced to 292 months in prison after being convicted of securities fraud, mail fraud, and three counts of wire fraud. The 24-year sentence is one of the longest terms for fraud in U.S. history, in part because of the large financial losses to thousands of investors. In addition to the jail time, Olis was fined $25,000. Olis, however, did not act alone in the concealment. Gene Foster and Helen Sharkey, both former Dynergy executives, pled guilty to conspiracy and aided in the investigation. They then entered into a plea bargain under which Foster and Sharkey were to receive sentences of up to 5 years in prison and $250,000 in fines.

(*Continued*)

EXHIBIT 6-1

RECENT MAJOR WHITE-COLLAR CRIME SENTENCES

---

[3]Coalition Against Insurance Fraud, "Consumer Information." Accessed May 10, 2010 at http://www.insurancefraud.org/fraud_backgrounder.htm.

Richard Scrushy, CEO of HealthSouth. Sentence: Almost 10 years
Richard Scrushy, the founder of HealthSouth, is no stranger to white-collar criminal allegations. After being acquitted of charges under the Sarbanes-Oxley Act for lack of evidence in 2005, Scrushy was indicted on new charges a mere four months later. The new charges were for bribery and mail fraud linked to former Alabama Governor Don Siegelman. The charges involved fraud through exchanging campaign funds for political favors.

Scrushy was ultimately found guilty by a federal jury in 2006 for bribery, mail fraud, and obstruction of justice. He was sentenced in 2007 to almost 10 years' imprisonment, in addition to having to pay a fine of $150,000 and an additional $267,000 in restitution to the United Way. Scrushy is currently in jail.

Walter Forbes, CEO of Cendant Corporation. Sentence: 17 years and 7 months
In 2004, Walter Forbes went on trial for fraudulently inflating the company's revenue by $500 million to increase its stock price. Forbes was charged with wire fraud, mail fraud, conspiracy, and securities fraud. In addition, Forbes was also accused of insider trading of $11 million in Cendant stock only weeks before the accounting scandal was discovered. The former vice president was also charged with similar crimes. The Cendant CFO testified against both the vice president and Forbes, saying that he was asked to be "creative" in reorganizing revenue.

Despite his persistent use of the "dumb CEO defense" (I did not know about the wrongdoing), Forbes was found guilty in his third trial, which lasted all of 17 days. In January 2007, Forbes was sentenced to 12 years and 7 months in federal prison. He was also required to pay $3.275 billion in restitution.

Kenneth Lay, CEO of Enron.
In 2004, Kenneth Lay went on trial, pleading not guilty to 11 felony counts, including wire fraud, bank fraud, securities fraud, and conspiracy, for his part in falsifying Enron's financial reports, and denying that he profited enormously from his fraudulent acts. The extent of the fraud was discovered when the energy company went bankrupt in late 2001. As a result of the accounting fraud, Enron's stock plummeted, leaving thousands of people with near-worthless stock, hitting retirement funds especially hard.

The Securities and Exchange Commission also filed a civil complaint against Lay, which could have led to more than $90 million in penalties and fines. Lay was accused of selling large amounts of stock at artificially high prices, resulting in an illegal profit of $90 million.

On May 25, 2006, Lay was found guilty of 10 of the 11 counts against him. Each count carried a maximum 5- to 10-year sentence, which would have amounted to 50 to 100 years maximum, with most commentators predicting a 20- to 30-year sentence. On July 5, 2006, however, Lay died of a heart attack before the scheduled date of his sentencing. Due to his death, the federal judge for the Fifth Circuit, pursuant to Fifth Circuit precedent, abated Lay's sentence. The abatement made it as if Lay had never been indicted.

Bernie Ebbers, CEO of WorldCom. Sentence: 25 years
In 2004, Bernie Ebbers, former CEO of the bankrupt phone company WorldCom, pleaded not guilty to three counts of fraud and conspiracy. The accounting fraud, which involved hiding expenses and inflating revenue reports, left $11 billion in debt at the time of the bankruptcy. The former CFO of WorldCom, Scott Sullivan, pleaded guilty to fraud and agreed to assist in the prosecution of Ebbers. Sullivan faced up to 25 years in prison for his role in the accounting scandal. In addition, MCI sued Ebbers to recover more than $400 million in loans that he took from WorldCom, now called MCI.

Bernie Ebbers, in one of the longest prison sentences given to a former CEO for white-collar crimes, was sentenced in 2005 to a 25-year prison term in a federal prison. Ebbers, 63 years old at the time of his sentencing, began serving his term in federal prison in 2006.

Dennis Kozlowski, CEO of Tyco International. Sentence: 8 years and 4 months to 25 years
In a second trial in early 2005 after a mistrial, Dennis Kozlowski faced charges of corruption and larceny for stealing more than $600 million from Tyco International and failing to pay more than $1 million in federal taxes. Kozlowski had Tyco pay for such over-the-top expenses as a $15,000 umbrella holder and a $2,200 garbage can. Kozlowski's sentence could have been up to 30 years in prison.

Kozlowski, as well as former Tyco CFO Mark Swartz, was sentenced to 8 years and 4 months to 25 years. Unlike other CEOs convicted for white-collar crimes, such as Bernie Ebbers, Kozlowski was convicted in state court. In addition to his prison sentence, to be served in a New York state prison, Kozlowski, with Swartz, was also ordered to pay $134 million to Tyco. In addition, Kozlowski was also fined an additional $70 million. Kozlowski is currently serving his term in prison.

EXHIBIT 6-1 CONTINUED

CRITICAL THINKING ABOUT THE LAW

Why should we be concerned about white-collar crime? You can use the following critical thinking questions to help guide your thinking about white-collar crime as you study this chapter.

1. As a future business manager, you may be forced to make tough decisions regarding white-collar crime. Imagine that you discover that one of your employees planned to offer a bribe to an agent from the Environmental Protection Agency to prevent your company from being fined. Although the result of the potential bribe could greatly benefit your company, you know that the bribe is illegal. What conflicting ethical norms are involved in your decision?

   Clue: Review the list of ethical norms offered in Chapter 1.

2. White-collar crime is typically not violent crime. Therefore, many people assume that street crime is more serious and should receive harsher punishment. Can you generate some reasons why that assumption is false? Why might white-collar crimes deserve more severe sentences?

   Clue: Reread the introductory paragraphs that provide information about white-collar crime. Why might a business manager deserve a more severe sentence than a young woman who commits a robbery? What are the consequences of both actions? Think about white-collar crime against this background as you study this chapter.

3. If a judge strongly valued justice, do you think he or she would give a lighter sentence to a business manager who embezzled $50,000 than to a person who robbed a bank of $50,000? Why?

   Clue: Think about the definitions of justice offered in Chapter 1.

---

become more publicized, people's attitudes toward corporations and white-collar crime are being affected.

The future manager must be prepared to respond to a growing lack of public confidence and avoid becoming a corporate criminal. He or she must find ways to develop a corporate climate that discourages, not encourages, the commission of white-collar crime. This chapter will help readers prepare to face the challenges posed by corporate crime. The first section defines crime and briefly explains criminal procedure. Next, the factors that distinguish corporate crime from street crime are discussed. The third section explains in detail some of the more common white-collar crimes. The fourth section introduces some ideas on how we can reduce the incidence of white-collar crime. The fifth and sixth sections discuss the federal and state responses to white-collar crime. The chapter closes with an overview of the international dimensions of white-collar crime.

Crime and Criminal Procedure

CRIME

Criminal law is designed to punish an offender for causing harm to the public health, safety, or morals. Criminal laws prohibit certain actions and specify the range of punishments for such conduct. The proscribed conduct generally includes a description of both a wrongful behavior (an act or failure to act where one has a duty to do so) and a wrongful intent or state of mind. The legal term for wrongful intent is *mens rea* (guilty mind). An extremely limited number of crimes do not require *mens rea*. These crimes are the "strict liability," or regulatory, crimes. They typically occur in heavily regulated industries and arise when a regulation has been violated. Regulatory crimes are created when the legislature decides that the need to protect the public outweighs the traditional requirement of *mens rea*. Because of the absence of the *mens rea* requirement for regulatory crimes, punishment for their violation is generally less severe than it is for wrongful behavior. In some states, punishment is limited to fines.

Crimes are generally classified as treason, felony, misdemeanor, or petty crime on the basis of the seriousness of the offense. *Treason* is engaging in war against the United States or giving aid or comfort to its enemies. **Felonies** include serious crimes such as murder or rape; felonies are punishable by death or imprisonment in a penitentiary. Defendants charged with a felony are entitled to a jury trial. **Misdemeanors**, which are considered less serious crimes, are punishable by a fine or by imprisonment of less than a year in a local jail. Examples of misdemeanors include assault (a threat to injure someone) and disorderly conduct. In most states, **petty crimes** are considered a subcategory of misdemeanors; they are usually punishable by a fine or incarceration for six months or less. A building code violation is an example of a petty crime. The statute defining the crime generally states whether it is a felony, misdemeanor, or petty crime. The more serious the offense, the greater the stigma that attaches to the criminal.

felony A serious crime that is punishable by death or imprisonment in a penitentiary.

misdemeanor A crime that is less serious than a felony and is punishable by fine or imprisonment in a local jail.

petty crime A minor crime punishable under federal statutes, by fine or incarceration of no more than six months.

CRIMINAL PROCEDURE

Criminal proceedings are initiated somewhat differently from civil proceedings. The procedures may vary slightly from state to state, but usually the case begins with an **arrest** of the defendant. The police must, in almost all cases, obtain an arrest warrant before arresting the defendant and taking him or her into custody. A magistrate (the lowest-ranking judicial official) will issue the arrest warrant when there is probable cause to believe that the suspect committed the crime. A *magistrate* is a public official who has the power to issue warrants; he or she is the lowest-ranking judicial official. **Probable cause** exists if it appears likely, from the available facts and circumstances, that the defendant committed the crime. An arrest may be made by a police officer without a warrant, but only if probable cause exists and there is no time to secure a warrant. An arrest without a warrant is most commonly made when police are called to the scene of a crime and catch the suspect committing the crime or fleeing from the scene.

arrest To seize and hold under the authority of the law.

probable cause The reasonable inference from the available facts and circumstances that the suspect committed the crime.

The Miranda Warnings. At the time of the arrest, the suspect must be informed of her or his legal rights. These rights are referred to as the **Miranda rights**, because they were developed in response to the Supreme Court's decision in *Miranda v. Arizona*.[4] If the defendant is not informed of these rights, any statements the defendant makes at the time of the arrest will be inadmissible at the defendant's trial. These rights are listed in Exhibit 6-2.

Miranda rights Certain legal rights—such as the right to remain silent to avoid self-incrimination and the right to an attorney—that a suspect must be immediately informed of upon arrest.

EXHIBIT 6-2

THE MIRANDA WARNINGS

Before any questioning by authorities, the following statements must be made to the defendant:

1. "You have the right to remain silent and refuse to answer any questions."
2. "Anything you say may be used against you in a court of law."
3. "You have the right to consult an attorney before speaking to the police and have an attorney present during any questioning now or in the future."
4. "If you cannot afford an attorney, one will be appointed for you before the questioning begins."
5. "If you do not have an attorney available, you have the right to remain silent until you have had an opportunity to consult with one."
6. "Now that I have advised you of your rights, are you willing to answer any questions without an attorney present?"

[4]384 U.S. 436 (1966).

Despite the courts' effort to create an "objective rule to give clear guidance to the police," many arrests and interrogations create significant questions about the application of the Miranda warnings. In 2004 alone, the Supreme Court issued three separate decisions clarifying the application and use of the Miranda warnings. In *United States v. Patane*,[5] the Court held that physical evidence found through statements made without receipt of the Miranda warnings were admissible in court so long as those statements were not forced by the police; the incriminating statements, however, would not be admissible.

In *Missouri v. Seibert*,[6] the Supreme Court found that a confession made after the Miranda warnings were given could not be admissible if the police first ask for the confession, then give the Miranda warnings and ask for the same confession. Delivering the opinion of the Court, Justice Souter wrote, "*Miranda* addressed interrogation practices... likely... to disable [an individual] from making a free and rational choice" about speaking, and held that a suspect must be "adequately and effectively" advised of the choice the Constitution guarantees. "The object of question-first is to render Miranda warnings ineffective by waiting for a particularly opportune time to give them, after the suspect has already confessed."

Finally, in *Yarborough v. Alvarado*,[7] the Court examined the ambiguity of when a person is "in custody" and, therefore, is entitled to the Miranda warnings. The "in custody" standard is whether a reasonable person would feel free to leave or end questioning. Such a standard, however, can be influenced by a person's age and education. Nevertheless, the Court held that maintaining a clear and objective standard for police is of utmost importance, and noted that considerations of age and education "could be viewed as creating a subjective inquiry." The Court found that the confession of guilt to police by Alvarado, age 17, during an interview was admissible even though he had not been read his Miranda warnings, because he was never "in custody."

Hundreds of cases have sought to clarify the Miranda warnings since they were first created in 1966 in *Miranda v. Arizona*.[8] The cases just discussed suggest the importance the judicial system places on informing suspects of their constitutional rights and privileges.

APPLYING THE FACTS OF THE CASE...

Adrienne was caught on camera robbing a gas station. Thus, the local police had probable cause and arrested her. Detective Joe walked up to Adrienne's house, cuffed her, told her she was being arrested and began to question her about her whereabouts on the night of the theft. What part of the arrest process is missing from this scenario? Why is it important?

Booking and First Appearance. After the defendant has been arrested, he or she is taken to the police station for *booking,* the filing of criminal charges against the defendant. The arresting officer then files a criminal complaint against the defendant. Shortly after the complaint is filed, the defendant makes his or her **first appearance** before a magistrate. At this time, the magistrate determines whether there was probable cause for the arrest. If there was not, the suspect is set free and the case is dismissed.

If the offense is a minor one, and the defendant pleads guilty, the magistrate may accept the guilty plea and sentence the defendant. Most defendants,

first appearance Appearance of the defendant before a magistrate, who determines whether there was probable cause for the arrest.

---
[5] 124 S. Ct. 2620 (2004).
[6] 124 S. Ct. 2601 (2004).
[7] 124 S. Ct. 2140 (2004).
[8] 384 U.S. 436 (1966).

however, maintain their innocence. The magistrate will make sure that the defendant has a lawyer; if the defendant is indigent, the court will appoint a lawyer for him or her. The magistrate also sets bail at this time. **Bail** is an amount of money that is paid to the court to ensure that the defendant will return for trial. In some cases, especially in white-collar crimes, if the magistrate believes that the defendant has such "ties to the community" that he or she will not try to flee the area to avoid prosecution, the defendant may be released without posting bail. In such cases, the defendant is said to be released "on his [or her] own recognizance."

bail An amount of money the defendant pays to the court upon release from custody as security that he or she will return for trial.

Information or Indictment. If the crime is a misdemeanor, the next step is the prosecutor's issuance of an **information**, a formal written accusation or charge. The information is usually issued only after the prosecutor has presented the facts to a magistrate who believes that the prosecution has sufficient grounds to bring the case.

information A formal written accusation in a misdemeanor case.

In felony cases, the process begins with the prosecutor (the prosecuting officer representing the United States or the state) presenting the facts surrounding the crime to a grand jury, a group of individuals under oath who determine whether to charge the defendant with a crime. The grand jury has the power to subpoena witnesses and require them to produce documents and tangible evidence. If the grand jury is convinced, by a preponderance of the evidence, that there is reason to believe the defendant may have committed the crime, an indictment (a formal, written accusation) is issued against the defendant. A grand jury does not make a finding of guilt; it simply decides whether there is enough evidence that the defendant committed the crime to justify bringing the defendant to trial. Government resources are limited, and the prosecution may not always believe it has sufficient evidence to prove a case beyond a reasonable doubt, so not every crime is prosecuted. Usually, the decision to seek an indictment depends on whether the prosecution believes it can get a conviction and whether the interests of justice would be served by prosecuting the crime.

At the federal level, almost all criminal prosecutions are initiated by the indictment process, and the decision on whether to prosecute is generally guided by the *Principles of Federal Prosecution*, published by the Justice Department in 1980. These principles state that the primary consideration is whether the existing admissible evidence is sufficient to obtain a conviction for a federal crime. Even if sufficient evidence does exist, the prosecutor's office might choose not to prosecute a crime if no substantial federal interest would be served by doing so, if the defendant could be efficiently prosecuted in another jurisdiction, or if an adequate noncriminal alternative to criminal prosecution exists. The factors influencing the substantiality of the federal interest are listed in Exhibit 6-3. The principles clearly recognize that other prosecutorial actions may offer fairer or more efficient ways to respond to the criminal conduct.

EXHIBIT 6-3

FACTORS FOR DETERMINING A SUBSTANTIAL FEDERAL INTEREST AND THE PRINCIPLES FOR FEDERAL PROSECUTION

1. Federal law enforcement priorities established by the Department of Justice
2. Deterrent effect
3. The subject's culpability
4. The subject's willingness to cooperate
5. The subject's personal circumstances
6. The probable sentence
7. The possibility of prosecution in another jurisdiction
8. Noncriminal alternatives to prosecution

Some alternatives might be to institute civil proceedings against the defendant or to refer the complaint to a licensing board or the professional organization to which the defendant belongs.

Another alternative to indictment is pretrial diversion (PTD). Pretrial diversion attempts to keep certain criminal offenders out of the traditional criminal justice system by channeling them into a program of supervision and services. A PTD participant signs an agreement with the government acknowledging responsibility for the act at issue but not admitting guilt. The participant agrees to be supervised by the U.S. Probation Office and comply with the terms established for the agreed-upon period of the agreement, up to 18 months. Terms, which vary according to the circumstances and the criminal activity, might include participating in community programs or paying restitution. If the participant complies with the agreement, the matter is closed. If not, he or she is then prosecuted.

After the indictment comes the **arraignment**, a time when the defendant appears in court and enters a plea of guilty or not guilty. A not-guilty plea entitles the defendant to a trial before a petit jury. If the defendant declines a jury trial, the case is heard by a judge alone, in a procedure called a *bench trial*.

A defendant may also enter a plea of **nolo contendere**. By making this plea, the defendant does not admit guilt but agrees not to contest the charges. The advantage of a nolo contendere plea over a plea of guilty is that the former cannot be used against the defendant in a civil suit.

arraignment Formal appearance of the defendant in court to answer the indictment by entering a plea of guilty or not guilty.

nolo contendere A plea of no contest that subjects the defendant to punishment but is not an admission of guilt.

plea bargaining The negotiation of an agreement between the defendant's attorney and the prosecutor, whereby the defendant pleads guilty to a certain charge or charges in exchange for the prosecution reducing the charges.

Plea Bargaining. At any time during the proceedings, the parties may engage in **plea bargaining**, which is a process of negotiation between the defense attorney and the public prosecutor or district attorney. The result of this process is that the defendant pleads guilty to a lesser offense, in exchange for which the prosecutor drops or reduces some of the initial charges. Plea bargaining benefits the criminal by eliminating the risk of a greater penalty. It benefits the prosecutor by giving her or him a sure conviction and reducing a typically overwhelming caseload. It saves both parties the time and expense of a trial.

Plea bargaining is used extensively for white-collar crimes, generally at a much earlier stage than for street crimes. In white-collar cases, plea bargaining often occurs even before the indictment. This process, as well as other modifications of criminal procedures in white-collar crime cases, helps to make the white-collar criminal seem less of a criminal, and thus reduces the likelihood of severe punishment.

Burden of Proof. If the case goes to trial, the burden of proof is usually on the prosecutor. The burden of proof has two aspects: the burden of production of evidence and the burden of persuasion. The prosecution bears the burden of production of evidence of all the elements of the crime. Thus, the prosecution must present physical evidence and testimony that prove all elements of the crime. The burden of producing evidence of any affirmative defenses (defenses in which the defendant admits to doing the act but claims some reason for not being held responsible, such as insanity, self-defense, intoxication, or coercion) lies with the defendant.

The prosecution also bears the burden of persuasion, meaning that the prosecutor must convince the jury beyond a reasonable doubt that the defendant committed the crime. In some states, a defendant who presents an affirmative defense must persuade the jury of the existence and appropriateness of the defense by a preponderance of the evidence, meaning that the defendant's lawyer must prove that it is more likely than not that the defense exists and is valid. In other states, the burden of persuasion lies with the prosecutor to show beyond a reasonable doubt that the defense does not exist or is invalid.

The actual trial itself is similar to a civil trial, and the role of the prosecutor or district attorney is similar to that of the plaintiff's attorney. One major difference, however, is that the defendant in a criminal case cannot be compelled to testify, and the finder of fact is not to hold the exercise of this right against the defendant. This right to not testify is guaranteed by the constitutional provision in the Fifth Amendment that no person "shall be compelled in any criminal case to be a witness against himself."

Defenses. Obviously, one of the most common defenses is that the defendant did not do the act in question. But even if the defendant did commit the act, a number of affirmative defenses might be raised to preclude the defendant from being convicted of the crime. Affirmative defenses may be thought of as excuses for otherwise unlawful conduct. Four of the most common are entrapment, insanity, duress, and mistake.

Entrapment occurs when the idea for the crime was not the defendant's but was, instead, put into the defendant's mind by a police officer or other government official. An extreme example is a case in which a government official first suggests to an employee that the employee could make good money by altering certain corporate records. The official then shows up at the employee's home at night with a key to the office where the records are kept and reminds the employee that the record keeper is on vacation that week. The official also reminds the employee that most of the other workers rarely stay late on Friday nights, so Friday after work might be an ideal time to get the books. If prosecuted for fraud, the employee could raise the defense of entrapment.

Entrapment is not always easy to prove. Police are allowed to set up legitimate "sting" operations to catch persons engaged in criminal activity. The key to a legitimate sting is that the defendant was "predisposed" to commit the crime; the officer did not put the idea in the defendant's head. If an officer dresses up like a prostitute and parades around in an area where prostitution is rampant, a potential customer who solicits sex could not raise the charge of entrapment against a charge of soliciting a prostitute. Most cases, however, fall between our two examples, so it is often difficult to predict whether the entrapment defense will be successful.

Insanity is one of the best-known criminal defenses, although it is not used nearly as frequently as its notoriety might imply. The **insanity defense** is used when a person's mental condition prevents him or her from understanding the wrongful nature of the act he or she committed or from distinguishing wrong from right.

Duress occurs when a person is forced to commit a wrongful act by a threat of immediate bodily harm or loss of life, and the affirmative **duress defense** can be used by a person who believes that he or she was forced to commit a crime. For example, if Sam holds a gun to Jim's head and tells him to forge his employer's signature on a company check or he will be shot, Jim can raise the defense of duress to a charge of forgery. This defense is generally not available to a charge of murder.

A **mistake-of-fact defense** may sometimes be raised when that mistake vitiates the criminal intent. For example, if Mary takes Karen's umbrella from a public umbrella rack, thinking it is her own, she can raise mistake as a defense to a charge of theft.

A mistake of law, however, is generally not a defense. A person could not, for example, fail to include payment received for a small job on his or her income tax return because of a mistaken belief that income for part-time work of less than $100 did not have to be reported.

Any defendant who does not prevail at the trial court can appeal the decision, just as in a civil case. The steps of a criminal action are set out in Exhibit 6-4.

---

entrapment An affirmative defense claiming that the idea for the crime did not originate with the defendant but was put into the defendant's mind by a police officer or other government official.

insanity defense An affirmative defense claiming that the defendant's mental condition precluded him or her from understanding the wrongful nature of the act committed or from distinguishing wrong from right in general.

duress defense An affirmative defense claiming that the defendant was forced to commit the wrongful act by threat of immediate bodily harm or loss of life.

mistake-of-fact defense An affirmative defense claiming that a mistake made by the defendant vitiates criminal intent.

EXHIBIT 6-4

STEPS OF A CRIMINAL PROSECUTION

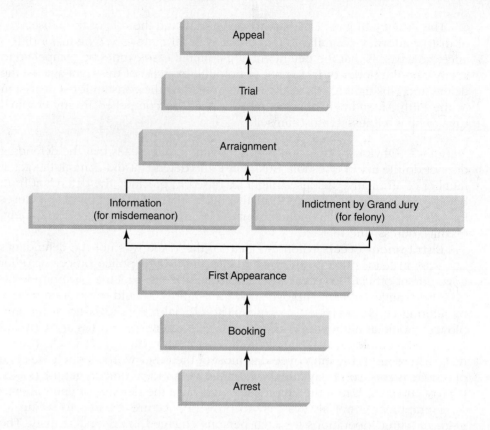

Distinguishing Features of White-Collar Crime

An initial problem with any discussion of white-collar crime is its definition. The term **white-collar crime** does not have a precise meaning. The term was first made popular in 1939 by sociologist Edwin Sutherland, who defined *white-collar crime* as "crime committed by a person of respectability and high social status in the course of his occupation." Traditionally, it has been the classification for those crimes committed in a commercial context by members of the professional and managerial classes. It includes such diverse acts as bribery of corporate or government officials and violations of federal regulations such as the Occupational Safety and Health Act and the Internal Revenue Service Code. In this book, we will use the traditional definition. For illustrations of some white-collar criminals and the crimes of which they were convicted, see Exhibit 6-1.

white-collar crime A crime committed in a commercial context by a member of the professional–managerial class.

THE CORPORATION AS A CRIMINAL

One of the distinguishing features of white-collar crime is that sometimes the "criminal" may be difficult to identify. In a street crime, the identity of the criminal is fairly clear: It is the person who committed the act. If a person hires another to commit a crime, the person doing the hiring is likewise guilty of a crime. In the case of white-collar crime, the crime is often committed on behalf of a corporation, which is as an artificial legal entity or an artificial person. An important question, then, is whether liability can be imposed on the corporation for the criminal acts committed by employees of the corporation on behalf of the corporation.

Initially, the courts said no. A corporation had no mind, so it could not have the mental state necessary to commit a crime. This rule was first eroded by the imposition of liability on corporations for so-called **strict liability offenses**,

strict liability offense An offense for which no state of mind or intent is required.

those for which no state of mind is required. These generally are cases in which corporate employees failed to take some action required by a regulation. For example, under most blue-sky laws (state securities regulations), it is a violation to file a false statement of a company's financial condition with a state's secretary of state. Filing a false statement is a crime, even if the corporate officer filing the statement believed it was true, as no state of mind is required to commit the crime.

The courts then began to impose liability on corporations for criminal acts of the employees by imputing the state of mind of the employee to the corporation. Today, as a general rule, the only crimes for which a corporation is not held liable are those that are punishable only by incarceration. Obviously, the rationale for this rule is that the punishment could not be carried out. Some states have eliminated this problem by passing a statute providing specific fines for corporations that commit offenses otherwise punishable by incarceration only.

Many corporate executives may not realize the extent to which a corporation today can be held liable for the acts of its employees. That liability can extend down to acts of even the lowest-level employees and even to acts in violation of corporate directives, as long as two conditions are met. First, the conduct must be within the scope of the employee agent's authority. Second, the action must have been undertaken, at least in part, to benefit the corporation. Some examples of employee actions for which corporations have been held liable include employees physically harming a customer for not paying his bill,[9] an employee sexually harassing another employee,[10] and a car salesman's obtaining automobile loans for the dealership's customers by misrepresenting financial data to the lending bank.[11] In a move indicating that the Department of Justice (DOJ) is serious about pursuing white-collar corporate criminals, in 2006, Deputy U.S. Attorney General Paul McNulty issued a memorandum stating that prosecuting corporate crimes was a priority of the DOJ. In addition, McNulty advised federal prosecutors not to stop with finding corporate liability, but also to look for individuals within the firm who should be punished for the firm's wrongdoings.

Many states have passed statutes imposing criminal liability on partnerships under the same circumstances as those under which liability is imposed on a corporation. In the absence of such a statute, liability is not imposed on the partnership because a partnership is not a legal person.

ARGUMENTS IN SUPPORT OF CORPORATE LIABILITY

Needless to say, there is no consensus about whether, as a matter of policy, corporations should be held criminally liable. Some of the arguments in favor of such liability include:

1. Imposing financial sanctions against the corporation will result in lower dividends for the shareholders, which, in turn, will prompt the shareholders to take a more active role in trying to make sure that the corporation behaves legally. They will carefully select directors who will scrupulously monitor corporate behavior and will express concern when anything appears to be unethical or illegal.

2. In situations in which the crime is an omission and the responsibility for performing the omitted duty is not clearly delegated to any specific party,

---

[9]*Crane Brothers, Inc. v. May,* 252 Ga. App. 690, 556 S.E.2d 865 (2001).
[10]*Pennsylvania State Police v. Suders,* 124 S. Ct. 2342 (2004).
[11]*Commonwealth v. Duddie Ford, Inc.,* 28 Mass. App. Ct. 426, 551 N.E.2d 1211 (1990).

the duty rests with no particular individual. Therefore, there is no one to blame, and the corporation cannot be held responsible. If no one, not even the corporation, is held liable, there is no incentive to obey the laws.

3. Closely related to reason 2 is the fact that there are a tremendous number of suspects in a corporation. Most enforcement agencies do not have the resources to investigate the large number of employees involved and to build cases against each. It is much easier and less expensive for the government to investigate and bring a case against the corporation as an entity.

4. The fact that many decisions are committee decisions—or else decisions made by an initial person or group and then approved by several tiers of management—again makes it hard to point the finger at one individual. Sometimes one individual is responsible for making an initial decision, and then someone else is responsible for implementing that decision.

5. A further reason is that corporate personnel are expendable. To lose a manager because of a conviction does not really harm the enterprise that profited from the wrongdoing. In fact, it allows the firm to externalize the costs of the criminal behavior; that is, to absolve itself of guilt in the eyes of the public. The manager takes the blame, and the corporation, which profited from the manager's illegal act, continues to thrive.

6. Even if one could impose liability on one or two individuals, it is unfair to single them out for punishment when the behavior in question probably resulted from a pattern of behavior common to the entire corporation.

7. Some people assume that the beneficiaries of crime committed on behalf of the corporation are the shareholders, because they may get higher dividends if corporate crime keeps costs lower. To fail to impose sanctions on the corporation would be to allow the shareholders to benefit from the illegal activity.

8. If an action is taken against the corporation, the criminal act will be linked to the corporation in the public's mind. In a market-oriented society such as ours, disclosure of full information about businesses is essential for consumers to make informed decisions about the types of firms with which they want to transact business.

ARGUMENTS IN OPPOSITION TO CORPORATE LIABILITY

The following are arguments against imposing liability on corporations:

1. Imposing fines on corporations is a waste of time and effort because the fines are never going to be severe enough to act as a deterrent. Even if more substantial fines were imposed, the firms would simply pass on the losses to consumers in the form of increased product prices. Thus, it would really be the consumers of the corporation's products who would be punished.

2. Some people believe that the shareholders' dividends may be reduced if fines are imposed on the corporation because the cost of the fines will reduce the profits available for dividend payments. Reducing dividends, it is argued, is unfair because in most corporations, the shareholders really do not have any power to control corporate behavior.

3. Because criminal prosecutions of corporations are not well publicized, they do not harm the corporation's public image. The corporations have enough money and public relations personnel to easily overcome any negative publicity with a well-run advertising campaign designed to polish their public

CHAPTER 6 • White-Collar Crime and the Business Community

image. For example, the prosecution of Revco of Ohio for defrauding Medicaid of hundreds of thousands of dollars resulted in only a temporary decline in the value of Revco's stock.[12]

IMPOSITION OF LIABILITY ON CORPORATE EXECUTIVES

Another potential candidate for liability in the case of white-collar crime is the corporate executive. Top-level corporate executives, as a group, have tremendous power through their control over national corporations. When these corporations earn record-setting profits, top-level executives rush forward to take credit for their corporations' successes. These same executives, however, do not rush forward to take responsibility for their corporations' criminal violations.

In fact, very few corporate officials are held liable by law enforcement for the actions of their companies.[13] This lack of liability is believed to occur because of the delegation of responsibility to lower tiers of management and reliance on unwritten orders, which frequently allow top-level management to protect itself from liability for the results of its policy decisions.

Traditionally, imposing liability on executives has been difficult because the criminal law usually requires an unlawful act to be accompanied by an unlawful intent (*mens rea*). In cases of violations of federal regulations, corporate executives often argue that they are not the ones directly responsible for filing the documents or conducting the studies. They certainly never explicitly ordered that such regulations be disregarded. In response to recent high-profile corporate scandals, Congress and the Securities and Exchange Commission have created more stringent certification requirements for CEOs, CFOs, and other corporate officials. For example, the Sarbanes-Oxley Act of 2002 holds high-ranking corporate officials responsible for the validity and accuracy of their companies' financial statements. Failure to comply with the financial statement certification or certification of false information is a corporate fraud under the act, punishable with fines that range from $1 million to $5 million and prison sentences from 10 to 20 years. This act is discussed in greater detail in Chapter 23.

As the following case shows, however, the courts are recognizing that corporate executives who have the power and authority to secure compliance with the law have an affirmative duty to do so. Failure to uphold that duty can lead to criminal sanctions.

CASE 6-1

United States v. Park
United States Supreme Court
421 U.S. 658 (1975)

Defendant Park, the president of a national food-chain corporation, was charged, along with the corporation, with violating the Federal Food, Drug, and Cosmetic Act by allowing food in the warehouse to be exposed to rodent contamination. Park had conceded that he was responsible for the sanitary conditions as part of his responsibility for the "entire operation," but claimed that he had turned the responsibility for sanitation over to dependable subordinates. He admitted at the trial that he had received a warning letter from the Food and Drug Administration regarding the unsanitary conditions at one of the company's warehouses.

---

[12]D. Vaughan, *Controlling Unlawful Organizational Behavior: Social Structure and Corporate Misconduct* (University of Chicago Press, 1983).

[13]Timothy P. Glynn, "Beyond 'Unlimiting' Shareholder Liability: Vicarious Tort Liability for Corporate Officers," 57 *Vand. L. Rev.* 329 (2004).

The trial court found the defendant guilty. The court of appeals reversed. The case was appealed to the U.S. Supreme Court.

Chief Justice Burger

The question presented was whether "the manager of a corporation, as well as the corporation itself, may be prosecuted under the Federal Food, Drug, and Cosmetic Act of 1938 for the introduction of misbranded and adulterated articles into interstate commerce." In *Dotterweich*, a jury had disagreed as to the corporation, a jobber purchasing drugs from manufacturers and shipping them in interstate commerce under its own label, but had convicted Dotterweich, the corporation's president and general manager.

In reversing the judgment of the Court of Appeals and reinstating Dotterweich's conviction, this Court looked to the purposes of the Act and noted that they "touch phases of the lives and health of people which, in the circumstances of modern industrialism, are largely beyond self-protection." It observed that the Act is of "a now familiar type" which "dispenses with the conventional requirement for criminal conduct—awareness of some wrongdoing. In the interest of the larger good it puts the burden of acting at hazard upon a person otherwise innocent but standing in responsible relation to a public danger."

Central to the Court's conclusion that individuals other than proprietors are subject to the criminal provisions of the Act was the reality that "the only way in which a corporation can act is through the individuals who act on its behalf." The Court also noted that corporate officers had been subject to criminal liability under the Federal Food and Drugs Act of 1906, and it observed that a contrary result under the 1938 legislation would be incompatible with the expressed intent of Congress to "enlarge and stiffen the penal net" and to discourage a view of the Act's criminal penalties as a "license fee for the conduct of an illegitimate business."

At the same time, however, the Court was aware of the concern which was the motivating factor in the Court of Appeals' decision, that literal enforcement "might operate too harshly by sweeping within its condemnation any person however remotely entangled in the proscribed shipment." A limiting principle, in the form of "settled doctrines of criminal law" defining those who "are responsible for the commission of a misdemeanor," was available. In this context, the Court concluded, those doctrines dictated that the offense was committed "by all who . . . have . . . a responsible share in the furtherance of the transaction which the statute outlaws."

The rule that corporate employees who have "a responsible share in the furtherance of the transaction which the statute outlaws" are subject to the criminal provisions of the Act was not formulated in a vacuum. Cases under the Federal Food and Drugs Act of 1906 reflected the view both that knowledge or intent were not required to be proved in prosecutions under its criminal provisions, and that responsible corporate agents could be subjected to the liability thereby imposed. Moreover, the principle had been recognized that a corporate agent, through whose act, default, or omission the corporation committed a crime, was himself guilty individually of that crime.

The rationale of the interpretation given the Act in *Dotterweich*, as holding criminally accountable the persons whose failure to exercise the authority and supervisory responsibility reposed in them by the business organization, resulted in the violation complained of, has been confirmed in our subsequent cases. Thus, the Court has reaffirmed the proposition that "the public interest in the purity of its food is so great as to warrant the imposition of the highest standard of care on distributors." In order to make "distributors of food the strictest censors of their merchandise," the Act punishes "neglect where the law requires care, and inaction where it imposes a duty." "The accused, if he does not will the violation, usually is in a position to prevent it with no more care than society might reasonably expect and no more exertion than it might reasonably extract from one who assumed his responsibilities."

Thus, *Dotterweich* and the cases which have followed reveal that in providing sanctions which reach and touch the individuals who execute the corporate mission—and this is by no means necessarily confined to a single corporate agent or employee—the Act imposes not only a positive duty to seek out and remedy violations when they occur but also, and primarily, a duty to implement measures that will ensure that violations will not occur. The requirements of foresight and vigilance imposed on responsible corporate agents are beyond question demanding, and perhaps onerous, but they are not more stringent than the public has a right to expect of those who voluntarily assume positions of authority in business enterprises whose services and products affect the health and well-being of the public that supports them.

The Act does not, as we observed in *Dotterweich*, make criminal liability turn on "awareness of some wrongdoing" or "conscious fraud." The duty imposed by Congress on responsible corporate agents is, we emphasize, one that requires the highest standard of foresight and vigilance, but the Act, in its criminal aspect, does not require that which is objectively impossible.

> *[I]t is equally clear that the Government established a prima facie case when it introduced evidence sufficient to warrant a finding by the trier of the facts that the defendant had, by reason of his position in the corporation, responsibility and authority either to prevent in the first instance, or promptly to correct, the violation complained of, and that he failed to do so. The failure thus to fulfill the duty imposed by the interaction of the corporate agent's authority and that statute furnishes a sufficient causal link. The considerations which prompted the imposition of this duty, and the scope of the duty, provide the measure of culpability.**

Reversed in favor of the Government.

---

*United States v. Park, United States Supreme Court 421 U.S. 658 (1975).

CRITICAL THINKING ABOUT THE LAW

The Court in Case 6-1 was guided by a fundamental principle that it did not explicitly state: To the extent that one has authority, he or she also has responsibility and can be liable for criminal action. This guiding principle played a significant role in the Court's justification (i.e., reasoning) for its decision.

Context was very important in the formulation of this principle, as well as in its application to Case 6-1. Key facts, primary ethical norms, and judicial precedent were important elements of this context. Consequently, the questions that follow focus on those aspects of Case 6-1.

1. What key fact was very important to the Court in its determination of Park's guilt?

    Clue: Think again about the Court's guiding principle in this case. You want to identify the key fact that allowed the Court to apply the principle to this particular case.

2. Precedent plays a crucial role in the Court's reasoning and, thus, in its decision. What key precedent in criminal law did the *Dotterweich* decision dispense with, thereby clearing the way for the guiding principle discussed previously to take on greater significance and make conviction in the present case possible?

    Clue: Reread the section in which Justice Burger discusses the *Dotterweich* decision.

Although the *Park* case demonstrates that corporate executives may be found guilty of committing a corporate crime, few are charged and even fewer are actually convicted. Even when they are convicted, harsh penalties are not likely to be imposed. Furthermore, even when executives do go to prison, because they are not seen as security risks they are often sent to prisons that some would argue are nicer than the motels many people can afford to stay in on vacations! See Exhibit 6-5 for a glimpse of some of these prisons where white-collar criminals are likely to end up.

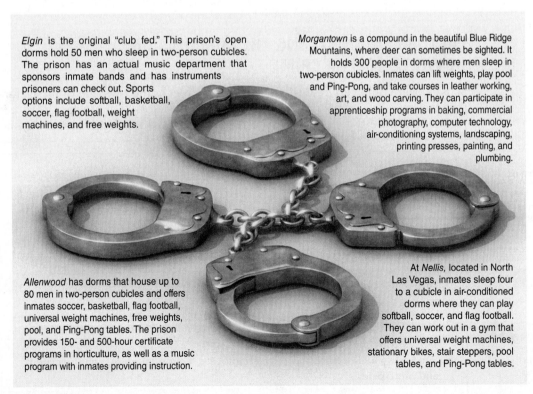

Elgin is the original "club fed." This prison's open dorms hold 50 men who sleep in two-person cubicles. The prison has an actual music department that sponsors inmate bands and has instruments prisoners can check out. Sports options include softball, basketball, soccer, flag football, weight machines, and free weights.

Morgantown is a compound in the beautiful Blue Ridge Mountains, where deer can sometimes be sighted. It holds 300 people in dorms where men sleep in two-person cubicles. Inmates can lift weights, play pool and Ping-Pong, and take courses in leather working, art, and wood carving. They can participate in apprenticeship programs in baking, commercial photography, computer technology, air-conditioning systems, landscaping, printing presses, painting, and plumbing.

Allenwood has dorms that house up to 80 men in two-person cubicles and offers inmates soccer, basketball, flag football, universal weight machines, free weights, pool, and Ping-Pong tables. The prison provides 150- and 500-hour certificate programs in horticulture, as well as a music program with inmates providing instruction.

At *Nellis*, located in North Las Vegas, inmates sleep four to a cubicle in air-conditioned dorms where they can play softball, soccer, and flag football. They can work out in a gym that offers universal weight machines, stationary bikes, stair steppers, pool tables, and Ping-Pong tables.

EXHIBIT 6-5

WHITE-COLLAR PRISON CAMPS

A primary reason for the limited number of convictions is the diffusion of responsibility. It is often difficult to establish who was responsible for the criminal act. Another problem related to corporate structure is that everyone usually has a specific job, and putting all of the pieces of the crime together is difficult. The reader should remember that the burden of proof is on the prosecution. Corporate executives also tend to have high-caliber counsel who specialize in defending white-collar criminals. These skilled attorneys, often paid for by the corporation or (when permitted by law) by an insurance carrier, usually get involved during the initial investigation of the case and perceive one of their key functions as keeping evidence out of the hands of the prosecutor. They often regard themselves as having lost the case if they do not prevent their client from being indicted or if they do not at least get the charge reduced to the lowest possible misdemeanor.[14] Another alleged reason for the limited number of convictions of corporate executives is that they are generally persons with a great deal of knowledge, including knowledge of inefficiencies and improprieties on the part of the government officials who are regulating them. It is argued that regulators are unlikely to press for prosecution when their own ineptness may be revealed in the process.

IMPOSITION OF LIABILITY ON LOWER-LEVEL CORPORATE CRIMINALS

Although much debate has been generated over the extent to which the corporation and its top executives should be held liable for crimes committed on behalf of the corporation, it is important to remember that lower-level and mid-level corporate employees can also be held liable for their individual criminal actions, and imposition of liability on the corporation does not in any way preclude imposition of liability on the individual actor as well. It is likewise no excuse on the part of employees that they were committing the wrongful act only because their employer instructed them to do so.

FACTORS ENCOURAGING THE COMMISSION OF WHITE-COLLAR CRIME

White-collar crime can be distinguished from street crime by some of the factors that facilitate its commission (Exhibit 6-6). Recognition of these factors is not meant to excuse this behavior; instead, it may be used to help devise ways to

EXHIBIT 6-6

FACTORS FACILITATING THE COMMISSION OF WHITE-COLLAR CRIMES

1. Societal stress on material success, without equal emphasis on means of achieving success.
2. Linkage of corporate rewards of salary and promotion to accomplishing short-term goals.
3. Groupthink.
4. Ease of rationalizing illegal behavior.
5. Dispersion of decision making.
6. Retention of status by persons convicted of white-collar crime.
7. The lack of an adversarial relationship between the corporation and government regulators.
8. Poor personnel policies that leave employees feeling insecure, unappreciated, and underpaid.

---

[14] K. Mann, *Defending White-Collar Crime* 10 (Yale University Press, 1985).

control corporate crime. As an informed corporate manager, your knowledge of these factors may help you to avoid the temptation to engage in criminal activities and to discourage others from doing so.

Initially, we must recognize that many people in our society value material success above all else. When the focus of our energies is on material success, we are much more willing to engage in illegal means to achieve our goal than we would be if our focus were on, for instance, ethical conduct. With the stress on success, the line between illegality and a shrewd business deal becomes blurred. The culture of some corporations creates an atmosphere in which corporate crime may thrive. For instance, if rewards such as salary and promotion are tied to meeting short-term goals, employees may use whatever means are available to help them achieve those goals.

Once illegal behavior is initiated, it tends to become institutionalized because of a phenomenon referred to by social psychologist Irving Janis as *groupthink*.[15] In groupthink, there is an implicit agreement not to bring up upsetting facts. In the corporation, where junior managers' success depends to some degree on the approval of senior managers, a junior manager would be extremely reluctant to criticize a senior manager's actions. One dramatic instance of this dynamic is the E. F. Hutton case, in which the practice of writing illegal checks spread throughout the company. Nobody wanted to bring up the upsetting fact that perhaps the practice was illegal. Instead, managers just went along.[16]

Another factor making white-collar crime easy to commit is the fact that decision making is often distributed among various individuals. Because responsibility is diffuse, individuals may feel only very limited personal responsibility for the results of their actions. This spreading of responsibility also results in an awareness that the likelihood of getting caught is small. Another factor related to the complex organizational structure of a corporation is that once a decision has been made, many people implement it. Thus, even if a manager has second thoughts about a decision, it is often too late to stop the process. Also, once a decision has been made, and when it is implemented by others, the decision maker feels limited responsibility.

The businesspersons who are unlucky enough to get caught do not automatically lose their status among their peers. Some, in fact, may be admired. Violations of the law are not necessarily violations of businesspersons' ethical codes. It is important to note that, unlike street criminals, who usually recognize that they are committing crimes, white-collar criminals are frequently regarded by themselves and their peers as respectable, law-abiding citizens.

In some instances, corporate crime is facilitated because the supposed adversarial relationship between the corporation and the government agency "watchdog" does not exist. Top corporate executives and high-level government officials may share similar values and lifestyles. Business managers often make career moves directly to a government agency regulating that business and then back to business. These factors may make some government officials reluctant to crack down on businesspersons, and businesspersons' awareness of this reluctance contributes to an environment in which white-collar crime is tolerated.

Finally, a corporation's personnel and operating procedures often encourage its employees to commit white-collar crimes. Although many of the previously discussed factors are not easily amenable to change, personnel and operating policies are under the direct control of management and, therefore, business managers concerned about white-collar crime can have an impact on the likelihood of its occurrence by carefully examining their corporate policies. For example, employees are much more likely to commit white-collar crimes when corporate policies lead to a lack of job security, inadequate pay, a lack

---

[15] D. Goleman, "Following the Leader," 85 *Science* 18 (Oct. 1985).
[16] *Id*.

of recognition for outstanding work, perceived inequities in employee salaries, poor promotion opportunities, inadequate expense accounts or unreasonable budget expectations, poor training, or poor communication practices.

SENTENCING OF WHITE-COLLAR CRIMINALS

Another feature that some would say distinguishes white-collar crime is the attitudes of the judges who hand down the sentences for these crimes. A public perception that judges were not giving long enough sentences to white-collar criminals and were not imposing large enough fines led to the adoption of the 1991 Sentencing Guidelines for use by federal judges. These guidelines are said to provide "just punishment, adequate deterrence, and incentives for organizations to maintain internal mechanisms for preventing, detecting, and reporting criminal conduct in all aspects of their activity."[17] Until early 2005, these guidelines were in fact mandatory restrictions on the judge's sentencing authority. Today, these guidelines are just what their name implies: guidelines.

Under these sentencing guidelines, a fine is a product of a "base fine" and a "culpability score." The base fine is the greatest of the company's gain, the victim's loss, or a dollar amount corresponding to an offense level. The culpability score provides a multiplier that is applied to the base fine. The culpability score is determined by looking at a chart of potential mitigating and aggravating factors. An example of an aggravating factor might be that high levels of management were aware of the criminal activity but did nothing to stop it. Mitigating factors would be having a meaningful compliance program in effect at the time of the offense and upper management's taking steps to remedy the harm, discipline the offender, and prevent a recurrence. Because of the difference that these aggravating and mitigating factors can have on the amount of the fine—a crime with a base fine of $5 million, for example, could be as low as $2 million or as high as $20 million, depending on the culpability score—supporters hope that the guidelines will not only result in fairer penalties but actually have a major impact on the way firms operate.

When the sentencing guidelines were established, there was a concern about the lack of prison time served by most white-collar criminals and the general judicial leniency toward white-collar crimes. Consequently, some confinement was mandated for almost all white-collar offenses. Sentences are determined in a way similar to fines, with a base sentence and culpability factor. Judges, however, were given discretion, under extraordinary circumstances, to modify the sentence and depart from the guidelines, and may impose alternative penalties (as described later in this section). Some of the factors that allow departure include "substantial cooperation" of the defendant; extraordinary effects of a prison sentence on third parties, including the defendant's family; an overstatement of loss from the crime; and diminished capacity.

Some critics of the guidelines argue that although prison sentences are required for a broad range of white-collar crimes, the sentences for these crimes are still much less than for street crimes. For example, the base sentence for an antitrust violation is as low as two to eight months.[18] During 2001, the average sentence in the 8,328 cases of fraud was 22.9 months.

The power of the sentencing guidelines began to crumble as a result of the 2004 U.S. Supreme Court decision in *Blakely v. Washington*,[19] in which the high court ruled that under the Sixth Amendment, juries, not judges, should

---

[17]C. C. Dow and R. J. Muehl, "Are Policies Keyed to New Sentencing Guidelines?" 35 *Securities Management* 98 (Nov. 1992).

[18]G. N. Racz, "Exploring Collateral Consequences: *Koon v. United States,* Third Party Departures from Federal Sentencing Guidelines," 72 *N.Y. Univ. L. Rev.* 1462 (1997).

[19]542 U.S. 296, 125 S. Ct. 2531 (2004).

determine the facts that increase sentences beyond guideline maximums. In that case, the facts admitted into evidence supported a maximum sentence of 53 months. The judge, however, imposed a sentence of 90 months after finding that the defendant had acted with deliberate cruelty, a statutorily enumerated ground for departing from the standard range. The Washington State Supreme Court had rejected the defendant's argument that the judge's acting on facts not admitted into evidence denied him his constitutional right to have a jury determine, beyond a reasonable doubt, all facts legally essential to his sentence. On appeal, the U.S. Supreme Court agreed with the defendant and overturned the state court's decision.

Although this case involved a state sentencing law, the decision immediately called into question the validity of the Federal Sentencing Guidelines. Within a month of the decision in *Blakely*, two circuit courts of appeal had ruled that the guidelines were unconstitutional, with the judge in one case stating that, "In order to comply with *Blakely* and the Sixth Amendment, the mandatory system of fixed rules calibrating sentences automatically to facts found by judges must be displaced by an indeterminate system in which the Federal Sentencing Guidelines in fact become 'guidelines' in the dictionary-definition sense."[20] He further said that instead of judges viewing the guidelines as mandatory, they should "view the guidelines in general as recommendations to be considered and then applied only if the judge believes they are appropriate and in the interests of justice in the particular case."[21]

A similar holding applied to the federal courts in the 2005 consolidated case of *United States v. Booker*[22] and *United States v. Fanfan*.[23] The key question in these cases was whether judges can increase criminal sentences based on arguments from prosecutors that juries never considered. The high court found that the mandatory guidelines are unconstitutional. The mandatory sentencing scheme of the guidelines when coupled with its reliance on judicial factfinding is incompatible with the Sixth Amendment. Although district courts still calculate sentencing ranges under the guidelines, these ranges are now considered merely advisory. Subsequent congressional action on the matter is anticipated.

Even before these rulings, judges had departed from the guidelines under extraordinary circumstances and, sometimes, also used alternative sentencing instead of imposing prison sentences or fines. One alternative to prison that judges favor for white-collar offenders is community service. Offenders have been assigned to perform such services as giving speeches about their wrongful acts to business and civic groups and working among the poor in drug rehabilitation clinics. A corporate criminal who dumped industrial waste into San Diego's sewers, claiming it was domestic sewage, received a $250 fine and three years' probation, with an unusual twist: For violating waste disposal laws, he was required to complete a hazardous materials course; perform 120 hours of volunteer work for the Veterans of Foreign Wars; and spend 40 hours at the city's pump station where waste is discharged by trucks into the city's sewage treatment system. The man was also ordered to inform the waste haulers who came into the pump station that he had falsified a report describing the type of waste he was discharging.[24]

Another popular alternative to prison is occupational disqualification. The white-collar criminal is prohibited for a specific period from engaging in an

---

[20] *United States v. Montgomery*, No. 03-5256, 2004 U.S. App. LEXIS 14384 (6th Cir. July 14, 2004).
[21] *Id.*
[22] No. 03-104 (2004).
[23] No. 03-105 (2004).
[24] K. Balint, "Falsifier on Waste," *San Diego Union-Tribune* (Aug. 21, 1993).

occupation in which she or he would be able to commit the same crime again. Policing such a prohibition is difficult, so if a corporation really wants the convicted employee's services, it can easily adjust the employee's job title to make it appear that the job is one the employee can legally hold.

An alternative touted as saving money for the taxpayers is house arrest or home confinement. The criminal is not allowed to leave home for the period of incarceration and is compelled to wear an unremovable sensor that allows government officials to detect his or her location at all times. In some cases, the defendant is allowed to serve his or her prison term on weekends.

Common White-Collar Crimes

Thus far, we have focused on the white-collar criminal, but any study of white-collar crime must include consideration of who the victims are and what some of the precise crimes are. The victims of white-collar crime are widespread, and they vary according to the precise crime committed. White-collar crimes may be committed against the public in general, as when environmental regulations are violated; against consumers, as when they are forced to pay higher prices because of violations of the antitrust laws or when they die because the products they purchased were made without undergoing the tests required by the Pure Food, Drug, and Cosmetic Act; against the taxpayers, as when income taxes are not paid; and against the corporation itself, as when employees steal from their employer. When the victims are the corporation the criminal works for, we sometimes refer to the crime as an *intrabusiness crime*. Estimates of the annual costs of intrabusiness crime range from $4 billion to $44 billion. In this section, we examine some of the more common white-collar crimes, noting their elements and their victims.

But before we examine common crimes, we want to emphasize how important it is to business managers to be aware of the legal definitions of particular white-collar crimes. Judges do not decide in favor of prosecutors when the actions of the defendant do not match the definition of the crime. Case 6-2 illustrates the importance of knowing the elements of possible crimes.

CASE 6-2

Sekhar v. United States
United States Supreme Court
133 S. Ct. 928

The Comptroller of the State of New York determines the Investment purchases made on behalf of the state pension system. In 2009, the Comptroller considered whether to invest in a fund managed by FA Technology. Based on its due diligence the General Counsel advised the Comptroller that such an investment would be unwise. The Comptroller followed this advice.

Four days later the General Counsel received an anonymous email warning the General Counsel that if he did not recommend the purchase from FA Technology, the General Counsel's extramarital affair would be disclosed. Similar emails were sent over the next several days. The emails were traced to the managing partner of FA Technology, Giridhar Sekhar, who admitted that he had sent the emails.

For the crime of extortion to have occurred, the law requires that the recommendation of the attorney had to be "property." Sekhar preferred to have the crime defined as "coercion" to avoid the more severe penalties attached to extortion. The United States claims that extortion is a subset of coercion, specifically coercion involving an economic element. The Hobbs Act is a Federal statute passed in 1948. The Court relied heavily for its decision on the language of that law. Both the federal district court and the Second Circuit Court of Appeals decided that the attorney's advice was a form of property that could be obtained by threats. Sekhar appealed this decision to the U.S. Supreme Court.

Justice Scalia

Petitioner was indicted for, and a jury convicted him of, attempted extortion, in violation of the Hobbs Act, 18 U.S.C. § 1951(a). That Act subjects a person to criminal liability if he "in any way or degree obstructs, delays, or affects commerce or the movement of any article or commodity in commerce, by robbery or extortion or attempts or conspires so to do." The Act defines "extortion" to mean "the obtaining of property from another, with his consent, induced by wrongful use of actual or threatened force, violence, or fear, or under color of official right." . . . On the verdict form, the jury was asked to specify the property that petitioner attempted to extort: (1) "the Commitment"; (2) "the Comptroller's approval of the Commitment"; or (3) "the General Counsel's recommendation to approve the Commitment." The jury chose only the third option.

The Court of Appeals for the Second Circuit affirmed the conviction. The court held that the general counsel "had a property right in rendering sound legal advice to the Comptroller and, specifically, to recommend—free from threats—whether the Comptroller should issue a Commitment for [the funds]." 683 F.3d, at 441. The court concluded that petitioner not only attempted to deprive the general counsel of his "property right," but that petitioner also "attempted to exercise that right by forcing the General Counsel to make a recommendation determined by [petitioner]."

Whether viewed from the standpoint of the common law, the text and genesis of the statute at issue here, or the jurisprudence of this Court's prior cases, what was charged in this case was not extortion.

"[W]here Congress borrows terms of art in which are accumulated the legal tradition and meaning of centuries of practice, it presumably knows and adopts the cluster of ideas that were attached to each borrowed word in the body of learning from which it was taken and the meaning its use will convey to the judicial mind unless otherwise instructed." *Morissette v. United States.*

Or as Justice Frankfurter colorfully put it, "if a word is obviously transplanted from another legal source, whether the common law or other legislation, it brings the old soil with it." . . . .

The Hobbs Act punishes "extortion," one of the oldest crimes in our legal tradition. The text of the statute at issue confirms that the alleged property here cannot be extorted. Enacted in 1946, the Hobbs Act defines its crime of "extortion" as "the obtaining of property from another, with his consent, induced by wrongful use of actual or threatened force, violence, or fear, or under color of official right." 18 U.S.C. § 1951(b)(2) (emphasis added). Obtaining property requires "not only the deprivation but also the acquisition of property." *Scheidler v. National Organization for Women, Inc.,* . . . . (citing *United States v. Enmons,* . . .). That is, it requires that the victim "part with" his property. The property extorted must therefore be transferable—that is, capable of passing from one person to another. The alleged property here lacks that defining feature.

Instead of defending the jury's description, the Government hinges its case on the general counsel's "intangible property right to give his disinterested legal opinion to his client free of improper outside interference." Brief for United States 39. But what, exactly, would the petitioner have obtained for himself? A right to give his own disinterested legal opinion to his own client free of improper interference? Or perhaps, a right to give the general counsel's disinterested legal opinion to the general counsel's client?

Either formulation sounds absurd, because it is. Clearly, the petitioner's goal was not to acquire the general counsel's "intangible property right to give disinterested legal advice." It was to force the general counsel to offer advice that accorded with petitioner's wishes. But again, that is coercion, not extortion. No fluent speaker of English would say that "petitioner obtained and exercised the general counsel's right to make a recommendation," any more than he would say that a person "obtained and exercised another's right to free speech." He would say that "petitioner forced the general counsel to make a particular recommendation," just as he would say that a person "forced another to make a statement." Adopting the Government's theory here would not only make nonsense of words; it would collapse the longstanding distinction between extortion and coercion and ignore Congress's choice to penalize one but not the other. That we cannot do.*

Judgment *reversed* in favor of the United States.

BRIBERY

Bribery is the offering, giving, soliciting, or receiving of money or any object of value for the purpose of influencing a person's action, especially a government official. The law against bribery is necessary to protect the integrity of the government and to ensure that the government functions fairly and efficiently. Bribery would include paying a judge to rule in favor of a party and giving a senator free use of your condominium if he or she votes for a particular piece of legislation. For example, Antonio Bras pled guilty to bribing Department of Public Works (DPW) officials in Washington, DC, to falsify records showing that Bras's company was providing more asphalt than it actually was, allowing Bras to overcharge the DPW. To enable his overcharging, Bras paid DPW officials to accept

bribery The offering, giving, soliciting, or receiving of money or any object of value for the purpose of influencing the judgment or conduct of a person in a position of trust, especially a government official.

---
*Sekhar v. United States, 133 S. Ct. 928, The Supreme Court of the United States.

falsified records. Bras was sentenced to three years and one month imprisonment, which was less than the maximum of five years he could have received.[25]

Some states broaden their definition of bribery to include certain payoffs in a commercial context. It is often considered bribery to offer to confer a benefit upon an employee or agent of another in an attempt to influence that agent's behavior on behalf of his or her employer or principal without that employer or principal's knowledge. Thus, an employee's offer to pay a contracting officer $5,000 in exchange for that officer's promise to purchase all the widgets the employer makes the next year would be considered a bribe in many states.

VIOLATIONS OF FEDERAL REGULATIONS

During the past few decades, regulatory agencies have been created and federal regulations enacted to control business. Analogous state regulations have also been enacted. Some primarily regulate economic matters; others are designed to protect the health, safety, and welfare of employees, consumers, and the public in general. You will study many of the regulations in the public law part of this book, Chapters 18 through 25.

Violation of any of these regulations constitutes a white-collar crime. The victims of these crimes vary according to the regulation that has been violated. For example, Occupational Safety and Health Act standards are established to protect the health and safety of workers. When these standards are violated, the violation is a criminal act that victimizes the employee, who is working under less safe or less healthful conditions than those required by law. When we examine these different regulations in later chapters of this book, the reader should consider who would be victimized by violations of each of the regulations. The reader who has a clear understanding of who might be hurt by such violations may be careful not to violate those regulations when he or she is a manager.

Violations of these regulations are often not perceived as criminal because they are frequently remedied outside the traditional courtroom setting. These regulations are often enforced by the appropriate regulatory agency through the issuance of a warning, a recall of defective products, or a *consent agreement* (a contract in which the violator agrees to cease engaging in the illegal activity). A cease-and-desist order may also be issued, ordering the corporation to cease violating the law and imposing a fine on the corporation for each day it violates the order. Warning letters are often the first approach of the regulatory agency, and the prudent businessperson should heed them. In cases of substantial violations, regulatory agencies may, of course, seek to impose fines on the corporation or manager responsible for the violation or may ask the court to impose a prison sentence on an offender.

The maximum monetary penalty that can be issued for violating federal regulations enacted to control business varies according to the regulation in question. For example, the maximum corporate penalty for violating the antitrust law is $100 million, a sum rarely awarded, which is larger than the previous maximum of $1 million but still not very large compared with the billions of dollars of assets and sales of some violators. For other acts, maximums are much lower. For example, $1,000 is the maximum for a first offense under the Pure Food, Drug, and Cosmetics Act, and $500,000 is the maximum for Occupational Health and Safety Act violations.

The maximum fines for individuals vary; a fine of $10,000 under acts such as the Pure Food, Drug, and Cosmetics Act and the Securities Exchange Act is typical. Maximum prison sentences usually range from six months to one year, with a few acts allowing up to five-year sentences. Maximum sentences are rarely imposed.

---

[25]482 F.3d 560 (D.C. Cir. 2007).

CRIMINAL FRAUD

Criminal fraud is a generic term that embraces a wide variety of means by which an individual intentionally uses some sort of misrepresentation to gain an advantage over another person. State fraud statutes vary, but most require proof of three elements: (1) an intent to defraud, (2) the commission of a fraudulent act, and (3) the accomplished fraud. It is very difficult to prove fraud, especially the first element: the intent to defraud. Some fraudulent acts that commonly occur in the corporate setting are:

criminal fraud Intentional use of some sort of misrepresentation to gain an advantage over another party.

1. Defalcation, the misappropriation of trust funds or money held in a fiduciary capacity.
2. False entry (entries), the making of an entry into the books of a bank or corporation that is designed to represent the existence of funds that do not exist.
3. False token, a false document or sign of existence used to perpetrate a fraud, such as making counterfeit money.
4. False pretenses, a designed misrepresentation of existing facts or conditions by which a person obtains another's money or goods, such as the writing of a worthless check.
5. Forgery, the material altering of anything in writing that, if genuine, might be the foundation of a legal liability.
6. Fraudulent concealment, the suppression of a material fact that the person is legally bound to disclose.

This list is by no means all-inclusive, but it demonstrates the broad variety of actions captured by the term *fraud*.

A person or corporation that uses the mail to execute a scheme or artifice to defraud the public out of money or property may be prosecuted under the federal law that prohibits mail fraud. Likewise, a person or corporation that uses the telephone, telegraph, television, radio, or other device to transmit a fraudulent message may be prosecuted for the federal crime of wire fraud. Mail fraud claims can be brought in a wide range of situations. Seeking greater deterrence of mail and wire fraud, Congress passed the White-Collar Crime Penalty Act of 2002 as part of the Sarbanes-Oxley Act. The act increased maximum prison sentences for mail and wire fraud from 5 years to a new maximum of 20 years.

As the following case indicates, the likelihood that a fraudulent scheme will succeed is inconsequential to the decision of guilt or innocence.

CASE 6-3

United States v. Gray
Eleventh Circuit Court of Appeals
367 F.3d 1263 (2004)

The appellant, Kevin Gray, was found guilty of mail fraud. He had attempted to convince a businessman, Frank Patti, who was on trial for tax evasion and faced substantial jail time, that for $85,000 he would bribe the jury, thus avoiding the threat of jail for the businessman. The methods used by the appellant were, however, at times hardly believable. As a result, the appellant seeks a judgment of acquittal on the ground that his fraudulent scheme was "so absurd" that a person of ordinary prudence would not have believed it; the scheme, therefore, fell outside the realm of conduct proscribed by the mail fraud statute.

Circuit Judge Tjoflat

On May 8, 2002, a federal grand jury in the Northern District of Florida returned an indictment charging the appellant on one count of mail fraud. He pled not guilty, and the case proceeded to trial before a jury. The jury, having received evidence establishing the facts set forth above, found the appellant guilty, and the district court sentenced him to prison for 28 months. This appeal followed.

The appellant's initial attack on his conviction is that the evidence was insufficient to make out a case of mail fraud. He argues that to prove the crime of mail fraud, the Government must establish that the defendant "intended to create a scheme 'reasonably calculated to deceive persons of ordinary prudence and comprehension.'" Additionally, it must show that the defendant took some action in furtherance of his scheme—to bring it to fruition—in the form of a material misrepresentation made to the would-be victim that "a reasonable person would have acted on." It is on this peg that the appellant hangs his hat, contending that a reasonable person would not have acted on his representations when considered as a whole. In bolstering his argument, he draws attention to his statement to Patti that $85,000 would be needed to bribe three of the jurors who would be trying his case: $35,000 for J-1, and $25,000 each for J-2 and J-3. The appellant contends that a reasonable person would know that since the pool from which these jurors would be selected would not be known until April 15—when the pool assembled at the courthouse for the trial—the representation had to be phony.

While it is true that statements like the one he cites would seem absurd or fanciful to a reasonable person, the mail fraud statute does not require that every representation a defendant utters while executing his scheme must be credible. Instead, the statute requires proof that the defendant's scheme to defraud involved the use of material, false representations or promises. The initial representations the appellant made to Patti satisfy this requirement.

In the letter to Patti, the appellant made a false promise: "we can assure you . . . no imprisonment but you must pay the agreed tax settlement issued by the court." In addition, he falsely represented that an undisclosed number of sympathizers—including "our mutual friend" and "our associates in Pensacola"—would work to extricate Patti from his legal predicament if the businessman would agree to follow certain instructions. True, the letter did not identify precisely how the writer and these sympathizers would help Patti, but this omission did not render the letter devoid of any material misrepresentations that were capable of prompting a reasonable person to act as Patti did. What the appellant overlooks is that the mail fraud statute "punishes unexecuted, as well as executed, schemes. This means that the government can convict a person for mail fraud even if his targeted victim never encountered the deception—or, if he encountered it, was not deceived." All that the Government needs to show to establish the *mens rea* element of the offense is that the defendant anticipated the intended victim's reliance, and the appellant's anticipation of Patti's reliance can be inferred from, among other things, the fact that he was prepared to call Patti at the pay phone at the time and location specified in the letter.

Because the letter received by Patti contained false material representations from the appellant as part of an effort to receive cash payments from the desperate businessman, the crime of mail fraud was complete when the appellant delivered the letter via FedEx to Patti.*

Affirmed in favor of the Prosecution.

CRITICAL THINKING ABOUT THE LAW

The defendant in this case seems to misunderstand the law. His contention is that his offer to bribe is an absurdity and that Patti should have recognized it as such. His logic is that there can be no fraud unless the person who is allegedly being helped by the fraud reasonably believes the promised act will occur. In other words, the more outrageous the promise, the more the person charged with fraud can escape liability. The court in this case makes it clear that it is the act of the person initiating the fraud that is the key to the offense, not the response of the person being defrauded.

1. Does the rule of law in mail fraud cases require any particular action or belief on the part of the person being defrauded?

    Clue: Study the quote in the next-to-last paragraph of the decision.

2. What ethical norm is being emphasized by the rule of law with respect to mail fraud?

    Clue: Look back at the list of alternative ethical norms and make a determination about which of them is being advanced by the rule of law as it applies to mail fraud.

Fraud can range from a single act that victimizes one individual to a long-term scheme that victimizes thousands. In the corporate setting, fraud is sometimes committed by managers to make themselves look better so that they can secure promotions at the expense of others who perhaps deserve the promotions. Fraudulent entries in corporate records may result in artificially inflating the purchase price of stock at the expense of its purchasers. Fraud may also

---
*United States v. Gray Eleventh Circuit, Court of Appeals 367 F.3d 1263 (2004).

be committed against the corporation, and thus against the shareholders, as when an employee on a bonus system fraudulently reports sales before they have been completed to collect an early bonus or "pads" his or her expense account. The more autonomy employees have, and the fewer people overseeing their actions, the greater the likelihood of their committing fraud.

Consumers may also be victims of corporate fraud, as when businesspersons make false representations in advertising and labeling. Consumers may also be victimized by the fraudulent substitution of inferior goods for higher-quality ones. These substitutions even have the potential to give rise to fraud claims.

Even student loans can provide the opportunity for fraud. In 2010, Rachel Yould, a former beauty queen, Stanford graduate, and Rhodes Scholar pled guilty to fraud charges after having defrauded lenders out of over $680,000 in fraudulent student loans. She borrowed the maximum amount of student loan money allowed, and then was able to obtain a new Social Security number through a domestic violence protection program, and used that new number to fraudulently obtain more student loan money, which she used to purchase a condominium, start a journal, and invest in the stock market.[26]

One of the fastest-growing forms of fraud over the last decade has been identity theft, whereby one's credit card, Social Security number, driver's license number, and other personal information are used for fraudulent purposes. According to Javelin Strategy & Research, a financial research company that has been tracking identity theft for the past decade, approximately 13.1 million Americans were victims of at least one identity theft in 2013, costing consumers more than $18 billion.[27] And in 2014, the IRS estimated that it mistakenly paid approximately $5 billion to identity thieves who filed fraudulent tax returns on behalf of unsuspecting citizens, and identified and stopped another $24.2 billion in attempted fraud.[28]

To combat this costly form of fraud, Congress passed the Identity Theft and Assumption Deterrence Act of 1998, making identity theft a federal felony punishable by up to 25 years in prison. The act also requires the FTC to help victims restore their credit. In addition, the FTC and other federal agencies suggest methods for reducing the likelihood of identity theft, including restricting the use of your Social Security number to occasions when it is legally required, reviewing your credit report at least twice a year, and proceeding with caution before giving out personal information over the Internet.

LARCENY

Another frequently occurring type of white-collar crime is larceny. **Larceny** is a matter of state criminal law, so the definition may vary slightly by state, but it can generally be defined as the secretive and wrongful taking and carrying away of the personal property of another with the intent to permanently deprive the rightful owner of its use or possession. The means of carrying out this crime is stealth: Larceny is not carried out by means of fear or force, which is the means of committing a robbery, and it is not carried out by means of false representation, which is one means of committing fraud. Larceny is commonly called theft by persons without legal training.

larceny The secretive and wrongful taking and carrying away of the personal property of another with the intent to permanently deprive the rightful owner of its use or possession.

---

[26]To read more about this white-collar crime, see Jeffrey Toobin, "The Scholar," *New Yorker Magazine* (Oct. 4, 2010), available at http://www.newyorker.com/reporting/2010/10/04/101004fa_fact_toobin.

[27]"Every Two Seconds Another American Becomes a Victim of Identity Theft," *Money* (Feb. 26, 2014), available at http://money.cnn.com/2014/02/06/pf/identity-fraud/.

[28]Kara Brandeisky and Susie Poppick, "How Identity Thieves Stole $5.2 Billion from the IRS," *Money* (Sept. 23, 2014), available at http://time.com/money/3419136/identity-theft-social-security-number-tax-return/.

Most states distinguish petty larceny from grand larceny, with the distinction based on the value of the item. Grand larceny involves items of higher value than those involved in petty larceny. It is usually considered a felony and, thus, is punishable by either a more severe fine or a longer term of imprisonment, or both.

In the corporate context, larceny generally involves employees taking the employer's property. Common instances of larceny include an employee's taking home stationery or supplies from the office.

EMBEZZLEMENT

embezzlement The wrongful conversion of the property of another by one who is lawfully in possession of that property.

Another white-collar offense is **embezzlement**. This crime is commonly defined as the wrongful conversion of the property of another by one who is lawfully in possession of that property. In some states, by statute, the crime may be committed only by certain classes of people, such as fiduciaries, attorneys, and public officials. As the reader might guess, larceny and embezzlement sometimes overlap. Like larceny, embezzlement is usually divided into degrees based on the value of the property embezzled. Some states also treat different kinds of embezzlers differently; that is, those embezzling from different types of institutions or those holding different types of positions may be distinguished.

COMPUTER CRIMES

As technology evolves, so do ways of committing corporate crime. With the arrival of the computer and the increasing automation of many facets of business, an area of crime has developed that society has not yet found an effective means of handling. No one knows the exact cost of computer crime each year, but estimates range from $300 million to $67 billion.

For the most part, computer crime, rather than being a new type of crime, is a means of making traditional crimes easier to commit. See Exhibit 6-7 for examples of some typical computer-crime techniques. Think about how many employees have access to a computer at work. When the number of individuals with home computers is added, there are myriad opportunities available for computer crime. Computer systems must now be protected from management, lower-level employees, and outsiders, sometimes known as *hackers*. With all these individuals having access to computers, a continuing increase in the amount of computer crime seems highly likely. Not only do computers make crime easier, but it appears that computers also make crime more profitable. According to federal officials, the average loss in a bank robbery is $10,000, and the average loss in a nonelectronic embezzlement is $23,500. But in a computer fraud, the average loss is $24,000.

Computer crimes are also not frequently prosecuted; some analysts estimate that less than 1 percent of those who engage in computer fraud are

EXHIBIT 6-7

COMMON COMPUTER-CRIME TECHNIQUES

 PIGGYBACKING—A nonauthorized person gaining access to a terminal when an authorized person failed to sign off, or an unauthorized person discovering an authorized user's password and signs on using that password

 IMPOSTER TERMINAL—Using a home computer with a telephone modem to gain access to a mainframe computer by cracking the password code and then using the computer free of charge

 TROJAN HORSE—Covertly placing instructions in a computer program that will generate unauthorized functions

 SALAMI SLICING—Stealing tiny amounts of money off large numbers of inputs (such as taking a penny off each entry) and transferring them into one's personal account

actually prosecuted. One reason these criminals are so successful is that many computer crimes are extremely difficult to detect. Even if the crime is detected, if the victim is a business, it will often not prosecute because it does not want its competitors to know that its system was vulnerable. A final problem with convicting people of computer crimes is that the crimes sometimes do not fit precisely within the statutory definitions of traditional crimes.

> **APPLYING THE LAW TO THE FACTS . . .**
>
> Let's say that Katie is a computer hacker, and hacks into the supposedly well-protected network system of a large corporation. Katie hacks into the corporation's financial accounts and withdraws a moderate sum of money. When the corporation's executives find out about the withdrawal, they fix the network so that Katie cannot reenter the network in the same way she did before, but they do not report the crime. Why might the corporation not report the crime?

Partially in response to this lack of adequate statutes under which to prosecute computer crime, Congress passed the Counterfeit Access Device and Computer Fraud and Abuse Act of 1984. The act not only imposes criminal sanctions, but also allows parties injured by a violation to bring a civil action to recover compensatory damages for the losses they incurred because of the violation. This act was subsequently amended by passage of the Computer Fraud and Abuse Act of 1986 (CFAA), which expanded the coverage of the original act. Congress continued to expand the scope of the law in 1989, 1990, 1994, 1996, and 2001. The most recent amendment to CFAA was passed as part of the USA PATRIOT Act to provide broader scope for prosecution of computer crimes. Under the latest version of the act, seven categories of activities are regulated:

1. The unauthorized use of or access to a computer to obtain classified military or foreign policy information with the intent to harm the United States or to benefit a foreign country
2. Accessing a protected computer (government computers, computers in financial institutions, or those used in interstate commerce) without authority or in excess of authority
3. The intentional, unauthorized access to a federal computer and the use, modification, destruction, or disclosure of data it contains or the prevention of authorized persons' use of such data
4. Accessing a protected computer without or in excess of authority with the intent to obtain something of value
5. Knowingly causing the transmission of a program, code, or command, and as a result causing damage to a protected computer
6. The fraudulent transfer of computer passwords or other similar data that could aid unauthorized access that either (a) affects interstate commerce or (b) permits access to a government computer
7. Transmitting in interstate or foreign commerce any threat that could cause damage to a protected computer with the intent to extort something of value

One important issue that had to be clarified under this act was whether the act would be violated if persons knowingly accessed a computer without permission but did so not knowing what damage they would cause by their action. This issue was settled by *U.S. v. Tappen*,[29] in which the court held that the intent goes to the intent to access the computers because Congress added

---
[29] Second Circuit Court of Appeals, 928 F.2d 504 (1991).

the intent element to ensure that people who inadvertently got into someone else's computer systems were not punished under the act.

Although it may seem as if the federal statute is fairly comprehensive, there are still a number of computer crimes that do not really violate that act. Those crimes must be prosecuted, if at all, under one of the state computer-crime statutes, which now exist in every state in some form, or under one of the traditional crime statutes.

As society attempts to find ways to respond to computer crimes, we have initially attempted to categorize the crimes. Following is just one way to categorize and think about these crimes.

Destruction of Data. Destruction of data is one of the biggest problems facing business today. A person with expertise in programming can create what is commonly referred to as a **virus**, a program designed to rearrange, replace, or destroy data. Once a virus is planted in a computer's instructions, it can spread to other systems or programs by rapidly copying itself. Hence, if a computer virus is not caught early, it can be extremely destructive.

A number of software programs have been developed to detect and destroy viruses. These programs, however, are reactive, not proactive. Every time a new type of virus is discovered, a new antiviral program must be developed. By 2006, there were more than 100,000 known viruses; hence, keeping antivirus software current is no easy task, though it is essential. In 2003, it was estimated that computer viruses and **worms** cost businesses more than $67 billion in damages.

Destruction of data may also be more limited. For example, a disgruntled employee who is fired might program his computer to destroy a section of data every time a file is saved. Before anyone realizes what has occurred, valuable data may be lost.

Unlawful Appropriation of Data or Services. Employees at work are often provided with expensive computer systems and software. They have access to vast amounts of data. When employees use their work computers or data accessed through these computers in a manner not authorized by their employer, they have engaged in theft of computer services or data.

Entering of Fraudulent Records or Data into a Computer System. Entering fraudulent records includes altering a person's credit rating electronically and breaking into a university's computer system and changing someone's course grades. For example, in 2007, 34 people, including former student employees, were charged with computer fraud and conspiracy charges regarding a six-year-long cash-for-grades scheme at Diablo Valley College in California. Allegedly, some of the students charged with the crimes illegally gained access to records of student grades and would take cash payments to alter grades for other students. By 2008, two of the ringleaders agreed to cooperate with prosecutors in exchange for sentences of one year in jail and four years' probation. Three others who were involved in changing grades were found guilty and charges against one were dropped because of a lack of evidence. The cases have not yet all been resolved. In May of 2010 another student implicated in the fraud was arrested, leaving four of those named in the scheme still at large.

Financial Crimes. Fraud, embezzlement, and larceny are now easier to accomplish by computer. An employee could electronically transfer ownership of funds from a corporate account to a personal account. Any object subject to lawful transfer electronically can also be stolen electronically.

Keeping these differences in mind, it is easy to see why, when fraud is suspected, the wise manager may wish to hire a fraud examiner, rather than simply rely on either an internal or external audit, to settle the question of whether fraud has occurred.

---

virus A computer program that destroys, damages, rearranges, or replaces computer data.

worm A program that travels from one computer to another but does not attach itself to the operating system of the computer it "infects." It differs from a "virus," which is also a migrating program but one that attaches itself to the operating system of any computer it enters and can infect any other computer that uses files from the infected computer.

Prevention of White-Collar Crime

We should all be interested in the prevention of white-collar crime because we are all its victims in more than one aspect of our lives. We are victims as consumers, shareholders, responsible employees, taxpayers, and citizens in general. The suggested ways for reducing white-collar crime are numerous and varied.

First, we will examine some of the ways in which corporate crime committed on behalf of the corporation may be prevented. One suggestion is to replace state chartering of corporations with federal chartering. Proponents argue that such chartering would prevent competition among states, which may lead to bribery and state officials' routinely overlooking corporate violations of the laws. As a part of federal chartering (or until then, by state law), corporations could be required to have outside directors (directors who are not also officers or managers). A counterargument to that suggestion is that outside directors do not have a significant impact on the behavior of management. They have neither the knowledge nor the interest to provide effective supervision.

An even more innovative suggestion, put forward by Christopher Stone,[30] is that each corporation doing more than a certain amount of business be required to have a general public director (GPD). The GPD would have an office at the corporation with a small staff, would be given access to corporate books and records, and would represent the public interest in the ongoing functions of the corporation. General public directors would sit on corporate committees and advise corporate officials about the legality of their activities. These public-interest watchdogs would be full-time workers paid from tax money. Stone further suggested that special public directors (SPDs), who would function like GPDs except that they would work in a special area such as workplace safety or antitrust, be assigned to corporations that have committed a series of violations in any specific area. SPDs would attempt to prevent further violations of a similar nature.

Two other proposals are to link the amounts of fines to the benefits obtained by the violations and to increase the amount of both corporate and individual

LINKING LAW AND BUSINESS

Accounting

When people initially think about how to detect fraud, they may think that fraud detection is the auditor's responsibility. If you remember what you learned about the role of an auditor from your accounting class, however, you would quickly recognize that although an auditor may discover something that tips him or her off that fraud has occurred, fraud detection is *not* the auditor's responsibility. It is useful, when thinking about fraud detection, to keep in mind the differences between the role of independent auditors and the role of fraud examiners, which are detailed in the following list:

- Auditors follow a program; fraud examiners look for the unusual.
- Auditors look for errors and omissions; fraud examiners look for oddities and exceptions.
- Auditors assess internal control risk; fraud examiners "think like a criminal" and look for holes in the controls.
- Auditors use the concept of materiality with amounts higher than a fraud examiner would use.
- Auditors usually start fresh with materiality each year; fraud examiners look at cumulative materiality.
- Auditors work with financial accounting and auditing logic; fraud examiners think about motive, opportunity, and integrity.

---

[30] C. Stone, *Where Law Ends: The Social Control of Corporations* 122, 183 (Harper & Row, 1976).

TECHNOLOGY AND THE LEGAL ENVIRONMENT

Are We Prepared to Fight Cybercrime?

After the worldwide attack by the Philippine "love bug" virus, and the difficult time that nation had in prosecuting the perpetrator of the cybercrime, McConnell International LLC, a global technology policy and management consulting firm, undertook a study to determine the readiness of most countries to successfully prosecute those who committed cybercrimes. Their conclusion was that in most countries around the world, existing laws are likely to be unenforceable against such crimes, meaning that businesses and governments must rely solely on technical measures to protect themselves from those who would steal, deny access to, or destroy valuable information. This unfortunate state of the law is very important because the number of cybercrimes seems to be continually escalating. For example, according to the Computer Emergency Response Team Coordination Center (CERT/CC), the number of reported incidences of security breaches in the first three quarters of 2000 rose by 54 percent over the total number of reported incidences in 1999 (see www.cert.org). What makes these figures all the more frightening, however, is that some corporations and governments do not want to admit being victims of cybercrimes, so the figures must be an understatement of the scope of the problem.

McConnell International's report analyzes the state of the law in 52 countries. It finds that only 10 of these nations have amended their laws to cover more than half of the kinds of crimes that must be addressed. Although many of the others have initiatives under way, most countries do not have laws that cover cybercrimes. For example, in many countries, laws that prohibit physical acts of trespassing or breaking and entering do not cover their "virtual" counterparts.

To prepare the report, McConnell International asked global technology policy officials in the 52 countries to provide copies of their laws that would be used to prosecute criminal acts involving both private- and public-sector computers. Countries that provided legislation were evaluated to determine whether their criminal statutes had been extended into cyberspace to cover 10 different types of cybercrime in four categories: data-related crimes, including interception, modification, and theft; network-related crimes, including interference and sabotage; crimes of access, including hacking and virus distribution; and associated computer-related crimes, including aiding and abetting cybercriminals, computer fraud, and computer forgery.

Thirty-three of the countries surveyed had not yet updated their laws to address any type of cybercrime. Of the remaining countries, 9 have enacted legislation to address 5 or fewer types of cybercrime, and 10 have updated their laws to prosecute 6 or more of the 10 types of cybercrime. Among those nations that had substantially or fully updated their laws were the United States, Canada, Japan, and the Philippines.

---

fines. More than 90 percent of the fines paid by corporations between 1975 and 1976 were less than $5,000. When imposed on a relatively small corporation, one with annual sales of $300 million, such fines are analogous to giving a person who earns $15,000 a year a two-and-a-half-cent fine.[31] Hence, fines as they currently exist do not really serve any deterrent function. Likewise, requiring greater and mandatory prison sentences for corporate executives who are found to have violated federal and state regulations might cause them to take these laws more seriously. Some also suggest elimination of the nolo contendere plea.

Because it is believed that the courts are unlikely to impose stiff fines on either corporations or convicted executives, the imposition of equity fines has been proposed. If a company is convicted of a white-collar crime, it would be forced to turn over a substantial block of its stock to a victim's compensation fund. This relinquishing of the company's stock would make its executives' holdings worth less. It would also provoke the ire of shareholders, whose holdings would be diluted, and might prompt them to call for the ouster of the responsible executives. Finally, it might put pressure on managers, who realize that the existence of a block of stock in one place would make a takeover attempt easier.

---

[31] G. Stricharchuk and A. Pasztor, "New Muscle in False Claims Act May Help in Combating Fraud against the Government," *WSJ* 19 (Dec. 19, 1986).

LINKING LAW AND BUSINESS

Management

Responsibility is defined as the obligation to complete specific activities. Typically, managers will assign individuals to particular positions in which they are entrusted with the responsibility of carrying out a task. Job activities are often divided by the functional similarity method, which is a basic method for separating duties in an organization. There are four interconnected steps in this method: (1) Examine organizational objectives; (2) delegate appropriate activities to meet established objectives; (3) design specific jobs for each activity; and (4) place individuals with responsibility for each specific job.

Another guide to the functional similarity method suggests that overlapping responsibility should be avoided. *Overlapping responsibility* refers to situations in which more than one person has responsibility for a specific task. When more than one employee is assigned to a certain task, there is a greater likelihood for employee conflicts and poorer working relationships.

There is, however, a disadvantage to avoiding overlapping responsibility. When individuals are given greater responsibility and autonomy, there is also a greater possibility for fraudulent activities. When there are fewer people working along with an individual, there may be a greater temptation to engage in fraud. Thus, managers should weigh the costs and risks associated with the level of responsibility given to employees before dividing responsibility for various activities.

It is also believed that regulations are often broken because of ineffective monitoring by agencies. One remedy, it is argued, is to increase the operating budgets of the regulatory agencies to allow them to hire more people to monitor corporations and to improve the training of regulatory agency employees. A related argument is that the regulations themselves are often vague and complex. Simplification of these laws would make them more understandable and easier to follow. It would also make violations of these laws easier to recognize and prosecute.

These ideas and suggestions are all beyond the direct control of most corporate managers. There are, however, some very practical things that managers can do to reduce the likelihood that their employees or companies will commit white-collar crimes. First, have a well-defined company code of ethics that the employees read and sign. Make sure that the employees understand that dishonest and unethical behavior is not acceptable. Second, provide a hotline for anonymous tips. Employees should be encouraged to use the hotline to report any instances of fraud on the part of other employees. Third, provide an employee assistance program. Often employees commit fraud and other white-collar crimes because they are having problems with substance abuse, gambling, or money management. If they can get assistance with these problems, it may prevent their trying to solve them by committing crimes against the company. Finally, conduct proactive fraud auditing.

Federal Laws Used in the Fight against White-Collar Crime

THE RACKETEER INFLUENCED AND CORRUPT ORGANIZATIONS ACT (RICO)

In the eyes of many plaintiffs' attorneys and prosecutors, a major weapon in the fight against white-collar crime can be found in Title IX of the Organized Crime Control Act of 1970,[32] the **Racketeer Influenced and Corrupt Organizations**

Racketeer Influenced and Corrupt Organizations Act (RICO) Federal statute that prohibits persons employed by or associated with an enterprise from engaging in a pattern of racketeering activity, which is broadly defined to include almost all white-collar crimes as well as acts of violence.

---

[32]18 U.S.C. § 1961 *et seq.*

Act, or RICO, as it is commonly called. This statute was originally enacted to help fight organized crime, but its application in the commercial context soon became apparent. In fact, a study by the Task Force on Civil RICO of the American Bar Association revealed that only about 9 percent of the RICO cases involved what is commonly considered organized crime; 37 percent of the cases involved common-law fraud in a commercial setting; and 40 percent involved securities fraud allegations.[33]

RICO is such a powerful statute because it allows any person whose business or property is injured by a violation of the statute to recover treble damages plus attorney's fees in a civil action. Hence, this is an instance in which a civil lawsuit may help prevent criminal action.

How does one violate RICO? RICO prohibits persons employed by or associated with an enterprise from engaging in a pattern of racketeering activity. Judicial interpretations have interpreted "pattern" to mean more than one act. Thus, RICO cannot be used against the one-time violator. What constitutes "racketeering activity," however, has been very broadly defined to include almost all criminal actions, including acts of violence, the provision of illegal goods and services, bribery, antitrust violations, securities violations, and fraud.

RICO has been used so successfully in white-collar crime prosecutions that many corporate and brokerage firm attorneys are making appeals to the legislature through the press to limit the application of RICO. They argue that the statute is unfair because it does not require that a defendant be convicted of the alleged criminal activity before a civil RICO suit can be brought. They argue that this lack of a requirement of a previous conviction leads to spurious lawsuits and encourages out-of-court settlements by intimidated legitimate businesspersons. Opponents also argue that the courts are, or will be, "flooded" with such lawsuits and that valuable resources will thus be wasted. They also argue that the law was designed to attack "organized crime," not employees of "legitimate businesses," against whom it is now being successfully used.

RICO has been used in a broad array of situations. For example, in 2015, RICO was used to attack alleged long-standing fraud and corruption in global soccer. According to federal prosecutors, in 2004, when FIFA, soccer's governing body, voted on where to hold soccer's World Cup, the vote of one of the governing body's members was purchased for $10 million. Initially, indictments against 14 soccer officials and marketing executives were handed down for numerous alleged illegal acts including $150 million in bribes. Prosecutors hope that RICO can help clean up the corruption that appears to have permeated soccer.[34]

Proponents of RICO argue that the law should continue in force as it is. They point out that fraud is a national problem, costing the nation more than $200 billion each year.[35] Proponents further believe that given our lack of success in prosecuting criminal fraud cases in the past, we should retain this tool, which may allow us to punish persons who have been able to escape criminal prosecution. Criminal activity in "legitimate" businesses is a major problem facing the country today, and curing it can only enhance the reputations of those in the business world. If a corporation and its officers are not engaging in illegal or quasi-legal activities, they have nothing to fear from RICO.

---

[33]Stricharchuk and Pasztor, *supra* note 26.

[34]Stephanie Clifford and Matt Apuzzo, "US Vows to Rid Global Soccer of Corruption," *New York Times*, May 25, 2015.

[35]Mann, *supra* note 13, at 86.

FALSE CLAIMS ACT

A largely ignored 141-year-old federal law has come back to life since 1986 and is now being used vigorously in the fight against white-collar crime. Under the False Claims Act, private citizens may sue employers on behalf of the government for fraud against the government. A successful party may receive 25 percent of the amount recovered if the government chooses to intervene in the action, or 30 percent if the government does not participate in the suit. Between the law's amendment in 1986 to encourage private whistleblowers and the end of 2009, the government has recovered $2,399,854,364 under this act.[36] As of September 30, 2009, the U.S. government had a total of 996 "qui tam" cases under investigation.[37]

The act also offers protection to persons using the law to sue their employer. An employer may be held liable to the employee for twice the amount of back pay plus special damages if found guilty of retaliation. Between the law's amendment in 1986 to encourage private whistleblowers and September 2006, the government paid whistleblowers a total of more than $1.6 billion.[38]

The Sarbanes-Oxley Act contains a provision that provides safeguards to protect whistleblowers, as does the Homeland Security Act of 2002. The new provisions were encouraged after whistleblowers received much attention with the rise in corporate fraud scandals. Perhaps no whistleblower is more notable in the business world than Sherron Watkins. As the vice president for corporate development at Enron, Watkins wrote several letters to chairman Kenneth Lay exposing corporate officials Andrew Fastow and Jeffery Skilling for committing the fraud. Although it was left for the courts to decide the facts of the Enron case and assign blame, Watkins played a significant role in exposing the company's failures. Because of the important role whistleblowers can play in revealing fraud or other illegal behavior, Congress considered the Congressional Whistleblower Protection Act of 2007, which would have expanded safeguards for federal employee whistleblowers who are fearful of future employment discrimination, and would have amended the Whistleblower Protection Act of 1989, which also protects federal employees who act as whistleblowers against the government. However, while the bill passed in the House, it failed in the Senate, and is not likely to be revived again in the near future.

The False Claims Act has also been used in a wide variety of circumstances. For example, in 2010, it was used to force Schwarz Pharma Inc. to pay $22 million to settle a claim that charged the company with marketing two unapproved drugs, Deponit and Hyoscyamine Sulfate Extended Release (Hyoscyamine Sulfate ER). The federal government received $12.24 million and the state received $9.76 million. Whistleblowers Constance and James Conrad received $1,836,575 from federal settlement, plus a share of the state settlement.[39] It has been used by an employee claiming that his employer, a school, helped its students fraudulently obtain millions of dollars in federal financial aid; by an employee who claimed that his employer, a defense contractor, knowingly manufactured defective parts for use in a guided missile system; and more recently by employees of hospitals, doctors, and nursing homes that overbilled Medicare claims. Claims related to health care fraud make up the biggest share of the claims, whereas the second greatest percentage of claims comes from fraud in defense procurement contracts.

---

[36]The False Claims Act Legal Center. Accessed May 1, 2000 at www.taf.org/statistics.htm.

[37]*Id.*

[38]Nolan Law Firm, "Qui Tam Frequently Asked Questions." Accessed February 22, 2008 at www.whistleblowerfirm.com/faq.html.

[39]http://www.taf.org/.

Because of Justice Department estimates that fraud costs the taxpayers as much as $100 billion a year,[40] there are many proponents of the use of this act. Its use is encouraged because it motivates persons in the best positions to be aware of fraud to report its occurrence. Employers who know that their employees may bring an action if asked to engage in fraud may be deterred from engaging in such acts. Nonetheless, proving that a party may legally bring a claim under the False Claims Act is often not an easy task.

Opponents of the use of the False Claims Act cite a variety of reasons. Some fear that frivolous or politically motivated lawsuits may be brought. Others say that the 60-day time period during which the Justice Department has to decide whether to join in the action is too brief and places an undue burden on the DOJ. During that limited time period, the government must decide whether to join in a complex suit that may take years to resolve.

In addition, the government can use, or choose not to use, the False Claims Act for political reasons. Despite the fact that more than a dozen False Claims Act lawsuits were filed by whistleblowers because of fraud in the Iraq reconstruction, the government has yet to join in on a single one of these suits. Fraud in Iraq reconstruction is estimated to be in the tens of millions of dollars, but the government has not helped out a single citizen who has filed suit. Some commentators have remarked that the government's refusal to sign on to a False Claims Act case frequently leads judges to doubt the validity of the claim, in that the government chose not to join the suit.[41]

For now, at least, use of the False Claims Act is increasing; but as the rising number of large judgments to whistleblowers keeps making headlines—at least a dozen whistleblowers thus far have collected more than $15 million each—more people are starting to question whether the whistleblower "rewards" are a wise use of taxpayers' money. In 2012, the top 30 settlements totaled $9,094,474,000. Some have proposed capping the amount a whistleblower can collect at $10 or $15 million. Those who have collected some of the awards above those amounts, however, have said that such a cap probably would have kept them from reporting the fraud. Because the majority of the awards are below $10 million, with the median award from 1986 to mid-2001 being around $150,000, it seems unlikely that amending the law to institute a cap will be a major item on the congressional agenda in the near future. And, as Kristin Amerling, President of Taxpayers Against Fraud says, "The False Claims Act is working exactly as it should, forging a public-private partnership that incentivizes integrity and discourages wrongdoing, even as it works to keep government and law enforcement more efficient."[42]

SARBANES-OXLEY ACT

The large number of high-profile companies engaged in fraudulent acts during the end of the 1990s and beginning of the 2000s led to the passage of the Sarbanes-Oxley Act of 2002. This act, fully discussed in Chapter 23, establishes new rules regarding corporate accounting, government oversight, and financial regulations. Although the act applies only to publicly held companies, many of the act's rules influence the behavior of private and nonprofit organizations. Specifically, the act seeks to eliminate the conflicts of interests that can lead to fraudulent activity. For example, the act establishes the Public Company Accounting

---

[40]H. Berleman, "A Few Big Penalties Make for a Record Year," *NLJ* (Oct. 24, 1994).

[41]Deborah Hastings, "Those Who Blow Whistle on Contractor Fraud in Iraq Face Penalties," *Findlaw Legal Headlines,* Aug. 27, 2007; retrieved Feb. 22, 2008, from news.findlaw.com/ap/o/51/08-26-2007/896d00a95e7ae2bf.html.

[42]Taxpayers Against Fraud Education Fund, "FY 2012 Is Record Year for FCA Recoveries," retrieved July 17, 2013 from http://www.taf.org/blog/fy-2012-record-year-fca-recoveries.

Oversight Board, which seeks to ensure proper accounting practices; mandates the separation of audit and nonaudit services; and requires corporate officials to certify their financial statements, which makes those officials responsible and liable for fraudulent statements. In addition, the Sarbanes-Oxley Act contains provisions that prevent publicly traded companies from firing, demoting, suspending, and otherwise discriminating against or harassing any employee who reports (blows the whistle on) corporate wrongdoing to the government.

WHISTLEBLOWER PROTECTION ACT

The Whistleblower Protection Act (WPA), similar to portions of the Sarbanes-Oxley Act, is intended to offer protections to those who act as whistleblowers. Whereas Sarbanes-Oxley protects private citizens, however, the WPA offers protections to federal employees who report illegal governmental activities. The WPA, originally enacted in 1989, offers protections to the vast majority of federal executive-branch employees. Congress explained the intent of the WPA as encouraging honest and responsible government by offering protections for those employees who know of governmental wrongdoing.

Generally, the WPA applies to present and past federal executive-branch employees, as well as applicants for such jobs. The WPA does not cover all of these federal executive employees, however. Positions that are related to policy making and policy determination, as well as positions related to foreign intelligence or counterintelligence, are not protected by the WPA. In addition, federal employees in the following areas are also not protected by the WPA: the postal service, the Government Accountability Office, the Federal Bureau of Investigation, the Central Intelligence Agency, the Defense Intelligence Agency, the National Security Agency, as well as several other agencies' employees.

Despite its good intentions, the WPA is not always effective at protecting and encouraging open government and citizen reporting of governmental wrongdoing. For example, as during 2010, stories of governmental wrongdoing and fraud in Iraq and Iraqi reconstruction contracts were starting to come to light. In its final report issued to Congress, the Commission on Wartime Contracting in Iraq and Afghanistan estimated that by 2011, as much as $60 billion had been lost to contractor fraud and waste in America's operations in Iraq and Afghanistan. Before the stories had started to make it into mainstream media, a number of governmental employees and officials, such as "Bunny" Greenhouse, were subjected to retaliatory actions for acting as whistleblowers. Bunny worked for the Army Corps of Engineers as a high-ranking civilian officer. In 2005, Bunny testified before Congress regarding widespread fraud in Iraq rebuilding contracts involving Halliburton. For her trouble, Bunny was demoted and moved to a different department where she now has no decision-making authority and is given little work to do.[43]

To a large extent, because of all the fraud in Iraq and Afghanistan, in September of 2012, the Non-Federal Employee Whistleblower Protection Act was introduced. The Act was designed to reduce fraud within the government and save taxpayer dollars by expanding the whistleblower-protections covering to federal contractors, subcontractors and grantees. If passed, it would have meant that any contractor, subcontractor, or grantee who spoke out against the misuse of federal funds could not be discharged, demoted, or otherwise discriminated against as reprisal for blowing the whistle.

While that bill did not pass, a comparable result was achieved by Section 828 of the National Defense Authorization Act of 2013. That section, beginning July 1, 2013, protects disclosures made by government contractors to any

---

[43]*Id.*

COMPARATIVE LAW CORNER

Whistleblowers in the United Kingdom

The United Kingdom, like the United States, is concerned about white-collar crime. One way to prevent or reduce white-collar crime is through whistleblowing. Although the United States has several federal laws that offer protections for whistleblowers, the primary protections for whistleblowers are through state laws. Conversely, the United Kingdom uses a national law to protect whistleblowers. The UK's whistleblower law is the Public Interest Disclosure Act of 1998 (PIDA). The United Kingdom's law differs in many ways from whistleblower laws in the United States.

First, the emphasis of the PIDA is fixing the problem within the company. The preferred (and most protected) disclosure is disclosure to a manager within the company. The basic form of the law is that a good-faith disclosure based on genuine concerns about crime, civil offenses, miscarriage of justice, danger to health and safety, or the environment, and the cover-up of any of these protects the whistleblower from firing or retaliation. The first step is often to disclose the possible violation to a manager in the company. Disclosure to someone at the company is not always advisable or possible, so whistleblowers have other protections as well.

Disclosures to the appropriate ministry are acceptable, when applicable, as well as disclosures to the appropriate regulatory agency, such as the Health and Safety Executive. Rather than just having a suspicion of wrongdoing, whistleblowers must believe that their allegations are substantially true before disclosing to a regulator. Wider disclosure is protected under certain circumstances. If the whistleblower reasonably believed that he or she would be victimized for disclosure to the company or regulator, reasonably believed that a cover-up was likely and there was no prescribed regulator, or had already raised the issue internally or with a regulator and no action had been taken, he or she would be protected for a wider disclosure.

At no point may whistleblowing be protected for the purposes of personal gain. Unlike in the United States, UK whistleblowers do not get monetary rewards. They may be protected and compensated if action is taken against them, but they do not receive anything for the act of whistleblowing.

Although there are differences in the approach to whistleblowing, the PIDA in the UK and various whistleblower protections in the United States share the same goal: to encourage people to come forward if they know about wrongdoing.

---

member of Congress, an Inspector General, the GAO, a contract oversight employee in an agency, authorized DOJ or law enforcement agencies, a court or grand jury, or a management official at the employing contractor with authority to investigate wrongdoing.

And on November 27, 2012, President Obama signed into law the Federal Whistleblower Enhancement Act, which had been unanimously passed into law by the House and Senate. Reforms of the new legislation include protecting federal employees (in addition to already-existing scenarios) from reprisal if they: are not the first person to disclose misconduct; disclose misconduct to coworkers or supervisors; disclose the consequences of a policy decision; or blow the whistle while carrying out their job duties.

State Laws Used in the Fight against White-Collar Crime

Some states have whistleblower statutes that protect employees who testify against their employers. Some state acts also allow government whistleblowers to bring actions against the state government, and a few—Texas, California, and Alaska—even allow government-employee whistleblowers to seek punitive damages. Although a whistleblower statute with unlimited potential for recovery may sound like a good idea, not everyone supports such laws.

Global Dimensions of White-Collar Crime

White-collar crime is not just a U.S. problem; it is a worldwide problem. Ironically, improvements in technology have increased the amount of white-collar crime on a worldwide scale by creating more opportunities for skilled employees to commit such crimes. For example, the availability of credit cards that can be used worldwide has increased the opportunities for credit card fraud.

When companies operate multinationally, they may be able to avoid regulation and escape the jurisdiction of any nation by not really basing their operations in any country. One example of a company that was able to successfully engage in criminal behavior for years without detection was Investor's Overseas Services (IOS). This company was much admired until its collapse in 1970. Using sales representatives recruited worldwide, IOS was able to persuade clients to invest $2.5 billion in a variety of mutual fund companies. Sales kept growing until the company finally went broke and the investors lost their money.

To some extent, IOS had been able to operate in a fraudulent manner because the company was not domiciled in any country or group of countries, and thus it was subject to no particular country's regulation. Managers registered and domiciled their funds wherever they could in order to avoid taxation and regulation. Consequently, they were able to do things that no company domiciled in a single country could do. The prudent manager should make sure that any multinational firm with which he or she does business is domiciled in some country.

Another factor leading to the commission of white-collar crime on an international scale is the unfortunate lack of cooperation among the police of different countries. This situation may be changing, however. In April 2001, the United States and 12 other countries agreed to start sharing confidential data about the complaints they receive from consumers, in a bid to crack down on cross-border Internet fraud. The FTC voted unanimously to begin pooling its U.S. complaints with those from other countries to create a single database, something an FTC spokesperson said would "greatly improve international law enforcement agencies' ability to address cross-border Internet fraud and deception."[44] The countries participating in the project, in addition to the United States, are Australia, Canada, Denmark, Finland, Hungary, Mexico, New Zealand, Norway, South Korea, Sweden, Switzerland, and the United Kingdom. Under the agreement, law enforcement agencies in each country will have access to the database through a single, password-protected website, the agency said. FTC officials said the information will tip them off to Internet scam artists.

Another important global initiative, the Convention on Cybercrime, is being undertaken by the United States, Canada, Japan, and the 43 members of the Council of Europe (COE). Initiated in 1997, after 27 drafts, a final version was agreed upon at the end of June 2001, paving the way for international rules governing copyright infringement, online fraud, child pornography, and hacking. The three main topics covered by the convention are harmonization of the national laws that define offenses, definition of investigation and prosecution procedures to cope with global networks, and establishment of a rapid and effective system of international cooperation. Full information about the treaty, including its text and a list of countries that have signed and ratified the treaty, can be found by doing a search at http://conventions.coe.int/.

The convention came into effect internationally on January 7, 2004, with the ratification by the fifth signing nation. As of December 2010, 46 nations had signed the convention, with 29 of the signators having ratified the convention. The United States officially ratified the Convention on Cybercrime on September 29, 2006, and the convention went into effect in the United States on January 1, 2007.

---

[44] "U.S., 12 Other Countries to Hit Internet Fraud," April 25, 2001. Available at news.findlaw.com/legalnews/s/20010424/techinternetfraud.html.

SUMMARY

Criminal law is that body of laws designed to punish persons who engage in activities that are harmful to the public health, safety, or welfare. A crime generally requires a wrongful act and a criminal intent.

Criminal procedure is similar to civil procedure, but there are some significant differences. A criminal prosecution begins with the issuance of an information by a magistrate or an indictment by a grand jury. The next step is the arraignment, which is followed by the trial, and then, in some cases, an appeal.

White-collar crimes—crimes committed in a commercial context—may be even more costly to society than street crimes are, but they are more difficult to prosecute and often carry relatively light sentences. Some of the more common white-collar crimes are larceny, the secretive and wrongful taking of another's property; embezzlement, the wrongful conversion of property of which one has lawful possession; and violations of federal regulations. Increasingly, computers are being used to commit white-collar crime, making the detection and prosecution of these crimes even more difficult.

Attempts are being made to fight white-collar crime on the federal level through such statutes as RICO and the False Claims Act. Some states have now passed whistleblower statutes to help fight white-collar crime.

When thinking about white-collar crime, it is important to remember that one of the drawbacks of increasing globalization is that it has led to increasing amounts of cross-border white-collar crime. Criminals have taken advantage of the lack of international cooperation among law enforcement officials.

REVIEW QUESTIONS

6-1 What is the purpose of criminal law?

6-2 Explain how crimes are classified.

6-3 Explain the basic procedural stages in a criminal prosecution.

6-4 State two alternative definitions of white-collar crime and give an example of one crime that fits under both definitions and another crime that would fit only one of the definitions.

6-5 Explain the rationale for imposing criminal liability on corporations.

6-6 Explain two sentencing alternatives to prison for white-collar criminals.

REVIEW PROBLEMS

6-7 Rawlsworth is an employee of General Sam Corporation. One of Rawlsworth's jobs is to monitor the amount of particular pollutants and to record the results on a form that is submitted to the Environmental Protection Agency. This self-monitoring is required by law to ensure that firms limit the amount of particular pollutants they discharge. One day the firm's equipment is malfunctioning, and so Rawlsworth records that the firm's discharge is in excess of the lawful amount. When Matheson, Rawlsworth's supervisor, sees what he has done, Matheson tells him to change the records to state that the firm is in compliance and, in the future, never to record such violations. When Rawlsworth calls the vice president to report this violation, he is told that the vice president does not take care of such matters and does not want to know about them. Rawlsworth then falsifies the records as instructed. When the falsification is discovered, who can be held criminally liable? Why?

6-8 Several corporations were convicted of violating the Sherman Act as a result of an unlawful agreement among their agents that the suppliers who supported an association to attract tourists would be given preferential treatment over those who did not contribute financially to the association. The corporations appealed on the grounds that the corporate agents involved were acting contrary to general corporate policy. Was the defense valid?

6-9 Defendant Laffal was the president of a corporation that operated a restaurant. It was alleged that prostitutes frequented the restaurant, picking up men there and returning them after a short time, thus making the restaurant an illegal "bawdy house" in violation of the state criminal law. Laffal argued that he could not be charged with operating a bawdy house because he was never present when any of the illegal acts took place and he did not even know they were going on. Was Laffal correct?

6-10 Evans was a loan officer for a bank and in this capacity had approved several loans to Docherty, all of which were legitimate and were repaid on time. Evans asked Docherty to apply for a loan from the bank for $2,000 and then to give the money to Evans, who would repay the loan. Evans explained that he could not obtain the loan himself because bank policy did not allow him to borrow from the bank. Docherty agreed. Was Evans's or Docherty's behavior illegal?

6-11 A defendant managed a corporation chartered for the purpose of "introducing people." He obtained a loan from a Mrs. Russ by telling her that he wanted the loan to build a theater on company property and that the loan would be secured by a mortgage on the property. The loan was not repaid; the mortgage was not given to the lender because the corporation owned no property. The defendant, in fact, simply deposited the money in the corporate account and used it to pay corporate debts. What crime, if any, did the defendant commit?

6-12 Jones worked for a small community college teaching business students how to set up inventories on various computer programs. The college had purchased the software and was licensed to use several copies of it for educational purposes. Jones started his own small business on the side and used the software for his own firm's inventory control. He saw his own use as "testing" the product to make sure he was teaching students to use a process that really worked. Is his behavior lawful or unlawful? Why?

CASE PROBLEMS

6-13 Teresa Chambers worked as the Chief of the United States Park Police from February 10, 2002 to July 9, 2004, when she was removed from her position. The United States Park Police is a component of the National Park Service, which is an agency within the Department of the Interior. In 2003, when the Office of Management and Budget decided against increasing the Park Police budget, Chambers spoke with a reporter from the *Washington Post* and a staffer for the United States House of Representatives Interior Appropriations Subcommittee. Chambers was placed on administrative leave when the staffer told her supervisor about the conversation and the *Washington Post* ran an article attributing several remarks to Chambers.

On December 17, 2003, the supervisor suggested removing Chambers from her position based on six charges of misconduct. The two relevant charges were: (2) making public remarks regarding security on the National Mall, in parks, and on parkways in the Washington, DC, metropolitan area; and (3) improperly disclosing budget deliberations to a *Washington Post* reporter. Chambers claims that charges 2 and 3 concern statements that are protected under the Whistleblower Act.

Charge 3 concerns improper disclosure of budget deliberations and was tied to Chambers's statement to the reporter from the *Washington Post* that "she said she has to cover a $12 million shortfall for this year and she asked for $8 million more for next year." Chambers contends that the disclosure was protected under the Whistleblower Protection Act as disclosing a substantial and specific danger to public safety. The agency contends that because the budget for the coming year had not been submitted to Congress, the disclosure was premature and in violation of agency protocol.

Charge 2 concerns statements that Chambers claims are related to issues of public safety. Chambers made statements about how traffic accidents have increased in an area that has only two officers instead of the recommended four and how there are not enough officials to protect the green areas around Washington, DC. How do you think the court should rule? *Chambers v. Department of the Interior,* 602 F.3d 1370; 2010 U.S. App. LEXIS 8209 (Fed. Cir. 2010).

6-14 Kyle Kimoto was president of Assail, Inc. Assail was a telemarketing firm responsible for marketing a financial package developed by another company that included a pay-as-you-go debit card. Assail employees would use call lists to choose individuals who had recently applied for a credit card and been rejected. The telemarketer would tell these individuals that they were now eligible for a Visa or Mastercard. If the individuals said they were interested, the telemarketer would place the call on hold to obtain "authorization." After letting a few minutes pass, the telemarketer would get back on the line to tell the individual that he or she had been approved. The telemarketer would tell the individuals that they would be charged a one-time processing fee of $159.95. The call was then transferred to another department to verify payment of the processing fee.

While the call was being transferred, the individual listened to a recording that informed him that the card was a pay-as-you-go MasterCard and there would be no credit on the card until a payment was made. However, any questions from the individual were answered in such a way as to support the assumption that this card would function as a typical credit card.

Defendant was indicted on one count of conspiracy, one count of mail fraud, and 12 counts of wire fraud. On appeal, Kimoto contends that as president he was in the business of selling pay-as-you-go credit cards and that the scripts used were not deceptive. Kimoto did not endorse employee statements indicating that the cards were credit cards. How do you think the court of appeals should rule? List the reasons you would use to support your position. *United States v. Kimoto,* 588 F.3d 464 (7th Cir. 2009).

6-15 Scott Bauknecht previously worked for Pontiac National Bank ("PNB"). In 2002, Bauknecht signed a confidentiality agreement with PNB where he agreed "not to disclose confidential customer information for a period of two years." PNB eventually changed its name to Freestar, and merged with First Financial Bank in 2011. In 2011, Bauknecht became a loan officer, as well as president of the bank. When PNB became Freestar, the bank had a policy forbidding release of confidential information any time during or after employment, as well as measures to require encryption of electronic devices.

In December of 2011, Bauknecht allegedly communicated with a competitor of Freestar, Ronald Minnaert, who was the president of Graymont Bank. Bauknecht spoke to Minnaert about the possibility of working for Graymont. Bauknecht then quit Freestar Bank and began working for Graymont Bank. Bauknecht continued to retain his previous customers after transitioning to employment at Graymont. Freestar Bank brought suit against Bauknecht, including breach of contract, breach of fiduciary duty, and misappropriation of trade secrets. The plaintiff also alleges that Bauknecht obtained customer information from a "Master Database." It was undisputed that both Bauknecht and Graymont Bank possessed a number of plaintiff's financial documents, including loan agreements. The plaintiff additionally sought summary judgment on a claim under the Computer Fraud and Abuse Act (CFAA) due to the defendant's accessing information from the "Master Database." How do you think the court ruled? What evidence would the plaintiff need to provide to satisfy a CFAA claim? *First Fin. Bank, N.A. v. Bauknecht,* 2014 U.S. Dist. LEXIS 151244 (C.D. Ill. 2014).

6-16 Van Chester Thompkins was arrested approximately a year after a Southfield, Michigan, shooting occurred in which one person was killed and another wounded. When he was taken into custody, a police officer gave him a written copy of the five Miranda warnings. To verify that the suspect understood the warnings, the officer asked him to read the fifth warning aloud, which he did. The suspect, upon request, then signed a form stating that he understood his rights. Officers then began questioning him, and at no point did he say he wanted to remain silent, although he was mostly silent throughout the almost three-hour interrogation, giving a few grunts, and brief responses such as "yes," "no," or "I don't know."

Almost three hours into the interrogation, the officer asked the suspect whether he believed in God, and the suspect said he did. And according to the police officer, his eyes started to tear. The officer then asked, "Do you pray to God?" and when the suspect said he did, the officer then asked, "Do you pray to God to forgive you for shooting that boy down?" The suspect again responded, "Yes," and looked away. He refused to make a written statement, and the interrogation ended about 15 minutes later.

At trial for the murder, the suspect sought to suppress the evidence of his statements on grounds that he had invoked his Fifth Amendment right to remain silent and his statements were involuntary. The trial court denied his motion and he was convicted of murder. After a number of appeals, the case ultimately was before the U.S. Supreme Court. How do you think the high court decided this case and why? *Berghuis v. Thompkins,* 130 S. Ct. 2250 (U.S. Ct. 2010).

6-17 Autohut was a used car business located in Englewood, Colorado. The business sold cars on a consignment basis and was not formed as a separate legal entity. Instead, Padilla, Mr. Lowell Andrews, and the Debtor in this case formed a partnership to run Autohut. This partnership agreement was not written. Autohut operated by "contacting individuals who were selling their cars on Craigslist" and offering "to sell their vehicles on consignment basis from Autohut's lot." The plaintiff, Chenaille, was a customer of Autohut who attempted to purchase a truck in 2011 from the business. The plaintiff customer paid the negotiated price to the business but never received the title to the truck. When the plaintiff returned to Autohut to discuss his receiving of the title for the truck, the business had shut down and the lot was empty. Plaintiff customer contacted the police and learned that many previous customers of Autohut had a similar experience.

The Colorado Motor Vehicle Dealer Board shut down Autohut's business and revoked its business license, but no charges were brought against Autohut or its owners. In the meantime, the original owner of the truck, a Mr. Jarred Johnson, had reported the vehicle as stolen, as he still retained the original title to the truck before he agreed to let Autohut sell the truck for him on the business's lot. When Mr. Johnson learned of the plaintiff's possession of the truck, he sued the plaintiff for return of the vehicle. The state court awarded ownership of the truck to Mr. Johnson, and the plaintiff was never refunded the money he paid Autohut for the truck. The plaintiff eventually filed suit for the debt owed to him from purchasing the truck. The plaintiff alleged that the debt owed to him met the requirements of embezzlement. How do you think the court ruled? Why? *Chenaille v. Palilla (In re Palilla),* 493 B.R. 248 (Bankr. D. Colo. 2013).

6-18 Betty Werner, the widow of a World War II veteran, had been receiving Veterans Administration (VA) Dependency and Indemnity Compensation (DIC) since the death of her husband in 1976. In 1995, Werner's son, Joel Ruben, had Werner admitted to an institution that would provide her primary care for severe mental problems. Ruben obtained a durable power of attorney, putting himself in charge of his mother's finances in order to provide for her care. Ruben redirected the VA's DIC payments made to his mother into his own personal accounts, and never actually used the money to pay for the care of his mother. When the VA began investigating the whereabouts of the payments made to Werner, Ruben admitted that he had used the money to pay off his own financial obligations and to care for his ill son. Of what crime is Ruben guilty? What evidence is necessary to prove these allegations? *United States v. Ruben,* 2006 U.S. App. LEXIS 12528 (2006).

6-19 Nayak paid doctors a kickback in order to get them to send their business to his surgery center. After learning of the kickback scheme, the government indicted Nayak. It later filed superseding information charging him with honest-services mail fraud. Although both the indictment and the superseding information alleged that Nayak intended "to defraud and to deprive patients of their right to honest services of their physicians" through his scheme, neither alleged that Nayak caused or intended to cause any sort of tangible harm to the patients in the form of higher costs or inferior care. In fact, the government later represented to the district court that the scheme did not cause patients any physical or monetary harm.

Nayak filed a motion to dismiss the mail fraud count, contending that the government needed to allege some form of actual or intended physical harm to the referring physicians' patients as an element of the crime. Why do you think Nayak lost his appeal? *Nayak v. United States,* 769 F.3d 968 (2014).

6-20 In 2004, James Clayton Terry was elected to serve as a member of the Lowndes County, Mississippi, Board of Supervisors, as the supervisor for District Four. As a member of the board, Terry was given a county vehicle and a Fuelman card, to be used for business purposes only. The county later began to receive complaints that Terry was making inappropriate purchases of gasoline and using the vehicle to take personal trips to casinos. Following an investigation, Terry was indicted for embezzlement.

The trial court found Terry guilty. On appeal, Terry argued that the conviction should be overturned because the state had failed to specify the dates during which the embezzlement supposedly occurred. The state argued that because the embezzlement was continuous, specific dates were not necessary. What exactly was embezzled, and should it matter whether the state alleged specific dates on which the embezzlement occurred? *Terry v. State,* 26 So.3d 378 (Miss. Appellate Court 2009).

6-21 Plaintiff Lisa Huff was injured by a falling tree near utility lines that were owned and operated by the defendant, Ohio Edison. The plaintiff alleges that the tree was located on property that was covered by an easement owned by Ohio Edison. When the plaintiff filed suit against the defendant, as well as First Energy Corporation and Asplundh Tree Expert Company, the state trial court granted summary judgment to the defendants, finding that there was no duty owed to the plaintiff. The court of appeals reversed this decision, finding that "there was a question of fact as to whether Huff had enforceable rights under the contract between Ohio Edison and Asplundh as a third-party beneficiary." The Ohio Supreme Court first denied discretionary review, and then granted jurisdiction and reversed the court of appeals decision, finding in favor of the defendant, Ohio Edison. On October 16, 2012, the plaintiffs filed suit alleging a conspiracy between judicial defendants and First Energy defendants to influence litigation, under the federal Racketeer Influenced and Corrupt Organizations Act (RICO). Did the court rule in favor of the plaintiff? Why or why not? Do you agree with the court's decision? *Huff v. First Energy Corp.,* 972 F. Supp. 2d 1018 (N.D. Ohio 2013).

THINKING CRITICALLY ABOUT RELEVANT LEGAL ISSUES

Allocating Blame for White-Collar Crime

Despite the prevalence of white-collar crime in the business world, it is wrong to try to pin the blame for a white-collar crime on just anyone. Our seeking someone to blame is exactly the motivation underlying the practice of pinning liability on managers for the wrongdoing of those employees below the managers. Just because a manager might be easier to identify and blame does not make it right to attach legal liability to the manager by virtue of his or her role as "manager."

The vast majority of crimes require a wrongful act, as well as a guilty mind, or *mens rea*. The problem with finding managers liable for the actions of those below them is that neither the element of wrongful act nor that of *mens rea* is met with respect to the manager. That is, just because a low-level employee commits a white-collar crime does not mean the manager participated or even intended for a crime to be committed. Liability imposed on managers for the wrongful acts of others overstretches the bounds of proper criminal liability. After all, is it right to legally hold parents responsible for the actions of their kids? What about holding teachers liable for an illegal act by one of their students? To hold a manager (or parent or teacher) accountable for the wrongful acts of another is absurd.

Another problem with imposing liability upon a manager for the actions of a lower-level employee is that the imposition of liability ignores the manager's other responsibilities. Managers are in charge of numerous people, as well as having their own work they need to complete. Accordingly, managers cannot be expected to keep an eye on every little action of all of the employees below them at all times. Even the most vigilant of managers has the potential to miss one act of wrongdoing by a lower-level employee. It seems wrong to hold this diligent manager legally liable because one employee was able to fool the manager and hide his wrongdoing from the manager.

Finally, businesses in general are harmed by the practice of holding managers liable for the actions of other employees. Businesses need qualified managers to ensure that everything runs smoothly. Holding managers liable for the actions of others, however, creates a disincentive for becoming a manager. If the practice of holding a manager liable for the white-collar crimes of lower-level employees continues, the number of competent, caring people who want to become managers will greatly decrease. This decrease could cost businesses by preventing the best people from rising to the top to be managers, because these people could fear the wrongful attribution of liability for the acts of another.

1. How would you frame the issue and conclusion of this essay?

2. Is there relevant missing information in the argument?

   Clue: What would you like to know before deciding whether the author is correct?

3. What ethical norm does the author appear to rely upon most heavily in making the argument?

4. Write an essay that someone who holds an opinion opposite to that of the essay author might write.

   Clue: What other ethical norms could influence an opinion on this issue?

ASSIGNMENT ON THE INTERNET

You have now learned several approaches to deter white-collar crime in this chapter, including federal laws such as RICO and the False Claims Act, individual state laws, and the proposed GPD. There are, however, many other independent steps a business can take to ensure that employees do not engage in illegal behavior. Use the Internet to find out what other businesses are doing to prevent white-collar crime. You can begin by visiting www.expertlaw.com and typing "corporate crime" in the search bar.

As a business manager, what additional deterrents to white-collar crime would you create or adopt? How does *United States v. Park*, discussed earlier in this chapter, affect your attitude toward the responsibility of a manager to prevent white-collar crime?

ON THE INTERNET

www.nw3c.org This site is the location of the National White Collar Crime Center, which provides investigative support services to fight against white-collar crimes.

www.usdoj.gov This is the website of the Department of Justice division that prosecutes fraud and white-collar crime. Click on the "Agencies" heading, then click on "Criminal Division" in the third row. Click on "About the Division," then "Organizations." Go to "List View," then click on "Fraud Section." Under the heading "Fraud Section," which should be about halfway down the page, click on "Website."

www.ic3.gov The Internet Crime Complaint Center is a joint endeavor by the FBI and the National White Collar Crime Center to create an easy-to-use forum for reporting Internet-based crimes.

FOR FUTURE READING

Blakely, G. Robert, and Michael Gerardi. "Eliminating Overlap or Creating a Gap? Judicial Interpretations of the Private Securities Litigation Reform Act of 1995 and RICO." 28 *Notre Dame Journal of Law, Ethics, and Public Policy* 435 (2014): 28.

Buell, Samuel W. "Is the White Collar Offender Privileged?" *Duke Law Journal* 63 (2014): 823.

De Geest, Gerrit. "The Rise of Carrots and the Decline of Sticks." *University of Chicago Law Review* 80 (2013): 341.

Earle, Beverley H., and Gerald A. Madek. "The Mirage of Whistleblower Protection under Sarbanes-Oxley: A Proposal for Change." *American Business Law Journal* 44 (2007): 1.

Friedrichs, David O. *Trusted Criminals: White Collar Crime in Contemporary Society* (3rd ed.). Belmont, CA: Thomson Higher Education, 2007.

Reiman, Jeffrey. *The Rich Get Richer and the Poor Get Prison: Ideology, Class, and Criminal Justice* (8th ed.). Boston: Allyn & Bacon, 2006.

Spalding, Jr., Albert D., and Mary Ashby Morrison. "Criminal Liability for Document Shredding after Arthur Anderson LLP." *American Business Law Journal* 43 (2006): 647.

Verstein, Andrew. "The Law as Violence: Violent White-Collar Crime." *Wake Forest Law Review* 49 (2014): 873.

CHAPTER SEVEN

Ethics, Social Responsibility, and the Business Manager

- DEFINITION OF BUSINESS ETHICS AND SOCIAL RESPONSIBILITY
- THEORIES OF ETHICAL THOUGHT
- CODES OF ETHICS
- SCHOOLS OF SOCIAL RESPONSIBILITY
- GLOBAL DIMENSIONS OF ETHICS AND SOCIAL RESPONSIBILITY

On December 2, 1984, in Bhopal, India, lethal methylisocyanate (MIC) gas leaked from a chemical plant owned by Union Carbide India Ltd., killing approximately 2,000 people and injuring thousands more, many of whom are still receiving treatment. Union Carbide's chairman, Warren Anderson, a lawyer, flew to India with a pledge of medical support and interim assistance totaling $7 million. He was arrested and deported from the country. Lawsuits on behalf of the deceased and injured were brought by U.S. law firms as well as by the government of India. Litigation continues in Indian courts.

The price of Union Carbide stock dropped from $48 to $33. In August 1984, GAF Corporation attempted to take over Union Carbide. Union Carbide successfully fought off the takeover attempt in 1985. On May 13, 1986, a federal court judge dismissed the personal injury and wrongful death actions, stating that the complaints should be more properly heard in a court in India. The judge attached certain conditions to the dismissal, one of which was that Union Carbide would have to agree to pay any damages awarded by an Indian court. The trial began in August 1988 in a New Delhi court amidst rumors that a former disgruntled employee had sabotaged Union Carbide's Bhopal plant, causing the gas leakage. In January 1988, Union Carbide shares traded on the New York Stock Exchange for $49, and the much leaner company was one of the 30 companies making up the composite Dow Jones Industrial Average. It might be helpful to read this case, set out in edited form in Chapter 8. For additional and updated facts through 2010, see the comments after Case 8-3.

The Bhopal incident in 1984, along with a stream of insider trading cases, as well "bank bailouts" and as a number of white-collar crime cases (highlighted in Chapter 6) have brought a heightened awareness of the need for debate as to whether the business community has a responsibility solely to shareholders or to other stakeholders as well. Such cases force us to ask ourselves, what should be the legal rules that businesses must obey in their daily operations? Additionally, there are ethical questions that force us to consider how we should behave if we are to live in a better world. *Business ethics* is the study of the moral practices of the firms that play such an important role in shaping that better world.

Below are results of a 2012 survey ranking countries by executives of companies as to the least corrupt or most corrupt.

Least Corrupt	*Most Corrupt*
New Zealand	Somalia
Denmark	Korea (North)
Finland	Myanmar
Sweden	Afghanistan
Singapore	Uzbekistan
Norway	Turkmenistan
Netherlands	Sudan
Australia	Iraq
Switzerland	Haiti
Canada	Venezuela
Luxembourg	Equatorial Guinea
Hong Kong	Burundi
Iceland	Libya
Germany	Democratic Republic of Congo
Japan	Chad
Austria	Yemen
Barbados	Kyrgyzstan
United Kingdom	Guinea
Belgium	Cambodia
Iceland	Zimbabwe
Bahamas	Paraguay
Chile	Papua Guinea
Qatar	Nepal
United States	Laos
France	Kenya
Santa Lucia	

Note: Who do you know who has a friend or relative in any of these countries?
Source: Corruption Perception Index, 2012, Transparency.org, C2012 Transparency International.

Whenever you wonder whether a business decision requires us to think about ethics, simply ask yourself, will this decision affect the quality of life of other people? If the answer is yes, the decision involves ethics. We think you will agree that business ethics is an extremely important aspect of our environment because almost all business decisions influence the quality of our lives.

This chapter presents material on business ethics in a neutral way. Readers are left to make their own choices about what part ethics should play in business decision making and about whether the business community, the trade groups that represent it, and individual managers should act in a "socially responsible" manner. This chapter includes (1) a broad definition of ethics and social responsibility; (2) some recognized theories of ethical thought and their application to business problems; (3) a discussion of individual, corporate, trade association, and professional ethical codes; and (4) schools of social responsibility as applied to business problems. The chapter ends with a brief discussion of some current trends in the area of ethics and social responsibility, as well as some proposals now being debated, which, if implemented, would change the structure of corporate governance.

CRITICAL THINKING ABOUT THE LAW

Business ethics is perhaps one of the most personal and emotional areas in business decision making. Business ethics can be confusing and complex because a right or wrong answer often does not exist. Because this area is so emotional and controversial, it is extremely important to use your critical thinking skills when responding to questions about business ethics. It would be very easy to make arguments based on your gut reaction to cases such as the Bhopal gas incident. You should, however, carefully use your critical thinking skills to draw an informed conclusion. The following questions can help you begin to understand the complexity surrounding business ethics.

1. As critical thinkers, you have learned that ambiguous words—words that have multiple possible meanings—can cause confusion in the legal environment. Perhaps the best example of ambiguity in the legal environment is the phrase social responsibility. What definitions of responsibility can you generate?

   Clue: Consider the Bhopal incident. Do you think Union Carbide would have the same definition of social responsibility as the families of the Indian accident victims?

2. It is common for individuals, businesses, judges, and juries each to use different meanings of the phrase social responsibility. Preferences for certain ethical norms might account for these different meanings. If executives of a company thought that security was extremely important, how might their definition of social responsibility be affected?

   Clue: Remember the definitions of security in Chapter 1. If Union Carbide valued security, how might the company treat the victims of the Bhopal incident?

3. Your friend discovers that you are taking a class on the legal environment of business. He says, "I'm extremely angry at the cigarette companies. They knew that cigarettes cause cancer. Don't those companies have a responsibility to protect us?" Because you are trained in critical thinking, you know that his question does not have a simple answer. Keeping your critical thinking skills in mind, how would you intelligently respond to his question?

   Clue: Consider the critical thinking questions about ambiguity, ethical norms, and missing information set out in Chapter 1 of this text.

Definition of Business Ethics and the Social Responsibility of Business

BUSINESS ETHICS

ethics The study of what makes up good and bad conduct, inclusive of related actions and values.

business ethics The study of what makes up good and bad conduct as related to business activities and values.

Ethics is the study of good and bad behavior. **Business ethics** is a subset of the study of ethics and is defined as the study of what makes up good and bad business conduct. This conduct occurs when the firm acts as an organization, as well as when individual managers make decisions inside the organization. For example, there may be differences between the way Warren Anderson personally looked at the Bhopal tragedy (a failure of the plant to implement company operating standards) and the way the corporation's board of directors and the chemical industry did (the Indian government allowed people to live too close to the plant). It is important to look at "business" ethics not as a single monolithic system, but from the perspective of individual managers, corporations, and industrywide ethical concerns. Each may view and judge a particular happening in a different way.

How these groups think depends on their ethical norms and on their philosophy or theory of ethics. To help you understand their thinking, we include a discussion of three schools of ethical thought. Individual managers, corporations, or industries may belong to any one of the schools, as each school has its advocates and refinements. In addition, each school attempts to explain why an action is right or wrong and how one knows it to be right or wrong.

THE SOCIAL RESPONSIBILITY OF BUSINESS

The **social responsibility** of business is defined as a concern by business about both its profit-seeking and its non-profit-seeking activities and their intended and unintended impacts on groups and individuals other than management or the owners of a corporation (e.g., consumers, environmentalists, and political groups). Since the late 1960s, an outcry has arisen for businesses to be more socially responsible. This expression of public concern has resulted in part from three factors:

social responsibility Concern of business entities about profit-seeking and non-profit-seeking activities and their unintended impact on others directly or indirectly involved.

1. *The complexity and interdependence of a postindustrial society.* No individual or business is an island. If a company builds a chemical plant in Bhopal, India, and its primary purpose is to make profits for its shareholders, can it be held responsible to the public that lives around the plant when there is a gas leak? The public is dependent on the firm's good conduct, and the firm is dependent on the public and its political representatives to supply labor, an adequate water supply, tax forgiveness, roads, and so on.

2. *Political influence that has translated public outcry for socially responsible conduct into government regulation.* Whether a malfunction occurs at a nuclear plant at Three Mile Island or a human disaster is caused by a gas leak in Bhopal, India, the political arm of government at all levels sees the solution as more regulation. This attitude pleases the government's constituents and makes its officials more electable.

3. *Philosophical differences about what the obligations of business should be.* Neoclassical economic theory would argue that the sole purpose of business is to make a profit for its investing shareholders, who in turn reinvest, creating expanded or new businesses that employ more people, thus creating a higher standard of living.

Different people hold different theories of social responsibility. Some argue for a managerial or coping approach; that is, "throw money" at the problem when it occurs, such as the Bhopal disaster, and it will go away. Others subscribe to a more encompassing theory of social responsibility, holding that business, like any other institution in our society (e.g., unions, churches), has a social responsibility not only to shareholders (or members or congregations) but also to diverse groups, such as consumers and political, ethnic, racial-group, and gender-oriented organizations. These and other schools of social responsibility are discussed later in this chapter.

The following case offers several possible theories of social responsibility as applied to a controversial factual situation.

CASE 7-1

In re Exxon Valdez
U.S. District Court, District of Alaska
296 F. Supp. 2d 107 (2004)

On Good Friday, March 24, 1989, the oil tanker *Exxon Valdez* was run aground on Bligh Reef in Prince William Sound, Alaska. On March 24, 1989, Joseph Hazelwood was in command of the *Exxon Valdez*. Defendant Exxon Shipping [Company] owned the *Exxon Valdez*. Exxon employed Captain Hazelwood, and kept him employed knowing that he had an alcohol problem. The captain had supposedly been rehabilitated, but Exxon knew better before March 24, 1989. Hazelwood had sought treatment for alcohol abuse in 1985 but had "fallen off the wagon" by the spring of 1986. Yet, Exxon continued to allow Hazelwood to command a supertanker carrying a hazardous cargo. Because Exxon did nothing despite its knowledge that Hazelwood was once again drinking, Captain Hazelwood was the person in charge of a vessel as long as three football fields and carrying 53 million gallons of crude oil. The best available estimate of the crude oil lost from the *Exxon Valdez* into Prince William Sound is about 11 million gallons. Commercial fisheries throughout this area were totally disrupted, with entire fisheries being closed for the 1989 season. Subsistence fishing by residents of Prince William Sound and Lower Cook Inlet villages was also disrupted. Shore-based businesses dependent upon the fishing industry were also disrupted as were the resources of cities such as Cordova. Exxon undertook a massive cleanup effort. Approximately $2.1 billion was ultimately spent in efforts to remove the spilled crude oil from the waters and beaches of Prince William Sound, Lower Cook Inlet, and Kodiak Island. Also, Exxon undertook a voluntary claims program, ultimately paying out $303 million, principally to fishermen whose livelihood was disrupted. [Lawsuits] (involving thousands of plaintiffs) were ultimately consolidated into this case.

The jury awarded a breathtaking $5 billion in punitive damages against Exxon. Exxon appealed the amount of punitive damages [to the U.S. Court of Appeals for the Ninth Circuit]. [T]he Ninth Circuit Court of Appeals in this case reiterated [that] the guideposts for use in determining whether punitive damages are grossly excessive [include] the reprehensibility of the defendant's conduct. The court of appeals remanded the case [and] unequivocally told this court that "[t]he $5 billion punitive damages award is too high" and "[i]t must be reduced."

Justice Holland

[T]he reprehensibility of the defendant's conduct is the most important indicium [indication] of the reasonableness of a punitive damages award. In determining whether a defendant's conduct is reprehensible, the court considers whether "The harm caused was physical as opposed to economic; the tortious conduct evinced an indifference to or a reckless disregard of the health or safety of others; the target of the conduct had financial vulnerability; the conduct involved repeated actions or was an isolated incident; and the harm was the result of intentional malice, trickery, or deceit, or mere accident."

The reprehensibility of a party's conduct, like truth and beauty, is subjective. One's view of the quality of an actor's conduct is the result of complex value judgments. The evaluation of a victim will vary considerably from that of a person not affected by an incident. Courts employ disinterested, unaffected lay jurors in the first instance to appraise the reprehensibility of a defendant's conduct. Here, the jury heard about what Exxon knew, and what its officers did and what they failed to do. Knowing what Exxon knew and did through its officers, the jury concluded that Exxon's conduct was highly reprehensible.

Punitive damages should reflect the enormity of the defendant's offense. Exxon's conduct did not simply cause economic harm to the plaintiffs. Exxon's decision to leave Captain Hazelwood in command of the *Exxon Valdez* demonstrated reckless disregard for a broad range of legitimate Alaska concerns: the livelihood, health, and safety of the residents of Prince William Sound, the crew of the *Exxon Valdez*, and others. Exxon's conduct targeted some financially vulnerable individuals, namely subsistence fishermen. Plaintiffs' harm was not the result of an isolated incident but was the result of Exxon's repeated decisions, over a period of approximately three years, to allow Captain Hazelwood to remain in command despite Exxon's knowledge that he was drinking and driving again. Exxon's bad conduct as to Captain Hazelwood and his operating of the *Exxon Valdez* was intentionally malicious.

Exxon's conduct was many degrees of magnitude more egregious [flagrant] [than defendant's conduct in other cases]. For approximately three years, Exxon management, with knowledge that Captain Hazelwood had fallen off the wagon, willfully permitted him to operate a fully loaded crude oil tanker in and out of Prince William Sound—a body of water which Exxon knew to be highly valuable for its fisheries resources. Exxon's argument that its conduct in permitting a relapsed alcoholic to operate an oil tanker should be characterized as less reprehensible than [in other cases] suggests that Exxon, even today, has not come to grips with the opprobrium [disgracefulness] which society rightly attaches to drunk driving. Based on the foregoing, the court finds Exxon's conduct highly reprehensible.

[T]he court reduces the punitive damages award to $4.5 billion as the means of resolving the conflict between its conclusion and the directions of the court of appeals.

[T]here is no just reason to delay entry of a final judgment in this case. The Court's judgment as to the $4.5 billion punitive damages award is deemed final.*

> **COMMENT:** In February 2008, the U.S. Supreme Court heard oral arguments on appeals from the U.S. Court of Appeals (D.C. Circuit), which had affirmed the U.S. District Court. The United States Supreme Court affirmed the lower courts (S. Ct. 2008).

---

*In re Exxon Valdez, U.S. District Court, District of Alaska 296 F. Supp. 2d 107 (2004).

Theories of Ethical Thought

CONSEQUENTIAL THEORIES

Ethicists, businesspeople, and workers who adhere to a consequential theory of ethics judge acts as ethically good or bad based on whether the acts have achieved their desired results. The actions of a business or any other societal unit are looked at as right or wrong only in terms of whether the results can be rationalized (Table 7-1).

This theory is best exemplified by the utilitarian school of thought, which is divided into two subschools: *act utilitarianism* and *rule utilitarianism*. In general, adherents of this school judge all conduct of individuals or businesses on whether that conduct brings net happiness or pleasure to a society. They judge an act ethically correct after adding up the risks (unhappiness) and the benefits (happiness) to society and obtaining a net outcome. For example, if it is necessary for a company to pay a bribe to a foreign official in order to get

TABLE 7-1

THEORIES OF ETHICAL THOUGHT

Consequential theories	Acts are judged good or bad based on whether the acts have achieved their desired results.
	Acts of the business community or any other social unit (e.g., government, school, fraternity, and sorority).
	Act and rule utilitarianisms are two subschools.
Deontological theories	Actions can be judged good or bad based on rules and principles that are applied universally.
Humanist theories	Actions are evaluated as good or bad depending on whether they contribute to improving inherent human capacities such as intelligence, wisdom, and self-restraint.

LINKING LAW AND BUSINESS

Business Ethics

Managers often attempt to encourage ethical practices in the workplace. A significant reason for managers' concern with ethics is to portray their organizations in a favorable light to consumers, investors, and employees. As a means of creating an ethical workplace, there are several methods that managers should implement that you may recall from your management class:

1. Create a code of ethics, which is a formal statement that acts as a guide for making decisions and actions within an organization. Distribution and continual improvement of the code of ethics are also important steps.
2. Establish a workplace or office for the sole purpose of overseeing organizational practices to determine if actions are ethical.
3. Conduct training programs to encourage ethical practices in the organization.
4. Minimize situations in which unethical behavior is common and create conditions in which people are likely to behave ethically. Two practices that often result in unethical behavior are to give unusually high rewards for good performance and uncommonly harsh punishments for poor performance. By eliminating these two causal factors of unethical behavior, managers are more likely to create conditions in which employees choose to behave ethically within the organization. Therefore, a manager's hope to represent the organization in a respectable manner may be achieved through the institution and implementation of these methods.

Source: S. Certo, *Modern Management* (Upper Saddle River, NJ: Prentice Hall, 2000), 66–69.

several billion dollars' worth of airplane contracts, utilitarians would argue, in general, that the payment is ethically correct because it will provide net happiness to society; that is, it will bring jobs and spending to the community where the airplane company is located. If the bribe is not paid, the contracts, jobs, and spending will go to a company somewhere else.

Act utilitarians determine if an action is right or wrong on the basis of whether that individual act (the payment of a bribe) alone brings net happiness to society, as opposed to whether other alternatives (e.g., not paying the bribe or allowing others to pay the bribe) would bring more or less net happiness. Rule utilitarians argue that an act (the payment of the bribe) is ethically right if the performance of similar acts by all similar agents (other contractors) would produce the best results in society or has done so in the past. Rule utilitarians take the position that whatever applicable rule has been established by political representatives must be followed and should serve as a standard in the evaluation of similar acts. If payment of bribes has been determined by the society to bring net happiness, and a rule allowing bribes exists, then rule utilitarians would allow the bribe. In contrast, the Foreign Corrupt Practices Act of 1977, as amended in 1988, which forbids paying bribes to foreign government officials to get business that would not have been obtained without such a payment, is an example of a standard that rule utilitarians would argue must be followed but that would lead to a different result. Hence, the act utilitarians might get the airplane plant, but the rule utilitarians, if they were following the Foreign Corrupt Practices Act, would not.

We must note that both act and rule utilitarians focus on the consequences of an act and not on the question of verifying whether an act is ethically good or bad.[1] Either one of these theories can be used by individuals or businesses to justify their actions.[2] Act utilitarians use the principle of utility (adding up the costs and benefits of an act to arrive at net happiness) to focus on an individual action at one point in time. Rule utilitarians believe that one should not consider the consequences of a single act in determining net happiness, but instead should focus on a general rule that exemplifies net happiness for the whole society. Case 7-1 illustrates a rule-utilitarian view of jurisprudence.

DEONTOLOGICAL THEORIES

Deontology is derived from the Greek word *deon,* meaning "duty." For advocates of deontology, rules and principles determine whether actions are ethically good or bad. The consequences of individual actions are not considered. The golden rule, "Do unto others as you would have them do unto you," is the hallmark of this theory.

Absolute deontology claims that actions can be judged ethically good or bad on the basis of absolute moral principles arrived at by human reason regardless of the consequences of an action, that is, regardless of whether there is net happiness.[3] Immanuel Kant (1724–1804) provided an example of an absolute moral principle in his widely studied "categorical imperative." He stated that a person ought to engage only in acts that he or she could see becoming a universal standard. For example, if a U.S. company bribes a foreign official to obtain a contract to build airplanes, then U.S. society and business should be willing to accept the principle that foreign multinationals will be morally free to bribe U.S.

---

[1] W. Lacroix, *Principles for Ethics in Business* (rev. ed.), 6 (Washington, DC: University Press, 1979).
[2] B. Brennan, "Amending the Foreign Corrupt Practices Act of 1977: Clarifying or Gutting a Law," *Journal of Legislation* 2 (1984), for an examination of the rule and act utilitarian schools of thought within the context of the proposed amendment of the 1977 Foreign Corrupt Practices Act.
[3] Lacroix, *supra* note 1, at 13.

government officials to obtain defense contracts. Of course, the reverse will be true if nonbribery statutes are adopted worldwide. Kant, as part of his statement of the categorical imperative, assumed that everyone is a rational being having free will, and he warned that one ought to "treat others as having intrinsic values in themselves, and not merely as a means to achieve one's end."[4] For deontologists such as Kant, ethical reasoning means adopting universal principles that are applied to everyone equally. Segregation of one ethnic or racial group is unethical because it denies the intrinsic value of each human being and thus violates a general universal principle.

HUMANIST THEORIES

A third school of thought, the humanist school, evaluates actions as ethically good or bad depending on what they contribute to improving inherent human capacities such as intelligence, wisdom, and self-restraint. Many natural law theorists (examined in Chapter 2) believe that humans would arrive by reason alone at standards of conduct that ultimately derive from a divine being or another ultimate source such as nature. For example, if a U.S. business participates in bribing a foreign official, it is not doing an act that improves inherent human capacities such as intelligence and wisdom; thus, the act is not ethical. In a situation that demanded choice, as well as the use of the intelligence and restraint that would prevent a violation of law (the Foreign Corrupt Practices Act of 1977), the particular business would have failed ethically as well as legally.

APPLYING THE LAW TO THE FACTS...

> Jordan is the CEO of a paper manufacturing company. Jordan's business advisors inform him that there is a cigarette corporation he could take over and he would then be able to increase his profits. Jordan rejects the idea, saying that manufacturing cigarettes would bring suffering to the public. What theory of ethical thought is guiding Jordan's actions?

Codes of Ethics

INDIVIDUAL CODES OF ETHICS

When examining business ethics, one must recognize that the corporations, partnerships, and other entities that make up the business community are a composite of individuals. If the readers of this book are asked where they obtained their ethical values, they might respond that their values come from parents, church, peers, teachers, brothers and sisters, or the environment. In any event, corporations and the culture of a corporation are greatly influenced by what ethical values individuals bring to them. Often, business managers are faced with a conflict between their individual ethical values and those of the corporation. For example, a father of three young children, who is divorced and their sole support, is asked by his supervisor to "slightly change" figures that will make the results of animal tests of a new drug look more favorable when reported to the Food and Drug Administration. His supervisor hints that if he fails to do so, he may be looking for another job. The individual is faced with a conflict in ethical values: individual values of honesty and humaneness toward potential users of the drug

---

[4] R. Wolff, ed., *Foundations of the Metaphysics of Moral Thought and Critical Essays* 44 (Bobbs-Merrill, 1964).

versus business values of profits, efficiency, loyalty to the corporation, and the need for a job. Which values should determine his actions?

Individual Ethical Codes versus Groupthink

On January 28, 1986, just 74 seconds into its launch, the space shuttle *Challenger* exploded, killing the first schoolteacher in space, Christa McAuliffe, and six other astronauts on board. A presidential commission set up to investigate the disaster found that faulty O-rings in the booster rockets were to blame. Two engineers testified before the commission that they had opposed the launch but were overruled by their immediate supervisor and other officials of the Thiokol Corporation that manufactured the booster rockets. The two engineers continued to warn of problems with the O-rings until the day before the launch. After the launch, one engineer was assigned to "special projects" for the firm. Another took leave and founded a consulting firm. The second schoolteacher in space, Barbara Morgan, was a backup to McAuliffe. She returned to teaching for 22 years after the *Challenger* incident until August 8, 2007, when she and six other astronauts were sent (successfully) to the International Space Station on the shuttle *Endeavor*.[a]

On September 11, 2001, two planes flew into the Twin Towers of the World Trade Center in New York City, one plane flew into the Pentagon in Washington, DC, and another flew into a field near Pittsburgh, Pennsylvania. Approximately 3,000 people were killed by terrorists flying the planes. Again, a presidential commission was set up. In 2004, the commission reported that the failure of intermediate-level employees to be heard within intelligence agencies, as well as the inability of agencies such as the Central Intelligence Agency (CIA), Defense Intelligence Agency (DIA), and National Security Agency (NSA) to bring early warning information forward to the decision makers (in the White House), was in part responsible for the events that took place. The CIA director resigned and other officials at some agencies retired. A new structure was set up for intelligence gathering in 2004, which allows a single individual to be responsible for intelligence provided to the president of the United States.

These factual situations are very different, but when reading the testimony presented to the presidential commissions, it appears that, in both cases, there were conflicts between individual ethical values and groupthink. *Groupthink*, as used here, is defined as a form of thinking that people engage in when they are involved in a cohesive in-group, striving for unanimity, which overrules a realistic appraisal of alternative courses of action. Groupthink refers to "a deterioration of mental efficiency, reality testing, and moral efficiency that results from in-group pressures."[b] For the engineers in the *Challenger* case and the middle-level managers of the intelligence agencies, the question always will remain: Were they part of a groupthink process that altered the outcome? Are there important factual differences in these cases: private-sector employment (Thiokol) as opposed to public-sector intelligence agencies? The rise of the groupthink process is currently popular. It is taught in colleges of business and in other academic sectors. For years, however, solitude had been associated with creativity. Great thinkers of our times (Moses, Buddha, writers of great books) have generally worked by themselves. A very quiet man named Steve Wozniak needed extra quietude to design the first calculators while working at Hewlett-Packard. HP made it easy for Wozniak to collaborate with his colleagues over coffee and donuts between 10 a.m. and 2 p.m. in a separate and easily accessible room. Mr. Wozniak later helped start Apple. The "New

---

[a]W. Leary, "Teacher Astronaut to Fly Decade after *Challenger*," *New York Times*, August 7, 2007, p. 11.

[b]Excerpt from Victims of Groupthink by Irving L. Janis. Published by Houghton Mifflin Harcourt, © 1972.

Groupthink Process" combines quietude of thinking and people engaged in a cohesive group striving for unanimity.

Groupthink, however, may be necessary in our society. Without it, how would we organize our corporations, the military, and government agencies? If we allowed everyone to think independently, would anyone follow orders in the military or build rocket boosters in industry? Also, people often do not think like whistleblowers (see discussion of the Sarbanes-Oxley Act later in this chapter and in Chapter 23) for fear of losing their status and their jobs, which are often necessary to support egos and families.

Before answering any of the questions posed here, return to Chapter 1 and review the eight steps in critical thinking outlined there.

CORPORATE CODES OF ETHICS

The total of individual employees' ethical values influences corporate conduct, especially in a corporation's early years. The activities during these years, in turn, form the basis of what constitutes a *corporate culture* or an environment for doing business. In a free-market society, values of productivity, efficiency, and profits become part of the culture of all companies. Some companies seek to generate productivity by cooperation between workers and management; others motivate through intense production goals that may bring about high labor turnover. Some companies have marketed their product through emphasis on quality and service; others emphasize beating the competition through lower prices.[5] Over time, these production and marketing emphases have evolved into what is called a *corporate culture,* often memorialized in corporate codes.

Since the mid-1960s, approximately 90 percent of all major corporations have adopted codes of conduct. In general, the codes apply to upper- and middle-level managers. They are usually implemented by a chief executive officer or a designated agent. They tend to establish sanctions for deviant behavior, ranging from personal reprimands that are placed in the employee's file, to dismissal. Some formal codes allow for due process hearings within the corporation, in which an employee accused of a violation is given a chance to defend himself or herself. With many employees bringing wrongful dismissal actions in courts of law, formal internal procedures are increasingly evolving to implement due process requirements. A study of corporate codes reveals that the actions most typically forbidden are:[6]

- Paying bribes to foreign government officials
- Fixing prices
- Giving gifts to customers or accepting gifts from suppliers
- Using insider information
- Revealing trade secrets

Corporate Ethics. *Internal Housecleaning* Following several financial scandals involving companies such as Enron, Martha Stewart Living, Inc., ImClone Systems, WorldCom, Inc. (now MCI), and Tyco International (see Chapter 6 for analysis of some of these), Congress passed the Sarbanes-Oxley Act in 2002. This act required publicly traded companies to set up confidential internal systems by April 2003 so that employees and others could have a method of reporting possible illegal or unethical auditing and accounting practices, as well as other problems such as sexual harassment.

---

[5]C. Power and D. Vogel, *Ethics in the Education of Business Managers* 6 (Hastings Center, 1980).

[6]K. Chatov, "What Corporate Ethics Statements Say," *California Management Review* 22: 206 (1980).

Web reporting systems such as Ethicspoint allow employees of companies to click an icon on their computers and be linked anonymously to the reporting services. Employees may report alleged unethical or illegal activity. The reporting system then alerts a management person or the audit committee of the board of directors to any possible problem. Other systems use a special hotline phone number (800 or 900). No system is perfect, but the key factor is that Sarbanes-Oxley has given legal impetus to "cleaning house" internally.

Whistleblowing protection under Section 806 of Sarbanes-Oxley[7] prohibits any publicly traded company from "discharging, demoting, suspending, threatening or otherwise discriminating against an employee who provides information to the government or assists in a government investigation regarding conduct that an employee believes may be a violation of the securities laws." As noted in Chapter 23, penalties are both civil and criminal in nature.

INDUSTRY CODES OF ETHICS

In addition to corporate ethical codes, industry codes exist, such as those of the National Association of Broadcasters or the National Association of Used Car Dealers. In most cases, these codes are rather general and contain either affirmative inspirational guidelines or a list of "shall-nots." A hybrid model including "dos and don'ts" generally addresses itself to subjects such as:[8]

- Honest and fair dealings with customers
- Acceptable levels of safety, efficacy, and cleanliness
- Nondeceptive advertising
- Maintenance of experienced and trained personnel, competent performance of services, and furnishing of quality products

Most trade associations were formed for the purpose of lobbying Congress, the executive branch, and the regulatory agencies, in addition to influencing elections through their political action committees (PACs). They have not generally been effective in monitoring violations of their own ethical codes. In light of the reasons for their existence and the fact that membership dues support their work, it is not likely that they will be very effective disciplinarians.

Some effective self-regulating mechanisms, however, do exist in industries. In Chapter 23, readers will see that self-regulating organizations (SROs) such as the National Association of Securities Dealers and the New York Stock Exchange have used the authority delegated to them by the Securities and Exchange Commission (SEC) in an extremely efficient manner. In addition, the Council of Better Business Bureaus, through its National Advertising Division (NAD), has provided empirical evidence that self-regulation can be effective. The NAD seeks to monitor and expose false advertising through its local bureaus and has done an effective job, receiving commendations from a leading consumer advocate, Ralph Nader.[9]

PROFESSIONAL CODES OF ETHICS

Within a corporation, managers often interact with individual employees who are subject to "professional" codes of conduct that may supersede corporate or industrywide codes in terms of what activities they can participate in and still

---

[7]H.R. 3782, signed into law by President George W. Bush on July 30, 2002, effective on August 29, 2002. Pub. L. No. 109–204; 15 U.S.C. § 78d (I)–(3), codified in Exchange Act § 4. See Chapter 23 for a full discussion.

[8]See R. Jacobs, "Vehicles for Self-Regulation Codes of Conduct, Credentialing and Standards," in *Self-Regulation*, Conference Proceedings (Washington, DC: Ethics Resource).

[9]See R. Tankersley, "Advertising: Regulation, Deregulation and Self-Regulation," in *Self-Regulation*, Conference Proceedings, *supra* note 7, at 45.

remain licensed professionals. For example, under the Model Code of Professional Responsibility, a lawyer must reveal the intention of his or her client to commit a crime and the information necessary to prevent the crime.[10] When a lawyer, a member of the law department of Airplane Corporation X, learns that his company deliberately intends to bribe a high-level foreign official in order to obtain an airplane contract, he may be forced, under the Model Code, to disclose this intention, because the planned bribe is a violation of the Foreign Corrupt Practices Act of 1977, as amended, an act that has criminal penalties. Failure to disclose could lead to suspension or disbarment by the lawyer's state bar. Management must be sensitive to this and to the several professional codes that exist (discussed later in this chapter).

Professionals is an often-overused term, referring to everything from masons to hair stylists to engineers, lawyers, and doctors. When discussing professions or professionals here, we mean a group that has the following characteristics:

- Mandatory university educational training before licensing, as well as continuing education requirements
- Licensing-examination requirements
- A set of written ethical standards that is recognized and continually enforced by the group
- A formal association or group that meets regularly
- An independent commitment to the public interest
- Formal recognition by the public as a professional group

Management must often interact with the professions outlined in the following paragraphs. Each of them has a separate code of conduct. An awareness of this fact may lead to a greater understanding of why each group acts as it does.

Accounting. The American Institute of Certified Public Accountants (AICPA) has promulgated a code of professional ethics and interpretive rules. The Institute of Internal Auditors has set out a code of ethics, as well as a Statement of Responsibilities of Internal Auditors. In addition, the Association of Government Accountants has promulgated a code of ethics.

Disciplinary procedures are set forth for both individuals and firms in the Code of Professional Ethics for Certified Public Accountants (CPAs). Membership in the AICPA is suspended without a hearing if a judgment of conviction is filed with the secretary of the institute as related to the following:[11]

- A felony as defined under any state law
- The willful failure to file an income tax return, which the CPA as an individual is required to file
- Filing a fraudulent return on the part of the CPA for his or her own return or that of a client
- Aiding in the preparation of a fraudulent income tax return of a client

The AICPA Division for CPA Firms is responsible for disciplining firms, as opposed to individuals. Through its SEC practice and its private company sections, this division requires member firms to (1) adhere to quality-control standards,

---

[10]See Model Code of Professional Responsibility DR 4-401(C) and Formal Op. 314 (1965).
[11]See AICPA Professional Standards, vol. 2, Disciplinary Suspensions and Termination of Membership Hearings, GL 730.01. See AICPA Professional Standards, vol. 2, Disciplinary Suspensions and Termination of Membership Hearings, GL 730.01. See AICPA Professional Standards, vol. 2, Disciplinary Suspensions and Termination of Membership Hearings, GL 730.01. The United States Public Company Accounting Oversight Board was established in 2012 to oversee accounting standards. Public No. 107-204 (Codified as Exchange Act Sec. 4, 15 U.S.C. Sec. 78(d).

(2) submit to peer review of their accounting and audit practices every three years, (3) ensure that all professionals participate in continuing education programs, and (4) maintain minimum amounts of liability insurance.

Accountants' ethical responsibility is reinforced by the Sarbanes-Oxley Act of 2002, which was passed by Congress following a series of financial scandals (see Chapters 6 and 23).[12] This act mandated the creation of a Public Company Accounting Oversight Board; it also included provisions requiring auditor independence. Under Section 802 of the act, accountants are required to maintain on file working papers relating to an audit or review for five years. A willful violation is subject to a fine, imprisonment for up to 10 years, or both.

Other statutory provisions affecting accountants include Sections 11 and 12(2) of the 1933 Securities Act, as well as Sections 10(b) and 18 of the 1934 Securities Exchange Act. The 1933 act deals with accountant liability for false statements or omission of a material fact in auditing financial statements required for registration of securities. A defense is due diligence and a reasonable belief that the work is complete.

Under Section 10(b) of the 1934 act, accountants are liable for false and misleading statements in reports required by the act (see Chapter 23). Willful violations bring criminal penalties. Additionally, provisions of the Internal Revenue Code provide (felony) criminal penalties for tax preparers who willfully prepare or assist in preparing a false return.[13] Tax preparers who negligently or willfully understate tax liability are also subject to criminal penalties. Furthermore, failure to provide a taxpayer with a copy of his or her return may subject a tax preparer to criminal penalties.[14]

Insurance and Finance. The American Society of Chartered Life Underwriters (ASCLU) adopted a Code of Ethics consisting of eight guides to professional conduct and six rules of professional conduct. The guides are broad in nature, whereas the rules are specific. Enforcement of the Code of Ethics is left primarily to local chapters. Discipline includes reprimand, censure, and dismissal. A local chapter can additionally recommend suspension or revocation to a national board. Very few disciplinary actions have been forthcoming.[15]

In addition, the Society of Chartered Property and Casualty Underwriters (CPCU) has a code of ethics consisting of seven "Specified Unethical Practices," as well as three "Unspecified Unethical Practices" of a more general nature. Upon receipt of a written and signed complaint, the president of the society appoints a three-member conference panel to hear the case. If a panel finds a member guilty of an unspecified unethical practice, the president directs the member to cease such action. If a member is found guilty of a specified unethical practice, the society's board of directors may reprimand or censure the violator or suspend or expel her or him from membership in the society.

Law. The American Bar Association's Model Rules of Professional Responsibility were submitted to the highest state courts and the District of Columbia for adoption, after the association's House of Delegates approved them in August 1983 (before then, the states had adopted the Model Code of Professional Responsibility). There are nine Canons of Professional Responsibility. From these are derived Ethical Considerations and Disciplinary Rules. The Model Rules set out a minimal level of conduct that is expected of an attorney. Violation of any of these rules may lead to warnings, reprimands, public censure, suspension, or

---

[12]H.R. 3762, signed into law by President George W. Bush on July 30, 2002, effective August 30, 2002.

[13]26 U.S.C. § 7208(2).

[14]26 U.S.C. § 7101(a)(36).

[15]See R. Horn, *On Professions, Professionals, and Professional Ethics* 74 (Malvern, PA: American Institute for Property and Liability Underwriters, 1978).

disbarment by the enforcement agency of the highest state court in which the attorney is admitted to practice. Most state bar disciplinary actions are published in state bar journals and local newspapers, so lawyers and the public in general are aware of attorneys who have been subject to disciplinary action.

The case excerpted here illustrates some legal problems surrounding professional ethical codes when they result in price-fixing.

Schools of Social Responsibility

Early in this chapter, the social responsibility of business was defined as a concern by business about both its profit and its nonprofit activities and their intended and unintended impact on others. As you will see, theories of ethics and schools of social responsibility are not necessarily mutually exclusive. For example, the primary purpose of a steel company is to make a profit for its individual and institutional shareholders. The unintended effects of this company's actions might be that the surrounding community has polluted waters, and homes are affected by ash that falls from the company's smokestack. Similarly, in the Union Carbide incident described at the beginning of this chapter, the purpose of Union Carbide India Ltd. was to make a profit for its shareholders. By doing so, it was able to employ people. The unintended effect of this activity was a gas leak that killed approximately 2,000 people and injured many more. The question in both of these cases is: What responsibility, if any, do firms have for the unintended effects of their profit-seeking activity? This section discusses five views of social responsibility that seek to answer that question: profit oriented, managerial, institutional, professional obligation, and regulation (Table 7-2). These schools reflect the ethical values or culture of a corporation. The reader should analyze each, realizing that the answer may not lie in any one.

PROFIT-ORIENTED SCHOOL

The profit-oriented school of social responsibility begins with a market-oriented concept of the firm that most readers were exposed to in their first or second course in economics. Holders of this theory argue that business entities are distinct organizations in our society and that their sole purpose is to increase profits

TABLE 7-2 SCHOOLS OF SOCIAL RESPONSIBILITY

Profit-oriented school	Business entities are distinct organizations in our society whose sole purpose is to increase profits for shareholders.
Managerial school	Advocates of this theory argue that business entities (particularly large ones) have a number of groups that they must deal with. They include not only stockholders but also employees, customers, activist groups, and government regulators, all of whom may make claims on the entities' resources.
Institutional school	Business entities have a responsibility to act in a manner that benefits all society.
Professional obligation school	Business managers and members of boards of directors must be certified as "professionals" before they assume managerial responsibilities. They must have a responsibility to the public interest beyond making profits. The Sarbanes-Oxley Act (see Chapters 6 and 23) may be leading in that direction.
Regulation school	All business units are accountable to elected officials. See the Sarbanes-Oxley Act in regard to dealing with independent financial audits (Chapter 23).

for shareholders. Businesses are to be judged solely on criteria of economic efficiency and how well they contribute to growth in productivity and technology. Corporate social responsibility is shown by managers who maximize profits for their shareholders, who, in turn, are able to reinvest such profits, providing for increased productivity, new employment opportunities, and increased consumption of goods.

Classical economists, who advocate this position, recognize that there will be unintended effects of such profit-seeking activities (*externalities*) that affect society and cannot be incorporated into or passed on in the price of output. They would argue that this is the "social cost" of doing business. Such social costs are a collective responsibility of the government. Individual businesses should not be expected to voluntarily incorporate in their product's price the cost of cleaning up water or air, because this incorporation will distort the market mechanism and the efficient use of resources. Profit-seeking advocates argue that, when government must act in a collective manner, it should act in a way that involves the least interference with the efficiency of the market system, preferably through direct taxation.

In summary, efforts at pollution control, upgrading minority workers, and bringing equality of payment to the workforce are all tasks of the government and not of the private sector, which is incapable of making such choices and is not elected in a democratic society to do so. Its sole responsibility is to seek profits for its shareholders. The following box represents an important set of issues.

"Old Joe Camel" was adopted by R. J. Reynolds (RJR) in 1913 as the symbol for the brand Camel. In late 1990, RJR revived Old Joe with a new look in the form of a cartoon meant to appeal to young smokers.

In December 1991, the *Journal of the American Medical Association* (*JAMA*) published three surveys that found the cartoon character Joe Camel reached children very effectively.[a] Of children between ages 3 and 6 who were surveyed, 51.1 percent recognized Old Joe Camel as being associated with Camel cigarettes. The 6-year-olds were as familiar with Joe Camel as they were with the Mickey Mouse logo for the Disney Channel.

An RJR spokeswoman claimed that "just because children can identify our logo doesn't mean they will use our product." Since the introduction of Joe Camel, however, Camel's share of the under-18 market climbed to 33 percent from 5 percent. Among 17- to 24-year-olds, Camel's market share climbed to 7.9 percent from 4.4 percent.

The Centers for Disease Control reported in March 1992 that smokers between ages 12 and 18 preferred Marlboro, Newport, or Camel cigarettes, the three brands with the most extensive advertising.[b]

Teenagers throughout the country were wearing Joe Camel T-shirts. Brown & Williamson, the producer of Kool cigarettes, began testing a cartoon character for its advertisements, a penguin wearing sunglasses and DayGlo sneakers. Company spokesman Joseph Helewicz stated that the advertisements were geared to smokers between 21 and 35 years old. Helewicz added that cartoon advertisements for adults were not new and cited the Pillsbury Doughboy and the Pink Panther as effective advertising images.

In mid-1992, then–Surgeon General Antonia Novello, along with the American Medical Association, began a campaign called "Dump the Hump" to pressure the tobacco industry to stop advertising campaigns that encourage kids to smoke. In 1993, the FTC staff recommended a ban on the Joe Camel advertisements.

---

[a]K. Deveny, "Joe Camel Ads Reach Children," *Wall Street Journal,* December 11, 1991, p. B-1.
[b]*Id.*

In 1994, then–Surgeon General Jocelyn Elders blamed the tobacco industry's $4 billion in advertisements for increased smoking rates among teens. RJR's tobacco division chief, James W. Johnston, responded, "I'll be damned if I'll pull the ads." RJR put together a team of lawyers and others it referred to as in-house censors to control Joe's influence. A campaign to have Joe wear a bandana was nixed, as was one for a punker Joe with pink hair.

In 1994, RJR's CEO James W. Johnston testified before a congressional panel on the Joe Camel controversy and stated, "We do not market to children and will not," and added, "We do not survey anyone under the age of 18."

Internal documents about targeting young people were damaging, though. A 1981 RJR internal memorandum on marketing surveys cautioned research personnel to tally underage smokers as "age 18." A 1981 Philip Morris internal document indicated that information about smoking habits in children as young as 15 was important, because "today's teenager is tomorrow's potential regular customer." Other Philip Morris documents from the 1980s expressed concern that Marlboro sales would soon decline because teenage smoking rates were falling.

A 1987 marketing survey in France and Canada by RJR before it launched the Joe Camel campaign showed that the cartoon image with its funny and humorous image of Joe Camel attracted attention. One 1987 internal document used the phrase *young adult smokers* and noted a campaign targeted at the competition's "male Marlboro smokers ages 13–24."

A 1997 survey of 534 teens by *USA Today* revealed the following:

Advertisement	*Have Seen Advertisement*	*Liked Advertisement*
Joe Camel	95%	65%
Marlboro Man	94%	44%
Budweiser Frogs	99%	92%

Marlboro was the brand smoked by most teens in the survey. The survey found that 28 percent of teens between ages 13 and 18 smoked—an increase of 4 percent since 1991. In 1987, Camels were the cigarette of choice for 3 percent of teenagers when Joe Camel debuted. By 1993, the figure had climbed to 16 percent.

In early 1990, the Federal Trade Commission (FTC) began an investigation of RJR and its Joe Camel advertisements to determine whether underage smokers were illegally targeted by the 10-year Joe Camel campaign. The FTC had dismissed a complaint in 1994 but did not have the benefits of the newly discovered internal memorandums.

In late 1997, RJR began phasing out Joe Camel. New Camel advertisements featured healthy-looking men and women in their twenties, in clubs and swimming pools, with just a dromedary logo somewhere in the advertisement. RJR also vowed not to feature the Joe Camel character on nontobacco items such as T-shirts. The cost of the abandonment was estimated at $250 million.

In 1996, Philip Morris proposed its own plan to halt youth smoking, which included no vending machine advertisements, no billboard advertisements, no tobacco advertisements in magazines with 25 percent or more youth subscribers, and limits on sponsorships to events (rodeos, motor sports) where 75 percent or more of attendees were adults.

In 1998, combined pressure from Congress, the state attorneys general, and ongoing class action suits produced what came to be known as "the tobacco settlement." In addition to payment of $206 billion, the tobacco settlement in all of its various forms bars outdoor advertising, the use of human images (Marlboro Man) and cartoon characters (Joe Camel), and vending-machine sales. This portion of the settlement was advocated by those who were concerned about teenagers and their attraction to cigarettes via these advertisements and cigarettes' availability in machines.

> **COMMENT:** Has this litigation and pressure from our elected officials at the state and federal level affected the smoking habits of teenagers today? Explain your answer, referring to Chapter 1 in this text. What relevant information is missing here?

MANAGERIAL SCHOOL

Advocates of the managerial school of social responsibility argue that businesses, particularly large institutions, have a number of interest groups or constituents both internally and externally that they must deal with regularly, not just stockholders and a board of directors. A business entity has employees, customers, suppliers, consumers, activist groups, government regulators, and others that influence decision making and the ability of the entity to make profits. In effect, modern managers must balance conflicting claims on their time and the company's resources. Employees want better wages, working conditions, and pensions; suppliers want prompt payment for their goods; and consumers want higher-quality goods at lower prices. These often-conflicting demands lead advocates of a managerial theory of social responsibility to argue that the firm must have the trust of all groups, both internal and external. Thus, it must have clear ethical standards and a sense of social responsibility for its unintended acts in order to maximize profits and to survive in both the short and long runs. A firm that seeks to maximize short-run profits and ignores the claims of groups, whether they be unions, consumer activists, or government regulators, will not be able to survive in the complex environment in which business operates.

APPLYING THE LAW TO THE FACTS . . .

Corrine was an executive at a company and was in charge of deciding whether to spend money to open more store locations. The additional locations would bring in more profits. However, Corrine was concerned about affording and protecting the high pay rates for employees, and making sure suppliers would be paid in a timely fashion. What school of responsibility is guiding Corrine's actions?

If one reviews the Union Carbide India Ltd. incident described earlier, it is clear that the explosion in Bhopal, India, had at least three consequences: (1) It precipitated an attempt by GAF to take over the company. (2) Union Carbide made a successful but costly attempt to fight off this takeover. (3) The value of the stock decreased and, thus, the investors suffered large losses. Advocates of managerial theory would point to the investors' trust in management's ability to deal with this disaster as being important to how the market evaluated Union Carbide's stock. They also would argue that when its Tylenol product was tampered with (poisoned), the management of Johnson & Johnson took decisive action and was thus perceived by investors and customers as being trustworthy.[16] As a result, Johnson & Johnson stock value recovered relatively quickly after the incident.

In the next case, note the conflicting claims of stakeholders.

INSTITUTIONAL SCHOOL

Advocates of an institutional school of social responsibility for business argue that business entities have a responsibility to act in a manner that benefits all of society, just as churches, unions, courts, universities, and governments have. Whether

---

[16]See M. Krikorian, "Ethical Conduct: An Aid to Management," address at Albion College, Albion, MI, April 16, 1985.

it is a sole proprietorship, a partnership, or a corporation, a business is a legal entity in our society that must be held responsible for its activities. Proponents of this theory argue that the same civil and criminal sanctions should be applied to business activities that injure the social fabric of a society (e.g., the pollution of water and air) as are applied to acts of individuals and of other institutions. When managers fail to deal adequately with "externalities," they should be held accountable not only to their boards of directors, but also to government enforcement authorities and individual citizens as well.

CASE 7-2

Cooper Industries v. Leatherman Tool Group, Inc.
United States Supreme Court
121 S. Ct. 1678 (2001)

Leatherman (plaintiff) sued Cooper (defendant) in federal district court for unfair competition. Leatherman Tool Group, Inc., manufactured and sold a multifunctional tool called the PST that improved on the classic Swiss Army knife. Leatherman dominated the market for multifunctional pocket tools.

In 1995, Cooper Industries, Inc., decided to design and sell a competing multifunctional tool under the name "ToolZall." Cooper introduced the ToolZall in August 1996 at the National Hardware Show in Chicago. At that show, Cooper used photographs in its posters, packaging, and advertising materials that purported to be a ToolZall but were actually of a modified PST. When those materials were prepared, the first of the ToolZalls had not yet been manufactured. A Cooper employee created a ToolZall "mock-up" by grinding the Leatherman trademark off a PST and substituting the unique fastenings that were to be used on the ToolZall. At least one of the photographs was retouched to remove a curved indentation where the Leatherman trademark had been. The photographs were used not only at the trade show, but also in marketing materials and catalogs used by Cooper's sales force throughout the United States.

The lower court found for Leatherman in the amount of $50,000 in compensatory damages and $4.5 million in punitive damages. Cooper appealed, but the court of appeals affirmed the lower court's decision, seeing no "abuse of discretion" by the lower court as to punitive damages. Cooper appealed to the U.S. Supreme Court, petitioning for a de novo review of the facts as to the size of the punitive damages.

Justice Stevens

Although compensatory damages and punitive damages are typically awarded at the same time by the same decision maker, they serve distinct purposes. The former are intended to redress the concrete loss that the plaintiff has suffered by reason of the defendant's wrongful conduct. The latter, which have been described as "quasi-criminal," operate as private fines intended to punish the defendant and to deter future wrongdoing. A jury's assessment of the extent of a plaintiff's injury is essentially a factual determination, whereas its imposition of punitive damages is an expression of its moral condemnation. The question [of] whether a fine is constitutionally excessive calls for the application of a constitutional standard to the facts of a particular case, and in this context de novo review of that question is appropriate.*

Reversed and *remanded* based on a de novo standard in favor of Cooper, to determin[e] whether the punitive damage award is excessive.

PROFESSIONAL OBLIGATION SCHOOL

Advocates of a professional obligation school of social responsibility state that business managers and members of boards of directors should be certified as "professionals" before they are allowed to assume managerial responsibility. In our discussion of professional ethical codes, we defined *professionals* as persons who are subject to (1) educational entrance requirements and continuing education standards, (2) licensing-examination requirements, (3) codes of conduct that are enforced, (4) a formal association that meets regularly, and (5) an independent commitment to the public interest. Advocates of a professional obligation theory argue that business directors and managers, like doctors and lawyers, have a responsibility to the public beyond merely making profits, and that the public

---
*Cooper Industries v. Leatherman Tool Group, Inc., United States Supreme Court 121 S. Ct. 1678 (2001).

must thus be able to be sure that these people are qualified to hold their positions. They should be licensed by meeting university requirements and passing a state or a national test. They should be subject to a disciplinary code that could involve revocation or suspension of their license to "practice the management of a business" if they are found by state or national boards to have failed to meet their codified responsibilities. Such responsibilities would include accountability for the unintended effects of their profit-making activities (externalities).

REGULATION SCHOOL

A regulation school of social responsibility sees all business units as accountable to elected public officials. Proponents of this theory argue that, because business managers are responsible only to a board of directors that represents shareholders, the corporation cannot be trusted to act in a socially responsible manner. If society is to be protected from the unintended effects of profit-making business activities (e.g., pollution, sex discrimination in the workplace, and injuries to workers), it is necessary for government to be involved.

The degree of government involvement is much debated by advocates of this theory. Some argue in the extreme for a socialist state. Others argue for government representatives on boards of directors, and still others argue that government should set up standards of socially responsible conduct for each industry. The last group advocates an annual process of reporting conduct, both socially responsible and otherwise, similar to the independent financial audits now required by the SEC of all publicly registered firms. The growth of ethics offices within corporations has played a role in dealing with ethical and legal problems. Sometimes these offices are mandated by courts when sentencing takes place in white-collar criminal cases. Often corporations set up such offices as preventive measures.

In Case 7-3 below there exists a relationship between law and ethics as Judge Lolley sets out in his opinion. Referring back to Chapter 1, explain the relationship.

CASE 7-3

Johnson Construction Co. v. Shaffer
Court of Appeal of Louisiana, Second Circuit
87 So. 3d 203 (2012)

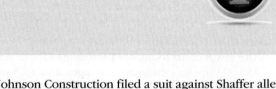

A truck owned by Johnson Construction Company needed repairs. John Robert Johnson, Jr., the company's president, took the truck with its attached 15-ton trailer to Bubba Shaffer, doing business as Shaffer's Auto and Diesel Repair. The truck was supposedly fixed, and Johnson paid the bill. The truck continued to leak oil and water. Johnson returned the truck to Shaffer, who again claimed to have fixed the problem. Johnson paid the second bill. The problems with the truck continued, however, so Johnson returned the truck and trailer a third time. Shaffer gave a verbal estimate of $1,000 for the repairs, but he ultimately sent an invoice for $5,863.49. Johnson offered to settle for $2,480, the amount of the initial estimate ($1,000) plus the costs of parts and shipping. Shaffer refused the offer and would not return Johnson's truck or trailer until full payment was made. Shaffer also charged Johnson a storage fee of $50 a day and 18 percent interest on the $5,863.49.

Johnson Construction filed a suit against Shaffer alleging unfair trade practices. The trial court determined that Shaffer had acted deceptively and wrongfully in maintaining possession of the trailer, on which no work had been performed. The trial court awarded Johnson $3,500 in general damages, plus $750 in attorneys' fees. Shaffer was awarded the initial estimate of $1,000 and appealed.

Judge Lolley

At the outset, we point out that Mr. Johnson maintained he had a verbal agreement with Bubba Shaffer, the owner of Shaffer's Auto Diesel and Repair, that the repairs to the truck would cost $1,000. Mr. Johnson also testified that he was not informed otherwise.

The existence or nonexistence of a contract is a question of fact, and the finder of fact's determination may not be set aside unless it is clearly wrong.

At the trial of the matter, the trial court was presented with testimony from Mr. Johnson, Mr. Shaffer, and Michael Louton, a mechanic employed by Shaffer. The trial court did not believe Mr. Johnson was informed of the cost for the additional work.

We cannot say that the trial court was clearly wrong in its determination. The trial court viewed Mr. Shaffer's testimony on the issue as "disingenuous" and we cannot see where that was an error.

So considering, we see no error in the trial court's characterization of Shaffer's actions with the trailer as holding "hostage in an effort to force payment for unauthorized repairs. Shaffer had no legal right to retain possession of the trailer. Thus, the trial court did not err in its determination that Shaffer's retention of Johnson Construction's trailer [for four years!] was a deceptive conversion of the trailer.

The state appellate court affirmed the judgment of the trial court in favor of Johnson Construction Company for $3,500, plus attorney fees and award of the original $10,000.*

Affirmed for the Plaintiff.

Global Dimensions of Ethics and Social Responsibility

CODE OF CONDUCT FOR TRANSNATIONAL CORPORATIONS

A United Nations effort to prevent misconduct by transnational corporations has been promulgated. Four objectives include:

1. *Respect for national sovereignty in countries where such companies operate.* Often transnational companies operate in developing nations where governments are less stable and more corrupt, making this goal very difficult to achieve.

2. *Adherence to sociocultural values.* The code seeks to prevent transnational companies from imposing value systems that are detrimental to those of the host country.

3. *Respect for human rights.* Companies should not discriminate on the basis of race, color, sex, religion, language, or political or other opinion. In developing nations, achievement of this goal is sometimes very difficult when the host country does discriminate on the basis of some of these factors.

4. *Abstention from corrupt practices.* Transnational corporations shall refrain from the offering, promising, or giving of any payment, gift, or other advantage to a public official or refrain from performing a duty in accordance with a business transaction.

Corruption is endemic to many developing countries (e.g., Nigeria or the People's Republic of China) and a way of doing business. The United States has set forth one approach (Foreign Corrupt Practices Act [FCPA]), and the Organization for Economic and Cultural Development another. See Chapter 23 for a discussion of the Foreign Corrupt Practices Act of 1977, as amended in 1988 and 1998, and the International Securities Enforcement Cooperation Act of 1990 (ISECA).

> **COMMENT:** Has the United States been successful in preventing corruption? If not, why not? Has the People's Republic been successful in its latest attempt to do away with corruption in many phases of its socialist system? Explain.
> Explain externalities that may exist when attempting to end corruption.

SUMMARY

We have sought to define ethics and social responsibility within the context of business associations. We examined consequential theories of ethics based on the consequences of the company's actions. Deontological schools of ethics, in contrast, are based on duties. Humanist theories of ethics evaluate actions as good or bad based on how the actions improved inherent human capacities.

*Court of Appeal of Louisiana, Second Circuit 87 So.3d 203 (2012).

This chapter also examined codes of ethics emanating from businesses and professions. It discussed five schools of social responsibility based on the unintended effects of corporate and human conduct. Finally, global dimensions of ethical and socially responsible conduct are highlighted, through an examination of the United Nations' Code of Conduct for Transnational Corporations.

REVIEW QUESTIONS

7-1 Define the deontological theory of ethics. On the basis of this theory and a reading of the Union Carbide case synopsis, do you think that Union Carbide acted ethically after the Bhopal incident? Explain.

7-2 Should corporations act in a socially responsible manner or solely for the profits owed to shareholders who invest in the corporation? Explain.

7-3 How are professional codes of ethics different from individual codes? Explain.

7-4 When someone tells you he or she is a "professional," what characteristics are expected of the individual and his or her profession? Explain.

7-5 What is meant by institutional school of responsibility? Explain.

REVIEW PROBLEMS

7-6 If corruption is endemic to many developing countries as well as developed countries, has the United States been successful in preventing corruption by legislation (e.g. the Foreign Corrupt Practices Act) and executive orders? Has the People's Republic of China been successful by using directives issued by President Xi Jinping? Explain. Critical thinking steps set out in Chapter 1 should be used in your answer.

7-7 Indra Wu, a sales rep at Rite Engineering, attended a trade show and conference at company expense. Main exhibitors donated prizes, which were awarded to attendees based on a drawing of free tickets given to all attendees upon registration. Wu won a $12,000 plasma television in one of the drawings. The winner's certificate included the winner's name with no mention of the company.

What should Wu do? What should Wu's supervisor do if she learns of the prize from someone other than Wu? Explain.

7-8 C, a student at University Z, discovers a method of bypassing the telephone system that allows him to make free telephone calls. C tells his roommate, D, about this method. How would you advise D to act, based on one of the ethical theories discussed in this chapter?

7-9 You are hiring a new manager for your department. You have several good applicants. Assume that the one who is best suited for the job in terms of training and experience has also been found to have done one of the things in the following list. How would knowing what this person has done affect your decision? Would your answer be the same regardless of which theory of ethical thought you applied? Examine each factual situation in the list and explain how your answer is affected.

1. The individual listed on his résumé that he had an MBA from Rutgers. He does not have an MBA.
2. The individual listed that his prior salary was $43,000. The prior salary was actually $40,000.

7-10 You are a purchasing manager for Alpha Corporation. You are responsible for buying two $1 million generators. Your company has a written policy prohibiting any company buyer from receiving a gratuity in excess of $50 and requiring that all gratuities be reported. The company has no policy regarding whistleblowing. A salesperson for a generator manufacturer offers to arrange it so that you can buy a $20,000 car for $7,000 from a third party. You decline the offer. Do you now report it to your superior? To the salesperson's superior? How would the various schools of social responsibility influence your decision? Pick one school and use its tenets to justify your decision. Explain.

7-11 You are a laboratory technician for the Standard Ethical Drug Company. You run tests on animals and prepare a summary that is then doctored by your superior to make a drug appear safe when in fact it is not. Your supervisor determines your salary and has significant influence on whether you keep your job. You are the sole source of support for your two children, have no close relative to help you, and are just barely making it financially. Jobs equivalent to yours are difficult to find. You are convinced that if the company markets the drug, the risk of cancer to the drug users will increase significantly. The drug provides significant relief for hemorrhoids. What will you do? Which school of social responsibility would be the basis for your decision? Explain.

CASE PROBLEMS

7-12 In 1999, Andrew Fastow, chief financial officer of Enron Corp., asked Merrill Lynch, an investment firm, to participate in a bogus sale of three barges so that Enron could record earnings of $12.5 million from the sale. Through a third entity, Fastow bought the barges back within six months and paid Merrill for its participation. Five Merrill employees were convicted of conspiracy to commit wire fraud, in part, on an "honest services" theory. Under this theory, an employee deprives his or her employer of "honest services" when the employee promotes his or her own interests, rather than the interests of the employer. Four of the employees appealed to the U.S. Court of Appeals for the Fifth Circuit, arguing that this charge did not apply to the conduct in which they engaged. The court agreed, reasoning that the barge deal was conducted to benefit Enron, not to enrich the Merrill employees at Enron's expense. Meanwhile, Kevin Howard, chief financial officer of Enron Broadband Services (EBS), engaged in "Project Braveheart," which enabled EBS to show earnings of $111 million in 2000 and 2001. Braveheart involved the sale of an interest in the future revenue of a video-on-demand venture to nCube, a small technology firm, which was paid for its help when EBS bought the interest back. Howard was convicted of wire fraud, in part, on the "honest services" theory. He filed a motion to vacate his conviction on the same basis that the Merrill employees had argued. Did Howard act unethically? Discuss. Explain with the school of ethical thought. *United States v. Howard*, 471 F. Supp. 2d 772 (S.D. Tex. 2007).

7-13 Steven Soderbergh is the Academy Award–winning director of *Erin Brockovich, Traffic,* and many other films. CleanFlicks, LLC, filed a suit in a federal district court against Soderbergh, 15 other directors, and the Directors Guild of America. The plaintiff asked the court to rule that it had the right to sell DVDs of the defendants' films altered without the defendants' consent to delete scenes of "sex, nudity, profanity and gory violence." CleanFlicks sold or rented the edited DVDs under the slogan "It's About Choice" to consumers, sometimes indirectly through retailers. It would not sell to retailers that made unauthorized copies of the edited films. The defendants, with DreamWorks, LLC and seven other movie studios that own the copyrights to the films, filed a counterclaim against CleanFlicks and others engaged in the same business, alleging copyright infringement. Those filing the counterclaim asked the court to enjoin (prevent) CleanFlicks and others from making and marketing altered versions of the films. *CleanFlicks of Colorado, LLC v. Soderbergh*, 433 F. Supp. 2d (D. Colo. 2006).

7-14 David Krasner, who worked for HSH Nordbank AG, complained that his supervisor, Roland Kiser, fostered an atmosphere of sexism that was demeaning to women. Among other things, Krasner claimed that career advancement was based on "sexual favoritism." He objected to Kiser's relationship with a female employee, Melissa Campfield, who was promoted before more qualified employees, including Krasner. How do a manager's attitudes and actions affect the workplace? *Krasner v. HSH Nordbank AG*, 680 FSupp.2d 502 (S.D.N.Y 2010) (see *Making Ethical Business Decisions*).

7-15 After the fall of the Soviet Union, the new government of Azerbaijan began converting certain state-controlled industries to private ownership. Ownership in these companies could be purchased through a voucher program. Frederic Bourke, Jr., and Viktor Kozeny wanted to purchase the Azerbaijani oil company SOCAR, but it was unclear whether the Azerbaijani president would allow SOCAR to be put up for sale. Kozeny met with one of the vice presidents of SOCAR (who was also the son of the president of Azerbaijan) and other Azerbaijani leaders to discuss the sale of SOCAR. To obtain their cooperation, Kozeny set up a series of parent and subsidiary companies through which the Azerbaijani leaders would eventually receive two-thirds of the SOCAR profits without ever investing any of their own funds. In return, the Azerbaijani leaders would attempt to use their influence to convince the president to put SOCAR up for sale. Assume that Bourke and Kozeny are operating out of a U.S. company. Discuss the ethics of this scheme, both in terms of the Foreign Corrupt Practices Act (FCPA) and as a general ethical issue. What duties did Kozeny have under the FCPA? *United States v. Kozeny*, 667 F.3d 122 (2d Cir. 2011) (see *Global Business Ethics*).

7-16 Charles Zandford was a securities broker for Prudential Securities, Inc., in Annapolis, Maryland. In 1987, he persuaded William Wood, an elderly man in poor health, to open a joint investment account for himself and his mentally retarded daughter. The stated investment objectives for the account were "safety of principal and income." The Woods gave Zandford discretion to manage their account and to engage in transactions for their benefit without prior approval. Relying on Zandford's promise to "conservatively invest" their money, the Woods entrusted him with $419,255. Zandford immediately began writing checks to himself on the account.

Paying the checks required selling securities in the account. Before William's death in 1991, all of the money was gone. Zandford was convicted of wire fraud and sentenced to more than four years in prison. The SEC filed a suit in a federal district court against Zandford, alleging in part misappropriation of $343,000 of the Woods' securities and seeking disgorgement of that amount.

Which theory of ethics did Zandford represent? Explain. Which theory of ethics did the SEC represent? Explain. *SEC v. Zandford,* 535 U.S. 813 (2002).

THINKING CRITICALLY ABOUT RELEVANT LEGAL ISSUES

Disbarment of Lawyers and Debarment of Officers and Directors of Corporations

Egil Krogh, Jr., was admitted [to practice] law in the state of Washington on September 20, 1968. On February 4, 1974, he was suspended as a result of his having been convicted of a felony. [Krogh now appeals the disciplinary board's decision to disbar him.]

The information referred to in the complaint charged that while the respondent was an officer and employee of the United States Government . . . and acting in his official capacity, in conjunction with others who were officials and employees of the United States Government, the defendant unlawfully, willfully and knowingly did combine, conspire, confederate and agree with his co-conspirators to injure, oppress, threaten and intimidate Dr. Lewis J. Fielding . . . in the free exercise and enjoyment of a right and privilege secured to him by the Constitution and laws of the United States, and to conceal such activities. It further charged that the co-conspirators did, without legal process, probable cause, search warrant or other lawful authority, enter the offices of Dr. Fielding in Los Angeles County, California, with the intent to search for, examine and photograph documents and records containing confidential information concerning Daniel Ellsberg, and thereby injure, oppress, threaten and intimidate Dr. Fielding in the free exercise and enjoyment of the right and privilege secured to him by the Fourth Amendment to the Constitution of the United States, to be secure in his person, house, papers and effects against unreasonable searches and seizures. . . . To all of these allegations, the respondent had pleaded guilty.

Both the hearing panel and the disciplinary board found that moral turpitude was an element of the crime of which respondent was convicted. The panel found that he has a spotless record except for the incident involved in these proceedings; that he is outstanding in character and ability; that his reputation is beyond reproach; that he acted, although mistakenly, out of a misguided loyalty to [President Nixon]; that the event was an isolated one, and that in all probability there would be no repetition of any such error on his part. The panel further found that the respondent had accepted responsibility and had made amends to the best of his ability; that he testified fully and candidly and that his attitude in the proceeding was excellent. The panel concluded that in this case which it found to be distinguishable from all other cases, the respondent apparently followed the order of a "somewhat distraught President of the United States" under the guise of national security to stop by all means further security leaks.

Th[e] rule [that attorneys are disbarred automatically when they are found guilty of a felony] still governs the disposition of such disciplinary proceedings in a number of jurisdictions. However, under our disciplinary rules, some flexibility is permitted, and the court retains its discretionary power to determine whether, on the facts of the particular case, the attorney should be disbarred.

We cannot accept the assumption that attorneys . . . can ordinarily be expected to abandon the principles which they have sworn to uphold, when asked to do so by a person who holds a constitutional office. Rather than being overawed by the authority of one who holds such an office . . . the attorney who is employed by such an officer should be the most keenly aware of the Constitution and all of its provisions, the most alert to discourage the abuse of power. In such a position those powers of discernment and reason, which he holds himself out as possessing, perform their most important function. If, when given a position of power himself, he forgets his oath to uphold the Constitution and laws of the land and instead [flouts] the constitutional rights of other citizens and holds himself above the law, can we say to the public that a person so weak in his dedication to constitutional principles is qualified to practice law?

That the reputation and honor of the bar have suffered severe damage as a result is now a matter of common knowledge. We find it difficult to believe that the respondent was not aware, when he authorized the burglary of Dr. Fielding's office, that if his conduct became known, it would reflect discredit upon his profession.

For the reasons set forth herein, we must conclude that the respondent, in spite of his many commendable qualities and achievements, has shown himself to be unfit to practice law.

The recommendation of the disciplinary board is *approved,* and the respondent's name shall be stricken from the roll of attorneys in this state. In answering these questions, refer to Chapter 1 of this text.

1. What ethical norm is central to the court's decision in this case?
2. What fact seems especially powerful in shaping the court's reasoning?
3. What reasons does the court provide for upholding the respondent's disbarment?
4. Outline the reasons why Egil Krogh, Jr., believed he should not be disbarred by the disciplinary board of the State of California.

Potential Debarment of Officers and Directors

In May of 2015 five of the world's largest banks pleaded guilty to an array of antitrust and fraud charges in a criminal proceeding. For most people, pleading guilty to these felonies would have landed them in jail for several years, loss of their job, and payment of a huge fine.

The Justice Department and the SEC agreed to $1 billion of fines and a negotiated plea agreement by which a 2012 non-prosecution agreement with one of the banks was torn up. Most of the banks' pleas were from holding companies.

1. Does the plea agreement meet the justice values set out in Chapter 1? Explain this from the point of view of the five corporations, and secondly, that of the Justice Department and the SEC.
2. Should the directors or officers of the companies be fined personally or go to jail? Remember, in a previous chapter we have learned that a corporation is considered a person at law. Should the officers and directors be debarred (not disbarred) by the SEC and be prevented from serving on boards or as officers of a publicly held corporation that is regulated by the agency forever? Would these steps bring justice to the case or are there externalities that have not been considered? For example, would this proposed punishment to five large companies affect the worldwide role of banks and other financial institutions?

Source: Based on B. Protess, and M. Corkery, "Five Big Banks to Plead Guilty to Felony Charges," *New York Times,* May 14, 2015, pp. B-1, B-5.

ASSIGNMENT ON THE INTERNET

This chapter introduces you to three theories of ethical thought and five schools of social responsibility. Explore how the three theories of ethical thought are put into practice. Using the Internet, find the code of ethics for a business or corporation that does business in your city or town. This site provides links to codes of ethics for hundreds of corporations: www.business-ethics.com.

Applying your critical thinking skills, determine if the chosen code of ethics relies more heavily on one theory of ethical thought than the others. Are there aspects of that business's code of ethics that you would like to see changed? Why?

ON THE INTERNET

www.business-ethics.com This site provides codes of ethics from hundreds of corporations and articles detailing current trends or methods of bringing ethics into the business environment.

www.legalethics.com This comprehensive ethics page provides numerous links to other valuable websites.

www.ethicsweb.ca This site contains links to numerous business ethics centers and institutions, as well as business ethics consultants. On the home page, click "businessethics.ca." Under the heading "Other Web-Based Resources," click on "Business Ethics Resources." From there, click on "Ethics Institutes and Organizations" for the list.

www.globalethics.org The website for the Global Ethics Institute provides information about business ethics from countries around the world.

FOR FUTURE READING

Besmer, Veronica. "Student Note: The Legal Character of Private Codes of Conduct: More Than Just a Pseudo-Formal Gloss on Corporate Social Responsibility." *Hastings Business Law Journal* 2 (2006): 279.

Epstein, Edwin. "Commentary: The Good Company: Rhetoric or Reality? Corporate Social Responsibility and Business Ethics Redux." *American Business Law Journal* 44 (2007): 207.

Jackall, Robert. *Moral Mazes: The World of Corporate Managers*. New York: Oxford University Press, 1998.

Lee, Ian B. "Is There a Cure for Corporate 'Psychopathy'?" *American Business Law Journal* 42 (2005): 65.

Ostas, Daniel T. "Deconstructing Corporate Social Responsibility: Insights from Legal and Economic Theory." *American Business Law Journal* 38 (2004): 261.

Yur, Mishima, Bernadine J. Dykes, Emily S. Block, and Timothy G. Pollock, "Why 'Good' Firms Do Bad Things: The Effect of High Aspirations, High Expectations, and Prominence on the Incidence of Corporate Illegality," *Academy of Management Journal*, 53 (2010): 701.

CHAPTER EIGHT

The International Legal Environment of Business

- DIMENSIONS OF THE INTERNATIONAL ENVIRONMENT OF BUSINESS
- METHODS OF ENGAGING IN INTERNATIONAL BUSINESS
- RISKS OF ENGAGING IN INTERNATIONAL BUSINESS
- LEGAL AND ECONOMIC INTEGRATION AS A MEANS OF ENCOURAGING INTERNATIONAL BUSINESS ACTIVITY
- GLOBAL DISPUTE RESOLUTION

At the outset of Chapter 2, we noted that U.S. managers can no longer afford to view their firms as doing business on a huge island between the Pacific and the Atlantic Oceans. Existing and pending multilateral trade agreements open vast opportunities to do business in Europe and Asia, throughout the Americas, and indeed throughout the world. If present and future U.S. managers do not become aware of these opportunities, as well as the attendant risks, they and their firms will be at a competitive disadvantage vis-à-vis foreign competitors from all over the world.

CRITICAL THINKING ABOUT THE LAW

This chapter (1) introduces the international environment of business; (2) sets forth the methods by which companies may engage in international business; (3) indicates the risks involved in such engagement; (4) describes organizations that work to bring down tariff barriers, and thus encourage companies of all nations to engage in international business; and (5) identifies the means by which disputes between companies doing business in the international arena are settled. Please note carefully that when we use the word companies in an international context, we are referring not only to private-sector firms, but also to nation-state subsidized entities and government agencies that act like private-sector companies.

Because of today's widespread international opportunities and advances in communication, business managers must be aware of the global legal environment of business. As you will soon learn, the political, economic, cultural, and legal dimensions are all important international business considerations. The following questions will help sharpen your critical thinking about the international legal environment of business.

1. Consider the number of countries that might participate in an international business agreement. Why might ambiguity be a particularly important concern in international business?

    Clue: Consider the variety of cultures as well as the differences in languages. How might these factors affect business agreements?

2. Why might the critical thinking questions about ethical norms and missing information be important for international businesses?

   Clue: Again, consider the variety of cultures involved in international business. Why might identifying the primary ethical norms of a culture be helpful?

3. What ethical norm might influence an entity's willingness to enter into agreements with foreign companies?

   Clue: How might international agreements differ from agreements between two U.S. companies?

Dimensions of the International Environment of Business

Doing international business has political, economic, cultural, and legal dimensions. Although this chapter emphasizes the legal dimensions of international business transactions, business managers need to be aware of those other important dimensions as well. (Ethical dimensions were examined in Chapter 7.)

POLITICAL DIMENSIONS

Managers of firms doing international business must deal with different types of governments, ranging from democracies to totalitarian states. They are concerned with the stability of these governments and with whether economic decisions are centralized or decentralized. In the Marxist form of government, such as that which existed in the former Soviet Union and in Eastern Europe until the 1990s, economic decisions were centralized, and there was political stability. This would seem to be an ideal environment in which to do business from a multinational business manager's perspective. But it was *not* ideal, because a centralized economy limits the supply of goods coming from outside a country, the price that can be charged for goods inside the country, and the amount of currency that can be taken out of the country by multinational businesses.

Despite the collapse of communism and the development of new political systems professing support of free enterprise in Eastern Europe and throughout the former Soviet Union, companies in the industrialized nations have delayed investing in some of these areas because they are uncertain of these areas' political stability and willingness to adhere to economic agreements. In the People's Republic of China (PRC or China), an early rush to invest was slowed by foreign companies' experiences with a seemingly capricious government. For example, McDonald's leased a prime location in Beijing from the centralized government but found itself ousted a few years later when the government revoked the lease to allow a department store to be built on that site. Moreover, doubts about the Chinese government's intention to honor its agreement with the British government—that Hong Kong would retain its separate political and economic status for 50 years after the 98-year British lease expired in 1997—led one long-time Hong Kong trading company, Jardine, to move its headquarters to the Bahamas. In 2010, China's monitoring of Google users' Internet communications prompted Google initially to move most of its activities to Hong Kong, a district now belonging to China. Three months later, China rejected the scheme. Google, instead of continuing to reroute queries to its Hong Kong engine, started sending visitors to a new "landing page" that linked to the Hong Kong website where users could perform searches beyond the reach of Chinese censors. Later in 2010, a compromise was reached. China allowed Google to operate in China if it tweaked its mainland search box to ask users if they still wanted to have communications sent to the Hong Kong site. Chinese users of Google thus had a choice, and both the Chinese government and Google saved face. Despite these political problems, China's growth rate in 2007–2009 proceeded at an annual gross domestic product (GDP) of 10–12 percent. Further, China has brought investment capital to Latin American, Asian, and African nations. It is searching for minerals and energy to develop its infrastructure. As this search continues, China's political influence spreads worldwide.

ECONOMIC DIMENSIONS

Every business manager should do a country analysis before deciding to do business in another nation-state. Such an analysis not only examines political variables, but also dissects a nation's economic performance as demonstrated by its rate of economic growth, inflation, budget, and trade balance. Four economic factors in particular affect business investment:

1. *Differences in size and economic growth rate* of various nation-states. For example, when McDonald's decided to engage in international business, the company initially located its restaurants only in countries that already had high growth rates. As more and more developing nations moved toward a market economy, McDonald's expanded into Russia, China, Brazil, Mexico, and other countries deemed to have potentially high growth rates.

2. *The impact of central planning versus a market economy* on the availability of supplies. When McDonald's went into Russia, it had to build its own food-processing center to be certain it would get the quality of beef it needed. Furthermore, because of distribution problems, it used its own trucks to move supplies.

3. *The availability of disposable income.* This is a tricky issue. Despite the fact that the price of a Big Mac, french fries, and a soft drink equals the average Russian worker's pay for four hours of work, McDonald's is serving thousands of customers a day at its Moscow restaurant.

4. *The existence of an appropriate transportation infrastructure.* Decent roads, railroads, and ports are needed to bring in supplies and then transport them within the host country. McDonald's experience in Russia is commonplace. Multinational businesses face transportation problems in many developing countries.

CULTURAL DIMENSIONS

Culture may be defined as learned norms of a society that are based on values, beliefs, and attitudes. For example, if people of the same area speak the same language (e.g., Spanish in most of Latin America, with the exception of Brazil and a few small nations), the area is often said to be *culturally homogeneous*. Religion is a strong builder of common values. In 1995, the Iranian government outlawed the selling and use of satellite communications in Iran on the grounds that they presented "decadent" Western values that were undermining Muslim religious values. Now, in 2010, it monitors Internet communications, which has led some U.S. congresspersons to seek a ban on all trade with Iran.

A failure to understand that some cultures are based on ascribed group membership (gender, family, age, or ethnic affiliation) rather than on acquired group membership (religious, political, professional, or other associations), as in the West, can lead to business mistakes. For example, gender- and family-based affiliations are very important in Saudi Arabia, where a strict interpretation of Islam prevents women from playing a major role in business. Most Saudi women who work hold jobs that demand little or no contact with men, such as teaching or acting as doctors only for women.

Another important cultural factor is the attitude toward work. Mediterranean and Latin American cultures base their group affiliation on family, and place more

culture Learned norms of society based on values and beliefs.

emphasis on leisure than on work. We often say that the Protestant ethic, stressing the virtues of hard work and thrift, is prevalent in Western and other industrialized nations. Yet the Germans work no more than 35 hours a week and take 28 days of paid vacation every year. The average hourly wage is higher in Germany than in the United States, and German workers' benefits far outpace those of U.S. workers.

Business managers must carefully consider language, religion, attitudes toward work and leisure, family versus individual reliance, and numerous other cultural values when planning to do business in another nation-state. They also need to find a method of reconciling cultural differences between people and companies from their own nation-state and those from the country in which they intend to do business.

CORRUPTION AND TRADE

The nature of trade between nations, between multinationals, and between multinationals and nation-states has led to global competition, and sometimes bribery, and thus corruption (see the "Corruption Perception Index" from Transparency International in the "Comparative Law Corner" feature later in this chapter). Attempts to lessen such bribery and corruption through bilateral and multilateral agreements have been led by the United States Foreign Corrupt Practices Act of 1977 (FCPA) and the Convention on Combating Bribery of Foreign Officials in International Business Transactions (CCBFOIBT) drafted by the Organization for Economic Cooperation and Development (OECD) and signed by 34 countries. The Convention adopts the standards of the FCPA. Chapter 22 provides additional details governing both acts.

In this text, Chapter 23, you will find a brief discussion of the FCPA provisions that forbid payments to foreign officials when those amounts are more than "grease payments." "Facilitating payments" made to obtain permits, licenses, or other official documents associated with contract performance, or movement of goods across a country, are considered lawful. The Justice Department, as well as other agencies and individuals, may enforce the FCPA. Activities that constitute a bribe are often the basis for legal action. The case excerpted here deals with this problem.

CASE 8-1

United States v. Kay
359 F.3d 738 (5th Cir. 2004)

David Kay (defendant) was an American citizen and a vice president for marketing of American Rice, Inc. (ARI), who was responsible for supervising sales and marketing in the Republic of Haiti. Douglas Murphy (defendant) was an American citizen and president of ARI.

Beginning in 1995 and continuing to about August 1999, Kay, Murphy, and other employees and officers of ARI paid bribes and authorized the payment of bribes to induce customs officials in Haiti to accept bills of lading and other documents that intentionally understated the true amount of rice that ARI shipped to Haiti for import, thus reducing the customs duties owed by ARI and RCH to the Haitian government.

In addition, beginning in 1998 and continuing to about August 1999, Kay and other employees and officers of ARI paid and authorized additional bribes to officials of other Haitian agencies to accept the false import documents and other documents that understated the true amount of rice being imported into and sold in Haiti, thereby reducing the amount of sales taxes paid to the Haitian government.

Kay directed employees of ARI to prepare two sets of shipping documents for each shipment of rice to Haiti, one that was accurate and another that falsely represented the weight and value of the rice being exported to Haiti.

Kay and Murphy agreed to pay and authorized the payment of bribes, calculated as a percentage of the value of the rice not reported on the false documents or in the form of a monthly retainer, to customs and tax officials of the Haitian government to induce these officials to accept the false documentation and to assess significantly lower customs duties and sales taxes than ARI would otherwise have been required to pay.

ARI, using official Haitian customs documents reflecting the amounts reported on the false shipping documents, reported only approximately 66 percent of the rice it sold in Haiti and thereby significantly reduced the amount of sales taxes it was required to pay to the Haitian government.

In 2001, a grand jury charged Kay with violating the FCPA and subsequently returned the indictment, which charged both Kay and Murphy with 12 counts of FCPA violations. Both Kay and Murphy moved to dismiss the indictment for failure to state an offense, arguing that obtaining favorable tax treatment did not fall within the FCPA definition of payments made to government officials in order to obtain business. The district court dismissed the indictment, and the United States of America appealed.

Justice Wiener

The principal dispute in this case is whether, if proved beyond a reasonable doubt, the conduct that the indictment ascribed to defendants in connection with the alleged bribery of Haitian officials to understate customs duties and sales taxes on rice shipped to Haiti to assist American Rice, Inc. in obtaining or retaining business was sufficient to constitute an offense under the FCPA. Underlying this question of sufficiency of the contents of the indictment is the preliminary task of ascertaining the scope of the FCPA, which in turn requires us to construe the statute.

Because an offense under the FCPA requires that the alleged bribery be committed for the purpose of inducing foreign officials to commit unlawful acts, the results of which will assist in obtaining or retaining business in their country, the questions before us in this appeal are (1) whether bribes to obtain illegal but favorable tax and customs treatment can ever come within the scope of the statute, and (2) if so, whether, in combination, there are minimally sufficient facts alleged in the indictment to inform the defendants regarding the nexus between, on the one hand, Haitian taxes avoided through bribery, and, on the other hand, assistance in getting or keeping some business or business opportunity in Haiti.

No one contends that the FCPA criminalizes every payment to a foreign official: It criminalizes only those payments that are intended to (1) influence a foreign official to act or make a decision in his official capacity, or (2) induce such an official to perform or refrain from performing some act in violation of his duty, or (3) secure some wrongful advantage to the payor. And even then, the FCPA criminalizes these kinds of payments only if the result they are intended to produce—their *quid pro quo*—will assist (or is intended to assist) the payor in efforts to get or keep some business for or with "any person."

Stated differently, how attenuated can the linkage be between the effects of that which is sought from the foreign official in consideration of a bribe (here, tax minimization) and the briber's goal of finding assistance or obtaining or retaining foreign business with or for some person, and still satisfy the business nexus element of the FCPA?

Invoking basic economic principles, the SEC reasoned in its amicus brief that securing reduced taxes and duties on imports through bribery enables ARI to reduce its cost of doing business, thereby giving it an "improper advantage" over actual or potential competitors, and enabling it to do more business, or remain in a market it might otherwise leave.

Section 78dd-1(b) excepts from the statutory scope "any facilitating or expediting payment to a foreign official . . . the purpose of which is to expedite or to service the performance of a routine governmental action by a foreign official. . . ." 15 U.S.C. § 78dd-1(b).

For purposes of deciding the instant appeal, the question nevertheless remains whether the Senate, and concomitantly Congress, intended this broader statutory scope to encompass the administration of tax, customs, and other laws and regulations affecting the revenue of foreign states. To reach this conclusion, we must ask whether Congress's remaining expressed desire to prohibit bribery aimed at getting assistance in retaining business or maintaining business opportunities was sufficiently broad to include bribes meant to affect the administration of revenue laws. When we do so, we conclude that the legislative intent was so broad.

Obviously, a commercial concern that bribes a foreign government official to award a construction, supply, or services contract violates the statute. Yet, there is little difference between this example and that of a corporation's lawfully obtaining a contract from an honest official or agency by submitting the lowest bid, and—either before or after doing so—bribing a different government official to reduce taxes and thereby ensure that the under-bid venture is nevertheless profitable. Avoiding or lowering taxes reduces operating costs and thus increases profit margins, thereby freeing up funds that the business is otherwise legally obligated to expend. And this, in turn, enables it to take any number of actions to the disadvantage of competitors. Bribing foreign officials to lower taxes and customs duties certainly can provide an unfair advantage over competitors and thereby be of assistance to the payor in obtaining or retaining business. This demonstrates that the question [of] whether the defendants' alleged payments constitute a violation of the FCPA truly turns on whether these bribes were intended to lower ARI's cost of doing business in Haiti enough to have a sufficient nexus to garnering business there or to maintaining or increasing business operations that ARI already had there, so as to come within the scope of the business nexus element as Congress used it in the FCPA. Answering this fact question, then, implicates a matter of proof and thus evidence.

Given the foregoing analysis of the statute's legislative history, we cannot hold as a matter of law that Congress meant to limit the FCPA's applicability to cover only bribes that lead directly to the award or renewal of contracts. Instead, we hold that Congress intended for the FCPA to apply broadly to payments intended to assist the payor, either directly or indirectly, in obtaining or retaining business for some person, and that bribes paid to foreign tax officials to secure illegally reduced customs and tax liability constitute a type of payment that can fall within this broad coverage. In 1977, Congress was motivated to prohibit rampant foreign bribery by domestic business entities, but nevertheless understood the pragmatic need to exclude innocuous grease payments from the scope of its proposals. The FCPA's legislative history instructs that Congress was concerned about both the kind of bribery that leads to discrete contractual arrangements and the kind that more generally helps a domestic payor obtain or retain business for some person in a foreign country; and that Congress was aware that this type includes illicit payments made to officials to obtain favorable but unlawful tax treatment.*

Reversed and *remanded* in favor of the United States.

CRITICAL THINKING ABOUT THE LAW

Congressional intent is a guiding principle of judicial interpretation. Here the court is asked to make a judgment about the scope of legislation. It answers that question by examining the purpose of the law and the applicability of that purpose to the facts of this case.

1. What is the difference between bribery and "innocuous grease payments?"
    Clue: For a payment to be innocuous, what effects would it have had to avoid?

2. What ethical norm is advanced by enforcing the statute in this case?
    Clue: How is fairness affected by permitting a firm to escape some of its tax liability?

COMPARATIVE LAW CORNER

Least Corrupt		Most Corrupt	
Denmark	9.4	Somalia	1.4
Finland	9.4	Myanmar	1.4
New Zealand	9.4	Iraq	1.5
Singapore	9.3	Haiti	1.6
Sweden	9.3	Uzbekistan	1.7
Iceland	9.2	Tonga	1.7
Netherlands	9.0	Sudan	1.8
Switzerland	9.0	Chad	1.8
Canada	8.7	Afghanistan	1.8
Norway	8.7	Laos	1.9
Australia	8.6	Guinea	1.9
Luxembourg	8.4	Equatorial Guinea	1.9
United Kingdom	8.4	Congo, Democratic Republic	1.9
Hong Kong	8.3	Venezuela	2.0
Austria	8.1	Turkmenistan	2.0
Germany	7.8	Papua New Guinea	2.0
Ireland	7.5	Central African Republic	2.0
Japan	7.5	Cambodia	2.0
France	7.3	Bangladesh	2.0
United States	7.2	Zimbabwe	2.1
Belgium	7.1	Tajikistan	2.1
Chile	7.0	Sierra Leone	2.1
		Liberia	2.1

Corruption generally discourages foreign investment, according to data published by Transparency International in its annual Corruption Perception Index (CPI). The CPI is determined by an annual survey of businesspeople, academicians, and analysts in each of 91 countries. The CPI, in its latest published data, lists

---

*United States v. Kay 359 F.3d 738 (5th Cir. 2004).

> Nigeria, Uganda, Indonesia, Bolivia, Kenya, Cameroon, and Russia as countries most prone to corruption. Finland, Denmark, New Zealand, Iceland, Singapore, Sweden, and Canada are perceived as having the least corruption. The United States ranks seventeenth.[a]
>
> When doing business with countries where corruption is rampant (e.g., a U.S. company trading oil equipment with Nigeria), it would behoove the business managers to learn what "facilitating payments" are lawful under U.S. law (FCPA), and what payments are legal under host-country laws (e.g., Nigerian statutes), if there exist such statutes. Even in countries that have statutes similar to those of the United States, it is also important to check with legal counsel to determine exceptions to host-countries' laws (e.g., when a U.S. company is trading oil equipment with Canada). In both cases, one should not presume that either Nigerian or Canadian laws are similar with regard to "facilitating payments" (grease payments) as set out in the FCPA.
>
> [a]*Corruption Perception Index,* 2009, transparency.org. © 2009 Transparency International EU.

LEGAL DIMENSIONS

When they venture into foreign territory, business managers have to be guided by the national legal system of their own country and that of the host country, and also by international law.

National Legal Systems. When deciding whether to do business in a certain country, business managers are advised to learn about the legal system of that country and its potential impact in such areas as contracts, investment, and corporate law. The five major families of law are (1) common law, (2) Romano-Germanic civil law, (3) Islamic law, (4) socialist law, and (5) Hindu law (Table 8-1).

The common-law family is most familiar to companies doing business in the United States, England, and 26 former British colonies. The source of law is primarily case law, and decisions rely heavily on case precedents. As statutory law has become more prominent in common-law countries, the courts' interpretation of laws made by legislative bodies and of regulations set forth by administrative agencies has substantially increased the body of common law.

Countries that follow the Romano-Germanic civil law (e.g., France, Germany, and Sweden) organize their legal systems around legal codes rather than around cases, regulations, and precedents, as do common-law countries. Thus, judges in civil-law countries of Europe, Latin America, and Asia resolve disputes primarily by reference to general provisions of codes and secondarily by reference to statutes passed by legislative bodies. As the body of written opinions in civil-law countries grows, and as they adopt computer-based case and statutory systems such as Westlaw and Lexis, however, the highest courts in these countries are taking greater note of case law in their decisions. Civil-law systems tend to put great emphasis on private law, that is, law that governs relationships between individuals and corporations or between individuals. Examples are the law of obligations, which includes common-law contracts, torts, and creditor–debtor relationships. In contrast to common-law systems, civil-law systems have an inferior public law: This is a law that governs the relationships between individuals and the state. In fact, their jurists are not extensively trained in such areas as criminal, administrative, and labor law.[1]

More than 1.3 billion Muslims in approximately 30 countries that are predominantly Muslim, as well as many more Muslims living in countries where Islam is a minority religion, are governed by Islamic law.[2] In many countries,

---

[1]See R. Davids and J. Brierly, *Major Legal Systems in the World Today* (Free Press, 1988) p. 437, and Schaffer, Richard, Augusti, Filiberio, and Dhooge, Lucien, *International Business Law and Its Environment* (Cengage Learning, 8th edition, 2015) p. 47.

[2]*Id.* at 437–38; *Id.* at 49.

TABLE 8-1

FAMILIES OF LAW

Family	Characteristics
Common law	Primary reliance is on case law and precedent instead of statutory law. Courts can declare statutory law unconstitutional.
Romano-Germanic	Primary reliance is on codes and statutory law rather than case law.
Civil law	In general, the high court cannot declare laws of parliament unconstitutional (an exception is the German Constitutional Court).
Islamic law	Derived from the *Shari'a*, a code of rules designed to govern the daily lives of all Muslims.
Socialist law	Based on the teachings of Karl Marx. No private property is recognized. Law encourages the collectivization of property and the means of production and seeks to guarantee national security. According to classical Marxist theory, both the law and the state will fade away as people are better educated to socialism and advance toward the ultimate stage of pure communism.
Hindu law	Derived from the *Sastras*. Hindu law governs the behavior of people in each caste (hereditary categories that restrict members' occupations and social associations). Primarily concerned with family matters and succession. Has been codified into India's national legal system.

Islamic law, as encoded in the Shari'a, exists alongside the secular law. In nations that have adopted Islamic law as their dominant legal system (e.g., Saudi Arabia), citizens must obey the Shari'a, and anyone who transgresses its rules is punished by a court. International business transactions are affected in many ways by Islamic law. For example, earning interest on money is forbidden (however, Islamic banks have found a way to work around this stricture: in lieu of paying interest on accounts, they pay each depositor a share of the profits made by the bank).

Socialist law systems are based on the teachings of Karl Marx and Vladimir Lenin (who was, incidentally, a lawyer). Right after the Bolshevik Revolution of 1917 in Russia, the Czarist legal system, which was based on the Romano-Germanic civil law, was replaced by a legal system consisting of People's Courts staffed by members of the Communist Party and peasant workers. By the early 1930s, this system had been replaced by a formal legal system with civil and criminal codes that has lasted to this day. The major goals of the Soviet legal system were to (1) encourage collectivization of the economy; (2) educate the masses as to the wisdom of socialist law; and (3) maintain national security.[3] Most property belonged to the state, particularly industrial and agricultural property. Personal (not private) property existed, but it could be used only for the satisfaction and needs of the individual, not for profit—which was referred to as "speculation" and was in violation of socialist law. Personal ownership ended either with the death of the individual or with revocation of the legal use and enjoyment of the property. Socialist law is designed to preserve the authority of the state over agricultural land and all means of production. It is still enforced in North Korea, Cuba, and to some degree, Libya, but the countries that made up the old Soviet Union and the East European bloc have been moving toward

---

[3]*Id.* at 437–38; *Id.* at 49.

Romano-Germanic civil-law systems and private-market economies in the past decade or so.

Hindu law, called *Dharmasastra*, is linked to the revelations of the Vedas, a collection of Indian religious songs and prayers believed to have been written between 100 BC and AD 300 or 400.[4] It is both personal and religious. Hindus are divided into social categories called *castes*, and the rules governing their behavior are set out in texts known as *Sastras*. The primary concerns of Hindu law are family matters and property succession. Four-fifths of all Hindus live in India; most of the remaining Hindus are spread throughout Southeast Asia and Africa, with smaller numbers living in Europe and the Americas. After gaining independence from England in 1950, India codified Hindu law. Today it plays a prominent role in Indian law alongside secular statutory law, which, especially in the areas of business and trade, uses legal terminology and concepts derived from common law. Both the Indian criminal and civil codes strongly reflect the British common-law tradition.[5] The civil code is particularly important today when issues involving outsourcing to and from India are discussed by the multinationals and governments involved.

SELECTED NATIONAL LEGAL SYSTEMS

Common Law	Romano-Germanic Civil Law	Islamic Law	Socialist Law
United States	Italy	Saudi Arabia	Russia
Canada	Japan	Kuwait	North Korea
Great Britain	Mexico	Abu Dhabi	Cuba
New Zealand	Poland	Iraq	China
Singapore	France	Indonesia	
Australia	Sweden	Bahrain	
	Finland	Libya	
	Germany	Algeria	

Comment: With the growth of the Internet the classification of each nation's law has become blurred. Further, former colonies, now independent, inherited legal principles and systems which are mixed with present-day principles, customs, and religious rites in many nations.

International Law. The law that governs the relationships between nation-states is known as **public international law**. **Private international law** governs the relationships between private parties involved in transactions across national borders. In most cases, the parties negotiate between themselves and set out their agreements in a written document. In some cases, however, nation-states subsidize the private parties or are signatories to the agreements negotiated by those parties. In such instances, the distinction between private and public international law is blurred.

The sources of international law can be found in (1) customs; (2) treaties between nations, particularly treaties of friendship and commerce; (3) judicial decisions of international courts, such as the International Court of Justice; (4) decisions of national and regional courts, such as the U.S. Supreme Court, the London Commercial Court, and the European Court of Justice; (5) scholarly writings; and (6) international organizations. These sources are discussed throughout this text.

public international law Law that governs the relationships between nations.

private international law Law that governs the relationships between private parties involved in transactions across borders.

---
[4]*Id.* at 176–79.
[5]*Id.* at 468–71.

APPLYING THE LAW TO THE FACTS . . .

Let's say that Sandra, a resident of Detroit, Michigan, is selling a car to Brianna, who lives in Detroit but is a citizen of The Bahamas. The two women draft and sign a contract for the sale of the vehicle. However, Brianna later finds out Sandra lied about the car and it was defective. Brianna wants to sue Sandra. What kind of international law would govern the conflict of the individuals from the two countries?

International business law includes laws governing (1) exit visas and work permits; (2) tax and antitrust matters and contracts; (3) patents, trademarks, and copyrights; and (4) bilateral treaties of commerce and friendship between nations and multilateral treaties of commerce such as the North American Free Trade Agreement (NAFTA), the European Union (EU), and the World Trade Organization (WTO). All are explored later in this chapter.

The U.S. Constitution grants the U.S. president the power to enter into treaties, with the advice and consent of the U.S. Senate (two-thirds must concur). The Constitution prohibits a state from entering into "any Treaty, Alliance or Confederation."[6] The U.S. Supreme Court, however, has allowed the states to enter into treaties that "do not encroach upon or impair the supremacy of the United States."[7] The states' power to enter into treaties is very limited, as indicated by the following case. This issue has gained some significance in today's world because states of the United States are presently seeking to enter into trade and other agreements with other countries, independent of the federal government.

CASE 8-2

Crosby v. National Foreign Trade Council
Supreme Court of the United States
530 U.S. 363 (2000)

In 1996, the Commonwealth of Massachusetts passed a law barring governmental entities in Massachusetts from buying goods or services from companies doing business with Burma (Myanmar). Subsequently, the U.S. Congress enacted federal legislation imposing mandatory and conditional sanctions on Burma. The Massachusetts law was inconsistent with the new federal legislation. The National Foreign Trade Council sued on behalf of its several members, claiming that the Massachusetts law unconstitutionally infringed on the federal foreign-affairs power, violated the Foreign Commerce Clause of the U.S. Constitution, and was preempted by the subsequent federal legislation. The district and appeals courts ruled in favor of the council, and the Commonwealth appealed.

Justice Souter

The Massachusetts law is preempted, and its application is unconstitutional under the Supremacy Clause of the U.S. Constitution. State law must yield to a congressional act if Congress intends to occupy the field, or to the extent of any conflict with a federal statute. This is the case even where the relevant congressional act lacks an express preemption provision. This Court will find preemption where it is impossible for a private party to comply with both state and federal law and where the state law is an obstacle to the accomplishment and execution of Congress's full purposes and objectives. In this case, the state act is an obstacle to the federal act's delegation of discretion to

---
[6] U.S. CONST. ART 1, § 10.
[7] *Virginia v. Tennessee,* 148 U.S. 503, 518 (1893).

the president of the United States to control economic sanctions against Burma. Within the sphere defined by Congress, the statute has given the President as much discretion to exercise economic leverage against Burma, with an eye toward national security, as law permits. It is implausible to think that Congress would have gone to such lengths to empower the President had it been willing to compromise his effectiveness by allowing state or local ordinances to blunt the consequences of his actions—exactly the effect of the state act.

In addition, the Massachusetts law interferes with Congress's intention to limit economic pressure against the Burmese Government to a specific range.... Finally, the Massachusetts law conflicts with the President's authority to speak for the United States among the world's nations to develop a comprehensive, multilateral Burma strategy. In this respect, the state act undermines the President's capacity for effective diplomacy.*

The Court *affirmed* the lower courts in favor of the defendant, National Foreign Trade Council.

Methods of Engaging in International Business

For purposes of this chapter, methods of engaging in international business are classified as (1) trade, (2) international licensing and franchising, and (3) foreign direct investment.

TRADE

We define **international trade** generally as exporting goods and services from a country and importing the same into a country. There are two traditional theories of trade relationships. The theory of *absolute advantage,* which is the older theory, states that an individual nation should concentrate on exporting the goods that it can produce most efficiently. For example, Sri Lanka (formerly Ceylon) produces tea more efficiently than most countries can, and thus any surplus in Sri Lanka's tea production should be exported to countries that produce tea less efficiently. The theory of *comparative advantage* arose out of the realization that a country did not have to have an absolute advantage in producing a good in order to export it efficiently; rather, it would contribute to global efficiency if it produced specialized products simply more efficiently than others did.

To illustrate this concept, let's assume that the best attorney in a small town is also the best legal secretary. Because this person can make more money as an attorney, it would be more efficient for her to devote her energy to working as a lawyer and to hire a legal secretary. Similarly, let's assume that the United States can produce both wheat and tea more efficiently than Sri Lanka can. Thus, the United States has an absolute advantage in its trade with Sri Lanka. Let us further assume that U.S. wheat production is comparatively greater than U.S. tea production vis-à-vis Sri Lanka. That is, by using the same amount of resources, the United States can produce two-and-a-half times as much wheat but only twice as much tea as Sri Lanka. The United States then has a comparative advantage in wheat over tea.[8]

In this simplified example, we made several assumptions: that only two countries and two commodities were involved; that transport costs in the two countries were about the same; that efficiency was the sole objective; and that political factors were not significant. In international trade, things are far more complex. Many nations and innumerable products are involved, and political factors are often more potent than economic considerations.

Trade is generally considered to be the least risky means of doing international business, because it demands little involvement with a foreign buyer or seller. For small and middle-sized firms, the first step toward involvement in international business is generally to hire an export management company, which

international trade The export of goods and services from a country, and the import of goods and services into a country.

---

[8]J. Daniels, L. Radebaugh, and D. Sullivan, *International Business: Environments and Operations* (10th ed.) 148, 149 (Upper Saddle River, NJ: Pearson Prentice Hall, 2004).

*Crosby v. National Foreign Trade Council, Supreme Court of the United States 530 U.S. 363 (2000).

LINKING LAW AND BUSINESS

Global Business

Your management class may have discussed the growing trend of globalization. One level of an organization's involvement in the international arena is the multinational corporation. There are three basic types of employees in multinational corporations: (1) expatriates—employees living and working in a country where they are not citizens; (2) host-country nationals—employees who live and work in a country where the international organization is headquartered; (3) third-country nationals—employees who are expatriates in a country (working in one country and having citizenship in another), while the international organization is located in another country. Typically, organizations with a global focus employ workers from all three categories. The use of host-country nationals, however, is increasing, considering the cost of training and relocating expatriates and third-country nationals. By hiring more host-country nationals, managers may spend less time and money training employees to adapt to new cultures, languages, and laws in foreign countries. In addition, managers may avoid potential problems related to sending employees to work in countries where they do not have citizenship or understand the culture; thus, managers may still obtain organizational objectives through cheaper and respectable means by hiring a greater number of host-country nationals.

Source: S. Certo, *Modern Management* (Upper Saddle River, NJ: Prentice Hall, 2000), pp. 78, 84–85.

---

is a company licensed to operate as the representative of many manufacturers with exportable products. These management companies are privately owned by citizens of various nation-states and have long-standing links to importers in many countries. They provide exporting firms with market research, identify potential buyers, and assist the firms in negotiating contracts.

Export trading companies, which are governed by the Export Trading Act in the United States, comprise those manufacturers and banks that either buy the products of a small business and resell them in another country or sell products of several companies on a commission basis. Small- and medium-sized exporting companies may also choose to retain foreign distributors, which purchase imported goods at a discount and resell them in the foreign or host country. Once a company has had some experience selling in other countries, it may decide to retain a foreign sales representative. Sales representatives differ from foreign distributors in that they do not take title to the goods being exported. Rather, they usually maintain a principal–agent relationship with the exporter.

INTERNATIONAL LICENSING AND FRANCHISING

international licensing
Contractual agreements by which a company (licensor) makes its intellectual property available to a foreign individual or company (licensee) for payment.

International licensing is a contractual agreement by which a company (*licensor*) makes its trade secrets, trademarks, patents, or copyrights (*intellectual property*) available to a foreign individual or company (*licensee*) in return either for royalties or for other compensation based on the volume of goods sold or a lump sum. All licensing agreements are subject to restrictions of the host country, which may include demands that its nationals be trained for management positions in the licensee company, that the host government receive a percentage of the gross profits, and that licensor technology be made available to all host-country nationals. Licensing agreements may differ vastly from country to country.

In the following case, the court was asked to resolve a conflict between two licensees of exclusive distribution rights to Russian films in the United States.

CASE 8-3

Russian Entertainment Wholesale, Inc. v. Close-Up International, Inc.
787 F. Supp. 2d 392 (2011)
United States District Court (E.D.N.Y.)

Two Russian film studios (the studios) granted rights to produce and distribute DVD versions of their films to multiple licensees. Each licensee received different limited exclusive rights. Krupny Plan, which could distribute the films only in the original Russian language, sublicensed its rights to the films for home use in the United States and Canada to Close-Up, a New York corporation. Ruscico could distribute multilingual versions of the same films that were dubbed or subtitled and sublicensed its rights to its distributor in the United States, Image. At the time of licensing, none of the parties considered that a viewer of the subtitled films could simply turn off the subtitles and hear the film in any of several languages, including Russian. None of the agreements had a requirement that the films prevent the disabling of subtitles. Close-Up brought this action against Ruscico and Image for damages from copyright infringement, claiming that it is the "exclusive" U.S. licensee of the Russian-language only versions of the films. The federal district court held for the defendants, and Close-Up appealed.

Judge Cogan

The Copyright Act establishes that the "legal or beneficial owner of an exclusive right under a copy-right" may bring suit for infringement under the act [citations omitted]. However, when this provision is invoked by an exclusive licensee, the licensee may seek relief from infringement only for the rights that the licensee has been exclusively licensed by the copyright holder. Plaintiff has shown that . . . it was the legal and beneficial licensee of the narrow right to reproduce and distribute *Russian-language-only* versions of the subject works. Therefore, even if plaintiff had a valid sublicense, plaintiff would still only have standing to sue for infringement of the narrow right to reproduce and distribute Russian-language-only DVDs. . . .

The evidence presented at trial proves that [the studios] elected to grant a "Russian language only" right to one licensee, and a separate "multilingual" right to another. The rights-holders did not consider sales of the multilingual DVDs manufactured by [the defendants] to violate the "Russian language only" license separately given to Krupny Plan. Instead, they considered the multilingual DVDs to be a distinct line of products, geared towards the separate non-Russian–speaking market.

Plaintiff has failed to put forth any evidence that defendants ever produced or distributed works that infringed plaintiff's limited rights in Russian-language–only DVDs . . . Instead, the evidence shows that all of the DVDs produced and distributed by defendants were multilingual DVDs, which [the studios] viewed as being distinct from the Russian-language-only DVDs that they had authorized Krupny Plan to reproduce and distribute. Plaintiff thus has failed to make out a claim for copyright infringement against any of the defendants.

Because there is no evidence that defendants reproduced or distributed DVD copies of the [films] that did not contain subtitles or dubbing in foreign languages, defendants' conduct was entirely within the scope of their rights . . .

Plaintiff next argues that paragraph 1.2.1 [of defendants' license], which states that "[r]eproduction of the Films in the original language without the accompaniment of the picture by sound and/or subtitles in a foreign language is a violation of the present Agreement," should be interpreted to mean that production of DVDs that *could be watched* in Russian without subtitles or dubbing was a violation of the agreement. However, plaintiff reads too much into this provision, which explicitly states that its purpose was to ensure that the DVDs produced by [the defendants] would be "multilingual versions." In this context, it is clear that paragraph 1.2.1 simply forbade [the defendants] from producing DVD copies . . . that did not include foreign subtitles or dubbing accompanying the films. Because all of the DVDs produced by defendants were multilingual versions that included subtitles in numerous foreign languages, defendants did not violate this provision of the agreement by producing DVDs that did not contain a disabling feature.*

The district court's opinion was *affirmed* in *Russian Entertainment Wholesale, Inc. v. Close-Up International, Inc.*, 482 Fed. Appx. 602 (2d Cir. 2012).

International franchising permits a licensee of a trademark to market the licensor's goods or services in a particular nation (e.g., Kentucky Fried Chicken franchises in China). Often companies franchise their trademark to avoid a nation-state's restrictions on foreign direct investment. Also, political instability is less likely to be a threat to investment when a local franchisee is running the business. Companies considering entering into an international franchise agreement should investigate bilateral treaties of friendship and commerce between

international franchising Contractual agreement whereby a company (licensor) permits another company (licensee) to market its trademarked goods or services in a particular nation.

---

*Russian Entertainment Wholesale, Inc. v. Close-Up International, Inc. 787 F. Supp. 2d 392 (2011) United States District Court (E.D.N.Y.).

the franchisor's nation and the franchisee's nation, as well as the business laws of the franchisee country.

In some instances, licensing and franchising negotiations are tense and drawn out because businesses in many industrialized nations are intent on protecting their intellectual property against "piracy" or are adamant about getting assurances that franchising agreements will be honored. These are major and legitimate concerns. For example, between 1994 and 1999, the United States threatened to impose sanctions against China because of that nation's sale of pirated U.S. goods and its failure to comply with international franchising requirements. A series of last-minute agreements encouraged by Chinese and U.S. businesses averted the sanctions, which would have proved expensive for private and public parties in both countries. By the year 2000, Congress and the president of the United States granted normal trade relations with China and left open the opportunity for the latter to join the WTO.[9] In June 2001, with a China–U.S. agreement on agriculture, a major barrier to entrance was overcome.

In November 2001, China and Taiwan entered the WTO after considerable negotiations between the Western nations over many issues. For example, at China's insistence, all membership documents refer to China as the People's Republic of China, whereas Taiwan is referred to as the Separate Customs Territory of Taiwan, Penghu, Kinmen, and Matsu. (The latter three islands are under the control of Taiwan.) Taiwan is not recognized as an independent nation by many, but as a territory belonging to the mainland. (See Chapter 25 for a discussion of franchising.)

FOREIGN DIRECT INVESTMENT

Direct investment in foreign nations is usually undertaken only by established multinational corporations. Foreign direct investment may take one of two forms: The multinational either creates a wholly or partially owned and controlled **foreign subsidiary** in the host country, or enters into a joint venture with an individual, corporation, or government agency of the host country. In both cases, the risk for the investing company is greater than the risk in international trade and international franchising and licensing, because serious amounts of capital are flowing to the host country that are subject to its government's restrictions and its domestic law.

foreign subsidiary A company that is wholly or partially owned and controlled in a company based in another country.

Large multinationals choose to create foreign subsidiaries for several reasons: (1) to expand their foreign markets; (2) to acquire foreign resources, including raw materials; (3) to improve their production efficiency; (4) to acquire knowledge; and (5) to be closer to their customers and competitors. Rarely do all these reasons pertain in a single instance. For example, U.S. companies have set up foreign subsidiaries in Mexico, Western Europe, Brazil, and India for quite different reasons. Mexico provided cheap labor and a location close to customers and suppliers for U.S. automobile manufacturers. In the case of Western Europe, the impetus was both a threat and an opportunity. The member nations of the EU have been moving to eliminate all trade barriers among themselves, but at the same time imposing stiffer tariffs on goods and services imported from non-EU countries. U.S. companies have been rushing to establish foreign subsidiaries in EU countries, not only to avoid being shut out of this huge and lucrative market, but also to expand sales among the EU's approximately 380 million people. Brazil is not only the largest potential market in Latin America, but it also offers low labor and transportation costs, making it ideal for U.S. automakers desiring to export to neighboring Latin American countries.

Union Carbide, Inc., a producer of chemicals and plastics, decided to establish a subsidiary in India, where cheap labor (including highly skilled chemists

---

[9]*See* "Backers Hope China Pact Will Promote Reform," *USA Today* 10 (Sept. 20, 2000).

and engineers) and low-cost transportation enabled the parent company to produce various materials cheaply and thus boost its bottom line. The Indian subsidiary turned out to be a very expensive investment for Union Carbide after the Bhopal disaster. The civil suit that resulted illustrates an issue that is often overlooked by managers of multinationals when setting up subsidiaries in foreign nation-states: Should a parent corporation be held liable for the activities of its foreign subsidiary? Although the case presented here is framed in a jurisdictional context (whether a U.S. court or an Indian court should hear the suit), bear in mind the issue of corporate parent liability as you read it.

CASE 8-4

In re Union Carbide Corp. Gas Plant Disaster v. Union Carbide Corp.
United States Court of Appeals
809 F.2d 195 (2d Cir. 1987)

The Government of India (GOI) and several private class action plaintiffs (Indian citizens) sued Union Carbide India Limited (UCIL) and the parent corporation, Union Carbide Corporation (UCC), for more than $1 billion after a disaster at a chemical plant operated by UCIL in 1984. There was a leak of the lethal gas methyl isocyanate from the plant on the night of December 2, 1984. The deadly chemicals were blown by wind over the adjacent city of Bhopal, resulting in the deaths of more than 2,000 persons and the injury of more than another 200,000 persons. UCIL is incorporated under the laws of India; 50.9 percent of the stock is owned by UCC, 22 percent is owned or controlled by the government of India, and the balance is owned by 23,500 Indian citizens. The federal district court (Judge Keenan) granted UCC's motion to dismiss the plaintiffs' action on the ground that Indian courts, not U.S. courts, were the appropriate forum for the suit. The plaintiffs appealed this decision.

Judge Mansfield

As the district court found, the record shows that the private interests of the respective parties weigh heavily in favor of dismissal on grounds of *forum non conveniens*. The many witnesses and sources of proof are almost entirely located in India, where the accident occurred, and could not be compelled to appear for trial in the United States. The Bhopal plant at the time of the accident was operated by some 193 Indian nationals, including the managers of seven operating units employed by the Agricultural Products Division of UCIL, who reported to Indian Works Managers in Bhopal. The plant was maintained by seven functional departments employing over 200 more Indian nationals. UCIL kept daily, weekly, and monthly records of plant operations and records of maintenance, as well as records of the plant's Quality Control, Purchasing, and Stores branches, all operated by Indian employees. The great majority of documents bearing on the design, safety, start-up, and operation of the plant, as well as the safety training of the plant's employees, is located in India. Proof to be offered at trial would be derived from interviews of these witnesses in India and study of the records located there to determine whether the accident was caused by negligence on the part of the management or employees in the operation of the plant, by fault in its design, or by sabotage. In short, India has greater ease of access to the proof than does the United States.

The plaintiffs seek to prove that the accident was caused by negligence on the part of UCC in originally contributing to the design of the plant and its provision for storage of excessive amounts of the gas at the plant. As Judge Keenan found, however, UCC's participation was limited and its involvement in plant operations terminated long before the accident. Under 1973 agreements negotiated at arm's length with UCIL, UCC did provide a summary "process design package" for construction of the plant and the services of some of its technicians to monitor the progress of UCIL in detailing the design and erecting the plant. However, the UOI controlled the terms of the agreements and precluded UCC from exercising any authority to "detail design, erect and commission the plant," which was done independently over the period from 1972 to 1980 by UCIL process design engineers who supervised, among many others, some 55 to 60 Indian engineers employed by the Bombay engineering firm of Humphreys and Glasgow. The preliminary process design information furnished by UCC could not have been used to construct the plant. Construction required the detailed process design and engineering data prepared by hundreds of Indian engineers, process designers, and subcontractors. During the ten years spent constructing the plant, the design and configuration underwent many changes.

In short, the plant has been constructed and managed by Indians in India. No Americans were employed at the plant at the time of the accident. In the five years from 1980 to 1984, although more than 1,000 Indians were employed at the plant, only one American was employed there and he left in 1982. No Americans visited the plant for more than one year prior to the accident, and during the five-year period before the accident the communications between the plant and the United States were almost nonexistent.

The vast majority of material witnesses and documentary proof bearing on causation of and liability for the accident is located in India, not the United States, and would be more accessible to an Indian court than to a United States court. The records are almost entirely in Hindi or other Indian languages, understandable to an Indian court without translation. The witnesses for the most part do not speak English but Indian languages understood by an Indian court but not by an American court. These witnesses could be required to appear in an Indian court but not in a court of the United States. India's interest is increased by the fact that it has for years treated UCIL as an Indian national, subjecting it to intensive regulations and governmental supervision of the construction, development, and operation of the Bhopal plant, its emissions, water and air pollution, and safety precautions. Numerous Indian government officials have regularly conducted on-site inspections of the plant and approved its machinery and equipment, including its facilities for storage of the lethal methyl isocyanate gas that escaped and caused the disaster giving rise to the claims. Thus India has considered the plant to be an Indian one and the disaster to be an Indian problem. It therefore has a deep interest in ensuring compliance with its safety standards.*

Affirmed in favor of Defendant, Union Carbide.

CRITICAL THINKING ABOUT THE LAW

Please refer to Case 8-4 and consider the following questions:

1. **Highlight the importance of facts in shaping a judicial opinion by writing an imaginary letter that, had it been introduced as evidence, would have greatly distressed Union Carbide Corporation (UCC).**

   Clue: Review the first part of the decision, in which Judge Mansfield discussed the extent of UCC's involvement in the plant where the accident occurred. What facts would counter his statement that the parent company had only "limited" involvement?

2. **Suppose a U.S. plant exploded, resulting in extensive deaths in the United States. Further, suppose that all the engineers who built the plant wrote and spoke German only. Could Judge Mansfield's decision be used as an analogy to seek dismissal of a negligence suit against the owners of the plant?**

   Clue: Review the discussion of the use of legal analogies in Chapter 1 and apply what you read to this question.

3. **What additional information, were it to surface, would strengthen Union Carbide's request for a dismissal of the case described?**

   Clue: Notice the wide assortment of facts that Judge Mansfield organized to support his decision.

COMMENT: The Bhopal victims filed their claims in U.S. courts against UCC because the parent company had more money than the subsidiary (UCIL). Also, suing the parent company made it more likely that the case would be heard in U.S. courts, which are considered to be far better forums for winning damages in personal injury actions than Indian courts are. After the lawsuits were removed to an Indian court, UCC agreed to pay $470 million to the Bhopal disaster victims. Union Carbide's stock substantially decreased in value, and UCC was threatened by a takeover (though the attempt was thwarted in 1985). More than half of UCC was subsequently sold or spun off, including the Indian subsidiary (UCIL). In 1989, the Indian Supreme Court ordered UCC to pay $470 million to compensate Bhopal victims; criminal charges against the company and its officials were dropped. About 12,000 people worked for UCC in 1995, in contrast to the 110,000 employed by the company a decade earlier. In 2001, Dow Chemical acquired Union Carbide (see *Wall Street Journal,* August 12, 2009, p. B-10).

On June 7, 2010, a district court in Bhopal found seven former Union Carbide India, Ltd. officials guilty of "causing death by negligence" for a gas leak

---

*In re Union Carbide Corp. Gas Plant Disaster v. Union Carbide Corp., United States Court of Appeals 809 F.2d 195 (2d Cir. 1987).

at the plant that killed approximately 3,000 people some 25 years before. This was the first criminal conviction. All convicted were Indian citizens. They were sentenced to 2 years in prison and fined 100,000 rupees ($2,130). The former Union Carbide subsidiary was convicted of the same charges and fined 5,000 rupees. All seven defendants were freed on bail. As of June, 2010, no cleanup has taken place in the affected area (see Lydia Polgreen and Hari Kumar, "8 Former Executives Guilty in '84 Bhopal Chemical Leak," *New York Times,* June 8, 2010, p. A-8; T. Lahiri, "Court Convicts Seven in Bhopal Gas Leak," *Wall Street Journal,* June 8, 2010, p. A-11).

Joint ventures, which involve a relationship between two or more corporations or between a foreign multinational and an agency of a host-country government or a host-country national, are usually set up for a specific undertaking over a limited period of time. Many developing countries (such as China) allow foreign investment only in the form of a joint venture between host-country nationals and the multinationals. Recently, three-way joint ventures have been established among United States–based multinationals (e.g., automobile companies such as Chrysler and General Motors), Japanese multinationals (e.g., Mitsubishi and Honda), and Chinese government agencies and Chinese nationals. Joint ventures are also used in host countries with fewer restrictions on foreign investment, often to spread the risk or to amass required investment sums that are too large for one corporation to raise by itself. Some of these joint ventures are private associations, with no host-government involvement.

joint venture Relationship between two or more persons or corporations or an association between a foreign multinational and an agency of the host-country government or a host-country national set up for a business undertaking for a limited time period.

Risks of Engaging in International Business

Unlike doing business in one's own country, the "rules of the game" are not always clear when engaging in business in a foreign country, particularly in what we have classified as middle- and low-income economies. Here we set out three primary risks that managers engaged in international business may face: (1) expropriation of private property by the host foreign nation, (2) the application of the sovereign immunity doctrine and the act-of-state doctrine to disputes between foreign states and U.S. firms, and (3) export and import controls.

EXPROPRIATION OF PRIVATE PROPERTY

Expropriation—the taking of private property by a host-country government for either political or economic reasons—is one of the greatest risks companies take when they engage in international business. Thus, it is essential for business managers to investigate the recent behavior of host-country government officials, particularly in countries that are moving from a centrally planned economy toward one that is market oriented (e.g., Russia and Eastern European nations).

expropriation The taking of private property by a host-country government for political or economic reasons.

One method of limiting risk in politically unstable countries is to concentrate on exports and imports (trade) and licensing and franchising. Another method is to take advantage of the low-cost insurance against expropriation offered by the Overseas Private Investment Corporation (OPIC). If a U.S. plant or other project is insured by OPIC and is expropriated, the U.S. firm receives compensation in return for assigning to OPIC the firm's claim against the host-country government.

Bilateral investment treaties (BITs), which are negotiated between two governments, obligate the host government to extend fair and nondiscriminatory treatment to investors from the other country. A BIT normally also includes a promise of prompt, adequate, and effective compensation in the event of expropriation or nationalization.

bilateral investment treaty (BIT) Treaty between two parties to outline conditions for investment in either country.

SOVEREIGN IMMUNITY DOCTRINE

sovereign immunity doctrine
States that a foreign-owned private property that has been expropriated is immune from the jurisdiction of courts in the owner's country.

Another risk for companies engaged in international business is the **sovereign immunity doctrine**, which allows a government expropriating foreign-owned private property to claim that it is immune from the jurisdiction of courts in the owner's country because it is a government rather than a private-sector entity. In these cases, the company whose property was expropriated often receives nothing because it cannot press its claims in its own country's courts, and courts in the host country are seldom amenable to such claims.

The sovereign immunity doctrine has been a highly controversial issue between the United States and certain foreign governments of developing nations. To give some protection to foreign businesses without impinging on the legitimate rights of other governments, the U.S. Congress in 1976 enacted the Foreign Sovereign Immunities Act (FSIA), which shields foreign governments from U.S. judicial review of their public, but not their private, acts. The FSIA grants foreign nations immunity from judicial review by U.S. courts unless they meet one of the FSIA's private exceptions. One such exception is the foreign government's involvement in "commercial activity." Case 8-5 clarifies the U.S. Supreme Court's definition of "commercial activity" under the FSIA. Note how the Court emphasizes the nature of the Nigerian government's action by asking whether it is the type of action a private party would engage in.

CASE 8-5

Keller v. Central Bank of Nigeria
United States Court of Appeals
277 F.3d 811 (6th Cir. 2002)

Prince Arthur Ossai, a government official in Nigeria, entered into a contract with Henry Keller (plaintiff), a sales representative for H.K. Enterprises, Inc., a Michigan-based manufacturer of medical equipment. They agreed that, among other things, Ossai would have an exclusive distribution right to sell H.K. products in Nigeria, which would buy $4.1 million of H.K. equipment for $6.63 million, plus a $7.65 million "licensing fee." Before the deal closed, though, Ossai demanded that $25.5 million on deposit in the Central Bank of Nigeria (CBN) be transferred into an account set up by Keller. CBN employees charged Keller $28,950 in fees for the transaction, but the funds were never transferred. Keller and H.K. filed a suit in a federal district court against the CBN and others, asserting in part a claim under the Racketeer Influenced and Corrupt Organizations Act (RICO). The defendants filed a motion to dismiss under the Foreign Sovereign Immunities Act. The court denied the motion, concluding that the claim fell within the FSIA's "commercial activity" exception. The defendants appealed to the U.S. Court of Appeals for the Sixth Circuit.

Justice Norris

[The defendants] claim that the illegality of the deal alleged precludes a finding that it is a commercial activity. The FSIA defines "commercial activity" as "either a regular course of commercial conduct or a particular commercial transaction or act." The commercial character of an activity shall be determined by reference to the nature of the course of conduct or particular transaction or act, rather than by reference to its purpose. [W]hen a foreign government acts, not as regulator of a market, but in the manner of a private player within it, the foreign sovereign's actions are commercial within the meaning of the FSIA.

In the instant case, the conduct was a deal to license and sell medical equipment, a type of activity done by private parties and not a "market regulator" function. The district court correctly concluded that this was a commercial activity, and that any fraud and bribery involved did not render the plan non-commercial.

Defendants claim that plaintiffs cannot establish another element of the commercial activity exception, namely, that there was a direct effect in the United States. [A]n effect is "direct" if it follows as an immediate consequence of the defendant's activity.

In this case, defendants agreed to pay but failed to transmit the promised funds to an account in a Cleveland bank. Other courts have found a direct effect when a defendant agrees to pay funds to an account in the United States and then fails to do so. The district court in the instant case correctly concluded, in accord with the other [courts], that defendant's failure to pay promised funds to a Cleveland account constituted a direct effect in the United States.*

Affirmed for the Plaintiff.

---

*Keller v. Central Bank of Nigeria., United States Court of Appeals 277 F.3d 811 (6th Cir. 2002).

CRITICAL THINKING ABOUT THE LAW

Context plays a vital role in any legal decision. The existence or nonexistence of certain events directly affects the court's verdict. In Case 8-5, the court applies the strictures of the FSIA to the specific facts of the case. If certain facts exist, the federal statute protects the plaintiff, and the court should appropriately reject the defendant's motion to dismiss. Otherwise, the CBN is immune, and the statute does not protect the plaintiff.

Understanding the facts is the starting point for legal analysis. The following questions encourage you to consider the significance of the facts in Case 8-5.

1. What facts are critical in the court's ruling in favor of the plaintiff?

   Clue: Reread the introductory paragraph.

2. Look at the facts you found. To illustrate the importance of context, which fact, if it had not been included in the case, might have resulted in the court's granting the defendant's motion to dismiss?

   Clue: Find the elements of the federal statute that the judge discusses and use these elements as a guide to highlight the most significant facts.

Act-of-State Doctrine

The **act-of-state doctrine** holds that each sovereign nation is bound to respect the independence of every other sovereign state and that the courts of one nation will not sit in judgment on the acts of the courts of another nation done within that nation's own sovereign territory. This doctrine, together with the sovereign immunity doctrine, substantially increases the risk of doing business in a foreign country. Like the sovereign immunity doctrine, the act-of-state doctrine includes some court-ordered exceptions, such as when the foreign government is acting in a commercial capacity or when it seeks to repudiate a commercial obligation.

Congress made it clear in 1964 that the act-of-state doctrine shall not be applied in cases in which property is confiscated in violation of international law, unless the president of the United States decides that the federal courts should apply it. As Case 8-6 demonstrates, the plaintiff has the burden of proving that the doctrine should not apply—that is, the courts should sit in judgment of public acts of a foreign government, in this case, a former government.

act-of-state doctrine A state that each nation is bound to respect the independence of another and the courts of one nation will not sit in judgment on the acts of the courts of another nation.

CASE 8-6

Linde v. Arab Bank, PLC
United States Court of Appeals
Second Circuit 706 F.3d 92 (2013)

Founded in 1930, Arab Bank is one of the largest financial institutions in the Middle East. Headquartered in Jordan, it serves clients in more than 500 branches in 30 countries, including branches in Australia, New York, and Switzerland. The bank is a major economic engine in Jordan and throughout the Middle East/Northern Africa, providing modern banking services and capital, and facilitating development and trade throughout the region. Victims of terrorist attacks that were committed in Israel between 1995 and 2004—during a period commonly referred to as the Second Intifada—filed a suit in a federal district court against Arab Bank, PLC, seeking damages under the Anti-Terrorism Act (ATA) and the Alien Tort Claims Act. According to plaintiffs, Arab Bank provided financial services and support to the terrorists. Over several years and despite multiple discovery orders, the bank failed to produce certain documents relevant to the case. As a result, the court issued an order imposing sanctions. Arab Bank appealed to the U.S. Court of Appeals for the Second Circuit, arguing that the order was an abuse of discretion.

Justice Carney

The Bank argues that the documents are covered by foreign bank secrecy laws such that their disclosure would subject the Bank to criminal prosecution and other penalties in several foreign jurisdictions. The sanctions order takes the form of a jury instruction that would permit—but not require—the jury to infer from the Bank's failure to produce these documents that the Bank provided financial services to designated foreign terrorist organizations, and did so knowingly.

The District Court carefully explained its decision to impose this sanction. It noted that many of the documents that plaintiffs had already obtained tended to support the inference that Arab Bank knew that its services benefitted terrorists. According to the District Court, these documents included documents from Arab Bank's Lebanon branch that suggested Arab Bank officials approved the transfer of funds into an account at that branch despite the fact that the transfers listed known terrorists as beneficiaries. As a consequence of Arab Bank's nondisclosure, the court reasoned, plaintiffs would be "hard-pressed to show that these transfers were not approved by mistake, but instead are representative of numerous other transfers to terrorists." The permissive inference instruction will, according to the District Court, "help to rectify this evidentiary imbalance."

Arab Bank argues that the District Court's decisions ordering production and imposing sanctions should be vacated because they offend international comity. This argument derives from the notion that the sanctions force foreign authorities either to waive enforcement of their bank secrecy laws or to enforce those laws, and in so doing create an allegedly devastating financial liabilities for the leading financial institution in their region. The Bank asserts, further, that international comity principles merit special weight here because the District Court's decisions affect the United States' interests in combating terrorism and pertain to a region of the world pivotal to United States foreign policy.

The [District] Court expressly noted that it had "considered the interests of the United States and the foreign jurisdictions whose foreign bank secrecy laws are at issue."

Additionally, international comity calls for more than an examination of only some of the interests of some foreign states. Rather, the concept of international comity requires a particularized analysis of the respective interests of the foreign nation and the requesting nation. In other words, the analysis invites a weighing of all of the relevant interests of all of the nations affected by the court's decision. The District Court recognized the legal conflict faced by Arab Bank and the comity interests implicated by the bank secrecy laws. But [the Court] also observed—and properly so—that Jordan and Lebanon have expressed a strong interest in deterring the financial support of terrorism, and that these interests have often outweighed the enforcement of bank secrecy laws, even in the view of the foreign states. Moreover, the District Court took into account the United States' interests in the effective prosecution of civil claims under that ATA [Anti-Terrorism Act]. This type of holistic, multi-factored analysis does not so obviously offend international comity.*

The U.S. Court of Appeals for the Second Circuit *affirmed* the lower court's decision and order.

EXPORT AND IMPORT CONTROLS

Export Controls. Export controls are usually applied by governments to militarily sensitive goods (e.g., computer hardware and software) to prevent unfriendly nations from obtaining these goods. In the United States, the Department of State, the Department of Commerce, and the Defense Department bear responsibility, under the Export Administration Act and the Arms Export Control Act, for authorizing the export of sensitive technology. Both criminal and administrative sanctions may be imposed on corporations and individuals who violate these laws.

Export controls often prevent U.S. companies from living up to negotiated contracts. Thus, they can damage the ability of U.S. firms to do business abroad.

Import Controls. Nations often set up import barriers to prevent foreign companies from destroying home industries. Two such controls are tariffs and quotas. For example, the United States has sought historically to protect its domestic automobile and textile industries, agriculture, and intellectual property (copyrights, patents, trademarks, and trade secrets). Intellectual property has become an extremely important U.S. export in recent years, and Washington has grown more determined than ever to prevent its being pirated. After several years of frustrating negotiations with the People's Republic of China, the U.S. government decided to threaten imposition of 100 percent tariffs on approximately $1 billion of Chinese imports in 1995, and again in 1996 and

---

*Linde v. Arab Bank, PLC, United States Court of Appeals, Second Circuit, 706 F.3d 92 (2013).

1997. In retaliation, the Chinese government has threatened several times to impose import controls on many U.S. goods. Washington took action only after documenting that hundreds of millions of dollars' worth of "pirated" computer software and products (including videodiscs, law books, and movies) was being produced for sale within China and for export to Southeast Asian nations in violation of the intellectual property laws of both China and the United States, as well as international law. The documentation showed that 29 factories owned by the state of Communist Party officials were producing pirated goods. A last-minute settlement in which the Chinese government pledged to honor intellectual property rights prevented a trade war that would have had negative implications for workers in import–export industries in both countries. American consumers would also have suffered because Chinese imports would have become twice as expensive had the 100 percent tariff taken effect—though the effect on consumers would have been offset by an increase in imports of the affected goods from other foreign countries (e.g., English and Japanese bikes would have replaced Chinese bikes in demand).

Another form of import control is the imposition of antidumping duties by two U.S. agencies, the International Trade Commission (ITC) and the International Trade Administration (ITA). The duties are levied against foreign entities that sell the same goods at lower prices in U.S. markets than in their own in order to obtain a larger share of the U.S. market (i.e., entities that practice "dumping").

Legal and Economic Integration as a Means of Encouraging International Business Activity

Table 8-2 summarizes a number of groups that have been formed to assist businesspeople in carrying out international transactions. These groups range from the WTO (formerly the General Agreement on Tariffs and Trade), which is attempting to reduce tariff barriers worldwide, to the proposed South American Common Market, which would form a duty-free zone for all the nations of South America. The most ambitious organization is the EU, which is in the process of forming a Western European political and economic community with a single currency and a common external tariff barrier toward nonmembers.

Multinational corporations are learning that doing international business is much easier when they are aware of the worldwide and regional groups listed in Table 8-2. We will describe three of these groups: the WTO, the EU, and NAFTA. We chose to examine these three because they represent three different philosophies and structures of legal and economic integration, not because we do not appreciate the major effects of other integrative groups outlined in the table.

THE WORLD TRADE ORGANIZATION

Purpose and Terms. On January 1, 1995, the 47-year-old General Agreement on Tariffs and Trade (GATT) organization was replaced by a new umbrella group, the World Trade Organization. The WTO has the power to enforce the new trade accord that evolved out of seven rounds of GATT negotiations, with more than 140 nations participating. All 144 signatories to this accord agreed to reduce their tariffs and subsidies by an average of one-third on most goods over the next decade, agricultural tariffs and subsidies included. Economists estimated that this trade pact would result in tariff reductions totaling $744 billion over the next 10 years.

Moreover, the accord, which the WTO will supervise, prohibits member countries from placing limits on the quantity of imports (*quotas*). For example, Japan had to end its ban on rice imports, and the United States had to end its import quotas on peanuts, dairy products, and textiles. Furthermore, the agreement

TABLE 8-2 LEGALLY AND ECONOMICALLY INTEGRATED INSTITUTIONS

Name	Members and Purpose
World Trade Organization (WTO)	Replaced General Agreement on Tariffs and Trade (GATT) in 1995 and the most-favored-nation clause with the normal trade relations principle. Composed of 144 member nations. Goal is to get the nations of the world to commit to the trade principles of nondiscrimination and reciprocity so that, when a trade treaty is negotiated between two members, the provisions of that bilateral treaty will be extended to all WTO members. All members are obligated to harmonize their trade laws or face sanctions. The WTO, through its arbitration tribunals, is to mediate disputes and recommend sanctions.
European Union (EU)	Composed of 28 European member states. Established as the European Economic Community (later called the European Community) by the Treaty of Rome in 1957. Goal is to establish an economic "common market" by eliminating customs duties and other quantitative restrictions on the import and export of goods and services among member states. In 1986, the treaty was amended by the SEA, providing for the abolition of all customs and technical barriers between nations by December 31, 1992. In 1991, the Maastricht Summit Treaty proposed monetary union, political union, and a "social dimension" (harmonizing labor and social security regulations) among EU members, although not all aspects were approved by all members. The treaty was subsequently amended by the treaties of Amsterdam (1997), Nice (2001), and Lisbon (2007). The Treaty of Lisbon entered into force in December of 2009 and is still effective today.
North American Free Trade Agreement (NAFTA)	The United States, Canada, and Mexico are members at present; Chile has been invited to join. NAFTA seeks to eliminate barriers to the flow of goods, services, and investments among member nations over a 15-year period, starting in 1994, the year of its ratification. Unlike the EU, NAFTA is not intended to create a common market. Whereas EU states have a common tariff barrier against non-EU states, members of NAFTA maintain their own individual tariff rates for goods and services coming from non-NAFTA countries.
Organization for Economic Cooperation and Development (OECD)	This organization was established in 1951 with Western European nations including Australia, New Zealand, United States, Canada, Japan, Russia, and some Eastern European nations as associate members. The OECD's original purpose was to promote economic growth after World War II. Today it recommends and evaluates options on environmental issues for its members and establishes guidelines for multinational corporations when operating in developed and developing countries.
European Free Trade Association (EFTA)	Founded in 1960 and originally composed of Finland, Sweden, Norway, Iceland, Liechtenstein, Switzerland, and Austria. EFTA has an intergovernmental council that negotiates treaties with the EU. Finland, Sweden, and Austria left the EFTA in 1995 and joined the EU. In the future, EFTA will have only minor significance.
Andean Common Market (ANCOM)	Composed of Bolivia, Venezuela, Colombia, Ecuador, and Peru. ANCOM seeks to integrate these nations politically and economically through a commission, the Juanta Andean Development Bank, and a Reserve Fund and a Court of Justice. Founded in 1969, achievement of its goals has been hampered by national interests.
Mercado Commun del Ser Mercosul (Mercosul)	Composed of Argentina, Brazil, Paraguay, and Uruguay. Mercosul's purposes are to reduce tariffs, eliminate nontariff barriers among members, and establish a common external tariff. The organization was founded in 1991, and these goals were to be met by December 31, 1994. For political reasons, they have not yet been fully met.
South American Common Market	A duty-free common market made up of countries in the ANCOM and Mercosul groups, created on January 1, 1995. On that date, tariffs were ended on 95 percent of goods traded among Brazil, Argentina, Paraguay, and Uruguay. All three nations adopted common external tariffs.
Asia-Pacific Economic Corporations (APEC)	Formed in 1989. A loosely organized group of 11 developed and developing Pacific Group nations, including Japan, China, United States, Canada, and New Zealand. APEC is not a trading bloc and has no structure except for a secretariat. An underlying Trans-Pacific Partnership (TPP) represents a proposed trade deal between the United States and 11 Asian-Pacific nations. If enacted, this TPP trade deal would affect nearly 1 billion people and 65% of global trade worldwide. It is opposed by trade unions, environmentalists, and political figures in all parties.

TABLE 8-2 CONTINUED

Name	Members and Purpose
	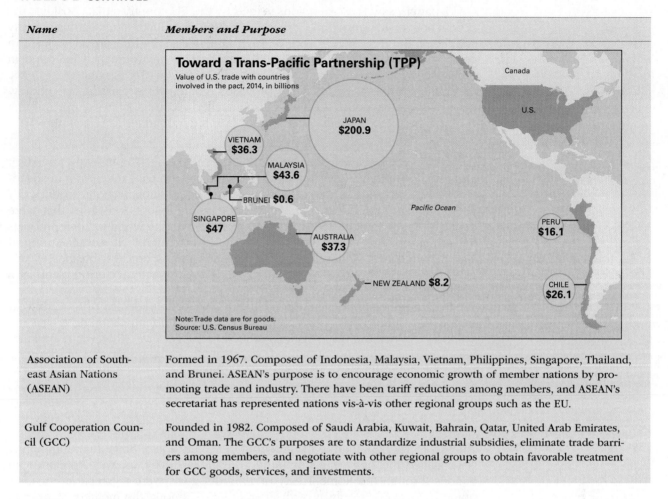
Association of Southeast Asian Nations (ASEAN)	Formed in 1967. Composed of Indonesia, Malaysia, Vietnam, Philippines, Singapore, Thailand, and Brunei. ASEAN's purpose is to encourage economic growth of member nations by promoting trade and industry. There have been tariff reductions among members, and ASEAN's secretariat has represented nations vis-à-vis other regional groups such as the EU.
Gulf Cooperation Council (GCC)	Founded in 1982. Composed of Saudi Arabia, Kuwait, Bahrain, Qatar, United Arab Emirates, and Oman. The GCC's purposes are to standardize industrial subsidies, eliminate trade barriers among members, and negotiate with other regional groups to obtain favorable treatment for GCC goods, services, and investments.

bans the practice of requiring high local content of materials for manufactured products such as cars. It also requires all signatory countries to protect patents, trademarks, copyrights, and trade secrets. (See Chapter 16 for a discussion of these topics, which fall under the umbrella of intellectual property.)

General Impact. The accord created the WTO, which consists of all nations whose governments approved and met the GATT—Uruguay Round Accord.[10] Each member state has one vote in the Ministerial Conference, with no single nation having a veto. The conference meets at least biannually, and a general council meets as needed. The conference may amend the charter created by the pact—in some cases, by a two-thirds vote; in others, by a three-fourths vote. Changes apply to all members, even those that voted *no*.

The WTO has been given the power to set up a powerful dispute-resolution system with three-person arbitration panels that follow a strict schedule for dispute-resolution decisions. WTO members may veto the findings of an arbitration panel. This is a matter of great concern to the United States, and it figured prominently in the House and Senate debates preceding approval of the WTO.

---

[10] Uruguay Round Amendments Act, Pub. L. No. 103–465, was approved by Congress on December 8, 1994. One hundred eight states signed the final act embodying the results of the Uruguay Round of multilateral trade negotiations. Bureau of National Affairs, *International Reporter* 11 (Apr. 20, 1994). Sixteen more states have joined since 1994.

U.S. farmers (and other groups), who had previously won decisions before GATT panels, saw these decisions vetoed by the EU countries that subsidize the production of soybeans and other agricultural products; they enthusiastically supported the new WTO process. Environmentalists and consumer groups, in contrast, feared that the WTO would overrule U.S. environmental laws and in other ways infringe on the sovereignty of the nation. To assuage these fears, the framers of the accord added a provision that allows any nation to withdraw upon six months' notice. Congress also attached a condition to its approval that calls for the establishment of a panel of U.S. federal judges to review the WTO panels' decisions.

Impact on Corporate Investment Decision Making. The WTO pact makes it less risky for multinationals to source parts—that is, to have them built in cheap-labor countries and then brought back to the multinational's home country for use in making a product (e.g., brakes for automobiles). Industries that were expected to shift quickly to buying parts from all over the world for their products include computers, telecommunications, and other high-tech manufacturing. It is anticipated that with a freer flow of goods across borders, businesses will gain increased economies of scale by building parts-manufacturing plants at a location that serves a wide market area.

Because the tariff cuts were to be introduced gradually over a six-year period and were to be finalized only after 10 years, the impact on trade was not immediate. Once the major nations ratified the WTO pact, however, companies began to make investment and employment decisions predicated on dramatic reductions in tariffs, subsidies, and other government-established deterrents to free trade.

THE EUROPEAN UNION

Purpose. The 28-member EU grew out of the European Economic Community (later called the European Community), established by six Western European nations through the Rome Treaty of 1957. Its goals were to create a "customs union" that would do away with internal tariffs among the member states and to create a uniform external tariff to be applied to all nonmembers. The EU is thoroughly committed to achieving the free movement of goods, services, capital, and people across borders (Exhibit 8-1).

The EU's ambitious plan to create an immense "common market" of 400 million people and $4 trillion worth of goods was greatly strengthened by the 1986 signing of the Single European Act (SEA), which set a deadline for economic integration of December 31, 1992, and instituted new voting requirements to make passage of EU legislation easier. A treaty that was proposed at the Maastricht Summit in 1991 was ratified in 1993. It provided for (1) monetary union through the creation of a single currency for the entire EU, (2) political union, and (3) a "social dimension" through the establishment of uniform labor and Social Security regulations. The leaders of the 12 nations that were members at the time agreed to the creation of a European Monetary Institute by 1994, a European Central Bank by 1998, and a uniform European currency unit (ECU) by 1999.

Three new members (Finland, Austria, and Sweden) joined the EU in 1994, bringing the total to 15 nations. In May 2004, 10 more nations joined the EU: Cyprus, Czech Republic, Estonia, Hungary, Latvia, Lithuania, Malta, Poland, Slovakia, and Slovenia (see Exhibit 8-2). Bulgaria, Croatia, and Romania were accepted into the EU in 2007.

Structure. The EU consists of (1) a Council of Ministers, (2) a Commission, (3) a Parliament (Assembly), (4) the European Court of Justice (with the addition of the Court of First Instance), and (5) the European Central Bank (ECB) and the European Monetary Union (EMU).

EXHIBIT 8-1

EUROPEAN UNION AND AN ATTEMPT TO CENTRALIZE POWER

EXHIBIT 8-2

FROM 15 TO 28: THE EU SPREADS TO THE EAST

Source: The Legal Environment of Business and On Line Commerce (5th ed.), by Henry Cheeseman (© 2007) p. 77, Ex. 5-2. Reproduced by permission of Pearson Education, Inc., Upper Saddle River, New Jersey.

After eight years of negotiations, the Lisbon Treaty became effective in December of 2009. The treaty seeks to give the European Union an international role beyond Europe. It creates a full-time presidency of a council of national leaders of the 27 member countries. The president is elected for a period of two and a half years. He (Herman Van Rompuy) has a full-time staff located in Brussels. The treaty also creates a new high representative for foreign and security policy. She (Catherine Ashton) will negotiate with the European Parliament and nations worldwide over major foreign policy issues, and will coordinate a diplomatic corps of approximately 5,000 diplomats, called the External Action Service, which is intended to supplement but not replace diplomats of the 27 individual member countries.

The treaty seeks to delineate the relationships between national legislatures of individual member countries and the EU Parliament. It will seek to implement a new voting system reflecting the population size of EU member countries. The treaty will use this new voting system to prevent a single member country from blocking a proposal such as an amendment to the constitution of the EU. Moreover, the treaty will give more power to the directly elected European Parliament.

The importance of the Treaty of Lisbon is that it replaces a failed attempt at a proposed constitution which was rejected by France and the Netherlands in referendums in 2005. To treaty supporters, it was important that a new voting system be established, as 12 new nations have been added since 2004. Critics argue that it was not needed because the new bureaucracy (located in Brussels, capital of the EU) encroaches upon the national sovereignty of member states.

1. *Council of Ministers.* The Council of Ministers is composed of one representative from each of the member nations. Its purposes are to coordinate the economic policies of member states and, more recently, to negotiate with nonmember states. In the past, the council generally rubber-stamped legislation proposed by the Commission, but this docility is less assured as the council begins to flex some of the authority granted it by the SEA and the Maastricht Treaty.

2. *Commission.* The Commission consists of 20 members who represent the EU, not national concerns. It is responsible for the EU's relations with international organizations such as the United Nations and the WTO. Member states are apportioned voting power in the Commission on the basis of their population and economic power. The Commission elects a president from among its members. Each Commission member supervises a functional area (e.g., agriculture or competition) that may be affected by several directorates. There are 22 directorates that are run by "supranational" civil servants called *director generals*. In theory, the directorates serve the Commission, but in fact the director generals often heavily influence legislation as it moves through the Commission.

Key Elements of the Treaty of Maastricht

Agreement	Goals and Stumbling Blocks
Monetary Union	The European Monetary Institute was created on January 1, 1994, and started operating on January 1, 1999.
	A single currency issued any day after January 1, 1999, is the goal of 25 nations that meet three standards: (1) Annual budget deficit cannot exceed the ceiling of 3 percent of gross domestic product; (2) the public debt limit for each country must not exceed 60 percent of gross domestic product; and (3) a country's inflation rate must be lower than 2.9 percent, based on a complex formula set out in 1997.
	Great Britain was allowed to opt out of the EC currency union until an unspecified date. It opposes monetary union for ideological reasons.
	Denmark was also allowed to opt out pending a referendum on the issue, a constitutional requirement. The Danish government backs monetary union.
Political Union	EC jurisdiction in areas including industrial affairs, health, education, trade, environment, energy, culture, tourism, and consumer and civil protection. Member states vote to implement decisions.
	Increased political cooperation under a new name—European Union. Permanent diplomatic network of senior political officials created in the EC capitals.
	Great Britain rejected EC-imposed labor legislation, forcing the removal of the so-called social chapter from the treaty. It will be implemented separately by the other 14 members, officials said.
Federalism	EC leaders dropped reference to an EC "with a federal goal." Instead, the political Union accord describes the community as "an ever-closer union in which political decisions have to be taken as near to the people as possible."
	Great Britain rejected the "federalism" concept as the embodiment of what it feels would be an encroaching EC superstate.
Foreign Affairs	EC states move toward a joint foreign policy, with most decisions requiring unanimity.
	Great Britain wanted the ability to opt out of any joint decision. How this provision will be interpreted by the various sides is yet to be determined.
Defense	The Western EU, a long-dormant group of nine EC states, will be revived to act as the EC's defense body but linked to the NATO alliance.
	Although France and Germany supported a greater military role for the Union, Great Britain, Italy, and others did not want to see NATO's influence diluted.
European Parliament	The 754 member EC assembly gets a modest say in shaping some EC legislation. Its new powers fall short of what the assembly had sought (i.e., an equitable sharing of the right to make EC laws with the EC governments).
	Great Britain and Denmark refused to grant the assembly broader powers.

3. *Parliament.* The Parliament (Assembly) is made up of representatives elected from each nation-state for a term set by the nation-state. The representatives come from most of the major European political factions (Socialists, Christian Democrats, Communists, Liberals, etc.), and each of the parties in the Parliament also exists in the member states. The Parliament elects a president to preside over its deliberations. The Parliament's general powers are to (1) serve as a consultative body to the Council, (2) refer matters affecting EU interests to the Commission or Council, (3) censure the Commission when necessary, (4) assent to trade agreements with countries outside the EU, (5) amend the EU budget, and (6) participate with the Commission in the legislative procedure.

4. *European Court of Justice.* The European Court of Justice performs the functions of arbiter and final decision maker in conflicts between EU law and individual member states. The national courts of member states are obligated to follow EU law and Court of Justice decisions. The Court of First Instance was established in 1989 to reduce the workload of the Court of Justice. It has jurisdiction over appeals of the Commission's decisions on mergers and acquisitions. It also sets the penalties for price-fixing when non-EU companies are involved.

5. *The European Central Bank and the European Monetary Union.* The Maastricht Treaty required all member states of the EU to converge their economic and monetary policies with the goal of creating a single currency, the euro. The criteria set forth required the harmonization of budget deficits, inflation levels, and long-term interest rates, and specific levels of currency inflations to achieve exchange-rate stability. The European Monetary Union was created in January 1999. By giving up their national currencies, members of the EMU have relinquished control of their exchange rates and monetary policy to an independent European Central Bank based in Frankfurt. The primary duty of the ECB is to maintain price stability, by lowering or raising interest rates. With the exception of Denmark and Great Britain, most members of the EMU have adopted the euro.

Impact.

Unity of Law. Agricultural, environmental, and labor legislation is being made uniform throughout the member nations, with allowances and subsidies for the poorer members. The national courts of member states are now following decisions of the European Court of Justice.

Economic Integration. The SEA and Maastricht Treaty have pushed the EU members to eliminate tariff and nontariff barriers among themselves. British and French differences over the creation of a single currency have forced the suspension of this goal.

Political Union. The political union envisioned by the Maastricht Treaty has been an elusive goal for the EU, because member states (and their citizens) have proved more reluctant to make the necessary compromises on national sovereignty than the treaty's architects anticipated. Nonetheless, the EU is the only regional organization that has in place the sophisticated structure required to make political union a realistic possibility.

The "Big Fat Greek" Bailout

On May 10, 2010, Euro-zone lenders (17 nations belonging to the European Union), the International Monetary Fund (IMF), and the European Central Bank decided to "bail out" the nearly bankrupt nation of Greece. They agreed to a nearly $1 trillion package of loans, guarantees, swaps, and bond purchases for use by Greece and other EU nations. The IMF contributed $318 billion. (It should be noted that the United States [taxpayer] had a role in this bailout because of its contributions to the IMF and its voting strength in that body. President Obama encouraged European leaders to act quickly. The Federal Reserve also provided credits for European banks that were lenders to Greece.) Given the existing precedents, the degree of international cooperation was surprising. As of May 2010, it seemed doubtful that the full $1 trillion would be needed, but the markets of Portugal, Spain, and Ireland, in the eyes of some, were similar to the situation in Greece.

Markets worldwide fell sharply following the bailout on May 9, 2010. Investors showed a lack of confidence in European markets and the euro. Similar to the U.S. bailout of banks and some other large business entities (insurance and automobile companies) in 2009, the EU sought to pledge support for debt markets in the Euro-zone combined with attempts to correct fiscal problems in certain countries. It may be easier to help individual banks and corporations in the United States than to bail out a sovereign country (or group of countries), having a single currency.

Why does a small country like Greece deserve a bailout, and why should it be able to shake the global financial system? Some thoughts follow:

- The problem may be larger than Greece. Many European nations, like Greece, have poor tax collection systems, and thus have problems servicing their sovereign debt. Spain, Portugal, Ireland, and Italy have similar problems. Spain may be too big to be bailed out.
- Prior to the establishment of the EU, problems with the countries of Europe would not have been as visible. If Greece had had its own currency (drachmas), it would have gone bankrupt, boosting both its exports and the cost of borrowing money, leaving it with less debt and perhaps more revenues and tax revenue to service the debt. Within the EU structure (with fixed exchange-rate systems like the euro), free-market signals are masked. Interest rates are held artificially low for reckless borrowers such as Greece, whose debt was largely held by private and state banks of Germany, France, and Spain. These nations played a role in bailing out Greece in 2010.
- The global economy may suffer deflationary effects. If enforceable austerity measures take place in countries that receive bailout money, millions of people (like residents in Greece) will end up living on less, resulting in less demand for worldwide natural resources, consumer goods, and housing components. The price of gold may initially jump as Europeans seek to protect a weakening euro.
- U.S. investors may be concerned about reduced earnings of multinationals that do business in EU countries. However, U.S. banks could gain from money that flees Europe, and U.S. consumer spending could benefit from weaker commodity prices.

   Update: As of July, 2013, Greece has been on an austerity program and received bailout money from the IMF, the ECB, Germany, and indirectly from the United States. With the election of a new government in 2015, a new liberal policy seeks to bring Greece away from an austerity program. What will result in Greece and the EU is unclear.

See Kopin Tan, "Bet on the Greenback to Beat the Euro," *Barron's*, May 17, 2010, F-19.

NORTH AMERICAN FREE TRADE AGREEMENT

Purpose. The NAFTA, ratified in 1994, sought to eliminate barriers to the flow of goods, services, and investments among Canada, the United States, and Mexico over a 15-year period. NAFTA envisioned a gradual phasing-out of these barriers, with the length of the phaseout varying from industry to industry. The ultimate goal was a totally free trade zone among the three member states, with eventual inclusion of other Central and Latin American countries. So far, the only country invited to join the founding members is Chile.

Structure. NAFTA is administered by a three-member Trade Commission, which oversees a Secretariat and arbitral panels.

Trade Commission. Staffed by trade ministers from each of the three nations, the Trade Commission meets once a year and makes its decisions by consensus. It supervises the implementation of the treaty and resolves disputes over interpretation. The daily operations of NAFTA are conducted by ad hoc working groups appointed by the three governments.

Secretariat. The permanent Secretariat is composed of national sections (departments) representing each member country. Its purposes are to provide technical support for the Trade Commission and to put together arbitral panels to resolve disputes between members.

Arbitral Panels. The treaty has detailed arbitration provisions for settling disputes, particularly those involving dumping of goods (selling a good in a member country at a lower price than at home) and interpretations of the treaty. Although the arbitration proceedings are designed especially to resolve disputes between member nations, the treaty encourages private parties to use them as well. If they do, they must agree to abide by the arbitral panel's decision.

Each arbitral panel has five members, chosen from a roster of 30 legal experts from NAFTA and non-NAFTA countries. Within 90 days, the panel will give the disputant countries a confidential report. Over the next 14 days, the disputants may present their comments on the report to the panel. Within 30 days of the issuance of the initial report, the arbitral panel must present its final report to the parties and to the Trade Commission, which publishes the report. The countries then have 30 days to resolve their dispute, or, if the panel has found one party wrong, the other may legally retaliate.

Impact. NAFTA has not only brought together three North American neighbors of different historical and cultural background, but it has also provided a model of economic integration for other countries in Central and Latin America.

The Asian Infrastructure Investment Bank and Development Bank

In April of 2015, China announced the formation of two new banks, attracting 46 countries, including Great Britain, Germany, Australia, and South Korea. The United States declined membership. The bank is to begin operation by the end of 2015. The Obama administration advised the allies not to join, fearing the undermining of the World Bank led by the United States and the Asian Development Bank led by Japan. Some nations feared that China's large state-owned enterprises will dominate and push for large infrastructure projects to the exclusion of competitors. The Chinese sought to allay such concerns, despite the fact that they will be the biggest shareholder. This is similar to the United States' interest in the World Bank.

The interim leader of the new Asian Investment Bank indicated that it will be "lean, clean, and green." It will seek to be corruption free. This will apparently seek to be in contrast to the World and Asian banks which many nations have criticized for their bureaucratic infrastructure. Both originated following World War II.

TECHNOLOGY AND THE LEGAL ENVIRONMENT

WTO Says U.S. Ban on Online Gambling Violates International Law

The island nation of Antigua and Barbuda (plaintiff) brought a case against the United States to a WTO panel. The nation licenses 19 companies that offer sports betting and casino games (e.g., blackjack) over the Internet. It argued that the United States is in violation of international law by prohibiting cross-border gambling operations via the Internet. The plaintiff argues that the U.S. trade policy does not prohibit cross-border gambling operations.

The WTO panel ruled in favor of Antigua and Barbuda. In its decision in 2004, the panel stated that U.S. policy prohibiting online gambling operations emanating from the plaintiff nation violates international law. WTO panels do not have to give reasons for their decisions.

Some important legal, political, and cultural considerations:

1. WTO panel decisions apply only to the set of facts and case before the panel. Internet gambling using credit cards takes place in the Caribbean, Costa Rica, Great Britain, and Canada. Although this case is not a precedent, the United States should expect more legal action, as several million customers are at stake, and revenues from offshore players are important to nation-states that operate Internet casinos.
2. Present federal law in the United States makes it illegal to bet over the Internet if not allowed by individual states. Although untested as legal theory, the Justice Department is seeking to crack down on broadcasters and print media that accept advertising from offshore Internet casinos, on the ground that they are aiding and abetting an illegal enterprise. The airwaves are controlled by the Federal Communications Commission. Some questions are being raised:

a. Is the lobbying of American gambling companies against Internet betting (emanating from abroad) the real reason for the U.S. government's stand? American companies have a lock on U.S. gambling and do not want to lose it to worldwide Internet casino interests.

b. On the basis of international trade law, countries allowing online casino gambling may seek to raise tariffs on services or goods of U.S. companies doing business within their jurisdiction (e.g., AT&T) as a way of retaliating against U.S. policy. Will this prompt U.S. companies to express their concerns to members of Congress? Does AT&T contribute to the reelection of members of Congress? Do gambling interests provide funds for reelection bids? Will there be a clash of interests?

Update: In June of 2013, the Justice Department indicated that *individual states* may sponsor games such as online poker with certain conditions attached.

It has had a significant impact on each country's exports and imports. It has served as an institution for arbitration of disputes. For example, one of the first arbitral panels was set up when the U.S. government filed a complaint on behalf of United Parcel Service (UPS) and UPS believed that it was being hampered by NAFTA government regulations in Mexico, which limit the size of delivery trucks to be used in delivering packages. The arbitral panel ruled in favor of UPS.

Global Dispute Resolution

Many times, when private or public parties enter into an international business agreement, they incorporate means for resolving future disputes (e.g., arbitration clauses) into the agreement. Another form of protection for firms doing business internationally is the insurance some nation-states offer domestic companies to encourage them to export (e.g., United States Overseas Private Investment Corporation). Still, the two methods used most frequently to resolve irreconcilable differences between parties involved in international transactions are arbitration and litigation.

ARBITRATION

Arbitration is a dispute resolution process whereby parties submit their disagreements to a private individual decision maker they have agreed on or to a panel of decision makers whose selection has been provided for in the contract the parties signed. Arbitration clauses in contracts involving international business transactions should meticulously stipulate what law will govern the arbitration, where and when the arbitration will take place, what language will be used, and how the expenses of arbitration will be shared. They should also stipulate a waiver of judicial (court) review by both parties to the dispute. All these matters should be carefully negotiated when the contract is being drafted.

Arbitration of disputes may also come about through treaties. For instance, the United Nations Convention on the Recognition of Foreign Arbitral Awards encourages the use of arbitration agreements and awards. The World Bank's International Center for the Settlement of Investment Disputes (ICSID), created in 1965 by treaty (the Washington Convention), provides arbitration rules as well as experienced arbitrators to disputants, and the International Chamber of Commerce offers a permanent arbitration tribunal. Finally, individual countries have arbitration associations that can provide experienced arbitrators to parties desiring assistance in settling their disputes.

LITIGATION

When contracts do not contain arbitration clauses and no other alternative (such as mediation or conciliation) is available, **litigation** may be the only way to resolve a dispute between parties. Some private international business contracts include a choice-of-forum clause so that the parties know which family of law is to be applied in case of a dispute and what nation's courts will be used. When negotiating contracts in the international arena, managers should make sure that choice-of-forum clauses are specific as to these questions. Either or both can make a major difference to the outcome. Because there is no single international court or legal system capable of resolving all commercial disputes between private parties, a choice-of-forum clause should be negotiated in all agreements involving major transactions. London's Commercial Court, established in 1895, is the most popular neutral forum for resolving commercial litigation, owing to its more than 100 years of experience.

Most of the international and regional organizations discussed in this chapter emphatically encourage the arbitration of private contractual disputes, because the arbitration process is a quicker and less public means of resolving disputes than litigation. In certain areas of the world (particularly the Far East), companies and governments strenuously seek to avoid litigation.

litigation A dispute resolution process going through the judicial system; a lawsuit.

APPLYING THE LAW TO THE FACTS...

Let's say that Jamal from Saudi Arabia and Annette from the United States have a contract stipulating that Jamal agrees to send Annette 15 pounds of Turkish coffee each week for six months. Jamal breaches the contract and Annette decides that she needs to take legal action. She is not sure what forum to use, but wants to choose the quickest option. What type of dispute resolution does she most likely attempt to pursue?

GLOBALIZATION: HURTS OR HELPS

In most of the chapters in this text, we have examined the globalization aspect of several areas of business ethics and law. We have attempted to present some factual bases for these discussions. In the following table, you will find a debate that is taking place all over the world. We invite you to participate in this sometimes vigorous discussion of whether globalization of business helps or hurts societies all over the world. It is skillfully summarized by Professor Murray Wiedenbaum in his text *Business and Government in the Global Marketplace*.

Pros	*Cons*
"Accelerates economic growth, increasing living standards"	"Generates widespread poverty in the pursuance of corporate greed"
"Offers consumers greater variety of products and at lower prices"	"Results in greater income inequality"
"Increases jobs and wages and improves working conditions"	"Moves jobs to low-wage factories that abuse workers' rights"
"Encourages a greater exchange of information and use of technology"	"Provides opportunity for criminal and terrorist groups to operate on a global scale"
"Provides wealth for environmental cleanup"	"Pollutes local environments that lack ecological standards"
"Helps developing nations and lifts millions out of poverty"	"Traps developing countries in high-debt loans"
"Extends economic and political freedoms"	"Threatens national sovereignty"
"Raises life expectancy, health standards, and literacy rates"	"Worsens public health and harms social fabrics of agricultural-based societies"

Source: M. Wiedenbaum, *Business and Government in the Global Marketplace*, 7th ed. (Upper Saddle River, NJ: Prentice Hall, 2004), p. 190.

SUMMARY

Although the emphasis in this book is on legal and ethical issues, managers whose companies are undertaking international business ventures need to consider the political, economic, cultural, and legal dimensions of the international environment of business. The major families of law are common law, which relies primarily on case law and precedent; civil law, which relies primarily on codes and statutory law; Islamic law, which relies on the Shari'a, a religious code of rules; socialist law, which is based on Marxism–Leninism and does not recognize private property; and Hindu law, which relies primarily on the Sastras, a religious code.

International law is divided into public international law, governing the relationships between nation-states, and private international law, governing the relationships between private parties involved in international transactions.

The major methods of engaging in international business are trade, international licensing and franchising, and foreign direct investment. The principal risk of engaging in international business are expropriation, the sovereign immunity doctrine and the act-of-state doctrine, export and import controls, and currency controls and fluctuations (particularly in developing nations).

World and regional integrative organizations, especially the WTO, the EU, and NAFTA, are making a strong impact on international business. Arbitration and litigation are the major methods of international dispute resolution.

REVIEW QUESTIONS

8-1 Contrast the Islamic family with the socialist-law family.

8-2 Which of the methods of engaging in international business discussed in this chapter is most risky for a foreign multinational company? Explain.

8-3 Define the following:
   a. expropriation of property
   b. doctrine of sovereign immunity
   c. act-of-state doctrine
   d. arbitration clause
   e. choice-of-forum clause

8-4 Why was the GATT Pact, creating the World Trade Organization, so important to doing international business? Explain.

8-5 Why is arbitration preferred to litigation as a means of resolving international business disputes? Explain.

REVIEW PROBLEMS

8-6 Royal Bed and Spring Company, a U.S. distributor of furniture products, entered into an exclusive distributorship agreement with a Brazilian manufacturer of furniture products. Under the terms of the contract, Royal Bed was to distribute in Puerto Rico the furniture products manufactured by Famossul in Brazil. The contract contained forum-selection and choice-of-law clauses, which designated the juridical district of Curitiba, State of Parana, Brazil, as the judicial forum and the Brazilian Civil Code as the law to be applied in the event of any dispute. Famossul terminated the exclusive distributorship and suspended the shipment of goods without just cause. As a matter of public policy, Puerto Rican law refuses to enforce forum-selection clauses specifying foreign venues. In what jurisdiction should Royal Bed bring suit? Explain.

8-7 X, a U.S. company, entered into a contract with C, a Swiss subsidiary of General Motors, to sell Chevrolet automobiles in Aruba. An arbitration clause in the parties' agreement provided that all disputes would be settled by arbitration in accordance with Aruban law. Aruba follows Dutch civil law. X argues that only U.S. law can apply, because the contract was made in the United States. Is X correct? Explain.

8-8 Zapata entered into a contract with a German corporation to use one of Zapata's oil-drilling rigs off the coast of Italy. The contract stated, "Any dispute arising must be treated before the London Court of Justice." A severe storm damaged the oil rig as it was being towed through the Gulf of Mexico. Zapata filed suit in federal district court. Does the U.S. court have jurisdiction to decide the dispute? What is the purpose behind a choice-of-forum clause, and should the clause be enforced?

8-9 The plaintiffs, who were descendants of victims of the Holocaust (the mass murder of 6 million Jews by the Nazis during World War II) in Europe filed a claim for breach of contract in the United States against an Italian insurance company, Assicurazioni Generali, S.P.A. (Generali). Before the Holocaust, the plaintiffs' ancestors had purchased insurance policies from Generali, but Generali refused to pay them benefits under the policies. Due to certain agreements among nations after World War II, such lawsuits could not be filed for many years. In 2000, however, the United States agreed that Germany could establish a foundation—the International Commission on Holocaust-Era Insurance Claims (ICHEIC)—that would compensate victims who had suffered losses at the hands of the Germans during the war.

Whenever a German company was sued in a U.S. court based on a Holocaust-era claim, the U.S. government informed the court that the matter should be referred to the ICHEIC as the exclusive forum and remedy for the resolution. There was no such agreement with Italy, however, so the federal district court dismissed the suit. The plaintiffs appealed. Did the plaintiffs have to take their claim to the ICHEIC rather than sue in a U.S. court? Why or why not?

8-10 Dr. Will Pirkey, a U.S. otolaryngologist, signed an employment contract in which he agreed to work for two years at the King Faisal Hospital in Saudi Arabia. Before his departure, Pirkey received his employment contract, which contained a clause providing that his agreement with the hospital would be construed in accordance with Saudi Arabian law. Because of the assassination of King Faisal and for other reasons, Pirkey did not go to Saudi Arabia as agreed and contested the choice-of-law provision of his employment contract as unconscionable. He argued that the laws of his home state (New York) should apply. Who won this case? Explain why.

8-11 U.S. Company (USC) owned a subsidiary in France that had a contract to deliver compressors for use in the Soviet natural gas pipeline then under construction. The U.S. government banned the export of goods to the Soviet Union by U.S. companies or U.S.-controlled foreign companies, and USC complied with the ban by ordering its French subsidiary to stop delivery of the compressors. The French government, however, ordered delivery. USC delivered the compressors. The U.S. government thereupon instituted a criminal action against USC. What is USC's defense? Explain.

CASE PROBLEMS

8-12 The Technology Incubation and Entrepreneurship Training Society (TIETS) entered into a joint-venture agreement with Mandana Farhang and M.A. Mobile to develop and market certain technology for commercial purposes. Farhang and M.A. Mobile filed a suit in a federal district court in California, where they both were based, alleging claims under the joint-venture agreement and a related nondisclosure agreement. The parties agreed that TIETS was a "foreign state" covered by the Foreign Sovereign Immunities Act (FSIA) because it was a part of the Indian government. Nevertheless, Farhang and M.A. Mobile argued that TIETS did not enjoy sovereign immunity because it had engaged in a commercial activity that has a direct effect in the United States. Could TIETS still be subject to the jurisdiction of U.S. courts under the commercial activities exception even though the joint venture was to take place outside the United States? If not, how? *Farhang v. Indian Institute of Technology*, 2012 WL113739 (N.D.Cal. 2012).

8-13 In 1954, the government of Bolivia began expropriating land from Francisco Loza for public projects, including an international airport. The government directed the payment of compensation in exchange for at least some of his land, but the government never paid the full amount. Decades later, Loza's heirs, Genoveva and Marcel Loza, who were both U.S. citizens, filed a suit in a federal district court in the United States against the government of Bolivia, seeking damages for the taking. Can the court exercise jurisdiction? Explain. *Santivanez v. Estado Plurinacional de Bolivia*, 2013 WL879983 (11th Cir. 2013).

8-14 In 1996, the International Trade Administration of the U.S. Department of Commerce assessed antidumping duties against Koyo Seiko Co., NTN Corp., on certain tapered roller bearings and their components imported from Japan. In assessing these duties, the ITA requested information from the makers about their home market sales. NTN responded in part that its figures should not include many sample and small-quantity sales, which were made to enable customers to decide whether to buy the products. NTN provided no evidence to support this assertion, however. In calculating the fair market value of the bearings in Japan, the ITA determined, among other things, that sample and small-quantity sales were within the makers' ordinary course of trade. Koyo and others appealed these assessments to the U.S. Court of International Trade, objecting in part to the ITA's inclusion of sample and small-quantity sales. Should the court order the ITA to recalculate its assessment on the basis of NTN's objection? Explain. *Koyo Seiko Co. v. United States,* 186 F. Supp. 2d 1332 (Court of International Trade [CIT] 2002).

THINKING CRITICALLY ABOUT RELEVANT LEGAL ISSUES

Margaux de Chien, a naturalized U.S. citizen, was born in France. She went to work for Vin Enterprises, a wholly owned subsidiary of a Delaware corporation licensed to do business in Colorado. The parent company, Spanish Wines Are Us, has its principal place of business in Spain. Margaux worked in the United States for two years and then was transferred, at her request, to the company headquarters in Spain. Unfortunately, Margaux did not get along well in Spain. It seems that she was always being harassed because of her strong feelings about the superiority of French versus Spanish wine and culture. After three years at Vin Enterprises in Spain, Margaux was terminated. Margaux filed a charge of discrimination with the Equal Employment Opportunity Commission, seeking relief under both Colorado law and Title VII of the Civil Rights Act of 1964 on the ground that Vin Enterprises discriminated against her and ultimately fired her because of her national origin. Vin Enterprises claimed that Title VII does not apply to the employment practices of U.S. employers who are subsidiaries of foreign-owned companies doing business in their home countries. What is the result? Discuss the legal and ethical aspects of this case using the critical thinking approach set out in Chapter 1 of this text.

ASSIGNMENT ON THE INTERNET

As this chapter demonstrates, there are many important issues to consider before engaging in international business. Go to www.lib.uchicago.edu and search for "Lyonette Louis-Jacques international law" (Louis-Jacques is a librarian and lecturer of law at the University of Chicago, and the author of the link you will be taken to). Under "catalog results," click on the first link. Click on "details" and copy and paste the URL given. Once here, pick a country you are not familiar with and research the issues you think most important to consider before doing business in that country. What are its cultural, economic, political, and legal dimensions? What are its trade laws? Does it belong to any international treaties or organizations?

If you cannot find all the information you need, make a list of detailed questions you would want answered. Finally, for each of the questions you researched, explain why that question was significant in your thinking.

ON THE INTERNET

www.asil.org/ Use this site, maintained by the American Society of International Law, to guide you in researching international law issues on the Internet. Highlight the dropdown menu "Resources/Pubs," then the subsection "Publications." From there, click on "American Journal of International Law."

www.washlaw.edu This page provides links to primary foreign and international legal resources, research aids, and sites useful for international business. Under the heading "International," click on "International Resources."

www.loc.gov/law/guide/nations.html This link to the Law Library of Congress's Nations of the World contains legal information for each country around the world. Under "Services," click on "Law Library." On the left-hand side of the page, click on "Research and Reports," then "read more" under "Guide to Law Online."

www.icj-cij.org This page contains information about and rulings of the International Court of Justice.

www.un.org This website of the United Nations, on international law, provides useful information, including treaties governing business transactions and trade law. Under the heading "What We Do," click on "Uphold International Law."

www.nafta-sec-alena.org The home page of the North American Free Trade Agreement contains many legal texts as well as methods for dispute settlement.

FOR FUTURE READING

Besmer, Veronica. "Student Note: The Legal Character of Private Codes of Conduct: More Than Just a Pseudo-Formal Gloss on Corporate Social Responsibility." *Hastings Business Law Journal* 21 (2006): 279.

Epstein, Edwin. "Commentary: The Good Company: Rhetoric or Reality? Corporate Social Responsibility and Business Ethics Redux." *American Business Law Journal* 44 (2007): 207.

Jackall, Robert. *Moral Mazes: The World of Corporate Managers*. New York: Oxford University Press, 1998.

Lee, Ian B. "Is There a Cure for Corporate 'Psychopathy'?" *American Business Law Journal* 42 (2005): 65.

Ostas, Daniel T. "Deconstructing Corporate Social Responsibility: Insights from Legal and Economic Theory." *American Business Law Journal* 38 (2004): 261.

PART TWO

Private Law and the Legal Environment of Business

Part Two explores areas of private law that affect the legal environment of business. It opens with a discussion of contract law, then proceeds to examine the law of torts, product liability law, property law, agency law, and finally, the law of business associations.

CHAPTER NINE

The Law of Contracts and Sales—I

- DEFINITION, SOURCES, AND CLASSIFICATIONS OF CONTRACT LAW
- ELEMENTS OF A LEGAL CONTRACT
- CONTRACTS THAT MUST BE IN WRITING
- PAROL EVIDENCE RULE
- THIRD-PARTY BENEFICIARY CONTRACTS AND ASSIGNMENT OF RIGHTS

It is a fundamental requirement of a free-enterprise economy that entities in the private sector and at all levels of government be able to enter into agreements that are enforceable by courts of law. Without the assurance that business agreements are legally enforceable, everyday commercial dealings would be difficult to carry out. Contract law has evolved to provide enterprises with the predictability and security they need to flourish and to produce quality products.

Contract law affects several other areas of law discussed in this book. When we take up the law of torts and product liability (Chapters 11 and 12), for instance, much of our discussion will concern breaches of the warranties of merchantability, fitness, or usefulness. In our review of the law of business associations (Chapters 16 and 17), we will examine contracts between principal and agents, employer and employees, and partners in partnership agreements. You will see when we analyze antitrust law (Chapter 24) that contracts that unreasonably restrain trade are prohibited. The basis for our discussion of labor law (Chapter 20) is collective bargaining agreements and what practices government regulation will tolerate being incorporated into those agreements. Finally, when we discuss the relationship between management and consumers (Chapter 25), the law of contracts will be our starting point. Contract law is thus immensely significant in the legal environment of business.

This chapter begins with a definition and classification of contract law. It analyzes the six elements of a contract and then explains which contracts must be in writing in order to be enforceable. The parol evidence rule, the nature of third-party beneficiary contracts, and the assignment of rights completes Chapter 9.

CRITICAL THINKING ABOUT THE LAW

Contract law promotes predictability in exchange. In other words, as a future business manager, you might enter into a contractual agreement in which you agree to pay a certain amount of money in exchange for another business's goods or services. Because contracts are legally enforceable, you are much more likely to receive these goods at the price on which you both agreed when there is an existing contract. Hence, there is greater predictability, and less risk, when a contract exists, because the possibility of a lawsuit deters businesses from deviating from the terms of a contract. Therefore, contracts keep businesses and individuals accountable to the agreements they make.

Predictability and accountability are only two reasons why contract law is so important, but certain primary ethical norms underlying these reasons have greater priority and influence than others. As you consider the benefits of contract law, this critical thinking exercise will urge you to think about the primary ethical norms that influence the law of contracts (see Chapter 1).

1. Which primary ethical norm would be most important to someone who viewed contract law as a crucial method of promoting predictability in exchange?

    Clue: This person might want to ensure that his business runs smoothly, even though some of his operations depend on another business or individual honoring the terms of the contract.

2. If someone were mostly concerned with contract law's function of keeping businesses and individuals accountable to their agreements, which primary ethical norm would this person value the most?

    Clue: This person might be fearful that other businesses or individuals might not perform as they agreed, thereby harming those who depend on their performance.

Definition, Sources, and Classifications of Contract Law

DEFINITION

A **contract** is generally defined as a legally enforceable exchange of promises or an exchange of a promise for an act that assures the parties to the agreement that their promises will be enforceable. Contract law brings predictability to the exchange. For example, if a corporation manufacturing video recorders enters into an agreement with a retailer to provide a fixed number of video recorders each month, the retailer knows that it can rely on the corporation's promise and advertise the availability of those video recorders to its customers, because if the manufacturer tries to renege on the agreement, its performance is enforceable in a court of law. Contracts are essential to the workings of a private-enterprise economy. They assist parties in the buying and selling of goods, and they make it possible to shift risks to parties more willing to bear them.

contract A legally enforceable exchange of promises or an exchange of a promise for an act.

SOURCES OF CONTRACT LAW

Contract law is grounded in the case law of the state and federal courts, as well as state and federal statutory law. Case law—or what is often known as the *common law* because it originated with the law of English courts—governs contracts dealing with real property, personal property, services, and employment contracts. Statutory law, particularly the Uniform Commercial Code (UCC), generally governs contracts for the sale of goods. The UCC has been adopted in whole or in part by all 50 states and the District of Columbia. This chapter integrates case law with the UCC.

Case Law. The law of contracts originated in judicial decisions in England and the United States. Later, states and the federal courts modified their case law through the use of statutory law. Nonetheless, the formation of contract law and its understanding are based on fundamental principles set out by the courts and, more recently, in the Restatement of the Law of Contracts. The Restatement summarizes contract principles as set out by legal scholars. Case law (or common law) applies to contracts that cover real property (land and anything attached permanently therein or thereto), personal property, services, and employment contracts.

Uniform Commercial Code. To obtain uniformity among state laws, particularly law as applicable to sales contracts, the National Conference of Commissioners on Uniform State Laws and the American Law Institute drafted a set of commercial laws applicable to all states. This effort was called the Uniform Commercial Code. Gradually, the states adopted the document in whole or in part. Thereafter, businesses had uniform requirements to expedite interstate contracts for the sale of goods.

In general, the requirements of Article 2 of the UCC regarding formation and performance of contracts are more liberal than those applied to contracts based on common-law principles. Particular differences are noted in sections of this chapter and Chapter 11. Article 2 seeks to govern the sale of goods in all states except Louisiana. A *sale* consists of the passing of title to goods from buyer to seller for a price. A contract for the sale of goods includes those for present and future sales. The code defines *goods* as tangible personal property. *Personal property* includes property other than real property (land and that attached thereto). When the UCC has not specifically modified the common law of contracts, common law applies.

Please note: Article 2 is continually being revised to provide coverage of contracts dealing with electronic data processing, licenses, and leases. Article 2A prescribes a set of uniform rules for the creation and enforcement of contracts for the sale of leases of goods (e.g., lease contracts for the sale of an automobile). In this case, a question of whether the case law or the UCC (Article 2) should govern.

CASE 9-1

Paramount Contracting Co. v. DPS Industries, Inc.
709 S.E.2d 288 (Ga. App. 2011)

Paramount, a civil engineering firm and general contractor, submitted a bid to perform the construction of runway improvements at the Atlanta Hartsfield-Jackson International Airport. Paramount included DPS's quote for supplying the fill dirt for the project in its bid. DPS's written quote described its work as "furnish[ing] and haul[ing]/deliver[ing] borrow dirt from DPS's location to the job site," and specifically excluded the provision of "traffic control, dust control, security and escort services" from the scope of work. The quote provides that the dirt would be delivered for a price of "$140/Truck Load."

After Paramount was awarded the airport project, it contacted DPS about the amount of dirt and number of trucks it would need for the airport project. DPS believed that the parties had a contract, and it sent a letter to Paramount confirming that it was "holding approximately 45,000 [cubic yards] of borrow dirt ready to be hauled in to your project once we receive [the] 10-day notice from you." Paramount did not respond.

Over the next two months, DPS sent other letters to Paramount about their agreement, but Paramount did not respond. After a meeting of executives from the two companies, Paramount sent the following:

[Y]ou insisted that we give commitment to you for buying the dirt before you will give us price [for other work]. This really was a surprise to us . . . Also please note that we have never committed to buy all the fill materials from you. In the last meeting you were informed that we intend to purchase some materials from you and it may be through other subcontractors. Our decisions will be conveyed to you as soon as possible.

Ultimately, Paramount bought the dirt it needed from another vendor. DPS sued Paramount for breach of contract. The jury found for DPS. Paramount appealed.

Judge Blackwell

The question of formation depends on [whether the contract is] governed by Article 2 of the Commercial Code or the common law. It is easier, generally speaking, to form a binding contract under Article 2 than under the common law. Article 2 applies only to contracts for the sale of goods, and it does not apply to contracts for the mere provision of services or labor. When a transaction involves both the sale of a good and the provision of services or labor, whether the transaction is governed by Article 2 depends upon the "predominant purpose" of the transaction. When the predominant element of a contract is the sale of goods, the contract is viewed as a sales contract and [Article 2] applies, even though a substantial amount of service is to be rendered in installing the goods.

DPS said that the parties contemplated only that DPS would sell and deliver dirt, and DPS urged that Article 2 applies because the sale of goods—the dirt that DPS offered to furnish Paramount—was the predominant purpose of the contemplated transaction. Paramount, on the other hand, said that the parties also contemplated that DPS would perform other tasks, such as placing and compacting dirt at the construction site.

It was for the jury to weigh the conflicting evidence, resolve this disputed issue of fact, and determine exactly what the contemplated transaction involved. So long as the evidence would permit a rational jury to resolve this issue in a way that would lead to a conclusion that the sale of goods was the predominant purpose of the contemplated transaction.

The evidence is consistent with the finding that the sale of dirt was the predominant purpose of the contemplated transaction. The DPS quote contains representations and warranties about the quality of the dirt. The pricing was based on the quantity of dirt furnished, not the miles driven or time spent to deliver the dirt. DPS did not provide a separate pricing for the sale of the dirt and its delivery. Paramount itself characterized the transaction repeatedly as one for the sale and furnishing of dirt, not the hauling of dirt.

Paramount relies on testimony that the costs of furnishing and hauling dirt would amount to approximately $70 or $80 per load, and that these costs would include both the expenses of operating a backhoe to load trucks and the expenses of operating the trucks. Paramount says that these cost factors all relate to hauling dirt and establish, therefore, that most of the transaction costs were related to hauling, not furnishing, the dirt.

Under Article 2, dirt is a "good" only if it is severed from the land by the seller, so the separation of dirt from the land is a necessary component of the sale of dirt, not its transportation after sale.

Finally, Paramount asks us to attribute the majority of the costs to hauling on the basis of the conventional wisdom that "dirt is cheap." We decline this invitation. Dirt might well be cheap, but we have no reason to believe that it is free and no basis for knowing just how cheap it is. We are not prepared to speculate that the commercial value of such dirt is negligible, much less to reverse a judgment entered on a jury verdict based on such speculation.*

Judgment *affirmed*.

CLASSIFICATIONS OF CONTRACTS

A contract is generally referred to as a binding set of promises (agreement) that courts will enforce. Terms that refer to types of contracts are sprinkled throughout this text. So that you will clearly understand what we are talking about, we define several classifications of contracts in this section.

Express and Implied Contracts. An **express contract** is an exchange of oral or written promises between parties, which are in fact enforceable in a court of law. Note that oral and written promises are equally enforceable. An **implied (implied-in-fact) contract** is established by the conduct of a party rather than by the party's written or spoken words. For example, if you go to the dentist in an emergency and have a tooth extracted, you and the dentist have an implied agreement or contract: She will extract your throbbing tooth in a professional manner and you will pay her for her service. The existence and content of an implied-in-fact contract are determined by the reasonable-person test: Would a reasonable person expect the conduct of these parties to constitute an enforceable contract? Additionally, see the following case for an examination of various types of contracts.

express contract An exchange of oral or written promises between parties, which are enforceable in a court of law.

implied (implied-in-fact) contract A contract that is established by the conduct of a party rather than by the party's written or spoken words.

---

*Paramount Contracting Co. v. DPS Industries, Inc. 709 S.E.2d 288 (Ga. App. 2011).

CASE 9-2

Pan Handle Realty, LLC v. Olins
Appellate Court of Connecticut
140 Conn. App. 556, 59A.3d 842 (2013)

The plaintiff is a Connecticut limited liability company (a form of business organization—see Chapter 17), which constructed a luxury home at 4 Pan Handle Lane in Westport [Connecticut] (the property). The defendant [Robert Olins] expressed an interest in leasing the property from the plaintiff for a period of one year. In pursuit of that interest, he submitted an application proposing to rent the property from the plaintiff at the rate of $12,000 per month, together with an accompanying financial statement. The plaintiff responded to the defendant's proposal by preparing a draft lease for his review, which the defendant promptly forwarded to his attorney.

On January 17, 2009, the defendant and his real estate agent, Laura Sydney, met with Irwin Stillman, then acting as the plaintiff's representative, to discuss the draft lease (January 17 meeting). At that meeting, the defendant and Irwin Stillman agreed to several revisions to the draft lease that had been proposed by the defendant's attorney, then incorporated the revisions into the lease and signed it. The resulting lease, which was dated January 19, 2009, specified a lump sum annual rent of $138,000. At the time of the signing, the defendant gave the plaintiff a postdated check for $138,000. The lease agreement required the plaintiff to make certain modifications to the property prior to the occupancy date, including the removal of all of the furnishings from the leased premises.

On January 21, 2009, the plaintiff's real estate broker informed it that, according to Sydney, the defendant planned to move into the property on January 28, 2009. The next day, the defendant requested information from the plaintiff for his renter's insurance policy, which the plaintiff duly provided. By that time, the plaintiff had also completed the modifications requested by the defendant at the January 17 meeting and agreed to in the lease agreement, including the removal of the furniture. The defendant's check, which was postdated January 26, 2009, was deposited by the plaintiff on that date.

The following day, however, Citibank advised the plaintiff that the defendant had issued a stop payment order on his postdated rental check and explained that the check would not be honored. The plaintiff subsequently received a letter from the defendant's attorney stating that "[the defendant] is unable to pursue any further interest in the property." Thereafter, the plaintiff made substantial efforts to secure a new tenant for the property, listing the property with a real estate broker, advertising its availability and expending $80,000 to restage it. Although, by these efforts, the plaintiff generated several offers to lease the property, was never able to find a qualified tenant, or, for that reason, to enter into an acceptable lease agreement with anyone for all or any part of the one year period of the defendant's January 19, 2009 lease.

Thereafter, on March 6, 2009, the plaintiff filed this action [in a Connecticut state court], alleging that the defendant had breached an enforceable lease agreement. The plaintiff further alleged that, despite its efforts to mitigate [lessen] its damages, it had sustained damages as a result of the defendant's breach, including unpaid rental payments it was to have received under the lease, brokerage commissions it incurred to rent the property again, and the cost of modifications to the property that were completed at the defendant's request.

The court issued a memorandum of decision resolving the merits of the case in favor of the plaintiff (May 11 decision). In that decision, the court found, more particularly, that the plaintiff had met its burden of proving that the parties had entered into an enforceable lease agreement, that the defendant had breached that agreement, and that the breach had caused the plaintiff damages in lost rent and utility bills incurred during the lease period. On the basis of these findings, the court awarded the plaintiff compensatory damages in the amount of $146,000—$138,000 in unpaid rent for the term of the lease and $8,000 in utility fees incurred by the plaintiff during the lease period, plus interest and attorney's fee.

Judge Sheldon

The defendant's claim on appeal is that the court improperly determined that the parties entered into a valid lease agreement. The defendant contends that because "material terms were still being negotiated and various issues were unresolved," there was no meeting of the minds, which is required to form a contract.

In order for an enforceable contract to exist, the court must find that the parties' minds had truly met. . . . If there has been a misunderstanding between the parties, or a misapprehension by one or both so that their minds have never met, no contract has been entered into by them and the court will not make for them a contract which they themselves did not make.

There was evidence in the record to support the court's finding that the parties entered into a valid lease agreement because there was a true meeting of the parties' minds as to the essential terms of the agreement. Prior to the January 17 meeting, the plaintiff had provided the defendant with a draft lease agreement, which the defendant had forwarded to his attorney for review. The defendant testified that at the January 17 meeting, he and the plaintiff's representative discussed the revisions proposed by the defendant's attorney, made the revisions and signed the lease.*

---
*Pan Handle Realty, LLC v. Olins Appellate Court of Connecticut 140 Conn. App. 556, 59A.3d 842 (2013).

Types of Contracts	
Express	An exchange between parties of oral or written promises that are enforceable
Implied	A contract established by the conduct of the parties
Unilateral	An exchange of a promise for an act
Bilateral	An exchange of one promise for another promise
Void	A contract that at its formation has an illegal object or serious defects
Voidable	A contract that gives one of the parties the option of withdrawing from the agreement
Valid	A contract that meets all the legal requirements for a fully enforceable contract
Executed	A contract the terms of which have been performed
Unenforceable	A contract that exists but cannot be enforced because of a valid defense
Quasi-contract	A court-imposed agreement to prevent the unjust enrichment of one party when the parties have not previously agreed to an enforceable contract

Unilateral and Bilateral Contracts. A **unilateral contract** is defined as an exchange of a promise for an act. For example, if City A promises to pay a reward of $5,000 to anyone who provides information leading to the arrest and conviction of the individual who robbed a local bank, the promise is accepted by the act of the person who provides the information. A **bilateral contract** involves the exchange of one promise for another promise. For example, Jones promises to pay Smith $5,000 for a piece of land in exchange for Smith's promise to deliver clear title and a deed at a later date. In the case below we see the importance of whether a unilateral contract exists.

unilateral contract An exchange of a promise for an act.

bilateral contract The exchange of one promise for another promise.

CASE 9-3

Audito v. City of Providence
United States District Court, District of Rhode Island
263 F. Supp. 2d 2358 (2003)

The city of Providence, Rhode Island, decided to hire a class of police officers in 2001. All had to graduate from the Providence Police Academy to be eligible. Two sessions were held. To be qualified, the applicants to the academy had to pass a series of tests and be deemed qualified after an interview. Those judged most qualified were sent a letter informing them that they had been selected to attend the academy if they completed a medical checkup and a psychological exam. The letter for the applicants to the 61st Academy, dated October 15, stated that it was a "conditional offer of employment."

Meanwhile, a new chief of police was appointed in Providence. He changed the selection process, which caused some who had received the letter to be rejected. Audito and 13 newly rejected applicants (who had completed the examination) sued the City of Providence in federal district court, seeking a halt to the 61st Academy unless they attended. The plaintiffs alleged in part that the city was in breach of its contract.

Justice Torres

[T]he October 15 letter *is a classic example of an offer to enter into a unilateral contract*. The October 15 letter expressly stated that it was a "conditional offer of employment" and the message that it conveyed was that the recipient would be admitted into the 61st Academy if he or she successfully completed the medical and psychological examinations, requirements that the city could not lawfully impose unless it was making a conditional offer of employment.

Moreover, the terms of that offer were perfectly consistent with what applicants had been told when they appeared [for their interviews]. At that time, [Police Major Dennis] Simoneau informed them that, if they "passed" the [interviews], they would be offered a place in the academy if they also passed the medical and psychological examinations.

The October 15 letter also was in marked contrast to notices sent to applicants by the city at earlier stages of the selection process. Those notices merely informed applicants that they had completed a step in the process and remained eligible to be considered for admission into the academy. Unlike the October 15 letter, the prior notices did not purport to extend a "conditional offer" of admission.

The plaintiffs accepted the city's offer of admission into the academy by satisfying the specified conditions. Each of the plaintiffs submitted to and passed lengthy and intrusive medical and psychological examinations. In addition, many of the plaintiffs, in reliance on the City's offer, jeopardized their standing with their existing employers by notifying the employers of their anticipated departure, and some plaintiffs passed up opportunities for other employment.

The city argues that there is no contract between the parties because the plaintiffs have no legally enforceable right to employment. The city correctly points out that, even if the plaintiffs graduate from the Academy and there are existing vacancies in the department, they would be required to serve a one-year probationary period during which they could be terminated without cause. That argument misses the point. The contract that the plaintiffs seek to enforce is not a contract that they will be appointed as permanent Providence police officers; rather, it is a contract that they would be admitted to the Academy if they passed the medical and psychological examinations.*

The federal district court ruled in favor of Audito et al.

CRITICAL THINKING ABOUT THE LAW

The offer of admission to the Police Academy under certain conditions does not provide a guarantee of an employment contract. It does, however, provide the opportunities promised if the applicants fulfill certain conditions.

1. When the new police chief changed the rules for employment, why was he not permitted to do so legally?

    Clue: The reasoning here is similar to the reasoning whenever one party to an agreement decides it does not like the contract it previously formed.

2. What is the relevant rule of law in this case?

    Clue: What requirements are imposed once one makes an offer to enter into a unilateral contract?

---

void contract A contract that at its formation has an illegal object or serious defects.

voidable contract A contract that gives one of the parties the option of withdrawing.

valid contract A contract that meets all the legal requirements for a fully enforceable contract.

executed contract A contract of which all the terms have been performed.

quasi-contract A court-imposed agreement to prevent the unjust enrichment of one party when the parties had not really agreed to an enforceable contract.

Void, Voidable, and Valid Contracts. A contract is **void** if at its formation its object is illegal or it has serious defects in its formation (e.g., fraud). If Jones promises to pay Smith $5,000 to kill Clark, the contract is void at its formation because killing another person without court sanction is illegal. A contract is **voidable** if one of the parties has the option of either withdrawing from the contract or enforcing it. If Jones, a 17-year-old in a state where the legal age for entering an enforceable contract is 18, executes an agreement with Smith, an adult, to buy a car, Jones can rescind (cancel) the contract before he is 18 or shortly thereafter. A **valid contract** is one that is not void, is enforceable, and meets the six requirements discussed later in this chapter.

Executed and Executory Contracts. An **executed contract** is one of which all the terms have been performed. In our earlier example, if Jones agrees to buy Smith's land for $5,000, and Smith delivers clear title and a deed and Jones gives Smith $5,000, the necessary terms (assuming no fraud) have been carried out or *performed*. In contrast, an *executory* contract is one of which all the terms have not been completed or performed. If Jones agrees to paint Smith's house for $2,500 and Smith promises to pay the $2,500 upon completion of the paint job, the contract remains executory until the house is completely painted. The importance of complete performance will be shown in Chapter 11's discussion of discharge and remedies for a breach of contract.

Quasi-Contract. A **quasi-contract** is a court-imposed agreement to prevent unjust enrichment of one party when the parties had not really agreed to an enforceable contract. For example, while visiting his neighbor, Johnson, Jones

---

*Audito v. City of Providence United States District Court, District of Rhode Island, 263 F.Supp.2d 2358 (2003).

sees a truck pull up at his own residence. Two people emerge and begin cutting his lawn and doing other landscaping work. Jones knows that neither he nor his wife contracted to have this work performed; nevertheless, he likes the job that the workers are doing, so he says nothing. When the landscapers finish, they put a bill in Jones's mailbox and drive off. It turns out that the landscapers made an honest mistake: They landscaped Jones's property when they were supposed to landscape Smith's. Jones refuses to pay the bill, arguing that he did not contract for this work. He even calls the landscapers unflattering names. The court orders Jones to pay, finding that he was unjustly enriched. (Jones would not have had to pay if he had not been in a position to correct the mistake before it took place. That is, Jones would not have had to pay had this mistaken landscaping occurred while he and his wife were vacationing in Paris.)

It should be noted that the receiver of the service must pay for the value of the service, not necessarily the contract price, only if he were in a position to stop the erroneous service.

Elements of a Legal Contract

A valid contract has six elements: (1) legal offer, (2) legal acceptance, (3) consideration, (4) genuine assent, (5) competent parties, and (6) a legal object. When these six elements are present, a legally enforceable contract usually exists.

LEGAL OFFER

The contractual process begins with a **legal offer**. "I will pay you $2,000 for your 1978 Cutlass," Smith says to Jones. Smith has initiated a possible contract and is known as the *offeror*. Jones is the *offeree*. For Smith's offer to be valid, by common-law principles, it must meet three requirements:

legal offer An offer that shows objective intent to enter into the contract, is definite, and is communicated to the offeree.

1. The offer must show objective intent to enter into the contract. The court will look at the words, conduct, writing, and, in some cases, deliberate omissions of the offer. The court will not concern itself with subjective measurements, such as what was in the person's mind at the time of entering into the contract. It simply asks whether a reasonable person who listened to Smith's statements would conclude that there was a serious intent to make an offer.

2. The offer must be definite; that is, there must be some reference to subject matter, quantity of items being offered, and price of the items. In Smith's offer, all three references are present. Article 2 of the UCC, because it is intended to govern daily transactions in goods, is less stringent. It allows the price and other terms—but not subject matter or quantity—to remain open or to be based on an industry standard of "reasonableness." Suppose, for example, that X offers to sell "20 widgets that are needed" to Y at a "reasonable price with specific terms to be negotiated." This is an example of the more open-ended approach to legal offers taken by UCC Section 2-204. The courts also weigh industry custom and prior dealings to determine whether the terms are definite.

3. The offer must be communicated to the party (offeree) intended by the offeror. Smith's offer was communicated directly to Jones, but what of an offer by Bank X to pay for information leading to the arrest and conviction of Y, a robber? Z does not know of the reward, but several days after it is offered, she sees Y running out of a store with something in his hand. Z apprehends Y. Should she get the reward from Bank X? The bank's offer of a reward was not communicated to her, so the technical requirements of contract law have not been met. Most state courts, however, and many state statutes, would allow Z to collect, for it is public policy to encourage citizens to assist in apprehending criminals.

The case below indicates that there is a lack of definiteness.

CASE 9-4

Baer v. Chase
United States Court of Appeals, Third Circuit
392 F.3d 809 (2004)

GREENBERG, Judge.

Chase, who originally was from New Jersey, but relocated to Los Angeles in 1971, is the creator, producer, writer and director of *The Sopranos*. Chase has numerous credits for other television productions as well. Chase had worked on a number of projects involving organized crime activities based in New Jersey, including a script for "a mob boss in therapy," a concept that, in part, would become the basis for *The Sopranos*.

In 1995, Chase was producing and directing a *Rockford Files* "movie-of-the-week" when he met Joseph Urbancyk who was working on the set as a camera operator and temporary director of photography.

Chase, Urbancyk, and Baer met for lunch on June 20, 1995, with Baer describing his experience as a prosecutor. Baer also pitched the idea to shoot "a film or television shows about the New Jersey Mafia." At that time Baer was unaware of Chase's previous work involving mob activity premised in New Jersey. At the lunch there was no reference to any payment that Chase might make to Baer for the latter's services.

In October 1995, Chase visited New Jersey for three days. During this "research visit" Baer arranged meetings for Chase with Detective Thomas Koczur, Detective Robert A. Jones, and Tony Spirito who provided Chase with information, material, and personal stories about their organized crime. Baer does not dispute that virtually all of the ideas and locations that he "contributed" to Chase existed in the public record.

After returning to Los Angeles, Chase sent Baer a copy of a draft of a *Sopranos* screenplay that he had written, which was dated December 20, 1995. Baer asserts that after he read it he called Chase and made various comments with regard to it. Baer claims that the two spoke at least four times during the following year and that he sent a letter to Chase dated February 10, 1997, discussing *The Sopranos* script.

Baer asserts that he and Chase orally agreed on three separate occasions that if the show became a success, Chase would "take care of" Baer, and "remunerate Baer in a manner commensurate to the true value of his services."

Baer claims that on each of these occasions the parties had the same conversation in which Chase offered to pay Baer, stating "you help me; I pay you." Baer always rejected Chase's offer, reasoning that Chase would be unable to pay him "for the true value of services Baer was rendering." Each time Baer rejected Chase's offer he did so with a counteroffer, "that I would perform the services while assuming the risk that if the show failed Chase would owe me nothing. If, however, the show succeeded he would remunerate me in a manner commensurate to the true value of my services." Baer acknowledges the counteroffer always was oral and did not include any fixed term of duration or price. In fact, Chase has not paid Baer for his services.

On or about May 15, 2002, Baer filed a complaint against Chase in [a federal] district court [claiming among other things] breach of implied contract. Eventually Chase brought a motion for summary judgment. Chase claimed that the alleged contract [was] too vague, ambiguous and lacking in essential terms to be enforced.

The district court granted Chase's motion.

Baer predicates [bases] his contract claim on this appeal on an implied-in-fact contract. The issue with respect to the implied-in-fact contract claim concerns whether Chase and Baer entered into an enforceable contract for services Baer rendered that aided in the creation and production of *The Sopranos*.

[A] contract arises from offer and acceptance, and must be sufficiently definite so that the performance to be rendered by each party can be ascertained with reasonable certainty. Therefore parties create an enforceable contract when they agree on its essential terms and manifest an intent that the terms bind them. *If parties to an agreement do not agree on one or more essential terms of the purported agreement, courts generally hold it to be unenforceable.*

[The] law deems the price term, *i.e.,* the amount of compensation, an essential term of any contract. An agreement lacking definiteness of price, however, is not unenforceable if the parties specify a practicable method by which they can determine the amount. However, *in the absence of an agreement as to the manner or method of determining compensation the purported agreement is invalid. Additionally, the duration of the contract is deemed an essential term and therefore any agreement must be sufficiently definitive to allow a court to determine the agreed upon length of the contractual relationship.*

The question with respect to Baer's contract claim, therefore, is whether his contract is enforceable in light of the traditional requirement of definitiveness. A contract may be expressed in writing, or orally, or in acts, or partly in one of these ways and partly in others. There is a point, however, at which interpretation becomes alteration. In this case, even when all of the parties' verbal and non-verbal actions are aggregated and viewed most favorably to Baer, we cannot find a contract that is distinct and definite enough to be enforceable.

Nothing in the record indicates that the parties agreed on how, how much, where, or for what period Chase would compensate Baer. The parties did not discuss who would determine the "true value" of Baer's services, when the "true value" would be calculated, or what variables would go into such a calculation. There was no discussion or agreement as to the meaning of "success" of *The Sopranos*. There was no discussion how "profits" were to be defined. There was no contemplation of dates of commencement or termination of the contract. And again, nothing in Baer's or Chase's conduct, or the surrounding circumstances of the relationship, shed light on, or answers, any of these questions. The district court was correct in its description of the contract between the parties: "The contract as articulated by the Plaintiff lacks essential terms, and is vague, indefinite and uncertain; no version of the alleged agreement contains sufficiently precise terms to constitute an enforceable contract." We therefore will affirm the district court's rejection of Baer's claim to recover under a theory of implied-in-fact contract.*

Affirmed for the Plaintiff.

Methods of Termination of an Offer. Generally, the common law provides five methods of terminating an offer.

1. *Lapse of time.* The failure of the offeree to respond within a reasonable time (e.g., 30 days) will cause an offer to lapse.
2. *Death of either party.* The death of the agent of a corporation, however, will not terminate the contract, because in most cases the company will continue.
3. *Destruction of the subject matter.* If the item contracted for cannot be replaced because of an accident or occurrence that is not the fault of the offeror, the offer may be terminated.
4. *Rejection by the offeree.* If the offeree does not accept the offer, it is terminated.
5. *Revocation by the offeror.* If the offeror withdraws the offer before the offeree accepts it, the offer is terminated.

The UCC differs somewhat from the common law in methods of termination of an offer. Here are some examples.

Rejection by the Offeree. At common law, a counteroffer by the offeree constitutes rejection (method 4 in the preceding list) and brings about a termination of the offer. For example, suppose Jones offers to sell his house for $200,000 and no more or less. If Smith offers Jones $185,000, Smith has terminated the original offer and now has set forth a counteroffer ($185,000), which Jones can either accept or reject.

UCC Section 2-207 allows for modification by offerees when dealing with the sale of goods. For example, in the case of nonmerchants, such as Smith and Jones, a counteroffer by the offeree (Smith) does not constitute a rejection, because there is still a clear intent to contract, but the additional term added by the offeree will not become part of the contract. For example, suppose Smith offers to sell his bicycle to Jones for $300. If Jones tells Smith that he will buy his bicycle for the amount of $300 if Smith paints it black, a contract exists even though the painting of the bicycle was not part of the original offer by Smith.

If both parties to a contract for goods are merchants, under Section 2-207, terms added to a contract by an offeree will become additional terms and part of an enforceable contract unless one of the following conditions exists: (1) the added terms are material to the contract; or (2) the offeror limited the term of the original offer to the offeree by placing it in writing; or (3) one of the parties objects to any added term within a reasonable period of time.

Revocation by the Offeror. At common law, the offer is terminated if the offeror notifies the offeree that the offer is no longer good before the offeree accepts it (*revocation*; method 5 in the preceding list). An offeree can forestall that type of termination by paying an offeror an amount of money to keep the offer open for a time. This tactic is called an *option,* and usually it will exist for 30 days. During this time, the offeror can neither sell the property to another nor revoke the offer. Under UCC Section 2-205, a firm offer made by a merchant

---

*Baer v. Chase, United States Court of Appeals, Third Circuit 392 F.3d 809 (2004).

in writing, and signed by the merchant with another, must be held open for a definite period (three months). The firm offer cannot be revoked, and no consideration is required (the offeree need not buy an option).

> ### APPLYING THE LAW TO THE FACTS...
>
> Linda offers Stacey a deal to purchase Linda's car for a great price. Stacey agrees, and puts down 5 percent of the price of the car, and says she will pay the rest in a month. Before the month is over, Linda changes her mind and tells Stacey that due to unforeseen circumstances, she cannot sell the car anymore. Stacey tells Linda she cannot change her mind like that. Is Stacey correct? Why or why not?

LEGAL ACCEPTANCE

legal acceptance An acceptance that shows objective intent to enter into the contract, that is communicated by proper means to the offeror, and that mirrors the terms of the offer.

Legal acceptance involves three requirements that must be met. For an acceptance to be valid,

1. an intent to accept must be shown by the offeree;
2. the intent must be communicated by proper means; and
3. the intent must satisfy, or "mirror," the terms of the offer.

Intent to Accept. There must be objective intent (words, conduct, or writing) similar to that required of a legal offer. If Jones offers to sell Smith his 1978 Cutlass for $2,000 and Smith responds by stating, "I'll think it over," there is no objective intent to accept because there exists no present commitment on the part of Smith. In general, silence does not constitute acceptance unless prior conduct of the parties indicates that they assume it does.

Communication of Acceptance. In general, any "reasonable means of communication" may be used in accepting an offer, and acceptances are generally binding upon the offeror when dispatched. Both industry custom and the subject matter will determine "reasonableness" in the eyes of a court. If, however, the offeror requires that acceptance be communicated only in a certain form (e.g., letter), any other form that is used by the offeree (e.g., telegram) will delay the effectiveness of the acceptance until it reaches the offeror. If the offer states that "acceptance must be by mail" and the mails are used, the acceptance is effective upon deposit at the post office. If a telegram is used instead of mail, the acceptance will not be effective until it reaches the offeror. The strictest interpretation of this rule (known as the "mailbox rule") states that any acceptance at variance with the terms of the offer cannot form a contract even when actually received by the offeror.

Determining whether there has been a valid acceptance is not always easy, as shown by the following classic case involving Pamela Lee Anderson, of *Baywatch* and (more recently) *Dancing with the Stars,* fame.

CASE 9-5

The Private Movie Company, Inc. v. Pamela Lee Anderson et al.
Superior Court of California, County of Los Angeles (1997)

The plaintiff, Private Movie Company (Efraim), sued the defendant, Pamela Lee Anderson (Lee), for $4.6 million, alleging that she breached both an oral and a written contract so that she could work on a different project. The plaintiff claimed that an oral contract existed on November 18, 1994, when the parties agreed on all of the principal terms of a "deal," at the conclusion of a "business meeting" at the offices of defendant's personal manager. The plaintiff claimed that a written contract was entered into on December 21, 1994, when the plaintiff's lawyer sent the defendant copies of a "long-form" contract. The plaintiff claimed that this contract was a written embodiment of the oral agreement reached on November 18, 1994.

The somewhat confusing facts that were testified to, and disputed, at trial made it difficult for the judge to determine whether a contract existed. The events began in October 1994, when plaintiff's attorney, Blaha, sent the plaintiff's script to the defendant's agent. After several conversations, an offer was also sent to her agent. At trial, Efraim testified that Lee had said she loved the script and the character, but she was concerned about the nudity and sexual content of the script. Efraim said that he told Lee that the script would be rewritten and he would do whatever she wished regarding the nudity.

On November 18, a business meeting was held by Efraim, his attorney, the defendant's agents (Joel and Stevens) and manager (Brody), and the director, to negotiate a contract. Those present at the meeting testified that agreement was reached on a specific makeup person, security, trailer to be provided for Lee, start date, expenses, and per diem. The issue of limiting the amount of nudity used in the theatrical trailer or any of the advertising material was raised, and apparently was resolved by an understanding that Brody (defendant's manager) would provide a list of dos and don'ts and that Private Movie would abide by them. The structure of the agreement was also discussed, with an understanding that there would be two contracts—an acting contract and a consulting contract—thereby allowing Private Movie to save money relating to payment of benefits. The issue of the sexual content or simulated sex in the movie script was not raised at the meeting, nor was the issue of any script rewrites brought up. At the end of the meeting, Efraim asked the defendant's agent whether the deal was closed if Lee's compensation was increased to $200,000. The agent said yes.

A few days later, Efraim had his attorney draft the agreement with the increased compensation. Several drafts were exchanged between the attorney and the defendant's agent, all containing the following nudity clause:

> Nudity. The parties hereto acknowledge that the Picture will include "nude and/or simulated sex scenes." Player has read the screenplay of the Picture prior to receipt of the Agreement and hereby consents to being photographed in such scenes, provided that such "nude and simulated sex scenes" will not be [handled] nor photographed in a manner different from what has been agreed to unless mutually approved by Artist and producer.

The rewritten script was sent to the defendant on December 27, 1994. The plaintiff's attorney testified that he called the defendant on December 29, 1994, and she said the script was great, but she wanted a different makeup artist and would split the difference in cost. The defendant testified that she recalled no such phone call. She said that she reviewed the script on January 1, 1995, saw that the simulated sex scenes remained, and called her manager to tell him she would not do the film.

The plaintiff found a less well-known actress to make the film and brought his action against the defendant.

Justice Horowitz

When the parties orally or in writing agree that the terms of a proposed contract are to be reduced to writing and signed by them before it is to be effective, there is no binding agreement until a written contract is signed. If the parties have orally agreed on the terms and conditions of a contract with the mutual intention that it shall thereupon become binding, but also agree that a formal written agreement to the same effect shall be prepared and signed, the oral agreement is binding regardless of whether it is subsequently reduced to writing.

Whether it is the intention of the parties that the agreement should be binding at once, or when later reduced to writing, or to a more formal writing, is an issue to be determined by reference to the words the parties used, as well as all of the surrounding facts and circumstances.

One of the essential elements to the existence of a contract is the consent of the parties. This consent must be freely given, mutual, and communicated by each party to the other.

Consent is not mutual unless the parties all agree upon the same thing in the same sense. Ordinarily, it is the outward expression of consent that is controlling. Mutual consent arises out of the reasonable meaning of the words and acts of the parties, and not from any secret or unexpressed intention or understanding. In determining if there was mutual consent, the Court considers not only the words and conduct of the parties, but also the circumstances under which the words are used and the conduct occurs.

Parties may engage in preliminary negotiations, oral or written, before reaching an agreement. These negotiations only result in a binding contract when all of the essential terms are definitely understood and agreed upon even though the parties intend that formal writing including all of these terms shall be signed later.

An acceptance of an offer must be absolute and unconditional. All of the terms of the offer must be accepted without change or condition. A change in the terms set forth in the offer, or a conditional acceptance, is a rejection of the offer.

Plaintiff has presented no testimony that Lee, on 11/18/94, the date [on] which Plaintiff alleged that an oral contract was created, personally agreed to perform in the movie *Hello, She Lied*; Plaintiff, therefore, has the burden of proving that Joel and/or Stevens, her "agent" and "manager," had the authority to bind her to an oral written contract.

The parties do not and did not agree on the definition of "simulated sex." Clearly the performance of simulated sexual scenes in the film was important and material to both Lee and Efraim. Efraim stated that he would abide by whatever Lee wanted in this regard.

Nudity and sexual content are material deal points that must be resolved before there can be a binding contract. An agreement concerning sexual content or simulated sex was not reached in this instance. Lee did not agree to the terms relating to simulated sex or to the script offered by the Plaintiff.

Plaintiff's letter of 1/13/95 to Lee claims she "agreed to perform simulated sex scenes, and the exact type of nudity had been agreed upon in detail." Efraim claimed in deposition that Lee agreed to perform simulated sex scenes and agreed to the draft contract to confirm that fact. Blaha testified that Paragraph 9 was a correct statement of the agreement. In deposition he stated it was a mistake. The rewritten script has three or four scenes that depict simulated sex. It is obvious that the "offer" made by Plaintiff concerning this issue was not complete and unqualified, nor was there any acceptance of this issue that was complete and unqualified.

Brody and Joel testified to their opinion that they thought they had "closed the deal" on 11/18/94 or shortly thereafter. Such perceptions have very little legal relevance. Brody testified that he had authority to negotiate this contract. Joel never spoke with Lee concerning the transaction and did not negotiate points such as script rewrite or sexual content.

Plaintiff has failed to prove by a preponderance of the evidence that Lee entered into an oral or written contract to perform in the movie Hello, She Lied.*

Judgment in favor of Defendant, Lee.

CRITICAL THINKING ABOUT THE LAW

We know that language is not usually clear. Words convey information but not always the information that the speaker or writer intends. Ambiguity characterizes those words and phrases that do not have a clear meaning. These ambiguous terms might result in another person's misinterpreting what the writer or speaker actually meant. In contract law, ambiguity could create problems between an offeror and offeree, as the two parties might not be in agreement on the same terms of the contract if the contract contains ambiguous language. In Case 9-5, the parties thought they understood each other. Key ambiguous phrases, however, created confusion in the contract negotiations and, consequently, raised concerns about whether there was actual consent by both parties.

As business managers, it is imperative that you demand clear definitions in the contracts that you offer and accept. The following questions pertaining to Case 9-5 prompt you to consider the importance of ambiguity in contract law.

1. What key ambiguous phrases did the court discuss?

    Clue: Find the legal term in dispute that the judge defined. Also, look for ambiguity in the specific elements of the contractual negotiations between the plaintiff and the defendant.

2. How did the ambiguity in the alleged contract affect the court's reasoning?

    Clue: Do you think the court would have ruled differently in Case 9-5 had the ambiguity not existed?

Satisfying, or "Mirroring," the Terms of the Offer. Under the common law, to be valid, the acceptance must satisfy, or "mirror," the terms of the offer. For example, if Jones offers to sell Smith his Cutlass for $2,000 and Smith responds by saying, "I'll give you $1,800," this is not a legal acceptance but a counteroffer by Smith, which then must be accepted by Jones in order for the terms of the counteroffer to be satisfied and a contract to arise. Under UCC Section 2-207, acceptance does not have to be a mirror image of the offer. Terms can be added to the contract without constituting a counteroffer if they meet one of the three conditions listed in the section on methods of termination of an offer.

Internet and E-Contracts: Acceptance Online. Parties now enter into many agreements online. Section 2-213 of the UCC deals with electronic communication of an acceptance by an offeree. This UCC section provides that "receipt of an electronic communication has a legal effect; it has that effect even though no individual is aware of its receipt," but "in itself does not establish that the content sent corresponds to the content received." Thus, receipt is required for acceptance by electronic communication, and receipt occurs when the email or other message arrives, even if the receiver does not know it has arrived. Also, the parties are left to use other means of proof to establish that all of the email

---

*The Private Movie Company, Inc. v. Pamela Lee Anderson et al. Superior Court of California, County of Los Angeles (1997).

or messages made it from one party to another. The company or offeror must list all the terms of the offer that the offeree is about to enter into. The offeree (buyer) must click on "I agree" or "I agree to the terms." Usually, terms set out by the offeror (seller) include cost, payment, warranties, arbitration provisions, and other substantive terms. As "click-on," "click-through," or "click-wrap" agreements have become customary in many industries, there is little dispute between parties as to the formation of a contract thereby.

CONSIDERATION

Consideration is defined as a bargained-for exchange of promises in which a legal detriment is suffered by the promisee. For example, Smith promises Jones that if she gives up her job with Stone Corporation, he will employ her at Brick Corporation. The two requirements of consideration are met: (1) Smith (promisor) has bargained for a return promise from Jones (promisee) that she will give up her job; (2) when Jones gives up her job, she has lost a legal right, the contractual right to her present job with Stone Corporation. The reader should note that legal detriment (giving up a legal right or refraining from exercising a legal right) must take place. Economic detriment is not necessary. For example, a student agrees not to go to any bars during fall semester in exchange for his mother's promise to give him $500. The student's giving up his right to go to bars is a legal detriment because he now cannot do something he previously could legally do.

consideration A bargained-for exchange of promises in which a legal detriment is suffered by the promisee.

Adequacy of Consideration. In general, the courts have not been concerned with the amount of consideration involved in a contract, especially in a business context. Even if one party makes a bad deal with another party—that is, if the consideration is inadequate—the courts will usually refuse to interfere. Unless a party can show fraud, duress, undue influence, or mistake, the court will not intervene on behalf of a plaintiff. However, sufficiency of consideration, as opposed to adequacy, will be examined by the court. Sufficiency of consideration requires both a bargained-for exchange of promises and legal detriment to the promisee.

Preexisting Duty Rule. In defining consideration, we said that a legal detriment to a promisee requires the giving up of a legal right or the refraining from exercising a right. Logically, the courts have then declared that if a party merely agrees to do what he or she is required to do, there exists no detriment to the promisee. For example, Smith contracted with Jones for Jones to build him a house by April 1, 1988, for $150,000. On February 1, 1988, Jones came to Smith and said that, because of the number of jobs he had, he would not be able to finish by April unless Smith agreed to a bonus of $10,000. Smith agreed to the bonus, and the house was completed by April 1. Smith then refused to pay the bonus, claiming that there was a preexisting duty on the part of Jones because he had a contractual duty to finish by April 1. Jones took him to court, but lost the suit because no consideration existed for the bonus agreement. There is an important exception to the preexisting duty rule: The UCC, which applies to the sale of goods, states that an agreement modifying the original contract needs no consideration to be binding.

Promises Enforceable without Consideration. The courts have enforced certain contracts when the requirements of consideration were not met, using the doctrine of promissory estoppel to do so. This doctrine requires (1) a promise justifiably relied on by the promisee, (2) substantial economic detriment to the promisee, and (3) an injustice that cannot be avoided except by enforcing the contract.

Consider this hypothetical example. An elderly couple pledged in writing to leave $1 million to their family church for a building fund if the church raised another $1 million. The church accepted the offer, raised the matching funds, and contracted with an architect and builder. The couple died and, in their will, left the money to another church. When the family church sued the deceased's estate on the basis of the promissory estoppel doctrine, the court awarded it the full amount pledged, even though a bargained-for exchange of promises did not exist. The family church justifiably relied upon the couple's promise, causing substantial economic injury to the church, and injustice could not be avoided in any other way.

Liquidated and Unliquidated Debts. A *liquidated* debt exists when there is no dispute about the amount or other terms of the debt. If A owes B and C $500,000, and B and C agree to accept $100,000 as settlement for the debt, they are not precluded from suing A later on for the balance. The courts reason that the first agreement by A to pay a particular amount ($500,000) to B and C was supported by consideration. The second agreement to pay $100,000 was not because A had a preexisting duty to pay $500,000, and there was, therefore, no legal detriment on A's part to support B and C's agreement to accept the lesser amount.

An *unliquidated* debt exists when there is a dispute between the parties as to the amount owed by the debtor. If there is an agreement similar to the preceding one, except that the amount A originally owed B and C is in dispute, the general rule is that consideration exists for the second agreement, and the creditors cannot come back and sue for the balance of what they thought they were owed. B and C would have no claim for the full $500,000, but would be limited to $100,000. The rationale is that new consideration was given for the second agreement. There exists a legal detriment because B and C are giving up a legal right to sue for an unspecified debt. The debtor is also giving up a legal right because there is uncertainty as to what he or she owes in an unliquidated debt situation.

Promises That Lack Consideration

Type of Consideration	*Description*
Illusory promises	A contract providing that only one of the parties need perform, only if he or she chooses to do so; the contract is not supported by consideration.
Moral obligation	Contracts based on love or affection lack consideration. A majority of the states hold that deathbed promises may constitute moral obligation but lack legal binding consideration.
Preexisting duty	A promise lacks consideration if a person promises to perform an act or do something she already has an obligation to do. For example, many states have statutes that prevent law enforcement officers from collecting rewards when apprehending a criminal who has a reward on his or her head. Also, as noted earlier, the original terms of a contract cannot be changed or modified unless unforeseen difficulties exist. Also, some exceptions are granted by the UCC.
Illegal consideration	A contract is not supported by consideration if the promise is supported by an illegal act. "I agree to pay you $10,000 if you burn my house down." Arson is unlawful and a promise to do such an act is unsupported by legal consideration.

GENUINE ASSENT

genuine assent Assent to a contract that is free of fraud, duress, undue influence, and mutual mistake.

When two parties enter into a legally enforceable contract, it is presumed that they have entered of their own free will and that the two parties understand the content of the contract in the same way. If fraud, duress, undue influence, or mutual mistake exists, **genuine assent**, or a "meeting of the minds," has not taken place, and grounds for rescission (cancellation) of the contract exist. Table 9-1 lists the factors that prevent genuine assent.

Fraud. **Fraud** consists of (1) a misrepresentation of a material (significant) fact, (2) made with intent to deceive the other party, (3) who reasonably relies upon the misrepresentation, (4) and as a result is injured. For example, Smith enters into a contract to sell a house to Jones. The house is 12 years old, and Smith knows that the basement is sinking. She fails to tell Jones. After Jones moves in, she finds that the house is sinking about two feet a year. In this case, there was a misrepresentation of a material fact, because Smith had a duty to disclose the fact that the house was sinking, but did not do so. Furthermore, there existed knowledge of the fact with intent to deceive. The law does not require that an evil motive exist, but only that the selling party (Smith) knew and recklessly disregarded the fact that the house was sinking. Reliance existed on the part of Jones, who thought the house was habitable, and of course injury to Jones took place because the house was not worth what she paid for it. The cost of preventing further sinking of the house would be part of the damages involved. Note that this example illustrates fraud based on a unique set of facts.

Duress. Another factor that prevents genuine assent of the parties is **duress**, defined as any wrongful act or threat that prevents a party from exercising free will when executing a contract. The state of mind of the party at the time of entering into the contract is important. If Smith, when executing a contract with Jones to sell a house, holds a gun on Jones and threatens to shoot Jones if he refuses to sign the contract, grounds exist for rescission of the contract. Duress is not limited to physical threats, however. Threats of economic ruin or public embarrassment also constitute duress. An illustration of duress is set forth in the case below.

TABLE 9-1
FACTORS PREVENTING GENUINE ASSENT

- Fraud
- Duress
- Undue influence
- Bilateral mistake
- Unilateral mistake

fraud Misrepresentation of a material fact made with intent to deceive the other party to a contract, who reasonably relied on the misrepresentation and was injured as a result. See also *criminal fraud*.

duress Any wrongful act or threat that prevents a party from exercising free will when executing a contract.

CASE 9-6

Stambovsky v. Ackley and Ellis Realty
Supreme Court, Appellate Division, State of New York
169 A.D.2d 254 (1991)

Plaintiff Stambovsky purchased a home from Ackley, who was represented by Ellis Realty. After entering the contract but before closing, Stambovsky learned that the house was said to be possessed by poltergeists, reportedly seen by Ackley and her family on numerous occasions over the previous nine years. As a resident of New York City, Stambovsky was unaware that apparitions seen by the Ackleys were reported in the *Reader's Digest* and the local press of Nyack, New York. In 1989, the house was also included on a five-home walking tour, in which the house was described as "a riverfront Victorian (with ghost)." Stambovsky brought an action for rescission of the contract, arguing that the reputation of the house impaired the present value of the property and its resale value. He argued that the failure of the Ackleys and Ellis Realty Company to disclose the nature of the house as haunted was fraudulent in nature. The defendants argued that the principle of caveat emptor (buyer beware) applied in the state of New York and they had no affirmative duty to disclose nonmaterial matters. They moved for dismissal. The lower court granted the dismissal. The plaintiff appealed.

Judge Rubin

While I agree with [the] Supreme Court [New York's trial court] that the real estate broker, as agent for the seller, is under no duty to disclose to a potential buyer the phantasmal reputation of the premises and that, in his pursuit of a legal remedy for fraudulent misrepresentation against the seller, plaintiff hasn't a ghost of a chance, I am nevertheless moved by the spirit of equity to allow the buyer to seek rescission of the contract of sale and recovery of his down payment. New York law fails to recognize any remedy for damages incurred as a result of the seller's mere silence, applying instead the strict rule of caveat emptor. Therefore, the theoretical basis for granting relief, even under the extraordinary facts of this case, is elusive if not ephemeral.

"Pity me not, but lend thy serious hearing to what I shall unfold." (William Shakespeare, *Hamlet*, Act I, Scene V [Ghost].)

From the perspective of a person in the position of plaintiff herein, a very practical problem arises with respect to the discovery of a paranormal phenomenon:

"Who you gonna call?" as the title song to the movie *Ghostbusters* asks. Applying the strict rule of caveat emptor to a contract involving a house possessed by poltergeists conjures up visions of a psychic or medium routinely accompanying the structural engineer and Terminix man on an inspection of every home subject to a contract of sale. It portends that the prudent attorney will establish an escrow account lest the subject of the transaction come back to haunt him and his client—or pray that his malpractice insurance coverage extends to supernatural disasters. In the interest of avoiding such untenable consequences, the notion that a haunting is a condition which can and should be ascertained upon reasonable inspection of the premises is a hobgoblin that should be exorcised from the body of legal precedent and laid quietly to rest.

The doctrine of caveat emptor requires that a buyer act prudently to assess the fitness and value of his purchase and operates to bar the purchaser who fails to exercise due care from seeking the equitable remedy of rescission. For the purposes of the instant motion to dismiss the action, the plaintiff is entitled to every favorable inference that may reasonably be drawn from the pleadings; specifically, in this instance, that he met his obligation to conduct an inspection of the premises and a search of available public records with most meticulous inspection and the search would not reveal the presence of poltergeists at the premises or unearth the property's ghoulish reputation in the community. Therefore, there is no sound policy reason to deny plaintiff relief for failing to discover a state of affairs that the most prudent purchaser would not be expected to even contemplate.

The case law in this jurisdiction dealing with the duty of a vendor of real property to disclose information to the buyer is distinguishable from the matter under review. The most salient distinction is that existing cases invariably deal with the physical condition of the premises and other factors affecting its operation. No case has been brought to this court's attention in which the property value was impaired as the result of the reputation created by information disseminated to the public by the seller (or, for that matter, as a result of possession by poltergeists). Where a condition that has been created by the seller materially impairs the value of the contract and is peculiarly within the knowledge of the seller or unlikely to be discovered by a prudent purchaser exercising due care with respect to the subject transaction, nondisclosure constitutes a basis for rescission as a matter of equity. Any other outcome places upon the buyer not merely the obligation to exercise care in his purchase but rather to be omniscient with respect to any fact that may affect the bargain. No practical purpose is served by imposing such a burden upon a purchaser. To the contrary, it encourages predatory business practice and offends the principle that equity will suffer no wrong to be without a remedy.

In the case at bar, defendant seller deliberately fostered the public belief that her home was possessed. Having undertaken to inform the public at large, to whom she has no legal relationship, about the supernatural occurrences on her property, she may be said to owe no less a duty to her contract vendee. It has been remarked that the occasional modern cases that permit a seller to take unfair advantage of a buyer's ignorance but has created and perpetuated a condition about which he is unlikely to even inquire, enforcement of the contract (in whole or in part) is offensive to the court's sense of equity. Application of the remedy of rescission, within the bounds of the narrow exception to the doctrine of caveat emptor set forth herein, is entirely appropriate to relieve the unwitting purchaser from the consequences of a most unnatural bargain.

Reversed in favor of Plaintiff, Stambovsky.

Dissenting Opinion

The parties herein were represented by counsel and dealt at arm's length. This is evidenced by the contract of sale, which contained various riders and a specific provision that all prior understandings and agreements between the parties were merged into the contract, that the contract completely expressed their full agreement and that neither had relied upon any statement by anyone else not set forth in the contract. There is no allegation that defendants, by some specific act, other than the failure to speak, deceived the plaintiff. Nevertheless, a cause of action may be sufficiently stated where there is a confidential or fiduciary relationship creating a duty to disclose, and there was a failure to disclose a material fact, calculated to induce a false belief. Plaintiff herein, however, has not alleged and there is no basis for concluding that a confidential or fiduciary relationship existed between these parties to an arm's length transaction such as to give rise to a duty to disclose. In addition, there is no allegation that defendants thwarted plaintiff's efforts to fulfill his responsibilities fixed by the doctrine of caveat emptor.*

CRITICAL THINKING ABOUT THE LAW

Although judges frequently rely on legal precedent, there is no fixed standard by which judges must give a certain weight to precedent. In other words, judges differ in the degree to which they show deference to legal precedent and legislative acts. Consequently, several judges viewing the same case and same set of facts could reach conflicting decisions. As you learned in Chapter 3 about the American legal system, judges work on the basis of different philosophies: some believe

---

*Stambovsky v. Ackley and Ellis Realty, Supreme Court, Appellate Division, State of New York 169 A.D.2d 254 (1991).

in judicial restraint and others believe in judicial activism. These underlying philosophies are in part responsible for the varying degrees of importance that judges give to the four primary ethical norms (see Chapter 1). In turn, the degree of importance that a judge attaches to each of these ethical norms plays a significant role in the shaping of a judge's reasoning and the court's conclusion. The following questions pertaining to Case 9-6 encourage you to consider the importance of primary ethical norms in a judge's reasoning.

1. Which primary ethical norm was guiding the judge's reasoning in the majority opinion?

    Clue: Consider the factors the judge discussed in favoring the plaintiff's interests over the defendant's interests.

2. To further illustrate the significance of ethical norms, to which ethical norm did the dissenting judge give highest priority?

    Clue: Similar to the first question, why do you think the judge elevated the defendant's interests over the plaintiff's interests?

3. What missing information would help you better evaluate the court's reasoning?

    Clue: Find the reasons the plaintiff provides for rescission of the contract. What omitted evidence could have made the plaintiff's case more convincing?

Undue Influence. If one party exerts mental coercion over the other party, there is **undue influence** and, therefore, no genuine assent. There are two court-established requirements for undue influence: (1) There must be a dominant–subservient relationship between the contrasting parties (e.g., a doctor and patient, lawyer and client, or any trusting relationship). (2) This dominant–subservient relationship must allow one party to influence the other in a mentally coercive way. An example of coercion that meets these requirements is a dying patient's contracting with a family doctor to sell his family land at an unreasonably low price in order to pay the doctor bills.

undue influence Mental coercion exerted by one party over the other party to the contract.

Mistake. A **mistake** also prevents a meeting of the minds, or genuine assent. If both parties made an error as to a material fact, a mutual, or *bilateral*, mistake has occurred, and as a general rule the courts will rescind such contracts. If, however, an error is made by only one party to the contract, a *unilateral* mistake has occurred, and the courts generally will not grant the mistaken party a rescission of the contract. An exception to this rule is made if the nonmistaken party knew or should have known of the mistake. For example, if five contractors bid on a $10 million hospital project and Smith's bid is $2 million below all other bids because of an accountant's error, the hospital should realize the error before accepting the bid, especially if Smith immediately notifies the hospital of the mathematical error and the contract is executory in nature. A mistake of future value or quality of an object is enforceable by either party; each party has assumed that there may be a change in value.

mistake Error as to material fact. A bilateral mistake is one made by both parties; a unilateral mistake is one made by only one party to the contract.

COMPETENT PARTIES

The fifth essential element of a legally enforceable contract is **competency** of the parties. A person is presumed to be competent at the time of entering into a contract, so most people who raise the defense of a lack of capacity must prove that, at the time the contract was entered into, the individual did not have the ability to understand the nature of the transaction and the consequences of entering into it. This defense often arises when contracts involve minors or insane or intoxicated persons.

competency A person's ability to understand the nature of the transaction and the consequences of entering into it at the time the contract was entered into.

Minors. A *minor* is a person under the legal age of majority. The states differ as to age of majority for entering into enforceable contracts. The age of majority for contractual capacity should not be confused with the age at which one can drink or can vote in state and federal elections. Contracts made by minors are voidable and can be disaffirmed by the minor at any time before the minor becomes of a majority age or shortly thereafter. If the minor fails to disaffirm a contract, he or she will be considered to have ratified (approved) it and is, thus, legally bound. For example, Smith, at age 17 years, 3 months, entered into a contract with Jones to buy the latter's automobile for $2,000. In the state in which the contract was executed, the majority age was 18. After using the automobile for 2 years and 1 month, Smith returned the auto and asked for his $2,000 back, minus depreciation on the car, because he was only 17 when he entered the contract and he claimed that, therefore, the contract was voidable. Smith was not allowed to disaffirm his contract because he had ratified the agreement by failing to disaffirm it before he turned age 18 or shortly thereafter.

A minor is generally liable for the reasonable value (not the contract or the market value) of necessaries (food, clothing, shelter), which enable the minor to live in a manner to which he or she is accustomed. To avoid the issues of what constitutes a *necessary*, prudent businesspeople check purchasers' ages very carefully and require a parent or guardian to cosign a loan when the borrower is a minor. For example, a student who is a minor for contractual purposes is not able to obtain a loan at a bank without a parent's signature, and merchants generally check to see if a charge card in the possession of a minor is issued in a parent's name.

APPLYING THE LAW TO THE FACTS...

Let's say that an 11-year-old signed her name to an online contract for a magazine subscription that would bill the address of the recipient at the time of shipment. Nowhere on the contract was the purchaser required to state his or her age. When the child receives the bill and the parents refuse to pay it for her, does the magazine company have any recourse against anyone? What if the online form said that the offer was limited to residents of the United States aged 18 or older?

Statutes That Make Minors Liable for Some Contracts

At common law, state statutes and some case law make minors liable for certain contracts. They cannot assent to the common-law doctrine of disaffirmance with regard to the following:

- Contracts to support children, especially when a minor is emancipated
- Contracts to enlist in the armed forces
- Contracts for life insurance
- Contracts for medical, surgical, and pregnancy care
- Contracts for health insurance
- Contracts for sports and entertainment when a court has approved the contracts

Insanity. If a person is adjudicated by a court of law to be insane, or is de facto (in fact) insane at the time of entering into a contract, the individual will be allowed to disaffirm a contract. Court-appointed guardians may also disaffirm such agreements.

Intoxication. A person who is intoxicated to the degree that understanding the nature of the contract and its consequences is impossible will be able to disaffirm a contract in all cases but those involving necessaries. Courts will look closely at the degree of intoxication. To disaffirm, the intoxicated individual must return the item bought. In the case of necessaries, the intoxicated individual is not allowed to disaffirm, but is held liable for the reasonable value of such items.

LEGAL OBJECT

The sixth necessary element of a contract is a **legal object**. This means that the subject matter of the agreement must be lawful. If it is not, the contract is void at its inception. Contracts that are in violation of state or federal statutes, as well as those in violation of case law, are void as a matter of public policy.

legal object Contract subject matter that is lawful under statutory and case law.

Statutory Law. State statutes that forbid wagering agreements (betting) and usurious (defined as exorbitant) finance charges or interest rates on loans, as well as those aimed at licensing and regulation, have been the subject of much adjudication. For example, if Smith practices law in a state without being admitted to practice before its highest state court, the courts will generally not enforce any contracts Smith made with clients for services rendered. Smith may also be subject to criminal charges. State statutes require licensing of nurses, doctors, accountants, real estate agents, electricians, and many other groups (the list varies from state to state), so as to protect the public from incompetents. Some opponents of these statutes argue that they were enacted at the behest of interest groups to decrease the supply of individuals in a particular profession or trade and, hence, prevent competition. Whatever their origin, however, courts will not enforce contracts made by unlicensed providers of services for which licensure is required.

Case Law. Often, when there is a question as to whether a contract has a lawful object, statutory law does not clearly indicate whether the contract is void or unenforceable. For example, agreements not to compete are found to be contrary to the public policy of fostering competition, and there are federal and state antitrust laws (statutes) making price-fixing between competitors illegal and void. However, when an otherwise lawful contract of employment contains a no-competition clause, pursuant to which an employee agrees not to become employed by a competitor of his or her employer, the courts will look at whether the restriction on the employee is for a reasonable time and area. In addition, it will look at the relative bargaining power of the employer and employee and the hardship on the employee who is contractually forbidden to

ONLINE FANTASY SPORTS: LUCK OR SKILL?

Online fantasy sports leagues are called "gambling," and millions play (estimates vary from 20 million to 40 million). Participants (owners) build teams based on real-life players (usually football). At the end of a week, performances of real-life players are translated into points and the points of all the players on a team are totaled at the end of a season. The winner may win money and prizes. Participants pay a fee to play and use the site facilities.

In some instances, money is paid into a "pot," and at the end of the (football) year, each owner of a fantasy team with the most points wins the "pot."

Are these games ones of chance (luck), or of skill? It could be argued that fantasy games are games of chance because events beyond the participants' control (e.g., injury to a key player such as a quarterback) could determine the outcome for a participant's team.

If you are a day trader in securities, you are betting on stocks that you buy in a one-day period. When you buy and sell such securities (stocks, bonds, commodities), are you gambling? You are in fact "betting" that the value of the securities will go up. Is this a game of chance, or of skill? How are these "games" different? What are the policy implications for our society, based on your answer? For example, is there an enforceable contract between fantasy-league participants? What if online "gambling" is declared illegal by Congress? *See Humphry v. Viacom,* 2001 WL 1797648 (D.N.J. 2007), which does not answer this question but raises some more questions with regard to fantasy games.

CASE 9-7

Brown & Brown, Inc. v. Johnson
New York Supreme Court, Appellate Division, Fourth Department
115 A.D.3d 162, 980 N.Y.S.2d 631(2014)

Brown & Brown, Inc., is a firm of insurance intermediaries—insurance agents, brokers, and consultants—in New York City. Brown hired Theresa Johnson to provide actuarial analysis. On Johnson's first day of work, she was asked to sign a nonsolicitaion covenant, which prohibited her from soliciting or servicing any of Brown's clients for two years after the termination of her employment. The covenant specified that if any of its provisions were declared unenforceable, they should be modified and then enforced "to the maximum extent possible." Less than five years later, Johnson's employment with Brown was terminated, and she went to work for Lawley Benefits Group, LLC. Brown filed a suit in a New York state court against Johnson, seeking to enforce the nonsolicitation covenant Johnson filed a motion to dismiss Brown's complaint. The court ruled in Brown's favor, and Johnson appealed.

Judge Whalen

A nonsolicitation covenant is overbroad and therefore unenforceable if it seeks to bar the employee from soliciting or providing services to clients with whom the employee never acquired a relationship through his or her employment. Here, the non-solicitation covenant purported to restrict Johnson from, inter alia [among other things], soliciting, diverting, servicing, or accepting, either directly or indirectly, "any insurance or bond business of any kind or character from any person, firm, corporation, or other entity that is a customer or account of the New York offices of the Company during the term of the Agreement" for two years following the termination of Johnson's employment, without regard to whether Johnson acquired a relationship with those clients. We conclude that the language of the non-solicitation covenant render it overbroad and unenforceable.

Plaintiffs contend that we nevertheless should partially enforce the covenant, in as much as plaintiffs seek to prevent Johnson from soliciting and servicing only those clients with whom Johnson actually developed a relationship during her employment with plaintiffs. We reject that contention. . . . Partial enforcement may be justified if the employer demonstrates an absence of overreaching, coercive use of dominant bargaining power, or other anti-competitive misconduct, but has in good faith sought to protect a legitimate business interest, consistent with reasonable standards of fair dealing. Factors weighing against partial enforcement are the imposition of the covenant in connection with hiring or continued employment[,] . . . the existence of coercion or a general plan of the employer to forestall competition, and the employer's knowledge that the covenant was overly broad. Here, it is undisputed that Johnson was not presented with the Agreement until her first day of work with plaintiffs, after Johnson already had left her previous employer. Plaintiffs have made no showing that, in exchange for signing the Agreement, Johnson received any benefit from plaintiffs beyond her continued employment.

. . . The fact that the Agreement contemplated partial enforcement does not require partial enforcement. . . . Allowing a former employer the benefit of partial enforcement of overly broad restrictive covenants simply because the applicable agreement contemplated partial enforcement would . . . enhance the risk that employers will use their superior bargaining position to impose unreasonable anti-competitive restrictions, uninhibited by the risk that a court will void the entire agreement, leaving the employee free of any restraint. In our view, the fact that the Agreement here contemplated partial enforcement does not demonstrate the absence of overreaching on plaintiffs' part, but, rather, demonstrates that plaintiffs imposed the covenant in bad faith, knowing full well that it was overbroad. We therefore conclude that the non-solicitation covenant should not be partially enforced.*

Affirmed for Defendant Johnson.

A state intermediate appellate court reversed the lower court's ruling and granted Johnson's motion to dismiss Brown's action with respect to the nonsolicitation covenant. The covenant was overbroad. Furthermore, it was not presented to Johnson until her first day of work, and she received no benefit for signing it beyond her continued employment.

---

*Brown & Brown, Inc. v. JohnsonMoore v. Midwest Distribution, Inc. New York Supreme Court, Appellate Division, Fourth Department 115 A.D.3d 162, 980 N.Y.S.2d 631(2014).

Contracts That Must Be in Writing

Most contracts need not be in writing. They are enforceable as long as the six elements of a contract exist. The statute of frauds, which originated in England in 1677, however, requires certain business contracts to be in writing. Originally, those contracts listed in this section, and some other nonbusiness contracts, were required to be in writing because they were thought to be the most likely situations in which perjury would occur. Today, each state requires by statute that various contracts be written in order to be enforceable.

In most states, the requirements for a written contract include some evidence of writing and the signature of the party being sued. The writing should reasonably outline the terms and state who are the parties to the agreement. Contracts governed by UCC Section 2-201 are not subject to these requirements. The party suing may have the only evidence of writing with his or her signature on it. Often, between merchants, confirmation memoranda summarize oral agreements and are satisfied by only one party. The nonsigning party must simply review the memorandum. The statute of frauds is satisfied if the nonsigning party agrees to its content.

The business-related contracts discussed in this section are those that most frequently fall within the statute of frauds. They therefore must be in writing to be enforceable.

CONTRACTS FOR THE SALE OF AN INTEREST IN LAND

An "interest in land" includes mortgages, easements (an *easement* is a contract that allows a party to cross your land, for example, with electrical wires), and, of course, the land itself and the buildings on it. Leases for longer than one year usually have to be in writing.

A notable exception to the requirement that contracts for sale of an interest in land must be written is *partial performance*. For example, if substantial improvements have been made on a piece of property by a lessee in reliance on an oral commitment by the lessor to sell, the oral contract will be enforced. Case 9-8 illustrates the importance of contracts performable within one year, and the concept of reasonableness.

CONTRACTS TO PAY THE DEBTS OF ANOTHER

If Smith promises to pay Jones's debt to the bank should Jones be unable to pay it, this contract must be in writing, under the statute of frauds. In this situation, Smith has secondary liability to the bank. If, however, Smith tells the bank that she will act as a surety for Jones's debt, Smith has a primary liability to the bank. This agreement is not within the statute of frauds and, therefore, may be enforced even if it is only oral.

In the first situation, Smith's promise was conditional in nature. That is, on condition that Jones does not pay, the bank may look to Smith, but first it must look to Jones. In the second situation, the bank may first look to Smith. It does not have to go to Jones at all because it has a guarantor or surety agreement with Smith.

CONTRACTS NOT PERFORMABLE IN ONE YEAR

A contract must be in writing if it specifies that it will last longer than one year (years run generally from the formation of the contract rather than the beginning of performance). For example, in most states, a baseball player who agrees to play for a team for three years at $2 million a year must sign a written contract in order for the agreement to be enforceable. If, however, no date is set in the contract for the completion of performance, the contract need not be in writing. For example, an agreement to provide help for a person until that person dies does not fall within the statute of frauds, and thus does not have to be in writing to be enforceable.

SALE OF GOODS OF $500 OR MORE

Under UCC Section 2-201, contracts for the sale of goods of $500 or more (generally the price of the goods is the determinant, not the value) fall within the statute of frauds and must be in writing to be enforceable. There are three exceptions to this rule: (1) one of the parties to a suit admits in writing or in court to the existence of an oral contract; (2) a buyer accepts and uses the goods; (3) the contract is between merchants, and the merchant who is sued received a written confirmation of the oral agreement and did not object within 10 days. In all of these instances, the oral contract will be enforced even if it is for goods worth $500 or more. An example below of the importance of the sale of goods ($500) rule is shown with the *Iacono* case.

CASE 9-8

Iacono v. Lyons
Court of Appeals of Texas
16 S.W.3d 82 (2000)

The plaintiff, Mary Iacono, and the defendant, Carolyn Lyons, had been friends for almost 35 years. In late 1996, the defendant invited the plaintiff to join her on a trip to Las Vegas, Nevada, for which the defendant paid. The plaintiff contended that she was invited to Las Vegas by the defendant because the defendant thought the plaintiff was lucky. Sometime before the trip, the plaintiff had a dream about winning on a Las Vegas slot machine. The plaintiff's dream convinced her to go to Las Vegas, and she accepted the defendant's offer to split "50-50" any gambling winnings. The defendant provided the plaintiff with money for the gambling.

The plaintiff and defendant started to gamble, but after losing $47, the defendant wanted to leave to see a show. The plaintiff begged the defendant to stay, and the defendant agreed on the condition that the defendant put the coins into the machines because doing so took the plaintiff, who suffers from advanced rheumatoid arthritis and was in a wheelchair, too long. The plaintiff agreed and took the defendant to a dollar slot machine that looked like the machine in her dream. The machine did not pay on the first try. The plaintiff then said, "Just one more time," and the defendant looked at the plaintiff and said, "This one's for you, Puddin." They hit the jackpot, and the slot machine paid $1,908,064. The defendant refused to share the winnings with the plaintiff and denied that they had an agreement to split any winnings. The defendant told Caesar's Palace that she was the sole winner and to pay her all the winnings.

The plaintiff sued the defendant for breach of contract. The defendant moved for summary judgment on the ground that any oral agreement was unenforceable under the statute of frauds. The trial court entered summary judgment in favor of the defendant. The plaintiff appealed.

Justice O'Connor

The defendant asserted the agreement . . . was unenforceable under the statute of frauds because it could not be performed within one year. There is no dispute that the winnings were to be paid over a period of 20 years.

The one year provision of the statute of frauds does not apply if the contract, from its terms, could possibly be performed within a year—however improbable performance within one year may be.

To determine the applicability of the statute of frauds with indefinite contracts, this Court may use any reasonably clear method of ascertaining the intended length of performance. The method is used to determine the parties' intentions at the time of contracting. The fact that the entire performance within one year is not required, or expected, will not bring an agreement within the statute.

Assuming without deciding that the parties agreed to share their gambling winnings, such an agreement possibly could have been performed within one year. For example, if the plaintiff and defendant had won $200, they probably would have received all the money in one pay-out and could have split the winnings immediately.

Therefore, the defendant was not entitled to summary judgment based on her affirmative defense of the statute of frauds.*

The Court of Appeals of Texas *reversed* and ruled in favor of the Defendant.

---

*Iacono v. Lyons, Court of Appeals of Texas 16 S.W.3d 82 (2000).

NONBUSINESS CONTRACTS

Nonbusiness-related contracts that must be in writing to be enforceable are (1) contracts in consideration of marriage and (2) contracts of an executor or administrator to answer for the debts of a deceased person.

Parol Evidence Rule

The **parol (oral) evidence rule** states that when parties have executed a written agreement, which is complete on its face, oral agreements made before or at the same time as the written agreement that vary, alter, or contradict the written agreement are invalid. Most state courts do not even allow such oral agreements to be introduced in evidence. For example, suppose that Smith enters into a written contract with Jones to sell him a two-year-old Chevy Citation for $5,000 and the contract terms state that "all warranties are excluded." At the time of signing, Smith orally tells Jones, "Don't worry, we'll warranty all parts and labor." This oral agreement made at the time of execution will not be allowed into evidence because it varies from the written agreement. Exceptions to the parol evidence rule are set out in Table 9-2.

Under UCC Section 2-202, written memoranda that are intended to be a final expression of the parties' agreement cannot be contradicted by prior or contemporaneous oral agreement, but may be explained or supplemented orally by course of dealing or usage of trade, by course of performance, or by evidence of consistent additional terms. This UCC rule allows the courts to admit into evidence oral testimony with regard to written agreements that would ordinarily be inadmissible under case law.

parol (oral) evidence rule When parties have executed a written agreement which is complete on its face, oral agreements made before or at the same time as the written agreement that vary, alter, or contradict the written agreement are invalid.

1. Oral agreements used to prove a subsequent modification of a written agreement are admissible.
2. Oral agreements to clear up ambiguity in the written agreement are admissible.
3. Oral agreements to prove fraud, mistake, illegality, duress, undue influence, or lack of capacity are admissible.
4. Oral agreements concerning collateral matters not germane to the written agreement are admissible.

TABLE 9-2

EXCEPTIONS TO THE PAROL EVIDENCE RULE

 LINKING LAW AND BUSINESS

Accounting

Land is treated specially, not just in the law but in other disciplines as well. As you may recall from your accounting class, *land* is classified as a long-term operational asset. One distinguishing characteristic of land is that it is not subject to depreciation or depletion. In other words, land is considered to have an infinite life because it is not destroyed by use. If an organization makes a basket purchase, or acquires several assets in a single transaction, the amount paid should be carefully divided between the land and the other assets on the organizational financial statements. Therefore, the balance sheet will reflect the nondepreciable nature of the land, while allowing noticeable depreciation or depletion of the other assets purchased.

Source: T. Edmonds, F. McNair, E. Milam, and P. Olds, *Fundamental Financial Accounting Concepts* (New York: McGraw-Hill, 2000), pp. 408–10.

Third-Party Beneficiary Contracts and Assignment of Rights

TYPES OF THIRD-PARTY BENEFICIARY CONTRACTS

So far, our discussion has focused on contracts between two parties (usually Smith and Jones). Two parties, however, may enter into a contract with the clear intent to benefit a third party; in these cases, there is a *third-party beneficiary contract*. The third party is not a party to the contract but merely a beneficiary of it. These contracts are divided into:

1. *Intended beneficiary contracts.* Two parties to a contract (promisor and promisee), either by words, writing, or actions, intend to bring benefits to a third party by virtue of an enforceable contract.
2. *Incidental beneficiary contracts.* There is no intent to bring benefits to the third party. The benefit is unintentional and the incidental beneficiary cannot sue to enforce such a contract.

donee–beneficiary contract A contract in which the promisee obtains a promise from the promisor to make a gift to a third party.

A **donee–beneficiary contract** exists when the purpose of the promisee in obtaining a promise from the promisor is to make a gift to a third person. For example, Liberty Insurance Company (promisor) promises to pay Smith (third party) a sum of $100,000 upon the death of Jones (promisee) in exchange for Jones's payment of a yearly premium. Under this third-party donee–beneficiary contract, Smith may sue Liberty Insurance Company if it fails to pay the $100,000 upon the death of Jones.

creditor–beneficiary contract A contract in which the promisee obtains a promise from the promisor to fulfill a legal obligation of the promisee to a third party.

A third-party **creditor–beneficiary contract** exists when the purpose of the promisee in requiring a promisor's performance to be made to a third person is to fulfill a legal obligation of the promisee to the third person. For example, Smith (promisee) works for Jones (promisor) in exchange for Jones's promise to pay Taylor $6,000 that Smith owes to Taylor. Under this third-party creditor–beneficiary contract, if Smith does the work and Jones refuses to pay, Taylor may sue both Jones and Smith.

Note that insurance contracts and all forms of creditor collection agreements are third-party beneficiary contracts. These types of agreements are obviously very important in our economy.

ASSIGNMENT OF RIGHTS

assignment The present transfer of an existing right.

An **assignment** is the present transfer of an existing right. Contracts between two parties may be assigned to a third party under certain conditions. Suppose that B, a manufacturing company, sells A, a retail company, 600 bicycles at $100 apiece on credit. A, known as the obligor-promisor, agrees to pay B, the obligee-promisee-assignor, $60,000. A does not pay, and therefore B has the right to sue A; but B has another choice: to assign this right to C, a collection agency, and receive immediate cash from C. In effect, B is assigning to C, known as the *assignee*, A's promise to pay in the future in exchange for receiving cash from C now (see Exhibit 9-1). C has the right to sue both A and B. That is, if the collection agency is unable to collect the money from the retailer A, it may sue not only retailer A but also the manufacturer, B, to recover the cash it advanced in anticipation of collecting the debt.

The conditions attached to most assignments are that unless the obligor (A) receives notice of assignment by the obligee-assignor (B), the obligor has no duty to the assignee (C). Once that notice is received, however, the assignee "stands in the shoes" of the obligee.

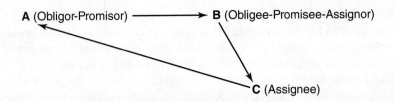

EXHIBIT 9-1

ASSIGNMENT OF CONTRACT RIGHTS

Certain classes of assignments are not recognized by law:

1. Assignments that materially change the duty of the obligor
2. Assignments forbidden by state statute
3. Any assignment forbidden by the original contract between the obligor-promisor and the obligee-promisee
4. Contracts for personal services (e.g., an artist's contract to paint an individual's portrait)

SUMMARY

A contract is defined as a legally enforceable exchange of promises. The sources of contract law are case law from state and federal courts and statutory law from the federal and state legislatures, particularly from the UCC. Contracts may be classified as express or implied; unilateral or bilateral; void, voidable, or valid; executed or executory; and quasi.

The six necessary elements of a legal contract are (1) a legal offer, (2) a legal acceptance, (3) a consideration, (4) genuine assent, (5) competent parties, and (6) a legal object. The UCC differs somewhat from the common law in its requirements for contracts. Table 9-3 outlines some of the differences.

TABLE 9-3

COMPARISON BETWEEN COMMON LAW AND THE UCC

Area of Comparison	Common Law	UCC
Contract application	Real property, services, and employment contracts	Contracts for the sale of goods
Requirements for offer	Includes subject matter, price, and quantity	Includes subject matter and quantity; may leave price and other terms open
Option agreements	Requires consideration for all option agreements	No consideration needed
Requirements for acceptance	Terms of acceptance are mirror image of offer	Mirror image not necessary; additional terms allowed if one of three requirements is met
Requirements for consideration	Consideration required for contract to be enforceable except under doctrine of promissory estoppel	Consideration not needed for modifications
Statute of frauds (contracts that must be in writing)	Real estate contracts, contracts not performable within one year	Sale of goods of $500 or more, with three exceptions

Some types of contracts fall within the statute of frauds and, therefore, must be in writing to be enforceable. The parol evidence rule invalidates most oral agreements made before or at the same time as a written contract if the oral agreements alter or contradict the terms of the written contract.

Third-party beneficiary contracts may be of either the donor–beneficiary or the creditor–beneficiary type. A contract made between two parties may be assigned to a third party under certain conditions.

REVIEW QUESTIONS

9-1 How does a contract based on common law differ from one based on the Uniform Commercial Code?

9-2 Explain the distinction between a void and a voidable contract; between an executed and an executory contract; between a unilateral and a bilateral contract.

9-3 Describe the three requirements for a valid acceptance.

9-4 Describe the three requirements for a valid offer.

9-5 Explain how an acceptance can be terminated.

9-6 Explain the difference between liquidated and unliquidated debts.

REVIEW PROBLEMS

9-7 Fisher, an employment agency, sued Catani, a minor, for breach of contract for the balance due the agency of $101.25 as a commission for finding Catani employment. The defendant disaffirmed his contract with the agency, while still a minor, two months after obtaining the job and one month after he quit. Can he disaffirm? Explain.

9-8 Robinson was employed as an assistant manager in Gallagher Drug Company Store. He was accused of theft and embezzlement, which he admitted to, and was fired. The following day, at company headquarters, he signed a contract promising to repay the company $2,000. Robinson made payments totaling $741, and then stopped. Gallagher sued for the balance. Robinson's defense was that he had signed the agreement under duress. What logical problem arises in Robinson's arguing that he became aware of the duress after he attended a support group for unemployed managers?

9-9 Osborne, a former chairman of the board of Locke Steel Company, entered into an agreement with Locke that, on retirement, he would hold himself available for consultation and would not work for any direct or indirect competitors of the company. In exchange for these promises, the company agreed to pay Osborne $15,000 a year for the rest of his life. After paying for two years, the company stopped payments when Osborne refused to consent to a modification of the agreement. Osborne sued. The defendant, Locke, argued that there was no consideration because the contract was based on past services, and thus there was no detriment to the promisee (Osborne). Who won? Explain.

9-10 Fisher, an inexperienced businessman, bought equipment and chinchillas from Division West Chinchilla in order to start a chinchilla ranch. Fisher got into the business because Division West had told him that chinchilla ranching was an "easy undertaking . . . and no special skills were required." Fisher lost money operating the ranch. He sued, claiming that he had relied on Division West's fraudulent representations. Who won? Explain.

9-11 William Story promised his nephew that he would pay the nephew $5,000 if he gave up "using tobacco, swearing, and playing cards and billiards until he was 21." The nephew did so and asked his uncle for the money. His uncle agreed to pay but died before he did so. The uncle's estate refused to pay, arguing that there was no consideration for the uncle's promise. Should the court uphold the agreement in this case? Why or why not?

CASE PROBLEMS

9-12 Wilcox hired Esprit Log and Timber Frame Homes to build a log house, which the Wilcoxes intended to sell. They paid Esprit $125,260 for materials and services. They eventually sold the home for $1,620,000 but sued Esprit due to construction delays. The logs were supposed to arrive at the construction site precut and predrilled, but that did not happen. So it took five extra months to build the house while the logs were cut and drilled one by one. The Wilcoxes claimed that the interest they paid on a loan for the extra construction time cost them about $200,000. The jury agreed and awarded them that much in damages, plus $250,000 in punitive damages and $20,000 in attorneys' fees. Esprit appealed, claiming that the evidence did not support the verdict because the Wilcoxes had sold the house for a good price. Is Esprit's argument credible? Why or why not? How should the court rule? *Esprit Log and Timber Frame Homes, Inc. v. Wilcox*, 302 Ga. App. 550, 691 S.E.2d 344 (2010).

9-13 Amber gave her boyfriend, Frederick, title to her house by quitclaim deed after he paid off her mortgage balance. The couple then separated, and Amber moved out. Two months later, they reconciled, but Amber refused to move back into the house with Frederick unless he granted her an undivided one-half interest in the house. The couple then signed a document making them "equal partners" in the house and providing for the disposition of the property in the event of a break-up. Is resuming a romantic relationship by moving back in with the other person adequate consideration to form a contract? *Willis v. Ormsby*, 966 N.E.2d 255 (Ohio 2012).

9-14 D.V.G. (a minor) was injured in a one-car auto accident in Hoover, Alabama. The vehicle was covered by an insurance policy issued by Nationwide Mutual Insurance Co. Stan Brobston, D.V.G.'s attorney, accepted that the settlement could be submitted to an Alabama state court for approval. D.V.G. died from injuries received in a second, unrelated auto accident. Nationwide argued that it was not bound to the settlement because a minor lacks the capacity to contract and so cannot enter into a binding settlement without court approval. Should Nationwide be bound to settlement? Why or why not? *Nationwide Mutual Insurance Co. v. Wood*, 402 So.3d 580 (Ala. 2013).

9-15 Brendan Coleman created and marketed Clinex, a software billing program. Later, Retina Consultants, P.C., a medical practice, hired Coleman as a software engineer. Together, they modified the Clinex program to create Clinex-RE. Coleman signed an agreement to the effect that he owned Clinex, Retina owned Clinex-RE, and he would not market Clinex in competition with Clinex-RE. After Coleman quit Retina, he withdrew funds from a Retina bank account and marketed both forms of the software to other medical practices. At trial, the court entered a judgment enjoining (preventing) Coleman from marketing the software that was in competition with the software he had developed for Retina Consultants. The court also obligated Coleman to return the funds taken from the company's bank account. Coleman appealed. *Coleman v. Retina Consultants, P.C.*, 286 Ga. 317, 687 S.E.2d 457 (2009).

9-16 The National Collegiate Athletic Association (NCAA) regulates intercollegiate amateur athletics among the more than 1,200 colleges and universities with which it contracts. Among other things, the NCAA maintains rules of eligibility for student participation in intercollegiate athletic events. Jeremy Bloom, a high-school football and track star, was recruited to play football at the University of Colorado (CU). Before enrolling, he competed in Olympic and professional World Cup skiing events, becoming the World Cup champion in freestyle moguls. During the Olympics, Bloom appeared on MTV and was offered other paid entertainment opportunities, including a chance to host a show on Nickelodeon. Bloom was also paid to endorse certain ski equipment and contracted to model clothing for Tommy Hilfiger. On Bloom's behalf, CU asked the NCAA to waive its rules restricting student-athlete endorsement and media activities. The NCAA refused, and Bloom quit the activities to play football for CU. He filed a suit in a Colorado state court against the NCAA, however, asserting breach of contract on the ground that NCAA rules permitted these activities if they were needed to support a professional athletic career. The NCAA responded that Bloom did not have standing to pursue this claim. What contract was allegedly breached in this case? Is Bloom a party to this contract? If not, is he a third-party beneficiary of it? Explain. *Bloom v. National Collegiate Athletic Association*, 93 P.3d 621 (Colo. Ct. App. 2004).

THINKING CRITICALLY ABOUT RELEVANT LEGAL ISSUES

Megawidget Corporation wants to buy 1,000 widget parts from Douglas Dealer. Megawidget's president, Mega Buyer, sends an email message to Douglas placing an order for the 1,000 widget parts at $10,000. Douglas emails back and clicks on a button on his computer screen, which is an icon of a little hand shaking. Mega receives the email and also clicks on the handshake icon on his computer. A week later, Mega finds out that he can buy the widget parts cheaper from Dealing Dan. Mega wants to rescind his contract with Douglas, and so claims that, under Article 2 of the UCC, his contract with Douglas had to be in writing to be valid; that is, the email was not a written contract and the "handshake" via icon was meaningless. Discuss the outcome of this contract dispute using a critical thinking approach.

1. What is the legal issue in this case? Explain.
2. What language in what section of Article 2 of the UCC is relevant in deciding this case? Is the language ambiguous? Explain.
3. What information seems to be missing from this case exposition? Explain.
4. What argument would Douglas use in upholding his position? Explain.

ON THE INTERNET

www.loc.gov From this site you can find links to a wealth of information about contracts.

www.lectlaw.com A site that provides multiple contract examples and forms. Under the heading "Some Main Rooms," click on "Free Legal Forms."

www.legaldocs.com This site provides information about how to prepare contracts and contains legal forms to use as templates for your own contracts.

lp.findlaw.com This site contains links to government agencies that regulate contracts. Click on "Corporate Counsel," then "Business Operations." Under the heading "Business Operations, click on "Commercial Contracts."

FOR FUTURE READING

Duxbury, Robert. *Contract Law (Nutshells)*. Sweet & Maxwell, 2006.

Friedman, Stephen E. "Improving the Rolling Contract." *American University Law Review* 56 (2006): 1.

Robertson, Andrew. "The Limits of Voluntariness in Contracts." *Melbourne University Law Review* 29 (2005): 179.

Wendt, John T., and Michael J. Garrison. "The Evolving Law of Employee Noncompete Agreements: Recent Trends and an Alternative Policy Approach." *American Business Law Journal* 45 (2008): 107.

CHAPTER TEN

The Law of Contracts and Sales—II

- METHODS OF DISCHARGING A CONTRACT
- REMEDIES FOR A BREACH OF CONTRACT
- E-CONTRACTS
- GLOBAL DIMENSIONS OF CONTRACT AND SALES LAW

In our discussion of the law of contracts in Chapter 9, we were concerned that agreements be enforceable. We noted that without enforceability of contract law by a court, there would be no predictability for enterprises that produce and sell goods. Without this there would be no security and financial stability for a firm. We would see the risk of loss increase and entrepreneurship decline if contracts were not enforceable at all levels of the manufacturing, marketing, and distribution of services and goods.

In this chapter we carefully examine the methods by which a contract can be discharged (particularly through performance), as well as the remedies that are possible for firms and individuals who are injured by a breach of contract for the sale of goods (Articles 2 and 2A of the Uniform Commercial Code [UCC]). E-contracts are highlighted and the concepts underlying them are distinguished from traditional contract principles. This chapter also addresses the international dimensions of contracts and sales agreements.

CRITICAL THINKING ABOUT THE LAW

In Chapter 9, we emphasized the importance of predictability and stability for those who enter into contracts. Nevertheless, we do not want parties to jump immediately to the conclusion that they have a contract every time they talk about an exchange. Instead, we want to make it possible for people to talk about an exchange without having actually made a commitment to the exchange. Why?

To aid your critical thinking about issues surrounding contract formation and performance, let's look at a fact pattern involving concert tickets.

Jennifer and Juan were recently involved in a breach-of-contract case. Juan had two extra tickets to a Garth Brooks concert, and he agreed to sell these tickets to Jennifer. After they had agreed on the price, Juan promised to give the tickets to Jennifer the next day. The next day, however, Jennifer did not want the tickets. Jennifer had discovered that it was an outdoor afternoon concert. Jennifer argued that she should not have to buy the tickets because she is allergic to sunlight and unable to spend any extended period of time outside. The judge ruled in favor of Jennifer.

1. What ethical norms seem to dominate the judge's thinking (see Chapter 1)?

    Clue: We have said that security is one reason for enforcing contracts. Review your list of ethical norms. Which norms seem to conflict with security in this case?

2. What missing information might be helpful in this case?

   Clue: To help you think about missing information, ask yourself the following question: Would the fact that Juan knew that Jennifer was allergic to sunlight affect your thinking about this case?

3. What ambiguous words might be troublesome in this case?

   Clue: Examine the reasoning that Jennifer uses to argue that she should be released from the contract.

Methods of Discharging a Contract

When a contract is terminated, it is said to be *discharged*. A contract may be discharged by performance (complete or substantial), mutual agreement, conditions precedent and subsequent, impossibility of performance, or commercial impracticability.

DISCHARGE BY PERFORMANCE

In most cases, parties to an agreement discharge their contractual obligation by doing what was required by the terms of the agreement. Many times, however, performance is substantial rather than complete. Traditional common law allowed suits for breach of contract if there was not **complete performance** of every detail of the contract. Today, however, the standard is **substantial performance**. This standard requires (1) completion of nearly all the terms of the agreement, (2) an honest effort to complete all the terms, and (3) no willful departure from the terms of the agreement.

Courts usually find substantial performance when there is only a minor breach of contract. For example, A, a contractor, agreed to build a house with five bedrooms for B. By the terms of the agreement, each of the rooms was to be painted blue. By mistake, one was painted pink, and B refused to pay A the $10,000 balance due on the house. The court awarded A $10,000 minus the cost of painting the wrongly painted room, finding that the departure from the contract terms was slight and unintentional and, therefore, insufficient for B to refuse to perform (pay) as agreed to in the contract.

If the breach is material, the injured party may terminate the contract and sue to recover damages. A *material breach* is substantial and, usually, intentional. Today, courts allow a party to "cure" a material breach if the time period within which a contract is supposed to be performed has not lapsed. In the following case both parties might be in breach of contract. For which party may the breach be material?

complete performance Completion of all the terms of the contract.

substantial performance Completion of nearly all the terms of the contract plus an honest effort to complete the rest of the terms, coupled with no willful departure from any of the terms.

CASE 10-1

Kohel v. Bergen Auto Enterprises, L.L.C.
Superior Court of New Jersey, Appellate Division
2013 WL 439970 (2013)

Per Curiam

On May 24, 2010, plaintiffs Marc and Bree Kohel entered into a sales contract with defendant Bergen Auto Enterprises, L.L.C. d/b/a Wayne Mazda Inc. (Wayne Mazda), for the purchase of a used 2009 Mazda. Plaintiffs agreed to pay $26,430.22 for the Mazda and were credited $7,000 as a trade-in for their 2005 Nissan Altima. As plaintiffs still owed $8,118.28 on their Nissan, Wayne Mazda assessed plaintiffs a net payoff of this amount and agreed to remit the balance due to satisfy the outstanding lien.

Plaintiffs took possession of the Mazda with temporary plates and left the Nissan with defendant. A few days later, a representative of defendant advised plaintiffs that the Nissan's vehicle identification (VIN) tag was missing. The representative claimed it was unable to sell the car and offered to rescind the transaction. Plaintiffs refused.

When the temporary plates on the Mazda expired on June 24, 2010, defendant refused to provide plaintiffs with the permanent plates they had paid for. In addition, defendant refused to pay off plaintiffs' outstanding loan on the Nissan as they had agreed. As a result, plaintiffs were required to make monthly payments on both the Nissan and the Mazda.

On July 28, 2010, plaintiffs filed a complaint in [a New Jersey state court] against Wayne Mazda.

On February 2, 2012, the court rendered an oral decision finding that there was a breach of contract by Wayne Mazda. On February 17, 2012, the court entered judgment in the amount of $5,405.17 in favor of the plaintiffs against Wayne Mazda. (The defendant appealed to a state intermediate appellate court.)

Defendant argued that plaintiffs' delivery of the Nissan without a VIN tag was itself a breach of the contract of sale and precludes a finding that defendant breached the contract. However, the trial court found that plaintiffs were not aware that the Nissan lacked a VIN tag when they offered it in trade. Moreover, defendant's representatives examined the car twice before accepting it in trade and did not notice the missing VIN until they took the car to an auction where they tried to sell it. There is a material distinction in the plaintiffs' conduct, which the court found unintentional, and defendant's refusal to release the permanent plate for which the plaintiffs had paid, an action the court concluded was done to maintain "leverage."

The evidence indicated that the problem with the missing VIN tag could be rectified. Marc Kohel applied and paid for a replacement VIN tag at Meadowlands [Nissan] for $35.31. While he initially made some calls to Meadowlands, he did not follow up in obtaining the VIN tag after the personnel at Wayne Mazda began refusing to take his calls.

The court concluded that "Wayne Mazda didn't handle this as adroitly (skillfully) as they could. . . ." Kevin DiPiano, identified in the complaint as the owner and/or CEO of Wayne Mazda, would not even take [the plaintiffs'] calls to discuss this matter. The court found:

> Mr. DiPiano could have been better businessman, could have been a little bit more compassionate or at least responsive, you know? He was not. He acted like he didn't care. That obviously went a long way to infuriate the plaintiffs. I don't blame them for being infuriated.

Here, plaintiffs attempted to remedy the VIN tag issue but this resolution was frustrated by defendant's unreasonable conduct. We thus reject defendant's argument that plaintiffs' failure to obtain the replacement VIN tag amounted to a repudiation of the contract.*

Affirmed, for the Plaintiffs Marc and Bree Kohel.

PERFORMANCE TO SATISFACTION OF ANOTHER

Contracts often state the completed work must personally satisfy one of the parties or a third person. When the subject matter of the contract is personal, the obligation is conditional, and performance must actually satisfy the party specified in the contract. For instance, contracts for portraits, works of art, and tailoring are considered personal because they involve matters of personal taste. Therefore, only personal satisfaction of the party fulfills the condition unless a court finds that the party is expressing dissatisfaction simply to avoid payment or otherwise is not acting in good faith.

Most other contracts need to be performed only to the satisfaction of a reasonable person unless they expressly state otherwise. When the subject matter of the contract is mechanical, courts are more likely to find the performing party has performed satisfactorily if a reasonable person would be satisfied with what was done. For example, Mason signs a contract with Jen to mount a new heat lamp pump on a concrete platform to her satisfaction. Such a contract normally need only be performed to the satisfaction of a reasonable person.

MATERIAL BREACH OF CONTRACT

A **breach of contract** is the nonperformance of a contractual duty. The breach is material when it is intentional and usually substantial.

breach of contract Failure of one of the parties to perform their obligations under the contract at the time performance is due.

Uniform Commercial Code and Performance and Convention on International Sale of Goods. The substantial performance doctrine does not apply to the sale of goods. The performance of a sale or lease contract under UCC Section 2-301 requires the seller or lessor to transfer and deliver what is known as *conforming goods* (perfect tender rule). The buyer or lessee must accept and pay for the conforming goods. The UCC states an exception to the perfect tender rule. If the goods or tender of delivery fail to conform to the contract in any respect, the buyer has the following options: (1) reject all the goods, (2) accept all that are tendered, or (3) accept any number of units

---

*Kohel v. Bergen Auto Enterprises, L.L.C. Superior Court of New Jersey, Appellate Division 2013 WL 439970 (2013).

the buyer chooses to and reject the rest. The buyer must generally give notice to the seller of any defect in the goods or tender of delivery and then allow the seller a reasonable time to cure the defect.

Seventy nations are signatories to the United Nations Convention on the International Sale of Goods (CISG), and goods are sold worldwide under the authority of the CISG. Under that convention, all four of the following conditions must be met to show an exception to the perfect tender rule: (1) the buyer has resorted to another remedy, such as avoidance or price reduction; (2) the seller failed to deliver or, in the case of nonconforming goods, the nonconformity was so serious that it constituted a fundamental breach; (3) the buyer gave timely notice to the seller that the goods were nonconforming; and (4) the buyer made a timely request that the seller provide substitute goods. As in the civil-law nations, the court may grant specific performance without regard to whether money damages are inadequate.

DISCHARGE BY MUTUAL AGREEMENT

After a contract is formed, the parties may agree that they should rescind (cancel) the contract because some unforeseen event took place that makes its fulfillment financially impracticable. For example, if A agrees to build B a house for $150,000, and then, while building the basement, runs into an unforeseen and uncorrectable erosion factor, both parties may want to cancel the agreement, with restitution to A for expenditures on the basement construction. The contract is then said to be discharged by mutual agreement.

Sometimes the parties wish to rescind an original agreement and substitute a new one for it. This type of discharge by mutual agreement is called *accord and satisfaction*. If the parties wish to substitute new parties for the original parties to the agreement, this is called *novation*. Note that novation does not change the contractual duties; it merely changes the parties that will perform those duties. Suppose a rock star (original party) is unable to perform at a concert because of illness, and a star of the same stature (substitute party) agrees to appear instead. If all parties, including the concert impresario, agree to the substitution, there is discharge by mutual agreement.

APPLYING THE LAW TO THE FACTS . . .

Bixley and Joe enter into an agreement whereby Joe agrees to buy Bixley's Cigar Shop. Bixley is planning to retire. As the date for the transfer draws closer, Bixley starts thinking that he really doesn't want to retire, so he calls Joe and asks whether Joe would be willing to reconsider the deal. Joe was actually starting to worry about whether he could run the operation, so he gladly agreed to just forget the deal. How has the contract been discharged? What if it were only Joe who got cold feet, and he found someone who was happy to buy the business according to the terms of the original sales contract? If Bixley agrees that Carley will replace Joe, then how would we describe the discharge of this contract?

DISCHARGE BY CONDITIONS PRECEDENT AND SUBSEQUENT

condition precedent A particular event that must take place to give rise to a duty of performance of a contract.

Condition Precedent. A **condition precedent** is a particular event that must take place in order to give rise to a duty of performance. If the event does not take place, the contract may be discharged. For example, when Smith enters into a contract with Jones to sell a piece of real estate, a clause in the agreement requires that title must be approved by Jones's attorney before closing and execution of the contract for sale. If Jones's attorney does not give this approval before the closing, then Jones is discharged from the contract. The following case illustrates discharge of a contract by a condition precedent.

CASE 10-2

Architectural Systems, Inc. v. Gilbane Building Co.
U.S. District Court, Maryland
760 F. Supp. 79 (1991)

Carley Capital Group (Carley) was the owner of a project in the city of Baltimore known as "Henderson's Wharf." The project was designed to convert warehouses into residential condominiums. On September 4, 1987, Carley hired Gilbane Building Company (Gilbane) to be the general contractor and construction manager for the project. Gilbane hired Architectural Systems, Inc. (ASI), as the subcontractor to perform drywall and acoustical tile work on the project. The subcontract included the following clause: "It is specifically understood and agreed that the payment to the trade contractor is dependent, as a condition precedent, upon the construction manager receiving contract payments from the owner."

Gilbane received periodic payments from Carley and paid ASI as work progressed. By late 1988, ASI had satisfactorily performed all of its obligations under the subcontract and submitted a final bill of $348,155 to Gilbane. Gilbane did not pay this bill because it had not received payment from Carley. On March 10, 1989, Carley filed for bankruptcy. ASI sued Gilbane, seeking payment.

Justice Young

ASI argues that it did not assume the credit risk simply by the inclusion of the statement "as a condition precedent" in the subcontract. It may not be sound business practice to accept such a business proposal but that is what occurred. The provision unambiguously declares that Gilbane is not obligated to pay ASI until it first received payment by the owner. The cause for the owner's nonpayment is not specifically addressed and could be due to a number of causes, including insolvency.

A provision that makes a receipt of payment by the general contractor a condition precedent to its obligation to pay the subcontractor transfers from the general contractor to the subcontractor the credit risk of non-payment by the owner for any reason (at least for any reason other than the general owner's own fault), including insolvency of the owner.*

Decision in favor of Gilbane.

Condition Subsequent. A **condition subsequent** is a particular future event that, when it follows the execution of a contract, terminates the contract. For example, a homeowner's insurance contract may discharge the insurer from responsibility for coverage in the event of an "act of war" (condition subsequent).

condition subsequent A particular event that, when it follows the execution of a contract, terminates the contract.

DISCHARGE BY IMPOSSIBILITY OF PERFORMANCE

In early common law, when disruptive unanticipated events (e.g., war) occurred after the parties entered into a contract, the contract was considered enforceable anyway. Thus, if a shipping line could not transport goods it had agreed to transport because of a wartime blockade, or if it could transport the goods but had to take a more expensive route to do so, it (or its insurance carrier) was required to pay damages or absorb the costs of the longer route. Courts today take the view that if an unforeseeable event makes a promisor's performance objectively impossible, the contract is discharged by **impossibility of performance**. *Objectively impossible* is defined as meaning that no person or company could legally or physically perform the contract.

impossibility of performance Situation in which the party cannot legally or physically perform the contract.

This defense to nonperformance is used most frequently in three circumstances. First is the death or illness of a promisor whose personal performance is required to fulfill the contract when no substitute is possible. For example, suppose that a world-renowned artist is commissioned to paint an individual's portrait. If the artist dies, the contract is discharged because there is no substitute for the artist. Second is a change of law making the promised performance illegal. For example, if A enters into a contract with B to sell B her home to be used for residential purposes, and subsequent to their agreement the property is zoned commercial, the contract will be discharged. The final circumstance is the destruction of the subject matter. If A enters into a contract with B to buy all

---

*Architectural Systems, Inc. v. Gilbane Building Co., U.S. District Court, Maryland 760 F. Supp. 79 (1991).

the hay in B's barn and the barn burns down with the hay in it before shipment takes place, the contract is discharged because performance is impossible.

DISCHARGE BY COMMERCIAL IMPRACTICABILITY

commercial impracticability Situation that makes performance of a contract unreasonably expensive, injurious, or costly to a party.

The courts have sought to enlarge the grounds for discharge of a contract by adding the concept of **commercial impracticability**, defined as a situation in which performance is impracticable because of unreasonable expense, injury, or loss to one party. In effect, a situation that was not foreseeable, or the nonoccurrence of which was assumed at the time the contract was executed, in fact occurs, making performance of the contract unreasonably expensive or injurious to a party. For example, a plastics manufacturer becomes extremely short of raw materials because of a war and an embargo on oil coming from the Middle East. The manufacturer's contracts with retailers for plastic goods will be discharged in most cases if the court finds that the manufacturer could not have anticipated the war and had no alternative source of materials costing about the same price.

CONTRACTS WITH THE GOVERNMENT AND THE SOVEREIGN ACTS DOCTRINE

When a party's performance is made illegal or impossible because of a new law, a performance contract is usually discharged, and damages are not awarded. But what happens when a party contracts with the government, and the government then promulgates a new law that makes its own performance impossible? If it no longer wishes to be bound by a contract, can the government discharge its obligations by changing the law to make performance illegal?

According to the *sovereign acts doctrine,* the government generally cannot be held liable for breach of contract due to legislative or executive acts. Because one Congress cannot bind a later Congress, the general rule is that subsequent acts of the government can discharge the government's preexisting contractual obligations.

This doctrine has limits, however. If Congress passes legislation deliberately targeting its existing contractual obligation, the defense otherwise provided by

LINKING LAW AND BUSINESS

Management and Production

Quality is defined as the extent to which a product functions as intended. The measure of excellence for a product is ranked primarily by the purchaser on the basis of certain characteristics and features, but the managers must oversee production processes to ensure that quality standards are being met. Managers are continuing to realize that an improvement in the quality of products leads to greater productivity for the organization. By emphasizing greater quality, a firm will probably spend less time and money on repairing defective products. Also, manufacturing quality products reduces the chance of production mistakes and inefficient use of materials.

One method of providing greater assurance that an organization is producing quality products is with statistical quality control. This is a process by which a certain percentage of products for inspection is determined to ensure that organizational standards for quality are met. Organizations that place a strong emphasis on quality, possibly by implementing an effective statistical quality control strategy, are less likely to be faced with having their goods rejected under the perfect tender rule or facing litigation for a breach of contract over defective products. Thus, a serious and consistent emphasis on quality reaps many benefits for an organization.

Source: Adapted from S. Certo, *Modern Management* (Upper Saddle River, NJ: Prentice Hall, 2000), pp. 445, 447. Reproduced by permission.

the sovereign acts doctrine is unavailable. The government it not prevented from changing the law, but it must pay damages for its legislatively chosen branch. On the other hand, if a new law of general application indirectly affects a government contract, making the government's performance impossible, the sovereign acts doctrine will protect the government in a subsequent suit for breach of contract.

The contract dispute in the following case arose out of the cancellation of a wedding reception due to a power failure. Is a power failure sufficient to invoke the doctrine of commercial impracticability?

CASE 10-3

Facto v. Pantagis
Superior Court of New Jersey, Appellate Division
915 A.2d 59 (2007)

Leo and Elizabeth Facto contracted with Snuffy Pantagis Enterprises, Inc., for the use of Pantagis Renaissance, a banquet hall in Scotch Plains, New Jersey, for a wedding reception in August 2002. The Factos paid the $10,578 price in advance. The contract excused Pantagis from performance "if it is prevented from doing so by an act of God (e.g., flood, power failure, etc.) or other unforeseen events or circumstances." Soon after the reception began, there was a power failure. The lights and the air-conditioning shut off. The band hired for the reception refused to play without electricity to power their instruments, and the lack of lighting prevented the photographer and videographer from taking pictures. The temperature was in the 90s, the humidity was high, and the guests quickly became uncomfortable. Three hours later, after a fight between a guest and a Pantagis employee, the emergency lights began to fade, and the police evacuated the hall. The Factos filed a suit in a New Jersey state court against Pantagis, alleging breach of contract, among other things. The Factos sought to recover their prepayment, plus amounts paid to the band, the photographer, and the videographer. The court concluded that Pantagis did not breach the contract and dismissed the complaint. The Factos appealed to a state intermediate appellate court.

Justice Skillman

Even if a contract does not expressly provide that a party will be relieved of the duty to perform if an unforeseen condition arises that makes performance impracticable, a court may relieve him of that duty if performance has unexpectedly become impracticable as a result of a supervening event. In deciding whether a party should be relieved of the duty to perform a contract, a court must determine whether the existence of a specific thing is necessary for the performance of a duty and its destruction or deterioration makes performance impractical. A power failure is the kind of unexpected occurrence that may relieve a party of the duty to perform if the availability of electricity is essential for satisfactory performance.

The Pantagis Renaissance contract provided: "Snuffy's will be excused from performance under this contract if it is prevented from doing so by an act of God (e.g., flood, power failure, etc.), or other unforeseen events or circumstances." Thus, the contract specifically identified a "power failure" as one of the circumstances that would excuse the Pantagis Renaissance's performance. We do not attribute any significance to the fact the clause refers to a power failure as an example of an "act of God." This term has been construed to refer not just to natural events such as storms but to comprehend all misfortunes and accidents arising from inevitable necessity which human prudence could not foresee or present. Furthermore, the clause in the Pantagis Renaissance contract excuses performance not only for "acts of God" but also "other unforeseen events or circumstances." Consequently, even if a power failure caused by circumstances other than a natural event were not considered to be an "act of God," it still would constitute *an unforeseen event or circumstance that would excuse performance.*

The fact that a power failure is not absolutely unforeseeable during the hot summer months does not preclude relief from the obligation to perform. Absolute unforeseeability of a condition is not a prerequisite to the defense of impracticability. The party seeking to be relieved of the duty to perform only needs to show that the destruction, or deterioration of a specific thing necessary for the performance of the contract makes performance impracticable. In this case, the Pantagis Renaissance sought to eliminate any possible doubt that the availability of electricity was a specific thing necessary for the wedding reception by specifically referring to a "power failure" as an example of an "act of God" that would excuse performance.

It is also clear that the Pantagis Renaissance was "prevented from" substantial performance of the contract. The power failure began less than forty five minutes after the start of the reception and continued until after it was scheduled to end. The lack of electricity prevented the band from playing, impeded the taking of pictures by the photographer and videographer and made it difficult for guests to see inside the banquet hall. Most significantly, the shutdown of the air-conditioning system made it unbearably hot shortly after the power failure began. It is also undisputed that the power failure was an area-wide event that was beyond the Pantagis Renaissance's control. These are precisely the kind of circumstances under which the parties agreed [in their contract] that the Pantagis Renaissance would be excused from performance.

Where one party to a contract is excused from performance as a result of an unforeseen event that makes performance impracticable, the other party is also generally excused from performance.

Therefore, the power failure that relieved the Pantagis Renaissance of the obligation to furnish plaintiffs with a wedding reception also relieved plaintiffs of the obligation to pay the contract price for the reception.

Nevertheless, since the Pantagis Renaissance partially performed the contract by starting the reception before the power failure, it is entitled to recover the value of the services it provided to plaintiffs.*

Reversed for Facto.

Remedies for a Breach of Contract

The fact that a court will enforce a contract does not mean that one party will automatically sue if the other breaches. Businesspeople need to consider several factors before they rush to file a lawsuit: (1) the likelihood of the suit's succeeding; (2) whether they wish to maintain a business relationship with the breaching party; (3) the possibility of arbitrating the dispute through a third party, thus avoiding litigation; and (4) the cost of arbitration or litigation as opposed to the revenues to be gained from enforcing the contract.

Remedies for a breach of contract are generally classified according to whether the plaintiff requests monetary damages ("legal" remedies) or nonmonetary damages (equitable remedies). Remedies for breach set out by the CISG include: (1) avoidance or cancellation of contract, (2) right to remedy or cure, (3) setting of additional time or extension, (4) price/reductions, (5) money damages, and (6) specific performance.

MONETARY DAMAGES ("LEGAL" REMEDIES)

monetary damages Dollar sums awarded for a breach of contract; "legal" remedies.

compensatory damages Monetary damages awarded for a breach of contract that results in higher costs or lost profits for the injured party.

Monetary damages (often referred to as *exemplary* damages) include compensatory, punitive, nominal, and liquidated damages.

Compensatory Damages. The purpose of **compensatory damages** is to place the injured (nonbreaching) party to a contract in the position in which that party would have been in had the terms of the contract been performed. For example, if a firm contracted to buy 8,000 widgets at $10 apiece to be delivered by August 15, the buyer has a right to go out and buy the widgets from another source if they are not delivered by the contract date. Suppose the widgets bought from the other source cost $10.50 apiece. In that case, the buyer can sue for the 50-cent difference in price per unit plus court costs. If the buyer cannot obtain widgets anywhere else, it can sue for lost profits.

The courts have set out three standards that the plaintiff-buyer must meet in order to recoup lost profits:

1. The plaintiff-buyer must show that it was reasonably foreseen by the defendant-seller that if it did not deliver the promised goods, the buyer would have no alternative source and, thus, would lose profits.

2. The plaintiff-buyer must show the amount of the damages with reasonable certainty; the buyer cannot just speculate about what this amount is.

3. The plaintiff-buyer must show that it did everything possible to mitigate the damage—that is, it looked for other possible sources of the goods.

In the case below the court is careful to award damages "naturally and proximately caused by the breach."

---

*Facto v. Pantagis, Superior Court of New Jersey, Appellate Division 915 A.2d 59 (2007).

CASE 10-4

Hallmark Cards, Inc. v. Murley
United States Court of Appeals, Eighth Circuit
703 F.3d 456 (2013)

Janet Murley served as Hallmark Cards, Inc.'s, vice president of marketing. In 2002, Hallmark eliminated her position as part of a corporate restructuring. Murley and Hallmark entered into a separation agreement under which she agreed not to work in the greeting card industry for eighteen months, disclose or use any confidential information, or retain any business records relating to Hallmark. In exchange, Hallmark offered Murley a $735,000 severance payment. After the non-compete agreement (see Chapter 9) expired, Murley accepted a consulting assignment with Recycled Paper Greetings (RPG) for $125,000. Murley disclosed confidential Hallmark information to RPG. On learning of the disclosure, Hallmark filed a suit in a federal district court against Murley, alleging breach of contract. A jury returned a verdict in Hallmark's favor and awarded $860,000 in damages, consisting of the $735,000 severance payment and the $125,000 Murley received from RPG. Murley appealed.

With respect to $735,000, Murley contends Hallmark was not entitled to a return of its full payment under the parties' separation agreement because Murley fulfilled several material terms of that agreement (e.g., the noncompete provision). Under the circumstances, we cannot characterize the jury's reimbursement of Hallmark's original payment under the separation agreement as grossly excessive or glaringly unwarranted by the evidence. Hallmark's terms under the separation agreement clearly indicated its priority in preserving confidentiality. At trial, Hallmark presented ample evidence that Murley not only retained but disclosed Hallmark's confidential materials to a competitor in violation of terms, and primary refund of its $735,000 is not against the weight of the evidence.

With respect to the remaining $125,000 of the jury award, Murley argues Hallmark can claim no entitlement to her compensation by RPG for consulting services unrelated to Hallmark. We agree. In an action for breach of contract, a plaintiff may recover the benefit of his or her bargain as well as damages naturally and proximately caused by the defendant at the time of the agreement. Moreover, the law cannot elevate the non-breaching party to a better position than she would have enjoyed had the contract been completed on both sides. By awarding Hallmark more than its $750,000 severance payment, the jury award placed Hallmark in a better position than it would find itself had Murley not breached the agreement. The jury's award of the $125,000 payment by RPG was, therefore, improper.

The U.S. Court of Appeals for the Eighth Circuit vacated the award of damages but otherwise affirmed the judgement in Hallmark's favor. The appellate court remanded the case to the lower court to reduce the award of damages to include only the amount of Hallmark's severance payment.

Punitive Damages. Damages in excess of compensatory damages that the court awards for the sole purpose of deterring the defendant and others from doing the same act again are known as **punitive damages**. They are infrequently awarded in contract cases.

Nominal Damages. Sometimes the court awards a very small sum (usually $1) in **nominal damages** to a party that is injured by a breach of contract but cannot show real damages. In these cases, the court generally enables the injured party to recover court costs, though not attorney's fees.

Liquidated Damages. **Liquidated damages** are usually set in a separate clause in the contract. The clause generally stipulates that the parties agree to pay so much a day for every day beyond a certain date that the contract is not completely performed. Liquidated damages clauses are frequently found in general contractors' agreements with individuals, corporations, and state or local agencies in situations in which it is essential that a building project be completed on time. Such clauses help the contracting parties avoid going to court and seeking a judicial determination of damages—with all the attendant delays and expenses.

The UCC (Sections 2-715[1] and 2A-504) permits parties to a sale or lease contract to establish in advance damages that will be paid upon a breach of contract. These liquidated damages will substitute for actual damages. The concept of liquidated damages is illustrated in Case 10-5.

punitive damages Monetary damages awarded in excess of compensatory damages for the sole purpose of deferring similar conduct in the future.

nominal damages Monetary damages of a very small amount (e.g., $1) awarded to a party that is injured by a breach of contract but cannot show real damages.

liquidated damages Monetary damages for nonperformance that are stipulated in a clause in the contract.

CASE 10-5

Arrowhead School District No. 75, Park County, Montana v. James A. Klyap, Jr.
Supreme Court of Montana
79 P.3d 250 (2003)

Arrowhead School District No. 75 is located in Park County, south of Livingston, Montana, and consists of one school, Arrowhead School for the 1997–98 school year, the School employed about 11 full-time teachers and several part-time teachers. During that school year, the School employed Klyap as a new teacher instructing mathematics, language arts, and physical education for the sixth, seventh, and eighth grades. In addition, Klyap, through his own initiative, helped start a sports program and coached flag football, basketball, and volleyball.

The School offered Klyap a contract for the 1998–99 school year in June 1998, which he accepted. This contract provided for a $20,500 salary and included a liquidated damages clause. The clause calculated liquidated damages as a percentage of annual salary determined by the date of breach; a breach of contract after July 20, 1998, required payment of 20 percent of salary as damages. Klyap also signed a notice indicating that he accepted responsibility for familiarizing himself with the information in the teacher's handbook that also included the liquidated damages clause.

On August 12, Klyap informed the School that he would not be returning for the 1998–99 school year even though classes were scheduled to start on August 26. The School then sought to enforce the liquidated damages clause in Klyap's teaching contract for the stipulated amount of $4,100.

After Klyap resigned, the School attempted to find another teacher to take Klyap's place. Although at the time Klyap was offered his contract the School had 80 potential applicants, only two viable applicants remained available. Right before classes started, the School was able to hire one of those applicants, a less experienced teacher, at a salary of $19,500.

The District Court determined that the clause was enforceable because the damages suffered by the School were impractical and extremely difficult to fix. Specifically, the court found that the School suffered damages because it had to spend additional time setting up an interview committee, conducting interviews, training the new, less experienced teacher, and reorganizing the sports program. The district court also found that such clauses are commonly used in Montana and that the School had routinely and equitably enforced the clause against other teachers. After concluding that the School took appropriate steps to mitigate its damages, the court awarded judgment in favor of the School in the amount of $4,100. Klyap appealed.

Justice Nelson

The fundamental tenet of modern contract law is freedom of contract; parties are free to agree mutually to terms governing their private conduct as long as those terms do not conflict with public laws. This tenet presumes that parties are in the best position to make decisions in their own interest. Normally, in the course of contract interpretation by a court, the court simply gives effect to the agreement between the parties in order to enforce the private law of the contract. When one party breaches the contract, judicial enforcement of the contract ensures [that] the nonbreaching party receives expectancy damages, compensation equal to what that party would receive if the contract were performed. By only awarding expectancy damages rather than additional damages intended to punish the breaching party for failure to perform the contract, court enforcement of private contracts supports the theory of efficient breach. In other words, if it is more efficient for a party to breach a contract and pay expectancy damages in order to enter a superior contract, courts will not interfere by requiring the breaching party to pay more than was due under their contract.

Liquidated damages are, in theory, an extension of these principles. Rather than wait until the occurrence of breach, the parties to a contract are free to agree in advance on a specific damage amount to be paid upon breach. This amount is intended to predetermine expectancy damages. Ideally, this predetermination is intended to make the agreement between the parties more efficient. Rather than requiring a postbreach inquiry into damages between the parties, the breaching party simply pays the nonbreaching party the stipulated amount. Further, in this way, liquidated damages clauses allow parties to estimate damages that are impractical or difficult to prove, as courts cannot enforce expectancy damages without sufficient proof.

After reviewing the facts of this case, we hold that while the 20% liquidated damages clause is definitely harsher than most, it is still within Klyap's reasonable expectations and is not unduly oppressive. First, as the School pointed out during testimony, at such a small school teachers are chosen in part depending on how their skills complement those of the other teachers. Therefore, finding someone who would provide services equivalent to Klyap at such a late date would be virtually impossible. [The anticipation of this] difficulty was borne out when only two applicants remained available and the School hired a teacher who was less experienced than Klyap. As a teacher, especially one with experience teaching at that very School, Klyap would have to be aware of the problem finding equivalent services would pose.

Second, besides the loss of equivalent services, the School lost time for preparation for other activities in order to attempt to find equivalent services. As the District Court noted, the School had to spend additional time setting up an interview committee and conducting interviews. Further, the new teacher missed all the staff development training earlier that year so individual training was required. And finally, because Klyap was essential to the sports program, the School had to spend additional time reorganizing the sports program as one sport had to be eliminated with Klyap's loss. These activities all took away from the other school and administrative duties that had been scheduled for that time.

Finally, although the School testified it had an intent to secure performance and avoid the above damages by reason of the clause, such an intent does not turn a liquidated damages clause into a penalty unless the amount is unreasonably large and therefore not within reasonable expectations.

Therefore, because as a teacher Klyap would know teachers are typically employed for an entire school year and would know how difficult it is to replace equivalent services at such a small rural school, it was within Klyap's reasonable expectations to agree to a contract with a 20% of salary liquidated damages provision for a departure so close to the start of the school year.

Accordingly, we hold the District Court correctly determined that the liquidated damages provision was enforceable.*

Affirmed for the School.

CRITICAL THINKING ABOUT THE LAW

This case highlights the court's flexibility. Ordinarily, when a person signs a contract with someone else, the courts will hold the parties to the terms of the agreement. In this case, there was no dispute, for instance, about the fact that a 20-percent-of-salary liquidated damages provision was in the contract between the plaintiff and the school district. Yet, as the reasoning in the case made clear, there are certain ground rules that must be satisfied before the court enforces the provisions of a contract.

1. What evidence would Klyap have had to possess for him to have prevailed in this case?

    Clue: Follow the court's reasoning step by step to see what the judge checked before agreeing to enforce the contract.

2. Why might the court have not enforced the contract provision in question had the liquidated damages provision been 40 percent, rather than 20 percent?

    Clue: Is it likely that Klyap would have agreed freely to such a provision?

EQUITABLE REMEDIES

When dollar damages are inadequate or impracticable as a remedy, the injured party may turn to nondollar or **equitable remedies**. Equitable remedies include rescission, reformation, specific performance, and injunction.

Rescission. **Rescission** is defined as the canceling of a contract. Plaintiffs who wish to be put back in the position they were in before entering into the contract often seek rescission. In cases of fraud, duress, mistake, or undue influence (discussed in Chapter 9), the courts will generally award rescission.

Reformation. The correction of terms in an agreement so that they reflect the true understanding of the parties is known as **reformation**. For example, if A enters into an agreement to sell B 10,000 widgets at $.50 per unit when the figure should have been $5.50 per unit, A may petition the court for reformation of the contract.

Specific Performance. A court order compelling a party to perform in such a way as to meet the terms of the contract calls for **specific performance**. Courts are reluctant to order specific performance unless (1) a unique object is the subject matter of the contract (e.g., an antique or artwork) or (2) real estate is involved. To obtain a specific performance order, the plaintiff must generally show that dollar damages (damages "at law") are inadequate to compensate for

equitable remedies Nonmonetary damages awarded for breach of contract when monetary damages would be inadequate or impracticable.

rescission Cancellation of a contract.

reformation Correction of terms in an agreement so that they reflect the true understanding of the parties.

specific performance A court order compelling a party to perform in such a way as to meet the terms of the contract.

---

*Arrowhead School District No. 75, Park County, Montana v. James A. Klyap, Jr., Supreme Court of Montana 79 P.3d 250 (2003).

injunction Temporary or permanent court order preventing a party to a contract from doing something.

the defendant's breach of contract. Even then, a court will often refuse to grant the order if it is incapable of supervising performance or is unwilling to do so. For these reasons, specific performance in contract cases is infrequently granted.

Injunctions. **Injunctions** are temporary (e.g., 30 days) or permanent orders of the court preventing a party to a contract from doing something. The plaintiff must show the court that dollar damages are inadequate and that irreparable harm will be done if the injunction is not granted. For instance, if an opera singer contracts with an opera company to sing exclusively for the company and then later decides to sing for other opera companies, the court may grant an injunction to prevent her from singing for the other companies. Injunctions, like specific performance, are seldom granted in contract law cases.

REMEDIES FOR BREACH OF A SALES CONTRACT (GOODS)

Remedies for the Seller. Under the UCC, remedies for the seller resulting from the buyer's breach include the following:

- The right to recover the purchase price if the seller is unable to sell or dispose of goods (Section 2-709[1])
- The right to recover damages if the buyer repudiates a contract or refuses to accept the goods (Section 2-708[1])

Remedies for the Buyer. If a seller breaches a sales contract by failing to deliver conforming goods or repudiating the contract prior to delivery, the buyer has a choice of remedies under the UCC:

- The right to obtain specific performance when the goods are unique or remedy at law is inadequate (Section 2-716[1])
- The right to recover damages after cancellation of the contract
- The right to reject the goods if the goods or tender fail to conform to the contract, or the right to keep some goods and reject the others
- The right to recover damages for accepted goods if the seller is notified of the breach within a reasonable time (Section 2-714[11])
- The right to revoke an acceptance of goods under certain circumstance (Section 2-608[1], [2])

In the case that follows, the issue is whether two years after the sale of goods is a reasonable period of time in which a buyer may discover a defect in goods and notify the seller.

CASE 10-6

Fitl v. Strek
Supreme Court of Nebraska
690 N.W.2d 625 (2005)

In September 1995, James Fitl attended a sports-card show in San Francisco, California, where he met Mark Strek, an exhibitor at the show, who was doing business as Star Cards of San Francisco. Later, on Strek's representation that a certain 1952 Mickey Mantle Topps baseball card was in near-mint condition, Fitl bought the card from Strek for $17,750. Strek delivered it to Fitl in Omaha, Nebraska, where Fitl placed it in a safe-deposit box. In May 1997, Fitl sent the card to Professional Sports Authenticators (PSA), a sports-card grading service. PSA told Fitl that the card was ungradable because it had been discolored and doctored. Fitl complained to Strek, who replied that Fitl should have initiated a return of the card within "a typical grade period for the unconditional return of a card . . . 7 days to 1 month" of its receipt. In August, Fitl sent the card to ASA Accugrade, Inc. (SAS), another grading service, for a second opinion on its value. ASA also concluded that the card had been refinished and trimmed. Fitl filed a suit in a Nebraska state court against Strek, seeking damages. The court awarded Fitl $17,750, plus his court costs. Strek appealed to the Nebraska Supreme Court.

Justice Wright

Strek claims that the lower court erred in determining that notification of the defective condition of the baseball card 2 years after the date of purchase was timely pursuant to [UCC] 2-607(3)(a).

The [trial] court found that Fitl had notified Strek within a reasonable time after discovery of the breach. Therefore, our review is whether the [trial] court's finding as to the reasonableness of the notice was clearly erroneous.

Section 2-607(3)(a) states, "Where a tender has been accepted the buyer must within a reasonable time after he discovers or should have discovered any breach notify the seller of breach or be barred from any remedy." [Under UCC1-204(2)] "[w]hat is a reasonable time for taking any action depends on the nature, purpose, and circumstances of such action."

The notice requirement set forth in Section 2-607(3)(a) serves three purposes. The most important one is to enable the seller to make efforts to cure the breach by making adjustments or replacements in order to minimize the buyer's damages and the seller's liability. A second policy is to provide the seller a reasonable opportunity to learn the facts so that he may adequately prepare for negotiation and defend himself in a suit. A third policy is the same as the policy behind statutes of limitation: to provide a seller with a terminal point in time for liability.

[A] party is justified in relying upon a representation made to the party as a positive statement of fact when an investigation would be required to ascertain its falsity. In order for Fitl to have determined that the baseball card had been altered, he would have been required to conduct an investigation. We find that he was not required to do so. Once Fitl learned that the baseball card had been altered, he gave notice to Strek.

One of the most important policies behind the notice requirement is to allow the seller to cure the breach by making adjustments or replacements to minimize the buyer's damages and the seller's liability. However, even if Fitl had learned immediately upon taking possession of the baseball card that it was not authentic and had notified Strek at that time, there is no evidence that Strek could have made any adjustment or taken any action that would have minimized his liability. In its altered condition, the baseball card was worthless.

Earlier notification would not have helped Strek prepare for negotiation or defend himself in a suit because the damage to Fitl could not be repaired. Thus, the policies behind the notice requirement, to allow the seller to correct a defect, to prepare for negotiation and litigation, and to protect against stale claims at a time beyond which an investigation can be completed, were not unfairly prejudiced by the lack of an earlier notice to Strek. Any problem Strek may have had with the party from whom he obtained the baseball card was a separate matter from his transaction with Fitl, and an investigation into the source of the altered card would not have minimized Fitl's damages.*

Affirmed in favor of Fitl.

APPLYING THE LAW TO THE FACTS...

Kallie operates a boutique and enters into a contract with JJ Scarves for the purchase of 25 print scarves that she plans to sell at her boutique. If the company fails to deliver the scarves at the agreed-upon date, and two weeks later sends her a letter saying they are sorry but they cannot fulfill the terms of the contract, what would her damages be? What if the opposite situation occurred and Kallie refused delivery of the scarves without even looking at them, saying that she was sorry, but her customers just didn't seem to purchase scarves, so she just couldn't accept the order? What would the company's damages be? In each situation, what would the nonbreaching party have to do to ensure they would be able to recover damages?

E-Contracts

In Chapter 9, and in this chapter, we have examined the traditional principles governing contracts for the sale of real and personal property, as well as the sale and lease of goods as defined by the UCC. E-contracts are now an everyday part of a cyberspace era. Although many of the traditional contract principles apply to contracts entered into online, they should include these minimum provisions:

- Remedies that are available to the buyer if any of the goods contracted for are defective.

---

*Fitl v. Strek, Supreme Court of Nebraska 690 N.W.2d 625 (2005).

- A statement of the seller's referral policy.
- A statement of how the goods are to be paid for.
- A forum selection clause, which indicates the location and/or forum where a dispute will be settled should one arise.
- A disclaimer-of-liability provision by the seller for certain uses of a good sold.
- The manner in which an offer can be accepted (e.g., by "click on"). Similarly, an acceptance may be made under traditional principles or by the provisions of Article 2 of the UCC (previously discussed in Chapter 9).
- Click-on terms that indicate agreement to the terms outlined in an offer.
- Browse-wrap terms that are enforceable (or binding) without the offeree's active consent, as with click-on terms.

E-SIGNATURES

In the year 2000, Congress enacted a law entitled the Electronic Signatures in Global and National Commerce Act (E-SIGN),[1] allowing consumers and businesses to sign contracts online and making e-signatures just as binding as ones in ink. Such contracts as those for bank loans and brokerage accounts may be entered into over the Internet 24 hours a day. This prevents delays that arise from the need for paper contracts to be written, mailed, signed, and then returned. With e-signatures, the cost of drawing up paperwork, mailing it, and storing agreements is eliminated in favor of electronic retention.

The law went into effect on October 1, 2000, but questions were raised as to whether people would be able to forge electronic signatures on everything from online purchases to credit card applications. The law does not specify what constitutes a digital signature. Possible requirements include (1) a password that must be entered into a form on a Web page, with the website having to confirm that it belongs to a certain person; or (2) the use of hardware such as thumbprint scanning devices that plug into personal computers and transmit the thumbprint over the Internet to a business, which would keep it on file for authentication purposes.

Laws governing e-signatures differ from state to state. Some states, such as California, prohibit documents from being signed with e-signatures, whereas others do not. In 1999, the National Conference of Commissioners on Uniform State Law and the American Law Institute set out the Uniform Electronic Transcription Act (UETA),[2] which sought to bring uniformity to this area of the law. The UETA, which has been adopted in whole or in part by more than 40 states, indicates that a signature may not be denied legal enforceability solely because of its electronic form.

THE UNIFORM COMPUTER INFORMATION TRANSACTION ACT

Prior to World War II, the common law of contracts was sufficient to handle most of the transactions in a mainly agricultural society. As the distribution and manufacturing of goods came to dominate commerce, the UCC was drafted by the National Conference of Commissioners on Uniform State Law, a consortium of lawyers, judges, businesspeople, and legal scholars. Individual states gradually adopted all or part of the UCC. For purposes of the law of contracts, Article 2 (Sales) and 2A (Leases) are highly significant, as shown in this and the preceding chapter.

As the world of computers and electronic commerce developed in the 1980s and 1990s, the common law of contracts and the UCC did not provide adequate guidelines, because the cyberspace economy is largely based on electronic

---

[1] 15 U.S.C. § 700 *et seq.*
[2] UETA §§ 102(8) AND (25).

contracts and the licensing of information. Questions developed as to how to enforce e-contracts, as well as what consumer protection should be provided. Therefore, in July 1999, the National Conference of Commissioners issued the Uniform Computer Information Transaction Act (UCITA).

Scope of the UCITA. It should be emphasized that the UCITA deals only with information that is electronically disseminated.[3] Under this act, a *computer information transaction* is an agreement to create, transfer, or license computer information. It does not cover licenses of information for traditional copyrighted materials such as books or magazines.

Many of the provisions of the UCITA are similar to Article 2 of the UCC. For example, a licensing agreement may be interpreted by the courts using the express terms of the agreement, as well as course of performance and usage. There are, however, several differences regarding licensing agreements under the UCITA:

- The party who sells the right to use a piece of software (licensor) can control the right of use by the buyer (licensee). In a mass market this is extremely important. An exclusive license means that for its duration, the licensor will not grant any other person rights to the same information. This is a matter of serious negotiation both nationally and internationally, with firms as well as with governments.

- If a contract requires a fee of more than $5,000, it is enforceable only if it is authenticated. To *authenticate* means to sign a contract or execute an electronic symbol, sound, or message attached to or linked with the record. Authentication may be attributed to a party's agent. Authentication may be proven if a party uses information or if he or she engages in operations that authenticate the record.

The Business Community: Criticisms of the UCITA. The UCITA would benefit software makers because it favors click-wrap agreements, and there would be uniformity to such agreements. This still leaves the question of enforcement of such agreements when they are between other business entities or governments.

Debate exists among those who want uniformity and critics who believe software makers would use uniformity to argue that they are not responsible for defect in the software they sell. For example, who would be held responsible for a software virus? Will consumers read the fine-print disclaimers? Some would argue that problems associated with e-contracts are similar to those encountered in dealing with contracts for real property, goods, and personal property.

Global Dimensions of Contract and Sales Law

As more nations in Europe, Latin America, and Asia have shifted toward market-oriented economies, international trade has increased, and, along with it, contracts implementing transactions between foreign entities (either governments or private companies) and U.S. companies have increased. International and regional treaties lowering or eliminating tariffs have hastened the trend to free trade. (See Chapter 8 for a detailed description of recent trade pacts.)

Given this accelerating tendency toward free trade, the United Nations Commission on International Trade Law drafted the Convention on Contracts for the International Sale of Goods[4] to provide uniformity in international transactions. The CISG covers all contracts for the sale of goods in countries that have ratified it. An estimated two-thirds of all international trade is conducted among

---

[3]UCITA § 102.
[4]15 U.S.C. App. (1997).

	Scope	Battle of the Forms	Warranties	Statute of Fraud
Common Law	1. Provision of services 2. Contracts for sale of land or securities 3. Loan agreements	Mirror image rule	Any express warranties made	1. Transfer of real estate 2. Contract cannot be performed within one year 3. Prenuptial agreement 4. Agreement to pay debt of another
UCITA	Computer information (including software, computer games, and online access)	Contract even if acceptance materially alters the offer	1. Warranty of noninterference and noninfringement 2. Implied warranties of merchantability of computer program, informational content, fitness for licensee's particular purpose and fitness for system integration	Contract for $5,000 or more
CISG	Sale of goods by merchants in merchants in different countries unless parties opt out	In practice, mirror image rule	1. Implied warranties of merchantability and fitness for a particular use 2. Any express warranties made	None
UCC	Sale of goods	If both are merchants additional terms are incorporated; if not, minor rules apply	Implies warranties for a particular purpose	Sale of 500 or more

the 70 nations that were signatories to the CISG as of April 2009.[5] Parties to a contract can choose to adhere to all or part of the CISG, or they may select other laws to govern their transactions.

On January 1, 1988, the CISG was approved as a treaty and incorporated into U.S. federal law. As a treaty, it overrides conflicting state laws dealing with contracts. Each of the 50 states is now examining conflicts between the Uniform Commercial Code (as adopted in the state) and the CISG, which supersedes it.

Some of the differences between the CISG and the UCC are highly significant. For example, under the CISG, a contract is formed when the seller (offeror) receives the acceptance from the offeree, whereas under the UCC, a contract is formed when the acceptance is mailed or otherwise transmitted. To take another example, under the CISG, a sales contract of any amount is enforceable if it is oral, whereas the UCC requires a written contract for a sale of goods of $500 or more.

Present and future business managers must become knowledgeable about these and other differences between the UCC and the CISG if they are to avoid costly and time-consuming litigation as international transactions in goods increase. You might want to review Chapter 8 at this point to refresh your memory about the methods and details of international transactions.

SUMMARY

Contracts are discharged by performance, mutual agreement, conditions precedent and subsequent, and sometimes through impossibility of performance. Remedies for breach of contract include dollar remedies such as lost profits, punitive, nominal, and liquidated damages. Often, when dollar damages are insufficient, the court will rely on equitable remedies (nondollar damages) such as rescission, restitution, specific performance, and reformation.

---

[5]See N. Karambelas, "Convention on International Sale of Goods: An International Law as Domestic Law," *Washington Lawyer*, April 2009, p. 27; R. Schauffer et al., *International Business Law* (9th edition, 2015) pp 90–92.

E-contracts require some minimum provisions that may differ slightly from common-law principles. With the help of statutory and case law, e-contracts are now becoming part of everyday business transactions. Contract laws became more uniform with the ratification by many nations of the Convention on Contracts for the International Sale of Goods. This has wide implications for the conduct of international transactions.

REVIEW QUESTIONS

10-1 List the criteria used by the courts in determining lost profits.

10-2 Why do courts rely on equitable remedies? Explain.

10-3 What is meant by "punitive damages?" Explain.

10-4 Explain what is meant by the "reformation" of a contract.

10-5 Why should a party who has not breached a contract be required to mitigate the damages of the breaching party?

10-6 What are some provisions that should be included in e-contracts?

REVIEW PROBLEMS

10-7 In 2003, Karen Pearson and Steve and Tara Carlson agreed to buy a 2004 Dynasty recreation vehicle (RV) from DeMartini's RV Sales in Grass Valley, California. On September 29, Pearson, the Carlsons, and DeMartini's signed a contract providing that "seller agrees to deliver the vehicle to you on the date this contract is signed." The buyers made a payment of $145,000 on the total price of $356,416 the next day, when they also signed a form acknowledging that the RV had been inspected and accepted. They agreed to return later to have the RV transported out of the state for delivery (to avoid paying state sales tax on the purchase). On October 7, Steve Carlson returned to DeMartini's to ride with the seller's driver to Nevada to consummate the out-of-state delivery. When the RV developed problems, Pearson and Carlson filed a suit in a federal district court against the RV's manufacturer, Monaco Coach Corp., alleging, in part, breach of warranty under state law. The applicable statute is expressly limited to goods sold in Nevada. How does the Uniform Commercial Code (UCC) define a sale? What does the UCC provide with respect to the passage of title? How do these provisions apply here? Discuss.

10-8 Internet Archive (IA) is devoted to preserving a record of resources on the Internet for future generations. IA uses the "Wayback Machine" to automatically browse websites and reproduce their contents in an archive. IA does not ask the owners' permission before copying the material but will remove it on request. Suzanne Shell, a resident of Colorado, owns a website that is dedicated to providing information to individuals accused of child abuse or neglect. The site warns, "IF YOU COPY OR DISTRIBUTE ANYTHING ON THIS SITE YOU ARE ENTERING INTO A CONTRACT." The terms, which can be accessed only by clicking on a link, include, among other charges, a fee of $5,000 for each page copied "in advance of printing." Neither the warning nor the terms require a user to indicate assent. When Shell discovered that the Wayback Machine had copied the contents of her site—approximately eighty-seven times between May 1999 and October 2004—she asked IA to remove the copies from its archive and pay her $100,000. IA removed the copies and filed a suit in a federal district court against Shell, who responded, in part, with a counterclaim for breach of contract. IA filed a motion to dismiss this claim. Did IA contract with Shell? Explain. *Internet Archive v. Shell*, 505 F. Supp. 2d 755 (D. Colo. 2007).

10-9 McDonald has contracted to purchase 500 pairs of shoes from Vetter. Vetter manufactures the shows and tenders delivery to McDonald. McDonald accepts the shipment. Later, on inspection, McDonald discovers that 10 pairs of shoes are poorly made and will have to be sold to customers as seconds. If McDonald decides to keep all 500 pairs of shoes, what remedies are available to her? Explain.

10-10 Kirk has contracted to deliver to Doolittle 1,000 cases of Wonder brand beans on or before October 1. Doolittle is to specify means of transportation 20 days prior to the date of shipment. Payment for the beans is to be made by Doolittle on tender of delivery. On September 10, Kirk prepares the 1,000 cases for shipment. Kirk asks Doolittle how he would like the goods to be shipped, but Doolittle does not respond. On September 21, Kirk, in writing, demands assurance that Doolittle will be able to pay on tender of the beans, and Kirk asks that the money be placed in escrow prior to October 1 in a bank in Doolittle's city named by Kirk. Doolittle does not respond to any of Kirk's requests, but on October 5 he wants to file suit against Kirk for breach of contract for failure to deliver the beans as agreed. Explain Kirk's liability for failure to tender delivery on October 1.

10-11 Julius W. Erving (Dr. J) entered into a four-year contract to play exclusively for the Virginia Squires of the American Basketball Association. After one year, he left the Squires to play for the Atlanta Hawks of the National Basketball Association. The contract signed with the Squires provided that the team could have his contract set aside for fraud. The Squires counterclaimed and asked for arbitration. Who won? Explain.

10-12 TWA had a sale/leaseback agreement with Connecticut National Bank. Because of the Gulf War, air travel was decreased and TWA was having trouble making its payments. Discuss the extent to which TWA could use commercial impracticability or impossibility as a defense for nonpayment.

CASE PROBLEMS

10-13 After submitting the high bid at a foreclosure sale, David Simard entered into a contract to purchase real property in Maryland for $192,000. Simard defaulted (failed to pay) on the contract. A state court ordered the property to be resold at Simard's expense, as required by state law. The property was to be resold for $163,000, but the second purchaser also defaulted on his contract. The court then ordered a second resale, resulting in a final price of $130,000. Assuming that Simard is liable for consequential damages, what is the extent of the liability? Is he liable for losses and expenses related to the first resale? If so, is he also liable for losses and expenses related to the second resale? Why or why not? *Burson v. Simard*, 35 A.3d 1154 (Md. 2012).

10-14 Cuesport Properties, LLC, sold a condominium in Anne Arundel County, Maryland, to Critical Developments, LLC. As a part of the sale, Cuesport agreed to build a wall between Critical Development's unit and an adjacent unit within thirty days of closing. If Cuesport failed to do so, it was to pay $126 per day until completion. This was an estimate of the amount of rent that Critical Developments would lose until the wall was finished and the unit could be rented. Actual damages were otherwise difficult to estimate at the time of the contract. The wall was built on time, but without a county building code. Critical Developments did not modify the wall to comply with the code until 260 days after the date of the contract deadline for completion of the wall. Does Cuesport have to pay Critical Developments $126 for each of the 260 days? Explain. *Cuesport Properties, LLC v. Critical Developments, LLC*, 209 Md. App. 607, 61 A.3d 91 (2013).

10-15 The Northeast Independent School District in Bexar County, Texas, hired STR Constructors, Ltd., to renovate a middle school. STR subcontracted the tile work in the school's kitchen to Newman Tile, Inc. (NTI). The project had already fallen behind schedule. As a result, STR allowed other workers to walk over and damage the newly installed tile before it had cured, forcing NTI to constantly redo its work. Despite NTI's requests for payment, STR remitted only half the amount due under their contract. When the school district refused to accept the kitchen, including the tile work, STR told NTI to quickly make the repairs. A week later, STR terminated their contract. Did STR breach the contract with NTI? Explain. *STR Constructors, Ltd. v. Newman Tile, Inc.*, S.W.3d, 2013 WI, 632969 (Tex. App.–El Paso 2013).

10-16 United Concrete purchased concrete from Red-D-Mix and later claimed that the concrete was defective because of issues with bleed water. Bleed water is excess water that seeps out of concrete after it has been poured; it rests on the surface and can weaken the concrete, leading to premature degeneration. United Concrete then sued Red-D-Mix for fraudulent representation under Wisconsin law. United Concrete claimed that a Red-D-Mix salesperson had assured United Concrete that the bleed water issue had been resolved. Red-D-Mix asserted in its defense that its salesperson's statement was mere puffery and therefore not actionable.

Were the statements by the Red-D-Mix salesperson puffery? Why or why not? Does it matter whether the Red-D-Mix salesperson had any reasonable basis for making the statements? How could United Concrete have avoided this lawsuit? *United Concrete & Construction, Inc. v. Red-D-Mix Concrete*, 836 N.W.2d 807 (Wis. 2013).

10-17 William West, an engineer, worked for Bechtel Corporation, an organization of about 160 engineering and construction companies, which is headquartered in San Francisco, California, and operates worldwide. Except for a two-month period in 1985, Bechtel employed West on long-term assignments or short-term projects for 30 years. In October 1997, West was offered a position on a project with Saudi Arabian Bechtel Co. (SABCO), which West understood would be for two years. In November, however, West was terminated for what he believed was his "age and lack of display of energy." After his return to California, West received numerous offers from Bechtel for work that suited his abilities and met his salary expectations, but he did not accept any of them and did not look for other work. Three months later, he filed a suit in a California state court against Bechtel, alleging in part breach of contract and seeking the salary he would have earned during the two years with SABCO. Bechtel responded in part that, even if there had been a breach, West had failed to mitigate his damages. Is Bechtel correct? *West v. Bechtel Corp.*, 117 Cal. Rptr. 2d 647 (2002).

THINKING CRITICALLY ABOUT RELEVANT LEGAL ISSUES

Judges should start offering punitive awards in breach-of-contract cases. Punitive damages are awarded for the sole purpose of deterring the defendant and others from doing the same act again. Our courtrooms are flooded with enough cases. Businesses know that they can get away with breaching their contracts. If we start imposing punitive damages, however, these businesses will alter their behavior.

This change would specifically affect those businesses that make it a practice to breach a contract whenever necessary. People are losing faith in the security of a contract. A survey of 100 first-year business students suggested that contracts are little respected and rarely followed in business.

1. What is the conclusion offered in this argument, and the reasons supporting the conclusion? Return to Chapter 1 of this text.

2. The author cites a survey of 100 first-year business students who suggested that contracts are "little respected and rarely followed in business." Do you see any problems with this evidence?

3. What additional information do you need to evaluate this argument?

4. What arguments could you make that are the opposite of those made by the author of this essay? Return to Chapter 1.

ASSIGNMENT ON THE INTERNET

This chapter briefly introduces the UCITA as a means of regulating the increasing volume of e-commerce and electronic contracts. Not everyone, however, is in favor of the UCITA. Using the Internet, search for articles and opinions that present multiple perspectives about how the UCITA influences business operations.

You can begin by visiting the Americans for Fair Electronic Commerce Transactions (AFECT) site (www.ucita.com). What arguments does it present against UCITA? Then visit http://www.repository.law.indiana.edu/cgi/viewcontent.cgi?article=1251&context=fclj to read an article containing arguments in support of UCITA.

Now draft a brief response to what you have read. Would you support or oppose wider enactment of UCITA? Why? Would you recommend that any changes be made if your state was considering adopting parts of UCITA? What would those changes be?

ON THE INTERNET

www.law.cornell.edu/ucc/2/overview.html This site contains the part of the Uniform Commercial Code relevant to contracts and sales.
www.lectlaw.com/files/bul08.htm This site provides more information on nonperformance and breach of contract.
www.cisg.law.pace.edu/cisg/text/treaty.html Here you can find the United Nations Convention on Contracts for the International Sale of Goods.
http://www.uniformlaws.org/ Go to this site and do a search for Uniform Electronic Transactions Act to find out more about this uniform law.

FOR FUTURE READING

DiMatteo, Larry. "A Theory of Efficient Penalty: Eliminating the Law of Liquidated Damages," *American Business Law Journal* 38 (2001): 633.

Eagan, William. "The Westinghouse Uranium Contracts." *American Business Law Journal* 18 (1980): 281.

Karambelas, Nicholas. "United Nations Convention on International Sale of Goods." *Washington Lawyer* (April 2009): 261–35.

Pryor, Jeremy. "Lost Profit or Lost Chance: Reconsidering the Measure of Recovery for Lost Profits in Breach of Contract Actions." *University Law Review* 19 (2006–2007): 561–80.

CHAPTER ELEVEN

The Law of Torts

- THE GOALS OF TORT LAW
- DAMAGES AVAILABLE IN TORT CASES
- CLASSIFICATIONS OF TORTS
- INTENTIONAL TORTS
- NEGLIGENT TORTS
- STRICT LIABILITY TORTS
- GLOBAL DIMENSIONS OF TORT LAW

We said in Chapter 2 that the law is divided into criminal law and civil law. The division, however, is not airtight. Although a given set of actions may constitute a crime or a wrong against the state and thus may give rise to a criminal prosecution, the same set of actions may also constitute a *tort,* a civil wrong that gives the injured party the right to bring a lawsuit against the wrongdoer to recover compensation for the injuries. We define a **tort** as an injury to another's person or property.

tort An injury to another's person or property; a civil wrong.

This chapter discusses torts as if they were the same across the country—and, in general, they are—but keep in mind that tort law is state law and so may vary somewhat from state to state. The total amount of tort litigation has been declining since 1996.[1] Even so, tort law is and will continue to be an important area of law and an essential subject for the student of the legal environment of business, because managers who do not have a basic understanding of tort law are placing themselves and the company's stakeholders at risk.

The Goals of Tort Law

Tort cases are commonly referred to as *personal injury cases,* although a tort case may involve harm solely to property. The primary goal of tort law is to compensate innocent persons who are injured or whose property is injured as a result of another's conduct, but tort law also fulfills other important societal goals. It discourages private retaliation by injured persons and their friends. It promotes citizens' sense of a just society by forcing responsible parties to pay for the injuries they have caused. Finally, it deters future wrongs because potential wrongdoers are aware that they will have to pay for the consequences of their harmful acts.

For example, if Sam takes Judy's car without her permission and wrecks it, he has committed a tort. If there were no tort law, she would get no compensation from Sam and would have to use her own money to have the car repaired or to buy a new one. She would feel that she lived in an unjust world. She might even be tempted to seek revenge against Sam by breaking his car window

---

[1]Ted Rohrlich, "America's Litigation Explosion Has Fizzled," *Los Angeles Times,* Feb. 1, 2001.

(or something else even more personal). Others, seeing what Sam has gotten away with, would be less likely to be careful with other people's property in the future. Because we have tort law, however, Judy can sue Sam and receive compensation from him for the damage he did to her car. She will then feel that she has received justice and will not be inclined to take any private retaliatory actions against Sam. Others, knowing that Sam had to pay for the harm he caused, may be deterred from committing torts themselves.

CRITICAL THINKING ABOUT THE LAW

Tort law allows compensation for individuals whose person or property has been injured. Applying some critical thinking questions to tort law can help you better understand this chapter.

1. As discussed in Chapter 1, courts have preferences for certain ethical norms. Using critical thinking skills will help us understand how those norms have shaped legal reasoning about tort law. It is quite possible for two judges hearing a tort case to disagree on a verdict. One reason for the different verdicts is their disagreement over which ethical norms are most important. What conflict of ethical norms is inherent in tort law?

    Clue: Think of the definitions of the primary ethical norms in Chapter 1. If a judge strongly values freedom, what ethical norm might conflict with the judge's loyalty to freedom? Why?

2. Loyalty to certain ethical norms will influence your attitude toward compensating injured individuals. Remember that the majority of civil jury trials involve torts. If you value efficiency, how might the large number of tort cases in the court system affect your thinking about tort law?

    Clue: Why would this large number of tort cases not trouble a person who values justice over efficiency?

3. One of the critical thinking skills you have learned to use is the identification of ambiguous words. Words with multiple possible meanings can result in different interpretations of a law. In tort law, this issue is especially important. Look at the definition of a tort. How does the definition of the word injury influence thinking about tort cases?

    Clue: Again, consider the number of tort cases in the courts. How would the number of court cases change if we loosely defined the word injury? What ethical norms would influence our definition of injury?

Damages Available in Tort Cases

The victim of a tort may sue the wrongdoer, known as the *tortfeasor,* and has the potential to recover from among three types of damages: compensatory, nominal, and punitive (Table 11-1). All three types of damages were defined and discussed in Chapter 10 in relation to breaches of contract. Here we describe their specific application to tort cases.

COMPENSATORY DAMAGES

The most common type of damages sought in tort cases are compensatory damages. *Compensatory damages* are designed to make the victim whole again, that

TABLE 11-1 TYPES OF TORT DAMAGES

Type	Purpose	Amount
Compensatory	To put the plaintiff in the position he or she would have been in had the tort never occurred	Sufficient to cover all losses caused by the tort, including compensation for pain and suffering
Nominal	To recognize that the plaintiff has been wronged	A nominal amount, usually $1–$5
Punitive	To punish the defendant	Determined by the severity of the wrongful conduct and the wealth of the defendant

is, to put the victim in the position he or she would have been in had the tort never taken place. They include compensation for all of the injuries that the tortfeasor caused to the victim and his or her property. Typical items covered by this class of damages are medical bills, lost wages, property repair bills, and compensation for pain and suffering. Note that attorneys' fees are not considered an item of compensatory damages, even though it would be virtually impossible for a victim to bring suit without the services of an attorney. Because the plaintiffs in personal injury cases usually must pay their attorneys by giving them a portion of the compensatory damages they are awarded, some people argue that compensatory damages do not fully compensate tort victims.

NOMINAL DAMAGES

Sometimes the plaintiff is unable to prove damages that would necessitate compensation. In such a case, the court may award the victim *nominal damages* (damages in name only). The sum of such awards is minuscule, usually $1, but recovery of nominal damages may be important because it allows the plaintiff to seek punitive damages. Punitive damages cannot be awarded alone; they must accompany an award of compensatory or nominal damages.

An illustration of an instance in which nominal damages were of utmost importance is the case of the death-row inmate in South Carolina who received nominal damages of only 10 cents from his case against prison guards.[2] This sum allowed the plaintiff—in this case, the inmate—to seek punitive damages. This award, however, was important for other reasons as well. For example, it may change the prison guards' behavior in the future. Also, because the inmate won, even though the damages were only nominal, perhaps the public will see the case as involving an important legal issue.

PUNITIVE DAMAGES

When the act of the tortfeasor is flagrant, unconscionable, or egregious, the court may award the victim *punitive damages*. These damages are designed not only to punish the tortfeasor for willfully engaging in extremely harmful conduct, but also to deter others from engaging in similar conduct. Punitive damages are considered by some legal scholars to be especially useful in deterring manufacturers from making unsafe products. If there were no possibility of incurring punitive damages, manufacturers might calculate how much money they would have to spend fighting and settling lawsuits resulting from the sale of a defective product and then calculate the cost of making a safer product. If it turned out to be cheaper to produce the defective product and compensate injured victims than to make a safer product, rational manufacturers would be likely to produce the defective product. The risk of incurring punitive damages, however, is often sufficient to convince manufacturers to produce the safe product.

Some people disagree with this reasoning. They claim that the costs of compensatory damages alone are a sufficient incentive to produce only safe products. They further argue that the almost unrestricted ability to award punitive damages gives juries too much power.

In recent years, there have been many attempts by insurance companies and tort reform groups to limit the amount of punitive damages that can be assessed. These advocates of tort reform have tried repeatedly to get the courts to strike down punitive damages as unconstitutional on the ground that such damages violate defendants' due process rights. This argument was unsuccessful until the 1994 case of *Honda Motor Co. v. Oberg*.[3] This case, in which the U.S. Supreme

---

[2]"Lawyer Gets $30,000 Fee Award for 10-Cent Win," *LWUSA* 95: 1037 (1995).
[3]114 S. Ct. 2331 (1994).

Court struck down a punitive damages award as being a violation of due process, was unusual in two respects. First, the punitive damages were more than 500 times the amount of the compensatory damages. Second, the state law had no provision for judicial review of the amount of the punitive damages award, whereas every other state allows such a review. It was Oregon's denial of judicial review of the amount of punitive damages that the high court said violated the Due Process Clause. Because of its unusual facts, *Honda v. Oberg* was not very instructive as to when punitive damages are so excessive as to violate due process. But in the 1995 case of *BMW v. Gore*,[4] the Supreme Court finally set forth a workable test. The so-called BMW guideposts require courts to look at three factors to determine whether a punitive damages award is so excessive as to violate due process: (1) the reprehensibility of the conduct, (2) the ratio of punitive damages to compensatory damages, and (3) comparable civil and criminal penalties for the same crime.

Then, in 2003, the U.S. Supreme Court reaffirmed and attempted to clarify the BMW rule when holding in *State Farm v. Campbell*[5] that a punitive damages award was grossly excessive and violated the Due Process Clause. In *Campbell*, the plaintiffs sued their auto insurance carriers alleging bad-faith failure to settle a claim against them within policy limits. The jury found in favor of the plaintiffs, awarding $2.6 million in compensatory damages. The jury further awarded $145 million in punitive damages, primarily because of evidence that State Farm's conduct was part of a long-standing pattern and practice of dishonest and fraudulent acts against policyholders.

The trial court reduced the punitive award to $25 million, but the Utah Supreme Court, in an extensive opinion evaluating the evidence, reinstated the jury's verdict. The U.S. Supreme Court granted review of whether the punitive damages award violated the standards of *BMW v. Gore* by a constitutionally excessive ratio to compensatory damages and by punishing the defendant in part for out-of-state conduct not directly affecting the plaintiffs.

The Supreme Court reiterated from *Gore v. BMW* that the most important indicium of the reasonableness of a punitive damages award is the degree of reprehensibility of the defendant's conduct, and noted that the reprehensibility of a defendant is determined by considering whether "the harm caused was physical as opposed to economic; the tortious conduct evinced an indifference to or a reckless disregard of the health or safety of others; the target of the conduct had financial vulnerability; the conduct involved repeated actions or was an isolated incident; and the harm was the result of intentional malice, trickery, or deceit, or mere accident."[6]

The Court also noted that punitive damages should be awarded only if the defendant's culpability, after having paid compensatory damages, is so reprehensible as to warrant the imposition of further sanctions to achieve punishment or deterrence.

In applying the first prong of the *BMW* case, the high court found that although the defendant's direct conduct against the plaintiff was reprehensible enough to warrant some modest punitive damages, the trial court erred in allowing the jury to consider the defendant's out-of-court conduct toward other policyholders, and without this improperly considered evidence of out-of-state conduct, the defendant's conduct in failing to settle the claim within its policy limits was not sufficiently reprehensible to warrant such a large punitive damages award.

Although the Court once again refused to draw a line as to what ratio of compensatory to punitive damages was acceptable, it did give some additional

---
[4] 116 S. Ct. 1589 (1995).
[5] 123 S. Ct. 1513 (2003).
[6] *Id.* at 1516.

guidance by saying that "[o]ur jurisprudence and the principles it has now established demonstrate . . . that, in practice, few awards exceeding a single-digit ratio between punitive and compensatory damages, to a significant degree, will satisfy due process."[7]*

Since *Gore*, some states have modified the standards they apply in their states to determine whether a punitive damages award violates due process. For example, notice how Florida has adopted a standard that is somewhat more favorable to defendants than the original Gore standards are.

CASE 11-1

Young v. Becker & Poliakoff
Court of Appeals of Florida, Fourth District
88 So.3d 1002 (2012)

Jacquelyn Young hired the law firm of Becker & Poliakoff to represent her in her federal employment discrimination lawsuit against her employer. The firm associate that filed the action made a mistake by attaching the wrong U.S. Equal Employment Opportunity Commission (EEOC) right-to-sue letter. The court dismissed the claims. The law firm did not try to re-file using the correct attachment, or try to dismiss the motion. Thirteen months later, the law firm informed Young that the claims had been dismissed, and that the firm was withdrawing from representing her further with the case.

Young argued that the firm had a conflict of interest when it continued to represent other employees of Young's employer, and when their settlement included a rule barring the firm from suing the employer in the future. Young believed that the firm had waited to pursue her case until its other case was settled. The jury determined that Becker & Poliakoff knew that the case had been dismissed, but withheld that information from Young so they could settle the other case and secure the $2.9 million fee and cost reimbursement in that case. The jury returned a verdict for Young of $394,000 in compensatory damages as a result of Becker & Poliakoff's breach of fiduciary duty. The total compensatory damages consisted of $144,000 in past lost wages and $250,000 in damages for "pain and suffering, mental anguish, or loss of dignity." However, the court reduced the punitive damages to $2 million, claiming that no evidence was presented to show that the firm could afford the award without facing bankruptcy. Both parties appealed.

Justice Taylor

We grant appellee's motion for rehearing, withdraw our previous opinion, and substitute the following in its place. No further motions for rehearing or clarification will be entertained. . . .

"Under Florida law, a trial court's determination of whether a damage award is excessive, requiring a remittitur or a new trial, is reviewed by an appellate court under an abuse of discretion standard." In ruling on a motion for remittitur, the trial court must evaluate the verdict in light of the evidence presented at trial. . . .

In evaluating a punitive damages award, the trial court must also determine whether the award comports with constitutional due process requirements. "The three criteria a punitive damages award must satisfy under Florida law to pass constitutional muster are: (1) 'the manifest weight of the evidence does not render the amount of punitive damages assessed out of all reasonable proportion to the malice, outrage, or wantonness of the tortious conduct;' (2) the award 'bears some relationship to the defendant's ability to pay and does not result in economic castigation or bankruptcy to the defendant;' and (3) a reasonable relationship exists between the compensatory and punitive amounts awarded."

In this case, the trial court found that the $4.5 million punitive damages award overcame the presumption of excessiveness under Section 768.73, Florida Statutes. The court, however, concluded that the award did not satisfy the criteria for constitutionality. Although the court found that the first and third criteria mentioned above were met because the award was proportional to reprehensible conduct of the defendant and bore a reasonable relationship between the compensatory and punitive amount awarded, it found that the award fell short on the second criteria; it was excessive because it was "too much for Defendant to bear without economic castigation or bankruptcy." As explained in the trial court's thorough and detailed order, this finding is supported by the record.

After noting that the jury apparently discredited evidence presented by the defense regarding Becker & Poliakoff's financial picture, the trial court turned to testimony of Young's financial expert, Dr. Pettingil, in determining that the $4.5 million punitive damages award would bankrupt Becker & Poliakoff. In short, the trial court found that Dr. Pettingil's opinion placed the law firm's net worth at $9.7 million to $11.1 million, and that "a $4.5 million punitive damages award constitutes forty percent of the net worth of the company." This amount,

---
[7]*Id.* at 1524.
*State Farm Mut. Automobile Ins. Co. v. Campbell, 538 U.S. 408 (2003), 123 S. Ct. 1513 (2003).

the court reasoned, was "too large" and exceeded "the highest amount that can be sustained based upon the evidence." Explaining how it arrived at the $2 million remittitur amount, the court stated the following:

The court finds that the maximum award that will not be excessive is $2 million which constitutes about 18%–20% of the firm's net worth. Dr. Pettingil's testimony establishes sufficient assets to bear this amount. His testimony established annual earnings of $675,000.00 per year increasing by 3% in 2010 and every year thereafter, $3 million per year in extraordinary compensation and a total of $1.5 million in retained earnings. Over 2009 and 2010 this would amount to assets exposable to collection of a punitive damage award of $6 million to $9 million, depending upon the extent of payment to officers' extraordinary compensation.

$2 million is as close to disgorging what the jury determined to be ill-gotten gains as Defendant's financial wealth will tolerate.

Contrary to Young's contention, the trial court did not improperly substitute its judgment for that of the jury, but instead properly exercised its discretion in reviewing the award upon the financial information in evidence. While a punitive damages award should be painful enough to provide some retribution and deterrence, it should not financially destroy a defendant. We, therefore, do not disturb the amount of the punitive damages ordered by remittitur; reasonable people could differ over this matter, and, therefore, no clear abuse of discretion is shown.

We further find any error in the trial court's ruling that prohibited Dr. Pettingil from testifying that an award of $10 million would not bankrupt Becker & Poliakoff to be harmless. Here, the witness was allowed to state his opinion concerning valuation of the firm's net worth and its financial ability to pay an award. And, even without hearing the witness's opinion as to whether an award of $10 million would bankrupt the firm, the jury still awarded $4.5 million in punitive damages—an amount the trial court found to be excessive in relation to the firm's net worth.

We reject Becker & Poliakoff's argument that the punitive damages award should have been set aside or remitted further. In connection with this argument, Becker & Poliakoff argues that the only "loss" cognizable in this case would have been loss of wages and that mental anguish damages were precluded by the impact rule. Assuming, without deciding, that the damages for mental anguish were not properly awardable as compensatory damages in this case, it is clear that the jury awarded at least some compensatory damages for breach of fiduciary duty. Thus, we need not consider whether punitive damages could have been awarded in this case in the absence of actual damages.*

Accordingly, we *affirm* the final judgment.

Although the *BMW v. Gore* guidelines have led to even more damages awards being overturned, most people fail to recognize that even before *BMW* and *Campbell*, most multimillion-dollar damages awards by juries making headlines and fueling the debate over limiting punitive damages were rarely paid and certainly not promptly paid. In September 1995, for example, a federal jury in Alaska slapped Exxon Corporation with the largest punitive damages award ever imposed on a corporation: $5 billion awarded to 32,000 people who were injured by the *Exxon Valdez* oil spill in March 1989. The award was 10 times the economic loss of $500 million suffered by the plaintiffs. Before the time for filing posttrial material had closed, Exxon had filed a total of 22 motions. Eventually, the appellate court ordered the trial judge to reduce the "excessive" award, so District Court Judge Holland reduced the punitive damages award to $4 billion—but in 2006, the appellate court said that the reduction was insufficient and further reduced the award to $2.5 billion. In late 2007, the U.S. Supreme Court refused to hear the final appeal of Exxon in this case. See Table 11-2 for some examples of appeals courts' reductions of extravagant punitive damages awarded by juries.

States tend to take a more active role in limiting punitive damages. For example, on February 2, 1994, the Texas Supreme Court handed down a decision in the case of *Transportation v. Moriel*[8] that is expected to make it more difficult for plaintiffs in that state to recover punitive damages, and many commentators believe that the ruling will influence decision makers in other states. Moriel suffered a broken pelvis and became impotent as a result of having a stack of countertops fall on him while he was working. He sued the insurance company when it delayed payment on some of his medical bills. The jury awarded Moriel $101,000 in compensatory damages, with $100,000 of that total being for mental anguish, and $1 million in punitive damages. The state court of appeals affirmed

---

[8] 879 S.W.2d 10 (Tex. 1994).

*Young v. Becker & Poliakoff 88 So.3d 1002 (2012).

TABLE 11-2 SOME MAJOR PUNITIVE DAMAGE AWARDS

Case	Jury Award	Ultimate Resolution
Geragos v. Borer	Attorney Geragos was defending Michael Jackson. Geragos chartered a private plane from XtraJet, Inc., to fly with Jackson from Las Vegas to Santa Barbara, so that Jackson could surrender for his arrest. Borer, the owner of XtraJet, installed hidden cameras on the plane, and then attempted to sell the recordings of Jackson and Geragos on their flight. When Geragos found out, he sued Borer, claiming, among other things, invasion of privacy, misappropriation of name and likeness, and unfair business practices. He received an award for $2.25 million in compensatory damages and $9 million in punitive damages.	On appeal in 2010, compensatory damages were reduced to $150,000 and punitive damages to $600,000. The judge found the punitive award to be so excessive under the circumstances as to violate due process. He believed that a ratio of no more than 4 to 1 was appropriate given that Borer's conduct was not as reprehensible as other forms of punishable conduct, in that he did not endanger anyone's health or safety, he did not target a financially vulnerable victim, and he had never engaged in similar misconduct in the past.
Frankson v. Browne & Williamson	A jury awarded the widow Gladys Frankson $350,000 in compensatory damages and $20 million in punitive damages following the death of her husband, who died from lung cancer caused by his using the defendant's cigarettes.	In June 2004, the Supreme Court of New York did not strictly follow the 4:1 ratio in *State Farm v. Campbell,* but it held that the punitive damages were still excessive and reduced the punitive damages award to $5 million if the plaintiff agreed to the new amounts. Otherwise, the judge directed a new trial on the issue of punitive damages.
Diamond Woodworks, Inc. v. Argonaut Insurance Co.	A jury awarded compensatory damages and $14 million in punitive damages to an employee who was denied insurance benefits after being injured at his place of employment.	The trial court reduced the punitive damages to $5.5 million, but the appellate court granted the defendant's motion for a new trial only if the defendant agreed to a remittitur of $1 million in punitive damages, in accordance with the 4:1 ratio established in *State Farm v. Campbell.*
Conroy v. Owens-Corning Fiberglass	A jury awarded $3.37 million in compensatory damages and $54 million in punitive damages to the families of three men who contracted mesothelioma from long-term workplace exposure to asbestos.	On appeal, punitive damages were reduced from $18.2 million per plaintiff to $1 million and one cent per plaintiff. The plaintiffs then settled for an undisclosed amount.
Liebeck v. McDonald's	A jury awarded Stella Liebeck $2.9 million in damages, including $2.7 million in punitive damages for extensive burns she received when she spilled hot coffee (170°F) on her legs. Jurors were influenced by McDonald's having known that prior customers had received severe burns from its coffee and its ongoing failure to warn customers about its unusually hot product.	The trial court reduced the award by 77 percent to $640,000. The parties subsequently settled the case for an undisclosed amount.

the verdict. The Texas Supreme Court struck down the punitive damages award, holding that an insurance company's refusal to pay a claim does not justify punitive damages unless the failure to pay was in bad faith and the insurer knew that its action would probably bring about extraordinary harm such as "death, grievous physical injury or genuine likelihood of financial catastrophe."

An example of a state law limiting punitive damages is that of Missouri, which limits punitive damages to five hundred thousand dollars or five times the net amount of the judgment awarded to the plaintiff against the defendant. Such limitations, however, do not apply if the state of Missouri is the plaintiff requesting the award of punitive damages, or the defendant pleads guilty to or is convicted of a felony arising out of the acts or omissions pled by the plaintiff. The restriction also doesn't apply to cases arising under a limited number of Missouri statutes.[9] Many pieces of legislation designed to reform tort law have been proposed at both the federal and the state levels. The majority of these proposals contained provisions limiting punitive damages, and many focused on medical malpractice. For example, in 1995, the proposed federal Common Sense Legal Reform Act contained a provision limiting punitive damages in certain types of tort cases—namely, torts involving defective products (so-called *product liability cases*, which are discussed in Chapter 12). This legislation would allow punitive damages in such cases only when the plaintiff could prove by clear and convincing evidence that the harm suffered was caused by "actual malice." Such damages would also be limited to $250,000 or three times the actual economic harm incurred by the plaintiff, whichever was greater. Part of this legislation, including the cap on punitive damages, was passed in 1996 but was vetoed by President Clinton. In 2011, HR 5, the **Help Efficient**, **Accessible**, **Low-cost**, **Timely Healthcare (HEALTH) Act of 2011** was proposed. The act would have imposed limits on medical malpractice litigation in state and federal courts by capping awards and attorney fees, modifying the statute of limitations, and eliminating joint and several liability.

Until 2005, reformers at the federal level had very little success. In February of 2005, however, in response to arguments that state tort laws lack uniformity, the Class Action Fairness Act, designed to transfer jurisdiction in large, multistate class action tort suits from state courts to federal courts, was signed into law. Because state courts commonly give larger awards in class action suits than do federal courts, the new legislation was seen as a way to reduce the awards in such cases.

According to the Federal Judicial Center, the law did have one very swift and certain impact on the courts: It sharply increased the number of class action suits filed in and removed to the federal courts.[10] Under the law, the federal courts have jurisdiction over class action cases in which (1) the aggregate value

COMPARATIVE LAW CORNER

Punitive Damages in Japan

Punitive damages have been a major source of contention in the United States. Tort reformists in the United States argue fervently for caps on the amount of punitive damages. Other groups argue that large punitive damages are necessary to discourage large corporations from committing torts, because compensatory damages are often less expensive than ceasing to commit torts. Very few people in the United States, however, argue for the complete elimination of punitive damages.

The view of punitive damages is very different in Japan. Japan's Supreme Court has ruled many times that punitive damages violate Japan's public policy. Recently, the Japanese legislature passed a law forbidding the acceptance of punitive damages in foreign courts as well. Although Japan's lack of punitive damages is not entirely unusual (many European countries do not have a system of punitive damages either), the reinforcement of this ban on punitive damages is atypical. Many of the other European countries that lack systems of punitive damages are moving toward having damages beyond compensation and acceptance of foreign awards of punitive damages.

Not everyone in Japan is against punitive damages, however. Some Japanese businesses have indicated that they would like some sort of punitive damages in cases of patent infringement. Most of the companies that favor an expansion of infringement damages are secondary industries, such as pharmaceuticals, rather than major industries such as automobiles. The major industries seem content with the damages system in its current form.

---

[9]Missouri Revised Statutes Chapter 510, Section 510-265, August 28, 2012.
[10]Maricia Coyle, "Class Action Changes Bring Quick Impact," *National Jaw Journal* 6 (Oct. 2, 2006).

LINKING LAW AND BUSINESS

Marketing

The establishment of torts for the purpose of deterring future crimes relates to a familiar concept in the field of marketing. This idea, *advertising,* is defined as the "presentation and promotion of ideas, goods, or services by an identified sponsor." Advertisers hope that what they are promoting will gain acceptance by the general public. Similarly, the law of torts is created to promote fair and just behavior among civilians. One intent of torts is that potential wrongdoers will refrain from injuring other persons or their property because of the consequences entailed in tort laws.

of the claims exceeds $5 million; (2) there are at least 100 class members; (3) any member of the plaintiff class is a citizen of a state different from any defendant; and (4) two-thirds or more of the class members and primary defendants are not members of the state in which the action was originally filed.

Tort reform advocates at the state level have been more successful thus far. Almost every state has passed some sort of tort reform legislation. Since 1986, 34 of these state tort reform laws limited punitive damages awards in some fashion.[11] Many of these reform efforts, however, have been struck down by the courts.

Classifications of Torts

There are three classifications of torts: intentional, negligent, and strict liability. The primary distinguishing feature among them is the degree of willfulness of the wrongful conduct. **Intentional torts** are those wherein the defendant took some purposeful action that he or she knew, or should have known, would harm the plaintiff. **Negligent torts** involve carelessness on the part of the defendant. Finally, **strict liability torts** involve inherently dangerous actions and impose liability on the defendant regardless of how careful he or she was. Defenses for the various categories of torts differ, as do the types of damages generally awarded (Table 11-3).

Intentional Torts

Intentional torts, the most "willful" torts, include a substantial number of carefully defined wrongful acts. What each of these acts has in common is the element of intent. *Intent* here does not mean a specific determination to cause harm to the plaintiff; rather, it means the determination to do a specific physical act that may lead to harming the plaintiff's person, property, or economic interests.

Intentional torts can be divided into three categories based on the interest being harmed: torts against persons, torts against property, and torts against economic interests. The following sections discuss a number of specific torts that fall into each category, along with the defenses to each.

INTENTIONAL TORTS AGAINST PERSONS

There are a number of torts against persons. We will discuss five of the most common ones: assault and battery, defamation, privacy torts, false imprisonment, and intentional infliction of emotional distress.

Assault and Battery. Torts against persons consist of harm to another's physical or mental integrity. One of the most common torts against the person is assault. An **assault** is the intentional placing of another in fear or apprehension of an

intentional tort A civil wrong that involves taking some purposeful action that the defendant knew, or should have known, would harm the person, property, or economic interests of the plaintiff.

negligent tort A civil wrong that involves a failure to meet the standard of care a reasonable person would meet and, because of that failure, harm to another results.

strict liability tort A civil wrong that involves taking action that is so inherently dangerous under the circumstances of its performance that no amount of due care can make it safe.

assault Intentional placing of a person in fear or apprehension of an immediate, offensive bodily contact.

---
[11] Congressional Budget Office, *The Effects of Tort Reform: Evidence from the States* (June 2004).

TABLE 11-3 CATEGORIES OF TORTS

Type	Description and Examples	Common Defenses	Type of Damages Usually Awarded
Intentional Torts	Purposeful action that results in harm	Specific to subtype	
Against persons	Assault and battery	• Self-defense • Defense of another • Defense of property	Compensatory damages for medical bills, lost wages, and pain and suffering
	Defamation	• Truth • Privilege, absolute (congressional and courtroom speech) • Privilege, conditional (speech concerning public figures or in employment context)	Compensatory damages for measurable financial losses
	Invasion of privacy	• Waiver by plaintiff of right to privacy	Compensatory damages for any resultant economic loss and pain and suffering
	False imprisonment	• Posted warnings of observation	Compensatory damages for treatment of physical injuries and lost time at work
	Intentional infliction of emotional distress	• Shopkeepers' privilege	Compensatory damages for the treatment of physical illness resulting from the emotional distress
Against property	Trespass to realty		Compensatory damages for harm caused to property and losses suffered by rightful owner
	Trespass to personalty		Compensatory damages for harm to the property
	Conversion		Compensatory damages for full value of converted item
Against economic interests	Disparagement	• Truth	Compensatory damages for actual economic loss
	Intentional interference with a contract	• No knowledge of contract	Compensatory damages for loss of expected benefits from the contract
	Unfair competition		Compensatory damages for lost profits
	Misappropriation	• Independent origination • Denial of discussion of idea	Compensatory damages for economic losses
Negligent Torts	Careless action that results in harm	• No duty • No breach of duty • No causation (actual or proximate) • No damages suffered by plaintiff • Contributory negligence by plaintiff • Pure comparative negligence • Modified comparative negligence	Compensatory damages for injuries, including medical bills, lost time from work, harm to property, and pain and suffering
Strict Liability Torts	Action that is so inherently dangerous that no amount of due care can make it safe	• Assumption of risk	Compensatory damages for personal injury and harm to property

immediate, offensive bodily contact. All of those elements must be present for an assault to exist. Thus, if the defendant pointed a gun at the plaintiff and threatened to shoot and the plaintiff believed the defendant would shoot, an assault would have taken place. If the plaintiff, however, thought that the defendant was joking when making the threat, there was no assault because there was no apprehension on the part of the plaintiff. Likewise, a threat to commit harm in a week is not an assault because there is no question of immediate bodily harm. A threat made with an unloaded gun, as long as the plaintiff does not know the defendant is incapable of carrying out the threat, however, is an assault.

An assault is frequently, but not always, followed by a **battery**, which is an intentional, unwanted, offensive bodily contact. Punching someone in the nose is a battery, whereas accidentally bumping into someone on a crowded street is not. The term *bodily contact* has been broadly interpreted to include such diverse situations as the defendant's using a projectile, such as a gun, to make physical contact with the plaintiff, and a defendant's pulling a chair out from under the plaintiff. A number of well-known figures, such as the boxer Mike Tyson, have been sued for battery.

battery Intentional, unwanted, and offensive bodily contact.

Defenses to Battery. The most common defense to a battery is self-defense. If one is attacked, one may repel the attacker—but with only that degree of force reasonably necessary to protect oneself. In most states, if a third person is in trouble, one may defend that person with the same degree of force that one would reasonably use to defend oneself, so long as the third party is unable to act in his or her own defense and there is a socially recognized duty to defend that person. This situation is often referred to as *defense of another*.

A third defense that may be raised against a charge of battery is defense of property. A person can use reasonable force to defend home and property from an intruder. Deadly force in defense of property, however, is rarely, if ever, considered justified.

Defamation. Another tort that most people have heard of is defamation. **Defamation** is the intentional publication (communication to a third party) of a false statement that is harmful to the plaintiff's reputation. If the defamation is published in a permanent form—for example, in a piece of writing or on television—the tort is called **libel**; if it is spoken, it is called **slander**.

defamation Intentional publication (communication to a third party) of a false statement that is harmful to the plaintiff's reputation.

libel Publication of a defamatory statement in permanent form.

slander Spoken defamatory statement.

Once a plaintiff proves the elements of a case of libel, "general" damages are presumed as a matter of law. These damages provide the plaintiff with compensation for harms that are hard to quantify but that would almost certainly arise from libel, such as feelings of humiliation and loss of standing in the community. In the case of slander, however, the plaintiff must prove "special" damages, which means that to recover damages, the plaintiff must demonstrate an actual monetary loss resulting or flowing from the slanderous statement.

There is an exception to this limitation on damages, however, and the exception is for statements that constitute *slander per se*. These are statements that are considered by their very nature to be so obviously harmful to a person that no proof of special damages is needed. Traditionally, statements are considered slander per se if they are statements that say (1) one has a loathsome communicable disease; (2) one has committed improprieties in the performance of his or her profession; (3) one has committed or been imprisoned for a serious crime; and (4) an unmarried female is not chaste.

One example of a libel case is the Warnaco claim against Calvin Klein. On the *Larry King Live* television show, Calvin Klein accused Warnaco, the company that manufactures Calvin Klein jeans and underwear, of making and selling substandard Calvin Klein products. In response, Warnaco brought a libel claim against Calvin Klein personally.[12]

---
[12]*National Law Journal* B9 (Feb 16, 1998).

There are limits on what is considered defamation. For instance, someone can say something that is potentially harmful to another's reputation but not suffer any consequences for saying it if the statement is merely one of opinion and not a statement of a fact. For example, in a 1997 case, Randolph Cook claimed to have had a past relationship with Oprah Winfrey. During that relationship, he claimed, Winfrey used cocaine regularly. Cook contacted several media outlets with his claim. After Winfrey heard about this, she called Cook a liar, both privately and publicly. Cook sued Winfrey for defamation, among other things, but because calling someone a liar is only an opinion, and one cannot be sued for stating one's opinion, the case was dismissed.[13] Likewise, it was not defamatory for employees of Apple Computer Company to refer to the famous astronomer Carl Sagan as "butthead astronomer."[14]

HYPOTHETICALLY SPEAKING

Suppose that. . . . Explain why defamation did or did not occur. If the facts were slightly differently in that . . . how would this change affect the validity of the defamation claim?

Defamation has become a little more confusing since people began communicating over the Internet. This medium of communication has generated two questions. First, when does a false statement made over this information network constitute defamation? Second, who can be held liable if defamation does exist? Both the legislatures and the courts have been grappling with these issues. The following case illustrates one court's approach.

CASE 11-2

Nemet Chevrolet, Ltd. v. Consumeraffairs.com, Inc.
United States Court of Appeals for the Fourth Circuit
591 F.3d 250 (2009)

The plaintiff, Nemet Chevrolet, Ltd., is in the business of selling and servicing automobiles. The defendant, Consumeraffairs.com, Inc., operates a website where consumers can comment on the quality of goods and services, including those at Nemet Chevrolet. Nemet felt that several of the postings on the defendant's website were false and harmful to its business reputation, so it filed suit alleging defamation. The defendant filed a motion to dismiss, under Federal Rule 12(b)(6), for failure to state a claim upon which relief could be granted. According to the defendant, the statements on its website are protected by the Communications Decency Act of 1996 (CDA), which prevents plaintiffs from holding Internet service providers liable for the publication of information created and developed by others. The district court granted the motion with leave to amend the complaint. The plaintiff amended, but the defendant filed another 12(b)(6) motion to dismiss. The district court again granted the dismissal. The plaintiff appealed.

Circuit Judge Agee

Recognizing that the Internet provided a valuable and increasingly utilized source of information for citizens, Congress carved out a sphere of immunity from state lawsuits for providers of interactive computer services to preserve the "vibrant and competitive free market" of ideas on the Internet. The CDA bars the institution of a "cause of action" or imposition of "liability" under "any State or local law that is inconsistent" with the terms of § 230. As relevant here, § 230 prohibits a "provider or user of an interactive computer service" from being held responsible "as the publisher or speaker of any information provided by another information content provider." Assuming a person meets the statutory definition of an "interactive computer service provider," the scope of § 230 immunity turns on whether that persons' actions also make it an "information content provider." The CDA defines an "information content provider" as "any person or entity that is responsible, in whole or in part, for the

---

[13] *Cook v. Winfrey*, 975 F. Supp. 1045 (N.D. Ill. 1977).

[14] *Sagan v. Apple Computer Co.*, 874 F. Supp. 1972 (C.D. Cal. 1994).

creation or development of information provided through the Internet or any other interactive computer service."

Taken together, these provisions bar state-law plaintiffs from holding interactive computer service providers legally responsible for information created and developed by third parties. Congress thus established a general rule that providers of interactive computer services are liable only for speech that is properly attributable to them. State-law plaintiffs may hold liable the person who creates or develops unlawful content, but not the interactive computer service provider who merely enables that content to be posted online.

To further the policies underlying the CDA, courts have generally accorded § 230 immunity a broad scope. This Circuit has recognized the "obvious chilling effect" the "specter of tort liability" would otherwise pose to interactive computer service providers given the "prolific" nature of speech on the Internet. Section 230 immunity, like other forms of immunity, is generally accorded effect at the first logical point in the litigation process. As we have often explained in the qualified immunity context, "immunity is an *immunity from suit* rather than a mere defense to liability" and "it is effectively lost if a case is erroneously permitted to go to trial." We thus aim to resolve the question of § 230 immunity at the earliest possible stage of the case because that immunity protects websites not only from "ultimate liability," but also from "having to fight costly and protracted legal battles."

Nemet does not dispute that Consumeraffairs.com is an interactive computer service provider under the CDA. What Nemet contends is that Consumeraffairs.com is also an information content provider as to the twenty posts and, therefore, cannot qualify for § 230 immunity. In other words, Nemet's argument is that its amended complaint pleads sufficient facts to show Consumeraffairs.com is an information content provider for purposes of denying statutory immunity to Consumeraffairs.com at this stage in the proceedings.

. . . We must determine . . . whether the facts pled by Nemet, as to the application of CDA immunity, make its claim that Consumeraffairs.com is an information content provider merely possible or whether Nemet has nudged that claim "across the line from conceivable to plausible."

In the amended complaint, Nemet recited the specific language from each customer about his or her automobile complaint for each of the twenty posts it claimed were defamatory. Then, Nemet pled as to each of the posts as follows:

> *Upon information and belief, Defendant participated in the preparation of this complaint by soliciting the complaint, steering the complaint into a specific category designed to attract attention by consumer class action lawyers, contacting the consumer to ask questions about the complaint and to help her draft or revise her complaint, and promising the consumer that she could obtain some financial recovery by joining a class action lawsuit. Defendant is therefore responsible, in whole or in part, for developing the substance and content of the false complaint . . . about the Plaintiffs.*

. . . In short, Nemet argues [that] the language . . . shows Consumeraffairs.com's culpability as an information content provider either through (1) the "structure and design of its website," or (2) its participation in "the preparation of" consumer complaints: *i.e.*, that Consumeraffairs.com "solicit[ed]" its customers' complaints, "steered" them into "specific categor[ies] designed to attract attention by consumer class action lawyers, contact[ed]" customers to ask "questions about" their complaints and to "help" them "draft or revise" their complaints, and "promis[ed]" customers would "obtain some financial recovery by joining a class action lawsuit."

We first examine the structure and design of the website argument, which encompasses all the facts pled in the Development Paragraph except for the claim Consumeraffairs.com asked questions and "help[ed] draft or revise her complaint." . . .

Even accepting as true all of the facts Nemet pled as to Consumeraffairs.com's liability for the structure and design of its website, the amended complaint "does not show, or even intimate," that Consumeraffairs.com contributed to the allegedly fraudulent nature of the comments at issue. Thus, . . . Nemet's pleading not only fails to show it is plausible that Consumeraffairs.com is an information content provider, but not that it is even a likely possibility.

We now turn to the remaining factual allegations, common to all twenty posts from the Development Paragraph, that Consumeraffairs.com is an information content provider because it contacted "the consumer to ask questions about the complaint and to help her draft or revise her complaint." Nemet fails to make any cognizable argument as to how a website operator who contacts a potential user with questions thus "develops" or "creates" the website content. Assuming it to be true that Consumeraffairs.com contacted the consumers to ask some unknown question, this bare allegation proves nothing as to Nemet's claim [that] Consumeraffairs.com is an information content provider.

The remaining claim, of revising or redrafting the consumer complaint, fares no better. Nemet has not pled what Consumeraffairs.com ostensibly revised or redrafted or how such affected the post. . . .

Moreover, in view of our decision in *Zeran*, Nemet was required to plead facts to show [that] any alleged drafting or revision by Consumeraffairs.com was something more than a website operator performs as part of its traditional editorial function. It has failed to plead any such facts. . . . § 230 forbids the imposition of publisher liability on a service provider for the exercise of its "editorial and self-regulatory functions."

We thus conclude that the Development Paragraph failed, as a matter of law, to state facts upon which it could be concluded that it was plausible that Consumeraffairs.com was an information content provider. Accordingly, as to the Development Paragraph, the district court did not err in granting the Rule 12(b)(6) motion to dismiss because Nemet failed to plead facts sufficient to show [that] Consumeraffairs.com was an information content provider and not covered by CDA immunity.

Even if the facts pled in the Development Paragraph are insufficient for Rule 12(b)(6) purposes, Nemet separately argues that as to eight of the twenty posts, the amended complaint pled other facts which show [that] Consumeraffairs.com is an information content provider. Thus, Nemet argues [that] the motion to dismiss should not have been granted as to these eight posts. . . . [A]s to the eight posts, Nemet pled as to each that "[b]ased upon the information provided in the post, [Nemet] could not determine which customer, if any, this post pertained to."

. . . Nemet's *sole* factual basis for the claim that Consumeraffairs.com is the author, and thus an information content provider not entitled to CDA immunity, is that Nemet cannot find the customer in *its* records based on the information in the post.

Because Nemet was unable to identify the authors of these comments based on "the date, model of car, and first name" recorded online, Nemet alleges that these comments were "fabricated" by Consumeraffairs.com "for the purpose of attracting other consumer complaints." But this is pure speculation and a conclusory allegation of an element of the immunity claim. . . . Nemet has not pled that Consumeraffairs.com created the allegedly defamatory eight posts based on any tangible fact, but *solely* because it [Nemet] can't find a similar name or vehicle of the time period in Nemet's business records. Of course, the post could be anonymous, falsified by the consumer, or simply missed by Nemet. There is nothing but Nemet's speculation which pleads Consumeraffairs.com's role as an actual author in the Fabrication Paragraph.

On appeal, Nemet argues that its supporting allegations nonetheless . . . [present] adequate facts that Consumeraffairs.com is the author of the eight posts, but each is meritless. These allegations include (1) that Nemet has an excellent professional reputation, (2) none of the consumer complaints at issue have been reported to or acted upon by the New York City Department of Consumer Affairs, (3) Consumeraffairs.com's sole source of income is advertising and this advertising is tied to its webpage content, and (4) some of the posts on Consumeraffairs.com's website appeared online after their listed creation date. Nemet's allegations in this regard do not allow us to draw any reasonable inferences that would aid the sufficiency of its amended complaint.

That Nemet may have an overall excellent professional reputation, earned in part from a paucity of complaints reported to New York City's Department of Consumer Affairs, does not allow us to reasonably infer that the particular instances of consumer dissatisfaction alleged on Consumeraffairs.com's website are false. Furthermore, Nemet's allegations in regard to the source of Consumeraffairs.com's revenue stream are irrelevant, as we have already established that Consumeraffairs.com's development of class-action lawsuits does not render it an information content provider with respect to the allegedly defamatory content of the posts at issue. Finally, the fact that some of these comments appeared on Consumeraffairs.com's website after their listed creation date does not reasonably suggest that they were fabricated by Consumeraffairs.com. Any number of reasons could cause such a delay, including Consumeraffairs.com's review for inappropriate content. . . .

Viewed in their best light, Nemet's well-pled allegations allow us to infer no more than "the mere possibility" that Consumeraffairs.com was responsible for the creation or development of the allegedly defamatory content at issue. Nemet has thus failed to nudge its claims that Consumeraffairs.com is an information content provider for any of the twenty posts across the line from the "conceivable to plausible." As a result, Consumeraffairs.com is entitled to § 230 immunity and the district court did not err by granting the motion to dismiss.*

Judgment *Affirmed*.

CRITICAL THINKING ABOUT THE LAW

In every legal case, there are at least two separate conclusions. The plaintiff believes that the court should rule one way, whereas the defendant thinks that the court should rule another. In Case 11-2, plaintiff Nemet provided one conclusion, but the court supported a conclusion more similar to Consumeraffairs.com's conclusion. The court's reasoning provides the answer for why the court reached its particular conclusion. The following questions address the court's reasoning.

1. Identify the court's conclusion in Case 11-2.

    Clue: Reread the final paragraph of the court's decision.

2. What are the reasons the court provides to support this conclusion?

    Clue: Look at the court's application of the Communications Decency Act.

3. To demonstrate the significance of primary ethical norms in court decisions such as this one, identify the ethical norm that would have reversed this decision.

    Clue: This norm is related to prioritizing the plaintiff's rights over those of the defendant in cases such as Case 11-2.

---

*Nemet Chevrolet, Ltd. v. Consumeraffairs.com, Inc. United States Court of Appeals for the Fourth Circuit 591 F.3d 250 (2009).

Defenses to Defamation. There are two primary types of defenses to a defamation action: truth and privilege. It is often stated that truth is an absolute defense. In other words, if I make an honest statement that harms the reputation of the defendant, there has been no defamation. For the ordinary plaintiff, however, a defendant cannot use the excuse that he or she thought the statement was true. Only when a possible privilege exists is the defendant's incorrect belief about the truth of the statement important.

Privilege is the second type of defense in a defamation action. Most privileges arise under certain circumstances in which our society has decided that encouraging people to speak is more important than protecting people's reputations.

There are two types of privilege: (1) absolute and (2) qualified or conditional. When an **absolute privilege** exists, one can make any statement, true or false, and cannot be sued for defamation. There are very few situations in which such a privilege exists. The Speech and Debate Clause of the U.S. Constitution gives an absolute privilege to individuals speaking on the House and Senate floors during congressional debate. This privilege encourages the most robust debate possible over potential legislation. Another absolute privilege arises in the courtroom during a trial.

absolute privilege The right to make any statement, true or false, about someone and not be held liable for defamation.

APPLYING THE LAW TO THE FACTS...

Consider a situation where Jerry and Melissa are witnesses in a trial, and during the trial, Jerry makes false statements about Melissa that cause harm to her reputation. After the trial, Melissa sues Jerry for defamation. Which defense to defamation would Jerry use to protect his speech? Would this defense work for him in light of the circumstances under which he made his statements? What if he had made the comments to Melissa in the hallway, and an unseen reporter had overheard them and published them in the newspaper without checking their accuracy? Is anyone liable now?

The other type of privilege is a qualified or conditional privilege. A **conditional privilege** provides that one will not be held liable for defamation unless the false statement was made with malice. *Malice* has a special meaning in a defamation case: it means knowledge of the falsity of the statement or reckless disregard for the truth. In other words, the defendant either knew that the statement was false or could easily have discovered whether it was false.

conditional privilege The right to make a false statement about someone and not be held liable for defamation provided the statement was made without malice.

The conditional privilege most often used is the public figure privilege. People in the public eye, such as politicians, often find themselves the victims of false rumors. When a defendant has made a false statement about a public figure—a person who has thrust herself or himself into the public eye and who generally has access to the media—the defendant will raise the public figure privilege as a defense to charges of defamation. If the defendant proves that the plaintiff is a public figure, the plaintiff will have to additionally prove that the defamation was made with malice (defined as knowledge of the falsity or reckless disregard for the truth) in order to recover for defamation.

The reason for this privilege to comment freely about public figures as long as statements are made without malice is to encourage open discussion about persons who have a significant impact on our lives. Also, because public figures generally have access to the media, they are in a position to defend themselves and, therefore, need less protection than an ordinary private citizen.

A libel or slander case brought by a public figure sometimes appears quite complex. First, the public figure plaintiff proves that the defendant made a false statement that harmed the plaintiff's reputation. Then the defendant must prove

EXHIBIT 11-1

THE SHIFTING BURDEN OF PROOF IN A DEFAMATION CASE

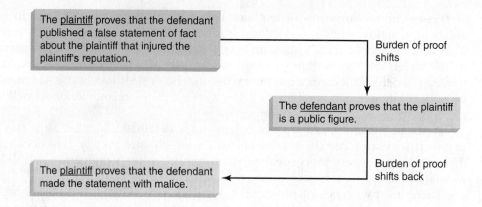

that the plaintiff is in fact a public figure. Then the burden of proof shifts back to the plaintiff, who must prove that the statements were made with malice (Exhibit 11-1).

There are two kinds of public figures: public figures for all purposes and public figures for a limited purpose. The *public figure for all purposes* was defined in the foregoing paragraph. Movie stars, musicians, and politicians fall into that category. The *public figure for a limited purpose* is a private figure who achieves substantial media attention for a specific activity. That person is then considered a public figure but only for matters related to that activity. For example, the leader of an antiabortion group would be considered a public figure for matters related to abortion. Thus, if the activist brought a defamation suit against a defendant who falsely stated that the activist had undergone three abortions as a teenager, the activist would have to prove that the defendant knew the statement was false, or acted recklessly, without even trying to check the veracity of the claim. In contrast, had the defendant claimed that the activist stole money from at least three former employers, no public figure privilege would arise, and it would not be necessary for the activist to prove that the claim had been made with malice.

Some people are trying to argue that the public figure privilege should also apply in another context: when the defamatory statement is published over the Internet. The rationale for this privilege is twofold. First, remember that part of the reason for the public figure privilege is that the public figure who has been defamed has access to the media and, therefore, has the ability to defend himself or herself. Likewise, when a person is defamed over the Internet, the defamed party can respond with a few keystrokes. Thus, there is less need for the stronger legal protection we ordinarily give to the private party. A second reason is that we want to encourage free expression and the exchange of ideas on the Internet. Requiring a plaintiff to prove malice would encourage such free discussion because people would not have to worry about making errors when they speak about others.

Another use of the conditional privilege arises with respect to job recommendations. To encourage employers to give honest assessments of their former employees, an employer who makes a false statement about a former worker can be held liable only if the statement is made with malice.

Privacy Torts. Although truth may be an absolute defense to defamation, one is not necessarily allowed to reveal everything one knows about another person. The recently developed tort of *invasion of privacy* is used to allow a person to keep private matters confidential. Just as defamation has two forms, libel and slander, the tort of invasion of privacy is really four distinct torts: (1) public disclosure of private facts, (2) false light, (3) appropriation, and (4) invasion of privacy.

Public disclosure of private facts occurs when the defendant makes public a fact about the plaintiff that the plaintiff is entitled to keep private. The disclosure must be unwarranted, and the plaintiff must not have waived his or her right to privacy. For example, if the defendant worked in a clinic and revealed the names of women who had obtained abortions at the clinic, the defendant would be liable for public disclosure of private facts.

False light occurs when you do not actually make a defamatory statement about someone, but by your actions you place the person in a false light. For example, a neighborhood newsletter publishes a story captioned "Gang Warfare Growing in Our Community," and between the caption and the article is an untitled photo of four girls sitting on the hood of a car. The photo is clear enough that the girls' identities are obvious. If these girls are not gang members, they have been placed in a false light and may sue the publisher. Often, illustrations in tabloids may lead to false-light claims.

Appropriation of a person's name for commercial gain occurs when a defendant uses another's name or likeness without that person's permission for commercial gain. This tort, for example, prohibits a cereal company from putting an athlete's picture on its cereal box without obtaining the athlete's permission.

The final privacy tort is **invasion of privacy**, which occurs when someone invades another's solitude, seclusion, or personal affairs when that person has the right to expect privacy. One example of invasion of privacy is wiretapping and using someone's password to gain access to the person's electronic mail messages. In another example, an owner of an ice skating rink who installed two-way mirrors in the women's dressing room would have committed an invasion of privacy, because the skaters should be able to expect a certain degree of privacy in a dressing room. Of course, the degree of privacy one may reasonably expect varies greatly. For example, if one is trying on clothes in a department store fitting room where signs are posted saying that the "area is under

public disclosure of private facts A privacy tort that consists of unwarranted disclosure of a private fact about a person.

false light A privacy tort that consists of intentionally taking actions that would lead observers to make false assumptions about the person.

appropriation A privacy tort that consists of using a person's name or likeness for commercial gain without the person's permission.

invasion of privacy A privacy tort that consists of encroaching on the solitude, seclusion, or personal affairs of someone who has the right to expect privacy.

TECHNOLOGY AND THE LEGAL ENVIRONMENT

CAN-SPAM: Putting Spam on the Stand

Spam, or unsolicited commercial email, constitutes more than one-half of all electronic mail traffic. The flooding of these unwanted messages prompted Congress to impose regulations on such messages. For instance, Congress passed the Controlling the Assault of Non-Solicited Pornography and Marketing Act of 2003 (the CAN-SPAM Act), which took effect in January 2004. The act states, "Most of these messages are fraudulent or deceptive in one or more respects." Consequently, Congress created three provisions for spammers. First, spammers must clearly label their messages as advertisements, avoiding misleading or untruthful subject lines that function simply to entice readers to view such messages. Second, spammers must provide a clear and convenient opt-out option in their messages, whereby recipients may reject future emails from these spammers. Third, spammers must send messages from legitimate return addresses, while also including the sender's postal address. These three restrictions on spammers, however, do not apply in situations in which a recipient has given prior affirmative consent to receive spam messages.

Congress created this federal act to preempt most state laws against spammers while making exceptions for state laws related to deceptive information in commercial electronic mail. The Federal Trade Commission, along with other federal and state agencies and attorneys general, can bring suit against spammers. Violations of the CAN-SPAM Act could result in civil and criminal penalties, including heavy fines and possible imprisonment.

observation to deter shoplifting," it would not be unreasonable for the store to have authorized security guards of the same sex as the dressing-room occupants observing the dressing rooms.

False Imprisonment. *False imprisonment* is the intentional restraint or confinement of a person against that person's will and without justification. The tort protects our freedom of movement. The confinement cannot be by moral force alone. There must be either physical restraint, such as locking a door; physical force, such as holding someone down; or threats of physical force.

Most cases of false imprisonment are brought against security guards and retailers. In fact, this tort is brought so frequently against retailers who have detained a person suspected of shoplifting that it has become known as the "shopkeepers' tort." In most states, retailers who detain suspected shoplifters for questioning are entitled to raise "the shopkeepers' privilege." Under this privilege, a merchant who has reason to believe that a person has shoplifted may detain the person for questioning about the incident. The detention must be conducted in a reasonable manner, and the suspect can be held for only a reasonable time.

Even if one is successful in bringing an action for false imprisonment, damages are often not easy to prove. Obviously, a person who is physically restrained might have medical bills for treatment of physical injuries, but most cases do not involve physical harm. Usually, plaintiffs ask for a monetary award to compensate them for time lost from work, pain and suffering from the mental distress, and humiliation.

Occasionally, however, as in the 2006 case of *Jackson v. Rich's*,[15] the store personnel's behavior leaves the jury no choice but to award the plaintiff a huge verdict. Jackson was leaving the store after buying some clothes for her terminally ill son when Rich's plainclothes detectives stopped her and told her that she had been caught shoplifting. Remembering the kidnapping and murder a few years earlier of a woman who had been abducted from the same parking lot, Jackson was terrified and offered to let the men search her bag. They said they could do that only in the store's detention room. She refused to go with them and attempted to use her cell phone to call 911. They took her phone and said they were the police. They handcuffed her and escorted her, crying, through the store, then handcuffed her to a bar in the store's detention center so that she could not move. She asked to call the police again, but they still refused. A manager then came in and examined her receipt and told the men that they could unhook her and let her go. He then apologized and offered her a discount on her purchase.

After two hours of deliberation, the jury awarded Jackson $1.2 million in compensatory damages for emotional distress. The plaintiff had testified that as a result of the incident, she no longer can go shopping alone, wear anything tight around her wrists, or be in close quarters. She gets frightened when anyone gets too close to her.

The jury felt that the security men's approach was overkill, when all they needed to do was ask to see what was in the bag, especially because she had offered to show it to them. According to store policy, they should not approach anyone unless they have shoplifting on tape. In Jackson's case, before leaving the store, she had seen a shirt that she thought might match the shorts she had just bought. She took the shorts out of the bag to compare them and then put the shorts back into the bag. The security officers saw her putting the shorts back into the bag and jumped to the conclusion that she had been shoplifting.

---

[15]Natalie White, "False Arrest for Shoplifting Yields $1.2 Million Verdict," *Lawyer's Weekly*, May 22, 2006, p. 10.

Before arguments for punitive damages could be made in the case, the defendants settled for a confidential amount.

Intentional Infliction of Emotional Distress. This tort arises when the defendant engages in outrageous, intentional conduct that is likely to cause extreme emotional distress to the party toward whom such conduct is directed. For example, a debt collector calls a debtor and tells the debtor that he is a police officer and he is sorry to inform the debtor that his wife has just been killed in an auto accident, and her last words to the medic at the scene of the crash were, "God must be punishing me for our not paying our debts." Such conduct would most likely be interpreted as the **intentional infliction of emotional distress**.

In most states, to recover damages for intentional infliction of emotional distress, the plaintiff must demonstrate some physical symptoms caused by his or her emotional distress. For example, in the preceding example, if the plaintiff had high blood pressure and after hearing the message had a heart attack, the heart attack would provide the necessary physical basis to prove his injury. Other physical symptoms commonly arising from emotional distress include headaches, a sudden onset of high blood pressure, hives, chills, inability to sleep, or inability to get out of bed.

Although some people argue that the requirement of physical harm puts an undue burden on the plaintiff, others fear that without the requirement of physical symptoms of harm, it would be too easy to successfully recover damages in a situation in which there is not any real harm. For example, critics point to the 1998 suit filed against Dennis Rodman of the Chicago Bulls for intentional infliction of emotional distress as an illustration of abuse of the tort. Rodman and a friend were playing craps at the Mirage Hotel in Las Vegas. While playing, Rodman allegedly rubbed the dealer's bald head for good luck. The dealer claimed that this act caused him "embarrassment, indignity, degradation, and anger." Because of the severe results of Rodman's head-rubbing, the dealer sought damages in excess of $10,000.[16]

intentional infliction of emotional distress Intentionally engaging in outrageous conduct that is likely to cause extreme emotional pain to the person toward whom the conduct is directed.

INTENTIONAL TORTS AGAINST PROPERTY

The second category of intentional torts involves damage to property. **Trespass to realty**, also called **trespass to real property**, occurs when a person intentionally enters the land of another or causes an object to be placed on the land of another without the landowner's permission. Trespass to realty also occurs when one originally enters another's land with permission, is told to leave, and yet remains on the land. It is no defense to argue that one did not know that the land belonged to another; the intent refers to intentionally being on that particular piece of land.

Trespass to personalty occurs when one intentionally interferes with another's use and enjoyment of his or her personal property. It is usually of short duration, but the trespasser is liable for any harm caused to the property or any loss suffered by the true owner as a result of the trespasser's having used the property.

Conversion is a more extreme wrong. It occurs when the defendant deprives the owner of his or her use and enjoyment of personal property. Traditionally, the tort required the defendant's permanent removal of the property from the owner's possession and control, such that the item could not be recovered or restored to its original condition. Today, however, a serious deprivation, even if not permanent, may constitute conversion. The plaintiff usually recovers damages for the full value of the converted item.

trespass to realty (trespass to real property) Intentionally entering the land of another or causing an object to be placed on the land of another without the landowner's permission.

trespass to personalty Intentionally exercising dominion and control over another's personal property.

conversion Intentional permanent removal of property from the rightful owner's possession and control.

---

[16]"Across the USA: Nevada," *Lawyers Weekly*. Accessed February 25, 2008. www.lawyersweekly.com.

If I take my neighbor's car for a drive without permission, but I return it unharmed before the owner knows I have it, I have committed trespass to personalty, but the true owner suffers no damages. If I take the car and hit a tree, damaging the bumper, before I return the car, I have again committed trespass to personalty and will be liable for the cost of repairing the car. If I take the car and sell it to a salvage firm that tears the car apart and sells its parts, I have committed conversion and will be liable for replacing the car.

INTENTIONAL TORTS AGAINST ECONOMIC INTERESTS

Torts against economic interests are the torts that most commonly arise within the business context. One such tort is **disparagement**.

To win a disparagement case, a plaintiff must prove four elements. First, the defendant made a false statement of a material fact about the plaintiff's business, product, or service. In general, the types of statements that are actionable are statements about the quality, honesty, or reputation of the business, as well as statements about the ownership of the business property. The second element is publication. Remember, *publication* in the context of any kind of defamation action means communication to a third party. So, if the defendant makes disparaging comments about the plaintiff's business in a public address to a consumer group or in an advertisement, the defendant has published the statement.

Third, there must be harm to the reputation of the business, product, or service. Finally, there must be actual economic loss as a result of the false statements. Proving the economic loss that provides a basis for compensatory damages is not always easy. Usually, damages will be based on a decrease in profits that can be linked to the publication of the false statement. An alternative, albeit a less common way to prove damages, is to demonstrate that the plaintiff had been negotiating a contract with a third party, but the third party lost interest shortly after publication of the false statement. The profits the plaintiff would have made on the contract would be the damages. Table 11-4 lists the elements of disparagement.

In 13 states, a closely related tort has been created: *food disparagement*. Dubbed "veggie libel" and "banana bills" by their critics, these laws provide ranchers and farmers a cause of action when someone spreads false information about the safety of a food product. The first major test of one of these laws came in a $6.7 million case filed by a rancher in a federal district court against talk-show host Oprah Winfrey and one of her guests. They were discussing the potential for U.S. cattle to contract mad cow disease, and, at one point, Oprah said that was it—the conversation had stopped her from ever eating a burger again. After the broadcast, which the show's producers said tried to show both sides of the issue, the price of cattle futures fell.

The Texas law at issue provides that anyone who says that a perishable food product is unsafe, knowing the statement is false, may be required to pay damages to the producer of the product. The defendants originally asked that the case be dismissed on the ground that the law unconstitutionally interferes with free speech. The judge dismissed the food-disparagement claims on the grounds that the cattlemen did not prove that "knowingly false" statements were made and that a perishable food was not involved. The jury then decided there was no case under traditional business disparagement law either.

Another tort against economic interests is the tort of intentional interference with a contract, a complex and difficult tort to prove. To prove the tort of **intentional interference with a contract**, the plaintiff must demonstrate that:

1. The plaintiff had a valid contract with a third party.
2. The defendant knew of the contract and its terms.
3. The defendant took action knowing that it was highly likely to cause the third party to breach the contract with the plaintiff.

---

disparagement Intentionally defaming a business product or service.

intentional interference with a contract Knowingly and successfully taking action for the purpose of enticing a third party to breach a valid contract with the plaintiff.

TABLE 11-4
ELEMENTS OF DISPARAGEMENT

1. A false statement of a material fact about the plaintiff's product or service
2. Publication
3. Damage to the reputation of the product or service
4. Economic loss

4. The defendant undertook the action for the purpose of causing the third party to breach the contract.
5. The third party did in fact breach the contract.
6. As a result of the breach, the plaintiff was injured.

Some of the most common cases concerning intentional interference with contracts in the business setting involve employers taking employees from another firm when they know that the employees have contracts for a set period of time. Luring an employee from a successful competitor is often a delicate situation. There is no problem if the employee does not have a contract for a fixed period of time, but if the employee is indeed bound by a contract of employment for a fixed term or by a contractual agreement not to work for a competitor for a set period of time, then pursuit of the employee opens a second employer with knowledge of the contract to liability.

A third tort against economic interest is **unfair competition**. Our legal system assumes that individuals go into business for the purpose of making a profit. Competition is supposed to drive inefficient firms out of business because the more efficient firms will be able to provide less expensive goods and services. For this system to work, however, firms must be in business to make a profit. Therefore, it is unlawful for a person to go into business for the purpose of causing a loss of business to another without regard for his or her own profit.

For example, assume that Mark wants to open a painting business but his father wants him to go to college. When Mark opens his business, his father starts a competing firm and is able to underbid every job his son bids because the father is willing to lose money. He just wants to force his son out of business. The father in this example is engaging in unfair competition.

Misappropriation is another tort against economic interest that is difficult to prove. **Misappropriation** occurs when a person presents an unsolicited idea for a product, service, or even method of marketing to a business with the expectation of compensation if the idea is used by the firm and the firm subsequently uses the idea without compensating the individual. The individual may then have the basis for an action for misappropriation.

The firm may always defend on the ground that it had already independently come up with the idea that the plaintiff proposed. The firm may also deny that the idea was even discussed. It is, therefore, extremely important that anyone offering an unsolicited idea to a firm have that idea and the offer to the firm documented.

> **unfair competition** Entering into business for the sole purpose of causing a loss of business to another firm.

> **misappropriation** Use of an unsolicited idea for a product, service, or marketing method without compensating the originator of the idea.

Negligent Torts

ELEMENTS OF NEGLIGENCE

The second classification of torts is negligent torts. **Negligence** results not from the willful wrongdoing of a party but from carelessness. A person is said to be *negligent* when her or his behavior falls below the standard of care necessary to protect others from an unreasonable risk of harm. To prove negligence, a plaintiff must establish four elements: (1) duty, (2) breach of duty, (3) causation, and (4) damages. Failure to establish any one of those elements precludes recovery by the plaintiff.

The first element to be proved is duty. The *duty* is the standard of care that the defendant owes the plaintiff. Under certain circumstances, a law establishes the duty of care for a particular party, but the courts generally use a "reasonable person" standard. Under this standard, the defendant must have exercised the degree of care and skill that a reasonable person would have exercised in similar circumstances to protect the plaintiff from an unreasonable risk of injury.

The reasonable-person standard is an objective standard; it is an illustration of how members of society would expect an individual to act in a certain

> **negligence** Failure to live up to the standard of care that a reasonable person would meet to protect others from an unreasonable risk of harm.

situation. Thus, the reasonable person is careful and wise. In negligence cases, a judge or jury must determine what the reasonable person would do in a similar situation and compare this standard to the actions of the individual in the case before it.

One of the reasons that a future business manager should be knowledgeable about duty of care is that courts generally expect businesses to meet a reasonable duty of care for customers who enter onto the businesses' property. Thus, businesses must warn customers about potential risks they might encounter while on the property, or even better, make sure that the property is safe for customers. Even if a business attempts to warn its customers about potential hazards, the business might still be considered negligent. For example, in a case decided in Los Angeles,[17] a woman sued the House of Blues restaurant because she tripped over lumber that was being stored on the front porch of the restaurant. Although the lumber was marked with yellow construction tape, the woman received $91,366 in damages.

The next element to be proved is a breach of duty. Once the plaintiff establishes the duty required of the defendant under the circumstances, the plaintiff must show that the defendant's conduct was not consistent with that duty. For example, a reasonable person does not leave a campfire burning unattended in the woods. A defendant who builds a campfire and then goes home without putting out the campfire has breached her or his duty of care to the owner of the campground and to other campers whose safety is endangered by the unguarded campfire.

The third element is causation. Causation is really two elements: actual cause and proximate cause. *Actual cause* is a factual matter of whether the defendant's conduct resulted in the plaintiff's injury. The breach of the duty must have resulted directly in the plaintiff's harm. To ascertain whether the breach of duty was the actual cause of the plaintiff's harm, one must ask, "If the defendant had obeyed his or her duty, would the plaintiff still have been injured?" If the answer is no, then the defendant's breach was the actual cause of the plaintiff's harm.

Proximate cause is a question of how far society wishes to extend liability. In the majority of states, *proximate cause* is defined as foreseeability. Proximate cause exists if both the plaintiff and the type of injury incurred by the plaintiff are foreseeable. For example, it is foreseeable that if a tire falls off a car, the car may run off the road and hit a pedestrian. It is not foreseeable that the pedestrian is carrying dynamite, which he will throw when he sees the car speeding toward him, causing the dynamite to explode, causing vibrations that shatter a window six blocks away, and causing glass shards to fly and cut a secretary. Neither the secretary nor the secretary's injury would be foreseeable, so the secretary would not succeed in a suit for negligence against the manufacturer of the car in most states, because of the lack of proximate cause. Proximate cause, however, would not prevent the pedestrian from suing in this example, because a pedestrian is a foreseeable victim when a car goes out of control.

In a minority of states, the courts do not differentiate between actual and proximate cause; once actual cause is proved, proximate cause is said to exist. Thus, in the minority of states, both the pedestrian and the secretary in the preceding example would be able to recover.

Damages, or compensable injury, are the final element. The defendant's action must have resulted in some harm to the plaintiff for which the plaintiff can be compensated. A party cannot bring an action in negligence seeking only nominal damages. One example of negligence involved the tragic death of R&B vocalist Aaliyah Dana Haughton, who died August 25, 2001, in an airplane accident, following the completion of her music video titled "Rock the

---

[17]*Haywood v. Baseline Construction Co.*, No. SC004942 (Los Angeles County Sup. Ct. 1999).

Boat."[18] Background Records, which had entered into a recording agreement with Aaliyah, brought suit against Instinct Productions, a company that produced Aaliyah's music video and made transportation arrangements for the filming. Background sued Instinct for negligence, claiming that Background and Instinct shared a long, trusting relationship, from which Instinct owed a duty to Background to provide safe transportation for Aaliyah. Background argued that Instinct breached this duty, causing foreseeable economic harm to Background, whose financial success depended primarily on Aaliyah. The trial court ruled in favor of Background, awarding it damages. On appeal, however, the case was reversed, and the appellate court labeled Background's lawsuit as frivolous.

In any negligence case, the plaintiff must show that the defendant owed a duty of care to the plaintiff and breached that duty, causing foreseeable harm to the plaintiff for which the plaintiff is seeking compensation. Place yourself in the plaintiff's position to see that proving negligence is often difficult. Frequently, direct proof of the defendant's negligent conduct does not exist because it was destroyed and there were no witnesses to the negligent act. To make it easier for plaintiffs to recover in negligence cases, most courts have adopted two doctrines that may apply in such situations: *res ipsa loquitur* and *negligence per se*.

Res ipsa loquitur literally means "the thing speaks for itself." The plaintiff uses this doctrine to allow the judge or jury to infer that the defendant's negligence was the cause of the plaintiff's harm when there is no direct evidence of the defendant's lack of due care. To establish *res ipsa loquitur* in most states, the plaintiff must demonstrate that:

1. The event was of a kind that ordinarily does not occur in the absence of negligence.
2. Other responsible causes, including the conduct of third parties and the plaintiff, have been sufficiently eliminated.
3. The indicated negligence is within the scope of the defendant's duty to the plaintiff.

Proof of these elements does not require a finding of negligence; it merely permits such a finding.

One of the earliest uses of *res ipsa loquitur* was the case of *Escola v. Coca-Cola*.[19] In this case, the plaintiff, a waitress, was injured when a bottle of Coca-Cola that she was removing from a case exploded in her hand. From the facts that (1) bottled soft drinks ordinarily do not spontaneously explode and (2) the bottles had been sitting in a case, undisturbed, in the restaurant for approximately 36 hours before the plaintiff simply removed the bottle from the case, the jury reasonably inferred that the defendant's negligence during filling of the bottle resulted in its explosion. The plaintiff, therefore, could recover without direct proof of the defendant's negligence. The doctrine has subsequently been used in numerous accident cases in which there was no direct evidence of negligence. Note that the jury does not *have* to infer negligence, but it may. The defendant's best response to the use of this doctrine is to try to demonstrate other possible and plausible causes of the accident.

Another doctrine that may aid the plaintiff is **negligence per se**. If a statute is enacted to prevent a certain type of harm and a defendant violates that statute, causing that type of harm to befall the plaintiff, the plaintiff may use proof of the violation of the statute as proof of negligence. For example, it is unlawful to sell certain types of glue to minors because they may inhale it to obtain a euphoric feeling. Such a use of the glue may lead to severe health problems or death. If

res ipsa loquitur Legal doctrine that allows a judge or a jury to infer negligence on the basis of the fact that accidents of the type that happened to the plaintiff generally do not occur in the absence of negligence on the part of someone in the defendant's position.

negligence per se Legal doctrine that says when a statute has been enacted to prevent a certain type of harm and the defendant violates that statute, causing that type of harm to befall the plaintiff, the plaintiff may use proof of the violation as proof of negligence.

---
[18]"Negligence Action Brought against Video Producer over Air Crash Death of Popular Singer Advances," *New York Law Journal* 2 (June 3, 2004).
[19]24 Cal. 2d 453, 150 P.2d 436 (Cal. 1944).

a retailer sold such glue to a minor who died from sniffing the glue, proof of the sale in violation of the statute establishes negligence per se by the retailer.

DEFENSES TO NEGLIGENCE

Although the courts have created the two foregoing doctrines to help plaintiffs establish their cases, the courts also accept certain defenses that will relieve a defendant from liability, even if the plaintiff has successfully established the elements of negligence.

Initially, all states made available a strong defense to negligence: **contributory negligence**. Under this defense, the defendant must prove that (1) the plaintiff did not exercise the degree of care that one would ordinarily exercise to protect oneself from an unreasonable risk of harm and (2) this failure contributed to causing the plaintiff's own harm. Proof of such contributory negligence is an absolute bar to recovery. In other words, once the defendant proves that the plaintiff was contributorily negligent, the defendant wins the lawsuit and will not have to pay any damages to the plaintiff. Because of the harshness of this defense, many states adopted the last-clear-chance doctrine (Exhibit 11-2). Under this doctrine, once the defendant establishes contributory negligence on the part of the plaintiff, the plaintiff may still recover by showing that the defendant had the last clear opportunity to avoid the accident that resulted in the plaintiff's loss.

The adoption of this doctrine, however, still left a lot of situations in which an extremely careless defendant caused a great deal of harm to a plaintiff who was barred from recovery because of minimal contributory negligence. Thus, today, most states have replaced the contributory negligence defense with either pure or modified **comparative negligence**. Under a pure comparative negligence defense, the court determines the percentage of fault of the defendant, and that is the percentage of damages for which the defendant is liable. Damages under modified comparative negligence are calculated in the same manner, except that the defendant must be more than 50 percent at fault before the plaintiff can recover. Twenty-eight states have modified comparative negligence, 13 have pure comparative negligence, and 9 have contributory negligence. Remember,

contributory negligence A defense to negligence that consists of proving that the plaintiff did not exercise the ordinary degree of care to protect against an unreasonable risk of harm and that this failure contributed to causing the plaintiff's harm.

comparative negligence A defense that allocates recovery based on percentage of fault allocated to plaintiff and defendant; available in either pure or modified form.

EXHIBIT 11-2

APPLICATION OF THE LAST-CLEAR-CHANCE DOCTRINE

The plaintiff establishes the negligence of the defendant by proving:
- duty,
- breach of duty,
- causation (actual and proximate), and
- damages.

↓

The defendant establishes contributory negligence of the plaintiff by proving that the plaintiff failed to exercise the degree of care one would ordinarily exercise to protect oneself from an unreasonable risk of harm.

↓

The plaintiff uses the last-clear-chance doctrine and proves that the defendant had the last chance to prevent the harm from occurring.

↓

Judgment for Plaintiff

every state adopts one of these three defenses. The parties do not get to pick from among them. If a party resides in a state that uses a defense that is not favorable to that party, however, he or she can always argue that the state should change its law to accept a different defense. For example, a plaintiff residing in a state that still allows the contributory negligence defense might try to argue that the state should follow the trend and modernize its law by moving to modified comparative negligence and abolishing the contributory negligence defense.

Another defense that may be used in a negligence case is **assumption of the risk**, in which the defendant must show that the plaintiff voluntarily and unreasonably encountered a known risk. To successfully use this defense, the defendant must establish that the harm suffered was indeed the risk assumed. For example, in one case, a plaintiff was using a grinding wheel while wearing only his eyeglasses, not the safety goggles provided by his employer to keep the pieces of stone chips and dust from flying into his eyes. The defective grinding wheel exploded into three pieces, and one piece flew into the plaintiff's eye, blinding him. When the plaintiff sued the defendant manufacturer, the defendant raised the defense of assumption of the risk. The court struck down that attempted use of the defense, noting that the wearing of safety goggles was not intended to prevent harm from exploding grinding wheels and that if any risk was assumed by the plaintiff, it was the risk of getting a small stone chip in his eye. As the plaintiff could not have known that the wheel would explode, he could not have assumed the risk.

assumption of the risk A defense to negligence based on showing that the plaintiff voluntarily and unreasonably encountered a known risk and that the harm the plaintiff suffered was the harm that was risked.

In the following case, the court discusses the applicability of both assumption of the risk and comparative negligence in the context of a rather unusual injury at a baseball game.

CASE 11-3

Coomer v. Kansas City Royals
Supreme Court of Missouri
437 S.W.3d 184

Coomer and his father were watching the Royals host the Detroit Tigers. Coomer and his father left their assigned seats early in the game and moved to empty seats six rows behind the visitors' dugout. Shortly after Coomer changed seats, Sluggerrr, the team mascot, began the "Hotdog Launch," a feature of every Royals home game since 2000. The launch occurs between innings, when Sluggerrr uses an air gun to shoot hotdogs from the roof of the visitors' dugout to fans seated beyond hand-tossing range, and tosses hotdogs by hand to the fans seated nearby when his assistant is loading the airgun. Sluggerrr generally tossed the hotdogs underhand while facing the fans but sometimes throws overhand, behind his back, and side-armed. Coomer and his father were seated approximately 15 to 20 feet from Sluggerrr, directly in his view. After employing his hotdog-shaped airgun to send hotdogs to distant fans, Sluggerrr began to toss hotdogs by hand to fans seated near Coomer. Coomer saw Sluggerrr turn away from the crowd as if to prepare for a behind-the-back throw, and turned to look at the scoreboard. A "split second later," he was hit in the face by a hotdog, and suffered a torn and detached retina that required surgery for a replacement lens.

Coomer sued for negligence, and the jury was given instruction for determining whether Coomer had assumed the risk of being injured by a flying hot dog, as well as for applying the comparative negligence. The jury found that Coomer was 100% responsible for his own injury, but it was not clear whether the jury had used comparative negligence or assumption of the risk to come to their decision in favor of the Royals. Coomer appealed on grounds of improper jury instructions.

En Banc

. . . Among the jury instructions was one asking the jury to decide whether the risk of being injured by Sluggerrr's hotdog toss is one of the inherent risks of watching a Royals home game that Coomer assumed merely by attending. Whether a particular risk is inherent in watching a sporting event is a question of law for the court, not a question of fact for the jury. This Court holds that the risk of being injured by Sluggerrr's hotdog toss is not one of the inherent risks of watching a Royals home game.

In the past, this Court has held that spectators cannot sue a baseball team for injuries caused when a ball or bat enters the stands. Such risks are an unavoidable—even desirable—part of the joy that comes with being close enough to the Great American Pastime to smell the new-mown grass, to hear the crack of 42 inches of solid ash meeting a 95–mph fastball, or to watch a diving third baseman turn a heart-rending triple into a soul-soaring double-play. The risk of being injured by Sluggerrr's hotdog toss, on the other hand, is not an unavoidable part of watching the Royals play baseball. That risk is no more inherent in watching a game of baseball than it is inherent in watching a rock concert, a monster truck rally, or any other assemblage where free food or T-shirts are tossed into the crowd to increase excitement and boost attendance.

. . . [T]he key inquiry here is whether the risk which led to plaintiff's injury involved some feature or aspect of the game which is inevitable or unavoidable in the actual playing of the game. . . . Can [this] be said about the antics of the mascot? We think not. Actually, the . . . person who dressed up as Sluggerrr recounted that there were occasional games played when he was not there. In view of this testimony, as a matter of law, we hold that the antics of the mascot are not an essential or integral part of the playing of a baseball game. In short, the game can be played in the absence of such antics.

. . . if Coomer was injured by a risk that is an inherent part of watching the Royals play baseball, the team had no duty to protect him and cannot be liable for his injuries. But, if Coomer's injury resulted from a risk that is not an inherent part of watching baseball in person—or if the negligence of the Royals altered or increased one of these inherent risks and caused Coomer's injury—the jury is entitled to hold the Royals liable for such negligence and, to the extent the reasonableness of Coomer's actions are in dispute, the jury must apportion fault between the parties using comparative fault principles. This approach has been used in Missouri and around the country.

. . . First, the Court holds that the evidence was sufficient to justify submitting Coomer's comparative fault to the jury. Coomer contends that, because he was "just sitting there," this cannot constitute negligence. The jury might reach that conclusion and, as a result, not attribute any percentage of the fault to Coomer. But that is not the only conclusion supported by this evidence. The evidence also was sufficient for the jury to find that Coomer acted unreasonably by: (a) watching Sluggerrr go into his leonine wind-up in preparation for a behind-the-back hotdog toss and then (b) choosing the precise moment that Sluggerrr was releasing the hotdog to let his gaze—and attention—wander elsewhere. The jury may find that this failure to keep a careful lookout, among other reasons, was sufficient to assess some percentage of fault to Coomer.

If this case is tried again on remand . . . If the jury finds that Sluggerrr failed to use reasonable care when he threw the hotdog at Coomer and injured him, it will assess a percentage of fault to the Royals . . . [I]f the evidence on retrial is the same as here, the jury may conclude that Coomer failed to use reasonable care to protect himself from Sluggerrr's negligence (failing to keep an adequate lookout or otherwise) and, on that basis, it could assess a percentage of fault to Coomer under a proper comparative fault instruction.*

Reversed, in favor of Coomer and remanded.

Although it involves neither contributory negligence nor assumption of the risk, the following case demonstrates an interesting defense where the argument was essentially that the accident was "unavoidable."

CASE 11-4

Venkateswarlu Thota and North Texas Cardiology Center v. Margaret Young
366 S.W.3d 678 (2012)

On March 10, 2005 William (Ronnie) Young died of leukemia. Preceeding his death, Ronnie additionally suffered from angina, hypertension, a rare blood disorder called polycythemia vera, and coronary artery disease. In 2001, Ronnie went to cardiologist Venkateswarlu Thota for chest pains. Medications did not work and finally Thota recommended a procedure called coronary angiography. After Thota performed the procedure, Ronnie went home. Ronnie later was rushed back to the hospital due to a tear in his artery from the surgery. Ronnie underwent many surgeries in the following months as a result of Thota's surgery. Eventually Ronnie died from leukemia that was caused by his struggle with polycythemia vera. Ronnie's wife brought a lawsuit against the hospital and Thota claiming Thota was negligent due to: 1. Not obtaining Ronnie's full medical history; 2. Not considering Ronnie's other medical problems might interfere with Thota's procedure; 3. Lacerating not only an artery but the wrong artery during the procedure; 4. Not seeing the artery tear before Ronnie was discharged; 5. Not being able to diagnose and treat the artery tear.

---

*Coomer v. Kansas City Royals, Supreme Court of Missouri. 437 S.W.3d 184.

Justice Green

Did the negligence, if any, of those named below, proximately cause the injury in question, if any?

"Negligence," when used with respect to the conduct of Venkat Thota, M.D., means failure to use ordinary care, that is, failing to do that which a cardiologist of ordinary prudence would have done under the same or similar circumstances or doing that which a cardiologist of ordinary prudence would not have done under the same or similar circumstances.

"Ordinary care," when used with respect to the conduct of Venkat Thota, M.D., means that degree of care that a cardiologist of ordinary prudence would use under the same or similar circumstances.

"Proximate cause," when used with respect to the conduct of Venkat Thota, M.D., means that cause which, in a natural and continuous sequence unbroken by any new and independent cause, produces an event, and without which cause such event would not have occurred. In order to be a proximate cause, the act or omission complained of must be such that a cardiologist using ordinary care would have foreseen that the event, or some similar event, might reasonably result therefrom. There may be more than one proximate cause of an event.

"New and independent cause," when used with respect to the conduct of Venkat Thota, M.D., means the act or omission of a separate and independent agency, not reasonably foreseeable by a cardiologist exercising ordinary care, that destroys the causal connection, if any, between the act or omission inquired about and the injury in question and thereby becomes the immediate cause of such injury.

"Negligence," when used with respect to the conduct of [Ronnie] Young means failure to use ordinary care, that is, failing to do that which a person of ordinary prudence would have done under the same or similar circumstances or doing that which a person of ordinary prudence would not have done under the same or similar circumstances.

"Ordinary care," when used with respect to the conduct of [Ronnie] Young means that degree of care that a person of ordinary prudence would use under the same or similar circumstances.

"Proximate cause," when used with respect to the conduct of [Ronnie] Young means that cause which, in a natural and continuous sequence, produces an event, and without which cause such event would not have occurred. In order to be a proximate cause, the act or omission complained of must be such that a person using ordinary care would have foreseen that the event, or some similar event, might reasonably result therefrom. There may be more than one proximate cause of an event.

We simply cannot determine, on this evidence, whether the jury properly found Dr. Thota not negligent, properly found that his negligence was excused based upon the unavoidable accident instruction, or improperly found that his negligence was excused based upon the new and independent cause instruction alone or combined with its improper finding of Ronnie's negligence.*

While Casteel's presumed harm analysis is necessary in instances where the appellate court cannot determine "whether the improperly submitted theories formed the sole basis for the jury's finding" because the broad-form question mixed valid and invalid theories of liability, or when the broad-form question commingled damage elements that are unsupported by legally sufficient evidence, an improper inferential rebuttal instruction and improper defensive theory of contributory negligence presented in a broad-form question with separate answer blanks in a single-theory-of-liability case does not prevent the harmed party from obtaining meaningful appellate review. When a trial court abuses its discretion by including erroneous charge questions or instructions in a single-theory-of-liability case, our traditional harmless error analysis applies and the appellate courts should review the entire record to determine whether the charge errors probably caused the rendition of an improper judgment.

Because we hold that Casteel's presumed harm analysis does not apply, we next consider whether, applying traditional harmless error analysis, the alleged charge errors constitute reversible error. We address Young's objections to the inclusion of Ronnie's contributory negligence and the instruction of new and independent cause in turn.

When charge questions are submitted in a manner that allows the appellate court to determine whether the verdict was actually based on a valid theory of liability, the error may be harmless. Young's argument that the inclusion of Ronnie's contributory negligence was harmful error fails for several reasons. First, Dr. Thota could only have been negligent in causing the tear in Ronnie's artery, and the jury failed to find that he was. The jury's finding as to Dr. Thota's non-negligence is entirely separate from its finding as to Ronnie's negligence. Perhaps the jury was confused about whether to find Ronnie negligent and, despite the unavoidable accident instruction, believed that they had to find someone negligent. Either way, any error associated with the inclusion of a jury question regarding Ronnie's negligence was harmless.

Moreover, when determining whether harm occurred, we consider the entire charge. Here, the clarifying instructions at the end of Question 1 made it clear that the jury could answer in any of the following combinations: (1) "Yes" to both Dr. Thota and Ronnie; (2) "No" to both; or (3) "Yes" to one and "No" to the other—the choice the jury ultimately made. The charge's definition of proximate cause also clearly informed the jury that "[t]here may be more than one proximate cause of an event." In light of the entire charge and the separate answer blanks for Dr. Thota and Ronnie, it is evident that the jury was well aware that its findings as to Dr. Thota's and Ronnie's negligence were separate and that there could be more than one proximate cause of an event.

At trial, Dr. Thota's and Dr. Doherty's testimony about Ronnie's medical reports conflicted. Dr. Doherty

---

*Venkateswarlu Thota, and North Texas Cardiology Center, v. Margaret YOUNG 366 S.W.3d 678 (2012).

testified that the standard of care for cardiac catheterization was to insert a needle and catheter into the right femoral artery below the inguinal ligament. In Dr. Doherty's opinion, Dr. Thota punctured Ronnie's artery at the wrong location, above the inguinal ligament and into the right external iliac artery. Dr. Doherty's opinion was based on Dr. Walker's report, the CT scan mentioned on Dr. Sudharshan's report, and the bleed in Ronnie's retroperitoneal cavity, which could occur when the puncture is too high, rather than the more visible femoral bleed that would occur if the puncture is in the femoral artery. In contrast, Dr. Thota claimed at trial that he did not breach the standard of care during Ronnie's catheterization procedure. He testified that he had no problems inserting the catheter and that he believed he entered the artery at the appropriate location. Dr. Thota stated that Dr. Sudharshan's finding that the puncture site was at "about the inguinal ligament," would indicate that the puncture site was correct. He further testified that Dr. Walker's report was ambiguous as to what he repaired and how far above or below the inguinal ligament the bleed originated. Also, Dr. Thota testified that a retroperitoneal bleed can occur with a femoral artery stick as well as an iliac artery stick and that, based on his review of the medical records and his own knowledge of the procedure, he met the standard of care.

Like many medical malpractice cases, this record contains conflicting expert opinions. The fact that Dr. Thota testified on his own behalf does not negate the weight that the jury could give to his testimony. Because of the conflicting testimony of Dr. Doherty and Dr. Thota, and because both testifying experts agreed that Ronnie was likely not bleeding upon his discharge from the hospital, the jury could have reasonably believed Dr. Thota's opinions and discounted Dr. Doherty's opinions. In circumstances where a reasonable jury could resolve conflicting evidence either way, we presume the jury did so in favor of the prevailing party.

Based on the conflicting evidence, the jury could have reasonably concluded that Dr. Thota did not breach the standard of care without reaching the issue of proximate cause. In that case, the jury would not have relied on the new and independent cause instruction because it pertains only to the proximate cause element. Thus, the record supports the jury's finding of no negligence as to Dr. Thota. Accordingly, our review of the entire record provides no clear indication that the new and independent cause instruction, if erroneous, probably caused the rendition of an improper verdict. We therefore conclude that any error in the trial court's submission of the new and independent cause instruction was harmless.

In sum, we hold that Young's timely and specific no-evidence objections were sufficient to preserve the disputed charge issues for appellate review. Because the trial court submitted a broad-form question on a single theory of liability that included separate answer blanks for Dr. Thota's and Ronnie's negligence, we hold that the court of appeals misapplied Casteel and its presumed harm analysis. Even assuming the trial court abused its discretion by including a question as to Ronnie's contributory negligence and an instruction on new and independent cause, for the reasons explained above, we hold that these alleged charge errors were harmless and did not probably cause the rendition of an improper judgment. Because Casteel's presumed harm analysis does not apply and any error in the disputed charge issues was harmless, we need not address Dr. Thota's remaining issues reversed and remanded.

Strict Liability Torts

A third type of tort is a strict liability tort. Under this theory, the defendant is engaged in an activity that is so inherently dangerous under the circumstances of its performance that no amount of due care can make it safe. The activity, however, does have some social utility, so we do not want to prohibit it entirely. Consequently, we allow people to engage in such activities, but hold them strictly liable for any damages caused by engaging in these activities. Inherently dangerous activities include blasting in a populated area and keeping nondomesticated animals. As you will see in Chapter 12, in today's society, strict liability has had perhaps its greatest impact on cases involving products that are considered unreasonably dangerous.

Global Dimensions of Tort Law

With the increasing globalization of business, it is becoming more common for citizens of foreign countries to temporarily reside in the United States, as well as for U.S. citizens to reside abroad for long periods of time. There are also a number of people with dual citizenship. It is, therefore, a realistic possibility that one might get a tort judgment in the United States and need to enforce that judgment in a foreign nation.

Although many European nations are signatories to treaties regarding enforcement of foreign judgments, the United States has not signed any such treaties. Therefore, the extent to which a U.S. judgment will be enforced in a foreign nation depends on that nation's laws. For example, some nations will review the judgment to ensure that it does not offend their country's notion of due process.

One area in which at least two nations have been unwilling to fully enforce U.S. judgments is with respect to punitive damages awards. Both a German federal court and an English court have ruled that punitive damages awards violate their nation's public policy interest in maintaining a purely compensatory tort system. They have, therefore, refused to enforce U.S. punitive damages awards. As international business and, thus, international litigation continue to grow, the U.S. business manager will have to become increasingly familiar with the policies of foreign courts.

SUMMARY

Tort law provides a means for an injured party to obtain compensation from the party whose actions caused the injury. Tort law provides three types of damages. Compensatory damages, which are the most common, are designed to put the plaintiff in the position he or she would have been in had the tort not occurred. Nominal damages, available only in intentional tort cases, are a minimal amount, such as $1, and signify that the defendant's behavior was wrongful but caused no harm. Punitive damages are assessed in addition to compensatory damages or nominal damages when the defendant's conduct is egregious. Punitive damages are designed primarily to punish the defendant and deter such conduct in the future.

Torts are classified as intentional, negligent, or strict liability, depending on the degree of willfulness required for the tort. The most willful are the intentional torts, which are further categorized by the interest that is injured. Intentional torts against the person include assault, battery, defamation, intentional infliction of emotional distress, false imprisonment, and the privacy torts. Intentional torts against property include trespass to realty, trespass to personalty, and conversion. Intentional torts against economic interests include disparagement, intentional interference with contractual relations, misappropriation, and unfair competition.

Negligence can be thought of as the tort of carelessness. To prove negligence, one must prove four elements: (1) duty of care, (2) breach of duty, (3) causation, and (4) damages. *Negligence per se* and *res ipsa loquitur* are two doctrines that may help the plaintiff prove negligence. Defenses to negligence include contributory negligence, modified and pure comparative negligence, and assumption of the risk.

Strict liability occurs when one causes injury to another by engaging in an unreasonably dangerous activity.

Given the globalization of business, it is becoming increasingly likely that one might need to enforce a tort judgment rendered in the United States in a foreign country. The enforceability of such a judgment depends on that foreign nation's laws; in several nations, including Germany and England, punitive damages awards will not be enforced.

REVIEW QUESTIONS

11-1 Evaluate the arguments for and against restricting the availability of punitive damages. Explain why you tend to agree more with one position than the other.

11-2 Distinguish intentional torts from negligent torts.

11-3 Define an assault and a battery, and explain how the two are related.

11-4 Explain why it is harder to win a defamation action if you are a public figure.

11-5 Explain the relationship between trespass to personalty and conversion.

11-6 Your state is proposing to pass a food-disparagement law. Construct the strongest arguments you can in support of and in opposition to such a law. Explain how emphasizing the importance of different ethical norms could lead to a different attitude toward the proposed law.

REVIEW PROBLEMS

11-7 Karen writes Bob a long letter in which she falsely accuses him of stealing her bike. Bob is outraged because no one has ever questioned his character in that way before. He is so incensed that he shows the letter to several colleagues, as well as to his boss.

A few weeks later, he applies for a promotion and is turned down. When he asks his boss why he lost the promotion, his boss, very reluctantly, says that a number of people were concerned about Bob's integrity in light of the recent accusations about his involvement in a bicycle theft. Bob sues Karen for defamation and intentional infliction of emotional distress. Why will he probably succeed or fail on each claim?

11-8 Madeline enters into a contract with Canyon Canoes to go on a weeklong canoe trip down a river. The contract states that, although the firm provides experienced guides and high-quality equipment, they are not insurers of the adventurers' safety. The firm cannot be responsible for harm resulting from ordinary dangers of outdoor activities. Madeline is injured when the Coleman stove she was provided with explodes. The explosion was caused by an inadequate repair that had been made by Canyon Canoes. The company raises the defense of assumption of the risk when she sues it for negligence. Is this a valid defense? Why or why not?

11-9 Action Advertising hired Alice Jones as an account executive. She signed an employment contract under which she agreed to work for the agency for a one-year term for an annual salary of $45,000. After the manager of Creative Ads saw an exceptional set of ads that Jones had created, he called Jones, asking her whether she would be interested in changing jobs. When Jones explained that she was bound by contract for six more months, the manager said that the contract was unenforceable and further offered to double her salary if she came to work for him, because he did not believe that she was being paid what she was worth. If Jones quits and goes to work for Creative, is there a tort? If so, what would the remedy be? If not, why not?

11-10 Sam was driving in excess of the speed limit and ran a red light at 11:00 p.m. He hit Suzanne's car, which was crossing the intersection when he ran the red light. Sam had not seen Suzanne's car because of his excess speed and also because she was driving a black car and had not turned on her headlights. Suzanne suffered extensive injuries and sued Sam for negligence. Detail the manner in which she tried to prove her case and describe how Sam attempted to defend himself. How do you think the court would resolve this dispute? How would the state in which the case arose affect the outcome?

11-11 Bill is having marital difficulties and has an affair with Sara, from whom he contracts herpes. Hoping to work out his marital problems, he does not inform his wife of his infection. Four years later, Bill and his wife, Eva, divorce. A month later, before she has had any relationships with other men, Eva discovers that she has herpes. Knowing she could have contracted the disease from only one person, she sues her husband for negligence, battery, and intentional infliction of emotional distress. What arguments would she make to support each of these causes of action? Explain how you believe the court would respond to each argument.

11-12 Devo Dynamite is imploding a building. Despite taking every known safety precaution and imploding the building at a time when the least traffic is likely to be in the area, the implosion is not perfect, and Ron, a passerby, is injured by a piece of flying debris. What tort may Devo be accused of committing? Explain why Ron is either likely or unlikely to be successful in his legal action.

CASE PROBLEMS

11-13 On January 22, 2007, Nyokia Stokes got into a conflict with another third grader at her school. That evening, the conflict transferred to the parents when the other girl's mother and a male companion went to Nyokia's home and threatened her. The police were called but deferred to the school. On January 23, 2007, a principal observed what looked like a fight among several adult women in the office of the school. The women were yelling, pulling out hair, and rolling around on the ground.

At the principal's urging, the police arrested the four women. After criminal charges were dropped, the four women brought a claim for false arrest against the principal and the Board of Education of the City of Chicago. The district court granted summary judgment to the defendants and plaintiffs appealed.

On appeal, the Stokes argue that because they were not the aggressors in the fight, the principal did not have probable cause to swear out the criminal complaints that caused their arrests. In this jurisdiction, probable cause exists if, at the time of the arrest, the circumstances known to the defendant are sufficient to cause a reasonable and prudent person to believe that the person committed an offense. Did the principal have probable cause? What reasons would you use to support your argument? *Stokes v. Board of Education of Chicago*, 599 F.3d 617 (7th Cir. 2010).

11-14 Oprah Winfrey was sued for, in pertinent part, intentional infliction of emotional distress after making statements about Lerato Nomvuyo Mzamane's performance as headmistress of the Oprah Winfrey Leadership Academy for Girls (OWLAG). OWLAG is a private academy in South Africa that provides education for children from impoverished families. After learning that one of the individuals working at the school had been abusive toward the students, Winfrey made statements to the effect that she had "lost confidence in [Mzamane's] ability to run the school," that Mzamane had failed to take student complaints "seriously," that the abusers thought they were "protected by [Mzamane]," and other similar statements.

Winfrey moved for summary judgment. How do you think the court ruled? Does it matter whether this is a jurisdiction that requires a physical injury to succeed on a claim for intentional infliction of emotional distress? Are there any other claims that the plaintiff could or may have pursued in this case? *Mzamane v. Winfrey*, 2010 U.S. Dist. LEXIS 23491 (2010).

11-15 In 2006, a 21-year-old girl named Christina Eilman was arrested at Midway Airport for causing a disturbance. Although she was mentally ill, a police officer took her from the airport and dropped her off in a neighborhood well known for being unsafe. Her family later pointed out that she was dressed in a way that attracted attention, she was unaware of where she was and mentally unable to appreciate the dangers surrounding her. Furthermore, she was a young white girl and her family argued that the officer dropped her off in a predominantly black, poverty stricken neighborhood where she stood out and could foreseeably be the victim of a crime. After being dropped off, Christina was sexually assaulted by a gang member and thrown out of a seven-story window. Her parents subsequently filed a lawsuit against the city and police officer for being negligence. How do you think the court decided? *Paine v. Cason* No. 10–1487 (2012).

11-16 Plaintiffs Mr. and Mrs. Rye were receiving medical care and family planning services at Women's Care Center of Memphis, MPLLC ("the Clinic"). Mrs. Rye had Rh negative blood, and was expected to receive an injection of RhoGAM from the Clinic during her time of pregnancy. The plaintiffs allege that Mrs. Rye did not receive this injection from her physician at the Clinic, and as a result, she became "Rh-sensitized" meaning that she had "antibodies in her body to Rh-positive blood." The plaintiffs asserted that this incident caused physical injury to Mrs. Rye, disruption of family planning, infliction of emotional distress, and future medical expenses. The plaintiffs filed a complaint alleging medical malpractice against the defendant. The defendants argued that the failure to provide Mrs. Rye with the Rh injection did not cause any harm to the patient or her baby, but did concede that it may disrupt future family planning. The trial court declined summary judgment for plaintiffs' physical injury claim. They granted summary judgment for the plaintiffs' claim of future medical expenses related to future pregnancy. The defendant appealed this decision. How do you think the court ruled on appeal? Why? *Rye v. Women's Care Ctr. of Memphis, Tenn.* App. LEXIS 131 (2014).

11-17 Defendant Jean Mincolla spread a rumor that Mark Yonaty was gay. Mincolla's alleged reasons were to cause Yonaty's girlfriend to believe the rumors and thus breakup with him. After Mincolla spread the rumor to close family friends and family members of the girlfriend, Yonaty decided to file a lawsuit against Mincolla for defamation. Yonaty argued that telling others that one is gay or bisexual was insulting and embarrassing enough to constitute defamation. On the other hand, Mincolla's side presented the argument that although Mincolla was trying to ruin Yonaty's relationship, ultimately being called gay or bisexual was not shameful enough to constitute defamation. What was the court's landmark decision in this case? *Yonaty v. Mincolla*, WL 1948006 (2012).

11-18 In New York, lawsuits involving injuries caused by domestic animals are decided by applying a theory of strict liability. The suing party must demonstrate that the owner of the animal was aware of the animal's tendencies to act dangerously. In 2007, Karen Hastings was driving her car down a road at 1:30 a.m. when she crashed her car into a cow that had wandered onto the dark road from Laurier Sauve's property. Hastings argued that an owner should be liable for a large animal escaping from its enclosure because there should be negligence claims available to injured persons affected by the wandering animals. Even if the animals aren't vicious, she pointed out that they were still dangerous when roaming free, seeing as she could not see the large animal on the dark road when she was driving in the middle of the night. The defendant argued that Hastings must prove that he was aware that the animal had a tendency or an inclination to cause harm to others, which he believed she couldn't. How did the court decide? *Hastings v. Sauve* 2012 NY Slip Op 02535 (2012).

11-19 Plaintiffs David Peshlakai and Darlene Thomas, husband and wife, were badly injured, and their daughters, Deshauna and Del Lynn Peshlakai, were killed, when their vehicle was crashed into by defendants James Ruiz and Gilbert Mendoza. Ruiz and Mendoza had very high blood alcohol levels at the time of the crash. According to the plaintiffs, Ruiz and Mendoza were continuously served alcohol by an Applebee's restaurant, regardless of their apparent drunkenness. The plaintiffs brought suit for wrongful death, personal injuries, and loss of consortium and other damages against Ruiz and Mendoza, as well as the franchisee responsible for the Applebee's Neighborhood Grill in Santa Fe, and Applebee's International, Inc. How do you think the court ruled? Why? *Peshlakai v. Ruiz*, 39 F. Supp. 3d 1264 (d.n.m. 2014).

11-20 Plaintiff Victor Restis was an entrepreneur for the shipping industry in Greece. His company, Enterprises Shipping and Trading s.a. ("EST"), was responsible for a fleet with a large commercial value. The plaintiff alleges that the defendant, a not-for-profit corporation named Uani, directed a campaign against the plaintiff which included defamatory publications, or a "name-and-shame campaign." Uani identifies as a corporation that "seeks to prevent Iran from fulfilling its ambition of obtaining nuclear weapons" and "engages in private sanction campaigns and legislative initiatives focused on ending corporate support of the Iranian regime." The plaintiff argues that the defendant's campaign falsely accused the plaintiff company of engaging in illicit business dealings with Iranian oil and shipping entities. The plaintiff brought suit against the defendant, alleging that the defendant tortuously interfered with the plaintiff's business relationships. How do you think the court ruled? Is the plaintiff required to show that the defendant acted solely out of malice? 2014 U.S. Dist. LEXIS *Restis v. Am. Coalition against Nuclear Iran, Inc.*, 2014 U.S. Dist. LEXIS 139402 (s.d.n.y. 2014).

THINKING CRITICALLY ABOUT RELEVANT LEGAL ISSUES

In 2004, at Governor Schwarzenegger's urging, California adopted a budget measure that involved taxing punitive damages awards 75 percent. What this measure did was to take a punitive damages award in a civil suit, give 25 percent of the award to the plaintiff, and then put the other 75 percent of the award in a state fund to be used for state needs. California was right to enact this legislation, and other states would be wise to do so as well. Governor Schwarzenegger has been an innovative leader, and even though he did not sign the bill to renew the legislation in 2006, he said that the idea was still good, but the legislation needed work before he would renew the law. Following the governor's lead, other states should enact similar legislation to help meet state budgetary needs.

The idea of taxing punitive damages and giving a large percentage of the money to the state is useful because so many parties benefit from the arrangement. The plaintiff still receives some of the punitive damages, the tortfeasor is still punished for the wrongdoing that sparked the civil suit, and the taxpayers benefit from the state's having more money to spend on its citizens. Besides, the purpose of punitive damages is to punish the tortfeasor, not to benefit the plaintiff, and California's law still allows tortfeasors to be punished for their egregious behavior. The punitive element is still there, but what is different is that more people benefit from the tortfeasor's wrongdoing, not just the one party that brought suit against the tortfeasor. The plaintiff is still compensated through compensatory damages, so she is not losing money she is owed. Rather, she no longer is awarded a windfall that was meant as punishment for another, not as reward for her injury.

Some critics argue that the tax is a bad idea because juries will be more likely to award higher punitive damages in order to bring in more tax revenue for the state. Higher damages, however, would help to further punish wrongdoers, as well as aid the state's needs. Also, the critics fail to account for the fact that the vast majority of punitive damage awards are greatly reduced on appeal; thus, any initial raise in the damages assessed would probably be reduced on appeal, essentially making no difference in the final amount of punitive damages against the tortfeasor. The policy should make no significant change in the amount of damages awarded but will instead benefit more people, which should be the goal of any good state policy.

1. What are the issue and conclusion of this essay?

2. Is significant information missing from the preceding argument?

Clue: What pieces of information would better aid you in deciding whether you agree with the author?

3. What ethical norm does the author appear to rely upon most in making the preceding argument?

4. Write an essay that someone who holds an opinion opposite to that of the essay the author might write.

Clue: What other ethical norms could influence an opinion on this issue?

ASSIGNMENT ON THE INTERNET

Now that you have a basic understanding of tort law, use the Internet and your critical thinking skills to evaluate arguments for tort reform. Find at least two websites, other than those that follow, with opposing views on the issue of tort reform. What is the primary argument of each one? What ethical norms support each argument?

After evaluating the arguments of others, formulate your own position about one aspect or area of tort reform. What ethical norms did you use in formulating your position?

The following links may be helpful in searching for other sites with positions on tort reform.

ON THE INTERNET

www.newsbatch.com News Batch is a website dedicated to helping explain policy issues in such a manner that the general public has a better chance of understanding and following policy debates. News Batch integrates many charts and graphs into its policy summaries in an attempt to facilitate understanding of important policy issues. News Batch also supplies multiple links to other pro and antitort reform sites. Simply click on "Tort Reform" on the right side of the page.

www.whatistortreform.com What Is Tort Reform is a website that helps you figure out exactly what "tort reform" is. The website contains links to other sites that help explain relevant topics related to the issue of tort reform.

http://www.tortdeform.com/ Tortdeform: the Civil Justice Defense Blog contains posts discussing the dangers of certain types of tort reform.

www.law.cornell.edu Use the Wex Legal Encyclopedia at this website to explore recent court cases involving torts.

lp.findlaw.com This site provides several links to tort-law resources, such as state and local laws, cases, and several databases. Click on "Corporate Counsel," then under the heading "Litigation and Disputes," click on "Civil Litigation."

FOR FUTURE READING

Cain, Terrence, "Third Party Funding of Personal Injury Tort Claims: Keep the Baby and Change the Bathwater." *Chicago-Kent Law Review* 89 (2014): 11.

Durkin, Chelsea Sage. "How Strong Stands the Federal Tort Claims Act Wall? The Effect of the Good Samaritan and Negligence Per Se Doctrines on Governmental Tort Liability." *Arizona State Law Journal* 39 (2007): 269.

Klutinoty, Maria C. "*Exxon Shipping Co. v. Baker*: Why the Supreme Court Missed the Boat on Punitive Damages." *Akron Law Review* 43 (2010): 203.

Lytton, Timothy D. "Clergy Sexual Abuse Litigation: The Policymaking Role of Tort Law." *Connecticut Law Review* 39 (2007): 809.

Oster, Jan. "Communication, Defamation and Liability of Intermediaries." *Legal Studies* 35 (2015): 348.

Angerer II, Ronald P. "Moving Beyond a Brick and Mortar Understanding of State Action: The Case for a More Majestic State Action Doctrine to Protect Employee Privacy in the Workplace." *Charlotte Law Review* 4 (2013): 1.

Steele, Jenny. *Tort Law: Text, Cases, and Materials*. New York: Oxford University Press, 2007.

CHAPTER TWELVE

Product and Service Liability Law

- THEORIES OF RECOVERY IN PRODUCT LIABILITY CASES
- MARKET SHARE LIABILITY
- SERVICE LIABILITY
- GLOBAL DIMENSIONS OF PRODUCT LIABILITY LAW

When consumers enter a store to purchase a product, they assume that the product will do the job the manufacturer claims it will do without injuring anyone, and the consumer may not be aware that each year more than 33.4 million injuries and around 28,200 deaths result from the use of products purchased in the United States.[1] Deaths, injuries, and property damage from consumer products incidents cost the nation more than $700 billion annually.[2] Estimates of the number of resultant product liability cases range as high as 1 million a year. Also, the verdicts for defective-product or product liability cases are increasing from year to year. The total of the five largest awards for product defect cases in 2009 was 52 percent larger than the total in 2008. In fact, the largest award from a 2009 product defect case amounted to around $300 million, from the Philip Morris tobacco case. Also, in 2008, only 1 of the 50 largest awards were the result of a verdict in a product defect case, but in 2009, 5 of the 50 largest judgments were awarded in product defect cases.[3]

Consequently, today's businessperson is likely to become involved in some aspect of product liability litigation. This chapter discusses the most significant aspects of this area of law, known as *product liability,* to help the student function as a prudent consumer and businessperson.

Product liability law developed out of tort law, discussed in Chapter 10. This chapter begins by introducing the three primary theories of recovery in product liability cases and the defenses raised in such cases. These sections are followed by an introduction to enterprise liability, a concept that has slightly broadened the potential reach of product liability cases. Closely related to product liability is service liability, discussed in the next-to-last section. The final section discusses global implications of product liability law.

---

[1] U.S. Department of Safety, http://www.yourlegalguide.com/defective-product-deaths/.
[2] U.S. Consumer Product Safety Commission, www.cpsc.gov/about/about.html (accessed July 27, 2007).
[3] John Cord, *Product Liability Statistics & Trends,* 2010. http://www.drugrecalllawyerblog.com/2010/01/product_liability_statistics_t.html.

CRITICAL THINKING ABOUT THE LAW

Manufacturers owe a certain responsibility to consumers. Consumers should be able to reasonably use a product without its causing harm to them or others. After you read the following scenario, answer the critical thinking questions that will enhance your thinking about product liability law.

Katherine purchased a can of hair spray from her local drugstore. When she removed the cap from the hair spray can, the can exploded in her hands. She suffered third-degree burns on her hands and face and was unable to work for three months.

Katherine sued the hair spray manufacturer after she discovered that another woman had suffered an identical accident when using the same brand of hair spray. The jury awarded Katherine $750,000 in compensatory damages.

1. Katherine's lawyer described a previous case in which an individual was injured because a product exploded. Two years earlier, a woman walking down a row of hair care products in a supermarket had been injured when three cans of hair spray spontaneously exploded. She lost her sight because of the explosion, and a jury awarded her $2.2 million in damages. Katherine's lawyer argued that because the previous woman had been compensated, Katherine should be awarded $2 million in damages for her injuries. Do you think the earlier case is similar enough to Katherine's case for Katherine to recover damages? Why?

   Clue: How are the cases similar and different? How does the fact that Katherine purchased the product affect your thinking about the earlier case?

2. The manufacturer argued that because it places a warning on the hair spray cans, it is free from responsibility for injury. The can states, "Warning: Flammable. Contents under pressure." The jury, however, ruled in favor of Katherine. What ethical norm seems to have shaped the jury's thought?

   Clue: Study the list of ethical norms in Chapter 1. The manufacturer argued that it should not have to assume responsibility because the can has a warning. What ethical norm is consistent with offering greater protection for the consumer?

3. What additional information about this case would make you more willing to state your own opinion about the situation?

   Clue: What information about the product would change your thinking about the responsibility of the manufacturer? For example, suppose that Katherine discovered that an identical accident had occurred with the same brand of hair spray. How might knowing the date that the similar accident occurred influence your thinking about Katherine's case?

Theories of Recovery in Product Liability Cases

Product liability law developed out of tort law. A glance at the three primary theories of recovery in product liability cases—negligence, breach of warranty, and strict product liability—reveals a relationship between product liability and tort law. A plaintiff usually brings an action alleging as many of these three grounds as possible.

NEGLIGENCE

Plaintiffs in product liability cases have traditionally used theories of negligence. To be successful, the plaintiff must prove the elements of negligence explained in Chapter 11: (1) that the defendant manufacturer owed a duty of care to the plaintiff, (2) that the defendant breached that duty of care, (3) that this breach of duty caused the plaintiff's injury, and (4) that the plaintiff suffered actual, compensable injury.

The Privity Limitation. An early problem with using negligence to recover for an injury caused by a defective product was establishing duty. Originally, a plaintiff who was not the purchaser of the defective product could not establish a duty of care and, thus, could not recover. This limitation was based on the concept of *privity*, which means that one is a party to a contract. In the earliest known product liability case, *Winterbottom v. Wright*,[4] the British court in 1842 established the rule that to recover for an injury caused by a defective product, the plaintiff must establish privity. In other words, before a manufacturer or seller of a defective good could owe a duty to the plaintiff, the plaintiff must have purchased that good directly from the defendant who manufactured it. Because plaintiffs rarely purchase goods from the manufacturer, few such suits were initially brought.

Gradually, especially in cases of defective food, courts began to eliminate the privity requirement, essentially abolishing it in the 1916 case of *MacPherson v. Buick Motor Co.*[5] In *MacPherson,* the court held the remote manufacturer of an automobile with a defective wheel liable to the plaintiff when the wheel broke and the plaintiff was injured. Judge Cardozo stated that the presence of a sale does not control the duty; if the elements of a product are such that it is harmful to individuals if negligently made, and if the manufacturer knows that the product will be used by someone other than the purchaser, then "irrespective of contract, the manufacturer of this thing is under a duty to make it carefully." The holding in the *MacPherson* case, which was quickly followed by similar holdings in other states, eliminated the privity requirement, thereby allowing a negligent manufacturer to be held responsible for a defective product that caused injuries to someone with whom the defendant manufacturer had no contract.

Eradication of the privity requirement and the subsequent increase in the liability of producers and sellers reflected a shift in social policy toward placing responsibility for injuries on those who market a product that could foreseeably cause harm if proper care were not taken in its design, manufacture, and labeling. Increasingly, the courts indicated that defendants should be responsible for their affirmative acts when they knew that such actions could cause harm to others. Also, because the manufacturer and seller derive economic benefits from the sale and use of the product, it seemed fair to impose liability on them if they earned profits from a defectively made product.

Thus, abolition of the privity limitation opened the door for negligence as a theory of liability when people were injured because of a product manufacturer's or seller's lack of care. A number of negligent acts or omissions typically give rise to negligence-based product liability actions; these are listed in Exhibit 12-1. We will discuss the most common ones: negligent failure to warn and negligent design.

Negligent Failure to Warn. Most of the negligence-based product liability actions result from a failure to warn or inadequate warning. To bring a successful negligence case for failure to warn, the plaintiff must demonstrate that the defendant knew or should have known that, without a warning, the product would be dangerous in its ordinary use or in any reasonably foreseeable use. In determining whether a reasonable manufacturer would have given a warning in a particular situation, the courts frequently consider the likelihood of the injury, the seriousness of the injury, and the ease of warning.

EXHIBIT 12-1

COMMON NEGLIGENT ACTIONS LEADING TO PRODUCT LIABILITY CASES

1. Negligent failure to warn
2. Negligent provision of an inadequate warning
3. Negligent design
4. Negligent manufacture
5. Negligent testing or failure to test
6. Negligent advertising

---
[4] 152 Eng. Rep. 402 (1842).
[5] 217 N.Y. 382, 111 N.E. 1050 (1916).

There is generally no duty to warn of dangers arising from unforeseeable misuses of a product or from obvious dangers. A producer, for example, need not give a warning that a sharp knife could cut someone. Similarly, some plaintiffs have argued that fast-food restaurants, like McDonald's, are liable to consumers for consumers' obesity-related health problems, because the restaurants failed to warn customers of the unhealthful attributes of fast food. In *Pelman v. McDonald's,* the plaintiff alleged that McDonald's failed to warn customers of the "ingredients, quantity, qualities and levels of cholesterol, fat, salt and sugar content and other ingredients in those products, and that a diet high in fat, salt, sugar and cholesterol could lead to obesity and health problems."[6] In his decision dismissing the plaintiff's claims against McDonald's, Judge Sweet specifically stated that "this opinion is guided by the principle that legal consequences should not attach to the consumption of hamburgers and other fast food fare unless consumers are unaware of the dangers of eating such food."* Because consumers know, or reasonably should know, the potential negative health effects of eating fast food, the plaintiff's claim was dismissed. But if future plaintiffs can allege that McDonald's food is dangerous in a manner not known to consumers, their claims may survive.

A defendant may give a warning in a manner not clearly calculated to reach those whom the defendant should expect to use the product. If the product is to be used by someone other than the original purchaser, the manufacturer is generally required to put some sort of warning on the product itself, not just in a manual that comes with the product. If children or those who are illiterate are likely to come into contact with the product and risk harm from its use, picture warnings may be required.

Products designed for intimate bodily use, especially drugs and cosmetics, often give rise to actions based on negligent failure to warn because the use of these products frequently causes adverse reactions. When a toxic or allergic reaction causes harm to the user of a cosmetic or an over-the-counter drug, many courts find that there is no duty to warn unless the plaintiff proves that (1) the product contained an ingredient to which an appreciable number of people would have an adverse reaction; (2) the defendant knew or should have known, in the exercise of ordinary care, that this was so; and (3) the plaintiff's reaction was due to his or her membership in this abnormal group.[7]

Other courts, however, determine negligence by looking at the particular circumstances of the case and by weighing the amount of danger to be avoided against the ease of warning. For example, in a 1995 case against McNeil Consumer Products Company, a jury awarded more than $8.8 million to a man who suffered permanent liver damage as a result of drinking a glass of wine with a Tylenol capsule. Although the corporation had known for years that combining a normal dose of Tylenol with a small amount of wine could cause massive liver damage in some people, the company failed to put a warning to that effect on the label. The jury did not accept the company's argument that the reaction was so rare that no warning was necessary.[8]

Marketing of prescription drugs is unique because the manufacturer almost never communicates directly with the user; instead, it communicates with the physician who prescribes the drug. In these cases, the courts generally hold that drug manufacturers have a duty to provide adequate warnings to physicians to enable them to decide whether to prescribe the drug or disclose the risk to the patient. The manufacturer must warn the physician of any chance of a serious adverse reaction, no matter how small. Prescription drugs are frequently the subject of product liability cases, as described in Exhibit 12-2.

---

[6]237 F. Supp. 2d 512 (S.D.N.Y. 2003).

[7]W. Page et al., *Prosser and Keeton on Torts* (5th ed.) (St. Paul, MN: West, 1984), 687.

[8]*Benedict v. McNeil Consumer Products Co.,* 1992 WL 729052 (L.R.P. Jury).

*McFat Litigation I–Pelman v. McDonald'sCorp.,237 F. Supp. 2d 512 (S.D.N.Y. 2003).

PRODUCT	CASE STATUS AND LEGAL CLAIMS
Avandia (prescription drug used to control blood sugar in Type II diabetics)	In May 2007, the *New England Journal of Medicine* released a study linking Avandia to a greatly increased risk of heart attack or heart-related death. The Food and Drug Administration (FDA) put out a safety alert and more research is being done on the safety of Avandia. The first law-suit as a result of this new information was filed in June 2007 and experts expect more to follow. As of May 2010, Avandia manufacturers made a $60 million settlement to end approximately 700 lawsuits; however, this was only the first settlement Avandia manufacturers are expected to make in regard to the drug's side effects.
Baycol (prescription drug to lower cholesterol)	Plaintiffs reportedly experienced rhabdomyolysis, a kidney disorder in which toxic muscle cells are released into the bloodstream. Patients can then develop fatal organ failure. Plaintiffs frequently bring claims of failure to warn or for a defectively designed drug. The manufacturer voluntarily removed Baycol from the market because of the legal claims it had spawned. As of January 2007, the court status update estimated that there were approximately 1,200 active cases. The status update also indicated that the manufacturer, Bayer, has settled 3,000-plus cases for more than $1.1 billion.
Fen-Phen (Redux) (drug to treat obesity)	Some patients experienced heart-valve disease after using Fen-Phen to lose weight. In January 2004, a $3.75 billion trust was created as a settlement between patients and the drug manufacturer, American Home Products, to compensate patients injured by Fen-Phen use. Under the settlement agreement, eligible patients may be entitled to compensation, diet drug prescription refunds, and echocardiography screenings.
Paxil (antidepressant) (similar claims have been brought regarding Zoloft, another antidepressant)	Patients taking Paxil reportedly had withdrawal reactions and problems: anxiety, agitation, confusion, dizziness, fatigue, headache, insomnia, irritability, nausea, palpitations, sweating, sleep disturbances, sensory disturbances, tremor, and vision distortion. As of April 2004, there were about 1,500 Paxil withdrawal plaintiffs in more than 30 states. These cases were consolidated into multidistrict litigation. Plaintiffs frequently bring the following claims: intentional misrepresentation, fraud, negligence, strict liability, and breach of warranty. Paxil has also been linked to increased suicide risk in teens and has faced many lawsuits on that front.
Prempro (prescription drug to relieve menopausal symptoms)	Researchers determined that women taking Prempro were more likely to suffer breast cancer, stroke, heart disease, blood clots, and dementia. After this research, the FDA approved new labels emphasizing these increased risks; however, Prempro still remains on the market. Approximately 6 million women had been taking Prempro before the researchers announced the increased health risks associated with Prempro use. The first of the lawsuits against the manufacturer was heard in August 2006 and several suits have been found for the plaintiffs since then, with millions in damages awarded. In one court case in 2010, a woman was awarded $9.54 million; another case in 2007 yielded the astounding verdict of $134 million. As of 2010, many cases are ongoing.
Vioxx (NSAID, COX-2 inhibitor)	Vioxx is a painkiller marketed to treat pain from osteoarthritis. Vioxx has been linked to increased risk of heart attacks and strokes among users. In 2004, the manufacturer pulled Vioxx from the market in response to results of an FDA study. As of July 2007, the manufacturer still faced more than 27,000 lawsuits. A $4.85 billion fund was created by Merck, the manufacturer, to cover those suits. Specifically, a $4 billion fund was created to cover those who had suffered heart attacks after using the drug, and another $850 million fund for those who suffered strokes as a result of using the drug. As of 2010, many lawsuits are still pending.

EXHIBIT 12-2

PRESCRIPTION DRUGS THAT HAVE LED TO PRODUCT LIABILITY CLAIMS

Zicam (over-the-counter nasal gel to remedy the common cold)	Plaintiffs contend that after using Zicam, they lost their sense of smell and taste. Plaintiffs argue that the Zicam manufacturer knew or should have known about the potential dangers associated with the use of nasal medications containing zinc. Use of zinc can cause nerve damage. Furthermore, plaintiffs argue that the manufacturer failed to provide sufficient warnings to the users of the products even though the side effects of zinc compounds have been documented. In January 2006, the manufacturer settled with 340 plaintiffs for $12 million.
Accutane (drug intended to treat severe acne)	Accutane is an acne medication that was used by millions of people to treat severe acne. In 2009, it was linked to the emergence of inflammatory bowel disease in users of the acne drug who had had no prior health problems related to the disease. On June 29, 2009, Hoffmann-La Roche announced a nationwide Accutane recall. Since the June 2009 Accutane recall, six court decisions have resulted in about $56 million in damages being paid to users who contracted inflammatory bowel disease from using Accutane by Roche.
Fosamax (anti-osteoporosis drug)	The *Journal of Oral and Maxillofacial Surgeons* released a report in 2004 that linked Fosamax to osteonecrosis of the jaw (ONJ). The FDA swiftly issued warnings about the drug, distributed by Merck. ONJ causes the decay and subsequent death of the bone matter associated with the jaw. As a result of this health defect, the drug lost its patent protection in 2008 and is no longer one of Merck's most financially successful drugs. Furthermore, Merck has set aside millions of dollars to battle dozens of lawsuits over the drug. The suits have resulted in mistrials, and both successes and defeats for Merck.
Zyprexa (drug to treat schizophrenia) Seroquel (antipsychotic drug from AstraZeneca)	In 2005, it was determined that Zyprexa and Seroquel led to severe weight gain in those who took the drugs. The pronounced effects of the drugs associated with patients' weight put patients in danger of contracting diabetes, among other health issues related to weight gain. Lawmakers claim that manufacturers refused to release prior knowledge of the weight-gain side effect and thus improperly marketed the drug. In 2005, the drug manufacturer Eli Lilly settled around 8,000 lawsuits, paying around $700 million to those patients affected by Zyprexa. AstraZeneca agreed to a $520 million settlement.
Ortho Evra (birth-control patch)	In 2006, clinical trial results that were released linked Ortho Evra to blood clots that could result in strokes. The drug, manufactured and distributed by Johnson & Johnson, comes in the form of a birth-control patch. Lawmakers claim that Johnson and Johnson had prior knowledge of this side effect yet did not release the information to the public and left it out of the drug's advertising. Johnson & Johnson settled in court for $1.25 million in 2007.
Yaz (birth control)	At least 40 lawsuits popped up in 2009 against Bayer Pharmaceuticals because of its Yaz birth-control drug. The lawsuits claimed that inadequate information about serious side effects was released to the public through the marketing of the product. These side effects include, but are not limited to, heart attacks, stroke, gallbladder disease, and sudden death. In fact, Yaz is the only birth control that contains both ethinyl estradiol and drospirenone, with the latter allegedly making the drug very dangerous. As of 2010, many of the lawsuits had been consolidated into large class action lawsuits and are pending in state courts from Florida to Ohio to New Jersey.

EXHIBIT 12-2 CONTINUED

Initially, almost all successful product liability actions based on negligence were for breach of the duty to warn. The range of successful actions was so limited because people believed that competition and an open market provided the best means for ensuring that products will have optimal safety features. Believers in the sanctity of the market feel that the manufacturer's job is to see that the purchaser is an informed purchaser and is not deceived about the safety of a product.[9]

Negligent Design. The foregoing attitude generally prevailed until approximately 1960, when the courts began, in a limited number of cases, to impose liability based on negligence in the sale of defectively designed products. Such liability is imposed only when a reasonable person would conclude that despite any warnings given with the product, the risk of harm outweighed the utility of the product as designed. Courts have found a wide variety of products to be negligently designed, including weed killers, gas stations, BB guns, airplanes, and traffic signs. One example of negligent design can be found in a 2010 case against Boston Scientific Corp.'s Guidant unit. The corporation was sued because it did not warn medical professionals and the United States Food and Drug Administration that some of the implantable heart defibrillators it was producing would short-circuit. The short-circuiting defects resulted in the deaths of many patients who had the medical device implanted. Furthermore, company officials had been aware of the defects for at least three years but refused to disclose the information. The corporation pled guilty and agreed to a settlement of $296 million. Thus, the design of the product was considered faulty, as the resulting deaths could have been prevented through the disclosure of the defects and spending resources on modifications and further testing.

In bringing an action for negligence in design, a plaintiff must generally prove that the product design (1) is inherently dangerous, (2) contains insufficient safety devices, or (3) consists of materials that do not satisfy standards acceptable in the trade.

Usually an action for product liability based on negligence is accompanied by a strict liability claim, which is easier to prove. With the growing use of strict liability and the broad range of defenses to negligence, negligence has become less important as a theory of liability.

Negligence per se. As stated in Chapter 11, violation of a statutory duty is considered *negligence per se*. This concept is used in negligence-based product liability cases.

When a statute establishes product standards, the manufacturer has a duty to meet those standards. A manufacturer that does not meet those standards has breached its duty of care. As long as the plaintiff can establish that the breach of the statutory duty caused injury, the plaintiff can recover under a theory of negligence per se.

Statutes that might be violated and lead to negligence per se actions include the Flammable Fabrics Act of 1953; the Food, Drug, and Cosmetics Act of 1938; and the Hazardous Substances Labeling Act of 1960.

APPLYING THE LAW TO THE FACTS...

Let's say that Rachel bought a sweater made of wool along with a mix of other materials. Rachel went to a bonfire with her friends and a small spark from the fire landed on her sweater, immediately causing it to burst into flames. If the manufacturer of Rachel's sweater did not comply with the duties laid out in the Flammable Fabrics Act, what action could Rachel take against the sweater manufacturer? What would she need to prove to win her case?

---

[9]R. Coase, "The Problem of Social Cost," *Law and Economics* 3: 1 (1960).

Defenses to a Negligence-Based Product Liability Action. All of the defenses discussed in the negligence section of Chapter 11 are available in product liability cases based on negligence. Remember that the plaintiff's own failure to act reasonably can provide a defense. Depending on the state in which the action is brought, the plaintiff's negligence will allow the defendant to raise the defense of contributory, modified comparative, or pure comparative negligence. If contributory negligence is proved, the plaintiff is barred from recovery. In a state where the defense of pure comparative negligence is allowed, the plaintiff can recover for only that portion of the harm attributable to the defendant's negligence. In a state that follows modified contributory negligence, the plaintiff can recover the percentage of harm caused by the defendant as long as the jury finds that the plaintiff's negligence was responsible for less than 50 percent of the harm. So, if a jury finds the defendant to be responsible for 60 percent of the plaintiff's harm, the plaintiff could recover nothing in a contributory negligence state, but could recover damages for 60 percent of his or her injuries in a modified or pure comparative negligence state. If the defendant were only 40 percent responsible, however, the plaintiff would be able to recover 40 percent of his or her injuries only in the pure comparative negligence state and nothing in the other two.

Another defense available in product liability cases based on negligence is assumption of the risk. A plaintiff is said to assume the risk when he or she voluntarily and unreasonably encounters a known danger. If the consumer knows that a defect exists but still proceeds unreasonably to make use of the product, he or she is said to have voluntarily assumed the risk of injury from the defect and cannot recover.

In deciding whether the plaintiff did indeed assume the risk, the trier of fact may consider such factors as the plaintiff's age, experience, knowledge, and understanding. The obviousness of the defect and the danger it poses are also relevant factors. If a plaintiff knows of a danger but does not fully appreciate the magnitude of the risk, the applicability of the defense is a question for the jury. In most cases, an employee using an unsafe machine at work is not presumed to have assumed the risk, because most courts recognize that the concept of voluntariness is an illusion in the workplace. Earlier, however, employees attempting to sue manufacturers of defective machines for injuries at work were defeated by this defense.

In many states, misuse of the product is raised as a defense in negligence-based product liability cases. To constitute a valid defense, such misuse must be unreasonable or unforeseeable. A defendant raising the defense of product misuse is really arguing that the harm was caused not by the defendant's negligence but by the plaintiff's failure to use the product in the manner in which it was designed and intended to be used.

Statutory defenses are also available to defendants. To ensure that there will be sufficient evidence from which a trier of fact can make a decision, states have **statutes of limitations** that limit the time within which all types of civil actions may be brought. In most states, the statute of limitations for tort actions, and thus for negligence-based product liability cases, is two to four years from the date of injury. Maine, however, has a six-year statutes of limitations. Kentucky, Louisiana, and Tennessee are the only states with one-year statutes of limitations. If the injured party is a minor, the statute of limitations does not start running until the injured party reaches age 18.

States also have **statutes of repose**, which bar actions arising more than a specified number of years after the product was purchased. Statutes of repose are usually much longer than statutes of limitations; they are often at least 10 years, and frequently are 25 or 50 years. Statutes of repose may seem unduly harsh on consumers who may be injured as a result of a latent defect in a product. In contrast, to make the manufacturer liable in perpetuity may be unduly

statute of limitations A statute that bars actions arising more than a specified number of years after the cause of the action arises.

statute of repose A statute that bars actions arising more than a specified number of years after the product was purchased.

state-of-the-art defense A product liability defense based on adherence to existing technologically feasible standards at the time the product was manufactured.

harsh on manufacturers and sellers, because of the resulting uncertainty about possible liability. Those who worry about the seller's liability, however, should remember that the older the product is, the more difficult it will be for the plaintiff to prove negligence on the part of the defendant.

A defendant may use the **state-of-the-art defense** to demonstrate that the alleged negligent behavior was reasonable, given the available scientific knowledge existing at the time the product was sold or produced. In a case based on the defendant's negligent defective design of a product, the state-of-the-art defense refers to the technological feasibility of producing a safer product at the time the product was manufactured. In cases of negligent failure to warn, the state-of-the-art defense refers to the scientific knowability of a risk associated with a product at the time of its production. This defense is valid in a negligence case because the focus is on the reasonableness of the defendant's conduct. Nevertheless, demonstrating that, given the state of scientific knowledge, there was no feasible way to make a safer product does not always preclude liability. The court may find that the defendant's conduct was still unreasonable because even in the product's technologically safest form, the risks posed by the defect in the design so outweighed the benefits of the product that the reasonable person would not have produced a product of that design.

An earlier section showed that failure to comply with a safety standard may lead to the imposition of liability. An interesting question is whether the converse is true: Does compliance with safety regulations constitute a defense? There is no clear answer to that question. Sometimes, however, compliance with federal laws may undergird a defense that use of state tort law is preempted by a federal statute designed to ensure the safety of a particular class of products. The following case illustrates one situation in which the Supreme Court had to determine whether a corporation could be found guilty of product liability after the company had complied with federal regulations, or whether the federal regulations preempted state product liability law.

CASE 12-1

Mutual Pharmaceutical Company, Inc. v. Bartlett
United States Supreme Court
133 S. Ct. 2466 (2013)

In 1978, the federal Food and Drug Administration (FDA), approved the nonsteroidal anti-inflammatory pain reliever sulindac, sold under the brand name Clinoril. The FDA also approved its labeling, which included warnings of potential side effects. When the company's patent expired, other companies, including Mutual Pharmaceutical Company, produced generic versions of the drug. As required by federal law, the companies selling the generic drugs used the exact same labels as approved for the original drug, without any modifications.

Karen Bartlett was prescribed Clinoril for shoulder pain, and the pharmacy filled the prescription with the generic form of the drug manufactured by Mutual Pharmaceutical. Shortly after starting to use the pain reliever, she developed an acute case of toxic epidermal necrolysis. As a consequence, she became severely disfigured and nearly blind. The label had not warned of the potential for toxic epidermal necrolysis, although subsequently, however, the FDA recommended changing all NSAID labeling to contain a more explicit toxic epidermal necrolysis warning.

Bartlett sued Mutual Pharmaceutical Company under the New Hampshire state design-defect law. A jury found in her favor and awarded her over $21 million in damages. The First Circuit Court of Appeals affirmed the decision, finding that neither the FDCA nor the FDA's regulations preempted respondent's design-defect claim. It distinguished *PLIVA, Inc. v. Mensing*, a previous Supreme Court case holding that failure-to-warn claims against generic manufacturers are preempted by the FDCA's prohibition on changes to generic drug labels, by stating that generic manufacturers facing design-defect claims could comply with both federal and state law simply by choosing not to make the drug at all.

Mutual Pharmaceuticals appealed to the U.S. Supreme Court.

Justice Alito

We must decide whether federal law preempts the New Hampshire design-defect claim under which respondent Karen Bartlett recovered damages from petitioner Mutual Pharmaceutical, the manufacturer of sulindac, a generic nonsteroidal anti-inflammatory drug (NSAID). New Hampshire law imposes a duty on manufacturers to ensure that the drugs they market are not unreasonably unsafe, and a drug's safety is evaluated by reference to both its chemical properties and the adequacy of its warnings.

Because Mutual was unable to change sulindac's composition as a matter of both federal law and basic chemistry, New Hampshire's design-defect cause of action effectively required Mutual to change sulindac's labeling to provide stronger warnings. But, as this Court recognized just two Terms ago in *PLIVA, Inc. v. Mensing*, 564 U.S. ___ (2011), federal law prohibits generic drug manufacturers from independently changing their drugs' labels. Accordingly, state law imposed a duty on Mutual not to comply with federal law. Under the Supremacy Clause, state laws that require a private party to violate federal law are preempted and, thus, are "without effect.". . .

Accordingly, we hold that state-law design-defect claims that turn on the adequacy of a drug's warnings are preempted by federal law under *PLIVA*.*

Reversed, in favor of Appellant, Mutual Pharmaceutical Company.

Each preemption case requires careful scrutiny of the purpose of the statute. Automobile manufacturers have attempted to use the preemption defense, but have met with limited success. For example, in 2000, a passenger who was wearing a seat belt was injured in an automobile accident, and sued Honda under state tort law, arguing that had the car had a driver's-side air bag in addition to the seat belt, she would not have been so seriously injured. The U.S. Supreme Court agreed that the state action was preempted by Honda's compliance with the Federal Motor Vehicle Safety Standard Act, which required auto manufacturers to equip some, but not all, of their vehicles with passive restraints.

Oil companies have had mixed results in cases in which they have argued that the Clean Air Act preempts their liability for methyl tertiary butyl ether (MTBE) water contamination. In an attempt to reduce air pollution in certain areas, Congress mandated that gasoline contain an oxygenate, which allows gasoline to burn more cleanly. Congress, however, did not require that a particular oxygenate be used. Most oil companies added the oxygenate MTBE to gasoline. Unfortunately, even very small amounts of MTBE can contaminate drinking water: It affects the smell and taste of water and may cause health problems. Public water utilities and private individuals, faced with millions of dollars in cleanup costs, began suing oil companies for damages to resolve the water contamination. In response, oil companies, raising the preemption defense, argue that because the government required them to add an oxygenate to gasoline, oil companies should not be responsible for the water contamination. Plaintiffs argue that oil companies chose to use MTBE; other oxygenates, such as ethanol, were available. Several courts have concluded that the Clean Air Act preempts oil companies' liability, but other courts have reached the opposite conclusion, emphasizing the fact that oil companies had a choice. Furthermore, these courts hold that the purpose of the Clean Air Act was to address air pollution, and the MTBE water contamination is a problem too far removed from the purposes of the Clean Air Act to preempt product liability claims associated with MTBE.

STRICT LIABILITY IN CONTRACT FOR BREACH OF WARRANTY

The Uniform Commercial Code (UCC) provides the basis for recovery against a manufacturer or seller on the basis of breach of warranty. A **warranty** is a guarantee or a binding promise. Warranties may be either **express** (clearly stated by the seller or manufacturer) or **implied** (automatically arising out of a transaction). Either type may give rise to liability (see Exhibit 12-3). Two types of implied warranties may provide the basis for a product liability action: warranty of merchantability and warranty of fitness for a particular purpose.

warranty A guarantee or binding promise that goods (products) meet certain standards of performance.

express warranty A warranty that is clearly stated by the seller or manufacturer.

implied warranty A warranty that automatically arises out of a transaction.

---

*Mutual Pharmaceutical Company, Inc. v. Barlett United States Supreme Court, 133 S. Ct. 2466 (2013).

EXHIBIT 12-3

WARRANTIES THAT MAY GIVE RISE TO LIABILITY

EXPRESS WARRANTIES	IMPLIED WARRANTIES
Written or oral description of the good	Of merchantability
Promise or affirmation of fact about the good	Of fitness for a particular purpose
Sample or model of the good	

implied warranty of merchantability A warranty that a good is reasonably fit for ordinary use.

Implied Warranty of Merchantability. The **implied warranty of merchantability** is a warranty or guarantee that the goods are reasonably fit for ordinary use. This warranty arises out of every sale, unless it is expressly and clearly excluded. According to the UCC, to meet the standard of merchantability, the goods

1. must pass without objection in the trade under the contract description;
2. must be of fair or average quality within the description;
3. must be fit for the ordinary purpose for which the goods are used;
4. must run, with variations permitted by agreement, of even kind, quality, and quantity within each unit and among all units involved;
5. must be adequately contained, packaged, and labeled as the agreement may require; and
6. must conform to any affirmations or promises made on the label or the container.[10]

If the product does not conform to those standards and, as a result of this nonconformity, the purchaser or his or her property is injured, the purchaser may recover for breach of implied warranty of merchantability. For example, some health care workers are using this theory to try to recover damages because they have developed latex protein toxic syndrome. When HIV rates began climbing, health workers uniformly used latex gloves to prevent transmission of the disease. Some health workers, however, started to develop an allergy to the proteins in the rubber latex. For some people, the allergy became so severe that they could not tolerate being in a room with a single latex balloon; consequently, they could no longer work in a profession that required them to wear latex gloves. These plaintiffs argue that glove manufacturers knew of the dangers of latex allergies but did not attempt to minimize the amount of protein in the gloves.

The UCC expressly provides that an injury to a person or property proximately caused by a breach of warranty is a recoverable type of consequential damage. The use of this warranty is limited, however, in two respects: (1) It is made only by one regularly engaged in the sale of that type of good, and (2) the seller may sometimes avoid liability by expressly disclaiming liability or by limiting liability to replacement of the defective goods. The UCC, however, has restricted applicability of the latter limitation by a provision declaring that a limitation of consequential damages for injury to a person in the case of consumer goods is prima facie "unconscionable." *Unconscionability* is a concept meaning gross unfairness. Under the UCC, unconscionable contract clauses are unenforceable. Thus, if a disclaimer is unconscionable, it will not be enforced.

Privity is not a problem in an action based on breach of warranty, because of UCC Section 2-318. This section allows states to adopt one of three alternatives to allow nonpurchasers to recover for breach of warranty. The most liberal alternative allows any person injured by the defective product to sue. One issue that frequently arises in product liability cases involving breach of the warranty

---
[10] Uniform Commercial Code, UCC § 2-314.

COMPARATIVE LAW CORNER

Warranties and Guarantees in the United Kingdom

Product liability laws in the United Kingdom are similar to many laws in the United States. There is a difference, however, in some of the language used to describe those laws. A consumer may sue under breach of implied or express warranty in the United States. In the United Kingdom, a *warranty* is a legally binding assurance that any manufacturing defects that appear in a product during a certain time period will be addressed. The consumer buys the warranty and it serves as an insurance policy for the product.

A guarantee in the United Kingdom is similar to an express warranty in the United States. A *guarantee* is a free promise made to the consumer by the producer to repair any manufacturing problems found within a certain amount of time. Guarantees are also legally binding.

A consumer who would sue under breach of implied warranty in the United States would sue under breach of contract under the Sale of Goods Act in the United Kingdom. If a product is not fit for its purpose or not of satisfactory quality, problems covered by implied warranties in the United States, the Sale of Goods Act would apply in the United Kingdom. These legal distinctions are important to know before doing business or purchasing goods in the United Kingdom.

---

of merchantability is whether this warranty is breached when the alleged breach arises from a naturally occurring characteristic of the product. This problem typically arises in cases involving food. Is it a breach of the warranty of merchantability when there is a bone in a fish fillet or a pit in an olive jar labeled "pitted olives?" The following case sets forth the two tests that are used in various jurisdictions.

CASE 12-2

Williams v. Braum Ice Cream Stores, Inc.
Oklahoma Court of Appeals
534 P.2d 700 (1974)

Plaintiff Williams purchased a cherry pecan ice cream cone from the defendant's shop. While eating the ice cream, she broke her tooth on a cherry pit that was in the ice cream. She sued defendant Braum Ice Cream Stores, Inc., for breach of implied warranty of merchantability. The trial court ruled in favor of the defendant, and the plaintiff appealed.

Judge Reynolds

There is a division of authority as to the test to be applied where injury is suffered from an object in food or drink sold to be consumed on or off the premises. Some courts hold there is no breach of implied warranty on the part of a restaurant if the object in the food was "natural" to the food served. These jurisdictions recognize that the vendor is held to impliedly warrant the fitness of food, or that he may be liable in negligence in failing to use ordinary care in its preparation, but deny recovery as a matter of law when the substance found in the food is natural to the ingredients of the type of food served. This rule, labeled the "foreign-natural test" by many jurists, is predicated on the view that the practical difficulties of separation of ingredients in the course of food preparation (bones from meat or fish, seeds from fruit, and nutshell from the nut meat) is a matter of common knowledge. Under this natural theory, there may be a recovery only if the object is "foreign" to the food served. How far can the "foreign-natural test" be expanded? How many bones from meat or fish, seeds from fruit, nutshells from the nut meat or other natural indigestible substances are unacceptable under the "foreign-natural test"?

The other line of authorities hold[s] that the test to be applied is what should "reasonably be expected" by a customer in the food sold to him.

[State law] provides in pertinent part as follows:

1. ... a warranty that the goods shall be merchantable is implied in a contract for their sale if the seller is a merchant with respect to goods of that kind. Under this section, the serving for value of food or drink to be consumed either on the premises or elsewhere is a sale.

2. Goods to be merchantable must be at least such as
   a. are fit for the ordinary purposes for which such goods are used; . . . In *Zabner v. Howard Johnson's Inc.* . . . the Court held:

   The "foreign-natural" test as applied as a matter of law by the trial court does not recommend itself to us as being logical or desirable. The reasoning applied in this test is fallacious because it assumes that all substances which are natural to the food in one stage or another of preparation are, in fact, anticipated by the average consumer in the final product served. . . .

Categorizing a substance as foreign or natural may have some importance in determining the degree of negligence of the processor of food, but it is not determinative of what is unfit or harmful in fact for human consumption. A nutshell natural to nut meat can cause as much harm as a foreign substance, such as a pebble, piece of wire, or glass. All are indigestible and likely to cause injury. Naturalness of the substance to any ingredients in the food served is important only in determining whether the consumer may reasonably expect to find such substance in the particular type of dish or style of food served.

The "reasonable expectation" test as applied to an action for breach of implied warranty is keyed to what is "reasonably" fit. If it is found that the pit of a cherry should be anticipated in cherry-pecan ice cream and guarded against by the consumer, then the ice cream was reasonably fit under the implied warranty.

In some instances, objects which are "natural" to the type of food but which are generally not found in the style of the food as prepared, are held to be the equivalent of a foreign substance.

We hold that the better legal theory to be applied in such cases is the "reasonable expectation" theory, rather than the "naturalness" theory as applied by the trial court. What should be reasonably expected by the consumer is a jury question, and the question of whether plaintiff acted in a reasonable manner in eating the ice cream cone is also a fact question to be decided by the jury.*

Reversed and *remanded* in favor of Plaintiff, Williams.

CRITICAL THINKING ABOUT THE LAW

The criteria selected are important in determining the outcome of a case. Put simply, depending on the court's selection from many possible criteria, it can reach multiple conclusions. Judging a case according to criteria X, Y, and Z can yield a vastly different decision than if the same case were judged according to criteria A, B, and C.

Case 12-2 illustrates the foregoing assertion. The trial court had made a legal decision based on criterion X, namely, the "foreign-natural" test. The appeals court, however, held that the trial court must redecide the case, this time on the basis of criterion Y, or the "reasonable expectation" test.

The critical thinking questions enable you to examine carefully the key differences between the two tests, including the possible implications. The larger project of the questions is to increase your awareness of the extent to which a legal decision is dependent upon the criteria chosen to reach that decision.

1. What is the fundamental difference between the natures of the two tests discussed by the court in Case 12-2?

   Clue: Reread the discussion of the two tests to formulate your answer.

2. Which of the two tests is more likely to yield ambiguous reasoning when applied?

   Clue: Refer to your answer in question 1.

implied warranty of fitness for a particular purpose A warranty that arises when the seller tells the consumer a good is fit for a specific use.

Implied Warranty of Fitness for a Particular Purpose. A second implied warranty that may be the basis for a product liability case is the **implied warranty of fitness for a particular purpose**, which arises when a seller knows that the purchaser wants to purchase a good for a particular use. The seller tells the consumer that the good can be used for that purpose, and the buyer reasonably relies on the seller's expertise and purchases the product. If the good cannot be used for that purpose, and as a result of the purchaser's attempt to use the good for that purpose, the consumer is injured, a product liability action for breach of warranty of fitness for a particular purpose is justified. For example, if a farmer needed oil for his irrigation engine and he went to a store and told the seller exactly what model irrigation engine he needed oil for, the seller would be creating an implied warranty of fitness for a particular purpose by picking up a can of oil, handing it to the farmer, and saying, "This is the product you need." If

---

*Williams v. Braum Ice Cream Store, Inc., Oklahoma Court of Appeals 534 P.2d 700 (1974).

the farmer purchases the recommended oil, uses it in the engine, and the engine explodes because the oil was not heavy enough, the seller would have breached the warranty of fitness for a particular purpose. If the farmer were injured by the explosion, he would be able to recover on the basis of breach of warranty.

Express Warranties. The seller or the manufacturer may also be held liable for breach of an *express warranty,* which is created by a seller in one of three ways: by describing the goods, by making a promise or affirmation of fact about the goods, or by providing a model or sample of the good. If the goods fail to meet the description, fail to do what the seller claimed they would do, or fail to be the same as the model or sample, the seller has breached an express warranty. For example, if a 200-pound man asks a seller whether a ladder will hold a 200-pound man without breaking, the seller who says that it will is affirming a fact and is thus expressly warranting that the ladder will hold a 200-pound man without breaking. If the purchaser takes the ladder home, climbs up on it, and it breaks under his weight, causing him to fall to the ground, he may bring a product liability action against the seller on the basis of breach of an express warranty.

A memorable example of an express warranty is the claim many companies made that their software was "Y2K compliant." When the nation was in fear of a possible chaotic result of the date change to the year 2000, many computer companies came out with "Year 2000–compliant" software. This claim, however, was usually written outside of the contract and, therefore, was an express warranty. Because businesses had the potential to lose a lot of money after Y2K, they felt that they needed someone to help reimburse what they might lose. Many decided that the easiest targets would be these "Y2K-safe" software companies, because the businesses could sue on the basis of breach of an express warranty.[11]

Defenses to Breach-of-Warranty Actions. Two common defenses used in cases based on breach of warranty arise from the UCC; they make sense in a commercial setting when a transaction is between two businesspeople, but they make little—if any—sense in the context of a consumer injury. Therefore, the courts have found ways to limit the use of these defenses in product liability cases in most states.

The first such defense is that the purchaser failed to give the seller notice within a reasonable time after he or she knew or should have known of the breach of warranty, as required by the UCC. Obviously, most consumers would not be aware of this rule and, as a result, many early breach-of-warranty cases were lost. Most courts today avoid this requirement by holding (1) that a long delay is reasonable under the particular circumstances, (2) that the section imposing the notice requirement was not intended to apply to personal injury situations, or (3) that the requirement is inapplicable between parties who have not dealt with each other, as when a consumer is suing a manufacturer.

The second defense is the existence of a **disclaimer**. A seller or manufacturer may relieve itself of liability for breach of warranty in advance through the use of disclaimers. The disclaimer may say (1) that no warranties are made (as is), (2) that the manufacturer or seller warrants only against certain consequences or defects, or (3) that liability is limited to repair, replacement, or return of the product price.

Again, these disclaimers make sense in a commercial context, but seem somewhat harsh in the case of consumer transactions. Thus, the courts do not look with favor on disclaimers. First, the disclaimer must be clear; in many cases, courts have rejected the defendant's use of a disclaimer as a defense on

disclaimer Disavowal of liability for breach of warranty by the manufacturer or seller of a good in advance of the sale of the good.

---

[11]J. L. Dam, "Can Business Sue for Cost of Fixing the 'Year 2000' Problem?" *Lawyers Weekly USA,* www.lawyersweekly.com.

the ground that the retail purchaser either did not see the disclaimer or did not understand it. Thus, a businessperson using disclaimers to limit liability must be sure that the disclaimers are very plainly stated on an integral part of the product or package that will not be removed before retail purchase by the consumer. In some cases, however, despite clear disclaimers, courts have held disclaimers to consumers invalid, stating either that these disclaimers are unenforceable adhesion contracts resulting from gross inequities of bargaining power and are, therefore, unenforceable; or that they are unconscionable and contrary to the policy of the law. The UCC, in fact, now contains a provision stating that a limitation of consequential damages for injury to a person in the case of consumer goods is prima facie unconscionable.

The statute of limitations may also be used defensively in a case based on strict liability for breach of warranty. Under the UCC, the statute of limitations runs four years from the date on which the cause of action arises. In an action based on breach of warranty, the cause of action, according to the UCC, arises at the time of the sale. This rule would severely limit actions for breach of warranty, as defects often do not cause harm immediately. Section 2-725(2) of the UCC, however, changes the time that the cause of action arises to the date when the breach of warranty is or should have been discovered when a warranty "explicitly extends to the future performance of the goods, and the discovery of the breach must await the time of performance." This section, when applicable, makes the statute of limitations less of a potential problem for the plaintiff. In a few states, courts have simply decided to apply the tort statute of limitations, as running from the date of injury or the date when the defect was or should have been discovered, to all product liability cases grounded in breach of warranty.

STRICT LIABILITY IN TORT

The third and most prevalent theory of product liability used during the past three decades is strict liability in tort, established in the 1963 case of *Greenman v. Yuba Power Products Co.*[12] and incorporated in Section 402A of the *Restatement (Second) of Torts*. This section reads as follows:

1. One who sells any product in a defective condition, unreasonably dangerous to the user or consumer or his family is subject to liability for physical harm thereby caused to the ultimate user or consumer, or to his property, if
   a. the seller is engaged in the business of selling such a product; and
   b. it is expected to and does reach the consumer or user without substantial change in the condition in which it was sold.
2. The rule stated in Subsection (1) applies although
   a. the seller has exercised all possible care in the preparation and sale of his product; and
   b. the user or consumer has not bought the product from or entered into any contractual relation with the seller.*

Under this theory, the manufacturer, distributor, or retailer may be held liable to any reasonably foreseeable injured party. Unlike causes of action based on negligence or, to a lesser degree, breach of warranty, product liability actions based on strict liability in tort focus on the product, not on the producer or seller. The degree of care exercised by the defendant is not an issue in these cases. The issue in such cases is whether the product was in a "defective condition, unreasonably dangerous" when sold. To succeed in a strict liability action, the plaintiff must prove that

---
[12] 59 Cal.2d 57 (1962).

*From Restatement (Second) of Torts Section 402A.

1. the product was defective when sold;
2. the defective condition rendered the product unreasonably dangerous; and
3. the product was the cause of the plaintiff's injury.

APPLYING THE LAW TO THE FACTS...

Let's say that Amy buys slippers from a slipper manufacturer named Slip-On. A year later she falls down the stairs in her home while wearing the slippers. Subsequently she brings a strict liability action against Slip-On. She proves that she fell down the stairs because her slippers were missing a grip on the sole that would allow for stable traction on all hard surfaces, thus rendering her slippers in their current state unsafe for walking on wood floors, tile, etc. Amy has proved two of the three prongs she needs to be successful against Slip-On. What prong is Amy missing?

The defect is usually the most difficult part of the case for the plaintiff to establish. A product may be defective because of (1) some flaw or abnormality in its construction or marketing that led to its being more dangerous than it otherwise would have been, (2) a failure by the manufacturer or seller to adequately warn of a risk or hazard associated with the product, or (3) a design that is defective. For example, in 1966, Mr. Dolinski purchased a bottle of Squirt from a vending machine and drank most of the contents. He soon felt ill and discovered a decomposed mouse and mouse feces at the bottom of the bottle. He suffered physical and mental distress and avoided soft drinks after this experience. Under strict liability in tort, he sued that bottle manufacturer and distributor, Shoshone Coca-Cola Bottling Company, and a jury awarded him $2,500. Moreover, this was the first case in which the Nevada state courts recognized the doctrine of strict liability.[13]

A defect in manufacture or marketing generally involves a specific product that does not meet the manufacturer's specifications. Proof of such a defect is generally provided in one or both of two ways: (1) experts testify as to the type of flaw that could have caused the accident that led to the plaintiff's injury; (2) evidence of the circumstances surrounding the accident led the jury to infer that the accident must have been caused by a defect in the product. Notice in the following case how the court makes an analogy to *res ipsa loquitur* when finding the existence of a defect caused by the defendant.

CASE 12-3

Welge v. Planters Lifesavers Co.
Court of Appeals for the Seventh Circuit
17 F.3d 209 (1994)

Richard Welge loved to sprinkle peanuts on his ice cream sundaes. Karen Godfrey, with whom Welge boarded, bought a 24-ounce vacuum-sealed plastic-capped jar of Planters peanuts for him at a Kmart. To obtain a $2 rebate from the maker of Alka-Seltzer, Godfrey needed proof of her purchase of the jar of peanuts. So, using an X-Acto knife, she removed the part of the label that contained the bar code. She then placed the jar on top of the refrigerator for Welge. About a week later, Welge removed the plastic seal from the jar, uncapped it, took some peanuts, replaced the cap, and returned the jar to the top of the refrigerator. A week after that, he took down the jar, removed the plastic cap, spilled some peanuts into his left hand to put on his sundae, and replaced the cap with his right hand—but as he pushed the cap down on the open jar,

---
[13]*Dolinski v. Shoshone Coca-Cola*, 82 Nev. 439, 420 P.2d 855 (1966).

the jar shattered. His hand, continuing in its downward motion, was severely cut and, he claimed, became permanently impaired.

Welge filed product liability actions against Kmart, which sold the jar of peanuts to Godfrey; Planters, which manufactured the product (filled the glass jar with peanuts and sealed and capped it); and Brockway, which manufactured the glass jar and sold it to Planters. After pretrial discovery, the defendants moved for summary judgment. The district judge granted the motion on the ground that the plaintiff had failed to exclude possible causes of the accident other than a defect introduced during the manufacturing process. The plaintiff appealed.

Justice Posner

No doubt there are men strong enough to shatter a thick glass jar with one blow. But Welge's testimony stands uncontradicted that he used no more than the normal force that one exerts in snapping a plastic lid onto a jar. So the jar must have been defective. No expert testimony and no fancy doctrine are required for such a conclusion. A nondefective jar does not shatter when normal force is used to clamp its plastic lid on. The question is when the defect was introduced. It could have been at any time from the manufacture of the glass jar by Brockway (for no one suggests that the defect might have been caused by something in the raw materials out of which the jar was made) to moments before the accident. But testimony by Welge and Godfrey . . . excludes all reasonable possibility that the defect was introduced into the jar after Godfrey plucked it from a shelf in the KMart store. From the shelf she put it in her shopping cart. The checker at the check-out counter scanned the bar code without banging the jar. She then placed the jar in a plastic bag. Godfrey carried the bag to her car and put it on the floor. She drove directly home, without incident. After the bar-code portion of the label was removed, the jar sat on top of the refrigerator except for the two times Welge removed it to take peanuts out of it. Throughout this process it was not, so far as anyone knows, jostled, dropped, bumped, or otherwise subjected to stress beyond what is to be expected in the ordinary use of the product. Chicago is not Los Angeles; there were no earthquakes. Chicago is not Amityville either; no supernatural interventions are alleged. So the defect must have been introduced earlier, when the jar was in the hands of the defendants.

But, they argue, this overlooks two things. One is that Karen Godfrey took a knife to the jar. And no doubt one can weaken a glass jar with a knife. But nothing is more common or, we should have thought, more harmless than to use a knife or a razor blade to remove a label from a jar or bottle. People do this all the time with the price labels on bottles of wine. Even though mishandling or misuse, by the consumer or by anyone else (other than the defendant itself), is a defense . . . to a products liability suit . . . and even if, as we greatly doubt, such normal mutilation as occurred in this case could be thought a species of mishandling or misuse, a defendant cannot defend against a products liability suit on the basis of a misuse that he invited. The Alka-Seltzer promotion to which Karen Godfrey was responding when she removed a portion of the label of the jar of Planters peanuts was in the KMart store. It was there, obviously, with KMart's permission. By the promotion KMart invited its peanut customers to remove a part of the label on each peanut jar bought, in order to be able to furnish the maker of Alka-Seltzer with proof of purchase. If one just wants to efface a label one can usually do that by scraping it off with a fingernail, but to remove the label intact requires the use of a knife or a razor blade. Invited misuse is no defense to a products liability claim. Invited misuse is not misuse. . . .

Even so, the defendants point out, it is always possible that the jar was damaged while it was sitting unattended on the top of the refrigerator, in which event they are not responsible. Only if it had been securely under lock and key when not being used could the plaintiff and Karen Godfrey be certain that nothing happened to damage it after she brought it home. That is true—there are no metaphysical certainties—but it leads nowhere. Elves may have played ninepins with the jar of peanuts while Welge and Godfrey were sleeping; but elves could remove a jar of peanuts from a locked cupboard. The plaintiff in a products liability suit is not required to exclude every possibility, however fantastic or remote, that the defect which led to the accident was caused by someone other than one of the defendants. The doctrine of *res ipsa loquitur* teaches that an accident that is unlikely to occur unless the defendant was negligent is itself circumstantial evidence that the defendant was negligent. The doctrine is not strictly applicable to a products liability case because[,] unlike an ordinary accident case[,] the defendant in a products case has parted with possession and control of the harmful object before the accident occurs. . . . But the doctrine merely instantiates the broader principle, which is as applicable to a products case as to any other tort case, that an accident can itself be evidence of liability. . . . If it is the kind of accident that would not have occurred but for a defect in the product, and if it is reasonably plain that the defect was not introduced after the product was sold, the accident is evidence that the product was defective when sold. The second condition (as well as the first) has been established here, at least to a probability sufficient to defeat a motion for summary judgment. Normal people do not lock up their jars and cans lest something happen to damage these containers while no one is looking. The probability of such damage is too remote. It is not only too remote to make a rational person take measures to prevent it; it is too remote to defeat a products liability suit should a container prove dangerously defective.

Of course, unlikely as it may seem that the defect was introduced into the jar after Karen Godfrey bought it if the plaintiff's testimony is believed, other evidence might make their testimony unworthy of belief—might even show, contrary to all the probabilities, that the knife or some mysterious night visitor caused the defect after all. The fragments of glass into which the jar shattered were preserved and were examined by experts for both sides. The experts agreed that the jar must have contained a defect but they could not find the fracture that had precipitated the shattering of the jar and they could not figure out when the defect that caused the fracture that caused the collapse of the jar had come into being. The defendants' experts could neither rule out, nor rule in, the possibility that the defect had been introduced at some stage of the manufacturing process. The plaintiff's expert noticed what he thought was a preexisting crack in one of the fragments, and he speculated that a similar crack might have caused the fracture that shattered the jar. This, the district judge ruled, was not enough.

But if the probability that the defect that caused the accident arose after Karen Godfrey bought the jar of Planters peanuts is very small—and on the present state of the record we are required to assume that it is—then the probability that the defect was introduced by one of the defendants is very high.

. . . The strict-liability element in modern products liability law comes precisely from the fact that a seller subject to that law is liable for defects in his product even if those defects were introduced, without the slightest fault of his own for failing to discover them, at some anterior stage of production. . . . So the fact that KMart sold a defective jar of peanuts to Karen Godfrey would be conclusive of KMart's liability, and since it is a large and solvent firm there would be no need for the plaintiff to look further for a tortfeasor. This point seems to have been more or less conceded by the defendants in the district court—the thrust of their defense was that the plaintiff had failed to show that the defect had been caused by any of them—though this leaves us mystified as to why the plaintiff bothered to name additional defendants.

. . . Evidence of KMart's care in handling peanut jars would be relevant only to whether the defect was introduced after sale; if it was introduced at any time before sale—if the jar was defective when KMart sold it—the source of the defect would be irrelevant to KMart's liability. In exactly the same way, Planters' liability would be unaffected by the fact, if it is a fact, that the defect was due to Brockway rather than to itself. To repeat an earlier and fundamental point, a seller who is subject to strict products liability is responsible for the consequences of selling a defective product even if the defect was introduced without any fault on his part by his supplier or by his supplier's supplier.

. . . Here we know to a virtual certainty (always assuming that the plaintiff's evidence is believed, which is a matter for the jury) that the accident was not due to mishandling after purchase but to a defect that had been introduced earlier.*

Reversed and *remanded* in favor of Plaintiff, Welge.

When a plaintiff is seeking recovery based on a design defect, he or she is not impugning just one item, but an entire product line. If a product is held to be defectively designed in one case, a manufacturer or seller may recognize that this particular case may stimulate a huge number of additional lawsuits. Thus, defendants are very concerned about the outcome of these cases. Therefore, the availability of this type of action has a greater impact on encouraging manufacturers to produce safe products than does the availability of any other type of product liability action.

For example, hundreds of claims, ranging from property damage to personal injury and wrongful death, have been brought in federal and state courts against Firestone and Ford for Firestone's ATX, ATX II, and Wilderness AT tires. Plaintiffs argue that the tires have a design defect that causes the treads to prematurely separate, leading to tire blowouts. Firestone issued a recall for these tires in August 2000, but plaintiffs argue that the recall was not sufficient to warn customers about the tire defects. Furthermore, plaintiffs allege that Firestone had knowledge of the defects earlier but failed to disclose this knowledge. In an attempt to encourage efficiency, the federal cases have been consolidated to the Southern District Court in Indiana, at least for the purposes of discovery.

Although all the states agree that manufacturers may not market defectively designed products, there is no uniform definition of a defective design. Two tests have evolved to determine whether a product is so defective as to be unreasonably dangerous. The first test, set out in the *Restatement (Second) of Torts,* is the consumer-expectations test. This test asks the question: Did the product meet the standards that would be expected by a reasonable consumer? Such a test relies on the experiences and expectations of the ordinary consumer and, thus, is not answered by the use of expert testimony about the merits of the design.

---

*Welge v. Planters Lifesavers Co., Court of Appeals for the Seventh Circuit 17 F.3d 209 (1994).

The second is the feasible alternatives test, sometimes referred to as the *risk-utility test*. Courts applying this test typically look at the following seven factors:

1. The usefulness and desirability of the product—its utility to the user and to the public as a whole.
2. The safety aspects of the product—the likelihood that it will cause injury, and the probable seriousness of the injury.
3. The availability of a substitute product which would meet the same need and not be as unsafe.
4. The manufacturer's ability to eliminate the unsafe character of the product without impairing its usefulness or making it too expensive to maintain its utility.
5. The user's ability to avoid danger by the exercise of care in the use of the product.
6. The user's anticipated awareness of the dangers inherent in the product and their avoidability, because of general public knowledge of the obvious condition of the product, or of the existence of suitable warnings or instructions.
7. The feasibility, on the part of the manufacturer, of spreading the loss by setting the price of the product or carrying liability insurance.

Impact of the Restatement (Third) of Torts. Even though elements of Section 402A of the *Restatement (Second) of Torts* have come to be adopted in all states, and it is generally considered to be the foundation of modern product liability law, a lot of dissatisfaction has centered on this provision. That dissatisfaction resulted in what may be the biggest change in product liability law since the passage of Section 402A: the adoption, on May 20, 1997, of the American Law Institute's *Restatement (Third) of Torts: Product Liability*, which is intended to replace Section 402A. Under the Restatement (Third), "one engaged in the business of selling or otherwise distributing products who sells or distributes a defective product is subject to liability for harm to persons or property caused by the defect." The seller's liability, however, is determined by a different standard, depending on which type of defect is involved: (1) a manufacturing defect, (2) a design defect, or (3) a defective warning.

When the defect is one in the manufacture, liability is strict. A *manufacturing defect* is said to exist when "the product departs from its intended design." The new rule imposes liability in such a case regardless of the care taken by the manufacturer.

The Restatement (Third) adopts a reasonableness standard for design defects. It states that a product is defective in design when the foreseeable risks of the harm posed by the product could have been reduced or avoided by the adoption of a reasonable alternative design by the seller and the omission of the alternative design renders the product not reasonably safe.*

402A. Under the Restatement (Third), anyone who sells or distributes a defective product can be held liable for damage to persons or property caused by the defect, but the extent of liability of the seller depends upon the type of defect that caused the harm.

When the defect is one in the manufacture, meaning that the product was not made the way it was designed to be made, liability is strict. The new rule imposes liability in such a case regardless of the care taken by the manufacturer.

The Restatement (Third) adopts a reasonableness standard for design defects. It states that a product is defective in design when the foreseeable risks of the harm posed by the product could have been reduced or avoided by the adoption of a reasonable alternative design by the seller and the omission of the

---

*Restatement (Third) of Torts: Product Liability.

alternative design renders the product not reasonably safe. Basically, the courts are to apply a risk-utility test, determining whether the value of the product outweighs the risks imposed by it.

If the product is defective because it does not contain an appropriate warning, the same reasonableness standard is imposed on the seller. If foreseeable risks could have been reduced by reasonable warnings or instructions by the seller, the failure to provide such information means that the seller can be liable because the product is unreasonably dangerous.

Defenses to a Strict Product Liability Action. Product misuse, discussed as a defense to a negligence-based action, is also available in a strict product liability case. Assumption of the risk is likewise sometimes raised as a defense in a strict liability action.

Controversy, however, has arisen over whether the state-of-the-art defense should be allowed in cases in which the cause of action is based on strict liability. In most strict liability cases, courts have rejected the use of this defense, stating that the issue is not what the producers knew at the time the product was produced, but whether the product was defectively dangerous. In the 1984 case of *Elmore v. Owens Illinois, Inc.*, a claim arose from the plaintiff's contracting asbestosis. The plaintiff's job required him to handle a product manufactured by the defendant that contained 15 percent asbestos. The Supreme Court of Missouri ruled that the state of the art of a product has no bearing on the outcome of a strict liability claim, because the issue is the defective condition of the product, not the manufacturer's knowledge, negligence, or fault.

The refusal of most courts to allow the state-of-the-art defense in strict liability cases makes sense if we consider the social policy reasons for imposing strict liability. One of the reasons for imposing strict liability is that the manufacturers or producers are best able to spread the cost of the risk; this risk-spreading function does not change with the availability of scientific knowledge.

The argument against this position, however, is equally compelling to some. If the manufacturer has indeed done everything as safely and carefully as available technology allows, it seems unfair to impose liability on the defendant. After all, how else could it have manufactured the product?

LIABILITY TO BYSTANDERS

Sometimes the person injured by the defective product is not a purchaser, nor even an owner, of the product. The question arises as to whether strict product liability can be used by someone other than the owner or user of the product. On the grounds that the bystander is in even greater need of protection from defective products that are dangerous, and because he or she can do less to protect himself or herself from them, many courts have extended liability to foreseeable bystanders.

Market Share Liability

The focus of this chapter has been on product liability cases in which the plaintiff knows who produced the defective product. But when injuries from a product show up 10 or 20 years after exposure to the product, even if the injury can be traced to the defective product, the plaintiffs cannot always trace the product to any particular manufacturer. If a number of manufacturers produced the same product, the plaintiff may have no idea whose product was used, and may even have used more than one manufacturer's product.

In this situation, the courts have to balance the interests of the plaintiffs in recovering for injuries caused by defective products against the manufacturers' interests in not being held liable for injuries caused by a product they did not

market share theory A theory of recovery in liability cases according to which damages are apportioned among all the manufacturers of a product based on their market share at the time the plaintiff's cause of action arose.

produce. The primary means used to resolve this dilemma today is the **market share theory**, created in 1980 by the California Supreme Court in the case of *Sindell v. Abbott Laboratories*.[14]

In *Sindell*, the plaintiffs' mothers had all taken a drug known as diethylstilbestrol (DES) during pregnancies that had occurred before the drug was banned. Because the drug had been produced 20 years before the plaintiffs suffered any effects from the drug their mothers had taken, it was impossible to trace the defective drug back to each manufacturer that had produced the drug that caused each individual's problems. To balance the competing interests of the victims, who had suffered injury from the drug, and the defendants, who did not want to be held liable for a drug they did not produce, the court allowed the plaintiffs to sue all of the manufacturers who had produced the drug at the time the plaintiffs' mothers used the drug. The judge then apportioned liability among the defendant-manufacturers on the basis of the share of the market they had held at the time the drug was produced.

Since *Sindell*, a number of other courts have applied and refined the market share theory, primarily in drug cases. One trial court judge in a Pennsylvania DES case laid out the factors that are generally necessary for applying market share liability: (1) all defendants are tortfeasors; (2) the allegedly harmful products are identical and share the same defective qualities; (3) plaintiff is unable, through no fault of his or her own, to identify which defendant caused his or her injury; and (4) the manufacturers of substantially all of the defective products in the relevant area and during the relevant time are named as defendants.[15]

Other courts have modified the market share approach of *Sindell*. Attempts to extend the theory to products other than drugs have not met with much success. For example, in 2003, the court refused to allow the city of Gary to rely on market share theory to prove damages in a negligence action against handgun manufacturers, wholesalers, and dealers, because of the wide mix of lawful and unlawful conditions and the many potentially intervening acts by nonparties.[16]

Service Liability

Along with the growth in lawsuits for defective products, there has also been an increase in the number of lawsuits brought for defective services. These actions are generally brought when someone or someone's property is harmed as a result of an inadequately performed service.

Unlike in the product liability area, strict liability has rarely been applied to services. The few cases in which a strict liability standard has been applied involved defendants that provided both a good and a service, such as a restaurant owner's serving spoiled food.

Most service liability cases involve services provided by professionals, such as doctors, lawyers, engineers, real estate appraisers, and accountants. Actions against these professionals are generally referred to as **malpractice suits** and are usually based on a theory of negligence, breach of contract, or fraud. Malpractice actions against professionals are increasing at an extremely rapid rate. For example, by the early 1970s, only about 700 legal malpractice decisions had been reported; today, that many legal malpractice cases are brought each year.

The businessperson, however, is most likely to become involved in a malpractice action involving accountant malpractice. The next section, therefore, explores the liability of accountants.

malpractice suits Service liability suits brought against professionals, usually based on a theory of negligence, breach of contract, or fraud.

---

[14] 607 P.2d 924 (Cal. 1980).

[15] *Erlich v. Abbott Laboratories*, 5 Phila. 249 (1981).

[16] *City of Gary ex rel. King v. Smith & Wesson Corp.*, 801 N.E.2d 1222, Prod.Liab.Rep. (CCH) ¶ 16,862 (Ind. 2003).

ACCOUNTANTS' LIABILITY

One group that has seen increasing liability is accountants. Much of their potential liability has come from the securities laws. Accountants' liability under these laws is discussed in Chapter 23.

Accountants' liability for malpractice generally arises in actions for negligence, fraud, or breach of contract. In a malpractice action based on negligence, the plaintiff must prove the same elements discussed in previous sections on negligence: duty, breach of duty, causation, and damages.

The accountant's duty is said to be that of using the degree of care, skill, judgment, and knowledge that can reasonably be expected of a member of the accounting profession. Two sets of standards have been developed by the American Institute of Certified Public Accountants, the professional accountants' association, that help determine reasonable care. A reasonable accountant should, at minimum, follow the Generally Accepted Accounting Principles (GAAP) and the Generally Accepted Auditing Standards (GAAS). These two codes provide standards against which to measure an accountant's practices.

A major issue in accounting malpractice is the question of to whom the accountant's duty is owed. States are not in agreement about to whom an accountant can be held liable. Of course, the accountant is always liable to his or her clients. Third parties who have relied on the accountant's work, however, present a problem.

States use three alternative rules to define the parameters of the accountant's liability to third parties. The first, and oldest, rule is often referred to as the **Ultramares Doctrine**. Under this rule, the accountant is liable only to those in a privity-of-contract relationship. In other words, only the party who contracted for the accountant's work may sue. For example, if a client contacted an accountant to prepare a statement that the accountant knew was going to be used to secure a loan from the First Founding Bank, First Founding could not sue the accountant for malpractice in a state that followed the Ultramares Doctrine, because there was no contractual relationship between First Founding and the accountant.

Ultramares Doctrine Rule making accountants liable only to those in a privity-of-contract relationship with the accountant.

A somewhat more liberal rule is found in Section 552 of the *Restatement (Second) of Torts*. This rule holds that accountants will be liable to a limited class of intended users of the information. Thus, the accountant owes a duty to the client and any class of persons the accountant knows is going to be receiving a copy of his or her work. Under this rule, First Founding could recover in the preceding example.

The broadest rule applies in an extremely limited number of states. The smallest minority of states holds the accountant liable to any reasonably foreseeable user of the statement the accountant prepares. Exhibit 12-4 illustrates the states that adhere to each rule.

The reasonably foreseeable user rule, which was adopted for accountants in the Florida State Supreme Court case of *First Florida Bank v. Max Mitchell & Co.*,[17] has been used to extend liability to third parties adversely affected by the performance of other professionals. For example, it was cited to justify allowing a condominium association to sue an engineer who had been retained to inspect buildings and to make structural reports before an apartment building was converted into a condominium; it was also cited to allow a real estate appraiser to be sued by a bank that relied on an inaccurate appraisal of the property.

Global Dimensions of Product Liability Law

Businesspersons are concerned with the transnational aspects of product liability law primarily in two situations: (1) when they sell an imported product that causes injury to a consumer in the United States, and (2) when they manufacture

---
[17] 588 So.2d (Fla. 1990).

EXHIBIT 12-4

LIABILITY OF ACCOUNTANTS TO THIRD PARTIES

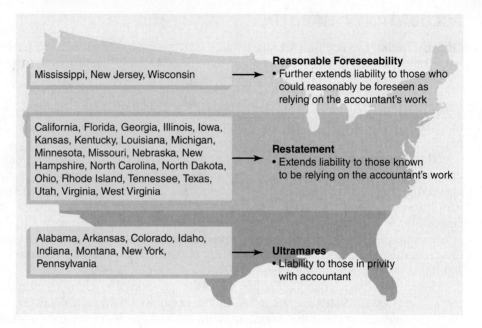

and export a product that causes harm to a consumer in a foreign country. In both instances, the U.S. corporation may be subject to liability for the injury.

Liability for a defective product that injures a consumer may be imposed on everyone in the chain of distribution of the product, from the retailer to the manufacturer. In about 80 percent of the cases, it is the manufacturer on whom plaintiffs tend to concentrate, because the manufacturer is usually responsible for the defect and has the greatest assets. If the manufacturer of a defective product is a company located in another country, a U.S. importer, wholesaler, distributor, or retailer may find itself liable for the injuries caused by a defective imported product. When a manufacturer is located in a foreign country, the plaintiff often simply sues only the retailer and the wholesaler. Because they do business in the state where the consumer lives, the court can easily assert personal jurisdiction over them. The foreign corporation may sometimes argue successfully that the corporation does not have enough minimum contacts with the state to allow the assertion of jurisdiction over the foreign manufacturer under the state's long-arm statute.

Even if the long-arm statute is satisfied, a potential problem arises in conjunction with service. Although the means of service acceptable in the United States are acceptable for serving corporations in most countries, the Hague Convention on the Service of Judicial and Extrajudicial Documents in Civil and Commercial Matters (adhered to by 28 countries, including most of the major trading partners of the United States) requires that the foreign defendant receive actual notice of the suit. This requirement is sometimes difficult to satisfy.

Still another consideration for the plaintiff is the collectability of the judgment. If the foreign defendant has no assets in the United States and refuses to pay, the plaintiff will be forced to ask the courts in the country where the manufacturer is located to execute a judgment against the defendant's assets there. With all of these potential problems resulting from an action against a foreign manufacturer, a plaintiff is very likely simply to sue those U.S. businesses in the chain of distribution. The prudent businessperson who sells foreign goods should be aware of this potential liability problem.

In the case of U.S. products sold abroad, U.S. manufacturers may be brought before the foreign courts. Since the late 1970s, European and other foreign countries have been adopting increasingly strict product liability rules, holding manufacturers, distributors, and retailers liable for injuries caused by defective products on theories similar to those used in the United States, such as breach of warranty, negligence, and strict liability.

Each country has its own set of rules, so the prudent businessperson will become familiar with the rules of the country to which he or she is exporting a product. New Zealand's Accident Compensation Act, for example, provides for almost automatic payment of compensation for pecuniary damages, such as medical expenses and lost wages, although it excludes most claims for pain and suffering.[18]

Foreign importers, retailers, and wholesalers of goods manufactured in the United States are not at all reluctant to join U.S. manufacturers in lawsuits so as to distribute the cost of the judgment. Moreover, once a plaintiff has obtained a judgment over a U.S. corporation in a foreign state, the U.S. courts will generally enforce the judgment rendered by a foreign court as long as the principles used to obtain jurisdiction over the person were reasonably similar to those accepted in the United States and the substantive law reasonably conforms to our sense of justice. Thus, it is extremely important that businesspersons remember that selling a product overseas does not mean freedom from product liability considerations. Product standards for goods sold overseas should be just as high as for goods sold domestically. In fact, extra precautions may have to be taken. For example, warning labels and instructions should always be printed in the languages spoken in the countries where the goods will be sold.

SUMMARY

Product liability law grew out of tort law and relies on basic tort theories. A product liability action can be based on negligence, breach of warranty, or strict product liability. The easiest of these to prove is strict product liability.

A product liability action may be brought by any party who is injured by a defective product, even if he or she did not purchase the product. An action based on strict product liability may even be brought by a bystander.

The defendant may be a retailer, distributor, or manufacturer. Sometimes, when the producer of the product cannot be clearly identified, as in the case of a drug, the theory of enterprise liability (market share theory) may be used to bring an action against all manufacturers of that product.

Service liability is analogous to product liability but is for defective services. Another difference between the two is that negligence is generally the only theory of liability available in a service liability case. Most service liability cases involve professional malpractice, such as accountant or medical malpractice.

Just because a U.S. corporation intends goods for sale overseas, the goods should not be less safe than those produced for American consumption. The manufacturer of a shoddy exported product may find itself defending an action brought in a foreign court. Conversely, an importer in the United States should be especially careful about inspecting the imported goods, because a U.S. plaintiff may not want to sue the foreign producer, leaving the U.S. importer as the primary defendant.

REVIEW QUESTIONS

12-1 Explain what privity is and what impact it had on the development of product liability law.

12-2 Explain the elements one would have to prove to bring a successful product liability case based on negligence.

12-3 Explain the defenses one can raise in a product liability case based on negligence.

12-4 Explain the various types of warranties that provide the basis for product liability cases based on breach of warranty.

12-5 Explain the difference between the foreign-natural and consumer-expectations tests.

12-6 Explain the defenses available in a breach-of-warranty case.

---

[18]G. Palmer, "Compensation for Personal Injury: A Requiem for the Common Law in New Zealand," *Journal of Compensation Law* 21: 1 (1972).

REVIEW PROBLEMS

12-7 Jack Clark was eating a chicken enchilada at Mexicali Rose restaurant when he swallowed a chicken bone. The bone lodged in his throat and had to be removed in the emergency room of a local hospital. What is the primary factor that will determine whether Mr. Clark's product liability lawsuit is successful?

12-8 Five people died of carbon monoxide poisoning from a gas heater that had been improperly installed in a cabin by the owner, who had not extended the vent pipe far enough above the roofline. The instruction manual stated that the pipe had to be vented outside, but did not specify how far outside the vent pipe should extend, other than having a drawing that showed it extending beyond the roofline. The manual also said, "Warning: To ensure compliance with local codes, have installed by a gas or utility inspector." Do the decedents' estates have a product liability action for failure to warn? Why or why not?

12-9 The plaintiff was injured when a fire extinguisher failed to work when it was needed to put out a fire. Could the defendant manufacturer of the fire extinguisher raise the defense of contributory negligence against the plaintiff if the plaintiff's negligence started the fire? Why or why not?

12-10 Mattie was injured when she lost control of the car she was driving because of a tire blowout. The tire was guaranteed by the manufacturer "against failure from blowouts." The guarantee also limited the manufacturer's liability to repair or replacement of any defective tire. Can Mattie sue under strict liability breach of warranty and recover damages for her injury? Why or why not?

12-11 Bob was waiting at the crosswalk for the light to turn green. As he stood there, a car that was stopped in the road next to him suddenly exploded, and Bob was injured by the blast. A defect in the engine caused the explosion. Will Bob be able to bring a strict product liability action against the manufacturer of the engine?

12-12 National Bank was deciding whether to loan money to Pateo Corporation. It asked Pateo to provide a copy of the company's most recent audit. While doing the audit, the auditors, Hamble & Humphries, failed to follow up on evidence indicating that one of the firm's managing partners might be siphoning money out of the corporation's funds. Relying on the audit, the bank made the loan. Six months later Pateo went into bankruptcy, primarily because one partner had stolen funds from the corporation and had then fled the country. Can the bank bring an action against Hamble & Humphries? Why is your answer dependent on the state in which the case arose?

CASE PROBLEMS

12-13 Thomas Woeste ordered raw oysters at Washington Platform Saloon and Restaurant. After eating the oysters, Woeste died as a result of contracting the bacterium *Vibrio vulnificus*. *Vibrio vulnificus* is a naturally occurring bacterium that grows in oysters harvested in warm water. It has minimal effects on healthy people, but can be dangerous to people with compromised immune systems. Woeste suffered from hepatitis C and cirrhosis of the liver, which made him susceptible to *Vibrio*. Woeste's estate claimed that Washington Platform was both negligent and strictly liable for not providing adequate warning about the dangers of raw oysters. Washington Platform made a motion for summary judgment on the ground that it did provide a warning against eating raw seafood on its menu and that Woeste did not read the warning before ordering oysters. The motion for summary judgment was granted for Washington Platform and Woeste appealed. Discuss the outcome of Woeste's appeal. *Woeste v. Washington Platform Saloon & Restaurant,* 836 N.E.2d 52 (Ohio Ct. App. 2005).

12-14 The plaintiff was partially paralyzed from a carotid artery tear that resulted from his being tackled during a high school football scrimmage. While he was on the field, his coaches removed his helmet, which was then lost. The plaintiff's mother filed suit against the helmet manufacturers, alleging that the helmet's liner and foam padding were defectively designed. The district court granted summary judgment to the defendants because the plaintiff could not produce the specific helmet at issue and thus could not prove the helmet was defective. The plaintiff appealed, arguing that the fact that she could not produce the specific helmet was irrelevant as she was arguing that all of the helmets were defective due to their design. Do you think the appellate court agreed that the specific helmet need not be produced? *A.K.W. v. Easton-Bell Sports, Inc. et al.,* No. 11-60293, 2011 U.S. App. LEXIS 21108 (5th Cir. 2011).

12-15 When the tire tread separated on a rear wheel of his truck, a driver lost control of his truck, which rolled over. The two passengers in the truck were killed, and the estate of one of the passengers brought suit against Cooper Tire, the manufacturer of the tire. The plaintiff argued that the tire was defective in design and had a manufacturing defect. During the discovery portion of the case, the plaintiff sought information regarding all tires manufactured by the defendant. He was trying to find information to show that Cooper Tire had notice of a tread separation problem in its other tires. Cooper Tire refused to produce that information, arguing that information regarding other tires that it manufactured but were not at issue in the case was irrelevant. What arguments, if any, could the plaintiff use to establish that the information regarding other tires is relevant? *Mario Alvarez v. Cooper Tire & Rubber Co.*, 75 So.3d 789 (Fla. Dist. Ct. App., 4th Dist. 2011).

12-16 The defendant company, Genetech, Inc., designed, manufactured, and sold a psoriasis medication called Raptiva. The psoriasis medication worked by suppressing T-cells to prevent their migration and contribution to psoriasis. Unfortunately, because T-cells also serve the purpose of fighting infections, the medication had the potential to cause life-threatening side effects. Following several reports of these life-threatening side effects, including one consumer who contracted a rare brain infection, Genetech voluntarily removed the product from the market in 2009. The plaintiff consumers claimed that the psoriasis medication caused several subsequent injuries, and as such, brought a product liability suit against Genetech. The district court granted Genetech's motion to dismiss, holding that Genetech was entitled to immunity under the Michigan Products Liability Act, and that the plaintiffs' claims that immunity did not apply to the defendant was preempted by federal law. The plaintiffs appealed. Do you think the court of appeals agreed that Genetech was immune to the product liability claims? Why or why not? *Marsh v. Genetech, Inc.*, 693 F.3d 546, U.S. App. LEXIS 18703 (2012).

12-17 Eighteen-year-old Brandon Patch was pitching in an American Legion baseball game when he was struck in the head by a batted ball hit by a batter using a model CB-13 aluminum bat made by Hillerich and Bradsby. Balls hit by aluminum bats travel at a much higher velocity than balls hit using a traditional wooden bat, thereby increasing the necessary reaction time of the infielders.

Brandon died from the injury he received when hit by the ball, so his parents sued the Hillerich and Bradsby Company for failure to warn Brandon of the alleged defect in the bat (the increased speed of balls hit by it). The company argued that there was no defect and that the company had no duty to warn a non-user of the bat.

The jury found a failure to warn on the part of the company and awarded the plaintiffs $850,000. Defendant appealed. How do you think the appellate court ruled, and why? *Patch v. Hillerich & Bradsby Co.*, 257 P.3d 214 (2011).

12-18 Danell Gomez had a surgical catheterization in 1999. A device known as an "Angio-Seal" was used to close the hole in Gomez's artery. The Angio-Seal is regulated by the FDA under the Federal Medical Device Amendments. The Angio-Seal uses a plug of collagen to seal the hole, and uses an anchor to prevent the collagen from entering the artery. Gomez alleges that the anchor failed to prevent the collagen from entering Gomez's artery and it formed a blockage in her femoral artery in her leg. Gomez underwent nine surgeries as a result of the blockage to her artery. She sued the manufacturer under Louisiana product liability law on several counts, but the defendant's motion for summary judgment on the ground of federal preemption was granted.

Gomez appealed; the Fifth Circuit Court of Appeals affirmed most of the preemption claims, but reversed and remanded the summary judgment for Gomez's claim of manufacturing defect. Gomez claimed that the Angio-Seal had been manufactured improperly and that the defect had caused the anchor to fail. Why would this claim not be preempted by the Federal Medical Device Amendments? *Gomez v. St. Jude Medical Daig Division, Inc.*, 442 F.3d 919 (5th Cir. 2006).

THINKING CRITICALLY ABOUT RELEVANT LEGAL ISSUES

Government Protection against Unsafe Products

The recent scares caused by toxic products imported from China should be of concern to all retailers. The problematic products have varied from toy trains to toothpaste to tires. In 2006, the United States imported almost $290 billion worth of goods from China. Because such a huge number of our products are made in China, these incidents are a major concern for both consumers and retailers. Under product liability laws, consumers can sue retailers for selling them a defective or dangerous product.

The ability of consumers to sue the retailers for a product that was defectively made in another country is very harmful to the retailers. The retailers either must perform many expensive checks on their merchandise, which is increasingly coming from other countries, such as China, or they just have to trust that the merchandise is being made responsibly. This dilemma is very dangerous for businesses and could mean major expenses either from increased merchandise testing or from product liability lawsuits. For example, Foreign Tire Sales, Inc. (FTS), is being held responsible for the recall of defectively made tires, which it says it cannot afford without going into bankruptcy. Even though FTS did not make the tires, and the designs that it provided to the manufacturer were safe, FTS is being punished because it sold the tires.

Rather than making businesses responsible, the government should institute more inspections of foreign merchandise to ensure that the public is being protected. In response to the poisonous toothpaste from China, the Food and Drug Administration (FDA) is checking all shipments of toothpaste for toxins. Rather than waiting until the problem emerges, the FDA and other government agencies should check foreign goods as a preventative measure. Although the government obviously cannot increase regulations or inspections in Chinese factories, it can perform more inspections and require more extensive documentation when the products come into our country.

More inspections and regulation by the government will help protect business owners from product liability lawsuits, because it will remove the defective products before they reach consumers. The government should be responsible for inspecting these products because it is better able to inspect all of the foreign products; it has the money, the manpower, and the systems already in place. If the government can afford to check every shipment of toothpaste now, it should be able to increase checks before the toxic product reaches consumers.

The government should take the responsibility of protecting its citizens and its businesses. The government should protect its citizens from purchasing unsafe merchandise and protect its businesses from lawsuits over unsafe foreign merchandise.

1. How would you frame the issue and conclusion of this essay?
2. What evidence is used to support the author's opinion?
3. What information is missing from this essay?
4. Write an essay on this topic from a point of view different from that of the essay author.

   Clue: How could the inclusion of any missing information change the conclusion of the essay?

ASSIGNMENT ON THE INTERNET

At the Legacy Tobacco Documents Library, **library.ucsf.edu**, you can find internal documents from the files of top tobacco companies. Within this site, you can find a summary chart of tobacco lawsuits by state. Click on "Search & Find" at the top of the page, then on "Archives & Online Collections." Once here, click on the heading "Manuscript Collections," then on "Tobacco Control Archives." Select one of the states to view the complaints filed in the case. Read the complaint to find out what claims (i.e., negligence, strict liability, breach of warranty) the plaintiff is bringing. Find out what type of relief the plaintiff is seeking (i.e., damages, injunction, etc.).

Earlier in this chapter, we discussed failure-to-warn claims against the fast-food industry, and we noted that Congress is considering banning all obesity-related claims against fast-food restaurants. Write a paper in which you compare the fast-food litigation to the tobacco litigation.

ON THE INTERNET

www.cpsc.gov This site is the home of the U.S. Consumer Product Safety Commission.
www.law.cornell.edu This site contains a brief overview of product liability and provides several links to liability law as well as recent product liability court cases. To access, click on "Wex legal encyclopedia" under the heading "Legal Resources." Then search for the term "products liability."
www.lawyersandsettlements.com A website with news on class action suits. Most class action suits are product liability cases.
www.lawprofessors.typepad.com This site is a blog written by law professors. Perform a search for "products liability" in the search bar and click on the first result of the search. This will take you to blogs written about product liability cases.

FOR FUTURE READING

Beck, Hugh C. "The Substantive Limits of Liability for Inaccurate Predictions." *American Business Law Journal* 44 (2007): 161.

Dorfman, Howard L., Vivian M. Quinn, and Elizabeth A. Brophy. "Presumption of Innocence: FDA's Authority to Regulate the Specifics of Prescription Drug Labeling and the Preemption Debate." *Food and Drug Law Journal* 61 (2006): 585.

Geistfield, Mark A. *Principles of Product Liability*. New York: Foundation Press, 2006.

Owen, David G., and Jerry J. Phillips. *Products Liability in a Nutshell*. St. Paul, MN: ThomsonWest, 2008.

Ramseyer, J. Mark. "Liability for Defective Products: Comparative Hypotheses and Evidence from Japan." *American Journal of Comparative Law* 61 (Summer 2013): 617.

Rushin, Stephen. "Warning Labels and FCC Regulations: The New Legal and Business Frontier for Cell Phone Manufacturers." *Berkeley Business Law Journal* 7 (2010): 150.

Wagner, Wendy. 2006. "When All Else Fails: Regulating Risky Products through Tort Litigation." *Georgetown Law Journal* 65 (2006): 693.

CHAPTER THIRTEEN

Law of Property: Real and Personal

- REAL PROPERTY
- INTERESTS IN REAL PROPERTY
- VOLUNTARY TRANSFER OF REAL PROPERTY
- INVOLUNTARY TRANSFER OF REAL PROPERTY
- RESTRICTIONS ON LAND USE
- PERSONAL PROPERTY
- GLOBAL DIMENSIONS OF PROPERTY LAW

property A bundle of rights, in relation to others, to possess, use, and dispose of a tangible or intangible object.

When people hear the word *property*, they generally think of physical objects: land, houses, cars. **Property**, however, is a bundle of rights and interests in relation to other persons with reference to a tangible or intangible object (Exhibit 13-1). The essence of the concept of property is that the state provides the mechanism to allow the owner to exclude other people.

By virtue of this right, persons with great amounts of property have an especially significant amount of power. Because possessing property facilitates the acquisition of even more property, the identification of those who possess a disproportionate amount of property rights provides insight into the dynamics of influence and authority in our society.

Property rights are not the same in every society, nor are they static. While reading this chapter, think about how property rights could be different and what impact that difference would have both on the legal environment of business and on society as a whole.

Because different types of property give their owners different rights, and because different bodies of law govern different types of property, we will discuss the three primary types of property in two separate chapters. Initially, this chapter focuses on real property, that is, land and anything permanently attached to it. The second half discusses personal property, both tangible (capable of being detected by the senses) and intangible (incapable of being detected by the senses). Chapter 14 shifts the focus to intellectual property, that is, things created primarily by mental rather than physical processes.

EXHIBIT 13-1

PROPERTY AS A BUNDLE OF RIGHTS

Property is often characterized as a bundle of rights. All of the rights in the bundle are not always held by one person. For example, when an owner leases a house to a tenant, the owner has transferred the rights to use and possess but has kept the rest of the rights.

Note: Notice that the rights of the holder decrease as one moves up the pyramid.

CRITICAL THINKING ABOUT THE LAW

Property is directly related to the power one has in society. Some individuals argue that the government should offer more protection for property owners. Others argue that in a fair country, citizens would have similar amounts of property because property provides a basis for so many other decisions. Property rights actually exist as a matter of degree. A property owner has some rights, but he or she must tolerate some restrictions on those rights. The following questions will help you think critically about the link between property rights and power.

1. If a group of politicians passed a law to increase protection of property rights, what ethical norm probably led to this legislation?

   Clue: Review the list of ethical norms in Chapter 1. Which ethical norm seems most likely to motivate increased protection of property?

2. If a group of radical politicians proposed a law that reduced personal property protection and redistributed some property rights to the poor, what ethical norm was probably behind this action?

   Clue: Again, review the list of ethical norms.

3. In light of the variation in rights attached to the concept of property in different countries, the idea of property has ambiguity as one of its primary characteristics. What could nations do to eliminate any ambiguity associated with saying "This property is mine"?

   Clue: Take another look at the definition of property rights.

Real Property

Real property is land and everything permanently attached to it. One's rights to a property depend on the type of interest one has in that property. The law provides the means to convey or transfer that interest. Although most conveyances are voluntary, the government may require involuntary conveyances to benefit the public and may place restrictions on the use of property to protect the public health, safety, and welfare.

real property Land and everything permanently attached to it.

DEFINITION OF REAL PROPERTY

Many disputes have arisen over whether certain items are real or personal property. The law says that an item that is attached to the land is a part of the realty. What does "attached" mean? Usually, an item is considered attached if its removal would cause damage to the property. Thus, built-in appliances are generally a part of the real property, whereas freestanding ones are not. Sometimes, however, an item is not really permanently affixed to the land, but its functioning is said to be essential to the functioning of the structure. In such cases, the courts usually find the item to be part of the real property.

Fixtures. An item that is initially a piece of personal property but is later attached permanently to the realty is known as a **fixture** and is treated as part of the realty. Thus, if a person rents another's property and installs a built-in microwave oven, the oven is a fixture and becomes part of the realty. The tenant may not remove the oven when he or she leaves. There are two exceptions to this rule.

First, the parties may agree that specific fixtures will be treated as personal property. To be enforceable, such an agreement must be in writing.

The second exception is for trade fixtures. A *trade fixture* is a piece of personal property that is affixed to realty in conjunction with the lease of a property for a business. When an entrepreneur opens an ice cream parlor in a leased building, the freezers he or she installs are trade fixtures. These are treated as personal property because of a presumption that neither party intends such fixtures to become a permanent part of the realty. The businessperson will need the freezers at any new location, and new business tenants will have different needs.

fixture An item that is initially a piece of personal property but is later attached permanently to the realty and is treated as part of the realty.

EXTENT OF OWNERSHIP

Any concept of land obviously includes the surface of the land, but legally, more is included. The landowner is entitled to the airspace above the land, extending to the atmosphere. Ownership of land also includes **water rights**, the legal ability to use water flowing across or underneath one's property.

Water rights, however, are somewhat restricted, in that one cannot divert water flowing across the property in such a manner as to deprive landowners downstream of the use of water from the stream.

Ownership of land usually also encompasses the land below the surface, including **mineral rights**, the legal ability to dig or mine the minerals from the earth. These mineral rights, however, may be sold or given to someone other than the person who owns the surface of the land.

water rights The legal ability to use water flowing across or underneath one's property.

mineral rights The legal ability to dig or mine the minerals from the earth below the surface of one's land.

Interests in Real Property

Not all interests in land are permanent. The duration of one's ownership depends on the type of estate one is said to hold. The estate that one has also determines what powers one has in regard to using the land. Exhibit 13-2 briefly summarizes these interests, which are described in detail in the following subsections.

FEE SIMPLE ABSOLUTE

The most complete estate is the fee simple absolute. When most people talk about owning property, they usually have in mind a fee simple absolute. If one has a **fee simple absolute**, one has all rights to own and possess that land. When the owner of a fee simple absolute interest dies, the interest passes to the owner's heirs.

fee simple absolute The right to own and possess the land against all others, without conditions.

CONDITIONAL ESTATE

The interest of a **conditional estate** is the same as that of a fee simple absolute, except that it is subject to a condition, the happening or nonhappening of which

conditional estate The right to own and possess the land, subject to a condition whose happening (or nonhappening) will terminate the estate.

EXHIBIT 13-2

ESTATES IN LAND

will terminate the interest. For example, Rose may own a farm subject to the condition that Rose never grow cotton. Once Rose grows cotton, the condition has occurred, and the land either reverts to the former owner or is transferred in accordance with the terms of the **deed** (the instrument that is used to convey real property). Conversely, a conditional estate could be set up so that the holder would own the farm as long as the primary crop planted every year was corn. Failure to meet the condition would terminate the estate.

deed Instrument of conveyance of property.

LIFE ESTATE

A **life estate** is the right to own and possess property until one dies. The use of a life estate may be more restricted than that of a fee simple absolute. The party who will take possession of the property upon the death of the holder of the life estate has an interest in making sure that the value of the property does not substantially decline as a result of neglect or abuse by the holder of the life estate. Thus, the life-estate holder is not allowed to waste the property, and cannot use the property in such a way as to destroy its value to future holders. Nor can the life tenant neglect to make necessary repairs to the property to prevent its destruction or deterioration.

life estate The right to own and possess the land until one dies.

FUTURE INTEREST

A person's present right to possession and ownership of land in the future is a **future interest** and usually exists in conjunction with a life estate or a conditional estate. For example, Sam owns a life estate in Blueberry Farm, and, upon Sam's death, fee simple absolute ownership of the land will pass to Jane. Jane has a future interest in Blueberry Farm. As a result of her interest, she may sue Sam to enjoin him from engaging in waste of the property.

future interest The present right to possess and own the land in the future.

APPLYING THE LAW TO THE FACTS...

Janice holds a life estate in the property known as Binghamton Farms. The property contains a three-story mansion that was built in the 1800s. She decides to tear down the mansion and replace it with a small cottage that would be easier to maintain. If the future interest holders hear of her plans and are opposed to them, do you think they will be able to stop her? Why or why not?

LEASEHOLD ESTATES

leasehold The right to possess property for an agreed-upon period of time stated in a lease.

lease The contract that transfers possessory interest in a property from the owner (lessor) to the tenant (lessee).

A **leasehold** is not an ownership interest. It is a possessory interest. One who has a leasehold is entitled to exclude all others, including the property owner under most circumstances, for the period of the lease and is entitled to use the property for any legal purpose that is not destructive of other occupiers' rights or prohibited by the terms of the lease. The **lease** is the contract that transfers the possessory interest. It generally specifies the property to be leased, the amount of the rent payments and when they are due, the duration of the leasehold, and any special duties or rights of either party. It is signed by both parties. The owner of the property is the *lessor*, or landlord. The holder of the lease is the *lessee*, or tenant.

Although the rights and obligations of the landlord and the tenant may be altered by the lease, some states have statutes requiring landlords to keep the premises in good repair and allowing tenants to withhold their rent if the landlord fails to do so. The landlord may enter the property only in an emergency, with permission of the tenant to make repairs, or with notice to the tenant near the end of the leasehold to show the property to a potential tenant. If the tenant fails to make the agreed-upon rental payments, the landlord may bring an action to evict the tenant.

Unless prohibited by the lease, a tenant may move out of the property and sublease it to another party. The initial tenant, however, still remains liable to the landlord for payment of rent due for the entire term of the lease.

EASEMENTS

easement An irrevocable right to use some portion of another's land for a specific purpose.

An **easement** is an irrevocable right to use some portion of another's land for a specific purpose. The party holding the easement does not own the land in question but has only the right to use it. Easements generally arise in one of three ways: express agreement, prescription, or necessity.

An easement by express agreement arises when the landowner expressly agrees to allow the holder of the easement to use the land in question for the agreed-upon purpose. For example, a utility company may have an easement to run power lines across one's property. The easement should be described on the deed to the property or recorded in the county office that keeps property records to protect the holder of the easement when the property is sold.

An easement by prescription arises under state law. When one openly uses a portion of another's property for a statutory period of time, an easement arises. In many states, the time period is 25 years.

An easement by necessity arises when a piece of property is divided and, as a result, one portion is landlocked. The owner of the landlocked portion has an easement to cross the other parcel for purposes of entrance to and exit from the land.

LICENSE

license A temporary, revocable right to be on someone else's property.

A **license** is a temporary, revocable right to be on someone else's property. When one opens a business, the public is given a license to enter the property to purchase the good or services provided by the business.

CO-OWNERSHIP

co-ownership Ownership of land by multiple persons or business organizations; all tenants have an equal right to occupy all of the property.

We have been referring to the holders of interest in land in the singular. Ownership of land may also be held by multiple persons, as well as by business organizations. Whenever there is ownership by multiple parties, it is important to know what type of ownership exists, because different forms confer different rights on the owners.

The three types of **co-ownership** are tenancy in common, joint tenancy, and tenancy by the entirety. Regardless of which type of tenancy exists, all

TABLE 13-1
JOINT OWNERSHIP

Type	Division of Ownership	Rights of Owners' Creditors	Ownership of Property upon Death
Tenancy in common	Equal or unequal shares	Can attach interest	Transferred to heirs
Joint tenancy	Equal shares	Can attach interest	Divided among other joint tenants
Tenancy by the entirety	Equal shares	Cannot attach interest	Goes to surviving spouse

tenants have the equal right to occupy all the property. Their other interests are described here and are summarized in Table 13-1.

Tenancy in common is the most common form of joint ownership. Owners may own unequal shares of the property, may sell their interest without consent of the other owners, and may have their interest attached by a creditor. Upon the death of the tenant in common, his or her heirs receive the property interest.

Under **joint tenancy**, all are co-owners of equal shares and may sell their shares without the consent of other owners. Their interest can be attached by creditors. Upon the death of a joint tenant, his or her interest is divided equally among the remaining joint owners.

Tenancy by the entirety exists only when co-owners are a married couple. One cannot sell his or her interest without the consent of the other, and creditors of only one cannot attach the property. Upon the death of one, full ownership of the property passes to the surviving spouse. Upon divorce, tenancy by the entirety automatically becomes tenancy in common.

CONDOMINIUMS AND COOPERATIVES

Two forms of co-ownership that have become increasingly popular over the past 20 years are condominiums and cooperatives.

Condominiums. In a condominium ownership interest, the owner acquires title to a "unit" within a building, along with an undivided interest in the land, buildings, and improvements of the "common areas" of the development. When the condominium is developed, a Declaration of Covenants, Conditions, and Restrictions (called *CC&Rs*) is filed. This document contains the architectural and use restrictions for the condominium development, assessments, and instructions for forming the condominium association that will manage the condominium. The association has the power, as provided in the CC&Rs, to levy assessments against the unit owners in order to manage, maintain, insure, repair, and replace the common areas. The association likewise has the authority to fine unit owners who do not comply with the CC&Rs. The following case illustrates the type of conflict that often arises out of this form of ownership.

tenancy in common Form of co-ownership of real property in which owners may have equal or unequal shares of the property, may sell their shares without the consent of the other owners, and may have their interest attached by creditors.

joint tenancy Form of co-ownership of real property in which all owners have equal shares in the property, may sell their shares without the consent of the other owners, and may have their interest attached by creditors.

tenancy by the entirety Form of co-ownership of real property, allowed only to married couples, in which one owner cannot sell without the consent of the other, and the creditors of only one owner cannot attach the property.

CASE 13-1

Plum Creek C.A. v. Oleg Borman
Appellate Court of Illinois, First District, Fifth Division
2013 Ill. App. Unpub. LEXIS 1582

A tenant in the Plum Creek Condominium community wanted laminate floors in his condominium. The request was granted in 2001, with the caveat that should noise complaints arise, the flooring would be carpeted at the expense of the tenant. Eight years later, complaints of excessive noise arose, which led to a fine of $500 being assessed. The fine would be suspended, however, if the tenant changed his flooring to carpet, as stated in the 2001 letter. The defendant refused to pay this fine or correct the issue, so the board issued a complaint in 2010 and attempted to obtain an injunction forcing the tenant to install carpeting. The court ruled in favor of the tenant, on the grounds that the board did not provide sufficient proof of a violated right, as the original 2001 letter spelled out the rights of both the board and the tenant.

Justice Palmer

The board argues that the circuit court erred in denying it permanent injunctive relief. A party seeking a permanent injunction must demonstrate "(1) a clearly ascertainable right in need of protection, (2) that he will suffer irreparable harm without protection of that right, (3) that there is no adequate remedy at law, and (4) that there is a substantial likelihood of success on the merits of the underlying action."

The court . . . reasoned that the "May 18, 2001, letter defined for the parties what their rights were" and granted the board the choice to request defendant "to carpet either wall to wall or area rug the floors in question" if noise complaints were registered. It found that the board made its choice in the September 30, 2009, letter when it told defendant that it felt that installing area rugs would eliminate the noise problem. The court held that, because defendant installed the area rugs in compliance with the two letters, the board, therefore, had no clear right in need of protection.

The board argues that it has a protectable interest in enforcing the covenant requiring unit owners to install wall to wall carpeting in their units . . . It asserts that the May 18, 2001, letter is not a contract between the parties and, moreover, does not bar the board . . . from reserving the right to have defendant install wall to wall carpeting if the area rugs failed to abate the noise . . .

In determining the intent of the parties, a court must consider the document as a whole and not focus on isolated portions of the document. . . . . If the language of a contract is clear and unambiguous, the intent of the parties must be determined solely from the language of the contract itself. . . .

The agreement . . . makes it clear that all unit owners must take any measures specified by the board in the rules in order to soundproof their unit. The rules establish that wall to wall carpeting is required to soundproof the units. The rules also establish a unit owner may install flooring other than wall to wall carpet but only upon the filing of a written application and obtaining written consent from the board, which will not be granted unless the owner can establish that adequate soundproofing will be provided with the alternative flooring.

Upon the parties' execution of the application, it became, by its own terms, "the entire Agreement between the parties." In the application, defendant specifically agreed "to comply with all Association Declarations, By-Laws, and Rules and Regulations in respect to this change and/or improvement." He also acknowledged that: "should complaints be registered by residents for noise, floor shall be carpeted." By its clear language, the application established that, if noise complaints were registered as a result of the laminate flooring, installation of wall to wall carpeting would be required . . .

Read together, the express language contained in the application, . . . even if the board approved a deviation from the wall to wall requirement, installation of wall to wall carpeting might still be required if noise complaints resulted from the alternative flooring. "When the language of a covenant is unambiguous, clear and specific, no room is left for interpretation or construction." . . . "Restrictions should be given the effect which the express language of the covenant authorizes."

It is clear that the stated purpose of the covenant, to soundproof the condominium building, remained intact even after defendant was given approval to install laminated floors in his unit. Throughout its dealings with defendant, the board consistently sought to ensure that defendant understood that the soundproofing requirements stated in the covenant were not waived and that wall to wall carpeting remained the abiding requirement, subject to modification by approved application only as long as no noise complaints resulting from the modification were registered.

The May 18, 2001, letter . . . specifically reminded defendant that "should complaints be registered by other residents because of the noise created by the hardwood flooring, you may be requested to carpet either wall to wall or area rug the floors in question."

Contrary to the circuit court's finding, the May 18, 2001, letter does not set forth the rights of the parties. It is not part of the agreement between the parties. First, the letter does not have the requisite offer and acceptance elements for an enforceable contract.

Second, the application, as agreed to by defendant, clearly provides that:

All verbal and written communication between the parties is expressed hereinabove, and no verbal understandings or agreements shall alter, change, or modify the terms and provisions of this Agreement, and the entire Agreement between the parties is expressed herein. No course of prior dealings between the parties and no usage of trade shall be relevant to supplement or explain any term used in this agreement. Further, this agreement shall not be modified or altered by subsequent course of performance between the parties.

Pursuant to this provision, the application is the entire agreement between the parties and the May 30, 2001, letter cannot alter the terms of that agreement. The letter is merely a written notification informing defendant that his application to deviate from the soundproofing covenant in the declaration and rules had been conditionally approved. Indeed, the letter restates the terms of the covenant, pointing out that "should complaints be registered by other residents because of the noise created by the hardwood flooring, you may be requested to carpet either wall to wall or area rug the floors in question."

For the same reasons, the September 30, 2009, letter is not part of the parties' agreement. It is merely a communication notifying defendant of the board's decision regarding the latest noise complaints resulting from his installation of laminate flooring and of the remedy required for this violation of the soundproofing covenant, a remedy that was expressly reserved by the May 30, 2001, letter.

The court found that (a) the "either/or" language of the May 2001 letter gave the board a choice to either require defendant to install area rugs or wall to wall carpet in the unit, but not both; and (b) the board made its choice in the September 30, 2009, letter when it required defendant to install area rugs in the unit. We disagree.

Further, the statement in the May 2001 letter that defendant might "be requested to carpet either wall to wall or area rug the floors in question" was merely a warning, setting forth the possible remedies should other residents complain regarding noise resulting from the laminate flooring. This caveat cannot be construed as limiting the board to only one of the two remedies. Lastly, the September 2009 letter . . . required defendant to comply with the soundproofing requirement by abating the noise resulting from his hardwood flooring but it gave him two options on how to accomplish this: (1) by installing area rugs in the unit, with the knowledge that, if the measure failed, defendant would be required to install wall to wall carpeting, or (2) by installing wall to wall carpeting in the unit without trying the area rug option first. Defendant chose the first option. The board did not make this choice for him.

Read together, the 2001 and 2009 letters cannot be interpreted as a waiver of the covenant requiring the installation of wall to wall carpeting. . . . In each letter, the board made clear that noise abatement remained its ultimate concern and that it would approve flooring options other than wall to wall carpeting only as long as no noise complaints were registered as a result of the other options.

The covenant at issue is a restrictive covenant. It unambiguously forces a unit owner to forgo installation of the flooring of his choice and to install wall to wall carpeting in order to soundproof the building. The application makes it very clear that the board intended to enforce the covenant, approving defendant's request to install laminate flooring or laminate flooring topped with area rugs only as long as no noise complaints resulted from his choice of flooring. If such complaints were registered, the board would require defendant to install wall to wall carpeting.

Defendant's failure to install wall to wall carpeting in his unit when noise complaints resulted from his choice of flooring is a breach of the covenant. "[T]he mere breach of a covenant is sufficient grounds to enjoin the violation.". . . . Therefore, defendant's breach is sufficient grounds to enjoin his violation. The board has a clear and ascertainable right to enforcement of the covenant that is in need of protection. Defendant breached that covenant. The board is, therefore, entitled to a permanent injunction requiring defendant to comply with the covenant by installing wall to wall carpet. Therefore, we reverse the court's decision denying the board's request for injunctive relief.*

Reversed in favor of Plum Creek C.A. and *remanded* to the circuit court for issuance of a permanent injunction.

CRITICAL THINKING ABOUT THE LAW

Despite the effort homeowners' associations put into specifying what can and cannot be done in individual housing units, disputes about whether particular actions by homeowners are in conformity with the rules governing the residents or the homeowners are common. The courts, as in this case, pay close attention to the provisions in the agreement signed by the homeowner at the time of purchase.

1. What reasons does the court use to determine that the injunction should be provided to require Mr. Borman to comply with the original agreement between him and the association?

    Clue: Look at the portion of the decision where the court discussed the importance of the initial agreement in determining rights and responsibilities between Borman and the association. Then study the portion of the decision where the court analyzes Borman's conduct with respect to the terms of the initial agreement.

2. What is the key section of the initial agreement that causes the decision to go against Mr. Borman's desire to be seen as being in compliance with the agreement with the homeowners' association?

    Clue: Ask yourself the following question: In light of the court's reasoning, what portion of the original agreement would Borman now wish were never included?

Cooperatives. Whereas condominiums often involve developments consisting of multiple buildings, each of which contains two or more units, cooperatives are more commonly used for ownership interests in apartment buildings. In a

---

*Plum Creek C.A v. Oleg Borman, Appellate Court of Illinois, First District, Fifth Division, 2013 Ill. App. Unpub. LEXIS 1582.

cooperative, the investor resident acquires stock in the corporation that owns the facility and receives a permanent lease on one unit of the facility. A board of directors is generally elected from among the unit owners to manage the facility. All unit owners are bound by rules established by the cooperative's board of directors. Commonly, the cooperative will provide that a member may be evicted for violation of the rules and, upon eviction, the cooperative will repurchase the evicted member's unit.

Voluntary Transfer of Real Property

The value of property is heightened by the owner's ability to transfer that property. In general, the owner may transfer the property to anyone for any amount of consideration or for no consideration. He or she may transfer all or any portion of the property.

To transfer property, however, the owner must follow the proper procedures. These are execution, delivery, acceptance, and (to protect the recipient of the property) recording (Exhibit 13-3). Unless something to the contrary is stated, it is presumed that a conveyance of ownership is the conveyance of a fee simple absolute.

EXECUTION

The first step in a voluntary transfer is the execution of a deed. The deed, as shown in Exhibit 13-4, is the instrument of conveyance. A properly drafted deed:

1. Identifies the *grantor* (the person conveying the property) and the *grantee* (the person receiving the property)
2. Contains words that express the grantor's intent to convey the property
3. Identifies the type and percentage of ownership
4. States the price paid for the property, if any
5. Contains a legal description of the physical boundaries of the property (not the street address)
6. Specifies any easements or restrictions on use of the land
7. Identifies any warranties or promises made by the grantor in conjunction with the conveyance

Once the properly drafted deed has been signed by the grantor and the grantee, it is said to have been *executed*. Many states require the signing of the deed to be witnessed or notarized (witnessed by an official of the state who certifies that she or he saw the parties sign the deed and was provided evidence that the signatories were who they purported to be).

EXHIBIT 13-3

STEPS IN A CONVEYANCE

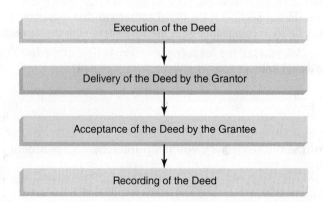

EXHIBIT 13-4

A SAMPLE GENERAL WARRANTY DEED

Form 1-A   8   Legal News, Toledo, Ohio

WARRANTY DEED

Received and Recorded at _____ M.   TRANSFERRED _____
                                                                           AUDITOR
                    RECORDER   PER _____

Know All Men By These Presents:

That   Sam Seller   the grantor ,

in consideration of $88,000

       to be   paid by   Betty Buyer   the grantee

whose present mail address is 2028 North Main, Bowling Green, Ohio

the receipt whereof is hereby acknowledged, do es   hereby BARGAIN, SELL and CONVEY to said Grantee

    and her   heirs, successors and assigns forever, the following described real estate, situate in the County of
                Wood   , State of Ohio: viz:

                                    Lot 57, plat 32 in
                                    Green Hills Subdivision

and all the estate, right, title and interest said grantor   has   or ought to have in and to said described premises,
together with the privileges and appurtenances to the same belonging, but subject to zoning ordinances, restrictions of
record and public utility or other easements of record.
    Grantor   acquired title to the above described premises by instrument recorded in Vol. **LXVI**   , Page **12**
    To Have and To Hold the same to the said Grantee   , and to   his   heirs, successors and assigns forever;
Grantor is   hereby covenanting that he is   the true and lawful owner   of said premises and he
    is   well seized of the same in fee simple, and ha s   good right and full power to bargain, sell and convey the same
in the manner aforesaid, and that the premises so conveyed are clear, free and unencumbered and that   he   will
warrant and defend the same against all claims whatsoever, except taxes and assessments due and payable .

    IN WITNESS WHEREOF, The said   Sam Seller   has

hereunto set his   hand   this   17   day of May   in the year of our Lord
One Thousand Nine Hundred and eighty-eight
        Signed, acknowledged and
        Delivered in the presence of

_____   _____

_____   _____

The State of Ohio,_____Wood_____County, ss.
        BE IT REMEMBERED, That on the   17th   day of May   in the year of our Lord
One Thousand Nine Hundred and eighty eight   , before me, the subscriber, a Notary Public within and for
said county, personally came Sam Seller

the grantor   in the above conveyance, and acknowledged the signing thereof to be his   voluntary act and deed,
for the purpose therein mentioned.
        IN WITNESS WHEREOF, I have hereunto subscribed my name and affixed my official seal on the day and
year aforesaid.

                                    Notary Public, ____Lucas____ County, Ohio.
    This Instrument Prepared By: __Andrea Attorney__   My Commission Expires 9-28-95

General Warranty Deed. Two basic types of deeds are normally used to transfer ownership of property. The first, a **general warranty deed**, illustrated in Exhibit 13-4, is preferred by grantees because it contains certain warranties or promises by the grantor. Although such covenants may vary slightly from state to state, they generally include the following:

1. The covenant of *seisin*—a promise that the grantor owns the interest that he or she is conveying

general warranty deed A deed that promises that the grantor owns the land and has the right to convey it and that the land has no encumbrances other than those stated in the deed.

2. The covenant of the right to convey—a promise that the grantor has the right to convey the property
3. The covenant against encumbrances—a promise that there are no mortgages or liens against the property that are not stated in the deed
4. The covenant for quiet enjoyment—a promise that the grantee will not be disturbed by anyone who has a better claim to title of the property and a promise to defend the grantee's title against such claims or to reimburse the grantee for any money spent in the defense or settlement of such claims
5. The covenant of further assurances—the promise that the grantor will provide the grantee with any additional documents that the grantee needs to perfect his or her title to the property

Quitclaim Deed. The other type of deed, which is more desirable from the grantor's perspective, is the **quitclaim deed**. With such a deed, the grantor simply transfers to the grantee the interest that the grantor owns in the property being conveyed. The grantor makes no additional covenants. Obviously, most grantees would be very reluctant to accept a quitclaim deed.

quitclaim deed A deed that simply transfers to the grantee the interest that the grantor owns in the property.

DELIVERY

Once executed, the deed must be delivered, or transferred, to the grantee with the intent of transferring ownership to the grantee. The delivery may be made directly to the grantee or to a third party who has been instructed to transfer the deed to the grantee.

ACCEPTANCE

The final essential step is acceptance by the grantee. This is the grantee's expression of intent to possess the property. Acceptance is presumed when the grantee retains possession of the deed.

RECORDING

Recording is not essential for the transfer of ownership, but it is so important to securing the grantee's rights to the property that it should always be a part of the process of conveyance. *Recording* refers to the filing of the deed (as well as any other documents related to realty, such as mortgages) with the appropriate county office. This office varies by state and may be the county clerk's office or the county recorder's office. Recording gives the world notice of the transfer. It is significant because in many states, if there are two deeds allegedly conveying the same piece of property, the owner of the property is the one whose deed was recorded first.

Involuntary Transfer of Real Property

Transfer of ownership interests in property may also occur without the owner's knowledge, or even against his or her will, by either adverse possession or condemnation.

ADVERSE POSSESSION

adverse possession Acquiring ownership of realty by openly treating it as one's own, with neither protest nor permission from the real owner, for a statutorily established period of time.

Most states provide that when a person openly treats realty as his or her own, without protest or permission from the real owner, for a statutorily established period of time, ownership is automatically vested in that person. Each state has its own exact requirements for **adverse possession**, but the necessary possession is often described as having to be actual (the party resides on or uses the land as would an owner), open (not secretive), and notorious (without the owner's permission). Some state laws require specific acts such as the payment of real estate taxes. Others require that the adverse possessor have taken possession of the land "under color of title," that is, thinking that he or she was the lawful possessor of the land.

CONDEMNATION

Condemnation is a process by which the government acquires the ownership of private property for a public use over the protest of the owner of the property. The property owner may have been contesting either the taking itself or the price that the government was willing to pay for the taking. Condemnation proceedings occur as a result of the government's exercise of its right of eminent domain, which was briefly discussed in Chapter 5. Remember from the discussion of the Takings Clause that the right of **eminent domain** is the constitutional right of the government to take private property, upon payment of just compensation, for a purpose that will benefit the general welfare of the citizens. This taking may be by any level of government or, in limited cases, by private companies fulfilling a public or governmental function.

Under eminent domain, the government offers to purchase the property. If the owner objects to the transfer or the price, the government institutes condemnation proceedings. If the court determines that the governmental purpose is legitimate, it then determines the fair market value of the property. Once this price has been paid, ownership is transferred to the government.

Sometimes the owner questions whether the taking is really for a public purpose. That question may be particularly difficult when the property to be acquired will be subsequently conveyed to another private individual, especially when the entity to whom the property is being transferred had previously attempted to purchase the property unsuccessfully. In the early 1980s, we started seeing an increasing number of cases in which city and state economic redevelopment agencies were using the power of eminent domain to take private property and transfer it to another private business for the ostensible public use of economic development or job creation.

The case that seemed to stimulate the increase in takings for economic redevelopment was the now-famous Michigan State Supreme Court case of *Poletown Neighborhood Council v. City of Detroit and the Detroit Economic Development Corporation*.[1] In that case, the Detroit Economic Development Corporation sought to acquire a large parcel of land on which members of the Poletown Neighborhood Council resided and had small businesses. Once the development agency had acquired the land using the city's eminent domain power, that property would be conveyed to General Motors to expand its plant. The plaintiffs, who did not want their community destroyed, sued the city and the development council on the ground that they were attempting to abuse their power of eminent domain to unconstitutionally take private property for a private use.

In decisions that surprised many legal scholars, the lower courts upheld the taking. The high court likewise held that the use of eminent domain was constitutional. The court said that the heart of this dispute was whether the proposed condemnation is for the primary benefit of the public or the private user, and it went on to say that once a legislature has declared that a particular government action meets a public need and serves a public purpose, that finding is entitled to deference from the court.

After this case, most courts appeared to give state agencies very broad latitude in cases involving takings for private development, finding public use when there is any plausible argument that the public will benefit from the new use.

condemnation The process whereby the government acquires the ownership of private property for a public use over the protest of the owner.

eminent domain The constitutional right of the government to take privately owned real property for a public purpose in exchange for just compensation to the owner.

---
[1]304 N.W.2d 455 (Mich. 1981).

Moreover, this broad latitude was not just given in cases where individuals' property was being taken. Increasingly, governments began using eminent domain to transfer property from one business to another.

As the number of these private property transfers continued to rise, a slowly increasing number of courts started backing off from giving so much deference to states in economic development situations. Finally, in the late summer of 2004, in *County of Wayne v. Hathcock et al.*, the Michigan Supreme Court overturned the *Poletown* decision that had been the stimulus for the use of eminent domain in so many of these economic redevelopment situations.[2]

In *Hathcock*, Wayne County, Michigan, was attempting to condemn private property to sell the property to a private business for the completion of the Pinnacle Project, which was supposed to bring hotels, conference centers, recreation areas, and other business to the metropolitan airport area. The property owners filed suit against Wayne County, claiming that the condemnation was an unfair taking because the Pinnacle Project did not constitute a public use. The trial court and the appeals court both ruled in favor of Wayne County, relying on *Poletown*. The property owners appealed to the Michigan Supreme Court, which used a test for public use set out in Justice Ryan's dissent in *Poletown*. The Pinnacle Project was not something that required government intervention for its existence, nor was it subject to public oversight once completed. Finally, the land to be condemned for the Pinnacle Project did not have to be condemned for an independent public good (such as cleaning up slums). The court determined that because the majority decision in *Poletown* did not have any basis in previous case law, *Poletown*'s broad construction of "public use" should not be applied. The Michigan Supreme Court found for Hathcock and the property owners, overturning *Poletown*. The influence of *Hathcock* ended up being limited, however, because soon after the decision, the U.S. Supreme Court finally decided to weigh in on the issue in the following case.

CASE 13-2

Susette Kelo et al., Petitioners v. City of New London, Connecticut et al.
Supreme Court of the United States 125 S. Ct. 2655 (2005)

The city of New London was faced with economic hard times and high unemployment, and the city's development corporation had come up with an integrated development plan that the city approved. The plan would allegedly create between 1,736 and 3,169 new jobs and generate a minimum of $680,544 in property tax revenues for the city. It would revitalize the city, including its downtown and waterfront areas. Using its development agent, the city purchased most of the property it needed for the project from willing sellers. A few property owners refused to sell, so the city initiated condemnation proceedings to take their property. The property owners argued that the taking of their properties would violate the public-use restriction in the Fifth Amendment's Takings Clause. The trial court granted a permanent restraining order prohibiting the taking of some of the properties, but allowing the taking of others. Relying heavily on the reasoning in *Poletown,* the high court in Connecticut ruled that the city could use its power of eminent domain to take the private property of a number of residents for the purpose of constructing a state park, new residences, research and development facilities, office space, and retail space. Kelo appealed.

Justice Stevens

Two polar propositions are perfectly clear. On the one hand, it has long been accepted that the sovereign may not take the property of A for the sole purpose of transferring it to another private party B, even though A is paid just compensation. On the other hand, it is equally clear that a State may transfer property from one private party to another if future "use by the public" is the purpose of the taking; the condemnation of land for a railroad with common-carrier duties is a familiar example.

As for the first proposition, the City would no doubt be forbidden from taking petitioners' land for the purpose of conferring a private benefit on a particular private party. . . . Nor would the City be allowed to take

---
[2]684 N.W.2d 765 (Mich. 2004).

property under the mere pretext of a public purpose, when its actual purpose was to bestow a private benefit. The takings before us, however, would be executed pursuant to a "carefully considered" development plan. The trial judge and all the members of the Supreme Court of Connecticut agreed that there was no evidence of an illegitimate purpose in this case. Therefore, the City's development plan was not adopted "to benefit a particular class of identifiable individuals."

On the other hand, this is not a case in which the City is planning to open the condemned land—at least not in its entirety—to use by the general public. Nor will the private lessees of the land in any sense be required to operate like common carriers, making their services available to all comers. But although such a projected use would be sufficient to satisfy the public use requirement, this "Court long ago rejected any literal requirement that condemned property be put into use for the general public." . . .

The disposition of this case therefore turns on the question [of] whether the City's development plan serves a "public purpose." Without exception, our cases have defined that concept broadly, reflecting our long-standing policy of deference to legislative judgments in this field. In *Berman v. Parker* (1954), this Court upheld a redevelopment plan targeting a blighted area of Washington, D.C., in which most of the housing for the area's 5,000 inhabitants was beyond repair. Under the plan, the area would be condemned and part of it utilized for the construction of streets, schools, and other public facilities. The remainder of the land would be leased or sold to private parties for the purpose of redevelopment, including the construction of low-cost housing . . . . The public use underlying the taking was unequivocally affirmed.

Viewed as a whole, our jurisprudence has recognized that the needs of society have varied between different parts of the Nation, just as they have evolved over time in response to changed circumstances . . . . For more than a century, our public use jurisprudence has wisely eschewed rigid formulas and intrusive scrutiny in favor of affording legislatures broad latitude in determining what public needs justify the use of the takings power.

IV

. . . [The city's] determination that the area was sufficiently distressed to justify a program of economic rejuvenation is entitled to our deference. The City has carefully formulated an economic development plan that it believes will provide appreciable benefits to the community, including—but by no means limited to—new jobs and increased tax revenue. . . . To effectuate this plan, the City has invoked a state statute that specifically authorizes the use of eminent domain to promote economic development. Given the comprehensive character of the plan, the thorough deliberation that preceded its adoption, and the limited scope of our review, it is appropriate for us, as it was in *Berman,* to resolve the challenges of the individual owners, not on a piecemeal basis, but rather in light of the entire plan. Because that plan unquestionably serves a public purpose, the takings challenged here satisfy the public use requirement of the Fifth Amendment.

. . . We emphasize that nothing in our opinion precludes any State from placing further restrictions on its exercise of the takings power. Indeed, many States already impose "public use" requirements that are stricter than the federal baseline. Some of these requirements have been established as a matter of state constitutional law, while others are expressed in state eminent domain statutes that carefully limit the grounds upon which takings may be exercised. This Court's authority, however, extends only to determining whether the City's proposed condemnations are for a "public use" within the meaning of the Fifth Amendment to the Federal Constitution.*

Reversed in favor of City of New London.

Outrage over the decision in this case led many states to pass legislation to protect private property rights and limit the government's ability to condemn property. In fact, shortly after *Kelo* was decided, a survey by the Saint Index, a polling organization specializing in land-use issues, found that 81 percent of Americans opposed the decision.[3] Since *Kelo,* more than 40 states have passed legislation limiting the government's ability to take private property and give it to private business.[4] Some of these laws have included tightening the definition of *public use* or closing loopholes such as vague definitions of "blight." In 2005, the House of Representatives introduced a bill to prevent "economic development" from being used to justify giving property to private business under eminent domain. It stalled in the Senate; a similar bill was introduced in the summer of 2007, but died without passage at the end of the year. In June of 2006, President Bush banned federal agencies from using eminent domain solely for "private

---

[3]"The Limits of Anti-*Kelo* Legislation: Reformers Are Trying to Outlaw Eminent Domain Abuse. But Will the Laws They're Passing Be Effective?" *Reason* (Aug. 1, 2007). Available at www.accessmylibrary.com/coms2/summary_0286-31855668_ITM.

[4]Castle Coalition, "50 State Report Card" (August 2007). Retrieved Feb. 29, 2008, from www.castlecoalition.org/pdf/publications/report_card/50_State_Report.pdf.

*Susette Kelo et al., Petitioners, v. City of New London, Connecticut, et al. Supreme Court of the United States 125 S. Ct. 2655 (2005). https://www.law.cornell.edu/supct/html/04-108.ZO.html.

development," but allowed takings for private owners who promise to use the land for both private and "public" development. New "anti-*Kelo* legislation" may still be introduced in the future.

Restrictions on Land Use

No one is allowed to use his or her land in a completely unrestricted manner. As previously indicated, the doctrine of waste prohibits some uses and abuses of land. There are other such restrictions, both voluntary and involuntary.

RESTRICTIVE COVENANTS

restrictive covenants Promises by the owner, generally included in the deed, to use or not to use the land in particular ways.

Parties may voluntarily enter into **restrictive covenants**, that is, promises to use or not to use their land in particular ways. These covenants are generally included in the deeds and are binding on the owners as long as the covenants are for lawful acts. For example, a restrictive covenant not to construct buildings higher than three stories would be lawful and enforceable, whereas a covenant never to convey property to minorities would be unlawful and, thus, unenforceable.

As with condominium and cooperative developments, restrictions may often be placed in CC&Rs, and may be enforced by a homeowners' association for the development in which one's house is located. It is important to read the restrictions contained in a deed and in the CC&Rs very carefully, because courts will generally enforce such restrictions and provisions unless they clearly violate public policy or state or federal law. Courts are not sympathetic to the argument that it would be very costly for an owner to comply with the CC&Rs, as a number of Florida homeowners found out in 2010, when a court ordered them to remove the docks they had constructed in violation of a community deed restriction. The homeowners had constructed docks on their properties in violation of a rule that prohibited the building of any new docks in the community. The court ordered the docks removed, as they were built in violation of the restriction, despite the facts that the owners had not read the deed restrictions and that one owner had spent $50,000 on his dock and another about $65,000.[5]

ZONING

zoning Government restrictions placed on the use of private property to ensure the orderly growth and development of a community and to protect the health, safety, and welfare of citizens.

Zoning is the restriction of the use of property to allow for the orderly growth and development of a community and to protect the health, safety, and welfare of its citizens. Zoning may restrict the type of use to which land may be put, such as residential, commercial, industrial, or agricultural. Zoning laws may also regulate land use in geographic areas on the basis of such factors as the intensity (single or multifamily dwellings), the size, or the placement of buildings.

When a community enacts new zoning ordinances, generally there is a public hearing on the proposed change in zoning. Often, the community allows a nonconforming use of a particular property to remain when the zoning of an area changes. This exception to the zoning law occurs when the property in question was being used for a purpose not allowed under the new zoning statutes, but because of the prior use, the owner is allowed to continue using the property in the nonconforming manner.

variance Permission given to a landowner to use a piece of his or her land in a manner prohibited by the zoning laws; generally granted to prevent undue hardship.

A landowner who wished to use her or his land in a manner prohibited by zoning laws may seek a variance from the appropriate governmental unit, usually a zoning board or a planning commission. A **variance** is permission to use a piece of land in a manner prohibited by the zoning laws. Variances are generally granted to prevent undue hardship.

Sometimes, persons negatively affected by a zoning law challenge that ordinance. Zoning is allowed under the *police power,* the power of the state to regulate to protect the health, safety, and welfare of the public. To be a valid exercise

---
[5] "HOA Rule Forces Pasco Homeowners to Dismantle Docks" (Feb. 16, 2010). Bay9News.com.

COMPARATIVE LAW CORNER

Eminent Domain in Germany

Germany, much like the United States, is a federal system and has both federal and state (*Lander*) regulations regarding eminent domain. The federal laws on eminent domain are very similar to the laws in the United States, with a few notable differences. Germany is a civil-law system, rather than a common-law system, so most of its rules on eminent domain are written into statutes, rather than determined through court cases.

The federal government may expropriate land for the public good as long as it provides compensation. Germany's Basic Law says: "Such compensation shall be determined by establishing an equitable balance between the public interest and the interests of those affected." Compensation in the German system is not based on fair market value alone, but must also take the interests of those affected into consideration. Germans may dispute the amount of compensation in the ordinary courts.

A taking requires that the government have the legal authority for the taking. The taking must also be appropriate for accomplishing the government's purpose, and done in the least intrusive way possible. This requirement means that the government cannot take property before it needs it, and the government can take only the property it needs. The taking must also pass a balancing test, showing that the public's interest is greater than the owner's interest. This last requirement is absent from the U.S. eminent domain laws.

Another difference is that if the property is no longer needed by the German government, it must revert to its former owner. In the United States, once the government has provided just compensation for the property, it belongs to the government.

The last major characteristic of eminent domain in Germany is currently being debated in the United States, as a result of the *Kelo* case. In Germany, the state cannot take property for purely private purposes (taking property and giving it to private business to spur the economy), but provision of economic benefits, such as job creation, can be an acceptable taking.

---

of such power, the zoning ordinance must not be arbitrary or unreasonable. An ordinance is unreasonable if (1) it encroaches on the private property rights of landowners without a substantial relationship to a legitimate government purpose, or (2) there is no reasonable relationship between the ends sought to be obtained and the means used to attain those ends.

Zoning is also unreasonable if it totally destroys the economic value of a property holder's land. In such cases, the zoning is really a constructive taking of the property, and the party whose land is so affected is entitled to just compensation. This type of challenge is frequently made, but is rarely successful.

It is important, when purchasing property, to make sure you understand how the land is zoned, and to make sure you understand exactly how the zoning is interpreted, as the following case illustrates. Notice that this case involves a zoning question that often arises near colleges.

CASE 13-3

Emine Bayram v. City of Binghamton and City of Binghamton Zoning Board of Appeals
New York Supreme Court
27 Misc.3d 1032, 899 N.Y.S.2d 566 (2010)

Bayram owns a single-family residence located in a R–1 Residential Single–Unit Dwelling District. On June 15, 2009, Bayram entered into a written agreement to lease the property to seven Binghamton University students, for a term of approximately two years commencing on August 15, 2009, and ending on August 30, 2011. On August 31, 2009, a complaint was received by the City of Binghamton that the property was being "rented to a frat house." Following investigation, the city issued a Notice of Violation, advising Bayram of the Department's determination that the seven individuals then residing at the property pursuant to the lease did not constitute a functional and factual family equivalent—as defined by the Zoning Ordinance—and, therefore, that their continued occupancy of the property would constitute a violation thereof.

Bayram sought a review by the Zoning Board, and after a public hearing, the Board affirmed its decision and ordered her to evict the tenants.

Justice Phillip R. Rumsey

. . . The interpretation by a zoning board of appeals of terms defined in its own zoning ordinance is entitled to great deference, and must be upheld if it is neither irrational nor unreasonable; judicial review is limited to an examination of whether there is a rational basis for the determination that is supported by evidence in the record.

Any dwelling unit located in a residential zoning district may be occupied only by a family or the equivalent of a family (City of Binghamton Zoning Ordinance § 410–26[D]). As relevant to resolution of the issue presented by this proceeding, a functional and factual family equivalent is, in turn, defined as:

A group of unrelated individuals living together and functioning together as a traditional family. In determining whether or not a group of unrelated individuals comprise [sic] a functional and factual family equivalent, a petition shall be presented before the Zoning Board of Appeals, who [sic] will consider, among other things, the following factors:

A. Whether the occupants share the entire dwelling unit or act as separate roomers.

B. Whether the household has stability akin to a permanent family structure. The criteria used to determine this test may include the following:

1. Length of stay together among the occupants in the current dwelling unit or other dwelling units.
2. The presence of minor, dependent children regularly residing in the household.
3. The presence of one individual acting as head of household.
4. Proof of sharing expenses for food, rent or ownership costs, utilities and other household expenses.
5. Common ownership of furniture and appliances among the members of the household.
6. Whether the household is a temporary living arrangement or a framework for transient living.
7. Whether the composition of the household changes from year to year or within the year.
8. Any other factor reasonably related to whether or not the group of persons is the functional equivalent of a family.

Based on a review of the ZBA's application of the foregoing criteria to the evidence in the record, it cannot be said that the ZBA's decision is irrational. The record of its deliberations . . . and its written decision dated January 12, 2010 . . . reflect that the ZBA considered each of the enumerated factors in reaching a determination that is supported by the evidence that was before it.

Following the specified procedure, the ZBA first concluded that the occupants do not share the entire dwelling unit, instead acting as separate roomers, based on evidence—including photographs—showing that each has a separate room with his own refrigerator, computer, and television, and that each resident has his own car.

. . . The ZBA's findings with respect to each factor, in support of its determination that the tenants' living arrangement lacks the requisite degree of stability, may be briefly summarized as follows. The ZBA found that the tenants' living arrangement is temporary, limited at most to the two-year lease term, based on evidence that the tenants are in the area only for the purpose of attending Binghamton University.

. . . There are no minor, dependent children residing at the property. The ZBA noted, appropriately, that this fact was not dispositive, and then proceeded to reject petitioner's contention that John Kim functions as a head of household. While recognizing that the lease designates Kim as "head of the family," the ZBA found little evidence that he functions in the role of a traditional head of household. It further determined that petitioner attempted to depict the tenants as a functional and factual family equivalent—in part by naming Kim as head of household—to circumvent the occupancy limitations imposed by the Zoning Ordinance.

The ZBA concluded that there was not substantial evidence of shared expenses. . . . [T]he Zoning Officer initially determined that each tenant separately paid his share of the rent directly to petitioner.

With respect to common ownership of furniture—all appliances being owned by the landlord—the ZBA decided that the evidence, including photographs, showed that most significant items, such as computers and televisions, were individually owned and not shared. In reaching its conclusion that the household is a temporary living arrangement or a framework for transient living, the ZBA noted that a lack of permanency was apparent in that none of the residents' automobiles was registered to the property address, none of their driver's licenses listed the property as a residence address, and none of the tenants was registered to vote in local elections.

Finally, after noting that there was no evidence to suggest a change of household composition during the lease term, the ZBA considered an additional factor, namely that the monthly rent of $2,800—which it believed is three to four times the monthly rent typically paid for a single-family residence in Binghamton—is indicative of a lease to seven individuals, rather than to a single family unit.

It is readily apparent from the foregoing summary of its deliberations that the ZBA's decision was rational in light of the evidence before it. . . . Thus, the decision of the ZBA must be upheld.*

Judgment *affirmed* in favor of the City.

---

*Emine Bayram v. City of Binghamton and City of Binghamton, Zoning Board of Appeals, New York Supreme Court 27 Misc.3d 1032, 899 N.Y.S.2d 566 (2010). http://law.justia.com/cases/new-york/other-courts/2010/2010-20116.html.

While the zoning board's decision was upheld in the foregoing case, and Bayram was prohibited from extending the lease or leasing to any other similarly situated group of college students in the future, the court did not require eviction of the current tenants before their lease was up because under the city's laws, eviction is authorized when the tenants are using the property for an illegal purpose. The fact that the lease violated the zoning ordinance did not establish that the tenants were engaging in criminal activity.

OTHER STATUTORY RESTRICTIONS ON LAND USE

In addition to zoning ordinances, states often use their police power to pass laws affecting individuals' use of their property. For example, some states have passed historic preservation statutes whereby certain buildings with historical importance are subject to certain restrictions. Owners of such structures may be required to keep the buildings in good repair or might be required to have any alterations to the buildings' facades approved prior to modification.

Like zoning laws, any laws restricting people's use of their property may be subject to constitutional challenge. In *Dolan v. City of Tigard*, the Court clarified and toughened its requirement that government planners produce specific justifications when they condition building permits on a party's rendering a portion of the property for public use.[6] Plaintiff Dolan sought a permit from defendant City of Tigard to expand her store and pave her parking lot. The city conditioned the grant of her permit on her dedication of a portion of her property for a public greenway to minimize flooding that would otherwise be likely to result from her construction and a pedestrian/bicycle pathway to decrease congestion in the business district. The U.S. Supreme Court ruled that the city had to show some evidence that the expansion of the permit would cause the problems that would be fixed by the new regulations, as well as that the regulations would fix those problems. Simply stating that a larger store *could* cause more congestion and that a bicycle path *could* help clear up that congestion was not enough.

The United States Supreme Court reinforced this decision in 2013, in the case of *Koontz v. St. John's Rangement District*.[7] In that case, Koontz purchased 14.9 acres of land, and applied for the necessary permits to develop 3.7 acres of wetlands. To mitigate the environmental effects of his proposed development, he offered to not develop his remaining 11 acres by deeding a conservation easement to the District for that land. The District denied his proposal and said they would grant him the permits and approve construction only if he (1) reduced the size of his development and, *inter alia*, deeded to the District a conservation easement on the resulting larger remainder of his property or (2) hired contractors to make improvements to District-owned wetlands several miles away. Koontz thought the demands were excessive and sued the District.

The trial court found the District's actions illegal because they failed to meet the requirement from *Dolan v. Tigard* that government may not condition the approval of a land-use permit on the owner's relinquishment of a portion of his property unless there is a nexus and rough proportionality between the government's demand and the effects of the proposed land use. The state's appellate court affirmed, but the Florida Supreme Court reversed. First, it held that petitioner's claim failed because, unlike in *Dolan*, the District *denied* the application. Second, the State Supreme Court held that a demand for money cannot give rise to a claim under *Dolan*.

The U.S. Supreme Court overruled the Florida State Supreme Court, holding that the government's demand for property from a land-use permit applicant must satisfy the *Dolan* requirements even when it denies the permit. They also

---
[6] 114 S. Ct. 2309 (1994).
[7] 133 S. Ct. 2586 (2013).

held that the government's demand for property from a land-use permit applicant must satisfy the *Dolan* requirements even when its demand is for money.

Personal Property

All property that is not real property is **personal property**. As previously explained, personal property may be either tangible or intangible. **Tangible property** includes movable items, such as furniture, cars, and even pets. **Intangibles** include such items as bank accounts, stocks, and insurance policies.

Because most intangibles (with the exception of some of those classified as intellectual property, discussed in Chapter 14) are evidenced by writings, most of the following discussion applies to both tangible and intangible property. The primary concerns that arise in conjunction with personal property involve (1) the means of acquiring ownership of the property and (2) the rights and duties arising out of a bailment. Both are discussed here in detail.

VOLUNTARY TRANSFER OF PERSONAL PROPERTY

The most common means by which personal property is acquired is by its voluntary transfer as a result of either a purchase or a gift. Ownership of property is referred to as **title**, and title to property passes when the parties so intend. When there is a purchase, the acquiring party gives some consideration to the seller in exchange for title to the property. In most cases, this transfer of ownership requires no formalities, although in a few cases changes of ownership must be registered with a government agency. The primary transfers requiring such formalities include sales of motor vehicles, watercraft, and airplanes. Transfer of such property requires that a certificate of title be signed by the seller, taken to the appropriate governmental agency, and then reissued in the name of the new owner.

Gifts are distinguished from purchases in that no consideration is given for a gift. As the reader knows, a promise to make a gift is, therefore, unenforceable. Once properly made, however, a gift is an irrevocable transfer.

For a valid gift to occur, three elements must be present (Exhibit 13-5). First, there must be a *delivery* of the gift. This delivery may be actual, that is, the physical presentation of the gift itself; or it may be constructive, that is, the delivery of an item that gives access to the gift or represents it, such as handing over the keys to a car. Second, the delivery must be made with **donative intent** to make a present, as opposed to future, gift. The donor makes the delivery with the purpose of turning over ownership at the time of delivery. The final element is *acceptance*, a willingness of the donee to take the gift from the donor. Usually, acceptance is not a problem, although a donee may not want to accept a gift because he or she does not want to feel obligated to the donor or because he or she believes that ownership of the gift may impose some unwanted legal liability.

INVOLUNTARY TRANSFERS OF PERSONAL PROPERTY

Involuntary transfers involve the transfer of ownership of property that has been abandoned, lost, or mislaid. The finder of such property may acquire ownership rights to such property through possession.

Abandoned property is property that the original owner has discarded. Anyone who finds that property becomes its owner by possessing it.

Lost property is property that the true owner has unknowingly or accidentally dropped or left somewhere. He or she has no way of knowing how to retrieve it. In most states, the finder of lost property has title to the lost good against all except the true owner.

---

personal property All property that is not real property; may be tangible or intangible.

tangible property Personal property that is material and movable (e.g., furniture).

intangible property Personal property that does not have a physical form and is usually evidenced in writings (e.g., an insurance policy).

title Ownership of property.

EXHIBIT 13-5

ELEMENTS OF A GIFT

- Delivery
- Donative Intent
- Acceptance

donative intent Intent to transfer ownership to another at the time the donor makes actual or constructive delivery of the gift to the donee.

Mislaid property differs from lost property in that the owner has intentionally placed the property somewhere but has forgotten its location. The person who owns the realty on which the mislaid property was placed has the right to hold the mislaid property. The reason is that it is likely that the true owner will return to the realty looking for the mislaid property.

In some states, the law requires that before becoming the owner of lost or mislaid property, a finder must place an ad in a newspaper that will give the true owner notice that the property has been found, or must leave the property with the police for a statutorily established reasonable period of time. Some state laws require both.

APPLYING THE LAW TO THE FACTS . . .

Let's say that Clara's father leaves some furniture he doesn't want by the road. Tim, seeing the abandoned furniture, picks it up and takes it home. When Clara finds out what happened to the furniture, she goes to retrieve it, telling Tim that she has more rights to the furniture because she is related to her father and Tim is a stranger with no rights to it. Is Clara legally correct? Why or why not?

BAILMENTS

A **bailment** of personal property is a special relationship in which one party, the *bailor*, transfers possession of personal property to another, the *bailee,* to be used by the bailee in an agreed-on manner for an agreed-on time period.

One example of a bailment is when a person leaves his or her coat in a coat-check room. The person hands the coat to the clerk and is given a ticket identifying the object of the bailment so that it can be reclaimed. The bailment may be gratuitous or for consideration. It may be to benefit the bailor, the bailee, or both. If the bailment is intended to benefit only the bailor, the bailee is liable for damage to the property caused by the bailee's gross negligence. If the bailment is solely for the bailee's benefit, then the bailee is responsible for harm to the property caused by even the slightest lack of due care on the part of the bailee. If the bailment is for the mutual benefit of both bailee and bailor, the bailee is liable for harm to the bailed property arising out of the bailee's ordinary or gross negligence. If the property is harmed by an unpreventable "act of God," there is no liability on the part of the bailee under any circumstances. These general rules notwithstanding, the parties to a bailment contract can limit or expand the liability of the bailee by contract. In general, conspicuous signs (e.g., "We are not responsible for lost items") have been held sufficient to limit liability.

If the bailee is to receive compensation for the bailment, the bailee may retain possession of the bailed property until payment is made. In most states, when the bailor refuses to provide the agreed-on compensation to the bailee, the property may ultimately be sold by the bailee after proper notice and a hearing. The proceeds are first used to pay the bailee and to cover the costs of the sale. The remaining proceeds then go to the bailor.

In addition to the duty to pay bailees their agreed-on compensation, a bailor must also warn the bailee of any hidden defects in the bailed goods. If such a warning is not given and the defect injures the bailee, the bailor is liable to the bailee for the injuries.

bailment A relationship in which one person (the bailor) transfers possession of personal property to another (the bailee) to be used in an agreed-on manner for an agreed-on period of time.

Global Dimensions of Property Law

U.S. citizens are increasing purchasing property abroad. There are some advantages, as well as disadvantages. It is certainly a way to diversify your investments; when the real estate market falls here, it isn't necessarily going to fall in China.

Another advantage may be that you may gain certain rights in the country where you buy property. For example, if you purchase property in Dubai, you will automatically receive a visa that will allow you to do business there.

Of course, there are dangers, too. For example, it is probably best to invest in a place where the currency is fairly stable, because a huge slide in the value of the currency of the nation where you purchased the property could lead to a significant decline in the value of your asset, especially if you are suddenly forced to sell and the currency has declined since you purchased the property.

Another factor to consider when investing in real estate in a foreign country is the political stability of the country. If you are buying in the European Union, you can probably count on the laws remaining fairly stable, but in a country with a less stable political system, a coup could result in a dramatic change in the laws of property ownership, which might even lead to your property being confiscated.

Regardless of where you choose to purchase property, however, before you invest, it is important to make sure you thoroughly understand the laws governing property ownership and transfer in that country because every country has its own set of laws. Do not assume that just because the law works one way in the United States, it is the same in any other nation. In fact, you should not even assume before checking that you can buy property in the country of your dreams because some nations place significant restrictions on the ability of foreigners to buy their property. For example, Mexico's constitution forbids foreigners from owning any real estate within 30 miles from the coastline or 60 miles from any U.S. border.

One final point to keep in mind when thinking about buying real estate in foreign countries is that U.S. mortgage companies generally will not finance purchases of foreign real estate, so you will need to obtain financing in the country where you are buying. Therefore, you also need to become familiar with the country's laws regarding lending. As the world becomes a smaller place, however, buying property in foreign countries will become increasingly common.

SUMMARY

Property is a bundle of rights in relation to a tangible object, the most significant of which is probably the right to exclude others. Property can be divided into three categories: real property, land and anything permanently attached to it; personal property, tangible movable objects and intangible objects; and intellectual property, property that is primarily the result of one's mental rather than physical efforts. Real property can be transferred voluntarily or involuntarily. Voluntary transfers include transfer by gift or sale. Adverse possession and condemnation are the two involuntary means. Personal property likewise can be transferred voluntarily through a gift or sale. It may also be transferred involuntarily if it is lost or mislaid.

REVIEW QUESTIONS

13-1 Explain why each of the following is or is not real property.
  a. A fence
  b. A tree
  c. A house trailer
  d. A built-in oven
  e. A refrigerator

13-2 Define the primary estates in land.

13-3 Explain the circumstances under which each of the following types of ownership would be most desirable. Give reasons for your responses.
  a. Joint tenancy
  b. Tenancy in common
  c. Tenancy by the entirety

13-4 Explain how the ownership of land may be transferred.

13-5 Explain how a general warranty deed differs from a quitclaim deed.

13-6 What are the similarities and differences between condemnation and adverse possession?

REVIEW PROBLEMS

13-7 Carlos owns a life estate in Morganberry Farms. The farm consists of 5 acres, with a house and landscaping located on 1/2 acre, and corn being raised on the other 4.5 acres. Carlos decides to cover one of the acres that was being used for corn with asphalt so that he can charge people to park on the property during the summer when local fairs and festivals are held in the area. The future owner of the estate hears of these plans and seeks an injunction prohibiting Carlos from asphalting over the acre. Why will the court either grant or deny the injunction?

13-8 The Hortons bought a home in a development that was subject to a set of CC&Rs, which included a restriction requiring homeowners to submit plans for changing their front landscaping for approval by the architectural control committee before any changes were made. Among the items included in a list of prohibited landscaping materials was artificial grass. The state was encouraging residents to save water by offering to pay residents $60 for every square foot of grass they replaced with either rock or artificial grass. The Hortons wished to take advantage of this option and submitted plans to replace their grass with artificial grass. When the architectural review committee refused to approve their plans, the Hortons decided to change their landscaping anyway. Are they within their legal rights to do so? What do you believe are the strongest arguments on each side?

13-9 The plaintiff owned land in an area zoned for buildable private parks. The city rezoned the land to allow only parks open to the public. This rezoning effectively prohibited the plaintiff from generating any sort of income from the land. Was the plaintiff correct in his contention that this constituted a deprivation of property without due process of law?

13-10 An athletic team wants to build a new sports arena in the middle of the downtown area. The team has bought enough land for the stadium, but needs more land to build an adjacent parking lot. The team asks the city to take the necessary property from the residents and transfer it to the team so they can build the arena. Would it be lawful for the city to do so? If so, how? If not, why not?

13-11 Judy worked the morning shift at Wild Oats. She bought a Pepsi but got too busy to drink it, so she left it on a shelf behind the counter to drink the next day. Cindy was working that afternoon. She got thirsty and asked her coworkers if the Pepsi belonged to any of them. When no one claimed it, she opened the bottle, saw some fine print on the underside of the cap, and read, "You have won a million dollars." When Judy found out what happened, she confronted Cindy and claimed the prize as hers. Explain who should receive the prize money and why.

13-12 Hallman spent the night at the New Colonial Hotel. On inquiry, he was told that the bellboy would take care of his car. The bellboy took the car to a nearby parking lot and left the car and keys with the lot attendant. He gave Hallman a claim check bearing the name of the lot and stamped "New Colonial." When Hallman went to pick up his car, the side window was broken and more than $500 worth of personal property was missing. Who was liable for the missing personalty and the damage to the car? Why?

CASE PROBLEMS

13-13 Six individuals living near a landfill sought recovery from defendant landfill company for substantial interference with the use and enjoyment of their property due to odors coming from the defendant landfill company. The plaintiffs alleged nuisance, trespass, and negligence claims. The jury awarded the plaintiffs actual and compensatory damages totaling $532,500 as well as $300,000 in punitive damages. The defendant landfill company appealed. How do you think the court ruled? Can invisible odors be recognized as a cause of action for a trespass claim? *Babb v. Lee County Landfill SC, LLC*, 405 S.C. 129 (S.C. 2013).

13-14 In 2011 the Nwankwos filed an application with the Design Committee of their residential subdivision to build an outdoor living space on their property. The Nwankwos wished for the outdoor living space to include a "gazebo attached to the house and additional landscaping," as well as a raised patio/fire pit. The Nwankwos's neighbors, the Baruks, filed a complaint to the president of the subdivision, expressing frustration that the Nwankwos's fire pit area was too close to their own property. The Design Committee of the Subdivision held a special meeting to discuss the construction plans of the Nwankwos, and eventually ordered the Nwankwos to cease building until a new construction application was approved by the Committee. The Nwankwos appealed this order, and the Subdivision's Association scheduled a hearing between the Nwankwos and the Baruks. In 2012, the Subdivision's Association reversed their decision and allowed the Nwankwos to complete their construction of the patio/fire pit. The Baruks then requested an appeal hearing, which was denied. The Baruks eventually filed suit against the Nwankwos and Subdivision's Association, alleging common-law nuisance and trespass. How did the court rule? What evidence would the Baruks need to provide to sustain their claims of common-law nuisance and trespass? *Baruk v. Heritage Club Homeowners' Ass'n*, 2014 Ohio App. LEXIS 1531 (2014).

13-15 The Parkowner has owned and operated Rancho de Calistoga Mobilehome Park, a 184 space mobile home park in the City of Calistoga since approximately 1977. In 1984, the City adopted the Ordinance, which was amended in 1993 and again in October 2007. The Ordinance provides, that "[i]t is necessary to protect mobile home homeowners and residents of mobile homes from unreasonable rent increases and at the same time recognize the rights of mobile home park owners to maintain their property and to receive just and reasonable return on their investments." The Ordinance authorizes a yearly increase in rent that equals the lesser of (1) 100 percent of the percent change in Consumer Price Index; or (2) 6 percent of the base rent. If a park owner wants to increase the rent above that amount, the Ordinance provides for an administrative process to consider such requests.

The Parkowner filed a claim arguing that the rent control ordinance constitutes a taking of his property without just compensation because it takes rental income he had counted on earning. Why do you believe the rent control ordinance does or does not constitute a taking? *Rancho de Calistoga v. City of Calistoga* (N.D. Cal.) 2012 WL 5636356.

13-16 Plaintiffs Kathryn and Jeremy Medlen owned a dog named Avery. The plaintiffs' pet escaped from the backyard and was obtained by the County's animal control agency. When the plaintiffs went to pick up their pet from the animal control shelter they did not have sufficient funds for the required pickup fee. The shelter held the plaintiffs' pet until the plaintiffs had the funds for pickup. The pet was then mistakenly placed on the shelter's euthanasia list, and was put to sleep. Plaintiffs sued the employee of the shelter for causing the pet's death and sought "sentimental or intrinsic value damages." The trial court dismissed the suit, and the plaintiffs appealed. The court of appeals reversed the trial court's decision, finding that pets were property in the eyes of the law. The employee of the shelter appealed this decision. How do you think the Supreme Court of Texas ruled? Why? *Strickland v. Medlen*, 397 S.W.3d 184 (Tex. 2013).

13-17 The plaintiff parked his motorcycle in the defendant's parking garage, from which it was allegedly stolen. The plaintiff had parked his motorcycle without assistance and retained his keys. The parties' relationship was governed by a written "Garage Agreement," which provided that the plaintiff "licensee" parked at his "own risk." The plaintiff sued to recover the value of his motorcycle. The trial court found that there was no bailment and dismissed the case. How do you think the appellate court ruled? *Ernest Burke, Plaintiff-Respondent v. Riverbay Corp.*, 2009 N.Y. Slip Op. 52386U, 2009 N.Y. Misc. LEXIS 3225.

THINKING CRITICALLY ABOUT RELEVANT LEGAL ISSUES

Eminent Domain and Democracy

A democracy is based on the idea that for many purposes, the best decision-making framework is to resolve difficult issues by following the wishes of the representatives of majority opinion. Therefore, when a duly elected group of County Commissioners or a local city council decide that a piece of land would be best used for the public purpose of expanding employment opportunities in the community, it would seem to follow that supporters of democracy would go along quietly with the decision.

But there are few legal issues that cause more conflict than a dispute over the use of eminent domain. The idea itself is not the problem. Almost all would agree that there are certain functions in the community (a school perhaps) that are so crucial to its ongoing growth and development that we need to be able to take land for a public purpose.

But what in the abstract seems so clear becomes a nightmare, when it is my or your property that is being taken. That property may have been owned and treasured by our family for decades; friendships and comfort zones with local businesses may have flowered as a result of numerous ongoing contacts. We do not want to give up any of these things that make our lives so comfortable.

In addition, when eminent domain is used, it is not just something abstract like the community that may benefit. Instead, specific people benefit. Regardless of how the eminent domain argument is made, those who will benefit from the construction and later uses of the land are special beneficiaries who have a vested interest in translating their personal interest into public-benefit arguments for the purpose of activating eminent domain. Those being forced to sell their land are quick to point out the private benefits flowing from this alleged public good.

Also lurking not so hidden in the background is a theme developed earlier in this chapter. The more powerful a group is, the more likely it is going to be the party pushing for the use of eminent domain. They, of course, word the issue in terms of the projected public benefit. But opponents wonder out loud whether they are not simply abusing their financial power to achieve yet one more advantage in a market economy where influence is often shaped by dollar votes.

Returning to the relationship between democracy and eminent domain, there are a number of questions that arise for a critical thinker pondering any eminent domain controversy.

1. What specific information would be needed to assist in making an intelligent decision about the issues raised in the potential tension between democracy and eminent domain?
2. What ambiguous words need to be clarified before an eminent domain controversy can be fairly decided?
3. How does this issue make it clear that the way a legal issue is framed or understood goes a long way toward telling us who will win the dispute?
4. How would you word the issue here if you were arguing on behalf of opponents to an eminent domain action?

   Clue: Think about what is required for a decision to be thoughtfully democratic? In other words, democracy relies on majority opinions, but some opinions are better formed than others. What would have to be true before you would support the decision of a majority of citizens?

ASSIGNMENT ON THE INTERNET

This chapter introduced you to government regulation and protection of various forms of property. One such regulation and protection used by the government is zoning. Use a legal research database such as **Findlaw.com** to find a recent court case in which a zoning law was challenged or where the plaintiff argued that a zoning law resulted in a taking of property.

Once you have found such a case, identify the court's reasoning and the ethical norms underlying that reasoning. Was the particular zoning law found to be unreasonable? Now compare the reasoning of your case to that of *Dolan v. City of Tigard*. Are the ethical norms different or the same?

ON THE INTERNET

https://www.justia.com/real-estate/ You can go to this site to find an overview of real estate law and research questions related to real estate law.
http://www.hg.org/realest.html This site contains information about real estate law, as well as tips for finding a real estate lawyer.

FOR FUTURE READING

Batches, Wayne. "Business Improvement Districts and the Constitution: The Troubling Necessity of Privatized Government for Urban Revitalization." *Hastings Constitutional Law Quarterly* 28 (2010): 91.

Davis, Seth. "Presidential Government and the Law on Property." *Wisconsin Law Review* (2014): 471.

Eagle, Steven J. "The Really New Property: A Skeptical Appraisal." *Indiana Law Review* 43 (2010): 1229.

Gatewood, Jace C. "The Evolution of the Right to Exclude—More Than a Property Right, A Privacy Right." *Mississippi College Law Review* 32 (2014): 447.

Marsh, Tanya D. "Sometimes Blackacre Is a Widget: Rethinking Commercial Real Estate Contract Remedies." *University of Nebraska Law Review* 88 (2010): 635.

Pomeroy, Chad J. "The Shape of Property." *Seton Hall Law Review* 44 (2014): 797.

Sirois, Lauren. "Recovering for the Loss of a Beloved Pet: Rethinking the Legal Classification of Companion Animals and the Requirements for Loss of Companionship Tort Damages." *University of Pennsylvania Law Review* 163 (2015): 1199.

Stern, James. "Property's Constitution, 101." *California Law Review* (April, 2013): 277.

CHAPTER FOURTEEN

Intellectual Property

- INTRODUCTION TO INTELLECTUAL PROPERTY
- TRADEMARKS
- PATENTS
- TRADE SECRETS
- COPYRIGHTS
- GLOBAL DIMENSIONS OF INTELLECTUAL PROPERTY LAW

Introduction to Intellectual Property

Intellectual property consists of the fruits of one's mind. The laws of intellectual property protect property that is primarily the result of mental creativity rather than physical effort. This category includes trademarks, patents, trade secrets, and copyrights, which are discussed in that order in this chapter.

Trademarks

A **trademark** is a distinctive mark, word, design, picture, or arrangement used by a seller in conjunction with a product that tends to cause the consumer to identify the product with the producer. Even the shape of a product or package may be a trademark if it is nonfunctional.

trademark A distinctive mark, word, design, picture, or arrangement used by the producer of a product that tends to cause consumers to identify the product with the producer.

CRITICAL THINKING ABOUT THE LAW

1. In Chapter 13 you studied real and personal property. Why might the legal rules for intellectual property be different in any fashion from the rules for real and personal property?

    Clue: Consider the different attitudes we might have toward the results from thinking and the output from physical exertion.

2. Some argue that intellectual property should be available for everyone once it is produced by someone's mind. What ethical norm underlies this attitude toward sharing mental output?

    Clue: Should a person benefit because her mind creates an idea that other minds, for some reason, did not produce?

3. If an idea becomes a legal source of great wealth for a person, what information about the "discovery" would cause you to be more supportive of protecting the property right to the fruits of the discovery?

    Clue: The play and movie *Amadeus* suggest that Mozart's brilliant compositions resulted more from his genius than from any hard work on his part.

TABLE 14-1
TYPES OF MARKS PROTECTED UNDER THE LANHAM ACT

1. *Product trademarks:* marks affixed to a good, its packaging, or its labeling.
2. *Service marks:* marks used in conjunction with a service.
3. *Collective marks:* marks identifying the producers as belonging to a larger group, such as a trade union.
4. *Certification marks:* marks licensed by a group that has established certain criteria for use of the mark, such as "UL Tested" or "Good Housekeeping Seal of Approval."

Even though the description of a trademark is very broad, there has still been substantial litigation over precisely what features can and cannot serve as a trademark. A sound (NBC's three chimes and Metro-Goldwyn-Mayer Corporation's roaring lion), a scent (plumeria blossoms on sewing thread and yarns), and even a color (green-gold of Qualitex dry-cleaning press pads) have all been found to constitute trademarks, because they distinguish the goods as unique and serve to identify their source.

Even phrases can be trademarked. For example, in 2014, Seattle Seahawks running back Marshawn Lynch filed a trademark for the phrase, "I'm just here so I won't get fined," the phrase he used to answer over 20 questions on Super Bowl XLIX media day. He trademarked the phrase to use on items in his Beast Mode apparel line. In 2013, he had trademarked the phrase "About that action BOSS," a phrase from his only Superbowl XLVII interview, to be used for a similar purpose.

A trademark used intrastate is protected under state common law. To be protected in interstate use, the trademark must be registered with the U.S. Patent Office under the Lanham Act of 1947. Several types of marks, listed in Table 14-1, are protected under this act.

If a mark is registered, the holder of the mark may recover damages from an infringer who uses that mark to pass off its own goods as those of the mark holder. The mark owner may also obtain an injunction prohibiting the infringer from using the mark. Only the latter remedy is available for an unregistered mark.

To register a mark with the Patent Office, one must submit a drawing of the mark and indicate when it was first used in interstate commerce and how it is used. The Patent Office conducts an investigation to verify these facts and will register a trademark as long as it is not generic, descriptive, immoral, deceptive, the name of a person whose permission has not been obtained, or substantially similar to another's trademark.

It is sometimes difficult to determine whether a trademark will be protected, and once the trademark is issued, it is not always easy to predict when a similar mark will be found to infringe upon the registered trademark. The following case demonstrates a typical analysis used in a trademark-infringement suit.

CASE 14-1

Toys "R" Us, Inc. v. Canarsie Kiddie Shop, Inc.
District Court of the Eastern District of New York
559 F. Supp. 1189 (1983)

Beginning in 1960, Plaintiff Toys "R" Us, Inc., sold children's clothes in stores across the country. The firm obtained a registered trademark and service mark for Toys "R" Us in 1961 and aggressively advertised and promoted its products using these marks. In the late 1970s, Defendant Canarsie Kiddie Shop, Inc., opened two children's clothing stores within two miles of a Toys "R" Us shop and contemplated opening a third. The owner of Canarsie Kiddie Shop, Inc., called the stores Kids "r" Us. He never attempted to register the name. Toys "R" Us sued for trademark infringement in the federal district court.

Judge Glasser

In assessing the likelihood of confusion and in balancing the equities, this Court must consider the now classic factors.

1. Strength of the Senior User's Mark

"[T]he term 'strength' as applied to trademarks refers to the distinctiveness of the mark, or more precisely, its tendency to identify goods sold under the mark as emanating from the particular, although possibly anonymous, source." A mark can fall into one of four general categories, which, in order of ascending strength, are: (1) generic; (2) descriptive; (3) suggestive; and (4) arbitrary or fanciful. The strength of a mark is generally dependent both on its place upon the scale and on whether it has acquired secondary meaning.

A generic term "refers, or has come to be understood as referring to the genus of which the particular product is a species." A generic term is entitled to no trademark protection whatsoever, since any manufacturer or seller has the right to call a product by its name.

A descriptive mark identifies a significant characteristic of the product, but is not the common name of the product. A mark is descriptive if it "informs the purchasing public of the characteristics, quality, functions, uses, ingredients, components, or other properties of a product, or conveys comparable information about a service." To achieve trademark protection a descriptive term must have attained secondary meaning, that is, it must have "become distinctive of the applicant's goods in commerce."

A suggestive mark is one that "requires imagination, thought, and perception to reach a conclusion as to the nature of the goods." These marks fall short of directly describing the qualities or functions of a particular product or service, but merely suggest such qualities. If a term is suggestive, it is entitled to protection without proof of secondary meaning.

Arbitrary or fanciful marks require no extended definition. They are marks which in no way describe or suggest the qualities of the product.

The Toys "R" Us mark is difficult to categorize.

The strength of the plaintiff's mark must be evaluated by examining the mark in its entirety.... I agree that the Toys "R" Us mark serves to describe the business of the plaintiff, and in this sense is merely descriptive. This descriptive quality, however, does require some "imagination, thought, and perception" on the part of the consumer since the plaintiff's mark, read quite literally, conveys the message, "we are toys," rather than "we sell toys."

Whether the "leap of imagination" required here is sufficient to render the mark suggestive rather than descriptive is a question with no clear-cut answer. Such an absolute categorization is not essential, however, since the defendants concede that through the plaintiff's marketing and advertising efforts the Toys "R" Us mark has acquired secondary meaning in the minds of the public, at least in relation to its sale of toys. Such secondary meaning assures that the plaintiff's mark is entitled to protection even if it is viewed as merely descriptive. Because I find that through the plaintiff's advertising and marketing efforts the plaintiff's mark has developed strong secondary meaning as a source of children's products, it is sufficient for purposes of this decision to note merely that the plaintiff's mark is one of medium strength, clearly entitled to protection, but falling short of the protection afforded an arbitrary or fanciful mark.

2. Degree of Similarity between the Two Marks

[T]he key inquiry is . . . whether a similarity exists which is likely to cause confusion. This test must be applied from the perspective of prospective purchasers. Thus, it must be determined whether "the impression which the infringing [mark] makes upon the consumer is such that he is likely to believe the product is from the same source as the one he knows under the trademark." In making this determination, it is the overall impression of the mark as a whole that must be considered.

Turning to the two marks involved here, various similarities and differences are readily apparent. The patent similarity between the marks is that they both employ the phrase, "R" Us. Further, both marks employ the letter "R" in place of the word "are," although the plaintiff's mark uses an inverted capitalized "R," while the defendants generally use a non-inverted lower case "r" for their mark.

The most glaring difference between the marks is that in one the phrase "R" Us is preceded by the word "Toys," while in the other it is preceded by the word "Kids." Other differences include the following: plaintiff's mark ends with an exclamation point, plaintiff frequently utilizes the image of a giraffe alongside its mark, plaintiff's mark is set forth in stylized lettering, usually multi-colored, and plaintiff frequently utilizes the words, "a children's bargain basement" under the logo in its advertising.

I attach no great significance to the minor lettering differences between the marks, or to the images or slogans usually accompanying the plaintiff's mark. . . . While the marks are clearly distinguishable when placed side by side, there are sufficiently strong similarities to create the possibility that some consumers might believe that the two marks emanated from the same source. The similarities in sound and association also create the possibility that some consumers might mistake one mark for the other when seeing or hearing the mark alone. The extent to which these possibilities are "likely" must be determined in the context of all the factors present here.

3. Proximity of the Products

Where the products in question are competitive, the likelihood of consumer confusion increases. . . . [B]oth plaintiff and defendants sell children's clothing; . . . the plaintiff and defendants currently are direct product competitors.

4. The Likelihood That Plaintiff Will "Bridge the Gap"

"[B]ridging the gap" refers to two distinct possibilities: first, that the senior user presently intends to expand his sales efforts to compete directly with the junior user, thus creating the likelihood that the two products will be directly competitive; second, that while there is no present intention to bridge the gap, consumers will assume otherwise and conclude, in this era of corporate diversification, that the parties are related companies. . . . I find both possibilities present here.

5. Evidence of Actual Confusion
Evidence of actual confusion is a strong indication that there is a likelihood of confusion. It is not, however, a prerequisite for the plaintiff to recover.

6. Junior User's Good Faith
The state of mind of the junior user is an important factor in striking the balance of the equities. In the instant case, Mr. Pomeranc asserted at trial that he did not recall whether he was aware of the plaintiff's mark when he chose to name his store Kids "r" Us in 1977.

I do not find this testimony to be credible. In view of the proximity of the stores, the overlapping of their products, and the strong advertising and marketing effort conducted by the plaintiff for a considerable amount of time prior to the defendants' adoption of the name Kids "r" Us, it is difficult to believe that the defendants were unaware of the plaintiff's use of the Toys "R" Us mark.

The defendants adopted the Kids "r" Us mark with knowledge of plaintiff's mark. A lack of good faith is relevant not only in balancing the equities, but also is a factor supporting a finding of a likelihood of confusion.

7. Quality of the Junior User's Product
If the junior user's product is of a low quality, the senior user's interest in avoiding any confusion is heightened. In the instant case, there is no suggestion that the defendants' products are inferior, and this factor therefore is not relevant.

8. Sophistication of the Purchasers
The level of sophistication of the average purchaser also bears on the likelihood of confusion. Every product, because of the type of buyer that it attracts, has its own distinct threshold for confusion of the source or origin.

The goods sold by both plaintiff and defendants are moderately priced clothing articles, which are not major expenditures for most purchasers. Consumers of such goods, therefore, do not exercise the same degree of care in buying as when purchasing more expensive items. Further, it may be that the consumers purchasing from the plaintiff and defendants are influenced in part by the desires of their children, for whom the products offered by plaintiff and defendants are meant.

9. Junior User's Goodwill
[A] powerful equitable argument against finding infringement is created when the junior user, through concurrent use of an identical trademark, develops goodwill in their [sic] mark. Defendants have not expended large sums advertising their store or promoting its name. Further, it appears that most of the defendants' customers are local "repeat shoppers," who come to the Kids "r" Us store primarily because of their own past experiences with it. In light of this lack of development of goodwill, I find that the defendants do not have a strong equitable interest in retaining the Kids "r" Us mark.

Conclusion on Likelihood of Confusion
[T]he defendants' use of the Kids "r" Us mark does create a likelihood of confusion for an appreciable number of consumers.

In reaching this determination, I place primary importance on the strong secondary meaning that the plaintiff has developed in its mark, the directly competitive nature of the products offered by the plaintiff and defendants, the plaintiff's substantially developed plans to open stores similar in format to those of the defendants', the lack of sophistication of the purchasers, the similarities between the marks, the defendants' lack of good faith in adopting the mark, and the limited goodwill the defendants have developed in their mark.*

Judgment for the Plaintiff, Toys "R" Us.

CRITICAL THINKING ABOUT THE LAW

Legal reasoning and decision making almost always entail a reliance on tradition. In Case 14-1, the court's deference to tradition is especially strong. In applying the "classic factors" to the case, the court judges the present case on the basis of the way in which an allegedly similar earlier case was judged.

The implications of relying on tradition in legal reasoning are quite significant. The questions that follow will help you to consider this significance more deeply.

1. To demonstrate your ability to recognize reasoning by analogy, identify the implicit analogy that pervades the court's reasoning in Case 14-1.

    Clue: Consider the source of the "classic factors."

2. What important piece of missing information hinders your ability to make a sound critical judgment about the appropriateness of the court's reasoning?

    Clue: Refer to your answer to Question 1; it is directly related to this question.

---

*Toys "R" Us, Inc., v. Canarsie Kiddie Shop, Inc., District Court of the Eastern District of New York 559 F. Supp. 1189 (1983).

The potential for consumer confusion seems to be a very important consideration in a trademark-infringement case. In the *Dentyne Ice* chewing gum case, the court ruled that there was little possibility for consumer confusion between "Dentyne Ice" and "Icebreakers" because Dentyne was a common household name; thus, Dentyne did not infringe on Nabisco's trademark.

APPLYING THE LAW TO THE FACTS . . .

Your roommate cannot find a summer job. He decides that he is going to make money instead by selling silkscreened T-shirts that have a picture of an upside down "swoosh" underneath the words "JUST DONE IT." He asked your opinion of his proposed business venture. What do you tell him?

TRADE DRESS

The term **trade dress** refers to the overall appearance and image of a product. Trade dress is entitled to the same protection as a trademark. To succeed on a claim of trade-dress infringement, a party must prove three elements: (1) the trade dress is primarily nonfunctional; (2) the trade dress is inherently distinctive or has acquired a secondary meaning; and (3) the alleged infringement creates a likelihood of confusion.

trade dress The overall appearance and image of a product that has acquired secondary meaning.

The main focus of a case of trade-dress infringement is usually on whether or not there is likely to be consumer confusion. For example, in a 1996 case, Tour 18, Limited, a golf course, copied golf holes from famous golf courses without permission of the course owners.[1] In copying a hole from one of the most famous courses in the country, Harbour Town Hole 19, the defendant even copied the Harbour Town Lighthouse, which is the distinctive feature of that hole. In its advertising, Tour 18 prominently featured pictures of this hole, including the lighthouse. The operator of the Harbour Town course sued Tour 18 for trade-dress infringement. The court found that there was infringement and made Tour 18 remove the lighthouse and disclaim in its advertising any affiliation with the owner of the Harbour Town course.

Trade-dress violations occur over a wide range of products. Two very different examples of trade-dress infringement include *Bubba's Bar-B-Q Oven v. Holland Co.*,[2] and *Two Pesos v. Taco Cabana*.[3] In the first case, Bubba's had almost exactly copied the physical appearance of Holland Company's very successful gas-fired barbecue grill, which the court found to be trade-dress infringement. In the second case, the court found that Taco Cabana's trade dress consisted of "a festive eating atmosphere having interior dining and patio areas decorated with artifacts, bright colors, paintings and murals"; "a patio that has interior and exterior areas with the interior patio capable of being sealed off from the outside patio by overhead garage doors"; a stepped exterior of the building that has "a festive and vivid color scheme using top border paint and neon stripes"; and "bright awnings and umbrellas."* When Two Pesos opened a series of competing Mexican restaurants that mimicked those features almost exactly, the court found Two Pesos guilty of trade-dress infringement.

A claim of trade-dress infringement is often accompanied by a claim of trademark infringement. For example, the Chippendale's name is protected as a trademark, and their dancers' costume—cuffs and a collar—is protected as trade dress. Chippendale's sued a European gaming machine manufacturer who had made and distributed "Chickendale's" slot machines featuring dancing chickens wearing cuffs and collars. The case settled in 2010 with Chippendale's receiving a cash settlement and the transfer of the mark Chickendale's to them.[4]

---

[1] *Pebble Beach Co. v. Tour 18, Ltd.* 942 F. Supp. 1513 (S.D. Tex. 1996).

[2] 175 F.3d 1013; Docket No. 98-1029 (4th Cir. 1999).

[3] 112 S. Ct. 2753 (1994).

[4] *Chippendales USA, LLC v. Atronic Americas, LLC,* Civ. No. 03-904 (D.N.J.), as reported at Chippendale's website. Accessed April 20, 2010 at http://www.chippendales.com/company/victories.php.

*Two Pesos, Inc. v. Taco Cabana, Inc. (91-971), 505 U.S. 763 (1992).

TECHNOLOGY AND THE LEGAL ENVIRONMENT

Trademarks and Domain Names

If a business has a very strong trademark, what better domain name to have than that trademark? Unfortunately, the same trademark may be owned by two companies selling noncompeting goods, yet there can be only one user of any single domain name.

Domain names are important because they are the way people and businesses are located on the Web. A domain name is made up of a series of domains separated by periods. Most websites have two domains. The first-level domain, the one that the address ends with, generally identifies the type of site. For example, if it is a government site, it will end in *gov*. An educational site will end in *edu,* a network site in *net,* an organization in *org,* and a business in *com* or *biz*. These top-level domain names are the same worldwide.

Until recently, these top-level domain names had to be in Latin language script. However, as of mid-2010, four countries (Egypt, the Russian Federation, Saudi Arabia, and the United Arab Emirates) have been authorized to use their own non-Latin language scripts in the top-level domain portion of their Internet address names.

The second-level domain is usually the name of whoever maintains the site. For a college, for example, it would be an abbreviation of the college, as in *bgsu*. Businesses generally use their firm name or some other trademark associated with their product, because that name will obviously make it easier for their customers to find them.

So, how does a firm go about securing a domain name that reflects its trademark? The Internet Corporation for Assigned Names and Numbers (ICANN) is the nonprofit corporation that is responsible for coordinating technical Internet functions, including the management of the domain name system. ICANN has accredited a number of companies, called *registrars,* to issue domain names to the public and place their owners' information on the registry. Network Solutions, Inc. (NSI), which is funded by the National Science Foundation, was the first registrar, but has now been joined by several other firms. A list of ICANN-accredited registrars can be found at **http://www.icann.org/en/registrars/accredited-list.html**.

Anyone seeking to register a domain name should contact one of these registrars and provide the necessary contact information so that a domain name can be issued. A registrant must now state in its application that the name will not infringe on anyone else's intellectual property rights, that the name is not being registered for an unlawful purpose, that statements in the registration agreement are complete and accurate, and that the domain name will not be knowingly used in violation of any law. Registrars have the flexibility to offer initial and renewal registrations in 1-year increments, for up to 10 years.

Domain names may be cancelled or transferred when a complaint is filed with ICANN and the dispute is resolved against a registrant in an administrative proceeding held in accordance with the Uniform Domain Name Dispute Resolution Policy. Claims subject to dispute resolution include allegations that your domain name is confusingly similar to a trademark or service mark in which the complainant has rights and that your domain name has been registered and is being used in bad faith.[5]

Some firms have tried to get the domain name they desire by going to another country. That alternative is certainly a possibility. Many countries, however, require that a firm be incorporated within their borders before it can gain the right to the domain name there. Also, an additional problem is that trademark law relating to domain names is even more unclear abroad.

For the new entrepreneur, the best advice is to try to simultaneously apply for federal trademark protection and register the domain name; for those not yet on the Web, the sooner you get your domain name, the more likely you are to get the name you want. Moreover, if you feel that your mark is being violated by another's domain name, you may want to sue them for infringement, because the unauthorized use of another's trademark in a domain name has been found to be illegal. You may, however, be in for quite a fight, because this is a new area of the law.

---

[5] *ICANN Uniform Domain Name Dispute Resolution Policy,* available at www.icann.org/udrp/udrp-policy-24oct99.htm.

FEDERAL TRADEMARK DILUTION ACT OF 1995

Under the Lanham Act, trademark owners were protected from the unauthorized use of their marks on only competing goods or related goods where the use might lead to consumer confusion. Consequently, a mark might be used without permission on completely unrelated goods, thereby potentially diminishing the value of the mark. In response to this problem, a number of states passed trademark dilution laws, which prohibited the use of "distinctive" or "famous" trademarks, such as McDonald's, even without a showing of consumer confusion.

In 1995, Congress made similar protections available at the federal level with the Federal Trademark Dilution Act (FTDA). Trademark dilution occurs through either "blurring" or "tarnishment." Blurring occurs when the distinctiveness of the famous mark is reduced by its association with a similar mark. For example, in 2011, the Trademark Trial and Appeal Board found that the use of the mark "JUST JESU IT" would cause dilution by blurring of Nike's "JUST DO IT" mark. Tarnishment occurs when someone uses a mark in a way that causes the famous mark to be linked with an inferior quality product or an unwholesome category of products. For example, using candyland.com for a child pornography site constituted dilution by tarnishment of Hasbro's trademarked "Candyland" for its children's board game.[6]

In one of the first cases decided under the FTDA, the court said that the protection available under this act extended not just to identical marks but also to similar marks. In that case, Ringling Brothers–Barnum & Bailey challenged Utah's use of the slogan "The Greatest Snow on Earth" as diluting its famous slogan, "The Greatest Show on Earth." In denying Utah's motion to dismiss because the slogans were not identical, the court said that the marks need not be identical to dilute a famous mark. In the case below the plaintiff alleged both trademark infringement and trademark dilution.

CASE 14-2

Victor Moseley and Cathy Moseley et al., *dba* Victor's Little Secret v. V Secret Catalogue, Inc. et al.
Supreme Court of the United States
537 U.S. 418, 123 S. Ct. 1115 (2003)

Petitioners, Victor and Cathy Moseley, own and operate an adult novelty store named "Victor's Secret" in a strip mall in Elizabethtown, Kentucky. An army colonel saw an advertisement for the store and thought petitioners were using a reputable trademark to promote unwholesome merchandise, so he sent a copy of the advertisement to respondents, the corporations that own the Victoria's Secret trademarks. These corporations operate more than 750 Victoria's Secret stores and spent more than $55 million advertising their brand in 1998.

Respondents asked petitioners to discontinue using the name, and they responded by changing the store's name to "Victor's Little Secret." Respondents then filed a lawsuit containing four separate claims for (1) trademark infringement, alleging that petitioners' use of their trade name was "likely to cause confusion and/or mistake"; (2) unfair competition, alleging misrepresentation; (3) "federal dilution" in violation of the FTDA; and (4) trademark infringement and unfair competition in violation of the common law of Kentucky. The dilution count was premised on the claim that petitioners' conduct was "likely to blur and erode the distinctiveness" and "tarnish the reputation" of the Victoria's Secret trademark.

Finding that the record contained no evidence of actual confusion between the parties' marks, the district court concluded that "no likelihood of confusion exists as a matter of law" and entered summary judgment for the Moseleys on the infringement and unfair competition claims. With respect to the FTDA claim, however, the court ruled for V Secret Catalogue, and the Moseleys appealed. The Sixth Circuit affirmed, finding that respondents' mark was "distinctive" and that the evidence established "dilution" even though no actual harm had been proved.

---

[6]*Hasbro Inc. v. Internet Entm't. Grp., Ltd.*

Justice Stevens

The Victoria's Secret mark is unquestionably valuable and petitioners have not challenged the conclusion that it qualifies as a "famous mark" within the meaning of the statute. Moreover, as we understand their submission, petitioners do not contend that the statutory protection is confined to identical uses of famous marks, or that the statute should be construed more narrowly in a case such as this.

The District Court's decision in this case rested on the conclusion that the name of petitioners' store "tarnished" the reputation of respondents' mark, and the Court of Appeals relied on both "tarnishment" and "blurring" to support its affirmance. Petitioners have not disputed the relevance of tarnishment, presumably because that concept was prominent in litigation brought under state antidilution statutes and because it was mentioned in the legislative history. . . . Indeed, the contrast between the state statutes, which expressly refer to both "injury to business reputation" and to "dilution of the distinctive quality of a trade name or trademark," and the federal statute which refers only to the latter, arguably supports a narrower reading of the FTDA.

The relevant text of the FTDA provides that "the owner of a famous mark" is entitled to injunctive relief against another person's commercial use of a mark or trade name if that use "causes dilution of the distinctive quality" of the famous mark. This text unambiguously requires a showing of actual dilution, rather than a likelihood of dilution.

This conclusion is fortified by the definition of the term "dilution" itself. That definition provides:

> "The term 'dilution' means the lessening of the capacity of a famous mark to identify and distinguish goods or services, regardless of the presence or absence of—
> "(1) competition between the owner of the famous mark and other parties, or
> "(2) likelihood of confusion, mistake, or deception."

The contrast between the initial reference to an actual "lessening of the capacity" of the mark, and the later reference to a "likelihood of confusion, mistake, or deception" in the second caveat confirms the conclusion that actual dilution must be established.

Of course, that does not mean that the consequences of dilution, such as an actual loss of sales or profits, must also be proved. . . . We do agree, however, with that court's conclusion that, at least where the marks at issue are not identical, the mere fact that consumers mentally associate the junior user's mark with a famous mark is not sufficient to establish actionable dilution. As the facts of that case demonstrate, such mental association will not necessarily reduce the capacity of the famous mark to identify the goods of its owner, the statutory requirement for dilution under the FTDA. For even though Utah drivers may be reminded of the circus when they see a license plate referring to the "greatest snow on earth," it by no means follows that they will associate "the greatest show on earth" with skiing or snow sports, or associate it less strongly or exclusively with the circus. "Blurring" is not a necessary consequence of mental association. (Nor, for that matter, is "tarnishing.")

The record in this case establishes that an army officer who saw the advertisement of the opening of a store named "Victor's Secret" did make the mental association with "Victoria's Secret," but it also shows that he did not therefore form any different impression of the store that his wife and daughter had patronized. There is a complete absence of evidence of any lessening of the capacity of the Victoria's Secret mark to identify and distinguish goods or services sold in Victoria's Secret stores or advertised in its catalogs. The officer was offended by the ad, but it did not change his conception of Victoria's Secret. His offense was directed entirely at petitioners, not at respondents. Moreover, the expert retained by respondents had nothing to say about the impact of petitioners' name on the strength of respondents' mark.

Noting that consumer surveys and other means of demonstrating actual dilution are expensive and often unreliable, respondents and their amici argue that evidence of an actual "lessening of the capacity of a famous mark to identify and distinguish goods or services" may be difficult to obtain. It may well be, however, that direct evidence of dilution such as consumer surveys will not be necessary if actual dilution can reliably be proved through circumstantial evidence—the obvious case is one where the junior and senior marks are identical. Whatever difficulties of proof may be entailed, they are not an acceptable reason for dispensing with proof of an essential element of a statutory violation. The evidence in the present record is not sufficient to support the summary judgment on the dilution count, and the judgment is therefore reversed.*

Judgment *reversed* for Petitioners.

APPLYING THE LAW TO THE FACTS . . .

Louis Vuitton Malletier is a well-known producer of luxury luggage, purses, leather goods, and accessories, and has several widely recognized registered trademarks for Louis Vuitton, the monogram LV, and designs that combine the LV mark with images of stars, flowers, and circles. They also have a limited line of luxury pet accessories, such as collars and leashes, that they sell at their upper-end boutiques, at prices ranging from $200 to $1,600. They sell no dog toys. Haute Diggity Dog produces and sells a line of pet chew toys and beds with names parodying elegant, high-end products with names

---

*Bilski v. Kappos, United States Supreme Court 130 S. Ct. 3218 (2010).

such as Chewy Vuitton and Dog Perignon. The chew toys are in the shape of the product they are parodying, so they sell a chewy plastic handbag with the monogram CV and the name Chewy Vuitton on it. These products are primarily sold in pet shops for around $20. If Louis Vuitton wanted to try to stop Haute Diggity Dog from selling these products, how would he attempt to do so? What defenses would Haute Diggity Dog raise? What do you think the outcome of this dispute would be?

Patents

Patents are used to protect inventions, which are defined by the patent office as including any new and useful process, machine, manufacture, or composition of matter, or any new and useful improvement thereof; designs for an article of manufacture; asexually reproduced plants; or living material invented by a person. For patent protection to be granted, certain criteria must be satisfied (Table 14-2). First, the object of the patent must be novel, or new. No one else must have previously made or published the plans for this object. The second criterion is that it be useful, unless it is a design patent. It must provide some utility to society. The final criterion is that it be nonobvious. The invention must not be one that the person of ordinary skill in the trade could have easily discovered.

While the criteria for a patent may seem straightforward, it is not always easy to determine whether something is patentable. One issue that has been very controversial since the 1990s is whether business methods should be patentable. The United States Supreme Court had the opportunity to make a definitive ruling on this matter in 2010, and while in the particular case they did not allow a patent on the business method, they did not categorically find that business methods could not be patented. The uncertainty of this area is clear by the fact that when the case was decided, while all the justices agreed that the subject matter of the case was not patentable, there were three separate opinions with no single opinion garnering a majority of the justices' support.

patent Grants the holder the exclusive right to produce, sell, and use a product, process, invention, machine, or asexually reproduced plant for 20 years.

TABLE 14-2

CRITERIA FOR A PATENT

- Novel
- Useful
- Nonobvious

CASE 14-3

Bilski v. Kappos
United States Supreme Court
130 S. Ct. 3218 (2010)

Petitioners filed an application to obtain a patent for a claimed invention that explains how commodities buyers and sellers in the energy market can protect, or hedge, against the risk of price changes. The key claims were claim 1, which describes a series of steps instructing how to hedge risk, and claim 4, which places the claim 1 concept into a simple mathematical formula. The remaining claims explain how claims 1 and 4 can be applied to allow energy suppliers and consumers to minimize the risks resulting from fluctuations in market demand. The patent examiner rejected the application on the grounds that the invention was not implemented on a specific apparatus, merely manipulated an abstract idea, and solved a purely mathematical problem.

The Board of Patent Appeals and Interferences agreed and affirmed. The Federal Circuit, in turn, affirmed. The en banc court rejected its prior test for determining whether a claimed invention was a patentable "process" under the Patent Act, i.e., whether the invention produced a "useful, concrete, and tangible result," and held instead that a claimed process is patent eligible if: (1) it is tied to a particular machine or apparatus, or (2) it transforms a particular article into a different state or thing. Concluding that this "machine-or-transformation test" is the sole test for determining patent eligibility of a "process" under § 101, the court applied the test and held that the application was not patent eligible. The U.S. Supreme Court agreed to hear the appeal.

Justice Kennedy

Section 101 specifies four independent categories of inventions or discoveries that are patent eligible: processes, machines, manufactures, and compositions of matter. In choosing such expansive terms, . . . Congress plainly contemplated that the patent laws would be given wide scope . . . Congress took this permissive approach to patent eligibility to ensure that "'ingenuity should receive a liberal encouragement.'"

This Court's precedents provide three specific exceptions to § 101's broad principles: "laws of nature, physical phenomena, and abstract ideas." While these exceptions are not required by the statutory text, these exceptions are consistent with the notion that a patentable process must be "new and useful." . . .

The § 101 eligibility inquiry is only a threshold test. Even if a claimed invention qualifies in one of the four categories, it must also satisfy "the conditions and requirements of this title. Those requirements include that the invention be novel, nonobvious, and fully and particularly described."

The present case involves an invention at issue that is claimed to be a "process" under § 101. Section 100(b) defines as a "process, art or method, and includes a new use of a known process, machine, manufacture, composition of matter, or material."

Under the Court of Appeals' formulation, an invention is a "process" only if: "(1) it is tied to a particular machine or apparatus, or (2) it transforms a particular article into a different state or thing." . . . This Court has "more than once cautioned that courts 'should not read into the patent laws limitations and conditions which the legislature has not expressed.'"

. . . Any suggestion in this Court's case law that the Patent Act's terms deviate from their ordinary meaning has only been an explanation for the exceptions for laws of nature, physical phenomena, and abstract ideas.

This Court has not indicated that the existence of these well-established exceptions gives the Judiciary *carte blanche* to impose other limitations that are inconsistent with the text and the statute's purpose and design. Concerns about attempts to call any form of human activity a "process" can be met by making sure the claim meets the requirements of § 101.

The Court of Appeals incorrectly concluded that this Court has endorsed the machine-or-transformation test as the exclusive test. . . .

This Court's precedents establish that the machine-or-transformation test is a useful and important clue, an investigative tool, for determining whether some claimed inventions are processes under § 101. The machine-or-transformation test is not the sole test for deciding whether an invention is a patent-eligible "process."

It is true that patents for inventions that did not satisfy the machine-or-transformation test were rarely granted in earlier eras, especially in the Industrial Age, . . . But times change. Technology and other innovations progress in unexpected ways. For example, it was once forcefully argued that until recent times, "well-established principles of patent law probably would have prevented the issuance of a valid patent on almost any conceivable computer program." A categorical rule denying patent protection for "inventions in areas not contemplated by Congress . . . would frustrate the purposes of the patent law."

The machine-or-transformation test may well provide a sufficient basis for evaluating processes similar to those in the Industrial Age—for example, inventions grounded in a physical or other tangible form. But there are reasons to doubt whether the test should be the sole criterion for determining the patentability of inventions in the Information Age. As numerous *amicus* briefs argue, the machine-or-transformation test would create uncertainty as to the patentability of software, advanced diagnostic medicine techniques, and inventions based on linear programming, data compression, and the manipulation of digital signals.

. . . In searching for a limiting principle, this Court's precedents on the unpatentability of abstract ideas provide useful tools. . . . [T]he Patent Act leaves open the possibility that there are at least some processes that can be fairly described as business methods that are within patentable subject matter under § 101.

Finally, even if a particular business method fits into the statutory definition of a "process," that does not mean that the application claiming that method should be granted. In order to receive patent protection, any claimed invention must be novel, § 102, nonobvious, § 103, and fully and particularly described, § 112. These limitations serve a critical role in adjusting the tension, ever present in patent law, between stimulating innovation by protecting inventors and impeding progress by granting patents when not justified by the statutory design.

Petitioners seek to patent both the concept of hedging risk and the application of that concept to energy markets. . . . Rather than adopting categorical rules that might have wide-ranging and unforeseen impacts, the Court resolves this case narrowly on the basis of this Court's decisions in Benson, Flook, and Diehr, which show that petitioners' claims are not patentable processes because they are attempts to patent abstract ideas. Indeed, all members of the Court agree that the patent application at issue here falls outside of § 101 because it claims an abstract idea.*

Affirmed.

As technology develops, the patent law faces increasing challenges. While the high court has upheld the patentability of a genetically engineered, living bacterium that is capable of breaking down crude oil,[7] in 2014, the Supreme Court held that a naturally occurring segment of DNA that has been isolated is not eligible for patent protection.[8]

---

[7]*Diamond v. Chakrabarty*, 447 U.S. 303 (1980).

[8]*Association for Molecular Pathology v. Myriad Genetics*, 133 S. Ct. 2107 (2014).

*Bilski v. Kappos, United States Supreme Court. 130 S. Ct. 3218 (2010). http://www.supremecourt.gov/opinions/09pdf/08-964.pdf.

Once a patent is issued, it gives its holder the exclusive right to produce, sell, and use the object of the patent for 20 years. The holder of the patent may license, or allow others to manufacture and sell, the patented object. In most cases, patents are licensed in exchange for the payment of *royalties,* a sum of money paid for each use of the patented process.

The only restriction on the patent holder is that he or she may not use the patent for an illegal purpose. The two most common illegal purposes would be tying arrangements and cross-licensing. A **tying arrangement** occurs when the patent holder issues a license to use the patented object only if the licensee agrees also to buy some nonpatented product from the holder. **Cross-licensing** occurs when two patent holders license each other to use their patents only on the condition that neither licenses anyone else to use his or her patent without the other's consent. Both of these activities are unlawful because they tend to reduce competition.

To obtain a patent, one generally contacts an attorney licensed to practice before the U.S. Patent Office. The attorney does a patent search to make sure that no other similar patent exists. If it does not, the attorney fills out a patent application and files it with the Patent Office. The Patent Office evaluates the application, and, if the object meets the criteria already described, a patent is issued.

In 2011, Congress passed the America Invents Act, which changed the U.S. patent system in a number of ways, the most important being how the patent office resolves a situation where the first to file for the patent was not the first to invent the object of the patent. Prior to the Act, the first to invent, assuming adequate records to substantiate the invention, would receive the patent, whereas now the patent goes to the first to file. This change brings us in line with what is done in Japan and Europe. The Act also made changes in the reexamination process to allow discovery requests during the process and makes it easier and less expensive for small businesses and entrepreneurs to file for patents.

Once the patent is issued, the holder may bring a patent-infringement suit in a federal court against anyone who uses, sells, or manufactures the patented invention without the permission of the patent holder. A successful action may result in an injunction prohibiting further use of the patented item by the infringer and also an award of damages. Sometimes, however, the result of the case is that the holder loses the patent. This loss would occur when the alleged infringer is able to prove that the Patent Office should not have issued the patent in the first place.

Unfortunately, despite the seemingly careful process for awarding patents, many patents are issued for ideas that are neither new nor nonobvious, or are overly broad, covering ordinary computing processes that should never have been patented. So-called patent trolls—companies that have no intention of ever using the patents they hold for any manufacturing or creative purpose—buy up these overly broad and vague patents, and then send threatening letters out to people and companies whom they argue are infringing on their patents. They threaten litigation if the party doesn't agree to pay them a licensing fee. Even though the trolls' claims are usually baseless, many companies or individuals will go ahead and pay the fee to avoid what they know will be costly and lengthy litigation.

One of these so-called trolls, a company called Lodsys, targets small app developers, telling them that their in-app purchasing technology, which is generally provided by Apple or Google, infringes on Lodsys's patent. The company has filed a lawsuit against small app developers. Apple has intervened in the case, arguing that the license it obtained from the former owner of the patent covered its app developers' use of the technology. Google has sought to have the patent reexamined on grounds that the patent isn't valid. But these actions take years, and no one knows how many small app developers are being threatened by Lodsys.[9]

> **tying arrangement** A restraint of trade wherein the seller permits a buyer to purchase one product or service only if the buyer agrees to purchase a second product or service. For example, a patent holder issues a license to use a patented object on condition that the licensee also agrees to buy nonpatented products from the patent holder.
>
> **cross-licensing** An illegal practice in which two patent holders license each other to use their patented objects only on condition that neither will license anyone else to use those patented objects without the other's consent.

---

[9]Patent Trolls, Electronic Freedom Foundation. Accessed April 15, 2015 at https://www.eff.org/issues/resources-patent-troll-victims.

There is widespread concern that these patent trolls are abusing the patent system and may be harming the very innovation the system is supposed to encourage. In fact, a study in 2014 revealed that roughly 67 percent of the patent claims filed that year were by nonpracticing entities, parties who buy up others' intellectual property and then sue anyone they think may be "infringing" on the patents.[10] President Obama issued five patent reform executive orders during 2014 in an attempt to address this issue by, among other strategies, requiring the Patent and Trademark Office to stop issuing overly broad patents and requiring patent applicants to provide more details about what invention they are claiming.[11] At the time this book went to press, Congress was considering bipartisan legislation designed to make it harder for these patent trolls to operate.[12]

A common dilemma facing an inventor is whether to protect an invention through patent or trade-secret law. If the inventor successfully patents and defends the patent, the patent holder has a guarantee of an exclusive monopoly on the use of the invention for 20 years, a substantial period of time. The problem for the inventor is that once this period is over, the patented good goes into the public domain and everyone has access to it. There is also the risk that the patent may be successfully challenged and the protection lost prematurely.

Trade-secret law, in contrast, could protect the invention in perpetuity. This method is described in the next section.

Trade Secrets

trade secret A process, product, method of operation, or compilation of information used in a business that is not known to the public and that may bestow a competitive advantage on the business.

A **trade secret** is a process, product, method of operation, or compilation of information that gives a businessperson an advantage over his or her competitors. Inventions and designs may also be considered trade secrets. A trade secret is protected by the common law from unlawful appropriation by competitors as long as it is kept secret and comprises elements not generally known in the trade. Businesses usually try to protect their trade secrets by having employees with access to trade secrets sign nondisclosure agreements.

Competitors may discover the "secret" by any lawful means, such as reverse engineering or by going on public tours of plants and observing the use of trade secrets. Discovery of the secret means there is no longer a trade secret to be protected.

Under common law, to enjoin a competitor from continuing the use of a trade secret, or to recover damages caused by the use of the secret, a plaintiff must prove that:

1. a trade secret actually existed;
2. the defendant acquired it through unlawful means, such as breaking into the plaintiff's business and stealing it or securing it through misuse of a confidential relationship with the plaintiff or one of the plaintiff's present or former employees; and
3. the defendant used the trade secret without the plaintiff's permission.

---

[10]Analee Newitz, "Google Attempts to Fight Patent Trolls with a Pretty Dubious Strategy." Accessed May 2, 2015 at http://gizmodo.com/google-attempts-to-fight-patent-trolls-by-buying-pat-1700413270.

[11]David Kravitz, "History Will Remember Obama as the Great Slayer of Patent Trolls," *Wired*. Accessed May 5, 2015 at http://www.reuters.com/article/2015/02/05/us-congress-patents-usa-idUSKBN0L92X720150205.

[12]Diane Bartz, "House Takes up Bill Designed to Fight Patent Trolls," *Reuters*. Accessed May 4, 2015 at http://www.reuters.com/article/2015/02/05/us-congress-patents-usa-idUSKBN0L92X720150205.

ECONOMIC ESPIONAGE ACT

Trade secrets are also protected under the Economic Espionage Act. Under Section 1832 of the Act, an individual who misappropriates a trade secret related to a product produced for interstate commerce, with the knowledge or intent that the misappropriation will harm the owner of the trade secret, may be subject to a prison term of up to 10 years; an organization convicted under the section may be fined up to $5 million. The misappropriation of trade secrets with the knowledge that the secret will benefit a foreign power may lead to individual prison sentences of up to 15 years per offense and fines for organizations of up to $10 million under Section 1831.

Copyrights

Copyrights protect the expression of creative ideas. They do not protect the ideas themselves but only their fixed form of expression. Copyrights protect a diverse range of creative works, such as books, periodicals, musical compositions, plays, motion pictures, sound recordings, lectures, works of art, and computer programs. Titles and short phrases may not be copyrighted.

There are three criteria for a work to be copyrightable (Table 14-3). First, it must be fixed, which means set out in a tangible medium of expression. Next, it must also be original. Finally, it must be creative.

A copyright arises under common law when the idea is expressed in tangible form, and is protected for the life of its creator plus 70 years, or, if the owner is a publisher, for 95 years after the date of publication or 120 years after creation, whichever occurs first.

copyright The exclusive legal right to reproduce, publish, and sell the fixed form of an expression of an original creative idea.

TABLE 14-3

CRITERIA FOR A COPYRIGHT

- Fixed form
- Original
- Creative

APPLYING THE LAW TO THE FACTS...

Lucky Break Wishbone designed wishbones using graphite electrodes to make them smooth and "attractive and sleek." They were thinner in the arms and more rounded on the edges than actual wishbones. The company then copyrighted, produced, and sold these wishbones. Sears subsequently started producing and selling similar wishbones. When Lucky Break sued them for copyright infringement, Sears claimed that the wishbones did not qualify for copyright protection. Who do you think should win? Why?

Under the common law of copyright, any infringer may be enjoined from reproducing a copyrighted work. For the creator to be able to sue the infringer to recover damages arising from the infringement, however, the copyrighted work must be registered. One may register a work by filing a form with the Registrar of Copyright and providing two copies of the copyrighted materials to the Library of Congress. Originally, publication of the work had to be accompanied by a notice of copyright, such as, in the case of printed works, for example, the word "copyright" and the symbol © or the abbreviation "copr.," followed by the first date of publication and the name of the copyright owner. While this notice is no longer required, it is still a good idea because it informs the public that the work is protected by copyright, identifies the copyright owner, and shows the year of first publication. It also ensures that in an infringement suit, the infringer cannot claim innocent infringement as a defense to reduce damages.

Statutory damages of up to $30,000 per infringement or $150,000 per willful infringement plus attorney fees are available only if the work was registered either prior to the infringement or within 90 days of the first publication of the work. Copyright holders may alternatively seek actual damages and the defendant's profits on the copyrighted works.

FAIR USE DOCTRINE

fair use doctrine A legal doctrine providing that a portion of a copyrighted work may be reproduced for purposes of "criticism, comment, news reporting, teaching (including multiple copies for classroom use), scholarships, and research."

A source of controversy involving copyrighted works is application of the **fair use doctrine**. This doctrine provides that a portion of a copyrighted work may be reproduced for purposes of "criticism, comment, news reporting, teaching (including multiple copies for classroom use), scholarships, and research." When deciding whether the use is fair, the court looks not only at the purpose of the use, but also at the amount of the work that is used and whether the use has any impact on the commercial value of the copyrighted work.

Under a broad conception of fair use, parody is also protected, but it is difficult to know when the author has taken so much of the original that the court will find infringement and not fair use. For example, in 2001, the estate of the author of *Gone with the Wind* sought an injunction to block the publication of *The Wind Done Gone*. The author of the second text said it was a parody of the former book and a portrayal of life in the Old South from a black point of view. The estate of the author of *Gone with the Wind,* however, claimed that the second book was an illegal sequel that infringed on the older work, citing the book's direct taking of characters, character traits, scenes, settings, physical descriptions, and plot from the copyrighted novel, as well as its appropriation of direct quotes from the novel.[13]

The district court initially granted the injunction, but on appeal, the higher court overturned the district court's decision, claiming that the injunction was an unlawful prior restraint of speech under the First Amendment. The court looked at the record and did not believe that the plaintiffs had made a strong enough case that (1) there was a substantial likelihood that the plaintiffs would prevail on the merits, (2) the plaintiffs would suffer irreparable harm if the injunction were not granted, (3) the threatened injury to the plaintiff outweighed that to the defendant, and (4) the granting of the injunction would be in the public interest, which are the necessary prerequisites for the granting of a preliminary injunction.

The Wind Done Gone went on sale on June 28, 2001, with the following words encircled on its cover: "An Unauthorized Parody." The estate of the author of *Gone with the Wind* eventually settled in 2002 with the author of *The Wind Done Gone* and her publishers. The appellate court accurately predicted that the plaintiffs would not find success on the merits. Had the plaintiffs succeeded, however, they would have been entitled to a permanent injunction and, most likely, damages in the amount of the profits made from the infringing work.

COPYRIGHTS IN THE DIGITAL AGE

The Copyright Act of 1976 was created well before the digital age. When that act was passed, the drafters did not even consider the potential implications of the Internet for copyright protection. However, as use of the Internet became widespread, copyrighted material immediately began to appear all over the Internet; copyrighted materials such as written works and pictures were transferred online and copied numerous times in violation of the spirit of copyright law. However, no laws existed to directly regulate such acts. Today, copyrighted material such as videos, movies, songs, and music videos are all frequently uploaded or transferred online, and our copyright laws have evolved to address these activities.

One of the first problems legislators had when addressing copyrighted material and the Internet was how to define whether a "copy" of something was created. Legislators determined that once material was downloaded into a computer's memory or RAM, the material was officially "copied."[14]

---

[13]*Mitchell & Joseph Reynolds Mitchell v. Houghton Mifflin Co.*, 58 U.S.P.Q.2d 1652 (2001).[9]
[14]*Peak Computer, Inc. v. MAI Systems Corp.*, 991 F.2d 511; 1993 U.S. App. LEXIS 7522 (9th Cir. 1993).

Another related issue is, "What constitutes the performance of copyrighted material?" This issue arises because the copyright holder has the exclusive right to perform the copyrighted work publicly. The United States Supreme Court addressed this issue in the following case.

CASE 14-4

American Broadcasting Company, Inc. et. al. v. Aereo, Inc.
United States Supreme Court
134 S. Ct. 2498 (2014)

Aereo developed a subscription service whereby subscribers, for a fee, could view broadcast television shows over the Internet or on their mobile devices. Aereo did not own the copyright in those works nor did it hold a license from the copyright owners to perform those works publicly.

Aereo's system was composed of servers, transcoders, and thousands of dime-sized antennas housed in a central warehouse. To watch a show currently being broadcast, the subscriber would go to Aereo's website and select that program. Then one of Aereo's servers would select an antenna for the use of the subscriber during that show. The server would save the data in a subscriber-specific folder on Aereo's hard drive, and once several seconds of programming had been saved, the server would stream the saved copy of the show to the subscriber over the Internet. Alternatively, the subscriber could request that the show be streamed at a later time.

The broadcast networks and cable companies who were required by law to pay a fee to rebroadcast network content believed Aereo was violating copyright law by broadcasting copyrighted materials, so they filed a lawsuit against Aereo, arguing that Aereo infringed their copyrighted material because Aereo's streams constituted public performances. The District Court denied their request for a preliminary injunction. Plaintiffs appealed, and the Second Circuit affirmed the lower court's decision, finding that Aereo's streams to subscribers were not "public performances," and thus did not constitute copyright infringement. Plaintiffs once again appealed.

Justice Breyer

. . . This case requires us to answer two questions: First, in operating in the manner described above, does Aereo "perform" at all? And second, if so, does Aereo do so "publicly"?

. . . In Aereo's view, it does not perform. It does no more than supply equipment that "emulate[s] the operation of a home antenna and digital video recorder (DVR). Like a home antenna and DVR, Aereo's equipment simply responds to its subscribers' directives. So it is only the subscribers who 'perform' when they use Aereo's equipment to stream television programs to themselves.". . . [W]hen read in light of its purpose, the Act is unmistakable: An entity that engages in activities like Aereo's performs.

History makes plain that one of Congress' primary purposes in amending the Copyright Act in 1976 was to overturn this Court's determination that community antenna television (CATV) systems (the precursors of modern cable systems) fell outside the Act's scope. . . . In 1976 Congress amended the Copyright Act . . . [and] enacted new language that erased the Court's line between broadcaster and viewer, in respect to "perform[ing]" a work. The amended statute clarifies that to "perform" an audiovisual work means "to show its images in any sequence or to make the sounds accompanying it audible.". . . Under this new language, both the broadcaster and the viewer of a television program "perform," because they both show the program's images and make audible the program's sounds.

. . . Congress also enacted the Transmit Clause, which specifies that an entity performs publicly when it "transmit[s] . . . a performance . . . to the public." . . . The Clause thus makes clear that an entity that acts like a CATV system itself performs, even if when doing so, it simply enhances viewers' ability to receive broadcast television signals. . . .

. . . Congress further created a new section of the Act to regulate cable companies' public performances of copyrighted works. . . . Section 111 creates a complex, highly detailed compulsory licensing scheme that sets out the conditions, including the payment of compulsory fees, under which cable systems may retransmit broadcasts . . .

Congress made these three changes to achieve a similar end: to bring the activities of cable systems within the scope of the Copyright Act. . . .

This history makes clear that Aereo is not simply an equipment provider. Rather, Aereo, and not just its subscribers, "perform[s]" (or "transmit[s]"). Aereo's activities are substantially similar to those of the CATV companies that Congress amended the Act to reach. . . . . We conclude that Aereo is not just an equipment supplier and that Aereo "perform[s].". . .

Next, we must consider whether Aereo performs petitioners' works "publicly," . . . . Under the Clause, an entity performs a work publicly when it "transmit[s] . . . a performance . . . of the work . . . to the public." . . . Aereo denies that it satisfies this definition. It reasons as follows: First, the "performance" it "transmit[s]" is the performance created by its act of transmitting. And second, because each of these performances is capable of being received by one and only one subscriber, Aereo transmits privately, not publicly. Even assuming Aereo's first argument is correct, its second does not follow. . . .

We assume arguendo that Aereo's first argument is correct. . . . So under our assumed definition, Aereo transmits a performance whenever its subscribers watch a program. But what about the Clause's further requirement that Aereo transmit a performance "to the public"?

. . . The fact that each transmission is to only one subscriber, in Aereo's view, means that it does not transmit a performance "to the public." In terms of the Act's purposes, these differences do not distinguish Aereo's system from cable systems, which do perform "publicly." . . . [T]he Clause suggests that an entity may transmit a performance through multiple, discrete transmissions. That is because one can "transmit" or "communicate" something through a set of actions.

The Transmit Clause must permit this interpretation, for it provides that one may transmit a performance to the public "whether the members of the public capable of receiving the performance . . . receive it . . . at the same time or at different times." § 101. Were the words "to transmit . . . a performance" limited to a single act of communication, members of the public could not receive the performance communicated "at different times." Therefore, in light of the purpose and text of the Clause, we conclude that when an entity communicates the same contemporaneously perceptible images and sounds to multiple people, it transmits a performance to them regardless of the number of discrete communications it makes. . . .

Moreover, the subscribers to whom Aereo transmits television programs constitute "the public." Aereo communicates the same contemporaneously perceptible images and sounds to a large number of people who are unrelated and unknown to each other.

In sum, having considered the details of Aereo's practices, we find them highly similar to those of the CATV systems. . . . And those are activities that the 1976 amendments sought to bring within the scope of the Copyright Act. Insofar as there are differences, those differences concern not the nature of the service that Aereo provides so much as the technological manner in which it provides the service. We conclude that those differences are not adequate to place Aereo's activities outside the scope of the Act. For these reasons, we conclude that Aereo "perform[s]" petitioners' copyrighted works "publicly," as those terms are defined by the Transmit Clause.*

Reversed, in favor of American Broadcasting Company, Inc. et.al.

Software and the Copyright Act of 1976. The first digitally related amendment to the Copyright Act of 1976 was for software. In 1980, the Computer Software Copyright Act was passed by Congress. This amendment was made in response to computer technology. The act declared that computer software was material that could now be protected with a copyright.[15] Subsequently, courts have determined that other parts of computer software that can be copyrighted are the language that humans can read and coded language, as well as the general framework or organization of computer software and programs.[16]

The No Electronic Theft Act of 1997. In 1997, President Clinton signed into law a law that would impose criminal penalties on individuals who intentionally distributed copies of copyrighted materials over the Internet without authorization from the copyright holder, regardless of whether such individuals received profits for these actions. Such an individual could be fined up to $250,000 and imprisoned for up to 5 years.

An interesting result of the No Electronic Theft (NET) Act was its effect on the application of fair use doctrine to materials copied on the Internet. This doctrine had formerly protected individuals who made copies for certain uses that were not for profit. The NET Act instead outlawed *any* distribution or copying of copyrighted material over the Internet regardless of profits or profit-making intent.

The Digital Millennium Copyright Act. In October 1998, the Digital Millennium Copyright Act (DMCA) was passed by Congress and signed into law by President Clinton. The DMCA, similar to the NET Act, aimed to protect copyrights in the digital age. Specifically, the DMCA[17] states that the act of evading anti-piracy

---

[15] Pub. L. No. 95-517 (1980), amendment to 17 U.S.C. §§ 101, 117.

[16] *Stern Electronics v. Kaufman*, 669 F.2d 852; 1982 U.S. App. LEXIS 22489 (2d Cir. 1982).

[17] Among other sites, the text of the DMCA is available from UCLA at http://www.gseis.ucla.edu/iclp/dmca1.htm.

*American Broadcasting Company, Inc., et. al. vs. Aereo, Inc. United States Supreme Court 134 S. Ct. 2498 (2014).

technologies included in software, including DVDs and CDs, is a crime. However, such activities are permitted if one is involved with and working on behalf of a nonprofit library, educational institution, archive, or other such entity. The DMCA also forbids creation or sale of a device that can crack codes to allow one to copy any kind of software. Certain exceptions apply to this provision, too, one example being encryption research.

Internet service providers were given some protection from copyright-infringement claims, because these providers simply transmit information and do not produce or intentionally distribute it. Colleges and universities are also protected if they act as service providers. However, service providers are required to take down or cease transmission of information if the information is reported to be an infringement of a copyright. Also, any website continually casting music over the Internet must have licensing agreements with the record companies or other copyright holder.

Finally, the DMCA demanded that the Register of Copyrights help Congress determine how long-distance education via the Internet can be appropriately conducted to avoid copyright infringement.

Every three years, the Librarian of Congress and the Copyright Office entertain proposed exemptions to the DMCA. From the consumer's standpoint, one of the more exciting requests for an exemption came from the Electronic Frontier Foundation, who requested that "jailbreaking," hacking any phone's operating system to allow consumers to run any app on the phone they choose. Apple opposed the requested exemption, arguing that the DMCA protects the copyrighted encryption built into the bootloader that starts up the iPhone OS operating system. The Copyright Office disagreed with Apple and found that a copyright owner might try to restrict the programs that can be run on a particular operating system, but copyright law is not the vehicle for imposing such restrictions.

In 2010, in addition to allowing the jailbreaking exemption, The Librarian of Congress and Copyright Office also allowed exemptions for the breaking of DVD encryption by professors, students, and documentary makers so the clips can be used for education and commentary.

File-Sharing Networks and Technologies. A major issue in the courts arose over file-sharing technologies. Basically, a type of software was created that enabled individuals to make the size of a music file extremely small and thus, able to be easily transmitted to someone else's computer. In other words, music files could be simply and rapidly distributed to many different people.

As a result of such technology, file-sharing websites popped up that facilitated this peer-to-peer networking. *Peer-to-peer networking* is made possible when many personal computers are connected to a network hosted online, and all of the files in those personal computers can be shared with others without the files actually being hosted on the Internet website.

Web hosts such as Napster began to facilitate peer-to-peer networks, which were violating music copyrights on a grand scale. Basically, such websites were facilitating the act of transferring copyrighted material from one individual to another without any permission from the copyright holder.

Eventually, in 2005, the Supreme Court determined that certain file-sharing Web hosts had the intent to violate copyright law, and therefore hosts of file-sharing and networking sites could be liable for the distribution of copyrighted material, even though the hosts were only indirectly liable for the actions of their users.

It was only a matter of time, as technology improved, before not just music files, but also movie files started to be copied and distributed without permission. The following case illustrates the legal system's response to technology designed to enable people to copy movies.

CASE 14-5

RealNetworks, Inc. v. DVD Control Copy Association, Inc. et al.
United States District Court for the Northern District of California
641 F. Supp. 2d 913 (2009)

RealNetworks is a media company that produces and sells software called RealDVD. The software enables people to make copies of DVDs they have purchased. In 2009, the company sued the DVD Control Copy Association of America and seven movie studios in pursuit of a legal declaration that its DVD copying software did not breach the DMCA. RealNetworks claimed that its software was only intended for people to make "backups" of DVDs that they had already purchased themselves.

The association and the movie studios disagreed, claiming that although the DMCA did permit RealNetworks' initial access to the copyrighted material, it did not protect the company's granting access to the material to additional individuals without any consent from the copyright owners. Furthermore, the software enables individuals to save copyrighted content to a computer hard drive or a portable drive, which deactivates any protection in the technology of the actual DVD.

Judge Marilyn Hall Patel

DMCA's anti-circumvention and anti-trafficking provisions establish "new grounds for liability in the context of the unauthorized access of copyrighted material." . . . There is nothing that limits the number of times a physical DVD can be copied using either [RealNetworks' products] Vegas or Facet, however. A DVD could be passed around a dormitory, office, or neighborhood and copied on any Facet box or any computer using Vegas. . . . Plaintiff must then show that Real [Network]'s RealDVD products are either: (a) primarily designed or produced for the purpose of circumventing technological measures that effectively control access to a copyrighted work; (b) have only a limited commercially significant purpose or use other than to circumvent such technological measures; or (c) marketed for use in circumventing such technological measures. These are disjunctive clauses. The court need look no further than the first enumerated condition to find that the Studios are likely to prevail on the Section 1201(a)(2) claim. As defined in Section 1201(a), "to 'circumvent a technological measure' means to descramble a scrambled work, to decrypt an encrypted work, or otherwise to avoid, bypass, remove, deactivate, or impair a technological measure, without the authority of the copyright owner." 17 U.S.C. § 1201(a)(3)(A). CSS is a "technological measure" that effectively controls access to copyrighted DVD content and RealDVD permits the access of that content without the authority of the copyright owner. RealDVD products are designed primarily for circumvention of that technology, as Real has admitted its intent upon initial development was to create a software product that copies DVDs to computer hard drives so that the user does not need the physical DVD to watch the content. This unauthorized access infringes the Studios' rights because it entails accessing content without the authority of the copyright owner, as discussed.*

Judgment for the Defendant.

The court agreed with the association and determined that the software did in fact violate the DMCA, for the reasons given in its opinion, and granted the association and studios a preliminary injunction that banned RealNetworks not only from selling the software, but also from licensing it to any other company. This decision effectively altered the digital technology and copyright laws for the future.

Global Dimensions of Intellectual Property Law

The primary international protection for intellectual property is offered through multilateral conventions and treaties. The major treaty on intellectual property is the Trade Related Aspects of International Property Rights (TRIPS), which is administered through the World Trade Organization (WTO). The WTO administers a multitude of trade treaties. On a regular basis, the member states convene to renegotiate the treaties. In the Uruguay Round of negotiation (1986–1994), intellectual property rights became a major issue for developed

---

*RealNetworks, Inc. v. DVD Control Copy Association, Inc., et al. United States District Court for the Northern District of California 641 F. Supp. 2d 913 (2009).

countries such as the United States and the European Union. TRIPS was the agreed-upon solution.

TRIPS is the most comprehensive intellectual property treaty to date. Previous agreements have included the Berne Convention of 1886, the Universal Copyright Convention, the Paris Convention of 1883, and the Patent Cooperation Treaty. These treaties covered various aspects of intellectual property, but they did not have universal membership, and not all parties had signed all versions of the various conventions. TRIPS must be signed and adhered to by all WTO members, numbered at 159 as of March 2013. There are also 25 observer countries that must begin negotiations to join the WTO within 5 years of becoming observers.

TRIPS begins by ensuring equal protection through its national treatment policy (domestic and foreign products must be treated the same) and its most-favored-nation policy (every member nation must be treated the same). TRIPS expands the copyright protection in the Berne Convention to include computer programs and rental rights, and continues to protect against unauthorized copying of written and recorded creations. TRIPS defines *trademark* and protects trademarks for both goods and services. TRIPS also has a protection category of geographical indications, to protect names such as "Champagne," and "Tequila," which refer to origin of the product and indicate special characteristics.

Under TRIPS, industrial designs must be protected for at least 10 years, and patents must be protected for at least 20 years. There are several exceptions to the patent requirements, allowing a country to deny a patent to prevent the commercial exploitation of something dangerous to public order or morality. Countries also have the right not to patent different processes, such as surgical methods. Furthermore, countries are also permitted to issue compulsory licenses to force production of a patented product while protecting the patent holder's rights.

The protection for integrated circuit layout designs has not yet come into force. Trade secrets and other undisclosed information are protected as long as a reasonable attempt was made to keep the information secret.

TRIPS allows governments the right to curb anticompetitive licensing contracts under certain circumstances. The governments must enforce the intellectual property rights with their national laws. TRIPS provides standards for intellectual property laws and punishments.

TRIPS entered into force on January 1, 1995. Member countries, based on their level of development, had different amounts of time to comply with TRIPS. Developed countries had a year to comply, whereas less-developed countries had 6 years to comply, and least-developed countries had 11 years to comply. Some of those deadlines have been extended. China has changed its intellectual property laws to come into compliance with TRIPS, although enforcement is still lacking. In summary, TRIPS provides strong protection of intellectual property rights through its near-universal membership and the WTO's ability to enforce treaties.

SUMMARY

The primary forms of protected intellectual property are trademarks, copyrights, patents, and trade secrets. Unlike most property, which is protected by state law, the first three forms of intellectual property are protected by federal statutes. These statutes set forth the criteria for establishment of a valid trademark, copyright, or patent, and permit the holders of intellectual property rights to sue for infringement or misuse of their protected property.

Intellectual property is protected internationally primarily through treaties such as TRIPS, which is enforced by the WTO.

REVIEW QUESTIONS

14-1 How does trademark infringement differ from trademark dilution?

14-2 Explain the different types of marks that are protected under the Lanham Act.

14-3 What factors would lead a person to choose patent protection over trade-secret protection, and vice versa?

14-4 Explain the four factors that are relevant to a determination of whether the fair use doctrine is available as a defense.

14-5 Explain the relationship between a tying arrangement and cross-licensing.

14-6 How is trade dress different from a trademark?

REVIEW PROBLEMS

14-7 Matt creates a small doughnut shop called DoGos. He uses the term *DoGo* surrounded by three concentric blue circles on all packaging. He starts selling his specialty DoGo doughnuts in local grocery stores, and soon develops a strong customer base, although he never sells the product outside of the state of Ohio. After five years of increasing business, he notices a sudden dropoff in sales. He soon discovers that a competitor is producing a similar product, but at a lower price, packaged in almost identical packaging, but with the name DoGoos inside the three concentric blue circles. Matt finds out who the producer of DoGoos is and sends him a request that he change his product name and packaging. Simon, the producer of DoGoos, refuses, saying that because Matt has no trademark, his signature packaging has no protection. Explain why Simon either is or is not correct.

14-8 Grover creates a machine that enables a manufacturer to make screws with 50 percent less waste than the industry norm. He patents his process and then decides to license the patent to Markham Manufacturing for a small royalty fee and a promise that Markham will buy a key ingredient needed to produce the screws from Grover. Is there anything wrong with this contract?

14-9 Plaintiffs created an original screenplay and copyrighted it before trying to pitch it to a major motion picture producer, who rejected the screenplay. That same company, five years later, produced a major motion picture with the same basic theme that the plaintiffs had pitched, but the plot, mood, setting, and pace of the two works were different. Do you think the plaintiffs had a case for copyright infringement?

14-10 Thrifty Inn decides to open a motel along an interstate that will provide cheap lodging. It calls the motel Sleep McCheap. McDonald's seeks to enjoin Thrifty Inns from using the name, claiming that it violates the McDonald's trademark as well as the McStop trademark that McDonald's has for its one-stop business that provides cheap food, cheap lodging, and cheap gas. Will the injunction be granted? Why or why not?

14-11 Amerec Corporation developed a secret, unpatented process for producing methanol and built a special plant where it was going to use this process. The Christophers were hired by a competitor of Amerec to take aerial photographs of the construction. Amerec sued the photographers for misappropriation of trade secrets. What defenses might the Christophers raise? Would these be successful?

14-12 Professor Kendall wants students to read three articles from a recently published journal. The professor, who is concerned about the students' expenditures for books, photocopies the articles and places them on reserve. The publisher of the journal sues the professor for copyright infringement. What defense will the professor raise? Would this defense be successful?

CASE PROBLEMS

14-13 Jerry Seinfeld's wife, Jessica Seinfeld, was sued for copyright infringement and plagiarism over her cookbook, *Deceptively Delicious*, by Missy Chase Lapine. Lapine is the author of *The Sneaky Chef*, a cookbook with a concept similar to Seinfeld's. Both cookbooks are based on sneaking healthy ingredients into food prepared for children so that children will inadvertently eat foods they otherwise would be opposed to. The goal of both cookbooks is for children to eat more healthfully. Lapine wrote her cookbook in 2007; Seinfeld's came out in 2008. Seinfeld argued that many cookbooks have a goal not only to help individuals eat healthier, but to get children to eat healthier.

Lapine later added a count to her complaint, stating that Jerry Seinfeld launched a verbal attack on her by calling her a "wacko" during a guest appearance on David Letterman's TV show. How do you think the judge ruled with respect to the plaintiff's claims of copyright infringement? Do you think the idea of sneaking healthy foods into children's meals is so obscure that Seinfeld was indeed guilty of plagiarism and copyright infringement? *Missy Chase Lapine v. Jessica Seinfeld et al.*, 2010 U.S. App. LEXIS 8778 (2d Cir., Apr. 28, 2010).

14-14 In 2009, Capitol Records sued Jammie Thomas-Rasset for damages after she was found liable for copyright violations when she illegally downloaded 24 songs off the Internet. Apparently, in addition to downloading the songs, Thomas-Rasset was involved in a peer-to-peer network and was additionally allowing Capitol's songs to be downloaded from her computer to other individuals' computers. In other words, Thomas-Rasset was distributing copyrighted material without permission from the holder of the copyrights. Ultimately, Capitol Records was suing for monetary damages. An award of $2 million (in response only to the 24 songs) was the expected decision of the court. However, Thomas-Rasset argued that $2 million was an excessive award for the copyright infringement of 24 songs. How do you believe this case was decided? Why do you think so? *Capitol Records Inc. et al. v. Jammie Thomas-Rasset*, 2010 U.S. Dist. LEXIS 5049 (D. Minn. 2010).

14-15 Five major media studios and entertainment companies ("the Studios") held copyrights on Disney Enterprises, Inc., Twentieth Century Fox Film Corporation, Universal City Studios Productions, LLLP, Columbia Pictures Industries, Inc., and Warner Bros. Entertainment, Inc. The Studios filed suit against Hotfile Corp. and one of its founders, Anton Titov, who were owners of an offshore technology company that provided online file storage devices. Plaintiff Studios alleged that the defendants' customers "abused its system by sharing licensed materials belonging to the Studios," and that the defendants were subsequently liable. The Studios argued that the defendants' website was designed to encourage users to engage in piracy, "complete with a system of financial incentives that fosters infringement." The defendants claimed to have been unaware of infringement that transpired on its system. How do you think the court ruled in this case? What evidence would the court need to rule in favor of the plaintiffs? *Disney Enters. v. Hotfile Corp.*, 2013 U.S. Dist. LEXIS 172339 (S.D. Fla. 2013).

14-16 Plaintiffs American Beverage Corporation ("ABC") and Pouch Pac Innovations, LLC ("PPi") brought suit against defendants Parrot Bay and Smirnoff brands for infringing the design and trade dress used for frozen cocktail mixtures produced by the plaintiffs. Specifically, plaintiffs allege that they patented a "pouch" design for frozen cocktails, which was later used by the defendants. The appearance of the plaintiffs' patented design is an "hourglass shape when viewed from the front, a wedge shape when viewed from the side, and a lenticular shape when viewed from the bottom." The defendants alleged that they sought a shape and size similar to that of the plaintiffs' "pouch" design because it is typical for companies in the beverage industry "to use the same or similar serving format" as their competitors. The plaintiffs' trade-dress infringement claim was brought pursuant to the Lanham Act. For the plaintiffs to obtain a preliminary injunction on a Lanham Act claim, what factors should the court have considered? Do you think the court ruled in favor of the plaintiffs? Why? *Am. Bev. Corp. v. Diageo N. Am., Inc.*, 936 F. Supp. 2d 555 (W.D. Pa. 2013).

14-17 In August 2000, plaintiff obtained a copyright registration for an original screenplay of fictional material entitled "The Challenge." She sent a copy to Disney, along with a cover letter containing additional story elements and asking the company to consider producing it. The company never returned her copy of the screenplay and never responded to her letter.

Subsequently, the plaintiff alleged that Disney's motion picture *Cars* infringed on her copyrighted script for The Challenge. The Challenge involved an off-road racer getting lost in the desert, and *Cars* involved a stock car getting separated from his friends, getting lost, and then getting stuck in a small town in the desert. Both drivers were young kids who were eventually mentored by older drivers. The driver in the former movie faces dangers like dehydration and exposure to the elements, whereas the other driver simply is stuck in a small town from which he wishes to leave. Thus the trial court found the plots are not significantly similar.

The works share themes of self-reliance and the importance of friendship and teamwork, themes often predominate stories of competition. The additional themes of "the importance of giving back to the community" and "important life lessons can come from unexpected people" are present in *Cars* but are absent from "The Challenge."

Both shows are set "in a desert" according to the plaintiff, although "The Challenge" is in the actual desert, whereas *Cars* is in a town surrounded by a desert. Further, outside of the scenes that include the desert, *Cars* is set at two large racetracks while "The Challenge" is set in a rough neighborhood, a school, and the main character's home and work. Thus, the trial court found no substantial similarity when comparing the settings in the two stories. Because the trial court found a lack of substantial similarity, they dismissed the case. Do you think the appellate court disagreed? Why or why not? *Campbell v. Walt Disney Co.*, 718 F. Supp. 2d 1108 (N.D. Cal. 2010).

14-18 In August 2014, plaintiff, who sold his photographic work online, sued defendants for copyright infringement. Specifically, the owner of the online photos alleged that the defendants had obtained those photos without the owner's consent, and placed one of the photographs on a commercial Internet website operated by the defendants. After the defendants ignored a letter by the plaintiff demanding that the online photos be taken down, the plaintiff filed this suit seeking $30,000 in damages. The owner also alleged that the defendants' conduct constituted a willful violation of his copyrights. The court considered the following factors in determining whether a damages remedy existed: the infringer's state of mind; the expenses saved by the infringer; the revenue lost by the copyright holder; the deterrent effect on the infringer and third parties; the infringer's cooperation in the case; and the attitude and conduct of the parties. Based on the allegations stated above, do you believe that these factors merit a damages reward of $30,000? In whose favor do you believe the court ruled? Explain your reasoning. *Oppenheimer v. Holt*, No. 1:14-CV-000208-MR, 2015 WL 2062189 (W.D.N.C. May 4, 2015).

THINKING CRITICALLY ABOUT RELEVANT LEGAL ISSUES

A Question under TRIPS

When TRIPS came into effect, many member states had to change their intellectual property laws. Additionally, any states joining the WTO had to comply with TRIPS, which prompted many other countries to change their intellectual property laws. Most developing countries were given a longer time period to come into compliance with TRIPS than was required for developed countries, but for many developing countries that time period has recently ended. One of these countries is India.

India's previous patent laws covered a manufacturing method, rather than the finished product, so Indian companies were taking patented products, making them a different way, and selling them as generics. This legal treatment was a major problem for pharmaceutical companies. India sold huge numbers of generic pharmaceuticals in Africa, primarily HIV/AIDS medications, denying the pharmaceutical companies their royalties.

It is very important for patents for HIV/AIDS medication to be upheld. Although it might at first seem unfair to charge more than a generic company would for a medication, it is crucial to realize that the generic company did not have to invest in the research and development of the drug. Although the drug may not cost as much as the name-brand price to produce, the research and development, clinical trials, and salaries of everyone involved in the process of creating new medication must be accounted for in the price of the medication. What is really unfair is allowing a generic company to take advantage of the hard work done by the original company.

Perhaps more importantly, patents provide incentive for pharmaceutical companies to continue to create new and better drugs. With the expansion of HIV/AIDS and the tolerance that patients build up to drugs, it is necessary for pharmaceutical companies to come up with new treatments. Without the incentive from patents, innovation would slow, a result more detrimental to the treatment of HIV/AIDS than paying the true cost for medications.

Enforcing patents on HIV/AIDS drugs fairly reimburses the pharmaceutical companies for their work in research and development and helps patients by ensuring innovations in the form of new medications.

1. What are the issues and conclusion of this essay?
2. What ethical norms drive the author's reasoning?
3. Ask and answer the critical thinking question that you believe reveals the main problem with the author's reasoning.
4. Write an essay about this issue with a different conclusion.

   Clue: What alternate definitions of the primary ethical norm might change the conclusion?

ASSIGNMENT ON THE INTERNET

To demonstrate your knowledge of the various types of marks, search the Web and find two examples of each of the following types of marks: product mark, collective mark, service mark, and certification mark. Provide the citation for the Web page, copy the mark, and explain how you know it fits under that category of marks.

ON THE INTERNET

www.copyright.gov The U.S. Copyright Office gives information about copyrights, as well as directions on how to copyright your own work.
www.uspto.gov This is the website of the U.S. Patent and Trademark Office.
www.wipo.int This is the home page of the World Intellectual Property Organization.
www.intelproplaw.com The Intellectual Property Law Server provides useful information on patent, trademark, and copyright intellectual property. The many links at this site contain laws, articles, and cases of interest.
www.icann.org This site is the home page for the Internet Corporation for Assigned Names and Numbers, where you can find information about domain name ownership.

FOR FUTURE READING

Christiansen, Linda. "Mickey Mouse Still Belongs to Disney: The Supreme Court Upholds Copyright Extension." *Journal of the Academy of Marketing Science* 32 (2004): 212.

Jacobs, Hannah. "Searching for Balance in the Aftermath of the 2006 Takings Initiatives." *Yale Law Journal* 116 (2007): 1518.

Hovenkamp, Herbert. "Symposium on Antitrust and IP: Consumer Welfare in Competition and Intellectual Property Law." *Competition Policy International* 9 (2013): 53.

Magid, Julie Manning, Anthony D. Cox, and Dena S. Cox. "Quantifying Brand Image: Empirical Evidence of Trademark Dilution." *American Business Law Journal* 43 (2006): 1.

Masur, Jonathon. "Raising the Stakes in Patent Cases." *Georgetown Law Review* 101 (2013): 637.

McJohn, Stephen. "Top Tens in 2012: Patent, Trademark, Copyright, and Trade Secret Cases." *Northwest Journal of Technology and Intellectual Property* 11 (2013).

Rogers, Eric, and Jeon Young. "Inhibiting Patent Trolling: A New Approach to Applying Rule 11." *Northwestern Journal of Technology and Intellectual Property* 12 (2014): 291.

Seaman, Christopher. "The Case Against Federalizing Trade Secrecy." *Virginia Law Review* 101 (2015): 317.

Serafino, Laurie. "Arguing for Protection of Data Stored in the Cloud." *Pennsylvania Lawyer* 35 (Sept./Oct. 2013): 28.

CHAPTER FIFTEEN

Agency Law

- DEFINITION AND TYPES OF AGENCY RELATIONSHIPS
- CREATION OF AN AGENCY RELATIONSHIP
- DUTIES OF AGENTS AND PRINCIPALS
- PRINCIPAL'S AND AGENT'S LIABILITY TO THIRD PARTIES
- TERMINATION OF THE PRINCIPAL–AGENT RELATIONSHIP
- GLOBAL DIMENSIONS OF AGENCY LAW

Agency law has become more prominent in our complex postindustrial society. In an earlier period, when business owners (principals) did most or all of their business with customers on a one-to-one basis, agents and third parties played a very small role in the business environment. As business entities became larger and conducted far-flung transactions within the United States and worldwide, businesspeople felt a need to hire domestic and foreign agents to represent their interests to potential customers.

This chapter examines (1) the definition and creation of an agency relationship, as well as employment relationships; (2) the rights and duties of agents and principals toward each other and toward third parties; (3) the law of contracts and torts as they affect agency relationships; and (4) the global dimensions of agency law.

CRITICAL THINKING ABOUT THE LAW

Agency law is based on a trusting relationship between two parties or individuals. Critical thinking skills, such as identifying ethical norms and missing information, are especially important in thinking about agency relationships.

1. Agency relationships are frequently used in modern society. Businesses increasingly rely on agents to conduct their business with customers. What ethical norms would businesses expect their agents to hold and act on in carrying out their responsibilities?

    Clue: Look at the list of ethical norms and ask yourself which would be most important to a business.

2. As you will soon learn, agency relationships can be formed by informal oral agreements or by formal written contracts. Judges are often faced with the task of determining whether an agency relationship exists when there is an informal oral agreement (and sometimes, even when there is a formal written contract) or whether an agency relationship exists after an agent has performed a task. What kinds of information would be especially important to a judge in deciding whether an agency relationship had been established?

    Clue: What would tend to happen if a business formed a relationship with an employee to carry out a task?

3. Consider the definitions of justice in Chapter 1. How might the fact that the agent has already performed the task affect the judge's decision?

   Clue: Do any of the definitions of justice give special guidance here?

Definition and Types of Agency Relationships

DEFINITION OF AGENCY

Agency is defined as a fiduciary relationship (one of trust and confidence) between two persons in which they mutually agree that one person (the *agent*) will act on behalf of the other (the *principal*) and be subject to the latter's control and consent. For example, a corporate officer who enters into contracts that legally bind the corporation is representing the corporation as an agent.

Provided they are able to understand the instructions of a principal, most people have the capacity to act as agents. For example, a minor can be employed by a principal to act as an agent in making an offer of employment to a third person. The minor's lack of capacity to enter a contract is immaterial here, because the contract is between the principal and the third party. (The minor's lack of contractual capacity, however, means that any agreement between the minor and the principal may be voided by the minor before he or she reaches majority age or shortly thereafter.)

TYPES OF AGENCY RELATIONSHIPS

Agency law is concerned with three types of relationships: principal–agent, employer–employee, and employer–independent contractor. These relationships are summarized in Table 15-1.

Principal–Agent. The **principal–agent relationship** (Exhibit 15-1) is a relationship in which the principal (usually an employer) hires an agent (employee) and gives the agent either expressed (express) or actual authority to act on the principal's behalf. **Expressed authority** arises from specific statements made by the principal to the agent. **Actual authority** includes expressed authority and implied authority—that authority customarily given to an agent in an industry, a trade, or a profession. The principal–agent relationship is the most basic type of agency relationship. It is, in fact, the foundation of the next two types of agency relationships we are about to discuss: that between employer and employee and that between principal (employer) and independent contractor.

Employer–Employee. The employer–employee relationship evolved out of the traditional master–servant relationship, in which a master employed a servant to perform services and the servant's conduct was subject to the master's physical control. Under the doctrine of ***respondeat superior***, the master was responsible for servants' acts that were within the scope of their employment. In the postindustrialized version of this arrangement, the **employer–employee relationship**, the employee is subject to the control of the employer and, in

agency A fiduciary relationship between two persons in which one (the agent) acts on behalf of, and is subject to the control of, the other (the principal).

principal–agent relationship Relationship in which the principal gives the agent expressed or actual authority to act on the former's behalf.

expressed authority Authority that arises from specific statements made by the principal (employer) to the agent (employee).

actual authority Includes expressed authority as well as implied authority, or that authority customarily given to an agent in an industry, a trade, or a profession.

respondeat superior Legal doctrine imposing liability on a principal for torts committed by an agent who is employed by the principal and subject to the principal's control.

employer–employee relationship Relationship in which an agent (employee) who works for pay and is subject to the control of the principal (employer) may enter into contractual relationships on the latter's behalf.

Principal–agent	Agent is hired by the principal to act on the principal's behalf subject to the latter's control and consent.
Employer–employee	Employer controls the employee's physical conduct and determines the details of performance.
Employer–independent	Employer does not control the details of performance or conduct of the independent contractor.

TABLE 15-1

TYPES OF AGENCY RELATIONSHIPS

EXHIBIT 15-1

PRINCIPAL–AGENCY RELATIONSHIP

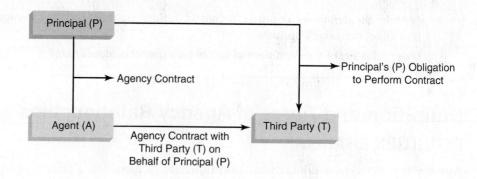

employer–independent contractor relationship
Relationship in which the agent (independent contractor) is hired by the principal (employer) to do a specific job but is not controlled with respect to physical conduct or details of work performance.

accordance with expressed or implied authority, may enter into contractual relationships on behalf of the employer. In general, any agent who works for pay is considered an employee, and any employee, unless specifically limited, who deals with a third party is given agent status. The employer–employee relationship is particularly important because it is the only agency relationship that is subject to workers' compensation, Social Security, and unemployment compensation laws (as well as numerous other state and federal laws). It is also important to distinguish the employer–employee relationship from an **employer–independent contractor relationship**, which normally is not subject to these statutes; this is an important distinction in the legal environment of business.

Employer–Independent Contractor. Independent contractors are hired by the principal (employer) to do a specific job but are not controlled with respect to their physical conduct or the details of their work performance. In this type of agency relationship, the employer is normally not required to make payments into workers' compensation pools or to withhold Social Security amounts and is not usually liable for torts committed by the independent contractor. (*Torts*, you will recall, are wrongful acts, other than breaches of contract, for which damages may be obtained.) For example, the employer of a taxi driver who is an independent contractor and owns his or her own cab is not liable, in most instances, for damages when the driver injures another person in an accident.

When called on to distinguish between an employee and an independent contractor, courts scrutinize the degree of control the employer has over the agent. Courts also consider factors such as (1) whether the hired persons had a distinct occupation or profession, (2) whether they supplied their own tools and equipment, (3) whether they were employed only for a specific time period, (4) whether they were paid hourly or on completion of a job, and (5) what degree of skill was required to do the job.

To escape paying unemployment insurance taxes as well as medical benefits, many employers in recent years have classified their workers as independent contractors or part-time workers. A common strategy in the highly competitive business environment of the 1990s was to "downsize" by cutting back on the number of employees in a corporation through early retirement plans and lay-offs and then to "restructure" by rehiring these people as "consultants" (independent contractors) to accomplish the same work they did as employees. This is essentially a cost-cutting measure. The "consultants" draw a per diem or lump sum instead of a salary for the job they are hired to perform, and the company saves on health care, unemployment compensation, and Social Security benefits. In turn, the independent contractors are free to work for others besides their old employer, perhaps while receiving a pension from the employer. The case that follows indicates the close scrutiny courts are giving to employers' classification of workers as independent contractors.

CHAPTER 15 • Agency Law

LINKING LAW AND BUSINESS

Finance

In finance, you have examined agency relationships and, particularly, costs associated with such relationships. For example, if both parties to a relationship are utility maximizers, there is reason to believe that the agent will not act in the best interest of the principal. The principal can limit divergences from its interest by establishing appropriate incentives for the agent and by monitoring costs of agents. In addition, in some situations, it pays the agent to expend resources (bonding costs) to make sure that the principal is not harmed.

The sum of agency costs includes:

1. monitoring expenditures by the principal;
2. bonding expenditures by the agent; and
3. residual loss.

In the following case, the court had to determine the status of an auto service company and its tow truck driver who assaulted the passenger of a vehicle the company had been hired to tow. Is the driver an employee or an independent contractor?

Source: M. Jensen and W. H. Meckling, "Agency Costs and the Theory of the Firm." *Journal of Financial Economics* 3 (1976): 79, 81.

CASE 15-1

Coker v. Pershad
Superior Court of New Jersey Appellate Division
2013 WL 1296271 (2013)

Background and Facts AAA North Jersey, Inc., contracted with Five Star Auto Service to perform towing and auto repair services for AAA. Terence Pershad, the driver of a tow truck for Five Star, responded to a call to AAA for assistance by the driver of a car involved in an accident in Hoboken, New Jersey. Pershad got into a fight with Nicholas Coker, a passenger in the car, and assaulted Coker with a knife. Coker filed a suit in a New Jersey state court against Pershad, Five Star, and AAA. The court determined that Pershad was Five Star's employee and that Five Star was an independent contractor, not AAA's employee. Thus, AAA was "not responsible for the alleged negligence of its independent contractor, defendant Five Star, in hiring Mr. Pershad." Five Star entered into a settlement with Coker. Coker appealed the ruling in AAA's favor.

The important difference between an employee and an independent contractor is that one who hires an independent contractor has no right of control over the manner in which the work is to be done.

Plaintiff [Coker] argues that AAA controlled the means and method of the work performed by Five Star. Factors [that] determine whether a principle maintains the right of control over an individual or a corporation claimed to be an independent contractor [include]:

(a) the extent of control which, by the agreement, the master may exercise over the details of the work;
(b) whether the one employed is engaged in a distinct occupation or business;
(c) the kind of occupation, with reference to whether, in the locality, the work is usually done under the direction of the employer or by a specialist without supervision;
(d) the skill required in a particular occupation;
(e) whether the employer or the workman supplies the tools;
(f) the length of time for which the person is employed.

Applying these factors to the facts of this case, it is clear that AAA did not control the manner and means of Five Star's work. The Agreement specifically stated that Five Star was an independent contractor. Five Star purchased its own trucks and any other necessary equipment. AAA assigned jobs to Five Star, and Five Star completed the work without any further supervision by AAA. Five Star chose the employees to send on towing calls and the trucks and equipment the employees would use.

Five Star was also in business for itself and performed auto repair services for principals and customers other than AAA. Five Star hired and fired its own employees.

Plaintiff also argues that Five Star should be considered to be controlled by AAA because "providing towing and other roadside assistance is arguably the focus of the regular business of AAA." [But] AAA is an automobile club that provides a wide variety of services to its members. It contracts with numerous service providers, such as gas stations, motels, and other businesses, to provide these services. Thus, AAA is not solely in the towing business.

AAA had used Five Star to provide towing services for approximately eight years, and there is nothing in the record to demonstrate that it lacked the skill needed to provide these services.

A state intermediate appellate court affirmed the lower court's ruling. AAA could not be held liable for the actions of Five Star, its independent contractor, because "AAA did not control the manner and means of Five Star's work."*

CRITICAL THINKING ABOUT THE LAW

When courts establish legal tests, they describe conditions that must be met for the test to be passed. In Case 15-1, the plaintiff hoped he could meet the standards of the common-law test attached to an agency relationship.

To meet this test, a large range of facts had to be present. If the facts all fell into place in a particular way, the plaintiff would prevail.

1. Identify specific facts from Case 15-1 that might permit the plaintiff to prevail.

   Clue: Go to the location in the decision where the court goes step-by-step through the test and pick out a few instances in which one part of the test was met because of a particular set of circumstances.

2. What "facts," had they existed, would have allowed the plaintiff to win this case?

   Clue: Read the court's statement of the necessary conditions for establishing an agency. Then imagine some "facts" that would satisfy those conditions.

"Temps": Employees or Independent Contractors?

In 1990, the Internal Revenue Service found that Microsoft Corporation had hundreds of workers who were actually employees of the company, as opposed to independent contractors, based on the control that Microsoft exercised over them. Instead of contesting the IRS ruling, which required the company to pay back payroll taxes, Microsoft required most of its former independent contractors to become associated with employment agencies and work for Microsoft as "temps." If they didn't, they were laid off. Those workers who didn't sign up with an agency sued Microsoft, claiming that they were employees of the company and were entitled to participate in the stock option plan. Microsoft argued that each of the workers had signed an independent contractor agreement and, thus, were not eligible for benefits. At the Ninth Circuit Court of Appeals level, the court held that the "temps" were "common-law employees" under agency law. Furthermore, the court stated that being employees of a temporary employment agency did not prevent the workers from being common-law employees of Microsoft at the same time.

Source: Vizcaino v. Microsoft, 173 F.3d 713 (9th Cir. 1999).

Creation of an Agency Relationship

Agency relationships can be created by informal oral agreements or formal written contracts. Under the statute of frauds (see Chapter 9), some states require that all agency contracts be in writing.

---

*Coker v. Pershad, Superior Court of New Jersey, Appellate Division 2013 WL 1296271 (2013).

TABLE 15-2 HOW AGENCY RELATIONSHIPS ARE CREATED

Agency Relationships	Method of Creation
Expressed agency or agency by agreement	Formed by detailed written or oral agreement. A power-of-attorney document is often used to establish this relationship.
Agency by implied authority	Formed in situations in which custom and circumstances determine the agent's authority to do business for the principal.
Agency through ratification	Formed when an unauthorized act of an agent is accepted by the principal.
Agency by estoppel or apparent authority	Formed when a principal leads a third party to believe that a certain person is acting as an agent for the principal. The principal is thereafter estopped from denying the agency relationship.

Most such relationships grow out of the consent of two parties—the principal indicating in some way that the agent should act on its behalf and the agent agreeing to do so. An agency relationship may be formed through (1) expressed agency, (2) implied authority, (3) ratification of the agent's previous acts by the principal, or (4) apparent agency (Table 15-2).

EXPRESSED AGENCY OR AGENCY BY AGREEMENT

An agency relationship formed through an oral or written agreement between the principal and agent is known as an **expressed agency** or **agency by agreement**. Under the statute of frauds, agency agreements that will last longer than one year must be in writing. An *exclusive agency contract* is a type of agreement whereby the principal agrees that it will not employ any other agent for a period of time or until a particular job is completed. If the principal fails to live up to this agreement, the agent may file a breach-of-contract suit to recover monetary damages plus court costs and attorney's fees.

A legal document used to establish an agency relationship is called a **power of attorney**. It gives the agent the authority to sign legal documents on behalf of the principal. Many states allow powers of attorney for both business and health care purposes. In the latter kind of power of attorney, the principal designates in writing an individual (agent) who will act for the principal in the event of a serious illness that renders the principal mentally incompetent to make decisions concerning his or her medical care. The agent is usually given the authority to make such decisions in consultation with physicians. The power of attorney for business purposes may be a general document that gives the agent broad powers to act on the principal's behalf, or it may be a more limited document that authorizes the agent to act only in matters enumerated in the agreement. Many people now incorporate by reference powers of attorney for both health and business purposes into their wills or trusts.

expressed agency (agency by agreement) Agency relationship formed through an oral or written agreement.

power of attorney An agency agreement used to give an agent authority to sign legal documents on behalf of the principal.

AGENCY BY IMPLIED AUTHORITY

When a principal and an agent create an agency relationship, they often do not set out every detail of the agent's authority in the written or oral agreement. In **agency by implied authority**, customs, circumstances, and the facts of a situation determine the authority of the agent to do business on behalf of the principal.

Unless the agency contract says otherwise, courts have generally allowed agents to:

1. Receive payments of money due the principal
2. Enter into contracts for incidentals

agency by implied authority Agency relationship in which customs and circumstances, rather than a detailed formal agreement, determine the agent's authority.

3. Employ or discharge employees
4. Buy equipment and supplies

In the following classic case, the court indicates the extent to which implied authority will be granted to an agent.

CASE 15-2

Penthouse International v. Barnes
United States Court of Appeals
792 F.2d 943 (9th Cir. 1986)

Priscilla Barnes was a hostess at a club in Hollywood, California, when she was approached by a freelance photographer (Dunas) who sold nude photographs to *Penthouse* magazine. Dunas was an independent contractor. He asked Barnes to pose nude. She agreed but did not want her actual name used. Dunas agreed to her terms, and Barnes signed a "Release, Authorization and Agreement Form" that gave Penthouse the right to "republish photographic pictures or portraits." Dunas added the term *AKA* (also known as) on the contract to indicate that the photographs would not be published under her actual name, but under a pseudonym. Penthouse did so in 1976.

Later, Barnes became a well-known television broadcaster for a station in Los Angeles. When Penthouse informed her in 1983 that it wished to republish her nude photograph, she threatened to sue, claiming that Dunas had implied agency to write the term *AKA* and that Penthouse was thus bound by his actions and representations. Penthouse requested a declaratory judgment from the federal district court allowing it to republish a nude photo of the defendant. The district court found for Barnes and issued an injunction against Penthouse. Penthouse appealed.

Judge Boochever

Under California law, questions regarding the existence of agency are questions of fact that we review for clear error. California Civil Code Section 2316 defines actual authority as "such as a principal intentionally confers upon the agent, or intentionally, or by want of ordinary care, allows the agent to believe himself to possess." At issue is whether Dunas contracted to act on behalf of Penthouse.

Penthouse instructed photographers "to get a signed model release and not to alter the release in any way, without our permission." However, Penthouse carried Dunas's name on its masthead, gave him blank Penthouse contracts, may have given him business cards, and had him present contracts to models. Thus, although the record conflicts as to whether Dunas was an actual agent of Penthouse, on review, we cannot find that the district court clearly erred in finding Dunas to be a Penthouse agent.

Having found that the district court did not err in characterizing Dunas as an agent, we next turn to whether Dunas was acting within the scope of his authority by modifying the contract. Barnes does not contend nor is there evidence that Dunas possessed express actual authority to modify the contract. Dunas, however, had implied actual authority. Implied actual authority requires that (1) Dunas believe that he was authorized to modify the contract based on Penthouse conduct known to him or (2) such a belief was reasonable.

Circumstantial evidence exists that Dunas placed "AKA" on other contracts with no objection from Penthouse. In June 1974, a year and a half before Barnes posed, contracts prepared by Dunas had "AKA" on them. Further, Penthouse internal memoranda reflect an understanding among Penthouse employees that "AKA" added to a contract meant that Penthouse was to associate a fictitious name with a woman's photograph. The evidence thus indicates that Dunas reasonably believed that he was authorized to add "AKA" and modify the contract to require that only a fictitious name be used.*

Affirmed in favor of Defendant, Barnes.

CRITICAL THINKING ABOUT THE LAW

One of the primary purposes of a judge's opinion is to explain the court's reasoning in a particular case. A judge's opinion is not arbitrary, in the sense that a judge must give due consideration to relevant facts and rules of law for any legal issue. From a judge's opinion, we are, therefore, able to know not only a judge's conclusion but also the

---

*Penthouse International v. Barnes, United States Court of Appeals 792 F.2d 943 (9th Cir. 1986).

reason the judge ruled for one party over another. These opinions provide the court's rationale in a particular case, which may later be used as precedent for subsequent cases that contain similar fact patterns. In Case 15-2, the judge provided several reasons to support the conclusion. The next two questions relate to the judge's reasoning in Case 15-2.

1. What reasons did the judge provide for ruling in favor of the defendant?

    Clue: To ensure that you have found a reason, ask yourself whether what you have listed answers this question: "Why did the court rule for the defendant?"

2. What aspects of the court's reasoning were particularly strong or weak? (Remember that just because reasons are given does not mean that these reasons are necessarily strong.)

    Clue: Reverse the roles in this case and assume that you are the plaintiff's lawyer. With which parts of the judge's opinion would you still disagree based on the court's reasoning? Would there be parts of the judge's reasoning with which, even though you were the opposing party, you would agree?

APPLYING THE LAW TO THE FACTS...

Let's say that Mandy and Adam would like to enter into an expressed agency or agency relationship. What document would these two need to facilitate such a relationship?

AGENCY THROUGH RATIFICATION BY PRINCIPAL

Agency by ratification occurs when a person misrepresents himself or herself as an agent and the principal accepts (*ratifies*) the unauthorized act. If the principal accepts the results of the agent's act, then the principal is bound, just as if he or she had authorized the individual to act as an agent. Two conditions are necessary for the ratification to be effective: (1) the principal must have full knowledge of the agent's action and (2) the existence of the principal must be clear to the third party at the time of the agent's unauthorized act. For example, Max Black tells his friend Mary Ann Jones, a realtor, that if she ever finds anyone who is ready, willing, and able to sell a certain kind of computer franchise store, he will buy it. Jones finds such a store, cannot reach Black, and has never been his agent, but she enters into the agreement with the franchisor anyway and signs the contract "Mary Ann Jones, agent for Max Black." Jones is not Black's agent. If Black, however, decides to honor the contract, he will have ratified an agency relationship, and Jones will most likely be due a commission.

agency by ratification Agency relationship in which an unauthorized agent commits the principal to an agreement and the principal later accepts the unauthorized agreement, thus ratifying the agency relationship.

AGENCY BY ESTOPPEL OR APPARENT AUTHORITY

If a third person is led by a principal to believe that a certain individual is the principal's agent, then there appears to be authority for the agent to act and the principal is **estopped** from denying that the individual is an agent. It is the writing, words, or acts, or some combination thereof, of the principal that create the **agency by estoppel**, or **apparent authority**, for the agent to act; the third party relies on the principal's conduct. Apparent authority is illustrated in the case below.

estoppel A legal bar to either alleging or denying a fact because of one's previous words or actions to the contrary.

agency by estoppel (apparent authority) An agency relationship in which the principal is estopped from denying that someone is the principal's agent after leading a third party to believe that the person is an agent.

CASE 15-3

Motorsport Marketing, Inc. v. Wiedmaier, Inc.
Missouri Court of Appeals
195 S.W.3d 492 (2006)

Wiedmaier, Inc., owns and operates a truck stop in Missouri. The owners are Marsha Wiedmaier and her husband (Jerry). Their son (Michael) had no interest in the company but worked for it as a fuel truck operator. In April 2003, Michael faxed a credit application to Motorsport's sales manager, Lesa James. Marsha signed the form as "Secretary-Owner" of Wiedmaier; after she signed, Michael added himself to the list of owners. A credit line was approved. Michael formed Extreme Diecast, LLC, and told Motorsport that it was a part of Wiedmaier. He then began ordering Motorsport merchandise. By early 2004, however, Michael had stopped making payments on the account, quit his job, and moved to Columbus, Ohio. Patrick Rainey, the president of Motorsport, contacted Marsha about the account, but she refused to pay. Motorsport filed a suit in a Missouri state court against Wiedmaier and others to collect the unpaid amount. The court entered a judgment in favor of Motorsport, assessing liability against the defendants for the outstanding balance of $93,388.58, plus $13,406.38 in interest and $25,165.93 in attorney's fees. The defendants appealed to the state intermediate appellate court.

Justice Howard

To establish apparent authority of a purported agent, Motorsport must show that

(1) the principal manifested his consent to the exercise of such authority or knowingly permitted the agent to assume the exercise of such authority; (2) the person relying on this exercise of authority knew of the facts and, acting in good faith, had reason to believe, and actually believed, the agent possessed such authority; and (3) the person relying on the appearance of authority changed his position and will be injured or suffer loss if the transaction executed by the agent does not bind the principal.

We find that Motorsport has shown that each of the criteria for establishing Michael's apparent agency has been satisfied. First, [t]he credit application constituted a direct communication from Wiedmaier, Inc. (through Marsha) to Motorsport causing Motorsport to reasonably believe that Michael had authority to act for Wiedmaier, Inc.

Second, Motorsport, relying on Michael's exercise of authority and acting in good faith, had reason to believe, and actually believe, that Michael possessed such authority. Motorsport received a credit application from Wiedmaier, Inc. signed by owner Marsha Wiedmaier, listing Michael as an owner. Motorsport had no reason to believe that Michael was not an owner of Wiedmaier or was otherwise unauthorized to act on Wiedmaier, Inc.'s behalf.

Wiedmaier, Inc. argues that even if Motorsport's reliance on Michael's apparent authority was reasonably prudent on April 10, 2003, when Michael submitted the credit application, such reliance could not have been and was not reasonably prudent from and after June 23, 2003. At that time, Michael personally made the first payment on the account with a check drawn on the account of Extreme Diecast. At the very least, Wiedmaier, Inc. argues, Motorsport had "red flags waving all around it suggesting that Michael was something other than the agent of Wiedmaier, Inc."

We find that this argument is without merit. It is a common practice for a truck stop to have a separate division with a separate name to handle its diecast and other related merchandise, and Michael represented that this is exactly what Extreme Diecast was. This evidence explains what Wiedmaier, Inc. characterizes as "red flags" concerning Michael's authority.

Third, Motorsport changed its position and will be injured or suffer loss if the transaction executed by Michael does not bind Wiedmaier, Inc. Motorsport extended credit to Wiedmaier, Inc. based on its interaction with Michael and based on its belief that it was dealing with Wiedmaier, Inc. Marsha Wiedmaier has refused to pay the account balance. If the transaction executed by Michael does not bind Wiedmaier, Inc., Motorsport will suffer the loss of the balance due on the account.*

Duties of Agents and Principals

PRINCIPAL'S DUTIES TO AGENT

Through an evolution of case law and scholarly writing, it is now generally recognized that a principal (employer) owes four duties to an agent (employee): (1) a duty of compensation, (2) a duty of reimbursement and indemnification,

---

*Motorsport Marketing, Inc. v. Wiedmaier, Inc., Missouri Court of Appeals 195 S.W.3d 492 (2006).
http://caselaw.findlaw.com/mo-court-of-appeals/1089412.html.

(3) a duty of cooperation, and (4) a duty to provide safe working conditions. These duties are part of a written contract or are implied by law.

Duty of Compensation. In the absence of a written agreement, it is implied that a principal will compensate the agent for services rendered, either when services are contracted for or when they are completed. If the parties cannot agree on the amount, courts will usually indicate that the compensation should be calculated on the basis of what is reasonable or customary. For example, lawyers and real estate brokers are sometimes paid on a **contingency fee** (commission) basis.

contingency fee Agent's compensation that consists of a percentage of the amount the agent secured for the principal in a business transaction.

Duty of Reimbursement and Indemnification. An **indemnity** is an obligation on the part of an individual (principal) to make good (or reimburse) another person (agent) against any losses incurred when the latter is acting on behalf of the former. All necessary expenses can be recouped by the agent, but the agent is not entitled to expenses arising out of tortious conduct (e.g., negligence) or unlawful activities (e.g., bribes paid to suppliers).

indemnity Obligation of the principal to reimburse the agent for any losses the agent incurs while acting on the principal's behalf.

Duty of Cooperation. The principal must do nothing to interfere with the reasonable conduct of an agent as agreed upon in an express or implied contract. For example, if a franchisor (principal) agrees with a franchisee (agent) that the latter has an exclusive right to sell a specific item within a geographic territory, the franchisor cannot legally compete with the franchisee by setting up another franchisee within that territory. If the franchisor does establish a competitive franchisee, the first franchisee can sue for breach of contract, asking for lost profit, court costs, and attorney's fees or for specific performance.

Duty to Provide Safe Working Conditions. The principal has a duty to its agent to provide safe working conditions. Federal legislation such as the Occupational Safety and Health Act (discussed in Chapter 18) set standards designed to create a safe working environment for employees. Employers who repeatedly violate those standards may be fined or imprisoned or both.

AGENT'S DUTIES TO PRINCIPAL

Just as a principal has legal duties to an agent, the agent has legal obligations toward the principal: loyalty, obedience, accounting, and performance.

Duty of Loyalty. Courts have often indicated that the agent's most important obligation is fiduciary. The fiduciary obligation includes loyalty to the principal— that is, acting on behalf of one principal only to avoid conflicts of interest, communicating all material information to the principal, and refraining from acting in a manner that is adverse to the principal's interest. See an example of the duty of loyalty in the case below.

CASE 15-4

Cousins v. Realty Ventures, Inc.
Court of Appeals of Louisiana
Fifth Circuit 844 So.2d 860 (2003)

Don Cousins acted as the representative/agent for Eagle Ventures, Inc., to find a real estate investment for the corporation to purchase. To do so, Cousins engaged the services of Leo Hodgins, a real estate agent/broker and owner of Realty Ventures, Inc. (RVI).

In March 1991, Leo Hodgins learned that 3330 Lake Villa Drive, an 8,000-square-foot office building in Metairie, Louisiana, owned by Westinghouse Credit Corporation, was being sold for $125,000. In June 1991, Leo Hodgins first brought 3330 Lake Villa Drive to Cousins's attention. Cousins asked Hodgins to submit an offer to Westinghouse on behalf of Eagle Ventures, Inc., for $90,000. Mr. Hodgins submitted the offer to Westinghouse. Westinghouse was unable to sell the property at the time because it was having difficulties with Tonti Management, which was managing some of its local commercial holdings.

In October 1991, Leo Hodgins resubmitted Eagle Ventures' June 1991 offer. Hodgins learned that Westinghouse was ready to sell 3330 Lake Villa, but only as part of a package deal with its neighboring property, 4141 Veterans Boulevard.

In April 1992, Mr. Hodgins and his brother, Paul, created a partnership, known as 4141 Vets Limited Partnership, to purchase the property. On May 7, 1992, Westinghouse sold 3330 Lake Villa to 4141 Vets Limited Partnership for $65,000 and sold 4141 Veterans to the same partnership for $355,000. Hodgins then offered to sell 3330 Lake Villa to Eagle Ventures for $175,000.

The plaintiff, Don Cousins, filed the instant suit against defendants, Realty Ventures, Inc. and others. The plaintiff's suit alleged that the defendants breached their fiduciary duties to him. The jury returned a verdict in favor of the plaintiff and awarded damages in the amount of $1,750,000. The defendants filed this appeal.

Justice Edwards

The precise duties of a real estate broker must be determined by an examination of the nature of the task the real estate agent undertakes to perform and the agreements he makes with the involved parties. In the instant case, Leo Hodgins accepted Mr. Cousins'[s] request to find commercial real estate for Eagle Ventures to purchase by turning his attention to 3330 Lake Villa. The common law sets forth the types of duties expected of a real estate agent. For example, a real estate agent may be found liable where he or she does not timely communicate an offer and that failure to communicate results in damages to the client. Similarly, a real estate broker has been found to have a duty to communicate to his principal all offers received and may be liable in damages for failure to do so. Moreover, a [R]ealtor has a duty to relay accurate information about property, a duty which extends to both vendor and purchaser, and may be held liable if such duty is breached.

Plaintiffs argued at trial that Leo Hodgins committed a breach of his fiduciary duty to them in several respects, [including] failing to communicate Westinghouse's response to their offer. Defendants argue [that] they owed no further duty to plaintiffs after the June purchase offer for 3330 Lake Villa was submitted to Westinghouse. Once Westinghouse responded that it could not sell the property due to the property management agreement, Leo Hodgins relayed this information to Don Cousins. Plaintiffs contend that the agent/client relationship persisted far beyond that point based on Leo Hodgins'[s] resubmission of Eagle Ventures' offer in October. They argue that the information regarding the package sale of 3330 Lake Villa and 4141 Veterans Boulevard constituted a counteroffer by Westinghouse that should have been communicated to Mr. Cousins.

In our opinion, as late as February or March of 1992, Mr. Cousins was still communicating with Leo Hodgins regarding the status of his offer and Leo Hodgins was still discussing Eagle Ventures' offer with Westinghouse as late as January 1992. We believe Mr. Cousins acted consistently with his belief that Leo Hodgins was acting as his agent with Westinghouse at that time, while Leo Hodgins did nothing to dispel that belief. During some of that time, Leo Hodgins was armed with the information that 3330 Lake Villa was only for sale as a package with 4141 Veterans Boulevard and was covertly planning to acquire the property himself. Hodgins never told Don Cousins or anyone else associated with Eagle Ventures about Westinghouse's decision to sell the properties together prior to March 1992. Leo Hodgins'[s] failure to communicate the package sale to plaintiffs the moment he learned of it constituted a breach of his fiduciary duties to them. Leo Hodgins'[s] duty was to give plaintiffs the information that Westinghouse rejected their offer for a single property sale and allow them to decide whether they wished to purchase both properties.*

Affirmed in favor of the Plaintiff, Cousins.

Duty of Obedience. The agent has the duty to follow all reasonable and lawful instructions of the principal. If the manufacturer of a drug to alleviate the symptoms of the common cold tells its advertising department, salespeople, and distributors to tout the product as a cure for the common cold, these agents have no duty to follow the instructions because they are in violation of Section 5 of the Federal Trade Commission Act (see Chapter 25).

Duty of Accounting. Whenever the principal requests an accounting of money or property, it is the duty of the agent to hand over the requested material. When courts have been asked to determine how the duty of accounting is met, their answers have varied from case to case, but the keeping of accurate books that can be viewed by the principal is the minimum activity expected.

Duty of Performance. In all agency relationships, courts have indicated that agents must use reasonable care and skill in performing their work. This is generally taken to mean that agents must live up to the standards of performance expected of people in their occupation.

*Cousins v. Realty Ventures, Inc. Court of Appeals of Louisiana, Fifth Circuit 844 So.2d 860 (2003). http://www.leagle.com/decision/20031704844So2d860_11636.

A bank acts as an agent for its customer when collecting an item presented for payment. Suppose that the item is an international check to be paid in currency subject to a variable rate of exchange. Does the bank have a duty to account for the difference between the amount paid to the customer and the amount collected on the check? That was a question in the following case.

CASE 15-5

Gossels v. Fleet National Bank
Appeals Court of Massachusetts
876 N.E.2d 872 (2007)

Peter Gossels received a check from the German government for 85,071.19 euros, drawn on Dresdner Bank of Germany. The check was a payment in reparation for the seizure by the Third Reich of property belonging to Gossels's family. On October 15, 1999, he took the check to a branch of Fleet National Bank in Boston and presented it to the international teller.

The teller failed to inform Gossels that Fleet paid international checks at a "retail exchange rate" several percentage points lower than the interbank "spot rate" for foreign currency.

As Gossels started to endorse the check, the international teller also incorrectly instructed Gossels not to endorse the check.

Fleet, as collecting bank, sent the check to Fleet's foreign correspondent bank in Germany, Deutsche Bank, which sent it to Dresdner Bank. After a delay to obtain Gossels's endorsement, Dresdner Bank debited the funds from the appropriate account and sent 85,071.19 euros to Deutsche Bank, which credited 84,971.19 euros to Fleet's account at Deutsche Bank (100 euros having been deducted as a collection fee).

On December 15, 1999, Gossels received notice from Fleet that it had credited his account with check proceeds in the amount of $81,754.77, which was based on the December 15, 1999, retail exchange rate offered by Fleet for 84,971.19 euros. The same number of euros would have been worth $88,616.45 based on the October 15, 1999, retail exchange rate offered by Fleet or $92,023.80 based on the October 15, 1999, spot rate offered by Dresdner Bank.

When Gossels learned of the different rates, he filed suit in a Massachusetts state court against Fleet. The court entered a judgment for Gossels on the ground of negligent misrepresentation. Both parties appealed to a state intermediate appellate court.

Justice Gelinas

Under [Section 4-201(a) of the Uniform Commercial Code,] Fleet became Gossels's agent when he passed the check to Fleet, and Fleet accepted the check, for collection.

As Gossels's agent and fiduciary, Fleet was obliged to disclose fully all facts material to the transaction. The principal has a right to be informed of all material facts known to the agent in reference to the transaction in which he is acting for his principal, and good faith requires a disclosure of such facts by the agent. Whenever facts known to the agent but not to the principal may affect the desires and conduct of the principal, the agent must communicate that information to the principal, particularly if the agent is engaging in any arrangement adverse to the principal's interest.

Fleet was required, as a fiduciary, to advise Gossels that it would pay his international check at a retail exchange rate that was several percentage points lower than the spot rate it received for foreign currency. Nowhere in the transaction did Fleet reveal that it would pay Gossels, its principal, a retail rate of exchange substantially less than the spot rate it obtained for the item, which former rate was set out daily on a "secret" rate sheet that bank employees were advised not to disclose to the public, and that it would profit by effectively keeping the difference.

By failing to disclose the rate, and by withholding the amount without giving Gossels an adequate explanation, Fleet further committed a breach of its fiduciary duty to give a complete accounting of the disposition of the funds. An agent or fiduciary is under a duty to keep and render accounts and, when called upon for an accounting, has the burden of proving that he properly disposed of funds which he is shown to have received for his principal.

If there is no prior agreement with regard to the profit, the agent must not only render an account of it, but must pay the funds over to the principal. Fleet was bound to account for the amounts it kept in the rate arbitrage [the profits it earned on the transaction], and to pay it over to Gossels.

The [trial] judge's finding that Fleet is liable to Gossels for negligent misrepresentation has not been shown to be erroneous. Fleet is also liable to Gossels for breach of fiduciary duty. The judgment is vacated and we remand the case to the trial court for entry of a single judgment in favor of Gossels on the aforementioned claims.*

Affirmed for Gossels.

---

*Gossels v. Fleet National Bank Appeals, Court of Massachusetts 876 N.E.2d 872 (2007).

Principal's and Agent's Liability to Third Parties

Principals are normally responsible for the acts of their agents if those acts come within the scope of the agents' employment. Principals, then, can be held liable when their agents enter contracts, commit torts within the scope of their employment, or commit crimes while acting on behalf of their principal (Table 15-3).

CONTRACTUAL LIABILITY

The purpose of a principal–agency relationship is to enable the principal to expand business through the use of agents who can negotiate agreements with third parties that the principal would find it difficult or impossible to contact. To do their job, therefore, agents must be authorized to enter into contracts. A principal's liability for an agent's contracts depends in large part on whether the principal is disclosed or undisclosed to the third party at the time the agreement is executed. In the following case, we have an agent for a disclosed principal. Usually, an agent (Taxman) does not have liability for nonperformance of a fully disclosed principal (Arctic). Case 15-6 focuses on a principal's liability for an agent's contract.

TABLE 15-3

DUTIES OF PRINCIPALS AND AGENTS

Principal's Duties to Agents	
Duty of cooperation	Must cooperate with and assist the agent in the performance of the agent's duties and the accomplishment of the agency.
Duty to provide safe working conditions	Must provide safe working conditions, warn agents of dangerous conditions, and repair and remedy unsafe conditions.
Duties of reimbursement and indemnification	Must reimburse an agent for all expenses paid that were authorized by the principal, within the scope of the agency, and necessary to discharge the agent's duties. Must indemnify the agent for any losses suffered because of the principal's acts.
Duty of compensation	Must pay the agent agreed-upon compensation. If there is no agreement, the principal must pay what is customary in the industry, or, if there is no custom, then the reasonable value of the services.
Agent's Duties to Principal	
Duty of performance	Performance of the lawful duties expressed in the agency contract with reasonable care, skill, and diligence.
Duty of notification	Duty to notify principal of any information the agent learns that is important to the agency. Information learned by the agent in the course of the agency is imputed to the principal.
Duty of loyalty	Duty not to act adversely to the interests of the principal. The most common breaches of loyalty are 1. *Self-dealing.* Agent cannot deal with the principal unless his or her position is disclosed and the principal agrees to deal with the agent. 2. *Usurping an opportunity.* Agent cannot usurp (take) an opportunity belonging to the principal as his or her own. 3. *Competing with the principal.* Agents are prohibited from competing with the principal during the course of an agency unless the principal agrees.
Duty of obedience	Must obey the lawful instructions of the principal during the performance of the agency.
Duty of accountability	Must maintain an accurate accounting of all transactions undertaken on the principal's behalf. A principal may demand an accounting from the agent at any time.

CASE 15-6

McBride v. Taxman Corp.
Court of Illinois, First District
765 N.E.2d 51 (2002)

Walgreens Company entered into a lease with Taxman Corporation (defendant) to operate a drugstore in Kedzie Plaza, a shopping center in Chicago, Illinois, owned by Kedzie Plaza Associates; Taxman was the center's property manager. The lease required the "landlord" to promptly remove snow and ice from the center's sidewalks. Taxman also signed, on behalf of Kedzie Associates, an agreement with Arctic Snow and Ice Control, Inc., to remove ice and snow from the sidewalks surrounding the Walgreens store. On January 27, 1996, Grace McBride, a Walgreens employee, slipped and fell on snow and ice outside the entrance to the store. McBride filed a suit in an Illinois state court against Taxman and others alleging, among other things, that Taxman had negligently failed to remove the accumulation of ice and snow. Taxman filed a motion for summary judgment in its favor, which the court granted. McBride appealed.

Justice Cerda

On October 10, 1995, Taxman signed, on behalf of the owner, Arctic's one-page "Snow Removal Proposal & Contract" (although dated August 7, 1995), for the term November 16, 1995 through April 16, 1996, for the shopping center where this Walgreens store was located.

Also on October 10, 1995, Arctic and Taxman signed a multi-page document dated October 3, 1995, that was apparently drafted by Taxman. The document was not given a title but contained several pages of terms concerning snow removal "per contract(s) attached."

Plaintiff argues that the contract between Taxman and Arctic created a duty of Taxman to remove ice and snow for the benefit of the plaintiff.

The Arctic proposal and contract was signed "Kedzie Associates by the Taxman." The Taxman-drafted portion of the contract contained a line above the signature of Taxman's director of property management stating "The Taxman Corporation, agent for per contracts attached." The latter document specifically stated that the contract was not an obligation of Taxman and that all liabilities were those of the owner and not Taxman. We conclude that Taxman was the management company for the property owner and entered into the two contracts for snow and ice removal only as the owner's agent.

Taxman did not assume a contractual obligation to remove snow or ice; it merely retained Arctic as a contractor on behalf of the owner.*

Affirmed in favor of Taxman.

E-COMMERCE: INTELLIGENT AGENTS

When discussing agency law in this chapter and contract law in Chapters 9 and 10, we have discussed relationships between human beings and entities represented by individual corporations, partnerships, or franchisors. Usually, the agent acted within the scope of her or his expressed or implied authority. Today, electronic agents (e-agents) such as computer programs are used in e-commerce to perform such tasks as searching the Web for the best price on a service or good.

As discussed in Chapter 10, "click-on" agreements often are part of e-commerce. The terms of such an agreement must be recognized by an intelligent agent. Failing to observe such terms by the agent (A) may bind a principal (P), especially in a lengthy franchise agreement. Also, it may be important that certain click-on agreements exempt third parties (T) from liability. To avoid such problems, click-on agreements are now being developed for online stores that are more easily understood by e-agents.

The Uniform Electronic Transaction Act (UETA) (previously discussed in Chapter 10) allows e-agents to bind principals. When an individual or a business entity places an order over the Internet, the principal or company that takes the order from an e-agent cannot disclaim the order.

---

*McBride v. Taxman Corp. Court of Illinois, First District 765 N.E.2d 51 (2002).

LIABILITY OF DISCLOSED, PARTIALLY DISCLOSED, AND UNDISCLOSED PRINCIPALS

disclosed principal One whose identity is known by the third party when the latter enters into an agreement negotiated by the agent.

partially disclosed principal One whose identity is not known to the third party at the time of the agreement, although the third party does know that the agent represents a principal.

A **disclosed principal** is one whose identity is known by the third party at the time he or she enters into the agreement; the third party is aware that the agent is acting on behalf of this principal. A **partially disclosed principal** is one whose identity is not known to the third party at the time of the agreement; the third party does know, however, that the agent represents a principal. In both of these cases, the agent has actual authority and, therefore, the principal is liable. For example, if Jones (agent) enters into an agreement with Thomas (third party) to buy a new car on behalf of Smith, Inc. (principal), Smith will be liable for the contract. Disclosure of a principal is usually shown in the agent's signature on the contract (e.g., Smith, Inc., by Jones, Agent).

LIABILITY OF UNDISCLOSED PRINCIPAL

undisclosed principal One whose identity and existence are unknown to the third party.

When the third party is aware of neither the identity of the principal nor the agency relationship, the agent may be liable to the third party if the principal does not perform the terms of the contract. **Undisclosed principal**–agent relationships are lawful and are often used by wealthy people to conceal their involvement in a negotiation. For example, Jones, a famous tycoon, hires Smith to go to an auction to bid on a castle. Jones wishes to remain undisclosed because he fears the other bidders will "bid up" the price if they know he is bidding. In this case, both Jones and Smith are liable to the seller whose castle is being auctioned.

TORT LIABILITY

Under the doctrine of *respondeat superior,* a principal (employer) may be liable for the intentional or negligent torts of his or her agent (employee) if such acts are committed within the scope of the agent's employment. If the unauthorized acts are committed outside the scope of employment, liability shifts to the agent. Thus, the crucial issue for the courts is whether the employee/agent was acting within the scope of employment when the tort took place.

Two criteria used by the courts to decide this issue are as follows: (1) Was the agent or employee acting in the principal's interest? (2) Was the agent or employee authorized to be in the particular place where he or she was at the time of commission of the tort? For example, Smith, an employee of Brennan Pizza Company, is authorized to deliver pizza in a vehicle owned by the company but not to make any stops in between delivering the pizza and returning the company vehicle to Brennan Pizza. Smith stops off at a local tavern after delivering the last pizza. Smith comes out of the tavern and hits a third party, Jones, who is walking across the street in a legal manner. If Jones sues the company and Smith, the court must decide these questions: Should the company be held liable? Should the agent, Smith, be held liable? Or should they be held liable jointly and severally? In most states, the agent would be liable given this factual situation because she was not authorized to be where she was at the time of the accident. When company rules against the tortious activity are in place, courts generally rule against the agent. Were the company and Smith found to be **jointly and severally liable**, the agent would have had to indemnify the principal for what the company has to pay. Practically speaking, though, an agent such as Smith usually does not have the money to pay either damages or indemnities. Under the "deep pockets" concept, Brennan Pizza Company alone would pay damages because it has the money to do so.

joint and several liability The legal principle that makes two or more people liable for a judgment, either as individuals or in any proportional combination. Under this principle, a person who is partially responsible for a tort can end up being completely liable for damages.

TORT LIABILITY AND NEGLIGENCE

In a series of cases, tort negligence in hiring and supervision has arisen. A school district's failure to do background checks before hiring teachers or bus drivers, when the potential employee's past offensive conduct toward a child could have been discovered (through computer checks), constitutes the tort of *negligent hiring*.

In the following, Case 15-7, the U.S. Supreme Court considered whether an employer can be held strictly liable for a supervisor's conduct (sexual harassment) which led to the latter's discharge.

Whether a real estate salesperson's actions in connection with certain real estate transactions fell within the salesperson's scope of employment was at issue in the following case.

CASE 15-7

Auer v. Paliath
Court of Appeals of Ohio, Second District
2013-Ohio-391, 986 N.E.2d 1052 (2013) [Judge], FROELICH

Torri Auer [a California resident] brought suit [in an Ohio State court] against real estate salesperson Jamie Paliath, real estate broker for Keller Williams Home Town Realty, and others based on alleged fraud by Paliath in the sale of several rental properties [in Dayton, Ohio] to Auer. After a jury trial . . . , Paliath was found liable to Torri Auer in the amount of $135,200 for fraud in the inducement of Auer's purchases of the properties.

Home Town Realty appeals from the trial court's judgment.

Under [Ohio Revised Code (R.C.) Section 4735.01] the term "real estate broker" includes "any person, partnership, association, limited liability company, limited liability partnership, or corporation who for another and who for a fee, commission, or other valuable consideration" engages in various activities regarding real estate, including selling, purchasing, leasing, renting, listing, auctioning, buying, managing, and advertising real estate. A real estate salesperson generally means "any person associated with a licensed real estate broker to do or to deal with any acts or transactions set out or comprehended by the definition of a real estate broker, for compensation or otherwise."

Under R.C. Section 4735.21, no real estate salesperson may collect any money in connection with any real estate transaction, except as in the name of and with the consent of the licensed real estate broker under whom the salesperson is licensed.

A real estate broker will be held vicariously liable for intentional torts committed by salesmen acting within the scope of their authority. Vicarious liability is appropriate because a real estate salesman has no independent status or right to conclude a sale and can only function through the broker with whom he is associated. A salesman is required to be under the supervision of a licensed broker in all of his activities related to real estate transactions.

When a real estate salesperson acts in the name of a real estate broker in connection with the type of real estate transaction for which he or she was hired and the broker collects a commission for the transaction, the salesperson's actions in connection with that real estate transaction are within the scope of the salesperson's employment, as a matter of law.

In this case, Paliath contracted with Home Town Realty as a real estate salesperson to assist clients with the purchase and sale of real estate. Paliath advised and assisted Auer in the purchase of the . . . properties, and her fraudulent conduct involved misrepresentations regarding those properties.

Reviewing the properties separately, the evidence at trial established that Home Town Realty was listed as the real estate broker on the purchase contract, the agency disclosure statement, and the settlement statement for the Belton Street sale. Home Town Realty received a commission check of $180 from the title company that conducted the closing. Based on this evidence, it was established, as a matter of law, that Paliath acted within the scope of her employment as a real estate salesperson with Home Town Realty in relation to Auer's purchase of the Belton property.

Similarly, Paliath's actions with respect to the 1111–1115 Richmond Avenue properties were taken as a real estate salesperson assisting Auer with the purchase of the properties. Home Town was listed as a broker on the purchase contract, the agency disclosure statement, and the settlement statement for 1111 Richmond Avenue, and it received a commission of $2,400 following the closing. . . . The evidence thus demonstrated, as a matter of law, that Paliath was acting in the scope of her employment regarding the sale of 1111 Richmond Avenue.*

The trial court's judgment will be *affirmed*.

---

*Auer v. Paliath, Court of Appeals of Ohio, Second District, 2013-Ohio-391, 986 N.E.2d 1052 (2013).

CRIMINAL LIABILITY

Principals are generally not liable for criminal acts (e.g., murder, rape, price-fixing, extortion) of their agents because it is difficult to show the intent of the principal that is required for liability for such crimes. This general rule, however, has two major exceptions: (1) when a principal participates directly in the agent's crime and (2) when the principal has reason to know that a violation of law by employees or agents is taking place (see *United States v. Park* in Chapter 6).

For an example of the first exception, assume that a refinery threatens to terminate Jones (employee) unless he agrees to set the prices of gasoline in a certain location and communicate those prices to all company-owned and independent dealers that use the refinery's gasoline. If Jones does what he is told and is caught, both the refinery and Jones will be prosecuted under the Sherman Act for price-fixing (see Chapter 25). Jones's defense that he was just an employee following his employer's orders when he fixed prices will not be accepted. In contrast, if Jones refuses to do what his employer is demanding and is discharged, he can sue the refinery on the basis of the tort of wrongful discharge.

APPLYING THE LAW TO THE FACTS...

Randall is an assistant plant manager. He was trying to negotiate a settlement with an EPA inspector for some violations of the Federal Water Pollution Control Act. In the course of trying to become friendly with the inspector in hopes of being able to negotiate a better settlement, he discovered that for his retirement, the inspector was trying to make some extra money through trading, which gave Randall an idea. He called the plant manager and asked whether it would be okay to let the inspector know that the firm was about to be bought out by a bigger company, on the grounds that the inspector might then want to delay settlement negotiations until the firm was in a better financial condition. Of course, the real motive for sharing the information was that if Randall gave the inspector a "tip" he could use to make a profitable stock purchase, the inspector might be willing to make a more favorable settlement with the company. The plant manager says that it is a good idea. The inspector uses the information to purchase a significant number of shares in the company. When an insider trading case is brought due to the use of the information by the inspector, can any liability be imposed as a result of the assistant manager's status as an agent of the corporation? Explain.

Termination of the Principal–Agent Relationship

The principal–agent relationship may be terminated by agreement or by operation of law.

TERMINATION BY AGREEMENT

Either of the parties may decide to terminate the principal–agent relationship. When such a relationship is terminated, all third parties who dealt with the agent should be given actual notice by the principal. Constructive notice may be given to others through advertisement in newspapers published in areas where the agent operated on behalf of the principal.

Some agency relationships are terminated by a lapse of time. For example, Jones agrees with Smith that Smith will serve as his agent for business purposes until May 20, 2007. On that day, Smith's actual or apparent authority lapses. Once again, the principal should notify all relevant third parties at the termination.

TERMINATION BY OPERATION OF LAW

When the principal–agent relationship is terminated by law, there is no requirement to give notification to relevant third parties. The five most common methods of termination by operation of law are (1) death of one of the parties, even if the other party is unaware of that death; (2) insanity of one of the parties; (3) bankruptcy of the principal; (4) impossibility of performance, which may come about through the destruction or loss of subject matter or a change in the law; and (5) an outbreak of war, particularly when the agent's country is at war with the principal's country. See the case below for an illustration of the termination of a contract.

CASE 15-8

Gaddy v. Douglass
Court of Appeals of South Carolina
597 S.E.2d 12 (2004)

Ms. M was born in 1918. After retiring, Ms. M returned to Fairfield, South Carolina, where she lived on her family farm with her brother, a dentist, until his death in the early 1980s. Ms. M never married. Dr. Gaddy was Ms. M's physician and a close family friend. The appellants are Ms. M's third cousins.

In 1988, Ms. M then executed a durable general power of attorney designating Dr. Gaddy as her attorney-in-fact. Concerns about Ms. M's progressively worsening mental condition prompted Dr. Gaddy to file the 1988 durable power of attorney in November 1995. Thereafter, Dr. Gaddy began to act as Ms. M's attorney-in-fact and assumed control of her finances, farm, and health care. His responsibilities included paying her bills, tilling her garden, repairing fences, and hiring caregivers.

In March 1996, Dr. Gaddy discovered that Ms. M had fallen in her home and fractured a vertebra. Ms. M was hospitalized for six weeks. During the hospitalization, Dr. Gaddy fumigated and cleaned her home, which had become flea-infested and unclean to the point where rat droppings were found in the house. Finding that Ms. M was not mentally competent to care for herself, he arranged for full-time caretakers to attend to her after she recovered from the injuries she sustained in her fall. He made improvements in her home, including plumbing repairs adapting a bathroom to make it safer for caretakers to bathe Ms. M, who was incapable of doing so unassisted. During Ms. M's hospitalization, neither of the appellants visited her in the hospital or sought to assist her in any manner.

Dr. Gaddy had Ms. M examined and evaluated by Dr. James E. Carnes, a neurologist, in December 1996. After examining Ms. M, Dr. Carnes found that she suffered from dementia and confirmed that she was unable to handle her own affairs. As Ms. M's Alzheimer's disease progressed and her faculties deteriorated, Dr. Gaddy managed her financial affairs, oversaw maintenance of her properties, and ensured that she received constant care, including food, clothing, bathing, and housekeeping.

Ms. M's long-standing distant relationship with some members of her family, including appellants, changed in March 1999. On March 12, 1999, the appellants visited Ms. M, and with the help of a disgruntled caretaker, took her to an appointment with Columbia attorney Douglas N. Truslow to "get rid of Dr. Gaddy." On the drive to Truslow's office, Heller had to remind Ms. M several times of their destination and purpose. At Truslow's office, Ms. M signed a document revoking the 1988 will and the 1988 durable power of attorney. She also signed a new durable power of attorney (1999 durable power of attorney) naming the appellants as her attorneys-in-fact. The appellants failed to disclose Ms. M's dementia to Truslow and to David Byrd, a witness to the execution of the March 12 document. Based on the revocation of the 1988 power of attorney and recently executed power of attorney, the appellants prohibited Dr. Gaddy from contacting Ms. M and threatened Dr. Gaddy with arrest if he tried to visit Ms. M.

On March 16, 1999, three days after Ms. M purportedly revoked the 1988 durable power of attorney and executed the 1999 durable power of attorney, Dr. Gaddy brought a legal action as her attorney-in-fact pursuant to the 1988 durable power of attorney. He alleged, among other things, that the purported revocation of the 1988 durable power of attorney and the execution of the 1999 durable power of attorney were invalid because "on March 12, 1999, the date on which Ms. M purportedly signed the 1999 power of attorney and the revocation, she was not mentally competent" due to "senile dementia of the Alzheimer's type." Medical testimony was presented from five physicians who had examined Ms. M. They concluded that Ms. M (1) was "unable to handle her financial affairs" and (2) would not "ever have moments of lucidity" to "understand legal documents."

The trial judge concluded that Ms. M lacked contractual capacity "from March 12, 1999 and continuously thereafter." As a result, he invalidated the 1999 revocation of the 1988 durable power of attorney and the 1999 durable power of attorney and declared valid the 1988 durable power of attorney. Finally, he awarded Dr. Gaddy litigation expenses to be paid from Ms. M's assets.

Justice Kittredge

Since 1986, the South Carolina Legislature has expressly authorized and sanctioned the use and efficacy of durable powers of attorneys.

Upon the execution of a durable power of attorney, the attorney-in-fact retains authority to act on the principal's behalf notwithstanding the subsequent physical disability or mental incompetence of the principal. To honor this unmistakable legislative intent, it is incumbent on courts to uphold a durable power of attorney unless the principal retains contractual capacity to revoke the then existing durable power of attorney or to execute a new power of attorney.

"In order to execute or revoke a valid power of attorney, the principal must possess contractual capacity." Contractual capacity is generally defined as a person's ability to understand in a meaningful way, at the time the contract is executed, the nature, scope, and effect of the contract. Where, as here, the mental condition of the principal is of a chronic nature, evidence of the principal's prior or subsequent condition is admissible as bearing upon his or her condition at the time the contract is executed.

Here, the credible medical testimony presented compellingly indicates that Ms. M suffered from at least moderate to severe dementia caused by Alzheimer's Disease, a chronic and permanent organic disease, on March 12, 1999. We are firmly persuaded that Ms. M's dementia, chronic and progressive in nature, clearly rendered her incapable of possessing contractual capacity to revoke the 1988 durable power of attorney or execute the 1999 power of attorney. We find this conclusion inescapable based on the record before us.

The very idea of a durable power of attorney is to protect the principal should he or she become incapacitated. This case is precisely the type of situation for which the durable power of attorney is intended.*

Affirmed for Plaintiff, Gaddy.

CRITICAL THINKING ABOUT THE LAW

The durable power of attorney exists to protect a person who may be unable to protect herself at some later point. In Case 15-8, the court was asked to decide whether the person who is allegedly being protected by a revocation of that power of attorney can revoke even when her mental condition suggests that she may not appreciate the meaning of such a revocation. The court was asked to decide who is legally permitted to care for her.

1. How did Dr. Gaddy's behavior during the period in which he so clearly held a durable power of attorney clearly strengthen his credibility before the court?

    Clue: How would he have behaved during that period of time had he not had Ms. M's best interests uppermost in his thinking?

2. What facts, had they been true, would have permitted the revocation of Dr. Gaddy's power of attorney?

    Clue: Review the court's analysis of why the revocation was invalid.

Global Dimensions of Agency Law

As global business by multinational, midsize, and even small corporations has expanded, many U.S. businesses have hired foreign agents to represent them abroad. With the lowering of tariff barriers through such treaties as the North American Free Trade Agreement and the World Trade Organization, foreign agents will become ever more essential to the movement of goods and services across national boundaries. Thus, it behooves managers in businesses of all sizes to become knowledgeable about the exporting companies, distributors, and sales agents they contract with and the rules that govern those agents in

---

*Gaddy v. Douglass, Court of Appeals of South Carolina 597 S.E.2d 12 (2004).

their own countries. We covered the general aspects of these treaties and the international legal environment of business in Chapter 8. Here we encapsulate some of the major differences between agency law and practice in the United States and in two importing trading partners of U.S. business: Japan and the European Union (EU). In hiring a representative for a firm abroad, it should always be determined whether the arrangement is a dependent agency where little discretion is left to an agent, as opposed to an independent agency where the time and work schedule are left to the representative.

JAPAN

In Japan, agents must disclose whom they are representing, and third parties often require proof that agents are acting on behalf of a particular principal. Unlike the United States, an agent in Japan must have expressed authority to act, and principals are held liable for acts of agents within their limited range.

EUROPEAN UNION

European Union principals are bound to act in good faith toward their agents. If there is no written contract, compensation is based on customary practice in the location where the agent is acting. Agents are also expected to act in good faith. All items in negotiations must be communicated to the principal, and the agent is bound to follow the principal's instructions quite literally. Most agency agreements are assumed to be for an indefinite time; thus, notice of one, two, and three months, respectively, is required to terminate a relationship of one, two, and three years' duration.

U.S. AGENTS ABROAD

From just these two examples, you can see how agency law elsewhere can differ significantly from agency law in the United States. Managers should carefully review a host country's laws on agency before carrying on business in that country.

There are special complications for managers when the law of the host country conflicts with the Foreign Corrupt Practices Act (FCPA) of 1977, as amended in 1988. Although the FCPA allows remuneration to lower-level foreign agents and officials to expedite the handling of goods that U.S. businesses are seeking to sell in another country, it strictly forbids payments to political officials beyond a certain level. Because companies are liable for both civil and criminal sanctions for violating the FCPA, managers must keep a close watch over the actions of hired foreign agents, as well as over the actions of their own employees abroad.

COMPARATIVE LAW CORNER

Many businesspeople doing business in Japan have particular difficulty with the authority level given to agents representing Japanese multinationals. As noted previously, agents must have expressed authority to represent Japanese companies, so such agents will likely have proof of some expressed (oral or written) authority. U.S. companies, in contrast, expect implied or apparent authority to be sufficient for agents. It would seem that U.S. company managers should be careful in reviewing Japanese law as well as U.S. law. Please note carefully the FCPA of 1977, as amended in 1988 and 1998. See the preceding section on "U.S. Agents Abroad," as well as details on the FCPA in Chapter 23.

SUMMARY

An agency relationship is a fiduciary relationship in which the agent acts on behalf of, and is subject to the control of, the principal. There are three types of such relationships: principal–agent, employer–employee, and employer–independent contractor. The four methods of creating an agency relationship are (1) through expressed agency (agency by agreement), (2) by implied authority, (3) through ratification by the principal, and (4) by estoppel or apparent authority.

The duties of the principal to the agent are compensation for services rendered, reimbursement for expenses and indemnification for losses, cooperation, and provision of safe working conditions. Those of the agent toward the principal are loyalty, obedience to reasonable and lawful instructions, an accurate accounting, and skillful and careful performance.

Principals are liable to third parties for contracts made by their agents on their behalf, torts committed by the agent within the scope of the agent's employment, and crimes committed by the agent at the principal's direction.

Agency relationships may be terminated by mutual consent or by operation of law.

Managers doing international business need to be aware of the sometimes significant differences between agency law in the United States and agency law in other countries.

REVIEW QUESTIONS

15-1 Explain the doctrine of *respondeat superior*.

15-2 Distinguish between actual and apparent authority.

15-3 Describe agency by implied authority. Explain.

15-4 Distinguish an agent from an independent contractor.

15-5 Is a principal responsible for all contracts entered into by an agent? Explain.

15-6 Agency through ratification is formed under what circumstance? Explain.

REVIEW PROBLEMS

15-7 Lei Chu-Ming is a purchasing agent-employee for the A & B Coal Supply partnership. Chu-Ming has authority to purchase the coal needed by A & B to satisfy the needs of its customers. While Chu-Ming is leaving a coal mine from which she just purchased a large quantity of coal, her car breaks down. She walks into a small roadside grocery store for help. While there, she runs into Darin Wilson, who owns 360 acres back in the mountains with all mineral rights. Wilson, in need of cash, offers to sell Chu-Ming the property for $1,500 per acre. On inspection of the property, Chu-Ming forms the opinion that the subsurface contains valuable coal deposits. Chu-Ming contracts to purchase the property for A & B Coal Supply, signing the contract "A & B Coal Supply, Lei Chu-Ming, agent." The closing date is August 1. Chu-Ming takes the contract to the partnership. The managing partner is furious, as A & B is not in the property business. Later, just before closing, both Wilson and the partnership learn that the value of the land is at least $15,000 per acre. Discuss the rights of A & B and Wilson concerning the land contract.

15-8 Peter hires Alice as an agent to sell a piece of property he owns. The price is to be at least $30,000. Alice discovers that the fair market value of Peter's property is at least $45,000 and could be higher because a shopping mall is going to be built nearby. Alice forms a real estate partnership with her cousin Carl. Then she prepares for Peter's signature a contract for the sale of the property to Carl for $32,000. Peter signs the contract. Just before closing and passage of title, Peter learns about the shopping mall and the increased fair market value of his property. Peter refuses to deed the property to Carl. Carl claims that Alice, as Peter's agent, solicited a price above that agreed upon when the agency was created and that the contract is therefore binding and enforceable. Discuss fully whether Peter is bound to this contract.

15-9 Owner orally authorized Agent to sell his house. Agent completed a sale of the house to Buyer. When Buyer attempted to enforce the contract against Owner, Buyer was told that the contract was not enforceable because Owner's agency relationship with Agent was not in writing. Must Owner have given Agent written authorization?

15-10 ABC Tire Corp. hires Arnez as a traveling salesperson and assigns him a geographic area and time schedule in which to solicit orders and service customers. Arnez is given a company car to use in covering the territory. One day Arnez decides to take his personal car to cover part of his territory. It is 11 a.m., and Arnez has just finished calling on all customers in the city of Tarrytown. His next appointment is at 2 p.m. in the city of Austex, 20 miles down the road. Arnez starts out for Austex, but halfway there he decides to visit a former college roommate who runs a farm 10 miles off the main highway. Arnez is enjoying his visit with his former roommate when he realizes that it is 1:45 p.m. and that he will be late for the appointment in Austex. Driving at a high speed down the country road to reach the main highway, Arnez crashes his car into a tractor, severely injuring Thomas, the driver of the tractor. Thomas claims that he can hold ABC Tire Corp. liable for his injuries. Discuss fully ABC's liability in this situation.

15-11 Julius and Olga Sylvester owned an unimproved piece of land near King of Prussia, Pennsylvania. They were approached by Beck, a real estate broker, who asked if they were willing to sell their land, stating that an oil company was interested in buying, renting, or leasing the property. The Sylvesters said that they were only interested in selling, and they authorized Beck to sell the property for $16,000. Several weeks later, Beck phoned the Sylvesters and offered to buy the property himself for $14,000. Olga asked, "What happened to the oil company?" and Beck responded, "They are not interested. You want too much money for it." The Sylvesters sold the property to Beck. A month later, Beck sold the property to Epstein for $25,000. When the Sylvesters learned that Beck had realized a huge profit in a quick resale of the property, they sued Beck, claiming that he owed them the $9,000 profit. Does Beck owe the Sylvesters the money? Why or why not?

15-12 Peter authorized Arnon, a grain broker, to buy at the market 20,000 bushels of wheat for Peter. At the time, Arnon had in storage 5,000 bushels belonging to Johnson, who had authorized Arnon to sell for him. Arnon also had 15,000 bushels that she owned. Arnon transferred these 20,000 bushels to Peter's name and charged Peter the current market price. Shortly thereafter, and before Peter had used or sold the wheat, the market price declined sharply. Peter refused to pay for the wheat and tried to cancel the contract. Can he cancel? Explain.

CASE PROBLEMS

15-13 Bailey, an owner-operator truck driver working as an independent contractor, was transporting goods for Norco when his truck was stolen. It was found in a lake with its diesel fuel contaminating the water. Bailey called the Norco dispatcher to alert her, and Protect Environmental Services later arrived on the scene with a contract to perform cleanup services. Bailey was told that the contract needed to be signed before cleanup could begin, and he testified that the Protect representative told him that the contract was between Protect and Norco. The Protect representative asked Bailey if he was "Mr. Norco," and when Bailey said "no," Protect testified that he told Bailey to contact Norco and "obtain approval" before cleanup started. Protect then said that Bailey made a phone call, and when the call ended, Bailey said that he had received authorization to sign the contract. Bailey signed the contract believing he had authority to do so.

Protect mailed and faxed a copy of the signed contract and the invoice to Norco. Norco never responded, nor did it pay the invoice. Norco never "assert[ed] Bailey's lack of authority as the basis for non-payment," nor did Norco ever dispute the bill or contract. Protect sued Norco for payment three months later.

What elements are required for finding of apparent authority? Do you think Protect had a reasonable basis to believe that Bailey had authority to sign the contract on Norco's behalf? How do you think the court treated Norco's silence? Would Norco's failure to dispute the contract and the invoice be considered a "manifestation of acceptance"? *Protect Environmental Services, Inc. v. Norco Corp.*, 403 S.W.3d 532 (Tex. App. 2013).

15-14 Hercules, Inc., is a large chemical corporation. Its operations in Brunswick, Georgia, extracts resins from three stumps, processes the resins into chemical compounds, and sells them to manufacturers. Hercules purchases tree stumps from various parties, including D. Hays Trucking, Inc. (Hays).

Hays owns its own equipment and delivery vehicles, hires its own truckers and other employees, pays for its employees' workers' compensation coverage, and withholds federal and state taxes from employees' paychecks. Hays directs the work of its employees who pull the stumps from the ground and the truckers who deliver the stumps to Hercules.

One night Mr. Hays was driving a tractor-trailer owned by D. Hays Trucking, Inc., from Alabama to the Hercules plant in Georgia. The truck was loaded with 80,000 pounds of pine stumps. Just before midnight, when he was 10 miles from the Hercules plant, Mr. Hays crashed the tractor-trailer into a car driven by Phyllis Lewis, killing her. Mr. Hays was driving the tractor-trailer approximately 10 to 15 miles per hour over the 65-mile-per-hour speed limit, and there were no skid marks from the tractor-trailer prior to the collision. Preston Lewis, the executor of the estate of Phyllis Lewis, brought suit in U.S district court against Mr. Hays; D. Hays Trucking, Inc.; and Hercules, Inc., to recover damages for negligence. Hercules made a motion for summary judgment, alleging that D. Hays Trucking, Inc., was an independent contractor and therefore Hercules could not be held liable for its negligence. *Lewis v. D. Hays Trucking, Inc.,* 701 F. Supp. 2d 1300, 2010 U.S Dist. LEXIS 28035.

15-15 Sam and Theresa Daigle decided to build a home in Cameron Parish, Louisiana. To obtain financing, they contacted Trinity United Mortgage Company. At a meeting with Joe Diez on Trinity's behalf, on July 18, 2001, the Daigles signed a temporary loan agreement with Union Planters Bank. Diez assured them that they did not need to make payments on this loan until their house was built and that permanent financing had been secured. Because the Daigles did not make payments on the Union loan, Trinity declined to make the permanent loan. Meanwhile, Diez left Trinity's employ. On November 1, the Daigles moved into their new house. They tried to contact Diez at Trinity but were told that he was unavailable and would get back to them. Three weeks later, Diez came to the Daigles's home and had them sign documents that they believed were to secure a permanent loan but were actually an application with Diez's new employer. Union filed a suit in a Louisiana state court against the Daigles for failing to pay on its loan. The Daigles paid Union, obtained permanent financing through another source, and filed a suit against Trinity to recover the cost. Who should have told the Daigles that Diez was no longer Trinity's agent? Could Trinity be liable to the Daigles on this basis? Explain. *Daigle v. Trinity United Mortgage, LLC,* 890 So.2d 583 (La. Ct. App. 2004).

15-16 Juanita Miller filed a complaint in an Indiana state court against Red Arrow Ventures, Ltd., Thomas Hayes, and Claudia Langman, alleging that they had breached their promise to make payments on a promissory note issued to Miller. The defendants denied this allegation and asserted a counterclaim against Miller. After a trial, the judge announced that although he would be ruling against the defendants, he had not yet determined what amount of damages would be awarded to Miller. Over the next three days, the parties' attorneys talked and agreed that the defendants would pay Miller $21,000. The attorneys exchanged correspondence acknowledging this settlement. When the defendants balked at paying this amount, the trial judge issued an order to enforce the settlement agreement. The defendants appealed to a state intermediate appellate court, arguing that they had not consented to the settlement agreement. What is the rule regarding the authority of an agent to agree to a settlement? How should the court apply the rule in this case? Why? *Red Arrow Ventures, Ltd. v. Miller,* 692 N.E.2d 939 (Ind. Ct. App. 1998).

THINKING CRITICALLY ABOUT RELEVANT LEGAL ISSUES

The only agency relationships that the courts should recognize are those created by agreement. Agency relationships are like contracts. There must be a meeting of the minds for a contract to be valid, and this standard also ought to apply to agency relationships.

Agency relationships created by implied authority, ratification, or estoppel all originate with either an ambiguous agreement or a false statement. In the first instance, the ambiguity surrounding the agreement prevents both parties from knowing if there has been a meeting of the minds. This situation may be likened to two people who speak different languages negotiating a contract. Each will never know if the other is willing to grant his or her desires.

The second scenario violates a notion that civilized human beings hold very dear: that we should tell the truth. By allowing agency relationships that began with one party's telling a lie, we encourage dishonesty. People are motivated to lie about their agency relationships out of hope that the other party will accept the lie. It's the same as telling all of your friends that you are engaged to someone to pressure that person into accepting your proposal for marriage.

Finally, by allowing such relationships the opportunity to become legally valid, we are merely opening the door for more lawsuits. Our courts' dockets are already full; we should work to reduce the court workload, not to increase it. To create a better society, we must treat agency relationships like other contracts.

The author of this passage relies heavily on analogies in his reasoning. Locate these analogies and evaluate how reasonable they are. (Remember, the way to evaluate an analogy is to state ways in which the two things being compared are similar and then to state ways in which they are different.)

1. What is the author's conclusion, and with what reasons does he support it?
2. Is there any additional information that would aid you in your task of evaluating this argument? Explain.
3. State an argument that is the opposite of the one set out by the author here.

ASSIGNMENT ON THE INTERNET

Using the first website that follows or another website of your choosing, find a recent case involving principal–agent law. Read the case and determine the issue being decided, the court's holding or conclusion, and the reasons cited for the holding. Next, identify the ideas discussed in this chapter that were used in the court's reasoning. Do you agree with the court's reasoning? Why or why not?

ON THE INTERNET

http://www.hcch.net/index_en.php?act=conventions.text&cid=89 This site contains an international treaty on agency law.

https://www.law.cornell.edu/topics/agency.html The Legal Information Institute's site on agency contains a brief overview and links to recent court cases involving agency law.

FOR FUTURE READING

DeMott, Deborah A. "When Is a Principal Charged with an Agent's Knowledge?" *Duke Journal of Comparative & International Law* 13 (2003): 291.

Jones, Aaron D. "Corporate Officer Wrongdoing and the Fiduciary Duties of Corporate Officers under Delaware Law." *American Business Law Journal* 44 (2007): 275.

CHAPTER SIXTEEN

Law and Business Associations—I

- FACTORS INFLUENCING A BUSINESS MANAGER'S CHOICE OF ORGANIZATIONAL FORM
- SOME COMMON FORMS OF BUSINESS ORGANIZATION IN THE UNITED STATES
- SPECIALIZED FORMS OF BUSINESS ASSOCIATIONS
- GLOBAL DIMENSIONS OF BUSINESS ASSOCIATIONS

The world of business is now more complex than ever before. Businesses vary greatly in size and organizational form. In Chapters 16 and 17, we outline the differences and the advantages and disadvantages of each form. See Table 16-1 for factors influencing a manager's choices of organizational form.

TABLE 16-1 COMPARISON OF ALTERNATIVE FORMS OF BUSINESS ORGANIZATION

Organizational Form	Tax Ramifications	Control Considerations	Liability	Ease and Expense of Formation	Transferability of Ownership Interests	Lifetime
Sole proprietorship	Profits taxed directly to proprietor as ordinary income and losses deducted by proprietor.	Sole proprietor has total control.	Sole proprietor has unlimited personal liability.	No formalities or expenses required other than those specific to the business to be operated.	Nontransferable.	Limited to life of proprietor.
General partnership	A federal income tax return must be filed for information only. Profits taxed to partners as ordinary income, and losses deducted by partners. Profits and losses shared equally unless changed by partnership agreement.	Each partner is entitled to equal control. Can be changed by partnership agreement.	Each partner has unlimited personal liability for debts of the partnership.	No formalities or expenses required other than those specific to the business to be operated. Written agreement advisable.	Nontransferable.	Limited to life of partners.

TABLE 16-1 CONTINUED

Organizational Form	Tax Ramifications	Control Considerations	Liability	Ease and Expense of Formation	Transferability of Ownership Interests	Lifetime
Limited partnership (limited liability partnership)	Same as general partnership.	Same as for general partners. Limited partners cannot take part in management.	Same as for general partners. Liability of limited partner is limited to his or her capital contribution.	Added expense and time required to draw up and file written partnership agreement. Failure to comply with formalities will result in loss of limited partnership.	Nontransferable.	Limited to life of general partners.
Public corporation	Profits taxed as income to corporation and again as income to owners when distributed as dividends.	Separation of ownership and control. No control over daily management decisions.	Liability limited to loss of capital contribution.	Expense and time required to comply with statutory formalities. Must receive charter from state; usually required to register and pay fees to operate in states other than state of incorporation.	Generally unlimited, except by shareholder agreement.	Unlimited.
Subchapter S corporation	Taxed as a partnership.	Separation of ownership and control. No control over daily management decisions.	Liability limited to loss of capital contribution.	Same as for public corporation; must follow IRS rules carefully or lose Subchapter S status.	Ownership interests limited to no more than 35 shareholders; all must be individuals, estates, or certain types of trusts, and no shareholder may be a nonresident alien. Shares may not be issued or transferred to more than 35 shareholders.	Unlimited.
Limited liability company	Taxed as a partnership.	Control over daily management decisions.	Liability limited to loss of capital contribution.	Same as for public corporation; must follow IRS rules carefully or lose limited liability status.	No limitation on number of members.	Unlimited.

This chapter outlines some of the common forms of business associations, including simple proprietorships, partnerships, and specialized associations (joint stock, joint venture, and franchises). In addition, global dimensions of some of these business associations will be examined here.

CRITICAL THINKING ABOUT THE LAW

We point out in this chapter that the organizational form of a business will determine the amount of regulation that business experiences. Furthermore, business owners will have different rights and responsibilities according to the organizational form. Although you will learn about the major organizational forms later in this chapter, you can prepare to use your critical thinking skills while you consider business associations by asking yourself the following questions.

1. This question is not a formal critical thinking question, but it lays the groundwork for critical thinking about forms of business organization. Think about the businesses with which you interact every day. For example, where do you buy groceries? Who cuts your hair? Where did you buy this book? Why should the community care whether these businesses are owned by a single person or by hundreds of shareholders?

    Clue: Ask yourself what the community expects from businesses and how these different forms of ownership affect the extent to which those expectations are fulfilled.

2. Joan wants to open a business that specializes in selling fine wines. She believes that as long as she follows the law of selling alcohol only to individuals 21 years of age or older, the government should not impose any other regulations on her business activity. What ethical norm seems to dominate Joan's thinking?

    Clue: Think about the list of ethical norms in Chapter 1. Which norm seems most consistent with little or no governmental regulation? Which ethical norm seems to conflict with the idea of little governmental regulation?

3. One of the factors that might influence business owners to choose one organizational form over another is the liability associated with that organizational form. For example, in a sole proprietorship, the owner is liable for all losses, whereas owners in a corporation have limited liability. Vanessa has a chance to be either a sole proprietor or a shareholder in a corporation. She chooses to become a sole proprietor, even though she considers this option to be riskier. Which ethical norms might be guiding her decision to become a sole proprietor?

    Clue: Look closely at the list of ethical norms. Which norms seem consistent with Vanessa's choice? Can you identify any other ethical norms that may be influencing her decision?

Factors Influencing a Business Manager's Choice of Organizational Form

There is no ideal form for a business venture. Each of the forms we discuss in this chapter has advantages and disadvantages. The entrepreneur, with the counsel of an attorney and an accountant or a tax expert, should carefully weigh the advantages and disadvantages of different organizational forms for the type of business the entrepreneur wishes to engage in. The principal factors influencing the choice of organizational form are (1) tax ramifications, (2) control considerations, (3) potential liability of the owner(s), (4) ease and expense of formation and operations, (5) transferability of ownership interests, and (6) projected life of the organization. Table 16-1 gives an overview of how these six factors differ for each organizational form.

Some Common Forms of Business Organization in the United States

Some common forms of business organization in the United States are the sole proprietorship, the partnership, the corporation, and the specialized businesses (joint stock, joint venture, and franchising). We examine each of these organizational forms in this chapter and Chapter 17.

SOLE PROPRIETORSHIPS

When a single proprietor seeks to go into business for himself or herself (alone), the person must examine the following: tax ramifications, control considerations,

TABLE 16-2 ALTERNATIVE FORMS OF BUSINESS ORGANIZATION

Sole proprietorship	Person going into business on his or her own; responsible for all profits and losses.
Partnership (general)	Voluntary association of two or more persons to carry on a business as co-owners for profit. A limited liability partnership is a voluntary association, as is a general partnership, but one or more partners contribute capital only, and those partners play no role in management. Their liability is limited to the amount of capital they contribute.
Corporation	A legal entity created by state law that raises capital by issuing stocks and bonds to investors, who become shareholders and owners of the corporation. Corporations can be classified as public or private and are further distinguished as multinational, professional, closely held, Subchapter S, or nonprofit.

liability, ease and expense of formation, transferability of ownership, and projected lifetime of the business. Each of these is outlined in Table 16-2.

The **sole proprietorship** is the easiest and least expensive way to create a business organization. After obtaining any licenses that are required by state and local law, an individual is in business. This is a distinct advantage over the other forms to be discussed in this chapter and in Chapter 17.

sole proprietorship
A business owned by one person who has sole control over management and profits.

A sole proprietor generally pays only personal income taxes (inclusive of Social Security and Medicare taxes) on profits. Sole proprietor's profits are reported on his or her personal income tax form. An individual can also set up retirement accounts for the owner and any employees.

A sole proprietor may be held personally liable for the debt of the business. Often other forms of business organizations are used to avoid this risk. The ease of transferring ownership of a single proprietorship is a potential advantage. Less documentation is needed than for transfer of other business forms and interests.

Some advantages (pros) and disadvantages (cons) of sole proprietorships over other forms of business organization are summarized here and in Table 16-1.

Advantages (Pros)	*Disadvantages (Cons)*
(1) Only one individual is involved—ease of formation.	(1) Personal liability for almost all assets, both personal and business.
(2) Total control belongs to one individual. The individual may dissolve the sole (single) proprietorship at any time.	(2) Individual pays all taxes on profits made by business. Ownership cannot be transferred.
(3) Sole proprietorships are subject to less government regulation at all levels.	(3) As a single proprietorship grows, it must gain maturity as to regulation, and management thereof, to be successful. Often, individuals lack the sophistication to make the next step and take the business to the next level.

CASE 16-1

Quality Car & Truck Leasing, Inc. v. Sark
Court of Appeals of Ohio, Fourth District
2013 WE 13959

Michael Sark operated a logging business as a sole proprietorship. To acquire equipment for the business, Sark and his wife, Paula, borrowed funds from Quality Car & Truck Leasing, Inc. When his business encountered financial difficulties, Sark became unable to pay his creditors, including Quality. The Sarks sold their house (valued at $203,500) to their son, Michael, Jr., for $1 but continued to live in it.

Three months later, Quality obtained a judgment in an Ohio state court against the Sarks for $150,481.85 and then filed a claim to set aside the transfer of the house to Michael, Jr., as a fraudulent conveyance. From a decision in Quality's favor, the Sarks appealed, arguing that they did not intend to defraud Quality and that they were not actually Quality's debtors.

Judge Kline

The trial court found that summary judgment was proper under [Ohio Revised Code (R.C.) Section 1336.04(A)(2)(a)]. That statute provides as follows:

> A transfer made or an obligation incurred by a debtor is fraudulent as to a creditor, whether the claim of the creditor arose before or after the transfer was made or the obligation was incurred, if the debtor made the transfer or incurred the obligation without receiving a reasonably equivalent value in exchange for the transfer or obligation, and the debtor was engaged or was about to engage in a business or a transaction for which the remaining assets of the debtor were unreasonably small in relation to the business or transaction.

The trial court found "that Michael Senior and Paula made a transfer without the exchange of reasonably equivalent value and that the debtor was engaged or was about to engage in a business . . . transaction for which the remaining assets of the debtor were unreasonably small in relation to the business or transaction."

The Sarks argue that the summary judgment was not proper because there is a genuine issue of material fact regarding whether they intended to defraud Quality Leasing. The Sarks argument fails because intent is not relevant to an analysis under R.C. Section 1336.04(A)(2)(a). A creditor does not need to show that a transfer was made with intent to defraud in order to prevail under R.C. Section 1336.04(A)(2)(a). Thus, the Sarks cannot defeat summary judgment by showing that they did not act with fraudulent intent when Michael Senior and Paula transferred the property to Michael Junior.

The Sarks also claimed that summary judgment was improper because there is an issue of fact regarding whether Michael Senior and Paula are actually Quality Leasing's debtors. Michael Senior apparently returned the equipment that secured the debts owed to Quality Leasing. According to the Sarks, Quality Leasing's appraisals of the equipment showed that the value of the equipment would be enough to satisfy the debts.

The Sarks' argument, however, does not address the fact that they are clearly judgment debtors to Quality Leasing and that the judgment has not been satisfied. The Sarks have not challenged the validity of the judgment against them nor have they shown that the judgment has been satisfied. Thus, there is no genuine issue of material fact regarding whether Paula and Michael Senior are debtors to Quality Leasing.

In conclusion, there is no genuine issue as to any material fact. Quality Leasing is entitled to judgment as a matter of law.*

State Court of Appeals *affirmed*.

GENERAL PARTNERSHIPS

partnership A voluntary association of two or more persons formed to carry on a business as co-owners for profit.

When two or more people wish to be involved in the ownership of a business, either a limited or a general partnership may be formed. Under the Uniform Partnership Act (UPA), the law that governs partnerships in most states, a **partnership** is defined as a voluntary association of two or more persons formed to carry on a business as co-owners for profit. As in a sole proprietorship, the profits of a partnership are taxed only as income to the partners. More recently, the Revised Uniform Partnership Act (RUPA) has been adopted in more than 35 states (others continue to follow the UPA). The RUPA states that a partnership need not dissolve just because a partner leaves (in contrast to the UPA), clarifies the fiduciary duties of partners, and establishes a formula for valuing a partnership interest during a buyout. It also provides greater protection for the limited liability partner (to be discussed later in this chapter).

general partnership A partnership in which management responsibilities and profits are divided (usually equally) among the partners and all partners have unlimited personal liability for the partnership's debts.

In a **general partnership**, all profits are usually divided equally among the partners, and all partners have unlimited personal liability for the partnership's debts. Also, management responsibilities may be, but are not always, divided equally. For example, if James and Carol decide to operate a florist shop as co-owners in a general partnership, each of them makes an initial contribution in the form of cash, realty, business supplies, or services. They take an equal role in the management of the business, with each expected to

---

*Quality Car & Truck Leasing, Inc. v. Sark Court of appeals of Ohio, Fourth District, 2013 WE 13959. https://www.supremecourt.ohio.gov/rod/docs/pdf/4/2013/2013-ohio-44.pdf.

contribute services. They split the profits equally (assuming that no other proportion has been specified in a written partnership agreement). At the end of the first year, they file a partnership return with the Internal Revenue Service (IRS) that shows the partnership's profit or loss. If the partnership makes a profit, James and Carol each pay income tax on half the profits; if the partnership incurs a loss, each deducts one-half of that loss from his or her ordinary income. The partnership itself pays no taxes. If the business fails, both James and Carol can be sued by creditors and forced to pay the partnership's debts out of their personal resources.

Creating a Partnership. In many partnerships, there is initially no written partnership agreement; the partners informally split the capital contribution and work between them. In the absence of a written agreement, the Revised Uniform Partnership Act of 1997, RUPA, controls. For example, the RUPA requires partners to share profits equally. Now, James and Carol from our preceding example may not have intended an equal distribution of profits when they went into business, and such a distribution may be unfair because Carol does 90 percent of the work and the less-energetic James does a mere 10 percent. However, if they have no written agreement, James can legally claim 50 percent of the profits. To avoid conflicts over management responsibilities, borrowing power, profit sharing, and other common bones of contention in a partnership, the partners should set out the rights and responsibilities of each partner in a written partnership agreement at the outset. Exhibit 16-1 lists some of the items that should be included in such an agreement.

Most of the RUPA consists of rules that will apply unless the partnership agreement states differently. This will force all partnerships to be thoughtfully and carefully created. All parties engaged in an existing partnership or in creating a new one should be careful to see if the RUPA has been adopted in the state where the partnership agreement is formed. As of this writing, the RUPA has been adopted by a majority of states, as well as the District of Columbia, Puerto Rico, and the U.S. Virgin Islands. If so, the RUPA will determine when a partnership exists based on three factors: (1) whether there exists a sharing of profits and losses, (2) a joint ownership of the business, and (3) an equal opportunity to participate in the management of the business. This test of partnership existence is illustrated by the next case.

EXHIBIT 16-1

ITEMS INCLUDED IN PARTNERSHIP AGREEMENTS

- Name and address of partnership
- Name and address of partners
- Purpose of partnership
- Duration of partnership
- Amount and type of investment of each partner (e.g., cash, realty, services)
- Loans to partnership
- How profits and losses are to be shared
- Management and voting power of each partner
- Method of settling disputes that should arise
- Cross-insurance of partners
- Duties of each partner
- How books are to be set up and maintained
- Banking arrangements—who has authority to deposit and withdraw
- Who has authority to borrow money in the name of the partnership
- Who does the hiring and firing of employees

CASE 16-2

In re KeyTronics
Supreme Court of Nebraska
744 N.W.2d 475 (2008)

In 1999, King was doing business under the name of "Washco" as a sole proprietorship engaged in selling, installing, and servicing car wash systems and accessories. King offered to his customers the "QuikPay" system, a cashless vending system for car washes that used a memory-chip key that interacted with a controller at the car wash. A cash value can be placed on the key, or the car wash usage recorded on the key can be billed monthly. Washco purchased QuikPay systems for resale from Datakey Electronics, Inc. (Datakey), but the arrangement was becoming unprofitable for Datakey, partly because the keys for QuikPay could only be obtained from an attendant. According to Glen Jennings, president of Datakey, because most car washes are unattended, this reliance on the presence of the car wash owner or employee was limiting the market for the product.

As QuikPay's largest distributor, King was aware that QuikPay's limitations made the product unattractive to many of his customers. King contacted Willson, an electronics technician and computer programmer, to see if Willson could develop a combined "key dispenser" and "revalue station" for the QuikPay system that would make the system self-service. King also asked Willson if he would design and install an interface between the QuikPay system and the car wash of one of King's customers; designing such an interface was beyond King's technical expertise. Willson individually designed and installed at least four specific customer interfaces that allowed King to sell the QuikPay system to those customers, but Willson was never paid for his work.

According to King, there was an oral agreement among himself, Willson, and Scott Gardeen (an employee of Datakey who was an original designer of QuikPay) to form a corporation whenever Willson developed the key dispenser-revalue station. The three parties met in the spring of 2002 to discuss a venture in which they would design and build the key dispenser-revalue station and sell it to Datakey. It was agreed that Willson would write the software and do the firmware, hardware, and any other electrical or software work; Gardeen would contribute his knowledge of the system and his contact with Datakey; and King would contribute financial resources and his experience and contacts as QuikPay's largest distributor. Together, Willson, King, and Gardeen came up with the name "Secure Data Systems" for their business. (They also discussed the fact that the entity's initials, "SDS," were the initials of their first names: Scott, Don, and Scott.) By the summer, Willson had built a handheld revalue station for a meeting with Jennings. Jennings indicated that if a final, marketable key dispenser-revalue station were developed, Datakey would be interested in a business relationship with Secure Data Systems.

Willson estimated that he had put at least 2,000 hours into QuikPay sales and maintenance and development of the key dispenser-revalue station. When Willson was asked why he invested his time and expertise into QuikPay without any remuneration, he explained, "That was my contribution to the company. I mean that was my piece." Willson contacted a law firm to draw up papers to formalize the partnership. These papers were never drafted. According to Willson, when he told King he was looking into creating a written agreement for their relationship, King "assured [him] that he was having his attorneys look at it." King and Willson had another meeting around the end of December and agreed to end their relationship and any joint QuikPay or key dispenser-revalue station activities. Approximately two weeks after this meeting, King called Willson and offered to compensate him for the time he had spent in maintaining or repairing QuikPay. Willson refused.

Willson brought an action for winding-up and an accounting, alleging formation of a partnership. King denied that they had formed a partnership. The trial court found that King and Willson had "pooled resources, money and labor," but found that no partnership existed because there was no "specific agreement." Alternatively, the trial court found that because King did not commit his preexisting business to any specifically formed partnership, the scope of the partnership did not encompass any activity garnering profits. Willson appealed the trial court's order.

Justice McCormack

This case is governed by the RUPA. Section [202(a)] of the Act defines that a partnership is formed by "the association of two or more persons to carry on as co-owners a business for profit" and explains that this is true "whether or not the persons intend to form a partnership." . . .

Obviously, the relationship between King and Willson is "of two or more persons." In addition, whether the business of QuikPay maintenance, or even the development of the never-produced key dispenser-revalue station, qualifies as a business "for profit" is not in issue. It is not essential that the business for which the association was formed ever actually be carried on, let alone that it earn a profit. Without Willson's technical assistance, King would have been unable to continue QuikPay's viability after Datakey abandoned the product. That King could have dealt with certain issues by hiring contractors or employees is irrelevant. He chose not to do so—presumably because the promise of the key dispenser-revalue station made a partnership relationship more worthwhile—and saved himself the expense of paying for this labor.

We also find that despite King's protestations to the contrary, the evidence shows that King and Willson shared control over QuikPay business. We note that control is "elusive because of the many gradations of control and because partners often delegate decision-making power." ... Still, Willson testified that he and King consulted with each other over what appropriate pricing would be as they picked up Datakey's equipment and customers.

Willson also testified that he had an agreement with King to share profits, although King denies this. Of the five indicia of co-ownership, profit sharing is possibly the most important, and the presence of profit sharing is singled out in [Section 202(c)(3)] as creating a rebuttable presumption of a partnership. However, what is essential to a partnership is not that profits actually be distributed, but, instead, that there be an interest in the profits. Willson's testimony that they agreed to share in the profits of the business is, in light of all the evidence, simply more credible than King's statement that compensation "was never discussed." And even King vaguely admits that they had an understanding to share profits of the key dispenser-revalue station, if that were developed. It seems reasonable to assume that this same understanding would apply to Willson as his participation and the scope of the venture expanded to encompass all QuikPay business.

We do not find any evidence that King and Willson had an agreement for loss sharing. But we find this of little import, since purported partners, expecting profits, often do not have any explicit understanding regarding loss sharing. Likewise, although King and Willson admittedly do not own any joint property, in an informal relationship, the parties may intend co-ownership of property but fail to attend to the formalities of title. Moreover, in this case, it is unclear that there is much QuikPay "property" at all.

We conclude that the objective, as well as subjective, indicia are sufficient to prove co-ownership of the business of selling, maintaining, and developing QuikPay. Having already concluded that there was an association for the same, we conclude that Willson proved that he and King had formed a partnership for the business of selling, maintaining, and developing QuikPay.*

Reversed and *remanded* for Willson and QuikPay.

Relationship between Partners. The RUPA states that each partner has a fiduciary relationship to the partnership and must act in good faith for the benefit of the partnership. In most general partnerships, each partner has one vote in decisions pertaining to the management of the business, although in some instances—such as a decision to merge with another partnership—a unanimous vote may be required. The fiduciary relationship (duty of loyalty) and a duty of good faith and fair dealings by partners are established by the RUPA, as demonstrated in the following case.

CASE 16-3

Enea v. Superior Court of Monterey County
Court of Appeals of California, Sixth Appellate District
34 Cal. Rptr. 3d 513 (2005)

In 1980, Defendants William Daniels and Claudia Daniels, and other family members, formed a general partnership known as 3-D. The partnership's sole asset was a building that had been converted from a residence to offices. A portion of the property had been rented since 1981, on a month-to-month basis, by the law practice of William Daniels (the firm's sole member). From time to time, the property was rented on similar arrangements to others, including defendant Claudia Daniels. The partnership agreement has as its principal purpose the ownership, leasing, and sale of the only partnership asset—the building. The partnership agreement contained no provision that the property would be leased for fair market value. The defendants asserted that there was no evidence of any agreement to maximize rental profits.

In 1993, plaintiff Benny Enea, a client of William Daniels, purchased a one-third interest in the partnership from William's brother, John P. Daniels. In 2001, however, the plaintiff questioned William Daniels about the rents being paid for the property, and in 2003, the plaintiff was "dissociated" from the partnership.

On August 6, 2003, the plaintiff brought an action for damages, alleging that the defendants had occupied the partnership property while paying significantly less than fair rental value, in breach of their fiduciary duty to the plaintiff. The trial court granted the defendants' motion for summary judgment, and the plaintiff appealed.

Justice Rushing

For present purposes it must be assumed that defendants in fact leased the property to themselves, or associated entities, at below-market rents. . . . Therefore the sole question presented is whether defendants were

---
*In re KeyTronics, Supreme Court of Nebraska 744 N.W.2d 475 (2008).

categorically entitled to lease partnership property to themselves, or associated entities (or for that matter, to anyone) at less than it could yield in the open market. Remarkably, we have found no case squarely addressing this precise question. We are satisfied, however, that the answer is a resounding "No."

The defining characteristic of a partnership is the combination of two or more persons to jointly conduct business. It is hornbook law that in forming such an arrangement the partners obligate themselves to share risks and benefits and to carry out the enterprise with the highest good faith toward one another—in short, with the loyalty and care of a fiduciary. "Partnership is a fiduciary relationship, and partners are held to the standards and duties of a trustee in their dealings with each other."

"[I]n all proceedings connected with the conduct of the partnership every partner is bound to act in the highest good faith to his copartner and may not obtain any advantage over him in the partnership affairs by the slightest misrepresentation, concealment, threat or adverse pressure of any kind." Or to put the point more succinctly, "Partnership is a fiduciary relationship, and partners may not take advantages for themselves at the expense of the partnership."

Here the facts as assumed by the parties and the trial court plainly depict defendants taking advantages for themselves from partnership property *at the expense of the partnership*. The advantage consisted of occupying partnership property at below-market rates, i.e., less than they would be required to pay to an independent landlord for equivalent premises. The cost to the partnership was the additional rent thereby rendered unavailable for collection from an independent tenant willing to pay the property's value.

Defendants . . . persuaded the trial court that they had no duty to collect market rents in the absence of a contract expressly requiring them to do so. This argument turns partnership law on its head. Nowhere does the law declare that partners owe each other only those duties they explicitly assume by contract. On the contrary, the fiduciary duties at issue here are *imposed by law,* and their breach sounds in tort.*

The trial court's order is set aside and the motion for summary judgment by the defendant is *denied*.

CRITICAL THINKING ABOUT THE LAW

The decision in this case clearly describes the purpose and expectations associated with a partnership. For the defendants to have succeeded in this case, partnership duties would have to consist only of those laid out in explicit contract form. The act of forming a partnership in and of itself, however, includes by law certain expectations.

1. What facts would have to have existed for the defendants to be granted a summary judgment?

    Clue: What does the court say at the start of its final paragraph?

2. The defendants are assuming that the duties of the partners to the business association are something about which the parties to the agreement must negotiate. What does the court mean in this regard when it says that the duties are "imposed by law"?

    Clue: Once a partnership agreement is signed, does the very fact of its existence as a partnership mean that certain duties exist?

Read Exhibit 16-1 carefully, for it will give you a good idea of the issues that should be resolved before people enter into a business partnership. The relationship between partners will go more smoothly if the parties agree beforehand on how much each will invest, the management duties each will undertake, methods of dispute resolution, banking arrangements and borrowing policies for the business, and how the books will be kept. See Exhibit 16-2 Model Partnership Agreement, which is a model form for a general partnership agreement. It does not contain all variables that might be wise to include for specific kinds of partnerships, but it is a useful starting point for drafting such an agreement. See Exhibit 16-3 for a summary of duties and rights between and among partners.

Terminating a Partnership. A partnership, unlike a corporation, does not have perpetual existence. It can "die" when partners leave the partnership, the partnership is merged with another business or goes bankrupt, or the partnership agreement expires.

---

*Enea v., Superior Court of Monterey County, Court of Appeals of California, Sixth Appellate District 34 Cal. Rptr. 3d 513 (2005). http://caselaw.findlaw.com/ca-court-of-appeal/1150082.html.

On its "deathbed," a partnership goes through a process called *dissolution and winding-up*. *Dissolution* prevents any new business from taking place after partners have initiated termination proceedings. **Winding-up** involves completing all unfinished transactions, paying off debts, dividing any remaining profits, and distributing assets. When winding-up is complete, the partnership's legal existence is terminated.

winding-up The process of completing all unfinished transactions, paying off outstanding debts, distributing assets, and dividing remaining profits after a partnership has been terminated or dissolved.

EXHIBIT 16-2

MODEL GENERAL PARTNERSHIP AGREEMENT FORM

MODEL GENERAL PARTNERSHIP AGREEMENT FORM
Kubasek-Brennan-Browne
PARTNERSHIP AGREEMENT

This agreement, made and entered into as of the [Date], by and among Kubasek-Brennan-Browne (referred to as "Partners").

WITNESSETH:

Whereas, the Parties hereto desire to form a General Partnership (hereinafter referred to as the "Partnership"), for the term and upon the conditions hereinafter set forth.

Now, therefore, in consideration of the mutual covenants hereinafter contained, it is agreed by and among the Parties hereto as follows:

Article I
BASIC STRUCTURE

Form. The Parties hereby form a General Partnership pursuant to the Laws of the State of Newgarth.

Name. The business of the Partnership shall be conducted under the name of Kubasek-Brennan-Browne.

Place of Business. The principal office and place of business of the Partnership shall be located at 2130 Foot Street, Justin, Newgarth, or such other place as the Partners may from time to time designate.

Term. The Partnership shall commence on [Date], and shall continue for [Number] years, unless earlier terminated in the following manner:
  (a) By the completion of the purpose intended, or
  (b) Pursuant to this Agreement, or
  (c) By applicable Newgarth law, or
  (d) By death, insanity, bankruptcy, retirement, withdrawal, resignation, expulsion, or disability of all of the then Partners.

Article II
MANAGEMENT

Managing Partners. The Managing Partner(s) shall be all partners.

Voting. All Managing Partner(s) shall have the right to vote as to the management and conduct of the business of the Partnership according to their then Percentage Share of [Capital/Income]. Except otherwise herein set forth a majority of such [Capital/Income] shall control.

Percentage Share of Profits. Distribution to the Partners of net operating profits of the Partnership shall be made quarterly in the percentage agreed upon (40%, 40%, 20%).

Article III
DISSOLUTION

Dissolutions. In the event that the Partnership shall hereafter be dissolved for any reason whatsoever, a full and general account of its assets, liabilities, and transactions shall at once be taken. Such assets may be sold and turned into cash as soon as possible and all debts and other amounts due the partnership collected. The proceeds thereof shall thereupon be applied as follows:

EXHIBIT 16-2

CONTINUED

(a) To discharge the debts and liabilities of the Partnership and the expenses of liquidation.

(b) To pay each Partner or his legal representative any unpaid salary, drawing account, interest or profits to which he shall then be entitled and in addition, to repay to any Partner his capital contributions in excess of his original capital contribution.

(c) To divide the surplus, if any, among the Partners or their representatives as follows: (1) First (to the extent of each Partner's then capital account) in proportion to their then capital accounts;

(2) Then according to each Partner's then Percentage Share of Capital/Income.

Right to Demand Property. Each partner shall have the right to demand and receive property in kind for his distribution.

Article IV
MISCELLANEOUS

Accounting Year, Books, Statements. The Partnership's fiscal year shall commence on January 1st of each year and shall end on December 31st of each year. Full and accurate books of account shall be kept at such place as the Managing Partner(s) may from time to time designate, showing the condition of the business and finances of the Partnership; and each Partner shall have access to such books of account and shall be entitled to examine them at any time during ordinary business hours.

Arbitration. Any controversy or claim arising out of or relating to this Agreement shall only be settled by arbitration in accordance with the rules of the American Arbitration Association's chosen Arbitrator, and shall be enforceable in any court having competent jurisdiction.

Witness	Partners
J. Foster	Kubasek-Brennan-Browne

EXHIBIT 16-3

RELATIONSHIP AMONG PARTNERS*

*The UPA and RUPA as interpreted by state case law have largely established these duties.

Duties among Partners
- *Fiduciary Duties*—include duty of loyalty, good faith, fairness, duty not to appropriate partnership opportunities, not to compete in some instances, not to have conflicts of interest, and not to reveal confidential information of the partnership.
- *Duty of Obedience*—duty to act in accord with partnership agreement.
- *Duty of Care*—duty to manage the affairs of the partnership without gross negligence, reckless, or intentional misconduct.

Rights among Partners
- *Partnership Property*—the right to possess and use for partnership purposes.
- *Transferability of Interest in Partnership Property*—the right to share of profits and losses of the partnership.
- *Management of Property*—partners have equal rights unless otherwise agreed on pursuant to the partnership agreement.
- *Information*—each partner has the right to information necessary to carry out rights of the partnership.

APPLYING THE LAW TO THE FACTS...

Mandy, Cassie, and Adam formed the MCA partnership to operate a bakery and signed a 10-year lease on a property where they planned to operate their business. After five years, the business is a failure and the partnership dissolves, but it does not have sufficient funds to pay money they owe for the remaining five years of the lease. From whom, if anyone, can the property owner collect the remaining money he is owed under the lease?

Partnerships: Advantages and Disadvantages

Advantages (Pros)	Disadvantages (Cons)
(1) Easily formed when contract is between two persons. Costs are relatively small.	(1) Most partnerships, unlike corporations, do not have perpetual existence. Partnerships can be dissolved at almost any time, not simply by virtue of partner death or withdrawal. A partner can dissociate himself or herself from a partnership by not carrying out the partnership business.
(2) No taxes; tax is usually not levied on the partnership itself, but on the profits and based on individual partners' conditions when the latter file taxes.	(2) Each partner has limited liability in most instances.
(3) Partnerships can operate in almost all states and are usually subject to less state and local regulations than are corporations.	(3) Partners are taxed on their share of partnership profit.
(4) A partnership can allow for a wide variety of profit-sharing and operational arrangements, as long as the operation or arrangement is not illegal.	(4) Partnership property is separate from the property of the individual partners, but the latter property may become partnership property if the instrument transferring an individual partner's property makes reference to the partnership.

LIMITED PARTNERSHIPS AND LIMITED LIABILITY LIMITED PARTNERSHIPS

Limited partnerships have at least one general partner and one limited partner and are easily identifiable because they must include the word *limited* (or an abbreviation of that word). Strictly following the statutory scheme for the formation of a limited partnership, general and limited partnerships function similarly (see Exhibit 16-4). The primary law governing limited partnerships is the Revised Uniform Limited Partnership Act (RULPA), which has been adopted by 48 states. The question of how much management activity a limited partner can engage in before losing the special status granted by statute is still unsettled.

Limited Liability Limited Partnership. A limited liability limited partnership (LLLP) differs and is distinguished from a limited partnership in that liability is the same for a general partner(s) as for a limited partner(s). The liability of each is limited to the amount of capital contributed. Some states provide by statute for LLLPs. Others may allow a limited partnership to register with the state (usually the secretary of state) as an LLLP.

A 2001 revision to the RULPA provides that an LLLP "means a limited partnership whose certificate of limited partnership states that the limited partnership is a limited liability limited partnership." Under the 2001 revision, a limited partner cannot be held liable for the partnership debts even if he or she participates in the management and control of the limited partnership.

limited partnership A partnership that has one general partner who is responsible for managing the business and one or more limited partners who invest in the partnership but do not participate in its management and whose liability is limited to the amount of capital they contribute.

APPLYING THE LAW TO THE FACTS...

Mandy, Cassie, and Adam would like to enter into a partnership. They all want to share equally in the profits, management, and risk. What type of partnership should they form? What if, after much discussion, Adam says that he really doesn't want to have any say in management and that he doesn't want the unlimited liability associated with a partnership but he'd still like to invest in his friends' business? What type of partnership should they form? How will creditors know that Adam doesn't have unlimited liability?

EXHIBIT 16-4

COMPARISON OF SOME FACTORS OF GENERAL PARTNERSHIPS AND LIMITED PARTNERSHIPS

Factors	General Partnerships	Limited Partnerships
(1) Creation	By contractual agreement between two or more persons to carry on a business as co-owners for profit.	Same, but most include one or more general partners and one or more limited partners on certificate filed with the state.
(2) Profits and losses	By agreement; or, in absence of an agreement, are shared equally	Profits are shared equally, as required by certificate, and losses are shared up to the amount of each limited partner's capital contribution.
(3) Liability	Unlimited personal liability.	Limited liability for general partners; for others, limited to amount of capital contribution.
(4) Capital contribution	Usually no minimum or mandatory amount, unless set by a partnership agreement.	Usually set by agreement.
(5) Management	Specified in the partnership agreement; or, in the absence of an agreement, each has an equal role.	General partners only. If limited partners have some voting right, they will generally be subject to liability. Sometimes limited partners act as agents.

Specialized Forms of Business Associations

Table 16-3 summarizes some common but more specialized forms of business associations used in the United States. These include joint stock companies, syndicates, joint ventures, and franchises.

JOINT STOCK COMPANY

A **joint stock company** is a partnership agreement in which members agree to stock ownership in exchange for partnership liability. Although members own shares that are transferable (as in a corporation), the joint stock company is

joint stock company A partnership agreement in which members of the company own shares that are transferable but all goods are held in the name of the members, who assume partnership liability.

TABLE 16-3 SPECIALIZED FORMS OF BUSINESS ASSOCIATIONS

Form	Description
Joint stock company	A partnership agreement in which individuals agree to take stock ownership while retaining partnership liability.
Syndicate	An investment group created primarily for the purpose of financing a purchase, usually a single transaction.
Joint venture	A partnership, an individual, or a corporation that pools labor and capital for a limited period of time.
Franchising	A method of marketing goods through a private agreement whereby the franchisor allows use of its trade name, trademark, or copyright in exchange for a percentage of the gross profits made by the franchisee. Federal and state laws govern franchising.

treated as a partnership because all goods are held in the name of the partners, who are held personally liable when sued successfully by a third party.

SYNDICATE

A **syndicate** is an investment group that makes a private agreement to come together for the purpose of financing a large commercial project (e.g., a hotel or a sports team) that the individual members (partnerships or corporations) could not finance alone. The advantage of a syndicate is that it can raise large amounts of capital quickly. The disadvantage is that if the project fails, the syndicate members may be held liable for a breach of the agreement by a third party.

syndicate An investment group that privately agrees to come together for the purpose of financing a large commercial project that none of the syndicate members could finance alone.

JOINT VENTURE

When individuals, partnerships, or corporations make a private agreement to finance, produce, and sell goods, securities, or commodities for a limited purpose and/or a limited time, they have formed a joint venture. Joint ventures are a popular way for developing nations (e.g., China) to attract foreign capital. Typically, two companies (e.g., Chrysler and Bank of America) join with foreign companies (e.g., Chinese state auto companies) to finance, produce, and market goods to be sold in the foreign nation and possibly in other nations. In 2010, Chinese state-owned companies backed by Chinese capital entered into joint ventures with Latin American and African nations as well as private-sector entities.

FRANCHISING

A **franchising** relationship is based on a private commercial agreement between the *franchisor*, who owns a trade name or trademark, and the *franchisee*, who sells or distributes goods using the trade name or trademark. It is a method of marketing goods or services.

The franchisee usually pays a percentage of the gross sales to the franchisor in exchange for use of the trademark name, construction of the building, and numerous other services. Usually, the franchisee is a local entrepreneur whom the franchisor supplies with goods to be sold under conditions set out in the agreement. Failure to meet such conditions (e.g., not keeping a fast-food restaurant clean) may lead to termination of the franchise agreement.

franchising A commercial agreement between a party that owns a trade name or trademark (the franchisor) and a party that sells or distributes goods or services using that trade name or trademark (the franchisee).

Laws Governing Franchising. When franchising exists based on a contractual relationship, traditional common law applies for nongoods. Article 2 of the UCC applies when the franchisor and franchisee are involved in a sales agreement. In addition, there are federal and state laws designed to protect franchisees from terminating franchise agreements without good cause. Other federal and state laws are discussed later in this section.

Federal Laws. The Franchising Rule promulgated by the Federal Trade Commission in 1979 requires a franchisor to disclose all material facts to a franchisee before entering into a franchise agreement. (See Chapter 25 for a discussion of this rule.)

State Laws. State deceptive trade practices acts and the Uniform Franchise Law set out by the National Conference of Commissioners on Uniform State Laws seek to make all states' franchise laws similar. (However, the latter must be adopted by a state's legislature before it is applicable in that state.) Further, state franchise administrators developed a Uniform Franchise Offering Circular (UFOC). It requires a franchisor to make specific presale disclosures to any prospective franchisee. Information that must be disclosed includes the franchisor's major business balance sheet and the franchisor's income statement for the preceding three years, material terms of the franchise agreement, restrictions on the franchisee's territory, grounds for termination of the franchise, and other relevant information. Prior

TERMS OF A FRANCHISE AGREEMENT

Generally, a franchise agreement is a standard form contract prepared by the franchisor. Franchise agreements cover the following topics:

1. Quality-control standards. The franchisor's most important assets are its name and reputation. The quality-control standards set out in the franchise agreement—such as the franchisor's right to make periodic inspections of the franchisee's premises and operations—are intended to protect these assets. Failure to meet the proper standards can result in loss of the franchise.

2. Training requirements. Franchisees and their personnel usually are required to attend training programs either on-site or at the franchisor's training facilities.

3. Covenant not to compete. Covenants not to compete prohibit franchisees from competing with the franchisor during a specific time and in a specified area after termination of the franchise. Unreasonable (overextensive) covenants not to compete are void.

4. Arbitration clause. Most franchise agreements contain an arbitration clause providing that any claim or controversy arising from the franchise agreement or an alleged breach thereof is subject to arbitration. The U.S. Supreme Court has held such clauses to be enforceable.

5. Other terms. Capital requirements include restrictions on the use of the franchisor's trade name, trademarks, and logo; standards of operation; duration of the franchise; recordkeeping requirements; sign requirements; hours of operation; prohibition on sale or assignment of the franchise; conditions for termination of the franchise; and other specific terms pertinent to the operation of the franchise and the protection of the parties' rights.

FRANCHISE FEES

Franchise fees payable by the franchisee are usually stipulated in the franchise agreement. The franchisor may require the franchisee to pay any or all of the following fees:

1. Initial license fee. A lump-sum payment for the privilege of being granted a franchise.

2. Royalty fee. A fee for the continued use of the franchisor's trade name, property, and assistance that is often computed as a percentage of the franchisee's gross sales.

3. Assessment fee. A fee for such things as advertising and promotional campaigns and administrative costs, billed either as a flat monthly or annual fee or as a percentage of gross sales.

4. Lease fees. Payment for any land or equipment leased from the franchisor, billed either as a flat monthly or annual fee or as a percentage of gross sales or other agreed-upon amount.

5. Cost of supplies. Payment for supplies purchased from the franchisor.

to execution of the franchise agreement, counsel for all parties should examine the UFOC and state and federal law to prevent action later on that leads to misunderstandings and costly litigation. The importance of good faith and fair dealing is shown in the following case.

CASE 16-4

Holiday Inn Franchising, Inc. v. Hotel Associates, Inc.
Court of Appeals of Arkansas
2011 Ark. App. 147, 382 S.W.3d 6 (2011)

Buddy House was in the construction business in Arkansas and Texas. For decades, he collaborated on projects with Holiday Inn Franchising, Inc. Their relationship was characterized by good faith—many projects were undertaken without written contracts. At Holiday Inn's request, House inspected a hotel in Wichita Falls, Texas, to estimate the cost of getting it into shape. Holiday Inn wanted House to renovate the hotel and operate it as a Holiday Inn. House estimated that recovering the cost of renovation would take him more than 10 years, so he asked for a franchise term longer than Holiday Inn's usual 10 years. Holiday Inn refused, but said that if the hotel was run "appropriately," the term would be extended at the end of 10 years. House bought the hotel, renovated it, and operated it as Hotel Associates, Inc. (HAI), generating substantial profits. He refused offers to sell it for as much as $15 million.

Before the 10 years had passed, Greg Aden, a Holiday Inn executive, developed a plan to license a different local hotel as a Holiday Inn instead of renewing House's franchise license. Aden stood to earn a commission from licensing the other hotel. No one informed House of Aden's plan. When the time came, HAI applied for an extension of its franchise, and Holiday Inn asked for major renovations. HAI spent $3 million to comply with this request. Holiday Inn did not renew HAI's license, however, but instead granted a franchise to the other hotel. HAI sold its hotel for $5 million and filed a suit in an Arkansas state court against Holiday Inn, asserting fraud. The court awarded HAI compensatory and punitive damages. Holiday Inn appealed.

Judge Abramson

Generally, a mere failure to volunteer information does not constitute fraud. But silence can amount to actionable fraud in some circumstances where the parties have a relation of trust or confidence, where there is inequality of condition and knowledge, or where there are other attendant circumstances.

In this case, substantial evidence supports the existence of a duty on Holiday Inn's part to disclose the Aden [plan] to HAI. Buddy House had a long-term relationship with Holiday Inn characterized by honesty, trust, and the free flow of pertinent information. He testified that [Holiday Inn's] assurances at the onset of licensure [the granting of the license] led him to believe that he would be relicensed after ten years if the hotel was operated appropriately. Yet, despite Holiday Inn's having provided such an assurance to House, it failed to apprise House of an internal business plan . . . that advocated licensure of another facility instead of the renewal of his license. A duty of disclosure may exist where information is peculiarly within the knowledge of one party and is of such a nature that the other party is justified in assuming its nonexistence. Given House's history with Holiday Inn and the assurance he received, we are convinced he was justified in assuming that no obstacles had arisen that jeopardized his relicensure.

Holiday Inn asserts that it would have provided Buddy House with the Aden [plan] if he asked for it. But, Holiday Inn cannot satisfactorily explain why House should have been charged with the responsibility of inquiring about a plan that he did not know existed. Moreover, several Holiday Inn personnel testified that Buddy House in fact should have been provided with the Aden plan. Aden himself stated that . . . House should have been given the plan. . . . In light of these circumstances, we see no ground for reversal on this aspect of HAI's cause of action for fraud.*

Affirmed by court of appeals.

Internet Franchising. With the advent of Internet marketing, the issue now has arisen as to whether a franchisee violates a territorial-area clause included in its franchise agreement when it uses an Internet website to market products worldwide. Both courts and arbitration panels have had difficulty dealing with this cyberspace issue. The franchisor believes that the exclusive territorial clause it grants the franchisee in the franchise agreement is lost if the franchisee can market goods from its website. The franchisee sees its website as a cheap and convenient method of marketing.

Global Dimensions of Business Associations

It is inevitable that the worldwide trend toward market-oriented economies will raise the demand for the investment capital and manufactured goods and services of the industrialized world. This certainty, together with the forging of international and regional agreements to lower or eliminate tariffs and other barriers to trade in recent years (see Chapter 8), is spurring many partnerships, single proprietorships, and specialized business associations that were once strictly domestic businesses to become transnational, multinational, or international buyers and sellers of goods and services.

In deciding whether to take the plunge into international waters, managers of small, midsize, and large businesses need to ask the following questions:

- Is there a demand for the product in the targeted country or countries?
- Are there legal obstacles in the targeted country that should be carefully considered?

---

*Holiday Inn Franchising, Inc. v. Hotel Associates, Inc. Court of Appeals of Arkansas, 2011 Ark. App. 147, 382 S.W.3d 6 (2011).

- Is managing an international business at a distance a realistic possibility for the firm?
- Will management be able to deal successfully with currency fluctuations?
- Is the risk of political interference by the target country's government too great to make doing business in that country worthwhile?
- Is there a serious risk of nonperformance, nonpayment, or loss of property or freight in the country or region the firm is considering entering?

If the responses to the preceding questions indicate that the firm should go ahead, managers need to determine the optimal level of global involvement by answering the following questions:

- Should the firm directly export to another firm in the target country, or should it hire an export trading company to market its products?
- Should the firm license the use of its products under an international licensing agreement? For example, *international franchising* is a form of licensing in which franchisees in the targeted countries are allowed to use the franchisor's name in exchange for a percentage of the gross profits. This specialized business form is the preferred way to go international among fast-food retailers such as McDonald's, Wendy's, and Pizza Hut. If the firm holds a patent, trademark, or copyright on its product, however, it must always be concerned about possible efforts by businesses or individuals in the target country to circumvent multilateral and bilateral international agreements.
- Should the firm go international by joint venture, merger, or acquisitions? These alternatives, which involve investment of large sums of capital, seem most appropriate for large multinational companies.

OUTSOURCING

The increase in outsourcing of manufacturing and service-sector jobs, despite political opposition (e.g., Democrats, unions, prisoners, and others), brought into focus a dilemma for private- and public-sector companies in developed nations worldwide. Treading the delicate balance between cost savings on goods and services and increased exports to fast-growing developing countries on the one hand and loss of jobs and "heat" from politicians on the other hand present problems for many private-sector companies. Candidates in the 2004 and 2008 presidential elections offered contrasting views on this subject.

Also, doing a cost-benefit analysis with regard to outsourcing does not produce many hard numbers. For example, if some jobs are outsourced to India, how is that balanced against Indians buying more U.S. services and U.S. citizens being able to buy cheaper goods manufactured in India? Most multinationals need to ask themselves many questions in regard to outsourcing, such as the types of employees and skill levels needed, the time required for employee training, and the effect of outsourcing on the process of delivering the services or manufacturing the goods. Legal questions also exist at the federal, state, and international levels.

Congress's grant of authority to the Obama administration to fast track the largest U.S. trade agreement ever entered into was passed in 2015. The trade deal included 12 nations from Asia, Latin America, inclusive of the United States and Canada. Some Democrats and Republicans opposed the president for fear that jobs would be "shipped out" to less developed countries where wages were noncompetitive with those in the United States. Members of Congress from districts with large union constituencies were sensitive to this problem. The argument(s) in favor of granting fast-track authority allowed only an up or down vote on the proposed bill (Trans-Pacific Partnership [TPP]), without any amendments on the floor of the U.S. Senate. Some of the same arguments, pro and con, regarding this bill were used when outsourcing was debated at the state and federal levels. The significance of this debate was that the trade bill would be a political and economic tool for the United States and Japan to compete with China (not included by choice). If enacted, TPP would affect nearly 1 billion people and 65 percent of all global trade.

SUMMARY

The sole proprietorship, the partnership, and the corporation are the three major forms of business organizations in the United States. (See Chapter 17 for a discussion of corporations.)

Factors influencing the choice of one of these three types of business associations include tax ramifications; control considerations; liability of owners; and (less significantly) ease and expense of ownership, transferability of properties, and the projected life of the business. The sole proprietorship allows the owner to have total control of management, assets, and profits; the owner also has unlimited personal liability. General partners usually exercise equal control over management and profits; they, too, incur unlimited personal liability. Limited partners forgo management control in return for limited liability. In both sole proprietorships and partnerships, profits are taxed as personal income to the owners.

Specialized forms of business associations are joint stock companies, syndicates, joint ventures, and franchising.

Business organizations of all forms should ask some basic questions before they enter global markets. Once they have decided to proceed, they need to decide the optimal level of involvement and marketing methods. Trade agreements such as the TTP need to be considered when forming an organization to do business in the United States.

REVIEW QUESTIONS

16-1 List the primary differences between a sole proprietorship and a general partnership.

16-2 Identify the factors that an entrepreneur should consider when selecting an organizational form for a business.

16-3 Describe the circumstances under which a partnership would offer greater tax advantages than would a single proprietorship. Explain.

16-4 What is one tax advantage of a general partnership? Explain.

16-5 Explain the difference between an LP and an LLP.

16-6 Why is a joint venture important for a new company? Explain.

REVIEW PROBLEMS

16-7 Daniel is the owner of a chain of shoe stores. He hires Rubya to be the manager of a new store opening in Grand Rapids, Michigan. Daniel, by written contract, agrees to pay Rubya a monthly salary and 20 percent of the profits. Without Daniel's knowledge, Rubya represents himself to Classen as Daniel's partner and shows Classen the agreement to share profits. Classen extends credit to Rubya. Ruyba defaults. Discuss whether Classen can hold Daniel liable as a partner.

16-8 National Foods, Inc., sells franchises to its fast-food restaurants, known as Chicky-D's. Under the franchise agreement, franchisees agree to hire and train employees based strictly on Chicky-D's standards. Chicky-D's regional supervisors are required to approve all job candidates before they are hired and all general policies affecting those employees. Chicky-D's reserves the right to terminate a franchise for violating the franchisor's rules. In practice, however, Chicky-D's regional supervisors routinely approve new employees and the policies of individual franchises. After several incidents of racist comments and conduct exhibited by Tim, a recently hired assistant manager at a Chicky-D's, Sharon, a counterperson at the restaurant, resigns. Sharon files a suit in a federal district court against National. National files a motion for summary judgment, arguing that it is not liable for harassment by franchise employees. Will the court grant National's motion? Why or why not?

16-9 Clark, who owned a vacant lot, and Bird, who was engaged in building houses, entered into an oral agreement by which Bird was to erect a house on the lot. Upon the sale of the house and lot, Bird was to have his money first. Clark was then to have the agreed value of the lot, and the profits were to be equally divided. Did a partnership exist? Explain.

16-10 Dunn and Welch both appeared to operate Ruidoso Downs Feed Concession. Dunn sought to obtain credit for Ruidoso Downs from Anderson Hay and Grain Company. Relying on Dunn's financial position, Anderson extended the credit. Dunn was the person who was responsible for making sure that the payments were made. When Ruidoso Downs could not pay Anderson, Anderson brought an action against Dunn, alleging that, as a partner, Dunn was personally liable for the debts the partnership could not repay. Dunn defended on the grounds that he was not a partner because there was no formal partnership agreement between him and Welch. Was Dunn liable? Explain.

16-11 Bron, Arthur, and David formed a partnership for the purpose of betting on boxing matches. Bron and Arthur would become friendly with various boxers and offer them bribes to lose certain bouts. David would then place large bets using money contributed by all three and would collect the winnings. After David had accumulated a large sum of money, Bron and Arthur demanded their share, but David refused to make any split. Can Bron and Arthur compel David to account for the profits of the partnership? Explain.

16-12 Hugo and Charles were brothers who did business as partners for several years. When Hugo died, Charles was appointed administrator of his estate. Tax returns disclosed that the partnership business continued to operate just as it had before Hugo's death and that Hugo's estate received the profits and was charged with the losses of the business. Did Charles have the authority to continue the partnership business after the dissolution of the partnership brought about by Hugo's death? Are the assets of Hugo's estate chargeable with the liabilities of the partnership incurred after Hugo's death? Explain.

CASE PROBLEMS

16-13 J.C., Inc., had a franchise agreement with McDonald's Corporation to operate McDonald's restaurants in Lancaster, Ohio. The agreement required J.C. to make monthly payments to McDonald's of certain percentages of the gross sales. If any payment was more than 30 days late, McDonald's had the right to terminate the franchise. The agreement also stated that even if McDonald's accepted a late payment, that would not "constitute a waiver of any subsequent breach." McDonald's sometimes accepted J.C.'s late payments, but when J.C. defaulted on the payments in July 2010, McDonald's gave notice of 30 days to comply or surrender possession of the restaurants. J.C. missed the deadline. McDonald's demanded that J.C. vacate the restaurants, but J.C. refused. McDonald's alleged that J.C. had violated the franchise agreement. J.C claimed that McDonald's had breached the implied covenant of good faith and fair dealing. Which party should prevail and why? *McDonald's Corp. v. C.B. Management Co.*, 13 F. Supp. 2d 705 (N.D.Ill. 1998).

16-14 Karl Horvath, Hein Rüsen, and Carl Thomas formed a partnership, HRT Enterprises, to buy a manufacturing plant. Rüsen and Thomas leased the plant to their own company, Merkur Steel. Merkur then sublet the premises to other companies owned by Rüsen and Thomas. The rent these companies paid to Merkur was higher than the rent Merkur paid to HRT. Rüsen and Thomas did not tell Horvath about the subleases. Did Rüsen and Thomas breach their fiduciary duties to HRT and Horvath? Discuss. *Horvath v. HRT Enterprises*, 489 Mich. App. 992, 800 N.W.2d 595 (2011).

16-15 George Oshana and GTO Investments, Inc., operated a Mobil gas station franchise in Itasca, Illinois. In 2010, Oshana and GTO became involved in a rental dispute with Buchanan Energy, to which Mobil had assigned the lease. In November 2011, Buchanan terminated the franchise because Oshana and GTO had failed to pay the rent. However, Oshana and GTO alleged that they were "ready, willing, and able to pay the rent" but that Buchanan failed to accept their electronic fund transfer. Have Oshana and GTO stated a claim for wrongful termination of their franchise? Why or why not? *Oshana v. Buchanan Energy*, 2012 WL 426921 (N.D.Ill. 2012).

16-16 In August 1998, Jea Yu contacted Cameron Eppler, president of Design 88, Ltd., to discuss developing a website that would cater to investors and provide services to its members for a fee. Yu and Patrick Connelly invited Eppler and Ha Tran, another member of Design 88, to a meeting to discuss the site. The parties agreed that Design 88 would perform certain Web design, implementation, and maintenance functions for 10 percent of the profits from the site, which would be called "The Underground Trader." They signed a "Master Partnership Agreement," which was later amended to include Power Uptik Productions, LLC (PUP). The parties often referred to themselves as partners. From Design 88's offices in Virginia, Design 88 designed and hosted the site and solicited members through Internet and national print campaigns. When relations among the parties soured, PUP withdrew. Design 88 filed a suit against PUP and the others. Did a partnership exist among these parties? Explain. *Design 88, Ltd. v. Power Uptik Productions, LLC*, 133 F. Supp. 2d 873 (W.D. Va. 2001).

16-17 On August 23, 1995, Climaco Guzman entered into a commercial janitorial services franchise agreement with Jan-Pro Cleaning Systems, Inc., in Rhode Island for a franchise fee of $3,285. In the agreement, Jan-Pro promised to furnish Guzman with "one (1) or more customer account(s) . . . amounting to $8,000.00 gross volume per year. . . . No portion of the franchise fee is refundable except and to the extent that the Franchisor, within 120 business days following the date of execution of the Franchise Agreement, fails to provide accounts." By February 19, Guzman had not received any accounts and demanded a full refund. Jan-Pro promised "accounts grossing $12,000 per year in income." Despite its assurances, Jan-Pro did not have the ability to furnish accounts that met the stated requirements. In September, Guzman filed a suit in a Rhode Island state court against Jan-Pro, alleging in part fraudulent misrepresentation. Should the court rule in Guzman's favor? Why or why not? *Guzman v. Jan-Pro Cleaning Systems, Inc.*, 839 A.2d 504 (R.I. 2003).

THINKING CRITICALLY ABOUT RELEVANT LEGAL ISSUES

Ed Johnson is a real estate manager and investor. Nearly 20 years ago, Ed embarked on a partnership with Jack Jones to improve and operate an office building in New Haven, Connecticut. The building and land are owned by James Jason. James gave Ed and Jack a 20-year lease; at the end of 20 years, the lease would terminate and the property would revert to James. Pursuant to their partnership agreement, Ed and Jack each provided 50 percent of the capital for improvements to the office space and received 50 percent of allocable net profits.

Although he no longer needed Jack's capital for the project, Ed suspected that Jack would be interested in participating in the mall development. However, it was not clear whether James made the offer to renew the lease solely to Ed or to the partnership. Because Jack had not been invited to dinner and his name had never been mentioned, Ed believed that the offer was made solely to him.

1. Does Ed have an ethical responsibility to inform Jack of the opportunity to renew the lease? Explain.

2. Does it matter that the renewal offer for the long-term lease was initially raised in a dinner conversation between Ed and James? Explain.

3. Should Ed be free to sever relations with Jack with regard to the property? Consider that Ed has managed the property and no longer needs Jack's capital. What competing values does his dilemma involve?

ASSIGNMENT ON THE INTERNET

This chapter introduces the common forms of business organization in the United States. Business structures outside the United States, however, are often very different. Use the Internet to determine some of the major differences in business organizations between the United States and the European Union. You can begin your search by visiting the Eurolegal Services website at www.eurolegal.org/webresources/eustartup.htm.

If you were looking to start a business in the European Union, what would you do differently than if you intended to start the business in the United States? The following websites may also assist you in your search.

ON THE INTERNET

www.irs.gov/faqs/faq-kw127.html The IRS provides information about LLCs at this site.
www.mycorporation.com/index.htm This is the site of a private firm that assists small companies and businesses in becoming incorporated.
www.llc-reporter.com/28.htm This site provides an analysis of the Uniform Limited Liability Company Act.

FOR FUTURE READING

Crouch, Holmes F. *Pros & Cons of LLCs: How to Shape a Limited Liability Company (LLC), Understand Its Rules, Prepare Tax Returns & Fend Off Con Artists* (Series 200: Investors & Businesses 2007). Allyeartax Guides.

Rothenberg, Robert, and Tatyana V. Melnikova. "Comparative Forms of Doing Business in Russia and New York State—Proprietorships, Partnerships, and Limited Partnerships." *American Business Law Journal* 40 (2004): 563.

CHAPTER SEVENTEEN

Law and Business Associations—II

- THE CORPORATION
- CLASSIFICATION OF CORPORATIONS
- CREATION OF CORPORATIONS
- FINANCING OF CORPORATIONS
- OPERATION OF CORPORATIONS
- LIMITED LIABILITY COMPANIES
- GLOBAL DIMENSION OF CORPORATIONS: A "BIG FAT GREEK" BAILOUT II AND III

The Corporation

The partnership is the most common form of business organization in the United States. The dominant business organizational form, however, is the **corporation**, a legal entity created by state law that raises capital by issuing **stock** to investors, who own the corporation. Although the corporation may have many owners, it is legally treated as a single person. Before we go into the laws governing the creation, financing, and operation of corporations, we'll explain how corporations are classified (see Table 17-1).

corporation An entity formed and authorized by state law to act as a single legal person and to raise capital by issuing stock to investors who are the owners of the corporation.

stock The capital that a corporation raises through the sale of shares that entitle the shareholders to certain rights of ownership.

CRITICAL THINKING ABOUT THE LAW

When we say that the corporation is treated legally as if it were a single person, we are making a statement that affects the nature of our democracy. A citizen is allowed to have freedom of speech and to participate fully in the political life of our country. Should we extend the same liberties to corporations?

Certainly, corporations are greatly affected by political actions. Therefore, would it not follow that they should be offered the opportunity to participate in the political life of the nation in the same manner as individual citizens?

But when they are allowed to participate fully, are they just like any other voter and political voice? Do they have characteristics that differentiate them from those of regular human citizens?

Suppose that Drill Baby Drill Oil Company, a Saudi Arabian–owned oil company incorporated in Delaware, wishes to spend $1 billion to elect the Republican candidate for president. Furthermore, suppose that this corporation has assets that exceed the gross national product of all but 12 nations.

1. As it spent money for political purposes to influence elections, would Drill Baby Drill, which is required by law to pursue profits, be likely to use its money to elect politicians who had the long-term health of the nation uppermost in their minds?

    Clue: Think about the ambiguity contained in the phrase "long-term health of the nation." Does that idea focus on the quality of our national environment? Or is it measured primarily in terms of how low our unemployment rate is? Or does it refer to some other quality or characteristic? Why would the answer matter?

2. Why does the extent to which the answer to the first question is "yes" influence the extent to which we might treat a corporation as just another citizen for purposes of the law?

   Clue: Suppose that the firm is a multinational corporation. How does that status, coupled with the firm's focus on profits, alter our understanding of what it means to be a citizen of a nation?

3. What are the inadequacies in seeing the corporation as identical to a person for political purposes?

   Clue: What evidence about corporate behavior in elections would shape your perception of whether a corporation is just another person?

TABLE 17-1 CLASSIFICATION OF CORPORATIONS

Closely held	Corporations held by a small group of people whose stock is not traded on national securities exchanges.
Publicly held	Corporations whose stock is traded on a national exchange (discussed in Chapter 23) and is held by numerous shareholders. This form is emphasized in this text.
Domestic, foreign, alien	A corporation is held to be domestic in the state in which it incorporates. If a corporation is formed in one state but does business in another, it is referred to in the other state as a *foreign* corporation. A corporation formed in one country but doing business in the United States is referred to in the United States as an *alien* corporation.
Multinational	A corporation that maintains worldwide distribution and manufacturing, trades on the exchanges of many nation-states, and has officers and managers who are citizens of many states.
Subchapter S	A corporation that organizes and operates as a domestic business corporation but for tax purposes is treated like a partnership.
Professional	A corporation that allows physicians, lawyers, dentists, accountants, and other professionals to incorporate to take advantage of health and pension-plan tax deductions.
Benefit corporation	A for-profit corporation that seeks to have a material positive impact on society and the environment. The purpose of this corporation is to benefit the public as a whole rather than increase long-term shareholder value.
Nonprofit	A corporation not formed for purposes of making a profit (e.g., charitable, religious, educational, and health care organizations).

Classification of Corporations

All corporations are broadly classified as either public or private. They are further differentiated as closely held, publicly held, multinational, Subchapter S, professional, or limited liability corporations. Not-for-profit (nonprofit) corporations also exist under state laws and qualify under federal and state tax laws.

CLOSELY HELD CORPORATION

closely held corporation
A corporation whose stock is not traded on the national securities exchanges but is privately held by a small group of people.

The greatest number of corporations in this country are private, or **closely held corporations**. The stock of these corporations is not traded on any of the national securities exchanges. Instead, it is usually held by a small group of people, often members of the same family or close friends, who serve as directors as well as officers and active managers of the corporation. When the corporation is formed, these original owners frequently enter into an agreement to restrict the sale of stock to the initial shareholders. By limiting ownership in this way, they retain control of the corporation.

PUBLICLY HELD CORPORATION

Those corporations whose stock is traded on the national securities exchanges are known as **publicly held corporations**. Although technically governed by the same rules as closely held corporations (except for securities law, which is discussed in Chapter 23), their operations are much different from those of closely held corporations. Publicly held corporations have numerous shareholders who are simply investors. Real control rests in the hands of the officers and managers, who may own some stock, although generally not a majority or controlling amount.

When the term *corporation* is used in this book, a publicly held corporation is meant, unless otherwise specified. Although some regulations apply only to publicly held corporations, in most instances, the same laws apply to both public and private corporations. The impact of such laws differs, though, depending on whether the affected corporation is public or private. The reason for our focus on public rather than private corporations is the same as our reason for emphasizing corporations rather than partnerships: impact on society. Public corporations are wealthier than any other form of business organization, and therefore, the impact of regulations on these corporations has the greatest effect on society.

publicly held corporation A corporation whose stock is traded on at least one national securities exchange.

MULTINATIONAL OR TRANSNATIONAL CORPORATION

This relatively new type of publicly held corporation now dominates the world economy. It is called a **multinational**, or **transnational**, **corporation** because it does not restrict its production to a single nation and generally maintains worldwide distribution sites. Its stock is usually traded on the securities exchanges of several nations, and its managers are often citizens of different countries. Through their tremendous wealth, power, and reach, multinationals have great influence on societies all over the world.

multinational (transnational) corporation A corporation whose production, distribution, ownership, and management span several nations.

SUBCHAPTER S CORPORATION

This type of closely held corporation is best described as a hybrid of the corporation and the partnership. The **Subchapter S corporation** is organized and operates as a regular business corporation, but for tax purposes, it is treated like a partnership. To qualify for Subchapter S treatment under the Internal Revenue Code (IRC), a domestic corporation must (1) have no more than 35 shareholders, all of whom are individuals, estates, or certain types of trusts and none of whom is a nonresident alien; (2) have only one class of stock outstanding; and (3) not be a member of an affiliated group of corporations. All shareholders must consent to the election of Subchapter S status.

Subchapter C corporations pay taxes on income generated by a business, and shareholders pay tax on the same income when it is distributed as dividends. Any corporation not meeting the requirements for an S corporation is automatically classified as a C corporation.

Subchapter S corporation A business that is organized like a corporation but under IRC, Subchapter S is treated like a partnership for tax purposes so long as it abides by certain restrictions pertaining to stock, shareholders, and affiliations.

Subchapter C corporation A corporation that organizes and operates as a domestic business but for tax purposes is treated like a partnership.

ROBS CORPORATION

A rollover business corporation (ROBS) is an e-corporation that the Internal Revenue Service (IRS) recognizes under certain conditions. The funds come from an individual's 401K plan.

PROFESSIONAL CORPORATION

The **professional corporation** is a fairly new form of business organization intended for doctors, lawyers, dentists, accountants, and other professionals who were once unable to incorporate legally. Most states now have passed statutes

professional corporation A corporation that is organized by doctors, dentists, lawyers, accountants, or other professionals and is specified in state statutes.

permitting specified professionals to incorporate so that they can take the tax advantages of deductions for health and pension plans that are allowable under the corporate form. In most states, the professional corporation differs from other corporations in that the owners are not accorded limited liability for professional acts. (It is generally considered contrary to public policy to grant professionals limited liability for their negligence.)

NONPROFIT CORPORATION

This type of corporation is formed for purposes other than making a profit. Some examples of nonprofit (not-for-profit) corporations include hospitals, educational institutions, charities, and religious groups. Most are private in nature, but not necessarily so. Nonprofit corporations are used by groups as a way to carry out transactions and own property without individuals (who belong to a group) being held liable. Often, in legal papers, they are referred to as *eleemosynary institutions* (private corporations created for charitable and benevolent purposes).

Creation of Corporations

Corporations are creatures of state, not federal, law. Each of the 50 states, as well as the District of Columbia, Puerto Rico, and Guam, has a general incorporation statute that stipulates the articles of incorporation to be used in that state. These articles, generally standardized forms, identify the name of the corporation, its registered address and resident agent, the general purpose of the business, the classes of stock to be issued by the corporation and their face value, and the names and mailing addresses of the incorporators. The articles, accompanied by the required fees, are filed with the secretary of state of the state of incorporation, who then issues a certificate of incorporation. Upon issuance of the certificate, the corporation holds its first board meeting, at which a board of directors is elected, bylaws are enacted, and corporate stock is issued. The bylaws are the governing regulations of the corporation and often are the basis for litigation when directors act on behalf of the shareholders of the corporation.

The Delaware Supreme Court is considered the most influential of all courts in the nation with regard to corporate governance and the chief arbiter of conflicts between corporations and between shareholders and a single corporation. More than 50 percent of *Fortune 100* companies are registered in Delaware. Anyone who merges, sues, or manages a Delaware corporation is subject to Delaware law as interpreted by its courts. This is why, fairly recently, there has been significant controversy surrounding the appointment and reappointment of Delaware State Supreme Court justices and the politicizing of appointments made by the commission that recommends candidates to the governor.

In creating a corporation, three activities are of importance:

1. Incorporation: (these activities, listed below, tend to vary among states)
   a. Selection of a state of incorporation. See the preference for Delaware above.
   b. Securing the corporate name. All states recognize the corporate names to include the word corporation (Corp.), Incorporated (Inc.), Limited (Ltd.), or Company (Co.). One should check whether a name will be deceptively similar, thus invoking a trade-name dispute.
   c. Preparation of the articles of incorporation should include the following (RMB CA2.3):
      i. Name
      ii. Number of shares

iii. Name and street address of the initial registered agent and the secretary of the state's office elected for incorporation

iv. Name and address of each incorporator

2. Corporate Powers: Upon creation, expressed powers of a corporation are found in the articles of incorporation, state laws of incorporation and operation, state and federal constitutions, and case law and statutes of the state and federal jurisdictions where the corporation is incorporated. The U.S. (federal) Constitution, state constitutions, state statutes, articles of incorporation, bylaws, and resolutions of the board of directors establish the order of priority the courts will follow in cases of conflict among these documents.

3. Piercing the Corporate Veil: Where a corporate entity is used to perpetrate a fraud or circumvent a law, courts ignore the corporate structure, pierce the corporate veil, and hold an individual shareholder personally liable (RMBCA 2.04).

In the case below, a corporation failed to pay its legal fees, so its attorney sued to hold the shareholders personally liable.

CASE 17–1

Brennan's Inc. v. Colbert
Court of Appeals of Louisiana
85 So.3d 787 (2012)

Pip, Jimmy, and Theodore Brennan are brothers and shareholders of Brennan's Inc., which owns and operates New Orleans' famous Brennan's Restaurant. In 1998, the Brennan brothers retained attorney Edward Colbert and his firm, Kenyon & Kenyon LLP, to represent Brennan's Inc., in a dispute with another family member. All bills were sent to Brennan's, Inc., and the payments came from the company's checking accounts.

As a close corporation, Brennan's, Inc., did not hold formal corporate meetings with agendas and minutes, but it did maintain corporate books, hold corporate bank accounts, and file corporate tax returns. In 2005, Brennan's, Inc., sued Colbert and his law firm for legal malpractice. In its answer, Kenyon & Kenyon demanded unpaid legal fees from both Brennan's, Inc., and the Brennan brothers personally. The trial court found that the Brennan brothers could not be held personally liable. Kenyon & Kenyon appealed. The law firm argued that the court should pierce the corporate veil because Brennan's, Inc., did not observe corporate formalities and the Brennan brothers did not honor their promise to pay their legal bills.

Judge Dysart

As a general rule, a corporation is a distinct legal entity, separate from the individuals who compose it, thus insulating the shareholders from personal liability.

These are limited exceptions where the court may ignore the corporate fiction and find the shareholders personally liable for the debts of a corporation. One of those exceptions is where the corporation is found to be the "alter ego" of the shareholder. It usually involves situations where fraud or deceit has been practiced by the shareholders through the corporation. Another basis is where the shareholders disregarded the corporate formalities to the extent that the corporation and the shareholder are no longer distinct entities.

Absent fraud, malfeasance, or criminal wrongdoing, courts have been reluctant to hold a shareholder personally liable, the totality of the circumstances is determinative.

The Kenyon firm was aware of the nature of the operation of Brennan's Inc., . . . prior to being retained. The client was Brennan's, Inc., bills were sent to Brennan's Inc., and payments were paid with checks from the Brennan's Inc., bank accounts. . . . The Kenyon firm acknowledged that Brennan's, Inc., acting through its shareholders, promised to make good on the debt.

There is no evidence that the Brennan brothers ever agreed to bind themselves personally of any debt incurred in connection with the legal services provided by the Kenyon firm. There is no written retention agreement between the corporation and the Kenyon firm, nor is there a written guaranty from any of the brothers.

The Kenyon firm admits that there is no requirement for small [close] corporations to operate with the formality usually expected of larger corporations. The Kenyon firm has failed to establish that the lack of corporate formalities, particularly meetings, agendas and minutes, is sufficient to pierce the corporation veil.

Brennan's, Inc., at all times since its inception has maintained corporate books, corporate bank accounts, and has filed corporate tax returns.

The Kenyon firm has not proven that any of the Brennan brothers made promises to pay the firm's bills without the intent to pay them. If a broken promise to pay was sufficient to establish fraud, then every lawsuit against a corporation for a debt would automatically allow for the piercing of the corporate veil. Clearly, a juridical entity such as a corporation can only speak through its shareholders.

The Louisiana appellate court held that Kenyon & Kenyon could not hold the Brennan brothers personally liable by piercing the corporate veil.*

Affirmed for the Brennans.

Two primary concerns of state corporate laws are the financing of the corporation and its operation. These two areas are necessarily entwined, but they are separated here for discussion purposes. Laws governing these two areas of corporate activity are primarily state regulations, although most states are guided by a modified version of the Revised Model Business Corporations Act (RMBCA).

Financing of Corporations

Financing refers to the acquisition of funds or capital for the operation or expansion of a corporation. Corporations generally engage in two types of financing: debt and equity. *Debt financing* may be described as the taking out of loans which generally involves the corporation's promise to pay the principal and interest at a stated time and rate. *Equity financing* is accomplished by the sale of ownership interests in the corporation.

DEBT FINANCING

notes Short-term loans.

bonds Long-term loans secured by a lien or mortgage on corporate assets.

debentures Unsecured long-term corporate loans.

A corporation can issue three primary types of debt instruments: notes, bonds, and debentures. **Notes** are short-term loans. **Bonds** are usually long-term loans secured by a lien or mortgage on corporate assets. **Debentures** are usually unsecured long-term corporate loans.

In all forms of debt financing, corporations incur a liability to the holder of the debt security. Periodic interest payments are generally required. Interest payments on debt securities are tax deductible, whereas dividend payments made to owners under equity financing are not. This difference in the tax treatment of the two types of financing is one reason for corporations' heavy reliance on debt financing. Even though debt financing offers distinct tax advantages to both the corporation and the investors, there is always the risk that a corporation which relies too much on debt financing will be deemed too thinly capitalized by the IRS and the Securities and Exchange Commission (SEC), which will then treat any loans made by shareholders as capital contributions.

The various types of bonds are discussed next.

Unsecured Bonds. Unsecured bonds, usually called debentures, have only the obligation of the corporation behind them. Thus, debenture holders are unsecured creditors that rank equally with other general creditors.

secured bond Type of bond secured by the issuer's pledge of a specific asset.

income bond A bond whose payment of interest is dependant on sufficient earnings.

participating bond A bond that pays both interest at a fixed rate and to dividends.

Secured Bonds. A secured creditor is a creditor whose claim not only is enforceable against the general assets of the corporation but also is a lien on specific property. Thus, **secured**, or mortgage, **bonds** provide the security of specific corporate property in addition to the general obligation of the corporation.

Income Bonds. Traditionally, debt securities bear a fixed interest rate that is payable without regard to the financial condition of the corporation. **Income bonds**, on the other hand, condition the payment of interest to some extent on corporate earnings. **Participating bonds** call for a stated percentage of return, regardless of earnings, with additional payments dependent on earnings.

---

*Brennan's Inc. v. Colbert Court of Appeal of Lousiana, 85 5o.3d 787 (2012).

Convertible Bonds. Usually at the option of the holder, **convertible bonds** may be exchanged, in a specified ratio, for other securities of the corporation.

Callable Bonds. Callable bonds are subject to a redemption provision that permits the corporation to redeem or call (pay off) all or part of the issue before maturity at a specified redemption pace.

AAA Bonds and Junk Bonds. These types of bonds connote the grade of investment bonds from low risk to high risk. Moody's Global Long Term Rating Scale rates bonds from a high of Aaa (highest quality subject to the lowest level of risk) to C (the lowest rated, typically in default, with little prospect for recovery of principal or interest).

> **convertible bonds** Bonds that can be exchanged at a fixed rate for another type of security.
>
> **callable bonds** A bond which the issuer has the right to redeem prior to its maturity date, under certain conditions.

EQUITY FINANCING

All business corporations must raise operating capital through the sale of stock, equity securities, or stock operations. *Shareholders*—persons who purchase shares of stock—generally acquire rights to control the corporation through voting; to receive income through dividends; and, upon dissolution of the corporation, to share in the net assets in direct proportion to the number of shares they own.

The number of shares of stock must be authorized in the corporation's articles of incorporation. All shares authorized by the articles need not be issued or sold to shareholders immediately, but no shares may be issued that are not authorized. Under the RMBCA, the articles of incorporation must authorize (1) one or more classes of stock that entitle their owners to unlimited voting rights and (2) one or more classes of stock (these may be the same classes as those with voting rights) that entitle their owners to receive the net assets of the corporation upon dissolution. This provision of the RMBCA ensures that there will be a class of shareholders with the power to elect directors and make other important decisions and a class of shareholders who will share in the residuary (remaining assets) of the corporation upon its termination.

Classes of Stock. Most states allow corporations to authorize and issue different classes of stock, with different rights attached to the different classes. The limitations and preferences of each class must be stated in the articles of incorporation. The three primary classes of stock are (1) common, (2) preferred, and (3) stock options.

Traditionally, **common stock** has carried with it the right to vote, the right to participate in income through dividends, and the right to participate in the net assets on liquidation. Common stock has no preferential rights (described in the next paragraphs); therefore, common stockholders bear the greatest risk of loss. If a corporation has only one class of stock, it is ordinarily assumed to be common stock.

Owners of **preferred stock** are given special preferences relating to either the payment of dividends or the distribution of assets. Most preferred stock is preferred as to dividends, meaning that in every year in which the corporation pays a dividend, the preferred shareholders are paid before the common shareholders at a rate stipulated in the articles of incorporation.

If the stock is *cumulative preferred*, the preferred shareholders do not lose their rights to a dividend during a year in which no dividends are paid. Rather, their rights to each year's unpaid dividends accumulate. Thus, during the next year in which dividends are paid, the preferred shareholders receive all past dividends that have accumulated plus the present year's dividends before any dividends are paid to the common shareholders.

If the stock is *participating preferred*, the preferred shareholders first receive their dividends at the preferred rate. The common shareholders receive dividends at the same rate. The remaining income is shared, on a pro rata basis, by the common and preferred stockholders.

> **common stock** A class of stock that entitles its owner to vote for the corporation's board of directors, receive dividends, and participate in the net assets upon liquidation of the corporation.
>
> **preferred stock** A class of stock that entitles its owner to special preferences relating to either dividends or the distribution of assets.

If the stock is *liquidation preferred*, upon liquidation of the corporation, preferred shareholders receive either the par value of their stock or a specified monetary amount before the common shareholders share pro rata in the remainder of the assets. Liquidation preferences may also be participating.

Finally, preferred stock may be *convertible*. At the holder's request, such stock may be exchanged for common stock at a stated ratio.

Preferred stock frequently has limited voting rights. It is also generally redeemable, meaning that the corporation has the right to exchange each preferred share for a prespecified monetary amount.

All these distinctions may become irrelevant as more states adopt the RMBCA. The RMBCA stipulates that the various classes of stock and the number of shares in each class that may be issued must be stated in the articles of incorporation, but it omits references to such classes of stock as "preferred" and "common." If only one class of stock is designated in the articles of incorporation, it is presumed that the shares confer both the right to vote and the right to participate in corporate assets. If more than one class of stock is authorized, either the distinguishing designation and preferences, limitations, and rights of those classes must be listed or the board of directors must be authorized to designate such features at a later date.

capital structure The percentage of each type of capital—debt, preferred stock, and common equity—used by a corporation.

stock warrant A document authorizing its holder to purchase a stated number of shares of stock at a stated price, usually for a stated period of time; may be freely traded.

stock option A stock warrant issued to employees; cannot be traded.

This broad flexibility given to the directors to affect the **capital structure** of the corporation may be desirable from management's perspective. It does not, however, benefit the shareholders, because it may dilute their interests. It is also contrary to present trends in securities regulation (discussed in Chapter 23).

Under the RMBCA, corporations can also issue rights to purchase a stated number of shares at a stated price, usually for a stated period of time. Legal documents, called **stock warrants**, that certify these rights may be freely traded. Employees may receive such rights as compensation in the form of **stock options**.

Stock Options and Government Regulation

Stock options were used as a way of providing incentives or bonuses for a company's top employees in the late 1800s and early 1900s. As publicly traded companies grew, tax laws promulgated by Congress in the 1950s allowed employees to pay a capital gains tax rate of 25 percent. By 1952, one-third of all New York Stock Exchange companies were using executive stock options.

In the 1960s and 1970s, Congress enacted a number of restrictions on stock options. In 1973, the Chicago Board of Trade opened the first public market for stockholders to trade options on the shares of public companies. In the 1980s, Congress enacted new laws that made stock options important. It restored the capital gains treatment for option profits and slashed further the capital gains tax rate, which encouraged companies to offer large stock options to key executives instead of increasing salaries. In 1992, the top five executives at the 1,500 largest U.S. corporations cashed in approximately $2.4 billion of options. During the 1990s, many companies gradually extended options to large numbers of employees as bonuses.

By 2002, between 20 and 25 percent of public companies made options available to many of their full-time employees. This was particularly important in the growth of high-tech companies where founders could not afford large salaries, and these stock options were frequently given as a substitute. Many people working for high-tech companies became wealthy overnight.[a]

In 2006, the SEC and the Justice Department began investigating the question of backdating stock options. Over 100 companies were served warrants under the 1934 Exchange Act. The SEC argued that backdating of stock options

---

[a]*Federal Securities Law Reports*, July 26, 2006; August 16, 2007.

is not, in and of itself, illegal if properly disclosed. Backdating increases stock options' potential value if the option grant is backdated to a time when the stock is at a lower market price. Many executives claimed that backdating was common in some industries as a way to obtain employment of talented people.

In August 2007, the former CEO of Brocade Communications, Gregory Reyes, was convicted on 10 counts that included securities fraud, mail fraud, falsifying books, and making false statements to auditors.[b]

This was the first U.S. Justice Department conviction based on backdating of stock options. It sent a message to executives about full disclosure under the 1934 Exchange Act. The government was able to prove that Reye knowingly manipulated an option grant date to defraud investors.

---

[b]See, generally, Joseph Blasi, Douglas Kruse, and Aaron Bernstein. *In the Company of Owners: The Truth About Stock Options* (New York: Basic Books, 2003).

When employees are granted the rights to purchase shares at a stated price, however, these rights cannot be traded. A major issue was whether director, officer, and employee stock options would become expense items under a Financial Accounting Standards Board (FASB) proposal. In December 2004, the FASB unanimously adopted the expense item rule, which was scheduled to take effect in July 2005. Further, the FASB told companies that if they expect to buy back shares of their own company stock in connection with a stock option plan, they must declare an estimate of how many shares they will buy back. Often, buyback programs are used to prevent stock options and the new shares resulting from the exercise of options from diluting stock prices and earnings per share.

CONSIDERATION

Stocks and warrants are issued in return for consideration—that is, something of value. That consideration cannot be less than the stated value of the shares. If a corporation issues shares for less than the stated value, the shareholder remains liable to the corporation for the difference between the stated value and the amount of consideration actually paid.

Traditionally, the minimum amount for which a share could be issued was called the **par value**. In general, this amount was so low (often $1) that the likelihood of a buyer not paying at least that much for the stock was slim. The total par value of all stock initially issued by a corporation was known as *stated capital*. Some states allowed the issuance of no-par stock, which is stock that does not have a stated par value, but when a corporation issued no-par stock, the board of directors still had to designate a stated value for the stock. The sum of the stated values was the stated capital of the corporation.

par value The nominal or face value of a stock or bond.

The RMBCA has done away with the terms *par*, *no par*, and *stated capital*. Now, before issuing any shares, the board of directors must determine that the amount of consideration received or about to be received is adequate. When the corporation receives the consideration for which the board of directors authorized the issuance of the shares, the shares are deemed fully paid.

Under the RMBCA and most current state laws, the consideration paid for the stock may be in the form of money, property, or past services. The RMBCA also allows for payment by promissory notes and agreements to provide future services.

In some states, problems can arise over the valuation of nonmonetary consideration. Most states use the good-faith rule, which presumes that the valuation of the property or services given as consideration for the stock was fair as long as it was honestly made—in other words, there was no fraud or bad faith on the part of the directors in making their valuation, and they exercised the degree of care that ordinarily prudent persons in their position would exercise.

Operation of Corporations

The question of how the corporation is financed can be answered relatively easily. As the previous section explained, the corporation is financed by debt and equity security holders. The answer to the question of who manages the corporation is not quite as simple. Even legal experts disagree to some extent over who actually manages the corporation, as well as who should manage it.

Three groups theoretically have a voice in the management of the corporation: the shareholders, the board of directors, and the corporate officers and managers. Formal responsibility for management of the corporation is vested in its board of directors, who are elected by the shareholders. These directors determine policy matters and appoint the officers who carry out those policies and manage the everyday affairs of the corporation. Exhibit 17-1 illustrates the division of responsibility in this corporate hierarchy of shareholders, board of directors, and officers and managers.

THE ROLE OF THE SHAREHOLDERS

The shareholders are the owners of the corporation, yet they have no direct control over its operation. They are not agents of the corporation and cannot act on its behalf. Their control of the corporation is limited to exercising their right to vote at shareholders' meetings and, through that voting, to select the board of directors who will set corporate policy. (See the *Wynn* case later in this chapter.)

Most corporations are obliged to hold an annual shareholders' meeting at a time specified in the corporate bylaws. In addition, special shareholders' meetings may be called by the board of directors, the holders of more than 5 percent of the shares are entitled to cast a vote at such meetings, or anyone else who is authorized to do so under the corporate bylaws.

proxy A document by which shareholders or a publicly held company can transfer its rights to vote at a shareholders' meeting to a second party.

Shareholders vote at the shareholders' meetings either in person or by a **proxy**, which is a written delegation of authority to cast one's votes. Most shareholders vote by proxy, and it is this process of proxy election that has led many people to question whether shareholders really have any say in operating the corporation.

The proxy election is usually run by a proxy committee of corporate executives, who, under the Security and Exchange Commission's proxy rules, must use a ballot form to solicit proxies. The form must state that the shares held by the shareholder will be voted in accordance with the way the shareholder marks the ballot. The shareholder has the option to indicate on the ballot that he or

EXHIBIT 17-1

THE CORPORATE HIERARCHY

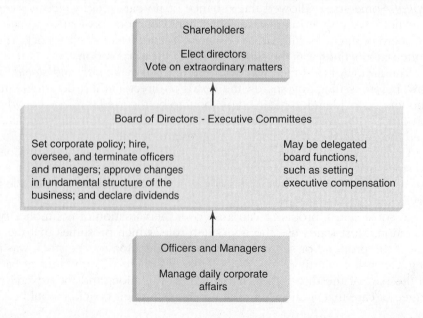

she wishes to allow the proxy committee to vote the shares in any way it sees fit. The proxy committee also sends all shareholders a statement of resolutions on which the shareholders are to vote, as well as a biographical sketch of each of the candidates for the board of directors.

This process sounds efficient, even benign. However, because shareholders of major corporations are scattered across the country and could not realistically attend a shareholders' meeting, the proxy process gives management effective control over the election. By placing on the ballot only the names of those candidates management wishes to see elected to the board of directors, management, in essence, selects the board. Although a shareholder can write in the name of another candidate, the cost of communicating with other shareholders makes the prospects for a write-in candidate quite slim.

Any shareholder may also engage in proxy solicitation. The costs of doing so, however, are almost prohibitive. In a fight between a shareholder and the corporate proxy committee for proxies, the corporate committee has access to corporate funds; corporate office materials such as paper, duplicating machinery, and postage accounts; shareholder contact information; corporate clerical personnel; and the corporate legal staff. The shareholders have only their personal funds.

In recent years, the proxy forms sent out by proxy committees of major corporations have sometimes contained resolutions submitted by politically active shareholders who understand how the corporate machinery operates. Most of these resolutions have sought to change the corporation's social policies; for example, resolutions seeking to prohibit the corporation from investing in countries that practice apartheid and from withholding information from shareholders regarding the environmental impact of the corporation's activities have been popular in the past decade or so. The public-interest proxy resolution shown in Exhibit 17-2 was proposed by a shareholder of General Motors in the proxy statement for the 1985 annual meeting. Its objective was to force General Motors to be politically neutral.

When these types of resolutions appear on proxy forms, the management of the corporation usually suggests that shareholders vote against them. Management also generally includes a strong argument against such resolutions on the proxy statement on which the proposal appears. Thus, even if a shareholder does get a resolution on the ballot, the chances of it passing are slim, although there are more frequent and more vigorous fights over policy resolutions than over the election of directors.

Recognizing that shareholder voting occurs primarily by proxy and, thus, that the shareholders' meeting really serves no purpose but to fulfill the demands of the law, Delaware decided to abolish the requirement that corporations hold an annual shareholders' meeting. Other states may soon follow Delaware's example.

EXHIBIT 17-2

SHAREHOLDER RESOLUTION

Resolved: That the stockholders of General Motors, assembled in annual meeting in person and by proxy, hereby recommended that the Corporation affirm the political non-partisanship of the Corporation. To this end, the following practices are to be avoided.

a. The handing of contribution cards of a single political party to an employee by a supervisor.

b. Requesting an employee to send a political contribution to an individual in the corporation for subsequent delivery as part of a group of contributions to a political party or fund-raising committee.

c. Requesting an employee to issue personal checks blank to payee for subsequent forwarding to a political party, committee, or candidate.

d. Using supervisory meetings to announce that contribution cards of one party are available and that anyone desiring cards of a different party will be supplied one on request to his supervisor.

e. Placing a preponderance of contribution cards of one party at mail station locations.

THE ROLE OF THE BOARD OF DIRECTORS

Most incorporation laws state that the corporation shall be managed by the board of directors. Although such a rule may reflect the behavior of the directors of a closely held corporation who are also its officers, it does not describe the behavior of the directors of publicly held corporations; nor does it describe the behavior that most people expect from directors of publicly held corporations.

Recognizing that most directors are not going to become involved in daily corporate affairs, the RMBCA in its statement on the role of the board of directors says merely that the corporation shall be managed "under the direction of" the board. This provision does seem to make it clear, however, that the directors are expected to function at least as overseers and policy makers. In this role, the board of directors generally must authorize or approve (1) the payment of dividends, changes in financing, and other capital changes; (2) the selection, supervision, and removal of officers and other executive personnel; (3) the determination of executive compensation and pension plans; (4) the adoption, amendment, or repeal of the corporate bylaws; and (5) the establishment of policy regarding products, services, and labor relations. Unfortunately, most corporate boards abdicate their policy-making function and merely rubber-stamp decisions already made by the officers and managers of the corporation. Rarely does a board challenge an action taken by its corporate officers (Exhibit 17-3).

EXHIBIT 17-3

CEO PAY: THE BENEFITS AND COSTS TO SHAREHOLDERS

Following the exposure of high salaries, and other financial problems, involving companies and individuals inclusive of Enron, WorldCom, Adelphia, and Richard Grasso (formerly of the New York Stock Exchange),[a] the compensation of officers of corporations has been spotlighted.

The decisions for compensation to be paid to a CEO, and other officers of a publicly held corporation, is usually made by the compensation committee of the board of directors. The board of directors is often made up of present and past CEOs, of corporations, nominated informally by the CEO of the company itself and voted on by the full board. Reforms on compensation, how committees are formed, and what members of the board are eligible have taken place.

Boards, representing shareholders, pay large sums to CEOs and other officers of a corporation for several reason as noted below.

- Some argue that CEOs should be paid based on stock market performance of the company. There are up and down years in the market, but generally qualified CEOs are needed for direction.
- Some argue that if CEOs own stock, they have the same interest as shareholders.
- Others doubt this analysis, arguing that compensation committees of board of directors seek to keep compensation packages similar to those in other corporations in the same sector.
- Compensation committees seek to increase compensation packages in order to keep CEOs from going to other companies, and often tie such packages to stock performance.
- Under rules adopted by the SEC in January, 2011, companies whose stock market value exceeds $75 million must give shareholders a chance at least once every three years to voice their approval or disapproval of how senior management is paid. The vote, if called for, is mandatory, but the result is advisory.[b]

[a]WSJ/Mercer, "CEO Compensation Survey," *Wall Street Journal*, April 12, 2003, R-6–10; and "Grasso Lawsuit Fallout Hits Other Executives," *Wall Street Journal*, May 25, 2003, C1, C6.
[b]Jason Swide, "A Chance to Veto a Bonus," *Wall Street Journal*, B1, Jan. 30, 2011.

The board, like the shareholders, must function as a group. Unlike shareholders, however, directors are not allowed to vote by proxy. Instead, they are supposed to vote in person at a formal directors' meeting. The rationale for this traditional restriction is that corporations derive benefits from the consultation, discussion, and collective judgment of their boards. Today, however, the laws in most states have been relaxed enough so that board members may act informally, without a meeting, when all the directors consent to such action in writing. Boards are often, however, faced with complex legal issues such as derivative shareholder litigation challenging board competence, expertise, and good faith, as in the following case.

CASE 17-2

In re Abbott Laboratories Derivative Shareholders Litigation
United States Court of Appeals for the Seventh Circuit
325 F.3d (7th Cir. 2003)

In 1999, Abbott Laboratories, an Illinois corporation, entered into a consent decree that required it to pay a $100 million fine, at the time the largest penalty ever imposed for a civil violation of Food and Drug Administration (FDA) regulations. The FDA also required Abbott to withdraw 125 types of medical diagnostic test kits, destroy certain inventory, and make a number of corrective changes in its manufacturing procedures after six years of quality control violations. Abbott shareholders brought a shareholder derivative suit alleging that the directors breached their fiduciary duties when they failed to take the necessary action to correct repeated noncompliance problems brought to Abbott's attention by the FDA in the period from 1993 until 1999.

The FDA not only had sent Abbott a Form 453 noting deviations from the requirements set forth in the FDA's "Current Good Manufacturing Practice" after each of its 13 inspections of Abbott's Abbott Park and North Chicago facilities but also sent four formal certified Warning Letters to Abbott. The first was sent to David Thompson, president of Abbott's Diagnostics Division (ADD), on October 20, 1993. The letter stated that the FDA had found adulterated in vitro diagnostic products and warned: "Failure to correct these deviations may result in regulatory action being initiated by the Food and Drug Administration without further notice. These actions include, but are not limited to, seizure, injunction, and/or civil penalties." A second Warning Letter was sent to Thompson on March 28, 1994, with a copy to Duane Burnham, Abbott's CEO and board chair.

On January 11, 1995, the *Wall Street Journal* reported that the FDA had uncovered a wide range of flaws in Abbott's quality assurance procedures used in assembling its diagnostic products. In July 1995, the FDA and Abbott entered into a Voluntary Compliance Plan to work together to correct Abbott's deficiencies. In February 1998, after finding continued deviations from the regulations, the FDA sent Abbott the equivalent of a Warning Letter closing out the plan. In 1999, the FDA sent the fourth and final Warning Letter to Miles White, a member of Abbott's board and the current CEO. White had replaced Burnham as CEO in April 1999.

The plaintiffs did not demand that the members of Abbott's board institute an action against themselves for breach of their fiduciary duties, arguing that such a demand would be futile. Under applicable law, such a demand is not required if the plaintiffs plead facts showing that the directors faced a substantial likelihood of liability for their actions. In particular, "demand can only be excused where facts are alleged with particularity which creates a reasonable doubt that the directors' action was entitled to the protections of the business judgment rule." The district court dismissed the complaint, and the plaintiffs appealed.

Justice Wood

Plaintiffs in *Abbott* allege facts that the directors were aware of known violations, providing evidence that there was direct knowledge through the Warning Letters and as members of the Audit Committee. Under proper corporate governance procedures—the existence of which is not contested by either party in *Abbott*—information of the violations would have been shared at the board meetings. In addition, plaintiffs have alleged that, as fiduciaries, the directors all signed the annual SEC forms which specifically addressed government regulation of Abbott's products. The *Abbott* case is clearly distinguished from the "unconsidered" inaction in *In re Caremark*.

. . . .

The district court noted, correctly, that the plaintiffs did not allege that Abbott's reporting system was inadequate. . . . Where there is a corporate governance structure in place, we must then assume the corporate governance procedures were followed and that the board knew of the problems and decided no action was required. . . .

. . . .

Delaware law states that director liability may arise from the breach of duty to exercise appropriate attention to potentially illegal corporate activities of from "an *unconsidered failure of the board to act* in circumstances in which due attention would, arguably, have prevented the loss." *In re Caremark* The court held that "a sustained or systematic failure of the board to exercise oversight will establish the lack of good faith that is a necessary condition to [director] liability."

Given the extensive paper trail in *Abbott* concerning the violation and the inferred awareness of the problems, the facts support a reasonable assumption that there was a "sustained and systematic failure of the board to exercise oversight," in this case intentional in that the directors knew of the violations of law, took no steps in an effort to prevent or remedy the situation, and that failure to take any action for such an inordinate amount of time resulted in substantial corporate losses, establishing a lack of good faith. We find that six years of noncompliance, inspections, 483s, Warning Letters, and notice in the press, all of which resulted in the largest civil fine ever imposed by the FDA and the destruction and suspension of products which accounted for approximately $250 million in corporate assets, indicate that the directors' decision to not act was not made in good faith and was contrary to the best interests of the company.

With respect to demand futility based on the directors' conscious inaction, we find that the plaintiffs have sufficiently pleaded allegations, if true, of a breach of the duty of good faith to reasonably conclude that the directors' actions fell outside the protection of the business judgment rule. . . .

The directors contend that they are not liable under Abbott's certificate of incorporation provision which exempts the directors from liability [for breach of the duty care]. Directors are not protected by that provision when a complaint alleges facts that infer a breach of loyalty or good faith. [*Eds.*: The court stated that the burden of establishing good faith rests with the director seeking protection under the exculpatory provision.]

Plaintiffs in *Abbott* accused the directors not only of gross negligence, but of intentional conduct in failing to address the federal violation problems, alleging "a conscious disregard of known risks, which conduct, if proven, cannot have been undertaken in good faith."*

Affirmed for the Plaintiff.

THE ROLE OF THE OFFICERS AND MANAGERS

The officers and managers are responsible for the actual management of corporate affairs. Unlike directors and shareholders, officers are agents of the corporation. Technically, they are appointed and supervised by the board of directors, but because the officers of the company control the proxy election, it is usually the directors who serve at the pleasure of the corporate officers.

Traditional statutes provided that a corporation must have certain officers, such as a president, a vice president, and a treasurer. The RMBCA, however, makes no such stipulations.

FIDUCIARY OBLIGATIONS OF DIRECTORS, OFFICERS, AND MANAGERS

As previously noted, the shareholders do not have any direct control over the corporation's operations. Whenever the property of one party is placed in the control of another, however, a fiduciary relationship exists between the two. Thus, certain obligations, called *fiduciary duties*, are placed on property holders to ensure that they will treat the property as carefully as if it were their own. Because the shareholders own the corporation (property) but its care is entrusted to the directors, officers, and managers, a fiduciary relationship exists between the shareholders and the officers, managers, and directors. The standards of conduct—and, thus, the fiduciary duties—imposed by the RMBCA on officers and directors of corporations are almost identical. Both are required to exercise their duties (1) in good faith, (2) with the care an ordinarily prudent person in a like position would exercise under similar circumstances, and (3) in a manner he or she reasonably believes to be in the best interests of the corporation.[1]

There is a potential breach of this duty to act in the best interests of the corporation when a corporate officer or director takes personal advantage of an opportunity that, in all fairness, should have belonged to the corporation. The following case illustrates the **corporate opportunity doctrine**.

corporate opportunity doctrine
A doctrine, established by case law, that says that corporate officers, directors, and agents cannot take personal advantage of an opportunity that in all fairness should have belonged to the corporation.

---
[1] RMBCA §§ 8.30, 8.42.

*In Re Abbott Laboratories Derivative Shareholders Litigation, United States Court of Appeals for the Seventh Circuit. 325 F.3d (7th Cir. 2003).

CASE 17-3

Beam v. Stewart
Court of Chancery of Delaware, New Castle
833 A.2d 961; affirmed, 845 A.2d 1040 (2003)

Monica A. Beam, a shareholder of Martha Stewart Living Omnimedia, Inc. (MSO), brought a derivative action against the defendants, all current directors and a former director of MSO, and against MSO as a nominal defendant. MSO is a Delaware corporation that operates in the publishing, television, merchandising, and Internet industries marketing products bearing the "Martha Stewart" brand name. Defendant Martha Stewart (Stewart) is a director of the company and its founder; chairman; chief executive officer; and by far, its majority shareholder, controlling roughly 94.4 percent of the shareholder vote. Stewart, a former stockbroker, has in the past 20 years become a household icon, known for her advice and expertise on virtually all aspects of cooking, decorating, entertaining, and household affairs generally.

The market for MSO products is uniquely tied to the personal image and reputation of its founder, Stewart. MSO retains "an exclusive, worldwide, perpetual royalty-free license to use [Stewart's] name, likeness, image, voice and signature for its products and services." In its initial public offering prospectus, MSO recognized that impairment of Stewart's services to the company, including the tarnishing of her public reputation, would have a material adverse effect on its business. In fact, under the terms of her employment agreement, Stewart may be terminated for gross misconduct or felony conviction that results in harm to MSO's business or reputation, but is permitted discretion over the management of her personal, financial, and legal affairs to the extent that Stewart's management of her own life does not compromise her ability to serve the company.

Stewart's alleged misadventures with ImClone arose in part out of a long-standing personal friendship with Samuel D. Waksal (Waksal). Waksal is the former chief executive officer of ImClone as well as a former suitor of Stewart's daughter. Waksal and Stewart have provided one another with reciprocal investment advice and assistance, and they share a stockbroker, Peter E. Bacanovic (Bacanovic) of Merrill Lynch. The speculative value of ImClone stock was tied directly to the likely success of its application for FDA approval to market the cancer-treatment drug Erbitux. On December 26, Waksal received information that the FDA was rejecting the application to market Erbitux. The following day, December 27, he tried to sell his own shares and tipped off his father and daughter to do the same. Stewart also sold her shares on December 27 (see Case 1-1). After the close of trading on December 28, ImClone publicly announced the rejection of its application to market Erbitux. The following day, the trading price closed slightly more than 20 percent lower than the closing price on the date that Stewart had sold her shares. By mid-2002, these events had attracted the interest of the *New York Times* and other news agencies, federal prosecutors, and a committee of the U.S. House of Representatives. Stewart's publicized attempts to quell any suspicion were ineffective at best because they were undermined by additional information as it came to light and by the other parties' accounts of the events. Ultimately, Stewart's prompt efforts to turn away unwanted media and investigative attention failed. Stewart eventually had to discontinue her regular guest appearances on CBS's *The Early Show* because of questioning during the show about her sale of ImClone shares. After barely two months of such adverse publicity, MSO's stock price had declined by slightly more than 65 percent. In January 2002, Stewart and the Martha Stewart Family Partnership sold 3 million shares of Class A stock to ValueAct, an investor group.

The complaint alleges that the director defendants breached their fiduciary duties by failing to ensure that Stewart would not conduct her personal, financial, and legal affairs in a manner that would harm the company, its intellectual property, or its business. It also alleges that Stewart breached her fiduciary duty of loyalty, usurping a corporate opportunity by selling large blocks of MSO stock.

Chancellor Chandler

The "duty to monitor" has been litigated in other circumstances, generally where directors were alleged to have been negligent in monitoring the activities of the corporation, activities that led to corporate liability. That the Company is "closely identified" with Stewart is conceded, but it does not necessarily follow that the Board is required to monitor, much less control, the way Stewart handles her *personal* financial and legal affairs.

Regardless of Stewart's importance to MSO, she is not the corporation. And it is unreasonable to impose a duty upon the Board to monitor Stewart's personal affairs because such a requirement is neither legitimate nor feasible. Monitoring Stewart by, for example, hiring a private detective to monitor her behavior is more likely to generate liability to Stewart under some tort theory than to protect the Company from a decline in its stock price as a result of harm to Stewart's public image.

[A] corporate officer or director may not take a business opportunity for his own if: (1) the corporation is financially able to exploit the opportunity; (2) the opportunity is within the corporation's line of business;

(3) the corporation has an interest or expectancy in the opportunity; and (4) by taking the opportunity for his own, the corporate fiduciary will thereby be placed in a position [inimical] to his duties to the corporation.

In this analysis, no single factor is dispositive. Instead the Court must balance all factors as they apply to a particular case. For purposes of the present motion, I assume that the sales of stock to ValueAct could be considered to be a "business opportunity." I now address each of the four factors articulated in *Broz*.

The amended complaint asserts that MSO was able to exploit this opportunity because the Company's certificate of incorporation had sufficient authorized, yet unissued, shares of Class A common stock to cover the sale to ValueAct. Defendants do not deny that the Company could have sold previously unissued shares to ValueAct. I therefore conclude that the first factor has been met.

An opportunity is within a corporation's line of business if it is an activity as to which [the corporation] has fundamental knowledge, practical experience and ability to pursue.

MSO is a consumer products company, not an investment company. Simply stated, selling stock is not the same line of business as selling advice to homemakers. . . . For the foregoing reasons, I therefore conclude that the sale of stock by Stewart was not within MSO's line of business.

A corporation has an interest or expectancy in an opportunity if there is "some tie between that property and the nature of the corporate business." . . . Here, plaintiff does not allege any facts that would imply that MSO was in need of additional capital, seeking additional capital, or even remotely interested in finding new investors.

The corporate opportunity doctrine is implicated only in cases where the fiduciary's seizure of an opportunity results in a conflict between the fiduciary's duties to the corporation and the self-interest of the director as actualized by the exploitation of the opportunity. Given that I have concluded that MSO had no interest or expectancy in the issuance of new stock to ValueAct, I fail to see, based on the allegations before me, how Stewart's . . . sales placed [her] in a position inimical to their duties to the Company. Were I to decide otherwise, directors of every Delaware corporation would be faced with the ever-present specter of suit for breach of their duty of loyalty if they sold stock in the company on whose Board they sit.

Additionally, Delaware courts have recognized a policy that allows officers and directors of corporations to buy and sell shares of that corporation at will so long as they act in good faith.

On balancing the four factors, I conclude that plaintiff has failed to plead facts sufficient to state a claim that Stewart . . . usurped a corporate opportunity for [herself] in violation of [her] fiduciary duty of loyalty to MSO. [This count] is dismissed in its entirety for failure to state a claim upon which relief can be granted.*

For Defendants.

LINKING LAW AND BUSINESS

Financial Accounting

As a consequence of the disadvantages involved with debt financing, users of financial statements use ratios to assess the risk. Using one ratio or even several ratios, however, is not the basis for effective analysis of financial statements. It is imperative that ratios be used concurrently with one another and with other information, including historical facts, because the usefulness of ratios depends on the likelihood that history will repeat itself. One example of a ratio that evaluates the risk involved in debt financing is the times-interest-earned ratio. This ratio is computed by dividing a firm's earnings before interest and taxes (EBIT) by the interest expense. The times-interest-earned ratio is used to determine how many times an organization can pay interest expenses using available earnings for interest payments. The higher the ratio, the more likely the firm will be able to make its interest payments. A firm's inability to make interest payments can lead to bankruptcy; therefore, a higher ratio indicates that there is less risk for the firm.

Source: T. Edmonds, F. McNair, E. Milam, and P. Olds, *Fundamental Financial Accounting Concepts* (New York: McGraw-Hill, 2000), 480–82.

Although an opportunity must be offered to the corporation, it may not necessarily be accepted by the corporation. Once an opportunity is rejected by a vote of the disinterested members of the board of directors, that opportunity no longer belongs to the corporation.

When the corporation cannot take advantage of an opportunity because of financial constraints, a director or an officer must do his or her best to secure

---

*Beam v. Stewart Court of Chancery of Delaware, New Castle 833 A.2d 961; affirmed, 845 A.2d 1040 (2003).

financing of the opportunity by the corporation, although neither need go so far as to lend the corporation money. In general, taking personal advantage of the opportunity when financing is clearly unavailable is allowed.

Another potentially troublesome situation occurs when an officer or a director, or a corporation in which the officer or director has an interest, enters into a transaction with the corporation. This problem, known as a **conflict of interest**, is specifically addressed by the RMBCA. The act provides that a transaction will not be voided because of a conflict of interest when any one of the following is true: (1) The material facts of the transaction and the director's interest were disclosed or known to the board of directors or a committee thereof, and the board authorized, approved, or ratified the transaction; (2) the material facts of the transaction and the director's interest were known or disclosed to the shareholders entitled to vote, and they approved, authorized, or ratified the transaction; or (3) the transaction was fair to the corporation. Use of the broad language "the transaction was fair" seems to give the utmost flexibility to directors. In many states that have not adopted the RMBCA, directors can protect themselves from conflict-of-interest charges through full disclosure and ratification by the board of directors.

conflict of interest A conflict that occurs when a corporate officer or director enters into a transaction with the corporation in which he or she has a personal interest.

We have said that officers and directors are required to exercise their duties in a manner they reasonably believe to be in the best interests of the corporation. Under the **business judgment rule**, the courts generally avoid second-guessing corporate executives and let stand any business decisions made in good faith that are uninfluenced by personal considerations.

In determining whether a director's conduct fulfills the duty of care, the courts are aided by the RMBCA, which states that a director is entitled to rely on information and reports provided or prepared by (1) officers or employees of the corporation whom the director reasonably believes are reliable and competent, (2) legal counsel or accountants in regard to matters that the director believes are within that professional's competence, or (3) a committee of directors of which the director is not a part if he or she reasonably believes that the committee merits confidence. The use courts make of the business judgment rule is illustrated in the following landmark case. Note carefully the court's emphasis on the financial expertise required when discussing the question of a valuation study.

business judgment rule A rule that says that corporate officers and directors are not liable for honest mistakes of business judgment.

CASE 17-4

Smith v. Van Gorkom
Delaware Supreme Court
488 A.2d 858 (1985)

In order to take advantage of a favorable tax situation, defendant Van Gorkom, chief executive of Trans Union Corporation (Trans Union or the Company), solicited a merger offer from Pritzker, an outside investor. Van Gorkom acted on his own and arbitrarily arrived at a buyout price of $55 per share. Without any investigation, the full Trans Union board accepted the offer informally. The offer was proposed two more times before its formal acceptance by the board. Plaintiff Smith and other shareholders brought suit, claiming that the board had failed to give due consideration to the offer. The trial court held that the shareholder vote approving the merger should not be set aside because the stockholders had been "fairly informed" by the board of directors before they voted on it.

The court also found that because the board had considered the offer three times before formally accepting it, the board had acted in an informed manner and was, therefore, entitled to the protection of the business judgment rule. The plaintiffs appealed.

Justice Horsey

On Friday, September 19, Van Gorkom called a special meeting of the Trans Union Board for noon the following day....

Van Gorkom began the Special Meeting of the Board with a twenty-minute oral presentation. Copies of the proposed Merger Agreement were delivered too late for study before or during the meeting. He reviewed the Company's ITC and depreciation problems and the efforts theretofore made to solve them. He discussed his initial meeting with Pritzker and his motivation in arranging that meeting. Van Gorkom did not disclose to the Board, however, the methodology by which he alone had arrived at the $55 figure, or the fact that he first proposed the $55 price in his negotiations with Pritzker.

Van Gorkom outlined the terms of the Pritzker offer as follows: Pritzker would pay $55 in cash for all outstanding shares of Trans Union stock, upon completion of which Trans Union would be merged into new T Company, a subsidiary wholly owned by Pritzker and formed to implement the merger; for a period of 90 days, Trans Union could receive, but could not actively solicit, competing offers; the offer had to be acted on by the next evening, Sunday, September 21; Trans Union could only furnish to competing bidders published information, and not proprietary information; the offer was subject to Pritzker obtaining the necessary financing by October 10, 1980; if the financing contingency were met or waived by Pritzker, Trans Union was required to sell to Pritzker one million newly issued shares of Trans Union at $38 per share.

The Board meeting of September 20 lasted about two hours. . . . The directors approved the proposed Merger Agreement.

On February 10, the stockholders of Trans Union approved the Pritzker merger proposal. Of the outstanding shares, 69.9% were voted in favor of the merger, 7.25% were voted against the merger, and 22.85% were not voted.

The determination of whether a business judgment is an informed one turns on whether the directors have informed themselves "prior to making a business decision of all material information reasonably available to them."

In the specific context of a proposed merger of domestic corporations, a director has a duty under 8 [Delaware Code §] 251(b), along with his fellow directors, to act in an informed and deliberate manner in determining whether to approve an agreement of merger before submitting the proposal to the stockholders. Certainly in the merger context, a director may not abdicate that duty by leaving to the shareholders alone the decision to approve or disapprove the agreement.

On the record before us, we must conclude that the Board of Directors did not reach an informed business judgment on September 20, 1980 in voting to "sell" the company for $55 per share pursuant to the Pritzker cash-out merger proposal. Our reasons, in summary, are as follows:

The directors (1) did not adequately inform themselves as to Van Gorkom's role in forcing the "sale" of the Company and in establishing the per share purchase price; (2) were uninformed as to the intrinsic value of the Company; and (3) given these circumstances, at a minimum, were grossly negligent in approving the "sale" of the Company upon two hours' consideration, without prior notice, and without the exigency of a crisis or emergency.

Without any documents before them concerning the proposed transaction, the members of the Board were required to rely entirely upon Van Gorkom's 20-minute oral presentation of the proposal. No written summary of the terms of the merger was presented; the directors were given no documentation to support the adequacy of [the] $55 price per share for sale of the Company; and the Board had before it nothing more than Van Gorkom's statement of his understanding of the substance of an agreement, which he admittedly had never read, nor which any member of the Board had ever seen.

There was no call by the Board, either on September 20 or thereafter, for any valuation study or documentation of the $55 price per share as a measure of the fair value of the Company in a cash-out context. It is undisputed that the major asset of Trans Union was its cash flow. Yet, at no time did the Board call for a valuation study taking into account that highly significant element of the Company's assets.

The record also establishes that the Board accepted without scrutiny Van Gorkom's representation as to the fairness of the $55 price per share for sale of the Company—a subject that the Board had never previously considered. The Board thereby failed to discover that Van Gorkom had suggested the $55 price to Pritzker and, most crucially, that Van Gorkom had arrived at the $55 figure based on calculations designed solely to determine the feasibility of a leveraged buyout. No questions were raised either as to the tax implications of a cash-out merger or how the price for the one million share option granted Pritzker was calculated.

We do not say that the Board of Directors was not entitled to give some credence to Van Gorkom's representation that $55 was an adequate or fair price. . . . The issue is whether the directors informed themselves as to all information that was reasonably available to them. Had they done so, they would have learned of the source and derivation of the $55 price and could not reasonably have relied thereupon in good faith.

The defendants ultimately relied on the stockholder vote of February 10 for exoneration. The defendants contend that the stockholders' "overwhelming" vote approving the Pritzker Merger Agreement had the legal effect of curing any failure of the Board to reach an informed business judgment in its approval of the merger.

The burden must fall on defendants who claim ratification based on shareholder vote to establish that the shareholder approval resulted from a fully informed electorate. On the record before us, it is clear that the Board failed to meet that burden.

To summarize: we hold that the directors of Trans Union breached their fiduciary duty to their stockholders (1) by their failure to inform themselves of all information reasonably available to them and relevant to their decision to recommend the Pritzker merger; and (2) by their failure to disclose all material information such as a reasonable stockholder would consider important in deciding whether to approve the Pritzker offer.

We hold, therefore, that the Trial Court committed reversible error in applying the business judgment rule in favor of the director defendants in this case.*

Reversed in favor of Plaintiff, Smith.

---

*Smith v. Van Gorkom, Delaware Supreme Court, 488 A.2d 858 (1985). http://businessentitiesonline.com/Smith%20v.%20Van%20Gorkom.pdf.

CRITICAL THINKING ABOUT THE LAW

The standards we use to make judgments are often ambiguous. For example, in judging someone's character, we might consider whether that person is fair, honest, and reasonable. To the extent that standards such as "fair" and "honest" and "reasonable" do not have universal meanings, they are ambiguous.

As you are well aware by this time, courts are not exempt from this tendency to use ambiguous standards in judging. The facts of a case are important in determining how ambiguous standards will be applied.

An important critical thinking skill is the ability to recognize ambiguous language in a court's opinion, for without this recognition, you are not prepared to make an informed decision about whether the court's application of standards was merited by the facts of the case. The questions that follow are intended to help you improve this critical thinking skill.

1. To demonstrate your ability to recognize ambiguous language, identify at least two examples of such language in the court's opinion in Case 17-4.

    Clue: Remember that adjectives are often ambiguous.

2. The court applies this ambiguous language in a manner unfavorable to the defendant. What reasons does the court provide for doing so?

    Clue: If the defendant had met the standards by which he was being judged, the decision would have been favorable to him. Another way of phrasing this question is: Why didn't the court find that he met these standards?

COMMENT: In *Cinerama Inc. v. Technicolor Inc.*, 663 A.2d 1156 (1995), the Supreme Court of Delaware distinguished the *Van Gorkom* case and held that the defendant (Technicolor) board of directors did not violate its duty of loyalty when, as interested directors, they participated in a unanimous vote to repeal the company's supermajority provisions. The court stated that an "entire fairness analysis is required" when considering how a board of directors discharges its fiduciary duties. The Delaware Supreme Court affirmed the chancery court's decision in favor of the defendants despite the fact that the interested directors had played a major part in negotiating the merger of their company without disclosing their material conflicts of interest to shareholders such as the plaintiff, Cinerama.

In *Brehm v. Eisner*, 746 A.2d 244 (Del. 2000), the issue was whether stockholders may impose personal liability on the directors of a Delaware corporation for lack of due care and for waste of corporate assets. The legal standards referring to corporate governance and compliance with statutory and case law were found not to have been violated. The appeals court affirmed the dismissal of the action of the lower court (chancery court). The appeals court stated that the lucrative buyout of the former president of Disney (Ovitz) did not constitute waste by the board. In the absence of fraud, disagreements were not grounds for imposing personal liability based on breach of fiduciary duty with regard to the board.

In January 2003, E. Norman Veasey, chief justice of the Delaware Supreme Court, indicated in a roundtable discussion on executive compensation at the University of Delaware[2] that corporate directors could be held personally liable if they failed to act in good faith and thus breached their fiduciary duty to shareholders.

Experts in the area of corporate governance seemed to interpret such comments as meaning that the independence of a board's compensation committee from the CEO will be looked at by the court carefully in terms of good faith, as

---

[2]T. Becker, "Delaware Justice Warns Boards of Liability for Executive Pay," *Wall Street Journal*, Jan. 6, 2003, A-4.

outlined in this chapter. The large number of companies incorporated in Delaware make such comments particularly important.

On March 2, 2004, the chairman and chief executive of Walt Disney Company (Michael Eisner) lost his position as chairman after an estimated 43 percent of voting shareholders declined to support his reelection. Other nominees to the board received fewer dissenting votes. Roy Disney, nephew of the founder, led the dissenting shareholders. Despite the no-confidence vote, Eisner still had the directors' support as CEO. In October of 2004, Eisner announced his resignation effective in 2005. In a related matter, the Court of Chancery of Delaware denied Roy Disney's request to remove the confidentiality tag from certain documents of the company. Roy Disney was seeking the removal of Eisner as CEO [857 A.2d 444 (Del. 2004)]. Eisner subsequently resigned at a later date.

> **APPLYING THE LAW TO THE FACTS...**
>
> Steve and Elaine Wynn were cofounders of Wynn Resorts (board) and members of the board of directors of this very large corporation until Friday, April 24, 2015, when Ms. Wynn was removed from the board by a vote of a majority of its members. Ms. Wynn was accused of improper activity. She lobbied unsuccessfully with large shareholders, who issued a warning that they would be watching the future activities of the board carefully.
>
> Three years before, Wynn Resorts had ousted the then-largest shareholder Kazuo Okada by removing him from the board after allegations that he was involved in corrupt activity. His 2010 stake in the company was redeemed at a 30 percent discount. As of this writing, Mr. Okada (once a close friend of Mr. Wynn) continues to fight to get back his shares while denying allegations of any wrongdoing.

In March 2015, Wynn Resorts' board decided not to renominate Ms. Wynn, who had been a director for more than 12 years when her term expired on April 24, 2015, which also was the date of the annual shareholders' meeting. Ms. Wynn nominated herself and began to campaign for a seat on the board touting her decade of experience and the fact that she was the company's third-largest shareholder and the sole female director. They alleged that Ms. Wynn had sold $10 million of shares through her personal foundation during a "blackout period" ahead of the company's earnings release, during which time directors are forbidden by company policy to sell common stock.

The company also accused Ms. Wynn of improper behavior regarding a land deal. Ms. Wynn participated in board meetings when these deals were discussed by the boards. She did not recuse herself during such discussions. Ultimately, her nephew was involved in a competing bid. A group, including Ms. Wynn's nephew, purchased the land. Ms. Wynn denied any involvement in the group's land purchase and indicated that the money she received from the shares she sold was to be used for charitable purposes.

Ms. Wynn seeks to dissolve a shareholder agreement with Mr. Wynn, putting her at odds with the board. The agreement puts Mr. Wynn in control of the company, even though he is the second-largest shareholder. It puts voting and selling restrictions on Ms. Wynn's shares. *Wynn v. Wynn Resorts*, Nev. E. Dist. 2014.

The Wynns were married, divorced, remarried, and divorced; both playing a role in building the company; and both were known to be billionaires. Were all the directors at the Wynn Resort meeting their fiduciary obligations outlined in this section? List the obligations and apply each one to the facts of the case. Which values are in conflict as outlined in Chapter 1?

Limited Liability Companies

The **limited liability company (LLC)** is a hybrid form of business organization. It allows entrepreneurs and small business owners to enjoy the same limited personal liability that shareholders in a corporation have while retaining the status of partners in a partnership. The LLC is federally taxed, not as a corporation, but as a partnership, so taxes are paid personally by members of the LLC. Limited liability corporation members share in the profits from the business and exercise management control without these actions affecting their profit share or limited liability status. The LLC differs from the Subchapter S corporation in that other corporations, partnerships, and foreign investors can be LLC members, and there is no limit on the number of members.

limited liability company (LLC) A hybrid corporation-partnership similar to the Subchapter S corporation but with far fewer restrictions.

THE UNIFORM LIMITED LIABILITY ACT

The National Conference of Commissioners on Uniform State Law (comprised of lawyers, academics, and judges) issued the Uniform Limited Liability Company Act (ULLCA). This act covers the formation, operation, and termination of LLCs. Each state through its legislature must determine whether the ULLCA will become state law. Forty-eight states have adopted all or part of this Uniform Act as a state statute.

APPLYING THE LAW TO THE FACTS...

Mark and Allen decide to form an LLC. Mark tells Allen that they probably won't be able to hire any employees in the near future because their profits will be low and the profits will be taxed like a corporation. Allen says that an LLC is not taxed like a corporation and that the members of an LLC pay taxes instead of profits being taxed. Who is correct?

LLC CHARACTERISTICS

The LLC is a separate legal person (or entity) under state law. It can sue or be sued, enforce contracts, and be found civilly and criminally liable for violations of law (ULLCA Section 201).

The owners of the LLC are called *members*. They are not personally liable to third parties for the debts, obligations, and liabilities of the LLC beyond an individual's capital contribution—that is, they are under a limited liability (ULLCA Section 303[9]). A member is personally liable for the debts of an LLC only if he or she agrees to become so in writing or if it is so stated in the articles of incorporation.

CREATING A LIMITED LIABILITY COMPANY

The following steps are important in creating an LLC:

- Choose a state where the LLC will be doing most of its business.
- Select a name. It is important to make sure that the state's statute governing the LLC is met and that the name is not already being used. The words *limited liability company* or the abbreviation *LLC* or other language approved by ULLCA Section 105(a) should be used, as adopted by the state statute. Reserve the name by filing the application with the secretary of state, following a determination of whether the proposed company name is federally trademarked by another company.
- Under the ULLCA, an LLC may be formed by one or more persons by delivering articles of organization to the secretary of state. The LLC comes into existence at the time of filing by the secretary (ULLCA Section 202).

DURATION OF THE LLC

Unless an LLC has specified otherwise in its articles of organization, its duration is unlimited or at will, meaning that it has no specific term.

FINANCING OF THE LLC

Members' contributions to the LLC may be in the form of money; real, personal, tangible, and intangible property (e.g., often a patent); or contracts for services, promissory notes, or other agreements to contribute (ULCC Section 401).

CONTROL CONSIDERATIONS

Members of an LLC should have an operating agreement that outlines responsibilities, voting rights, and the way the LLC is to be managed, at minimum. Often, outside management companies are hired. Experienced counsel and accountants should be involved in the drafting of such an agreement.

TAX RAMIFICATIONS

As noted previously, the LLC does not pay taxes. Income and losses are passed through to members to be reported on individual tax returns. Accountants and tax lawyers should be hired to set up this "flow-through" treatment on tax returns.

Some advantages and disadvantages of LLCs over other forms of business organizations are summarized here and in Table 16-1 of the previous chapter.

Advantages	*Disadvantages*
1. Members are not personally liable for the LLC's debts and obligations. Loss is limited to personal investment of capital.	1. State statutes are not uniform.
2. The LLC has continuity beyond the death of a member or members.	2. A foreign LLC (an LLC in a state not originally chartered within the state) may not be treated the same.
3. The LLC can include foreign investors, with some limitations.	3. In a member-managed LLC, all decisions are made by a majority, which often causes delays in decision making or derivative lawsuits filed by dissenting members.
4. The LLC can be taxed as a partnership (two or more members) or a corporation.	

The case below explores the fiduciary duties imposed by the terms of an LLC operating agreement.

CASE 17-5

Gatz Properties, LLC v. Auriga Capital Corporation
Supreme Court of Delaware
59 A.3d 1206 (Del. 2015)

In 1997, Gatz Properties, LLC, and Auriga Capital Corp., along with other minority investors, formed Peconic Bay, LLC, to lease and develop a golf course on property owned by the Gatz family. Under Peconic Bay's operating agreement, Gatz Properties was the manager of Peconic Bay, and William Gatz (Gatz) managed and controlled Gatz Properties. The Peconic Bay agreement provided that certain major decisions required the approval of 66 2/3 percent of the Class A membership interests and 51 percent of the Class B interests. The Gatz family controlled 85 percent of the Class A interests and 52% of the Class B interests.

The property was leased to Peconic Bay under a long-term lease. Peconic Bay borrowed $6 million for improvements and entered into a sublease with American Golf Corp. to operate the course. The operation was never profitable. Knowing that American Golf would terminate the sublease, Gatz had the property appraised. The improved land was valued at $10.1 million, but the development value of the vacant land was $15 million.

A third party made an unsolicited offer to buy the property. Even though Gatz was not cooperative, the buyer submitted two offers, both of which Gatz disclosed to the members. The members rejected both. One member asked Gatz to inquire whether the potential buyer would offer $6 million. Gatz allegedly told the buyer that the offer had to be "well north of $6 million." The bidder was still interested and attempted to work out satisfactory price terms, but again, Gatz did not cooperate. Meanwhile, Gatz made his own buyout offer to the Peconic Bay minority investors after misrepresenting to them the terms the outside buyer was willing to negotiate. After all the minority investors (except one) rejected the offer, Gatz had the property reappraised, but he did not provide the appraiser with the information necessary to make an accurate appraisal. The appraised leasehold value was $2.8 million if the property was used as a public course and $3.9 million if it was used as a private course.

Gatz then offered to buy 25 percent of the minority members' capital account balances. He also retained legal counsel, who advised the minority members that the majority members had the right to vote out the minority members "so long as fair price is paid for the interests of the minority members." Counsel further advised that in light of the existing debt and the most recent appraisal, "that value is, at best, zero."

Shortly thereafter, Gatz formally proposed to sell the property at auction and told the minority members that Gatz Properties would bid. The proposal was approved because the Gatz family had majority voting power. Gatz used an auction company, with no experience in golf courses, that "specialized in 'debt-related'" sales. He ended up being the only bidder. He purchased the property for $50,000 and assumed the outstanding debt; the minority investors received $20,985 in the aggregate.

The minority members later instituted suit, and the Delaware Court of Chancery awarded them almost $800,000 in damages as well as all of their attorney fees. Gatz appealed.

Percurian

The pivotal legal issue presented . . . is whether Gatz owned contractually agreed-to fiduciary duties to Peconic Bay and its minority investors. Resolving that issue requires us to interpret Section 15 of the LLC Agreement, which both sides agree is controlling. Section 15 pertinently provides that:

> Neither the Manager nor any other Member shall be entitled to cause the Company to enter into any additional agreements with affiliates on terms and conditions which are less favorable to the Company than the terms and conditions of similar agreements which could then be entered into with arms-length third parties, without the consent of a majority of the non-affiliated Members . . .

The Court Chancery determined that Section 15 imposed fiduciary duties in transactions between the LLC and affiliated persons. We agree. [W]e construe its operative language as an explicit contractual assumption by the contracting parties of an obligation subjecting the manager. . . .

. . . There having been no majority-of-the-minority approving vote in this case, the burden of establishing the fairness of the transaction fell upon Gatz. That burden Gatz could easily have avoided. If (counterfactually) Gatz had conditioned the transaction upon the approval of an informed majority of the nonaffiliated members, the sale of Peconic Bay would not have been subject to, or reviewed under, the contracted-for entire fairness standard.

Entire fairness review normally encompasses two prongs, fair dealing and fair price. "However, . . . [a]ll aspects of the issue must be examined as a whole since the question is one of entire fairness," [The Court of Chancery] also properly considered the "fairness" of how Gatz dealt with the minority "because the extent to which the process from the behavior one would expect in an arms-length deal bears importantly on the price determination." The court further held that "in order to take cover under the contractual safe harbor of Section 15, Gatz bears the burden to show that he paid a fair price to acquire Peconic Bay." We agree.

The trial judge found facts . . . that firmly support his conclusion that Gatz breached his contracted-for duty to the LLC's minority members. [T]he court found that "Peconic Bay was worth more than what Gatz paid." . . . The Court . . . also properly relied on Auriga's expert witness's discounted cash flow analysis, which valued Peconic Bay at approximately $8.9 million.

. . . [T]he court relied on the fact that Gatz had rebuffed [the outside buyer's] interest in discussing a deal "well north of $6 million" . . .

. . . [T]he Court of Chancery did not "view the Auction process as generating a price indicative of what Peconic Bay would fetch in a true arms-length negotiation." Indeed, the court found, the Auction was a "sham", "the culmination of Gatz's bad faith efforts to squeeze out the Minority Members." . . . "Gatz manufactured a situation of distress to allow himself to purchase Peconic Bay at a fire sale price at a distress sale."

. . . Section 16 permits both exculpation and indemnification of Peconic Bay's manager in specified circumstances. Gatz was not entitled to exculpation because the Court of Chancery properly found that he had acted in bad faith and had made willful misrepresentation in the course of breaching his contracted-for fiduciary duty. Consequently, Section 16 of the LLC Agreement provides no safe harbor.

Further, the court correctly found that Gatz's offer to Peconic Bay's minority members "contained incomplete and misleading information [about the outsider offer]." Gatz "intentionally [misled] the Minority Members when accurate information concerning third-party offers would have been material to their decision whether to accept Gatz's own offer."

The trial court demanded that if Gatz had engaged with [the outside buyer] . . . as Gatz's contracted-for entire fairness duty required, Peconic Bay could probably have been sold at a price that returned to the minority investors both their initial capital ($725,000) plus 10% aggregate return ($72,000) . . .

The damages award was based on conscience and reason, and we uphold it.

"Under the American Rule, absent express statutory language to the contrary, each party is normally obliged to pay only his or her own attorneys' fees." . . .

Our courts have, however, recognized bad faith litigation conduct as a valid exception to that rule.

. . . The court did not abuse its discretion in awarding attorneys' fees.*

The Supreme Court *affirmed* the judgment in favor of the minority investors, finding the manager had breached his fiduciary duties.

Global Dimensions of Corporations: A "Big Fat Greek" Bailout II and III

Before examining this material, you would do well to go back and review the facts of the bailout of Greece set out in Chapter 8.

Following the threat of bankruptcy to Greece and later Spain and Portugal in 2010, the question became who held the debt (approximately $2.6 trillion) issued by these nations. It appeared that no one knew the full story, and because of this mystery, the private- and public-sector banks and other institutions of Europe and the United States stopped lending money to one another. Interbank loans came to a halt. There was a call for transparency from governments and other institutions around the world—including China, which holds a large amount of U.S. debt. Approximately 2.2 trillion euros of debt were issued by Spanish, Portuguese, and Greek institutions as of June 2010. The exposure of foreign banks and other institutions was largest in France, Germany, Britain, the Netherlands, Italy, and Belgium; U.S.-based institutions had the smallest exposure. Estimates were vague because little was publicly known as to which banks held what amounts. Government regulators of the European Union (EU) had only incomplete information, as did those of individual countries. In Europe, bank secrecy has traditionally been part of many countries' business culture.

The European Central Bank (ECB) joined with the International Monetary Fund (IMF) to offer $1 trillion in loan guarantees to Europe's banks. The ECB began buying government bonds for the first time ever to prevent a debt sell-off of Greek, Spanish, and other sovereign debt. Through June 2010, this "bailout," along with severe austerity measures, helped calm the waters to a degree. Nonetheless, major indexes in Europe and the United States fell. The euro fell dramatically, and the U.S. dollar became a safe haven. Investors remained skeptical that Greek institutions and others in Europe could repay their debts. As the euro fell, there was no indication that transparency had arrived or would come soon enough. Would the "bailout" be sufficient? Would it help as the growth of European economies faltered? Some skeptics also wondered whether the United States would fall back into a second recession, after seemingly recovering (at least somewhat) from the one in 2008. By September 2013, there had been some recovery in banking institutions (in Southern Europe), but the recovery was very tentative.

SUMMARY

The classification of corporations includes (1) closely and publicly held, (2) domestic and foreign, (3) multinational, (4) Subchapters S and C, (5) professional, and (6) nonprofit. The policy implications of, and the role of multinational corporations in, actions taken by national and supranational entities were discussed briefly.

---

*Gatz Properties, LLC v. Aurigo Capital Clorporation Supreme Court of Delaware, 59 A. 3d 1206 (Del.2015). http://courts.delaware.gov/opinions/download.aspx?ID=180430.

The nature of corporations was examined with regard to their creation, financing, and operation. This chapter also reviewed the advantages and disadvantages of limited liability companies. Finally, the global dimensions of both private- and public-sector corporations set out in 2010 were related to the context of a European economic crisis. Similar arguments were made in the Greek bank bailouts in 2012. This could be called "Greek" bailout III.

REVIEW QUESTIONS

17-1 Explain the difference between notes and debentures.

17-2 Distinguish between the various types of bonds set out in this chapter.

17-3 Define a closely held corporation and distinguish it from a publicly held corporation.

17-4 Describe the circumstances under which a partnership would offer greater tax advantages than a corporation.

17-5 Explain how corporations provide limited liability to their owners.

17-6 What is one tax advantage of a limited liability company?

REVIEW PROBLEMS

17-7 Thomas Persson and Jon Nokes founded Smart Inventions, Inc., to market household consumer products. The success of their first product, the Smart Mop, continued with later products, which were sold through infomercials and other means. Persson and Nokes were the firm's officers and equal shareholders. Nokes was in charge of the day to day operations. By 1998, they had become dissatisfied with each other's efforts. Nokes represented the firm as financially "dying," "in grim state, . . . worse than ever," and offered to buy all of Persson's shares for $1.6 million. Persson accepted. On the day they signed the agreement to transfer the shares, Smart Inventions began marketing a new product—the Tap Light. It was an instant success, generating millions of dollars in revenues. In negotiating with Persson, Nokes had intentionally kept the Tap Light a secret. Persson sued Smart Inventions, asserting fraud and other claims. Under what principle might Smart Inventions be liable for Noke's fraud? Is Smart Inventions liable for Nokes's fraud? Is Smart Inventions liable in this case? Explain. *Persson v. Smart Inventions, Inc.*, 125 Cal. App.4th 1141, 23 Cal.Rptr.3d 335 (2 Dist. 2005).

17-8 Mike Lyons incorporated Lyons Concrete, Inc., in Montana but did not file its first annual report, so the state involuntarily dissolved the firm in 1996. Unaware of the dissolution, Lyons continued to do business as Lyons Concrete. In 2003, he signed a written contract with William Weimar's property in Lake County for $19,810. Weimar was in a rush to complete the entire project, and he and Lyons orally agreed to additional work on a time-and-materials basis. When scheduling conflicts arose, Weimar had his own employees set some of the forms, which proved deficient. Weimar also directed Lyons to pour concrete in the rain, which undercut its quality. In mid-project, Lyons submitted an invoice for $14,389, which Weimar paid. After the work was complete, Lyons sent Weimar an invoice for $25,731, but he refused to pay, claiming that the $14,389 covered everything. To recover the unpaid amount, Lyons filed a mechanic's lien as "Mike Lyons d/b/a Lyons Concrete, Inc." against Weimar's property. Weimar filed a suit in a Montana state court to strike the lien, and Lyons filed a counterclaim to reassert it. Who won? Explain.

17-9 Allan Jones sold a ski shop franchise to Edward Hamilton. Although Mr. Jones did not contribute equity to the business or share in the profits, he did give Mr. Hamilton advice and share his experience to help him get started. Most of Mr. Hamilton's capital came in the form of a loan from Union Bank. When Mr. Hamilton failed to pay, Union Bank sued Mr. Jones for payment under the theory that Mr. Jones was a partner by implication or estoppel. Was he? Explain.

17-10 Heritage Hills (a land development firm) was organized on July 2, 1975, as a limited partnership, but the partnership agreement was never properly filed. Heritage Hills went bankrupt, and the bankruptcy trustee sought to recover the debts owed by the partnership from the limited partners. Can he? Explain.

CASE PROBLEMS

17-11 Vince Lutriario claims that the defendant, A World of Pets and Supplies (A World of Pets), sold him a dog infected with Giardia. Lutriario sought to recover the price of the dog and the costs he incurred curing the dog. The marshal who went to A World of Pets was told that the store no longer existed and was now World of Pups. Lutriario had previously talked to the owner of A World of Pets (also the owner of the World of Pups) and had arranged a payment plan to be reimbursed. Now, however, the defendant sought an order vacating the judgment that led to that plan and claimed that World of Pups was distinct and separate from the original defendant, A World of Pets. Despite the name change, there appeared to be continuity not only of management and owner but also of physical location, assets, and general business operation. Was there a *de facto* . . . between the two "Worlds"? *Lutriario v. A World of Pets and Supplies, Ltd.*, 907 N.Y.S.2d 101 (Civ. Ct. 2010).

17-12 Mark Burnett and Kamran Pourgol were the only shareholders in a corporation that built and sold a house. When the buyers discovered that the house exceeded the amount of square footage allowed by the building permit, Pourgol agreed to renovate the house to conform to the permit. No work was done, however, and Burnett filed a suit against Pourgol. Burnett claimed that without his knowledge, Pourgol had submitted incorrect plans to obtain the building permit, misrepresented the extent of the renovation, and failed to fix the house. Was Pourgol guilty of misconduct? If so, how might the situation have been avoided? Discuss. *Burnett v. Pourgol*, 83 A.D.3d 756, 921 N.Y.S.2d 280 (2 Dept. 2011).

17-13 In 1997, Leon Greenblatt, Andrew Jahelka, and Richard Nichols incorporated Loop Corp. with only $1,000 of capital. Three years later, Banco Panamericano, Inc., which was run entirely by Greenblatt and owned by a Greenblatt family trust, extended a large line of credit to Loop. Loop's subsidiaries then participated in the credit, giving $3 million to Loop while acquiring a security interest in Loop itself. Loop then opened an account with Wachovia Securities, LLC, to buy stock. Values plummeted, Loop owed Wachovia $1.89 million. Loop also defaulted on its loan from Banco, but Banco agreed to lend Loop millions of dollars more. Rather than repay Wachovia with the influx of funds, Loop gave the funds to closely related entities and "compensated" Nichols and Jahelka without issuing any W-2 forms (forms reporting compensation to the Internal Revenue Service [IRS]). The evidence also showed that Loop made loans to other related entities; shared office space, equipment, and telephone and fax numbers with related entities; failed to file its tax returns on time (or sometimes at all); and failed to follow its own bylaws. In a lawsuit brought by Wachovia, can the court hold Greenblatt, Jahelka, and Nichols personally liable by piercing the corporate veil? Why or Why not? *Wachovia Securities, LLC v. Banco Panmericano, Inc.*, 674 F.3d 743 (9th Cir. 2012).

17-14 Harry Hoaas and Larry Griffiths were shareholders in Grand Casino, Inc., which owned and operated a casino in Watertown, South Dakota. Griffiths owned 51 percent of the stock, and Hoaas owned 49 percent. Hoaas managed the casino, which Griffiths typically visited once a week. At the end of 1997, an accounting showed that the cash on hand was less than the amount posted in the casino's books. In October 1999, Griffiths did a complete audit. Hoaas was unable to account for $135,500 in missing cash. Griffiths then kept all of the casino's most recent profits, including Hoaas's $9,447.20 share, and, without telling Hoaas, sold the casino for $100,000 and kept all of the proceeds. Hoaas filed a suit in a South Dakota state court against Griffiths, asserting, among other things, a breach of fiduciary duty. Griffiths countered with evidence of Hoaas's misappropriation of corporate cash. What duties did these parties owe each other? Did either Griffiths or Hoaas or both of them breach those duties? Explain. *Hoaas v. Griffiths*, 314 N.W.2d 61 (2006).

THINKING CRITICALLY ABOUT RELEVANT LEGAL ISSUES

Corporations should be outlawed. They have harmed our society in far more ways than they have helped it. After all, we have corporations to thank for all of the sweatshops, old-boy networks, and underpaid workers. And let's not forget about the pollution. It's always those big chemical corporations and industrial plants that are filling the air we breathe and the water we drink with pollutants. It's a miracle that our life expectancy is more than 50 years!

Just look at what corporations today do. They put poisonous products on the market and then use million-dollar marketing schemes to trick the public into buying them. Take cigarettes, for example. For decades, the big tobacco corporations have been lying to the public and paying off our government officials. They've invested millions in covering up the fact that cigarettes kill. Hundreds of thousands of innocent Americans have died because of the lengths to which these corporations have gone just to make money.

In addition, corporations don't know how to treat their workers. A recent survey showed that workers at corporations were five times as unhappy as were sole proprietors. Another study demonstrated that 87 percent of on-the-job accidents happened to people who worked at corporations. If we'd just get rid of these corporations, Americans would be much safer at work.

We don't have these kinds of problems with partnerships or sole proprietorships. When is the last time you heard of a small mom-and-pop store that severely polluted the environment or killed people with its products?

Furthermore, the capitalist ideas that our economy is built upon do not support corporations. Capitalism is based on the idea that all competing businesses will be mom-and-pop stores. Even the founder of capitalism stated that he didn't trust businessmen and that in order for his ideas to work, the businesses would have to be small. Today, corporations just seek to widen the guilt between the rich and the poor. Academic institutions do the same with their high tuition. We should have free schools like Saudi Arabia.

Our founding fathers would probably roll over in their graves if they knew that our country is being run by these large corporations. It's time that we return to the ideas on which our economy is based. All businesses should be either sole proprietorships or partnerships. The time has come for corporations to join the ranks of the dinosaurs.

1. What reasons does the author give for his conclusion?
2. What important information is omitted from the author's reasoning?
3. Create some arguments in opposition to the ones the author sets out in the essay.

ASSIGNMENT ON THE INTERNET

This chapter describes the common corporate forms of business organization in the United States. Business structures outside the United States, however, are often very different. Use the Internet to determine some of the major differences in corporate business organization between the United States and the European Union. You can begin your search by visiting Europa's Starting Up Web page at europa.eu/. If you were looking to start a business in the European Union, what would you do differently from starting the business in the United States? You may also be able to find other websites to assist you in your search.

ON THE INTERNET

www.irs.gov The IRS provides information about LLCs at this site. Perform a search for "LLC" in the search bar, then click on the result titles "Limited Liability Company."

www.mycorporation.com This is the site of a private firm that assists small companies and businesses in becoming incorporated.

www.llc-reporter.com This site provides an analysis of the ULLCA.

www.hg.org This site provides links to useful information on corporate law and recent court cases that decide issues of corporate law.

www.lexfori.net This site provides information about starting a corporation in France and describes the various structures that corporations can take in that country.

FOR FUTURE READING

Chandler, Martin. "Delimiting Liability for South Carolina Limited Liability Corporations: When Can an LLC Manager Be Personally Liable for Tortious Interference? *South Carolina Law Review* 64 (2013): 801–832.

Crouch, Holmes F. *The Pros & Cons of LLCs: How to Shape a Limited Liability Company (LLC), Understand Its Rules, Prepare Tax Returns & Fend Off Con Artists* (Series 200: Investors & Businesses 2007). Allyeartax Guides.

Dhir, Aaron A. "Realigning the Corporate Building Blocks: Shareholder Proposals as a Vehicle for Achieving Corporate Social and Human Rights Accountability." *American Business Law Journal* 43 (2006): 365.

Jones, Aaron D. "Corporate Officer Wrongdoing and the Fiduciary Duties of Corporate Officers under Delaware Law." *American Business Law Journal* 44 (2007): 475–520.

Miller, Sandra K., Penelope Sue Greenberg, and Ralph H. Greenberg. "An Empirical Glimpse into Limited Liability Companies: Assessing the Need to Protect Minority Investors." *American Business Law Journal* 43 (2006): 609.

PART THREE

Public Law and the Legal Environment of Business

Part Three focuses on the public laws that regulate the legal environment of business. Because most public laws governing the legal environment are administered and created by administrative agencies, this part opens with a chapter introducing administrative agencies. After this foundational chapter, we examine laws affecting employees in the workplace, laws governing employee benefits, labor–management relationships, and employment discrimination. The focus then shifts to laws governing the physical environment, securities, antitrust, and consumer protection, as well as debtor–creditor relationships.

CHAPTER EIGHTEEN

The Law of Administrative Agencies

- INTRODUCTION TO ADMINISTRATIVE LAW AND ADMINISTRATIVE AGENCIES
- CREATION OF ADMINISTRATIVE AGENCIES
- FUNCTIONS OF ADMINISTRATIVE AGENCIES
- LIMITATIONS ON ADMINISTRATIVE AGENCIES' POWERS
- STATE AND LOCAL ADMINISTRATIVE AGENCIES
- GLOBAL DIMENSIONS OF ADMINISTRATIVE AGENCIES

The first two federal administrative agencies—the Interstate Commerce Commission (ICC) (no longer in existence) and the Federal Trade Commission (FTC)—were created by Congress in the late nineteenth century and early twentieth century during an era of reform. Congress thought that the anticompetitive conduct of railroads and other corporations could best be controlled by separate administrative agencies with defined statutory mandates. In another era of reform following the stock market crash of 1929 and the beginning of the Great Depression, Congress saw a need for additional agencies to assist a free-market economy and to act in the public interest—hence, the creation of such agencies as the Securities and Exchange Commission (SEC), the National Labor Relations Board (NLRB), and the Federal Communications Commission (FCC). From time to time since then, new administrative agencies have been established, until today there are some 76 federal administrative agencies now functioning and affecting nearly every aspect of life in the United States.[1] With the passage of the Dodd-Frank Wall Street Reform and Consumer Protection Act[2] (Dodd-Frank) in 2010, the usage of federal regulatory agencies to implement federal statutes passed by Congress continued. Additional agencies have been added by Dodd-Frank (see Chapter 23).

For example, the clothes we wear are subject to regulation by the Consumer Product Safety Commission (CPSC), the cars we drive are subject to regulation by the National Highway Safety Transportation Board (NHSTB) and the

---

[1] Office of the Federal Register, *General Index: Code of Federal Regulations* (rev. ed.) (Washington, DC: National Archives, General Services Administration, Jan. 1, 2006).
[2] Public Law 111-203, 124 Stat. 1376 (2010).

Environmental Protection Agency (EPA), and the television we watch and the out-of-state telephone calls we make are subject to regulation by the FCC.

Because government agencies have such a powerful effect on business, businesspeople need to understand the laws (regulations) these agencies are empowered to create and the rules of procedure they use to make such laws. The impact of administrative agency regulations on business and society is a primary focus of this text. Although we emphasize the role of federal administrative agencies, each of the 50 states, the District of Columbia, and Puerto Rico, as well as counties, cities, and some towns, have administrative agencies that also regulate business and societal conduct. In this chapter, we define administrative law and administrative agencies and discuss the reasons for their growth, how they were created, and their functions. We also explore the federal administrative agencies' relationship to the executive, legislative, and judicial branches of government, as well as the institutions and laws that limit the power of administrative agencies. We end this chapter with a brief consideration of the global dimensions of administrative agencies.

Introduction to Administrative Law and Administrative Agencies

ADMINISTRATIVE LAW

For the purposes of this text, **administrative law** is defined broadly as any rule (statute or regulation) that affects, directly or indirectly, an administrative agency. These rules may be procedural or substantive, and they may come from the legislative, executive, or judicial branch of government or from the agencies themselves. Such rules may be promulgated at the federal, state, or local levels. A **procedural rule** generally has an impact on the internal processes by which the agencies function or prescribes methods of enforcing rights. For example, under the Administrative Procedure Act (APA), a federal administrative agency must give adequate notice to all parties involved in an agency hearing. A **substantive rule** defines the rights of parties. An example is an act of Congress that forbids the FTC from applying the antitrust laws to all the Coca-Cola bottlers in the United States. In this instance, the rights and regulations of both the FTC and the Coca-Cola bottlers were defined by Congress.

administrative law Any rule (statute or regulation) that directly or indirectly affects an administrative agency.

procedural rule A rule that governs the internal processes of an administrative agency.

substantive rule A rule that creates, defines, or regulates the legal rights of administrative agencies and the parties they regulate.

CRITICAL THINKING ABOUT THE LAW

As a future business leader, you will certainly encounter many governmental regulations. Congress created administrative agencies, in part, because it could not hope to address the enormous variety and number of concerns that are now covered by administrative agencies. Although you will not learn about every administrative agency in this chapter, you can jump-start your thinking about administrative agencies by answering these critical thinking questions.

1. Your roommate states that people do not have to follow the regulations passed by administrative agencies because these regulations are not laws. She argues that only Congress can make laws. Which critical thinking question could be applied to settle this disagreement?

    Clue: Do you and your roommate agree on the meaning of the words she is using?

2. Some individuals may argue that the creation of regulations by administrative agencies promotes unfair restrictions on business. What ethical norm seems to be behind this thought?

    Clue: If you want fewer restrictions from the government, what ethical norm is influencing your thought? What ethical norm seems to conflict with the wish for fewer governmental regulations?

3. Congress assumes that the administrative agencies will address problems effectively in their respective areas. For example, the EPA ensures compliance with environmental laws. If Matt makes the assumption that environmental problems are so complex and widespread that the EPA could not hope to make a difference, what conclusion do you think Matt would draw regarding administrative agencies?

Clue: Think about a contrary assumption. If Matt assumed that the administrative agencies were effective, would he be more likely to support the regulations passed by the various agencies?

ADMINISTRATIVE AGENCIES

administrative agency Any body that is created by the legislative branch to carry out specific duties.

An **administrative agency** is any body that is created by the legislative branch (e.g., Congress, a state legislature, or a city council) to carry out specific duties. Some agencies are not situated wholly in the legislative, executive, or judicial branch of government. Instead, they may have legislative power to make rules for an entire industry, judicial power to adjudicate (decide) individual cases, and executive power to investigate corporate misconduct. Examples of such independent federal administrative agencies are the EPA, the FCC, and the FTC; at the state level, examples are public utilities commissions and building authorities; at the city level, examples are city planning commissions and tax appeals boards.

independent administrative agency An agency whose appointed heads and members serve for fixed terms and cannot be removed by the president except for reasons defined by Congress.

Types. Administrative agencies are generally classified as independent or executive (see Table 18-1). **Independent administrative agencies**, such as the FTC and the SEC, are usually headed by a board of commissioners, who are appointed for a specific term of years by the president with the advice and consent of the Senate. A commissioner can be removed before serving out a full term only for causes defined by Congress, not at the whim of the president—which is why these agencies are called *independent*. Generally, the majority of a board must be of the sitting president's party. This is dictated by the Administrative Procedure Act of 1946.

executive administrative agency An agency located within a department of the executive branch of government; heads and appointed members serve at the pleasure of the president.

Executive administrative agencies are generally located within departments of the executive branch of government. For example, the Occupational Safety and Health Administration (OSHA) is located in the Department of Labor, and the National Transportation Safety Board (NTSB) is in the Department of Transportation. Heads and members of these boards have no fixed term of office. They serve at the pleasure of the president, meaning that they can be removed from their positions by the chief executive at any time.

TABLE 18-1 SELECTED FEDERAL ADMINISTRATIVE AGENCIES

Independent Agencies	Executive Agencies
Commodity Futures Trading Commission (CFTC)	Federal Deposit Insurance Corporation (FDIC)
Consumer Product Safety Commission (CPSC)	General Services Administration (GSA)
Equal Employment Opportunity Commission (EEOC)	International Development Corporation Agency (IDCA)
Federal Communications Commission (FCC)	National Aeronautics and Space Administration (NASA)
Federal Trade Commission (FTC)	National Science Foundation (NSF)
National Labor Relations Board (NLRB)	Occupational Safety and Health Administration (OSHA)
National Transportation Safety Board (NTSB)	Office of Personnel Management (OPM)
Nuclear Regulatory Commission (NRC)	Small Business Administration (SBA)
Securities and Exchange Commission (SEC)	Veterans Administration (VA)
	Department of Homeland Security (DHS)

Reasons for Growth. Administrative agencies have proliferated rapidly since the late 1890s for the following reasons:

1. *Flexibility.* Unlike the court proceedings presented in Chapter 3, administrative agency hearings are not governed by strict rules of evidence. For example, hearsay rules are waived in most cases.
2. *Need for expertise.* The staff of each of the agencies has technical expertise in a relatively narrow area, gained from concentrating on that area over the years. It would be impossible, for example, for 435 members of the House of Representatives and 100 senators to regulate the television, radio, and satellite communications systems of the United States on a daily basis. Only the FCC staff has that expertise.
3. *Prevention of overcrowding in courts.* If administrative agencies did not exist, our highly complex, often litigious society would have to seek redress of grievances through the federal and state court systems. As explained in Chapter 4, both corporations and individuals are already seeking alternatives to the overburdened court system.
4. *Expeditious solutions to national problems.* After the 1929 stock market crash and the ensuing Depression, Congress sought to give investors confidence in the securities markets by creating the SEC in 1934. The SEC was intended to be a "watchdog" agency that would ensure full disclosure of material information to the investing public and prevent a repeat of the fraudulent practices that marked the freewheeling 1920s. When the public became concerned about the deterioration of the nation's water, land, and air, Congress created the EPA to implement air, water, and waste regulations.

All these reasons and more are why administrative agencies exist. These reasons, however, are frequently challenged by proponents of deregulation—or no regulation—of industry.

The debate between advocates of returning to a period in our history when market forces were the sole regulators of business conduct and champions of administrative agency regulation is highlighted throughout Part III of this text. We also now have advocates of reregulation in areas such as the power sector. With rising prices for electricity, for both homes and businesses, calls for reregulation are often heard. Administrative agencies are becoming more important again. This is especially true today, because the current administration's philosophy includes a belief that federal regulation is necessary to solve national problems.

Creation of Administrative Agencies

Congress creates federal administrative agencies through statutes called **enabling legislation**. In general, an enabling statute delegates to the agency congressional **legislative power** for the purpose of serving the "public interest, convenience, and necessity." Armed with this mandate, the administrative agency can issue rules that control individual and business behavior. In many instances, such rules carry criminal as well as civil penalties. In Chapter 23, you will see how the SEC, using its mandate under the 1933 and 1934 Securities Acts, can both fine and criminally prosecute individuals involved in insider trading. The enabling statute also delegates **executive power** to the agency to investigate potential violations of rules or statutes. Chapter 23 sets out the wide-ranging investigative powers of the SEC staff. Finally, the enabling statutes delegate **judicial power** to the agency to settle or adjudicate any disputes it may have with businesses or individuals. For example, the SEC, using its congressional mandate under the 1933 and 1934 Securities Acts, has prescribed rules governing the issuance of, and trading in, securities by businesses as well as by brokers and underwriters. Administrative

---

enabling legislation Legislation that grants lawful power to an administrative agency to issue rules, investigate potential violations of rules or statutes, and adjudicate disputes.

legislative power The power delegated by Congress to an administrative agency to make rules that must be adhered to by individuals and businesses regulated by the agency; these rules have the force of law.

executive power The power delegated by Congress to an administrative agency to investigate whether the rules enacted by the agency have been properly followed by businesses and individuals.

judicial power The power delegated by Congress to an administrative agency to adjudicate cases through an administrative proceeding; includes the power to issue a complaint, have a hearing held by an administrative law judge, and issue either an initial decision or a recommended decision to the head(s) of an agency.

law judges are assigned to the SEC adjudicate cases in which individuals or corporations may have violated the rules.

Because the framers of the U.S. Constitution carefully separated the legislative (Article I), executive (Article II), and judicial powers of government into three distinct branches, some people complain that allowing administrative agencies (Article III) to exercise all three powers violates the spirit of the Constitution. Critics go so far as to state that these agencies constitute a "fourth branch of government." In 1995, the House of Representatives and the Senate overwhelmingly passed a bill that required all administrative agencies (both executive and independent) to do a cost-benefit analysis of any proposed regulation that would cost the economy more than $25 million. All agencies also have to identify possible alternatives to the proposed regulation that would require no government action, as well as varying actions customized for different regions of the country, and "the use of market-based mechanisms." The Office of Management and Budget (OMB) reviews all proposed rules judged to be "major." Because the OMB (located in the executive branch) recommends annual budgets to Congress for each administrative agency, it has influence over all rulemaking.

Functions of Administrative Agencies

Administrative agencies perform the following functions: (1) rulemaking; (2) adjudication of individual cases brought before administrative law judges by agency staff; and (3) administrative activities, which include (a) informal advising of individual businesses and consumers, (b) preparation of reports and performance of studies of industries and consumer activities, and (c) issuance of guidelines for the business community and others as to what activities are legal in the eyes of agency staff.

RULEMAKING

We said that administrative agencies are authorized to perform the legislative function of making rules or regulations by virtue of their enabling statutes. For example, the enabling statute of the OSHA gave the secretary of labor authority to set "mandatory safety and health standards applicable to businesses affecting interstate commerce." The secretary was also given the power to "prescribe such rules and regulations that he may deem necessary to carry out the responsibilities under this act." In some cases, the procedures for implementing the rulemaking function are spelled out in the enabling act. If they are not, agencies follow the three major rulemaking models—formal, informal, and hybrid—outlined in the **Administrative Procedure Act (APA)** of 1946.

Administrative Procedure Act (APA) Law that establishes the standards and procedures that federal administrative agencies must follow in their rulemaking and adjudicative functions.

Formal Rulemaking. Section 553(c) of the APA requires formal rulemaking when an enabling statute or other legislation states that all regulations or rules must be enacted by an agency as part of a formal hearing process that includes a complete transcript. This procedure provides for (1) an agency notice of proposed rulemaking to the public in the *Federal Register*; (2) a public hearing at which witnesses give testimony on the pros and cons of the proposed rule, each witness is cross-examined, and the rules of evidence are applied; and (3) the making and publication of formal findings by the agency. On the basis of these findings, an agency may or may not promulgate a regulation. Because of the expense and time involved in creating a formal transcript and record, most enabling statutes do not require agencies to go through a formal rulemaking procedure when promulgating regulations.

Informal Rulemaking. As provided by Section 553 of the APA, informal rulemaking applies in all situations in which the agency's enabling legislation

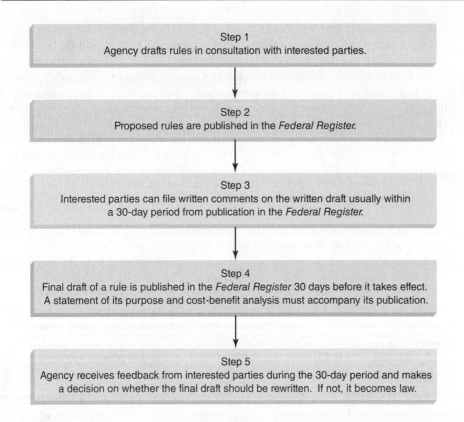

EXHIBIT 18-1

STEPS IN THE INFORMAL RULEMAKING PROCESS

or other congressional directives do not require another form. The APA requires that the agency (1) give prior notice of the proposed rule by publishing it in the *Federal Register*; (2) provide an opportunity for all interested parties to submit written comments; and (3) publish the final rule, with a statement of its basis and purpose, in the *Federal Register*. Exhibit 18-1 lays out the five-step process for promulgating a rule according to the informal rulemaking model. Executive and independent agencies are required to set out a cost-benefit analysis in Step 1 of the process.

Informal rulemaking is the model most often used by administrative agencies because it is efficient in terms of time and cost. No formal public hearing is required, and no formal record needs to be established, as in formal rulemaking. Parties opposed to a particular rule arrived at through informal rulemaking, however, often seek to persuade the appellate courts that the agency in question did not take important factors into account when the rule was being made.

Hybrid Rulemaking. Interested parties often complained that informal rulemaking gave them little opportunity to be heard other than in writing. Both Congress and the executive branch wanted administrative agencies to do a cost-benefit analysis of proposed regulations. Out of this input, from the public and two branches of government, came hybrid rulemaking, which combines some of the aspects of formal and informal rulemaking. This model requires the agency to give notice of a proposed regulation, set a period for public comments, hold a public hearing, and have a cost-benefit analysis done by an independent executive agency.

Exempted Rulemaking. Section 553 of the APA allows the agencies to decide whether there will be public participation in rulemaking proceedings relating to the "military or foreign affairs" and "agency management or personnel," as well as in proceedings relating to "public property, loans, grants, benefits, or contracts" of an agency. Public notice and comment are also not required when the agency is making interpretive rules or general statements of policy.

It is generally conceded that proceedings dealing with military and foreign affairs often require speed and secrecy, both of which are incompatible with public notice and hearings.

Judicial Review of Rulemaking. After a regulation is promulgated by an administrative agency and is published in the *Federal Register*, it generally becomes law. Appellate courts have accepted agency-promulgated regulations as law unless a business or other affected groups or individuals can show that:

1. the congressional delegation of legislative authority in the enabling act was unconstitutional because it was too vague and not limited;

2. an agency action violated a constitutional standard, such as the right to be free from unreasonable searches and seizures under the Fourth Amendment (e.g., if an agency such as OSHA promulgated a rule that allowed its inspectors to search a business property at any time without its owner's permission and without an administrative search warrant, that rule would be in violation of the Fourth Amendment); and

3. the act of an agency was beyond the scope of power granted to it by Congress in its enabling legislation.

Judicial review of administrative agency action provides a check against agency excesses that can prove very costly to the business community. The landmark case below illustrates this power of judicial review of administrative agencies.

CASE 18-1

City of Arlington v. Federal Communications Commission
United States Supreme Court
133 S. Ct. 1863 (2013)

State and local zoning authorities must approve applications by wireless telecommunications network companies to build new towers, as well as approve applications to place new antennas on existing towers. The Telecommunications Act of 1996 "imposed specific limitations on the traditional authority of state and local governments to regulate the location, construction, and modification" of wireless telecommunications networks. These limitations are incorporated into the Communications Act of 1934, which "empowers the Federal Communications Commission [FCC] to 'prescribe such rules and regulations as may be necessary in the public interest to carry out its provisions.'" One such provision requires state and local governments to act "within a reasonable period of time" after an application for a wireless tower or antenna site is filed.

An organization representing various wireless service providers petitioned the FCC to clarify the meaning of "reasonable period of time." In late 2009, the FCC, "relying on its broad statutory authority," issued a declaratory ruling determining that "a reasonable period of time" is presumptively (but rebuttably) 90 days when the application requests permission to place a new antenna on an existing tower and 150 days for all other applications. Certain state and local governments argue that the FCC did not have authority to issue such a declaratory ruling because the FCC did not have authority to "interpret ambiguous provisions" of the act and because two of the act's clauses (a saving clause and a judicial review provision) "together display[ed] a congressional intent to withhold from the [FCC] authority to interpret" the phrase "reasonable period of time." The cities of Arlington and San Antonio, Texas, then petitioned for review of the declaratory order in the U.S. Court of Appeals for the Fifth Circuit.

The Fifth Circuit held that the framework articulated in *Chevron U.S.A. Inc. v. National Resources Defense Council, Inc.*, governed the threshold question of whether the FCC had the statutory authority to adopt the 90- and 150-day time frames. In *Chevron*, the U.S. Supreme Court held that ambiguities in statutes should be resolved "within the bounds of reasonable interpretation, not by the courts but by the [applicable] administering agency." After concluding that certain language in the act was "ambiguous," the court held that the FCC had permissibly construed the statute to determine its statutory authority. The Plaintiff cities appealed.

Justice Scalia

Chevron is rooted in a background presumption of congressional intent: namely, "that Congress, when it left ambiguity in a statute" administered by an agency, "understood that the ambiguity would be resolved, first and foremost, by the agency, and desired the agency (rather than the courts) to possess whatever degree of discretion the ambiguity allows." *Chevron* thus provides a stable background rule against which Congress can legislate: Statutory ambiguities will be resolved, within the bounds of reasonable interpretation, not by the courts but by the administering agency. Congress knows to speak in plain terms when it wishes to circumscribe, and in capacious terms when it wishes to enlarge, agency discretion.

The question here is whether a court must defer under *Chevron* to an agency's interpretation of a statutory ambiguity that concerns the scope of the agency's statutory authority (i.e., its jurisdiction). The argument against deference rests on the premise that there exist two distinct classes of agency interpretations: Some interpretations—the big, important ones, presumably—define the agency's "jurisdiction." Others—humdrum, run-of-the-mill stuff—are simply applications of jurisdiction the agency plainly has. That premise is false, because the distinction between "jurisdictional" and "nonjurisdictional" interpretations is a mirage. No matter how it is framed, the question a court faces when confronted with an agency's interpretation of a statute it administers is always, simply, *whether the agency has stayed within the bounds of its statutory authority*.

In sum, judges should not waste their time in the mental acrobatics needed to decide whether an agency's interpretation of a statutory provision is "jurisdictional" or "nonjurisdictional." Once those labels are sheared away, it becomes clear that the question in every case is, simply, whether the statutory text foreclosed the agency's assertion of authority, or not . . . The federal judge as haruspex, sifting the entrails of vast statutory schemes to divine whether a particular agency interpretation qualifies as "jurisdictional," is not engaged in reasoned decision-making.

. . . If "the agency's answer is based on a permissible construction of the statute," that is the end of the matter.*

The judgment of the Court of Appeals was *affirmed*.

ADJUDICATION

In carrying out its adjudicative function in individual cases, as opposed to rulemaking for whole industries, an administrative agency usually pursues a four-step process. After receiving a complaint alleging violation of an administrative law, the agency notifies the party against whom the complaint is made and conducts an investigation into the merits of the complaint. If the agency staff finds that the complaint has merit, the agency next negotiates with the party to see if it can get the party to stop the violation voluntarily. If negotiation is unsuccessful, the third step is to file a complaint with an administrative law judge (ALJ). Step 4 consists of a hearing and decision by the ALJ. The party may appeal the ALJ's decision to the full commission or agency head and ultimately to a federal court of appeals and the U.S. Supreme Court.

All these steps are guided by the APA, which sets out minimum procedural standards for administrative agency adjudication. Enabling statutes that create agencies often add other procedural requirements. Finally, case law arising out of appeals of agency decisions to the U.S. circuit courts of appeal and the U.S. Supreme Court provides further guidelines for agencies in carrying out their adjudicative function. In the following detailed description of the four-step adjudicative process for federal administrative agencies, we use the FTC as a representative agency. You will find it easier to follow our discussion if you look first at the organizational outline of the FTC provided in Exhibit 18-2 and the summary of adjudication and judicial review of agency decision making given in Exhibit 18-3.

---

*City of Arlington v. Federal Communication Commission, United States Supreme Court 133 S.Ct. 1863 (201).

EXHIBIT 18-2 FEDERAL TRADE COMMISSION

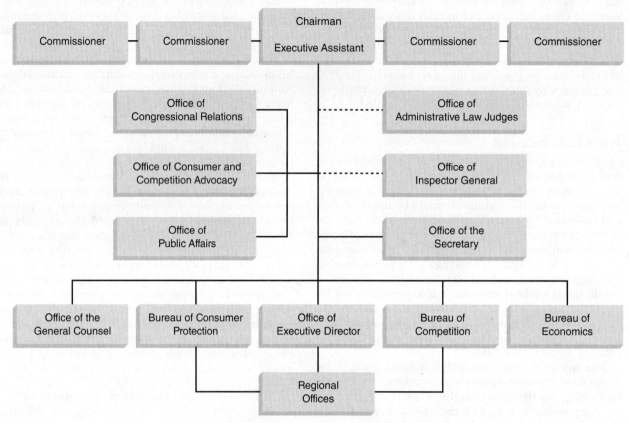

Source: U.S. Government Manual 2002–2003. Washington, DC: Office of the Federal Register, 2003, 722.

APPLYING THE LAW TO THE FACTS . . .

The EPA is informed of a company dumping waste into Lake Erie, so it contacts the company and attempts to get the company to stop dumping waste into the lake. The company ignores all messages from the EPA. Finally, the EPA files a complaint with an administrative law judge (ALJ). A hearing almost immediately follows. What step did the EPA mistakenly leave out in this scenario?

Investigation and Complaint. The FTC, which includes the Bureau of Competition and the Bureau of Consumer Protection, is obliged to conduct an investigation whenever it receives a complaint from other government agencies, competitors, or consumers. For example, upon receiving a complaint about a mouthwash product that is advertised as killing germs and protecting people from sore throats, the commission examines the product to see if the statement has any scientific validity. Should the commission's staff find that the advertising is "deceptive" or "unfair" within the meaning of Section 5 of the Federal Trade Commission Act, it will seek to stop the advertising campaign in one of two ways:

1. *Voluntary compliance.* The staff will ask the corporation to stop the advertising campaign voluntarily. Usually, no penalty is assessed if the company agrees to do this.

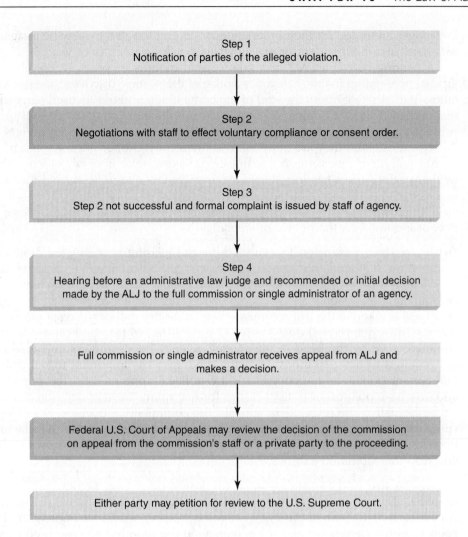

EXHIBIT 18-3

ADJUDICATION AND JUDICIAL REVIEW OF AGENCY DECISION MAKING

2. *Consent order.* If voluntary compliance is not obtained, the staff notifies the mouthwash company that it has 10 days to enter into a consent order; otherwise, the staff will issue a formal complaint.

Most cases are closed at this stage because, under a **consent order**, the company does not have to admit that it was deceptive or unfair in its advertising; it only has to promise that it will not do such unlawful advertising again and agree to the remedy the commission imposes. The latter may be some form of corrective advertising that tells the public that the mouthwash does not kill germs. A consent order helps the commission staff to obtain a binding cease-and-desist order with limited effort and time. It also benefits the company, because by agreeing to a consent order, the company avoids both an admission of guilt and the cost of litigation and shareholder and consumer lawsuits that might ensue if the next steps in the adjudication process—a formal complaint and a hearing by an ALJ—resulted in an adverse decision for the company.

Formal Complaint and Hearing. If the case is not settled by voluntary compliance or a consent order, the commission's staff, usually through the FTC's Office of the General Counsel (see Exhibit 18-2), will issue a formal complaint listing the charges against the mouthwash company and will request that certain penalties be assessed by the ALJ. **Administrative law judges (ALJs)**, who number approximately 1,150, are selected on the basis of a merit examination and are assigned to specific administrative agencies. They usually come from within the federal administrative bureaucracy and are given life tenure. Administrative

consent order An agreement by a business to stop an activity that an administrative agency alleges to be unlawful and to accept the remedy the agency imposes; no admission of guilt is necessary.

administrative law judge (ALJ) A judge, selected on the basis of a merit exam, who is assigned to a specific administrative agency.

law judges are noted for their independence, even though they may be assigned to a particular independent or executive agency for a number of years.[3]

A hearing before an ALJ may take several months or years. It resembles a judicial proceeding in that it includes notice to the parties, discovery, the presentation of evidence by both the staff of the commission and the accused party (the respondent), direct examination and cross-examination of the witnesses, and presentation of motions and arguments to the ALJ. There is, however, no jury at these hearings, and they are more informal than court proceedings. For instance, an ALJ will often intervene to ask questions and to take note of evidence that neither of the parties has introduced. At times in fact, the ALJ becomes a severe questioner of both parties, especially in hearings involving disability and welfare claims. Thus, adjudicative proceedings are less adversarial and more investigative, or inquisitorial, than court proceedings.

Initial or Recommended Decision. After the hearing is completed, both the commission staff and the respondent submit proposed findings of facts and conclusions of law. Under the APA, the ALJ must then prepare an initial or recommended decision. An initial decision becomes the final agency action unless an appeal is taken to the full commission by either the staff or the respondent. In contrast, a recommended decision is not final; it has to be acted on by the full commission or by the head of the agency. Agency heads and commissions are not required to defer to the ALJ's factual findings. It is important to remember that commissioners and agency heads are political appointees of the president and may have political or policy reasons for overruling an ALJ's decision. Should either the staff or the respondent appeal a full commission's decision to a federal circuit court of appeals, the court is likely to give deference to the ALJ's factual findings, because the ALJ is the person who actually heard the witnesses testify and read the submitted exhibits.

Appeal to the Full Commission. If the losing party (the agency staff or the respondent) does not agree with the ALJ's decision, it may appeal to the full commission in the case of the FTC or to the head of an executive department (or agency) in the case of an executive agency. In the mouthwash case used as an example, a majority of the commission members must rule in favor of one of the parties on the basis of a **preponderance of the evidence** standard (51 percent or more). The APA requires that the commission state factual, legal, and policy bases for its decision. This requirement makes the agency responsible for its decision both to the public and to the courts that may later review it.

preponderance of the evidence A legal standard whereby a bare majority (51 percent) of the evidence is sufficient to justify a ruling.

Judicial Review of Adjudicative Proceedings. If the party that loses at the full-commission or agency-head level in an adjudicative proceeding wishes to appeal, it must file a motion for appeal with the federal circuit court of appeals that has jurisdiction in the case. Briefs are filed by both parties, and the court hears oral argument. The court also reviews the whole record, including the ALJ's findings, in the case. It does not review the commission's factual findings as long as they are supported by substantial evidence in the record. (The substantial evidence rule requires that the court find that a reasonable person, after reviewing the record, would make the same findings the agency did.) Rather, it reviews the commission's legal findings to ensure that (1) it acted in a constitutionally approved way, (2) it acted within the scope of its jurisdiction as outlined by the enabling statute, and (3) it followed proper statutory procedures and did not act in an arbitrary or capricious manner.[4] The following case considers the application of the arbitrary and capricious standard.

---

[3] See "Administrative Law Judges Are Washington's Potent Hybrids," *New York Times*, Dec. 3, 1980 and "Symposium: Administrative Law Judges," *Western New England Law Review* 6: 1 (1984).

[4] Administrative Procedure Act of 1946 (APA), 5 U.S.C. §§ 551–706; 5 U.S.C. § 706(2)(A).

CASE 18-2

Fox Television Stations, Inc. v. Federal Communications Commission
United States Court of Appeals
489 F.3d 444 (2d Cir. 2007)

The Federal Communications Commission's (FCC's) policing of "indecent" speech stems from 18 U.S.C. Section 1464, which provides that "[w]hoever utters any obscene, indecent, or profane language by means of radio communication shall be fined or imprisoned not more than two years, or both." The FCC first exercised its statutory authority to sanction indecent (but nonobscene) speech in 1975, when it found Pacifica Foundation's radio broadcast of comedian George Carlin's "Filthy Words" monologue indecent.

Under the FCC's definition, *indecent speech* is language that describes, in terms patently offensive as measured by contemporary community standards for the broadcast medium, sexual or excretory activities and organs.

During [a] January 19, 2003, live broadcast of the Golden Globe Awards, musician Bono stated in his acceptance speech: "[T]his is really, really, brilliant. Really, really, great ****" (expletive included in the original broadcast).

On a complaint about the broadcast by individuals associated with the Parents Television Council, the FCC held that any use of any variant of "the F-Word" inherently has sexual connotation and therefore falls within the scope of the indecency definition. The Commission found that use of the word was fleeting and isolated irrelevant, and it overruled all prior decisions in which fleeting use of an expletive was held not indecent.

On February 21, 2006, the Commission found Fox Television Stations, Inc.'s broadcast of the 2002 Billboard Music Awards and Fox's broadcast of the 2003 Billboard Music Awards indecent and profane. During the 2002 broadcast, Cher stated: "People have been telling me I'm on the way out every year, right? So f*** 'em."

Fox filed a petition for review of the FCC's order in the U.S. Court of Appeals for the Second Circuit.

Justice Pooler

Agencies are of course free to revise their rules and policies. Such a change, however, must provide a reasoned analysis for departing from prior precedent. When an agency reverses its course, a court must satisfy itself that the agency knows it is changing course, has given sound reasons for the change, and has shown that the rule is consistent with the law that gives the agency its authority to act. In addition, the agency must consider reasonably obvious alternatives and, if it rejects those alternatives, it must give reasons for the rejection. The agency must explain why the original reasons for adopting the rule or policy are no longer dispositive [a deciding factor].

The primary reason for the crackdown on fleeting expletives advanced by the FCC is the so-called "first blow" theory. Indecent material on the airwaves enters into the privacy of the home uninvited and without warning. To say that one may avoid further offense by turning off the [television or] radio when he hears indecent language is like saying that the remedy for an assault is to run away after the first blow.

We cannot accept this argument as a reasoned basis justifying the Commission's new rule. First, the Commission provides no reasonable explanation for why it has changed its perception that a fleeting expletive was not a harmful "first blow" for the nearly thirty years between [the decisions in Pacifica's case] and Golden Globes. More problematic, however, is that the "first blow" theory bears no rational connection to the Commission's actual policy regarding fleeting expletives.

A re-broadcast of precisely the same offending clips from the two Billboard Music Award programs for the purpose of providing background information on this case would not result in any action by the FCC.

The Order makes passing reference to other reasons that purportedly support its change in policy, none of which we find sufficient. For instance, the Commission states that even non-literal uses of expletives fall within its indecency definition because it is "difficult (if not impossible) to distinguish whether a word is being used as an expletive or as a literal description of sexual or excretory functions." This defies any commonsense understanding of these words, which, as the general public well knows are often used in everyday conversation without any "sexual or excretory" meaning. Even the top leaders of our government have used variants of these expletives in a manner that no reasonable person would believe referenced "sexual or excretory organs or activities." [The court proceeded to recount examples of when President Bush and Vice President Cheney used the questionable words in public.]

Accordingly, we find that the FCC's new policy regarding "fleeting expletives" fails to provide a reasoned analysis justifying its departure from the agency's established practice. For this reason, Fox's petition for review is granted.*

Affirmed for Fox Television.

---

*Fox Television Stations, Inc. v. Federal Communications Commission, United States Court of Appeals. 489 F.3d 444 (2d Cir. 2007).

ADMINISTRATIVE ACTIVITIES

In addition to rulemaking and adjudication, executive and independent agencies perform a variety of tasks that are less well known but equally important to the average individual or business. The most significant of these are the following:

1. Advising businesses and individuals concerning what an agency considers legal and not legal. The antitrust merger guidelines we discuss in Chapter 24 are an example of an attempt by the Justice Department and the FTC to advise all interested parties about what conduct will be considered violations of Section 7 of the Clayton Act. More generally, lawyers representing interested parties meet daily with agency officials to receive informal comments or advice.

2. Conducting studies of industry and markets. Agencies such as the FTC, OSHA, and the FDA carry out studies to determine the level of economic concentration in an industry, dangerous products in the workplace, and the harmful effects of legal drugs.

3. Providing information to the general public on myriad matters by answering telephone calls, distributing pamphlets, and holding seminars.

4. Licensing of businesses in certain areas, such as radio and television stations (FCC).

5. Managing property. The General Services Administration (GSA) is the largest landlord in the country. It buys, sells, and leases all property used by the U.S. government.

Limitations on Administrative Agencies' Powers

STATUTORY LIMITATIONS

Certain federal statutes, summarized in Table 18-2, limit the power of administrative agencies and their officials. We have already discussed the APA. You should carefully review the brief descriptions of the other statutes listed in the table. It is important that you know, both as a future business manager and as an individual citizen, their major provisions. For instance, under the Federal Register Act of 1933, the Federal Privacy Act of 1974, and the Freedom of Information Act of 1966 as amended in 1974 and 1976, the decision-making processes of administrative agencies are open to the public. This legislation prevents secret, arbitrary, or capricious activity by the "fourth branch of government." Also, as you have seen, judicial review of administrative agencies' rulemaking and adjudication functions serves a similar purpose. Note also that private citizens have a means of relief against improper acts by employees of federal administrative agencies through the Federal Tort Claims Act of 1946, which forces agencies to waive sovereign immunity for their tortious actions and those of their employees. Tortious actions under this act include assault, battery, abuse of prosecution, false arrest, and trespass. For example, if an inspector from the EPA illegally enters a business property after being told to leave, the inspector, as well as the agency, may be held liable.

We said earlier that agencies were exempted from holding open hearings in certain circumstances, chiefly when proceedings concern military matters or foreign affairs. Some agencies have tried to stretch the exemption to cover proceedings in other "sensitive" matters.

INSTITUTIONAL LIMITATIONS

Executive Branch. The power of administrative agencies is limited by the executive branch through (1) the power of the president to appoint the heads

TABLE 18-2 FEDERAL STATUTES LIMITING ADMINISTRATIVE AGENCIES' AUTHORITY

Statute	Summary of Provisions
Federal Register of Act of 1993	Created the Federal Register system, which mandates the publication of all notices of federal agency meetings, proposed regulations, and final regulations in the *Federal Register*. The Federal Register system includes the *Government Manual*, which lists information, updated yearly, about each administrative agency, and the *Code of Federal Regulations (CFR)*, which codifies regulations promulgated by agencies of the federal government.
Freedom of Information Act of 1966 (FOIA)	Requires each agency to publish in the *Federal Register* places where the public can get information from the agency, procedural and substantive rules and regulations, and policy statements. Also, the FOIA requires each agency to make available for copying on request such items as staff manuals, staff instruction orders, and adjudicated opinions, as well as interpretations of policy statements. Nine exceptions enable an agency to deny an FOIA request by the public, a business, or other groups.
Government in Sunshine Act of 1976 (Sunshine Act)	Requires each agency headed by a collegiate body to hold every portion of a business meeting open to public attendance. A *collegiate body* exists if the agency is headed by two or more individuals, the majority of whom are appointed by the president and confirmed by the Senate.
Federal Privacy Act of 1974 (FPA)	Prevents an agency from disclosing any record in a system of records, by any means of communication, to any person or agency without the written authority of the individual. Eleven exceptions to the statute allow information to be released by the agency without the consent of individuals. Some exceptions are (1) to meet an FOIA request, (2) for use by the Selective Service System, (3) for use by another federal agency in civil or criminal law enforcement, (4) for use by a committee of the Congress, or (5) to meet a court order. Also, under the FPA, an individual may obtain information and correct errors in his or her record.
Administrative Procedure Act (APA)	The APA requires that all federal administrative agencies follow certain uniform procedures when performing their rulemaking and adjudicative functions.
Federal Tort Claims Act of 1946 (FTCA)	Requires the federal government to waive sovereign immunity and to assume liability for the tortious acts of its employees if nondiscretionary functions are being carried out by the employee.
Small Business Regulatory Enforcement Fairness Act of 1996 (SBREFA)	Requires analysis that measures the cost that a proposed rule would impose on a business (particularly a small business). Congress may review a proposed regulation for 60 days. The SBREFA helps enforce the Regulatory Flexibility Act to make sure federal agencies seek to reduce the impact of new regulations on small businesses.
Congressional Review Act of 1996	This law, in effect, gives Congress a "veto power" over every single regulation that agencies pass. Under this statute, a regulation cannot take effect until at least 60 working days have passed since the regulation was promulgated. If, during the 60-day period, a majority of the members of Congress pass a resolution of disapproval of the rule and either the president signs it or Congress overrules his veto of it, the regulation is nullified.

of the agencies, (2) the power of the OMB to recommend a fiscal-year budget for each agency, and (3) presidential executive orders.

The president not only appoints the head of each administrative agency but also designates some lower-level heads of departments and divisions that do not come under the federal civil service system. Naturally, presidential appointees tend to have the same philosophical bent as the chief executive and are often of the same party. In this way, the president gains some influence over both independent and executive agencies.

Presidents exercise even greater influence over executive agencies through the budget process and executive orders. In 1981, for instance, President Reagan signed Executive Order 12291, which requires executive agencies to perform a cost-benefit analysis before promulgating a major federal regulation. A *major federal regulation* is a regulation that will cost businesses $100 million or more to comply with. Another example, this one concerning the budget process, Executive Order 12498, signed in 1985 (also by President Reagan), extended the OMB's powers so that it now has authority over "pre-rulemaking action" by executive agencies. This executive order requires civilian government agencies to submit a Draft Regulatory Program listing all pre-rulemaking and other significant actions they intend to take in a fiscal year. These Draft Regulatory Programs become part of the administration's Regulatory Program. Once that program is published by the OMB, no agency may deviate from the plan without approval from the OMB unless forced to do so by the courts.

Note that these executive orders affect executive administrative agencies. However, independent administrative agencies have been requested to comply voluntarily with these orders, and some have done so. A bill passed by the House of Representatives and the Senate in 1995, and discussed in this chapter under "Creation of Administrative Agencies," applies to both executive and independent administrative agencies.

Legislative Branch. Congress limits the authority of administrative agencies through its (1) oversight power, (2) investigative power, (3) power to terminate an agency, and (4) power to advise on and consent to presidential nominations for heads of administrative agencies.

When Congress creates an agency, it delegates to that agency its own legislative power over a narrow area of commerce (e.g., human rights). Each year, through one of its oversight committees, it determines whether the agency has been carrying out its mandated function. Suppose, for example, that the House Energy and Commerce Committee's Subcommittee on Consumer Finance and Telecommunications finds that the SEC is not enforcing laws against insider trading and fraud. The full committee will investigate and, if it finds dereliction, will order the SEC to enforce the laws as it is charged to do.

The greatest legislative limitation on agency power, however, lies in Congress's right to approve or disapprove an agency budget submitted by the executive branch (the OMB). If Congress disagrees with the agency's actions, it can slash the budget or refuse to budget the agency at all. The latter action, of course, will shut down the agency. In contrast, if Congress believes that the executive branch is shortchanging an agency for some reason, it can raise that agency's budget above the amount proposed by the OMB.

Judicial Branch. The courts can curb administrative agencies' rulemaking and adjudicative excesses by reversing or modifying such actions, as explained earlier in this chapter. You might want to go back to that portion of the chapter at this point and reread the standards used by the courts in reviewing these agency functions.

State and Local Administrative Agencies

Each of the 50 states, the District of Columbia, and territories such as Puerto Rico and Guam have created state and local administrative agencies to carry out tasks assigned to them by their legislative bodies. Most have utilities commissions (or the equivalent) that regulate local and in-state telephone rates and are similar to the Federal Communications Commission, which regulates telephone rates for calls between states (interstate calls). Agencies that regulate state-chartered banks, workers' compensation, state universities, and state taxes are common at the state level and are assigned duties by the state legislature. At the city and county level, real estate planning boards, zoning commissions, and supervisory boards are just a few of the administrative agencies that have a profound effect on the life of all citizens; future business leaders should not overlook these when determining where their companies will be located.

When federal and state agency laws conflict, the Supremacy Clause of Article VI of the U.S. Constitution plays a significant role, as shown in the following case.

CASE 18-3

Vonage Holdings Corp. v. Minnesota Public Utilities Commission
U.S. District Court of Minnesota
290 F. Supp. 2d 993 (2003); aff'd 394 F.3d 568 (2004)

Vonage Holdings Corporation markets and sells Vonage DigitalVoice, a service that permits voice communication via a high-speed (broadband) Internet connection. Vonage's service uses a technology called Voice over Internet Protocol (VoIP), which allows customers to place and receive voice transmissions routed over the Internet.

Traditional telephone companies use circuit-switched technology. Voice communication using the Internet has been called Internet protocol (IP) telephony, and rather than using circuit switching, it utilizes "packet switching," a process of breaking down data into packets of digital bits and transmitting them over the Internet.

Vonage has approximately 500 customers with billing addresses in Minnesota. The Minnesota Department of Commerce (MDOC) investigated Vonage's services and on July 15, 2003, filed a complaint with the Minnesota Public Utilities Commission (MPUC). The complaint alleged that Vonage failed to obtain a proper certificate of authority required to provide telephone service in Minnesota.

Vonage then moved to dismiss the MDOC complaint. The MPUC concluded that Vonage was required to comply with Minnesota statutes and rules regarding the offering of telephone service. Vonage then filed a complaint seeking an injunction.

Justice Davis

The issue before the Court is whether Vonage may be regulated under [a] Minnesota law that requires telephone companies to obtain certification authorizing them to provide telephone service. Vonage asserts that the Communications Act of 1934, as amended by the Communications Act of 1996, preempts the state authority upon which the MPUC's order relies. Vonage asserts that its services are "information services," which are not subject to regulation, rather than "telecommunications services," which may be regulated.

The Supremacy Clause of Article VI of the Constitution empowers Congress to preempt state law. Preemption occurs when (1) Congress enacts a federal statute that expresses its clear intent to preempt state law; (2) there is a conflict between federal and state law; (3) compliance with both federal and state law is in effect physically impossible; (4) federal law contains an implicit barrier to state regulation; (5) comprehensive congressional legislation occupies the entire field of regulation; or (6) state law is an obstacle to the accomplishment and execution of the full objectives of Congress. *Moreover, a federal agency acting within the scope of its congressionally delegated authority may preempt state regulation.*

Examining the statutory language of the Communications Act, the Court concludes that the VoIP service provided by Vonage constitutes an information service because it offers the "capability for generating, acquiring, storing, transforming, processing, retrieving, utilizing, or making available information via telecommunications." Vonage's services are closely tied to the provision of telecommunications services as defined by Congress, the courts, and the [Federal Communications Commission (FCC)], but this Court finds that Vonage *uses* telecommunications services, rather than provides them.

The Court acknowledges the attractiveness of the MPUC's simplistic "quacks like a duck" argument, essentially holding that because Vonage's customers make phone calls, Vonage's services must be telecommunications services. However, this simplifies the issue to the detriment of an accurate understanding of this complex question. The Court must follow the statutory intent expressed by Congress, and interpreted by the FCC. Short of explicit statutory language, the Court can find no stronger guidance for determining that Vonage's service is an information service.

Where federal policy is to encourage certain conduct, state law discouraging that conduct must be preempted.*

For the Plaintiff, injunction granted.

---

*Vonage Holdings Corp. v. Minnesota Public Utilities Commission, U.S. District Court of Minnesota 290 F. Supp. 2d 993 (2003); aff'd 394 F.3d 568 (2004).

Global Dimensions of Administrative Agencies

In the United Kingdom, the Financial Services Authority (FSA) regulates the banking, securities, commodities futures, and insurance industries. The FSA is an independent, nongovernmental body whose board of directors is nominally appointed by the Crown. In contrast, in the United States, banks are regulated by the Federal Reserve Board, the Comptroller of the Currency, and state bank regulators; securities firms are regulated by the SEC, state securities commissions, and the National Association of Securities Dealers; commodities futures are regulated by the Commodities Futures Trading Commission; and insurance is regulated by state insurance commissions.

The FSA's self-avowed goals are to (1) maintain confidence in the British financial system, (2) promote public understanding of that system, (3) secure the right degree of protection for customers, and (4) help reduce financial crime. Like its American counterparts, the FSA oversees transactions, demands ethical and legal conduct from firms, and sets standards. Unlike the American system, which utilizes government funding, the FSA charges all firms it regulates annual licensing fees and thus is privately funded. This idea is to allow the FSA to act independently by removing all subjectivity, such as governmental wishes. The concentration of power in the FSA theoretically allows it to better regulate the country's banking and trading exchanges because it does not have to coordinate with other bodies that may have diverging goals and interests.

SUMMARY

Administrative law is defined broadly as any rule (statute or regulation) that directly or indirectly affects an administrative agency. The APA provides procedural guidelines for federal agencies; these guidelines are often copied, in whole or in part, by state and local administrative agencies.

The major functions of administrative agencies are rulemaking, adjudication, and the carrying out of numerous administrative activities. The executive, judicial, and legislative branches of government limit the power of federal agencies in numerous ways. In addition, several federal statutes limit the authority of administrative agencies.

Federal administrative agencies meet with their counterparts in other nations and enter into international agreements that aid the enforcement powers of U.S. agencies.

REVIEW QUESTIONS

18-1 Does Congress usually create administrative agencies? Explain.

18-2 What are the two major functions of administrative agencies? Explain each function.

18-3 Explain the distinction between executive administrative agencies and independent administrative agencies.

18-4 How do the courts check the power of administrative agencies?

18-5 Why do administrative agencies grow over time?

18-6 Distinguish between formal and informal rulemaking.

REVIEW PROBLEMS

18-7 For decades, the FTC resolved fair trade and advertising disputes through individual adjudications. In the 1960s, the FTC began promulgating rules that defined fair and unfair trade practices. In cases involving violations of these rules, the due process rights of participants were more limited and did not include cross-examination. Although anyone charged with violating a rule would receive a full adjudication, the legitimacy of the rule itself could not be challenged in the adjudication. Furthermore, a party charged with violating a rule was almost certain to lose the adjudication. Affected parties complained to a court, arguing that their rights before the FTC were unduly limited by the new rules. What would the court examine to determine whether to uphold the new rules?

18-8 Assume that the Food and Drug Administration (FDA), using proper procedures, adopts a rule describing its future investigations. This new rule covers all future circumstances in which the FDA wants to regulate food additives without giving food companies an opportunity to cross-examine witnesses. Later, the FDA wants to regulate methyl isocyanate, a food additive. The FDA undertakes an informal rulemaking procedure, without cross-examination, and regulates methyl isocyanate. Producers protest, saying that the FDA promised them the opportunity for cross-examination and that it is free to withdraw the promise made in its new rule. If the producers challenge the FDA in court, on what basis would the court rule in their favor? Explain.

18-9 Holabird & Root hired the Chicago law firm Sabo & Zahn to bring a collection action against a local real estate developer, Horwitz Matthews, Inc. (HM). Sabo won a $150,000 judgment against HM on behalf of Holabird and then began enforcing proceedings to collect on it. Responding to a citation to discover assets, HM provided Sabo with tax returns that were subject to a confidentiality agreement barring the attorneys from disclosing the information derived from the returns to anyone outside the firm. Despite the agreement, the Sabo attorneys told about 40 of HM's associates and investors by mail that HM had apportioned itself a greater percentage of some partnership business than it was entitled to and that investors' losses had been underreported. HM sued Sabo and Holabird, claiming it was vicariously liable. Is a law firm an agent of its clients, or is it an independent contractor? Should a client be liable for torts committed by its law firm? Explain.

18-10 In 1969, the secretary of transportation approved a plan to extend an interstate highway through Overton Park in Memphis, Tennessee. A group of environmentalists petitioned the courts, seeking to enjoin the Department of Transportation (DOT) from financing the project. They argued that under the Federal Aid Highway Act, federal funds could not be used for highway construction through a public park if a "feasible and prudent alternative" existed. DOT personnel produced affidavits indicating that the secretary had considered other alternatives and argued that unless there was substantial evidence to the contrary, the secretary's decision should be upheld by the reviewing court. Who won this case and why?

18-11 The FTC instituted proceedings against the Soft Drink Bottling Company, alleging violations of laws prohibiting unfair methods of competition. The complaint against Soft Drink Bottling challenged the validity of exclusive bottling agreements between the company and franchised bottlers, who had to agree not to sell the company's products outside a designated territory. Soft Drink Bottling asked the FTC to include the 513 bottlers in the case. The FTC refused to do so on the grounds that inclusion of so many bottlers would make the case unmanageable—although it said that any bottlers who wished to could intervene in the case. Soft Drink Bottling decided to appeal the decision to a federal court. Did the court entertain the action? Why or why not?

18-12 One of the hottest business issues today is the outrageously lavish pay of many chief executive officers (CEOs) vis-à-vis corporate profits. In 1990, for example, the CEO of United Airlines got $18.3 million—1,200 times what a new flight attendant makes—even though United's profits had fallen by 71 percent that year. The ICC would like to regulate CEO pay. Does it have the authority to do so? Explain.

CASE PROBLEMS

18-13 After Dave Conley died of lung cancer, his widow filed for benefits under the Black Lung Benefits Act. To qualify for benefits under the act, exposure to coal dust must have been a substantial contributing factor to a person's death. Conley had been a coal miner, but he had also been a longtime smoker. At the benefits hearing, a physician testified that coal dust was a substantial factor in Conley's death. No evidence was presented to support this conclusion, however. The administrative law judge awarded benefits. On appeal, should a court defer to this decision? Discuss. *Conley v. National Mines Corp.*, 595 F.3d 297 (6th Cir. 2010).

18-14 Michael Manin, an airline pilot, was twice convicted of disorderly conduct, a minor misdemeanor. To renew his flight certification with the NTSB, Manin filed an application that asked about his criminal history. He did not disclose his two convictions. When they came to light more than 10 years later, Manin argued that he did not know that he was required to report convictions for minor misdemeanors. The NTSB's policy was to consider an applicant's understanding of the information a question sought before determining whether an answer was false. But without explanation, the agency departed from this policy, refused to consider Manin's argument, and revoked his certification. Was this action arbitrary or capricious? Explain. *Manin v. National Transportation Safety Board*, 627 F.3d 1239 (D.C. Cir. 2011).

18-15 To ensure highway safety and protect drivers' health, Congress charged federal agencies with regulating the hours of service of commercial motor vehicle operators. Between 1940 and 2003, the regulations that applied to long-haul truck drivers were mostly unchanged. (Long-haul drivers operate beyond a 150-mile radius of their base.) In 2003, the Federal Motor Carrier Safety Administration (FMCSA) revised the regulations significantly, increasing the number of daily and weekly hours that drivers could work. The agency had not considered the impact of the changes on the health of the drivers, however, and the revisions were overturned. The FMCSA then issued a notice that it would reconsider the revisions and opened them up for public comment. The agency analyzed the costs of the options and concluded that the safety benefits of not increasing the hours were less than the economic costs. In 2005, the agency issued a rule that was nearly identical to the 2003 version. Public Citizen, Inc., and others, including the Owner-Operator Independent Drivers Association, asked the U.S. Court of Appeals for the District of Columbia Circuit to review the 2005 rule as it applied to long-haul drivers. *Owner-Operator Independent Drivers Association, Inc. v. Federal Motor Carrier Safety Administration*, 494 F.3d 188 (D.C. Cir. 2007). Should the court rule in favor of the Independent Drivers? Why or why not?

18-16 OSHA is part of the U.S. Department of Labor. OSHA issued a "directive" under which each employer in selected industries was to be inspected unless it adopted a "Comprehensive Compliance Program"—a safety and health program designed to meet standards that in some respects exceeded those otherwise required by law. The Chamber of Commerce of the United States objected to the directive and filed a petition for review with the U.S. Court of Appeals for the District of Columbia Circuit. The Chamber claimed, in part, that OSHA did not use proper rulemaking procedures in issuing the directive. OSHA argued that it was not required to follow those procedures because the directive itself was a "rule of procedure." OSHA claimed that the rule did not alter the rights or interests of parties, "although it may alter the manner in which the parties present themselves or their viewpoints to the agency." What are the steps of the most commonly used rulemaking procedure? Which steps are missing in this case? *Chamber of Commerce of the United States v. U.S. Department of Labor*, 74 F.3d 206 (D.C. Cir. 1999).

18-17 In 2007, the U.S. Supreme Court decided that under the Clean Air Act, the EPA had the right to regulate industry emissions of heat-trapping gases. Subsequently, challengers, including individuals doubtful of the existence of global warming, and representatives from certain industries brought the EPA to court, arguing against the EPA's claim that human health and well-being are imperiled by greenhouse gases. Specifically, the opponents challenged EPA studies that indicate the danger to human health. Ultimately, many of the EPA's rules and regulations rely on that finding. One opponent of the EPA—Sullivan, the Alaska Attorney General—said that Alaska had "important interests that will be impaired by the endangerment finding." Specifically, he argued that his state could not handle the encumbrance of some of the EPA's regulations regarding greenhouse gases. This case was a 2010 consolidation of 16 separate cases challenging the EPA. What main issues does this case come down to? How did the court decide? *Coalition for Responsible Regulation, Inc. et al. v. EPA*, Case No. 09-1322 (D.C. Cir. 2010).

18-18 Lion Raisins, Inc., is a family-owned business that grows and markets raisins to private enterprises. In May 1999, a USDA investigation reported that Lion appeared to have falsified inspectors' signatures, given false moisture content, and changed the grade of raisins on three USDA raisin certificates issued between 1996 and 1998. Lion was subsequently awarded five more USDA contracts. In January 2001, however, the USDA awarded these contracts to other bidders and, on the basis of the May 1999 report, suspended Lion from participating in government contracts for one year. Lion filed a suit in the U.S. Court of Federal Claims against the USDA, seeking, in part, lost profits on the school lunch contracts on the grounds that the USDA's suspension was arbitrary and capricious. Who won? Explain. *Lion Raisins, Inc. v. United States*, 51 Fed. Cl. 238 (2001).

THINKING CRITICALLY ABOUT RELEVANT LEGAL ISSUES

For our administrative agencies to be more effective, we ought to require them to use formal rulemaking. To continue to allow agencies to use the informal rulemaking process is to undermine the democratic principles we hold so dear.

Representative democracy is based on the notion that our elected officials will represent the views of the people. When the ideas and opinions of the people are not expressed, or are expressed but then ignored, the ideal of representative democracy is not realized. In such instances, our form of government is more like an autocracy because the government is able to rule at its whim, free of responsibility to the citizens.

Formal rulemaking requires a full public hearing in which the testimony is printed in an official transcript. It also requires the publication of formal findings. These features ensure that the agency will listen to the opinions of the people. In addition, the printed records of testimony and factual findings guarantee the people an opportunity to make sure the agency's reasoning in each decision is sound.

Informal rulemaking does not provide this guarantee that the agency's officials will be responsive to the public's desires. Informal rulemaking requires only the publication of the proposed rule in the *Federal Register*. Afterward, interested parties are able to submit written arguments for or against the rule. How can we be sure, however, that the agency will adequately consider this written testimony? If the head of the agency disagrees with what you write, he may tell you that your letter "must have gotten lost in the mail." Clearly, it is too easy for agency officials to ignore the public in the process of informal rulemaking.

Many people argue that we should continue to allow informal rulemaking because it is more convenient for the agencies compared with formal rulemaking. Well, I suppose it would also be more convenient for the heads of the agencies to make the decision or rule by themselves. (Refer to Chapter 1 when answering the following questions.)

1. What primary ethical value is least important to the author of this essay? Explain.

2. What are some problems with the evidence the author gives to support her arguments? Explain.

3. Give some arguments that are opposite of those made by the author in this essay. Explain your choices.

ASSIGNMENT ON THE INTERNET

This chapter introduced administrative law and several administrative agencies. There are, however, many agencies not discussed that play a significant role in creating regulations. Use the Internet to discover three federal administrative agencies not discussed in this chapter. Explain the purpose of each agency and state at least two recent rules or regulations it has issued. The websites listed in the following section may be of use in locating the many federal administrative agencies.

ON THE INTERNET

www.whitehouse.gov This site, maintained by the White House, contains links to federal agencies and commissions. Highlight "1600 Penn," then go to "Federal Agencies and Commissions" under the heading "Our Government."

www.americanbar.org/aba.html Visit this site if you have ideas about how to improve the way administrative agencies cooperate. Highlight "ABA Groups," then click on Sections. Then click on "Administrative Law and Regulatory Practice."

www.law.cornell.edu The Legal Information Institute's site on administrative agencies includes recent administrative law judicial decisions. Click on "Wex Legal Encyclopedia," then "Browse." Scroll until you find "Administrative Law."

www.constitution.org This site contains a discussion of the constitutional issues related to administrative activities. Perform a search for "administrative sanctions." DO NOT click on the links at the top of the page (Sanctions, Common Law, etc.) as these are ads.

www.oalj.dol.gov This is the site of the Office of Administrative Law Judges. Click on "Agencies," then on "Office of Administrative Law Judges (OALJ)."

FOR FUTURE READING

Fox, William F. "Understanding Administrative Law." Lexis Nexis (2012).

Garry, Patrick M. "Accommodating the Administrative State: The Interrelationship between the Chevron and Nondelegation Doctrines." *Arizona State Law Journal* 38 (2006): 921.

Hall, Daniel E. *Administrative Law: Bureaucracy in a Democracy* (5th ed.). New Jersey: Prentice Hall, 2011.

Lubbers, Jeffrey S. *A Guide to Administrative Rulemaking*. Chicago: American Bar Association, 2006.

CHAPTER NINETEEN

The Employment Relationship and Immigration Laws

- WAGE AND HOUR LAWS
- UNEMPLOYMENT COMPENSATION
- CONSOLIDATED OMNIBUS BUDGET RECONCILIATION ACT OF 1985
- WORKERS' COMPENSATION LAWS
- THE FAMILY AND MEDICAL LEAVE ACT OF 1993
- THE OCCUPATIONAL SAFETY AND HEALTH ACT OF 1970
- EMPLOYEE PRIVACY RIGHTS
- IMMIGRATION LAW
- GLOBAL DIMENSIONS OF THE EMPLOYMENT RELATIONSHIP

One way to think about the employment relationship is that it is a contractual relationship between the employer and the employee: The employer agrees to pay the employee a certain amount of money in exchange for the employee's agreement to render specific services. Early in our history, the employer and employee were free to determine all the conditions of their employment relationship. Today, however, both the federal and state governments specify many of the conditions under which that relationship exists.

This chapter explains many of the conditions that the government imposes. The first five sections focus on the laws that affect employee wages and benefits: wage and hour laws; unemployment compensation legislation; the Consolidated Omnibus Budget Reconciliation Act of 1985; state workers' compensation laws; and the Family and Medical Leave Act. The sixth and seventh sections shift the focus to protection of worker safety and health and privacy rights. The ninth section is a new section discussing an area of growing importance: immigration law. The final section discusses the international dimensions of the employment relationship.

CRITICAL THINKING ABOUT THE LAW

Employers are required by law to provide specific employment conditions for their employees. For example, minimum wage and hours laws help to ensure that the worker will not be required to work an extraordinary number of hours for little pay.

Why do we have laws to protect workers? One reason for these laws is the belief that employees cannot protect themselves in the workplace. Because they have very little power over the employer, we have traditionally offered protection for employees to prevent their exploitation. After you read the following case example and answer the questions, you will have sharpened your critical thinking about employee benefits.

Mike works full time at a large factory. As he is preparing to go home after his shift, he sees his boss yelling angrily at the television in the executive office. The news station has just reported congressional passage of a new law that requires employers to provide paid leave for the birth of a child; both mothers and fathers will receive the paid leave. Mike's boss thinks the new law is an outrageous restriction on employers. Mike, however, welcomes the new law because his wife is pregnant; he will be happy to stay home with his wife when she has the baby.

1. Mike's opinion and his employer's opinion about the new law obviously conflict. Their respective ethical preferences are one important cause of their disagreement. What ethical norm seems to be dominating the employer's thought?

    Clue: Why is the employer upset? Review your list of ethical norms; try to match one of the norms to the reason the employer is angry.

2. Mike's boss makes the following argument: "It is ridiculous that the government would make me pay my employees to sit at home! They are my employees. I should get to decide if employees should be allowed to leave for the birth of a child." Do you see any problems with this argument? Explain.

    Clue: How does the employer's argument conflict with the reason for offering employee benefits?

3. Before you make a judgment about the worth of the new law, is there any additional information that would help your thinking? Explain.

    Clue: What information seems to be missing? Several crucial pieces of information about the paid leave are not offered in the news report.

Wage and Hour Laws

Employers no longer have complete freedom to pay workers any wages they choose or to require them to work any number of hours. Several federal laws impose minimum wage and hour requirements. For example, the Davis–Bacon Act[1] requires that contractors and subcontractors working on government projects pay the "prevailing wage." The most pervasive law regulating wages and hours, however, is the Fair Labor Standards Act (FLSA),[2] which covers all employers engaged in interstate commerce or the production of goods for interstate commerce.

One of the most significant aspects of the FLSA is its requirement that a minimum wage of a specified amount be paid to all employees in covered industries. This specified amount is periodically raised by Congress to compensate for increases in the cost of living caused by inflation. The most recent increase in the minimum wage was to $7.25 on July 24, 2009.

Employees who receive a significant portion of their income from tips are entitled to a lower minimum wage from their employer. An employer may pay a tipped employee not less than $2.13 an hour in direct wages if that amount plus the tips received equal at least the federal minimum wage, the employee retains all tips, and the employee customarily and regularly receives more than

---

[1] 40 U.S.C. §§ 276a–276a-5 (1998).
[2] 29 U.S.C. §§ 201–260 (1998).

$30 a month in tips. If an employee's tips combined with the employer's direct wages of at least $2.13 an hour do not equal the federal minimum hourly wage, the employer must make up the difference.

The FLSA also requires that employees who work more than 40 hours a week be paid no less than one and one-half times their regular wage for all the hours beyond the standard 40-hour workweek. Four categories of employees, however, are excluded from this provision of the law: executives, administrative employees, professional employees, and outside salespersons. To prevent employers from taking advantage of these exemptions, however, the act generally requires that these employees earn at least a minimum amount of income and spend a certain amount of time engaged in specified activities before they can fall into each of those exempted categories.

A seemingly unending list of well-known companies appear to have attempted to avoid paying overtime, often by miscategorizing nonmanagerial employees as managers. For example, in 2002, Starbucks and UPS collectively agreed to pay $36 million in damages to settle claims by groups of workers who had been misclassified as "managers." CVS agreed to pay an undisclosed sum to settle claims of pharmacists that they were not salaried, exempt employees, and PETCO agreed to pay its managers and assistant managers in California up to $25 million to settle their claims that they were not exempt from being paid overtime pay.[3] In 2007, Computer Sciences Corporation paid a $24 million settlement to its technical support workers, who claimed that they had not received proper overtime wages. The technical support workers represented about 40 percent of the total staff at the company.[4] In 2007, Wal-Mart, the world's largest employer, turned itself in to the Department of Labor (DOL) for unpaid overtime; Wal-Mart agreed to pay $33 million in back wages and unpaid overtime.[5]

In 2012, the U.S. Supreme Court decided a major lawsuit interpreting the outside salespersons exception in a manner favorable to the pharmaceuticals industry. Two employees filed a lawsuit against GlaxoSmithKline, arguing that they had not been paid for the 10 to 20 hours a week they worked, on average, outside the normal business day. Although the Labor Department initially held that the exemption did not apply when the workers were visiting doctors' offices to promote the drugs they were selling, in a 5–4 decision, the high court ruled that pharmaceutical companies do not have to pay overtime to sales reps who visit doctors' offices because they are outside salespersons exempt from FLSA.[6]

When the workday starts and ends can also lead to FLSA litigation. For example, in 2011, Tyson Foods settled a lawsuit by agreeing to pay more than 17,000 workers employed at 41 poultry plants in 12 states an average of $1,000 apiece in back pay as part of a $32 million settlement. The workers had sued Tyson for violating the Fair Labor Standards Act by failing to pay them for time spent putting on and taking off gear they were required to wear to protect themselves and the poultry.[7]

---

[3]Connie Robbins Gentry, "Off the Clock," *Chain Store Age* 79; WL 13301250 (Feb. 21, 2003).

[4]Ed Frauenheim, "CSC Agrees to Overtime Settlement," CNET News.com (Mar. 25, 2005). Accessed March 3, 2008 at www.news.com/CSC+agrees+to+overtime+settlement/2100-1022_3-5638162.html.

[5]Marcus Kabel, "Wal-Mart to Pay More Than $33 Million for Overtime Violations," Associated Press (Jan. 25, 2007).

[6]*Christopher v. SmithKline Beecham Corp.*, 132 S. Ct. 2156 (2012).

[7]Steven Greenhouse, "Georgia: Tyson Agrees to $32 Million Settlement," *New York Times* (Sept. 19, 2012). Available at www.nytimes.com/2011/09/20/us/georgia-tyson-agrees-to-32-million-settlement.html.

During a yearlong period ending September 30, 2014, a record high of 8,160 FLSA lawsuits were filed.[8] This number topped the previous year's then-record high of 7,500 cases. If we looked at the past two decades, we would see a fairly consistent increase every year in the number of FLSA lawsuits filed, with the most significant increases beginning around 2003. Most of the cases are for alleged misclassification of employees, uncompensated "work" performed off the clock, and miscalculation of overtime pay for nonexempt workers. Technology has also now blurred the line between when you are working and when you are not.

In what some had perceived as an attempt to quell the amount of litigation under the FLSA, in 2004, the Bush administration made substantial changes to the DOL overtime regulations. In August 2004, the new DOL rules went into effect, redefining eligibility requirements for overtime protection. Under these new rules, workers earning less than $23,660 a year were guaranteed overtime protection, which is estimated to affect 1.3 million workers. Other changes to the rules, however, remove overtime protection from an estimated 4 to 6 million workers. One such change redefined "executive" to include those workers whose job is primarily manual or routine labor but who also perform supervision. The new definition of *executive* removed overtime protection from 1.4 million supervisors. Other key definitions, such as those of "professional" and "administrative," were changed, resulting in a net loss of workers with overtime protection. Because fewer employees were guaranteed overtime protection, many predicted that the number of overtime-related lawsuits would drop. Obviously, this drop did not materialize, as litigation in this area continues to increase.

The Wage and Hour Division (WHD) of the DOL is responsible for enforcing the FLSA, and even though the agency tries to assist as many workers as possible, it does not have the resources to help every worker who has a legitimate complaint. However, under a program instituted in late 2010, when FLSA complainants are informed that the WHD will not be able to pursue their complaints, they will be given a toll-free number to contact the ABA-Approved Attorney Referral System. If complainants choose to call the toll-free number, they will be advised of the ABA-Approved Lawyer Referral and Information Service providers in their area. The complainant may then contact one of the providers and determine whether to retain a qualified private-sector lawyer.

To ensure that employers comply with the FLSA, the law imposes recordkeeping requirements on employers. For each nonexempt employee, the employer must maintain the following records: (1) employee's full name and Social Security number; (2) address, including zip code; (3) birth date if under 18; (4) sex and occupation; (5) time and day of week when employee's workweek begins; (6) hours worked each day; (7) total hours worked each workweek; (8) basis on which employee's wages are paid (e.g., "$6 an hour," "$220 a week," "piecework"); (9) regular hourly pay rate; (10) total daily or weekly straight-time earnings; (11) total overtime earnings for the workweek; (12) all additions to or deductions from the employee's wages; (13) total wages paid each pay period; and (14) date of payment and the pay period covered by each payment.[9]

Finally, to ensure that employees become aware of their rights, the employer must post in a conspicuous place a poster similar to the one shown in Exhibit 19-1.

Unemployment Compensation

Having a minimum wage provides a great deal of security for employees when they are working, but what happens to those who lose their jobs? In 1935,

---

[8]Ben James, "FLSA, FMLA Lawsuits Soaring, New Statistics Show," *Law* 360 (March 11, 2015). Available at http://www.law360.com/articles/630168/flsa-fmla-lawsuits-soaring-new-statistics-show.

[9]Accessed March 3, 2008, at www.dol.gov/esa/regs/compliance/whd/whdfs21.htm.

EMPLOYEE RIGHTS
UNDER THE FAIR LABOR STANDARDS ACT
THE UNITED STATES DEPARTMENT OF LABOR WAGE AND HOUR DIVISION

FEDERAL MINIMUM WAGE
$7.25 PER HOUR
BEGINNING JULY 24, 2009

OVERTIME PAY At least 1½ times your regular rate of pay for all hours worked over 40 in a workweek.

CHILD LABOR An employee must be at least **16** years old to work in most non-farm jobs and at least **18** to work in non-farm jobs declared hazardous by the Secretary of Labor.

Youths **14** and **15** years old may work outside school hours in various non-manufacturing, non-mining, non-hazardous jobs under the following conditions:

No more than
- **3** hours on a school day or **18** hours in a school week;
- **8** hours on a non-school day or **40** hours in a non-school week.

Also, work may not begin before **7 a.m.** or end after **7 p.m.**, except from June 1 through Labor Day, when evening hours are extended to **9 p.m.** Different rules apply in agricultural employment.

TIP CREDIT Employers of "tipped employees" must pay a cash wage of at least $2.13 per hour if they claim a tip credit against their minimum wage obligation. If an employee's tips combined with the employer's cash wage of at least $2.13 per hour do not equal the minimum hourly wage, the employer must make up the difference. Certain other conditions must also be met.

ENFORCEMENT The Department of Labor may recover back wages either administratively or through court action, for the employees that have been underpaid in violation of the law. Violations may result in civil or criminal action.

Employers may be assessed civil money penalties of up to $1,100 for each willful or repeated violation of the minimum wage or overtime pay provisions of the law and up to $11,000 for each employee who is the subject of a violation of the Act's child labor provisions. In addition, a civil money penalty of up to $50,000 may be assessed for each child labor violation that causes the death or serious injury of any minor employee, and such assessments may be doubled, up to $100,000, when the violations are determined to be willful or repeated. The law also prohibits discriminating against or discharging workers who file a complaint or participate in any proceeding under the Act.

ADDITIONAL INFORMATION
- Certain occupations and establishments are exempt from the minimum wage and/or overtime pay provisions.
- Special provisions apply to workers in American Samoa and the Commonwealth of the Northern Mariana Islands.
- Some state laws provide greater employee protections; employers must comply with both.
- The law requires employers to display this poster where employees can readily see it.
- Employees under 20 years of age may be paid $4.25 per hour during their first 90 consecutive calendar days of employment with an employer.
- Certain full-time students, student learners, apprentices, and workers with disabilities may be paid less than the minimum wage under special certificates issued by the Department of Labor.

For additional information:
1-866-4-USWAGE
(1-866-487-9243)   TTY: 1-877-889-5627

U.S. Wage and Hour Division

WWW.WAGEHOUR.DOL.GOV

U.S. Department of Labor | Wage and Hour Division

WHD Publication 1088 (Revised July 2009)

EXHIBIT 19-1
YOUR RIGHTS UNDER THE FAIR LABOR STANDARDS ACT POSTER

COMPARATIVE LAW CORNER

Ireland and the Minimum Wage

Ireland, like the United States and most industrialized countries, has a national minimum wage. However, Ireland's approach to the minimum wage is very different from the approach in the United States. Ireland has different stages of minimum wage. In 2015, in the United States, the national minimum wage was $7.25 per hour, with a few exceptions for tipped workers such as waiters and waitresses. The national minimum wage in Ireland in 2015 for an experienced adult worker was €8.65 per working hour. In Ireland, you must be an experienced adult worker to receive the national minimum wage.

So who qualifies as an experienced adult worker in Ireland? Anyone who is not under the age of 18, is in his or her first two years of working after turning 18, or is undergoing training covered under the specific statute is considered an *experienced adult worker*. A worker under the age of 18 must be paid at least €6.06 per working hour. A worker in his or her first year of employment after turning 18 must be paid at least €6.92 per working hour, and a worker in his or her second year of employment after turning 18 must be paid at least €7.79 per working hour. Ireland's minimum wage law assumes that a worker learns more and becomes more valuable with experience and therefore rewards more experienced workers with a higher rate of pay.

Although many businesses in the United States might like a minimum wage law similar to Ireland's, the size difference between the two countries makes such a step improbable. The more complex a law is, the more complex the enforcement tends to be. Ireland has about 2 million workers, compared with more than 150 million workers in the United States. Ireland's small size allows efficient application and administration of its more complex minimum wage law.

Congress passed the Federal Unemployment Tax Act,[10] which created a state system that provides unemployment compensation to qualified employees who lose their jobs. Under this act, employers pay taxes to the states. These tax dollars are deposited into the federal government's Unemployment Insurance Fund.

Each state has an account from which to access the money in the fund. States then set up their own system of allocating these funds, determining such matters as how the amount of compensation is determined and how long it can be collected. Eligibility requirements must also be set, with most states requiring, at minimum, an applicant not to have been fired for just cause or not to have quit voluntarily. Some states' eligibility requirements are more generous than others. For example, although states generally require that employees did not voluntarily quit, some will still allow an employee to receive unemployment if he or she quit because of a compelling reason, as in the 2000 case of *Beachem v. Unemployment Compensation Board of Review*.[11] In that case, Mr. Beachem was a single parent of a son who began having emotional and behavioral problems. The son was living with Beachem's mother in Alabama at the time, and Beachem quit his job to move to Alabama to care for the boy after exhausting all possible child care options. He found a job in Alabama but was laid off after a month, too soon to collect unemployment compensation in that state. So he filed for benefits in Pennsylvania. The court found that "a cause of necessitous and compelling nature exists where there are circumstances that force one to terminate his employment that are real and substantial and would compel a reasonable person under those circumstances to act in the same manner,"[12] and allowed him to receive benefits.

---

[10] 26 U.S.C. §§ 3301–3310.

[11] *Beachem v. Unemployment Compensation Board of Review*, 2000 WL 1357484 (Pa. Commw. Ct. 2000).

[12] *Id.*

CASE 19-1

Cassandra Jenkins v. American Express Financial Corp.
Supreme Court of Minnesota
721 N.W.2d 286 (2006)

Cassandra Jenkins worked as an insurance specialist for American Express Financial Corporation. After being convicted of assaulting a nurse in 2004, she was sentenced to 30 days in jail, with work-release privileges scheduled to begin on April 18, 2004. Prior to serving her sentence, Jenkins notified her supervisor, Joel Hansen, of her conviction, sentence, and work-release privileges. Hansen said that Jenkins could continue her employment while on work release. When Jenkins reported to the workhouse on April 18, 2004, she discovered that her employer had not verified her employment. She attempted to contact Hansen multiple times, but he never verified her employment and then terminated her employment for absenteeism on April 26. Jenkins filed for unemployment benefits but was denied because the department adjudicator determined that Jenkins had been discharged by her employer for misconduct. Jenkins appealed, but the unemployment law judge (ULJ) upheld the denial of unemployment benefits. Jenkins appealed to the court of appeals, which affirmed the department's decision. Jenkins appealed to the Minnesota Supreme Court.

Justice Meyer

Whether an employee has engaged in conduct that disqualifies him from unemployment benefits is a mixed question of fact and law. Specifically, the determination of whether an employee was properly disqualified from receipt of unemployment compensation benefits is a question of law on which we are free to exercise our independent judgment.

An otherwise eligible employee will be disqualified from the receipt of unemployment benefits for a variety of reasons, including a discharge for employment misconduct. Employment misconduct is defined as "any intentional, negligent, or indifferent conduct, on the job or off the job (1) that evinces a serious violation of the standards of behavior the employer has the right to reasonably expect of the employee, or (2) that demonstrates a substantial lack of concern for the employment."

Absence from work under circumstances within the control of the employee, including incarceration following a conviction for a crime, has been determined to be misconduct sufficient to deny benefits. Importantly, though, we declined to adopt a rule that absenteeism resulting from incarceration was misconduct as a matter of law. Instead, we directed the agency to base its determinations "upon the facts in each particular case, [leaving] the commissioner . . . with the responsibility of finding the facts as to 'good cause' and 'fault' within the intent and purpose of the act."

We turn to the facts of this case to determine whether the employee engaged in misconduct. The first statutory definition of misconduct is conduct "that evinces a serious violation of the standards of behavior the employer has the right to reasonably expect of the employee." This definition is an objective determination: was the employer's expectation for the employee reasonable under the circumstances? The court of appeals applied this definition and determined that because the absenteeism resulted from Jenkins' criminal conviction it was misconduct because American Express was not obligated to verify Jenkins' employment. The employer in this case does not argue that Jenkins' off-the-job behavior violated its standard of behavior for its employees. Rather, the employer contends that it was the simple fact of Jenkins' failure to report to work that violated the employer's reasonable expectation.

But the facts here lead us to conclude that it was unreasonable for the employer to expect Jenkins to report to work by April 26: the employer allowed Jenkins to continue working between the time of her conviction in March and the time she reported to the workhouse in April; the employer knew in advance of April 18 that Jenkins would be able to participate in a work-release program if the employer verified her employment; the employer told Jenkins that employees in the past had been allowed to participate in work-release programs; the employer told Jenkins that she would be able to continue working while she served her sentence; Jenkins and others provided the employer with the name and phone number of the person to contact to verify Jenkins' employment; and the employer failed to verify Jenkins' employment despite good faith efforts on the part of Jenkins and others to obtain the verification.

Jenkins' case is distinguishable from cases in which absenteeism due to incarceration was found to be misconduct: *Grushus v. Minnesota Mining and Mfg. Co.*; *Smith v. American Indian Chemical Dependency Diversion Project*; and *Smith v. Industrial Claim Appeals Office of the State of Colorado*. In all three of those cases, the claimant simply failed to show up at work because he had been incarcerated. In both of the *Smith* cases, the claimant did not contact his employer until after he had missed work because he had been incarcerated. The *Grushus* case also involved deception to the employer as to the reasons for the claimant's inability to return to work.

For the above reasons, we conclude that the first statutory definition of misconduct relied on by the court of appeals is not satisfied.

The second statutory definition of misconduct is whether the employee "demonstrate[d] a substantial lack of concern for [her] employment." Committing a crime that results in a period of incarceration may be evidence that an employee lacked concern for her employment. In this case, however, there is substantial evidence that Jenkins' inability to report to work was not caused by a substantial lack of concern for her employment. The record establishes that Jenkins made diligent efforts to report to work. She informed her employer of her conviction and the availability of work release to allow her to continue her employment; she obtained a verbal assurance from the employer that the employer would cooperate with the work-release program; as soon as she became aware that she would not be able to report to work because her employment had not been verified, she made every effort to contact the employer; and she provided her employer with the necessary information to permit her release from the workhouse. The evidence as a whole amply demonstrates that she engaged in significant attempts to report for work and continue her employment. Her conduct does not demonstrate a substantial lack of concern for her employment.

We hold that under the facts presented, Jenkins' absence from work was not misconduct that disqualified her from receiving unemployment benefits.*

Reversed in favor of the Appellant, Cassandra Jenkins.

Consolidated Omnibus Budget Reconciliation Act of 1985

The Consolidated Omnibus Budget Reconciliation Act of 1985 (COBRA) ensures that employees who lose their jobs or have their hours reduced to a level at which they are no longer eligible to receive medical, dental, or optical benefits can continue receiving benefits for themselves and their dependents under the employer's policy. By paying the premiums for the policy, plus up to a 2 percent administration fee, employees can maintain the coverage for up to 18 months, or 29 months for a disabled worker. Employees have 60 days after their coverage would ordinarily terminate to decide whether to maintain the coverage.

This obligation does not arise if the employee was fired for gross misconduct or if the employer decides to eliminate the benefit for all current employees. Employers who refuse to comply with the law may be required to pay up to 10 percent of the annual cost of the group plan or $500,000, whichever is less.

Workers' Compensation Laws

Unlike many other laws affecting the employment relationship, workers' compensation legislation is purely state law. Therefore, our coverage of this topic must be rather generalized. Prudent businesspeople will familiarize themselves with the workers' compensation statutes of the states within which their companies operate.

COVERAGE

workers' compensation laws State laws that provide financial compensation to covered employees or their dependents when employees are injured on the job.

Workers' compensation laws provide financial compensation to employees or their dependents when the covered employee of a covered employer is injured on the job. For administrative convenience, most states exclude certain types of businesses and small firms from coverage. A few states also allow employers to "opt out" of the system. These states may likewise give the employee the opportunity to reject coverage.

Workers' compensation is said to be "no fault" because recovery does not depend on showing that the injury was caused by an error of the employer. Think of workers' compensation as analogous to insurance: The employer pays premiums based on the frequency of accidents in the employer's business, and the employees receive insurance-like benefits if injured.

---

*Cassandra Jenkins v. American Express Financial Corp., Supreme Court of Minnesota 721 N.W.2d 286 (2006).

To recover workers' compensation benefits, the injured party must demonstrate that he or she is an employee as opposed to being an independent contractor. This distinction, discussed in Chapter 15, is based on the degree of control the employer can exert over the worker: The greater the degree of control, the more likely the party will be considered an employee. Factors showing employer control include the employer's dictating how the job is to be done, providing the tools to do the job, and setting the worker's schedule. In contrast, in employer–independent contractor relationships, the employer generally specifies the task to be accomplished but has no control over how the task is done. A broker hired by a firm to sell a piece of property is an example of an independent contractor.

The employee must also establish that the injury occurred on the job, meaning that it must have taken place during the time and within the scope of the claimant's employment. Once an employee is on company property, the courts generally find that the employee was on the job, a finding based on the application of the so-called premises rule.

More difficult, however, is the situation of the employee who is injured on the way to or from work. If the employee works fixed hours at a fixed location, injuries on the way to or from work are generally not compensable, but some states establish exceptions to this rule. For example, the *special-hazards exception* applies when a necessary means of access to the employer's premises presents a special risk, even if the hazardous area is beyond the control of the employer. For example, if an employee must make a left-hand turn across a busy thoroughfare to enter the company parking lot, this situation has been held to be a risk of employment; therefore, employees involved in accidents while making the left-hand turn into their employer's parking lot have been allowed compensation under this exception.

Another exception is when an employee is requested to run an errand for the employer on the way to or from work. Compensation is usually allowed for injuries sustained during the course of running the errand.

Sometimes, as a consequence of the job, an employee is forced to temporarily stay away from home. What if the employee is injured while away from home? In some states, reasonable injuries suffered while away from work are covered. In a New York case, a typist was required to travel to Canada to transcribe depositions. While showering in her hotel, she fell and injured herself. She filed a successful workers' compensation claim.

The elements necessary for recovery under workers' compensation laws are summarized in Exhibit 19-2.

APPLYING THE LAW TO THE FACTS...

Andrea worked at a steel mill that employed 500 workers. While she was working one day, the mill owner was having construction done on the ceiling of the mill. A lighting unit fell and injured Andrea so that she couldn't work. Why do you think Andrea's injury will or will not be covered by workers' compensation? How, if at all, would your answer be different if, at the time the light unit fell on her, Andrea was in the women's locker room at the mill, changing out of her protective coveralls at the end of her shift?

EXHIBIT 19-2

ELEMENTS REQUIRED FOR A SUCCESSFUL WORKERS' COMPENSATION CLAIM

1. Claimant is an employee, as opposed to an independent contractor.
2. Both claimant and employer are covered by the state workers' compensation statute.
3. The injury occurred while the claimant was on the job and acting within the scope of the claimant's employment.

RECOVERABLE BENEFITS

The amounts and types of benefits recoverable under workers' compensation are specified by each state's relevant statute. Most statutes cover medical, hospital, and rehabilitation expenses. Some statutes also cover "other treatment" or "appliances" necessary to treat workers for work-connected injuries.

Compensation under state statutes also generally includes payment for lost wages. When employees become disabled as a result of their injury, most states have a schedule that determines the amount of compensation for the disability, as well as compensation schedules for loss of body parts. Thus, an employee who lost a toe in an industrial accident, even though not disabled as a result, would be entitled to some compensation for the loss.

THE CLAIMS PROCESS

Most workers' compensation claims are handled by a state agency responsible for administering the compensation fund and adjudicating claims. In general, an employee who is injured fills out a claim form and files it with the responsible administrative agency. A representative of the agency then verifies the claim with the employer. If the employer does not contest the claim—and most do not—an employee of the bureau (usually called a "claims examiner") investigates the claim and determines the proper payment.

If either the amount or the validity of the claim is contested, there is an informal hearing before a regional office of the agency. Most states provide for an appeals process within the agency. A dissatisfied party who has exhausted the administrative appeals process may appeal to the state trial court of general jurisdiction.

BENEFITS OF THE WORKERS' COMPENSATION SYSTEM

Workers' compensation systems are often touted as benefiting workers. In many ways, they do. Before we had workers' compensation statutes, an employee who was injured on the job could successfully gain compensation from the employer only if he or she was able to prove that the injury was caused by the employer's negligence. Such proof was often difficult to produce.

Recovery was further restricted by the availability of powerful defenses. If the employee's own negligence contributed to the injury, the employee could not recover because of the defense of contributory negligence. Under the fellow servant rule, an employee could not recover if the act of another employee caused or contributed to the injury. In some cases, the courts would say that if the claimant employee engaged in work knowing that it presented a particular safety risk, that employee had assumed the risk of injury and, therefore, the employer was not liable.

Overall then, workers' compensation helps employees because it has removed from them the burden of having to prove employer negligence and has made all the defenses just mentioned irrelevant to compensation claims. Today, employees can obtain compensation in situations in which formerly they could not have recovered. The system also helps employees because they do not need to hire an attorney to recover. Many injured workers have never used attorneys and would not know how to seek legal help. They might also fear retaliation from their employers for bringing suit.

Certain aspects of workers' compensation, however, hurt employees and help employers. In most states, employees who are covered by workers' compensation statutes do not have the right to bring personal injury lawsuits against their employers. Because the payment schedule adopted under any state law is a political compromise heavily influenced by business lobbies, the amount of compensation given is minimal—generally far less than the amount a party could recover through a successful lawsuit against the employer.

Some people argue that claimants are not completely compensated under most payment schedules and that workers' compensation laws compel employees to give up the chance of a large (or full) recovery in exchange for certain, but minimal, recovery. Because employers simply make regular payments to the workers' compensation fund, the costs of occupational injuries are a routine part of their operating expenses. Companies do not have to worry about incurring substantial unanticipated losses; at most, they may find that their premiums increase if the number of claims filed against them goes up.

Some people argue further that by reducing potential costs for workplace injuries in this manner, workers' compensation has made employers less careful about employee safety than they would be if they had to fear huge damage awards.

The Family and Medical Leave Act of 1993

On August 5, 1993, the **Family and Medical Leave Act (FMLA)** went into effect. This law was designed to guarantee that workers facing an unexpected medical catastrophe or the birth or adoption of a child would be able to take the needed time off from work. The act was hailed by its supporters as a "breakthrough" in U.S. law, but it was denigrated by its opponents as an unwieldy encumbrance upon business. So far, the jury is still out on the effectiveness of this act.

Family and Medical Leave Act (FMLA) A law designed to guarantee that workers facing a medical catastrophe or certain specified family responsibilities will be able to take needed time off from work without pay but without losing medical benefits or their jobs.

MAJOR PROVISIONS

The FMLA is a highly complex piece of legislation containing 6 titles divided into 26 sections. The regulations, designed to guide the implementation of the act, are eight times longer than the statute itself! No wonder many employers were still not in full compliance with the act a year after it became effective.

The act covers all public employers and private employers of 50 or more employees. Covered employers must formulate a family leave policy and revise employee handbooks accordingly. The policy must provide all eligible employees with up to 12 weeks of leave during any 12-month period for any of the following family-related occurrences:

- The birth of a child
- The adoption of a child
- The placement of a foster child in the employee's care
- The care of a seriously ill spouse (including same-sex spouses in those states that recognize same-sex marriage), parent, or child
- A serious health condition that renders the employee unable to perform any of the essential functions of his or her job

To exercise the rights under the FMLA, an employee whose need for a leave is foreseeable must advise the employer of that need at least 30 days prior to the anticipated date on which the leave should begin, or as soon as practicable. A typical foreseeable leave would be for the birth of a child.

If the leave is unforeseeable, notice must be given as soon as practicable. "As soon as practicable" is defined as within one or two business days from when the need for the leave becomes known.

The act does not provide a clear definition of exactly what type of notice is necessary. At minimum, the employee must inform the employer of why the employee needs the leave and, if possible, the length of time needed. The request need not specifically mention the FMLA.

Courts look at the facts of each case individually to determine whether notice was sufficient.

Upon termination of their leaves, employees must be restored to the same position or one that involves substantially equivalent skills, effort, responsibility, and authority. Also, although the leave itself can be unpaid, the employer must continue health insurance benefits during the leave period. The employer may also require an employee to substitute paid time off for unpaid leave. For example, an employee who has 4 weeks of accrued sick leave and 2 weeks of vacation and wishes to take a 12-week leave for the birth of a new baby may be required to take the paid vacation and sick leave plus 6 weeks of unpaid leave.

Two of the most contentious issues under the FMLA are (1) what constitutes proper notice and (2) what constitutes a medical condition that is serious enough to invoke the protections of the act. In *Miller v. AT&T Corp.*, the decision hinged on determining whether ongoing treatment for the flu constituted a serious medical condition.[13] FMLA regulations say that ordinarily, the flu does not constitute a serious medical condition that is eligible for FMLA leave. The court in the *Miller* case examined the specific facts of the case and determined that Miller's illness was not an ordinary case of the flu and, therefore, did qualify for FMLA leave.

The following case demonstrates one way the court has attempted to help clarify the issue of what constitutes a serious medical condition.

CASE 19-2

Jeffrey Bonkowski v. Oberg Industries, Inc.
United States Court of Appeals for the Third Circuit
2015 U.S. App. LEXIS 8492

Bonkowski was a wirecut operator and machinist for Oberg who suffered from a number of health conditions, including an aortic bicuspid and diabetes. He testified in court that on November 14, 2011, he met with two supervisors to discuss his recent suspension for allegedly sleeping on the job. He began feeling shortness of breath, pain, and dizziness and received permission from them to clock out early and continue the meeting the next day. He clocked out at 5:18 p.m. He went to the hospital that evening, but wasn't admitted to the hospital until just after midnight on the morning of the 15th. He underwent testing that revealed no problems and was released the evening of the 15th with instructions to follow up with his personal physician and a cardiologist, but with no activity restrictions. He received a doctor's note stating that "Jeff was hospitalized and is excused from work."

On November 16, the head of Oberg's human resources department notified Bonkowski that his employment was terminated because he had walked off the job on November 14. Bonkowski filed an FMLA action against Oberg, alleging two causes of action under the FMLA: (1) Oberg retaliated against him for exercising his FMLA rights, and (2) Oberg interfered with his FMLA rights.

After discovery, Oberg filed a motion for summary judgment, which the District Court granted, finding that there was no way Bonkowski had a serious medical condition entitling him to leave under FMLA and, therefore, that he could not have been retaliated against under FMLA. Bonkowski appealed.

Circuit Judge Cowen

. . . In this appeal, the Court must interpret a Department of Labor regulation—which states in relevant part that "[i]npatient care means an overnight stay in a hospital, hospice, or residential medical care facility." We conclude that "an overnight stay" means a stay in a hospital, hospice, or residential medical care facility for a substantial period of time from one calendar day to the next calendar day as measured by the individual's time of admission and his or her time of discharge. Because Bonkowski was admitted and discharged on the same calendar day, we will affirm the District Court's order . . .

. . . Bonkowski "does not dispute that if he was not qualified for leave . . . i.e., if he did not have a serious health condition, his claims fail as a matter of law." . . .

. . . "[T]he term 'serious health condition' means an illness, injury, impairment, or physical condition that involves (A) inpatient care in a hospital, hospice, or

---
[13]*Miller v. AT&T Corp.*, 250 F.3d 820 (4th Cir. 2001).

residential medical care facility; or (B) continuing treatment by a health care provider." In turn, "[t]he FMLA's legislative history noted that '[t]he definition of serious health condition' . . . is broad and intended to cover various types of physical and mental conditions.". . .

. . . . In the preamble to its 1995 rulemaking promulgating final FMLA regulations, the DOL observed that "[t]his scant statutory definition [of a "serious health condition"] is further clarified by the legislative history.". . . Specifically, "[t]he congressional reports did indicate that the term was not intended to cover short-term conditions for which treatment and recovery are very brief, as Congress expected that such conditions would be covered by even the most modest of employer sick leave policies."

The DOL has adopted regulations that define the various terms incorporated into the FMLA's definition of a "serious health condition." Both the parties and the District Court appear to turn to the current version of these DOL regulations, [under which] "Serious health condition," provides that, ". . . serious health condition entitling an employee to FMLA leave means an illness, injury, impairment or physical or mental condition that involves inpatient care. . . . " Inpatient care means an overnight stay in a hospital, hospice, or residential medical care facility . . .

. . . It is our responsibility to interpret this regulation defining the statutory terms "inpatient care" as "an overnight stay." The District Court and the parties have proffered three basic approaches . . . (1) the District Court's "sunset-sunrise" approach; (2) the "totality of the circumstances" approach offered by Bonkowski; and (3) Oberg's "calendar day" approach. Specifically, the District Court relied on dictionary definitions of "overnight," "duration," and "night" to conclude that "an 'overnight' stay at a hospital is a stay from sunset on one day to sunrise the next day." Bonkowski argues that "[t]he totality of the circumstances demonstrate a genuine issue of material fact regarding whether Mr. Bonkowski stayed overnight at a hospital." . . . In addition to defending the District Court's "sunset-sunrise" definition, Oberg contends that, at a minimum, the term "an overnight stay" refers to a stay from one calendar day to the next calendar day as measured by the inpatient's admission and discharge times.

This Court ultimately agrees with the interpretation proffered by Oberg—although with one major modification. We believe that "an overnight stay" means a stay in a hospital, hospice, or residential medical care facility for a substantial period of time from one calendar day to the next calendar day as measured by the individual's time of admission and his or her time of discharge.

. . . We believe that any kind of "totality of the circumstances" approach would make it more difficult for both employers and employees to predict whether a specific set of circumstances rises to the level of "an overnight stay". . . and lead to additional litigation in the future with possibly inconsistent results.

Having considered and rejected both the "sunset-sunrise" definition as well as an open-ended "totality of the circumstances" approach, we conclude that "an overnight stay" . . . means a stay in a hospital, hospice, or residential medical care facility for a substantial period of time from one calendar day to the next calendar day as measured by the individual's time of admission and time of discharge.

While he was not admitted until shortly after midnight on November 15, 2011, Bonkowski testified at his deposition that, when he was being wheeled into Butler Memorial Hospital, he saw a clock showing that "it was a few minutes before 12:00." He [therefore takes issue with Oberg's position that a patient's stay . . . should be measured from the moment the individual was admitted. According to Bonkowski, it would be absurd (and contrary to the remedial purpose of the FMLA) to exclude from the definition of "an overnight stay" an individual who arrived at the hospital at 9:00 p.m. on November 14, 2011, was admitted at 12:01 a.m. on November 15, 2011, and was finally discharged at 11:59 p.m. on November 15, 2011. However, as Oberg points out, the Second Circuit has specifically addressed the admission concept under a similar statutory and regulatory scheme.

. . . [W]e believe it is appropriate to follow the Second Circuit's example. We accordingly conclude that "an overnight stay" . . . is triggered by the individual's admission—and not his or her arrival at the hospital. After all, both the Medicare and FMLA schemes incorporate the same basic notion of inpatient care. . . . After all, the fact that an individual is sitting in a hospital emergency or waiting room does not necessarily indicate that his or her condition constitutes more than a short-term medical problem that would generally be covered by the employer's sick leave policy. . . . The time of admission also provides a relatively straightforward and objective criterion to apply . . . In the end, the time of admission—whether considered under the auspices of the FMLA or the Medicare Act—represents a bright-line rule that targets the persons that Congress . . . intended to protect.

Like the time of admission, a "calendar day" interpretation constitutes an objective "bright-line" criterion for deciding whether the individual's time in the hospital rises to the level of "an overnight stay". . . . This should help to simplify any disputes arising out of the regulation's "overnight stay" language (and perhaps even help to deter future disputes and FMLA violations because a bright-line interpretation should put employers [and their employees] on notice of when exactly an employee is entitled to leave under the FMLA) . . .

Although we largely adopt Oberg's reading, we do so with one significant modification. The Court agrees with Bonkowski that it would be absurd to read the terms "an overnight stay" to include an employee who was admitted at 11:59 p.m. on one calendar day and discharged at 1:00 a.m. on the next calendar day. Accordingly, the individual must stay for a substantial period of time in the hospital, hospice, or residential medical facility (as measured by his or her time of admission and time of discharge). Under the circumstances, a minimum of eight hours would seem to be an appropriate period of time. However, because we need not decide this issue to resolve this dispute, we leave this issue of the requisite length of time for another day. It is uncontested that Butler Memorial Hospital formally admitted and discharged Bonkowski on November 15, 2011. Under our "calendar day" approach, the time Bonkowski spent in the hospital did not rise to the level of "an overnight stay" . . . because he did not stay in the hospital from one calendar day to the next calendar day as measured by his time of admission and time of discharge.*

Affirmed, in favor of Oberg.

CRITICAL THINKING ABOUT THE LAW

This decision highlights the importance of a person carefully reading a statute before determining whether he or she has a basis for a legal remedy. The judges of the 3rd Circuit Court of Appeals leaned on the specific words of the FMLA to decide the meaning of the key language.

1. What wording in the FMLA was at issue in this case? As is typical in any case, certain elements of a statute or a precedent are especially significant in shaping the court's decision. Each side is struggling to get the judges to see his or her interpretation of the words as the best meaning in terms of the intent of the legislature and the effect on public policy.

    Clue: Return to the section of the decision where the judges discuss the importance of the length of time a person must spend in the hospital to trigger application of the FMLA.

2. Statutes are enacted in an effort to advance specific ethical norms. This case presents an unusually clear instance where the ethical norms are made explicit. To what ethical norms does the FMLA pay allegiance?

    Clue: Early in the decision, the judges review the purposes the FMLA was designed to fulfill. Translate those purposes into the language of the ethical norms.

REMEDIES FOR VIOLATIONS OF THE FMLA

If an employer fails to comply with the FMLA, the penalties can be substantial. The plaintiff may recover damages for unpaid wages or salary, lost benefits, denied compensation, and actual monetary losses up to an amount equivalent to the employee's wages for 12 weeks, as well as attorney's fees and court costs. If the plaintiff can prove bad faith on the part of the employer, double damages may be awarded. An employee may also be entitled to reinstatement or promotion.

When employees prevail in FMLA trials, they generally are awarded a monetary amount double their actual lost wages and certain incurred expenses, such as payments made for COBRA insurance coverage. One of the largest jury awards under the FMLA thus far has been $2,227,241.48, awarded in March 2008 to Nicholas Lore from Chase Manhattan Mortgage Corporation.[14] Lore was a regional manager for Chase, earning $600,000 a year. He requested time off to respond to a serious medical condition, undiagnosed pain in his groin and knee that prevented him from sleeping more than a few hours, and was told that the company needed to hire another manager to take over some of his duties. Several months later, no replacement had been found. He approached his supervisor again about taking time off to have the problem diagnosed and treated and the next day was informed by the supervisor's assistant that his resignation was accepted, even though he had simply sought a leave. The reason the jury verdict was so high was that Lore was such a highly paid employee.

---

[14] www.customwebexpress.com/bflaw/sub/Atlanta-Employment-Attorneys.jsp;jsessionid=4C8167966EC444BDF7DAE6B9C882F19B.

*Jeffrey Bonkowski v. Oberg Industries, Inc. United States Court Of Appeals For The Third Circuit 2015 U.S. App. Lexis 8492.

While his jury verdict was just over $2 million, Lore's ultimate award was over $6 million because of the FMLA's provision of a doubling of the jury award, plus interest.

In 2008, the FMLA was amended to provide benefits for family members of military personnel, adding a 12-week "exigency leave" for the child, spouse, or parent of National Guard and Reserves service members to respond to exigencies related to a call to active duty, and a 26-week "military caregiver's leave" for family members of service members of the armed forces, National Guard, and Reserves seriously injured while on active duty.

In 2009, the Department of Defense Authorization Law for fiscal year 2010 expanded the military caregiver's leave to include veterans with injuries such as posttraumatic stress disorder that would not manifest itself until after the service member was no longer on active duty, as well as to service members with preexisting medical conditions that were made worse by incidents that occurred during their service. The amendments also included special eligibility provisions for airline flight crew employees (and a unique way of calculating their benefits) in recognition of their unique scheduling requirements. These changes will give airline employees greater access to the benefits of the FMLA. The new rules for both military and airline employees were not effective until 2013.

THE FUTURE OF THE FMLA

Writing in 2013, on the 20th anniversary of the FMLA, a commentator opined that the FMLA has established a norm that employees do not have to risk job loss because of a leave necessitated by a short-term health crisis or the new obligations imposed by giving birth to a child. It has also demonstrated that early fears of the burden on employers of implementing the act were grossly exaggerated, as employers now report that they are administering the act with little cost or trouble. However, the FMLA does not meet the standards of other industrialized nations. Nearly 40 percent of American employees are not covered by the act because their employers do not employ 50 or more persons or employees do not work enough hours in a week to be covered.

The DOL also released a survey that same year showing that the law has had a positive effect on working families without imposing an undue burden on employers. The study shows that employers generally find it easy to comply with the law, and misuse of the FMLA by workers is rare. The vast majority of employers, 91 percent, report that complying with the FMLA has no noticeable effect or has a positive effect on business operations such as employee absenteeism, turnover, and morale. Finally, the survey revealed that 90 percent of workers return to their employer after FMLA leave, showing little risk to businesses that investment in a worker will be lost as a result of leave granted under the act.

The unpaid nature of the leave, however, is still seen as problematic by some. Many of those eligible for leave cannot afford to take it. A 2000 survey administered by the DOL showed that 3.5 million employees needed FMLA leave but did not take it, and their primary reason for not taking it was that they could not afford to do so. Ninety percent of those said that they would have taken leave if some of it had been paid. Another problem seems to be that although one part of the impetus for the act was to get men more involved in parenting, very few men take advantage of the act for the birth or adoption of a child, and when they do, they typically take less than a week of time.

Despite these imperfections with the FMLA, it does not seem as if any changes will be coming in the near future. There have been a number of attempts to amend the legislation and make it applicable to firms with 25 or more employees, but all such attempts have failed to garner enough votes.

A few states, however, have chosen to expand coverage. For example, California's SDI program provides a partial replacement wage for workers who have a non-work-related illness, injury, or medically disabling condition, including disability resulting from childbirth or pregnancy. California law also makes it illegal for an employer to refuse to grant an employee's request to transfer to a less strenuous or less hazardous job during pregnancy as long as the request can be reasonably accommodated. California's law also extends parental and family leave rights to workers caring for a domestic partner or child of a domestic partner. For more information about what individual states are doing, you can read "Expecting Better: A State-by-State Analysis of Laws That Help New Parents," listed in the "On the Internet" section at the end of this chapter.

The Occupational Safety and Health Act of 1970

Employees worry about more than compensation. They also want to work in a safe environment. Although working can be hazardous to one's health and safety, not all jobs are equally hazardous. The most dangerous industry is mining, followed by construction, agriculture, and transportation. The safest industries are the service industries. Fortunately, workplaces are getting safer. Exhibit 19-3 illustrates the declining rates of occupational injuries in the U.S. workplace.

Occupational Safety and Health Act (OSH Act)
A regulatory act designed to provide a workplace free from recognized hazards that are likely to cause death or serious harm to employees.

The primary regulatory measure designed to provide a safer workplace is the **Occupational Safety and Health Act (OSH Act)** of 1970. The OSH Act requires every employer to "furnish to each of his employees . . . employment . . . free from recognized hazards that are likely to cause death or serious physical harm. . . ." To ensure that this objective will be met, Congress, under the OSH Act, authorized the creation of three agencies: the Occupational Safety and Health Administration (OSHA), the National Institute for Occupational Safety and Health (NIOSH), and the Occupational Safety and Health Review Commission (OSHRC).

Occupational Safety and Health Administration (OSHA) The agency responsible, under the OSH Act, for setting and enforcing standards for occupational health and safety.

OCCUPATIONAL SAFETY AND HEALTH ADMINISTRATION

The most important of these agencies is the **Occupational Safety and Health Administration (OSHA)**. It has both a standard-setting and an enforcement role. In addition, it undertakes educational programs among employers and employees. In the years since OSHA's establishment in 1971, workplace fatalities

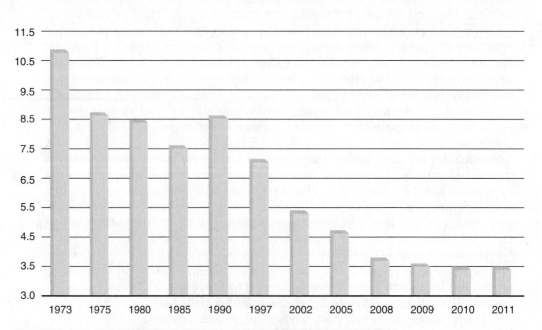

EXHIBIT 19-3 **OCCUPATIONAL INJURY AND ILLNESS INCIDENCE RATES**

have been cut by over 60 percent and occupational injury and illness rates by over 40 percent even though the American workforce has more than doubled.

Standard Setting. OSHA sets the standards for occupational health and safety in the United States. These standards are frequently opposed by labor and management, but for different reasons. In general, labor organizations criticize them for insufficiently protecting employees' health and safety, whereas employer groups claim that they are unnecessarily stringent and too costly.

In establishing health standards, OSHA has used the following four-step process since 1981:

1. The agency asks whether the hazard presents a "significant risk" that warrants intervention.
2. If it does, OSHA decides whether regulatory action can reduce the risk.
3. If it can, the agency establishes a standard to reduce the risk "to the extent feasible," taking into account both technological and economic feasibility.
4. OSHA then analyzes the cost effectiveness of various implementation options to determine which will achieve its goals most efficiently.[15]

Enforcement. OSHA is charged with enforcing the OSH Act through unannounced inspections and the levying of fines against violators. The goals of OSHA inspections are to find and correct existing hazards and to encourage employers to eliminate hazards before inspection.

OSHA conducts several different types of inspections. In order of priority, these are imminent danger inspections, when OSHA learns of a hazard that can be expected to cause physical harm or death; catastrophe and fatality investigations, whenever an accident hospitalizes five or more workers or causes a death; employee complaints, when an employee alleges a violation and requests an inspection; special inspection programs, including those aimed at certain hazards or industries; and programmed or random inspections. Follow-up inspections may also be conducted at any time.

OSHA's budget strongly affects the number of inspectors the agency has and, hence, the number of inspections it can perform. The agency's budget obviously fluctuates annually. For fiscal year 2009, its budget was roughly $515 million; the agency was estimated to have conducted 39,004 workplace inspections that year. During fiscal year 2008, OSHA's budget was $486 million and the agency conducted 38,591 workplace inspections.

Inspection Procedure. The safety and health compliance officer conducting the inspection arrives at the plant, usually unannounced; presents his or her credentials; and asks to meet with the person in charge and a union representative to explain the purpose of the visit. The inspector then asks to see any relevant records. Since 1981, the policy of OSHA has been for the inspector in a safety inspection to calculate from these employer records the average lost-workday rate of injury. If it is lower than the average for the industry, the inspection is ended. If it is higher, the inspector walks through the workplace, taking notes, pictures, and exposure samples when relevant. An employer, an employee representative, or both may accompany the inspector on this tour. After the inspection, the inspector discusses any apparent violations with the employer. Any citations are usually mailed to the employer at a later date.

As noted in Chapter 5, an employer has the constitutional right to refuse to allow an inspection if the inspector does not have a warrant. If the employer does refuse to allow an inspection, however, the OSHA representative generally just goes to court and obtains a warrant.

---

[15]*Preventing Illness and Injury in the Workplace* (Washington, DC: Office of Technology Assessment, 1985), p. 16.

Penalties. When violations are found, compliance inspectors may issue citations for violations. These violations fall into three categories: willful or repeat, serious, and nonserious. The maximum fine for serious and nonserious violations is $7,000 per day of noncompliance, whereas the maximum for a willful or repeat violation is up to $70,000 per violation. If a willful violation results in the death of a worker, criminal penalties may be imposed. For a first conviction, a fine of up to $10,000 and six months in jail are possible. For subsequent convictions, the penalty may be up to $20,000 and one year in jail.

Penalties can mount when a firm has multiple violations. In 2006, the John J. Steuby Company was fined $788,000 for more than 50 violations. Twelve violations were "willful," 37 were "serious," 1 violation was "repeat," and 3 were "nonserious."[16] In 2007, after the investigation of an employee death, OSHA proposed a $2.78 million fine for the Cintas Corporation.[17] The high fine was based on 42 citations designated as "willful," 1 "repeat" violation, and 3 "serious" violations. In 2010, VT Halter Marine, Inc., a shipbuilder, was fined $1,322,000 following a November 2009 explosion and fire that killed two workers and seriously injured two others.[18] The high fine resulted from the court's finding of 17 "willful" and 11 "serious" violations.

On October 30, 2009, OSHA fined BP $87 million, the largest fine in OSHA history, for its failure to correct hundreds of safety violations discovered after a 2005 explosion at its Texas City, Texas, oil refineries. The fine was four times as large as any previous OSHA sanction.[19]

And although no fine has come close to that since 2009, significant fines are still issued every year. For example, in 2011, OSHA issued a number of six- and seven-figure citations.[20] By January of 2014, the number of companies facing total OSHA fines above $100,000 had tripled since 2010.[21]

Often, when a firm is cited for multiple violations, the case will end with not only a fine but also an agreement to improve working conditions significantly. For example, in 2013, as a result of several OSHA citations, Wal-Mart settled by paying a penalty of $190,000 and agreeing to make significant improvements in safety conditions at all of its 2,857 retail and Sam's Club stores.[22]

In addition to citing companies for safety violations, OSHA, through its voluntary participation program (VPP), recognizes companies that have created a particularly safe workplace by designating them as "Star" worksites. VPP sets criteria-based performance standards, and when a company applies to the program, there is a rigorous on-site evaluation of the applicant. These designated "Star" sites, which have put into practice outstanding safety and health programs, are not targeted by OSHA for regularly scheduled programmed inspections. These sites, however, tend to have fewer employee illnesses and injuries.

---

[16]"John J. Steuby Company Cited for Alleged Workplace Safety and Health Violations; OSHA Proposes $788,000 in Penalties," OSHA National News Release (Jan. 12, 2006).

[17]"U.S. Department of Labor's OSHA Proposes $2.78 Million Fine against Cintas Corp. Following Tulsa, Okla. Employee Death in Industrial Dryer," OSHA National News Release (Aug. 16, 2007).

[18]"U.S. Department of Labor's OSHA Cites VT Halter Marine More Than $1.3 Million for Willfully Exposing Workers to Toxic Vapors in a Confined Space," OSHA National News Release (May 19, 2010).

[19]Steven Greenhouse, "BP Faces Record Fine for '05 Refinery Explosion," *New York Times*, Oct. 30, 2009. Accessed June 10, 2010 at www.nytimes.com/2009/10/30/business/30labor.html.

[20]Fred Hosier, "Top 10 OSHA Fines of 2011," Safety and OSHA News. Available at www.safetynewsalert.com/top-10-osha-fines-of-2011/.

[21]Polytron, "Are More Costly OSHA Citations Coming in 2014?" http://polytron.com/blog/are-more-costly-osha-citations-coming-in-2014.

[22]Fred Hosier, "Wal-Mart Enters Corporate-Wide Settlement with OSHA, Pays $190K Fine," Safety and OSHA News. Available at www.safetynewsalert.com/wal-mart-enters-corporate-wide-settlement-with-osha-pays-190k-fine/.

Two other categories of workplaces are also recognized. "Merit" workplaces have developed and implemented good safety and health management systems, but must take additional steps to reach "Star" quality.

"Demonstration" workplaces operate effective safety and health management systems that are different from current VPP requirements. These sites allow OSHA to test the efficacy of different approaches.

The program began in 1982 and is highly selective. As of August, 2015, 2,200 workplaces were participating in the program at one of the three levels.[23] An illustration of a company receiving recognition as a "Star" site is Parsons Corporation, a California-based company with seven active OSHA VPP sites. The company employs 11,500 employees worldwide, including 3,500 contractors. Because of the size of the company, it is able to influence the health and safety of numerous employees worldwide.[24]

Every year, to give employers an idea of what kind of actions leads to citations, OSHA publishes a list of the top 10 violations. For example, in 2014, the top 10 list was led by fall protection violations (7,516 citations), followed by mislabeling and the absence of safety data sheets (6,148 citations) and scaffolding violations (4,968 citations). Additional information about the most frequent violations can be found on the OSHA website.

Public Education. Since 1975, OSHA has provided free, confidential, on-site consultations with employers seeking to bring their businesses up to OSHA standards. Any unsafe or unhealthful working conditions discovered during consultations are not reported to the OSHA inspection staff as long as the firm commits to correcting any serious job safety and health hazards found during the consultation. The steps of a typical consultation are found in Exhibit 19-4. OSHA also sponsors a Targeted Training Program, which provides short-term grants to employers, employees, and nonprofit organizations to address health and safety issues about which OSHA has particular concerns.

State Plans. Under the present OSH Act, a state may regulate its own workplaces if its program provides for the establishment and enforcement of standards that will be at least as effective as the federal standards. As of September 2015, 22 states and territories had OSHA-approved programs.[25]

OCCUPATIONAL SAFETY AND HEALTH REVIEW COMMISSION

The **Occupational Safety and Health Review Commission (OSHRC)** is an independent review body before which an employer can contest the issuance of a citation by OSHA, the amount of the penalty, or the time within which abatement is expected. A hearing is conducted before an administrative law judge, whose ruling becomes OSHRC's final order 30 days after it is issued, provided it is not contested. Within the 30-day period, any party may request a review of the decision by OSHRC, which is automatically granted. A final order is issued after the review is completed. The final order may be appealed to a circuit court of appeals.

NATIONAL INSTITUTE FOR OCCUPATIONAL SAFETY AND HEALTH

The **National Institute for Occupational Safety and Health (NIOSH)** was established under the OSH Act to identify occupational health and safety

Occupational Safety and Health Review Commission (OSHRC) An independent body that reviews OSHA citations, penalties, and abatement periods when they are contested by an employer.

National Institute for Occupational Safety and Health (NIOSH) A research facility established by the OSH Act to identify occupational health and safety problems; develop controls to prevent occupational accidents and diseases; and disseminate its findings, particularly to OSHA.

---

[23]U.S. Department of Labor, OSHA, Current Federal and State Plan Sites, https://www.osha.gov/dcsp/vpp/sitebystate.htm.

[24]"Parsons Corporation Becomes Newest Member of OSHA's VPP Corporate Pilot," OSHA Trade News Release (Oct. 31, 2008). Available at www.osha.gov/pls/oshaweb/owadisp.show_document?p_table=NEWS_RELEASES&p_id=16886.

[25]United States Dept. of Labor, OSHA, Frequently Asked Questions. Available at https://www.osha.gov/dcsp/osp/.

EXHIBIT 19-4

STEPS OF A CONSULTATION ADAPTED FROM THE CONSULTATION PROGRAM FOR SMALL BUSINESSES, DECEMBER 9, 2004

A total of 4.7 million nonfatal injuries and illnesses were reported in private industry workplaces during 2002, resulting in a rate of 5.3 cases per 100 equivalent full-time workers in 2002–2003 = 5.3.

For the most part, that rate has continued to fall, going from a rate of 4.9 per hundred workers in 2004 to 4.6 in 2005, 4.4 in 2006, 4.2 in 2007, and 3.9 in 2008. By 2011, the drop was even more significant, with roughly 3.0 million nonfatal workplace injuries and illnesses reported by private industry employers that year, resulting in an incidence rate of 3.5 cases per 100 equivalent full-time workers, according to the Survey of Occupational Injuries and Illnesses (SOII) conducted by the U.S. Bureau of Labor Statistics. The rate both in 2013 and 2014 was a record low of 3.3.

Walk-through

Together, the employer and consultant will examine conditions in the workplace. OSHA strongly encourages maximum employee participation in the walk-through. Talking with employees during the walk-through helps the consultant identify and judge the nature and extent of specific hazards.

The consultant studies the entire workplace or the specific operations designated by the employer and discusses the applicable OSHA standards. Consultants also will point out other safety or health risks that might not be cited under OSHA standards, but still pose safety or health risks to employees. They may suggest other measures to prevent future hazardous situations.

A comprehensive consultation also includes (1) appraisal of all mechanical and environmental hazards and physical work practices, (2) appraisal of the present job safety and health program or establishment of one, (3) a conference with management on findings, (4) a written report of recommendations and agreements, and (5) training and assistance with implementing recommendations.

Opening Conference

Upon arrival at the site, the consultant will first meet with the employer in an opening conference to briefly review the consultant's role and the obligations of the employer.

Initiation

The consulting process is initiated by a confidential phone call or letter to OSHA. A consultant will discuss the employer's specific needs and set up a visit date.

Abatement and Follow-through

Following the closing conference, the consultant sends the employer a detailed written report explaining the findings and confirming any abatement periods agreed upon. Consultants may also contact the employer from time to time to check the firm's progress. Ultimately, OSHA requires hazard abatement so that each consultation visit achieves its objective—effective employee protection. If you fail to eliminate or control identified serious hazards (or an imminent danger) according to the plan and within the limits agreed upon or an agreed-upon extension, the situation must be referred from consultation to an OSHA enforcement office for appropriate action, although these situations almost never arise.

Closing Conference

The consultant then reviews detailed findings with the employer in a closing conference. Problems, possible solutions, and abatement periods to eliminate or control any serious hazards identified during the walk-through are discussed. In rare instances, the consultant may find an "imminent danger" situation during the walk-through. If so, immediate action must be taken to protect all employees. In certain other situations that would be judged a "serious violation" under OSHA criteria, the employer and the consultant are required to develop and agree to a reasonable plan and schedule to eliminate or control that hazard. The consultants will offer general approaches and options to the employer, and may also suggest other sources for technical help.

problems, to develop controls to prevent occupational accidents and diseases, and to disseminate its findings. This body functions primarily as a research facility that attempts to make its findings on hazards available to those who can use them. Although independent of OSHA, NIOSH provides OSHA with information on which the agency bases many of its "criteria documents" (recommendations for health and safety standards that contain supporting evidence and bibliographical references).

IMPLEMENTATION OF THE OSH ACT

OSHA, NIOSH, and OSHRC work together to implement the OSH Act. By taking advantage of the assistance offered by these agencies and by meeting the standards that OSHA has established, businesspeople can make their workplaces safer and more healthful. It may even be in their interests to do so, for often the increased costs of providing a safer workplace are more than offset by the benefits employers derive from reduced absenteeism, fewer workers' compensation claims, and less work time lost because of accident investigations.

Employee Privacy Rights

Although employees are rightly concerned about safety, they are also increasingly concerned about privacy on the job. Privacy issues arise in a broad array of contexts, including the hiring process, monitoring of employee performance on the job, and the use of new technology.

ELECTRONIC MONITORING AND COMMUNICATION

Increasing use of technology in the workplace has raised a number of new privacy issues. When, for example, can employers monitor employees' telephone conversations, read their email, or listen to their voice mail? The legal issues related to workplace privacy involve the common-law tort of invasion of privacy (discussed in Chapter 12) and the Omnibus Crime Control and Safe Streets Act of 1968,[26] as amended by the Electronic Communications Privacy Act (ECPA) of 1986.[27]

The Omnibus Crime Control Act prohibits employers from listening to the private telephone conversations of employees and disclosing the contents of those conversations. Employers are allowed to ban personal calls and monitor

LINKING LAW AND BUSINESS

Management

In your management class, you may have learned about several theories of motivation related to human needs. The most widely known and accepted theory is Maslow's hierarchy of needs. This concept, developed by Abraham Maslow, includes five fundamental human needs: physiological; security or safety; social; esteem; and self-actualization. As previously mentioned, the OSH Act of 1970 has promoted safer working environments via the aid of the three agencies established by Congress. Therefore, one of the five needs in Maslow's hierarchy, specifically safety needs, has been endorsed by congressional legislation. Because the statutes under the OSH Act encourage employers to create safer workplaces and better satisfy safety needs, managers have a more probable means of motivating employees to perform to the best of their ability to accomplish a particular task. Thus, advocating safer workplaces could potentially result in motivated employees, which is an essential element in attaining organizational goals.

S. Certo, *Modern Management* (Upper Saddle River, NJ: Prentice Hall, 2000), 354, 358–59.

---

[26] 42 U.S.C. § 2210 *et seq.*
[27] 18 U.S.C. §§ 2510–2521.

calls for compliance, but once they determine that a call is personal, they are not allowed to continue listening to the conversation. Violators may be subject to fines of up to $10,000.

The ECPA extended employees' privacy rights to electronic forms of communication, including email and cellular telephones. The act prohibits the intentional interception of electronic communications or the intentional disclosure or use of the information obtained through such interception. The act, however, includes a "business-extension exemption" that allows employers to monitor employees' telephone conversations in the ordinary course of their employment, as long as the employer does not continue to listen to conversations once it recognizes that they are of a personal nature. Employers often need to do such monitoring to improve job performance and to offer some protection to employees from harassing calls from the public. A second exception arises when the employees consent to the monitoring of their conversations.

The key question in cases involving employer monitoring and interception of employee communications via email, telephone, or voice mail has been whether the employee had a reasonable expectation of privacy with respect to the communication in question. Therefore, many employment law specialists recommend that to minimize the likelihood of being sued by employees for invasion of privacy and to keep good morale in the workplace, employers should have written employee privacy policies that are explained to employees and printed in employee handbooks. But even a good privacy policy may not be sufficient to keep an employer from being sued.

Over the past few years, however, as social media has become more popular, another aspect of the EPCA, the Stored Communications Act (SCA), has taken on more significance. The SCA, designed to protect the privacy of stored electronic communications, provided the basis for holding an employer liable when a jury found no invasion of privacy, discussed in Case 19-3.

CASE 19-3

**Brian Pietrylo and Doreen Marino, Plaintiffs v.
Hillstone Restaurant Group *dba* Houston's, Defendant**
United States District Court, D. New Jersey
Civil Case No. 06–5754, September 25, 2009

Two restaurant employees, Pietrylo and Marino, set up a password-protected group on MySpace called the Spec-Tator, where past and present employees could vent about everything going on in their workplace without having to worry about "outside eyes prying." No managers were invited to join. A manager heard about Spec-Tator and got one of the employees to give him her password to the group. He used her information to go onto the Spec-Tator several times. Even though the company had no policy prohibiting employees from using social media to comment on its business and workplace issues, because of what he read on the site, the manager fired the employees who had started Spec-Tator.

The plaintiffs sued for, among other claims, violations of the federal Stored Communications Act (the SCA) and invasion of privacy. The jury returned a verdict in favor of Plaintiffs on the SCA claims, finding that Defendant had, through its managers, knowingly or intentionally or purposefully accessed the Spec-Tator without authorization on five occasions, and further found that Defendant had acted maliciously. The jury awarded $2,500 and $903 in compensatory damages to Pietrylo and Marino, respectively. They also were awarded punitive damages, which, by stipulation of the parties, equaled four times the amount of compensatory damages awarded by the jury. The jury found that there was no violation of the plaintiffs' right to privacy.

Defendants filed a motion for judgment in their favor as a matter of law, or, in the alternative, a new trial.

HOCHBERG, District Judge.

III. Discussion

A. Defendant's Statutory Liability

[T]o prevail on their statutory claims, Plaintiffs were required to offer sufficient evidence to allow the jury to conclude that Houston's managers knowingly, intentionally, or purposefully accessed the Spec-Tator without authorization. According to the SCA, if access to the Spec-Tator was authorized "by a user of that service with respect to a communication of or intended for that user," there is no statutory violation. . . . Defendant now argues (1) that there was no evidence that an invited member of the chat group did not authorize the Houston's managers to use her password to access the Spec-Tator; and (2) that Plaintiffs presented no evidence that Houston's managers possessed the requisite state of mind for their actions to constitute violations of the statutes at issue.

. . . Defendant emphasizes that St. Jean (the key witness on the issue of authorization) was undisputedly an authorized user of the Spec-Tator and showed the website to Rodriguez on her own. Defendant further emphasizes that St. Jean willingly provided her MySpace.com log-in information to Anton and Marano (indirectly through Rodriguez) without indicating any reservations to them. This evidence, Defendant argues, indicates that St. Jean authorized Anton, Marano, and other Houston's employees to access the Spec-Tator and precludes a finding that Houston's managers accessed the Spec-Tator knowing that such access was unauthorized.

In response, Plaintiffs argue that there is ample evidence in the record from which a reasonable jury could find, and indeed did find, that Defendant violated the federal SCA. . . . Plaintiffs emphasize that . . . the jury was required to rely on the testimony and demeanor of the witnesses. St. Jean's testimony, in particular, was critical, in establishing that she had not authorized access to the Spec-Tator. St. Jean testified that she felt she had to give her password to Anton because she worked at Houston's and for Anton.

She further testified that she would not have given Anton her password if he had not been a manager, and that she would not have given her information to other co-workers. Furthermore, when asked whether she felt that something would happen to her if she did not give Anton her password, she answered "I felt that I probably would have gotten in trouble." . . . the jury could reasonably infer from such testimony that St. Jean's purported "authorization" was coerced or provided under pressure. As a result, this testimony provided a basis for the jury to infer that Houston's accessing of the Spec-Tator was not, in fact, authorized.

In addition, St. Jean's testimony, taken in conjunction with the testimony provided by Anton and Marano, provided evidence upon which the jury could reasonably infer that Houston's managers acted with the state of mind proscribed by the federal and state statutes. Evidence was presented that Houston's managers accessed the Spec-Tator on several different occasions, even though it was clear on the website that the Spec-Tator was intended to be private and only accessible to invited members. A reasonable jury could conclude that these repeated visits to the Spec-Tator were intentional or purposeful, as opposed to accidental, and that Houston's managers knew that they were not authorized to access the contents of the Spec-Tator from the manner and means that Anton and Marano used to get access to the password-protected MySpace page of St. Jean.

Evidence was presented that Houston's managers continued to access the Spec-Tator after realizing that St. Jean had reservations about having provided her log-in information after being asked for it by Anton while at work. Robert Marano, for example, testified that he knew that "[St. Jean] was very uneasy with the fact that she had given me and the rest of the managers her password," and that she was worried about the consequences of having provided such information. . . . On the basis of such testimony, the jury could reasonably infer that Houston's managers knew that they were not actually authorized to continue accessing the Spec-Tator through St. Jean's MySpace password, but continued to do so anyway.

The jury found that Houston's managers unlawfully accessed the Spec-Tator five separate times, further supporting their finding that they accessed it without authorization and not by mistake or accident. This Court does not re-weigh the evidence or reevaluate the credibility of the witnesses. This Court does not find that the jury's verdict resulted in a miscarriage of justice, nor that it cries out to be overturned nor that it shocks the conscience, because the verdict was based in large part upon credibility determinations made by the jury and because such determinations are fully within the province of the jury. Accordingly, the Court finds that there was sufficient evidence upon which the jury could find a verdict for Plaintiffs on their statutory claims and denies Defendant's motion for judgment as a matter of law and Defendant's motion for a new trial.

C. Punitive Damages

Finally, Defendant seeks to overturn the jury's determination that Plaintiffs are entitled to punitive damages, arguing that there was no evidentiary support for the finding that Houston's access of the Spec-Tator was malicious. This argument is again based on Defendant's view of Karen St. Jean's testimony, as set forth above, and Defendant's fervent belief that the Houston's managers pursued access to the Spec-Tator for legitimate business reasons.

Punitive damages are available under the federal Stored Communications Act "[i]f the violation is willful or intentional." . . . After the jury made its finding on the substantive counts, the jury was instructed to answer these predicate questions only if they had found Houston's liable on the substantive count. The two predicate questions asked the jury whether they found the conduct of Houston's to be "wanton or willful," or "malicious."

Although Houston's certainly does have a right and obligation to protect its employees and managers from harassment or humiliation, and to protect the core values of the restaurant, the jury's findings indicate that the jury did not believe that the method used by Houston's to protect those values was proper conduct. . . . The jury was not persuaded by the well-tried case of Houston's regarding its claimed lack of knowledge of "non-authorization." . . . [T]he jury had sufficient evidence from which it could reasonably infer that Houston's acted maliciously in repeatedly accessing the Spec-Tator via St. Jean's password. . . . The jury could well have inferred that the manner and means used by Anton, as St. Jean's boss, to ask her for her private password would have given Anton knowledge that the access was not authorized. . . . In addition, St. Jean testified that she was not informed that Marano would further convey her personal log-in information to other members of Houston's management. Nor was St. Jean informed that Anton, Marano, or others would repeatedly access the site. No evidence was presented that St. Jean was asked for authorization for such repeated use of her password to access the Spec-Tator nor was she asked for "blanket" authorization for access. The jury's inference from such evidence that Houston's acted maliciously was neither unreasonable nor a miscarriage of justice. . . .

The parties' stipulation to a modest amount of punitive damages if either predicate question was answered in the affirmative does not cry out as unjust. In light of the small sum of punitive damages agreed by stipulation and looking at the evidence presented at trial in its entirety, together with the reasonable inferences that the jury could draw from that evidence, this Court cannot say that there was insufficient evidence to support the jury's answer to the predicate question that led to the stipulated amount of punitive damages.*

Judgment for the Plaintiff,
as Defendant's motions are denied.

Privacy policies should cover matters from employer surveillance policies to control of and access to medical and personnel records, to drug testing, and to all issues unique to the electronic workplace. Exhibit 19-5 provides a sample of a good email policy.

A related privacy concern in the workplace is the use of surveillance cameras. Again, the courts fall back on the reasonable expectation of privacy. Closed-circuit cameras are increasingly being used to combat employee theft, to monitor employee performance, and to safeguard employees. Clearly visible cameras are generally not challenged. As one arbitrator said, "Since one of the supervisor's jobs is to observe employees at work, such supervision cannot be said to interfere with an employee's right to privacy, even if it is done by a camera." This logic, however, did not hold in a recent case involving a very well-known employer. During the night shift at Wal-Mart, a visible surveillance camera recorded four employees eating candy and nuts from damaged incoming boxes (which was against store policy). When Wal-Mart fired them for theft, the four employees sued the corporation, and won $5 million each. The employees received the money in part because of "slander" and "emotional distress," but also, more relevantly, because of Wal-Mart's eavesdropping. So sometimes employers can get into trouble for taking action based on what is seen on a surveillance camera—even if it is in plain sight.[28]

Hidden cameras may generally be placed in areas where employees do not have a reasonable expectation of privacy, such as hallways. Installation of video cameras in bathrooms, however, may prompt claims of invasion of privacy. In cases in which a plant is unionized, the courts may consider the installation of a closed-circuit television monitoring system to be an issue the firm must bargain over with the union, because it changes working conditions.

HYPOTHETICALLY SPEAKING

Let's assume that Doug was working in an office and went to the bathroom. He went to the bathroom to fix his shirt, but noticed a hidden camera recording the more public area of the bathroom in front of the mirror. He complained to his employers, but his employers stated that the camera was not an invasion of privacy because it was aimed at a public area of the room. How much leeway are employers given to video record "public" areas of the premises of a company?

---

[28]"Wal-Mart Employees Win $20 Million after Being Fired for Eating Company Candy." Accessed March 5, 2008 at www.lawyersweekly.com.

*Brian Pietrylo and Doreen Marino, Plaintiffs, v. Hillstone Restaurant Group d/b/a Houston's, Defendant. United States District Court, D. New Jersey Civil Case No.06–5754, September 25, 2009.

EXHIBIT 19-5

EMAIL POLICY AND EMPLOYEE ACKNOWLEDGMENT

ELECTRONIC MAIL POLICY

* Email is the property of the company and should be used solely for work-related purposes.

* Employees are prohibited from sending messages that are harassing, intimidating, offensive, or discriminatory.

* Each employee will be given a password to access email. Your password is personal and should not be shared with anyone else. However, the company retains a copy of all passwords and has a right to access email at any time for any reason without notice to the employee. The employee has NO expectation of privacy or confidentiality in the email system.

* The employee must sign and return an Acknowledgment and Consent form indicating receipt and acceptance of our company's policy.

ACKNOWLEDGMENT

I understand that the company's electronic mail and voice mail systems (herein together referred to as "the company's systems") are company property and are to be used for company business. I understand that [excessive] use of the company's systems for the conduct of personal business is strictly prohibited.

I understand that the company reserves the right to access, review, and disclose information obtained through the company's systems at any time, with or without advance notice to me and with or without my consent. I also understand that I am required to notify my supervisor and the company's Security Department if I become aware of any misuse of the company's systems.

I confirm that I have read this employee acknowledgment and have had an opportunity to ask questions about it. I also agree to abide by the terms of the company's policy in this regard a copy of which has been provided to me.

AGREED TO THIS _____ DAY OF _____, 19_____.

_____

Witness _____ Employee Signature _____

DRUG TESTING

In the interests of safety and to increase productivity, employers have become increasingly interested in using drug testing for employees. Employer interest in drug testing was also encouraged by Congress in 1988, when it passed the Drug-Free Workplace Act, which required employers who receive federal aid or do business of $25,000 or more with the federal government to develop an antidrug policy for employees, provide drug-free awareness programs for them, make them aware of assistance programs for those with drug problems, and warn them of penalties for violating the company's drug-free policies.

Employers, however, must be careful when establishing drug-testing policies, because employees may challenge such policies on the grounds that they violate state or federal constitutional rights, common-law privacy rights, or state or local drug-testing laws. Because of the variation among states regarding drug testing, an employer must examine very closely both the statutory and case law in its state before implementing a drug-testing policy. With that caution in mind, however, some broad generalizations can be made.

Drug testing generally arises in four different contexts, and standards may be different for each. The most common drug testing is preemployment testing, allowed by most states. Sometimes, a prospective employee will try to sue the employer for requiring preemployment drug testing, but as long as the drug tests do not violate discrimination laws, the tests are usually accepted. In one case, a man failed his preemployment drug test and sued the physicians because he thought they must have made a mistake. The drug test came up positive, and he claimed that he had never taken drugs. He did not win his case, however, because the physicians were hired by the employer and not by the plaintiff; hence, the physicians owed the plaintiff no duty, as they had no contract with him. Interestingly, however, in a case the following year in a different state, the state supreme court imposed liability on a drug-testing firm whose negligent testing as part of an employer's substance abuse testing program led to the plaintiff's termination.[29] The employer could not be sued because it was allowed to reject a prospective employee on any grounds, as long as the employer was not violating discrimination laws.[30]

Most states also recognize the legitimacy of drug testing as a part of a periodic physical examination, assuming that proper standards for drug testing are followed. The third context in which drug testing may be used is when the employer has a reasonable suspicion that an employee may be under the influence of drugs in the workplace. "Reasonable suspicion" generally requires the employer to have some evidence, such as an unexplainable drop in performance, to justify the tests. If the employer does have such sound evidence, the test is generally upheld. The most controversial area is that of random drug testing, on which the courts are split. Random drug testing is most likely to be upheld when an employee may pose a safety risk to others by doing a job while he or she is under the influence of drugs and when employees have been informed in advance in writing of the random drug-testing policy.

Private-sector employers have much greater freedom to test for drugs than do public-sector employers. Whereas private employers may be restricted only by state constitutions and state and local laws, federal entities are also restricted by the Constitution. The exception would, in the end, swallow the rule.

OTHER TESTING

In recent years, the number of companies that use some form of preemployment testing has increased significantly. The time and costs of preemployment testing are offset in the long run because businesses are able to reduce turnover rates and low productivity by carefully screening applicants before they are offered a job. In addition, preemployment testing often saves businesses money by paring down the applicant pool efficiently with some assurance that applicants are treated consistently.

Businesses, however, must be careful to ensure that their tests do not violate discrimination laws or the Americans with Disabilities Act. Tests should measure only what is necessary for the performance of a job. For example, a typing test that screens out those applicants who cannot use their hands is acceptable if the

---

[29] *Ney v. Axelrod*, 723 A.2d 719 (Pa. 1999).

[30] *Duncan v. Afton, Inc.*, 1999 WL 1073434 (Wyo. 1999).

position is for a typist. Tests, however, should seek to accommodate those with disabilities if their disability does not impair the performance of a job.

Of the many forms of preemployment testing, polygraph testing has stimulated the greatest amount of legal action. In 1988, Congress passed the Employee Polygraph Protection Act. This law prevents employers from using lie-detector tests while screening job applicants. It also prohibits administering lie-detector tests randomly, but it does allow their use in specific instances in which there has been an economic injury to the employer's business. The act also provides an exception to allow private security companies and those selling controlled substances to use lie-detector testing of applicants and current employees. The Labor Department may seek fines of up to $10,000 against firms that violate the act.

Skills testing, such as a typing test or a professional exam, is widely used and generally legal so long as the tests examine skills necessary for the job. Much less used are aptitude tests that seek to evaluate applicants' general abilities. Aptitude tests often leave the employer open to discrimination lawsuits if the tests are not carefully written to elicit specific information necessary to job performance.

Another common type of test is a medical test, used because the applicant must be in good health to ensure safety and good job performance. For example, airline pilots must undergo regular medical examinations to ensure their health and the safety of their passengers.

Public-sector employees are granted an additional level of protection from discriminatory testing. Federal employees are protected under the Uniform Guidelines on Employee Selection Procedures of 1978. These procedures are evaluated and enforced by the federal Equal Employment Opportunity Commission.

Immigration Law

The United States has always been a nation of immigrants, and until 1921, it placed few restrictions on immigration. Starting in 1921, however, the United States began imposing quotas on the number of immigrants it would accept annually from each foreign country. It is estimated that in 2012, there were 11.2 million unauthorized or illegal immigrants in this country, and 8.1 million of them were either working or looking for work.[31] Because approximately 5 percent of the workforce consists of undocumented workers and illegal immigrants in a business's workforce can lead to liability for the firm, today's business manager must understand the laws governing immigration.

IMMIGRATION REFORM AND CONTROL ACT OF 1986

The Immigration Reform and Control Act of 1986 (IRCS), administered by the U.S. Citizenship and Immigration Services (USCIS), requires employers to hire only those people who can legally work in this country—U.S. citizens and noncitizen residents who are authorized to work in the country. To ensure that only these individuals are employed, employers are required to verify the eligibility of all job applicants. Verification is done by having the applicant complete an eligibility form, also known as Form I-9, for at least three years.

Form I-9 has three sections. The first section, filled out by the applicant, asks for name, other names used, address, date of birth, U.S. Social Security number, and email address and telephone number (optional). It also asks the applicant to check whether he or she is (1) a citizen of the United States, (2) a noncitizen

---

[31]Jens Manuel Krogstad and Jeffrey S. Passel, "5 Facts about Illegal Immigration," U.S., Pew Research Center. Accessed May 10, 2015 at www.pewresearch.org/fact-tank/2014/11/18/5-facts-about-illegal-immigration-in-the-u-s/.

national of the United States, (3) a lawful permanent resident, or (4) an alien authorized to work. Aliens authorized to work here must then record the date their employment authorization ends and their Alien Registration Number/USCIS Number. The applicant signs the form and gives it to the employer.

Once the applicant has accepted the job, the employer completes Section 2 by examining the verifying documents to make sure they are genuine and recording the necessary information from the documents on the form, recording information about when and where the employee will be working, and signing and dating the attestation on the form. Although it is not necessary to photocopy the documents, it is probably a good idea to do so in case the business is inspected. For a list of verifying documents, see Exhibit 19-6.

All employees must complete these I-9 forms because an employer can be fined up to $16,000 and imprisoned up to six months for each undocumented worker they have hired. Even minor paperwork offenses can end up being costly; failing to comply with Form I-9 requirements imposes a minimum of $110 for each form and a maximum of $1,100 for each form for all offenses. Some of the potential problems employers must guard against include failing to keep an I-9 for the required amount of time, failing to destroy a form in a timely manner, failing to complete necessary portions of the forms, failing to obtain an employee's signature or attestation date, failing to record acceptable documents relied upon by the employer in hiring the employee, failing to include dates of rehire, failing to reverify the employee's employment eligibility, and failing to comply with the appropriate deadlines for completion of the form.

U.S. Immigration and Customs Enforcement's (ICE) Office of Homeland Security Investigations (HSI) enforces the law and engages in worksite enforcement, conducting what have become known as "ICE raids." Once of the more publicized enforcement actions was against Abercrombie & Fitch, which resulted in a settlement of $1,047,110 for violations of the Immigration and Nationality Act related to an employer's obligation to verify the employment eligibility of its workers.[32]

Since 2009, ICE has been focusing on workplace enforcement actions, believing that employment of undocumented workers is a big cause of illegal immigration. So if we reduce the number of jobs available for illegal immigrants, it should slow the number of individuals attempting to enter this country illegally. In fiscal year 2013, ICE made 452 criminal arrests tied to worksite enforcement investigations and debarred 277 business and individuals for administrative and criminal violations. It also served 3,127 Notices of Inspection and 637 Final Orders, totaling $15,808,365 in administrative fines.[33]

AUTHORIZED NONCITIZEN WORKERS

As mentioned above, some noncitizens are authorized to work in this country on both a permanent and temporary basis. Those authorized on a permanent basis are green card holders—permanent residents with rights to enter, exit, work, and live in the United States for their entire life. An unlimited number of green cards are available for immediate relatives (such as spouses and unmarried children under the age of 21) whose relatives with U.S. citizenship petition for them. Also, 480,000 green cards are available for other family members each year, but these are competitive and the waiting time varies from 4 to 24 years in the various family categories.

A total of 140,000 green cards are offered each year to people whose job skills are needed in the United States. Ordinarily, a job offer also is required, and

---

[32] www.ice.gov/news/releases/abercrombie-and-fitch-fined-after-i-9-audit.

[33] Worksite Enforcement, U.S. Immigration and Customs Enforcement. Available at www.ice.gov/factsheets/worksite.

EXHIBIT 19-6 VERIFYING DOCUMENTS

LISTS OF ACCEPTABLE DOCUMENTS
All documents must be UNEXPIRED

Employees may present one selection from List A
or a combination of one selection from List B and one selection from List C.

LIST A Documents That Establish Both Identity and Employment Authorization	LIST B Documents That Establish Identify	LIST C Documents That Establish Employment Authorization
1. U.S. Passport or U.S. Passport Card	1. Driver's license or ID card issued by a State or outlying possession of the United States provided it contains a photograph or information such as name, date of birth, gender, height, eye color, and address	1. A Social Security Account Number card, unless the card includes one of the following restrictions: (1) NOT VALID FOR EMPLOYMENT (2) VALID FOR WORK ONLY WITH INS AUTHORIZATION (3) VALID FOR WORK ONLY WITH DHS AUTHORIZATION
2. Permanent Resident Card or Alien Registration Receipt Card (Form I-551)		
3. Foreign passport that contains a temporary I-551 stamp or temporary I-551 printed notation on a machine-readable immigrant visa		
4. Employment Authorization Document that contains a photograph (Form I-766)	2. ID card issued by federal, state, or local government agencies or entities, provided it contains a photograph or information such as name, date of birth, gender, height, eye color, and address	2. Certification of Birth Abroad issued by the Department of State (Form FS-545)
5. For a nonimmigrant alien authorized to work for a specific employer because of his or her status: a. Foreign passport; and b. Form I-94 or Form I-94A that has the following: (1) The same name as the passport; and (2) An endorsement of the alien's nonimmigrant status as long as that period of endorsement has not yet expired and the proposed employment is not in conflict with any restrictions or limitations identified on the form.	3. School ID card with a photograph	3. Certification of Report of Birth issued by the Department of State (Form DS-1350)
	4. Voter's registration card	4. Original or certified copy of birth certificate issued by a State, county, municipal authority, or territory of the United States bearing an official seal
	5. U.S. Military card or draft record	
	6. Military dependent's ID card	
	7. U.S. Coast Guard Merchant Mariner Card	
	8. Native American tribal document	5. Native American tribal document
	9. Driver's license issued by a Canadian government authority	6. U.S. Citizen ID Card (Form I-197)
	For persons under age 18 who are unable to present a document listed above:	7. Identification Card for Use of Resident Citizen in the United States (Form I-179)
6. Passport from the Federated States of Micronesia (FSM) or the Republic of the Marshall Islands (RMI) with Form I-94 or Form I-94A indicating nonimmigrant admission under the Compact of Free Association Between the United States and the FSM or RMI	10. School record or report card	8. Employment authorization document issued by the Department of Homeland Security
	11. Clinic, doctor, or hospital record	
	12. Day-care or nursery school record	

the employer must prove that it has recruited for the job and not found any willing, able, and qualified U.S. workers to hire instead of the immigrant. Because of annual limits, preferences are given to different categories of applicants, and applicants often wait years for an available green card. The top category includes (1) persons of extraordinary ability in the arts, the sciences, education, business, or athletics; (2) outstanding professors and researchers; and (3) managers and executives of multinational companies. The employers generally apply for these noncitizens' green cards.

Two primary visas are important to employers—the H-1B visa and the EB-1 visa. The H-1B visa is issued to an employee in a professional or specialty occupation (generally one requiring a bachelor's degree or higher) when the employer can show difficulty in recruiting qualified workers in the United States.

The visa is generally good for three years and can be renewed for an additional three years. During that time, the employer can sponsor the visa holder for a green card, and if it is granted, the visa holder becomes a permanent resident. If not, the worker must return to his or her home country. When a worker has an H-1B visa, immediate family members may accompany him or her to the United States on an H-4 visa. The annual cap on H-1B visas is currently set at 65,000, so competition for them is strong.

The EB-1, or employment-based, first-preference visa, is granted to a person who (1) has extraordinary ability in the sciences, arts, education, business, or athletics through sustained national or international acclaim, (2) is an outstanding professor or researcher, or (3) is a multinational executive or manager. Each occupational category has certain requirements that must be met. People in the first category apply for the visa themselves; the employer applies for the visa for those in the other two categories. An employee with an EB-1 visa can apply for U.S. citizenship, usually within five years. Fewer than 50,000 of these visas are issued every year, so competition for them is strong.

In addition to these two visas, a number of other temporary work visa programs are available for workers in different fields. For example, the H-2A program allows U.S. employers who meet specific regulatory requirements to bring foreign nationals to the United States to fill temporary agricultural jobs when there are not enough U.S. workers who are able, willing, qualified, and available to do the temporary work. Requirements for these visas and applications can be found on the USCIS website.

Global Dimensions of the Employment Relationship

Although some of the benefits described in this chapter may seem significant, laws in other nations ensure much greater benefits for workers. One area in which other nations provide significantly greater benefits is parental leave. Workers in the United States are guaranteed up to 12 weeks of unpaid leave; most European countries require a guaranteed paid leave. For example, in France, Austria, and Finland, paid parental leave begins 6 weeks before childbirth; it extends to 10 weeks after birth in France, 8 weeks in Austria, and almost 1 year in Finland. French working mothers are then entitled to an additional (unpaid) job-protected leave until their children reach age 3. Altogether, more than 120 nations require paid maternity leave, with the Czech Republic providing 24 weeks of paid leave. One of the reasons such extensive benefits can be offered in those nations, however, is that the tax-funded social insurance/social security systems provide most of the money for the benefits.

Another area in which employees have fewer rights in the United States is that of vacation time. The United States, unlike most European nations, does not mandate a minimum amount of annual vacation time for employees. But in Ireland, for example, the Holiday Act of 1973 guarantees every worker, regardless of how long he or she has been with a company, three weeks of paid vacation time and nine additional days off for public holidays. Luxembourg also has tremendous holiday benefits: Irrespective of the employee's years of service, he or she is given 25 days of holiday, 12 of which must be taken in succession. In addition, Luxembourg workers are given time off for 10 public holidays.

To try to ensure that workers around the world receive the rights and benefits described in this chapter, the United Nations Commission on Human Rights developed the International Covenant on Economic, Social, and Cultural Rights. This covenant does not really bind any employers, but the Commission exerts influence by documenting and publicizing where the terms of the covenant are not followed.

SUMMARY

The employment relationship today is still a contract between employer and employee, but the national and state governments specify certain parameters of this relationship. Laws that affect wages and hours include the Davis–Bacon Act, the FLSA, and the Federal Unemployment Tax Act.

The FMLA ensures that workers will be able to take necessary time off from work when they or members of their family suffer from a serious medical condition. COBRA ensures that if an employee loses his or her job, he or she will be able to purchase health insurance for up to 18 months. The OSH Act tries to secure safe working conditions for employees, but if they are injured on the job, workers' compensation laws provide benefits to compensate them for the injury or disability they received.

A final concern of workers and employers is the extent of worker privacy rights. Federal employees have more rights in this area than do private employees, because the public employer must comply with the Fourth Amendment; nevertheless, private employers must still be sure that they do not violate state and federal laws designed to protect worker privacy. The prudent employer today will have a written privacy policy that clearly sets out when employees should and should not have a reasonable expectation of privacy.

A comparison of other countries' employee benefits with those in the United States shows that the United States offers less protection to employees with respect to family leave benefits and less vacation.

REVIEW QUESTIONS

19-1 What benefits do employees receive under the FLSA?

19-2 Explain how workers' compensation laws benefit both employers and employees but in different ways.

19-3 What requirements are imposed on an employer by the Family and Medical Leave Act?

19-4 What remedies can an employee seek if an employer violates the Family and Medical Leave Act?

19-5 Explain the relationship between OSHA and NIOSH.

19-6 Explain the principles you want to keep in mind when drafting a firm's employee privacy policy.

REVIEW PROBLEMS

19-7 In early December, Decco Manufacturing Corporation underwent a major downsizing and laid off 40 percent of its employees. Robert Banks was not laid off, but he thought he probably would be laid off in the near future unless business got dramatically better. In fact, his supervisor told him that he would probably be laid off in early January. Robert decided that if he was going to get laid off, he would rather have the time between jobs over the holidays. So he quit and filed for unemployment compensation. Should he be able to collect workers' compensation?

19-8 Nellie Mandle worked as a machine operator. She had been told on numerous occasions that if her press ever jammed, she was not to stick her hand in to unjam it; rather, she was to use a special safety fork with a long handle that would allow her to unjam the machine without inserting any part of her hand into the press. Her machine jammed, and she looked around for the safety fork. Realizing that the worker on the previous shift must have removed the fork, she reached inside the machine to unjam it. Before she could remove her hand, the machine cycled, catching her hand and injuring it severely. She filed a workers' compensation claim, which her employer contested because she caused the injury by disobeying the safety rules. Evaluate the employer's argument.

19-9 Caroline Williams works for a firm that has a written policy prohibiting the use of company telephones for personal use. Employees have been told that the company randomly monitors telephone conversations to enforce this policy. Caroline uses a company telephone to call her doctor to find out the results of a blood test she had to determine whether she had contracted a sexually transmitted disease. Her employer intercepted the phone call and, once he heard her question, stayed on the line to find out the results of her test. Discuss why you believe the employer's behavior is lawful or unlawful. If unlawful, what penalty should he receive?

19-10 Michael Meuter was an employee in a hospital emergency room. He used an extension phone to call one of the workers in the pharmacy to order some drugs for the emergency room. After placing his request, he started to complain to the pharmacy worker about his supervisor, calling the supervisor a number of offensive names. Unknown to Michael, his supervisor was listening to the conversation. Was the supervisor's listening to the conversation lawful? Explain.

19-11 Ginny Morris applied for a job as an executive assistant to the president of a software firm. She received high evaluations from those on the hiring committee and was told that she looked like an excellent candidate for the job. But before a final decision could be made, she would have to take a drug test so that the firm could be confident that she did not use illegal drugs and a lie-detector test to ensure that she could be entrusted with trade secrets. Discuss whether you believe there are any problems with the firm's requests.

CASE PROBLEMS

19-12 Shallene Alayon, who was employed by Urban Management Corporation as a services coordinator, filed for workers' compensation benefits. Alayon claimed that she slipped and fell at work and, as such, her injuries warranted her receiving workers' compensation benefits. However, Urban Management asserts that Alayon received non-work-related injuries in a motor vehicle accident in 1999 and thus argues that her injuries are attributable to preexisting conditions. The Labor and Industrial Relations Appeals Board held that Urban Management may be liable for Alayon's injury; however, "the nature and extent of the injury must be determined by the Director of the Department of Labor and Industrial Relations." Alayon appealed this ruling. What reasoning do you believe the director of the Department of Labor and Industrial Relations used to come to a conclusion on Alayon's eligibility for benefits? *Alayon v. Urban Mgmt. Corp.*, 2014 Haw. App. LEXIS 581 (2014).

19-13 Maria Escriba was an employee of Foster Poultry Farms, Inc. When Escriba was fired in 2007 for failing to comply with the company's "three day no-show, no-call rule," Escriba filed suit under the Family and Medical Leave Act. Escriba's failure to comply with the company's no-show rule followed an already approved leave to take care of her sick father in Guatemala. The employer argued that although Escriba had a qualifying reason to take leave, she denied having this time off count as FMLA leave and instead pursued it as vacation time. The lower court concluded that the case at hand was a classic "he said, she said case." Both parties moved for judgment as a matter of law. The district court denied Escriba's motion, concluding that substantial evidence supported the jury's finding that Escriba unequivocally declined to take FMLA leave. What evidence did the district court rely on when ruling in favor of Foster Poultry Farms, Inc.? How do you think the appeals court ruled? Why? *Escriba v. Foster Poultry Farms, Inc.*, 743 F.3d 1236 (9th Cir. 2014).

19-14 Plaintiff Wolfe worked for defendant company Tobacco Express II, Inc. After Wolfe had worked for the defendant company for some time, the defendant claimed that cash and inventory were missing. The company then required Wolfe and other employees to take a polygraph test. Wolfe refused. Plaintiff alleged that he then reported back to work, where he was terminated. Applying the Employee Polygraph Protection Act, Wolfe brought a claim against the defendant company. The defendant company brought counterclaims against Wolfe for conversion of merchandise. How do you think the court ruled? Why? *Wolfe v. Tobacco Express II, Inc.*, 26 F. Supp. 3d 560 (2014).

19-15 Yatram Indergit worked for Rite Aid as a store manager from 1979 to 2007. Indergit's duties at Rite Aid included disciplining employees, interviewing employees, training employees, scheduling employee hours, performing various office duties, running cash registers, and stocking shelves. When Indergit's employment was terminated in 2007, he filed suit against Rite Aid for, among other things, failure to pay overtime as required by the FLSA. The defendants contended that Indergit's employment fell within the executive exemption to the overtime requirement. Indergit contended that the majority of his job responsibilities were the same as those performed by nonexempt employees. Moreover, Indergit argued that his job duties were primarily nonexempt, as shown by evidence that nonexempt employees previously performed his duties. Rite Aid moved for summary judgment, arguing that there was no genuine issue of material fact that could be heard by the trier of fact. Is there an issue of material fact? What evidence would be necessary to decide whether Indergit's job responsibilities made him exempt from the overtime requirements? *Indergit v. Rite Aid Corp.*, 2010 U.S. Dist. LEXIS 32322 (2010).

19-16 Iliana Rodriguez worked at the University of Miami Hospital when she took a medical leave of absence. Current law states that employees may return to the position they had prior to their leave, unless the employer can make a substantial argument for the change in position being due to factors other than the leave. When Rodriguez returned to work, the original plan was for her to return to her original position with a plan in place to improve her work. However, when Rodriguez returned to work, she was placed in a different position that she did not want and argued that before she left, she was told she would stay at her original position. Her employers argued that even though there was a potential improvement plan, their meeting before she left was a prior indication that she was not performing well in her position and that was the reason for her moving to a new position after her leave. Rodriquez sued the hospital. How do you think the court decided? *Rodriguez v. University of Miami Hospital*, 11th Cir., No. 11-15206 (2012).

19-17 Olofsson was a delivery driver for the Mission Linen Supply. In 2004, he went to Sweden for about five weeks (nonvaction time) to visit his parents. The plant manager, Jack Anderson, Sr., had said Olofsson could go if he filled out the proper slip. Olofsson filled out the slip and went through the proper company procedures with the payroll clerk, Ruth Clark. Once the paper was filled out with Clark's help, it had to be authorized by the company. After Olofsson returned from his trip, his elderly mother informed him that she would be undergoing extensive back surgery and asked him to come back to take care of her during the recovery process. Olofsson said that Anderson (an area manager) said he could go if he filled out another form and turned in a doctor's note. However, Clark told him that Anderson was not authorized to grant him medical leave and they had to wait for a response from HR. She also said that the doctor's note had no letterhead and did not look as if it had come from a legitimate medical establishment. Clark then gave Olofsson a government form that the doctor filled out and faxed back. Olofsson went to Sweden, and after he was there, executives informed him that he had not worked enough hours that year to qualify for medical leave and thus he was terminated. Olofsson sued for wrongful termination in that even though he didn't work enough hours that year and his papers had never been authorized by executives, he was led to believe that he could take the medical leave. How do you think the court decided? *Olofsson v. Mission Linen Supply*, Cal. Rptr.3d (2012).

THINKING CRITICALLY ABOUT RELEVANT LEGAL ISSUES

For years, courts, employers, and employees have struggled with the issue of employee drug testing. Employers want their employees to be drug free, whereas employees do not want to submit to an invasive and embarrassing procedure. Courts want to uphold employees' right to privacy while allowing employers to run a safe, drug-free workplace. This issue may finally have a solution.

A drug test has been developed that uses a swab of saliva rather than urine or blood. It is being implemented in several states already, including Georgia and Hawaii. This test gives much faster results than urine and blood tests, and some experts believe that it might be more effective than urine or blood tests. A result of negative or nonnegative can be returned within 10 minutes, and employees who test negative can begin or return to work immediately. This faster process alleviates the stress and anxiety of employees who must be tested for drugs.

The major benefit of the new saliva test is that it is noninvasive. Employees do not have to suffer the embarrassment of having to "pee in a cup" or the painful experience of having blood drawn. Giving a swab of saliva is easy. This drug test protects employees from having to undergo an invasive or embarrassing test, while allowing employers to keep their workplaces drug free.

Allowing employers to drug test their employees is very important in keeping the workplace safe and ethical. Employees who use drugs while on the job are a danger to themselves and to others and could be responsible for lawsuits against an employer. Also, employees who break the law by taking illegal drugs might engage in other illegal activity, which could also be damaging to the employer. Allowing drug testing is the best way for an employer to keep the workplace drug free. Random drug testing has been legally questionable because of the employees' right to privacy. The saliva drug test eliminates the concerns with the right to privacy and is a major step in helping keep the workplace safe and drug free.

1. How would you express the issue and conclusion?
2. Are any of the terms the author uses to make her point ambiguous?

Clue: How is the new drug test better than the old drug tests?

3. What missing information would help your evaluation of this argument?
4. Write a short essay written by someone with a different opinion about this topic.

Clue: How could different definitions of ambiguous words in this essay change the conclusion?

ASSIGNMENT ON THE INTERNET

Laws that govern employment relationships indicate the ethical norms a particular country or state wishes to advance. This can be seen no more clearly than in cases of family or medical leave. Visit the website for the Clearinghouse on International Development in Child, Youth and Family Policies (www.childpolicyintl.org) and compare the family and medical leave policies of the United States with those of other industrialized countries around the world. Write a paper in which you address the following questions: How does the United States compare on a world scale with regard to employee benefits for family leave? What does this comparison say about the ethical norms of the United States in areas of employment relationships?

ON THE INTERNET

ebn.benefitnews.com This site provides news about employee benefits for employers, employees, and advisers. The site also offers an email newsletter.

www.dol.gov At this site, the home page of the DOL, you can find all the information you need as an employer to make sure you are in compliance with the FLSA.

www.osha.gov The Web page of OSHA provides information about worker health and safety, including how to file a complaint. It also provides the text of the Occupational Safety and Health Act of 1970, as well as OSHA standards, regulations, and directives.

hr.ucdavis.edu/ This site contains a document entitled "Family and Medical Leave . . . What Every Supervisor Should Know," which provides a good overview of the FMLA and illustrates its implementation by a public employer. To access this document, search "family and medical leave" in the search bar on the home page.

www.ohchr.org/ At this page, you can find the International Covenant on Economic, Social and Cultural Rights by performing a search on the home page.

http://www.law.du.edu/index.php To learn more about privacy issues related to employment, go to this site and search for "Privacy Foundation."

www.dol.gov This website provides an interactive map detailing the minimum wage laws in U.S. states and territories. To access this map, click on "Agencies" and then find "Wage and Hour Division (WHD)."

www.law.cornell.edu This website, run by Cornell Law School, provides links to the employment and labor laws in each state. Use this site to look up the employment laws in your state. Under the heading "Legal Resources," go to "Wex Legal Encyclopedia," then search for "labor and employment laws."

http://www.nationalpartnership.org site/DocServer/Expecting_Better_Report.pdf?docID=10301 Expecting Better: A State-by-State Analysis of Laws That Help New Parents. This page discusses what individual states do to help new parents. Click on the search button (here, a magnifying glass) and search "Expecting Better." Find the most recent edition of the Expecting Better document for your reading.

FOR FUTURE READING

Estreicher, Samuel, and Jeffrey M. Hirsch. "Comparative Wrongful Dismissal Law: Reassessing American Exceptionalism." *North Carolina Law Review* 92 (2014): 343.

Fineman, Jonathan. "The Vulnerable Subject at Work: A New Perspective on the Employment At-Will Debate." *Southwestern Law Review* 43 (2013): 275.

Gaudet, Dennis. "It's Time to Fix Our Workers' Compensation System." *New Hampshire Business Review* 37 (2015): 11.

Helleck, Adam M., Amy Rohde Leslie, and Sharla J. Frost. "Welding Fumes: A Review of the History, Workplace Standards, Research, and Litigation from the 1920s to Present for Welding Fumes in General and Manganese Dust/Fumes." *South Texas Law Review* 48 (2006): 527.

Hornung, Meir S. "Think Before You Type: A Look at Email Privacy in the Workplace." *Fordham Journal of Corporate & Financial Law* 11 (2005): 115.

Pryal, Katie R. "Keep Your Personal Email Personal." *Women in Higher Education* 24 (2015): 16.

Tomassetti, Julia. "The Contracting/Producing Ambiguity and the Collapse of the Means/Ends Distinction in Employment." *South Carolina Law Review* 66 (2014): 315.

CHAPTER TWENTY

Laws Governing Labor–Management Relations

- STRUCTURE OF THE PRIMARY U.S. LABOR LEGISLATION AND THE MECHANISMS FOR ITS ENFORCEMENT
- LABOR ORGANIZING
- THE COLLECTIVE BARGAINING PROCESS
- STRIKES, BOYCOTTS, AND PICKETING
- GLOBAL DIMENSIONS OF LABOR–MANAGEMENT RELATIONS

In the early 1800s, labor unions were very rare. Despite the mistreatment of workers during the Industrial Revolution, most attempts to organize workers during the nineteenth and early twentieth centuries were treated by the courts as criminal conspiracies. Finally, in the economic chaos of the Great Depression, Congress enacted laws giving employees the right to organize and to bargain collectively over wages and terms and conditions of employment.

Union strength has fluctuated in the years since unions were legalized. More than a third of U.S. workers were organized in the post–World War II period. By 1983, however, only 20.1 percent of workers were unionized, and by 2006, the percentage had fallen to 12.0 percent, or 15.4 million workers. However, in 2007, union membership increased to 12.1 percent, the first time union "density" had increased in several years.[1] But by 2014, the percentage of workers who were unionized had fallen to 11 percent, with 35.7 percent of public employees being unionized, compared with 6.6 percent of private-sector workers.[2] Many attribute this falling rate of unionization to the passage of tough anti-union laws in such states as Wisconsin and Michigan. Not surprisingly, the greatest drop in public employee union membership in 2012 was in Wisconsin.

Moreover, in 2012, the median weekly earnings of workers who were represented by unions was $970, compared with $763 per week for those not represented by unions.[3] Not all occupations are equally organized. Exhibit 20-1 shows the percentages of workers organized by occupational group in 2011. Interestingly, even though the number of union members has fallen dramatically in the past 30 years, a poll taken by the Pew Research Center in May 2015 showed that more respondents (48 percent) viewed unions favorably than

---

[1] U.S. Department of Labor, Union Members Summary. Accessed January 2, 2011 at www.bls.gov/news.release/union2.nr0.htm.
[2] BLS News Release, Union Members 20152012, USDL-15-0072, January 23, 2015.
[3] Ibid.

EXHIBIT 20-1 PERCENTAGES OF WORKERS ORGANIZED BY OCCUPATIONAL GROUP IN 2011

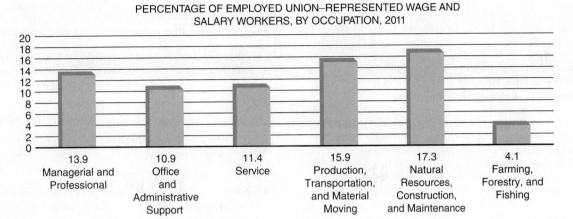

unfavorably (39 percent). Support for unions was strongest in the Midwest (55 percent) and weakest in the South (41 percent).[4]

Although organized workers are still not in as powerful a position as their employers, they are distinctly better off than they were during most of our nation's existence. The primary basis for their improved status is the National Labor Relations Act (NLRA), which was passed in 1935 and is the focus of this chapter.

The first section outlines the structure and enforcement of the NLRA and the Landrum–Griffin Act. The next three sections discuss areas of labor–management relations governed by the NLRA: organizing; collective bargaining; and the collective activities of striking, picketing, and boycotting. This chapter concludes with a consideration of the international dimensions of labor law.

Before you read about our system of labor laws, examine Table 20-1, which summarizes the two conflicting views of the role of unions in economics and society. Whether someone thinks labor laws should strengthen or restrain labor organizations depends largely on which of the two "faces" of unions he or she thinks is "prettier."

CRITICAL THINKING ABOUT THE LAW

Many people hold strong feelings about unions. Workers who belong to unions often view unions as positive forces that work to their benefit. In contrast, employers are often suspicious of the activity of unions. As you consider the aspects of labor law in this chapter, be aware of the role of biases in complex legal issues. The following critical thinking questions will help you consider the role of biases and ethical norms in labor legislation.

1. In the language of ethical norms, what function do unions serve for workers?

    Clue: Remember the list of ethical norms and reread the beginning paragraphs of this chapter. How do labor unions help workers? Can you match this answer to an ethical norm?

2. What role do you think personal ethical norms should play when someone thinks about labor legislation?

    Clue: Could paying attention to these ethical norms benefit workers or employers in any way?

3. For days, your coworkers have been excitedly talking about plans to unionize. You are unsure if you will join the union. One of your coworkers argues that joining the union will help you get a raise. What questions do you have for your coworker about missing information?

    Clue: Think about any possible costs associated with getting a raise.

---

[4] J. Lewis, "Americans Divided on Decline in Union Membership, According to Opinion Poll," May 15, 2015. Available at www.lexology.com/library/detail.aspx?g=809fbccb-edb2-46c1-8fb1-bb041ff3d3f7&utm_source=Lexology+Daily+Newsfeed&utm_medium=HTML+email+-+Body+-+General+section&utm_campaign=Lexology+subscriber+daily+feed&utm_content=Lexology+Daily+Newsfeed+2015-05-19&utm_term=.

TABLE 20-1 THE TWO FACES OF UNIONS

Collective Face	*Monopoly Face*
Unions primarily provide a collective voice through which workers can express their job-related concerns.	Unions are institutions that primarily serve to raise wages above competitive levels.
Unions increase efficiency because unionized firms have lower employee turnover rates, so the employer spends less money and time training new employees.	Unions decrease efficiency by securing unmerited wage increases for their workers, thereby causing a misallocation of resources.
Because unions usually negotiate contracts that base wage increases primarily on seniority, older workers are more likely to help newer ones and a more cooperative workplace will exist, thereby increasing efficiency.	Unions decrease efficiency by causing strikes that result in lost production and obtaining special contract provisions that reduce productivity.
Unions decrease inequality of wage distribution within the firm because they try to raise the wages of below-average workers to match those of average workers; for solidarity purposes, unions have to try to make the wages of those in the bargaining unit more equal.	Unions increase the existing inequality of wage distribution by providing higher wages for unionized workers at the expense of nonunionized workers.
Unions are democratic institutions, representing the interests of workers in general in the political process.	Unions gain their power through coercion and the threat of physical violence, and they use that power to lobby for legislation to restrict competition in their respective industries.

Source: Adapted from R. B. Freeman and J. L. Medoff, *What Do Unions Do?* (New York: Basic Books, 1994).

Structure of the Primary U.S. Labor Legislation and the Mechanisms for Its Enforcement

Three major pieces of legislation govern labor–management relations in the United States today: the Wagner Act of 1935, the Taft–Hartley Act of 1947, and the Landrum–Griffin Act of 1959 (the last is also cited as the Labor–Management Reporting and Disclosure Act, or LMRDA). The Taft–Hartley Act amended the Wagner Act, and they are jointly referred to as the NLRA. In this section, we briefly describe the primary features of each of these acts and discuss the situations in which the business manager will most likely need an understanding of these laws.

THE WAGNER ACT OF 1935

Wagner Act Guarantees the rights of workers to organize and bargain collectively and forbids employers from engaging in specified unfair labor practices. Also called the National Labor Relations Act (NLRA).

collective bargaining Negotiations between an employer and a union primarily over wages, hours, and terms and conditions of employment.

The **Wagner Act** (also called the National Labor Relations Act, or NLRA) was the first major piece of federal legislation adopted explicitly to encourage the formation of labor unions. Many supporters of this act recognized that a number of labor problems were caused by gross inequality of bargaining power between employers and employees. They hoped that the Wagner Act would bring about industrial peace and raise the standard of living of U.S. workers. The act was to accomplish those goals by facilitating the formation of labor unions as a powerful collective voice for employees and by providing for **collective bargaining** between employers and unions as a means of obtaining the peaceful settlement of labor disputes. The key section of the Wagner Act is Section 7. This section provides that:

> [e]mployees shall have the right to self-organization, to join, form or assist labor organizations, to bargain collectively through representatives of their own choosing, and to engage in concerted activities for the purpose of collective bargaining or other mutual aid and protection.*

---

*Wagner Act Section 7, 29 U.S. Code § 157.

TABLE 20-2 EMPLOYER UNFAIR LABOR PRACTICES

Section	Prohibited Practice
8(a)(1)	Interference with employees' Section 7 rights
8(a)(2)	Employer-dominated unions
8(a)(3)	Discrimination by employers in hiring, firing, and other employment matters because of union activity
8(a)(4)	Retaliation against an employee who testifies or makes charges before the National Labor Relations Board
8(a)(5)	Failure to engage in good-faith collective bargaining with duly certified unions

Employees' Section 7 rights are protected through Section 8(a) of the act, which prohibits specific "employer unfair labor practices." These practices are delineated in Table 20-2. Section 9 of the act sets forth the procedures, including the secret ballot election, by which the exclusive employee-bargaining-unit representative (union representative) is chosen.

The final important portion of the Wagner Act authorized an administrative agency, the National Labor Relations Board (NLRB), to interpret and enforce the act. It also provided for judicial review in designated federal courts of appeal.

THE TAFT–HARTLEY ACT OF 1947

The passage of the Wagner Act led to a growth in unionization and an increase in workers' power. Given that they had had almost no power before, any power workers obtained was bound to look like a dramatic increase. Thus, the public's perception of union power may have been greater than the actual power of unions. At any rate, this perception led to the passage of the **Taft–Hartley Act**, which was designed to curtail the powers that unions appeared to have acquired under the Wagner Act.

Section 8(b) of the Taft–Hartley Act, titled "Union Unfair Labor Practices," bars unions from engaging in certain specified activities (see Table 20-3). The act also (1) amended Section 7 of the Wagner Act to include the right of employees to refrain from engaging in collective activity, (2) made collective bargaining agreements enforceable in federal district courts, and (3) provided a civil damages remedy for parties injured by certain prohibited union activities.

Taft–Hartley Act Bars unions from engaging in specified unfair labor practices, makes collective bargaining agreements enforceable in U.S. district courts, and provides a civil damages remedy for parties injured by certain prohibited union activities.

TABLE 20-3 UNION UNFAIR LABOR PRACTICES

Section	Prohibited Practice
8(b)(1)	Restraining or coercing employees in the exercise of their Section 7 rights
8(b)(2)	Forcing the employer to discriminate against employees on the basis of union or anti-union activity
8(b)(3)	Failing to engage in good-faith collective bargaining with the employer
8(b)(4)	Striking, picketing, and engaging in secondary boycotts for illegal purposes
8(b)(5)	Charging excessive union dues or initiation fees in a union shop
8(b)(6)	Featherbedding (charging employers for services not performed)
8(b)(7)	Picketing for recognition or to force collective bargaining under certain circumstances

THE LANDRUM–GRIFFIN ACT OF 1959

Landrum–Griffin Act Governs the internal operation of labor unions.

The final major piece of labor legislation is the **Landrum–Griffin Act**, which primarily governs the internal operations of labor unions. Passage of this act was prompted by congressional hearings that uncovered evidence of looting of union treasuries by some powerful union officials and of corrupt, undemocratic practices in some labor unions. The act requires certain financial disclosures by unions and establishes civil and criminal penalties for financial abuses by union officials. It also includes a section, "Labor's Bill of Rights," that gives employees protection against their own unions. The rights established by the Landrum–Griffin Act are summarized in Table 20-4.

THE NATIONAL LABOR RELATIONS BOARD

National Labor Relations Board (NLRB) The administrative agency set up to interpret and enforce the Wagner Act (NLRA).

Structure. The **National Labor Relations Board (NLRB)**, as stated earlier, is the administrative agency responsible for the interpretation and enforcement of the NLRA. Its structure is diagrammed in Exhibit 20-2. The NLRB's three primary functions are as follows:

1. Monitoring the conduct of the employer and the union during an election to determine whether workers want to be represented by a union
2. Preventing and remedying unfair labor practices by employers or unions
3. Establishing rules and regulations interpreting the acts

The NLRB is composed of five members, each appointed by the president with the advice and consent of the Senate. Members serve staggered five-year terms. The board meets in Washington, DC. Three-member panels decide routine cases involving disputes between employees, union, and employer, but the entire board may hear significant cases.

The general counsel of the NLRB, also appointed by the president with the advice and consent of the Senate, oversees the investigation and prosecution of unfair practice charges before the board. If a board decision is subsequently

TABLE 20-4

EMPLOYEE RIGHTS UNDER THE LANDRUM–GRIFFIN ACT

Section	Right
101(A)(1)	*Equal Rights.* Every union member has an equal right to nominate candidates, to vote in elections, and to attend and fully participate in membership meetings, subject to the organization's reasonable constitution and bylaws.
101(A)(2)	*Freedom of Speech and Assembly.* Members have the right to meet freely with one another at any time and to express any views about the labor organization, candidates for office, or business affairs at organization meetings, subject to reasonable rules pertinent to conduct of meetings.
101(A)(3)	*Dues, Initiation Fees, and Assessments.* Increases in local union dues, initiation fees, or assessments must be voted on by a majority of the members through secret ballot.
101(A)(4)	*Protection of Right to Sue.* Labor organizations cannot prohibit members from bringing any legal actions, including those against the organization. Organizations may require that members first exhaust reasonable hearing procedures established by the organization.
101(A)(5)	*Safeguards against Improper Discipline.* No member may be fined or otherwise disciplined except for nonpayment of dues without being (1) served with written notice of specific charges, (2) given a reasonable time to prepare a defense, and (3) afforded a full and fair hearing.

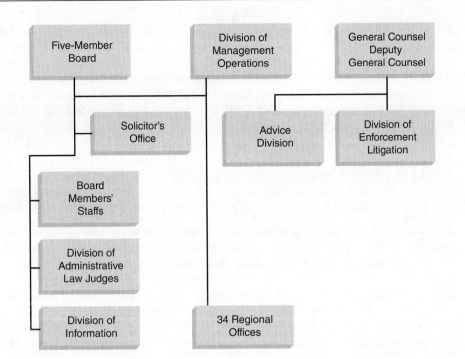

EXHIBIT 20-2

THE NATIONAL LABOR RELATIONS BOARD

challenged in court, it is the general counsel who represents the board before the appellate court.

There are too many cases for the board and general counsel to handle each one personally, so most cases are handled by 34 regional offices, located in major cities across the country. These regional offices are headed by a regional director, who is appointed and overseen by the general counsel. The regional director and his or her staff are directly responsible for investigating charges of unfair labor practices, which they prosecute before administrative law judges (ALJs). They are also responsible for conducting representation elections, in which employees of a firm decide whether they wish to be represented by a union.

Jurisdiction. Just as a civil court must have jurisdiction over the parties before it, the NLRB must have jurisdiction over the parties before it in both representation and unfair labor practice cases. The basis for NLRB jurisdiction is found in the NLRA, under which Congress granted jurisdiction to the NLRB over any business "affecting commerce," with certain specific exceptions. Any employer or employee not covered by the NLRA need not abide by its provisions. (Noncovered employees and employers, however, may be covered by state labor laws.) Employees specifically omitted from NLRA coverage are those who work in federal, state, and local government; employees in the transportation industry and those covered by the Railway Labor Act; independent contractors; agricultural workers; household domestics; and persons employed by a spouse or parent.

Also excluded from NLRB jurisdiction are supervisors, managerial employees, and confidential employees. Much litigation has arisen over disputed definitions of *managerial employee* and *supervisor*. One of the employee groups that is facing difficulty today over the issue of whether they are supervisors is nurses. Thus far, the NLRB and the courts have been carefully examining the duties of the nurses in each case, with the nurse's eligibility depending on the circumstances. It was thought for a while that the NLRB had finally come up with a workable test for what constituted "independent judgment," which characterizes an employee as a supervisor. The board had said that employees do not exercise "independent judgment" when they exercise "ordinary professional or technical judgment in directing less-skilled employees to deliver services in accordance with employer-specified standards." The U.S. Supreme Court, however, rejected that test in 2001,[5] in a decision many commentators believe may make it more difficult for health care personnel and other professionals to organize.

Just because Congress has granted the NLRB the authority to act in a given case does not mean that the board *will* act. The board does not have unlimited funds. Consequently, the NLRB has established its own set of guidelines, which it uses to determine whether it will exercise jurisdiction over an employer. These guidelines are designed basically to determine whether a firm does a significant amount of business and, thus, has enough employees to justify the expenditure of NLRB resources. The guidelines are established industry by industry (e.g., a transit system must have a total annual business volume of at least $250,000).

Procedures in Representation Cases. An important function of the NLRB and the general counsel is to ensure that employees will be uncoerced in choosing a bargaining representative or in choosing not to be represented by a union. On April 14, 2015, new procedures for representation cases went into effect. According to the NLRB, these minor changes will "remove unnecessary barriers to fair and expeditious resolution of representation cases, simplify representation-case procedures, codify best practices, and make them more transparent and uniform across regions." Under NLRB procedures, which are illustrated in Exhibit 20-3, a petition for a representation election is initially filed,

EXHIBIT 20-3

STEPS IN A REPRESENTATION PROCEEDING

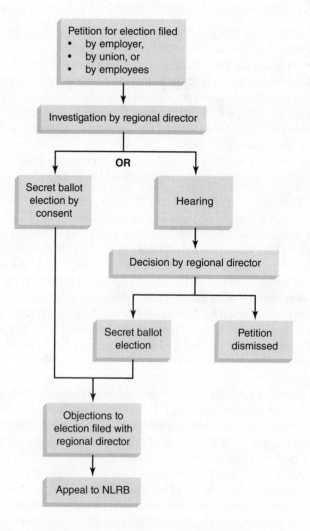

---

[5]*NLRB v. Kentucky River Community Care, Inc.*, 532 U.S. 706 (2001).

preferably electronically, with the regional director (1) by the union, when it can demonstrate that it has the support of more than 30 percent of the employees it seeks to represent (known as "majority support"); (2) by the employer, when two or more unions are claiming to be the exclusive representative of the employees or when one union claims to have majority support; or (3) by the employees themselves. The union demonstrates its support by submitting authorization cards or a signature sheet. Each employee signature indicates that the employee gives the union the authority to act as his or her exclusive bargaining representative. When the petition and supporting documents are filed with the NLRB, they must simultaneously be served on all other parties affected by the petition.

Within two days of receipt of the notice of filing, the employer must post a Notice of Petition for Election and distribute the notice electronically if the employer customarily communicates with its employees electronically.

The regional director then conducts an investigation to determine whether the employer is under the jurisdiction of the NLRB, whether the group of employees the union is seeking to represent is covered by the NLRA, whether the group of employees seeking representation is an appropriate bargaining unit, and whether there is sufficient support (30 percent) for the union.

Within two days of receiving the Notice of Petition and Notice of Hearing from the NLRB, the employer is required to post a Notice of Petition for Election in conspicuous places, including all places where notices to employees are customarily posted. If the employer customarily communicates with employees in the petitioned-for unit through electronic means, the employer must also distribute the Notice of Petition for Election electronically.

When the petition is filed, it is reviewed in the regional office for sufficiency and the showing of interest. If these findings are affirmative, the petition will be assigned a case number, and letters will be sent to the parties notifying them of the board agent who will be handling the case and providing a notice of the date of the representation hearing. The hearing is normally set eight days from the date the notice is received, as well as the time by which the Statement of Position must be received by the board and the petitioning party, which is usually noon the day before the hearing.

The Statement of Position from the employer should include commerce and other information that will facilitate entry into election agreements or streamline the pre-election hearing if the parties are unable to enter into an election agreement. It also must include a list of the full names, work locations, shifts, and job classifications of all individuals in the proposed unit. An employer who contends that the proposed unit is not appropriate and wants to correct it must make a separate list of the same information for all individuals it contends must be added to the proposed unit, as well as the names of those it believes should be excluded.

At any point prior to the hearing, all parties may enter into an agreement consenting to an election; agreeing on an appropriate unit; and agreeing on the method, date, time, and place of a secret ballot election that will be conducted by an NLRB agent. If no agreement is reached, the regional office will hold a hearing to receive evidence on whether a question of representation exists and an election should be held. An affirmative decision results in an order for an election. Within two business days of the order for an election or approval of an agreement for an election, the employer must provide to the regional director and the parties named in the election agreement a list of the full names, work locations, shifts, job classifications, and contact information (including home addresses, available personal email addresses, and available home and cell phone numbers) of all eligible voters. An election by secret ballot will then be conducted by a representative of the regional office. In fiscal year 2009, the NLRB conducted 1,619 representation elections, and workers chose union representation in 63.8 percent of the elections.[6]

After the election, the losing party may file objections to the outcome of the election with the regional director, who either orders a new election or certifies the results. The decision may be appealed to the board.

Procedures in Unfair Labor Practice Cases. A second important function of the NLRB is to prevent and remedy unfair labor practices by both employers and employees. An unfair labor practice charge is initiated when an aggrieved employee, union, or employer files an unfair labor practice charge with the appropriate regional office. (A sample charge is pictured in Exhibit 20-4.) In

EXHIBIT 20-4

FORM FOR FILING AN UNFAIR LABOR PRACTICE CHARGE

UNITED STATES OF AMERICA
NATIONAL LABOR RELATIONS BOARD
CHARGE AGAINST EMPLOYER

DO NOT WRITE IN THIS SPACE	
Case	Date Filed

INSTRUCTIONS
File an original and 4 copies of this charge with NLRB Regional Director for the region in which the alleged unfair labor practice occurred or is occurring.

1 EMPLOYER AGAINST WHOM CHARGE IS BROUGHT

a Name of Employer		b Number of workers employed
c Address, street, city, state, zip code	d Employer Representative	e Telephone No.
f Type of Establishment: factory, mine, wholesaler, etc.	g Identify principal product or service	

h The above-named employer has engaged in and is engaging in unfair labor practices within the meaning of section 8 (a) subsections ________ (1) and first subsections of the National Labor Relations Act and these unfair labor practices are unfair practices affecting commerce within the meaning of the Act

2 Basis of the Charge (set forth a clear and concise statement of the facts constituting the alleged unfair labor practices)

By the above and other acts, the above-named employer has interfered with, restrained, and coerced employees in the exercise of the rights guaranteed in Section 7 of the Act

3 Full name of the party filing charge (labor organization: give full name including local name and number)

4a Address (street and number, city, state and ZIP code)	4b Telephone No.

5 Full name of national or international labor organization which is an affiliate or constituent unit (to be filled in when charge is filed by a labor organization)

6 DECLARATION

I declare that I have read the above charge and that the statements are true to the best of my knowledge and belief.

B ________________________ ________________
Signature of representative making charge (title if any)

Address ________________________ ________________ ________
Telephone No. (date)

WILLFUL FALSE STATEMENTS ON THIS CHARGE CAN BE PUNISHED BY FINE AND IMPRISONMENT (U.S. CODE, TITLE 18, SECTION 1001)

[6]NLRB, *Seventy-Fourth Annual Report of the NLRB for the Fiscal Year Ended September 30, 2009*, 1. Accessed June 4, 2010 at www.nlrb.gov/sites/default/files/attachments/basic-page/node-1677/nlrb2009.pdf.

fiscal year 2009, 22,943 unfair labor charges were filed with the NLRB.[7] After regional office employees, called *field examiners*, investigate the charge, the regional director decides whether to issue a complaint. If a complaint is issued, an attorney from the regional office tries to resolve the complaint informally.

If informal negotiations are unsuccessful, the regional office attorney prosecutes the case before an ALJ. In fiscal year 2009, NLRB administrative law judges issued 190 decisions.[8] The ALJ issues an order recommending a remedy or suggesting a dismissal, in either case stating the rationale and evidence for the decision. If no party objects to the decision within 20 days, it automatically becomes a final order of the NLRB. If an order is issued and any party fails to abide by it, the board must petition a U.S. court of appeals for an enforcement order. The procedure used in unfair labor practice cases is illustrated in Exhibit 20-5.

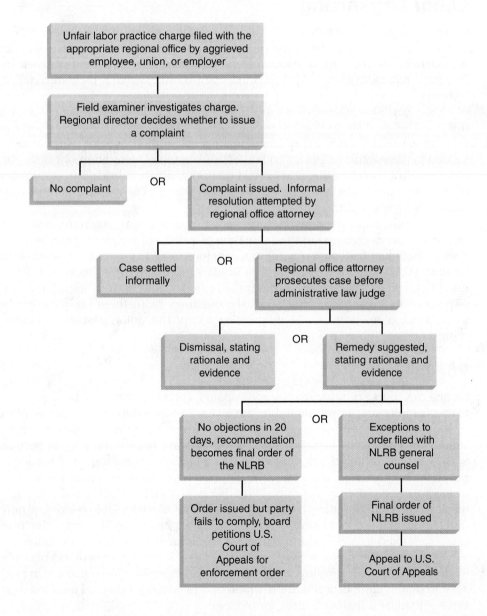

EXHIBIT 20-5

PROCEDURES USED IN AN UNFAIR LABOR PRACTICE CASE

---

[7]*Id.*, at 1.

[8]*Id.*, at 2.

Appeal to the board is available at times. If the regional director refuses to issue a complaint, the charging party may appeal to the general counsel in Washington, DC. Such appeals are almost always denied. Alternatively, a party dissatisfied with the ALJ's decision may file an appeal, called an *exception* to the recommended order, with the board in Washington, DC. Briefs are then filed, and in rare instances, oral arguments are heard. The board then issues its final order. In fiscal year 2012, the board issued 341 decisions in unfair labor practice cases.[9]

No one is required to honor an NLRB order, because the board has no contempt-of-court powers. If its order is not followed, the NLRB brings an enforcement proceeding in a circuit court of appeals, asking the court to order the parties to abide by its order. A dissatisfied party may appeal to a circuit court of appeals.

Labor Organizing

In the 1936 novel *In Dubious Battle*, John Steinbeck graphically described the extreme hardships faced by union organizers just after the passage of the Wagner Act. Derided as communist sympathizers, they were often run out of town by company representatives and sometimes even by the workers they were trying to help.

Many small-town law enforcement officers were indebted to business during this era. The most protection they were willing to offer organizers was to advise them to get out of town. Many labor organizers lost their lives or were severely injured in these early unionization battles, and victories were often not clear-cut. Gradually, the violence directed against labor organizers subsided as companies realized that the NLRA would not be repealed and the courts were going to enforce employees' right to organize.

Today, dissatisfied employees can contact a national union representing other employees engaged in the same type of work. For example, employees of a shop that manufactures parts for auto engines would contact the United Auto Workers (UAW) union. The union then sends a representative to meet with interested employees and explains what unionization would do for them. If the employees want to pursue unionization, the organizer helps them run a campaign to convince a majority of the workers to accept the union as their exclusive representative. Table 20-5 lists some of the largest unions in the United States.

BOARD RULES

During the course of this organizing campaign, certain activities of both employers and employees are prohibited by the NLRA and by "board rules," that is, rules of conduct developed over the years by the NLRB in a number of cases. The constraints on employers' behavior under the NLRA are found primarily in Section 8(a)(1), which prohibits interference with employees exercising their Section 7 rights. It is important to distinguish conduct that constitutes an unfair labor practice from violations of board rules. Why? Because a violation of board rules may result in the NLRB's setting aside the results of an election and ordering a new one, whereas the commission of an unfair labor practice by the employer may cause the board to ignore the election results altogether and order the employer to bargain with the union without a new election. The latter remedy occurs only in cases in which the employer's conduct was so egregious as to make it impossible to hold a fair election and the union had previously collected authorization cards signed by a majority of the employees.

---

[9]NLRB, Board Decisions Issued. Accessed August 10, 2013 at www.nlrb.gov/news-outreach/graphs-data/decisions/board-decisions-issued.

TABLE 20-5 SOME OF THE LARGEST UNIONS IN THE UNITED STATES AS OF 2013

Union Name	Approximate Number of Members
National Education Association (NEA)	3,000,000[10]
Service Employees International Union (SEIU)	2,100,000[11]
American Federation of State, County, and Municipal Employees (AFSCME)	1,600,000[12]
American Federation of Teachers (AFT)	1,500,000[13]
International Brotherhood of Teamsters (IBT)	1,400,000[14]
United Food and Commercial Workers International Union (UFCW)	1,300.000[15]
United Steelworkers	1,200,000[16]
Communications Workers of America	700,000[17]

Board rules are designed to guarantee a fair election. One very important rule, the **24-hour rule**, prohibits both union representatives and employers from making speeches to "captive audiences" of employees within 24 hours of a representation election. A captive audience exists when the employees have no choice but to listen to the speech.

Another important board rule requires employers to file with the regional director, within seven days after an election order is issued, a list of the names and addresses of all employees eligible to vote. This list, known as the *Excelsior list* (after the case that created it), is then made available to the union or union organizers by the regional director.

It is essential that employers be aware of these and other board rules, because a violation of the rules may result in the setting aside of an election, even when the behavior does not constitute an unfair labor practice. All unfair labor practices, whether by employers or employees, are also considered violations of the board's election rules.

24-hour rule Prohibits both union representatives and employers from making speeches to "captive audiences" of employees within 24 hours of a representative election.

UNFAIR LABOR PRACTICES BY EMPLOYERS

In fiscal year 2009, the majority of claims of unfair labor practices by employers (8,723 claims) alleged that the employer refused to bargain.[18] The second-largest category of unfair labor practices alleged that workers were illegally discharged or that the employer engaged in other illegal discrimination against employees (6,411 charges).

---

[10]National Education Association, www.nea.org/home/2580.htm.
[11]SEIU.org, www.seiu.org.
[12]AFSME, www.afscme.org/union/about.
[13]AFT, www.aft.org/about.
[14]Teamsters, teamster.org/content/fast-facts.
[15]UFCW, www.ufcw.org/about.
[16]United Steel Workers, www.usw.org.
[17]Communications Workers of America, www.cwa-union.org/about.
[18]NLRB, *Seventy-Fourth Annual Report of the NLRB for the Fiscal Year Ended September 30, 2009*, 1. Accessed June 4, 2010 at www.nlrb.gov/sites/default/files/attachments/basic-page/node-1677/nlrb2009.pdf.

Interference with Organizing. Section 8(a)(1) of the NLRA prohibits employer interference with, restraint of, or coercion of employees in the exercise of their Section 7 rights. It is sometimes difficult for a businessperson to know when his or her speech or conduct rises to the level of coercion, restraint, or interference. To make the issue even more complicated, Section 8(c) expressly provides that the expression of a view, an argument, or an opinion is not evidence of an unfair labor practice as long as it does not contain any threats of reprisals or promises of benefits.

As these sections have been interpreted since 1969, employers are allowed to communicate to employees their general views on unions or their specific views on a particular union, even to the point of predicting the impact that unionization would have on the company, so long as these statements do not amount to threats of reprisals or promises of benefits. To stay within these bounds, an employer must phrase any predictions carefully and make sure they are based on objective facts; in essence, the consequences that the employer predicts must be outside the control of the employer. Threats of reprisals that constitute unfair labor practices include threats to close a plant if the employees organize; threats to discharge union sympathizers; and threats to discontinue present employee benefits, such as coffee breaks or employee discounts. An example of a promise of benefits that would be an unfair labor practice is the announcement of a new employee profit-sharing plan a few days before the election.

APPLYING THE LAW TO THE FACTS . . .

Mike was working at a factory that was in the midst of an organizing campaign. Mike had been openly active in the campaign, trying to persuade workers to support the union any chance he got. His supervisor cautioned Mike that he didn't think the campaign would succeed, and because he liked Mike, he wanted to warn him that if the campaign failed, some members of the management team would remember who the union supporters were and would take that into account in their future decision making. Did the manager do anything wrong? What if he had said that once the organizing campaign failed, those employees who had opposed the union would receive preferential treatment to reward them for their devotion to the company?

No-solicitation rules may also constitute an employer unfair labor practice because they interfere with communications among employees. To exercise their Section 7 rights, employees must be able to communicate with one another. Today, email provides the easiest way for employees to communicate with each other when organizing. In one of its most significant rulings in 2014, Purple Communications, the NLRB held that if an employer allows an employee to use its email system at work, use of that employer-provided email system "for statutorily protected communications on nonworking time must presumptively be permitted." While employers may be able to restrict or prohibit the use of the systems when necessary to "maintain production and discipline," the burden will be on an employer to establish why such a prohibition or restriction is necessary. If employers allow employees to use a company email account for nonwork matters, the company cannot prevent employees from using it for organizing purposes.

Businesspeople must understand what types of organizing behavior can lawfully be prohibited and what prohibitions constitute unfair labor practices. Understandably, employers do not want employees to use work time or company property to organize, and in general, employers may prohibit union solicitation and the distribution of literature during work time. During nonwork time such as lunch and coffee breaks, employers may prohibit organizing activity on company property only for legitimate safety or efficiency reasons and only if the restraint is not manifestly intended to thwart organizing efforts. The burden of proof is on the employer to demonstrate these safety or efficiency concerns.

It is not always easy to predict when employer conduct is going to be found unlawful, because in many cases, the court is trying to strike a balance between the employer's private property rights and the employees' rights to communicate with one another during an organizing campaign. For example, it is not unlawful for an employer to refuse to allow workers to post organizing notices on a company bulletin board on which employees had previously been allowed to post only "for sale" cards. Only when an employer has opened up bulletin boards for all employee postings and then disallows postings related to organizing have courts found a violation of the NLRA.

Nonemployee organizers have fewer rights than employee organizers. As long as they have some way to communicate with employees (and they do in almost all cases now, because they are entitled to the Excelsior list of names and addresses of employees), nonemployee organizers may be prohibited from entering the employer's property, including private parking lots. The courts have held that in the case of nonemployee organizers, the employer's private property rights will be protected.

Domination or Support of Labor Organizing. In addition to the NLRA's all-encompassing Section 8(a)(1), other subsections of 8(a) set forth specific behaviors that constitute unfair labor practices; a violation of any of these subsections is simultaneously a violation of Section 8(a)(1). For example, under Section 8(a)(2), an employer cannot dominate, support, or interfere with a labor organization. Thus, in response to an organizing campaign by one union, the employer cannot aid some of its employees in contacting a different union to compete for the right to represent the workers at that plant. Nor can the employer play any role in establishing or operating any committee or other organization designed to represent or aid employees in their dealings with the employer over wages, rates of pay, or other terms and conditions of employment.

The employer must also be careful about voluntarily recognizing a union claiming to represent a majority of the employees. If the employer recognizes a union that does not represent the majority of employees, that is a violation of Section 8(a)(2), even if the employer was acting in good faith.

A union dominated by an employer will be *disestablished*; that is, it may never again represent those employees. A union unlawfully supported by an employer will be *decertified*; that is, it will not be able to represent the employees until it has been certified as a result of a new representation election monitored by the NLRB.

One of the major concerns facing labor relations specialists at present is the impact of Section 8 on some of the more cooperative labor–management programs that employers are trying to put in place. Many commentators from both business and academia have cited the traditional adversarial relationship between labor and management as being at least partially responsible for productivity problems in many industries.[19] Consequently, U.S. management has been experimenting with programs to increase cooperation between labor and management. There is strong evidence that these programs do boost productivity.[20] Many of them, however, have been found to constitute unfair labor practices.

Labor–management committees are one such program. They provide a forum in which workers can communicate directly with upper management. Worker participants on these committees are either elected by their fellow

labor–management committee A forum in which workers communicate directly with upper management. May be illegal under the NLRA if the committee has an impact on working conditions, unless all workers in a bargaining unit or a plant participate or the employees on the committee are carrying out a traditional management function.

---

[19]C. Farrell and M. J. Mandell, "Industrial Policy," *Business Week* 70, no. 75 (Apr. 6, 1992); B. Childs, "United Motors: An American Success Story," *Labor Law Journal* 40: 453 (1989).

[20]B. Childs, "United Motors: An American Success Story," *Labor Law Journal* 40: 453 (1989).

workers or appointed by management. They usually serve for a limited time, such as six months, to ensure maximum participation.

Under a literal interpretation of the NLRA, most participatory committees would appear to constitute labor organizations. In fact, in the leading U.S. Supreme Court case on this issue, *NLRB v. Cabot Carbon Co.*,[21] the Court found that "employee committees" established by the employer to allow employees to discuss with managers such issues as safety, increased efficiency, and grievances at nonunion plants and departments are labor organizations. Numerous decisions since *Cabot* have followed this strict interpretation. The dilemma for management is that these committees are rather useless unless they are allowed to discuss issues that have an impact on working conditions, but having such discussions makes them illegal employer-dominated labor organizations under the NLRA.

There are narrow exceptions to the strict interpretation the Supreme Court set forth in *Cabot*. The first applies when all of the workers in a bargaining unit or plant participate in the program. In that situation, the committee does not "represent" the employees because it comprises all the employees; if it does not represent employees, it cannot be a labor organization and, hence, cannot be construed as an employer-dominated union.

The second exception involves a situation in which employees carry out a traditional management function. In such cases, the employee group no longer "deals with" management because it is performing the delegated function itself.

Remember, even if a labor–management committee is found to constitute a labor organization, to be unlawful, the committee must be dominated or supported by the employer. The traditional test[22] asks whether the committee is structurally independent of management. Most participatory committees have some minimal association with management that would render them unfair under this strict test. Again, however, some circuit courts are using two factors to minimize the impact of this holding on participatory programs. First, the court asks whether the employers had good motives in establishing the plan. Second, the court asks whether the employees are satisfied with it.

Applying this two-part analysis, the court may then distinguish illegal domination and support from legal cooperation.[23] In the case in which this two-part test was initially set forth, the court focused on the NLRA's goal, which is to protect employee free choice, and said that employees should be free to enter into cooperative arrangements with their employers as long as the employer does not try to use the plan to interfere with free choice.

Discrimination Based on Union Activity. An employer that discriminates against employees because of their union activity is in violation of Section 8(a)(3). An ambiguous situation arises when a marginal employee who is also an organizer for a union is fired. The NLRB will have to determine whether the firing was motivated by the employee's poor performance or by the employee's union activity. Discharge of even the most strident union activist is legal as long as the primary motivation for the firing was poor performance rather than union activity. In such cases, courts look at such factors as how others who engaged in similar misconduct have been treated by the employer.

Firing is the ultimate form of discrimination, but other forms of discrimination, such as reducing break time and being treated unfavorably in job and overtime assignments, also constitute violations of Section 8(a)(3). The following case examines a situation in which employers allegedly fired a number of employees for their union activity.

---

[21] 360 U.S. 203 (1959).

[22] This test was established in *NLRB v. Newport News Shipbuilding & Drydock Co.*, 308 U.S. 241 (1939).

[23] *Chicago Rawhide v. NLRB*, 221 F.2d 165 (7th Cir. 1955).

CASE 20-1

Gaetano & Associates, Inc. v. National Labor Relations Board
United States Court of Appeals for the Second Circuit
2006 U.S. App. LEXIS 12436 (2006)

Gaetano & Associates Inc. (the Company) is the owner and developer of properties in New York City. The Company is its own general contractor and tries to perform as much of the construction work as possible. As part of two joint renovation projects in New York, the Company hired a number of carpenters. Representatives of the Carpenters' Union began organizing efforts in early April 2003. The Company was aware of the union organizing activity even before a petition for an election was filed by either union.

A union representative called the Company on April 16, 2003, and spoke to William Gaetano, the owner, stating that his union represented the carpenters. This was followed up by a letter dated April 16 advising that the union claimed to represent a majority of the carpenters employed by the Company. Also on April 16, 2003, the Carpenters' Union filed a petition for an election. The NLRB's regional office immediately faxed a copy of this petition, along with a notice that a representation hearing would take place on April 25, 2003.

At the end of the working day on April 16, 2003, the owner laid off a large number of the carpenters. On the same day, the Company began replacing some of the fired employees. Furthermore, after the April 16 layoff, the Company entered into two subcontracts to install windows and sheetrock. With two minor exceptions, the subcontracted work could easily have been performed by the fired carpenters. The workers and the union brought charges against the Company. The ALJ found in favor of the carpenters, and the NLRB affirmed.

En Banc

Our review of NLRB orders is limited. "We must enforce the Board's order where its legal conclusions are reasonably based, and its factual findings are supported by substantial evidence on the record as a whole." . . . Moreover, we accept an ALJ's credibility determinations, as adopted by the NLRB, unless the testimony is "incredible or flatly contradicted by undisputed documentary testimony."

Mass Lay-Offs

The ALJ found, and the Board agreed, that the Company's April 16, 2003 mass lay-off of carpenters was motivated by anti-union animus and that the company would not have made the same decision absent the concerted activity and thus that the Company had violated NLRA § 8(a)(1) and (3). Contrary to the Company's argument, temporal proximity can be a sufficient basis from which to infer anti-union animus as a matter of law. Here, the lay-off occurred at the end of the day on April 16, 2003, in the middle of the workweek on the same day that union representative Byron Schuler called the Company to inform it that its carpenters sought to be represented by his union and the NLRB faxed the Company the union's petition for a representation election. Immediately after laying off a significant number of its carpenters on April 16, 2003, the Company hired two carpenters to work at the main site and hired three more carpenters to work at that site a week later. The ALJ credited the testimony of a number of employees that at least 20 percent of rough carpentry remained at the time of the lay-off. The testimony is neither so incredible as to defy the laws of nature nor contradicted by documentary evidence, and therefore we do not disturb the ALJ's finding.

The Company argues further that the Board erred in finding it liable for an unfair labor practice under the two-part test of *Wright Line*, . . . because its actions were taken for economic reasons. It relies on cases in which we found, on the basis of testimony and corroborating documentary evidence, substantial support for the employer's proffered business justification for lay-offs that occurred in close temporal proximity to protected activity. The Company's reliance is misplaced, however, because it has failed to produce any documentation supporting its proffered economic justifications, and the ALJ found the testimony of the Company's principals inconsistent with its actions in hiring additional carpenters in April and May. In these circumstances, we find no error in the Board's rejection of the Company's affirmative defense.

Anti-Union Animus in Sub-Contracting

The Company next contends that the Board erred in finding that anti-union animus was a substantially motivating factor in the Company's decision to subcontract the window installation and related work because it relied on circumstantial evidence and the fact that the Company had committed other unfair labor practices. We disagree.

First, there is no prohibition on the Board's consideration of circumstantial evidence. Second, the Board only noted the existence of other unfair labor practices in concluding that the circumstances of the outsourcing gave rise to an inference of anti-union animus. There is substantial evidence supporting the Board's conclusion; namely, that the Company's decision to subcontract the window installation was a departure from its policy, adopted in 1998, to use its employees to perform as much of the construction work as possible; that the Company knew of the employees' union activity; and that the Company's numerous unfair labor practices in response to the union campaign gave rise to the inference that the decision to subcontract soon after the employees engaged in protected activity was motivated by anti-union animus.*

Affirmed in favor of NLRB.

UNFAIR LABOR PRACTICES BY EMPLOYEES

Unfair labor practices by employees are less common in organizing campaigns than are unfair labor practices by employers, perhaps because when a union is trying to gain representational status, it generally does not have enough power to engage in such practices. The sections of the NLRA most applicable to unions during the organizing period are 8(b)(1), which prohibits restraint or coercion of employees in the exercise of their Section 7 rights; 8(b)(2), which prohibits forcing the employer to discriminate, encourage, or discourage union activity; and 8(b)(7), which prohibits picketing for recognition when another union has been certified or when the picketing union has lost an election within the past year.

A more common situation where a union is likely to be charged with an unfair labor practice is when it operates a hiring hall. The following case is one such illustration.

CASE 20-2

Laborers' International Union of North America, Local 872, AFL–CIO, *and* Stephanie Shelby
Case 28–CB–065507 (May 3, 2013)

Respondent operated an exclusive hiring hall that referred employees to multiple employers in and around Las Vegas, Nevada. Shelby was a union member who used the hiring hall to obtain employment. Over the year preceding October 4, she was concerned about her proper place on the union's out-of-work list, which the union's dispatchers used to make job referrals. She also had concerns about her "skill sheet," which listed her certified skills and referral qualifications, and had made several unsuccessful attempts to resolve these concerns during visits to the hiring hall in August and September.

On September 16, she had a heated confrontation at the hiring hall with dispatcher Rocio Lucero over her alleged failure to attend a required roll call, which had led to her lower placement on the out-of-work list. On October 3, she attended another roll call and attempted to upgrade her skill set, but was told she needed to bring in the proper documentation. She returned the next day with "transcript sheets," instead of the required "certification." When Lucero told her that the transcript sheets were not equivalent to the necessary certification, she started yelling and swearing. Lucero told her to leave or the police would be called, but she still refused to leave. So Lucero called the hiring hall manager, Taylor, who was upstairs. He came down and told Shelby that she was "86'ed" and had to leave. She kept screaming, but backed away toward the door. The manager then called the police on his cell phone, and she backed out onto the parking lot about 40 feet from the door, still screaming and swearing.

The police arrived, handcuffed Shelby, and asked the manager whether he wanted to "trespass" Shelby. He said he did. Police then gave him a card to read, which told Shelby that she was to leave the premises or be charged with trespassing, a misdemeanor, and would be charged with trespassing if she returned. She left shortly thereafter.

The day after the incident, she called the dispatch office and left a message for Lucero and the manager, apologizing for her behavior. Between the incident and the Board hearing, she returned to the hall three times, with a police escort, and there was no trouble with any of her visits.

She filed an unfair labor practice charge with the Board, alleging that the Respondent had breached its duty of fair representation by having her removed from the hiring hall on October 4 and by placing her under a trespass order that required her to have a police escort anytime she returned to the hall. The administrative law judge found that her initial removal from the hiring hall was justified and therefore not an unfair labor practice. However, he found that the imposition of the continuing trespass order was a breach of the Respondent's duty of fair representation. This appeal followed.

---

*Gaetano & Associates, Inc. v. National Labor Relations Board, United States Court of Appeals for the Second Circuit, 2006 U.S. App. LEXIS 12436 (2006).

Chairman Pearce and Members Griffin and Block

It is well established that a union's duty of fair representation extends to its operation of an exclusive hiring hall, and that where a union "causes, attempts to cause, or prevents an employee from being hired or otherwise impairs the job status of an employee," the Board draws an inference of unlawful coercion. The union may overcome that inference by demonstrating that its actions were justified.

Here, there is no dispute that on October 4, Shelby, at least initially, was seeking to enhance her prospects of being hired by augmenting her skill sheet at the hiring hall. The Respondent's ejection of Shelby from the hiring hall at least temporarily impaired her ability to achieve that objective. We nevertheless agree with the judge that the Respondent's action was justified by Shelby's conduct that day.

. . . As the judge found, her "loss of temper and her inability to control her actions [led] to her continued string of epithets directed at Respondent's agents"; when Taylor asked her to calm down, "she refused until the police finally arrived." The judge also found from the credited testimony that although expletives were common at the hiring hall, it was not common for a member to curse directly at a dispatcher in a personal confrontation. In a business office setting (as opposed to a dockside or construction site), the combination of Shelby's tirade of continuous screaming, her repeated use of expletives, and her persistent refusal either to "calm down" or to leave justified the Respondent's decision to remove her from the property at that time. Accordingly, we agree with the judge that the Respondent's conduct on October 4 did not violate the Act.

B. *Maintenance of "Trespass" Status after October 4*

. . . Although this is a close issue, we disagree with the judge's finding of a violation for the following reasons. The judge's finding is premised, mistakenly, on the notion that the Respondent imposed the police-escort requirement on Shelby. In fact, the police effectively placed Shelby in an ongoing "trespass" status on October 4 when, before acting on the Respondent's lawful request to have Shelby removed from the property that day, they instructed Taylor to read a card (which they provided him) to that effect to Shelby. Taylor did no more than was necessary to obtain Shelby's immediate removal; nor did he take any further action to keep the resulting access restriction in place. In short, the police-escort requirement was a direct consequence of the Respondent's lawful removal of Shelby . . .

Further . . . we are not persuaded that Shelby's apology triggered an affirmative duty on the part of the Respondent to seek dissolution of the escort requirement. Although Shelby called on October 5 to apologize in some manner for her misbehavior, she apparently made no attempt to speak directly to Taylor. Nor did she convey any clear commitment not to behave in a similar manner in the future. Equally, if not more important, Shelby never asked the Respondent for assistance in ending the escort requirement.

Finally, the record does not establish . . . that the escort requirement actually impaired Shelby's employment in any way. She was able to maintain her place on the out-of-work list during the period that the escort requirement was in place.

In addition, the escort requirement did not prevent her from being hired by the two employers who contacted the hiring hall to request her by name in November and December. There is no allegation that she suffered any resulting loss of pay or benefits.

In these circumstances, we find that the Respondent did not have an affirmative duty to seek to lift the trespass order or the escort requirement. Accordingly, we find that the Respondent's failure to do so did not violate Section 8(b)(1)(A); nor was the Respondent's behavior "so far outside a 'wide range of reasonableness' as to be irrational" under the duty of fair representation.*

Affirmed in part. *Reversed* in part in favor of Respondent.

CRITICAL THINKING ABOUT THE LAW

This case reminds us that the duty of fair representation does not mean that the union is required to support a member with regard to whatever demands a member may make. You can get a sense of what unfair representation in this case would have been by studying the Board's discussion of the possible harms to Ms. Shelby that she did not experience as a result of the continuation of the trespass order.

1. What would have had to have happened to Ms. Shelby as a result of the continuation of the trespassing order for the Board to have been more impressed by her request for a legal remedy?

    Clue: Toward the end of the decision, the Board discusses several things that did not happen to Ms. Shelby as a result of the trespass order.

2. Are there other possible harms to Ms. Shelby that the Board might have considered before deciding that the continuation of the trespassing order did not constitute unfair representation? Explain.

    Clue: If a union member must come to the hiring hall with a police escort, might an employer understandably be hesitant to contact that employee as a prospective hire?

---

*Laborers' International Union of North America, Local 872, AFL–CIO and Stephanie Shelby Case 28–CB–065507 (May 3, 2013).

LINKING LAW AND BUSINESS

Management

In your management class, you may have discussed several different approaches to management. One type is the behavioral approach, which emphasizes an increase in productivity based on a better understanding of people. Proponents of this approach believe that a better understanding of human behavior by the managers, while adapting an organization to its workers, will lead to greater organizational success. The behavioral approach to management highlights the human relations movement, which places a strong emphasis on the people-oriented aspects of a firm. The human relations movement involves the observation of relations among people in organizations to determine the relational impact on a firm's success. By developing a better understanding of human behavior, managers become equipped to engage in healthier relations with their employees. Therefore, the overall efficiency of the organization may increase. As already mentioned in this section, adversarial relations between a company's management and employees may, in contrast, lead to productivity problems. Programs such as quality circles, autonomous and semiautonomous work groups, and labor–management committees are several means of enhancing organizational productivity by improving employer–employee relations. Thus, a lawful use of these programs may result in better relational interactions with employers and employees, which could boost organizational productivity.

Source: Adapted from S. Certo, *Modern Management* (Upper Saddle River, NJ: Prentice Hall, 2000), 32–33. Reproduced by permission.

ORGANIZING THE APPROPRIATE UNIT

The purpose of an all organizing activity is for the union to gain the right to be the exclusive representative of employees in negotiations with the employer over wages, hours, and terms and conditions of employment. Because the potential power of the union clashes with the employer's desire to maintain control over the workplace, organizational campaigns can become very heated. In any organizing campaign, the first step for the union is to gain the support of a substantial number of the members (generally 30 percent) of an appropriate bargaining unit so that a board-run election may be ordered.

As mentioned previously, an important question in this initial stage is: What is the **appropriate bargaining unit**? The appropriate unit, as defined by the NLRA, is one that can "ensure the employees the fullest freedom in exercising the rights guaranteed by the Act." In making such a determination, the regional director of the NLRB examines a number of alternatives. An entire plant may be an appropriate unit; so may a single department of highly skilled employees; so may all the employees of a single employer located at more than one facility (e.g., all employees of a group of retail stores located in a metropolitan area).

In determining whether a proposed bargaining unit is appropriate, the NLRB considers primarily whether there is a mutuality of interest among the proposed members of the unit. All proposed members should have similar skills, wages, hours, and working conditions, for only then is it possible for a union to look out for the interests of all members. Other factors considered include the desires of the employees, the extent of organization, and the history of collective bargaining of the employer and of the industry in which the employer operates. In the following case, the NLRB considers a situation in which the employer wants to include individuals in the bargaining unit that the petitioners for the union do not wish to include. Notice that unlike the other cases in this chapter, the text of this case is the NLRB decision rather than a court decision.

appropriate bargaining unit May be an entire plant, a single department, or all employees of a single employer, as long as there is a mutuality of interest among the proposed members of the unit.

CASE 20–3

Specialty Healthcare and Rehabilitation Center of Mobile *and* United Steelworkers, District 9, Petitioner
National Labor Relations Board
Case 15–RC–8773 (August 26, 2011)*

The Employer operates a nursing home and rehabilitation center in Mobile, Alabama, with no history of collective bargaining. The Petitioner sought to represent a unit of 53 CNAs, who must be certified by the State of Alabama. The certification requires that an individual complete 16 hours of classroom training and 72 hours of general education. The certification course includes the basic components of caring for geriatric and incapacitated patients, such as bathing, dressing, and feeding. CNAs are required to attend specialized training on a periodic basis to maintain their certification. The Employer believed that the only appropriate unit consists of its approximately 86 nonsupervisory, nonprofessional service and maintenance employees such as cooks, supply clerks and data entry clerks, along with the CNAs. CNAs['] wages all start at $8.50 an hour, whereas all the other employees the employer wanted to group with them had starting wages ranging from $7 to $10 per hour.

The Regional Director found that a petitioned-for bargaining unit of certified nursing assistants (CNAs) was appropriate under a traditional community-of-interest analysis. The Employer, however, contended that the only appropriate unit containing the CNAs consists of the CNAs plus all other nonprofessional service and maintenance employees at its facility. The employer requested a review of the decision. The employer argued that because CNAs in acute care facilities were included in bargaining units with all nonprofessional and maintenance employees, they should be similarly grouped in nursing homes.

Chairman Liebman and Members Becker, Pearce, and Hayes

. . . For our purposes here, the critical fact about the Board's acute care hospital unit rule is that by its express terms it does not apply to this case or to nursing homes generally, and no party contends otherwise. . . . The rule also expressly provides that "[t]he Board will determine appropriate units in other health care facilities . . . by adjudication." . . . [W]e have decided to overrule *Park Manor* and to apply our traditional community of interest standards in this case and others like it. . . .

The traditional community-of-interest test is intended, as the Act requires, to assure employees the "fullest freedom in exercising the rights guaranteed by th[e] Act," rather than to satisfy an abstract notion of the most appropriate unit, and is thus pragmatic. In addition, it has always been informed by empirical knowledge acquired by the Board about the industry and workplace at issue.

Our determination of whether a proposed unit is an appropriate unit must be guided by the principles of unit determination drawn from the language of the statute . . . The existing presumptions are thus consistent with the statutory requirement that the proposed unit need only be an appropriate unit.

As the Supreme Court has recognized, Section 9(a), "read in light of the policy of the Act, implies that the initiative in selecting an appropriate unit resides with the employees." . . . The Board has construed that statutory first step in the representation case process to permit the petitioner to describe the unit within which "a substantial number of employees . . . wish to be represented." Procedurally, the Board examines the petitioned-for unit first. If that unit is an appropriate unit, the Board proceeds no further. As the Board recently explained, "the Board looks first to the unit sought by the petitioner, and if it is an appropriate unit, the Board's inquiry ends." Here, of course, the employees have proposed a unit consisting of a set of employees who are clearly identifiable as a group: all employees in the CNA classification.

The Act further declares in Section 9(b) that "[t]he Board shall decide in each case whether, in order to assure to employees the fullest freedom in exercising the rights guaranteed by this Act, the unit appropriate for the purposes of collective bargaining shall be the employer unit, craft unit, plant unit, or subdivision thereof." . . . The Board has historically honored this statutory command by holding that the petitioner's desire concerning the unit "is always a relevant consideration." . . . We thus consider the employees' wishes, as expressed in the petition, a factor, although not a determinative factor here.

We proceed, then, to determine if the employees' proposed unit consisting of all CNAs is "a unit" appropriate for the purposes of collective bargaining under Section 9(a). . . . Again, the Supreme Court has recognized that the language of Section 9(a) "suggests that employees may seek to organize 'a unit' that is 'appropriate'—not necessarily *the* single most appropriate unit."

In making the determination of whether the proposed unit is an appropriate unit, the Board's "focus

---
*Available at lawyersusaonline.com/wp-files/pdfs-5/specialty-healthcare-and-rehabilitation-center-of-mobile.pdf.

is on whether the employees share a 'community of interest.'" . . . In determining whether employees in a proposed unit share a community of interest, the Board examines: [W]hether the employees are organized into a separate department; have distinct skills and training; have distinct job functions and perform distinct work, including inquiry into the amount and type of job overlap between classifications; are functionally integrated with the Employer's other employees; have frequent contact with other employees; interchange with other employees; have distinct terms and conditions of employment; and are separately supervised.

Here, employees in the proposed unit clearly (and undisputedly) share a community of interest. The Regional Director so concluded based on the CNAs' "[d]istinct training, certification, supervision, uniforms, pay rates, work assignments, shifts, and work areas." The CNAs, of course, all occupy the same job classification. The CNAs in the Employer's nursing department are unlike all the other employees the Employer would include in the unit. Thus, they wear distinctive nursing uniforms unlike all the other employees, most of whom wear no uniform at all. Because they are in the nursing department, the CNAs' immediate and intermediate supervision (by LPNs and RNs) is separate and distinct from all other employees. The primary duty of the CNAs, unlike all the other employees, is the direct, hands-on care of facility residents. As a consequence, CNAs at this facility and nationwide experience unique risks and are subject to unique requirements. Only CNAs are routinely exposed to blood and other bodily fluids. Only CNAs routinely perform the physically demanding tasks of assisting residents with repositioning and ambulation. There is no evidence of significant functional interchange or overlapping job duties. Finally, the Regional Director correctly found "no evidence" of transfers into the CNA position from the other job classifications and only one such transfer out of the CNA position.

Applying traditional community of interest factors to these facts, we have little difficulty in concluding that the petitioned-for unit is an appropriate unit.

. . . Because a proposed unit need only be an appropriate unit and need not be the only or the most appropriate unit, it follows inescapably that demonstrating that another unit containing the employees in the proposed unit plus others is appropriate, or even that it is more appropriate, is not sufficient to demonstrate that the proposed unit is inappropriate. . . . "[I]t is not enough for the employer to suggest a more appropriate unit; it must 'show that the Board's unit is clearly inappropriate.'" . . . The fact that a proposed unit is small is not alone a relevant consideration, much less a sufficient ground for finding a unit in which employees share a community of interest nevertheless inappropriate. A cohesive unit—one relatively free of conflicts of interest—serves the Act's purpose of effective collective bargaining, and prevents a minority interest group from being submerged in an overly large unit. . . .

We therefore take this opportunity to make clear that, when employees or a labor organization petition for an election in a unit of employees who are readily identifiable as a group (based on job classifications, departments, functions, work locations, skills, or similar factors), and the Board finds that the employees in the group share a community of interest after considering the traditional criteria, the Board will find the petitioned-for unit to be an appropriate unit, despite a contention that employees in the unit could be placed in a larger unit which would also be appropriate or even more appropriate, unless the party so contending demonstrates that employees in the larger unit share an overwhelming community of interest with those in the petitioned-for unit.

We set out a clear test—using a formulation drawn from Board precedent and endorsed by the District of Columbia Circuit—for those cases in which an employer contends that a proposed bargaining unit is inappropriate because it excludes certain employees. In such cases, the employer must show that the excluded employees share an "overwhelming community of interest" with the petitioned-for employees.*

Affirmed, in favor of the Petitioning Employees.

    The appropriateness of the proposed bargaining unit is the first issue that the staff of the regional office determines when it receives a petition for a representation election. Once that issue has been resolved, employer and union representatives try to reach an agreement on such matters as the time and place of the election, standards for eligibility to vote, rules of conduct during the election, and the means for handling challenges to the outcome of the election. If the parties cannot reach an agreement, the NLRB regional director determines these matters and orders an election.

    If the union obtains signed authorization cards from more than 50 percent of the appropriate employee unit, it may ask the employer to recognize the union on the basis of this showing of majority support alone. Realizing that it is futile to try to prevent the union from representing its employees, the employer may decide that it would ultimately be beneficial to recognize the union and begin the bargaining process on an amicable note. Such behavior is risky, however,

---

*Specialty Healthcare and Rehabilitation Center of Mobile and United Steelworkers, District 9, Petitioner. National Labor Relations Board, Case 15–RC–8773 (August 26, 2011).

because it may constitute a violation of Section 8(a)(1), which prohibits employers from interfering with employees' Section 7 right of free choice. In other cases, the employer may wish to avoid the risk of violating Section 8(a)(2), which prohibits employer-dominated unions and may, therefore, request that the union file a petition for certification. Having a board-run election to ensure that there indeed is majority support protects the employer.

If a union receives a majority of the votes and the election results are not challenged, the board will certify the union as the exclusive bargaining representative of the employees of that unit. If two or more unions are seeking to represent employees and neither of the unions or "no union" receives a majority of the votes, there will be a runoff election between the choices that got the greatest and second-greatest number of votes. Once a valid representation election has been held and there has been either a certification of a representative union or a majority vote for no union, there cannot be another election for one year. Nor can there be an election during the term of a collective bargaining agreement, unless either the union is defunct or there is such a division in the ranks of the union that it is unable or unwilling to represent the employees.

The Collective Bargaining Process

Shortly after a union has been *certified*, or recognized, the collective bargaining process begins. Both the employer and the bargaining-unit representative are required by the NLRA to bargain collectively in good faith with respect to wages, hours, and other terms and conditions of employment. Note that the requirement is only to bargain in good faith, not to reach an agreement. The board has no power to order the parties to accept any contract provision; it can only order them to bargain.

To a great extent, **good-faith bargaining** is defined procedurally. Under Section 8(d), the parties must (1) meet at reasonable times and confer in good faith; (2) sign a written agreement if one is reached; (3) when intent on terminating or modifying an existing contract, give 60 days' notice to the other party with an offer to confer over proposals and give 30 days' notice to the federal or state mediation services in the event of a pending dispute over the new agreement; and (4) neither strike nor engage in a lockout during the 60-day notice period.

good-faith bargaining Following procedural standards laid out in Section 8 of the NLRA; failure to bargain in good faith, by either the employer or the union, is an unfair labor practice.

Failure of the employer to bargain in good faith is an unfair labor practice under Section 8(a)(5). Employers violate this section not only by disregarding proper procedural standards but also by assuming a take-it-or-leave-it attitude. Hence, if the employer takes a position and says that it will alter the position only if new information shows that its proposal contains incorrect assumptions, this is not good-faith bargaining. Employers who refuse to provide the union with relevant information that it requests and needs to represent the employees in the bargaining process responsibly are also engaging in an unfair labor practice. Relevant information includes job descriptions, time-study data, financial data supporting a company claim that it is unable to meet union demands, and competitive wage data to support a company claim that the union is demanding noncompetitive wage rates.

Taking unilateral action on a matter subject to bargaining is also an unfair employer labor practice under Section 8(a)(5). One example is giving employees a raise or additional benefits during the term of a collective bargaining agreement without first consulting the union. This behavior would have the effect of undermining the union as a bargaining representative and, thus, would be unlawful.

Because bargaining is meant to secure benefits for employees, there are fewer cases of union refusals to bargain, but failure of a union to bargain in good faith is a violation of Section 8(b)(3). Thus, unions may not violate any of the procedural requirements already delineated, nor may they refuse to sign a contract after an agreement has been reached or insist on bargaining for clauses that fall outside the scope of mandatory bargaining.

SUBJECTS OF BARGAINING

All subjects of bargaining are either mandatory or permissive. Many people mistakenly assume that wages are the only real issue that unions bargain over. As Table 20-6 illustrates, unions have been successful in negotiating higher wages for workers in most professions. The scope of bargaining items, however, is more expansive than merely wages. **Mandatory subjects of collective bargaining** are those subjects over which the parties *must* bargain: rates of pay, wages, hours of employment, and other terms and conditions of employment. Failure to bargain over these subjects constitutes an unfair labor practice. All other bargaining subjects are **permissive** and need not be bargained over. Management decisions concerning the commitment of capital and the basic scope of the enterprise, for instance, are not primarily about conditions of employment and, thus, are not mandatory. Inclusion of a permissive subject in the bargaining process in one year, even if that results in its inclusion in a collective bargaining agreement, does not make that subject an issue of mandatory bargaining in any future contract. The only subjects that cannot be included in the bargaining process are illegal terms, such as a contract clause that would require unlawful discrimination by the employer.

Unions have traditionally tried to expand the scope of mandatory items. Mandatory items concerning wages include piece rates, shift differentials, incentives, severance pay, holiday pay, vacation pay, profit sharing, stock option plans, and hours (including overtime provisions). Mandatory items concerning conditions of employment include layoff and recall provisions, seniority systems, promotion policies, no-strike and no-lockout clauses, grievance procedures, and work rules. In a case that generated a great deal of interest, the U.S. Supreme Court held that even changes in prices offered in a company cafeteria were subject to mandatory bargaining.[24]

mandatory subjects of collective bargaining Subjects over which the parties must bargain, including rates of pay, wages, hours of employment, and other terms and conditions of employment.

permissive subjects of collective bargaining Subjects that are not primarily about conditions of employment and, therefore, need not be bargained over.

Strikes, Boycotts, and Picketing

In 2012, there were 19 major strikes and lockouts involving 1,000 or more workers and lasting at least one shift.[25] Major work stoppages may occur through worker-initiated strikes or employer-initiated lockouts. These major work stoppages idled 148,000 workers and led to 1.13 million idle workdays.[26] The prudent businessperson needs to understand the NLRA as it applies not only to labor organizing and collective bargaining but also to three other common occurrences in the labor–management relations context: strikes, picketing, and boycotts.

TABLE 20-6 UNION AND NONUNION WAGES BY OCCUPATION, 2011

Occupational Group	*Union*	*Nonunion*
All workers	$23.04	$19.84
Management, professional, and related	$32.95	$35.70
Natural resources, construction, and maintenance	$29.69	$18.71
Production, transportation, and material moving	$21.78	$14.40
Sales and office	$16.60	$15.98
Service	$16.17	$10.16

Source: Union and Nonunion Wages, December 2011, Bureau of Labor Statistics, www.bls.gov/opub/ted/2013/ted_20130513.htm.

---

[24] *Ford Motor Co. v. National Labor Relations Board*, 441 U.S. 488 (1979).

[25] *Major Work Stoppages Summary*, www.bls.gov/news.release/wkstp.nr0.htm.

[26] *Id.*

COMPARATIVE LAW CORNER

Unions in Sweden

In the United States, we generally consider labor unions to be something of the past. The United States' big push to unionize came in the early 1900s, in response to worker abuse and health and safety issues. Some people see unions as unnecessary today, given the laws that protect workers' rights. Now, 12.3 percent of U.S. workers are unionized.[a] The U.S. view of unions, however, is not prevalent around the world. Sweden has very high union membership; approximately 80 percent of the workforce is unionized.

Swedish trade unions are very powerful in collective bargaining. For example, Sweden does not have a national minimum wage, but their workers are well compensated and get regular pay increases to account for inflation. Their trade unions, through collective bargaining, set a guiding wage for different business sectors. Although the company-level contracts have primacy, most workers do not have to settle for a wage below the guiding wage, because they know they can get a higher wage at another company.

Despite the high union concentration, Sweden has one of the lowest strike rates in Europe, about 10 strikes of any kind per year, whereas the United States records about 20 "major stoppages" a year. Perhaps because unions are such a force in Sweden, there is a much bigger push for cooperation between unions and business during collective bargaining.

Labor unions in Sweden are set up a little differently from those in the United States. Swedish unions do not have minimum membership requirements or registration requirements. Consequently, there are fewer barriers to union membership. A union is simply a group of workers who wish to associate to protect workers' rights. They must create bylaws and a board of directors to carry out their aim of protecting workers' rights. The few legal requirements regarding unions make it much easier for Swedes to join unions. There is also less antipathy toward workers' unionizing in Sweden compared with the United States. Unions are so widespread in Sweden that it is not unusual or unexpected for workers to want to join a union.

[a]*Union Members Summary*, www.bls.gov/news.release/union2.nr0.htm.

STRIKES

A **strike**, simply defined, is a temporary, concerted withdrawal of labor. It is the ultimate weapon used by employees to secure recognition or to gain favorable terms in the collective bargaining process because it can be so costly to employers. For example, beginning in 2007 and into 2008, television and film writers went on a 100-day strike in an attempt to receive pay raises. The strike caused a loss of approximately $250 billion to the Los Angeles economy.

Not all strikes, however, are legal, and employees engaging in certain types of legal strikes may still lose their jobs as a consequence of striking. The type of strike that one is engaged in is determined both by the purpose of the strike and by the methods used by the strikers.

Lawful Strikes. A lawful **economic strike** is a nonviolent work stoppage for the purpose of obtaining better terms and conditions of employment under a collective bargaining agreement. Because this type of strike is a protected activity, strikers are entitled to return to their jobs once the strike is over. Employers are, however, allowed to fill economic strikers' jobs while the strike is taking place, and if permanent replacements are hired, strikers are not entitled to return to their jobs. This ability of the employer to replace economic strikers permanently tends to make the strike a less potent weapon than it first appears.

Because of the ability of employers to permanently replace workers engaged in an economic strike if they first hire permanent replacement workers, unions have fought to try to get Congress to pass legislation prohibiting the use of permanent replacement workers. Workers did achieve a minor victory in 1995, when President Clinton issued an executive order prohibiting federal contractors

strike A temporary, concerted withdrawal of labor.

economic strike A nonviolent work stoppage for the purpose of obtaining better terms and conditions of employment under a collective bargaining agreement.

with government contracts worth over $100,000 from hiring permanent replacements for strikers. If any such firm does hire replacement workers, the labor secretary is to notify the head of any agencies that have contracts with such firms, and the contracts are to be terminated and no contracts are to be made with said firms in the future.

Even if replaced, however, economic strikers still are entitled to vote in representational elections at their former place of employment within one year of their replacement or until they find "regular and substantially similar employment" elsewhere, whichever comes first. Any permanently replaced economic striker is also entitled to be rehired when job vacancies arise at the former place of employment.

Employees may also lawfully engage in a strike over employer unfair labor practices. An **unfair labor practice strike** is a nonviolent work stoppage in protest against an employer's committing an unfair labor practice. Employees engaged in such a strike are entitled to return to their jobs at the end of the strike. If the employer fires such strikers, they will be able to sue for reinstatement and back pay for the time during which they were unlawfully prohibited from working.

unfair labor practice strike A nonviolent work stoppage for the purpose of protesting an employer's commission of an unfair labor practice.

APPLYING THE LAW TO THE FACTS . . .

At the factory where Amy worked, a number of employees were called "managers" but were not salaried and did not receive overtime. She and her coworkers believed that the company was misclassifying these workers to avoid paying required overtime to them, so they went on strike. The employer hired permanent workers to replace them. Amy believed that because the workers were on strike due to unfair labor practices, she was entitled to get her job back. Her employer stated that the company was legally entitled to hire replacement workers and turn them into permanent employees. Who is right in this instance?

Unlawful Strikes. Strikes are unlawful when either their means or their purpose is unlawful. Strikes with unlawful means include (1) sit-down strikes, wherein employees remain on the job but cease working; (2) partial strikes, wherein only some of the workers leave their jobs; and (3) wildcat strikes, which are not authorized by the parent union and are frequently in violation of the collective bargaining agreement. Strikes that include acts of violence or blockading of exits or entrances of a plant are also strikes with unlawful means.

Strikes with an unlawful purpose include *jurisdictional strikes*, which are work stoppages for the purpose of forcing an employer to resolve a dispute between two unions. Jurisdictional strikes are most common in the construction industry, where two unions often disagree over which trade (and, thus, which union's members) is entitled to do a particular type of work on a given project.

Also unlawful are a number of strikes that fall into the category of secondary strikes. A *secondary strike* occurs when the unionized workers of one employer go on strike to force their employer to exert pressure on another employer with which the union has a dispute.

When employees engage in a strike with unlawful means or an unlawful purpose, they are not legally protected and, therefore, may be discharged by their employer. For example, in 2003, the NLRB decided that a strike by an employer's security guards was not protected because the union failed to take reasonable precautions to protect the employer's operations from foreseeable dangers resulting from security guards' work stoppage. Because the strike was not protected, the employer's termination of the security guards was lawful.[27]

---

[27]NLRB, *Sixty-Eighth Annual Report of the NLRB for the Fiscal Year Ended September 30, 2003*, 23. Available at www.nlrb.gov/sites/default/files/attachments/basic-page/node-1677/nlrb2009.pdf.

BOYCOTTS

A **boycott** is a refusal to deal with, purchase goods from, or work for a business. Like a strike, it is a means used to prohibit a company from carrying on its business so that it will accede to union demands. Primary boycotts are legal; that is, a union may boycott an employer with which the union is directly engaged in a labor dispute.

Secondary boycotts, like secondary strikes, are illegal. A **secondary boycott** occurs when unionized employees who have a labor dispute with their employer boycott another employer to force it to cease doing business with their employer.

One type of secondary boycott is legal under the NLRA in the construction and garment industries. That is the "hot cargo agreement," an agreement between the union and the employer that union members need not handle nonunion goods and that the employer will not deal with nonunion employers. This type of secondary boycott, however, is considered an unfair labor practice under Section 8(b) in all other industries.

boycott A refusal to deal with, purchase goods from, or work for a business.

secondary boycott A boycott against one employer to force it to cease doing business with another employer with which the union has a dispute.

PICKETING

Picketing is the stationing of individuals outside an employer's place of business for the purpose of informing passersby of the facts of a labor dispute. Picketing usually accompanies a strike, but it may occur alone, especially when employees want to continue to work to draw a paycheck.

Just as there are numerous types of strikes and boycotts, there are multiple types of picketing with different degrees of protection. **Informational picketing**, designed to truthfully inform the public of a labor dispute between an employer and employees, is protected. This protection may be lost, however, if the picketing has the effect of stopping deliveries and services to the employer. Picketing designed to secure a stoppage of service to the employer is called *signal picketing* and is not protected.

Jurisdictional picketing, like jurisdictional strikes, occurs when two unions are in dispute over which union's workers are entitled to do a particular job. If one union pickets because work was assigned to the other union's members, this action is unlawful; jurisdictional disputes are resolved by the NLRB under an expedited procedure, so there is no need to take coercive action against the employer. The other union or the employer may secure an injunction to preserve the status quo until the board resolves the dispute.

Organizational or recognitional picketing, which is designed to force the employer to recognize and bargain with an uncertified union, is illegal when (1) another union has already been recognized as the exclusive representative of the employees and the employer and employees are operating under a valid collective bargaining agreement negotiated by that union, (2) there has been a valid representation election within the past year, or (3) the union has been picketing for longer than 30 days without filing a petition for a representation election.

When (1) and (2) are not applicable, a union may picket up to 30 days while attempting to secure signed authorization cards from more than 30 percent of the employees. Once the appropriate signatures have been obtained, the union may then file its petition for recognition and continue picketing to inform the public of the company's refusal to recognize the union.

Picketing, like striking, may be illegal because of its means. Violent picketing is, of course, unlawful. So is *massed picketing*, though this type of unlawful activity is somewhat more difficult to define. It is said to exist when the pickets are so massed (physically arranged) as to be coercive or to block entrances and exits. In cases interpreting this term, the courts have tended to find unlawful massed picketing when there have been so many picketers before a gateway to a plant that free entry or exit is made difficult or almost impossible.

picketing The stationing of individuals outside an employer's place of business to inform passersby of the fact of a labor dispute.

informational picketing Picketing designed to truthfully inform the public of a labor dispute.

jurisdictional picketing Picketing by one union to protest an assignment of jobs to another union's members.

organizational (recognitional) picketing Picketing designed to force the employer to recognize and bargain with an uncertified union.

Global Dimensions of Labor–Management Relations

Many U.S. corporations have moved their operations overseas to obtain cheaper labor costs. Our laws permit this, and a corporation's foreign operations are not subject to U.S. labor laws; however, most countries have labor laws of their own to which U.S. companies are subject. Likewise, foreign companies with plants in the United States must comply with our labor laws.

Many scholars argue that it would be desirable to have uniform labor laws across the world, or at least among all the industrialized nations. Their reasons differ. Some believe that certain inherent rights of workers should be protected regardless of where they live and work. Others believe that uniformity would make it easier for multinationals that have a presence in many countries because then their labor practices could be uniform throughout their operations.

The likelihood of a worldwide uniform labor law is almost nonexistent, because different countries' leaders have very different philosophies about the purpose of labor law. There are a few similarities, however, among most countries' labor laws. For example, in almost all countries, a worker has the right to refuse to perform unsafe work.[28]

The International Labor Organization (ILO), to which 174 nations now send government, management, and labor delegates, has attempted to create some uniformity among the labor laws of member nations. The ILO formulates conventions and recommendations for labor legislation that can be adopted by all countries. These include minimum specifications and often provisions for national or traditional variations on those basics. Enforcement procedures are generally left to the individual nations.[29] The ILO has promulgated 182 conventions and 190 recommendations; there have been over 6,600 ratifications of these.[30] Although there are still more differences than similarities in labor laws from country to country, the ILO has stimulated some harmonization of those laws.

SUMMARY

The NLRA is the primary piece of legislation governing labor–management relations. Section 7 of the NLRA sets forth the rights of employees, and Section 8(a) identifies specific employer behaviors, called unfair labor practices, that are prohibited. Employee unfair labor practices are set out in Section 8(b). The Landrum–Griffin Act was passed to ensure proper internal governance of labor organizations.

The administrative agency responsible for oversight and enforcement of the NLRA is the NLRB. The board is primarily responsible for ensuring that organizing campaigns are conducted fairly and that neither employers nor employees commit unfair labor practices.

Prospects for a worldwide uniform labor law are negligible, but the ILO is attempting to create some harmony among the labor laws of member nations.

REVIEW QUESTIONS

20-1 Explain why some people support unions whereas others oppose them.

20-2 Explain the relationship between Section 7 and Section 8 of the NLRA.

20-3 Explain why someone would argue that the existence of Section 8(a)(1) makes Sections 8(a)2–5 unnecessary.

---

[28] M. Lennards, "The Right to Refuse Unsafe Work," *Comparative Labor Law Journal* 4: 217 (1981).
[29] International Labor Organization (ILO), *The ILO: What It Is, What It Does* (Feb. 10, 2000).
[30] *Id.*

20-4 Describe the roles of the NLRA and its general counsel.

20-5 What is the difference between violating a board rule and committing an unfair labor practice?

20-6 What advice would you give an employer that wants to adopt some form of employee participation program but is concerned about the legality of such programs?

REVIEW PROBLEMS

20-7 A national union wanted to organize the employees of Dexter Thread Mills. The company parking lot was adjacent to a public highway, separated from the highway by a 10-foot-wide grassy public easement. The union sought to distribute handbills in the parking lot; the company sought to exclude the union from the lot. Was the union allowed to distribute the handbills on the company lot? Why or why not?

20-8 Otis Elevator was acquired by United Technologies (UT) in 1975. A review of Otis's operations showed its technology to be outdated. The company's products were poorly engineered and were losing money. Its production and research facilities were scattered across the United States with many duplications of work. Research, in particular, was done at two New Jersey facilities, one of which was outdated. UT did all of its research at a major research and development center in Connecticut; some research for Otis was also done there. UT decided to transfer Otis research from the two New Jersey locations to an expanded facility in Connecticut to strengthen the overall research effort and to allow Otis to redesign its product. The union representing Otis's employees alleged that UT had engaged in an unfair labor practice by refusing to bargain with the union over its decision to relocate the work. Was UT's refusal to bargain over this decision a violation of Section 8(a)(5)? Explain.

20-9 The clerks at Raley's were represented by the Independent Drug Clerks Association (IDCA). When the Retail Clerks Union (Retail Clerks) began a campaign to oust IDCA, Raley's maintained a neutral posture. Retail Clerks picketed the store in July. Could Raley's obtain an injunction prohibiting the picketing? Explain.

20-10 The workers at the Big R Restaurant were trying unsuccessfully to negotiate a new contract. On Monday, all of the bussers called in sick. On Tuesday, all the cooks called in sick. On Wednesday, all the servers called in sick. Was there anything unlawful about the employees' behavior? If so, what recourse does the employer have?

20-11 Ajax Manufacturing Company was unionized in 1998 and was operating under a contract negotiated at the beginning of that year. Several new workers were hired near the end of the year, and they thought a stronger union was needed. They began picketing the employer to recognize a different union as the representative of the workers, or at least to have a decertification election. Is their picketing legal? Why or why not?

CASE PROBLEMS

20-12 Costco, a large retailer similar to Wal-Mart, placed a controversial statement in Costco's employee handbook. According to the handbook, employees were not to make disparaging comments about other coworkers or the company on any sort of social media. A union representing the employees brought suit against Costco. The union alleged that Costco's prohibitions on social media use constituted a violation of § 8(a)(1) of the NLRA. The court thought that by prohibiting disparaging comments about the company online, employees might believe they would also not be allowed to discuss any sort of union. What concept is Costco omitting from its business practices in this case? How do you think the court ruled? Why? *Costco v. United Food and Commercial Workers, No. 34-CA-012421* (National Labor Relations Board, 2012).

20-13 Several bargaining unit employees of American Crystal Sugar's (ACS) facilities were participating in contract negotiations with ACS for a successor agreement. ACS made a final offer, which was rejected by the employees' union. ACS then locked out its bargaining-unit employees and transitioned to using replacement workers. When the unit employees applied for unemployment compensation, it was determined that the employees did not qualify for compensation because they were "unemployed due to a labor dispute." The unit employees appealed the decision, but the benefits denial was affirmed. The unit employees then requested that Job Service review the denial. This request was denied. Finally, the employees petitioned the district court for review of the benefits denial. Do you think the district court considered the lockout a "labor dispute" and thus affirmed the benefits denial? Why or why not? *Olson v. Job Serv. N.D.*, 826 N.W.2d 36 (2013).

20-14 The union was attempting to organize the Aladdin Hotel and Casino in Las Vegas. One afternoon when a human resources manager overheard an organizer asking hotel workers to sign union cards during the lunch break in the employee dining room, the manager interrupted the conversation and told the workers that they should make sure they understood all the facts before signing the cards. A discussion about unionization of approximately eight minutes' duration then took place. The union filed a complaint with the NLRB alleging that management was engaging in illegal surveillance in violation of the NLRA. The ALJ ruled in favor of the union, and the company appealed. What do you believe the court held on appeal? Why? *Local Joint Executive Board of Las Vegas v. NLRB*, 515 F.3d 942 2008 WL 216935 (9th Cir. 2008).

20-15 Sacred Heart is an acute care hospital in Spokane, Washington. The Washington State Nurses Association (WSNA) is a union that represents approximately 1,200 registered nurses employed at Sacred Heart. In the fall of 2003, WSNA and Sacred Heart began negotiations for a new collective bargaining agreement (CBA) to replace the then-existing agreement, set to expire in January 2004.

During the negotiations, nurses at Sacred Heart wore a number of union buttons without incident. The buttons read "Together Everyone Achieves More," "WSNA SHMC RNs Remember 98," "Staffing Crisis-Nursing Shortage-Medical Errors-Real Solutions," and "RNs Demand Safe Staffing." Several months after the nurses began wearing these buttons, the hospital issued a memorandum banning the nurses from wearing the "RNs Demand Safe Staffing" buttons. The hospital stated that it was concerned about the potential for patients to interpret buttons incorrectly and fear that they would receive inadequate treatment. The hospital claimed that its ban was designed to minimize or eliminate the negative impact on patients.

Shortly after the memorandum was issued, WSNA filed an unfair labor practice charge with the NLRB. There was an evidentiary hearing, and the ALJ issued a decision concluding that Sacred Heart had engaged in an unfair labor practice by enforcing the button prohibition. Shortly thereafter, a divided three-member panel of the Board reversed, finding that Sacred Heart had demonstrated that the message would disturb patients. Given your understanding of the NLRA, did Sacred Heart commit an unfair labor practice? Why or why not? *Washington State Nurses Association v. NLRB*, 526 F.3d 577 (8th Cir. 2008).

20-16 The International Association of Firefighters Local Chapter 230 is a union for the benefit of public service officials such as firefighters and police officers in the City of San Jose. Local 230 ran a pension benefits program for retired public service employees. The union was attempting to bargain with the employer of the public service employees, the City of San Jose, and had two proposals regarding the Local 230 pension package. The first proposal involved the makeup of the board in charge of the pension program. Local 230 wished to have an equal number of firefighters and policemen on the board. The second proposal consisted of planning how to track certain characteristics of program participants to determine their eligibility for the program. The City of San Jose refused to bargain with Local 230, arguing that the two proposals did not fall under mandatory subjects of collective bargaining. A new city ordinance had already placed equal numbers of firefighters and policemen on the pension board. Does the second proposal fall under permissive subjects of collective bargaining? Why or why not? *City of San Jose v. International Association of Firefighters Local 230*, No. CV075858 (Cal. App. 6th Dist. 2013), unpublished.

20-17 The Mahoning County Board of Developmental Disabilities (MCBDD), an employer, and MCBDD employees held a collective bargaining agreement that contained a grievance arbitration procedure. In 2007, the employee union filed a notice to begin negotiations for a new contract. When MCBDD later held a board meeting, the union members were found to be picketing outside the building where the meeting was being held, holding signs that related to the contract negotiations. MCBDD filed an unfair labor practice charge alleging that the union violated the 10-day notice requirement for picketing. The State Employee Relations Board found probable cause that the union did commit an unfair labor practice charge. The union appealed this decision, asserting that the notice requirement is a "restraint of its right to picket and is therefore unconstitutional." How do you think the court of appeals ruled? Why? *Mahoning Educ. Ass'n. of Developmental Disabilities v. State Empl. Rels. Bd.*, 137 Ohio St. 3d 257 (2013).

20-18 Egg Harbor Township authorized the construction of a community center, which was controlled by the Township's Project Labor Agreement (PLA). All contractors working on this community center project were required to become signatories to the PLA. Sambe was assigned as general contractor and, thus, was a signatory to the PLA. Sambe then subcontracted the project's roofing work to Donnelly. Later, a dispute arose when Donnelly selected several carpenters who were not signatories to the PLA to work on the roofing. Another signatory to the PLA, "Sheet Metal," opposed the hiring of the carpenters and informed Donnelly that the carpenters could not complete the project. The carpenters then threatened to picket if Donnelly reassigned the roofing work to Sheet Metal.

Sheet Metal attempted to resolve the dispute by pushing for an arbitration hearing. However, the carpenters continued to assert that they would picket the project if the work was reassigned. Donnelly then filed an unfair labor practice charge, alleging that the carpenters violated the NLRA by threatening to picket. The NLRB ended up awarding the disputed roofing work to the carpenters. What do you think the Board's reasoning was that led to this conclusion? *Sheet Metal Workers Int'l Ass'n Local Union No. 27 v. E.P. Donnelly, Inc.*, 737 F.3d 879 (3rd Cir. 2013).

THINKING CRITICALLY ABOUT RELEVANT LEGAL ISSUES

Despite the claim that a strike is the "ultimate weapon" for workers, strikes are never in the interest of workers. Strikes are an outdated tool that cannot work now and were never all that effective. It does not make sense for workers to rush to an ineffective extreme and go on strike. If workers stopped to think about their interests, they would realize that strikes do not help them achieve their goals.

The point of engaging in a strike is for workers to put pressure on their employers to get an economic benefit from their actions. If the workers are on strike because they want more money, how can deliberately missing work help them reach their goals? If the workers' claim is that they need more money, missing work, and thus losing out on pay, does not seem like it will accomplish the workers' goal. Even if the workers manage to get a raise by striking, the raise would have to account for what the workers "need" in addition to making up for the lost wages during the strike. It seems very unlikely that any strike will be effective enough to make up for lost wages during the strike. Accordingly, the strike will not serve the interest of the workers. If the strike is for some other sort of benefits, the gained benefits would still have to offset the lost wages to be rational, and once again, the likelihood of the benefits accounting for the lost work in addition to the benefits originally sought seems highly unlikely.

Not only is it illogical to strike to get money or benefits, but strikes are also very risky for workers. Employers can temporarily replace workers during strikes. With a substitute workforce, the employer will not necessarily feel pressure from the strike. What replacements mean is that the workers on strike do not get paid and the work still gets done, so the employer does not have incentive to give in to the striking workers' demands. In addition, the employer can simply hire permanent replacements. Permanent replacements mean that after the strike, the striking workers are not entitled to get their old jobs back. Clearly, it is not in the interest of workers to strike for economic reasons and end up losing their jobs because of it. The risk is too high to make strikes in the interest of the workers. Even if the employer negotiates with the workers because of the strike, the whole process is likely to foster resentment, which will not create a good working environment for anyone, further making strikes not in the interest of workers.

Another reason strikes are not in the interest of workers is that strikes are likely to breed contempt for the workers, as opposed to creating sympathy. Strikes do not affect just the workers and employer. Rather, others are harmed by strikes. For example, in 2005, when New York City transit workers went on strike, the masses of people who rely on public transportation in New York City were forced to walk, take taxis, or find other means of transport about the island. In addition, other workers who rely on the products or services produced by the striking workers have their work affected. As the other workers who are not involved in the strike feel the effects of the strike, they are more likely to resent the striking workers than they are to put pressure on the strikers' employer to end the strike. Being despised is typically not in anyone's interest.

1. How would you word the issue and conclusion of this essay?

2. Is relevant information missing from the argument? Explain.

   Clue: What would you like to know before deciding whether the author is correct?

3. Does the argument contain significant ambiguity?

   Clue: What words or phrases could have multiple meanings, where changing the meaning either strengthens or weakens the argument?

4. Write an essay that someone who holds an opinion opposite that of the essay's author might write.

   Clue: What other ethical norms could influence an opinion about this issue?

ASSIGNMENT ON THE INTERNET

You have learned about many laws and regulations that govern labor–management relations. Many of those laws and regulations are used in the NLRB's decisions. Visit the NLRB's website for recent decisions www.nlrb.gov and read at least two decisions. Then answer the questions in the next paragraph.

What information from this chapter helps you better understand the decisions? What reasons were given for the decisions? Did the Board have to resolve any ambiguity in the law to decide each of the cases? Finally, what ethical norms support the decisions?

ON THE INTERNET

www.nlrb.gov This site is the home page of the NLRB.

www.aflcio.org Find out about the AFL-CIO at this site.

www.albany.edu/history/LaborAudio This site contains a number of audio recordings, many historical, detailing firsthand accounts of the labor and industrial history in the United States.

ctb.ku.edu/en/ This website contains information on why groups strike, as well as the steps typically taken to organize an effective strike.

www.dol.gov The home page for the Department of Labor provides a wealth of useful information and links. This is a good place to begin research in areas of employment law, especially the Bureau of Labor Statistics.

www.gklaw.com This site contains information regarding a 2006 NLRB decision that attempts to define who counts as a supervisor. In addition, the website contains links to useful information for business managers regarding labor topics.

FOR FUTURE READING

Arthurs, Harry. "Reconciling Differences Differently: Reflections on Labor Law and Worker Voice after Collective Bargaining." *Comparative Labor Law & Policy Journal* 28 (2007): 155.

Brice, Roger, Samuel Fifer, and Gregory Naron. "Social Media in the Workplace: The NLRB Speaks." *Intellectual Property and Technology Law Journal.* 24, no. 10 (2012): 13–17.

DeMaria, Alfred T. "Two Companies Ordered to Bargain with Unions." *Management Report* 38 (2015): 2.

Dray, Phillip. *There Is Power in a Union: The Epic Story of Labor in America.* Doubleday, 2010.

Estlund, Cynthia. "Labor Law Reform Again? Reframing Labor Law as a Regulatory Project." *New York University Journal of Legislation and Public Policy* 16 (2013).

Estreicher, Samuel. "Trade Unionism Under Globalization: The Demise of Voluntarism?" *St. Louis University Law Journal* 54, no. 2 (2010): 415–426.

Fudge, Judy. "Trade Unions, Democracy, and Power." *International Journal of Law in Context* 7, no. 1 (2011): 95–105.

Gould, William B., IV. "Labor Law Beyond U.S. Borders: Does What Happens Outside of America Stay Outside of America?" *Stanford Law and Policy Review* 21 (2011): 401.

Kroncke, Jedidiah J. "Property Rights, Labor Rights and Democratization: Lessons from China and Experimental Authoritarians." *New York University Journal of International Law Politics* 46, no. 1 (2013).

LaJeunesse, Raymond J., Jr. "The Controversial 'Card-Check' Bill, Stalled in the United States Congress, Presents Serious Legal and Policy Issues." *Texas Review of Law and Politics* 14 (2010): 209.

MacDonald, Alexander T. "Permanent Replacements: Organized Labor's Fall, Employment Law's (Incomplete) Rise, and the Way Forward." *Idaho Law Review* 50 (2013): 19–52.

Mitchel, Simon. "Can Collective Bargaining Deliver Decent Work?" *New Zealand Journal of Employment Relations* 39 (2015): 15.

CHAPTER TWENTY ONE

Employment Discrimination

- THE EMPLOYMENT-AT-WILL DOCTRINE
- CONSTITUTIONAL PROVISIONS
- THE CIVIL RIGHTS ACTS OF 1866 AND 1871
- THE EQUAL PAY ACT OF 1963
- THE CIVIL RIGHTS ACT OF 1964, AS AMENDED (TITLE VII), AND THE CIVIL RIGHTS ACT OF 1991
- THE AGE DISCRIMINATION IN EMPLOYMENT ACT OF 1967
- THE REHABILITATION ACT OF 1973
- THE AMERICANS WITH DISABILITIES ACT OF 1991
- AFFIRMATIVE ACTION
- GLOBAL DIMENSIONS OF EMPLOYMENT DISCRIMINATION LEGISLATION

Being an employer was so much easier 100 years ago. Managers could use almost any criteria for hiring, promoting, and firing employees. Today, employers' decision-making powers are restricted by both federal and state laws, many of which are discussed in this chapter.

The right of the employer to terminate an employment relationship was originally governed almost exclusively by the employment-at-will doctrine, discussed in the first section of this chapter. The second section discusses the constitutional provisions that affect an employer's ability to hire and fire workers.

The following six sections discuss each of the major pieces of federal legislation designed to prohibit discrimination in employment; these acts are discussed in the order of their enactment. The ninth section discusses the increasingly controversial subject of affirmative action. Global dimensions of employment discrimination are discussed in the final section.

CRITICAL THINKING ABOUT THE LAW

You will soon be a businessperson and may be responsible for hiring, promoting, and firing people. When you hold this position, you need to be aware of federal and state laws that prohibit discrimination in employment. Why do you think the government has prohibited discrimination in employment? What ethical norm does the government emphasize by prohibiting discrimination in employment? The government seems to emphasize justice, in the sense that it wants all human beings to be treated equally, regardless of class, race, gender, age, and so on. Reading the following case example and answering the critical thinking questions will sharpen your thinking about laws prohibiting employment discrimination.

Tom, Jonathan, and Bob were hired to work as executive secretaries at a major corporation. The other secretaries for the corporation were surprised that three men were hired, because no man had ever before been hired as a secretary at the corporation. All secretaries were required to type 20 five-page reports each day in addition to completing work for their respective departments. After the male secretaries had been working at the corporation for approximately one month, they received pay raises. None of the female secretaries received raises. When the women asked the manager why the male secretaries had received raises, the manager claimed that the men were performing extra duties and consequently received raises.

1. The manager claimed that the men received raises because they were performing extra duties. Can you identify any potential problems in the manager's response?

   Clue: What words or phrases are ambiguous in the manager's response?

2. The female secretaries decided to bring a suit against the corporation. They claimed that they did not receive raises because of their gender. Assume that you are a lawyer and the female secretaries have come to you with their complaint. After talking with the secretaries, you realize that you need some additional information. What additional information might be helpful in this case?

   Clue: The female secretaries claimed that the male secretaries received raises because they are male. Can you think of any alternative reasons why the men might have received raises?

3. You discover only one case regarding equal pay that was decided in your district. In this case, both men and women performed hard labor in a factory, but only men received offers to work during the third shift. Those employees who worked the third shift received an additional $30 per hour. The women in this factory claimed that they were not asked to work the third shift because of their gender. The factory argued that the women who worked at the factory were not physically strong enough to endure the work of the third shift. The court ruled in favor of the women. Do you think that you should use this case as an analogy? Why or why not?

   Clue: How are the two cases similar? How are they different?

The Employment-at-Will Doctrine

employment-at-will doctrine
A contract of employment for an indeterminate term is terminable at will by either the employer or the employee; the traditional American rule governing employer–employee relations.

In all industrial democracies except the United States, workers are protected by law from unjust termination. The traditional "American rule" of employment—the **employment-at-will doctrine**—has been that a contract of employment for an indeterminate term is terminable at will by either party. Thus, an employee who did not have a contract for a specific length of time could be terminated at any time, without notice, for any reason. For example, Melissa Nelson, a dental assistant, was fired from her job because her employer found Ms. Nelson "irresistible" and feared he might try to have an extramarital affair with her.[1]

This doctrine has been justified by the right of the employer to control its property and on the grounds that it is fair because both employer and employee have the equal right to terminate the relationship. Some question the latter justification because the employer usually can replace a terminated employee, whereas it is not equally easy for the employee to find a new job. Thus, the employment-at-will doctrine places the employer in a position to treat employees arbitrarily.

The doctrine has been slowly restricted by state and federal legislation, as well as by changes in the common law. One of the first laws to restrict the employer's right to freely terminate employees was the National Labor Relations Act (discussed in Chapter 20), which has reduced the number of employees covered by the employment-at-will doctrine. This reduction has occurred because the act gives employees the right to enter into collective bargaining agreements, which usually restrict the employer's ability to terminate employees except for "just cause." Employees covered by these agreements are thus no longer "at-will" employees.

---

[1] *Nelson v. James H. Knight DDS, P.C.*, 834 N.W.2d 64, 66 (Iowa 2013).

The doctrine has also been restricted by common-law and state statutory exceptions, which fall into three categories: implied contract, violations of public policy, and implied covenant of good faith and fair dealing. In some states, the courts find that an implied contract may arise from statements made by the employer in advertising the position or including them in an employment manual. For example, sometimes a company provides an employment manual delineating the grounds for termination but not containing any provision for termination "at will." Under such circumstances, if the court finds that the employee reasonably relied on the manual, the court will not apply the employment-at-will doctrine and will allow termination only for the reasons stated in the manual. Thirty-seven states and the District of Columbia recognize this exception.

The **public policy exception** prohibits terminations that contravene established public policy. "Public policy" varies from state to state, but some of the terminations commonly deemed unlawful include dismissals based on actions "in the public interest," such as participation in environmental or consumer protection activities, and dismissals resulting from whistleblowing. Many states have also cut away at the employment-at-will doctrine with laws that specifically prohibit the termination of employees in retaliation for such diverse activities as serving jury duty, performing military service, filing for or testifying at hearings for workers' compensation claims, whistleblowing, and refusing to take lie-detector tests. A total of 43 states accept the public policy exception.

Eleven states recognize the **implied covenant of good faith and fair dealing exception**. This theory holds that every employment contract, even an unwritten one, contains an implicit understanding that the parties will deal fairly with one another. Because there is no clear agreement on what constitutes "fair treatment" of an employee, this theory is not often used.

Many federal laws also restrict the employment-at-will doctrine. Employees cannot be fired for filing a complaint, testifying, or causing a hearing to be instituted regarding the payment of the minimum wage, equal pay, or overtime. Pursuit of a discrimination claim is likewise statutorily protected.

The doctrine of employment-at-will, however, still exists and is strongly adhered to in many states. So although the doctrine is being cut back and business managers of the future therefore cannot rely on its continued availability, it may be a long time before the doctrine is no longer applicable. As its applicability varies from state to state, however, familiarity with the parameters of the doctrine in one's own state is extremely important. Table 21-1 breaks down which states accept each of the three major exceptions.

public policy exception An exception to the employment-at-will doctrine that makes it unlawful to dismiss an employee for taking certain actions in the public interest.

implied covenant of good faith and fair dealing exception An exception to the employment-at-will doctrine, based on the theory that every employment contract, even an unwritten one, contains the implicit understanding that the parties will deal fairly with each other.

TABLE 21-1 EXCEPTIONS TO THE EMPLOYMENT-AT-WILL DOCTRINE

Public Policy Exception	Implied Contract Exception	Good Faith and Fair Dealing Exception
Alaska, Arizona, Arkansas, California, Colorado, Connecticut, Delaware, District of Columbia, Hawaii, Idaho, Illinois, Indiana, Iowa, Kansas, Kentucky, Maryland, Massachusetts, Michigan, Minnesota, Mississippi, Missouri, Montana, Nevada, New Hampshire, New Jersey, New Mexico, North Carolina, North Dakota, Ohio, Oklahoma, Oregon, Pennsylvania, South Carolina, South Dakota, Tennessee, Texas, Utah, Vermont, Virginia, Washington, West Virginia, Wisconsin, Wyoming	Alabama, Alaska, Arizona, Arkansas, California, Colorado, Connecticut, District of Columbia, Hawaii, Idaho, Illinois, Iowa, Kansas, Kentucky, Maine, Maryland, Michigan, Minnesota, Mississippi, Nebraska, Nevada, New Hampshire, New Jersey, New Mexico, New York, North Dakota, Ohio, Oklahoma, Oregon, South Carolina, South Dakota, Tennessee, Utah, Vermont, Washington, West Virginia, Wisconsin, Wyoming	Alabama, Alaska, Arizona, California, Delaware, Idaho, Massachusetts, Montana, Nevada, Utah, Wyoming

TABLE 21-2 FEDERAL STATUTES PROHIBITING DISCRIMINATION IN EMPLOYMENT

Law	Prohibited Conduct	Remedies
Civil Rights Acts of 1866 and 1871, codified as 42 U.S.C. §§ 1981 and 1982	Discrimination on the basis of *race* and *ethnicity*	Compensatory damages, including several years of back pay, punitive damages, attorney's fees, court costs, and court orders
Equal Pay Act of 1963	Wage discrimination on the basis of *sex*	Back pay, liquidated damages equivalent to back pay (if defendant was not acting in good faith), attorney's fees, and court costs
Civil Rights Acts of 1964 (Title VII) and 1991	Discrimination in terms and conditions of employment on the basis of *race, color, religion, sex,* or *national origin*	Back pay for up to two years; remedial seniority; compensatory damages; punitive damages (may be limited due to class); attorney's fees, court costs; and court orders for whatever actions are appropriate including reinstatement and affirmative action
Age Discrimination in Employment Act of 1969	Discrimination in terms and conditions of employment on the basis of *age* when the affected individual is age 40 or older	Back pay, liquidated damages equal to back pay (if defendant acted willfully), attorney's fees, court costs, and appropriate court orders including reinstatement
Rehabilitation Act of 1973	Discrimination by government or governmental contractor on the basis of a *handicap*	Back pay, attorney's fees, court costs, and court orders for appropriate affirmative action
Americans with Disabilities Act of 1991	Discrimination in employment on the basis of a *disability*	Hiring, promotion, reinstatement, back pay, reasonable accommodation, compensatory damages, and punitive damages

Constitutional Provisions

The beginnings of antidiscrimination law can be traced back to three constitutional provisions: the Fifth Amendment, which states that no person may be deprived of life, liberty, or property without due process of law; the Thirteenth Amendment, which abolished slavery; and the Fourteenth Amendment, which granted former slaves all the rights and privileges of citizenship and guaranteed the equal protection of the law to all persons. These provisions alone, however, were not sufficient to prohibit the unequal treatment of citizens on the basis of their race, sex, age, religion, and national origin. Congress needed to enact major legislation to bring about a reduction in discrimination. These laws, referred to as *civil rights laws* and *antidiscrimination laws*, are summarized in Table 21-2 and discussed in detail in the following sections.

The first major civil rights act was passed immediately after the Civil War: the **Civil Rights Act of 1866** (42 U.S.C. Section 1981). This act was designed to effectuate the Thirteenth Amendment and guarantees that all persons in the United States have the same right to make and enforce contracts and have full and equal benefit of the law. The **Civil Rights Act of 1871** (42 U.S.C. Section 1982) prohibited discrimination by state and local governments. Initially used only when there was state action, today these acts are also used against purely private discrimination, especially in employment.

The Civil Rights Acts of 1866 and 1871

APPLICABILITY OF THE ACTS

Initially, the civil rights acts were interpreted very narrowly to prohibit discrimination based only on race. For several years, circuit courts of appeals were split

---

Civil Rights Act of 1866
Statute guaranteeing that all persons in the United States have the same right to make and enforce contracts and have the full and equal benefit of the law.

Civil Rights Act of 1871
Statute that prohibits discrimination by state and local governments.

as to how race is defined. In June 1986, the U.S. Supreme Court resolved that issue by holding that both an Arabic and Jewish individual were protected by the Civil Rights Act of 1866. Justice White, writing the majority opinion in *Saint Francis College et al. v. Majid Ghaidan Al-Khazraji,*[2] said that it was clear from the legislative history that the act was intended to protect from discrimination "identifiable classes of persons who are subjected to intentional discrimination solely because of their ancestry or ethnic characteristics, even if those individuals would be considered part of the Caucasian race today." Thus, today these laws have a broader application.

Remedies. The acts themselves do not have specific provisions for remedies. A wide variety of both legal remedies (money damages) and equitable remedies (court orders) have been awarded under these statutes. The courts are free under these acts to award compensatory damages, that is, damages designed to make the plaintiff "whole" again, which may amount to several years of back pay. The courts may also award punitive damages, an amount intended to penalize the defendant for wrongful conduct. Finally, the courts may require the defendant to pay the plaintiff's attorney's fees.

Procedural Limitation. Unlike most antidiscrimination laws, the Civil Rights Acts of 1866 and 1871 do not require the plaintiff to first attempt to resolve the discrimination problem through administrative procedures. The plaintiff simply files the action in federal district court within the time limit prescribed by the state statute of limitations, requesting a jury trial if one is desired. Often, a claim under the 1866 or 1871 Civil Rights Act will be added to a claim under another antidiscrimination statute.

The Equal Pay Act of 1963

The next major piece of federal legislation to address the problem of discrimination was the **Equal Pay Act of 1963**, an amendment to the Fair Labor Standards Act. Enacted at a time when the average wages of women were less than 60 percent of those of men, the act was designed with a very narrow focus: to prevent wage discrimination based on sex within a business establishment. It was designed primarily to remedy the situations in which women, working alongside men or replacing men, were being paid lower wages for doing substantially the same job.

As stated in 29 U.S.C. Section 206(d)(1), the act prohibits any employer from discriminating within any "establishment"

> *between employees on the basis of sex by paying wages to employees in such establishment at a rate less than the rate at which he pays wages to employees of the opposite sex . . . for equal work on jobs the performance of which requires equal skill, effort, and responsibility, and which are performed under similar working conditions, except where payment is made pursuant to (i) a seniority system; (ii) a merit system; (iii) a system which measures earnings by quantity or quality of production; or (iv) differential based on any factor other than sex.**

Equal Pay Act of 1963
Statute that prohibits wage discrimination based on sex.

In the typical Equal Pay Act case, the burden of proof is initially on the plaintiff to show that the defendant-employer pays unequal wages to men and women for doing equal work at the same establishment. Two questions necessarily arise: What is equal work? What is an establishment?

---
[2] 483 U.S. 1011 (1987). The accompanying case, filed by a Jewish plaintiff, was *Shaare Tefila Congregation et al. v. John William Cobb et al.,* 481 U.S. 615 (1987).
*U.S.C. Section 206(d)(1), United States Constitution.

EQUAL WORK

The courts have interpreted *equal* to mean substantially the same in terms of all four factors listed in the act: skill, effort, responsibility, and working conditions. If the employer varies the actual job duties affecting any one of those factors, there is no violation of the act. For example, if jobs are equal in skill and working conditions but one requires greater effort whereas the other requires greater responsibility, the jobs are not equal. Obviously, a sophisticated employer could easily vary at least one duty and then pay men and women different wages or salaries.

Skill is defined as experience, education, training, and ability required to do the job. *Effort* refers to physical or mental exertion needed for performance of the job. *Responsibility* is measured by the economic and social consequences that would result from a failure of the employee to perform the job duties in question. *Similar working conditions* refers to the safety hazards, physical surroundings, and hours of employment. An employer, however, is entitled to pay a shift premium to employees working different shifts, as long as the employer does not use sex as a basis for determining who is entitled to work the higher-paying shifts.

Extra Duties. Sometimes, employers try to justify pay inequities on the grounds that employees of one sex are given extra duties that justify their extra pay. The courts scrutinize these duties very closely. The duties are sufficient to preclude a finding of equal work only if:

1. the duties are actually performed by those receiving the extra pay;
2. the duties regularly constitute a significant portion of the employee's job;
3. the duties are substantial, as opposed to inconsequential;
4. additional duties of a comparable nature are not imposed on workers of the opposite sex; and
5. the extra duties are commensurate with the pay differential.

In some jurisdictions, the additional duties must also be available on a nondiscriminatory basis.

Establishments. One business location is obviously an establishment, but if an employer has several locations, they may all be considered part of the same establishment on the basis of an analysis of the company's labor relations policy. The greater the degree of centralized authority for hiring, firing, wage setting, and other human resource matters, the more likely the courts are to find multiple locations to be a single establishment. The more freedom each facility has to determine its own human resource policies, the more likely the court will find it to be independent of other facilities.

DEFENSES

Once an employee establishes that an employer is paying different wages to employees of different sexes doing substantially equal work, there are certain defenses the employer can raise. These are, in essence, legal justifications for paying unequal wages to men and women.

The first defense that an employer may use is that the pay differential is based on one of the four statutory exceptions found in the Bennett Amendment to the Equal Pay Act. If the wage differential is based on one of these four factors, the differential is justified and the employer is not in violation of the act. The four factors are:

1. A bona fide seniority system
2. A bona fide merit system
3. A pay system based on quality or quantity of output
4. Factors other than sex

The first three factors are fairly straightforward. Seniority-, merit-, and productivity-based wage systems must be enacted in good faith and must be applied to both men and women. As minimal evidence of good faith, any such system should be written down.

The fourth factor presents greater problems. Circumstances such as greater availability of females and their willingness to work for lower wages do not constitute "factors other than sex."

One frequently litigated factor is training programs. A training program that requires trainees to rotate through jobs that are normally paid lower wages will be upheld as long as it is a bona fide training program and not a sham for paying members of one sex higher wages for doing the same job. The court will look at each case individually, but factors that would lead to a training program's being found bona fide include a written description of the training program that is available to employees, nondiscriminatory access to the program for members of both sexes, and demonstrated awareness of the availability of the program by employees of both sexes.

REMEDIES

An employer found to have violated the act cannot remedy the violation by reducing the higher-paid workers' wages or by transferring those of one sex to another job so that they are no longer doing equal work.

A person who has been subjected to an Equal Pay Act violation may bring a private action under Section 16(b) of the act and recover back pay in the amount of the differential paid to members of the opposite sex. If the employer did not act in good faith in paying the discriminatory wage rates, the court will also award the plaintiff damages in an additional amount equal to the back pay. A successful plaintiff is also entitled to attorney's fees.

The Civil Rights Act of 1964, as Amended (Title VII), and the Civil Rights Act of 1991

The year after it passed the Equal Pay Act, Congress passed the Civil Rights Act of 1964. Title VII of this act is the most common basis for lawsuits premised on discrimination, because it covers a broader area of potential claimants than does either of the statutes that were discussed previously. **Title VII** prohibits employers from (1) hiring, firing, or otherwise discriminating in terms and conditions of employment and (2) segregating employees in a manner that would affect their employment opportunities on the basis of their race, color, religion, sex, or national origin. These five categories are known as *protected classes*.

Title VII Statute that prohibits discrimination in hiring, firing, or other terms and conditions of employment on the basis of race, color, religion, sex, or national origin.

Today's business manager must be familiar with Title VII, because the number of claims filed under the act is significant. According to the Equal Employment Opportunity Commission (EEOC), the total number of charges filed in 2014 was 88,778.[3] That number is lower than the 93,727 claims filed in 2013, making 2014 the fourth year in a row that the total number of claims filed has declined.

APPLICABILITY OF THE ACT

Employers covered by Title VII include only those who have 15 or more employees, that year or last, for 20 consecutive weeks and are engaged in a business that affects interstate commerce. In 1994, the term *employer* was broadened to include

---

[3]EEOC, Charge Statistics FY 1997 through 2014. Available at www.eeoc.gov/eeoc/statistics/enforcement/charges.cfm.

the U.S. government, corporations owned by the government, and agencies of the District of Columbia. The act also covers Indian tribes, private clubs, unions, and employment agencies.

In addition to prohibiting discrimination by covered employers, unions, and employment agencies, the act also imposes recordkeeping and reporting requirements on these parties. Covered parties must maintain all records regarding employment opportunities for at least six months. Such records include job applications, notices for job openings, and records of layoffs. If an employment discrimination charge is filed against an employer, such records must be kept until the case is concluded. EEO-1 forms (forms containing information concerning the number of minorities in various job classifications) must be filed annually with the EEOC by employers of more than 100 workers. A copy of this form is shown in Exhibit 21-1. Finally, each covered employer must display a summary of the relevant portions of Title VII where the employees can see it. The notice must be printed in a language that the employees can read.

PROOF IN EMPLOYMENT DISCRIMINATION CASES

The burden of proof in a discrimination case is initially on the plaintiff. He or she attempts to establish discrimination in one of three ways: (1) disparate treatment, (2) disparate impact, or (3) harassment.

disparate treatment Occurs when the employer treats one employee less favorably than another because of that employee's color, race, religion, sex, or national origin.

Disparate Treatment. **Disparate treatment** occurs when one individual is treated less favorably than another because of color, race, religion, sex, or national origin. The key in such cases is proving the employer's unlawful discriminatory motive. This process is referred to as *building a* prima facie *case*.

The plaintiff must establish the following set of facts: (1) The plaintiff is within one of the protected classes, (2) he or she applied for a job for which the employer was seeking applicants for hire or promotion, (3) the plaintiff possessed the minimum qualifications to perform that job, (4) the plaintiff was denied the job or promotion, and (5) the employer continued to look for someone to fill the position.

Once the plaintiff establishes these facts, the burden shifts to the defendant to articulate legitimate and nondiscriminatory business reasons for rejecting the plaintiff. Such reasons for a failure to promote, for instance, might include a poor work record or excessive absenteeism. If the employer meets this burden, the plaintiff must then demonstrate that the reasons the defendant offered were just a pretext for a real discriminatory motive. In other words, the alleged reason was not the real reason; it was just put forth because it sounded good. One way in which the plaintiff can demonstrate pretext is by showing that the criteria used to reject the plaintiff were not applied to others in the same situation. Introducing past discriminatory policies would also be relevant, as would statistics showing a general practice of discrimination by the defendant. At the pretext stage, the issue of proving an employer's intent to discriminate appears first and is usually the key to the plaintiff's winning or losing the case. Exhibit 21-2 shows how the burden of proof shifts in a disparate treatment case.

disparate impact Occurs when the employer's facially neutral policy or practice has a discriminatory effect on employees who belong to a protected class.

Disparate Impact. As complex as disparate treatment cases are, disparate impact cases are even more difficult to establish. **Disparate impact** cases arise when a plaintiff attempts to establish that an employer's facially neutral employment policy or practice has a discriminatory effect or impact on a protected class. In other words, a requirement of the policy or practice applies to everyone equally, but in application, it disproportionately limits employment opportunities for a particular protected class.

EQUAL EMPLOYMENT OPPORTUNITY

EMPLOYER INFORMATION REPORT EEO-1
1994

Joint Reporting Committee
- Equal Employment Opportunity Commission
- Office of Federal Contract Compliance Programs (Labor)

1 OF 1

RETURN COMPLETED REPORT TO:
THE JOINT REPORTING COMMITTEE
P.O. BOX 779
NORFOLK, VA 23501

PHONE: (804) 461-1213

Section A—TYPE OF REPORT
Refer to instructions for number and types of reports to be filed.

1. Indicate by marking in the appropriate box the type of reporting unit for which this copy of the form is submitted (MARK ONLY ONE BOX).

   (1) ☐ Single-establishment Employer Report

   Multi-establishment Employer:
   (2) ☐ Consolidated Report (Required)
   (3) ☐ Headquarters Unit Report (Required)
   (4) ☐ Individual Establishment Report (Submit one for each establishment with 50 or more employees)
   (5) ☐ Special Report

2. Total number of reports being filed by this Company (Answers on Consolidated Report only) _____

Section B—COMPANY IDENTIFICATION (To be answered by all employers)

OFFICE USE ONLY

1. Parent Company
   a. Name of parent company (owns or controls establishment in item 2) omit if same as label

   Address (Number and street)

City or town	State	ZIP code

2. Establishment for which this report is filed. (Omit if same as label)
   a. Name of establishment

Address (Number and street)	City or town	County	State	ZIP code

   b. Employer identification No. (IRS 9-DIGIT TAX NUMBER)

   c. Was an EEO-1 report filed for this establishment last year?   Yes ☐   No ☐

Section C—EMPLOYERS WHO ARE REQUIRED TO FILE (To be answered by all employers)

☐ Yes ☐ No 1. Does the entire company have at least 100 employees in the payroll period for which you are reporting?

☐ Yes ☐ No 2. Is your company affiliated through common ownership and/or centralized management with other entities in an enterprise with a total employment of 100 or more?

☐ Yes ☐ No 3. Does the company or any of its establishments (a) have 50 or more employees AND (b) is not exempt as provided by 41 CFR 60-1.5. AND either (1) is a prime government contractor or first-tier subcontractor, and has a contract, subcontract, or purchase order amounting to $50,000 or more, or (2) serves as a depository of Government funds in any amount or is a financial institution which is an issuing and paying agent for U.S. Savings Bonds and Savings Notes?

→ If the response to question C-3 is yes, please enter your Dun and Bradstreet identification number (if you have one):

NOTE: If the answer is yes to questions 1, 2, or 3, complete the entire form, otherwise skip to Section G.

NSN 7540-00-180-6384

(Continued)

EXHIBIT 21-1

EEO-1 FORM

Source: United States Department of Labour.

100 Page 2

Section D—EMPLOYMENT DATA

Employment at this establishment—Report all permanent full-time and part-time employees including apprentices and on-the-job trainees unless specifically excluded as set forth in the instructions. Enter the appropriate figures on all lines and in all columns. Blank spaces will be considered as zeros.

JOB CATEGORIES		NUMBER OF EMPLOYEES									
		OVERALL TOTALS (SUM OF COL. B THRU K)	MALE					FEMALE			
			WHITE (NOT OF HISPANIC ORIGIN)	BLACK (NOT OF HISPANIC ORIGIN)	HISPANIC	ASIAN OR PACIFIC ISLANDER	AMERICAN INDIAN OR ALASKAN NATIVE	WHITE (NOT OF HISPANIC ORIGIN)	BLACK (NOT OF HISPANIC ORIGIN)	HISPANIC	ASIAN OR PACIFIC ISLANDER	AMERICAN INDIAN OR ALASKAN NATIVE
		A	B	C	D	E	F	G	H	I	J	K
Officials and Managers	1											
Professionals	2											
Technicians	3											
Sales Workers	4											
Office and Clerical	5											
Craft Workers (Skilled)	6											
Operatives (Semi-Skilled)	7											
Laborers (Unskilled)	8											
Service Workers	9											
TOTAL	10											
Total employment reported in previous EEO—1 report	11											

NOTE: Omit questions 1 and 2 on the Consolidated Report.
1. Date(s) of payroll period used: 2. Does this establishment employ apprentices:
   1 ☐ Yes   2 ☐ No

Section E—ESTABLISHMENT INFORMATION *(Omit on the Consolidated Report)*

1. What is the major activity of this establishment? (Be specific, i.e., manufacturing steel castings, retail grocer, wholesale plumbing supplies, title insurance, etc.) Include the specific type of product or type of service provided, as well as the principal business or industrial activity.

OFFICE USE ONLY

Section F—REMARKS

Use this item to give any identification data appearing on last report which differs from that given above, explain major changes in composition or reporting units and other pertinent information.

Section G—CERTIFICATION *(See instructions G)*

Check One
1 ☐ All reports are accurate and were prepared in accordance with the instructions (check on consolidated only).
2 ☐ This report is accurate and was prepared in accordance with the instructions.

Name of Certifying Official	Title	Signature	Date
Name of person to contact regarding this report (Type or print)	Address (Number and Street)		
Title	City and State	ZIP Code	Telephone Number (Including Area Code)	Extension

All reports and information obtained from individual reports will be kept confidential as required by Section 709(e) of Title VII.
WILLFULLY FALSE STATEMENTS ON THIS REPORT ARE PUNISHABLE BY LAW, U.S. CODE, TITLE 18, SECTION 1001.

EXHIBIT 21-1
CONTINUED

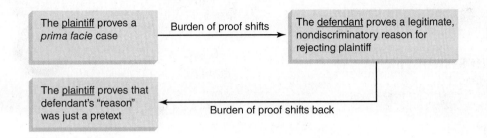

EXHIBIT 21-2

THE SHIFTING BURDEN OF PROOF IN A DISPARATE TREATMENT CASE

To establish a case of discrimination based on disparate impact, the plaintiff must first establish statistically that the rule disproportionately restricts employment opportunities for a protected class. The burden of proof then shifts to the defendant to demonstrate that the practice or policy is a business necessity. The plaintiff, at this point, can still recover by proving that the "necessity" was promulgated as a pretext for discrimination.

The first two steps for proving a prima facie case of disparate impact were laid out in *Griggs v. Duke Power Co.*[4] In that case, the employer-defendant required all applicants to have a high school diploma and a successful score on a professionally recognized intelligence test for all jobs except that of laborer. By establishing these criteria, the employer proposed to upgrade the quality of its workforce.

The plaintiff demonstrated the discriminatory impact by showing that 34 percent of the white males in the state had high school diplomas whereas only 12 percent of the black males did and by introducing evidence from an EEOC study showing that 58 percent of the whites compared with 6 percent of the blacks had passed tests similar to the one given by the defendant. The defendant could show no business-related justification for either employment policy, so the plaintiff was successful. Not all employees of Duke Power needed to be smart or have high school diplomas. After all, when does a student in high school learn how to install power lines or repair company vehicles? A high IQ or a high school or college diploma may be necessary for some jobs, but not for all jobs at Duke Power.

Harassment. The third way to prove discrimination is to demonstrate harassment. Harassment is a relatively new basis for a discrimination claim; it first developed in the context of discrimination based on sex and then evolved to become applicable to other protected classes.

The definition of **sexual harassment** stated in the EEOC Guidelines and accepted by the U.S. Supreme Court is "unwelcome sexual advances, requests for sexual favors, and other verbal or physical conduct of a sexual nature" that implicitly or explicitly make submission a term or condition of employment, make employment decisions related to the individual dependent on submission to or rejection of such conduct, or have the purpose or effect of creating an intimidating, hostile, or offensive environment.

The courts have recognized two distinct forms of sexual harassment. The first, *quid pro quo*, occurs when a supervisor makes sexual demands on someone of the opposite sex and this demand is reasonably perceived as a term or condition of employment. The basis for this rule is that similar demands would not be made by the supervisor on someone of the same sex.

The second form of sexual harassment involves the creation of a hostile environment. Case 21-1 demonstrates the standards used by the U.S. Supreme Court to determine whether an employer's conduct has indeed created a hostile work environment.

sexual harassment
Unwelcome sexual advances, requests for sexual favors, and other verbal or physical conduct of a sexual nature that explicitly or implicitly make submission a term or condition of employment or creates an intimidating, hostile, or offensive environment.

---
[4]401 U.S. 424 (1971).

CASE 21-1

Teresa Harris v. Forklift Systems, Inc.
United States Supreme Court
510 U.S. 17 (1994)

Plaintiff Harris was a manager for Defendant Forklift Systems, Inc. During her tenure at Forklift Systems, Plaintiff Harris was repeatedly insulted by the defendant's president and, because of her gender, subjected to sexual innuendos. Numerous times in front of others, the president told Harris, "You're just a woman. What do you know?" He sometimes asked Harris and other female employees to remove coins from his pockets and made suggestive comments about their clothes. He suggested to Harris in front of others that they negotiate her salary at the Holiday Inn. When Harris complained, he said he would stop, but he did not; so she quit and filed an action against the defendant for creating an abusive work environment based on her sex.

The district court found in favor of the defendant, holding that some of the comments were offensive to the reasonable woman but were not so serious as to affect Harris's psychological well-being severely or to interfere with her work performance. The court of appeals affirmed. Plaintiff Harris appealed to the U.S. Supreme Court.

Justice O'Connor

In this case we consider the definition of a discriminatorily "abusive work environment" (a "hostile work environment") under Title VII.

Title VII of the Civil Rights Act of 1964 makes it "an unlawful employment practice for an employer . . . to discriminate against any individual with respect to his compensation, terms, conditions, or privileges of employment, because of such individual's race, color, religion, sex, or national origin." . . . [T]his language "is not limited to 'economic' or 'tangible' discrimination. The phrase 'terms, conditions, or privileges of employment' evinces a congressional intent 'to strike at the entire spectrum of disparate treatment of men and women' in employment," which includes requiring people to work in a discriminatorily hostile or abusive environment. When the workplace is permeated with "discriminatory intimidation, ridicule, and insult," that is "sufficiently severe or pervasive to alter the conditions of the victim's employment and create an abusive working environment."

This standard, which we reaffirm today, takes a middle path between making actionable any conduct that is merely offensive and requiring the conduct to cause a tangible psychological injury. As we pointed out in *Meritor*, "mere utterance of an 'epithet which engenders offensive feelings in a employee,' does not sufficiently affect conditions of employment to implicate Title VII. Conduct that is not severe or pervasive enough to create an objectively hostile or abusive work environment—an environment that a reasonable person would find hostile or abusive"—is beyond Title VII's purview. Likewise, if the victim does not subjectively perceive the environment to be abusive, the conduct has not actually altered the conditions of the victim's employment, and there is no Title VII violation.

But Title VII comes into play before the harassing conduct leads to a nervous breakdown. A discriminatorily abusive work environment, even one that does not seriously affect employees' psychological well-being, can and often will detract from employees' job performance, discourage employees from remaining on the job, or keep them from advancing in their careers. Moreover, even without regard to these tangible effects, the very fact that the discriminatory conduct was so severe or pervasive that it created a work environment abusive to employees because of their race, gender, religion, or national origin offends Title VII's broad rule of workplace equality. The appalling conduct alleged in *Meritor*, and the reference in that case to environments "so heavily polluted with discrimination as to destroy completely the emotional and psychological stability of minority group workers," merely present some especially egregious examples of harassment. They do not mark the boundary of what is actionable.

We therefore believe the District Court erred in relying on whether the conduct "seriously affected plaintiff's psychological well-being" or led her to "suffer injury." Such an inquiry may needlessly focus the fact-finder's attention on concrete psychological harm, an element Title VII does not require. Certainly Title VII bars conduct that would seriously affect a reasonable person's psychological well-being, but the statute is not limited to such conduct. So long as the environment would reasonably be perceived, and is perceived, as hostile or abusive, there is no need for it also to be psychologically injurious.

This is not, and by its nature cannot be, a mathematically precise test. But we can say that whether an environment is "hostile" or "abusive" can be determined only by looking at all the circumstances. These may include the frequency of the discriminatory conduct; its severity; whether it is physically threatening or humiliating, or a mere offensive utterance; and whether it unreasonably interferes with an employee's work performance. The effect on the employee's psychological well-being is, of course, relevant to determining whether the plaintiff actually found the environment abusive. But while psychological harm, like any other relevant factor, may be taken into account, no single factor is required.*

Reversed and *remanded* in favor of Plaintiff, Harris.

---

*Teresa Harris v. Forklift Systems, Inc., United States Supreme Court, 510 U.S. 17 (1994).

CRITICAL THINKING ABOUT THE LAW

As was previously touched upon, the judiciary most often operates in relationship to shades of gray and not to the black and white between which those shades lie. The Court's decision in Case 21-1, in large part dependent on its determination of a definition, illustrates this point.

The Court's primary test was to decide what constitutes an "abusive work environment," the second type of sexual harassment actionable under Title VII. Deciding on such a definition is not as easy as going to a legal dictionary and looking up "abusive work environment." The Court had to interpret the meaning of such an environment, and important to this interpretation were legal precedent, ambiguity, and primary ethical norms.

Hence, the questions that follow will aid in thinking critically about these factors influential in the Court's interpretation.

1. What ambiguous language did the Court leave undefined in Case 21-1?

    Clue: To find this answer, look at the Court's definition of an "objectively hostile work environment." As always, remember that ambiguities are most often adjectives.

2. In her discussion of the precedent, Justice O'Connor made it clear that the district court misinterpreted the decision in rendering its decision. Contrary to the district court's decision, the existence of which key fact was not necessary for the Court to find the defendant guilty of sexual harassment?

    Clue: Revisit the paragraph discussing the district court's dismissal of Harris's claim. On what basis was this dismissal made? This is the key fact the existence of which the Supreme Court found unnecessary for judgment in favor of the plaintiff.

Since *Meritor,* conflicting lower-court decisions have created confusion in the area of sexual harassment. It appeared that in a quid pro quo case, a company was liable regardless of its knowledge, but in a hostile environment case, a company could not be held liable without direct knowledge of the situation. Another question was whether there could be recovery when only empty threats were made.

For example, in *Jones v. Clinton,*[5] the district court judge threw out Jones's sexual harassment case against the president because Jones had no clear and tangible job detriment (necessary to establish a quid pro quo case), and she was not subject to a hostile environment when the totality of the circumstances was viewed. Even if the allegations were true, the contacts did not constitute "the kind of pervasive, intimidating, abusive conduct"[6] necessary for a hostile environment.

The U.S. Supreme Court attempted to clarify these issues in *Ellerth v. Burlington.*[7] Ellerth was subjected to a litany of dirty jokes and sexual innuendos from her boss. He propositioned her and threatened to make her life miserable if she refused him. She refused him without reprisals and was even promoted. She did not complain about harassment but quit after a year because she could not stand the threats and innuendos.

In a decision that offered something to both plaintiffs and defendants, the high court ruled that

> an employer is subject to vicarious liability to a victimized employee for an actionable hostile environment created by a supervisor with immediate (or successively higher) authority over the employee. When no tangible employment action is taken, a defending employer may raise an affirmative defense to liability [by showing that] (a) the employer exercised reasonable care to prevent and correct promptly any sexually harassing behavior, and (b) the plaintiff employee

---

[5] No. LR-C-94-290 (E.D. Ark. 1998).
[6] *Id.*
[7] 118 S. Ct. 2275 (1998).

unreasonably failed to take advantage of any preventive or corrective opportunities provided by the employer or to avoid harm otherwise. No affirmative defense is available, however, when the supervisor's harassment culminates in a tangible employment action.[8*]

The Court then remanded the case to the lower court for a new trial.

Under limited circumstances, employers may be held liable for harassment of their employees by nonemployees: If an employer knows that a customer is harassing an employee but does nothing to remedy the situation, the employer may be liable. For example, in *Lockhard v. Pizza Hut Inc.*,[9] the franchise was held liable when the company failed to take any steps to stop the harassment of a waitress by two male customers.

Same-Sex Harassment. Initially, same-sex harassment did not constitute sexual harassment. In the first appellate case on this issue, a male employee sued his employer for sexual harassment, alleging that on several occasions, his male supervisor had approached him from behind and grabbed his crotch.[10] The court of appeals affirmed the trial court's dismissal of the claim on the grounds that no prima facie case had been established. The court said that Title VII addressed gender discrimination, and harassment by a male supervisor of a male employee did not constitute sexual harassment, regardless of the sexual overtones of the harassment.

However, the circuit courts soon became split on whether one could be sexually harassed by a person of the same sex. The U.S. Supreme Court finally rendered a definitive answer to that issue in the case of *Joseph Oncale v. Sundowner Offshore Services*,[11] with its holding that "nothing in Title VII necessarily bars a claim of discrimination 'because of . . . sex' merely because the plaintiff and the defendant are of the same sex."[12] As long as the discrimination was because of the victim's sex, it was actionable.

Hostile Environment Extended. Hostile environment cases have also been used in cases of discrimination based on religion, race, and even age.[13] For example, in one case,[14] Hispanic and black corrections workers demonstrated that a hostile work environment existed by proving that they had been subjected to continuing verbal abuse and racial harassment by coworkers and that the county sheriff's department had done nothing to prevent the abuse. The white employees had continually used racial epithets and posted racially offensive materials on bulletin boards, such as a picture of a black man with a noose around his neck, cartoons favorably portraying the Ku Klux Klan, and a "black officers' study guide" consisting of children's puzzles. White officers once dressed a Hispanic inmate in a straw hat, sheet, and sign that said "spic." Such activities were found by the court to constitute a hostile work environment.

A New Limitation on the Employer's Liability. As explained previously, to win a lawsuit for harassment by customers, the employee must show that the employer knew of the harassment and did nothing to stop it. Similarly, if the harassment is by a coworker, the employee must show that the employer is

---

[8]*Id.*

[9]162 F.3d 1062 (10th Cir. 1998).

[10]*Garcia v. Elf Atochem*, 28 F.3d 466 (5th Cir. 1994).

[11]118 S. Ct. 998 (1998).

[12]*Id.*

[13]*Crawford v. Medina General Hospital*, 96 F.3d 830 (6th Cir. 1996).

[14]*Snell v. Suffolk County*, 782 F.2d 1094 (1986).

*Burlington Industries, Inc. v. Ellerth, 118 S. Ct. 2257 (June 26, 1998).

negligent in responding to complaints about harassment. But to win a lawsuit for harassment by a supervisor, the employer does not have to be negligent because Title VII imputes the supervisor's acts to the employer. In 2013, in the following case, the U.S. Supreme Court made it more difficult for plaintiffs to win harassment cases by limiting the definition of who can be considered a supervisor. Like so many other cases, this one was a 5–4 decision.

CASE 21-2

Vance v. Ball State University
United States Supreme Court
133 S. Ct. 2423 (2013)

Maetta Vance claimed that Saundra Davis, a catering specialist, had made Vance's life at work contentious through physical acts and racial harassment. Vance sued her employer, Ball State University, for workplace harassment by a supervisor. Vance argued that Davis was a supervisor, whereas Ball State claimed that Davis was not actually Vance's supervisor. The District Court and Court of Appeals for the 7th Circuit determined that Davis was not Vance's supervisor because Davis did not have the power to direct the terms and conditions of Vance's employment and granted summary judgment to Ball State. The 7th Circuit agreed that Davis was not a supervisor and, therefore, the university could not be held vicariously liable. Vance appealed to the U.S. Supreme Court.

Justice Alito

In this case, we decide . . . who qualifies as a "supervisor" in a case in which an employee asserts a Title VII claim for workplace harassment?

Under Title VII, an employer's liability for such harassment may depend on the status of the harasser. If the harassing employee is the victim's co-worker, the employer is liable only if it was negligent in controlling working conditions. In cases in which the harasser is a "supervisor," however, different rules apply. If the supervisor's harassment culminates in a tangible employment action, the employer is strictly liable. But if no tangible employment action is taken, the employer may escape liability by establishing, as an affirmative defense, that (1) the employer exercised reasonable care to prevent and correct any harassing behavior and (2) that the plaintiff unreasonably failed to take advantage of the preventive or corrective opportunities that the employer provided. . . . Under this framework, therefore, it matters whether a harasser is a "supervisor" or simply a co-worker . . .

. . . For present purposes, the only relevant incidents concern Vance's interactions with a fellow BSU employee, Saundra Davis. During the time in question, Davis, a white woman, was employed as a catering specialist in the Banquet and Catering division. The parties vigorously dispute the precise nature and scope of Davis's duties, but they agree that Davis did not have the power to hire, fire, demote, promote, transfer, or discipline Vance. . . . We hold that an employer may be vicariously liable for an employee's unlawful harassment only when the employer has empowered that employee to take tangible employment actions against the victim, *i.e.*, to effect a "significant change in employment status, such as hiring, firing, failing to promote, reassignment with significantly different responsibilities, or a decision causing a significant change in benefits."*

Affirmed, in favor of Defendant Ball State University.

RETALIATION

In addition to protecting an employee from discrimination, Title VII also makes it unlawful for an employer to retaliate against an employee who has reported or otherwise complained about discrimination under the act. To prove a prima facie retaliation claim against an employer, the employee must show three things:

1. The employee engaged in a protected activity under the act, such as complaining of, participating in an investigation of, or filing a formal charge about perceived discrimination under Title VII or refusing to participate in conduct the employee reasonably believed was unlawful under Title VII.

---

*Vance v. Ball State University, United States Supreme Court.

2. The employee was fired, demoted, or suffered some other adverse employment action.

3. The employee's protected activity was the cause of or the determining factor for the adverse employment action.

Until the U.S. Supreme Court decided the case of *University of Texas Southwestern Medical Center v. Nassar*,[15] the third factor was easier to prove because the protected activity could be simply a motivating factor in the decision, which is the standard for a discrimination case. But in this recent case, the court tightened the standard from a motivating factor to "the determinative factor." In this 5–4 decision, Justice Kennedy said that it was important to have the proper causation standard in retaliation cases because the number of such cases filed with the EEOC had nearly doubled in the last 15 years, rising to more than 31,000 in 2012. During fiscal year 2014, that number had risen to 37,955.[16]

In 2015, the 4th Circuit Court of Appeals handed down a decision that some corporate counsel fear will increase the number of retaliation cases even more because it will expand the number of situations wherein a retaliation case will make it to the jury.

CASE 21-3

Reya C. Boyer-Liberto v. Fontainbleu Corporation
United States Court of Appeals for the 4th Circuit
2015 U.S. App. LEXIS 7557 (2015)

Reya C. Boyer-Liberto, an African American woman, sued her former employer for racial discrimination and retaliation, in violation of Title VII of the Civil Rights Act of 1964 and 42 U.S.C. § 1981. She based her racial discrimination claim on a hostile work environment allegedly created by two conversations she had with a coworker about an incident that occurred on September 14, 2010. During the conversations, which took place on two consecutive days, the coworker twice called Liberto a "porch monkey." And she grounds her retaliation claim on the termination of her employment after she complained about the statements.

The district court granted the defendants' motion for summary judgment, concluding that the conduct was too isolated to support either of Liberto's claims. Liberto appealed. In a 2–1 decision, a panel of the 4th Circuit affirmed the lower court's decision, but its decision was vacated by the 4th Circuit's granting Liberto a rehearing en banc.

Circuit Judge King

. . . [W]e now vacate the judgment of the district court and remand for further proceedings on Liberto's claims. In so doing, we underscore the Supreme Court's pronouncement in *Faragher v. City of Boca Raton* . . . that an isolated incident of harassment, if extremely serious, can create a hostile work environment. We also recognize that an employee is protected from retaliation when she reports an isolated incident of harassment that is physically threatening or humiliating, even if a hostile work environment is not engendered by that incident alone . . .

. . . The district court then invoked *Jordan* for the proposition that an "isolated racist comment" is "a far cry from . . . an environment of crude and racist conditions so severe or pervasive that they alter[] the conditions of [plaintiff's] employment." . . . In concomitantly rejecting Liberto's retaliation claims, the court

---

[15] 133 S. Ct. 2517 (2013).

[16] U.S. Equal Employment Opportunity Commission, EEOC Releases Fiscal Year 2014 Enforcement and Litigation Data, February 4, 2015. Available at www1.eeoc.gov/eeoc/newsroom/release/2-4-15.cfm.

again looked to *Jordan* and ruled that "no objectively reasonable person could have believed that the [plaintiff's work environment] was, or was soon going to be, infected by severe or pervasive racist, threatening, or humiliating harassment."

The panel decision was unanimous that the defendants were properly awarded summary judgment on Liberto's hostile work environment claims, in that Clubb's "use of [the term 'porch monkey'] twice in a period of two days in discussions about a single incident, was not, as a matter of law, so severe or pervasive as to change the terms and conditions of Liberto's employment." The panel observed that Liberto had "not pointed to any Fourth Circuit case, nor could she, finding the presence of a hostile work environment based on a single incident." . . .

The panel was split, however, with respect to Liberto's retaliation claims. The opinion of the panel majority validated the district court's summary judgment award on those claims, explaining that, "if no objectively reasonable juror could have found the presence of a hostile work environment . . . it stands to reason that Liberto also could not have had an objectively reasonable belief that a hostile work environment existed." Although the panel majority allowed that an "employee's opposition may be protected before the hostile environment has fully taken form," the majority faulted Liberto for failing to "present any indicators that the situation at the Clarion would have ripened into a hostile work environment."

In thus vacating the summary judgment award on Liberto's hostile work environment claims, we identify this as the type of case contemplated in *Faragher* where the harassment, though perhaps "isolated," can properly be deemed to be "extremely serious." We reject, however, any notion that our prior decisions . . . were meant to require more than a single incident of harassment in every viable hostile work environment case.

. . . Turning to Liberto's retaliation claims, Title VII proscribes discrimination against an employee because, in relevant part, she "has opposed any practice made an unlawful employment practice by this subchapter." Employees engage in protected oppositional activity when they "complain to their superiors about suspected violations of Title VII." . . . To establish a prima facie case of retaliation in contravention of Title VII, a plaintiff must prove "(1) that she engaged in a protected activity," as well as "(2) that her employer took an adverse employment action against her," and "(3) that there was a causal link between the two events.". . . . A prima facie retaliation claim under 42 U.S.C. § 1981 has the same elements.

. . . In the context of element one of a retaliation claim, an employee is protected when she opposes "not only . . . employment actions actually unlawful under Title VII but also employment actions [she] reasonably believes to be unlawful." . . . The Title VII violation may be complete, or it may be in progress. . . . "*Navy Federal* holds that an employee seeking protection from retaliation must have an objectively reasonable belief in light of all the circumstances that a Title VII violation has happened or is in progress." . . . In other words, an employee is protected from retaliation when she opposes a hostile work environment that, although not fully formed, is in progress.

The panel majority in *Jordan* ruled that, where an employee has complained to his employer of an isolated incident of harassment insufficient to create a hostile work environment, the employee cannot have possessed a reasonable belief that a Title VII violation was in progress, absent evidence "that a plan was in motion to create such an environment" or "that such an environment was [otherwise] likely to occur." . . . We reject that aspect of *Jordan* today, however, for several reasons. First of all, the *Jordan* standard "imagines a fanciful world where bigots announce their intentions to repeatedly belittle racial minorities at the outset, and it ignores the possibility that a hostile work environment could evolve without some specific intention to alter the working conditions of African-Americans through racial harassment."

The *Jordan* standard also is at odds with the hope and expectation that employees will report harassment early, before it rises to the level of a hostile environment. Where the harasser is her supervisor and no tangible employment action has been taken, the victim is compelled by the *Ellerth/Faragher* defense to make an internal complaint, i.e., "to take advantage of any preventive or corrective opportunities provided by the employer." . . .

Similarly, the victim of a co-worker's harassment is prudent to alert her employer in order to ensure that, if the harassment continues, she can establish the negligence necessary to impute liability. . . . The reporting obligation is essential to accomplishing Title VII's "primary objective," which is "not to provide redress but to avoid harm." . . . But rather than encourage the early reporting vital to achieving Title VII's goal of avoiding harm, the *Jordan* standard deters harassment victims from speaking up by depriving them of their statutory entitlement to protection from retaliation. Such a lack of protection is no inconsequential matter, for "fear of retaliation is the leading reason why people stay silent instead of voicing their concerns about bias and discrimination." . . .

The question, then, becomes this: What is the proper standard for determining whether an employee who reports an isolated incident of harassment has a reasonable belief that she is opposing a hostile work environment in progress? We conclude that, when assessing the reasonableness of an employee's belief that a hostile environment is occurring based on an isolated incident, the focus should be on the severity of the harassment. . . . That assessment thus involves factors used to judge whether a workplace is sufficiently hostile or abusive for purposes of a hostile environment claim— specifically, whether the discriminatory conduct "is physically threatening or humiliating, or a mere offensive utterance." . . . Of course, a single offensive utterance . . . generally will not create a hostile environment without significant repetition or an escalation in the harassment's severity. . . . But an isolated incident that is physically threatening or humiliating will be closer—even if not equal—to the type of conduct actionable on its own because it is "extremely serious.". . .

Accordingly, as relevant here, an employee will have a reasonable belief that a hostile work environment is occurring based on an isolated incident if that harassment is physically threatening or humiliating. This standard is consistent not only with *Clark County*, but also with other Supreme Court precedent, including *Crawford* and *Burlington Northern*. That is so because it protects an employee like *Jordan* who promptly speaks up "to attack the racist cancer in his workplace," rather than "remain[ing] silent" and "thereby allowing [discriminatory] conduct to continue unchallenged," while "forfeiting any judicial remedy he might have." . . . In sum, under the standard that we adopt today with guidance from the Supreme Court, an employee is protected from retaliation for opposing an isolated incident of harassment when she reasonably believes that a hostile work environment is in progress, with no requirement for additional evidence that a plan is in motion to create such an environment or that such an environment is likely to occur. The employee will have a reasonable belief that a hostile environment is occurring if the isolated incident is physically threatening or humiliating. . . . Because the defendants contested Liberto's retaliation claims on the lone ground that she did not engage in a protected activity, our analysis is limited to whether a jury could find that Liberto reasonably believed there was a hostile work environment in progress when she reported Clubb's use of the "porch monkey" slur. Applying the standard that we adopt today, the answer plainly is "yes." As we recognized in analyzing Liberto's hostile work environment claims, "porch monkey" is a racial epithet that is not just humiliating, but "degrading and humiliating in the extreme." . . . Indeed, we determined that a reasonable jury could find that Clubb's two uses of "porch monkey" were serious enough to engender a hostile environment. We must further conclude, therefore, in the context of the retaliation claims, that Liberto has made the lesser showing that the harassment was sufficiently severe to render reasonable her belief that a hostile environment was occurring. Accordingly, we vacate the summary judgment award on Liberto's retaliation claims, in addition to her hostile work environment claims. We also underscore that, on remand, a jury would be entitled to simultaneously reject the hostile work environment claims on the ground that Clubb's conduct was not sufficiently serious to amount to a hostile environment, but award relief on the retaliation claims by finding that Clubb's conduct was severe enough to give Liberto a reasonable belief that a hostile environment, although not fully formed, was in progress.

. . . Contrary to the dissent, we seek to promote the hope and expectation—ingrained in our civil rights laws and the Supreme Court decisions interpreting them—that employees will report harassment early, so that their employers can stop it before it rises to the level of a hostile environment. Employers are powerless in that regard only if they are unaware that harassment is occurring. But employees will understandably be wary of reporting abuse for fear of retribution. Under today's decision, employees who reasonably perceive an incident to be physically threatening or humiliating do not have to wait for further harassment before they can seek help from their employers without exposing themselves to retaliation.*

Reversed, in favor of Petitioner Liberto.

STATUTORY DEFENSES

The three most important defenses available to defendants in Title VII cases are bona fide occupational qualification (BFOQ), merit, and seniority. These defenses are raised by the defendant after the plaintiff has established a prima facie case of discrimination based on disparate treatment, disparate impact, or a pattern or practice of discrimination.

Bona Fide Occupational Qualification. The BFOQ defense allows an employer to discriminate in hiring on the basis of sex, religion, or national origin when such a characteristic is necessary to the performance of the job. Race or color cannot be a BFOQ. Such necessity must be based on actual qualifications, not on stereotypes about one group's abilities. Being a male cannot be a BFOQ for a job because it is a dirty or "strenuous" job, although there may be a valid requirement that an applicant be able to lift a certain amount of weight if such lifting is a part of the job. A BFOQ does not arise because an employer's customers would prefer to be served by someone of a particular gender or national origin; nor does inconvenience to the employer, such as having to provide two sets of restroom facilities, make a classification a BFOQ.

---

*Reya C. Boyer-Liberto v. Fontainebleau Corporation. United States Court of Appeals for the Fourth Circuit. 2015 U.S. App. Lexis 7557 (2015).

COMPARATIVE LAW CORNER

Sexual Harassment in France

The French deal with the problem of sexual harassment in employment very differently from Americans. In the United States, sexual harassment is a civil offense and can receive compensatory and punitive damages. In France, sexual harassment is instead part of the criminal code. Part of this difference has to do with a difference in the definitions of sexual harassment. The United States recognizes both quid pro quo and hostile work environment sexual harassment, whereas the French recognize only quid pro quo. Sexual harassment in France is defined as "[t]he fact of harassing anyone using orders, threats or constraint, in order to obtain favors of a sexual nature, by a person abusing the authority that functions confer on him. . . ." With this definition, it makes sense that the French consider sexual harassment a criminal offense. The French do not recognize the idea of a hostile work environment, and it is considered somewhat normal for male employees to comment on the attractiveness of female employees at work.

The French sexual harassment law also differs significantly from American law in its method of enforcement. Women in France are responsible for filing their own claims with the court, and the punishment their harasser can receive is limited to one year in jail or a fine. Also, French companies are not seen as responsible for the behavior of their employees, so if a supervisor sexually harasses a female subordinate, the woman cannot claim damages from the company. Her charges will be filed only against the supervisor who sexually harassed her.

Merit. Most merit claims involve the use of tests. Using a professionally developed ability test, which is not designed, intended, or used to discriminate, is legal. Such tests may have an adverse impact on a class, but do not violate the act as long as they are manifestly related to job performance. The *Uniform Guidelines on Employee Selection Procedures* (UGESP) have, since 1978, contained the policy of all governmental agencies charged with enforcing civil rights, and they provide guidance to employers and other interested persons about when ability tests are valid and job related. Under these guidelines, tests must be validated in accordance with standards established by the American Psychological Association.

Acceptable validation includes (1) *criterion-related validity*, which is the statistical relationship between test scores and objective criteria of job performance; (2) *content validity*, which isolates some skill used on the job and directly tests that skill; and (3) *construct validity*, wherein a psychological trait needed to perform the job is measured. A test that required a secretary to type would be content valid. A test of patience for a teacher would be construct valid.

Seniority Systems. A final statutory defense, available under Section 703(h), is a bona fide seniority system. A seniority system, in which employees are given preferential treatment based on their length of service, may perpetuate discrimination that occurred in the past. Nonetheless, such systems are considered bona fide and thus are not unlawful if (1) the system applies equally to all persons, (2) the seniority units follow industry practices, (3) the seniority system did not have its genesis in discrimination, and (4) the system is maintained free of any illegal discriminatory purpose.

Mixed Motives. One problem with discrimination cases is proving that the plaintiff's membership in a protected class is the reason for unfair treatment. In the 1991 act, Congress addressed the concept of a "mixed motives" case (i.e., a case in which the plaintiff proves that being a member of a protected class was one reason for the unfair treatment, but the defendant also proves that it also had a legal reason). If the court determines that the defendant had mixed motives, the verdict is for the plaintiff, but the court decides whether the plaintiff is entitled to damages based on the weight of the two motives.

PROTECTED CLASSES

Five classes are protected under Title VII. Unique problems have arisen with regard to each of them.

Race and Color. A primary goal of Title VII was to remedy the discrimination in employment to which blacks had long been subjected. The act, however, also contains a proviso stating that nothing in the act requires that preferential treatment based on an imbalance between their representation in the employer's workplace and their representation in the population at large be given to any protected class. This proviso paved the way for questions about "reverse discrimination," or discrimination against whites, as a result of employers' attempts to create a racially balanced workforce. (This issue is discussed later in the section on affirmative action.)

National Origin. The act prohibits discrimination based on national origin, not on alienage (citizenship of a country other than the United States). Thus, an employer can refuse to hire non–U.S. citizens. This prohibition applies even to owners of foreign corporations who have established firms in the United States. In the absence of a treaty between the United States and the foreign state authorizing such conduct, a corporation cannot discriminate in favor of those born in a foreign state.

Since the terrorist attack on the World Trade Center on September 11, 2001, there has been a significant increase in charges based on national origin by individuals who are, or are perceived as being, Arab or South Asian. Many of these claims are combined with claims of discrimination based on religion. Many are based on harassment. For example, two California auto dealers agreed to pay seven Afghan workers $550,000 to settle their complaint of harassment based on national origin and religion. The workers alleged that they were called everything from "camel jockeys" to "bin Laden's gang." One of the women with an Arabic name was asked to call herself by an American name, such as Sara.[17]

During fiscal year 2014, the EEOC received 9,579 charges of national-origin discrimination and resolved 9,768, recovering $31.4 million for the charging parties. Interestingly, these numbers represented a decline from 2013, when the number of such charges was 10,642, with 11,307 resolved (some carried over from the previous year) for a total recovery of $35.3 million.[18]

Religion. Under Title VII, employers cannot discriminate against employees on the basis of religion. Although an exception has been made allowing religious corporations, associations, and societies to discriminate in their employment practices on the basis of religion, they may not discriminate on the basis of any other protected class. In fiscal year 2013, the EEOC received 3,721 charges of religious discrimination and resolved 3,865 such charges, recovering $11.2 million.[19] In 2010, the number of charges received was 3,549, with 3,575 being resolved, generating $8.7 million for claimants.[20]

Employers are required to make reasonable accommodation to their employees' religious needs, as long as such accommodation does not place an undue hardship on the employer or other employees. For example, an employer has a dress code that prohibits clerical workers visible to the public from wearing hats or scarves. A Muslim worker requests that she be granted an exemption from the dress code so that she may wear the *hijab* (head scarf) in conformance with her

---

[17]Bob Egelko, "Two Auto Dealers Agree to Settle Suit with Afghan Workers," *San Francisco Chronicle*, B7 (Apr. 7, 2004).

[18]U.S. Equal Employment Opportunity Commission, National Origin-Based Charges FY 1997—FY 2014. Accessed April 15, 2015 at www.eeoc.gov/eeoc/statistics/enforcement/origin.cfm.

[19]U.S. Equal Employment Opportunity Commission, Religion-Based Charges FY 1997–FY 2014. Accessed April 15, 2015 at www.eeoc.gov/eeoc/statistics/enforcement/religion.cfm.

[20]Ibid.

Muslim beliefs. Her exemption would be a reasonable accommodation. Flexible scheduling, voluntary substitutions or swaps, job reassignments, and lateral transfers are other examples of reasonable accommodations to an employee's religious beliefs. Courts will examine the requested accommodation very carefully to ensure that it does not place an undue burden on the workplace. For example, the reasonableness of accommodating an employee's request not to work on Saturday would depend on the availability of other workers who would willingly work that day.

APPLYING THE LAW TO THE FACTS . . .

Let's say that Talal needs a day off to observe his religious holiday. However, he gave his employer only two days' notice and no other employee can cover for him at the last minute. Talal argues that forcing him to work on a religious holiday is religious discrimination. His employers argue that prohibiting him from engaging in a religious practice that would interfere with his work and impose hardship on his company is legal. Who is correct in this case?

As mentioned previously, since September 11, 2001, the number of charges of religious discrimination by individuals who are, or are perceived to be, Muslim or Sikh has increased. From September 11, 2000 to September 11, 2001, 323 charges based on "religion-Muslim" were filed with the EEOC. The following year, 706 similar charges were filed. In 2003, the EEOC settled one of the largest workplace discrimination suits against Muslims. In that case, four Muslim Pakistani machine operators alleged that their employer, Stockton Steel, routinely gave them the worst jobs, ridiculed their daily prayers, and called them "camel jockey" and "raghead." The four workers shared a $1.1 million settlement.[21] At the time of the settlement, then EEOC Commissioner Steven Miller expressed hope that such cases would sensitize employers to issues of religious and ethnic discrimination.[22] Since the 9/11 attacks, the EEOC has been attempting to reach out to Arab and Muslim groups to explain what illegal discrimination is and what actions they can take to enforce their rights. The percentage of religious discrimination suits by those whose religion is Muslim has fallen from its high of 28 percent in 2002 to 20 percent in 2012, although during that time period, the number of overall charges of religious discrimination continued to grow.[23]

In 2007, Bilan Nur, a Muslim woman, won an award of $287,000 for religious discrimination. Nur had requested permission to wear a head covering during the holiday of Ramadan, a deviation from her employer's dress code. Her employer, Alamo Rent-a-Car, refused to allow her to wear the head scarf in front of customers while she worked at the front counter. Nur wore the head scarf while at the front counter in violation of the dress code. Alamo sent Nur home several times and eventually fired her for wearing the head scarf. The EEOC brought a case against Alamo on behalf of Nur, and her award included $21,640 in back pay, $16,000 in compensatory damages, and $250,000 in punitive damages.[24] The EEOC stated that it hoped the large punitive damages would send a message to employers that religious discrimination would not be tolerated.[25]

---

[21]Marjorie Valbrun, "U.S. Battles Bias against Arabs and Muslims in the Workplace," *The Asian Wall Street Journal*, A6 (Apr. 14, 2003).

[22]*Id.*

[23]U.S. Equal Employment Opportunity Commission, Religion-based charges filed from 10/10/2000 through 9/30/2011. Showing percentage filed on the basis of religion—Muslim. Accessed April 15, 2015 at www.eeoc.gov/eeoc/events/9-11-11_religion_charges.cfm.

[24]Kevin D. Kelly, "Jury Awards $287,000 to Muslim Employee Denied a Religious Accommodation." Accessed March 12, 2008 at www.lexology.com/library/detail.aspx?g=43ea1ef7-353c-4134-b6e5-fe408356149a&l=6G99TH2.

[25]*Id.*

Sex. Under Title VII, *sex* is interpreted as referring only to gender and not to sexual preferences. Hence, homosexuals and transsexuals are not protected under the act. It would, however, be sex discrimination to fire male homosexuals while retaining female homosexuals.

Also, as you may recall from the earlier discussion of the case of *Joseph Oncale v. Sundowner Offshore Services*,[26] the U.S. Supreme Court has held that Title VII prohibits same-sex harassment regardless of the harasser's sexual orientation as long as the discrimination is tied to some kind of gender discrimination.[27]

While the *Oncale* case did not explicitly extend Title VII protection to discrimination based on a person's sexual orientation, the EEOC has muddied the waters a bit by holding that discrimination claims based on gender identity are cognizable under Title VII.

In *Macy v. Dept. of Justice*,[28] a case some have described as groundbreaking, the EEOC held that a complaint of discrimination based on "gender identity, change of sex, and/or transgender status" *is* cognizable under Title VII. In that case, Mia Macy completed a telephone interview for a position with a federal agency while she presented as a man and was told the position was hers barring any issues with her background check. While her background check was being done, she told the agency that she was in the process of transitioning from a male to a female. Five days later, Macy was told that the position was no longer available; another person was hired for the position soon thereafter. Macy filed her original claim with the EEOC, and the agency separated her claims into two distinct claims—one for discrimination based on "sex" and one for discrimination based on "sex stereotyping," "gender transition/change of sex," and/or "gender identity." The agency further indicated that the gender identity stereotyping aspect of her claim would be processed outside the EEOC's standard Title VII adjudication process.

Macy appealed to the full commission to have both aspects of her claim handled through the normal Title VII process. On appeal, the EEOC concluded that each of these formulations of Macy's claims was merely different ways of stating the same claim for discrimination "based on . . . sex," which clearly was cognizable under Title VII.

The EEOC further stated, "[a]s used in Title VII, the term 'sex' 'encompasses both sex—that is, the biological differences between men and women—and gender.' As the 11th Circuit noted . . . Title VII barred 'not just discrimination because of biological sex, but also gender stereotyping—failing to act and appear according to expectations defined by gender.' As such, the terms 'gender' and 'sex' are often used interchangeably to describe the discrimination prohibited by Title VII. That Title VII's prohibition on sex discrimination proscribes gender discrimination, and not just discrimination on the basis of biological sex, is important. If Title VII proscribed only discrimination on the basis of biological sex, the only gender-based disparate treatment would be when an employer prefers a man over a woman, or vice versa. But the statute's protections sweep far broader than that, in part because the term 'gender' encompasses not only a person's biological sex but also the cultural and social aspects associated with masculinity and femininity."

In September 2014, the EEOC filed two lawsuits in federal court challenging transgender discrimination. The first alleged that Lakeland Eye Clinic, a Florida-based organization of health care professionals, discriminated based on sex in violation of federal law by firing an employee because she was transgender, because she was transitioning from male to female, and/or because she did not conform to the employer's gender-based expectations, preferences, or

---

[26] 523 U.S.75 (1988).

[27] Ibid.

[28] EEOC Appeal No. 0120120821, 2012 WL 1435995 (E.E.O.C.) (April 20, 2012).

stereotypes.[29] The second suit, *EEOC v. R.G. & G.R. Harris Funeral Homes, Inc.*, found that Harris violated Title VII by firing the funeral director because of her transgender status, because of her gender transition, and/or because the firing was based on gender-based stereotypes.[30]

Those cases were filed just two months after President Obama issued Executive Order 13672, prohibiting federal contractors from discriminating against workers based on their sexual orientation or gender identity.

As noted earlier, sexual harassment is addressed by Title VII's prohibition against discrimination based on sex. Although sexual harassment cases were not filed in large numbers immediately after the passage of Title VII, the number of such cases filed has increased tremendously since law professor Anita Hill captivated the nation in late 1991 by testifying before Congress about the harassment to which she was subjected by U.S. Supreme Court nominee Clarence Thomas. According to the EEOC, 9,953 sexual harassment complaints were filed in the year ending in October 1992, an increase of 2,564 over the previous year. In fiscal year 2011, 11,364 sexual harassment complaints were filed; 11,717 were filed in 2010. The EEOC recovered $48.4 million for successful claimants in 2010 and $52.3 million in 2011.[31] Not all the sexual harassment charges are filed by women; in 2010, 16.4 percent of those charges were filed by males.[32]

These sex discrimination cases can be quite costly. For example, it cost Morgan Stanley $54 million to settle a sex discrimination case brought by 67 female officers and women eligible for officer promotions.[33] The women had alleged workplace discrimination in promotions, assignments, and compensation, along with a hostile work environment. Although management admitted no guilt, it agreed to set up mechanisms to prevent sex discrimination. Thus, it is important that businesspeople be able to recognize sexual harassment and prevent its occurrence in the workplace. Exhibit 21-3 provides some suggestions on how managers can avoid liability for sexual harassment.

---

1. Senior management must make clear its position that sexual harassment in any form will not be tolerated.
2. Have an explicit written policy on sexual harassment that is widely disseminated in the workplace and given to every new employee.
3. Make sure employees know what is, and is not, sexual harassment.
4. Provide a gender-neutral training program on sexual harassment for all employees.
5. Establish an efficient system for investigating charges of sexual harassment and punishing violators.
6. Make sure that complaints are to be filed with a neutral party, not with the employee's supervisor.
7. Thoroughly investigate and resolve every complaint, punishing every violation appropriately. If no violation is found, explain to the complainant why there was no violation.

EXHIBIT 21-3

TIPS FOR AVOIDING SEXUAL HARASSMENT CHARGES

Source: Adapted from K. Swisher, "Corporations Are Seeing the Light on Harassment," *Washington Post National Weekly Edition*, February 14–20, 1994, 21.

---

[29] "EEOC Sues Lakeland Eye Clinic for Sex Discrimination Against Transgender Employee." EEOC Press Release, September 25, 2014. Available at www.eeoc.gov/eeoc/newsroom/release/9-25-14e.cfm.

[30] Civ. No. E.D. Mich. 2:14-cv-13710-SFC-DRG.

[31] U.S. Equal Employment Opportunity Commission, Sexual Harassment Charges: EEOC & FEPAs Combined: FY 1997–FY 2011. Available at www.eeoc.gov/eeoc/statistics/enforcement/sexual_harassment.cfm.

[32] U.S. Equal Employment Opportunity Commission, Sexual harassment charges: EEOC & FEPAs Combined: FY 1997–FY 2011." Accessed December 31, 2010 at www.eeoc.gov/eeoc/statistics/enforcement/sexual_harassment.cfm.

[33] "EEOC and Morgan Stanley Announce Settlement of Sex Discrimination Lawsuit." EEOC Press Release, July 12, 2004. Accessed March 19, 2008 at www.eeoc.gov/press/7-12-04.html.

Pregnancy Discrimination Act. After a U.S. Supreme Court ruling that discrimination on the basis of pregnancy was not discrimination on the basis of sex under Title VII,[34] Congress amended the law by passing the Pregnancy Discrimination Act (PDA), which specifies that discrimination based on pregnancy is sex discrimination and that pregnancy must be treated the same as any other disability, except that abortions for any purpose other than saving the mother's life may be excluded from the company's medical benefits. The U.S. Supreme Court has concluded that Congress intended the PDA to be "a floor beneath which pregnancy disability benefits may not drop—not a ceiling above which they may not rise."[35] Consequently, the high court held that a California statute requiring unpaid maternity leave for pregnant women and reinstatement after the birth of the child was constitutional because the intent of the law was to make women in the workplace equal, not to give them favored treatment.[36]

In the summer of 2001, the PDA became the basis for the first ruling on the employment discrimination issue of gender equity in drug coverage. In a class action lawsuit against Bartell Drug Company, a Seattle judge ruled that the drugstore chain discriminated against women when it excluded prescription contraceptives from its employee health plan.[37] Granting summary judgment to the plaintiff, the judge said, "Male and female employees have different sex-based disability and health care needs, and the law is no longer blind to the fact that only women can get pregnant, bear children, or use prescription contraception."[38]

In 2015, an important case interpreting the PDA was handed down by the U.S. Supreme Court. In *Young v. UPS*,[39] the high court established a new test for determining when an employer's failure to accommodate a pregnant woman constitutes a violation of the act. According to that test, a pregnant employee denied accommodation can establish a "prima facie" claim of pregnancy discrimination where she shows that the employer did accommodate others "similar in their ability or inability to work." To avoid liability, the employer must demonstrate that its denial of the accommodation was based on a legitimate, nondiscriminatory reason. However, the Supreme Court's majority substantially limited the employer's ability to satisfy this burden by stating that the employer's proffered reason normally cannot consist of a claim that it was more expensive or less convenient to add pregnant women to the category of those whom the employer accommodated.

ENFORCEMENT PROCEDURES

Enforcement of Title VII is a very complicated procedure and is full of pitfalls. Failure to follow the proper procedures within the appropriate time framework may result in the plaintiff's losing her or his right to file a lawsuit under Title VII. An overview of these procedures is provided in Exhibit 21-4.

The Charge. The first step in initiation of an action under Title VII is the aggrieved party's filing a charge with the state agency responsible for enforcing fair employment laws (a state EEOC) or, if no such agency exists, with the federal EEOC. A *charge* is a sworn statement that sets out the name of the charging party, the name(s) of the defendant(s), and the nature of the discriminatory act. In states

---

[34]*General Electric Co. v. Gilbert*, 429 U.S. 125 (1976).

[35]*California Federal Savings & Loan Association et al. v. Department of Fair Employment & Housing et al.*, 479 U.S. 272 (1987).

[36]*Id.*

[37]*Erickson v. Bartell Drug Co.*, 141 F. Supp. 2d 1266 (W.D. Wash. 2001).

[38]*Id.*

[39]575 U.S. _ (2015).

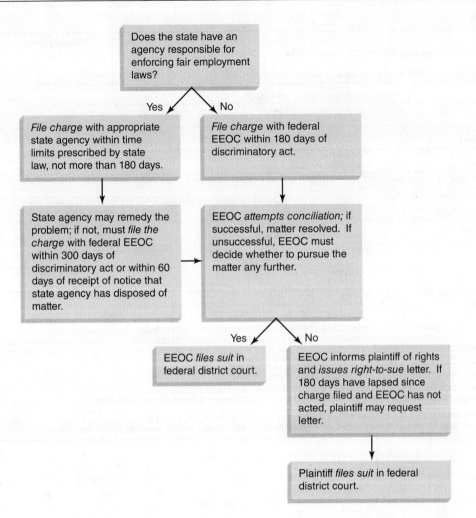

EXHIBIT 21-4

ANATOMY OF A TITLE VII CASE

that do not have state EEOCs, the aggrieved party must file the charge with the federal EEOC within 180 days of the alleged discriminatory act. In states that do have such agencies, the charge must be filed either with the federal EEOC within 180 days of the discriminatory act or with the appropriate state agency within the time limits prescribed by local law, which cannot be more than 180 days. If initially filed with the local agency, the charge must be filed with the federal EEOC within 300 days of the discriminatory act or within 60 days of receipt of notice that the state agency has disposed of the matter, whichever comes first. Exhibit 21-5 shows a typical charge.

Conciliation and Filing Suit. Once the EEOC receives the charge, it must notify the alleged violator of the charge within 10 days. After such notification, the EEOC investigates the matter in an attempt to ascertain whether there is "reasonable cause" to believe that a violation has occurred. If the EEOC does find such reasonable cause, it attempts to eliminate the discriminatory practice through conciliation. If unsuccessful, the EEOC may file suit against the alleged discriminator in federal district court.

If the EEOC decides not to sue, it notifies the plaintiff of his or her right to file an action and issues the plaintiff a right-to-sue letter. The plaintiff must have this letter to file a private action. The letter may be requested any time after 180 days have elapsed since the filing of the charge. As long as the requisite time period has passed, the EEOC will issue the right-to-sue letter regardless of whether the EEOC members find a reasonable basis to believe that the defendant engaged in discriminatory behavior.

CHARGE OF DISCRIMINATION	AGENCY	CHARGE NUMBER
This form is affected by the Privacy Act of 1974; See Privacy Act Statement before completing this form.	[X] FEPA [X] EEOC	

OHIO CIVIL RIGHTS COMMISSION _____ and EEOC
State or local Agency, if any

Name (Indicate Mr., Ms., Mrs.) Ms. Nellie Baldwin	Home Telephone (Include Area Code) (419) 863-4125	
STREET ADDRESS 826 Potter Road	CITY, STATE AND ZIP CODE Toledo, Ohio 43602	DATE OF BIRTH 11/10/56

NAMED IS THE EMPLOYER, LABOR ORGANIZATION, EMPLOYMENT AGENCY APPRENTICESHIP COMMITTEE, STATE OR LOCAL GOVERNMENT AGENCY WHO DISCRIMINATED AGAINST ME *(If more than one list below.)*

Name Mancum Manufacturers	NUMBER OF EMPLOYEES, MEMBERS +15	TELEPHONE (Include Area Code) (419) 693-8296
STREET ADDRESS 896 Lewis Ave.	CITY, STATE AND ZIP CODE Toledo, Ohio 43605	COUNTY Lucas
Name		TELEPHONE NUMBER (Include Area Code)
STREET ADDRESS	CITY, STATE AND ZIP CODE	COUNTY

CAUSE OF DISCRIMINATION BASED ON (Check appropriate boxes) [ ] RACE [ ] COLOR [X] SEX Female [ ] RELIGION [ ] NATIONAL ORIGIN [ ] RETALIATION [ ] AGE [ ] DISABILITY [ ] OTHER (Specify)	DATE DISCRIMINATION TOOK PLACE / / 02/05/93 [ ] CONTINUING ACTION

THE PARTICULARS ARE *(If additional space is needed, attach extra sheet(s)):*

1. I was employed by Canfield for 2 years as a machine operator general.
2. An opening for machine operator specialist, a higher position, was posted.
3. I applied for the position along with 4 other males and 2 females.
4. All applicants took a dexterity test.
5. I received the highest score on the test, but a male who scored second highest was promoted.
6. I was told that the posted job was "better suited for a male," but that with my test score I would be first in line when a more appropriate opening arose.

CXM/IFL:bd

[X] I also want this charge filed with the EEOC. I will advise the agencies if I change my address or telephone number and I will cooperate fully with them in the processing of my charge in accordance with their procedures.	Notary - (When necessary for State and Local Requirements) I swear or affirm that I have read the above charge and that it is true to the best of my knowledge, information and belief.
I declare under penalty of perjury that the foregoing is true and correct.	SIGNATURE OF COMPLAINANT *Ms. Nellie Baldwin*
Date Charging Party (Signature) EEOC TEST FORM 5 (09/01/91)	SUBSCRIBED AND SWORN TO BEFORE ME THIS DATE (Day, month, and year)

EXHIBIT 21-5

A TYPICAL CHARGE OF DISCRIMINATION FILED WITH THE EEOC

Source: United States Department of Labour.

REMEDIES

The plaintiff bringing a Title VII action can seek both equitable and legal remedies. The courts have broad discretion to order "such affirmative action as may be appropriate."[40] Under this broad guideline, courts have ordered parties to engage in diverse activities ranging from publicizing their commitment to minority hiring to establishing special training programs for minorities.

In general, a successful plaintiff is able to recover back pay for up to two years from the time of the discriminatory act. *Back pay* is the difference between the amount of pay received after the discriminatory act and the amount of pay that would have been received had there been no discrimination. For example, if two years before the case came to trial the defendant refused a promotion to a plaintiff on the basis of her sex and the job for which she was rejected paid $100 more per week than her current job, she would be entitled to recover back pay in the amount of $100 multiplied by 104. (If the salary rose at regular increments, these are also included.) The same basic calculations are used when plaintiffs were not hired because of discrimination. Such plaintiffs are entitled to the back wages they would have received minus any actual earnings during that time. Defendants may also exclude wages for any period during which the plaintiff would have been unable to work.

That same plaintiff may also receive remedial seniority dating back to the time the plaintiff was discriminated against.

The most significant impact of the 1991 Civil Rights Act resulted from its changes to the availability of compensatory and punitive damages. Under the new act, plaintiffs discriminated against because of race (and those discriminated against on the basis of sex, disability, religion, or national origin) may recover both compensatory damages, including those for pain and suffering, and punitive damages. In cases based on discrimination other than race, however, punitive damages are capped at $300,000 for employers of more than 500 employees, $100,000 for firms with 101 to 200 employees, and $50,000 for firms with 100 or fewer employees.

Attorney's fees are ordinarily awarded to a successful plaintiff in Title VII cases. They are denied only when special circumstances would render the award unjust. In those rare instances in which the courts determine that the plaintiff's action was frivolous or unreasonable or was without foundation, the courts may use their discretion to award attorney's fees to the prevailing defendant.

LILLY LEDBETTER FAIR PAY ACT OF 2009

In many cases, determining when a cause of action accrued can play a vital role in disposition of the case. Prior to 2007, the EEOC supported the position that every time an individual received a paycheck of a discriminatory amount, a new discriminatory compensation action arose. After every paycheck, an individual had 180 days to file a claim. In 2007, the Supreme Court decided, in *Ledbetter v. Goodyear Tire & Rubber Co.*,[41] that a compensation discrimination charge must be filed within 180 days of a discriminatory pay-setting decision. In other words, after an individual received the first discriminatory paycheck, she or he had 180 days to file a claim; subsequent paychecks no longer gave rise to new causes of action. Two years after the Court's decision in *Ledbetter*, President Obama signed the Lilly Ledbetter Fair Pay Act of 2009. That act, which explicitly recognizes the importance of protecting individuals who are victims of wage discrimination, restores the pre-*Ledbetter* policy that each paycheck gives rise to a new cause of action.

---

[40] § 706(a).
[41] 550 U.S. 618 (2007).

LINKING LAW AND BUSINESS

Management

Perhaps you learned in your organizational behavior or management class about *biculturalism*. This term refers to instances in which individuals of a particular racial or ethnic minority class have been socialized in two cultures—the dominant culture and the individual's ethnic or racial culture. Living in two cultures often increases stress, which is referred to as *bicultural stress*. Two general characteristics of bicultural stress are (1) role conflict—the conflict that exists when an individual fills two competing roles due to his or her dual cultural membership and (2) role overload—the excess expectations that result from living in two cultures. The intensity of these problems tends to increase for women of color, because of the negative dynamics directed toward both women and minorities. Hiring minorities can pose adaptation problems in the workplace for some managers. Accustomed to the cultural norms of the majority, some managers may be insensitive to the bicultural stress with which minorities are often burdened. In addition, managers may not realize that employees usually do not set aside their values and lifestyle preferences while at work. Therefore, it is important for managers to recognize differences and respond in ways that increase productivity without discriminating. This shift in management philosophy may include diversity training for managers and other employees to help them raise behavioral awareness, recognize biases and stereotypes, avoid assumptions, and modify policies. Therefore, an acute sensitivity to differences in the workplace may result in a friendlier environment where productivity is increased.

On January 29, 2009, President Obama signed the first bill of his presidency into law, which altered the measures set out in the Civil Rights Act of 1964 and overturned the Supreme Court's decision in the Ledbetter case. The bill was called the Lilly Ledbetter Fair Pay Act. After the Supreme Court's opinion in Ledbetter's case in 2007, Congress introduced legislation that would allow an employee six months to sue after every paycheck received. However, President George Bush opposed the legislation arguing that it would incite too many lawsuits. The legislation was put on hold until two years later when Congress, under President Obama, passed the measure, and President Obama then signed.

The Age Discrimination in Employment Act of 1967

Age Discrimination in Employment Act of 1967 (ADEA) Statute that prohibits employers from refusing to hire, discharging, or discriminating against people in terms or conditions of employment on the basis of age.

Our society does not revere age. Older employees detract from a firm's "youthful" image and are expensive. They have accumulated raises over the years and thus earn more than younger employees. They have pension benefits, which the employer will have to pay when they retire. They are sometimes viewed as rigid and unwilling to learn new technology. Thus, it is understandable that firms may attempt to discriminate against older employees. The **Age Discrimination in Employment Act of 1967 (ADEA)** was enacted to prohibit employers from refusing to hire, discharging, or discriminating in terms and conditions of employment on the basis of age. The language describing the prohibited conduct is virtually the same as that of Title VII, except that a person's being age 40 or older is the prohibited basis for discrimination.

Although the motivation for the ADEA was to prevent the unfair treatment of older people in the workplace, after the legislation had been in place for several years, some began to question whether the law also prohibited giving older workers more favorable treatment. In 2004, the U.S. Supreme Court decided that issue in *General Dynamics Land Systems, Inc. v. Dennis Cline et al.*[42] In *General Dynamics,*

---
[42] 540 U.S. 581 (2004).

present and former employees of General Dynamics brought suit under the ADEA. General Dynamics had instituted a policy effectively eliminating a retiree health insurance benefits program for workers under the age of 50. Those employees who were 50 or older at the time the policy was enacted would still be eligible for benefits, but others would not. The Supreme Court held that discrimination against "the relatively young" was beyond the scope of the protection offered by the ADEA. According to the Court's interpretation, the ADEA was designed to protect a "relatively old worker from discrimination that works to the advantage of the relatively young."[43] General Dynamics' policy did not violate the ADEA.

As the U.S. economy started a downward turn in late 2000, which continued through 2001, age discrimination claims began to increase. Charges of age discrimination filed with the EEOC rose from roughly 14,000 in fiscal year 1999 to 16,000 in 2000 and continued to increase to a peak of 22,778 in fiscal year 2009. In 2014, 20,588 charges of age discrimination were filed and the EEOC secured $77.7 million in benefits for aggrieved individuals.[44]

APPLICABILITY OF THE STATUTE

The ADEA applies to employers having 20 or more employees in an industry that affects interstate commerce. It also applies to employment agencies and to unions that have at least 25 members or operate a hiring hall. As a result of a Supreme Court ruling in *Kimel v. Florida Board of Regents*,[45] however, the act does not apply to state employers.

PROVING AGE DISCRIMINATION

Discrimination under the ADEA may be proved in the same ways that discrimination is proved under Title VII: by the plaintiff's showing disparate treatment or disparate impact. Most of the ADEA cases today involve termination. To prove a prima facie case of age discrimination involving a termination, the plaintiff must establish facts sufficient to create a reasonable inference that age was a determining factor in the termination. The plaintiff raises this inference by showing that he or she (1) belongs to the statutorily protected age group (age 40 or older), (2) was qualified for the position held, and (3) was terminated under circumstances giving rise to an inference of discrimination.

Until 1996, the plaintiff also had to demonstrate that he or she was replaced by someone outside the protected class. In *O'Connor v. Consolidated Caterers Corp.*,[46] however, the U.S. Supreme Court held that replacement by someone outside the protected class was not a necessity as long as evidence showed that the termination was based on age.

APPLYING THE LAW TO THE FACTS...

Carla worked as a secretary in a law firm. One day she was let go, but two other secretaries who were retained by the firm were under 40, while she was over 40. Carla said that because she had worked at the firm the longest, she deserved to keep her job and the firm keeping the secretaries younger than 40 proved age discrimination. Would Carla need any more evidence to prove her case? How might Carla be confused about the purpose of the ADEA?

---

[43]*Id.*

[44]U.S. Equal Employment Opportunity Commission, Age Discrimination in Employment Act (includes concurrent charges with Title VII, ADA and EPA) FY 1997–FY 2014. Retrieved May 15, 2010, from www.eeoc.gov/eeoc/statistics/enforcement/adea.cfm.

[45]120 S. Ct. 631 (2001).

[46]529 U.S. 62 (2000).

If the plaintiff establishes these three facts, the burden of proof then shifts to the defendant to prove that there was a legitimate, nondiscriminatory reason for the discharge. If the employer meets this standard, the plaintiff may recover only if he or she can show by a preponderance of the evidence that the employer's alleged legitimate reason is really a pretext for a discriminatory reason.

Initially, circuit courts were split on the evidentiary standard to which an age discrimination plaintiff must be held. Some courts have relied only on a pretext standard, as described earlier, whereas others have required a plaintiff to show direct, not just inferential, proof of discrimination (known as "pretext plus").[47] To resolve this circuit court confusion, the Supreme Court agreed to hear the case, *Reeves v. Sanderson Plumbing Products, Inc.*,[48] filed by a former employee who raised issues of age discrimination under the ADEA. The Supreme Court held that when a plaintiff establishes a prima facie case of age discrimination and subsequently provides sufficient evidence of pretext by the employer, a trier of fact can find unlawful discrimination without additional, independent evidence of discrimination. Therefore, "pretext plus" is no longer necessary. Case 21-4 demonstrates how the courts have applied the pretext standard.

CASE 21-4

Jones v. National American University
Eighth Circuit Court of Appeals
608 F.3d 1093 (2010)

Kathy Jones had been an employee at National American University (NAU) since 1998, when she was hired as a part-time corporate liaison at the university's Rapid City, South Dakota, campus. Later that year, Jones became a full-time admissions representative. In 2004, the director of admissions position at the Rapid City campus became available. Jones, then age 56, applied for the position. NAU formed a four-person committee to make the hiring decision. Individuals who had recently been denied a position as vice president of admissions for the university's online program were notified of the opening.

After receiving initial applications, the applicant pool was narrowed to six of the prior vice-president candidates and Jones. After phone interviews, the pool was narrowed to two of the former applicants and Jones. Each candidate attended an in-person interview. The position was offered to both of the prior vice-president candidates, both of whom rejected it. Jones was never offered the position, but was asked to serve as interim director until a candidate could be found. As part of her interim director duties, Jones helped to interview candidates for admissions positions. Following the interview of a 50-year-old candidate, one of the members of the hiring committee said, "I'm not sure we want a grandpa working with our college students." Jones never reported this comment. The open directorial position was eventually offered to a 34-year-old candidate for an admissions representative position. Jones resigned and filed discrimination charges with the EEOC.

After trial, the jury found that NAU had discriminated against Jones and that its conduct had been willful. The district court denied NAU's motion for judgment as matter of law and its motion for a new trial. Judgment was entered for $35,130 in damages, as well as attorney's fees and costs. NAU appealed.

Judge Murphy

NAU does not contest that Jones satisfied her burden of presenting a prima facie claim of age discrimination. Instead the university asserts that after it rebutted Jones's prima facie case by providing a legitimate, non-discriminatory reason for the failure to promote—specifically Jones's lack of management experience—she failed to establish that NAU's proffered reason was pretext.

An employee can prove that her employer's articulated justification for an adverse employment action is pretext "either directly by persuading the court that a discriminatory reason more likely motivated the employer or indirectly by showing that the employer's proffered explanation is unworthy of credence." Pretext may be shown with evidence that "the employer's

---
[47]M. Coyle, "How to Judge Age Bias," *National Law Journal* A10 (Mar. 20, 2000).
[48]530 S. Ct. 2097 (2000).

reason for the [adverse employment decision] has changed substantially over time."

Viewing the evidence in the light most favorable to Jones, we conclude that she presented sufficient evidence for the jury to conclude that NAU's proffered reason for the failure to promote was a pretext for age discrimination. Jones presented evidence that between the time of its EEOC charge response and the trial, NAU shifted its reasons for failure to promote her to the director position. NAU's response to the EEOC charge provided that throughout her employment, "Ms. Jones struggled with her performance. She consistently received moderate to low scores on her semiannual reviews. . . . She has consistently mediocre performance." By contrast, at trial NAU asserted that its primary reason for not promoting Jones was her lack of managerial and marketing experience. The university did not present evidence at trial that Jones was deficient in her performance.

Jones also presented evidence to dispute each of NAU's proffered reasons for their failure to promote her to the director position. She established that she was the only candidate considered who had the three years' recruiting experience listed as required in one job posting, and preferred in the other. She also presented evidence that Beck lacked the extensive management experience that the hiring committee asserted had been their primary qualification. She presented evidence that she had received consistently positive reviews and performance awards, and that she had a good relationship with her colleagues. Finally, Jones testified about the two age-related comments made by Buckles: (1) that he wasn't sure he wanted a "grandpa" working with the college kids, and (2) that Beck was a better long-term choice for the director position while Jones would have been the better short-term choice.

Given the benefit of all reasonable inferences from the evidence, Jones presented sufficient evidence at trial for the jury to determine that NAU's proffered reasons for the failure to promote were pretext for intentional age discrimination.

NAU alternatively argues that it is entitled to judgment as a matter of law under what it terms "the honest belief doctrine." See *Scroggins v. Univ. of Minn.*, 221 F.3d 1042, 1045 (8th Cir. 2000); *McNary v. Schreiber Foods, Inc.*, 535 F.3d 765, 769-70 (8th Cir. 2008). Relying on *Scroggins* and *McNary*, NAU asserts that "what ultimately matters is whether NAU's hiring committee established its honest belief in the nondiscriminatory facts that led to its decision." NAU's reliance on these cases is misplaced. In both *Scroggins* and *McNary*, the employer prevailed in a discrimination claim because the employee plaintiff failed to present any evidence to contradict the employer's asserted reason for the adverse employment decision. Thus, the employee failed to present evidence showing that the employer's proffered reason was mere pretext for discrimination. See *Scroggins*, 221 F.3d at 1045 (concluding that the employee had presented "no evidence suggesting anything other than the [employer's] honest belief"); *McNary*, 535 F.3d at 770 (same). By contrast, Jones has presented evidence sufficient to support a jury finding that NAU's alleged "honest belief" was pretext. The district court properly denied judgment as a matter of law to the university.*

Affirmed in favor of Plaintiff, Jones.

CRITICAL THINKING ABOUT THE LAW

So much about legal reasoning depends on taking a close look at analogies. No set of facts is ever exactly like another. When citing precedents, however, each party hopes that the facts in certain cases are similar enough in significant ways to cause the courts to select their cited cases as the more relevant ones in each case.

The questions here focus on the quality of the precedent cited by the defendant in Case 21-4.

1. What brought about and led to their being cited as authority by the defendant?

    Clue: Who won? Why?

2. What caused Judge Murphy to reject the analogies of *Scroggins* and *McNary* as binding in Case 21-4?

    Clue: Check her discussion of *Scroggins* and *McNary* to find how she distinguished *Scroggins* and *McNary* from the *Reeves* case.

STATUTORY DEFENSES

Bona Fide Occupational Qualification. A number of statutory defenses are available to an employer in an age discrimination case. The first is the bona fide occupational qualification, which requires the defendant to establish that he or she must hire employees of only a certain age to safely and efficiently operate the business in question. The courts generally scrutinize very carefully any attempt to demonstrate that age is a BFOQ.

---

*Jones v. National American University, Eighth Circuit Court of Appeals 608 F.3d 1093 (2010).

One example of an employer's successful use of this defense is *Hodgson v. Greyhound Lines, Inc.*,[49] wherein the employer refused to hire applicants aged 35 or older. Greyhound demonstrated that its safest drivers were those between the ages of 50 and 55, with 16 to 20 years of experience driving for Greyhound. Greyhound argued that this combination of age and experience could never be reached by those who were hired at age 35 or older. Therefore, to ensure the safest drivers, it should be allowed to hire only applicants younger than 35. The court accepted the employer's rationale. Although safety considerations are important, to use them in establishing age as a BFOQ, the employer must indeed prove, as did the defendant in the *Greyhound* case, that safety is related to age.

Other Defenses. As under Title VII, decisions premised on the operation of a bona fide seniority system are not unlawfully discriminatory despite any discriminatory impact. Likewise, employment decisions may also be based on "reasonable factors other than age."

Executive Exemption. In addition, termination of an older employee may be legal because of the **executive exemption**. Under this exemption, an individual may be mandatorily retired after age 65 if (1) he or she has been employed as a bona fide executive for at least two years immediately before retirement and (2) upon retirement, he or she is entitled to nonforfeitable annual retirement benefits of at least $44,000.

executive exemption
Exemption to the ADEA that allows mandatory retirement of executives at age 65.

After-Acquired Evidence of Employee Misconduct. An important issue, not just for the ADEA but also for other employment discrimination claims, is whether an employer can use evidence of an employee's misconduct discovered after a charge has been brought to defeat that charge. In *McKennon v. Nashville Banner*,[50] a unanimous Supreme Court decided that issue in a manner that pleased lawyers who represented both businesses and plaintiffs.

In *Nashville Banner*, the plaintiff had feared being fired by the company because of age, so she copied confidential documents to use (if needed) in her subsequent lawsuit. She was, in fact, fired, and filed a discrimination claim. The employer subsequently discovered that she had copied the documents and argued that her ADEA action should be dismissed because she would have been fired anyway had the firm known that she had copied the documents. The circuit court held that she deserved to be fired because of her misconduct and, therefore, she could not sue for discrimination.

The Supreme Court overruled the circuit court and held that after-acquired knowledge of misconduct will not bar a discrimination action. The Supreme Court did not, however, believe that such conduct should be totally irrelevant. If the defendant can prove that the misconduct was substantial enough to have warranted termination of the employee, then reinstatement will not be required. The amount of back pay required will also be reduced. The employee's back pay will be calculated from the date of the unlawful discharge until the date the evidence of misconduct was discovered. Thus, the after-acquired evidence may be used to reduce, but not completely bar, an action for discrimination.

ENFORCEMENT PROCEDURES

Enforcement of ADEA is similar to the enforcement of Title VII. The victim of age discrimination may file a charge with the appropriate state agency or with the EEOC within 180 days of the act. If a charge has been filed with the state agency, an EEOC charge must be filed within 300 days of the discrimination or within 30 days of receiving notice of the termination of state proceedings, whichever comes first. The charge

---
[49] 499 F.2d 859 (7th Cir. 1974).
[50] 513 U.S. 352 (1995).

must identify the defendant and specify the nature of the discriminatory act. Upon receipt of a charge, the EEOC must notify the accused and attempt to conciliate the matter. If conciliation fails, the EEOC may then bring a civil action against the violator.

A party who does not plan to file a private action may choose to file a charge with the EEOC only; if a party wishes to file a private civil action, complaints must be filed with both the appropriate state agency and the EEOC. If these complaints are filed within the appropriate time limits, a party then has three years from the date of the discriminatory act within which to file a private action under ADEA, assuming that the alleged discriminatory act was willful. If the alleged discrimination is purportedly unwillful, the party has two years within which he or she must file the private action. The party, however, must wait 60 days from the date of the filing of the complaints with both the EEOC and the state agency before filing the lawsuit. If the EEOC or the state agency files an action on the matter during that time, the plaintiff is precluded from filing suit.

REMEDIES UNDER ADEA

A successful ADEA plaintiff is entitled to back pay for up to two years. In addition, in a private action, a plaintiff may be able to recover liquidated damages in an amount equal to the back pay recovered if the plaintiff can prove that the employer acted willfully. *Willfully* means that the employer was substantially aware of the possibility that it was in violation of the ADEA but did not attempt to ascertain the legality of its actions. If liquidated damages are not granted, the plaintiff is generally entitled to interest on the back pay; interest is not awarded when liquidated damages have been granted. Compensatory damages for items such as mental distress from the discrimination are occasionally, but rarely, awarded by a few courts. Likewise, punitive damages are rarely awarded.

The Rehabilitation Act of 1973

In 1973, Congress broadened the class of individuals protected against discrimination to include the handicapped by passing the **Rehabilitation Act of 1973**, an act designed to protect the handicapped from discrimination in employment and to help them secure rehabilitation, training, access to public buildings, and all benefits of covered programs that might otherwise be denied them because of their handicap. It also requires that covered employers have a qualified affirmative action program for hiring and promoting the handicapped. A *handicapped individual*, for purposes of the act, is defined as one who has a "physical or mental impairment, which substantially limits one or more of such person's major life activities"[51] or who has a record of such impairment. Even people who are falsely regarded as having such an impairment are protected. The major provisions of this act are outlined in Table 21-3.

This act applies only to the federal government and employers that have contracts with the federal government, so its impact is relatively limited. However, in 1991, the Americans with Disabilities Act (ADA) was passed, which extends similar prohibitions against discrimination to private-sector employers that do not have federal contracts. Because of its broader impact, the ADA is discussed in greater detail in the next section; bear in mind that the principles discussed with respect to the ADA apply to the Rehabilitation Act as well.

It is important to remember that neither the Rehabilitation Act nor the ADA requires any employer to hire an unqualified individual. The acts require only the hiring of an individual with a disability who, with reasonable accommodation for his or her disability, can perform the job at the minimum level of

Rehabilitation Act of 1973 Prohibits discrimination in employment against otherwise qualified persons who have a handicap. Applies only to the federal government, employers that have contracts with the federal government, and parties who administer programs that receive federal financial assistance.

---

[51] 31 U.S.C. § 706(b).

TABLE 21-3 SUMMARY OF MAJOR PROVISIONS OF THE REHABILITATION ACT

Section	Potential Defendant	Prohibited Conduct	Required Conduct
501	Federal departments and agencies	Cannot discriminate against otherwise qualified workers because of a handicap	Prepare and implement an affirmative action plan for hiring and promoting the handicapped
502	Federal agencies entering into contracts with private employers for property or services and the private employers entering into these contracts	Private party with government contract cannot discriminate against otherwise qualified workers because of a handicap	Contracts must contain a clause requiring the private employer to take affirmative action in hiring and promoting the handicapped and not to discriminate against them
504	Parties who administer programs receiving federal assistance	Discrimination by those administering programs is prohibited	

productivity that would be expected of an individual with no disability. Nor does this act or the ADA prohibit an employer from terminating an employee whose disability does in fact prevent him or her from doing the job.

The Americans with Disabilities Act of 1991

Americans with Disabilities Act of 1991 (ADA) Statute requiring that employers make reasonable accommodations to the known disabilities of an otherwise qualified job applicant or employee with a disability, unless the necessary accommodation would impose an undue burden on the employer's business.

Like the Rehabilitation Act, the **Americans with Disabilities Act of 1991 (ADA)** is intended to prevent employers from discriminating against employees and applicants with disabilities by requiring employers to make reasonable accommodations to the known physical or mental disabilities of an otherwise qualified person with a disability, unless the necessary accommodation would impose an undue burden on the employer's business.

The ADA now covers all employers of 15 persons or more, which includes approximately 660,000 businesses, so its impact has the potential to be significant. Some fear that because of Amendments to the ADA passed in 2008 and described below, the impact of the act may become even more significant.

COVERED INDIVIDUALS

The definition of an *individual with a disability*, for purposes of the act, is essentially the same as the definition of a *handicapped individual* in the Rehabilitation Act. A *disability* is "(1) a physical or mental impairment which substantially limits one or more of the major life activities of such individual, (2) a record of such impairment, or (3) being regarded as having such an impairment."[52] A wide variety of impairments are captured under such a definition. Individuals suffering from diseases such as cancer, epilepsy, and heart disease are included, as are those who are blind or deaf. Those who are infected with the human immunodeficiency virus (HIV) but are not yet symptomatic are covered,[53] as are those whose past records may harm them. For example, persons who suffer from alcoholism but are not currently drinking or persons who are former drug addicts are protected. However, an employee who is currently a substance abuser and whose abuse would affect job performance is not protected.

Initially, both the EEOC and the U.S. Supreme Court interpreted both the "substantial impairment" requirement fairly strictly and considered a fairly limited number of activities to be "major life activities." But in 2008, deciding

---
[52] 42 U.S.C. § 12102(2). Title 42—The Public Health and Welfare.
[53] 1998 WL 332958.

that the rulings of the Supreme Court and the EEOC were too restrictive and were not in keeping with the spirit of the ADA, Congress passed the Americans with Disabilities Act Amendments Act of 2008 (ADAAA) to remedy the situation.

The ADAAA now provides that "major life activities include, but are not limited to, caring for oneself, performing manual tasks, seeing, hearing, eating, sleeping, walking, standing, lifting, bending, speaking, breathing, learning, reading, concentrating, thinking, communicating, and working." The ADAAA also added as a major life activity "the operation of a major bodily function, including, but not limited to, functions of the immune system, normal cell growth, digestive, bowel, bladder, neurological, brain, respiratory, circulatory, endocrine, and reproductive functions," thereby expanding the number of people who could potentially be protected under the act.

One other change by the ADAAA that could increase the number of claimants is its overturning previous interpretation of the act that held that when determining whether a person was disabled, mitigating actions were taken into account. Thus, a person who was hard of hearing but who could hear when wearing his hearing aids was not considered disabled. The ADAAA now prohibits the use of mitigating measures in evaluating whether a person has a disability, except for the use of glasses or corrective lenses if they fully correct the vision problem.

Initially, there was not a huge spike in claims. The law took effect January 1, 2009, and the EEOC reported 21,451 claims filed in fiscal year 2009, which is an increase of almost 2,000 claims over the 19,453 claims filed in 2008, the same as the increase in the number of claims between 2007 and 2008, when there was no change in the law. However, in fiscal year 2010, the number of disability claims rose more significantly to 25,165 claims, an increase of almost 4,000 claims, whereas in the prior three years, the increases were around 2,000 claims.[54]

Employers often find it difficult to know how the ADA applies to those who have mental disabilities. Under the ADA, employers are not only forbidden from discriminating against persons with mental disabilities but also must make reasonable accommodations for them unless such accommodations could cause undue hardship. Typical accommodations include providing a private office, flexible work schedule, restructured job, or time off for treatment.

Another type of disability that employers must be aware of is an intellectual disability. According to the EEOC, roughly 1 percent of Americans have an intellectual disability, and only about 31 percent of these individuals are employed, even though a much greater percentage would like to work.

An individual is considered to have an intellectual disability when (1) the person's intellectual functioning level (IQ) is below 70–75; (2) the person has significant limitations in adaptive skill areas as expressed in conceptual, social, and practical adaptive skills; and (3) the disability originated before the age of 18. "Adaptive skill areas" refers to basic skills needed for everyday life and includes communication, self-care, home living, social skills, leisure, health and safety, self-direction, functional academics (reading, writing, basic math), and work.

Another difficulty employers often have with the act is determining whether the accommodation an employee requests is reasonable. Case 21-5 is an example of such a difficult determination.

---

[54]U.S. Equal Employment Opportunity Commission, Americans with Disabilities Act of 1990 (ADA) charges (includes concurrent charges with Title VII, ADEA, and EPA) FY 1997–FY2014. Retrieved January 1, 2011, from www.eeoc.gov/eeoc/statistics/enforcement/ada-charges.cfm.

CASE 21-5

McMillan v. City of New York
United States Court of Appeals for the Second Circuit
2013 U.S. App. LEXIS 4454 (2013)

Rodney McMillan had schizophrenia, which was treated with medication regularly. Despite this condition, McMillan held employment as a case manager for the HRA Community Alternative Systems Agency (CASA). His job duties included conducting annual home visits, processing social assessments, recertifying clients' Medicaid eligibility, and making referrals. At McMillan's place of employment, there was a "flex-time policy" that allowed employees to arrive at work anytime between 9 a.m. and 10 a.m. More specifically, employees were not considered "late" to the office unless they arrived after 10:15 a.m. Further, an employee's tardiness could be approved or disapproved by that person's supervisor.

McMillan testified in court that he usually woke up around 7 or 7:30 a.m. each morning, but his medications for schizophrenia made him "drowsy and sluggish." As a result, McMillan often arrived late to work, around 11 a.m. The Defendant City of New York did not contest that the reason for McMillan's frequent tardiness was the treatment for his disability. For a period of time, McMillan's tardiness was continuously approved by his supervisor. However, in 2008, McMillan's supervisor Loshun Thornton, refused to approve any more of McMillan's late arrivals. Later, on May 8, 2009, McMillan was fined for eight days' pay for his late arrivals, and in 2010, the City brought charges of "misconduct and/or incompetence" against McMillan.

Because of his issues at work, on March 23 and April 22, 2010, McMillan requested accommodations for his disabilities, including a later flex start time that would permit him to arrive at work between 10 a.m. and 11 a.m. These requests were sent to Donald Lemons, Deputy Director of HRA's Equal Employment Opportunity Office for evaluation, where they were denied. The Office did not approve McMillan's requests for accommodations because there would be no one to supervise McMillan after 6 p.m. if he "flexed" his hours to arrive later in the morning.

McMillan contended that the response to his request was insufficient and subsequently brought suit against the City of New York, alleging violations of the ADA. The district court granted summary judgment for the City and dismissed all of McMillan's claims. McMillan appealed.

Judge Walker

One of the central goals of the Americans with Disabilities Act of 1990 ("ADA") is to ensure that, if reasonably practicable, individuals are able to obtain and maintain employment without regard to whether they have a disability. To accomplish this goal, the ADA requires that employers provide reasonable accommodations to qualified individuals. This case highlights the importance of conducting a fact-specific analysis in ADA claims.

It is undisputed that Rodney McMillan's severe disability requires treatment that prevents him from arriving to work at a consistent time each day. In many, if not most, employment contexts, a timely arrival is an essential function of the position, and a plaintiff's inability to arrive on time would result in his failure to establish a fundamental element of a prima facie case of employment discrimination. But if we draw all reasonable inferences in McMillan's favor—as we must at summary judgment—it is not evident that a timely arrival at work is an essential function of McMillan's job, provided that he is able to offset the time missed due to tardiness with additional hours worked to complete the actual essential functions of his job.

In our view, the United States District Court for the Southern District of New York did not conduct a sufficiently detailed analysis of the facts that tend to undermine the City's claim that a specific arrival time is an essential function of McMillan's position before granting summary judgment for the City.

If a plaintiff suggests plausible accommodations, the burden of proof shifts to the defendant to demonstrate that such accommodations would present undue hardships and would therefore be unreasonable. An "undue hardship" is "an action requiring significant difficulty or expense."

The City already has a policy of allowing employees to "bank" any hours they work in excess of seven hours per day and apply banked time against late arrivals, provided that those late arrivals are approved. Because there is no evidence that pre-approving McMillan's tardiness would constitute an undue burden on the City, the question is whether McMillan would be able to bank sufficient time to cover his late arrivals.

The district court correctly concluded that assigning a supervisor to work past 6:00 p.m. would constitute an undue hardship. However, McMillan was presumably unsupervised when he made home visits for his clients or when he worked past 7:00 p.m. It is unclear from this record whether his home visits or after-hours work was supervised and, if not, whether McMillan could bank these unsupervised hours.

Even if McMillan could not bank post-6:00 p.m. time, he also states that he would be willing to work through his one-hour lunch. The City has a policy, based on a collective bargaining agreement, of not allowing employees to work through lunch unless they receive advance approval. The district court concluded, without further explanation, that "plaintiff's proposed accommodation could not have been accommodated without undue hardship." We disagree. On the limited record before us, such pre-approval does not strike us as "requiring significant difficulty or expense."

Additionally, although the parties do not discuss this in their briefs, it might be the case that on some days McMillan would be able to arrive (relatively) early. If he also worked through lunch or stayed through 6:00 p.m. on those days, he would be able to bank that time against future tardiness as well.

On the present record, we cannot find as a matter of law that McMillan's suggested accommodations would constitute undue hardships to the City and are therefore unreasonable. Accordingly, McMillan states a prima facie case of discrimination based on his disability and, at least with regard to his late arrivals, a prima facie case for failure to provide accommodations.

For the reasons given above, on this record, we cannot conclude that a reasonable juror would find McMillan's claims to be without merit. If the factual record is developed further, some or all of McMillan's claims may not survive summary judgment. On the record before us, however, dismissal is premature.*

Vacated and *Remanded*.

CRITICAL THINKING ABOUT THE LAW

1. What facts need to be discovered before a decision is reached with respect to this case once it is returned to the district court?

    Clue: Review the decision and make note of each of the facts the 2nd Circuit Court of Appeals mentions that would have been relevant to the decision.

2. The Appeals Court makes it clear that the key to establishing reasonableness is dependent on facts specific to the case. In addition to the facts that the appeals court mentions need to be discovered, what other fact would be relevant to considering the reasonableness of the accommodation sought by the plaintiff?

    Clue: Do you think it is relevant to examine the disciplinary record of the plaintiff as it pertains to other situations where he was unsupervised? Would it affect the reasonableness of the accommodation if the record of the office was checked and it was found that other employees are permitted to work without supervision? Explain.

A major difficulty employers face under the ADA is ensuring that they do not violate the law during the interview process. The EEOC issued guidelines to help employers comply with the law. The guidelines emphasize that employers' questions must be designed to focus on whether a potential employee can do the job, not on the disability, but it is often difficult to know when a question violates the act. Exhibit 21-6 provides examples of acceptable and unacceptable questions drawn from the EEOC's guidelines.

You May Ask	Do Not Ask
• Can you perform the functions of this job (essential and/or marginal), with or without reasonable accommodations? • Can you meet the attendance requirements of this job? • Do you illegally use drugs? Have you used illegal drugs in the past two years? • Do you have a cold? How did you break your leg? • How much do you weigh? Do you regularly eat three meals a day?	• Do you have a disability that would interfere with your ability to perform the job? • How many days were you sick last year? • What prescription drugs are you currently taking? • Do you have AIDS? Do you have asthma? • How much alcohol do you drink each week? Have you ever been treated for alcohol problems?

EXHIBIT 21-6

INTERVIEWING POTENTIAL EMPLOYEES WITHOUT VIOLATING THE ADA

Source: Adapted from EEOC's "Enforcement Guidance on Pre-Employment Disability—Related Inquiries and Medical Examinations Under the Americans with Disabilities Act" (EEOC, 1998).

---

*McMillan v. City of New York, United States Court of Appeals for the Second Circuit 2013, U.S. App. LEXIS 4454 (2013).

It is well worth the prudent employer's time to study these guidelines, because the liability for violating the rules for job interviews can be substantial. For example, one job applicant who was asked about his disability during an interview was awarded $15,000 in compensatory damages and $30,000 in punitive damages.[55] The plaintiff was partially disfigured, partially deaf, and partially blind as a result of two brain tumor operations. The plaintiff brought up the disability himself to explain a gap in his work record, but the interviewers told him they felt uncomfortable with his disability and asked him to make them feel more comfortable by describing the condition and its treatment. They also asked whether managers or customers had a problem with him because of his disability.

ENFORCEMENT PROCEDURES

The ADA is enforced by the EEOC in the same way Title VII is enforced. To bring a successful claim under the ADA, the plaintiff must show that he or she (1) had a disability, (2) was otherwise qualified for the job, and (3) was excluded from the job solely because of that disability.

REMEDIES

Remedies are likewise similar to those available under Title VII. A successful plaintiff may recover reinstatement, back pay, and injunctive relief. In cases of intentional discrimination, limited compensatory and punitive damages are also available. An employer who has repeatedly violated the act may be subject to fines of up to $100,000.

Affirmative Action

affirmative action plans
Programs adopted by employers to increase the representation of women and minorities in their workforces.

reverse discrimination
Discrimination in favor of members of groups that have been previously discriminated against; claim usually is raised by white males.

One of the most controversial workplace issues of the past two decades has been the legitimacy of **affirmative action plans**. Ever since employers began to try to create balanced workforces by focusing on increasing their employment of minorities, there have been cries that such actions constitute **reverse discrimination**, which is a violation of the Equal Protection Clause of the Fourteenth Amendment.

Many of the significant cases challenging affirmative action plans have arisen in contexts other than private employment. Other areas in which these programs have been challenged include school admissions policies and government policies to set aside contracts for minority businesses. Table 21-4 summarizes the major affirmative action cases. A close reading of the cases reveals the increasing scrutiny the courts have come to apply to affirmative action policies, and it now appears that any affirmative action plan that can withstand constitutional muster must (1) attempt to remedy past discrimination, (2) not use quotas or preferences, and (3) end or change once it has met its goal of remedying past discrimination. This standard was set forth by the Supreme Court in *Adarand Constructors, Inc. v. Pena*, a case challenging a federal affirmative action program (see Table 21-4).

In 1997, many interested observers hoped the U.S. Supreme Court would hand down a definitive decision regarding affirmative action cases in the employment setting, as the high court agreed to hear the case of *Taxman v. Board of Education* described in Table 21-4. The parties, however, settled the case before it went to trial.

The next major affirmative action case arose in the area of university admissions. In *Grutter v. Bollinger*,[56] admissions policies of the University of Michigan

---

[55] *EEOC v. Community Coffee Co.*, No. H-94-1061 (S.D. Tex. 1995).
[56] 288 F.3d 732 (6th Cir. 2002), *aff'd*, 539 U.S. 306 (2003).

TABLE 21-4 MAJOR REVERSE DISCRIMINATION CASES

Case	Alleged Discriminatory Action	Outcome
Regents of the University of California v. Bakke, 438 U.S. 265 (1978)	The school's special admissions policy reserved 16 out of the 100 available seats for minority applicants. Bakke was denied admission while minorities with lower test scores were admitted.	Although race could be one of a number of factors considered by a school in passing on applications, this special admissions policy was illegal because a classification that benefits victims of a victimized group at the expense of innocent individuals is constitutional only where proof of past discrimination exists.
United Steelworkers v. Weber, 443 U.S. 193 (1979)	The employer and union entered into a voluntary agreement that half the openings in a skilled craft training program would go to blacks until the rough proportion of blacks in the program was equal to that of blacks in the labor force. A white male who would have been admitted to the training program absent the plan challenged the plan.	The Court said it was clear that Congress did not intend to wholly prohibit private and voluntary affirmative action. To be valid, such plans must not unnecessarily trammel the rights of whites, should be temporary in nature, and should be customized to solve the past proven pattern of discrimination.
Johnson v. Santa Clara County Transportation Agency, 480 U.S. 616 (1987)	The County affirmative action plan authorized the agency to consider applicant's sex as a relevant factor when making promotion decisions for job classifications in which women have traditionally been underrepresented.	The Court held that the plan represented a moderate, flexible, case-by-case approach to gradually effecting improvement of the representation of women and minorities in traditionally underrepresented positions. The Court emphasized that the agency had identified a conspicuous imbalance in representation, that no slots were set aside for women or minorities, and that no quotas were established. Race or sex could just be one of several factors considered.
Adarand Constructors, Inc. v. Pena, 515 U.S. 200 (1995)	Plaintiff submitted the lowest bid for a government contract, but the contract was awarded to a Hispanic firm submitting a higher bid. The job was sent to the Hispanic firm in accordance with a government program giving 5 percent of all highway construction projects to disadvantaged construction firms.	In a landmark decision, the Supreme Court held that any federal, state, or local affirmative action program that uses racial or ethnic classifications as a basis for making decisions is subject to strict scrutiny by the courts. This level of scrutiny can be met only when (1) the program attempts to remedy past discrimination, (2) does not use quotas or preferences, and (3) will be ended or changed once it has met its goal of remedying past discrimination.
Hopwood v. State of Texas, 84 F.3d 720 (5th Cir. 1996)	Two white law school applicants were denied admission to the University of Texas Law School because of the school's affirmative action program. That program allowed admissions officials to take race and other factors into account when admitting students.	The Court of Appeals for the 5th Circuit held that the program violated the equal protection clause because it discriminated in favor of minorities. The U.S. Supreme Court refused to hear the case.
Taxman v. Board of Education of the Township of Piscataway 91 F.3d 1547 (3rd Cir. 1996)	The Board of Education wanted to eliminate one teaching position at Piscataway High School. A black female and white female had the same seniority and qualifications. Because minority teachers were underrepresented in the school, the board chose to lay off the white teacher to promote racial diversity.	Taxman challenged the policy as violative of Title VII. The trial court granted summary judgment in her favor. The circuit court of appeals affirmed, awarding her complete back pay. The defendants appealed to the U.S. Supreme Court, but the case was settled prior to the hearing before the high court.

TABLE 21-4 CONTINUED

Case	Alleged Discriminatory Action	Outcome
Jennifer Johnson v. Board of Regents of the University of Georgia, 263 F.3d 1234; 2001 WL 967756 (11th Cir. 2001)	The University of Georgia had an admissions policy that awarded a fixed numerical bonus to nonwhite and male applicants that it did not give to white and female applicants. The three plaintiffs, white females who were denied admission to the University of Georgia, filed an action arguing that the use of race violated the Equal Protection Clause, among other claims.	The district court found in favor of the plaintiffs and entered summary judgment in their favor. The defendants appealed on the issue of preferential treatment based on race. The circuit court said that it did not need to address the issue of whether student body diversity is a sufficiently compelling interest to withstand the strict scrutiny that the court must apply to government decision making based on race. Even if it were a compelling interest, a policy that mechanically awards an arbitrary diversity bonus to every nonwhite applicant at a decisive stage in admissions, and severely limits the range of other factors relevant to diversity that may be considered at the stage, is not narrowly tailored to achieve that interest. The policy, therefore, violates the Equal Protection Clause of the Fourteenth Amendment.
Grutter v. Bollinger, 288 F.3d 732 (6th Cir. 2002), *aff'd*, 539 U.S. 306 (2003)	A white female was denied admission to the University of Michigan Law School despite a 3.8 GPA and a 161 LSAT score. She argued that the school's admissions policy, which focused on applicants' academic ability coupled with a flexible assessment of their talents, experience, and potential to contribute to the learning environment as well as the life and diversity of the law school, resulted in her being discriminated against on the basis of race.	After a 15-day bench trial, the federal district court found that Michigan's use of race as a factor in admissions decisions was unlawful. The 6th Circuit Court of Appeals reversed, finding the use of race to be narrowly tailored because it was used only as a potential plus factor. The U.S. Supreme Court affirmed, agreeing that the Equal Protection Clause does not prohibit the narrowly tailored use of race in admissions decisions to further a compelling interest in obtaining the educational benefits that flow from a diverse student body.
Parents Involved in Community Schools v. Seattle School District No. 1, 551 U.S. 701 (2007)	Parents in Louisville and Seattle filed suit arguing that the secondary schools' admissions plans used race as a factor in violation of the Equal Protection Clause. The schools voluntarily adopted policies designed to assign students to schools so as to counteract segregated housing patterns. The students were classified as "white" or "nonwhite" and then race was used as a tiebreaker when students were being assigned to oversubscribed schools.	In an opinion written by Chief Justice Roberts, the Court held that the policy was not narrowly tailored. However, on the issue of diversity as a compelling state interest, the Court was intensely divided. Ultimately, the plurality opinion provides little guidance as to whether the Court would hold that diversity is a compelling state interest in secondary education.

Law School were challenged by Grutter, a white Michigan resident with a 3.8 GPA and 161 LSAT score, who was denied admission. The school followed an official admissions policy seeking to achieve student body diversity through compliance with *University of California v. Bakke*. Focusing on students' academic abilities, coupled with a flexible assessment of their talents, experiences, and potential, the policy required admissions officials to evaluate each applicant based on all the information available in the file, including a personal statement, letters of recommendation, an essay describing how the applicant would contribute to law school life and diversity, and the applicant's undergraduate grade point average (GPA) and Law School Admissions Test (LSAT) score. In addition, officials were required to look beyond grades and scores to so-called "soft" variables such as recommenders' enthusiasm, the quality of the undergraduate institution and the applicant's essay, and the areas and difficulty of undergraduate course selection. The policy did not define diversity solely in terms of racial and ethnic status and did not restrict the types of diversity contributions eligible for "substantial weight," but it did reaffirm the law school's commitment to diversity with special reference to the inclusion of African American, Hispanic, and Native American students, who otherwise might not be represented in the student body in meaningful numbers. By enrolling a "critical mass" of underrepresented minority students, the policy sought to ensure their ability to contribute to the law school's character and to the legal profession.

In 2014, in the case of *Schuette v. Coalition to Defend Affirmative Action*, the high court upheld a Michigan law that banned affirmative action in public education and in state employment and contracting. Although the court did not reexamine the constitutionality of affirmative action, six justices agreed that states could end racial preferences without violating the U.S. Constitution.

TECHNOLOGY AND THE LEGAL ENVIRONMENT

The Internet as a Public Accommodation?

As e-commerce and the use of the Internet become more common, the courts are likely to begin to treat the Internet as a public accommodation, meaning that it would be subject to the ADA, which states that "[n]o individual shall be discriminated against on the basis of disability in the full and equal enjoyment of the goods, services, facilities, privileges, advantages, or accommodations of any place of public accommodation by any person who owns, leases to, or operates a place of public accommodation."

No one yet knows the full implications of treating the Internet as a public accommodation, but a lawsuit filed by the National Federation of the Blind (NFB) against America Online (AOL) may help give us an idea of what some disabled individuals could expect.

In November 1999, the NFB sued AOL, alleging that the company's software does not work with other software required to translate computer signals into Braille or synthesized speech. By failing to remove communication barriers presented by its designs, and thus denying the blind independent access to its service, AOL is alleged to be violating the ADA.

In July 2000, AOL and the NFB reached an agreement. The NFB suspended the lawsuit against AOL, and AOL promised to have appropriate software by April 2001. The lawsuit was subsequently dropped. Regulations tacked on to Section 508 of the Rehabilitation Act, which went into effect in December 2000, also assisted the blind. Under these regulations, federal agencies must construct and design their websites using applications and technologies to make site information available to all users.

Grutter alleged that she was rejected because the law school used race as a "predominant" factor, giving applicants belonging to certain minority groups a significantly greater chance of admission than students with similar credentials from disfavored racial groups, and that respondents had no compelling interest to justify that use of race. The district court found the law school's use of race as an admissions factor unlawful. On appeal, the 6th Circuit reversed, holding that Justice Powell's opinion in *Bakke* was binding precedent establishing diversity as a compelling state interest and that the law school's use of race was narrowly tailored because race was merely a "potential 'plus' factor" and because the law school's program was virtually identical to the Harvard admissions program described approvingly by Justice Powell and appended to his *Bakke* opinion.

The U.S. Supreme Court upheld Michigan's policy, stating: "All government racial classifications must be analyzed by a reviewing court under strict scrutiny. But not all such uses are invalidated by strict scrutiny. . . . Race-based action necessary to further a compelling governmental interest does not violate the Equal Protection Clause so long as it is narrowly tailored to further that interest."[57]* The Court reaffirmed Justice Powell's view that student body diversity is a compelling state interest that can justify using race in university admissions and deferred to the law school's educational judgment that diversity is essential to its educational mission.

The Court recognized that attaining a diverse student body was at the heart of the law school's proper institutional mission and noted that its "good faith" is "presumed," absent "a showing to the contrary." The justices noted that enrolling a "critical mass" of minority students simply to ensure some specified percentage of a particular group merely because of its race or ethnic origin would be patently unconstitutional, but the law school justified its critical-mass concept by reference to the substantial, important, and laudable educational benefits that diversity is designed to produce, including cross-racial understanding and the breaking down of racial stereotypes. The justices also noted that the law school's position was bolstered by numerous expert studies and reports showing that such diversity promotes learning outcomes and better prepares students for an increasingly diverse workforce for society and for the legal profession.

In the high court's eyes, the law school's admissions program was a narrowly tailored plan, which meant that it "did not insulate each category with certain desired qualifications from competition with all other applicants." Instead, it considered race or ethnicity only as a "'plus' in a particular applicant's file" and was flexible enough to ensure that each applicant was evaluated as an individual and not in a way that made race or ethnicity the defining feature of the application. Finally, the policy was limited in time.

This case has provided some guidance to those in higher education, but the Supreme Court's most recent ruling on affirmative action has muddied the waters. While still recognizing *Grutter* as valid in higher education, in *Parents Involved in Community Schools v. Seattle School District No. 1*, the Court found that assigning K–12 students in such a way as to keep the schools racially diverse may not fall under the *Grutter* precedent, because it did not demonstrate a benefit gained by forcing diversity on schools. The most important lesson for employers to draw from these cases is that any policy based on race will need to adhere very closely to carefully set guidelines, especially in showing the necessity of using race to get a specific benefit.

While school districts and universities look to these court cases for guidance, employers have the guidance of the EEOC to look to when setting up affirmative action policies. The EEOC has issued guidelines in an attempt to help employers set up valid affirmative action plans. According to these guidelines, Title VII is not violated if (1) the employer has a reasonable basis for determining that an affirmative action plan is appropriate and (2) the affirmative action plan is reasonable. Quotas, however, are specifically outlawed by the 1991 Civil Rights Act amendments.

Global Dimensions of Employment Discrimination Legislation

With many U.S. firms having operations overseas, the question of the extent to which U.S. laws prohibiting discrimination apply to foreign countries naturally arises. The Civil Rights Act of 1991 extended the protections of Title VII and the

---

[57]*Id.*

*Grutter v. Bollinger (02-241) 539 U.S. 306 (2003) 288 F.3d 732.

ADA to U.S. citizens working abroad for U.S. employers. Amendments to the ADEA in 1984 had already extended that act's protection in a similar manner. The provisions of these acts also apply to foreign corporations controlled by a U.S. employer.

It is not always easy to determine whether a multinational corporation will be considered "American" enough to be covered by these acts. According to guidelines issued by the EEOC in October 1993, the EEOC will initially look at where the company is incorporated, but will often have to look at other factors as well. These other factors must also be considered when the employer is not incorporated, as, for example, in the case of an accounting partnership. Some of these additional factors include the company's principal place of business, the nationality of the controlling shareholders, and the nationality and location of management. No one factor is considered determinative, and the greater the number of factors linking the employer to the United States, the more likely the employer is to be considered "American" for purposes of being covered by Title VII and the ADEA.

LINKING LAW AND BUSINESS

Management

Affirmative action plans promote greater diversity in workplaces. Despite the controversial issues related to these programs, diversity itself can be advantageous to organizations. In your management class, you probably discussed some of the advantages of diversity. First, group decisions that include contributions from diverse employees are advantageous because a greater assortment of ideas for dealing with work issues may have gone into those decisions. Second, diversity may also enhance a firm's credibility with its customers, in the sense that the firm is portrayed and perceived as more able to identify with customers of various backgrounds. Third, diversity can encourage greater creativity and innovation in organizations. Fourth, diversity also tends to promote a more flexible organizational structure that is beneficial when a firm is faced with a need to change. Therefore, diversity, if properly managed, could be beneficial to a firm.

Sources: S. Certo, *Modern Management* (Upper Saddle River, NJ: Prentice Hall, 2000), 529–30; S. Robbins, *Organizational Behavior* (Upper Saddle River, NJ: Prentice Hall, 2001), 14.

In determining whether a foreign corporation is controlled by a U.S. employer, the EEOC again looks at a broad range of factors. Some of these factors include the interrelation of operations, common management, centralized labor relations, and common ownership or financial control over the two entities. However, a corporation that is clearly a foreign corporation and is not controlled by a U.S. entity is not subject to U.S. equal employment laws. An employer may also violate the ADA and Title VII if compliance with either law would constitute an illegal action in the foreign country in which the corporation is operating.

SUMMARY

During the early years of our nation's history, the employment-at-will doctrine governed the employment relationship. Under this doctrine, an employee without a contract for a set period of time could be fired at any time for any reason. The doctrine has been gradually eroded, and most states today recognize at least one of three exceptions to the employment-at-will doctrine: the public policy exception, the implied contract exception, and the implied covenant of good faith and fair dealing exception.

Civil rights laws have also eroded the employer's ability to hire and fire at will. This chapter examined those laws in the order in which they were enacted. The Civil Rights Act of 1866 prohibits employers from discriminating against individuals because of their race.

The Equal Pay Act of 1963 prohibits employers from paying male and female employees doing the same job different wages because of their sex.

Title VII prohibits employers from discriminating in terms and conditions of employment on the basis of race, color, national origin, religion, and sex. This act was amended by the PDA, which essentially requires employers not to discriminate against pregnant women and to treat pregnancy like any other temporary disability. Title VII was also amended by the Civil Rights Act of 1991, which expanded the remedies available under Title VII.

The ADEA prohibits discrimination based on age against persons aged 40 or over. Enforcement of the ADEA is similar to enforcement of Title VII.

The Rehabilitation Act requires federal agencies, employers that have contracts with the federal government, and parties who receive any type of federal funds not to discriminate against persons with handicaps. The ADA extended the basic protections of the Rehabilitation Act to private employers, requiring them to reasonably accommodate persons with disabilities.

Employers locating overseas must remember that they can no longer avoid Title VII and the ADEA simply by leaving the country. U.S. corporations operating in foreign nations, as well as foreign companies controlled by U.S. corporations, must follow the Title VII requirements.

REVIEW QUESTIONS

21-1 Explain the employment-at-will doctrine. Discuss why some people prefer the complete abolition of this doctrine, whereas others feel saddened by its gradual demise.

21-2 Explain why each of the following sets of jobs would or would not be considered equal under the Equal Pay Act:
   a. Male stewards and female stewardesses on continental air flights
   b. Male checkers of narcotics and female checkers of nonnarcotic drugs at a pharmacy
   c. Male tailors and female seamstresses

21-3 Explain the following aspects of the Equal Pay Act:
   a. Its purpose
   b. The remedies available under the act
   c. The defenses available to employers

21-4 Explain the following aspects of Title VII:
   a. Its purpose
   b. The remedies available under the act
   c. The defenses available to employers

21-5 Explain two significant ways in which the Civil Rights Act of 1991 has changed the application of Title VII.

21-6 What constitutes "reasonable accommodation" under the Rehabilitation Act and the ADA?

REVIEW PROBLEMS

21-7 The City of Los Angeles provided equal monthly retirement benefits for men and women of the same age, seniority, and salary. The benefits were partially paid for by employee contributions and partially by employer contributions. Because women, on average, live longer than men, the city required women to make contributions to the retirement fund that were 14.84 percent higher than those made by men. Was this a violation of the Civil Rights Act? Why or why not?

21-8 JoAnn, Ann, and Bryon were all laboratory analysts, performing standardized chemical tests on various materials. JoAnn was hired first, with no previous experience, and was trained on the job by the supervisor. She later trained Ann. When Bryon was hired, he was trained by the supervisor with the assistance of the two women. All initially worked the same shift and received the same pay. Then Bryon received a 5-cent-per-hour raise and was to work a swing shift every other two weeks.

Was his higher wage a violation of the Equal Pay Act? Explain.

21-9 Administrators of an Ohio Christian school refused to renew a teacher's contract after she became pregnant, on the basis of its belief that "a mother's place is in the home." When she filed sex discrimination charges under the state civil rights statute, she was fired. Was this termination unlawful? Explain.

21-10 Ellen's immediate supervisor repeatedly required her to have "closed door" meetings with him, in violation of company policy. As a consequence, rumors began to spread that the two were having an office romance, although the meetings in fact involved her boss's attempt to convince Ellen to loan him money, again in violation of company policy. When Ellen asked her immediate supervisor to try to stop the rumors, he said that he found them somewhat amusing and refused to do anything to stop them. As a consequence of the rumors, she began to be treated as an "outcast" by her coworkers and received low evaluations from other supervisors in the areas of "integrity" and "interpersonal relations." She was passed over for two promotions for which she had applied. She filed an action against her employer on the grounds that her supervisor had created a hostile environment by his refusal to stop the rumors. Do you believe she has a valid claim under Title VII? Why or why not? Are there any other causes of action she might raise? Explain.

21-11 A U.S. citizen was working at a multinational company's Zaire facility. The employer was incorporated in the state of Louisiana. When the employee was terminated, allegedly because of his age, he sought recovery under the federal ADEA as well as the Louisiana Age Discrimination in Employment law. The employer argued that its overseas operations were not subject to the federal ADEA. Was the employer correct? Explain.

21-12 Davis, D'Elea, and Sims were former heroin or narcotics addicts. Davis and Sims were told by the city director that they could not be hired by the city because of their former habit. D'Elea was rejected from a city CETA program because of his former habit. The three sued the city, alleging that drug addiction was a handicap under the Rehabilitation Act of 1973 and that the city's refusal to hire them was therefore unlawful under this act. Were they correct in their contention? Explain.

CASE PROBLEMS

21-13 Plaintiff Michael Matanic was offered a process engineer position with American Tool & Mold, Inc. (ATM). The offer for the position required the plaintiff to complete medical testing to determine his ability to perform the responsibilities of the position. ATM also required that all prospective employees be able to lift 35 pounds. During the medical testing for employment, prospective employees had to complete a "Back History" form, which specifically inquires whether an employee's back injury was a workers' compensation injury.

The plaintiff's completed "Back History" form revealed that he had suffered an injury to his spine, after which he had had multiple surgeries. After ATM was made aware of this past injury, they did not require completion of the back screen and did not assess whether he could lift 35 pounds. Those who administered the medical testing then provided the plaintiff with a statement stating that he was "recommended not fit for employment/work at this time because: More medical information needed, specifically, old records of back surgery. . . ." The plaintiff then provided ATM with the appropriate medical records and was further examined by a physician who provided the plaintiff with a Worker Status Report. This report was provided to ATM.

ATM still concluded that the plaintiff had failed to provide "appropriate medical documentation from the surgical team that performed the procedure . . . stating that he has no permanent restriction in order to proceed with the clearance for the position with ATM." The plaintiff was then terminated from his conditional employment. The plaintiff filed a charge of discrimination with the EEOC, and the EEOC found reasonable cause to believe that ATM had violated the ADA. The EEOC motioned for summary judgment. How do you think the court ruled? Why? *EEOC v. Am. Tool & Mold, Inc.*, 21 F. Supp. 3d 1268 (2014).

21-14 Plaintiff Jean Robert Paul was hired by defendant employer as a case manager at a residential facility for people needing various social services. The plaintiff was a 58-year-old Haitian man. Paul shared a work space with four people and frequently did not get along with them. Significant problems among the plaintiff and his coworkers began when the plaintiff complained that two of his coworkers told him that he was "old, Haitian, and that he can't be a boss" and excluded the plaintiff from a daytime boat excursion with clients, stating that "the trip is for young people."

The plaintiff alleged that his boss's interventions were not satisfactory and that harassment in the workplace continued. The plaintiff also alleged that the defendant retaliated against him for complaining about the harassment and discrimination in the workplace. The plaintiff ended up filing Title VII and ADEA claims, alleging that the defendant harassed him and discriminated against him on the basis of age and nationality. The defendant motioned for summary judgment. What evidence would the court need to conclude that this action could sustain a prima facie Title VII or ADEA hostile work environment claim? How do you think the court ruled? *Paul v. Postgraduate Ctr. for Mental Health*, 2015 U.S. Dist. LEXIS 42944 (2015).

21-15 Plaintiff Kargbo was a 52-year-old service coordinator for defendant Mendelsohn, a manager of social and health care services for senior citizens. The plaintiff completed five weeks of training before starting full time. The plaintiff alleges that during this training, the defendant stated, "I don't believe you are the right man for this job. You are 52 years old. This job is normally for young college graduates." The plaintiff further alleges that the defendant employer treated the plaintiff poorly throughout the beginning of his employment.

During the plaintiff's three-month evaluation, the defendant listed the plaintiff's performance as "unsatisfactory in the categories of effective communication and learning orientation." The defendant later wrote an interoffice memorandum highlighting several complaints about the plaintiff's work performance. Finally, the defendant submitted a form recommending that the plaintiff be terminated. The plaintiff brought claims under Title VII of the Civil Rights Act and the Age Employment Discrimination Act related to his termination of employment. The defendant moved for summary judgment on all claims. How do you think the court ruled? Why? *Kargbo v. Phila. Corp. for Aging*, 16 F. Supp. 3d 512 (2014).

21-16 Katharine Richardson was hired by Friendly's Ice Cream Corporation (Friendly's) as an assistant manager of its Ellsworth, Maine, store in 2000. Between 2000 and 2006, Richardson performed both administrative and manual tasks as a part of her job. In January 2006, Richardson began to experience severe pain in her right shoulder. The pain was caused by the manual tasks that she had been performing at work. The company sent Richardson to see a physician, who diagnosed Richardson with shoulder impingement syndrome. The physician recommended that Richardson stop doing the manual tasks she had been doing at Friendly's. Richardson continued to work until September 2006, when she took a leave of absence to undergo shoulder surgery. After the surgery, physicians indicated that Richardson would still be unable to perform manual tasks at Friendly's. When Richardson did not recover as quickly as anticipated, the company terminated her employment, explaining that she was disabled and had exceeded the leave guaranteed her by the Family and Medical Leave Act. Friendly's moved for and won summary judgment after the close of discovery. Richardson appealed, arguing that she was discharged because of her disability. Friendly's argued that Richardson was no longer qualified for the position because she could not perform the essential functions of the position with or without reasonable accommodation. What was the essential function of Richardson's position? How do you think the appellate court ruled? *Richardson v. Friendly Ice Cream Corp.*, 594 F.3d 69 (1st Cir. 2010).

21-17 Susann Bashir worked at Southwestern Bell, a division of AT&T, from 1999 until 2010, when she was fired. In 2005, Bashir had converted to Islam. Several of her coworkers and two of her supervisors made degrading comments to her concerning her traditional hijab head covering. Bashir's supervisor referred to her as "one of those bomb-people" and was told repeatedly to remove her "hat thing." One of Bashir's supervisors allegedly tried to pull Bashir's hijab off of her head. Southwestern Bell countered that there was no evidence that such actions had occurred. AT&T asserted its dedication to maintaining an inclusive workplace. Does this case constitute discrimination? If so, how? What act protects Bashir if discrimination is present? *Bashir v. S.W. Bell Tel. Co.*, No. 1016-CV38690 (Mo., Jackson Co. Cir. May 3, 2012).

21-18 A.D.P. began working for the company that would eventually become ExxonMobil in 1978. She performed her job very well and was subsequently promoted to a position of senior research associate in 2005. Shortly after being promoted, A.D.P.'s husband passed away. A.D.P. became depressed. However, her work performance did not change and was still as strong as ever. A.D.P. voluntarily informed another employee of ExxonMobil that A.D.P. was going to seek treatment for her alcoholism. Upon returning to work, A.D.P. was subjected to random breathalyzer tests in keeping with company policy, even though A.D.P. had never been documented as being under the influence of alcohol at her job. A.D.P. passed the first nine breath tests that were administered to her. The tenth test estimated her blood alcohol content at 0.047 and 0.043. A.D.P. was then fired by ExxonMobil. A.D.P. brought suit against ExxonMobil for discrimination on the grounds of her disability. The appellate court found in favor of the plaintiff, A.D.P., after a state court had dismissed the case. How is A.D.P. covered by the Americans with Disabilities Act if she is not handicapped? Were there grounds for breathalyzer testing of A.D.P. based on her work performance? Explain. Why do you think the appellate court found in her favor? *A.D.P. v. ExxonMobil Research and Engineering*, 54 A.3d 813 (N.J. Super. App. 2012).

THINKING CRITICALLY ABOUT RELEVANT LEGAL ISSUES

Currently, one of the biggest issues surrounding discrimination is whether sexual orientation should be a protected characteristic along with age, race, color, gender, national origin, religion, and pregnancy. Proponents of this legislation often present the issue as one of fairness. All of these other groups get protection, so we also deserve protection, they plead. Supporters claim that gay people can be legally fired for being gay in many states, suggesting that being legally fired for a characteristic is inherently unfair.

Unfortunately, the supporters of this legislation do not understand the at-will doctrine of employment that is standard (with some exceptions) in the United States. Unless an employee is part of a collective bargaining agreement or under contract, an employer can fire him or her for anything at any time (with some exceptions, such as discrimination laws and whistleblower laws). Employees under collective bargaining agreements and contracts must be fired for "just cause"—meaning there has to be a good reason, such as sleeping on the job, for firing the employee. Anybody else can be fired for just about anything.

Supporters of adding sexual orientation to the list of protected characteristics do not seem to recognize that there are hundreds of unprotected characteristics. Employers can fire employees for coming to work with purple hair, even if the purple hair has no effect on employee productivity. Many proponents of this legislation argue that being gay is not a choice, just like race or gender, so it should be protected. There are many characteristics about which people have no choice that are not protected. In every state, at-will employees can be fired for having annoying voices. Unless the voice is a consequence of that employee's race or gender (such as finding that Latino voices or female voices are annoying), having an annoying voice is an unprotected and unchosen characteristic.

Moreover, if sexual orientation is added to the list of protected characteristics, it would essentially take away the rights of an already protected characteristic—religion. Many Americans object to homosexuality on religious grounds. Forcing a religious employer to violate his religious principles by hiring homosexual employees is discrimination against religious employers! It just wouldn't be fair to give sexual orientation protected status.

1. How would you frame the issue and the conclusion of this essay?

2. The writer gives several reasons to support her conclusion. Identify the reasons and describe the reasoning.

   Clue: The reasoning is the logic that ties the reasons to the conclusion. Ask yourself: "How does saying 'I like cake' lead to 'Let's go get cake'"?

3. How appropriate are the analogies used in this argument?

4. Write a short essay approaching the issue from a viewpoint different from the author's.

   Clue: What analogies could you use in making your case?

ASSIGNMENT ON THE INTERNET

Affirmative action remains a contentious issue in areas of employment, school admissions, and government policies. Yet, many companies use a form of affirmative action to create a diverse workforce. Using the Internet, find a company with an affirmative action policy and review the policy.

Using the test set forth in *Adarand Constructors, Inc. v. Pena* and EEOC guidelines, determine whether the affirmative action policy would withstand a legal challenge. Explain. Also make a list of information not available to you on the company's website that would assist you in better determining the legality of its affirmative action policy.

ON THE INTERNET

www.policyalmanac.org This site is a useful resource for research on affirmative action.

www.eeoc.gov The home page of the U.S. Equal Employment Opportunity Commission provides numerous links to helpful information, including statistics, laws, and regulations as well as how to file a charge.

www.ada.gov The ADA home page provides numerous resources for employers trying to comply with the ADA.

www.law.cornell.edu Here is a page that gives you information about discrimination law, as well as allows you to search for statutes and cases related to employment discrimination.

www.dol.gov This Department of Labor website provides labor policy information and an employment law guide.

www.discriminationattorney.com This site contains information about discrimination laws, exemplary cases, and articles for use by employers and employees.

www.ohchr.org This United Nations website provides information on all of the international treaties and conventions concerning discrimination.

FOR FUTURE READING

American Bar Association. *Guide to Workplace Law* (2nd ed.). New York: Random House Reference, 2006.

Areheart, Bradley A. "The Anticlassification Turn in Employment Discrimination Law." *Alabama Law Review* 63 (2012): 955.

Beiner, Theresa M. "The Many Lanes Out of Court: Against Privatization of Employment Discrimination Disputes." *Maryland Law Review* 73 (2014): 837.

Byrd, Robert C., and John D. Knopf. "Do Disability Laws Impair Firm Performance?" *American Business Law Journal* 47 (2010): 145.

Choi, Victor, Kendrick Kleiner, and Brian Kleiner. "New Developments Concerning Age Discrimination in the Workplace." *Franklin Business and Law Journal* no. 1 (2011): 63–71.

Clarke, Jessica A. "Beyond Equality? Against the Universal Turn in Workplace Protections." *Indiana Law Journal* 86, no. 4 (2011): 1219–1287.

Corbett, William R. "The Ugly Truth about Appearance Discrimination and the Beauty of Our Employment Discrimination Law." *Duke Journal of Gender, Law & Policy* 14 (2007): 153.

Corbett, William R. "What Is Troubling About the Tortification of Employment Discrimination Law?" *Ohio State Law Journal* 75 (2014): 1027.

Flake, Dallan F. "Image Is Everything: Corporate Branding and Religious Accommodation in the Workplace." *University of Pennsylvania Law Review* 163 (2015): 699.

King, Nancy J., Sukanya Pillay, and Gail A. Lasprogata. "Workplace Privacy and Discrimination Issues Related to Genetic Data: A Comparative Law Study of the European Union and the United States." *American Business Law Journal* 43 (2006): 79.

Prenkert, Jamie Darin, and Julie Magrid Manning. "A Hobson's Choice Model for Religious Accommodation." *American Business Law Journal* 43 (2006): 467.

Reghabi, Nedda. "A Balancing Act for Businesses: Transsexual Employees, Other Employees, and Customers." *Arizona State Law Journal* 43 (2011): 1047.

Solotoff, Ari B. "*Trott v. H.D. Goodall Hospital*: When Analyzing Employment Discrimination Cases Under Maine Law, Should Maine Courts Continue to Apply the McDonnell Douglas Analysis at the Summary Judgment Stage?" *Maine Law Review* 66 (2014): 571.

Sperino, Sandra F. "A Modern Theory of Direct Corporate Liability for Title VII." *Alabama Law Review* 61 (2010): 773.

Zellers, Victoria, and Stephen L. Bowers. "Critical Update on LGBT Rights in the Workplace." *Employee Benefit Plan Review* 69 (2015): 5.

CHAPTER TWENTY TWO

Environmental Law

- ALTERNATIVE APPROACHES TO ENVIRONMENTAL PROTECTION
- THE ENVIRONMENTAL PROTECTION AGENCY
- THE NATIONAL ENVIRONMENTAL POLICY ACT OF 1970
- REGULATING WATER QUALITY
- REGULATING AIR QUALITY
- REGULATING HAZARDOUS WASTE AND TOXIC SUBSTANCES
- THE POLLUTION PREVENTION ACT OF 1990
- GLOBAL DIMENSIONS OF ENVIRONMENTAL REGULATION

As previous chapters have demonstrated, this country has often turned to the government to solve problems created by business enterprises. Early in the history of our nation, people recognized that certain problems, such as monopolization and labor strife, were national in scope and required a national solution.

Unfortunately, we did not exercise the same degree of foresight in thinking about protecting our physical environment. We looked at our smokestack industries with pride and saw them as symbols of our great productivity and technological advances. People did not fully appreciate that the billowing smoke was making the air less healthful to breathe and that the industrial sewage dumped into rivers was killing or contaminating many forms of aquatic life. The demands placed on nature to serve as a garbage disposal grew ever greater.

Some people eventually started to realize that pollution was a negative externality. It was a cost of the product not paid for by the manufacturers in their costs of production or by consumers in the purchase price. Rather, its costs were being imposed on the community, as community members were forced to breathe dirty air and to fish and swim in impure water. People who had the misfortune of living in industrialized areas were paying even higher costs than were people in rural areas through pollution-related diseases and discomfort. These costs not only were being borne by those who did not use or manufacture the products whose production caused the pollution but also, in many cases, were higher than the cost of preventing the pollution in the first place.

During the late 1960s, environmental problems became a major national concern, which led to the enactment of legislation to protect the environment and clean up existing problems. This chapter first examines alternatives to the regulatory approach for solving pollution problems and examines the primary agency responsible for enforcing environmental laws, the Environmental Protection Agency. Next we discuss the primary direct regulations designed to protect

Alternative Approaches to Environmental Protection

TORT LAW

nuisance An unreasonable interference with someone else's use and enjoyment of his or her land.

Torts are injuries to one's person or property. Pollution injures citizens and their property. Our first attempts to regulate pollution were through the use of tort law, in particular, through the use of the tort of nuisance. A **nuisance** is an unreasonable interference with someone else's use and enjoyment of his or her land. If a factory were emitting black particles that settled on a person's property every day, depositing a layer of dirt on everything in the vicinity, that person might bring an action based on nuisance. He or she would be asking the court to enjoin the emission of the particulates. Before the tort of nuisance was used in attempts to stop pollution, an injunction was always granted when a nuisance was found. Nuisance, therefore, would appear to be the perfect solution to the problem of pollution. The following classic case, however, demonstrates why actions claiming the tort of nuisance are ineffective.

CASE 22-1

Boomer et al. v. Atlantic Cement Co.
New York State Court of Appeals
257 N.E.2d 870 (1970)

Defendant Atlantic Cement Company operated a large cement plant that emitted considerable amounts of dirt and smoke into the air. These emissions, combined with vibrations from the plant, caused damage to the plaintiffs, Boomer and other owners of property located close to the plant. The plaintiffs brought a nuisance action against the defendant, seeking an injunction. The trial court ruled in favor of the defendants; it found a nuisance but denied plaintiffs the injunction they sought. The plaintiffs appealed to the intermediate appellate court, and the judgment of the trial court was affirmed in favor of the defendant. The plaintiffs then appealed to the state's highest appellate court.

Judge Bergan

[T]here is now before the court private litigation in which individual property owners have sought specific relief from a single plant operation. The threshold question raised on this appeal is whether the court should resolve the litigation between the parties now before it as equitably as seems possible, or whether, seeking promotion of the general public welfare, it should channel private litigation into broad public objectives.

A court performs its essential function when it decides the rights of parties before it. Its decision of private controversies may sometimes greatly affect public issues. Large questions of law are often resolved by the manner in which private litigation is decided. It is a rare exercise of judicial power to use a decision in private litigation as a purposeful mechanism to achieve direct public objectives greatly beyond the rights and interests before the court.

Effective control of air pollution is a problem presently far from solution even with the full public and financial powers of government. In large measure adequate technical procedures are yet to be developed and some that appear possible may be economically impracticable.

It seems apparent that the amelioration of air pollution will depend on technical research in great depth, on a carefully balanced consideration of the economic impact of close regulation, and on the actual effect on public health. It is likely to require massive public expenditure and to demand more than any local community can accomplish and to depend on regional and interstate controls.

A court should not try to do this on its own as a by-product of private litigation and it seems manifest that the judicial establishment is neither equipped in the limited nature of any judgment it can pronounce nor prepared to lay down and implement an effective policy for the elimination of air pollution. This is an area beyond the circumference of one private lawsuit. It is a direct responsibility for government and should not thus be undertaken as an incident to solving a dispute between property owners and a single cement plant—one of many—in the Hudson River Valley.

The cement-making operations of defendant have been found by the Court at Special Term to have damaged the nearby properties of plaintiffs in these two actions. That court accordingly found defendant maintained a nuisance and this has been affirmed at the Appellate Division. The total damage to plaintiffs' properties is, however, relatively small in comparison with the value of defendant's operation and with the consequences of the injunction which plaintiffs seek.

The ground for the denial of injunction, notwithstanding the finding both that there is a nuisance and that plaintiffs have been damaged substantially, is the large disparity in economic consequences of the nuisance and of the injunction.

[T]o grant the injunction unless defendant pays plaintiffs such permanent damages as may be fixed by the court seems to do justice between the contending parties. All of the attributions of economic loss to the properties on which plaintiffs' complaints are based will have been redressed.

The nuisance complained of by these plaintiffs may have other public or private consequences, but these particular parties are the only ones who have sought remedies and the judgment proposed will fully redress them. The limitation of relief granted is a limitation only within the four corners of these actions and does not foreclose public health or other public agencies from seeking proper relief in a proper court.

It seems reasonable to think that the risk of being required to pay permanent damages to injured property owners by cement plant owners would itself be a reasonably effective spur to research for improved techniques to minimize nuisance.

The damage base here suggested is consistent with the general rule in those nuisance cases where damages are allowed. "Where a nuisance is of such a permanent and unabatable character that a single recovery can be had, including the whole damage past and future resulting therefrom, there can be but one recovery." It has been said that permanent damages are allowed where the loss recoverable would obviously be small compared with the cost of removal of the nuisance.

Thus, it seems fair to both sides to grant permanent damages to plaintiffs which will terminate this private litigation.*

Reversed in favor of Plaintiff, Boomer.

CRITICAL THINKING ABOUT THE LAW

In Case 22-1, the New York Court of Appeals became the third court to find the Atlantic Cement Company guilty of committing a nuisance against the plaintiff Boomer. At the same time, the state's highest court also became the third court not to grant an injunction to halt the cement company's pollution.

At first glance, the finding of the court and its subsequent decision seem to contradict one another. A closer look at the case, however, reveals that Judge Bergan, in delivering the decision, qualified when a nuisance warrants an injunction. The questions that follow will help you identify this qualification and determine the primary ethical norm to which such a qualification is tied.

1. To demonstrate your ability to follow legal reasoning, in your own words, run down the court's reasoning for its decision.

    Clue: Do not be too narrow here. You want to identify (1) why the court granted damages to the plaintiff and (2) why the court did not order an injunction.

2. The court argued that granting the plaintiff monetary damages should promote more environmentally friendly practices on the part of businesses, because they would develop technologies to avoid having to pay damages. What assumption did the court make in this reasoning?

    Clue: Reread the court's reasoning. This assumption is related to the quantitative relationship between the damages imposed on businesses for polluting and the economic benefits of polluting for businesses.

In *Boomer*, the plaintiffs technically "won" the case because they were granted a greater remedy than the lower courts had granted; they were granted an injunction in the event the defendant failed to pay permanent damages within a set period of time. They did not, however, achieve their objective, which was to eliminate the nuisance through receipt of an injunction, the traditional remedy in a nuisance action. Thus, in *Boomer v. Atlantic Cement Co.*, the court decided that before it would apply the traditional nuisance remedy to stop the pollution, it would weigh the harms resulting from the injunction against the benefits. Because of a lack of scientific knowledge, judges at that time did not

---

*Boomer et al. v. Atlantic Cement Co. New York State Court of Appeals, 257 N.E.2d 870 (1970).

see the true costs that the polluting behavior was imposing on the community. Thus, a major problem with using nuisance laws to stop pollution is that the courts will not necessarily use their authority to issue an injunction to stop the polluting behavior even when they find that a nuisance exists. Nuisance actions can be and are used, but they are used primarily as a way for plaintiffs injured by pollution to recover damages for their losses.

Negligence, an Alternative Tort Solution. Negligence is also used at times in the fight against pollution. Plaintiffs must establish the elements of negligence as described in Chapter 11: duty, breach of duty, causation, and damage. Negligence would most often be used in a case in which a defendant's polluting behavior harmed a plaintiff. For example, if a defendant buried hazardous waste in the ground and the waste seeped down into the water table, contaminating the plaintiff's well water and injuring the plaintiff, the plaintiff might bring a negligence action.

Negligence actions involving hazardous materials are often difficult to prosecute successfully, primarily because many of the pollutants do not cause immediate harm. By the time the harm occurs, it is often difficult to link the damage to the defendant's release of the material, making the element of causation extremely difficult to prove. The availability of defenses such as contributory or comparative negligence, as well as assumption of the risk, helps weaken the effectiveness of this tort. It also shares with nuisance the attribute of being reactive rather than preventing pollution in the first place.

The primary method of controlling pollution today is through direct regulation. Before we discuss the regulatory approach, though, some additional alternatives to regulation should be considered.

GOVERNMENT SUBSIDIES APPROACH

One such approach is the use of government subsidies. Under a subsidy system, the government pays polluters to reduce their emissions. Some subsidies that could be used are tax breaks, low-interest loans, and grants for the purchase and installation of pollution-control devices. The primary problem with this approach is that when a subsidy is for less than 100 percent of the cost, the firm that limits its pollutants must still pay the difference between the actual cost and the subsidy, a cost not borne by its competitors.

EMISSION CHARGES APPROACH

Another approach is simply to charge the polluter a flat fee on every unit of pollutant discharged. Each rational polluter would theoretically reduce pollution to the point at which the cost of reducing one more unit of pollutant is greater than the emission fee. The larger the fee for each unit, the greater the motivation of firms to reduce their emissions. Difficulties in monitoring every discharge of the pollutant and in calculating the amount that should be assessed for each unit of the various pollutants are major problems with this approach. A final problem with this approach is that it may amount to licensing a continuing wrong. Some firms might simply pay the charges and continue to emit pollutants that would be difficult to clean up even with the fees collected.

MARKETABLE DISCHARGE PERMITS APPROACH

Discharge permits provide a similar approach to pollution control. The government would sell permits for the discharge of various pollutants. These pollutants could be discharged only if the polluter had the appropriate permit. Polluters would be encouraged to reduce their emissions because this reduction would enable them to sell their permits. This approach is currently being attempted on a limited scale to reduce emissions of one significant air pollutant, sulfur.

From the perspective of people wishing to reduce the total amount of pollution emitted into the environment, the primary advantage that this system offers over a system of charges is that the government actually limits the total amount of pollution through the permits; no permits will be issued once a certain amount of emissions has been authorized. To reduce pollution, the government can simply reduce the number of permits that it issues. Again, however, there is the problem of monitoring the pollution sources.

DIRECT REGULATION APPROACH

Direct regulation is the primary device currently used for environmental protection. During the late 1970s, a comprehensive set of regulations designed to protect the environment and specifically to improve air and water quality was adopted. These regulations established specific limits on the amount of pollutants that could be discharged.

One issue that must be determined when direct regulations are to be used is whether the standards set by the regulations are "technology forcing" or "technology driven." So-called **technology-forcing standards** are set primarily on the basis of health considerations, with the assumption that once standards have been established, the industries will be forced to develop the technology needed to meet the standards. **Technology-driven standards**, in contrast, try to achieve the greatest improvements possible with existing levels of technology. Most of the early environmental regulations in this country were technology forcing. In some instances, this approach was highly successful, and impressive technological gains were made. In other instances, sufficient technology had not yet been developed, and we were unable to meet some rather lofty goals.

Environmental regulations are enforced primarily by administrative agencies. The judiciary is available as a last resort to ensure that these agencies fulfill their obligations under the law. Because the administrative agencies are staffed by presidential appointment, the attitude of the chief executive has a substantial impact on an agency's behavior. Under different administrations, federal environmental regulations have been enforced with varying degrees of vigor.

The remainder of this chapter focuses primarily on direct regulation as a means of protecting the environment, because despite some minor changes in some of the environmental laws, direct regulation is still the primary means of protecting the environment. We will first examine the Environmental Protection Agency, which has primary responsibility for enforcing the direct regulations.

technology-forcing standards Standards of pollution control set primarily on the basis of health considerations, with the assumption that once regulators have set the standards, industry will be forced to develop the technology needed to meet them.

technology-driven standards Standards that take account of existing levels of technology and require the best control system possible given the limits of that technology.

The Environmental Protection Agency

Like other areas of administrative law, environmental law is primarily made up of regulations passed by a federal agency operating under the guidance of congressional mandates. The primary agency responsible for passage and enforcement of these regulations is the **Environmental Protection Agency (EPA)**.

The EPA is one of the largest federal agencies, having approximately 17,384 employees as of the year 2011. The agency was created by executive order in 1972 to mount an integrated, coordinated attack on pollution in the areas of air, water, solid waste, pesticides, radiation, and toxic substances—a rather substantial mandate for any agency. The reason for placing control of all types of environmental problems within one agency was to ensure that the attack on pollution would be integrated. In other words, Congress wanted to be certain that we would not have a regulation reducing air pollution that simply led to increased water pollution. Unfortunately, such integration did not occur. Within the agency, separate offices were established for each of the areas of pollution, and there was very little interaction between them.

Environmental Protection Agency (EPA) The federal agency charged with the responsibility of conducting an integrated, coordinated attack on all forms of pollution of the environment.

Recognizing the inefficiency of the EPA's organizational structure, in 1993, then EPA administrator Carol Browner took one of the first major steps toward trying to make the agency one with a truly integrated focus. She moved all enforcement actions from the various program offices into one main enforcement office, the Office of Compliance, which has as its primary focus "providing industry with coherent information about compliance requirements." The office is divided into groups of regulators that focus on separate sectors of the economy: energy and transportation, agriculture, and manufacturing. Browner also created a new Office of Regulatory Enforcement to take on the tough responsibility of deciding which polluters would be taken to court.[1]

One area of special concern to business managers, especially since 1990, has been the EPA's use of criminal sanctions, including incarceration, to enforce environmental laws. These cases are not actually tried by the EPA; rather, they are passed on by the EPA to the Justice Department with a recommendation for prosecution.

Since 1994, the agency has been operating under a policy statement issued to guide its special agents in their enforcement activities. Under this policy, the agents are to look for "significant environmental harm" and "culpable conduct." To satisfy the second criterion, the EPA looks for a "history of repeated violations," "concealment of misconduct," "falsification of required records," "tampering with monitoring or control equipment," and "failing to obtain required licenses or permits."[2]

By issuing this policy, the EPA is trying to put firms on notice as to when their conduct is clearly unacceptable and may subject them to criminal liability. The policy also reflects the EPA's intent to target the worst violators and make examples of them, hoping that such prosecutions will have a deterrent effect.

The EPA's *Final Policy on Penalty Reductions* encourages firms to engage in environmental self-auditing. If a firm can demonstrate that it discovered a violation and moved to correct it, the EPA will seek to reduce the penalty for the violation. Of course, the firm that engages in a self-audit, discovers a violation, and chooses not to change the harmful practice is setting itself up as a candidate for criminal prosecution. See Exhibit 22-1 for the elements of a successful environmental auditing program.

EXHIBIT 22-1

ELEMENTS OF A SUCCESSFUL AUDITING PROGRAM

- Explicit senior management support for environmental auditing and the willingness to follow up on the findings
- An environmental auditing function independent of audited activities
- Adequate auditor training and staffing
- An explicit audit program, with objectives, scope, resources, and frequency
- A process that collects, analyzes, interprets, and documents information sufficient to achieve audit objectives
- A process that includes specific procedures to promptly prepare candid, clear, and appropriate written reports on audit findings, corrective actions, and schedules for implementation
- A process that includes quality assurance procedures to verify the accuracy and thoroughness of such audits

---

[1] P. Wallach and D. Levin, "Using Government's Guidance to Structure Compliance Plan," *National Law Journal*, S10 (Aug. 30, 1993).
[2] E. Devaney, *The Exercise of Investigative Discretion* (American Law Institute, 1995).

The National Environmental Policy Act of 1970

One of the first major environmental laws passed in this nation set forth our country's policy for protecting the environment. This act, the National Environmental Policy Act of 1970 (NEPA), is regarded by many as the country's most influential piece of environmental legislation.

The NEPA is also viewed as an extremely powerful piece of legislation, because its primary purpose and effect have been to reform the process by which regulatory agencies make decisions. Title II of the NEPA requires the preparation of an **Environmental Impact Statement (EIS)** for every major legislative proposal or agency action that would have a significant impact on the quality of the human environment. A substantial number of these statements are filed every year and are the basis of a significant amount of litigation.

> **Environmental Impact Statement (EIS)** A statement that must be prepared for every major federal activity that would significantly affect the quality of the human environment.

THRESHOLD CONSIDERATIONS

An EIS is required when three elements are present. First, the action in question must be federal, such as the grant of a license, the making of a loan, or the lease of property by a federal agency. Second, the proposed activity must be major, that is, requiring a substantial commitment of resources. Finally, the proposed activity must have a significant impact on the human environment.

CONTENT OF THE EIS

Once an agency has determined that an EIS is necessary, it must gather the information necessary to prepare the document. The NEPA requires that an EIS include a detailed statement of

1. the environmental impact of the proposed action;
2. any adverse environmental effects that cannot be avoided should the proposal be implemented;
3. alternatives to the proposed action;

COMPARATIVE LAW CORNER

Pollution Controls in Japan

Japan's first pollution legislation was passed in 1970, protecting air, water, and other areas. Instead of using a system like that of the United States, in which a national agency (the EPA) performs checks and assessments, Japan addressed the problem from inside the industries themselves.

Japan's solution was to require certain industries to have personnel specifically in charge of making sure the company was following environmental laws. Any company in one of the following industries is covered under this regulation: manufacturing, electric power supply, gas supply, or heat supply, which has facilities that generate soot, dust, noise, polluted water, or vibration. Larger companies are required to have three levels of pollution control personnel. At the highest level is the pollution control supervisor, who supervises and manages the work relating to control of pollution in factories. A higher-level manager, such as the factory manager, is suitable and may fulfill this role. Below the supervisor is the senior pollution control manager, who assists the pollution control supervisor and directs the pollution control managers. At the lowest level are the pollution control managers, who actually do the inspections and make sure everything is up to environmental standards in their facility type.

4. the relationship between local short-term uses of the human environment and the maintenance and enhancement of long-term productivity; and

5. any irreversible and irretrievable commitments of resources that would be involved in the proposed activity should it be implemented.

A continuing problem under the act, however, is interpreting what is meant by *environmental impacts*. Clearly, they extend beyond the immediate effects on the natural environment; in some cases, they have been held to include noise, increased traffic and congestion, the overburdening of public facilities such as sewage and mass transportation systems, increased crime, increased availability of illegal drugs, and (in a small number of cases) damage to the psychological health of those affected by the agency action. Other cases, however, have not allowed all such damages. For example, the loss of business profits resulting from a proposed agency action has not been considered an environmental impact.

The following case illustrates how difficult it sometimes is for the court to determine a significant environmental impact that requires the filing of an EIS.

Another problem regarding the scope of the EIS pertains to the requirement of a detailed statement of alternatives to the proposed actions. What alternatives must be discussed, and how detailed must the discussion be? In general, any reasonable alternatives, including taking no action, must be discussed. The more likely the alternative is to be implemented, the more detailed the statement must be.

CASE 22-2

Brodsky v. United States Nuclear Regulatory Commission
United States Court of Appeals for the Second District
2013 U.S. App. LEXIS 339 (2013)

Richard L. Brodsky, a New York State Assemblyman, asserted, among other claims, that the Nuclear Regulatory Commission (NRC) erred in not producing an environmental impact statement (EIS) under the NEPA. The plaintiff claimed that the defendant's production of an environmental assessment (EA) and a finding of no significant impact (FONSI) were inadequate. According to the defendant, the environmental assessment looked at an increase in fire safety risk and every other adverse environmental effect. In the end, the U.S. District Court for the Southern District of New York concluded that the defendant agency's environmental assessment satisfied its minimal burden to justify foregoing the environmental impact statement and granted the defendant summary judgment. The plaintiff appealed.

Judge Sack
The Need for an Environmental Impact Statement under NEPA

Plaintiffs contend that the NRC erred in failing to produce an environmental impact statement ("EIS") under NEPA, instead producing only an environmental assessment ("EA") and a finding of no significant impact ("FONSI"). We disagree:

"Judicial review of agency decisions regarding whether an EIS is needed is essentially procedural," and "the decision not to prepare an EIS is left to the informed discretion of the agency proposing the action." "[A] reviewing court must ensure that [the agency] has taken a 'hard look' at the environmental consequences and assess whether the agency has convincingly documented its determination of no significant impact."

The NRC's EA and FONSI satisfy the agency's minimal burden to justify foregoing the EIS. The EA contains extended discussion of why the exemption does not create any fire safety risk, examines whether this exemption would have any other adverse environmental effect, and considers the alternative of not granting the exemption (and thereby requiring compliance). The NRC was not required to say more.

We have considered plaintiffs' remaining arguments and, with the exception of the public participation challenge under NEPA addressed in our related opinion issued today, conclude they are without merit. The judgment of the district court is therefore AFFIRMED IN PART in accordance with this order.*

Affirmed in part in favor of Nuclear Regulatory Commission.

---

*Brodsky v. United States Nuclear Regulatory Commission, United States Court of Appeals for the Second District, 2013 U.S. App. LEXIS 339 (2013).

CRITICAL THINKING ABOUT THE LAW

1. Reasons or facts by themselves do not necessarily lead to one and only one decision. In this case, for instance, could you make the case that the court strains to find on behalf of the Nuclear Regulatory Commission? Explain.

   Could the same evidence have been used to overturn the original decision? Why or why not?

2. Is "significant impact" ambiguous? In other words, is it reasonable to wonder just what that term means in this instance? Explain.

EFFECTIVENESS OF THE EIS PROCESS

The EIS requirement has clearly changed the process of agency decision making, but many wonder whether the requirement has improved the quality of that decision making.

Now that this umbrella environmental act has been discussed, we will examine some of the specific laws designed to protect various aspects of the environment. The focus will initially be on protecting the quality of the nation's water.

Regulating Water Quality

Water pollution is controlled today primarily by two pieces of legislation: the Federal Water Pollution Control Act (FWPCA; also called the Clean Water Act) and the Safe Drinking Water Act (SDWA). The first concentrates on the quality of water in our waterways; the second ensures that the water we drink is not harmful to our health. (Some people say that the former law protects the environment from humans, whereas the latter protects humans from the environment.)

THE FEDERAL WATER POLLUTION CONTROL ACT

When Congress passed the 1972 amendments to the FWPCA, it established two goals: (1) "fishable" and "swimmable" waters by 1983 and (2) the total elimination of pollutant discharges into navigable waters by 1985. These goals were to be achieved through a system of permits and effluent discharge limitations. Obviously, these goals were not attained. Many argue that no one really expected their attainment. They did, however, set a high goal toward which we could aspire.

Point-Source Effluent Limitations. One of the primary tools for meeting the goals of the 1972 FWPCA amendments was the establishment and enforcement of point-source effluent limitations. **Point sources** are distinct places from which pollutants can be discharged into water. Factories, refineries, and sewage treatment facilities are a few examples of point sources. *Effluents* are the outflows from a specific source. **Effluent limitations**, therefore, are the maximum allowable amounts of pollutants that can be discharged from a source within a given time period. Different limitations were established for different pollutants.

Under the National Pollutant Discharge Elimination System (NPDES), every point source that discharges pollutants must obtain a discharge permit from the EPA or from the state if the state has an EPA-approved plan at least as strict as the federal standards. The permits specify the types and amounts of effluent discharges allowed. The discharger is required to monitor its discharges continually and report any excess discharges to either the state or federal EPA. Discharges without a permit or in amounts in excess of those allowed by the permit may result in the imposition of criminal penalties. Enforcement of the act is left primarily to the states when those states have an approved program for regulation.

point sources Distinct places from which pollutants are discharged into water, such as paper mills, electric utility plants, sewage treatment facilities, and factories.

effluent limitations Maximum allowable amounts of pollutants that can be discharged from a point source within a given time period.

CASE 22-3

Los Angeles County Flood Control District v. Natural Resources Defense Council, Inc. et al.
United States Supreme Court
133 S. Ct. 710 (2013)

Petitioner Los Angeles County Flood Control District operates a "municipal separate storm sewer system" (MS4), a drainage system that collects, transports, and discharges storm water. Because storm water is often heavily polluted, the CWA and its implementing regulations require certain MS4 operators to obtain an NPDES permit before discharging storm water into navigable waters. The District has such a permit for its MS4. Respondents Natural Resources Defense Council, Inc. (NRDC) and Santa Monica Baykeeper filed a citizen suit against the District and others under § 505 of the CWA, alleging, among other things, that water-quality measurements from monitoring stations within the Los Angeles and San Gabriel Rivers demonstrated that the District was violating the terms of its permit.

The District Court granted summary judgment to the District on these claims, concluding that the record was insufficient to warrant a finding that the MS4 had discharged storm water containing the standards-exceeding pollutants detected at the downstream monitoring stations. The 9th Circuit reversed in relevant part. The court held that the District was liable for the discharge of pollutants that, in the court's view, occurred when the polluted water detected at the monitoring stations flowed out of the concrete-lined portions of the rivers, where the monitoring stations are located, into lower unlined portions of the same rivers.

Justice Ginsburg

The Court granted review in this case limited to a single question: Under the Clean Water Act . . . does the flow of water out of a concrete channel within a river rank as a "discharge of a pollutant"? In this Court, the parties and the United States as *amicus curiae* agree that the answer to this question is "no." They base this accord on *South Fla. Water Management Dist. v. Miccosukee Tribe*, . . . in which we accepted that pumping polluted water from one part of a water body into another part of the same body is not a discharge of pollutants under the CWA. Adhering to the view we took in *Miccosukee*, we hold that the parties correctly answered the sole question presented in the negative. The decision in this suit rendered by the Court of Appeals for the Ninth Circuit is inconsistent with our determination. We therefore reverse that court's judgment.

. . . [W]e held in *Miccosukee* that the transfer of polluted water between "two parts of the same water body" does not constitute a discharge of pollutants under the CWA. . . . We derived that determination from the CWA's text, which defines the term "discharge of a pollutant" to mean "any *addition* of any pollutant to navigable waters from any point source." . . . Under a common understanding of the meaning of the word "add," no pollutants are "added" to a water body when water is merely transferred between different portions of that water body. . . ."

In *Miccosukee*, polluted water was removed from a canal, transported through a pump station, and then deposited into a nearby reservoir. . . . We held that this water transfer would count as a discharge of pollutants under the CWA only if the canal and the reservoir were "meaningfully distinct water bodies." . . . no discharge of pollutants occurs when water, rather than being removed and then returned to a water body, simply flows from one portion of the water body to another. We hold, therefore, that the flow of water from an improved portion of a navigable waterway into an unimproved portion of the very same waterway does not qualify as a discharge of pollutants under the CWA. . . .*

Reversed in favor of Petitioner, Los Angeles Country Flood Control District.

Permissible discharge limits under the discharge system are based on technological standards. Most sources today must use the best available control technology (BACT). All new sources must meet this standard, but some existing facilities are allowed to meet a slightly lower standard, best practicable control technology (or BPCT). The EPA issues regulations explaining which equipment meets these standards.

---

*Los Angeles County Flood Control District v. Natural Resources Defense Council, Inc., et al. United States Supreme Court (2013).

THE SAFE DRINKING WATER ACT

The FWPCA ensures that the waterways are clean, but "clean" does not necessarily mean "fit to drink." The SDWA, therefore, sets standards for drinking water supplied by a *public water supply system*, which is defined by the act as a water supply system that has at least 15 service connections or serves 25 or more persons.

The SDWA requires the EPA to establish two levels of drinking water standards for potential drinking water contaminants. Primary standards are to protect human health, and secondary standards are to protect the aesthetic quality of drinking water.

Primary standards are based on maximum contaminant level goals (MCLGs) and maximum contaminant levels (MCLs) for all contaminants that have the potential to have an adverse effect on human health. MCLGs are the levels at which there are no potential adverse health effects. These are unenforceable, health-based goals; they are the high standards to which we aspire. The MCLs are the enforceable standards. They are developed from the MCLGs but also take into account the feasibility and cost of meeting the standard. By 1991, the EPA was to have set MCLs for 108 of the hundreds of contaminants found in our drinking water and MCLs for 25 more contaminants every three years thereafter. These goals were not met, and the 1996 amendments to the SDWA gave the EPA more flexibility in setting standards so that the agency could focus first on setting standards for the contaminants that posed the greatest potential health hazards.

Keeping up with the ever-increasing MCLs is a difficult task for public drinking water suppliers. Monitoring these systems is also a chore. Most states do monthly monitoring. Violations may be punished by administrative fines or orders. The 1996 amendments also imposed a "right to know" provision, requiring drinking water suppliers to provide every household with annual reports on water contaminants and the health problems they may cause.

Regulating Air Quality

A second major environmental concern is protecting the quality of the air. To that end, Congress enacted the Clean Air Act in 1970. Although air quality continues to improve, the EPA estimated that in 2006, more than 60 percent of Americans lived in areas that did not meet the ambient air quality standards for at least one of six major conventional air pollutants: carbon monoxide, lead, nitrogen oxides, suspended particulates, ozone, and sulfur dioxide.[3]

Table 22-1 illustrates some of the most common health problems caused by these pollutants. In addition to these enumerated health problems, nitrogen oxides and sulfur dioxide contribute to the formation of acid rain, which defaces buildings and causes the pH levels of lakes to reach such low levels that most plants and animals can no longer survive in them. These pollutants, frequently referred to as *criteria pollutants*, have been regulated primarily through national air quality standards.

TABLE 22-1 AIR POLLUTANTS AND ASSOCIATED HEALTH PROBLEMS

Pollutant	Associated Problems
Carbon monoxide	Angina, impaired vision, poor coordination, lack of alertness
Lead	Neurological system and kidney damage
Nitrogen oxides	Lung and respiratory tract damage
Ozone	Eye irritation, increased nasal congestion, reduction of lung function, reduced resistance to infection
Sulfur dioxide	Lung and respiratory tract damage

---

[3]EPA, Basic Information. Accessed March 15, 2008 at www.epa.gov/airtrends/sixpoll.html.

Although the EPA is authorized to regulate air quality, environmentalists and others do not always believe that the EPA does its job effectively. In *Massachusetts v. Environmental Protection Agency*, ultimately heard by the Supreme Court in 2007, the state of Massachusetts and a number of environmental organizations challenged the EPA's refusal to regulate greenhouse gas emissions from motor vehicles. The high court ordered the EPA to determine whether greenhouse gases did indeed endanger human health. The EPA subsequently made an endangerment finding, which paved the way for the current mobile source performance standards that regulate tailpipe emissions of greenhouse gases and mileage requirements. In 2010, the EPA and the National Highway Transportation and Safety Administration enacted the Tailpipe Rule, which required passenger cars, light-duty trucks, and medium-duty passenger vehicles to meet a 35-mile-per-gallon standard for model years 2012 through 2016. In 2012, a final rule was promulgated establishing greenhouse gas emissions standards for model years 2017 through 2025.

THE NATIONAL AMBIENT AIR QUALITY STANDARDS

National Ambient Air Quality Standards (NAAQS) A two-tiered set of standards developed for the chief conventional air pollutants: primary standards designed to protect public health and secondary standards designed to protect public welfare.

The **National Ambient Air Quality Standards (NAAQS)** provide the focal point for air pollution control. The administrator of the EPA establishes primary and secondary NAAQS for criteria pollutants. Primary standards are standards that the administrator determines are necessary to protect the public health, including an adequate margin of safety. Secondary standards are more stringent, as they are the standards that would protect the public welfare (crops, buildings, and animals) from any known or anticipated adverse effect associated with the air pollutant for which the standard is being established. Currently, the primary and secondary standards are the same for all criteria pollutants except sulfur dioxide. The administrator of the EPA retains the authority to establish new primary and secondary standards if scientific evidence indicates that the present standards are inadequate or that such standards must be set for currently unregulated pollutants.

state implementation plan (SIP) A plan required of every state that explains how the state will meet federal air pollution standards.

Once each of the NAAQS is established, each state has nine months to establish a **state implementation plan (SIP)** that explains how the state is going to ensure that the pollutants in the air within a state's boundaries will be kept from exceeding the NAAQS. Primary NAAQS must be achieved within three years of the creation of a SIP, and secondary standards are to be met within a reasonable time. The administrator of the EPA has to approve all SIPs. When a SIP is found to be inadequate, the administrator has the power to amend it or send it back to the state for revision.

In the 1990 Clean Air Act Amendments, Congress specifically addressed those areas of the country that had not yet met the NAAQS, the so-called *nonattainment areas*. Such areas are classified into five categories ranging from "marginal" to "extreme," depending on how far out of compliance they are. New deadlines for meeting the primary standard for ozone were set, ranging from 5 to 20 years. Nonattainment areas also must establish or upgrade vehicle inspection and maintenance programs.

Because emissions from upwind areas may travel and pollute downwind areas, when states develop their SIPs, they are required to take into account the effect of pollution in their state on downwind areas and eliminate those amounts of pollutants that can contribute to nonattainment in those areas. In 2014, the United States Supreme Court, in *EPA v. EME Home City Generation, L.P.*,[4] upheld the EPA's Cross-State Air Pollution Transport Rule, the rule that mandates that sulfur dioxide and nitrogen oxide emissions from upwind states do not contribute significantly to nonattainment in downwind states.

---

[4] 134 S. Ct. 1584 (2014).

NEW SOURCE REVIEW

As part of the 1977 Clean Air Act Amendments, Congress established the New Source Review (NSR) program, which regulates criteria pollutants and ensures acceptable levels of NAAQS by mandating the installation of new pollution control technology in new or modified stationary sources. In 2002, it was estimated that the NSR regulated more than 17,000 stationary sources, such as power plants, oil refineries, and chemical factories. Consequently, many view the NSR as a key provision in the Clean Air Act, as it removes millions of tons of sulfur dioxide, nitrogen oxides, and mercury from the air each year.[5]

The NSR program, however, can be changed, and some view proposed changes as further rollbacks of long-standing environmental protections. One proposed change to the NSR would have allowed significant maintenance, upgrades, and expansions to occur without requiring new pollution controls as long as the costs of the modifications did not exceed 20 percent of the cost of the entire "process unit." Under this proposed rule, major utility plant changes that cost millions of dollars and increase pollution by thousands of tons could be defined as "routine maintenance" and thus be exempt from Clean Air Act protections. Environmental groups expressed strong opposition to this proposal, arguing that it would substantially harm the quality of the air, increase respiratory ailments such as asthma, and cause thousands of premature deaths. In 2004, the EPA reported that more than 100 million people in the United States breathe unhealthy levels of particulates emitted from stationary sources. Citing the widespread health effects of increased particulate matter in the air, environmentalist groups sued to stop implementation of the changes. In March 2006, the D.C. Circuit Court sided with the environmentalists and unanimously ruled to invalidate the NSR rule changes. However, future attempts to change the rule are still possible.

THE ACID RAIN CONTROL PROGRAM

One of the major air quality problems facing the United States, as well as other countries, is acid rain. Roughly 75 percent of **acid rain** is caused by emissions of sulfur dioxide and nitrogen oxides from the burning of fossil fuels by electric utilities. The 1990 Clean Air Act Amendments included an innovative approach to controlling sulfur dioxide emissions.

acid rain Precipitation with a high acidic content (pH level of less than 5) caused by atmospheric pollutants.

Under the 1990 Clean Air Act Amendments, Congress required the EPA to establish an emissions trading program that would significantly cut sulfur dioxide emissions. Under the program, the EPA auctioned a given number of sulfur dioxide allowances each year. A holder could emit one ton of sulfur dioxide for each allowance. Firms holding allowances would be able to use the allowances to emit pollutants, "bank" their allowances for the next year, or sell their allowances to other firms. The purpose of the program was to reduce total emissions in the most efficient way possible. Those firms for which emission reduction was the cheapest would reduce their emissions extensively, whereas those for which emission reduction would be extremely expensive would find it more efficient to buy allowances. Total emissions would fall because every succeeding year, the number of allowances issued would be reduced, but the firms actually reducing their emissions would be the ones whose emissions could be reduced at the lowest cost.

On March 29, 1993, the first auction of EPA pollution allowances was held. More than 150,000 allowances were sold, with each allowance permitting the emission of one ton of sulfur dioxide. Prices for each allowance ranged from

---

[5]EPA, *New Source Review, Report to the President* (June 2002). Available at www.epa.gov/nsr/documents/nsr_report_to_president.pdf.

$122 to $450. Utilities were given a fixed amount of allowances and could bid for others at the auction. Some environmental groups also participated in the auction, buying allowances to retire unused to help clean the air.

By 1995, after three years of program operation, the price of the allowances had fallen to less than $140 per ton. In 1998, a total of 150,000 allowances were offered for use that year at a price ranging from $115.01 to $228.92. In 2007, only 125,000 allowances were auctioned, at an average price of $444.39. By 2008, total sulfur dioxide emissions from regulated sources were down to 7.6 million tons, exceeding the program's long-term goal of 9.5 million tons long before the 2010 deadline. This program is often cited as a model for achieving cost-effective pollution reduction; consequently, many people are looking at emissions trading as a possible way to meet the worldwide problem of too many harmful greenhouse gases.[6]

CLIMATE CHANGE

Global climate change is the term increasingly being used by scientists and environmentalists to refer to the process in which Earth's climate changes in response to greenhouse gases and other pollutants. *Global climate change* is preferred to "global warming" because the process is complex and involves many more changes than simply an increase in Earth's temperature. Environmentalists and scientists argue that global climate change is a matter of extreme concern because as Earth's temperature rises, a number of events are likely to happen. First, the polar ice caps, as well as glaciers in general, will melt. In fact, it appears that glacial loss has already begun. One report indicated that the Arctic Sea ice cover had decreased in 2007 to a drastically new low; another record low was reached in 2012, although there appeared to be a slight rebound in 2013.[7]

The melting ice caps will release formerly frozen water, which will raise ocean levels.[8] Higher ocean water levels means that low-lying coastal areas will begin to be flooded.[9] In addition, the release of cold, formerly frozen water will mix with the warmer sea water, which will produce more storms. Further climate changes could follow, and many species of animals could be in danger of extinction from loss of habitat, change of climate, or loss of a different species that served as a food source.[10]

In general, most scientists and environmentalists recognize the existence of global climate change and the negative effects of greenhouse gases. In reaction to such dramatic changes in temperatures across the globe, for the first time since 1990, the federal U.S. Department of Agriculture has updated the map of planting zones in the United States, in preparation for the "warmer 21st century." With May 2014 through April 2015 being the warmest 12-month period among all months in the 136-year period of modern meteorological records, evidence of the increased greenhouse effect and emission of global gases seems to be growing. In 2007, 158 UN member countries held a weeklong conference in Bali to discuss strategies for reducing greenhouse gas emissions.[11] The Bali meetings,

---

[6]EPA, 2007 EPA Allowance Auction Results. Accessed March 15, 2008 at www.epa.gov /airmarkets/trading/2007/07summary.html.

[7]"Arctic Sea Ice Cover at Record Low," CNN, September 11, 2007. Accessed March 15, 2008 at www.cnn.com/2007/TECH/science/09/11/arctic.ice.cover/index.html?iref=mpstoryview; "Polar Ice Cap Melt Not as Great This Summer." Accessed September 20, 2013 at www.earthweek.com /2013/ew130920/ew130920a.html.

[8]Pew Center on Global Climate Change, The Basics. Accessed March 15, 2008 at www .pewclimate.org/global-warming-basics.

[9]*Id.*

[10]*Id.*

[11]Andrew Revkin, "Voices on Bali, and Beyond," *New York Times*, December 6, 2007. Accessed January 2, 2011 at dotearth.blogs.nytimes.com/2007/12/16/voices-on-bali-and-beyond/?scp =4&sq=bali%20climate%20conference&st=cse.

which ended with general agreement among the 158 countries, were intended to create a continued strategy for reducing greenhouse gases starting in 2012 when the first commitment period of the Kyoto Protocol expires.

In 2009, the Copenhagen Agreement was reached and contained promises of major emitting countries to cut carbon and develop a monitoring system to track success or failure. Industrialized countries also agreed to contribute $30 billion in near-term climate aid while raising $100 billion annually by 2020 for vulnerable nations. However, no firm allowances were set. In 2010, countries met once more and agreed to the Cancun Agreement, which fleshed out some of the details of the Copenhagen Agreement and bound countries to keep temperature rise below 2 degrees Celsius above pre-industrial levels, but no other firm commitments resulted, and parties agreed that they need to continue working to resolve the problems that are resulting from climate change.[12]

Since that agreement, annual climate conferences have continued to be held, as nation-states that signed the Kyoto Protocol continue to search for ways to fight climate change. During the meetings since 2010, the long-term goal has been a universal UN treaty on climate change by 2015, which would enter force by 2020. During the Bonn meeting in 2013, participants focused on how to transform the world's energy systems quickly enough toward low-carbon energy, including renewable energy, energy efficiency, and the consideration of carbon capture and storage, while also making significant strides toward the universal treaty extending the Kyoto Protocol. Unfortunately, the United States never joined the Kyoto Protocol and Canada pulled out of the agreement in 2011; Japan, New Zealand, and Russia subsequently pulled out and announced that they would not be signing up to a new interim commitment to the treaty. As a result, even if parties can agree on an extension of the treaty, the extension would cover only about 15 percent of the world's emissions of greenhouse gases.

On May 15, 2015, leaders from 12 states and provinces in 7 countries, collectively representing more than 100 million people, signed an agreement to limit the increase in the global average temperature to below 2 degrees Celsius, the warming threshold at which scientists say that there will likely be catastrophic climate disruptions. The agreement, called "Under 2 MOU," was created to provide a template for the world's nations to follow as work continued toward an international agreement to reduce greenhouse gas emissions ahead of 2015's United Nations Climate Change Conference in Paris.

The signatories committed to reduce greenhouse gas emissions to 80 to 95 percent below 1990 levels by 2050 or to achieve a per capita annual emission target of less than 2 metric tons by 2050. The targets allow each government to tailor emission reduction plans to fit regional needs. Parties also committed to recruiting additional global partners before the United Nations Climate Change Conference in December 2015.[13] As this book went to press before the Paris meeting, it is too early to know the effect this treaty will have, but many are hopeful that the Paris meeting will result in a new global climate change agreement.

Regulating Hazardous Waste and Toxic Substances

Most of us want to enjoy the products that technology has developed, but what price are we willing to pay for these amenities?

---

[12] Lisa Friedman, "A Near-Consensus Decision Keeps U.N. Climate Process Alive and Moving Ahead," *New York Times*, December 13, 2010. Accessed January 1, 2011 at www.nytimes.com/cwire/2010/12/13/13climatewire-a-near-consensus-decision-keeps-un-climate-p-77618.html?pagewanted=1&sq=bali%20climate%20conference&st=cse&scp=1.

[13] Governor Brown, "International Leaders Form Historic Partnership to Fight Climate Change." Accessed May 15, 2015 at under2mou.org/?page_id=447.

COMPARATIVE LAW CORNER

Solar Energy in Germany

Germany has determined that the fossil fuel-based energy system is not sustainable; thus, it has adopted an impressive new policy, called *Energiewende*, that has as its goal the generation of 80 percent of the nation's electricity from renewable energy sources by 2050. To attain this goal, Germany has had to move quickly. By the end of 2012, Germany had installed considerably more solar power capacity per capita than any other country, and in the first quarter of 2014, renewable energy sources met a record 27 percent of the country's electricity demand due to additional installations and favorable weather. It remains to be seen whether Germany can meet its goals, but the huge growth in renewable energy in the country is a good illustration of how governmental policy can be effective in improving environmental conditions. Germans view their approach as an important step in fighting global climate change.

Solar panels on house after house in Germany are helping the nation work toward accomplishing its goal of developing a sustainable economy.

Until the mid-1970s, most people were content to take advantage of newly available products without giving much thought to the by-products resulting from their manufacture. Most businesspeople were primarily concerned about creating new products and using new technology to increase production and profits. Then came a growing awareness of the potential health and environmental risks posed by the waste created in the production process. In addition to the problems created by waste, some of the new products themselves (and their newly created chemical components) were proving to be harmful.

The potential health risks from these chemicals and wastes include a plethora of cancers, respiratory ailments, skin diseases, and birth defects. Environmental risks include not only pollution of the air and water but also unexpected explosions and soil contamination. Species of plants and animals may be threatened with extinction.

During the mid-1970s, Congress began to take a closer look at regulating waste and toxic materials. One of the problems that regulators face in this area, however, is a lack of scientific knowledge concerning the impact of many chemicals on human health. We know that exposure to many chemicals causes cancer in laboratory animals. We are unable, however, to ascertain the impact of each increment of exposure. For example, we know that saccharin in some quantity can cause cancer in humans, but we do not know what quantity or whether especially sensitive persons may be affected by substantially smaller amounts. Congress has responded to these and related problems in a variety of ways.

Four primary acts are designed to control hazardous waste and toxic substances: (1) the Resource Conservation and Recovery Act of 1976; (2) the Comprehensive Environmental Response, Compensation, and Liability Act of 1980; (3) the Toxic Substances Control Act of 1979; and (4) the Federal Insecticide, Fungicide, and Rodenticide Act of 1972.

THE RESOURCE CONSERVATION AND RECOVERY ACT OF 1976

The Resource Conservation and Recovery Act of 1976 (RCRA) regulates both hazardous and nonhazardous waste, with the primary emphasis on control of hazardous waste. The focus of the act is on the treatment, storage, and disposal of hazardous waste (see Exhibit 22-2). The reason for this focus was the belief that it was not necessarily the creation of waste that was the problem, but rather the improper disposal of such waste. Also, it was hoped that making firms pay the true costs of safe disposal would provide the financial incentive for them to generate less waste.

The Manifest Program. The best-known component of the RCRA is its **manifest program**, which is designed to provide "cradle-to-grave" regulation of hazardous waste. A waste may be considered hazardous and, thus, fall under the manifest program in one of three ways. First, it may be listed by the EPA as a hazardous waste. Second, the generator may choose to designate the waste as hazardous. Finally, according to the RCRA, a **hazardous waste** may be "garbage, refuse, or sludge or any other waste material that has any one of the four defining characteristics: ignitability, corrosivity, reactivity, or toxicity."

Once a waste is designated as hazardous, it falls under RCRA's manifest program. Under this program, generators of hazardous waste must maintain records called *manifests*. These manifests list what amount and type of waste is produced, how it is to be transported, and how it will ultimately be disposed of. Some wastes cannot be disposed of in landfills at all. Others must receive chemical or biological treatment to reduce toxicity or to stabilize them before they can be deposited in landfills. If the waste is transported to a landfill, both the transporter and the owner of the disposal site must certify their respective sections of the manifest and return it to the creator of the waste. The purpose of these manifests is to provide a record of the location and amount of all hazardous wastes and to ensure that such waste will be properly transported and disposed of. Exhibit 22-3 shows the hazardous waste manifest trail. An electronic

manifest program A program that attempts to see that hazardous wastes are properly transported to disposal facilities licensed by the EPA so that the agency will have an accurate record (manifest) of the location and amount of all hazardous wastes.

hazardous waste Any waste material that is ignitable, corrosive, reactive, or toxic when ingested or absorbed.

According to the Resource Conservation and Recovery Act of 1976 (RCRA) and the Hazardous and Solid Waste Amendments of 1984 (RCRA Amendments), a hazardous waste may be "garbage, refuse, or sludge or any other waste material" that exhibits one or more of the following characteristics:

- Ignitability
- Corrosivity
- Reactivity (unstable under normal conditions and capable of posing dangers)
- Toxicity (harmful or fatal when ingested or absorbed)

Improperly handled, hazardous wastes can contaminate surface waters and groundwater, release toxic vapors into the air, or cause other dangerous situations, such as explosions.

EXHIBIT 22-2

WHAT IS A HAZARDOUS WASTE?

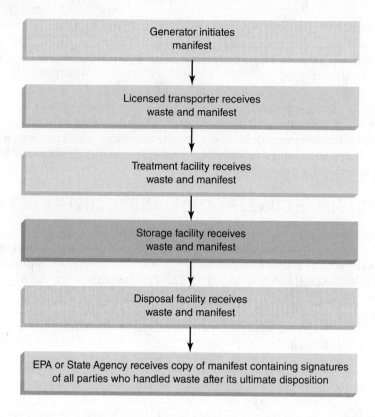

EXHIBIT 22-3

THE HAZARDOUS WASTE MANIFEST TRAIL

Source: EPA, *Environmental Programs and Challenges: EPA Updates* (Washington, DC: EPA, August 1988), 88.

hazardous waste manifest trail is currently being developed to increase the safety of hazardous waste disposal.

All firms involved in the transportation and disposal of hazardous waste must be certified by the EPA in accordance with standards established under RCRA. Every year, approximately 12 million tons of hazardous waste are transported for treatment, storage, or disposal.

RCRA Amendments of 1984 and 1986. Congress amended RCRA in 1984 and 1986. The primary effect of the amendments was to make landfills (or hazardous waste dumps) a last resort for the disposal of many types of waste. Advanced treatment, recycling, incineration, and other forms of hazardous waste treatment are all assumed to be preferable to land disposal. Some wastes were banned entirely from landfill disposal.

The 1986 amendment requires that companies report the amount of hazardous chemicals they release into the environment each year. From 1997 to 2001, RCRA reported a decrease in overall chemical emissions each year, but in 2004, toxic chemical emissions increased 5 percent over the previous year; this included a 3.4 percent increase in lead emissions and a 10 percent increase in mercury emissions. Environmental groups blame the lax standards of the Bush administration for the increase in toxic chemicals released into the environment.[14]

Enforcement of RCRA. RCRA is enforced by the EPA. States, however, may set up their own programs as long as these programs are at least as stringent as the federal program. The EPA gives any state that has taken the responsibility for regulating its hazardous wastes the first opportunity to prosecute violators. This procedure is consistent with the EPA's enforcement of other environmental laws.

If the state fails to act within 30 days, the EPA takes action to enforce the state's requirements. The EPA may issue informal warnings; seek temporary or permanent injunctions with criminal penalties of up to $50,000 per day of

---

[14] Juliet Eilperin, "Toxic Emissions Rising, EPA Says," *Washington Post* (June 23, 2004), A-2.

violation, civil penalties of up to $25,000 per violation, or both; or impose other penalties that the EPA administrator finds appropriate.

THE COMPREHENSIVE ENVIRONMENTAL RESPONSE, COMPENSATION, AND LIABILITY ACT OF 1980, AS AMENDED BY THE SUPERFUND AMENDMENT AND REAUTHORIZATION ACT OF 1986

If the manifest program is followed, waste will be disposed of properly and there will be no more contaminated waste sites. Before RCRA was enacted, however, there was extensive unregulated dumping. Something had to be done to take care of cleaning up the sites created by improper disposal. Exhibit 22-4 shows some of the risks posed by these sites.

To alleviate the problems created by improper waste disposal, the Comprehensive Environmental Response, Compensation, and Liability Act of 1980 (CERCLA) authorized the creation of the **Superfund**, primarily from taxes on corporations in industries that create significant amounts of hazardous waste. Additional funding also came from appropriations from the general fund; fines, penalties, and recoveries from responsible parties; and interest accrued on the balance of the fund. The money in Superfund was then used by the EPA or state and local governments to cover the cost of cleaning up leaks from hazardous waste disposal sites when their owners could not be located or were unable or unwilling to pay for the cleanup. Superfund also provides money for emergency responses to hazardous waste spills other than oil spills. When an owner is found after a cleanup or was initially unwilling to pay, the EPA may sue to recover the costs of the cleanup.

Under CERCLA, liability for cleanup extends beyond the immediate owner. So-called *potentially responsible parties* who also may be held liable include (1) present owners or operators of a facility where hazardous materials are stored, (2) owners or operators at the time the waste was deposited there, (3) the hazardous waste generators, and (4) those who transported hazardous waste to the site.

Successful actions under CERCLA to recover costs have been less frequent than originally hoped. The fund was intended to be self-replenishing but has not been. Thus, CERCLA was amended in late 1986 by the Superfund Amendment and Reauthorization Act of 1986 (SARA). These amendments provided more stringent cleanup requirements and increased Superfund's funding to $8.5 billion, to be generated primarily by taxes on petroleum, chemical feedstocks, imported chemical derivatives, and a new "environmental tax" on corporations. Additional money was to come from general revenues, recoveries, and interest.

The future of Superfund, however, remains in question. The taxes on chemical and petroleum companies used to support Superfund cleanups expired in

Superfund A fund authorized by CERCLA to cover the costs of cleaning up hazardous waste disposal sites whose owners cannot be found or are unwilling or unable to pay for the cleanup.

- Contaminated air
- Direct contact with hazardous waste
- Contaminated drinking water
- Ecological damage
- Fire or explosion hazard
- Exposure through food web
- Contaminated groundwater
- Contaminated soil
- Contaminated surface water

EXHIBIT 22-4

ENVIRONMENTAL OR PUBLIC HEALTH THREATS REQUIRING SUPERFUND EMERGENCY ACTIONS

Source: Office of Emergency and Remedial Response (Superfund), U.S. EPA, reprinted in *Environmental Programs and Challenges: EPA Updates* (EPA, August 1988), 96.

1995 and require reauthorization from Congress. Consequently, in 2003, the fund was depleted of any money from the chemical or petroleum industries, shifting the cleanup burden primarily to the taxpayers through allocations by Congress from the general fund.[15] Because Superfund is no longer funded by polluter and industry dollars, the completion of Superfund cleanups has declined significantly because the fund is now primarily dependent on annual allocations from the general fund. Environmentalists warn that without a renewal of the tax on chemical and petroleum industries, Superfund will not be able to treat many of the nation's most polluted sites, and in 2010, the EPA seemed to support their position, estimating that the cost of cleanup was increasing beyond the current funding needed for Superfund sites.

A temporary upswing in cleanups did occur as a result of a $600 million allocation to the Superfund from the American Recovery and Reinvestment Act of 2009 (the federal "stimulus" designed to create jobs and improve the economy). The beginning of construction at 26 new Superfund sites increased operations at another 25 ongoing cleanups, and completion of projects at over 20 sites as a consequence of this injection of money, has been cited by some as evidence for the reinstitution of the tax on the chemical and petroleum industries.[16]

To see whether any of these Superfund sites are near you, go to www.epa.gov/superfund/sites/index.htm and use the EPA's interactive map of Superfund sites.

APPLYING THE LAW TO THE FACTS . . .

Bob owns a rather large piece of property that is in an area zoned for industrial use. He sells the property for a price that is significantly below market value. When his friend Rodney asks why he unloaded the property so quickly for such a low price, he explains that when he first purchased the property, some people were paying the former owner to allow them to dump stuff in an unused area in the back. He continued to pick up extra cash that way for a couple of years, but then decided that the stuff was starting to smell. So he was going to cover it all up so that he could put a building on top of the area. He never did construct the building, but he did level the land so that the area was ready to be used for a building. Then he worried that some of what was dumped might be hazardous waste and he could have actually owned a future Superfund site. By selling the property before it was declared a Superfund site, he believed he could save himself quite a bit of money since the new owner would be the one who was liable. Is Bob correct? Why or why not?

THE TOXIC SUBSTANCES CONTROL ACT OF 1979

Toxic substances are integral parts of some products that we use every day. Neither RCRA nor CERCLA regulates these substances. The Toxic Substances Control Act (TSCA) attempts to fill this regulatory gap. It attempts to ensure that the least amount of damage will be done to human health and the environment while allowing the greatest possible use of these substances.

---

[15]General Accounting Office, *Superfund Program: Current Status and Future Fiscal Challenges*, GAO/RECD-03-850, July 2003.

[16]Braunson Virjee, "Stimulating the Future of Superfund: Why the American Recovery and Reinvestment Act Calls for a Reinstatement of the Superfund Tax to Polluted Sites in Urban Environments," *Sustainable Development Law & Policy* 11, no. 1, article 12. Available at http://digitalcommons.wcl.american.edu/sdlp/vol11/iss1/12.

The term *toxic substances* has not been clearly defined by Congress. By reviewing the types of substances regulated under TSCA, however, one would probably conclude that a **toxic substance** is any chemical or mixture whose manufacture, processing, distribution, use, or disposal may present an unreasonable risk of harm to human health or the environment. This broad definition encompasses a large number of substances. Thus, control of these substances is a major undertaking.

The primary impact of TSCA comes from its procedure for evaluating the environmental impact of all chemicals, except those regulated under other acts. Under TSCA, every manufacturer of a new chemical must give the EPA a pre-manufacturing notice (PMN) at least 90 days before the first use of the substance in commerce. The PMN contains all the available data and test results showing the risk posed by the chemical, although no amount or type of data is specifically required. The EPA then determines whether the substance presents an unreasonable risk to health or whether further testing is required to establish the substance's safety. The manufacture of the product is banned when the risk of harm is unacceptable. If the EPA determines that the supplied data is insufficient to determine that the product is safe, then more testing can be required, and a manufacturer of the product must wait until the tests have been satisfactorily completed. Otherwise, manufacturing may begin as scheduled.

toxic substance Any chemical or mixture whose manufacture, processing, distribution, use, or disposal presents an unreasonable risk of harm to human health or the environment.

APPLYING THE LAW TO THE FACTS . . .

Belinda wants to start importing a chemical for use in a new product she hopes to manufacture, but she is concerned because not much research has been published about the safety of the chemical. In fact, there has been only one blind-reviewed study of side effects of the chemical, and that one was inconclusive. "Don't worry, her friend reassures her. All you have to do when filling out the PMN is submit what data are available. If there are no data showing that the chemical presents an unreasonable risk, then the EPA has to approve the chemical." Is her friend right? Why or why not?

THE FEDERAL INSECTICIDE, FUNGICIDE, AND RODENTICIDE ACT OF 1972

One category of toxic substances that has been singled out for special regulatory treatment is **pesticides**, which are defined as substances designed to prevent, destroy, repel, or mitigate any pest or to be used as a plant regulator or a defoliant. Insecticides, fungicides, and rodenticides are all forms of pesticides.

Pesticides are obviously highly important to us. Their use results in increased crop yields. Some pesticides kill disease-carrying insects. Others eradicate pests, such as mosquitoes, that simply cause us discomfort. Yet, pesticides have harmful side effects; they may cause damage to all species of life. A pesticide that does not degrade quickly may be consumed along with the crops on which it was used, potentially harming the consumer's health. The pesticide may get washed into a stream to contaminate aquatic life and animals that drink from the stream. Once the pesticide gets into the food chain, it may do inestimable harm.

In 1972, FIFRA created the registration system that is used to control pesticide use. To be sold in the United States, a pesticide must be registered and properly labeled. A pesticide will be registered when (1) its composition warrants the claims made for it; (2) its label complies with the act; and (3) the manufacturer provides data to demonstrate that the pesticide can perform its intended function, when used in accordance with commonly accepted practice, without presenting unreasonable risks to human health or the environment.

pesticide Any substance designed to prevent, destroy, repel, or mitigate any pest or to be used as a plant regulator or defoliant.

A pesticide with general use registration can be sold without any restrictions. A restricted use registration will be granted if the pesticide will not cause an unreasonable risk only if its use is restricted in some manner. Typical restrictions include allowing the pesticide to be used only by certified applicators or allowing it to be sold only during certain times of the year or only in certain regions of the country or only in certain quantities.

Registration is good for five years, at which time the manufacturer must apply for a new registration. If at any time prior to the end of the registration period the EPA obtains evidence that a pesticide poses a risk to human health or the environment, the agency may institute proceedings to cancel or suspend the registration.

The EPA believes that progress under FIFRA has been significant, although there are critics of the act. In fiscal year 2006 alone, 297 product registrations were canceled as a result of FIFRA.[17]

Pesticide Tolerances in Food. Under the Federal Food, Drug, and Cosmetic Act (FFDCA), the EPA establishes legally permissible maximum amounts of pesticide residues in processed food or in animal products such as meat or milk, as well as on food crops such as apples and tomatoes. Before a pesticide can be registered, an applicant must obtain a tolerance for that pesticide. To obtain the tolerance, the applicant must provide evidence of the level of residue likely to result and data to establish safe residue levels. Under the 1996 Food Quality Protection Act, a *safe residue level* is a level at which there is a "reasonable certainty of no harm" from exposure to the pesticide. The law also requires distribution of a brochure on the health effects of pesticides.

The Pollution Prevention Act of 1990

Tremendous gains have been made through the laws described in the preceding sections, but it has become more costly to get increasingly smaller reductions of pollutants. Whereas initially a $1 million expenditure on end-pipe controls might have reduced emissions by 80 percent, today that same investment is likely to result in only a 5 percent reduction.

Recognition of this decline in the effectiveness of direct regulation and the consequent need to look for alternative approaches to pollution problems led to passage of the Pollution Prevention Act of 1990, in which Congress set forth the following policy:

> *Pollution should be prevented or reduced at the source whenever feasible; pollution that cannot be prevented should be recycled in an environmentally safe manner, whenever feasible; pollution that cannot be prevented or recycled should be treated in an environmentally safe manner whenever feasible; and disposal or other release into the environment should be employed only as a last resort and should be conducted in an environmentally safe manner.**

The government's role in encouraging this policy is one of providing a "carrot" as opposed to the "stick" of direct end-pipe regulations. The federal government is providing states with matching funds under the act for programs to promote the use of source reduction techniques for business. A clearinghouse has been established to compile the data generated by the grants and to serve as a center for source reduction technology transfer.

---

[17]Environmental Protection Agency, Board of Scientific Counselors, National Center for Environmental Research (NCER) Standing Subcommittee-2007, *Federal Register* (Aug. 22, 2007). Accessed March 15, 2008 at www.epa.gov/fedrgstr/EPA-PEST/2007/August/Day-22/p16560.pdf.

*Pollution Prevention Act of 1990, United States Environmental Protection Agency.

BUSINESS ASPECTS OF VOLUNTARY POLLUTION PREVENTION

Despite the voluntary nature of actions under this act, pollution prevention is becoming an important concept in business today. Chemical companies, for example, are beginning to see waste as avoidable and inefficient and are looking for ways to change their production processes to reduce the amount of waste they create. Examples abound of firms that are jumping on the pollution prevention bandwagon. For example, DuPont, America's largest producer of chemicals, voluntarily cut its greenhouse gas emissions by more than 50 percent from its 1991 levels.

Whether it is the increasing cost of waste disposal, a fear of stricter direct regulations, public pressure for firms to be "greener," or the federal government's new emphasis on pollution prevention, firms are changing their attitudes toward the environment. Whether this trend toward voluntary source reduction will continue remains to be seen.

SUSTAINABLE DEVELOPMENT

Along with a shift toward pollution prevention, we have seen a shift from concern about simply being in compliance with environmental regulations toward a concern about sustainability or sustainable development. *Sustainable development*, a term coined in 1987, refers to development that meets the needs of the current generation without compromising the needs of future generations.

Sustainability, however, goes beyond just environmental matters. Firms talk about maintaining their "triple bottom line," which refers to looking not just at profit, but at profit, people, and the planet. In other words, firms concerned about sustainable development want to evaluate their performance in broader terms than just making a profit; they also evaluate treating employees fairly and reinvesting in the communities in which they live or are located, as well as minimizing the firm's ecological impact. Triple-bottom-line accounting attempts to describe the social and environmental impact of an organization's activities, in a measurable way, in relation to the firm's economic performance, although at this time, the idea is still not well developed or widely used.

Global Dimensions of Environmental Regulation

THE NEED FOR INTERNATIONAL COOPERATION

In most areas of regulation, the United States first enacted national legislation and only later, if at all, considered the worldwide implications of the problem that the law was enacted to resolve. Nevertheless, its first major piece of environmental legislation, the NEPA, addressed the global nature of environmental problems. The act instructed the federal government to recognize the worldwide and long-range character of environmental problems and, when consistent with the foreign policy of the United States, lend appropriate support to initiatives, resolutions, and programs designed to maximize international cooperation in anticipating and preventing a decline in the quality of the world environment.

THE TRANSNATIONAL NATURE OF POLLUTION

International cooperation on environmental matters is essential because environmental problems do not respect national borders. There are three primary means by which environmental problems originating in one area of the globe affect

LINKING LAW AND BUSINESS

Marketing and Management

In your marketing or management class, you may have learned about environmental sustainability, which is a management approach that focuses on sustaining the environment and still generating profits for a firm. As firms advance toward environmental sustainability, there are four levels that companies examine to gauge their progress.

The first and most basic level is pollution prevention. As already discussed in the previous section, this involves the prevention or reduction of waste before it is created. Companies that are highlighting pollution prevention often use "green marketing" plans by developing environmentally friendly packaging, better pollution controls, and ecologically safer products.

The second level in environmental sustainability is product stewardship, which extends the focus from production creation to the entire product life cycle. At this level, firms often implement design for environment (DFE) policies that consider future consequences of the firm's products. Consequently, firms are taking measures to find more efficient ways of recovering, reusing, or recycling their products.

The third level is new environmental technologies. Because some companies that have already progressed in pollution prevention and product stewardship are limited by available technologies, new technologies are sometimes needed to meet their environmental goals.

The fourth level of environmental sustainability is sustainability vision, in which organizations develop a guide for their firms' future methods of environmental responsibility. This vision provides a framework for pollution control, product stewardship, and environmental technology.

By focusing on these four levels of environmental sustainability, there is a greater likelihood that firms using this management approach will promote the goals of environmentalists, which, it is hoped, will result in a greener and safer planet. In addition, these organizations will potentially be at less risk of litigation for unsafe practices and will, therefore, maintain a more positive image with the general public.

Source: P. Kotler and G. Armstrong, *Principles of Marketing*, 12th ed. (Upper Saddle River, NJ: Prentice Hall, 2008), 582–85.

---

other areas: (1) movement of air in prevailing wind patterns, (2) movement of water through ocean currents, and (3) active and passive migration of numerous species of plants and animals.

Scientists have discovered that air tends to circulate within one of three regional areas, or *belts*, that circle the globe north and south of the equator. For example, between the latitudes 30°N and 60°N of the equator, the prevailing air currents are the westerly winds. Thus, the air between these latitudes circulates in a westerly direction all around the globe, remaining primarily within those latitudes.

The United States and China both have much of their land masses within these two latitudes. As a result, pollutants emitted into the air in the United States may be carried by these westerly winds to China, just as pollutants emitted into the air anywhere between 30°N and 60°N of the equator anywhere in the world may ultimately end up in the air above the United States. Consequently, the United States could have extremely strict air pollution laws, yet still have polluted air as a result of other countries' emissions. Likewise, our failure to enact adequate air pollution control laws can adversely affect air quality in other countries. Canada, for instance, attributes some of its pollution problems to the failure of the United States to enact stricter control on sulfur dioxide emissions.

A similar situation exists with respect to the flow of water, except that the regions are not as clearly defined. All ocean currents ultimately connect with one another, so a pollutant discharged into any body of water that flows into an ocean may end up having a negative impact on water quality hundreds of miles away from the country in which it was dumped.

The migration of birds and animals also spreads pollutants. Many animals, such as geese, whales, salmon, seals, and whooping cranes, travel across national borders seasonally. If an animal ingests a hazardous chemical in one country, travels to another country, and is eaten by an animal in that country, that pollutant has now been inserted into the food web in the second country.

THE GLOBAL COMMONS

Another closely related reason for international cooperation on environmental matters is that many of the planet's resources, such as the oceans, are within no country's borders and are, therefore, available for everyone's use. Because of this availability, these resources are often called the *global commons*. Because everyone has access to them, they are susceptible to exploitation and overuse. Cooperation to protect these global resources is the only way to preserve them.

PRIMARY RESPONSES OF THE UNITED STATES

The United States has played a role in establishing global environmental policies in four primary ways: (1) research, (2) conferences, (3) treaties, and (4) economic aid. Unfortunately, to date, these responses have not been extremely successful, nor has there been a major commitment of U.S. resources to the resolution of transnational environmental problems.

Research. Research, the results of which are shared with other nations, is the typical U.S. response to international environmental issues. For example, in response to international concerns about changes in environmental conditions, the U.S. government sponsors research in universities and in federal laboratories by various governmental agencies. Some critics argue that we need to commit more money to research. Others claim that we use research as an excuse for inaction. Many environmentalists view a "commitment to research" as a stalling technique to prevent the imposition of needed controls.

Conferences. Conferences to discuss specific transnational environmental problems are often held; many are arranged through the United Nations. The first such conference was the United Nations Conference on the Human Environment, held in Stockholm in 1972. Similarly, in 1992, delegates from more than 120 nations met in Rio de Janeiro for the United Nations Conference on Environment and Development, commonly referred to as the Rio Earth Summit. Marking the 10-year anniversary of the Rio Earth Summit, more than 100 heads of state from around the world gathered for the Johannesburg Earth Summit during the summer of 2002 to discuss global climate change and sustainable development. These conferences serve primarily to promote an understanding of the global implications of environmental problems. Often, these conferences lead to the negotiation of treaties designed to help resolve environmental problems.

Treaties. *Treaties* are written agreements between two or more nations that specify how particular issues are to be resolved. The process of accepting a treaty varies from country to country. In the United States, a treaty must be negotiated and signed by a representative of the executive branch, generally the president. Then it must be approved by two-thirds of the U.S. Senate. Implementation of a treaty generally requires the passage of federal legislation that translates the objectives of the treaty into laws.

The United States has entered into numerous bilateral (signed by only two nations) and multilateral (signed by more than two nations) treaties, sometimes called *conventions*, in the area of environmental protection. One of the more successful multilateral treaties the United States has signed is the Montreal Protocol.

Originally signed by 24 nations and the European Community on September 16, 1987, the Montreal Protocol on Substances That Deplete the Ozone Layer ultimately led to an elimination of the production of ozone-depleting chlorofluorocarbons (CFCs) by January 1, 1996. A series of summits concerning the problem of ozone depletion has taken place since that initial meeting, and nations continue to amend the treaty to restrict production of more ozone-destroying compounds as our understanding of these chemicals grows.

One of the problems with treaties, however, is that they are unenforceable when the signatories decide not to obey them any longer. Many include clauses that allow a nation to withdraw from a contract or to cease abiding by particular terms after giving notice of its intent to the other parties to the treaty.

More recently, other methods have been used to foster international environmental action. Trade agreements have started to incorporate provisions regarding environmental protection. The North American Free Trade Agreement (NAFTA), for example, included a side agreement on the environment. Although it has been called the most environmentally sensitive trade agreement ever, there is concern that this agreement may ultimately result in a lessening of environmental protection.

Aid. A final way in which the United States affects environmental policy worldwide is by the judicious use of foreign aid, either financing pollution control projects or giving economic aid for a particular project only when certain environmentally sound conditions have been met. Some aid is also given in the form of technical assistance and training. For example, the U.S. Soil Conservation Service (SCS) provides technical assistance in soil and water conservation to many Latin American and African countries. The SCS also teaches conservation techniques to students from these countries.

SUMMARY

There are many ways a nation can protect its environment. Some of these methods include tort law, subsidies, discharge permits, emission charges, and direct regulation. Beginning in 1970, with the passage of the NEPA, our nation began a course of environmental protection based primarily on specific direct regulations.

The FWPCA established a discharge permit system designed to make the waterways fishable and swimmable. The SDWA sets standards to make our drinking water safe. The Clean Air Act, as amended several times, establishes the NAAQS, standards designed to ensure that conventional air pollutants do not pose a risk to human health or the environment. This act also establishes standards for toxic air pollutants.

Hazardous wastes and toxic substances are regulated primarily by four pieces of legislation. The RCRA sets standards for waste disposal sites and establishes the manifest system for the tracking of hazardous wastes from creation to disposal. CERCLA, as amended by SARA, provided funding and a mechanism for cleaning up hazardous waste sites. The TSCA provided a mechanism for testing new chemicals to ensure that they do not pose unreasonable risks before being used in commerce. Finally, FIFRA established a procedure for the regulation of pesticides through a registration system.

The newest trend in the environmental area is toward pollution prevention. This trend is encouraged by the Pollution Prevention Control Act of 1990.

Solving environmental problems requires cooperation among all nations. Four ways in which the United States works to solve these problems on a global scale are through shared research, conferences, treaties, and aid.

REVIEW QUESTIONS

22-1 Explain the common-law methods of resolving pollution problems and evaluate their effectiveness.

22-2 Explain the circumstances under which an environmental impact statement must be filed and describe the statement's required content.

22-3 Explain how emission charges and discharge permits could be used to help control pollution.

22-4 Present the arguments of those who would abolish the use of the EIS. How would you evaluate those criticisms?

22-5 Describe the structure of the amended FWPCA and explain how each element of the act is designed to further the goals of the FWPCA.

22-6 Compare the structure of the FWPCA with that of the Clean Air Act.

REVIEW PROBLEMS

22-7 The defendant operated a mining company. Because he used improper drainage techniques, drainage of pollutants from his mining operation contaminated the private water supplies of the plaintiff property owners located downstream from him. What legal theories would the plaintiffs use to sue the defendant? Would the plaintiffs be likely to win their lawsuit? Why or why not?

22-8 The lead industry challenged the EPA's establishment of a primary air quality standard for lead that incorporated an "adequate margin of safety." In setting the standard, the EPA had not considered the feasibility or the cost of meeting the standards. Must the EPA take such factors into consideration in setting primary air quality standards?

22-9 Ohio's SIP was submitted to the EPA. Approval of a portion of the plan was denied because it was not adequate to ensure the attainment and maintenance of the primary standard for photochemical oxidants in the Cincinnati area. The EPA supplemented the Ohio plan with a provision requiring a vehicle inspection and registration procedure for the Cincinnati area. Cincinnati set up the requisite inspection facilities but refused to withhold registration from those vehicles failing the inspection. The EPA sought an injunction ordering Ohio to implement "as written" the inspection and registration procedure described in the plan. Was the injunction granted? Explain.

22-10 The Idaho EPA, in developing its SIP, determined that the maximum sulfur dioxide emissions that could be captured from zinc smelters with the currently available technology was 72 percent. The state consequently adopted that standard for zinc smelters under the SIP. The federal EPA refused to accept that part of the SIP and promulgated an 82 percent standard. Did the federal EPA have authority to make such a change in the SIP? Why or why not?

22-11 As a cleaning agent in its production process, Kantrell Corporation uses about 50 gallons a day of a highly corrosive acid. It collects the used acid and funnels it through a pipe into a pond located on company property; the pond was dug to serve as a place in which to dispose of the acid and other wastes that could not be incinerated or recycled. Is Kantrell violating any federal environmental regulations? Explain.

22-12 The defendant operated a plant that had refined coal tar for 55 years. It had disposed of its wastes on the site. After the plant closed, the land was purchased by a municipal housing authority. The wastes buried on the site leaked into the groundwater, contaminating the drinking water of nearby cities. The state and the municipalities spent considerable sums of money cleaning up the site. The U.S. government joined the suit, seeking to hold the defendant liable under CERCLA. Was the defendant responsible even though it no longer owned the dump site? Explain.

CASE PROBLEMS

22-13 A nonprofit conservation organization, Western Watersheds Project, filed a lawsuit challenging the Bureau of Land Management's (BLM) decision to grant a 10-year grazing permit for four federal public land allotments for violating the NEPA. In deciding to grant the permits, the BLM had filed an EA, but had not included in the EA a discussion of the "no action" alternative. The organization argued that the EA was defective without a discussion of the effects of the no action alternative. The BLM believed that it did not need to discuss the "no action" alternative because it was relying on the previously prepared Resource Management Plan to determine that elimination of grazing was not viable or was not a necessary option. The BLM noted that it had also examined two other alternatives. The District Court granted the BLM summary judgment. How do you believe the Court of Appeals ruled in this case? Why? *Western Watersheds Project v. Bureau of Land Management*, 721 F.3e 1264 (2013) CA 10 (Wyo.), 2013 WL3801818.

22-14 An insurer, Chubb Custom Insurance Company, sought recovery of insurance payments made to its insured, the defendant, Space Systems/Loral, Inc., for environmental response costs incurred from cleaning up pollutants. The plaintiff insurer sought this recovery by asserting claims of CERCLA. The district court held that the insurer lacked standing to sue because it had not become statutorily liable for response costs under CERCLA. The district court dismissed the insurer's claim of CERCLA, and the plaintiff insurer appealed. Do you agree with the decision of the district court? Why or why not? Do you think the appellate court affirmed or reversed the decision of the district court? Why? *Chubb Custom Ins. Co. v. Space Systems/Loral, Inc.*, 788 F. Supp. 2d 1017 (9th Cir. 2013).

22-15 From 1979 to 2005, AVX Corporation leased 27 acres of property, referred to as "Horry Land Property." The plaintiff, AVX Corporation, claimed that the U.S. operations at the airfield during World War II caused TCE contamination on all of the real estate parcels that the airfield formerly encompassed, including the Horry Land Property. AVX Corporation sued the United States under CERCLA to recover costs it incurred after cleaning up the Horry Land Property in Myrtle Beach, South Carolina. The United States then filed a counterclaim for equitable contribution under CERCLA. The district court concluded that the United States did not contribute to any contamination on the property. The plaintiff, AVX Corporation, appealed. How do you think the appellate court ruled? Why? What evidence do you think the appellate court would need to reverse the decision of the district court and conclude that the United States did contribute to the contamination on the property? *AVX Corp. v. United States of America*, 2013 U.S. App. LEXIS 2762 (2013).

22-16 Sonat operated a natural gas company that included multiple compressor stations along a pipeline that spanned from Texas to Georgia. Many of the compressor stations included mercury metering. Sonat also used the lubricating oil Pydraul in the compressor engines at the stations. In 1989, Sonat discovered that this lubricating oil contained the toxic chemical polychlorinated biphenyl (PCB). Sonat tried to find solutions to remove Pydraul from the compressor stations. It was further discovered that Sonat's mercury-metering stations were discharging mercury along the pipeline. In 1992, the EPA notified outside gas pipeline companies that Sonat's mercury meters were leaking. Sonat conducted multiple remedial activities at its various mercury stations to prevent migration of mercury into groundwater. The business then received umbrella and excess-liability insurance policies from LMI Insurers. LMI later argued that Sonat's cleanup costs were not "damages" that LMI was obligated to pay because Sonat's cleanup projects were an "internal business decision and not the result of any compulsory process by a court or a state or federal agency." However, in court, LMI was required to pay Sonat's environmental remediation costs because the Toxic Substances Control Act required Sonat to report contaminations to the EPA and Sonat's remedial cleanup costs constituted "damages." The insurers appealed. How do you think the appellate court ruled? Why? *Certain Underwriters at Lloyd's v. Southern Natural Gas Co.*, 142 So.3d 436 (Ala. 2013).

22-17 In 2007, it was found that Title II of the Environmental Protection Act (EPA) "authorized EPA to regulate greenhouse gas emissions from new motor vehicles if the Agency formed a judgment that such emissions contribute to climate change." Subsequently, the EPA began heavily regulating greenhouse gas emissions. In addition to regulating emissions from new motor vehicles, the EPA also began making stationary sources of greenhouse gases, such as factories and power plants, subject to the Act's "Prevention of Significant Deterioration" (PSD) provisions, based on the *potential* of stationary sources to emit greenhouse gases. Numerous parties, including several states, challenged the EPA's greenhouse gas-related actions by filing for petitions for review in the D.C. Circuit. The case eventually went to the U.S. Supreme Court, where it would decide if it was permissible for the EPA to determine whether its new motor vehicle greenhouse gas regulations could lead to requirements under the act for stationary sources of greenhouse gases. How do you think the Court ruled? Why? *Util. Air Regulatory Group v. EPA*, 134 S. Ct. 2427 (U.S. 2014).

22-18 Plaintiff Violet Gallagher filed a complaint against defendant East Buffalo Township for alleged discharge of "turbid, malodorous garbage laden water" onto her property and into the Susquehanna River due to the Township's storm water management system. The plaintiff alleged that these actions violated the federal Clean Water Act, as well as Pennsylvania's Stormwater Management Act. What evidence would the plaintiff need to provide for the court to find that the federal Clean Air Act was violated? How do you think the court ruled? Why? *Gallagher v. E. Buffalo Twp.*, 2013 U.S. Dist. LEXIS 123626 (M.D. Pa. 2013).

THINKING CRITICALLY ABOUT RELEVANT LEGAL ISSUES

As the issue of global climate change comes to the fore in international discussions and as countries attempt to find solutions for climate change, attention frequently turns to the Kyoto Protocol. By June 2007, 17 countries had signed and ratified the Kyoto Protocol. By 2011, the number of parties to the Kyoto Protocol had risen to 192. Notably absent from the list of countries is the United States, which has stated that it will not ratify the Protocol. Although some environmentalists argue that the United States should ratify, the United States is correct to refuse.

The Kyoto Protocol, although well intentioned, is doomed to fail. An analysis indicates that its goals are ineffective. Although countries that ratify the Kyoto Protocol agree to reduce their greenhouse gas emissions to pre-1990 levels, only 35 countries have agreed to cap their greenhouse gas emissions. Agreeing to a cap is not part of the treaty. Also, the Protocol exempts developing nations and instead requires developed nations to limit their greenhouse gas emissions. By not requiring developing countries to limit their emissions, those who created the treaty have permitted these countries to continue to pollute at high volumes, thus offsetting any efforts taken by developed countries.

What further makes the Kyoto Protocol ineffective is that developing countries are excluded and China is counted as a developing country. By not having to reduce its emissions, China will continue to pollute in large quantities, preventing any hope of curbing global emissions. In addition, it is unfair that China and other developing countries can pollute at will, thus avoiding engaging in costly emission reduction strategies. The ability to avoid paying to reduce emissions gives China an unfair advantage on the global market, as it can produce and sell products cheaper compared with developed countries that need to pay for emissions-reducing technology.

There is another irony in the Kyoto Protocol that China helps to exemplify. If developed nations lower their demand for fossil fuels in an attempt to reduce greenhouse gas emissions, this reduction will lower the price of fossil fuels. As fossil fuels become cheaper, developing countries, especially China, will increase their use of cheap fossil fuels, which will produce even more greenhouse gases. Exempting China means that the Protocol cannot work.

But the main reason the United States shouldn't ratify the treaty is that it doesn't need to. In 2012, the United States became the first industrialized nation to meet the original 2012 target for CO_2 reductions. Signatories to the agreement are now being asked to commit to a new goal—to reduce GHG emissions by at least 18 percent below 1990 levels in the eight-year period from 2013 to 2020. If past behavior is a good predictor of future behavior, the United States doesn't need to ratify any treaty; it most likely will continue reducing its emissions and meet this goal, too. The current efforts of the United States are more than enough to try to address the problem of global climate change, thus making ratifying the Kyoto Protocol unnecessary.

1. What are the issue and conclusion in this essay?

2. Does the argument contain significant ambiguity in the reasoning? Explain.

   Clue: What words or phrases could have multiple meanings?

3. Ask and answer the critical thinking question that you believe reveals the main problem with the author's reasoning in this essay. Explain why the question you asked is particularly harmful to the author's argument.

4. Write an essay from the viewpoint of someone who holds a different opinion from that of the essay author.

   Clue: What other ethical norms could influence an opinion about this issue?

ASSIGNMENT ON THE INTERNET

Environmental protection is often a slow and arduous process because many enforcement efforts end up in the court system. Using Internet sites such as LexisNexis and **www.law.cornell.edu/wex/environmental_law**, find a recent court case involving a Superfund cleanup site.

What was disputed in the case? What reasons and/or laws were cited in the court's decision? How does the information in this chapter better assist you in understanding the court's decision? If cleanup was done, who was found responsible for the cleanup of the site? Finally, do you agree with the ethical norms that underlie the court's decision? Why or why not?

ON THE INTERNET

www.epa.gov The EPA home page is a source of valuable information about the main agency responsible for protecting the environment.

www.unep.org This is the home page of the United Nations Environmental Programme.

sedac.ciesin.columbia.edu/entri/index.jsp This site provides lists of environmental treaties and resource indicators.

www.epa.gov/superfund Information about the EPA's Superfund program can be found on this site.

www.nrdc.org The National Resource Defense Council works to prevent negative externalities that harm the environment.

www.eere.energy.gov The U.S. Department of Energy provides information regarding renewable energy sources and explains how this type of energy can help combat pollution and global climate change.

http://newsroom.unfccc.int/ Go to this site to find the most current information about UN action with respect to climate change.

FOR FUTURE READING

Frederickson, Robert. "A Green Bird in the Hand: An Example of Environmental Regulations Operating to Stifle Environmentally Conscious Industry." *Boston College Environmental Affairs Law Review* 34 (2007): 303.

Inman, Kelly. "Recent Development: The Symbolic Copenhagen Accord Falls Short of Goals." *University of Baltimore Journal of Environmental Law* 17 (2010): 219.

LeBel, Mark E. "Lack of Judicial CAIR: Chevron Deference and Market-Based Environmental Regulations." *New York University Environmental Law Journal* 20 (2013): 277.

Malloy, Thomas F. "The Social Construction of Regulation: Lessons from the War Against Command and Control." *Buffalo Law Review* 58 (2010): 267.

Palassis, Stathis N. "Beyond the Global Summits: Reflecting on the Environmental Principles of Sustainable Development." *Colorado Journal of International Environmental Law and Policy* 22 (2011): 41.

Shufelt, Jennie. "New York's CO_2 Cap-And-Trade Program: Regulating Climate Change Without Climate Change Legislation." *Albany Law Review* 73 (2010): 1583.

Thompson, Aselda. "Comment: Exposing a Gap in CERCLA Case Law: Is There a Right to Recover Costs Following Compliance with an Administrative Order after *Atlantic* and *Aviall*?" *Houston Law Review* 46 (2010): 1679.

CHAPTER TWENTY THREE

Rules Governing the Issuance and Trading of Securities

- INTRODUCTION TO THE REGULATION OF SECURITIES
- DODD-FRANK WALL STREET REFORM AND CONSUMER PROTECTION ACT OF 2010
- THE SARBANES-OXLEY ACT OF 2002
- THE SECURITIES ACT OF 1933
- THE SECURITIES EXCHANGE ACT OF 1934
- STATE SECURITIES LAWS
- E-COMMERCE, ONLINE SECURITIES DISCLOSURE, AND FRAUD REGULATION
- GLOBAL DIMENSIONS OF RULES GOVERNING THE ISSUANCE AND TRADING OF SECURITIES

In Chapter 17, we said that the corporation was the dominant form of business organization in the United States—and the most regulated. Two of the most strongly regulated aspects of corporate business are the issuance and trading of securities. Corporate securities—stocks and bonds—are used to raise capital for the corporation. They are also used by individuals and institutional investors to accumulate wealth. In the case of individuals, this wealth is often passed on to heirs, who use it to accumulate more wealth. Thus, securities provide a means for one generation in a family to "do better" than the preceding generation. Securities also provide a means for financing pension funds and insurance plans through institutional investment.

Securities holders are powerful determinants of trends in business: If an individual company, industry, or segment of the economy is not growing and paying a good rate of return, investors will switch their funds to another company, industry, or segment in expectation of better returns. Securities holders (or their proxies) elect the board of directors of a corporation, who, in turn, select the officers who manage the daily operations of a corporation. Finally, securities holders' ability to bring lawsuits helps keep officers and directors honest in their use of investors' funds.

Because of their importance to the operation of our free enterprise society and because of the ease with which they can be manipulated, securities have been regulated by governments for nearly a century. This chapter chiefly examines the role of the federal government in regulating securities. We introduce the subject with a brief history of securities regulation that contains a summary of the most important federal legislation. We then turn to the creation, function, and structure of the Securities and Exchange Commission (SEC). In a survey of major and representative securities legislation, we examine the provisions of the Dodd-Frank Act of 2010 and the Sarbanes-Oxley Act of 2002. Both the Securities Act of 1933, which governs the issuance of securities and outlines the registration requirements for both securities and transactions (and the allowable exemptions from those requirements), and the Securities Exchange Act of 1934, which governs trading in securities, are discussed. We then examine the state securities laws and online securities disclosure and fraud regulations. We end with a discussion of the global dimensions of the 1933 and 1934 securities acts; the Foreign Corrupt Practices Act, as amended in 1988; and the Convention on Combating Bribery of Foreign Public Officials in International Business Transactions.

CRITICAL THINKING ABOUT THE LAW

Because issuers can manipulate securities easily, federal and state governments have strongly regulated the issuance and trading of securities. Studying the following case example and answering some critical thinking questions about it will help you better appreciate the need for regulation of securities.

Jessica received a phone call from a man claiming to represent Buy-It-Here, a corporation that was relocating to Jessica's town. The man stated that the corporation was planning to issue new securities, and he was extending this offer to residents in Jessica's town. He claimed that Buy-It-Here would easily double its profits within six months. The man said that if Jessica sent $3,000, he would buy stock in Buy-It-Here for Jessica. Jessica sent the money; two weeks later, she discovered that Buy-It-Here was in the process of filing for bankruptcy.

1. This case is an example of the need for government regulation. We want the government to protect citizens from cases such as Jessica's buying stock in a bankrupt company. If we want governmental protection from potentially shady businesses, what ethical norm are we emphasizing?

   Clue: Put yourself in Jessica's place. Why would you want governmental protection? Now match your answer to an ethical norm listed in Chapter 1. Think about which ethical norm businesses would emphasize.

2. Jessica wants to sue Buy-It-Here for misrepresentation. Before she brings her case, what additional information do you think Jessica should discover?

   Clue: What additional information do you want to know about the case? Even without having extensive knowledge about securities, you can identify areas in which you might need more information about Jessica's case. For example, pay close attention to the role of the telephone caller.

3. Jessica did some research about securities cases in her state. She discovered a case in which a woman named Andrea Stevenson had purchased $100,000 worth of stock from a stockbroker. The company went bankrupt three months later. The stockbroker had known that the company was suffering financial problems but had said nothing to Andrea. The jury in this case found in favor of Andrea. Jessica wants to use Andrea's case as an analogy in her lawsuit. Do you think that Andrea's case is an appropriate analogy? Why or why not?

   Clue: What are the similarities between the cases? How are the cases different? Are these differences so significant that they overwhelm the similarities?

Introduction to the Regulation of Securities

Securities have no value in and of themselves. They are not like most goods produced or consumed (e.g., television sets or toys), which are easily regulated in terms of their hazards or merchantability. Because they are paper, they can be produced in unlimited numbers and can be manipulated easily by their issuers.

The first attempt to regulate securities in the United States was made by the state of Kansas in 1912. When other states followed the Kansas legislature's example, corporations played off one state against another by limiting their securities sales to states that had less stringent regulations. Despite the corporations' ability to thwart state efforts at regulation rather easily, there was strong resistance to the idea of federal regulation in Congress. It was not until after the collapse of the stock market in 1929 and the free fall of stock prices on the New York Stock Exchange (NYSE)—when the Dow Jones Industrial Average registered an 89 percent decline between 1929 and 1933—that Congress finally acted.

SUMMARY OF FEDERAL SECURITIES LEGISLATION

The following legislation, enacted by Congress since 1933, provides the framework for the federal regulation of securities. It is also important to note that this legislation is the basis (enabling act) for rulemaking by the SEC. Congressional legislation is emphasized here, but it is important to remember that SEC rulemaking may be equally significant in the long term. (You will remember that we discussed rulemaking for federal agencies in Chapter 18.)

- The Securities Act of 1933 (also known as the Securities Act or the 1933 Act) regulates the initial offering of securities by public corporations by prohibiting an offer or a sale of securities not registered with the Securities and Exchange Commission. The 1933 Act sets forth certain exemptions from the registration process, as well as penalties for violations of the act. This act is examined in detail in this chapter. Both the 1933 and 1934 Acts have been amended by Congress and the SEC rulemaking process, much of which is summarized in the following pages.

- The Securities Exchange Act of 1934 (also known as the Exchange Act) regulates the trading in securities once they are issued. It requires brokers and dealers who trade in securities to register with the Securities and Exchange Commission, the regulatory body created to enforce both the 1933 and 1934 Acts. The Exchange Act is also examined in detail in this chapter.

- The Public Utility Holding Company Act of 1935 requires public utility and holding companies to register with the SEC and to disclose their financial organization, structure, and operating process.

- The Trust Indenture Act of 1939 regulates the public issuance of bonds and other debt securities in excess of $5 million. This act imposes standards for trustees to follow to ensure that bondholders are protected.

- The Investment Company Act (ICA) of 1940, as amended in 1970 and 1975, gives the SEC authority to regulate the structure and operation of public investment companies that invest in and trade in securities. No private causes of action are created by this law. A company is an *investment company* under this act if it invests or trades in securities and if more than 40 percent of its assets are "investment securities" (which are all corporate securities and securities invested in subsidiaries). Accompanying legislation, entitled the *Investment Advisers Act of 1940*, authorizes the SEC to regulate persons and firms that give investment advice to clients. This act requires the registration of all such individuals or firms and contains antifraud provisions that seek to protect broker-dealers' clients.

- The Securities Investor Protection Act (SIPA) of 1970 established the nonprofit Securities Investor Protection Corporation (SIPC) and gave it authority to supervise the liquidation of brokerage firms that are in financial trouble, as well as to protect investors from losses up to $500,000 due to the financial failure of a brokerage firm. The SIPC does not have the monitoring and "bailout" functions that the Federal Deposit Insurance Corporation (FDIC) has in banking; it only supervises the liquidation of an already financially troubled brokerage firm through an appointed trustee.

- Chapter 11 of the Bankruptcy Abuse Prevention and Consumer Protection Act of 2005 gives the SEC the authority to render advice when certain debtor corporations have filed for reorganization.

- The Foreign Corrupt Practices Act (FCPA) of 1977, as amended in 1988, prohibits the direct or indirect giving of "anything of value" to a foreign official for the purpose of influencing that official's actions. The FCPA sets out an intent or "knowing" standard of liability for corporate management. It requires all companies (whether doing business abroad or not) to set up a system of internal controls to provide reasonable assurance that the company's records "accurately and fairly reflect" its transactions. The FCPA is discussed in detail in the last section of this chapter.

- The International Securities Enforcement Cooperation Act (ISECA) of 1990 clarifies the SEC's authority to provide securities regulators of other governments with documents and information and exempts from Freedom of Information Act disclosure requirements all documents given to the SEC by foreign regulators. The ISECA also authorizes the SEC to impose administrative sanctions on securities buyers and dealers who have engaged in illegal activities in foreign countries. Finally, it authorizes the SEC to investigate violations of the securities law set out in the act that occur in foreign countries. The ISECA is also discussed in the last section of this chapter.

- The Market Reform Act of 1990 authorizes the SEC to regulate trading practices during periods of extreme volatility. For example, the SEC can take such emergency action as suspending trading when computer program–driven trading forces the Dow Jones Industrial Average to rise or fall sharply within a short time period.

- The Securities Enforcement Remedies and Penny Stock Reform Act of 1990 (the 1990 Remedies Act) gives the SEC powerful new means for policing the securities industry: cease-and-desist powers and the power to impose substantial monetary penalties (up to $650,000) in administrative proceedings. The 1990 Remedies Act also gives the SEC and the federal courts the following powers over anyone who violates federal securities law:

    1. The imposition of monetary penalties by a federal court for a violation of the securities law on petition by the SEC.
    2. The power of the federal courts to bar anyone who has violated the fraud provisions of the federal securities laws from ever serving as an officer or a director of a publicly held firm.
    3. The power of the SEC to issue permanent cease-and-desist orders against "any person who is violating, has violated, or is about to violate" any provision of a federal securities law.

    This act arms the SEC with some of the most sweeping enforcement powers ever given to a single administrative agency other than criminal enforcement agencies such as the Justice Department.

- The Private Securities Litigation Reform Act of 1995 (Reform Act) provides a safe harbor from liability for companies that make statements to the public and investors about risk factors that may occur in the future.

- The Securities Litigation Uniform Standards Act of 1998 sets national standards for securities class action lawsuits involving nationally traded securities. This act amends the 1933 and 1934 Acts and prohibits any private class action suits in state or federal court alleging (1) any untrue statement or omission in connection with the purchase or sale of a covered security or (2) the defendant's use of any manipulation or deceptive device in connection with the transaction.

- The Sarbanes-Oxley Act of 2002 amends the 1933 and 1934 Acts. Sarbanes-Oxley includes provisions dealing with corporate governance, financial regulation, criminal penalties, and corporate responsibility, all of which are discussed in detail in this chapter. The Credit Rating Agency Reform Act of 2006 creates a new regulatory system by which the SEC identifies and oversees five nationally recognized agencies that issue credit ratings.

- The Dodd-Frank Act of 2010, a wide-ranging reform of regulatory actions that seeks to prevent the recurrence of a major financial catastrophe such as the one that occurred in 2008.

For your convenience, some of the federal securities legislation is summarized in Table 23-1.

TABLE 23-1 SUMMARY OF THE MAJOR FEDERAL SECURITIES LEGISLATION

Federal Securities Legislation	Purpose
Securities Act of 1933	Regulates generally the issuance of securities.
Securities Exchange Act of 1934	SEC regulates trading in securities.
Public Utility Holding Company Act of 1935	SEC regulates public utility and holding companies through registration and disclosure processes.
Trust Indenture Act of 1939	SEC regulates the public issuance of bonds and other debt securities.
Investment Company Act of 1940	SEC regulates the structure and operation of public investment companies.
Securities Investor Protection Act of 1970	Securities Investor Protection Corporation supervises the liquidation of financially troubled brokerage firms.
Foreign Corrupt Practices Act of 1977, as amended in 1988	Prohibits the payment of anything of value to influence foreign officials' actions.
International Securities Enforcement Corporation Act of 1990	SEC has authority to provide securities regulators of other governments with information on alleged violators of securities law in the United States and abroad.
Market Reform Act of 1990	SEC regulates the trading practices during periods of extreme volatility.
Securities Enforcement Remedies and Penny Reform Act of 1990	Requires more stringent regulation of broker-dealers who recommend penny-stock transactions to customers.
Securities Enforcement Remedies and Penny Reform Act of 1991	SEC regulates the securities industry through cease-and-desist powers and threat of substantial monetary penalties.
Private Securities Litigation Reform Act of 1995	SEC provides a safe harbor from liability for companies that make statements to the public or investors about risk factors that may occur in the future.
Securities Litigation Uniform Standards Act of 1998	Sets national standards for securities class action lawsuits involving nationally traded securities. Amends the 1933 and 1934 Acts and prohibits any private class action suit in state or federal court alleging (1) any untrue statement or omission in connection with the purchase or sale of a covered security or (2) that the defendant used manipulation or a deceptive device in connection with the transaction.
Sarbanes-Oxley Act of 2002	Amends the 1933 and 1934 Acts and other federal statutes. Includes provisions dealing with corporate governance, financial regulation, criminal penalties, and corporate responsibility.
Bankruptcy Abuse Prevention and Consumer Protection Act of 2005	SEC has authority to advise debtor corporations that have filed for reorganization.
Credit Rating Agency Reform Act of 2006	
Dodd-Frank Act of 2010	Creates a registration process through the SEC for rating agencies wishing to become nationally recognized. Congress sought to meet the need to increase the number of agencies from the five established by Section 15E of the 1934 Act. Seeks to amend several statutes and the regulatory process involving the SEC and other federal agencies of the federal government. Statute was passed following a major economic downturn (recession) in 2008.

THE SECURITIES AND EXCHANGE COMMISSION

Securities and Exchange Commission (SEC) The federal administrative agency charged with overall responsibility for the regulation of securities, including ensuring that investors receive "full and fair" disclosure of all material facts with regard to any public offering of securities. It has wide enforcement powers to protect investors against price manipulation, insider trading, and other dishonest dealings.

Creation and Function. The **Securities and Exchange Commission (SEC)** was created under the Securities Exchange Act of 1934 for the purpose of ensuring that investors receive "full and fair" disclosure of all material facts with regard to any public offering of securities. The SEC is not charged with evaluating the worth of a public offering of securities by a corporation (e.g., determining whether the offering is speculative); it is concerned only with whether potential investors are provided with adequate information to make investment decisions. To this end, the commission was given the power to set up and enforce proper registration regulations for securities, as well as to prevent fraud in the registration and trading of securities.

Structure. Exhibit 23-1 lays out the structure of the SEC. It has five commissioners (inclusive of the chairman), who are appointed by the president with the advice and consent of the Senate; each serves for a period of five years, and no more than three commissioners can be of the same political party. The SEC, based in Washington, DC, has 11 regional offices across the United States. There are five divisions: Corporation Finance, Market Regulation, Enforcement, Corporate Regulation, and Investment Management. (Note in Exhibit 23-1 that in addition to the 5 major divisions, there are several other important offices.)

Division of Corporation Finance. The Division of Corporation Finance is responsible for establishing and overseeing adherence to standards of financial reporting and disclosure for all companies that fall under SEC jurisdiction, as well as for setting and administering the disclosure requirements prescribed by the 1933 and the 1934 Securities Acts, the Public Utility Holding Company Act, and the Investment Company Act. This division reviews all registration statements, prospectuses, and quarterly and annual reports of corporations, as well as their proxy statements. Its importance in offering informal advisory opinions to issuers (corporations about to make a public offering of stock) cannot be overemphasized. Accountants, lawyers, financial officers, and underwriters all rely heavily on advice from this division.

Division of Trading and Markets. This SEC division regulates the national security exchanges (such as the NYSE), as well as broker-dealers registered under the Investment Advisers Act of 1940. Through ongoing surveillance of both the exchanges and broker-dealers, the Division of Market Regulation seeks to discourage manipulation or fraud in the issuance, sale, or purchase of securities. It can recommend to the full commission the suspension of an exchange for up to one year, as well as the suspension or permanent prohibition of a broker or dealer because of certain types of conduct. In addition, the division provides valuable informal advice to investors, issuers, and others on securities statutes that come within the SEC's jurisdiction.

Division of Enforcement. The Division of Enforcement is responsible for the review and supervision of all enforcement activities recommended by the SEC's other divisions and regional offices. It also supervises investigations and the initiation of injunctive actions.

Division of Economic and Risk Analysis. This division integrates financial economics and rigorous data analytics into the core mission of the SEC. It is involved across the entire range of SEC activities, including policy-making, rule-making, enforcement, and examination.

Division of Investment Management. This SEC division administers the ICA of 1940 and the Investment Advisers Act of 1940. All investigations arising under these acts dealing with issuers and dealers are carried out by this Division of Investment Management.

CHAPTER 23 • Rules Governing the Issuance and Trading of Securities

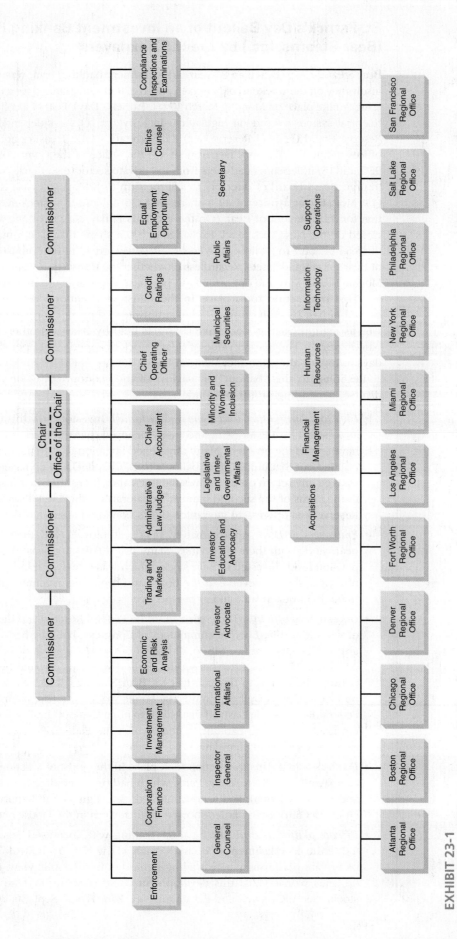

EXHIBIT 23-1
THE SECURITIES AND EXCHANGE COMMISSION

St. Patrick's Day Bailout of an Investment Banking Firm (Bear Stearns, Inc.) by the U.S. Taxpayers

Bear Stearns, Inc. (Bear), an 85-year-old investment banking firm, was headed for insolvency on the weekend of March 15 and 16, 2008, and planned for a bankruptcy filing to take place on Monday, March 17 (St. Patrick's Day). Fear of a collapse of the financial system led federal regulators—inclusive of (1) the independent Federal Reserve Board (Federal Reserve), (2) the Secretary of the Treasury and his many offices within the Treasury Department, (3) the Office of the Comptroller, (4) the SEC, and (5) independent advisers from the private sector (e.g., Black Rock, Inc.)—to urge Bear's board of directors to sell the firm to JPMorgan Chase & Company (J.P. Morgan), at a price of $2 a share, or $236 million, in a stock-swap transaction for 39.9 percent of Bear. (On the previous Friday, March 14, the stock value closed on the New York Stock Exchange at $30 a share, that is, at a market value of $3.54 billion.) In addition, the Federal Reserve agreed to fund up to $30 billion of Bear's nonliquid assets. Regardless of whether the transaction went through, J.P. Morgan would have the opportunity to purchase the headquarters of Bear.

This transaction took place in the midst of a nationwide credit crunch caused in part by cash outflows from subprime and prime mortgage holders, as well as margin calls on derivative contracts held by Bear and other investment banking firms. The market value of Bear's stock dropped to $11 per share in the days following the announcement on March 17.

Response to this "bailout" or "savior" of the economy (U.S. or world) was diverse, depending on the responder:

- *Investors (individual and some institutional).* Investors saw this transaction as a "steal" by J.P. Morgan, and many believed that the market itself would have solved the problem. Many threatened litigation to stop the bailout. They pointed out that unlike other bailouts (e.g., Chrysler), the taxpayers were not assured any return on their investment. Further, there was no transparency as to the terms of the secured interest (collateral) for the $30 billion the Federal Reserve was offering to guarantee Bear's nonliquid assets.

- *Employees of Bear.* Approximately 14,000 employees saw their jobs disappear, along with their life savings. Many had 401(k) funds as well as private pension funds invested in Bear stock, which was now worth little. After years of loyalty, they believed that the board and senior management of Bear had "sold them out" from a moral perspective.

- *Government officials.* The chairman of the Federal Reserve and the chairman of the SEC testified before committees of Congress that they had varying degrees of advance notice (48 to 72 hours) of the seriousness of Bear Stearns' problems. Their response (as set out earlier) was thus dictated by this short period. The failure to find a buyer for Bear Stearns could have led to a run on investment as well as commercial banks worldwide. The chairman of the Federal Reserve emphasized that this was not a "bailout" but rather an action required to save the banking system, which the Federal Reserve is directed to do by its enabling legislation.

- *Political actors.* This transaction took place in the midst of a national primary campaign by the Democratic Party, which had two candidates (Hillary Clinton and Barack Obama), and a noncontested campaign by the Republican Party (John McCain). Both parties showed their concern in the House and Senate.

- *Private platforms and "dark pools."* Trading with increased regulation has led to the establishment of bourses (such as the New York Stock Exchange) or private platforms where trading is hidden from public view. Regulators are thus worried that this development could obscure the true price of a stock. (See J. Creswell and P. Latman, *New York Times*, Sept. 30, 2010, F-6.)

CRITICAL THINKING ABOUT THE LAW

1. What values were in conflict for the parties to the St. Patrick's Day bailout described previously?

    Clue: The parties included, among others, the Federal Reserve, the U.S. Secretary of the Treasury and the Treasury Department, the SEC, and the president of the United States, as well as employees of Bear Stearns. Chapter 1 discusses the values involved in answering this question.

2. Should federal and state "bailouts" of private-sector firms such as Bear Stearns take place as a matter of general principle? Why or why not?

    Clue: What values are in conflict for federal and state governments? What about taxpayers—or are they represented? Politicians? Lobbyists?

Electronic Media, the Age of the Internet, and SEC Internal Functions

Through internal rulings, the SEC has recognized that the use of "electronic media . . . enhances the efficiency of the securities market by allowing for the rapid dissemination of information to investors and financial markets in a more cost-efficient, widespread and equitable manner than traditional paper methods." The SEC has provided interpretive guidance for the use of electronic media for the delivery of information required by the federal securities law. The SEC has defined *electronic media* to include audiotapes, videotapes, CD-ROM, email, bulletin boards, Internet websites, and computer networks. Further, securities regulators have authorized the use of social media sites such as Twitter and Facebook for communications by companies to investors and shareholders (April 2, 2013, *New York Times*, B-1).

The SEC has established the EDGAR (electronic data gathering, analysis, and retrieval) computer system, which performs automated collection, validation, indexing, acceptance, and dissemination of reports required to be filed with the SEC. The SEC requires all domestic companies to make their filings on EDGAR, except those exempted for hardship. EDGAR filings are posted at the SEC website 24 hours after the date of filing.

Dodd-Frank Wall Street Reform and Consumer Protection Act of 2010

In July 2010, Congress passed a bill that wrought a massive overhaul of federal government financial regulations and seemed to affect every sector of the economy. This piece of legislation became known as the Dodd-Frank bill (statute)[1] so named after its major sponsors: Senator Dodd (D-Conn.), who was chairman of the Senate Banking Committee, and Congressman Frank (D-Mass.), who was chairman of the House Financial Affairs Committee. This bill passed Congress in the midst of a recession and after a collapse of the financial markets in 2008. It sought to respond to the major causes of the financial crises. Actions taken by Congress, outlined in the following subsection, sought to prevent problems similar to those faced by the nation in the period between 2008 and 2010. Nonetheless, a 2011 survey of 94 fund advisers concluded that three-quarters of those surveyed indicated that Dodd-Frank regulations did not change their way of doing business. (See Walt King, Professor, St. Thomas University, Minneapolis, reprinted in parts, *Bloomberg*

---

[1] Pub. L. No. 111–203 (2010).

Newsweek, October 22, 2013.) As of June 2014, four years after passage of the Act, 208 of 398 proposed rules missed their deadlines. Industry groups continued to lobby to defeat the Act while seeking to tailor the proposed rules to their interests. Some rules are being challenged in court.[2]

OVERSIGHT OF FINANCIAL PROBLEMS BY REGULATORY AGENCIES

- A new Financial Stability Oversight Council (Council) was established by the Dodd-Frank Act. The council is made up of heads of major regulatory agencies (e.g., Treasury, SEC, FDIC, the Federal Reserve). The council will identify banks or nonbanks that pose a threat to the financial system. The Fed, with the approval of the council, will have the power to break up large firms. It could also require such firms to increase their reserves against future losses.
- The Fed was to be subject to oversight by the Government Accountability Office for a short period during 2008, particularly as to its loans via the discount window.
- Hedge funds larger than $100 million must register with the SEC and provide some information as to trades and their individual portfolios.
- The Office of Thrift Supervision will be absorbed into the Office of the Comptroller of the Currency.

RISK TAKING BY LARGE BANKS AND NONBANKS

Bank holding companies (e.g., Citigroup and Bear Stearns) participated in speculative trades involving mortgage-backed securities and other financial instruments (e.g., derivatives). When these speculative bets went under, the institutions involved could not sell the assets involved, which thus became known as "toxic" assets. The federal government had to spend billions in taxpayer money to bail out these companies. They are presently repaying these loans (at least in part), plus interest and/or preferred shares, to the federal government.

The Dodd-Frank legislation was also intended to prevent FDIC-insured institutions from making speculative trades and to require these entities to sell their interests in hedge funds and private equity funds; only 3 percent of their capital could remain invested in such funds. Investment banks also had to set aside reserves to cover losses. Originators of mortgage securities must hold 5 percent of the credit risk, thus retaining an interest in the performance of the securities. For reasons other than speculation, banks will be allowed to trade in a "proprietary" manner. Banks can also continue to buy or sell from their own accounts to hedge against other investments.

EXECUTIVE COMPENSATION

Compensation to executives of the largest financial firms was based on quarterly earnings. Earnings increased as these firms sold mortgage-backed securities and derivatives—until the housing "bubble burst." When subprime mortgages began to fail, the federal government had to bail out the large financial institutions that had speculated heavily in these instruments and derivatives based on them. Anger over the enormous compensation paid to executives of these institutions became a major public issue, as taxpayers saw their "bailout" tax dollars apparently being used to reward executives for serious mismanagement and poor performance.

---

[2]Dodd-Frank Progress Report, June 2014, Davis Polk at 2.

The Dodd-Frank Act did some things to deal with executive compensation:

- Shareholders were allowed a nonbinding vote on executive compensation, as directed by the SEC.
- Only independent directors of a company could sit on compensation committees of the board.
- Companies would be required to take back compensation if it was based on accounting statements that were later found to be inaccurate.

TOO BIG TO FAIL

Nonbank financial companies such as insurance giant AIG could not be legally shut down during the 2008 crisis. The government bailed them out, believing that their bankruptcy would bring about the collapse of the financial system both in the United States and markets worldwide.

The new statute gave the FDIC authority to shut down banks and nonbank financial firms. Taxpayers initially would foot the bill for liquidation, but the money was eventually to be returned to the federal coffers from shareholders and unsecured creditors. Further, the statute ordered an increase in the reserve ratio of the FDIC, but specified that small depository institutions (those with less than $10 billion in consolidated assets) were exempt from making such increases.

A fund of $11 billion was initially established within the Troubled Asset Relief Program (TARP) to cover the costs of shutting down companies. In theory, the government could then shut down huge companies without the taxpayers having to bail them out.

CREDIT RATING AGENCIES

Credit rating agencies (such as Moody's and Standard & Poor's) evaluated and rated billions in mortgage securities; both the private sector and governments at all levels relied on these ratings. These agencies were paid by the same companies that were issuing and trading in mortgage securities and other forms of debt (and thus had a vital interest in positive ratings). When the housing market crashed, many of the rating agencies sought to downgrade the ratings they had given mortgage securities and other assets.

- Despite what appeared to be a conflict of interest, Congress could not agree on a format to replace the ratings agencies. The Dodd-Frank legislation was intended to make it easier to sue credit rating agencies. In addition, this statute eliminated any federal requirement that banks and other investors rely on ratings set out by these agencies.
- The legislation orders the SEC to study ways to eliminate ratings shopping by issuers.
- It allows the SEC to deregister ratings agencies that have a bad record of violating financial regulations.
- All ratings agencies now have to disclose how they arrive at ratings and how they comply with conflict-of-interest regulations.

DERIVATIVES

Derivatives are synthetic securities that are dependent upon the movement of underlying variables (e.g., interest rates, commodity prices, and security indexes). They are used largely as hedges against risk; often, they are a form of insurance. Many times derivatives are negotiated privately between companies.

For example, company X agrees to make a number of payments to company Y, which, in turn, will pay up if a bond issuer Z defaults. When the terms of derivatives are negotiated privately, they are more difficult for regulators to track and assess for risk. They represent a market of approximately $600 trillion worldwide. Derivatives played a huge role in the fall of AIG and the large government bailout that ensued.

The Dodd-Frank statute sought to standardize derivatives traded on exchanges to increase transparency. Derivatives must now be routed through a clearinghouse to ensure that companies using them post collateral (margins). Banks will have to spin off their riskier derivatives and trade them through a subsidiary. Those derivatives will include any that deal in energy, mortgages, credit-default swaps, commodities, and agriculture. Banks can continue to trade derivatives in-house based on interest rates and foreign exchanges and for purposes of hedging risk. The Commodity Futures Trading Commission (CFTC) and the SEC will be the chief regulators of the derivatives market, both drafting and interpreting the regulations. There continues to be some debate over rules set forth to regulate derivatives and swaps.

CONSUMER PROTECTION

When Dodd-Frank was written, no regulatory authority had the sole responsibility for protecting consumers from predatory lenders. None of the regulatory agencies considered consumer protection their number-one priority. Mortgage brokers steered huge numbers of home buyers into subprime mortgages, often without much attention to the buyers' ability to pay based on their income. When the credit markets froze up and the 2008 recession came on, these consumers were the first to suffer.

The Dodd-Frank Act developed a new independent regulatory agency, called the Consumer Financial Protection Bureau (Bureau), which was originally located within the Federal Reserve Board (later it was moved to the Treasury Department). The head of the Bureau is appointed by the president for a five-year term. The Bureau is guaranteed a percentage of the annual Fed's operating expenses. Initially, the formula agreed upon will bring in $500 million annually, although the Bureau may request another $200 million yearly. Staff for this independent agency was drawn initially from several federal agencies, including the Federal Reserve, Federal Trade Commission, Federal Deposit Insurance Corporation, Department of Housing and Urban Development, and National Credit Union. The purpose of the Bureau is to police the financial markets on behalf of savers and borrowers. It is charged with regulating such firms as:

- Banks that issue consumer loans, checking accounts, and/or credit cards
- Mortgage lenders, services, brokers, appraisers, and settlement firms
- Credit counseling firms
- Debt collectors and consumer reporting agencies
- Private-sector student loan companies

EXEMPTIONS

Under Dodd-Frank, Congress has exempted auto dealers from the new agency's jurisdiction, even though they originate nearly 80 percent of all auto loans. Also, 99 percent of the nation's 7,939 banks (as of this writing) and thrifts (those with less than $10 billion in assets) will not fall under the Bureau's rules. These banks will instead be examined by traditional regulators, although the Bureau's rules will be enforced by such examiners. These exemptions at the federal level are a result of strong lobbying at the national and local levels.

The Dodd-Frank statute exempts payday lenders and check-cashing firms as well as auto dealers, leaving these entities to local and state regulation. It also failed to deal with Fannie Mae and Freddie Mac, the mortgage bodies that were responsible for approximately 90 percent of the subprime mortgages that gave rise to the need for this statute. The Dodd-Frank Act also provided for limited regulation of asset management and mutual fund companies.

REGULATION OF THE REGULATORS BY A COURT OF LAW

As with all statutes passed by Congress, the regulating agencies charged with carrying out the Dodd-Frank law are important to its actual enforcement. (See the discussion on rulemaking in Chapter 18 of this text.) With this particular statute, some 15 separate agencies have been involved in rulemaking and enforcement. Some of these agencies include the Federal Reserve Board, the SEC, the Treasury Department, the Financial Stability Oversight Council, the FDIC, the Commodities Future Trading Commission, the FTC, the OCC, and the Office of Financial Research.

Following the completion of administrative agency rulemaking, there will ordinarily be appeals by those affected. The federal courts of appeals normally hear these cases (see Chapter 18 on judicial review of rulemaking).

The Sarbanes-Oxley Act of 2002

Following financial and accounting scandals involving Martha Stewart Living, Inc., Tyco International, Inc., Enron Corporation, and others, Congress passed a bipartisan measure in 2002 sponsored by Senator Paul Sarbanes (D-Md.) and Representative Michael Oxley (R-Ohio) and signed into law by President Bush.[3] The act requires a new approach to corporate governance. Chief executive officers (CEOs) and chief financial officers (CFOs) must now certify that statements and reports are accurate, under pain of imprisonment if intent to mislead can be shown (Exhibit 23-2). The Public Company Accounting Oversight Board (PCAOB) was established to regulate accounting firms.[4] The SEC was given

EXHIBIT 23-2

STATEMENT UNDER OATH OF PRINCIPAL EXECUTIVE OFFICER AND PRINCIPAL FINANCIAL OFFICER REGARDING FACTS AND CIRCUMSTANCES RELATING TO EXCHANGE ACT FILINGS

[Name of principal executive officer or principal financial officer], states and attests that:
- to the best of my knowledge, based upon a review of the covered reports of [company name], and, except as corrected or supplemented in a subsequent covered report;
- no covered report contained an untrue statement of a material fact as of the end of the period covered by such report (or, in the case of a report on Form 8-K or definitive proxy materials, as of the date on which it was filed); and
- no covered report omitted to state a material fact necessary to make the statements in the covered report, in light of the circumstances under which they were made, not misleading as of the end of the period covered by such report (or, in the case of a report on Form 8-K or definitive proxy materials, as of the date on which it was filed).

---

[3]H.R. 3762. The act became effective on August 29, 2002; Pub. L. No. 107–204 (codified as Exchange Act § 4), 15 U.S.C. § 78(d)–3. See Greg Ip, "Maybe U.S. Markets Are Still Supreme: Study Finds No Proof that Sarbanes-Oxley Tarnishes the Allure," *Wall Street Journal,* C-1 (Apr. 27, 2007).

[4]In *Free Enterprise Fund v. Public Company Accounting Oversight Board,* 129 S. Ct. 2378 (2009), the board's membership rules were found to be constitutionally wanting in that members could be removed only for good cause. The Supreme Court said that this arrangement violated the separation of powers doctrine and the need of the president to manage the executive branch. The Court ruled 5–4 that the SEC will be able to remove members of the PCAOB at will. However, the Court unanimously held that the Sarbanes-Oxley Act remained fully operational as law.

new, expansive powers regarding private civil actions, as well as administrative actions. Some of the provisions of Sarbanes-Oxley are outlined in the following subsections. The SEC, using its rulemaking power, is responsible for implementing these provisions.

CORPORATE ACCOUNTABILITY

Sarbanes-Oxley requires CEOs and CFOs to certify financial reports. Officers must forfeit profits and bonuses if earnings are restated by a company due to securities fraud. Companies are required to disclose material changes in their financial condition immediately. Section 404 of this Act has been the source of much criticism from the business community, especially with regard to its impact on smaller companies. The paperwork is extensive. Management must assure the SEC of the effectiveness of internal controls for financial reporting purposes. Disclosure is required by Sections 404, 406, and 407 of this Act.

NEW ACCOUNTING REGULATIONS

A five-member board with legislative and disciplinary power was established: the Public Company Accounting Oversight Board. A majority of the board is independent from publicly held accounting companies. The board is funded by publicly held companies overseen by the SEC.

The act prohibits auditors (accounting firms) from offering nine specific types of consulting services to their corporate clients.

CRIMINAL PENALTIES

- The maximum penalty for securities fraud was raised to 25 years.
- A new crime was created under this act for destruction, alteration, or fabrication of records; the maximum penalty permitted under the act is 20 years imprisonment.
- Penalties are increased for CEOs or CFOs who knowingly certify a report that does not meet the requirements of this act; they are now subject to $1 million in fines and up to 5 years in prison. If officers "willfully" certify a noncomplying report, the penalty may be up to $5 million in fines or 20 years in prison or both.
- Under this act, penalties for mail and wire fraud are raised to 20 years; for defrauding pension funds, up to 10 years.

Other Sarbanes-Oxley provisions include:

- Lengthening of the statute of limitations for securities fraud to five years or two years from discovery.
- Protection for whistleblowers who report wrongdoing to employers or participate in a government investigation involving a potential securities violation.
- Preventing officials who are facing court judgments based on fraud charges from using bankruptcy laws to escape liability.
- Prohibiting certain loans to directors and officers if the loans come from public and private companies that are filing initial public offerings (IPOs). Arranging, receiving, or maintaining personal loans, except consumer or housing loans, is forbidden under this act.

The Securities Act of 1933

In the depths of the Great Depression, Congress enacted this first piece of federal legislation regulating securities. Its major purpose, as we have said, was to ensure full disclosure on new issues of securities.

DEFINITION OF A SECURITY

When most people use the word *securities*, they mean stocks or bonds that are held personally or as part of a group in a pension fund or a mutual fund. Congress, the SEC, and the courts, however, have gone far beyond this simple meaning in defining securities. Section 2(1) of the 1933 Act defines the term **security** as:

> any note, stock, treasury stock, bond, debenture, evidence of indebtedness, certificate of interest or participation in any profit sharing agreement, collateral trust certificate, reorganization certificate or subscription, transferable share, investment contract, voting trust certificate, certificate of deposit for a security, fractional undivided interest in oil, gas or other mineral rights, or, in general, any interest or instrument commonly known as a security.*

security A stock, a bond, or any other instrument of interest that represents an investment in a common enterprise with reasonable expectations of profits that are derived solely from the efforts of those other than the investor.

The words "or, in general, any interest or instrument commonly known as a security" have led to various interpretations by the SEC and the courts of what constitutes a security. In the landmark case of *SEC v. Howey*,[5] the Supreme Court sought to discover the economic realities behind the façade or form of a transaction and set out specific criteria for the courts to use in defining a security. In *Howey*, the Court held that the sale to the public of rows of orange trees, with a service contract under which the Howey Company cultivated, harvested, and marketed the oranges, constituted a security within the meaning of Section 2(1) of the 1933 Act. Its decision was based on three elements or characteristics: (1) There existed a contract or scheme whereby an individual invested money in a common enterprise, (2) the investors had reasonable expectations of profits, and (3) the profits were derived solely from the efforts of persons other than the investors. These criteria are examined in detail here because they have been the basis of considerable litigation.

Common Enterprise. The first element of the *Howey* test has been interpreted by most courts as requiring investors to share in a single pool of assets so that the fortunes of a single investor are dependent on those of the other investors. For example, commodities accounts involving commodities brokers' discretion have been held to be "securities" on the grounds that "the fortunes of all investors are inextricably tied" to the success of the trading enterprise.

Reasonable Expectations of Profit. The second element of the *Howey* test requires that the investor enter the transaction with a clear expectation of making a profit on the money invested. The U.S. Supreme Court has held that neither an interest in a noncontributory, compulsory pension plan nor stock purchases by residents in a low-rent cooperative constitute securities within the definition of *Howey*. In the case involving the pension plan,[6] the Court stated that the employee expected funds for his pension to come primarily from contributions made by the employer rather than from returns on the assets of the pension plan fund. Similarly, in the low-rent housing case,[7] the Court decided that shares purchased solely to acquire a low-cost place to live were not bought with a reasonable expectation of profit.

Profits Derived Solely from the Efforts of Others. The third element of the *Howey* test requires that profits come "solely" from the efforts of people other than the investors. The word *solely* was interpreted to mean that the investors can exert "some efforts" in bringing other investors into a pyramid sales scheme, but that the "undeniably significant ones" must be the efforts of management, not the investors.

The following case sets out a summary of a U.S. Supreme Court decision on what constitutes a security.

---

[5] 328 U.S. 293 (1946).

[6] *International Brotherhood of Teamsters, Chauffeurs, Warehousers, & Helpers of America v. Daniel*, 439 U.S. 551 (1979).

[7] *United Housing Foundation, Inc. v. SEC*, 423 U.S. 884 (1975).

*Section 2(1) of the 1933 Securities Act, U.S. Securities and Exchange Commission.

CASE 23-1

Securities and Exchange Commission v. Edwards
United States Supreme Court
540 U.S. 389 (2004)

Charles Edwards, the CEO and sole shareholder of ETS Payphones, Inc., offered the public investment opportunities in pay phones. The arrangement involved an investor paying $7,000 to own a pay phone. Each investor was offered $82 per month under a leaseback and management arrangement with ETS. The investors also were to recoup their $7,000 investment at the end of five years. ETS did not generate enough revenue to pay its investors, so it filed for bankruptcy. The SEC sued ETS for civil damages arising from alleged violations of federal securities laws. The SEC won at the trial level. The district judge ruled that pay phone leaseback and management agreements were investment contracts covered by federal securities laws. The 11th Circuit Court of Appeals reversed this judgment and ruled in favor of ETS. The SEC was granted certiorari to have the Supreme Court review the definition and application of the term *security*.

Justice O'Connor

Congress's purpose in enacting the securities laws was to regulate investments, in whatever form they are made and by whatever name they are called. To that end, it enacted a broad definition of *security*, sufficient to encompass virtually any instrument that might be sold as an investment, *investment contract* is not itself defined.

The test for whether a particular scheme is an investment contract was established in our decision in *SEC v. W. J. Howey Co.*, 66 S. Ct. 1100 (1946). We look to whether the scheme involves an investment of money in a common enterprise with profits to come solely from the efforts of others. This definition embodies a flexible rather than a static principle, one that is capable of adaptation to meet the countless and variable schemes devised by those who seek the use of the money of others on the promise of profits. . . .

There is no reason to distinguish between promises of fixed returns and promises of variable returns for purposes of the test. . . . In both cases, the investing public is attracted by representations of investment income, as purchasers were in this case by ETS's invitation to watch the profits add up. Moreover, investments pitched as low-risk (such as those offering a "guaranteed" fixed return) are particularly attractive to individuals more vulnerable to investment fraud, including older and less sophisticated investors. Under the reading respondent advances, unscrupulous marketers of investments could evade the securities laws by picking a rate of return to promise. We will not read into the securities laws a limitation not compelled by the language that would so undermine the laws' purposes.

Respondent protests that including investment schemes promising a fixed return among investment contracts conflicts with our precedent. We disagree.

Given that respondent's position is supported neither by the purposes of the securities laws nor by our precedents, it is no surprise that the SEC has consistently taken the opposite position, and maintained that a promise of a fixed return does not preclude a scheme from being an investment contract. It has done so in formal adjudications and in enforcement actions.

The Eleventh Circuit's perfunctory alternative holding, that respondent's scheme falls outside the definition because purchasers had a contractual entitlement to a return, is incorrect and inconsistent with our precedent. We are considering investment contracts. The fact that investors have bargained for a return on their investment does not mean that the return is not also expected to come solely from the efforts of others. Any other conclusion would conflict with our holding that an investment contract was offered in *Howey* itself.

We hold that an investment scheme promising a fixed rate of return can be an *investment contract* and thus a *security* subject to the federal securities laws.*

Reversed and *remanded*, for the SEC.

REGISTRATION OF SECURITIES UNDER THE 1933 ACT

Purpose and Goals. The 1933 Act requires the registration of nonexempt securities, as defined by Section 2(1), for the purpose of full disclosure so that potential investors can make informed decisions on whether to buy a proposed public offering of stock. As we noted earlier in this chapter, the 1933 Act does not authorize the SEC or any other agency to decide whether the offering is meritorious and should be sold to the public.

Registration Statement and Process. Section 5 of the 1933 Act requires that to serve the goals of disclosure, a registration statement consist of two parts—the

---

*Securities and Exchange Commission v. Edwards, United States Supreme Court, 540 U.S. 389 (2004).

prospectus and a "Part II" information statement—to be filed with the SEC before any security can be sold to the public. The registration statement provides (1) material information about the business and property of the issuer; (2) description of the significant provisions of the offering; (3) the use to be made of the funds garnered by the offering and the risks involved for investors; (4) the managerial experience, history, and remuneration of the principals, including pensions or stock options; (5) financial statements certified by public accountants attesting to the firm's financial health; and (6) pending lawsuits. The prospectus must be given to every prospective buyer of the securities. Part II is a longer, more detailed statement than the prospectus. It is not given to prospective buyers, but is open for public inspection at the SEC.

prospectus The first part of the registration statement the SEC requires from issuers of new securities. It contains material information about the business and its management, the offering itself, the use to be made of the funds obtained, and certain financial statements.

Disclosure. Issuers may use the detailed form (Form S-1). Effective December 4, 2005, the SEC amended the disclosure requirement noted here to recognize four categories of issuers:

1. A *nonreporting* issuer that is not required to file reports under the 1934 Act. It must use Form S-1, which it did not have to do previously.
2. An *unseasoned issuer* is an issuer that has reported continuously under the 1934 Act for at least three years. Such an issuer must use Form S-1, but is permitted to disclose less detailed information and to incorporate some information by reference to reports filed under the 1934 Act.
3. A *seasoned issuer* is an issuer that has filed continuously under the 1934 Act for at least one year and has a minimum market value of publicly held voting and nonvoting stock of $75 million. Such an issuer is permitted to use Form S-03, thus disclosing even less detail in the 1933 Act registration and incorporating even more information by reference to 1934 Act reports.
4. A *well-known seasoned issuer* is an issuer that has filed continuously for at least one year under the 1934 Act.

During the registration process, the SEC generally bans public statements by some issuers, other than those contained in the registration statement, until the effective date of registration. There are three important stages in this process: prefiling, waiting, and posteffective periods. They are summarized in Table 23-2.

Prefiling Period. Section 5(c) of the 1933 Act prohibits any offer to sell or buy securities before a registration statement is filed. The key question here is what constitutes an "offer." Section 2(3) of the act exempts from the definition any preliminary agreements or negotiations between the issuer and the underwriters or among the underwriters themselves. **Underwriters** are investment banking firms that purchase a securities issue from the issuing corporation with a view to eventually selling the securities to brokerage houses, which, in turn, sell them to the public. These underwriters—such as Goldman Sachs, Kidder Peabody, and First Boston—may arrange for distribution of the public offering of securities, but they cannot make offerings or sales to dealers or the public at this time. During the

underwriter Investment banking firm that agrees to purchase a securities issue from the issuer, usually on a fixed date at a fixed price, with a view to eventually selling the securities to brokers, who, in turn, sell them to the public.

TABLE 23-2 STAGES IN THE SECURITIES REGISTRATION PROCESS

Stage	Prohibitions
1. Prefiling period	No offer to sell or buy securities may be made before a registration statement is filed.
2. Waiting period	SEC rules allow oral offers during this period but no sales. A "red herring" prospectus that disavows any attempt to offer or sell securities may be published.
3. Posteffective period	Registration generally becomes effective 20 days after the registration statement is filed, although effective registration may be accelerated or postponed by the SEC. Offer and sale of securities are permitted thereafter.

prefiling period, the SEC regulations also forbid sales efforts in the form of speeches or advertising by the issuer that seeks to "hype" the offering or the issuer's business. However, a press release setting forth the details of the proposed offering and the issuer's name, without mentioning the underwriters, is generally permitted.

Waiting Period. In the interim between the filing and the time when registration becomes effective, SEC rules allow oral offers but not sales. The SEC examines the prospectus for completeness during this period. SEC rules permit the publication of a written preliminary, or **red herring**, prospectus that summarizes the registration but disavows in red print (hence, its name) any attempt to offer or sell securities. Notices of underwriters containing certain information about the proposed issue are also allowed to appear in newspapers during this period, but such notices must be bordered in black and specify that they are not offers to sell or solicitations to buy securities.

red herring A preliminary prospectus that contains most of the information that will appear in the final prospectus, except for the price of the securities. The "red herring" prospectus may be distributed to potential buyers during the waiting period, but no sales may be finalized during this period.

Posteffective Period. The third stage in the process is called the *posteffective period* because the registration statement usually becomes effective 20 days after it is filed, although sometimes the SEC accelerates or postpones registration for some reason. Underwriters and lenders can begin to offer and sell securities after the 20 days or upon commission approval, whichever comes first.

Under Section 8 of the 1933 Act, the SEC may issue a "refusal order" or "stop order," which prevents a registration statement from becoming effective or suspends its effectiveness, if the staff discovers a misstatement or an omission of a material fact in the statement. Stop orders are reserved for the most serious cases. In general, the issuers are forewarned by the SEC in informal **letters of comment or deficiency** before a stop order is put out, so they have the opportunity to make the necessary revisions. The commission may shorten the usual 20-day period between registration and effectiveness if the issuer is willing to make the modifications requested by the SEC staff. This procedure, in fact, is the present trend.

letter of comment or deficiency Informal letter issued by the SEC indicating what corrections must be made in a registration statement for it to become effective.

The 1933 Act requires that a prospectus be issued upon every sale of a security in interstate commerce except sales by anyone who is not an "issuer, underwriter or dealer." If a prospectus is delivered more than 9 months after the effective date of registration, it must be updated so that the information is not more than 16 months old. The burden is on the dealer to update all material information about the issuer that is not in the prospectus. Dealers who fail to do so risk civil liability under Sections 12(1) and 12(2) of the 1933 Act.

Communications. The December 2005 revisions brought flexibility to rules regarding written communication by issuers before and during registration of securities. This flexibility depended on certain characteristics of the issuer, including (1) the type of issuer, (2) the issuer's history of reporting, and (3) the issuer's market capitalization. These new rules created a type of written communication called a "free-writing prospectus," which is any written offer, including electronic communication (as defined previously) other than a prospectus required by statute. The new rules provide that:

- Well-known seasoned issuers may engage at any time in oral and written communications, including a free-writing prospectus, subject to certain conditions.
- All reporting issuers (unseasoned issuers, seasoned issuers, and well-known seasoned issuers) may at any time continue to publish regularly released factual business information and forward-looking information (predictions).
- Nonreporting issuers may at any time continue to publish factual business information that is regularly released and intended for use by persons other than in their capacity as investors or potential investors.
- Communications by issuers more than 30 days before filing a registration statement are permitted so long as they do not refer to a securities offering that is the subject of a registration statement.*

---

*Securities Act of 1933, U.S. Securities and Exchange Commission.

All issuers may use a free-writing prospectus after the filing of the registration statement, subject to certain conditions.

Shelf Registration. Traditionally, the marketing of securities has taken place through underwriters who buy or offer to buy securities and then employ dealers across the United States to sell them to the general public. With Rule 415, the SEC has established a procedure, called **shelf registration**, that allows a large corporation to file a registration statement for securities it may wish to sell over a period of time rather than immediately. Once the securities are registered, the corporation can place them on the "shelf" for future sale and need not register them again. It can then sell these securities when it needs capital and when the marketplace indicators are favorable. A company that files a shelf-registration statement must file periodic amendments with the SEC if any fundamental changes occur in its activities that would be material to the average prudent investor's decision to invest in its stock.

shelf registration Procedure whereby a large corporation can file a registration statement for securities it wishes to sell over a period of time rather than immediately.

"Fictional Filings" with the SEC

Like an extraordinary whimsical tale, *Universal Express* (not American Express), a small company of alleged postal stores, was able to lose money faster than it issued news releases. The SEC filed suit for fraud against Universal (Company) in 2007, because the company continued to issue billions of unregistered shares following the issuance of news releases. The unregistered shares were used to finance the company and its officers. In 2004, a federal district court in New York ruled that the company and its officers had violated the securities laws and ordered them to pay $21.9 million. The CEO, Richard Altomare, was barred from being an officer or a director of any public company.

Despite the court's order, Universal continued to issue news releases forecasting $9 million in annual revenues from 9,000 private postal stores. Judge Lynch of the federal district court ruled that there was no evidence the company had any such network of stores. As of the first quarterly report in 2007, Universal said that the 9,000 stores were "members" of its network regardless of the findings of the federal judge. In its suit, the SEC alleged that the company had issued 500 million unregistered shares over 33 months in violation of the 1933 Securities Act and other federal statutes (the basis for the original SEC suit). The company claimed that its old stock issues (before 2004) were allowed by a bankruptcy court ruling. Judge Lynch found such claims to be baseless and dismissed Universal's justification.

As of 2007, Universal Express continued to trade billions of shares weekly in an over-the-counter penny-stock bulletin board in California. Shares have never sold at more than $0.40 a share. In 2006, Universal lost $18.9 million on revenue of $1.1 million. Altomare acted as the sole member of Universal's board of directors and received a salary in 2006 of $650,000 (paid for with the sale of unregistered stock). As of 2007, news releases continue to be issued, claiming a "network of postal stores" in the United States producing annual revenues of $9 million.[a]

In April of 2008, Altomare was found in contempt of court for failing to make payments on the judgment issued against him, and in May of that year he was sent to federal prison for 80 days. In 2010, he wrote a book proclaiming his innocence and arguing that he was prosecuted for being a whistleblower against the SEC.

Meanwhile, in 2011, a federal judge ruled that two Florida penny-stock traders who issued unregistered shares of the defunct Universal Express must pay fines of $14 million and $5.3 million.[b]

---

[a]*SEC v. Universal Express, Inc.* (SDNY), reported at No. 2267, § 94165 (2007).
[b]"Universal Express Penny Stock Traders Must Pay $14 M," by Brian Bandell. Published by South Florida Business Journal, © 2011. Available at www.bizjournals.com/southflorida/news/2011/09/19/universal-express-penny-stock-traders.html.

SECURITIES AND TRANSACTIONS EXEMPT FROM REGISTRATION UNDER THE 1933 ACT

Section 5 of the 1933 Act requires registrations of any sale by any person of any security unless specifically exempted by the 1933 Act. The cost of the registration process, in terms of hiring lawyers, accountants, underwriters, and other financial experts, makes it appealing for a firm to put a transaction together in such a way as not to fall within the definition of a security. If that is impossible, firms often attempt to meet the requirements of one of the following four classes of exemptions to the registration process (summarized in Table 23-3).

Private Placement Exemptions. Section 4(2) of the 1933 Act exempts from registration transactions by an issuer that do not involve any public offering. Behind this exemption is the theory that institutional investors have the sophisticated knowledge necessary to evaluate the information contained in a private placement and, thus, unlike the average investor, do not need to be protected by the registration process set out in the 1933 Act. The private placement exemption is often used in stock option plans, in which a corporation issues securities to its own employees for the purpose of increasing productivity or retaining top-level managers. Because various courts had different views on what factual situations qualified for the private placement exemption, the SEC published Rule 146, which seeks to clarify the criteria used by the commission in allowing this exemption.

The issuer should follow the statutory guidelines (rules) listed here.

1. The number of purchasers of the company's (issuer's) securities should not exceed 35. If a single purchaser buys more than $150,000, that purchaser will not be counted among the 35.

2. Each purchaser must have access to the same kind of information that would be available if the issuer had registered the securities.

3. The issuer can sell only to purchasers who it has reason to believe are capable of evaluating the risks and benefits of investment and are able to bear those risks or to purchasers who have the services of a representative with the knowledge and experience to evaluate the risks for them.

4. The issuer may not advertise the securities or solicit public customers.

5. The issuer must take precautions to prevent the resale of securities issued under a private placement exemption.

TABLE 23-3 EXEMPTIONS FROM THE REGISTRATION PROCESS UNDER THE 1933 SECURITIES ACT

Exemptions	Definition
Private placement	Transactions by an issuing company not involving any public offering. Usually, the transaction involves sophisticated investors with enough knowledge to evaluate information given them (e.g., stock option plans for top-level management).
Intrastate offering	Any security or part of an offering offered or sold to persons resident within a single state or territory.
Small business	Section 3(b) of the 1933 Act allows the SEC to exempt offerings not exceeding $5 million. Regulations A and D promulgated by the SEC define the type of investors and the amount of securities that are exempt within a certain time period.
Other offering exemptions	By virtue of the 1933 Act, exemptions are allowed for transactions by any person other than an issuer, an underwriter, or a dealer. Also, government securities (federal, state, or municipal bonds) are exempt. Also exempt are securities issued by banks, charitable organizations, and savings and loans institutions.

Intrastate Offering Exemption. Section 3 of the 1933 Act provides an exemption for any "security which is part of an issue offered or sold to persons resident within a single state or territory, where the issuer of such security is a resident and doing business within, or, if a corporation, incorporated by, or doing business within such a state." To qualify for this exemption, an issuer must meet the strictly interpreted doing-business-within-a-state requirement: The issuer must be a resident of the state and "do business" solely with (i.e., offer securities to) people who live within the state.

Courts have interpreted Section 3 very strictly. One federal court ruled that a company incorporated in the state of California and making an offering of common stock solely to residents of California did not qualify for the intrastate exemption because it advertised in the *Los Angeles Times*, a newspaper sold by mail to residents of other states. Another factor in the court's decision in this case was that 20 percent of the proceeds from the securities sale were to be used to refurbish a hotel in Las Vegas, Nevada.[8]

After that decision, the SEC issued Rule 147, which sets standards for the intrastate exemption by defining important terms in Section 3 of the 1933 Act. For example, an issuer is "doing business within" a state if (1) it receives at least 80 percent of its gross revenue from within the state, (2) at least 80 percent of its assets are within the state, (3) it intends to use 80 percent of the net proceeds of the offering within the state, and (4) its principal office is located in the state. Rule 147 is also concerned with whether the offering has "come to rest" within a state or whether it is the beginning of an interstate distribution. An offering is considered intrastate only if no resales are made to nonresidents of the state for at least nine months after the initial distribution of securities is completed.

Small Business Exemptions. Section 3(b) of the 1933 Act authorizes the SEC, by use of its rulemaking power, to exempt offerings not exceeding $5 million when it finds registration unnecessary. Under this authority, the commission has promulgated Regulations A and D.

Regulation A exempts *small public offerings* made by an issuer, defined as offerings not exceeding $5 million over a 12-month period. The issuer must file an "offerings" and a "notification circular" with an SEC regional office 10 days before each proposed offering. The circular contains information similar to that required for a 1933 Act registration prospectus, but in less detail, and the accompanying financial statements may be unaudited. It should be noted that for these small business offerings, the SEC staff follows the same "letter of comment" procedure associated with registration statements; thus, a Regulation A filing may be delayed. Regulation A circulars do not give rise to civil liability under Section 11 of the 1933 Act (discussed later in this chapter), but they do make an issuer liable under Section 12(2) for misstatements or omissions (also discussed later in this chapter). The advantages of this regulation for small businesses are that the preparation of forms is simpler and less costly and the SEC staff can usually act more quickly.

Regulation D, which includes Rules 501–506, attempts to implement Section 3 of the 1933 Act. Rule 501 defines an *accredited investor* as a bank; an insurance or investment company; an employee benefit plan; a business development company; a charitable or educational institution (with assets of $5 million or more); any director, officer, or general partner of an issuer; any person with a net worth of $1 million or more; or any person with an annual income of more than $200,000. This definition is important because an accredited investor, as defined by Rule 501, is not likely to need the protection of the 1933 Act's registration process.

---

[8] *SEC v. Trustee Showboat*, 157 F. Supp. 824 (S.D. Cal. 1957).

noninvestment company
A company whose primary business is not in investing or trading in securities.

Rule 504 allows any **noninvestment company** (one whose primary business is not investing or trading in securities) to sell up to $1 million worth of securities in a 12-month period to any number of purchasers, accredited or nonaccredited, without furnishing any information to the purchaser. This $1 million maximum, however, is reduced by the amount of securities sold under any other exemption.

Rule 505 allows any private noninvestment company to sell up to $5 million of securities in a 12-month period to any number of accredited investors (as previously defined) and to up to 35 nonaccredited purchasers. Sales to nonaccredited purchasers are subject to certain restrictions concerning the manner of offering—for example, no public advertising is allowed—and resale of the securities.

Rule 506 allows an issuer to sell an unlimited number of securities to any number of accredited investors and to up to 35 nonaccredited purchasers. The issuer, however, must have reason to believe that each nonaccredited purchaser or representative has enough knowledge or experience in business to be able to evaluate the merits and risks of the prospective investment. Again, certain resale restrictions are attached to offerings made under this rule, as well as a prohibition against advertising. Rule 506 seeks to clarify Section 4(2) of the 1933 Act, dealing with private placement exemptions, as already discussed.

Other Offering Exemptions. Section 4(2) of the 1933 Act allows exemptions for "transactions by any person other than an issuer, underwriter or dealer." Because Sections 4(3) and 4(4) allow qualified exemptions for dealers and brokers, the issuer and the underwriters become the only ones not exempted. SEC Rules 144 and 144a define the conditions under which a person is not an underwriter and is not involved in selling securities.

Exempt Securities. Government securities issued or regulated by agencies other than the SEC are exempt from the 1933 Act. For example, debt issued by or guaranteed by federal, state, or local governments, as well as securities issued by banks, religious and charitable organizations, savings and loan associations, and common carriers under the Interstate Commerce Commission, are exempt. These securities usually fall under the jurisdiction of other federal agencies, such as the Federal Reserve System or the Federal Home Loan Board, or of state or local agencies.

The collapse of the Penn Central Railroad in 1970 and the default of the cities of Cleveland and New York on municipal bonds led Congress and the SEC to reexamine certain exemptions with a view to eliminating them. In fact, the Railroad Revitalization Act of 1976 eliminated the 1933 Act exemption for securities issued by railroads (other than trust certificates for certain equipment), and 1975 amendments to the securities acts now require firms that deal solely in state and local government securities to register with the commission and to adhere to rules laid down by the Municipal Securities Rulemaking Board.

Other exempt securities are issued in a corporate reorganization or bankruptcy and securities issued in stock dividends or stock splits.

RESALE RESTRICTIONS

Restrictions are placed on the resale of securities issued for investment purposes pursuant to intrastate, private placement, or small business exemptions.

- Rule 147 states that securities sold pursuant to an intrastate offering exemption (mentioned previously) cannot be sold to nonresidents for nine months.

- Rule 144 states that securities sold pursuant to the private placement or small business exemptions must be held one year from the date the securities are sold.
- Rule 144(a) permits "qualified institutional investors" (institutions that own and invest $100 million in securities, such as banks, insurance companies, and investment companies) to buy unregistered securities without being subject to the holding period of Rule 144. This rule seeks to permit foreign issuers to raise capital in this country from sophisticated investors without registration process disclosures. This also seeks to create a domestic market for unregistered securities.
- Regulation S and Rule 144(a) have attempted to expand the private placement market. See the section in this chapter on the "Global Dimensions of Rules Governing the Issuance and Trading of Securities."

Exempt Transactions for Issuers under the 1933 Securities Act

Exemption	Price Limitation	Limitations on Purchasers	Resales
Regulation A	$5 million	None	Unrestricted
Intrastate Rule 147	None	Intrastate only	Only to residents before nine months
Rule 506	None	Unlimited accredited; 35 unaccredited	Restricted
Rule 505	$5 million	Unlimited accredited; 35 unaccredited	Restricted

LIABILITY, REMEDIES, AND DEFENSES UNDER THE 1933 SECURITIES ACT

Private Remedies. The 1933 Act provides remedies for individuals who have been victims of (1) misrepresentations in a registration statement, (2) an issuer's failure to file a registration statement with the SEC, or (3) misrepresentation or fraud in the sale of securities. Each is examined here, along with some affirmative defenses.

Misrepresentations in a Registration Statement. Section 11 of the 1933 Act imposes liability for certain untruths or omissions in a registration statement. Section 11 allows a right of action to "any person acquiring such a security" who can show (1) a material misstatement or omission in a registration statement and (2) monetary damages. The term *material* is defined by SEC Rule 405 as pertaining to matters "of which an average prudent investor ought reasonably to be informed before purchasing the security registered." In addition, the issuer's omission of such facts as might cause investors to change their minds about investing in a particular security are considered material omissions for the purposes of Section 11. These facts include an impending bankruptcy, new government regulations that may be costly to the company, and the impending conviction and sentencing of the company's top executives for numerous violations of the FCPA of 1977 (discussed later in this chapter).

Three affirmative defenses are available to defendants:

1. The purchaser (plaintiff) knew of the omission or untruth.
2. The decline in value of the security resulted from causes other than the misstatement or omission in the registration statement.
3. The statement was prepared with the due diligence expected of each defendant.

The following case examines alleged omissions of material information. The defendants contended that the omissions were nonmaterial, and due diligence was exercised.

CASE 23-2

Litwin v. Blackstone Group, LP
United States Court of Appeals, Second Circuit
634 F.3d 706 (2011)

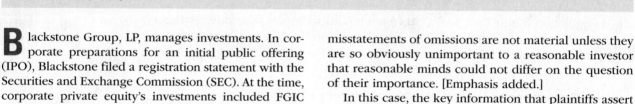

Blackstone Group, LP, manages investments. In corporate preparations for an initial public offering (IPO), Blackstone filed a registration statement with the Securities and Exchange Commission (SEC). At the time, corporate private equity's investments included FGIC Corporation and Freescale Semiconductor, Inc. FGIC insured investments in subprime mortgages. Before the IPO, FGIC's customers began to suffer large losses. By the time of the IPO, this situation was generating substantial losses for FGIC and, in turn, for Blackstone.

Meanwhile, Freescale had recently lost an exclusive contract to make wireless 3G chipsets for Motorola, Inc. (its largest customer). Blackstone's registration statement did not mention the impact on its revenue of the investments in FGIC and Freescale. Martin Litwin and others who invested in the IPO filed a suit in a federal district court against Blackstone and its officers, alleging material omissions from the statement. Blackstone filed a motion of dismiss, which the court granted. The plaintiffs appealed.

Justice Straud

Materiality is inherently fact-specific finding that is satisfied when a plaintiff alleges a statement or omission that a reasonable investor would have considered significant in making investment decisions.

However, it is not necessary to assert that the investor would have acted differently if an accurate disclosure was made. Rather, when a district court is presented with a motion [to dismiss,] a complaint may not properly be dismissed on the ground that the alleged misstatements of omissions are not material unless they are so obviously unimportant to a reasonable investor that reasonable minds could not differ on the question of their importance. [Emphasis added.]

In this case, the key information that plaintiffs assert should have been disclosed is whether, and to what extent, the particular known trend, event, or uncertainty might have been reasonably expected to materially affect Blackstone's investments. Plaintiffs are not seeking the disclosure of the mere fact of Blackstone's investment in FGIC, of the downward trend in the real estate market, or of Freescale's loss of its exclusive contract with Motorola. Rather, plaintiff claims that Blackstone was required to disclose the manner in which those then-known trends, events, or uncertainties might reasonably be expected to materially impact Blackstone's future revenues.

The question, of course, is whether a loss in a particular investment's values will merely affect revenues, because it will almost certainly have some effect. *The relevant question is whether Blackstone reasonably expects the impact to be material.* [Emphasis added.]

Because Blackstone's Corporate Private Equity segment plays such an important role in Blackstone's business and provides value to all of its other asset management and financials related to that segment that Blackstone reasonably expects will have a material adverse effect on its future revenues. Therefore, the alleged omissions related to FGIC and Freescale were plausibly material.*

The U.S. Court of Appeals for the 2nd Circuit vacated the lower court's dismissal and *remanded* the case.

due diligence defense
An affirmative defense raised in lawsuits charging misrepresentation in a registration statement. It is based on the defendant's claim to have had reasonable grounds to believe that all statements in the registration statement were true and no omission of material fact had been made. This defense is not available to the issuer of the security.

Whereas others associated with the company can raise the **due diligence defense**, the issuing company cannot. Section 11(a) is very specific about what other individuals may be held jointly or severally liable in addition to the issuing company:

1. Every person who signed the registration statement (Section 16 of the 1933 Act requires signing by the issuer, the issuing company's CEO, the company's financial and accounting officers, and a majority of the company's board of directors)
2. All directors
3. Accountants, appraisers, engineers, and other experts who consented to being named as having prepared all or part of the registration statement
4. Every underwriter of the securities

---

*Litwin v. Blackstone Groups, LP United States Court of Appeals, Second Circuit, 634 F.3d 706 (2011).

It should be noted that there are two exceptions to Section 11 liability:

1. An expert is liable only for the misstatements or omissions in the portion of the registration statement that the expert prepared or certified.
2. An underwriter is liable only for the aggregate public offering portion of the securities it underwrote.

Section 11 liability has made such a strong impact that today virtually all professionals and experts involved in the preparation of a registration statement make precise agreements concerning the assignment of responsibility for that statement. Failure to grasp the import of Section 11 and related sections of the 1933 and 1934 Acts can lead to loss of reputation and employment by businesspersons and professionals. The first case brought under Section 11[9] sent tremors through Wall Street, the accounting profession, and outside directors. In that case, the court evaluated each defendant's plea of due diligence on the basis of each individual's relationship to the corporation and expected knowledge of registration requirements.

Failure to File a Registration Statement. Failure to file a registration statement with the SEC when selling a nonexempt security is the second basis for a private action by the purchaser for rescission (cancellation of the sale). Section 12(1) of the 1933 Act provides that any person who sells a security in violation of Section 5 (which you recall from the discussion of the registration statement) is liable to the purchaser to refund the full purchase price. A purchaser whose investment has decreased in value may recover the full purchase price without showing a misstatement or fraud if the seller is unable to meet the conditions of one of the exemptions discussed earlier. In short, a business that fails to file a registration statement because of a mistaken assumption that it has qualified for one of the exemptions could be making a very expensive mistake.

Misrepresentation or Fraud in the Sale of a Security. A third basis for a private action is misrepresentation in the sale of a security, as defined by Section 12(2) of the 1933 Act, which holds liable any person who offers or sells securities by means of any written or oral statement that misstates a material fact or omits a material fact that is necessary to make the statement truthful. Unlike Section 11, Section 12(2) is applicable whether or not the security is subject to the registration provisions of the 1933 Act, provided there is use of the mails or other facilities in interstate commerce. The persons liable are only those from whom the purchaser bought the security. For example, under Section 12(2), a purchaser who bought the security from an underwriter, a dealer, or a broker cannot sue the issuer unless able to show that the issuer was "a substantial factor in causing the transaction to take place." A further requirement is that the purchaser must prove that the sale was made "by means of" the misleading communications. The defense usually raised by sellers in such suits is that they did not know, and using reasonable care could not have known, of the untruth or omission at the time the statement was made.

Fraud in the sale of a security is covered by Section 17(a) of the 1933 Act, which imposes criminal, and possibly civil, liability on anyone who aids and abets any fraud in connection with the offer or sale of a security. Violators may be penalized by fines of up to $10,000, imprisonment up to 5 years, or both. Section 12(a) (2) and Section 17(a) of the 1933 Act set out antifraud enforcement mechanisms. Rule 10(b)-5 applies to the issuance or sales of securities under the 1934 Act and other securities acts, even those exempted by the 1933 Act.

---

[9]*Escott v. Barchris Construction Corp.*, 283 F. Supp. 643 (S.D.N.Y. 1968).

Governmental Remedies. When a staff investigation uncovers evidence of a violation of the securities laws, the SEC can (1) take administrative action, (2) take injunctive action, or (3) recommend a criminal prosecution to the Justice Department.

Administrative Action. Upon receiving information of a possible violation of the 1933 Act, the SEC staff undertakes an informal inquiry. This involves interviewing witnesses but generally does not involve issuing subpoenas. If the staff uncovers evidence of a possible violation of a securities act, it may order an administrative hearing before an administrative law judge (ALJ) or ask the full commission for a formal order of investigation. A formal investigation is usually conducted in private under SEC rules. A witness compelled to testify or to produce evidence may be represented by counsel, but no other witness or counsel may be present during the testimony. A witness may be denied a copy of the transcript of his or her testimony for good cause, although the witness is allowed to inspect the transcript.

Witnesses at a private SEC investigation do not enjoy the ordinary exercise of Fourth, Fifth, and Sixth Amendment rights. For example, Fourth Amendment rights are limited because the securities industry is subject to pervasive government regulation, and those going into it know this in advance. Fifth Amendment rights are limited because the production of records related to a business may be compelled despite a claim of self-incrimination. (See the sections on the Fourth and Fifth Amendments in Chapter 5.) As for the Sixth Amendment, in a private investigation, the SEC is not required to notify the targets of the investigation, nor do such targets have a right to appear before the staff or the full commission to defend themselves against charges. The wide scope of SEC powers in these nonpublic investigations was reinforced when the U.S. Supreme Court upheld a lower court's decision to deny injunctive relief with regard to subpoenas directed at plaintiffs in an SEC private investigation.[10] For some of the constitutional reasons noted here, the SEC is being challenged in a 2015 Georgia case.[11]

An administrative proceeding may be ordered by the full commission if the SEC staff uncovers evidence of a violation of the securities laws. This proceeding before an ALJ can be brought only against a person or firm that is registered with the commission (an investment company, a dealer, or a broker). The ALJ has the power to impose sanctions, including censure, revocation of registration, and limitations on the person's or the firm's activities or practice.

In addition, after a hearing, the full commission may issue a stop order to suspend a registration statement found to contain a material misstatement or omission. If the statement is later amended, the stop order will be lifted. As mentioned earlier in the chapter, stop orders are usually reserved for the most serious cases. The SEC more frequently uses letters of deficiency to obtain corrections to registration statements. The remedies available under the 1991 Remedies Act, discussed earlier in this chapter under "Summary of Federal Securities Legislation," apply here as well.

Injunctive Action. The SEC may commence an injunctive action when there is a "reasonable likelihood of further violation in the future" or when a defendant is considered a "continuing menace" to the public. For example, under the 1933 Act, the SEC may go to court to seek an injunction to prevent a party from using the interstate mails to sell a nonexempt security. Violation of an injunctive order may give rise to a contempt citation. Also, parties under such an order are disqualified from receiving an exemption under Regulation A (the small business exemption). Again, the remedies available under the 1991 Remedies Act apply here.

---

[10] *SEC v. Jerry T. O'Brien, Inc. et al.*, 467 U.S. 735 (1984).

[11] *Hill v. EC*, N. District (Georgia), Case No. 1.2015cvo1802, May 19, 2015.

Criminal Penalties. Willful violations of the securities acts and the rules and regulations promulgated pursuant to those acts are subject to criminal penalties. Anyone convicted of willfully omitting a material fact or making an untrue statement in connection with the offering or sale of a security can be fined up to $1 million for each offense or imprisoned for up to 5 years or both. The SEC does not prosecute criminal cases itself, but instead refers them to the Justice Department.

The Securities Exchange Act of 1934

One year after passing the Securities Act of 1933, Congress crafted this second extremely important piece of securities legislation to come out of the Great Depression. More comprehensive than the 1933 Act, it had two major purposes: to regulate trading in securities and to establish the SEC to oversee all securities regulations and bar the kind of large market manipulations that had characterized the 1920s and previous boom periods.

REGISTRATION OF SECURITIES ISSUERS, BROKERS, AND DEALERS

Registration of Securities Issuers. Section 12 of the 1934 Securities Exchange Act requires every issuer of debt and equity securities to register with both the SEC and the national exchange on which its securities are to be traded. Congress extended this requirement to all corporations that (1) have assets of more than $10 million, (2) have a class of equity securities with more than 500 shareholders, and (3) are involved in interstate commerce. Registration becomes effective within 60 days after filing unless the SEC accelerates the process. Such companies are referred to as "Section 12" companies.

The commission has devised forms to ensure that potential investors will have updated information on all registrants whose securities are being traded on the national exchanges. Thus, registrants are required to file annual reports (Form 10-K) and quarterly reports (Form 10-Q) as well as SEC-requested current reports (Form 8-K). This last form must be filed within 15 days of the request, which is usually made in response to a perceived material change in the corporation's position (e.g., a potential merger or bankruptcy) that the commission's staff believes a prudent investor should know about. It should be noted that the Sarbanes-Oxley Act, discussed earlier in this chapter, should be reviewed for all requirements regarding CEOs and CFOs of issuing companies. Further, the accounting requirements of the FCPA of 1977 set out at the end of this chapter should be reviewed as to company officers' duties.

In a proposed Codification of the Federal Securities Law (CFSL), the American Law Institute has sought to streamline the registration process under the 1933 and 1934 Acts by requiring single-issuance registration under the 1933 Act and an annual company "offering statement" for securities traded on a national exchange under the 1934 Act. At present, Section 22 of the Exchange Act makes a registering company liable for civil damages to securities purchasers who can show that they relied on a misleading statement contained in any of the SEC-required reports.

Registration of Brokers and Dealers. Brokers and dealers are required to register with the SEC under the Exchange Act unless exempted. A **dealer**, as defined by the 1934 Act, is a "person engaged in the business of buying and selling securities for his own account," whereas a **broker** is a person engaged in the business of "effectuating transactions in securities for the account of others." The convenient term *broker-dealer* is used throughout to refer to all those who trade in securities; the specific term *broker* or *dealer* is used when only one type of trader is meant.

dealer A person engaged in the business of buying and selling securities for his or her own account.

broker A person engaged in the business of buying and selling securities for others' accounts.

Broker-dealers must meet a financial responsibility standard that is based on a net capital formula; a minimum capital of $25,000 is required in most cases. Brokers are obliged to segregate customer funds and securities.

Analysts or "Cheerleaders"? Conflicts of Interest

In July 2001, to prevent a potential conflict of interest, Merrill Lynch barred its stock analysts from investing in stock they had researched. This was a reaction to several events:

- Individuals and members of Congress had lost confidence in analysts, particularly in an economy and market that had been in a turndown for 18 months. Never before had so many individual investors actually invested in stocks and bonds and seen their paper wealth grow and then fall. The "party" appeared to be over for a time.

- Federal investigators were alleging the manipulation of Internet stock IPOs and the taking of kickbacks by investment bankers in July 2001.

- The same trading companies that had investment banking divisions floating new issuances of securities hired securities analysts to rank the securities of companies for which they were raising. The individual investor had begun to realize that there existed no "Chinese wall" between stock analysts and investment bankers in the same brokerage firm. In fact, some companies gave stock analysts bonuses when they helped encourage investment banking business by giving stocks high ratings. For example, one study showed that bullish ratings were so meaningless that "sell" ratings were less than 2 percent of all ratings shown. In many cases, stocks fell as much as 90 percent from their high before analysts removed their "buy" ratings.[a] This could be called "cheerleading."

- Money managers who ran large mutual funds with money from IRAs, Keoghs, 401(k)s, and 403(b)s became skeptical of analysts as investors lost confidence in the mutual funds and their managers. These investors, who tended to be passive in nature and dependent on the fund managers and the analysts, saw their retirement funds dwindling rapidly (and in some instances, disappearing almost entirely).

Although this assessment appeared gloomy, many investors argued that analysts should be encouraged to own stocks in the companies on which they do research, believing that they should "put their money where their mouth is." Some argued that the very purpose of the securities laws is full disclosure and that all analysts should be forced to disclose what holdings they have and to advise investors when they have a potential conflict of interest.

---

[a]"Stock Analysts Get Overall Rap for Deceiving Investors," *USA Today*, July 5, 2001, 10A.

The *Securities Investment Protection Act* (SIPA) provides a basis for indemnifying the customers of a brokerage firm that becomes insolvent: All registered brokers must contribute to a SIPA fund managed by the SIPC, a nonprofit corporation whose functions are to liquidate an insolvent brokerage firm and to protect customer investments up to a maximum of $500,000. Upon application to the SEC, the SIPC can borrow up to $1 billion from the U.S. Treasury to supplement the fund when necessary.

Under Section 15(b) of the Exchange Act, which contains the antifraud provisions, the SEC may revoke or suspend a broker-dealer's registration or may censure a broker-dealer. (Municipal securities dealers and investment advisers are subject to similar penalties.) In general, the commission takes such actions against broker-dealers either for putting enhancement of their personal worth ahead of their professional obligation to their customers—conflict of interest—or for trading in or recommending certain securities without having reliable information about the company. Broker-dealers are liable to both government and private action for failing to disclose conflicts of interest. When even the potential for such a conflict exists, a broker must supply a customer with written confirmation of each transaction, including full disclosure of whom the broker is representing in the transaction.

APPLYING THE LAW TO THE FACTS...

Assume that Rogers owned 2.1 million shares in WorldCom, a firm that collapsed after the revelation of accounting fraud. He claimed that he ordered his broker-dealer to sell his stock when it was worth $92 a share, but his broker told him not to sell because research reports showed that the company would continue to do well. In fact, the broker knew that WorldCom was grossly overvalued, but he promoted the stock anyway to retain WorldCom's investment banking business. What is problematic about the broker's behavior? What liability is he potentially subject to?

Monday, October 23, 2000, was an important day for securities analysts: That was when Regulation Fair Disclosure (FD) became effective. This rule required companies to publicize all potentially market-moving data at the time the data become available. No longer could such data be made available only to certain analysts in a securities firm before being given to the public at large. Analysts traditionally followed one industry and were in frequent contact by phone or email with its CFOs, investor relations officials, and often CEOs. By gaining bits of information from several companies in an industry, they were able to provide earning forecasts and then determine buy, sell, and hold ratings for the trading company they were employed by.

For example, Hallie Frobose, an analyst for 17 years for Brennan and Kubasek Company, may have concentrated on companies that were involved in lumber and forest products. By making telephone calls in the pre-Regulation FD days, she could obtain financial factors off the record for hundreds of variables that might affect earnings expectations for a major company (e.g., Lumber Pacific). By constructing a model with all the important variables and checking them with officers of the lumber company, recommendations could be made to Brennan and Kubasek's large investors; this was no longer the case for Frobose or her company after October 2000.

Those opposed to Regulation FD argue that it has a "chilling effect" on analysts and contacts they once had with corporate managers and has led to a decrease in the predictability of earnings reports such as Frobose's. Market inefficiency is a result in the eyes of many.

Those favoring Regulation FD argue that the major purpose of our securities laws is full disclosure and that both large and small investors should have access to the same data, resulting in a "level playing field."

DISCLOSURE: COMPENSATION

In discussing registration of securities and securities issuers, it is important to note that in 2006, pursuant to its mandate under the Exchange Act of 1934, the SEC set out rules requiring clearer and more complete disclosure of compensation paid to directors, the principal executives, and the three most highly paid executive officers. The issuer (usually a company) must disclose executives' compensation over the last three years, including salary, bonuses, a dollar value of stock and option awards, amount of compensation over nonequity plans, annual changes in present value of accumulated pension benefits, and all other compensation including perquisites. This type of disclosure is a start toward meeting some of the political arguments made by unions and other groups that compensation has not been fully disclosed to shareholders.

SECURITIES MARKETS

Earlier in this chapter, we defined a security as a stock or bond or any other instrument or interest that represents an investment in a common enterprise with reasonable expectations of profits derived solely from the efforts of people other than the investors. A security can also be considered as a form of currency that, once issued, can be traded for other securities on what is called a *securities market*. The concern here is with the markets for stocks and how they are regulated under the Exchange Act.

There are generally two types of markets in stocks: exchange markets and over-the-counter (OTC) markets. The **exchange market** provides for the buying and selling of securities within a physical facility such as the NYSE or regional exchanges such as the Boston, Detroit, Midwest (Chicago), Pacific Coast (Los Angeles and San Francisco), and Philadelphia exchanges. (This, however, is changing as computer and Internet systems are increasingly the medium through which stock sales or trades are made.) These exchanges traditionally prescribed not only the number and the qualifications of their broker-members but also the commissions they could charge. In 1975, commissions were deregulated by the SEC; since then, brokers have been free to set the commissions they charge their customers. Brokers do not trade directly on an exchange market, but rather transmit a customer's order to a registered specialist in a stock, who buys and sells that security for his or her own account on the floor of the exchange. The NYSE has now become a publicly traded company. In 2007, the SEC approved rule changes by the NYSE relating to the combination of Euronext and NYSE Group. Euronext owns five European exchanges. The combined company now competes with other national and international exchanges for securities business. The SEC will continue to regulate the NYSE Group in the same manner as noted here.

The **over-the-counter (OTC) market** has no physical facility—computers, telephones, and other forms of communication link OTC members—and no qualifications for membership. Its commissions have always been determined by the law of supply and demand. OTC firms serve as dealers or market makers in stocks and deal directly with the public.

BATS (Better Alternative Trading System) has been open since 2006. It operates the third-largest electronic exchange in the United States. It has been "stealing" business from its older rivals. Its founder is Steve Ratterman. It handles about 12 percent of the stock traded on public markets.

Today, the National Association of Securities Dealers (NASD) and the exchanges help the SEC to regulate the securities market. In enacting the Securities Exchange Act in 1934, Congress recognized that the stock exchanges had been regulating their members for 140 years and did not seek to dismantle their self-regulatory mechanisms. Rather, it superimposed the SEC on already existing self-regulatory bodies by requiring every "national securities exchange" to register with the SEC. Under Section 6(b) of the Exchange Act, an exchange cannot be registered unless the SEC determines that its rules are designed "to prevent fraudulent and manipulative acts and practices" and to discipline its members for any violations of its rules or the securities laws. Both the NYSE and the NASD have promulgated rules relating to stock transactions and qualifications for those participating in such transactions. In general, these rules are enforced by the self-regulating bodies.

To clarify the SEC's role, Congress amended the Exchange Act in 1975 to give the SEC explicit authority over all self-regulating organizations (SROs). Any exchange or OTC rule change now requires advance approval from the SEC.

---

exchange market A securities market that provides a physical facility for the buying and selling of stocks and prescribes the number and qualifications of its broker-members. These brokers buy and sell stocks through the exchange's registered specialists, who are dealers on the floor of the exchange.

over-the-counter (OTC) market A securities market that has no physical facility and no membership qualifications and whose broker-dealers are market makers who buy and sell stocks directly from the public.

The commission also has reviewing power over all disciplinary actions taken by the SROs. Moreover, as mentioned earlier, the 1975 amendments eliminated the power of exchanges to fix minimum commission rates.

The movement by exchanges to "go public" (sell shares and, thus, ownership rights in the exchanges) has changed their nonregulatory aspects as traded companies. The regulatory function continues through SEC oversight. The Financial Industry Regulatory Authority (FINRA) was created as an SRO in 2007 when the NASD merged with the New York Stock Exchange's regulatory arm. FINRA may soon be overseeing both brokers and financial advisers.

PROXY SOLICITATIONS

Procedural and Substantive Rules. Section 14 of the Exchange Act and the accompanying SEC regulations set forth the ground rules governing proxy solicitations by inside management, dissident shareholders, and potential acquirers of a company. You will remember from the discussion in Chapter 17 that *proxies* are documents by which the shareholders of a publicly registered company designate another individual or institution to vote their shares at a shareholders' meeting. They are often used by inside management to defeat proposals by dissident shareholders or to prevent a takeover by a "hostile" company. The real significance of the proxy solicitation process, however, is that it may result in materially changing the direction of the corporation without its owners' (the shareholders') awareness. Because very few individual shareholders (under 1 percent) attend annual shareholders' meetings, proxy voting is management's major instrument for electing the directors and setting the policy it wants.

Against this background, Congress enacted Section 14 of the Exchange Act—the section known as the Williams Act—making it unlawful for a company to solicit proxies in "contravention of such rules and regulations as the Commission [SEC] may prescribe as necessary or appropriate in the public interest or for the protection of investors." With this broad statutory authority, the SEC has promulgated rules and regulations that require all companies registered under the securities acts to file proxy statements with the commission 10 days before mailing them to shareholders. During this 10-day period, the SEC staff comments on the statements and sometimes asks for changes, usually because it believes that not all material information has been included. Under its Rule 22, the commission requires proxy statements to carry several items of information, ranging from a notice on the revocability of proxies to a notification of the interest that the soliciting individuals or institutions have in the subject matter to be voted on. The purpose of this procedure is to make sure that shareholders have full disclosure on a matter before they agree to any grant of their proxy. The SEC also requires companies to send shareholders a form on which they can mark their approval or disapproval of the subject matter to be voted on. If a proxy is solicited for electing new directors, the shareholders must receive an annual report of the corporation as well.

Shareholder Proposals. If a shareholder of a registered issuing company wishes to place an item on the agenda, Rule 14(a)(8) requires that management be notified in a timely way before a regular shareholder meeting or a special meeting. Once notified, management must include the proposal (200 words or fewer) in the proxy statement it sends to all the shareholders. Management may also include its own view on the proposal. Shareholder proposals in recent years have included prohibitions against discrimination, pollution, dumping of wastes, "golden parachutes," "poison pills," and "greenmail" (the last three topics are discussed later in this section under the heading "Remedies and Defensive Strategies"). In a sense, proxy solicitation became a form of shareholder democracy—one that corporate management believed was getting out of hand. After vigorous debate by all interested groups, the SEC amended Rule 14(a) in 1983 to allow management to exclude a shareholder proposal if:

1. under the particular state law governing the corporation, the proposal would be unlawful if agreed to by the directors;
2. it involves a personal grievance;
3. it is related to ordinary operational business functions;
4. it is a matter not significantly related to the company's business (the commission has defined this criterion as matters accounting for less than 5 percent of the assets, earnings, and sales of a company);
5. the stockholder making the proposal has not owned more than $1,000 worth of stock or 1 percent of the shares outstanding for a period of 1 year or more (although several shareholders may accumulate shares to meet this criterion); and
6. the shareholder proposal received less than 5 percent of the votes when submitted in a previous year.

Furthermore, shareholders are limited to one proposal per annual company meeting.

If management excludes a proposal, it must explain why, and the shareholder may then appeal to the SEC. The SEC staff decides whether the proposal should be placed on the agenda for the next annual meeting. The Dodd-Frank Act (examined earlier in this chapter) gives shareholders a nonbinding vote on executive compensation as directed by the SEC (see Chapter 18).

Proxy Contests. Proxy contests normally come about when an insurgent group of shareholders seeks to elect its own slate of candidates to the board of directors to replace management's slate. Both insurgent shareholders and management may seek shareholder proxies in this contest.

The SEC has set out specific rules governing disclosure by insurgents and management and the rights of each. An information statement must be filed by the insurgents, disclosing all participants in their group and the background of each, including past employment and any criminal history. In 2010 the SEC set out the "proxy access" role, which requires companies to include the names of all board nominees (even those not backed by the company), directly on the standard ballots distributed before shareholder annual meetings. To win the right to nominate, the investor or group of investors must own at least 3 percent of the company's stock and have held shares for a minimum of three years. If dissident shareholders wish to oust board members, current shareholders must foot the bill for preparation and mailing of the official proxy for themselves and the dissidents.

Both criminal and civil liability attach to a company that sends a misleading proxy statement to its shareholders. The civil liability is based on a negligence standard (preponderance of the evidence). The SEC may, through injunctive relief, prevent the solicitation of proxies or may declare an election of directors, based on misleading proxies, to be invalid. Under the Insider Trader Sanctions Act of 1984 (discussed later in this chapter), the commission may also institute criminal and administrative proceedings against a company. Furthermore, persons who rely on a misleading proxy statement to buy or sell securities have the right to institute a private action, as do insurgent shareholders in a proxy fight.

TENDER OFFERS AND TAKEOVER BIDS

A series of hostile takeovers in the 1990s and creative defensive strategies used by management of some targeted companies, renewed concerned parties' interest in the regulation of tender offers and takeover bids.

In a takeover bid, the acquiring company or individual, using a public **tender offer**, seeks to purchase a controlling interest (51 percent) in another company—the target company—which would lead to a takeover of that company's board of directors and management. The acquiring company makes this public offer in such national newspapers as the *Wall Street Journal* or the *New York Times* to company shareholders, requesting that they tender their shares for cash or for the acquiring company's securities or for both, usually at a price exceeding that quoted for the shares on a national exchange. Because of abuses in the 1960s, when shareholders frequently were given only a short time to make up their minds and, thus, could not properly evaluate tender offers, Congress enacted legislation to give shareholders more information and a longer period to make a decision. This legislation became Sections 13 and 14 of the Securities Exchange Act.

tender offer A public offer by an individual or a corporation made directly to the shareholders of another corporation in an effort to acquire the targeted corporation at a specific price.

Rules Governing Tender Offers. Sections 13 and 14 together constitute the regulatory framework for tender offers. Section 13 requires any person (or group) that acquires more than 5 percent of any class of registered securities to file within 10 days a statement with both the issuer (the target company) and the SEC. This statement must set forth (1) the background of the acquiring person or group, (2) the source of the funds used to acquire the 5 percent, (3) the purpose(s) of the acquisition of the stock, (4) the number of shares presently owned, (5) any relevant contracts with the target company, and (6) plans of the person or group for the targeted company.

Section 14 provides that no one may make a tender offer that results in ownership of more than 5 percent of a class of registered securities unless that person or group files with the SEC and with each offeree a statement containing information similar to that required by Section 13. It also restricts the terms of the offer, particularly the right of withdrawal by the offerer and extensions or changes in the offer. The "best price" rule applies during tender offers but only to consideration paid for past or future services.

The SEC has issued detailed rules concerning Section 14. For example, even if the offer is a **hostile bid**—meaning that the management of the target company opposes it—the target company must either mail the tender offer to all shareholders or promptly forward a list of the shareholders to the tender offerer. Management must also, within 10 days of receiving a tender offer, state whether it opposes or favors it or lacks enough information to make a judgment. SEC rules also compel management to file a form called Schedule 14-9. Schedule 14-9 requires top managers to (1) disclose whether they intend to hold their shares in the company or tender them to the offerer; (2) describe any agreements they may have made with the tender offerer; and (3) disclose, if the tender offer is hostile, whether they have engaged in any negotiations with a friendly or "white knight" company.

hostile bid A tender offer that is opposed by the management of the target company.

Section 14 of the 1934 Act and SEC rules require that a tender offer be open for at least 20 days so that shareholders will have a reasonable amount of time to consider it. The SEC has also set out certain withdrawal rights for shareholders who have tendered their shares.

REMEDIES AND DEFENSIVE STRATEGIES

Remedies. Section 14(e) (known as the Williams Act) makes it a criminal offense to make an untrue or misleading statement or to engage in fraudulent acts or deceptive practices in connection with a tender offer. The emphasis here is on intent to deceive. Shareholders of a targeted company can bring civil actions under Section 14(e) for violations of Sections 13(d) and 14(b) if they can show that they have been injured because they relied on fraudulent statements in the tender offer. In addition, under the Insider Trader Sanctions Act (discussed

CASE 23-3

Barbara Schreiber v. Burlington Northern, Inc.
United States Supreme Court
472 U.S. 1 (1985)

Petitioner Schreiber, on behalf of herself and other shareholders of El Paso Gas Company, sued Respondent Burlington Northern, claiming that the company had violated Section 14(e) of the Securities Exchange Act of 1934. In December 1982, Burlington issued a hostile tender offer for El Paso Gas Company. Burlington did not accept the shares tendered by a majority of shareholders of El Paso but instead rescinded the December offer and substituted another offer for El Paso in January. The rescission of the first tender offer resulted in a smaller payment per share to El Paso shareholders who retendered after the January offer. The petitioners claimed that Burlington's withdrawal of the December tender offer and the substitution of the January offer were a "manipulative" distortion of the market for El Paso stock and a violation of Section 14(e). The respondent argued that "manipulative" acts under 14(e) require misrepresentation or nondisclosure and that no such acts had taken place in this case. Therefore, the respondent moved for dismissal of the case based on failure to state a cause of action. The federal district court granted the motion for dismissal. The court of appeals affirmed. Schreiber appealed to the U.S. Supreme Court.

Chief Justice Burger

We are asked in this case to interpret Section 14(e) of the Securities Exchange Act. The starting point is the language of the statute. Section 14(e) provides:

> It shall be unlawful for any person to make any untrue statement of a material fact or omit to state any material fact necessary in order to make the statements made, in the light of the circumstances under which they are made, not misleading, or to engage in any fraudulent, deceptive or manipulative acts or practices, in connection with any tender offer or request or invitation for tenders, or any solicitation of security holders in opposition to or in favor of any such offer, request, or invitation. The Commission shall, for the purposes of this subsection, by rules and regulations define, and prescribe means reasonably designed to prevent, such acts and practices as are fraudulent, deceptive, or manipulative.

Our conclusion that "manipulative" acts under Section 14(e) require misrepresentation or nondisclosure is buttressed by the purpose and legislative history of the provision. Section 14(e) was originally added to the Securities Exchange Act as part of the Williams Act.

It is clear that Congress relied primarily on disclosure to implement the purpose of the Williams Act. Senator Williams, the bill's Senate sponsor, stated in the debate:

> Today, the public shareholder in deciding whether to accept or reject a tender offer possesses limited information. No matter what he does, he acts without adequate knowledge to enable him to decide rationally what is the best course of action. This is precisely the dilemma which our securities laws are designed to prevent.

The expressed legislative intent was to preserve a neutral setting in which the contenders could fully present their arguments. To implement this objective, the Williams Act added Sections 13(d), 13(e), 14(e), and 14(f) to the Securities Exchange Act. Some relate to disclosure; Sections 13(d), 14(d), and 14(f) all add specific registration and disclosure provisions. Others—Sections 13(e) and 14(d)—require or prohibit certain acts so that investors will possess additional time within which to take advantage of the disclosed information.

To adopt the reading of the term "manipulative" urged by petitioner would not only be unwarranted in light of the legislative purpose but would be at odds with it. Inviting judges to read the term "manipulative" with their own sense of what constitutes "unfair" or "artificial" conduct would inject uncertainty into the tender offer process. An essential piece of information—whether the court would deem the fully disclosed actions of one side or the other to be "manipulative"—would not be available until after the tender offer had closed.

This uncertainty would directly contradict the expressed Congressional desire to give investors full information.

Congress's consistent emphasis on disclosure persuades us that it intended takeover contests to be addressed to shareholders. In pursuit of this goal, Congress, consistent with the core mechanism of the Securities Exchange Act, created sweeping disclosure requirements and narrow substantive safeguards. The same Congress that placed such emphasis on shareholder choice would not at the same time have required judges to oversee tender offers for substantive fairness.

We hold that the term "manipulative" as used in Section 14(e) requires misrepresentation or nondisclosure. It connotes "conduct designed to deceive or defraud investors by controlling or artificially affecting the price of securities." *Ernst & Ernst v. Hochfelder*, 425 U.S., at 199. Without misrepresentation or nondisclosure, Section 14(e) has not been violated.

Applying that definition to this case, we hold that the actions of respondents were not manipulative. The amended complaint fails to allege that the cancellation of the first tender offer was accompanied by any misrepresentation, nondisclosure, or deception.*

Affirmed in favor of Defendant, Burlington Northern.

CRITICAL THINKING ABOUT THE LAW

Sometimes ambiguity is present in the court's own reasoning; for example, a judge might argue that a "reasonable" person would not be offended by sexual advances made by a fellow employee. At other times, the court must interpret ambiguity in congressional legislation to make a legal judgment.

Case 23-3 deals with the second of those judicial confrontations with ambiguity. It is important to be aware not only of the Court's interpretation of an ambiguity but also of the evidence it selects to support that interpretation. The very fact that an important term is ambiguous means that there might be other legitimate interpretations; thus, in judging whether you agree with the particular interpretation at hand, you must evaluate the evidence presented for it. It is also important to recognize the primary ethical norm that informed the Court's interpretation. The following questions address those considerations.

1. What legislative ambiguity was the Court dealing with in Case 23-3?

   Clue: The meaning of this term is the central issue of the case.

2. Specifically, to what evidence did the Court refer to support its own interpretation of the ambiguity?

   Clue: Reread the paragraph immediately following the quotation from Section 14(e).

3. In supporting its strict interpretation of legislative ambiguity, the Court stated that Congress intended to leave issues of fairness up to shareholders and not judges, making full disclosure the most important consideration. In this prioritization of the liberty (of shareholders) over potentially more just outcomes (allowing judges to decide fairness), one might argue that the primary ethical norm of liberty drove the Court's reasoning. What other primary ethical norm is implicit in this prioritization?

   Clue: Consider the primary ethical norm that would be damaged if judges decided fairness (especially with the inevitable increase in court cases).

Defensive Strategies. The business judgment rule, which was discussed in Chapter 17, is based primarily on the 50 states' case law and the Revised Model Business Corporations Act. It has traditionally allowed wide latitude to the managers of targeted companies, as long as they act in good faith in the best interests of shareholders, do not waste the corporate assets, and do not enter into conflict-of-interest situations.

Here is a list of defensive strategies that managements of targeted companies have used to repel hostile takeovers in recent years.

- Awarding large compensation packages (golden parachutes) to target-company management when a takeover is rumored.

- Issuing new classes of securities before or during a takeover battle that require a tender offerer to pay much more than the market price for the stock (poison pill).

---

*Barbara Schreiber v. Burlington Northern, Inc., United States Supreme Court 472 U.S. 1 (1985).

- Buying out a "hostile" shareholder at a price far above the current market price of the target company's stock in exchange for the hostile shareholder's agreement not to buy more shares for a period of time (greenmail). Congress has now eliminated this defense by legislation.
- Writing supermajority requirements for merger approval into the bylaws and articles of incorporation (porcupine provisions).
- Issuing Treasury shares (stock that was repurchased by the issuing corporation) to friendly parties.
- Moving to states with strong antitakeover (shark repellent) laws.
- Bankrupting the company (scorched-earth policy).
- Prevailing upon another company or individual (a white knight) to buy out the hostile bidder to prevent the undesirable takeover.

Competition between State Legislatures

Antitakeover Regulations

With more than 40 states having statutes that seek to regulate takeovers, it is wise to note that target companies often have sought protection from takeovers by lobbying with state legislatures for detailed regulatory or antitakeover statutes. There are several types of state statutes:

- Statutes similar to the Williams Act (see Section 14 of the 1934 Act).
- Statutes that allow the state legislature to review the merits of a tender offer and/or the adequacy of disclosure. Many of these exempt tender offers that are supported by the target company's management. Strong lobbying is often involved here.
- Statutes that require "fair prices." Acquirers must pay all shareholders the highest price paid to any shareholders.
- Statutes that prohibit transactions with an acquirer for a specified period of time after a change in control unless disinterested shareholders approve.

This has led to some debate as to whether such statutes are in the best interest of shareholders (e.g., pension funds, insurance companies, and large institutional shareholders) or voters. Such statutes may be in the short-term interest of a company located in state X, but may be contrary to the best interests of the population of the state when company Y is looking for a place to locate. There are arguments from both sides as to whether such regulatory statutes are helpful when states are competing to obtain corporate opportunities and the jobs that go with them.

SECURITIES FRAUD

The courts have had a difficult time defining *securities fraud*. An appellate court once stated:

> *Fraud is infinite, and were a Court of Equity once to lay down rules, how far they would go, and no further, in extending their relief against it, or to define strictly the species or evidence of it, the jurisdiction would be cramped, and perpetually eluded by new schemes which the fertility of man's invention would contrive.**

This is the philosophical position that has been adopted by the SEC: There cannot be a law against every type of fraud imaginable. Instead, the SEC staff has sought to use Section 10(b) of the Securities Act broadly, going beyond its exact language to develop a "fraud-on-the-market" theory that does not require the investor-plaintiff ever to have relied on false documents or specific acts, but only on the integrity of the market and a fair stock price.

*From *A History of the Court of Chancery* by Joseph Parkes. Published by Longman, Rees, Orme, Brown, and Green, 1828.

Section 10(b) of the Securities Exchange Act. One of the purposes of the Securities Exchange Act of 1934 was to ensure the full disclosure of all material information to potential investors. Full disclosure enables the market mechanism to operate efficiently and ensures that consumers are provided with a fair price for securities. Section 10(b) prohibits the use of the mails or other facilities (e.g., truck or car and satellite or data transmission) in interstate commerce

> *in connection with the purchase or sale of any security, any manipulative or deceptive device or contrivance in contravention of such rules and regulations as the Commission may prescribe as necessary or appropriate in the public interest or for the protection of investors.**

This broad statutory language signals a congressional intent to cover all possible forms of fraud. To that end,

> *it shall be unlawful for any person, directly or indirectly, by the use of any means or instrumentality of interstate commerce, or of the mails, or of any facility of any national securities exchange, (1) to employ any device, scheme, or artifice to defraud, (2) to make any untrue statement of a material fact necessary in order to make the statements made, in the light of circumstances under which they were made, not misleading or (3) to engage in any act, practice, or course of business which operates or would operate as a fraud or deceit upon any person, in connection with the purchase or sale of any security.*

A private party's standing to sue under Section 10(b) and associated SEC Rule 10(b)-5 has been upheld in cases in which manipulative or deceptive acts were committed in connection with the purchase or sale of securities. The question of standing that the Supreme Court answers again in the following case is whether the corporation (Matrixx) has made a disclosure of material fact in a timely manner as required by Rule 10(b)-5.

CASE 23-4

Siracusano v. Matrixx Initiatives, Inc.
131 S. Ct. 1309 (2011)

Matrixx Initiatives, Inc., is pharmaceutical company that sells Zicam through its wholly owned subsidiary Zicam, LLC. One of its products, responsible for 70 percent of its sales, is Zicam Cold Remedy, a homeopathic product marketed as stopping or minimizing cold symptoms. In December 1999, Matrixx began to receive questions from physicians whose patients were developing anosmia (loss of the sense of smell). Researchers at medical facilities contacted Matrixx in 2002 to offer access to studies showing that zinc sulfate (present in Zicam) was linked to anosmia.

During this time, Matrixx's public disclosures did not discuss these inquiries and studies. In fact, on October 22, 2003, Matrixx issued an optimistic press release announcing that its net sales for the third quarter of 2003 had increased by 163 percent over the third quarter of 2002. On an October 23, 2003, earnings conference call, executives for Matrixx expressed their "enthusiasm for the most recently completed quarter" and "optimis[m] about the future." At one point during the call, Zicam executives were asked to "make any comment on the litigation MTXX or officers are involved in, or whether or not there is any SEC [Securities and Exchange Commission] investigation." They replied that "[t]he officers of this company are not involved in any litigation" and that they were not aware of any SEC investigation. In fact, a lawsuit alleging that Zicam caused anosmia had already been filed at this time.

By January 30, 2004, the FDA was "looking into complaints that an over-the-counter common-cold medicine manufactured by a unit of Matrixx Initiatives, Inc. may be causing some users to lose their sense of smell." Matrixx's stock declined after this report, "falling from $13.55 per share on January 30, 2004, to $11.97 per share on February 2, 2004."

NECA-IBEW Pension Fund and James Siracusano (plaintiffs/appellants) brought a class action suit against Matrixx and three Matrixx executives (Appellees), alleging a violation of the Securities Exchange Act of 1934 by their failure to disclose material

---

*Securities Exchange Act of 1934. Section 10(b) of the Securities Exchange Act.

information regarding problems with Zicam. The district court granted Matrixx's motion to dismiss the complaint. The court of appeals reversed. The shareholders appealed.

Justice Sotomayor

To prevail on a § 10(b) claim, a plaintiff must show that the defendant made a statement that was "misleading as to a material fact."

Matrixx urges us to adopt a bright-line rule that reports of adverse events associated with a pharmaceutical company's products cannot be material absent a sufficient number of such reports to establish a statistically significant risk that the product is in fact causing the events. Absent statistical significance, Matrixx argues, adverse event reports provide only "anecdotal" evidence that "the user of a drug experienced an adverse event at some point during or following the use of that drug." Accordingly, it contends, reasonable investors would not consider significant because only then do they "reflect a scientifically reliable basis for inferring a potential causal link between product use and the adverse event."

A lack of statistically significant data does not mean that medical experts have no reliable basis for inferring a causal link between a drug and adverse events. As Matrixx itself concedes, medical experts rely on other evidence to establish an inference of causation.

The FDA similarly does not limit the evidence it considers for purposes of assessing causation and taking regulatory action to statistically significant data. For example, the FDA requires manufacturers of over-the-counter drugs to revise their labeling "to include a warning as soon as there is reasonable evidence of an association of serious hazard with a drug; a causal relationship need not have been proved."

This case proves the point. In 2009, the FDA issued a warning letter to Matrixx stating that "[a] significant and growing body of evidence substantiates that the Zicam Cold Remedy intranasal products may pose a serious risk to consumers who use them." The letter cited as evidence 130 reports of anosmia the FDA had received few reports of anosmia associated with other intranasal cold remedies, and "evidence in the published scientific literature that various salts of zinc can damage olfactory function in animals and humans." It did not cite statistically significant data. Given that medical professionals and regulators act on the basis of evidence of causation that is not statistically significant, it stands to reason that in certain cases reasonable investors would as well.

The information provided to Matrixx by medical experts revealed a plausible causal relationship between Zicam Cold Remedy and anosmia. Consumers likely would have viewed the risk associated with Zicam (possible loss of smell) as substantially out-weighing the benefit of using the product (alleviating cold symptoms), particularly in light of the existence of many alternative products on the market. Viewing the allegations of the complaint as a whole, the complaint alleges facts suggesting a significant risk to the commercial viability of Matrixx's leading product. It is substantially likely that a reasonable investor would have viewed this information "'as having significantly altered the "total mix" of information made available'"

Matrixx elected not to disclose the reports of adverse events not because it believed they were meaningless but because it understood their likely effect on the market. "[A] reasonable person" would deem the inference that Matrixx acted with deliberate recklessness (or even intent). We conclude, in agreement with the Court of Appeals, that respondents have adequately pleased [materially and] scienter.*

Affirmed.

CRITICAL THINKING ABOUT THE LAW

1. Based on this decision, if you had been the executives at Matrixx, what would you have disclosed and when would you have disclosed it?

2. Why is it relevant that consumers would be affected by the studies, whether significant or not?

3. By 2009, the FDA required that Zicam be removed from stores and warned consumers about the risk of losing their sense of smell. What happens to the company as a result? What will be the impact on Matrixx investors? What will their damages be?

The use of Section 10(b) and Rule 10(b)-5 has been controversial in three major types of securities fraud cases: insider trading, misstatements by corporate management, and mismanagement of a corporation (Table 23-4). After each of these areas is explored in turn, a new concept will be mentioned that shareholder suits based on fraud have been invoking: fraud-on-the-market theory.

---

*Siracusano v. Matrixx Initiatives, Inc. 131 S. Ct. 1309 (2011).

TABLE 23-4

SECURITIES FRAUD UNDER SECTION 10(b) OF THE SECURITIES EXCHANGE ACT OF 1934

Activity	Definition
Insider trading	The use of nonpublic information received from a corporate source by an individual(s) who has a fiduciary obligation to shareholders and potential investors and who benefits from trading on such information.
Misstatement of corporation	Any report, release, or financial statement or any other statement that is released by an officer, a director, or an employee of a corporation in connection with the purchase or sale of a security that shows an intent to mislead shareholders or potential investors.
Corporate mismanagement	Any transaction involving the purchase or sale of a security in which there is fraud based on an action of management. The plaintiff must be either a purchaser or a seller of securities in such a transaction.

Insider Trading and Section 10(b)5-1 and 10(b)5-2 of the Securities Act. **Insider trading** is the use of material, nonpublic information received from a corporate source by an individual who has a fiduciary obligation to shareholders and potential investors and who benefits from trading on such information. The SEC adopted new guidelines as provided here because federal appellate courts have disagreed on the definition. For example, a trader must now be "aware" of nonpublic information when making the purchase or sale of a security. In 2015, the 2nd Circuit Court of Appeals refused a rehearing of its December 2014 decision. In the 2014 decision, the court set out its original standard for judging insider trading cases but added to it that prosecutors must prove that traders knew the person who would provide the tip and the latter had gained some tangible reward for doing so. Friendship of career advice alone did not count as a benefit or reward. This case has broad implications for those previously convicted of insider trading who have argued that the addition to the standard should be considered in their cases. This case, if adopted by other circuit courts of appeal, could stimulate Congress to act on legislation that would more clearly define insider trading, so that we would no longer need to rely on the case by case approach presently used by courts.

Insiders have been found by the courts to be (1) officers and directors of a corporation; (2) partners in investment banking and brokerage firms; (3) attorneys in a retained law firm; (4) underwriters and broker-dealers; (5) financial reporters; and (6) in a unique case, an employee of a financial printing firm that printed documents for a tender offer. (See Exhibit 23-3 for a look at Wall Street's army of insiders, from the general to the grunts.)

There is no statutory definition of insider trading—only case law.

insider trading The use of material, nonpublic information received from a corporate source by someone who has a fiduciary obligation to shareholders and potential investors and who benefits from trading on such information.

"Holy Toledo": Securities Fraud May Be Easy

On April 29, 1999, Martin Frankel, a high school graduate and a native of Toledo, Ohio, absconded with a reputed $335 million. In Mr. Frankel's capacity as founder of "Thunor Trust," a string of eight insurance companies in several southern states had provided him with large sums of money to invest. The con was uncovered when the Franklin American Corporation, owner of these insurance companies, introduced Frankel as a prominent bond trader on April 28, 1999. After failing to attend a meeting with regulators, Frankel wired $334.6 million of Franklin American's money to a foreign bank account. In

EXHIBIT 23-3 INSIDER TRADERS

Insiders: Who Are They?
The Key Inside Players

These are the ones directly responsible for setting up the mergers and acquisitions that will have a significant effect on stock.

The CEO   Vice Chairman   Board of Directors   General Counsel

A Host of Other Related Insiders

A host of people outside the firm have information about key transactions long before the public knows.

Research Analysts from Investment Banks

Merger and Acquisition Teams

Law Firms

Public Relations

Proxy Solicitors

Secretaries

Friends and Relatives of Key Players who get information from the players.

---

addition, he burned all the documents at his Greenwich, Connecticut, mansion, creating a suspicious fire that led to a police search. He then fled to Europe, ending up in jail in Germany. After a period of six months, he was extradited to the United States and was convicted of fraud charges after pleading guilty. Five states sought roughly $215 million, which Frankel was charged with stealing. Approximately $9 million was received at an auction of 822 diamonds seized from Frankel. On December 10, 2004, he was sentenced to a federal prison for 16 years, 6 months.

Unlike convicted felons Ivan Boesky and Michael Milken, Frankel was not a high-profile individual. The use of aliases and the pretension of being a multimillionaire were all that was required. The fact that he used state-regulated insurance companies in this case may have helped the scheme. The SEC and the Justice Department did not become involved until he fled the country.

Sources: L. Mergener, "Frankel Gets 16 Years for Fraud," *The Blade*, December 11, 2004, 1; A. Cowan, "Onetime Fugitive Gets 17 Years for Looting Insurers," *New York Times*, December 11, 2004, PB-3; L. Vellequette, "Frankel to Be Subject of CNBC Show," *The Blade*, March 18, 2008, 27.

The expansion of targets in insider-trading cases, from management and corporate directors to a *Wall Street Journal* reporter and a printer employee, has resulted from the SEC enforcement staff's determination that to provide full disclosure in the marketplace for potential investors, it had to extend its jurisdiction over tippers (insiders) and tippees (those who receive tips from insiders).

APPLYING THE LAW TO THE FACTS...

Dallas was a CEO of a pharmaceuticals company, and he purchased significant amounts of stock in the company for his children over the years. When news that a new drug manufactured by the company was likely to get FDA approval was leaked by an unknown source, the value of the stock skyrocketed. Unfortunately, the FDA ultimately rejected the drug's application. Dallas got early notice of the FDA's planned rejection and immediately called all his children and told them to sell their stock, without telling them why. "Trust me," he said. They all sold immediately, and one daughter also told her boyfriend to sell his stock on her dad's advice, which he did. Is anyone in this scenario in trouble? If so, why? What potential penalties might he or she face?

Misstatements of Corporations and Section 10(b). The second area of controversy regarding Section 10(b) application involves statements by corporate executives. Any report, release, or financial statement or any other statement that sets forth material information (information that would affect the judgment of the average prudent investor) falls within Section 10(b).

Whereas Sections 13 and 14 of the Exchange Act (discussed earlier) apply only to reports, proxy statements, and other documents filed by a company registered with the SEC and a national exchange, Rule 10(b)-5 applies to any statement made by any issuer, registered or not. To be considered a securities fraud, corporate misstatements must meet two requirements: (1) They must be issued "in connection with the purchase or sale of any security," and (2) there must be a showing of **scienter** (intent). As you read the following case, try to determine how closely the defendants met those requirements.

scienter Knowledge that a representation is false.

CASE 23-5

Securities and Exchange Commission v. Texas Gulf Sulphur Co.
United States Court of Appeals
401 F.2d 833 (2nd Cir. 1968)

The SEC (plaintiff) brought an action against the Texas Gulf Sulphur Company (TGS) and 13 of its directors, officers, and employees (defendants) for violation of Section 10(b) of the Exchange Act and SEC Rule 10(b)-5, seeking an injunction against further misleading press releases and requesting rescission of the defendants' purchases and stock options. On June 6, 1963, TGS had acquired an option to buy 160 acres of land in Timmons, Ontario. On November 11, 1963, preliminary drilling indicated that there would be major copper and zinc finds. TGS acquired the land and resumed drilling on March 31, 1964, and by April 8, it was evident that there were substantial copper and zinc deposits. On April 9, Toronto and New York newspapers reported that TGS had discovered "one of the largest copper deposits in America." On April 12, TGS's management said that the rumors of a major find were without factual basis. At 10 a.m. on April 14, the board of directors authorized the issuance of a statement confirming the copper and zinc finds and announcing the discovery of silver deposits as well. On April 20, the NYSE announced that it "was barring stop orders [orders to brokers to buy a stock if its price rises to a certain level to lock in profits in case of a sharp rally in that stock] in Texas Gulf Sulphur" because of the extreme volatility in the trading of the stock.

Approximately one month later, rumors circulated about insider trading. It was later found that when drilling began on November 12, 1963, TGS's directors, officers, and employees owned only 1,135 shares of stock in the company and had no calls (options to purchase shares at a fixed price). By March 31, 1964, when drilling resumed, insiders (tippers) and their tippees had acquired an additional 7,100 shares and 12,300 calls. On February 20, 1964, TGS had issued stock options to three officers and two other employees as part of a compensation package.

From April 9, 1964 to April 14, 1964, when the confirmatory press release was issued, 10 insiders and their tippees made estimated profits of $273,892 on the purchase of their shares or calls of TGS stock. The federal district court dismissed charges against all but two defendants. Those defendants, Clayton and Crawford, appealed, and the SEC appealed from the part of the district court decision that had dismissed the complaint against TGS and the nine other individual defendants.

Judge Waterman

Rule 10(b)-5 was promulgated pursuant to the grant of authority given the SEC by Congress in Section 10(b) of the Securities Exchange Act of 1934. By that Act Congress proposed to prevent inequitable and unfair practices and to ensure fairness in securities transactions generally, whether conducted face-to-face, over the counter, or on exchanges. The Act and the Rule apply to the transactions here, all of which were consummated on exchanges.

The essence of the Rule is that anyone who, trading for his own account in the securities of a corporation, has "access, directly or indirectly, to information intended to be available only for a corporate purpose and not for the personal benefit of anyone" may not take "advantage of such information knowing it is unavailable to those with whom he is dealing," i.e., the investing public. Insiders, as directors or management officers, are, of course, by this Rule, precluded from so unfairly dealing, but the Rule is also applicable to one possessing the information who may not be strictly termed an "insider" within the means of Sec. 10(b) of the Act. Thus, anyone in possession of material inside information must either disclose it to the investing public, or, if he is disabled from disclosing it in order to protect a corporate confidence, or he chooses not to do so, must abstain from trading in or recommending the securities concerned while such insider information remains undisclosed. So, it is here no justification for insider activity that disclosure was forbidden by the legitimate corporate objective of acquiring options to purchase the land surrounding the exploration site; if the information was, as the SEC contends, material, its possessors should have kept out of the market until disclosure was accomplished.

As we stated in *List v. Fashion Park, Inc.*, "The basic test of materiality is whether a reasonable man would attach importance in determining his choice of action in the transaction in question." This, of course, encompasses any fact "which in reasonable and objective contemplation might affect the value of the corporation's stock or securities." Such a fact is a material fact and must be effectively disclosed to the investing public prior to the commencement of insider trading in the corporation's securities. The speculators and chartists of Wall and Bay Streets are also "reasonable" investors entitled to the same legal protection afforded conservative traders. Thus, material facts include not only information disclosing the earnings and distributions of a company but also those facts which affect the probable future of the company and those which may affect the desire of investors to buy, sell, or hold the company's securities.

The core of Rule 10(b)-5 is the implementation of the Congressional purpose that all investors should have equal access to the rewards of participation in securities transactions. It was the intent of Congress that all members of the investing public should be subject to identical market risks—which market risks include, of course, the risk that one's evaluative capacity or one's capital available to put at risk may exceed another's capacity or capital. The insiders here were not trading on an equal footing with the outside investors. They alone were in a position to evaluate the probability and magnitude of what seemed from the outset to be a major ore strike; they alone could invest safely, secure in the expectation that the price of TGS stock would rise substantially in the event such a major strike should materialize, but would decline little, if at all, in the event of failure, for the public, ignorant at the outset of the favorable probabilities, would likewise be unaware of the unproductive exploration, and the additional exploration costs would not significantly affect TGS market prices. Such inequities based upon unequal access to knowledge should not be shrugged off as inevitable in our way of life, or, in view of the congressional concern in the area, remain uncorrected.

We hold, therefore, that all transactions in TGS stock or calls by individuals apprised of the drilling results of K-55-1 were made in violation of Rule 10(b)-5.*

Reversed and *remanded* in favor of Plaintiff, SEC.

LINKING LAW AND BUSINESS

Economics: Efficient Markets

In microeconomics, you learned that the law of supply and demand brings about equilibrium price levels, assuming a free flow of information and mobility of resources. These factors will create efficient markets.

Microeconomics presents an important link to securities law. When plaintiffs argue a fraud-on-the-market theory, they are arguing that there is a distortion or omission of information and, thus, the purchase or sale of the security is subject to fraud and a violation of securities law.

---

*The Wharf (Holdings) Limited v. United International Holdings, Inc., United States Supreme Court 121 S. Ct. 1776 (2001).

Corporate Mismanagement and Section 10(b). The third controversial area of securities fraud under Section 10(b) is corporate mismanagement. Suits alleging corporate mismanagement and fraud brought by minority shareholders in class action or derivative suits must prove three elements: (1) that the transaction being attacked (e.g., the sale of a controlling stock interest in a corporation at a premium) involves the purchase or sale of securities, (2) that the alleged fraud is in connection with a purchase or sale, and (3) that the plaintiff is either a purchaser or a seller of securities in the transaction involved. The *Hochfelder* case, referred to in the *Schreiber* case excerpted earlier in this chapter, is an example of fraud perpetrated on shareholders by management. Other cases alleging fraud dealing with reorganizations and mergers have been brought, but since the mid-1970s, the Supreme Court has been reluctant to allow cases brought under Section 10(b) to preempt state laws and, thus, has made plaintiffs meet all three elements in an exacting manner.

Fraud-on-the-Market Theory and Section 10(b). The Supreme Court has attached stringent criteria to all private-party actions brought under Section 10(b)-5 and SEC Rule 10(b)-5. Defrauded investors generally need to show that their losses resulted from specific conduct of the company or its employees or agents and that they relied on specific misstatements, omissions, or fraudulent actions in making investment decisions. More recently, shareholders have used an efficient-market concept as the basis for suits claiming fraud. That is, they have alleged that they relied on the integrity of an efficient market to assimilate all information about a company and to reflect this information in a fair price for securities. The plaintiff in such a suit argues that when a company makes fraudulent disclosures or omissions, it distorts the information flow to the market and thus fixes the price of the company's securities too high, in violation of Section 10(b) and Rule 10(b)-5. This fraud-on-the-market theory assumes that the market price reflects all known material information. In a landmark decision (*United States v. James O'Hagan*, 117 S. Ct. 2199 [1997]) the U.S. Supreme Court upheld this theory.

Insider Trading Bills: Is Congress Corrupt?

The Senate bill states in part: "No member of Congress and no employee of Congress shall use any non-public information derived from the individual's position as a member of Congress or employee of Congress or gained from performance of the individual's duties" for personal benefit.

For many years, a separate industry, known as "political intelligence," was practiced by political insiders connected with hedge funds, mutual funds, and other investors. When bills were passed by members of the House and Senate, powerful new tools were added by which prosecutors could pursue public corruption cases. Numerous amendments were passed. For the first time, firms that collect "political intelligence" must register and report their activities as lobbyists. All such activities obtained from Congress and federal agencies that are used to influence or guide investment decisions must be reported.

LIABILITY AND REMEDIES UNDER THE 1934 EXCHANGE ACT

Criminal Penalties. Violations of Section 10(b) and Rule 10(b) may lead individuals to be fined up to $5 million or imprisoned up to 20 years, or both under the Sarbanes-Oxley Act as discussed earlier in this chapter. A partnership or corporation may be fined $25 million for a proven willful violation. The violator may be imprisoned for 25 years in addition to being fined. The SEC must refer all criminal actions to the Justice Department.

SEC Action. Under the Insider Trading Sanctions Act of 1984[12] and the Insider Trading and Securities Enforcement Act of 1988, the SEC may bring a civil suit against anyone violating or aiding or abetting a violation of the 1934 Act or an SEC rule by purchasing or selling a security while in possession of material, nonpublic information. Violations must occur on or through a national securities exchange or from or through a broker or dealer. If the defendant is found liable, the court may impose a fine in an amount triple (treble) the profits that were gained illegally.

The Insider Trading and Securities Fraud Enforcement Act of 1988 enlarged the class of people who may be subject to civil liability for insider trading. Also, bonus payments can be given to anyone providing information leading to the prosecution of insider-trading violations.[13]

Private Actions. Private parties may sue violators under Rule 10-5 and Section 10(b). Potential violators include accountants, attorneys, and others who aid and abet violations of Section 10(b). As noted later in this chapter, a corporation can bring an action to recover short-swing profits under Section 16(b). Under Section 10(b), a private plaintiff may seek rescission of a securities contract or recover damages for breach by disgorgement of illegal profits gained. In the following case, the U.S. Supreme Court emphasizes the role of a private plaintiff in actions for securities fraud violations.

CASE 23-6

The Wharf (Holdings) Limited v. United International Holdings, Inc.
United States Supreme Court
121 S. Ct. 1776 (2001)

United International Holdings, Inc., sued The Wharf (Holdings) Limited in federal district court for securities fraud for violating Section 10(b) and Rule 10(6)-5 of the 1934 Exchange Act.

The Wharf (Holdings) Limited is a Hong Kong firm that was interested in obtaining a license to operate a cable television system in Hong Kong. In 1991, the Hong Kong government announced that it would accept bids for the award of an exclusive license to operate a cable television system in Hong Kong. Wharf decided to find a business partner with cable system experience. Wharf located United International Holdings, Inc., a Colorado-based company with substantial experience in operating cable television systems. Wharf orally agreed to grant United an option to buy 10 percent of the stock of the new Hong Kong cable system if Wharf was awarded the license.

In May 1993, Hong Kong awarded the cable franchise to Wharf. When United raised $66 million and tried to exercise its option to invest 10 percent in the new cable company, Wharf refused to permit United to buy any of the new company's stock. Documents and other evidence showed that at the time Wharf orally granted United the 10 percent stock option, it had not intended ever to sell United any stock in the new venture. The jury held for United and awarded it $67 million in compensatory damages and $58.5 million in punitive damages against Wharf. The court of appeals affirmed. The U.S. Supreme Court granted review.

Justice Breyer

Wharf points out that its agreement to grant United an option to purchase shares in the cable system was an oral agreement. And it says that Section 10(b) does not cover oral contracts of sale. There is no convincing reason to interpret the Act to exclude oral contracts as a class. The Act itself says that it applies to "any contract" for the purchase or sale of a security. Oral contracts for the sale of securities are sufficiently common that the Uniform Commercial Code and statutes of frauds in every State now consider them enforceable. To sell an option while secretly intending not to permit the option's exercise is misleading, because a buyer normally presumes good faith. Since Wharf did not intend to honor the option, the option was, unbeknownst to United, valueless.*

Affirmed for Plaintiff, United.

---
[12] 15 U.S.C. § 78u(d)(2)(A).
[13] 15 U.S.C. § 78u-1.
*The Wharf (Holdings) Limited v. United International Holdings, Inc., United States Supreme Court 121 S. Ct. 1776 (2001).

SHORT-SWING PROFITS

Purpose and Coverage. Section 16(b) of the Exchange Act seeks to further the goal of complete disclosure of insider trading by requiring directors, officers, and owners of more than 10 percent of a class of stock of a registered company to file regular reports with the SEC and the exchanges on which the stock trades. Directors and officers must file an initial statement of their holdings in the company when they take office, and the others must file when they come to own more than 10 percent. A follow-up statement is due monthly if they change their holdings in any manner. Any profits made by a director or an officer or a 10 percent beneficial owner as a result of buying and selling the securities within a 6-month period—known as **short-swing profits**—are presumed to be based on insider information. A plaintiff does not have to show that these insiders had access to, relied on, or took advantage of any insider information. In 1991, the SEC adopted new rules relating to Section 16 that (1) created a new form (Form 5) that must be filed by all insiders within 45 days of the end of the issuer's calendar year, (2) waive liability for insiders for transactions that occur within six months of becoming an insider, (3) make the acquisition of a derivative security (e.g., warrant) fall under Section 16, and (4) define an *officer* under Section 16 as a person who has a policy function (e.g., CEO, president, and vice president). Now, those company officers who handle day-to-day operations do not fall under Section 16.

short-swing profits Profits made by directors, officers, or owners of 10 percent or more of the securities of a corporation as a result of buying and selling the securities within a six-month period.

COMPARATIVE LAW CORNER

Insider Trading Worldwide

Over the past 20 years, insider trading has been vigorously prosecuted. From 1995 to 2005, the stock markets of Britain, Germany, France, Italy, and Switzerland had only 19 criminal convictions. In the United States, the total for Manhattan alone during the same time period was 46 convictions. Britain had only three in this time frame. France had none. From a comparative viewpoint, why the differences in convictions, and what are the implications for people doing business in the United States as opposed to other nations?

The United States has uniform requirements for disclosure for primary offerings and proxy voting. Practices outside the United States are not uniform.[a] Further, the regulatory framework set out by the United States is in sharp contrast to the European and Japanese frameworks. Enforcement is weak with regard to insider trading in many countries.

Some Comparative Results

From a comparative viewpoint, it is very difficult for U.S. companies and U.S. citizens to believe that laws dealing with insider trading are not the same in civil law, socialist, or emerging nations' legal systems. Often, they know only vaguely about their own laws dealing with insider trading (unless advised by specialized counsel), much less about those of other nations. They are shocked when they are cited for violations of many securities laws outlined in this chapter (e.g., the FCPA of 1977 as amended in 1988 and 1998) or of laws of other nations.

Even more striking is the globalization of insider trading when it is alleged to have taken place in another country that has a weak insider-trading regulatory structure (e.g., Pakistan). When a foreign citizen takes part in insider trading on a U.S. exchange following a tip by another foreign citizen, both are stunned when the U.S. Justice Department charges them with insider trading for which criminal penalties are possible.[b] The SEC and Justice Department have entered into separate agreements with several nations for assistance in enforcing U.S. securities laws.

[a]L. Thomas M. deMercedi, "Insider Trading Can Now Touch Many Corners of the World," *New York Times*, August 4, 2007, C-1.
[b]H. Immons and K. Bennhold, "Call for Foreigners to Have a Say on U.S. Market Rules," *New York Times*, August 29, 2007, C-1.

Liability. Suits based on short-swing profits seek to force the insiders to return the profits to the corporation. Only the issuers—meaning the directors and officers of the corporation—and the shareholders have standing to sue. Officers and directors generally do not sue other officers and directors, so virtually all suits are brought by other shareholders. The expense of such litigation, however, makes use of this enforcement action infrequent.

State Securities Laws

State securities laws, often referred to as *"blue sky" laws*, regulate securities purchased and sold in intrastate commerce. Securities are regulated (concurrently) by both state and federal laws. State laws require that securities be registered (or qualified) with both federal and state authorities. A state official or office is usually so designated. State disclosure requirements and antifraud provisions are similar to each other and to their federal counterparts, such as the Securities Exchange Act of 1934, SEC Rule 10(b), and Section 10(b).

The Uniform Securities Act has been adopted in part by some states to bring uniformity to state security laws. The 1995 National Securities Markets Improvement Act limited regulation of investment companies to the SEC and did away with some state authority in this area.

E-Commerce, Online Securities Disclosure, and Fraud Regulation

MARKETPLACE OF SECURITIES

The Internet is now used daily to register securities. Online IPOs are a frequent occurrence that has brought efficiency (in economic terms) to the marketplace of securities. Small companies in particular use the Internet to avoid paying commissions to brokers or underwriters. Regulation A, discussed in this chapter, allows a simple method of registration.

In addition to using the Internet to provide information (e.g., 10-K) to the SEC as required by the 1933 and 1934 Acts, as well as others, potential investors receive information more rapidly and are willing to take advantage of this to purchase or sell securities. Online filing of many documents with the SEC is now routine for most large corporations, which benefit from the additional time given to them that did not exist when they had to use snail mail.

Furthermore, investors and companies can take advantage of the EDGAR database, which includes proxy statements, annual corporate reports, and a multitude of other documents that are filed with the commission. All of this allows the SEC to accomplish its goal of full disclosure. (See Exhibit 23-1 for where EDGAR is located in the structure of the SEC.)

E-COMMERCE AND FRAUD IN THE MARKETPLACE

The SEC continues to deal with fraud in the marketplace via the Internet. Chat rooms in particular have become cyberworld locations for violations of the 1933 and 1934 Securities Acts. Often, a stock price is "pumped up" as a result of information obtained in chat rooms. After it is "pumped," it is quickly "dumped." The SEC has been tracing such actions that seek to manipulate stock prices in violation of the 1934 Exchange Act. As noted in the discussion of the Sarbanes-Oxley Act, penalties can be costly, not only in dollar terms but also in possible prison time.

The list of millions of credit card names and addresses of customers of Target and other companies raised the stakes for the SEC, the FTC, and other government and nongovernment entities that sought to combat such activities.

Global Dimensions of Rules Governing the Issuance and Trading of Securities

The growing internationalization of money and securities markets has made it important to understand the transnational reach of U.S. securities regulations. Both the 1933 Act and the Exchange Act speak of the use of "facilities or instrumentation in interstate commerce." *Interstate commerce* is defined in the 1933 Act to include "commerce between any foreign country and the United States."

This section discusses (1) legislation prohibiting certain forms of bribery and money laundering overseas by U.S.-based corporations and (2) legislation governing foreign securities sold in the United States. You will notice throughout the discussion that provisions of the 1933 Act and the 1934 Exchange Act overlap.

LEGISLATION PROHIBITING BRIBERY AND MONEY LAUNDERING OVERSEAS

The Foreign Corrupt Practices Act of 1977, as Amended in 1988 and 1998. In the course of investigating illegal corporate payments made to President Nixon's 1972 reelection campaign, the SEC staff came across information showing that hundreds of corporations had also made questionable payments to foreign political parties, heads of state, and individuals to obtain business they would not otherwise have gained. The companies argued that these payments were not illegal under U.S. law and that they were necessary to compete with foreign state-owned and state-operated enterprises and with state-subsidized multinationals. The SEC, however, considered this information material under both the 1933 and 1934 securities acts because it affected the integrity of management and the records of the corporations involved. In 1974 and 1975, the SEC allowed approximately 435 companies to enter into consent orders whereby the companies did not admit to making illegal payments but agreed to report such payments to the SEC in the future. The SEC also urged Congress to enact legislation prohibiting bribery overseas by U.S. corporations.

The FCPA of 1977, as amended in 1988 and 1998, seeks to meet this goal. It applies to companies registered under the Securities Acts of 1933 and 1934 and to all other domestic concerns, whether they do business abroad or not. Its antibribery provisions prohibit all domestic firms from offering or authorizing a "corrupt" payment to a foreign official, a foreign political party, or a foreign political candidate to induce the recipient to act, or to refrain from acting, so that a U.S. corporation can obtain business it would not ordinarily get without the payment. The standard of criminal conduct to which corporate officials and employees are held is "knowing." If such a payment is known to violate the FCPA, the corporation can be fined up to $2 million, and its officers, directors, stockholders, employees, and U.S. agents can be fined up to $100,000 and imprisoned for up to 5 years. In addition to prohibiting the payment of bribes, the FCPA also bans the "offer" or "promise" of "anything of value," even if the offer or promise is never consummated. "Facilitating or expediting payments" to ensure routine governmental action is not prohibited.

The FCPA's accounting provisions, enacted as amendments to Section 13(b) of the Exchange Act, apply only to registered nonexempt companies. They require that companies make and keep records and accounts in "reasonable detail" that "accurately and fairly" reflect transactions. Also, companies are required to maintain systems that provide "reasonable assurance" that transactions have been recorded in accordance with generally accepted accounting principles.

The FCPA is jointly enforced by the Justice Department and the SEC. The SEC can investigate and bring civil charges under the act's bribery provisions, but it refers criminal cases to the Justice Department for prosecution. The Justice

Department can bring both civil and criminal charges against alleged violators of the FCPA. The SEC is charged with enforcement of the accounting provisions and can bring both civil actions and administrative proceedings.

Convention on Combating Bribery of Foreign Officials in International Business Transactions. The Convention on Combating Bribery of Foreign Officials in International Business Transactions (CCBFOIBT) was signed in December 1997 by 34 countries after debate by the Organization for Economic Cooperation and Development (OECD). Signatories are required to criminalize bribery of foreign officials, eliminate the tax deductibility of bribes, and subject companies to wider disclosure. Russia (an observer) and China are not signatories; the 34 signatories include the United States, Canada, Japan, and Germany. The treaty creates one loophole: "grease payments." It is acknowledged that these facilitating payments are the cost of doing business and can be paid to low-level officials.

The major change for the United States was the need to amend the FCPA to cover foreign subsidiaries of U.S. companies whose activities have a nexus with interstate or foreign commerce. Under current U.S. law, such subsidiaries are not subject to the FCPA.

The International Securities Enforcement Cooperation Act of 1990. After it was found that many insider traders in the United States were holding secret accounts in Switzerland, the SEC in June 1982 entered into a memorandum of understanding with the Swiss government that established a procedure for processing SEC requests for information about Swiss bank clients suspected of insider trading. As reports of overseas "money laundering" of profits made from insider trading in the United States mounted throughout the late 1980s, the SEC encouraged Congress to clarify the commission's authority to act, not only against those who were sheltering their profits from illegal insider trading in foreign countries but also against others who were violating U.S. securities laws abroad.

In 1990, Congress passed the International Securities Enforcement Cooperation Act (ISECA). The most important provisions of this act are as follows:

1. It provides for giving foreign regulators U.S. government documents and information needed to trace laundered money and those suspected of doing the laundering.

2. It exempts from the Freedom of Information Act (FOIA) disclosure requirements documents given to the SEC by foreign regulators. Without this exemption, foreign regulators would be reluctant to provide U.S. regulators with information and alleged violators could obtain information too easily.

3. It gives the SEC authority to impose administrative sanctions on buyers and dealers who engage in activities that are illegal under U.S. law while they are in foreign countries.

4. It authorizes the SEC to investigate violations of all U.S. securities laws that occur in foreign countries.

LEGISLATION GOVERNING FOREIGN SECURITIES SOLD IN THE UNITED STATES

Schedule B of the Securities Act of 1933 sets forth disclosure requirements for initial offerings by foreign issuers of stock on U.S. exchanges. Foreign issuers are entitled to some of the same exemptions in this area as domestic issuers, except that exemptions under Regulation A are granted only to U.S. and Canadian issuers. Also, the SEC has special registration forms for initial foreign offerings. In Section 12(g)(3) of the Exchange Act, Congress gave the SEC power to exempt foreign issuers whose securities are traded on U.S. exchanges or the OTC markets from certain registration requirements if the commission believes that such action would be in the public interest. Under SEC Rule 12(g)(3)-2, the securities of a foreign issuer are exempt from annual and current reports if the issuer or its government furnishes the SEC with annual information material to investors, which is made public in the issuer's own country. In 1983, the commission published a list of exemptions for foreign-issued securities and adopted regulations that generally require foreign securities registered under the Exchange Act also to be quoted on the National Association of Securities Dealers Authorized Quotations (NASDAQ).

REGULATIONS AND OFFSHORE TRANSACTIONS

Regulation S governs offers and sales of any securities made in *offshore transactions*. Such *transactions* are defined as those in which no offer is made to a person in the United States and (1) the buyer is outside the United States at the time the buy order is originated or (2) the transaction is executed in, on, or through the facilities of a designated offshore securities market. No directed selling efforts may be made in the United States.

Regulation S allows U.S. companies to offer securities abroad with some certainty that such securities will be exempt from federal securities registration. The combination of Rule 144A and Regulation S has expanded the private placement market by increasing the liquidity of private placement securities.

SUMMARY

The SEC is the federal agency responsible for overseeing the securities markets and enforcing federal securities legislation.

Several pieces of legislation provide the framework for the federal regulation of securities issuance and trading, but the most important are the Securities Act of 1933 and the Securities Exchange Act of 1934. The most recent additions are the Dodd-Frank Act and the Sarbanes-Oxley Act, which amend both the 1933 and 1934 Acts.

The Securities Act of 1933 seeks to ensure that investors receive full and fair disclosure of all material information about a new stock issue. It prescribes a three-stage registration process for new securities: prefiling, filing, and postfiling. Several types of securities are exempt from registration: principally private placements, intrastate offerings, and small business offerings.

The Securities Exchange Act of 1934 governs six areas of securities trading: the registration of securities issuers and broker-dealers, securities markets, proxy solicitations, tender offers and takeover bids, securities fraud, and short-swing profits. Provisions of the act dealing with securities fraud address insider trading, misstatements by corporate management, and mismanagement of a corporation.

The e-commerce world has made online investing via the Internet a common occurrence. The SEC, with the help of EDGAR, has sought to modernize the securities regulatory system and assist both investors and issuing companies.

The increasing internationalization of securities markets has led Congress and the SEC to extend the reach of U.S. securities regulations through specific agreements with foreign governments and through provisions of the FCPA and the ISECA.

REVIEW QUESTIONS

23-1 Do members of Congress presently have to disclose any illegal insider trading? Explain.

23-2 Under the proxy rules, when may the management of a registered issuing company exclude a shareholder proposal from the agenda of an annual meeting? Explain.

23-3 What criteria do the courts use to determine whether an instrument or transaction will be called a security? Explain.

23-4 Which securities must be registered under the 1933 Act? Which must be registered under the 1934 Act? Explain.

23-5 Explain the following statement: The 1934 securities law imposes liability for making a "material" statement or omission in proxy documents.

REVIEW PROBLEMS

23-6 Livingston had worked for Merrill Lynch for 20 years as a securities sales representative (account executive). In January 1972, he and 47 other account executives were given the honorary title of "vice president" because of their outstanding sales records. None of their duties changed, however, and they never attended a meeting of the board of directors. In November and December 1972, Livingston sold and repurchased the same number of shares of Merrill Lynch, making a profit of $14,836.37. Merrill Lynch sued Livingston for recovery of the profits, claiming that he had violated Section 16(b) of the Securities and Exchange Act of 1934. Livingston denied such charges. Who won this case? Why?

23-7 Estrada Hermanos, Inc., a corporation incorporated and doing business in Florida, decides to sell $1 million of its common stock to the public. The stock will be sold only within the state of Florida. Jose Estrada, the chair of the board, says that the offering need not be registered with the SEC. Is he right? Explain.

23-8 The boards of directors of DuMont Corp. and Epsot, Inc., agreed to enter into a friendly merger, with DuMont to be the surviving entity. The stock of both corporations was listed on a national stock exchange. In connection with the merger, both corporations distributed to their shareholders proxy statements seeking approval of the proposed merger. About three weeks after the merger was consummated, the price of DuMont stock fell from $25 to $13 as a result of the discovery that Epsot had entered into several unprofitable long-term contracts two months before the merger had been proposed. The contracts will result in substantial losses from Epsot's operations for at least the next four years. The existence and effect of these contracts, although known to both corporations at the time of the proposed merger, were not disclosed in the proxy statements of either corporation. Can the shareholders of DuMont recover in a suit against DuMont under the 1934 Act? Explain.

23-9 Schlitz Brewing Company failed to disclose on its registration statement, as well as in its periodic reports to the SEC, certain kickback payments that it was making to retailers to encourage them to sell Schlitz products, as well as the fact that the company had been convicted of violating a Spanish tax law. The SEC claimed that the failure to include such information was a violation of the antifraud provisions of the 1933 and 1934 Acts because it was material. Schlitz claimed that the information was not material because the kickbacks represented only $3 million, a tiny sum compared with the company's $1 billion in revenues. Was the information material, and was its omission thus a violation of Section 10(b) of the 1934 Act and SEC Rule 10(b)-5? Explain.

23-10 International Mining Exchange and a person named Parker sold a "Gold Tax Shelter Investment Program." Anyone who wished to invest had to write a check payable to an individual designated by International Mining and sign certain papers. Investors acquired a leasehold interest in a gold mine with proven reserves, and they agreed to allow International Mining to arrange for sale options to purchase the gold that would be mined. In effect, investors received the right to profits from the gold mined, in addition to a tax deduction based on the cost of developing the mine. The SEC claimed that this transaction involved "securities" and thus was not exempt from registration under the 1933 Act. International Mining claimed that this transaction did not fall within the definition of a security. What was the result? Explain.

23-11 Continental, a manufacturer of cigarettes, sold to a group of 38 investors bonds with warrants attached to purchase common stock. The sales took place in a high-pressure atmosphere in a room with phones ringing and new orders apparently coming in. Each investor signed an agreement that he or she had received written information about the corporation, and each testified to having access to additional information if requested. The SEC brought an action claiming that Continental was in violation of the registration provisions of the 1933 Act for selling unregistered nonexempt securities. Continental argued that it qualified for a private placement exemption. What was the result? Explain.

CASE PROBLEMS

23-12 To comply with accounting principles, a company that engages in software development must either "expense" the cost (record it immediately on the company's financial statement) or "capitalize" it (record it as a cost incurred in increments over time). If the project is in the pre- or postdevelopment stage, the cost must be expensed. Otherwise, it may be capitalized. Capitalizing a cost makes a company look more profitable in the short term. Digimarc Corp., which provides secure personal identification documents, announced that it had improperly capitalized software development costs over at least the previous 18 months. The errors had resulted in $2.7 million in overstated earnings, requiring a restatement of prior financial statements. Zucco Partners, LLC, which had bought Digimarc stock during the relevant period, filed a suit in a federal district court against the firm. Zucco claimed that it could show that there had been disagreements within Digimarc over its accounting. Is this sufficient to establish a violation of SEC Rule 10(b)-5? Why or why not? *Zucco Partners, LLC v. Digimarc Corp.*, 552 F.3d 981 (9th Cir. 2009).

23-13 Jabil Circuit, Inc., is a publicly traded electronics and technology company headquartered in St. Petersburg, Florida. In 2008, a group of shareholders who had owned Jabil stocks from 2001 to 2007 sued the company and its auditors, directors, and officers for insider trading. Jabil's compensation for executives included stock options. Some stock options were backdated to a time when the stock price was lower, making the options worth more to certain company executives. Backdating is legal as long as it is reported, but Jabil did not report that backdating had occurred. Thus, expenses were understated and net income was overstated by millions of dollars. The shareholders claimed that by rigging the value of the stock options through backdating, the executives had engaged in insider trading. The shareholders also asserted that there was a pattern of stock sales by executives before unfavorable news about the company was reported to the public. The shareholders, however, had no specific information about these stock trades or when (or even if) an executive was aware of any accounting errors related to backdating. Were the shareholders' allegations sufficient to assert that insider trading had occurred under SEC Rule 10(b)-5? Why or why not? *Edward J. Goodman Life Income Trust v. Jabil Circuit, Inc.*, 594 F.3d 783 (11th Cir. 2010).

23-14 Northstar Financial invested some of the funds of its investors with Schwab Total Bond Market Fund, a fund registered with the SEC under the Investment Company Act. In compliance with the Act, Schwab stated its investment policies in its registration statement. In violation of the Act, it changed its investment strategy from what was stated in its registration statement. Because of this change in investment strategy, the company lost significant money when the housing market imploded. So Northstar Financial filed a class action suit on behalf of its investors against the Schwab Fund for violating the Investment Company Act. The Fund filed a motion to dismiss the suit on grounds that no private action was available under the Act, and the District Court rejected the motion. The Schwab Fund appealed the dismissal. How do you think the Court of Appeals ruled? Why? *Northstar Financial Advisors v. Schwab Investments*, 615 F.3d 1106 (2010 WL 3169400, 9th Cir. 2010).

THINKING CRITICALLY ABOUT RELEVANT LEGAL ISSUES

Our laws are far too easy on those who commit securities fraud. Ten thousand dollars and five years in prison are not stiff enough penalties for those who violate the public's trust in the stock market and thereby undermine our economy.

Because our economy is based on capitalism, our businesses are dependent upon outside sources of funding. If a company wishes to expand or purchase another company, it needs access to outside funds. In addition, people need to make money, and their prospects in today's job market are uncertain. Thus, people need to supplement their income. For this reason, they are willing to "loan" businesses money to expand by purchasing securities.

When people commit securities fraud, they cause two undesirable outcomes. First of all, they cause investors to lose their money. Many of these investors are counting on their investments to enable them to survive after retirement. By committing fraud, white-collar criminals wipe out the investors' savings and force them to work extra years. In this sense, these corporate con artists not only are taking money from the investors but also are taking away years of their lives.

Second, when members of the public hear about acts of securities fraud, they become afraid to invest. When people fail to invest, businesses are unable to raise the capital they need. This lack of capital may cause them to lay off workers, close down divisions, or possibly close altogether. At any rate, lack of capital makes our economy run less efficiently, and when that happens, we all suffer.

Clearly, securities fraud isn't a small crime committed against a business. Instead, its effects are felt by companies, investors, and even workers all across the country. Because acts of securities fraud have such far-reaching effects, those who commit them must be subject to stiffer punishment.

1. What reasons does the author give for harshly punishing those who use securities to defraud people?
2. What evidence does the author provide to support these reasons?
3. What information would be helpful for you to have in evaluating the worth of the author's claims? (Refer to your answer to question 2.)
4. Which words or phrases in this essay are especially ambiguous?
5. Set out some arguments for the hypothesis "Our laws are far too hard on those who commit securities fraud."

ASSIGNMENT ON THE INTERNET

The SEC brings civil lawsuits against individuals who are accused of violating one or more securities laws, regulations, or rules. Using the Internet, visit the Securities and Exchange Commission's litigation website (www.sec.gov) and read a brief or press release of a recent case. What rule or regulation was violated? What did the individual do that violated the rule or regulation? Finally, how does examining the reasoning in a court's decision about securities law help you to better understand the chapter you just read? To access the briefs and press releases, highlight "Enforcement" and then click on "Litigation Releases."

ON THE INTERNET

securities.stanford.edu The Securities Class Action Clearinghouse provides a wealth of information about federal securities litigation, including cases, statutes, reports, and settlements.

www.sec.gov This is the home page of the Securities and Exchange Commission.

www.nasaa.org From this page, find out about the North American Securities Administrators Association, an organization devoted to investor protection.

www.seclaw.com This site provides securities statutes, rules, and regulations at both state and federal levels. Click on the heading "SEC Rules" to find more information.

www.law.cornell.edu/topics/securities.html This site contains a basic overview of securities law as well as recent judicial decisions about securities law. Click on "Wex legal encyclopedia," then "Browse." Under "S," scroll until you find "Securities law history."

lp.findlaw.com This site contains numerous links to securities laws and securities fraud information. In the "Quick Links" tabs at the top of the page, click on "Corporate Counsel." Once there, click on "Finance," then on "Securities."

FOR FUTURE READING

Casenote: "Securities Law—The Implied Private Right of Action Under Rule 10b-5 Does Not Extend Liability to Aiders and Abettors." *Cumberland Law Review* 42 (2011): 425.

Miller, Sandra K., Penelope Sue Greenberg, and Ralph H. Greenberg. "An Empirical Glimpse into Limited Liability Companies: Assessing the Need to Protect Minority Investors." *American Business Law Journal* 43 (2006): 609.

Palmiter, Alan. *Securities Regulation: Examples & Explanations*. New York: Wolters Kluwer, 2011.

Soderquist, Larry D., and Theresa A. Gabaldon. *Securities Law (Concepts and Insights)*. Mineola, NY: Foundation Press, 2006.

CHAPTER TWENTY FOUR

Antitrust Laws

- INTRODUCTION TO ANTITRUST LAW
- ENFORCEMENT OF AND EXEMPTIONS FROM THE ANTITRUST LAWS
- THE SHERMAN ACT OF 1890
- THE CLAYTON ACT OF 1914
- OTHER ANTITRUST STATUTES
- GLOBAL DIMENSIONS OF ANTITRUST STATUTES

There is disagreement in many areas of public law between those who believe that business conduct should be disciplined through government regulation and those who favor the marketplace and economic-efficiency criteria as the sole instruments of business discipline. Nowhere is this struggle sharper than in the area of antitrust law.

American attitudes toward government restraints in the area of contracts originated in the English common law, which traditionally upheld the freedom of the individual to contract. U.S. courts generally refused to interfere with commercial agreements: Price-fixing and horizontal and vertical territorial divisions of markets were considered part of the business environment and, hence, legal. This laissez-faire approach was accepted up to the second half of the nineteenth century, when the economic might of huge monopolies stirred Congress to enact the Interstate Commerce Act of 1887 and the Sherman Act of 1890. Despite the advance of technology and 117 years of knowledge and court decisions, the goals of antitrust laws continue to be debated.

This chapter begins with an introduction to the meaning of antitrust and a summary of the federal antitrust statutes. It then discusses enforcement of the antitrust laws and exemptions made to those laws. Next it examines the types of business conduct that are forbidden by the Sherman Act, as well as the Clayton Act, the Federal Trade Commission Act, and the Bank Merger Act of 1966. Because these acts have affected, directly and indirectly, almost every business and political institution in American society and carry criminal and/or civil penalties much of the chapter focuses on dissecting them. Finally, it examines the global dimensions of antitrust policy.

CRITICAL THINKING ABOUT THE LAW

Antitrust law is full of controversial cases. Should the government restrict businesses? If so, to what extent? Should we prevent businesses from creating monopolies? These questions are addressed in a variety of laws regarding antitrust policy. A fairly recent antitrust action by the government involved Microsoft, which was charged with unfair monopolistic practices. The government claimed that Microsoft unfairly restricted its competitors by forcing computer manufacturers to ship the Microsoft Internet Explorer browser along with Windows 95. Answering the following questions about Microsoft can help you think critically about antitrust law.

1. Before you can critically evaluate claims, you need to identify the reasons and conclusion. To get in the habit of paying attention to reasons, try to generate reasons for and against governmental intervention in Microsoft's practices.

    Clue: Reread the introduction. Why would Microsoft want to be free of governmental intervention? Why would the government want to regulate Microsoft?

2. Your roommate makes the following statement: "Businesses have to comply with far too many regulations. They should just be free to make their own rules. Businesses that aren't fair to the public will not be successful. The government shouldn't regulate Microsoft." How would you respond to your roommate?

    Clue: Even though you have not yet read this chapter, you can evaluate your roommate's statement. Do you see any problems with this statement?

3. Microsoft argues that the Internet Explorer browser is simply part of its Windows operating system. Furthermore, Microsoft claims that it is serving its customers by including the Internet Explorer. Customers don't have to worry about finding or installing an additional Web browser. Thus, the government is essentially hurting the public by regulating Microsoft. Are you persuaded by Microsoft's argument? Why or why not?

    Clue: What information might be missing from Microsoft's argument? What more would you like to know about the Web browser industry?

Introduction to Antitrust Law

A DEFINITION OF ANTITRUST

trust A business arrangement in which owners of stocks in several companies place their securities with trustees, who jointly manage the companies and pay out a specific share of their earnings to the securities holders.

Trusts were originally business arrangements in which owners of stocks in several companies placed their securities in the hands of trustees, who controlled and managed the companies. The securities owners, in return, received certificates that gave each a specified share of the earnings of the jointly managed companies. The trust device itself was not—and is not today—illegal. In the late 1880s and 1890s, however, trusts were used by a few companies to buy up or drive out of business many small companies in a single industry. Standard Oil Company, for example, used this process to monopolize the oil industry. Large trusts used unscrupulous methods of competition—such as bribery, setting up bogus companies, and harassing small companies with lawsuits—to gain monopolistic profits. Magazine and newspaper exposes of scandalous transactions involving trusts shook the public's confidence in unregulated markets. Against this background, the Sherman Act was enacted in 1890. Because it was aimed at monopolies that called themselves trusts, it was called an *antitrust* statute.

LAW AND ECONOMICS: SETTING AND ENFORCING ANTITRUST POLICY

The formulation and enforcement of antitrust policy have been substantially affected by the disciplines of law and economics. Moreover, there is a strong difference of opinion about antitrust law between lawyers and economists who favor some government regulation of business and those who want to see deregulation or, more radically, no regulation at all. These two approaches to antitrust policy are known, respectively, as the Harvard School and the Chicago School, after the universities where many of their proponents have taught and written.

Chicago School An approach to antitrust policy that is based solely on the goal of economic efficiency or the maximization of consumer welfare.

The **Chicago School** (a market or efficiency approach to antitrust policy) argues that antitrust decisions should be based solely on the criterion of economic efficiency—that is, the maximization of consumer welfare, which may be defined as improving the allocation of scarce resources without in some way decreasing productive efficiencies. In this chapter's discussion of antitrust goals, you will find that one of these goals is the "promotion of the maximization of consumer welfare using market principles and efficiency criteria"; however, the Chicago School argues that unless efficiency is the sole criterion for antitrust policy making, consumers will not be able to obtain goods at the lowest price possible and U.S.-based multinationals will not be able to compete with foreign multinationals. Adherents of the Chicago School would like to see antitrust statutes enforced less strictly, especially in the areas of vertical price and territorial restraints, and would decriminalize many antitrust offenses. Finally, proponents of this approach believe that large size in U.S. business is far from bad, considering that competitive foreign firms are big and are sometimes aided by their governments as well. You will see the Chicago approach at work when you read the GTE Sylvania case later in this chapter.

The **Harvard School** (a structural approach to antitrust policy) favors the preservation of an economy characterized by many buyers and sellers with little domination by anyone. Adherents of this approach condemn the accumulation of economic power because they believe that it leads to substantial political power at the federal, state, and local levels as politicians are "bought" by the holders of economic power. The resulting concentration of economic and political power allows a small elite to dominate society and to dictate the closing of plants, downsizing, and the loss of jobs from a community. The Harvard School's position on the creation and enforcement of antitrust policy is embodied in all four of the antitrust goals discussed next.

Harvard School An approach to antitrust policy that is based on the desirability of preserving competition to prevent the accumulation of economic and political power, the dislocation of labor, and market inefficiency.

Try not to take sides on this issue until you have read and critically analyzed this chapter. Whatever you decide, it is important that you understand from the outset of your study of antitrust policy that the political, economic, and judicial systems of this country are profoundly affected by these two opposing schools of thought (Table 24-1). Also, it might be interesting to see how the courts have adopted some criteria belonging to both schools.

However, because uncertainty about the legality of joint ventures seemed to discourage their use for joint research and development, Congress passed the National Cooperation Research Act to facilitate such applications. The Act provides that the courts must judge joint ventures in the research and development of new technology under the rule of reason test and that treble damages do not apply to ventures formed in violation of Section 1 if those forming the venture have notified the Justice Department and the FTC of their intent to form the joint venture.

TABLE 24-1

CHICAGO AND HARVARD SCHOOLS' APPROACHES TO ANTITRUST POLICY

Chicago School	*Harvard School*
1. Sole criterion for formulating antitrust policy is efficiency: the maximization of consumer welfare.	1. Several criteria, including (a) preservation of many buyers and sellers in the economy, (b) prevention of concentration of political and economic power, (c) preservation of local control of business and prevention of dislocation of labor markets, and (d) efficiency of markets.
2. Enforce the antitrust statutes rigorously and increase the criminal penalties in most areas of antitrust.	2. Decriminalize many offenses, including vertical restraints of trade and monopolies.
3. Encourage joint ventures between the United States and foreign multinationals without requiring government approval.	3. Allow joint ventures but retain strict oversight by the Justice Department and Federal Trade Commission to prevent worldwide concentration and division of global markets by multinationals.

CASE 24-1

American Needle, Inc. v. National Football League
Supreme Court of the United States, 2010
130 S. Ct. 2201

Originally organized in 1920, the National Football League (NFL) is an unincorporated association that encompasses 32 separately owned professional football teams. Each team has its own name, colors, and logo and owns related property and marketing trademarked items such as caps and jerseys. In 1963, the teams formed National Football League Properties (NFLP) to develop, license, and market their intellectual property. Most, but not all, of the substantial revenues generated by NFLP have either been given to charity or shared equally among the teams. However, the teams are able to and have at times sought to withdraw from this arrangement.

Between 1963 and 2000, NFLP granted nonexclusive licenses to a number of vendors, permitting them to manufacture and sell apparel bearing team insignias. American Needle, Inc., was one of those licensees. In December 2000, the teams voted to authorize NFLP to grant an exclusive 10-year license to manufacture and sell trademarked headwear for all 32 teams. It thereafter declined to renew American Needle's nonexclusive license.

American Needle filed this action in the Northern District of Illinois, alleging that the agreements between the NFL, its teams, NFLP, and Reebok violated Sections 1 and 2 of the Sherman Act. In their answer to the complaint, the defendants asserted that the teams, the NFL, and NFLP were incapable of conspiring within the meaning of Section 1 "because they are a single economic enterprise, at least with respect to the conduct challenged." The District Court granted summary judgment for the NFL on this question. The Court of Appeals for the 7th Circuit affirmed.

Justice Stevens

As the case comes to us, we have only a narrow issue to decide: whether the NFL respondents are capable of engaging in a "contract, combination . . ., or conspiracy" as defined by § 1."

We have long held that concerted action under § 1 does not turn simply on whether the parties involved are legally distinct entities. Instead, we have eschewed such formalistic distinctions in favor of a functional consideration of how the parties involved in the alleged anticompetitive conduct actually operate.

Conversely, there is not necessarily concerted action simply because more than one legally district entity is involved.

The key is whether the alleged "contact, combination . . ., or conspiracy" is concerted action—that is, whether it joins together separate decision-makers. The relevant inquiry, therefore, is whether there is a "contract, combination . . . or conspiracy" amongst "separate economic actors pursuing separate economic interests," such that the agreement "deprives the marketplace of independent centers of decision-making," and therefore of "diversity of entrepreneurial interests." Thus, while the president and vice president of a firm could (and regularly do) act in combination, their joint action generally is not the sort of "combination" that § 1 is intended to cover. Such agreements might be described as "really unilateral behavior flowing from decisions of a single enterprise." Nor, for this reason, does § 1 cover "internally coordinated conduct of a corporation and one of its unincorporated divisions," because "[a] division within a corporate structure pursues the common interests of the whole," and therefore "coordination between a corporation and its division does not represent a sudden joining of two independent sources of economic power previously pursuing separate interests." Nor, for the same reasons, is "the coordinated activity of a parent and its wholly owned subsidiary" covered. Nor, however, is it determinative that two legally distinct entities have organized themselves under a single umbrella or into a structured joint venture. The question is whether the agreement joins together "independent centers of decision-making." If it does, the entities are capable of conspiring under § 1, and the court must decide whether the restraint of trade is an unreasonable and therefore illegal one.

The NFL teams do not possess either the unitary decision-making quality of the single aggregation or economic power characteristics of independent action. Each of the teams is a substantial, independently owned, and an independently managed business. "[T]heir general corporate actions are guided or determined" by "separate corporate consciousnesses," and "[t]heir objectives are" not "common." The teams compete with one another, not only on the playing field, but to attract fans, for gate receipts and for contracts with managerial and playing personnel.

Directly relevant to this case, the teams compete in the market for intellectual property. To a firm making hats, the Saints and the Colts are two potentially competing suppliers of valuable trademarks. When each NFL team licenses its intellectual property, it is not pursuing the "common interests of the whole" league but is instead pursuing the interests of each "corporation itself"; teams are acting as "separate economic actors pursuing separate economic interests," and each team is to license their separately owned trademarks collectively and to only one vendor are decisions that "depriv[e] the marketplace of independent centers of decision-making," and therefore of actual or potential completion.

Although NFL teams have common interests such as promoting the NFL brand, they are still separate, profit maximizing entities, and their interests in licensing team trademarks are not necessarily aligned.

The question is whether NFLP decisions can constitute concerted activity covered by § 1. This is so both because NFLP is a separate corporation with its own management and because the record indicated that most of the revenues generated by NFLP are shared by the teams on an equal basis. Nevertheless we think it clear that for the same reasons the 32 teams' conduct is covered by § 1, NFLP's actions also are subject to § 1, at least with regards to its marketing of property owned by the separate teams. NFLP's licensing decisions are made by the 32 potential competitors, and each of them actually owns its share of the jointly managed assets. Apart from their agreement to cooperate in exploiting those assets, including their decisions relating to purchases of apparel and headwear, to the sale of such items, and to the granting of licenses to use its trademarks, 32 teams operating independently through the vehicle of the NFLP, are not like the components of a single firm that act to maximize the firm's profits. The teams remain separately controlled, potential competitors with economic interests that are distinct from NFLP's financial well-being. Unlike typical decisions by corporate shareholders, NFLP licensing decisions involve more than a mere majority of shareholders. And each team's decision reflects not only an interest in NFLP's profits. The 32 teams capture individual economic benefits separate and apart from the NFLP profits as a result of the decisions they make for the NFLP. NFLP's decisions thus affect each team's profits from licensing its own intellectual property. "Although the business interests of" the teams "will often coincide commonality of interest exists in every carrel." In making the relevant licensing decisions, NFLP is therefore "an instrumentality" of the teams. If the fact that potential competitors shared in profits or losses from a venture meant that the venture was immune from § 1, then any cartel "could evade the antitrust law simply by creating a 'joint venture' to serve as the exclusive seller of their competing products." However, competitors "cannot simply get around" antitrust liability by acting "through a third-party intermediary of 'joint venture'."

The fact that NFL teams share an interest in making the entire league successful and profitable, and that they must cooperate in the production and scheduling of games, provides a perfectly sensible justification for making a host of collective decisions. But the conduct at issue in this case is still concerted activity under the Sherman Act that is subject to § 1 analysis. When "restraints on completion are essential if the product is to be available at all," per se rules of illegality are inapplicable, and instead the restrain must be judged according to the flexible Rule of Reason. And depending upon the concerted activity in question, the Rule of Reason may not require a detailed analysis; it "can sometimes be applied in the twinkling of an eye."

Other features of the NFL may also save agreements amongst the teams. We have recognized, for example, "that the interest in maintaining a competitive balance" among "athletic teams is legitimate and important." While that same interest applies to the teams in the NFL, it does not justify treating them as a single entity for § 1 purposes when it comes to the marketing of the teams' individually owned intellectual property. It is, however, unquestionably an interest that may well justify a variety of collective decisions made by the teams. What role it properly plays in applying the Rule of Reason to the allegations in this case is a matter to be considered on remand.*

Reversed and *remanded*.

GOALS OF THE ANTITRUST STATUTES

A century of debate by lawyers, economists, and others has not produced a real consensus on the goals of the antitrust statutes. Nonetheless, these four goals can be derived from the study of antitrust legislation and case law:

1. *The preservation of small businesses and an economy characterized by many sellers competing with one another.* Proponents of this goal would break up large corporations such as General Motors (GM), International Business Machines (IBM), and Microsoft.

2. *The prevention of concentration of political and economic power in the hands of a few sellers in each industry.* Proponents of this goal argue that there is a direct correlation between large corporations, economic power, and control of the political process. They point to 1980, 1984, and 2000 postpresidential election analyses indicating that well-financed political action committees (PACs) controlled by big businesses had a tremendous effect on the elections' outcomes.

3. *The preservation of local control of business and protection against the effects of labor dislocation.* The advocates of this antitrust goal argue that when large companies are allowed to merge, fix prices, and participate in joint ventures, jobs are lost and plants are shut down in some areas. The consequences are a dislocation of labor and a decline in local and state economies as their tax bases shrink because people are moving elsewhere in pursuit of jobs.

---

*American Needle, Inc. v. National Football League, 2010, 130 S. Ct. 2201.

4. *The promotion of the maximization of consumer welfare using market principles and efficiency criteria.* Advocates of this goal define *consumer welfare* as an improvement in the allocation of resources without an impairment to productive efficiency. In effect, the proponents of this goal argue that by encouraging the allocation of resources in an efficient manner, antitrust enforcement can make sure that consumers will be provided goods at the lowest possible prices.

Some of these goals conflict.[1] For example, the Harvard School proponents of goal 1 (preserving small businesses) are criticized by adherents of the Chicago School, who favor only goal 4 (consumer welfare maximization), because they believe that large firms are needed to manufacture goods at the lowest cost per unit. They point out that until Henry Ford introduced the assembly-line production of automobiles, few people could afford cars. Many small businesses are economically inefficient, they say, and attempts to preserve them through antitrust policy will be underwritten by consumers in the form of higher prices. The Chicago School also insists that if U.S. manufacturers are not allowed to merge and participate in joint ventures, they will be unable to compete with large foreign multinationals and foreign companies owned or subsidized by their governments. One of the reasons the Justice Department moved to dismiss an antitrust suit against the International Business Machines Corporation in 1982 was that the computer market had become international in character since the original government complaint against IBM in 1972. If IBM had been broken up, it would not have been able to compete with large foreign multinationals either in the United States or in other countries.

A summary of the major federal antitrust statutes is provided in Table 24-2.

TABLE 24-2

SUMMARY OF MAJOR FEDERAL ANTITRUST LAWS

Act	Provisions
Sherman Act of 1890	
Section 1	Makes illegal every contract, combination, or conspiracy in restraint of trade; felony offense punishable by a fine up to $100 million per corporation and up to $1 million per individual; a person may be imprisoned up to 10 years, fined, or both.
Section 2	Forbids monopolizing, attempts to monopolize, or conspiracies to monopolize; penalties are the same as for Section 1.
Clayton Act of 1914	
Section 2	Forbids discrimination in price between different purchasers of goods of like grade and quality where the effect may be to lessen competition or to tend to create a monopoly in any line of commerce, or to injure, destroy, or prevent competition with the seller, the buyer, or either's customers.
Section 3	Forbids selling or leasing goods on the condition that the buyer or lessee shall not use or deal in goods sold or leased by the seller's or lessor's competitor, where the effect of such an agreement may be to substantially lessen competition or to tend to create a monopoly. In effect, this section outlaws exclusive dealing and tying arrangements.
Section 7	Forbids unlawful selling of corporate assets or stock mergers when the effect may be to substantially lessen competition or to tend to create a monopoly.
Federal Trade Commission Act of 1914	Forbids unfair methods of competition in commerce and unfair or deceptive acts in commerce.

---

[1]To examine four conflicting opinions on the goals of antitrust, see T. Calvani, "Consumer Welfare Is Prime Objective of Antitrust," *Legal Times* 4 (Dec. 24–31, 1984); B. Brennan, "A Legal-Economic Dichotomy: Contribution to Failure in Regulatory Policy," *American Business Law Journal* 4: 52 (1976); W. Cann Jr., "The New Merger Guidelines: Is the Justice Department Enforcing the Law?" *American Business Law Journal* 21: 12–13 (1983); R. Bork, *The Antitrust Paradox* 90–91, 104, 108 (1978).

Enforcement of and Exemptions from the Antitrust Laws

ENFORCEMENT

Enforcement of the antitrust laws is carried out in both the public and private sectors. The Department of Justice and the Federal Trade Commission (FTC) (see Table 24-3) are primarily responsible for enforcement in the public sector, whereas any individual or business entity in the private sector that establishes that it has been directly injured by illegal business conduct may bring an action under the federal statutes outlined here (as well as under state antitrust statutes). Table 24-3 shows which parties have enforcement powers for the three major federal antitrust statutes.

Public Enforcement. The Antitrust Division of the Justice Department exclusively enforces the Sherman Act and has concurrent jurisdiction with the FTC to enforce the Clayton Act. The FTC has exclusive jurisdiction to enforce the Federal Trade Commission Act (FTCA).

Usually, the Justice Department files civil suits in a federal district court. The remedy requested is ordinarily an injunction to prevent a particular action from occurring, along with a specific order requiring the business to change its conduct or operation. Most of the time, the defending parties, because of the cost of litigation and the attendant bad publicity, choose not to fight the case and instead enter into a consent decree (consent order) with the Justice Department, which binds them to stop the activity complained of (e.g., attempting to manipulate a market). As you know from the discussion of administrative agencies in Chapter 18, entering into a consent order does not involve admitting to any guilt or liability. The federal district court must approve the consent order.

For serious violations of the Sherman Act (e.g., price-fixing among competitors), the Justice Department may bring a criminal action. A corporation convicted of criminal conduct under the act faces a fine of up to $100 million for each offense; individual officers and employees who are convicted face a maximum $1 million fine for each offense or up to 10 years in jail or both. Nolo contendere pleas are often negotiated between the Justice Department and corporate or individual criminal defendants. This plea of no contest subjects the defendant to a lesser punishment than would result from conviction at a trial. Although technically not an admission of guilt, a nolo contendere plea is treated as such by a judge. Like a consent decree, it must be approved by the court.

Both consent decrees in a civil action and nolo contendere pleas in a criminal action are often entered into by defendants to avoid the cost of litigation and publicity. Another advantage of these decrees and pleas for defendants is that they cannot be used as a basis for shareholder derivative or indemnity suits. For the Justice Department, such decrees and pleas save time and taxpayers' money.

TABLE 24-3 PARTIES THAT ENFORCE THE FEDERAL ANTITRUST LAWS

	Sherman Act	*Clayton Act*	*Federal Trade Commission Act*
Justice Department	Civil and criminal enforcement powers	Civil enforcement power	No power to enforce
Federal Trade Commission	No power to enforce civil or criminal	Civil enforcement power	Civil enforcement power
Private parties	Power to enforce civil litigation	Power to enforce civil litigation	No power to enforce

In December 1997, at the urging of the Justice Department, a federal judge issued a temporary restraining order to prevent Microsoft from allegedly violating a 1995 court order by forcing computer makers to install its Internet browser software along with its Windows 95 operating system. Several days later, the federal court issued a contempt order advising that the company was making a mockery of the court's order. Microsoft appealed in January 1998. Violation of antitrust laws was at issue. On another front, 18 states have brought a separate antitrust action (later joined with the federal government), which has wider implications than the mere tying of a browser system to Windows 95. In this case, the 18 state attorneys general sought to show an effort by Microsoft to eliminate competition under Clayton Act Section 7 and to fix prices under Section 1 of the Sherman Act, as well as several other anticompetitive practices to be presented in this chapter. This action by the states was a continuation of their joint efforts to pursue cases against the tobacco companies, telemarketing advertisers, and environmental violators, all in the name of consumer protection. For elected attorneys general, these are popular cases to be litigated.[2] In 2004, all the states except Massachusetts settled with Microsoft (some contested it). Also, in 2004, the D.C. Circuit Court of Appeals ruled against Massachusetts and affirmed the settlement.

The settlement allowed Microsoft to remain one company under certain conditions:

- Microsoft may not retaliate against a computer maker in any way, including raising prices or withholding support for dealing with Microsoft's competitors.
- Microsoft must establish a schedule of fixed prices.
- Other computer makers (IBM, Gateway, and Dell) were allowed to install non-Microsoft products and desktop shortcuts of any size or shape on the computers.
- Microsoft had to reveal previously confidential programming interfaces that its products rely on to link to Windows code.

Furthermore, Microsoft was not to retaliate against other companies because their products compete with other Microsoft applications.

The FTC can bring only civil actions, which are usually argued before an administrative law judge (ALJ). The ALJ makes findings of fact and recommends action to the full five-member commission, which may issue a cease-and-desist order. The defendant has the option of appealing such an order to a U.S. Court of Appeals and further to the Supreme Court, but usually such cease-and-desist orders are negotiated by the parties before a hearing by the ALJ and are approved by the commission. Failure to abide by a cease-and-desist order carries a penalty of $10,000 a day for each day the defendant is not in compliance.

Private Enforcement. Section 4 of the Clayton Act says that

any person who shall be injured in his person or in his business or property by reason of anything forbidden in the antitrust laws may sue [and] . . . shall recover threefold the damages by him sustained and the cost of suit including a reasonable attorney's fee.*

This section provides the incentive for private enforcement of our antitrust laws, because it requires the court to triple the amount of damages awarded to a plaintiff by a jury or by a judge. It also awards reasonable attorney's fees to the plaintiff's attorney.

Private actions can be brought by individuals or businesses against perceived violators of the antitrust laws. In recent years, approximately 90 percent of all antitrust claims were brought by private-party plaintiffs. Moreover, when a small company such as Microwave Communication, Inc. (MCI) sues a large company

---

[2]See J. Marttott, "13 States Planning Broader Suits against Microsoft," *New York Times* (Apr. 30, 1988).
*Private Enforcement, Section 4 of Clayton Act.

such as American Telephone and Telegraph (AT&T) for antitrust violations, victory has the double advantage of enhancing its cash flow and showing bond-rating agencies and investors that it is a viable entity able to take on a big company.

Individuals in a class action suit and state attorneys general in *parens patriae* actions on behalf of their citizenry can also bring suits. In a **class action suit**, one member of a group of plaintiffs injured by an antitrust violation (e.g., price-fixing, which results in higher prices for direct purchasers) institutes an action on behalf of the entire group. This kind of suit is particularly useful when the amount of each individual claim is small. Similar to class actions are ***parens patriae* suits**, which are usually brought by a state attorney general on behalf of purchasers and taxpayers in a state (previously discussed in relation to the action against Microsoft and actions against tobacco companies by attorneys general of several states).

class action suit A lawsuit brought by a member of a group of persons on behalf of all members of the group.

parens patriae suit A lawsuit brought by a state attorney general on behalf of the citizenry of that state.

EXEMPTIONS

Several activities and industries are fully or partially exempt from the antitrust statutes (Table 24-4). These exemptions are based on federally enacted statutes or case law of the courts. When exemptions are granted by statute, they are largely the result of successful lobbying of Congress by an industry. Soft-drink franchisers, for example, lobbied successfully in 1980 to obtain a limited exemption from the antitrust statutes, and shipping lines received a similar exemption in 1984.

TABLE 24-4 SELECT ACTIVITIES EXEMPT UNDER U.S. ANTITRUST LAW

Activity	Basis for Exemption and Examples
Regulated industries	Transportation, electric, gas, telephone, and securities.
Labor union activities	Collective bargaining.
Intrastate activities	Intrastate telephone calls are regulated by state public utility commissions.
Agricultural activities	Farmers may belong to cooperatives that legally set prices.
Baseball	The U.S. Supreme Court declared baseball a sport, not a trade. No other professional sport has been exempted by Congress or the courts. [*Federal Baseball of Baltimore v. National League*, 259 U.S. 200 (1992).]
Activities falling within the "state action" doctrine	In *Parker v. Brown*, 317 U.S. 341 (1943), the U.S. Supreme Court held a state marketing program that was clearly anticompetitive to be exempt from the federal antitrust statutes, because it obtained its authority from a "clearly articulated legislative command of the state." The Court looks at the degree of involvement before exempting any activity under this doctrine.
Cities', towns', and villages' activities	The Local Government Antitrust Act of 1984 prohibits monetary recovery under the federal antitrust laws from any of these local subdivisions or from local officials, agents, or employees.
Export activities	The Webb-Pomerene Trade Act of 1918 and the Export Trading Company Act of 1982 made the formation of selling cooperatives of U.S. exporters exempt. Also, the Joint Venture Trading Act of 1983 exempted certain joint ventures of competing companies when seeking to compete with foreign companies that are private and/or state controlled. Approval of the Justice Department is required. The Shipping Act of 1984 allows shipping lines to enter into joint ventures and to participate in international shipping conferences that set worldwide rates and divide routes and shipments.
Oil marketing	The Interstate Oil Compact of 1935 allows states to set quotas on oil to the market in interstate commerce.
Joint efforts by businesses to seek government action	Such actions are exempt unless it is an attempt to make anticompetitive use of governmental processes. See *Eastern Railroad v. Noer Main Freight*, 365 U.S. 127 (1961).

The Sherman Act of 1890

The Sherman Act of 1890 is intended to prevent control of markets by any one powerful entity. In other words, it is designed to thwart anticompetitive behavior. Sections 1 and 2 of the act, covered in this section, profoundly affect decisions and behaviors of business managers. The Sherman Act is also an important tool to protect consumers from a number of activities discussed in this chapter.

SECTION 1: COMBINATIONS AND RESTRAINTS OF TRADE

Section 1 of the Sherman Act reads:

> *Every contract, combination in the form of trust or otherwise, or conspiracy, in restraint of trade or commerce among the several States, or with foreign nations, is declared to be illegal. Every person who shall make any contract or engage in any combination or conspiracy hereby declared to be illegal shall be deemed guilty of a felony, and, on conviction thereof shall be punished by fine not exceeding ten million dollars if a corporation, or, if any other person, three hundred fifty thousand dollars or by imprisonment not exceeding three years, or by both.**

Sherman Act Federal statutes that make illegal every combination, contract, or conspiracy that is an unreasonable restraint of trade when this concerted action involves interstate commerce.

The **Sherman Act** requires three elements for a violation: (1) a combination, contract, or conspiracy; (2) a restraint of trade that is unreasonable; and (3) a restraint that is involved in interstate, as opposed to intrastate, commerce. We examine the third element in Chapter 5 in the discussion of the effects of the Commerce Clause on business. Here, we analyze the first two elements.

Combination, Contract, or Conspiracy. The Sherman Act requires a contract, combination, or conspiracy, so more than one person must be involved (one cannot make a contract or conspire or combine with oneself). Just as an offeror and an offeree are necessary parties to a contract, there must be coconspirators in a conspiracy. In other words, there must be concerted action (action taken together) by two or more individuals or business entities. In antitrust language, there must be an "agreement" or **collusion**. Such an agreement can be expressed in writing or orally, or it can be implied, as established by circumstantial evidence such as trends toward uniformity in pricing in an industry and opportunities to conspire.

collusion Concerted action by two or more individuals or business entities in violation of the Sherman Act.

It is in cases of implied agreements established by circumstantial evidence that the courts deal with two major problems: (1) whether there can be an intra-enterprise conspiracy that violates the Sherman Act and (2) what actions constitute conscious parallelism as opposed to price-fixing. **Conscious parallelism** exists when identical actions (usually price increases) are taken independently but nearly simultaneously by two or more leading companies in an industry and thus have the apparent effect of having arisen from a conspiracy.

conscious parallelism Identical actions (usually price increases) that are taken independently but nearly simultaneously by two or more leading companies in an industry.

The courts' solution to the first problem has generally been that there can be no conspiracy between two divisions, departments, or subsidiaries of the same corporation. In answer to the second problem, the courts have generally held that if there is supportable evidence of conscious parallelism, as opposed to an expressed or implied agreement among competitors, no antitrust violation exists.

restraint of trade Action that interferes with the economic law of supply and demand.

Restraints of Trade. The second element required to prove a violation of Section 1 of the Sherman Act is that there be a **restraint of trade** and that this restraint be "unreasonable" as defined by the courts. In enacting the Sherman Act, Congress gave no indication of whether it meant all restraints or just some.

rule-of-reason standard A legal standard holding that only unreasonable restraints of trade violate Section 1 of the Sherman Act. If the court determines that an action's anticompetitive effects outweigh its procompetitive effects, the restraint is unreasonable.

Rule-of-Reason Standard. Taking its direction from the English courts, the U.S. Supreme Court adopted a **rule-of-reason standard**. Over time, the Court has followed certain indices, laid out by Justice Louis D. Brandeis in 1918, to determine whether a specific business activity is an unreasonable restraint of trade:

---

*The Sherman Antitrust Act (1890), 15 U.S. Code § 1, Trusts, etc., in restraint of trade illegal; penalty.

1. The nature and purpose of the restraint
2. The scope of the restraint
3. Its effect on the business and on competitors
4. Its intent

When using a rule-of-reason standard, the U.S. Supreme Court terms a restraint reasonable, and therefore legal, if it has a procompetitive purpose and its effect does not go beyond that purpose. A restraint is unreasonable and unlawful if it allows the parties to substitute themselves and their judgment for the laws of supply and demand. Restraints that are judged by a rule-of-reason standard to determine whether they are in violation of Section 1 of the Sherman Act (as seen in the Microsoft litigation discussed in this chapter) include some tying arrangements, activities of trade and athletic associations, some exclusive-dealing arrangements, nonprice vertical restraints, and some franchising arrangements. Most of these are discussed later in this chapter.

Per Se Standard. Over time, the courts have judged certain business activities and arrangements facially so anticompetitive in nature that they have seen no need to listen to any procompetitive economic justifications. This **per se standard** favors the plaintiff because all that has to be proved is that the restraint (e.g., price-fixing) took place; the only defense possible is that the activity did not occur. Because the present Supreme Court, however, began looking at the economic impact of previous per se rulings, the number of restraints of trade in this category has fallen. Restraints that are judged by the per se standard, and are therefore automatically in violation of Section 1 of the Sherman Act, include horizontal price-fixing, some tying arrangements, some divisions of markets, and group boycotts. Table 24-5 compares the types of activities that come under each of these standards.

Discussed next are the various activities that are considered restraints of trade under two headings: horizontal restraints and vertical restraints.

per se standard A legal standard applicable to restraints of trade that are inherently anticompetitive. Because such restraints are automatically in violation of Section 1 of the Sherman Act, courts do not balance procompetitive and anticompetitive effects in such cases.

Horizontal Restraints. **Horizontal restraints of trade** take place between competitors at the same level of the marketing structure. Three types of activities are considered horizontal restraints: horizontal price-fixing, horizontal divisions of markets, and horizontal boycotts.

horizontal restraint of trade Restraint of trade that occurs between competitors at the same level of the marketing structure.

Horizontal Price-Fixing. Suppose that competitors X, Y, and Z are the only manufacturers of a certain heat tape used in the construction of office buildings. They agree to take turns bidding on specific jobs, thus eliminating competition and keeping prices at a certain level. From the viewpoint of these companies, this is a way to make high profits, stay in business, and keep their employees working. The Supreme Court, however, views such **horizontal price-fixing**

horizontal price-fixing Collusion between two or more competitors to set prices for a product or service, directly or indirectly.

Per Se Standard	*Rule-of-Reason Standard*
Price-fixing—horizontal agreements	Restrictive covenant in a sale or employment; vertical price and nonprice restraints
Group boycotts	Location and resale restraints by manufacturer on some tying arrangements
Some tying arrangements	Exchange of information
Some divisions of markets	Joint research and development ventures; some horizontal price tampering based on unique nature of industry

TABLE 24-5

SHERMAN ACT SECTION 1 ACTIVITIES JUDGED BY THE PER SE STANDARD AND THE RULE-OF-REASON STANDARD

as a per se illegal restraint because it interferes with the price mechanism—that is, the law of supply and demand—which requires that competing sellers make decisions about prices on their own without agreement or collusion, either expressed or implied. In addition to direct price-fixing agreements, the Court has struck down such indirect price-fixing arrangements as minimum fee schedules for lawyers and engineers, exchanges of price information among groups of competitors, and agreements between competitors about terms of credit when these become part of the overall price structure in an industry.

The case that follows illustrates a possible agreement among competitors that may affect prices. Note carefully that this conduct (conscious parallelism) is carefully examined by the court to make sure it does not in fact constitute illegal price-fixing.

CASE 24-2

Williamson Oil Co. v. Philip Morris, USA
United States Court of Appeals for the Eleventh Circuit
346 F.3d 1287 (2003)

Between 1993 and 2000, Philip Morris (PM), R.J. Reynolds (RJR), Brown & Williamson (B&W), and Lorillard (the manufacturers) produced more than 97 percent of the cigarettes sold in the United States. During the early 1990s, as a price gap widened between premium brands such as Marlboro and Camel and discount brands such as Basic and Doral, some "premium smokers" began to shift to nonpremium brands. By 1993, nonpremium brands had captured more than 40 percent of the U.S. market. Although this trend benefited RJR and B&W, it was undesirable for premium-intensive manufacturers such as PM and Lorillard. PM then began to look for ways to reverse the trend toward discount cigarettes. In 1993, PM announced that it was cutting the retail price of Marlboro cigarettes, the single best-selling brand in America, by 40 cents per pack and foregoing price increases on other premium brands "for the foreseeable future." This price cut was followed by price cuts on PM's other premium brands. PM's price cuts set off a price war, as RJR, B&W, and Lorillard matched PM's retail price reductions, which cut into the market share held by the discount brands.

Subsequently, however, RJR announced that it would no longer sacrifice profitability for market share and increased the price of its premium and discount brands. The other manufacturers matched the increases within a couple of weeks. Eleven more parallel increases occurred between May 1995 and January 2000.

A class of several hundred cigarette wholesalers (plaintiffs) sued the manufacturers, alleging that they had conspired between 1993 and 2000 to fix cigarette prices at unnaturally high levels, which resulted in wholesale list-price overcharges of nearly $12 billion. The district court entered summary judgment in favor of the manufacturers after concluding that the wholesalers had failed to demonstrate the existence of a "plus factor." The court went on to state that even if the class had shown that a plus factor was present, the manufacturers had rebutted the inference of collusion because the economic realities of the 1990s cigarette market made the class's conspiracy theory untenable. The district court characterized the manufacturers' pricing behavior as nothing more than "conscious parallelism," a perfectly legal phenomenon often associated with oligopolistic industries. The wholesalers appealed.

Justice Marcus

[T]he distinctive characteristic of oligopoly is recognized interdependence among the leading firms: the profit-maximizing choice of price and output for one depends on the choices made by others.

When they are the product of a rational, independent calculus by each member of the oligopoly, as opposed to collusion, these types of synchronous actions have become known as "conscious parallelism." The Court has defined this phenomenon as the process, not in itself unlawful, by which firms in a concentrated market might in effect share monopoly power, setting their prices at a profit-maximizing, supra-competitive level by recognizing their share of economic interests and their interdependence with respect to price and output decisions.

As numerous courts have recognized, it often is difficult to determine which of these situations—illegal price-fixing or conscious parallelism—is present in a given case.

[P]rice-fixing plaintiffs are relegated to relying on indirect means of proof. The problem with this reliance on circumstantial evidence, however, is that such evidence is by its nature ambiguous, and necessarily requires the drawing of one or more inferences in order to substantiate claims of illegal conspiracy.

"[T]o survive a motion for summary judgment . . . a plaintiff seeking damages for [collusive price-fixing] . . . must present evidence that tends to exclude the possibility that the alleged conspirators acted independently." Evidence that does not support the existence of a price-fixing conspiracy any more strongly than it supports conscious parallelism is insufficient to survive a defendant's summary judgment motion.

In applying these principles, we have fashioned a test under which price fixing plaintiffs must demonstrate the existence of "plus factors" that remove their evidence from the realm of equipoise and render that evidence more probative of conspiracy than of conscious parallelism.

[T]he district court delineated distinct factors that appellants had denominated "plus factors." These are: "(1) signaling of intentions; (2) permanent allocations programs; (3) monitoring of sales; (4) actions taken contrary to economic self-interest."

[W]e are satisfied that none of the actions on which appellants' arguments are based rise to the level of plus factors. . . . Indeed, when all of appellees' actions are considered together, the class has established nothing more than that the tobacco industry is a classic oligopoly, replete with consciously parallel pricing behavior, and that its members act as such.*

Affirmed the District Court's grant of summary judgment for the Defendants. The suit was *dismissed*.

CRITICAL THINKING ABOUT THE LAW

1. What reasons would the plaintiff have had to advance to avoid summary judgment against it?

    Clue: How can a plaintiff show that more than conscious parallelism is the explanation for the imitative pricing behavior?

2. What is the difference between "signaling of intentions" and being the first oligopolist to increase prices?

    Clue: In that the first may be illegal and the second is not, the difference may be one of degree.

How "per se" is the per se rule in the area of price-fixing? For example, it is clear that certain forms of price-fixing are per se legal by statute. Before deregulation of long-distance telephone service in the early 1980s, prices for this service were set by the Federal Communications Commission and, therefore, were exempt from the Sherman Act. Similarly, before deregulation of the trucking industry, trucking rates were set by the Interstate Commerce Commission and, therefore, were exempt. In contrast, organizations of engineers, lawyers, and doctors have been found guilty of per se illegal price-fixing when they set minimum or maximum schedules of rates or recommend certain minimum prices. There is no bright line in this area of antitrust law showing which pricing activities will be judged per se illegal in the future.

Horizontal Division of Markets. Territorial division, customer allocation, and product-line division of markets between competitors have traditionally been deemed illegal per se. The **horizontal division of markets** is considered particularly dangerous to a free-market economy because it eliminates all forms of competition, in contrast to price-fixing, which eliminates only price competition. "Naked" horizontal agreements to divide markets, customers, or product lines have had no redeeming value in the eyes of the courts.

horizontal division of markets Collusion between two or more competitors to divide markets, customers, or product lines among themselves.

More recently, however, it has been argued that price-fixing and division-of-market agreements that are part of cooperative productive activity (as opposed to "naked" restraints) are economically efficient and therefore desirable. Thus, the advocates of these "more-than-naked horizontal agreements" contend that a rule-of-reason standard should guide the courts.

An example of price-fixing and division of markets that has been long accepted is a law partnership. Lawyers who would ordinarily compete with each other eliminate competition by signing a partnership agreement that restricts work

---

*Williamson Oil Co. v. Philip Morris, USA United States Court of Appeals for the Eleventh Circuit 346 F.3d 1287 (2003).

output to their specialization (market division) and that sets the fees to be charged by partners and their associates (price-fixing). In effect, an integrated economic unit fixes prices and divides markets (output) internally so that the partnership may operate more efficiently in competing externally with other law firms.

Horizontal Boycotts. Trade associations frequently promulgate rules among their memberships that amount to concerted refusals to deal with members who do not follow the association's regulations. This activity constitutes a **horizontal boycott** and is per se illegal because it takes away the freedom of other members to interact with the boycotted members and, in many instances, lessens the ability of a boycotted member to compete.

horizontal boycott
A concerted refusal by a trade association to deal with members that do not follow the association's regulations.

Many professional associations have rules that contain sanctions for violations, ranging from reprimands to suspension or expulsion from the association.

In this era of deregulation and the resulting indirect encouragement of industry self-regulation, certain antitrust cases have become extremely important. If an industry's own rules are arbitrary and capricious or lacking in due process, the courts will generally not uphold them under the rule of reason. For example, when the New York Stock Exchange, without a hearing, ordered all exchange members to withdraw wire connection with a nonmember broker, the Supreme Court held that

> *concerted termination of trade relations, which would ordinarily constitute an illegal boycott, might be exempt from the antitrust law as a result of the duty of self-regulation imposed on the Exchange [by Congress and the Securities and Exchange Commission], but only if fair procedures were followed, including notice and hearing.*[3]

In another case,[4] however, the Court found no unreasonable restraint of trade arising out of the program established by the National Sanitation Foundation (NSF) for testing products and issuing a seal of approval for products that complied with the NSF's promulgated standards, which were strictly enforced among manufacturers. The Court has indicated that when an alleged boycott of an unapproved manufacturer takes place, the plaintiff must show either that it was discriminated against vis-à-vis its competitors or that it was subjected to anticompetitive conduct.

Clearly, the courts will continue to watch self-regulatory associations carefully for due process and reasonable conduct. In July 1985, the Supreme Court ruled that a wholesale purchasing cooperative's expulsion of a member without notice, a hearing, or an opportunity to challenge the decision could not be conclusively presumed to be a per se violation of Section 1 of the Sherman Act.[5] The Court remanded the case to the federal district court, directing that a rule-of-reason approach be used to determine whether the cooperative had the market power to exclude competitors and whether the expulsion of a member was likely to have an anticompetitive effect.

In the case of noncommercial refusals to deal, it is clear that the courts will not apply the per se rule. For example, when the National Organization of Women (NOW) organized a boycott of convention facilities in all states that had refused to endorse the proposed Equal Rights Amendment to the U.S. Constitution, Missouri sued NOW, claiming that the organization was in violation of Section 1 of the Sherman Act. The circuit court of appeals stated that the Sherman Act was not applicable in this case because the boycott had a noncommercial goal, namely, to influence legislation in the political arena.[6] The court used

---

[3]*Silver v. New York Stock Exchange*, 373 U.S. 341 (1965).
[4]*Eliason Corp. v. National Sanitation Foundation*, 614 F.2d 126 (6th Cir. 1980).
[5]*Northwest Wholesale Stationers, Inc. v. Pacific Stationery & Printing Co.*, 472 U.S. 284 (1985).
[6]*Missouri v. NOW, Inc.*, 620 F.2d 1301 (8th Cir. 1980), *cert. denied*, 449 U.S. 8412 (1981).

the rule-of-reason standard to determine whether the group's purpose was truly noncommercial in nature.

A conflict between constitutional principles—such as the right to free speech and to petition one's government under the First Amendment of the Constitution—and enforcement of the Sherman Act against boycotts or refusals to deal must often be resolved by the courts. Usually, as in the NOW case, constitutional principles have prevailed when noncommercial groups have been involved.

Vertical Restraints. Those restraints agreed to between individuals or corporations at different levels of the manufacturing and distribution process are called **vertical restraints of trade**. For example, manufacturers and retailers, as well as franchisors and franchisees, are often involved in the following types of vertical restraints: resale price maintenance (price-fixing), territorial and customer restrictions, tying agreements, and exclusive-dealing contracts. Although the latter two restraints involve violations of Section 3 of the Clayton Act, the courts have also condemned such actions under Section 1 of the Sherman Act.

> **vertical restraint of trade**
> Restraint that occurs between individuals or corporations at different levels of the marketing structure.

As we examine these vertical restraints, you should focus on two policy implications for business managers:

1. What effect do court decisions have on intrabrand competition (retailers competing with one another in selling the same manufacturer's brand) and interbrand competition (competition between different manufacturers of a similar product when sold at the retail level)?

2. Are the courts moving in the direction of judging such restraints by a per se or a rule-of-reason standard?

Vertical Price-Fixing (Resale Price Maintenance). When a manufacturer sells to a retailer, the manufacturer may attempt to specify what price it expects the retailer to charge for the product or, at least, a minimum price. **Vertical price-fixing** agreements of this type have traditionally been judged per se illegal by the courts if a "contract, combination, or conspiracy" exists under Section 1 of the Sherman Act. This has been true regardless of whether the manufacturer coerced the retailer into entering the agreement (by refusing to supply the product) or the retailer entered the agreement voluntarily.

> **vertical price-fixing**
> Stipulation by a manufacturer to a retailer to which it sells products as to what price the retailer must charge for those products.

The courts' major concern in this area has been whether the retailer made the pricing decision independently or by agreement with the manufacturer. For example, many manufacturers offer suggested prices to their retailers in the form of price lists. The question often before the court is what type of surveillance the manufacturer uses to coerce an initial agreement or to gain compliance. If the prices are truly just suggestions, how many times has the retailer deviated from such prices, and what has been the manufacturer's response? The courts have scrutinized manufacturers' responses carefully, particularly when there is evidence of a manufacturer's refusal to deal. The courts have generally agreed that a manufacturer, on its own initiative, can announce in advance an intention not to deal with an individual retailer who does not sell the manufacturer's product at a specific price. No agreement is involved in these unilateral cases. The court, however, will infer an agreement, and thus per se illegal price-fixing, if the manufacturer refuses to deal with a retailer who fails to adhere to a resale price and then reinstates the retailer when it agrees to conform. In the following case excerpt, the issue is whether the setting of a minimum resale price was a per se violation of the Sherman Act or a resale activity to be judged by a rule of reason.

CASE 24-3

Leegin Creative Leather Products, Inc. v. PSKS, Inc., *dba* Kay's Kloset, Kay's Shoes
United States Supreme Court
127 S. Ct. 2705 (2007)

Given its policy of refusing to sell to retailers that discount its goods below suggested prices, the petitioner (Leegin) stopped selling to the respondent's (PSKS) store. PSKS filed suit, alleging, among other things, that Leegin violated the antitrust laws by entering into vertical agreements with its retailers to set minimum resale prices. The district court excluded expert testimony about the procompetitive effects of Leegin's pricing policy on the ground that *Dr. Miles Medical Co. v. John D. Park & Sons Co.*[7] makes it per se illegal under Section 1 of the Sherman Act for a manufacturer and its distributor to agree on the minimum price the distributor can charge for the manufacturer's goods. At trial, PSKS alleged that Leegin and its retailers had agreed to fix prices, but Leegin argued that its pricing policy was lawful under Section 1 of the Sherman Act. The jury found for PSKS. On appeal, the 5th Circuit declined to apply the rule of reason to Leegin's vertical price-fixing agreements and affirmed, finding that the *Dr. Miles* per se rule rendered irrelevant any procompetitive justifications for Leegin's policy.

Justice Kennedy

In *Dr. Miles Medical Co. v. John D. Park & Sons Co.*, the Court reestablished the rule that it is *per se* illegal, under § 1 of the Sherman Act, for a manufacturer to agree with its distributor to set the minimum price the distributor can charge for the manufacturer's good.

Section 1 of the Sherman Act prohibits "[e]very contract, combination in the form of trust or otherwise, or conspiracy, in restraint of trade or commerce among the several States." The Court has never "taken a literal approach to [its] language." Rather, the Court has repeated time and again that § 1 "outlaw[s] only unreasonable restraints."

The rule of reason is the accepted standard for testing whether a practice restrains trade in violation of § 1. Under this rule, the factfinder weighs all of the circumstances of a case in deciding whether a restrictive practice should be prohibited as imposing an unreasonable restraint on competition. Appropriate factors to take into account include "specific information about the relevant business" and "the restraint's history, nature, and effect." Whether the businesses involved have market power is a further, significant consideration. In its design and function the rule distinguishes between restraints with anticompetitive effect that are harmful to the consumer and restraints stimulating competition that are in the consumer's best interest.

Resort to *per se* rules is confined to restraints "that would always or almost always tend to restrict competition and decrease output." To justify a *per se* prohibition, a restraint must have "manifestly anticompetitive" effects, and lack any redeeming virtue.

As a consequence, the *per se* rule is appropriate only after courts have had considerable experience with the type of restraint at issue and only if courts can predict with confidence that it would be invalidated in all or almost all instances under the rule of reason. It should come as no surprise, then, that we have expressed reluctance to adopt *per se* rules with regard to restraints imposed in the context of business relationships where the economic impact of certain practices is not immediately obvious.

The Court has interpreted *Dr. Miles Medical Co. v. John D. Park & Sons Co.* as establishing a *per se* rule against a vertical agreement between a manufacturer and its distributor to set minimum resale prices. In *Dr. Miles* the plaintiff, a manufacturer of medicines, sold its products only to distributors who agreed to resell them at set prices. The Court found the manufacturer's control of resale prices to be unlawful. It relied on the common-law rule that "a general restraint upon alienation is ordinarily invalid." The Court then explained that the agreements would advantage the distributors, not the manufacturer, and were analogous to a combination among competing distributors, which the law treated as void.

The reasoning of the Court's more recent jurisprudence has rejected the rationales on which *Dr. Miles* was based. By relying on the common-law rule against restraints on alienation, the Court justified its decision based on "formalistic" legal doctrine rather than "demonstrable economic effect." The Court in *Dr. Miles* relied on a treatise published in 1628 but failed to discuss in detail the business reasons that would motivate a manufacturer situated in 1911 to make use of vertical price restraints. Yet the Sherman Act's use of "restraint of trade" invokes the common law itself . . . not merely the static content that the common law had assigned to the term in 1890. The general restraint on alienation, especially in the age when then Justice Hughes used the term, tended to evoke policy concerns extraneous to the question that controls here. Usually associated with land, not chattels, the rule arose from restrictions removing real property from the stream of commerce for generations. The Court should be cautious about putting dispositive weight on doctrines from antiquity but of slight relevance. We reaffirm that "the state of the common law 400 or even 100 years ago is irrelevant to the issue before us: the effect of the antitrust laws upon vertical distributional restraints in the American economy today."

---
[7]220 U.S. 373, 31 S. Ct. 376 (1911).

Dr. Miles, furthermore, treated vertical agreements a manufacturer makes with its distributors as analogous to a horizontal combination among competing distributors. In later cases, however, the Court rejected the approach of reliance on rules governing horizontal restraints when defining rules applicable to vertical ones. Our recent cases formulate antitrust principles in accordance with the appreciated differences in economic effect between vertical and horizontal agreements, differences the *Dr. Miles* Court failed to consider.*

The judgment of the Court of Appeals is *reversed* in favor of Leegin (Appellant, Plaintiff).

Is Price-Fixing Legal?

Consequences of a Landmark Decision

In June 2007, the U.S. Supreme Court overruled with *Leegin* a near-century-old decision (*Dr. Miles*) that had made price-fixing (minimum resale maintenance) per se illegal. Henceforth, such activity will be judged by a rule of reason. See Table 24-5 for a brief outline of some activities that are judged by either standard.

The impact of this decision raises some questions. (Remember that you just read only an excerpt of the case.)

- Is vertical price-fixing legal, or can such activity be judged illegal by a court of law based on this decision?
- Why did the U.S. Supreme Court change its mind in this case? Do you need more information to answer this question? Does the concept of stare decisis no longer have any meaning for the U.S. Supreme Court?
- Why did the four dissenting judges reach a different decision? What ethical standards are involved here?

What is the global impact of this decision? Assume that high courts in other common-law and some civil-law countries follow U.S. Supreme Court decisions; assume that the United States and foreign multinationals doing business in the United States and in other nations follow U.S. Supreme Court rulings. What approach should they take to this decision to bring company policy into accord with the *Leegin* case—or should they?

Vertical Territorial and Customer Restraints. The restraints used by a manufacturer to limit the territory in which a retailer may sell the company's product and to restrict the number of retailer-owned stores, as well as the customers a retailer can serve in a location, are classified as **nonprice vertical restraints**. Lawyers, economists, and scholars in many disciplines have studied nonprice vertical restraints in great depth.

Those urging that a rule-of-reason standard be applied to such territorial restraints argue that they encourage economic efficiencies and thus provide for spirited interbrand competition. Vertical restraints allow a manufacturer to concentrate its advertising and distributional programs on one or two retailers in a location, making it better able to compete at the retail level with different brand manufacturers of the same product. Customer restrictions are also beneficial, in this view, because they enable a manufacturer to give better service and to cut out the costs of distributors' and retailers' services. For example, a manufacturer may reserve certain large commercial customers for itself, selling directly to them in large quantities and disallowing retailer involvement with them.

Those arguing that a per se standard should be applied to vertical territorial restraints suggest not only that intrabrand competition is enhanced by this approach but also that customers are better able to compare the prices charged by different retailers selling the same brand. They argue that the elimination of intrabrand competition reduces the number of sellers of a leading brand in a market and increases overall market concentration in the product.

nonprice vertical restraint Restraint used by a manufacturer to limit the territory in which a retailer may sell the manufacturer's products and the number of stores the retailer can operate, as well as the customers the retailer can serve in a location.

---

*Leegin Creative Leather Products, Inc. v. PSKS, Inc., dba Kay's Kloset, Kay's Shoes, United States Supreme Court 127 S. Ct. 2705 (2007).

In the following landmark opinion, the U.S. Supreme Court changed the standard for judging vertical territorial and customer restrictions from the per se to the rule-of-reason standard. This was just 10 years after it had gone in the opposite direction in another case.[8]

CASE 24-4

Continental TV, Inc. v. GTE Sylvania
United States Supreme Court 433 U.S. 36 (1977)

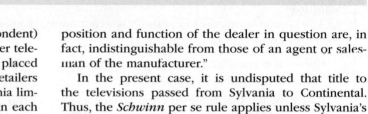

Before 1962, GTE Sylvania (the plaintiff-respondent) found that it was losing market share to other television manufacturers, so it adopted a plan that placed both territorial and customer restrictions on its retailers and phased out its wholesale distributors. Sylvania limited the number of retailers selling its product in each area and designated the location within each area where the stores could be located. When Sylvania became unhappy with its sales in San Francisco, it established another retailer besides Continental (the defendant-appellant) to carry its product. Continental protested, canceled a large order of Sylvania televisions, and ordered a competitor's product. Continental then requested permission to open another store in Sacramento. Sylvania opposed such an opening, claiming that it would be in violation of Continental's franchise agreement. When Continental advised Sylvania that it was nevertheless going to open in the new location, Sylvania cut Continental's credit line, and Continental, in turn, withheld all payments on inventory owed to the manufacturer's credit company. Sylvania terminated the franchise and sued for the money owed and the Sylvania merchandise in the hands of the defendant. Continental filed a cross-claim, alleging that Sylvania had violated Section 1 of the Sherman Act with its restriction on the locations of the retailers that could sell its product. The district court found in favor of Continental on the cross-claim. The U.S. Court of Appeals reversed for Sylvania. Continental appealed to the U.S. Supreme Court.

Justice Powell

The Court [in *Schwinn* (1967)] proceeded to articulate the following "bright line" per se rule of illegality for vertical restrictions. "Under the Sherman Act, it is unreasonable for a manufacturer to seek to restrict and confine areas or persons with whom an article may be treated after the manufacturer has parted with dominion over it." But the Court expressly stated that the rule of reason governs when "the manufacturer retains title, dominion, and risk with respect to the product and the position and function of the dealer in question are, in fact, indistinguishable from those of an agent or salesman of the manufacturer."

In the present case, it is undisputed that title to the televisions passed from Sylvania to Continental. Thus, the *Schwinn* per se rule applies unless Sylvania's restriction on locations falls outside *Schwinn*'s prohibition against a manufacturer attempting to restrict a "retailer's freedom as to where and to whom it will resell the products."

Sylvania argues that if *Schwinn* cannot be distinguished, it should be reconsidered. Although *Schwinn* is supported by the principle of stare decisis, we are convinced that the need for clarification in this area justified reconsideration. Since its announcement, *Schwinn* has been the subject of continuing controversy and confusion, both in the scholarly journals and in the federal courts. The great weight of scholarly opinion has been critical of the decision, and a number of the federal courts confronted with analogous vertical restrictions have sought to limit its reach. In our view, the experience of the past 10 years should be brought to bear on this subject of considerable commercial importance.

In essence, the issue before us is whether *Schwinn*'s per se rule can be justified under the demanding standards of *Northern Pac. R. Co.* (1958). The Court's refusal to endorse a per se standard in *White Motor Co.* (1963) was based on its uncertainty as to whether vertical restrictions satisfied those standards. Addressing this question for the first time, the Court stated:

> We need to know more than we do about the actual impact of these arrangements on competition to decide whether they have such a "pernicious effect on competition and lack . . . any redeeming virtue" and therefore should be classified as per se violations of the Sherman Act.

Only four years later, the Court in *Schwinn* announced its sweeping per se rule without even a reference to *Northern Pac. R. Co.* and with no explanation of its sudden change in position.

---

[8]*United States v. Schwinn & Co.*, 388 U.S. 365 (1967).

The question remains whether the per se rule stated in *Schwinn* should be expanded to include nonsale transactions or abandoned in favor of a return to the rule of reason. We have found no persuasive support for expanding the rule. As noted above, the *Schwinn* Court recognized the undesirability of "prohibit[ing] all vertical restrictions of territory and all franchising." And even Continental does not urge us to hold that all such restrictions are per se illegal.

We revert to the standard articulated in *Northern Pac. R. Co.*, and reiterated in *White Motor Co.*, for determining whether vertical restriction must be "conclusively presumed to be unreasonable and therefore illegal without elaborate inquiry as to the precise harm they have caused or the business excuse for their use." Such restrictions, in varying forms, are widely used in our free market economy. As indicated above, there is substantial scholarly and judicial authority supporting their economic utility. There is relatively little authority to the contrary. Certainly, there has been no showing in this case, either generally or with respect to Sylvania's agreements, that vertical restrictions have or are likely to have a "pernicious effect on competition" or that they "lack . . . any redeeming virtue." Accordingly, we conclude that the per se rule stated in *Schwinn* must be overruled. In so holding, we do not foreclose the possibility that particular applications of vertical restrictions might justify per se prohibition under *Northern Pac. R. Co.* But we do make clear that departure from the rule-of-reason standard must be based upon demonstrable economic effect rather than—as in *Schwinn*—upon formalistic line drawing.

In sum, we conclude that the appropriate decision is to return to the rule of reason that governed vertical restrictions prior to *Schwinn*.*

Affirmed in favor of Plaintiff, Sylvania.

CRITICAL THINKING ABOUT THE LAW

The Supreme Court's decision in Case 24-4 had a significant impact on the standard by which vertical territorial and consumer restraints are judged. The Supreme Court overturned the per se standard set out in the *Schwinn* case. The Court's decision, however, is not arbitrary. The Justices attempted to provide sound reasoning for their decision to move away from the precedent established in the case—but before you can evaluate the soundness of the Supreme Court's argument, you must first find the argument. The following questions highlight the crucial steps prior to evaluation of an argument: finding the issue, conclusion, and reasons.

1. State the issue in Case 24-4 in question form. What is the Court's conclusion?

    Clue: Remember that the issue is the primary question that a court is addressing. The conclusion is the court's response to the issue.

2. What reasons does the Court provide for its conclusion?

    Clue: In Case 24-4, the reasons answer this question: "Why did the Court overturn the per se standard?"

Tying Arrangements. As discussed previously in the enforcement section of this chapter (and the Microsoft case), in a *tying arrangement*, a single party agrees to sell a product or service (tying product; e.g., Windows 95) on the condition that the other party agrees to buy a second (tied) product service (e.g., Internet browser). For example, if a company owns a patent on a tabulating machine (tying product), it will attempt to get its customers to buy only tabulating cards that it produced (tied product), or if a franchisor owns a trademark symbol such as golden arches (tied product), it will seek to get its franchisees to use only products with the designated trademark symbol on them or products approved or manufactured by the franchisor (tying products). As noted in Table 24-2, tying arrangements are violations of Section 3 of the Clayton Act; however, that section of the act applies only to tying arrangements involving tangible commodities. Therefore, actions are frequently brought under Section 1 of the Sherman Act when either the tying or tied products involve services or real property.

Tying arrangements have generally been adjudged per se illegal if the manufacturer of the tying product has a monopoly on the tying product either by virtue of a patent or as a result of a natural monopoly situation. If the tying arrangement does not exist in a monopoly situation, it may still be an illegal vertical restraint of trade if the following three conditions are present:

---

*Continental TV, Inc. v. GTE Sylvania, United States Supreme Court, 433 U.S. 36 (1977).

1. *The manufacturer or seller of the tying product has sufficient economic power to lessen competition in the market of the tied product.* For example, if the owner A of a patent on salt-dispensing machines (tying product) leases the machines only to companies or individuals B who agree to buy salt (tied product) from A, such an agreement may be considered per se illegal because it limits the sellers of the tied product (salt) from competing vigorously in the salt market. If, however, there are similar salt-dispensing-machine manufacturers and lessors that the lessees B can buy from, it is clear that A will not be able to lessen competition in the salt market.

2. *A substantial amount of interstate commerce is affected.* If the tying agreement between the manufacturer and the lessor of the salt machines has little impact on the market of the tied product (salt), the courts will not consider this agreement to be per se illegal and, using a rule-of-reason approach, will dismiss the case.

3. *Two separate products or services are involved.* Some franchisors have argued successfully that their trademark (e.g., the golden arches) and their products and services (building, equipment, and service contract) are one and the same package rather than separate products, and thus no tying arrangement exists.

Exclusive-Dealing Contracts. Agreements between manufacturers and retailers (dealers) or between franchisors and franchisees, which require the second party to sell and promote only the brand of goods supplied by the first party, are known as **exclusive-dealing contracts**. For example, the Standard Oil Company of California had exclusive-dealing contracts with independent stations in seven Western states that required the stations to buy all their oil and other petroleum products from Standard Oil. Sales under that exclusive-dealing contract involved approximately 7 percent of all sales of such products in the seven states. Using a comparative substantiability test (one comparing the effect of such agreements on competing sellers of petroleum products in the geographic area), the U.S. Supreme Court found a violation of Section 3 of the Clayton Act.[9]

exclusive-dealing contract
Agreement in which one party requires another party to sell and promote only the brand of goods supplied by the first party.

Since that case, the Court has generally followed a rule-of-reason approach in cases involving exclusive-dealing agreements. Such agreements are found to be illegal when they foreclose a substantial portion of a relevant market. The Court has found that legitimate business reasons for exclusive-dealing contracts exist in certain industries. For example, it ruled that an exclusive-dealing contract between an electrical utility and a coal supplier extending 20 years was lawful because it had procompetitive effects.[10] The contract assured the utility and its customers a regular supply of coal at a reasonable fixed rate and allowed the coal company to better plan its production and employment needs over a long period; in turn, it was able to offer the utility a lower price.

SECTION 2: MONOPOLIES

Section 2 of the Sherman Act reads:

> *Every person who shall monopolize, or attempt to monopolize, or combine or conspire with any other person or persons to monopolize any part of trade or commerce among the several states, or with foreign nations, shall be deemed guilty of a felony, and on conviction thereof shall be punished by a fine not exceeding one million dollars if a corporation or, if any other person, one hundred thousand dollars, or by imprisonment not exceeding three years, or by both.*

---

[9]*Standard Oil Co. of California v. United States*, 337 U.S. 293 (1949).
[10]*Tampa Electric Co. v. Nashville Coal Co.*, 365 U.S. 320 (1961).
*The Sherman Act (1890)—15 U.S. Code § 2—Monopolizing trade a felony; penalty.

Section 2, therefore, prohibits monopolization, attempts to monopolize, and conspiracies to monopolize. Each of these prohibitions is examined in this section. In reading this material, keep the following four factors in mind:

1. One of the purposes of antitrust law, as stated earlier in this chapter, is to promote a competitive model. Such a model traditionally assumes the existence of many buyers and sellers who have equal access to information about the marketplace and labor that is mobile.

2. In framing Section 2 of the Sherman Act, Congress was vague about what it meant by a **monopoly**. Therefore, it has been up to the courts to define the concept case by case, sometimes with the aid of economic analysis.

3. Some claim that U.S. corporations need to be large to compete with state-owned and state-supported foreign multinationals.

4. Large companies that have attained their monopolistic positions through innovation and research, leading to patents, may be forced in some cases to share the results of their efforts with competitors to avoid bringing down on themselves a Section 2 enforcement and possible penalties (fines, imprisonment, or both).

monopoly An economic market situation in which a single business has the power to fix the price of goods or services.

Monopolization. The U.S. Supreme Court has developed three criteria, or steps, to determine whether a firm has attained a monopolistic position and is misusing its power in violation of Section 2 of the Sherman Act:

1. It determines the relevant product and geographic markets within which the alleged monopolist operates.
2. It determines whether the defendant has overwhelming power in the relevant markets.
3. It examines whether there is an intent on the part of the alleged monopolist to monopolize.

CASE 24-5

E.I. DuPont de Nemours and Co. v. Kolon Industries
United States Court of Appeals, Fourth Circuit
37 F.3d 435 (2011)

DuPont was founded in 1802 as a gunpowder manufacturer. Today, it operates in 90 countries in a broad range of industries from agriculture to nutrition to home construction. It recently made a major investment in "para-aramid," a biodegradable ingredient used in cosmetics, liquid detergents, and antifreeze.

DuPont manufactures and sells para-aramid fiber, which is a complex synthetic fiber used to make body armor, fiber-optic cables, and tires, among other things. Although several companies around the world manufacture this fiber, only three sell into the U.S. market—DuPont (based in the United States), Teijin (based in the Netherlands), and Kolon Industries, Inc. (based in Korea). DuPont is the industry leader, producing more than 70 percent of all para-aramid fibers purchased in the United States. In 2009, DuPont brought a lawsuit against Kolon for misappropriation of trade secrets.

Kolon counterclaimed that DuPont had monopolized and attempted to monopolize the para-aramid market in violation of Section 2 of the Sherman Act. Kolon claimed that DuPont had illegally used multiyear supply agreements for all of its high-volume para-aramid fiber customers. Under the agreements, the customers were required to purchase between 80 and 100 percent of their para-aramid needs from DuPont. Kolon alleged that those agreements removed substantial commercial opportunities from competition and limited other para-aramid fiber producers' ability to compete. On DuPont's motion, a federal district court dismissed Kolon's counterclaim, finding that Kolon had failed to sufficiently plead (demonstrate) unlawful exclusionary conduct. Kolon appealed to the U.S. Court of Appeals for the 4th Circuit.

To prove a Section 2 monopolization offense, a plaintiff must establish two elements: (1) possession of monopoly power and (2) willful acquisition of maintenance of that power—as opposed to simply superior products of historic accidents. An attempted monopolization offense consists of (1) the use of anticompetitive conduct (2) with specific intent to monopolize and (3) a dangerous probability of success.

To run afoul of Section 2, a defendant must be guilty of illegal conduct "to foreclose competition, to gain a competitive advantage, or to destroy a competitor." Conduct that might otherwise be lawful may be impermissibly exclusionary under antitrust law when practiced by a monopolist's behavior. And although not per se illegal, exclusive dealing arrangements can constitute an improper means of acquiring or maintaining a monopoly.

Here, the district court assumed that Kolon adequately pled possession of monopoly power. That assumption was correct, given that Kolon pled, among other things, that numerous barriers to entry into the U.S. para-aramid fiber market exist and supply is low; DuPont has long dominated the U.S. para-aramid fiber market; (and DuPont currently) controls over 70 percent of that market, [that is,] that "DuPont's market share remains greater than 70 percent of all sales by purchase volume of para-aramid fiber in the United States."

Kolon complained that "because DuPont's supply contracts severely restricted access to customers and preclude effective competition, DuPont's conduct has had a direct, substantial, and adverse effect on competition. And DuPont's anticompetetive conduct has allowed it to control output and increase prices for para-aramid fiber in the United States." And "by . . . excluding Kolon from competition for these customers significantly increase and supply is low has constrained the only potential entrant to the United States in decades from effectively entering the market, reducing if not practically eliminating additional competition, as well as preserving and growing DuPont's monopoly position." These allegations are sufficient to withstand a motion to dismiss.

The relevant product market includes all products that, although produced by different firms, have identical attributes, such as sugar. It also includes products that are reasonably interchangeable for the purpose for which they are produced. Products will be considered interchangeable if consumers treat them as acceptable substitutes.

Establishing the relevant product market is often the key issue in monopolization cases because the way the market is defined may determine whether a firm has monopoly power.

1. The possession of monopoly power in the relevant market.
2. "The willful acquisition or maintenance of the power as distinguished from growth or development as a consequence of a superior product, business acumen, or historic accident."

To establish a violation of Section 2, a plaintiff must prove both of these elements—monopoly power and an intent to monopolize.

The Sherman Act does not define monopoly. In economic theory, monopoly refers to control of a specific market by a single entity. It is well established in antitrust law, however, that a firm may be a monopolist even though it is not the sole seller in a market.

In addition, size alone does not determine whether a firm is a monopoly. A "mom and pop" grocery located in the isolated town of Happy Camp, Idaho, is a monopolist if it is the only grocery serving that particular market. Size in relation to the market is what matters because monopoly involves the power to affect prices.

Monopoly power may be proved by direct evidence that the firm used its power to control prices and restrict output. Usually, though, there is not enough evidence to show that the firm intentionally controlled prices, so the plaintiff has to offer indirect, or circumstantial, evidence of monopoly power indirectly, the plaintiff must show that the firm has a dominant share of the relevant market and that there are significant barriers for the new competitors entering the market.*

Relevant Product and Geographic Markets. Markets are divided into product and geographic markets. How the courts determine the boundaries of those markets helps decide what market share a company has and, thus, its market power.

The courts have generally defined the relevant product market as that in which the company alleged to be a monopolist can raise or lower prices with relative independence of the forces of supply and demand. In a monopoly situation, the courts look to the concept of **cross-elasticity of demand or substitutability**.

Cross-elasticity of demand measures the impact that upward and downward changes in price have on the demand for the product. If cross-elasticity of demand is positive, an increase in price of the alleged monopolistic product will result in consumers switching to a substitute product. For example, in a landmark case,[11] the government charged the DuPont Company with monopolizing the cellophane industry because it produced 75 percent of all the cellophane

cross-elasticity of demand or substitutability If an increase in the price of one product leads consumers to purchase another product, the two products are substitutable and there is said to be cross-elasticity of demand.

---

[11]*United States v. DuPont Co.*, 351 U.S. 3717 (1956).

*E.I DuPont de Nemours and Co. v. Kolon Industries, United States Court of Appeals, Fourth Circuit, 37 E.3d 435 (2011).

sold in the United States. DuPont argued that cellophane was not the correct product market because there were many substitutes for cellophane; rather, flexible packaging materials was the correct product market to consider in this case. If the court agreed, DuPont would not be a monopolist because cellophane constituted only 25 percent of the flexible packaging materials market. The U.S. Supreme Court did rule in favor of DuPont, on the basis of the availability of substitutes and the high elasticity of demand for cellophane. The Court noted that a slight increase in the price of cellophane caused many customers to switch to other flexible wrapping materials, which showed that there was a positive cross-elasticity of demand and that DuPont lacked monopoly status.

The courts generally have defined the geographic market as the area where the defendant's firm competes head-on with others in the previously determined relevant product market. Usually, geographic markets are stipulated (agreed to) by the plaintiff and the defendant as regional, national, or international. An exception, however, occurred in a leading monopoly case decided by the Supreme Court.[12] In this case, the parties argued over whether the products that were sold were at a regional market level (Grinnell) or a national level (United States). The products in which Grinnell had ownership interests included tires, sprinklers, plumbing supplies, and burglar alarm systems, together called accredited central station protection services.

The case that follows illustrates a question often faced by business leaders and lawyers. What is the "relevant market"?

CASE 24-6

Newcal Industries, Inc. v. Ikon Office Solutions
United States Court of Appeals, Ninth Circuit
513 F.3d 1038 (2008)

Newcal Industries (Plaintiff-Appellant) and Ikon Office Solutions (IKON) Defendant-Respondent) compete in the brand-name copier equipment-leasing market for commercial customers and in the provision of service.

When a lease approaches its term, these companies compete for the lease of upgraded copier equipment. When a service contract approaches its term, these companies also compete to buy out the service contract in order to provide another one. Newcal alleged that IKON "tricked" its customers by amending its lease agreements and service contracts without disclosing that such amendments would lengthen the terms of the original agreements. The purpose of these contract extensions was to shield IKON customers from competition in the aftermarkets for upgraded copier equipment and service agreements. When IKON succeeded in extending the terms of the original contract, it was able to raise that contract's value. Consequently, Newcal and other competitors had to pay higher prices to buy out such contracts in the aftermarkets for upgraded equipment and services. Newcal brought claims under the Sherman Act, alleging antitrust violations. The District Court held that Newcal had failed to allege a legally recognizable "relevant market" under the Sherman Act. Newcal appealed.

Justice Thomas

First and foremost, the relevant market must be a product market. The consumers do not define the boundaries of the market; the products or producers do. Second, the market must encompass the product at issue as well as all economic substitutes for the product. As the Supreme Court has instructed, "The outer boundaries of a product market are determined by the reasonable interchangeability of use between the product itself and substitutes for it." As such, the relevant market must include all economic substitutes; it is legally permissible to premise antitrust allegations on a submarket. That is, an antitrust claim may, under certain circumstances, allege restraints of trade within or monopolization of a small part of the general market of substitutable products. In order to establish the existence of a legally cognizable submarket, the plaintiff must be able to show (but need not necessarily

---
[12]*United States v. Grinnell Co.*, 384 U.S. 563 (1966).

establish in the complaint) that the alleged submarket is economically distinct from the general product market. In [another case], the Supreme Court listed several "practical indicia" of an economically distinct submarket: "industry or public recognition of the submarket as a separate economic entity, the product's peculiar characteristics and uses, unique production facilities, distinct customers, distinct prices, sensitivity to price changes, and specialized vendors."

First, the law permits an antitrust claimant to restrict the relevant market to a single brand of the product at issue. Second, the law prohibits an antitrust claimant from resting on market power that arises solely from contractual rights that consumers knowingly and voluntarily gave to the defendant. Third, in determining whether the defendant's market power falls in the category of contractually-created market power or in the category of economic market power, the law permits an inquiry into whether a consumer's selection of a particular brand in the competitive market is the functional equivalent of a contractual commitment, giving that brand an agreed-upon right to monopolize its consumers in an aftermarket. The law permits an inquiry into whether consumers entered into such "contracts" knowing that they were agreeing to such a commitment.

The relevance of this point to the legal viability of Newcal's market definition may not be intuitively obvious, but it is nevertheless significant. IKON has a contractually-created monopoly over services provided under original IKON contracts. That contractually-created monopoly then gives IKON a unique relationship with those consumers, and the contractual relationship gives IKON a unique position in the wholly derivative aftermarket for replacement equipment and lease-end services. The allegation here is that IKON is exploiting its unique position—its unique contractual relationship—to gain monopoly power in a derivative aftermarket in which its power is not contractually mandated.

This case is not a case in which the alleged market power flows from contractual exclusivity. IKON is not simply enforcing a contractual provision that gives it the exclusive right to provide replacement equipment and lease-end services. Rather, it is leveraging a special relationship with its contracting partners to restrain trade in a wholly derivative aftermarket. We therefore reverse the district court's holding that Newcal's complaint is legally invalid.

That holding, however, does not quite end the matter. In considering the legal validity of Newcal's alleged market, we must also determine whether IKON customers constitute a cognizable subset of the aftermarket, such that they qualify as a submarket. That is, we have thus far concluded only that there is no per se rule against recognizing contractually-created submarkets and that such submarkets are potentially viable when the market at issue is a wholly derivative aftermarket. A submarket must bear the "practical indicia" [indicators] of an independent economic entity in order to qualify as a cognizable submarket. In this case, Newcal's complaint sufficiently alleges that IKON customers constitute a submarket according to all of those practical indicia.*

Reversed and *remanded.*

APPLYING THE LAW TO THE FACTS...

A natural gas company began providing services in about one-third of the states in the United States. The company provided natural gas services for the same prices as other natural gas providers. For unknown reasons, other natural gas providers began to go out of business. Finally, the natural gas company was dominating the natural gas markets in many states. Does this company meet the three-prong test for determining whether a company is a monopoly?

Overwhelming Power in the Market. Once the relevant markets have been determined, the alleged monopolist's market power to control prices and to exclude fringe competition is significant. The courts are interested in whether the defendant has overwhelming market power, not absolute power, because there are usually small competitors that produce poor substitutes. The courts ask: Do the pricing and output of the alleged monopolist control the conduct of the few competitors in the industry?

To answer that question, the courts have traditionally looked at five factors: market share, the size of other firms in the industry or market, the pricing structure of the market, entry barriers, and the unique nature of the industry. In the case of *Aspen Skiing Company v. Aspen Highlands Skiing Corp.*,[13] the U.S.

---

[13]388 U.S. 365 (1967).

*Newcal Industries, Inc. v. Ikon Office Solutions, United States Court of Appeals, Ninth Circuit, 513 F.3d 1038 (2008).

Supreme Court stated that in viewing market power, it would look not only at the market share that the alleged monopolist held but also at whether the power was acquired and maintained through predatory conduct that would be illegal or as a result of a superior product, business acumen, or historical accident. This approach is fair, because a corporation attaining monopoly power through innovation and research could easily be punished instead of rewarded if market share were the sole measure of overwhelming market power. A company would then have little incentive to compete to gain a market share in excess of 50 percent, because if it did gain that much (or perhaps even less), it would risk being charged as a monopolist under Section 2 of the Sherman Act.

Intent to Monopolize. After defining the relevant product and geographic markets and determining whether the company has overwhelming market power, the courts must decide if the company has a general intent to monopolize the market. This step is significant because having overwhelming power by virtue of being "big" in the market is not enough to make a firm liable for a Section 2 violation. The courts will look at specific conduct that tends to show intent, such as attempts to exclude competitors or to raise barriers to entry. In particular, they look at the foreseeable consequences of an alleged monopolist's actions: Would these actions naturally lead to a monopoly position? For example, the Aluminum Company of America (ALCOA) anticipated every demand increase and expanded its output in the aluminum industry. It was thus able to exclude competitors from the aluminum ingot market by lowering prices. These generally would be good business practices if ALCOA had not been judged to have overwhelming market power in the relevant product market. The courts draw a fine line between a monopoly gained by innovation, patents, and business acumen and one attained by conduct whose foreseeable consequence is the reinforcement of a monopoly position. The first position is gained in a passive manner; the second, in an active manner that shows intent to monopolize. Note carefully the court's analysis outlined in the case that follows.

The following case includes an allegation of a violation of Section 2 of the Sherman Act involving monopolization.

CASE 24-7

United States v. Microsoft Corporation
United States Court of Appeals for the District of Columbia Circuit
253 F.3d 34 (2001)

The authors recommend a close reading of the facts of *United States v. Microsoft* set out earlier in this chapter.

Section 2 of the Sherman Act makes it unlawful for a firm to "monopolize." The offense of monopolization has two elements: (1) the possession of monopoly power in the relevant market and (2) the willful acquisition or maintenance of that power as distinguished from growth or development as a consequence of a superior product, business acumen, or historic accident.

The district court found that Microsoft possessed monopoly power in the market for Intel-compatible PC operating systems. Focusing primarily on Microsoft's efforts to suppress Netscape Navigator's threat to its operating systems monopoly, the court also found that Microsoft maintained its power not through competition on the merits, but through unlawful means. Microsoft challenged both conclusions on appeal.

Per Curiam (by the Whole Court of Appeals)

We begin by considering whether Microsoft possesses monopoly power and finding that it does, we turn to the question [of] whether it maintained this power through anticompetitive means. Agreeing with the District Court that the company behaved anticompetitively and that these actions contributed to the maintenance of its monopoly power, we affirm the court's finding of liability for monopolization.

Monopoly Power

While merely possessing monopoly power is not itself an antitrust violation, it is a necessary element of a monopolization charge. The Supreme Court has defined monopoly power as the power to control prices or exclude competition. More precisely, a firm is a monopolist if it can profitably raise prices substantially above the competitive level[;] where [there is] evidence that a firm has in fact probably done so, the existence of monopoly power is clear. Because such direct proof is only rarely available, courts more typically examine market structure in search of circumstantial evidence of monopoly power. Under this structural approach monopoly power may be inferred from a firm's possession of a dominant share of a relevant market that is protected by entry barriers.

"Entry barriers" are factors (such as certain regulatory requirements) that prevent new rivals from timely responding to an increase in price above the competitive level.

The District Court considered these structural factors and concluded that Microsoft possesses monopoly power in a relevant market. Defining the market as Intel-compatible PC operating systems, the District Court found that Microsoft has a greater than 95 percent share. It also found the company's market position protected by a substantial entry barrier.

Microsoft argues that the District Court incorrectly defined the relevant market. It also claims that there is no barrier to entry in that market. Alternatively, Microsoft argues that because the software industry is uniquely dynamic, direct proof, rather than circumstantial evidence, more appropriately indicates whether it possesses monopoly power. Rejecting each argument, we uphold the District Court's finding of monopoly power in its entirety.

Microsoft's pattern of exclusionary conduct could only be rational if the firm knew that it possessed monopoly power. It is to that conduct that we now turn.

Provisions in Microsoft's agreements licensing Windows to [computer makers] reduce usage share of Netscape's browser and, hence, protect Microsoft's operating system monopoly.

Therefore, Microsoft's efforts to gain market share in one market (browsers) served to meet the threat to Microsoft's monopoly in another market (operating systems) by keeping rival browsers from gaining the critical mass of users necessary to attract developer attention away from Windows as the platform for software development.

We conclude that Microsoft's commingling of browser and nonbrowser code has an anticompetitive effect; the commingling deters computer makers from pre-installing rival browsers, thereby reducing the rivals' usage share and, hence, developers' interest in rivals.

By ensuring that the majority of all [ISP] subscribers are offered [Internet Explorer] either as the default browser or as the only browser, Microsoft's deals with the [ISP] clearly have a significant effect in preserving its monopoly.

Microsoft's exclusive deals with the [Internet software vendors] had a substantial effect in further foreclosing rival browsers from the market.*

> Judgment in favor of the United States (Plaintiff) affirming the U.S. District Court decision that Microsoft did possess and maintain monopoly power in the market for Intel-compatible operating systems. An appellate court *reversed* other holdings of the district court and *remanded* these matters for further proceedings.

COMMENT: The European Court of First Instance upheld a $600 million fine against Microsoft in September 2007. The fine had been levied by the European Commission (see Chapter 8 for the European Union court structure). A spokesperson for the U.S. Justice Department expressed regret at the European court's opinion and indicated that such a decision might limit innovation on the part of other multinational companies such as Microsoft. In April 2015, Google made the same threat involving innovation with the European Union's antitrust action hanging over its head. Is it real or imagined?

Attempt to Monopolize. Section 2 of the Sherman Act forbids not only monopolization but also attempts to monopolize, because the drafters of the section were concerned about the damage that efforts to attain a monopoly could inflict on an industry even if such efforts failed. So great was their concern, in fact, that the penalties are the same for both monopolization and attempts to monopolize. Case law indicates that after determining the relevant geographic and product markets, the courts look for one or some combination of three factors when a firm is charged with an attempt to monopolize: specific intent, predatory conduct, and a dangerous probability of success. We will discuss the first two; the third is self-explanatory.

---

*United States v. Microsoft Corporation, United States Court of Appeals for the District of Columbia Circuit, 253 F.3d 34 (2001). http://www.justice.gov/atr/case-document/opinion-1#f.

Specific intent is shown by bringing forth evidence that a firm has engaged in predatory or anticompetitive conduct aimed at a stated or potential competitor.

Predatory conduct includes (1) stealing trade secrets, (2) interfering unlawfully in requirement contracts that third parties have with other competitors, and (3) attempting to destroy the reputation of a competitor through defamatory actions. Recently, the courts have added **predatory pricing**—pricing below average variable cost (or, in some cases, below average total cost)—to this list, on the grounds that when a company is pricing below average variable cost, it is not seeking to maximize profits but is intending to drive a competitor out of business.

See *Matsushita Electric Industrial Co. v. Zenith Radio Corp.*,[14] in which the U.S. Supreme Court accepted the argument that predatory pricing (predation) at some point is an irrational strategy. Debate exists as to whether predatory pricing may exist at below average variable cost, below average total cost, above average total cost, or a quantitative rule that forbids increasing output by a monopolist when a new entrant comes into the market.

predatory pricing Pricing below the average variable cost to drive out competition.

The Clayton Act of 1914

The Clayton Act was enacted in 1914 after a major debate in the presidential campaign of 1912. The Supreme Court had ruled in 1911 that only restraints that were unreasonable by their nature or in their effect could be declared unlawful under the Sherman Act. This ruling left much room for interpretation by federal judges as well as by Justice Department prosecutors. Democratic candidate Woodrow Wilson argued during the presidential campaign that the Supreme Court was hostile to the antitrust laws and that businesspeople needed guidance as to what specific practices were illegal. He urged the creation of an agency to investigate trade practices and to advise businesspeople about what actions were and were not lawful. Upon election, Wilson proposed a bill that, after strenuous debate and a good deal of compromise in Congress, was enacted into law as the **Clayton Act** of 1914. It declared the following acts to be illegal under certain circumstances:

1. Price discrimination (Section 2)
2. Tying arrangements and exclusive-dealing contracts (Section 3)
3. Corporate mergers and acquisitions that tend to lessen competition or to create a monopoly (Section 7)
4. Interlocking directorates (Section 8)

Clayton Act Prohibits price discrimination, tying, exclusive-dealing arrangements, and corporate mergers that substantially lessen competition or tend to create a monopoly in interstate commerce.

At the same time, Congress passed the FTCA of 1914, setting up the FTC and giving it authority to police these and other "unfair or deceptive acts or practices affecting interstate commerce."

SECTION 2: PRICE DISCRIMINATION

Section 2 of the Clayton Act (as amended in 1936 by the Robinson–Patman Act) prohibits each of the business activities set out in Table 24-6. As you read the following paragraphs, pay attention to the italicized words, because they have been the source of litigation and acceptable defenses to the charges raised in that litigation.

Section 2(a) of the Clayton Act prohibits discrimination in price by seller between two purchasers of a commodity of like grade and quality, in interstate commerce, and resulting in injury to competition. Each of these elements must be proved by a plaintiff in any action brought under Section 2(a). The following discussion dissects these elements one by one.

---

[14] 475 U.S. 574 (1986).

TABLE 24-6

SUMMARY OF PROVISIONS OF THE CLAYTON ACT AS AMENDED BY THE ROBINSON–PATMAN ACT

Section	Action(s) Prohibited	Defense
2(a)	Discrimination in price by seller between two purchasers of a commodity of like grade and quality where effect may be to substantially lessen competition or tend to create a monopoly.	Cost justification or a good-faith attempt to meet equally low prices of competitors.
2(c)	Fictitious brokerage payments (or discounts where services not rendered).	None.
2(d)	Payments for promotions or allowances for promotional services by seller unless made available to all buyers on proportionately equal terms.	Meeting competition.
2(e)	Promotional services by seller unless provided to all buyers on proportionately equal terms.	Meeting competition for seller.
2(f)	Inducing to discriminate in price or knowingly receiving the benefits of such discrimination.	Cost justification.

price discrimination A price differential that is below the average variable cost for the seller; considered predatory, and therefore illegal, under the Clayton Act.

- *Price.* Section 2(a) forbids direct or indirect discrimination in price. **Price discrimination** is deemed by most courts and scholars to be a price differential that is below the average variable cost for the seller and, thus, is "predatory" and illegal. An example of indirect price discrimination is a seller giving a preferred buyer a 60-day option to purchase a product at the present price, while giving another purchaser only a 30-day option. The courts have ruled that this situation constitutes price discrimination under Section 2(a).
- *Sales.* There must be two actual sales (not leases or consignments) by a single seller that are close in time. Assume that seller A offers to sell to B a widget for $1.00 and then sells the widget to C for $0.95. If B charges price discrimination, that claim will not be upheld because there was no sale between A and B, but merely an offer to sell. A sale exists only when there is an enforceable contract.
- *Commodities.* Commodities are movable or tangible properties (e.g., milk or bicycle tires). Services and other intangibles are not covered by Section 2(a).
- *Like Grade and Quality.* The commodities must be of similar grade and quality; they need not be exactly the same. For example, price differences in milk cartons that are slightly different in size do fall under Section 2(a) jurisdiction. However, differences in price between car tires and bicycle tires are differences in prices of commodities of different grade and quality, so they do not fall under Section 2(a) jurisdiction.
- *Interstate Commerce.* The sales must occur in interstate commerce. If the two sales by a single seller to two purchasers take place in intrastate commerce, the Clayton Act, being a federal statute, does not apply.
- *Competitive Injury.* Finally, the plaintiff must show that the price discrimination caused competitive injury, which under the Clayton Act is price discrimination that substantially lessens competition, tends to create a monopoly, or injures, destroys, or prevents competition with the person or firm that knowingly receives the benefits of discrimination.

Injury to competition includes the following:

primary-line injury A form of price discrimination in which a seller attempts to put a local competitive firm out of business by lowering its prices only in the region where the local firm sells its products.

1. **Primary-line injury** (at the seller level) occurs when a seller cuts prices in one geographic area to drive out a local competitor.

2. **Secondary-line injury** (at the buyer level) occurs when competitors of one of the buyers are injured because the seller sold to that one buyer at a lower price than it sold to the others. The buyer that received the lower wholesale price can then undersell the other buyers, which may substantially lessen competition.

3. **Tertiary-line injury** (at the retailer level) occurs when a discriminatory price is passed along from a secondary-line buyer to a retailer. Retailers that get the benefit of a seller's lower price to a buyer will be able to undersell their competitors.

secondary-line injury A form of price discrimination in which a seller offers a discriminatory price to one buyer but not to another buyer.

tertiary-line injury A form of price discrimination in which a discriminatory price is passed along from a secondary-line party to a favored party at the next level of distribution.

The Meeting-the-Competition Defense. Section 2(b) of the Clayton Act allows a seller to discriminate in price if that seller is able to show that the lower price "was made in good faith to meet an equally low price of a competitor." The seller can discriminate to meet the competition but not to "bury" or "beat" the competition. The breadth of the meeting-the-competition defense has long been debated.

SECTION 3: TYING ARRANGEMENTS AND EXCLUSIVE-DEALING CONTRACTS

Section 3 of the Clayton Act reads:

> ... [I]t shall be unlawful for any person engaged in commerce, in the course of such commerce, to lease or make a sale or contract for sale of goods, wares, merchandise, machinery, supplies, or other commodities, whether patented or unpatented for use, consumption or resale within the United States or any territory thereof or the District of Columbia or any insular possession or other place under the jurisdiction of the United States, or fix a price charged therefore, or discount from or rebate upon, such price, on the condition, agreement or understanding that the lessee or purchaser thereof shall not use or deal in the goods, wares, merchandise, machinery, supplies, or other commodities of a competitor or competitors of the lessor or seller, where the effect of such lease, sale, or contract for sale or such condition, agreement or understanding may be to substantially lessen competition or tend to create a monopoly in any line of commerce.*

This is the section of the act on which courts have generally relied in cases concerning tying arrangements and exclusive-dealing contracts. Tying arrangements and exclusive-dealing contracts may not be per se illegal in a particular instance, despite past treatment of them as per se illegal in other instances by the courts.

SECTION 7: MERGERS AND ACQUISITIONS

Section 7 of the Clayton Act reads:

> [N]o corporation engaged in commerce shall acquire, directly or indirectly, the whole or any part of the stock or other share capital and no corporation subject to the jurisdiction of the Federal Trade Commission shall acquire the whole or any part of the assets of another corporation engaged also in commerce, where in any line of commerce in any section of the country, the effect of such acquisition may be substantially to lessen competition, or to tend to create a monopoly.†

The purpose of Section 7 of the Clayton Act, as amended in 1950, is to prohibit anticompetitive mergers and acquisitions that tend to lessen competition at their incipiency—that is, in the words of Justice Brennan, "to arrest apprehended

---

*Section 3 of Clayton act—15 U.S. Code § 14—Sale, etc., on agreement not to use goods of competitor.

†Section 7 of the Clayton act.

consequences of intercorporate relationships before those relationships [can] work their evil, which may be at or any time after the acquisition."[15]

The language of the statute has led to controversy and considerable litigation, especially because the business world went on a **merger** binge in the early 1980s. There were more than 2,000 mergers each year from 1983 through 1986, and some of this country's largest corporations were involved in the deal-making. For example, in 1984, Chevron purchased Gulf Oil for $13.2 billion, and Texaco bought Getty for $10.1 billion. The emphasis in 1983 and 1984 was on large oil company acquisitions, but 1985 and 1986 saw acquisitions by companies in the manufacturing, technology, and service areas of the economy as well. Although the 1980s is the decade associated with big deal-making, the merger frenzy continued into the 1990s and finally decreased in 2001. In 2014, more mergers and acquisitions took place than had occurred in years. The tally sent the M&A volume over the $3 trillion mark for the year. Stocks were soaring and debt was cheap for the acquiring companies.

merger One company's acquisition of another company's assets or stock in such a way that the second company is absorbed by the first.

Reasons for the Increase in Mergers. Mergers are a method of external growth as opposed to internal corporate expansion. They may take place for one or any combination of the following reasons:

1. *Undervalued assets.* It is cheaper for a company such as GM to buy Electronic Data Systems (EDS) and Hughes Aircraft to obtain computer capabilities, a computer transmission network, and telecommunications capabilities than to borrow money and expand internally in those areas. In the opinion of GM and its investment banking advisers, both EDS and Hughes Aircraft were undervalued stocks in the marketplace and, therefore, a "good buy."

2. *Diversification.* During a recession (e.g., 1981–1983 and 2008–2010), when stocks are generally underpriced, companies may seek to diversify—that is, to reduce their risks in one industry's business cycle by investing in another industry. U.S. Steel's acquisition of Marathon Oil Company was an attempt at diversification by a steel company hit hard by recession and foreign imports.

3. *Tax credits for research and development.* Between the middle of 1981 and the end of 1985, the Internal Revenue Code allowed a 25 percent tax credit for increases in research capabilities acquired through mergers.

4. *Economies of scale.* A merger often brings about greater efficiency and lower unit costs, particularly in research and development and in manufacturing.

5. *The philosophy that "bigness" is not "bad."* This flexible approach to mergers was embodied in the Justice Department's Merger Guidelines in the period 2001–2008.[16]

Criteria for Determining the Legality of Mergers under Section 7. The U.S. Supreme Court, the lower federal courts, the Justice Department, and the FTC use the following criteria, or steps, to decide on the legality of a merger:

1. Relevant product and geographic markets

2. Probable impact of the merger on competition in the relevant product and geographic markets

Relevant Product and Geographic Markets. The earlier discussion of monopolies said that how courts determine the boundaries of the product and geographic markets helps them decide what market share a company has and,

---

[15] *United States v. E.I. du Pont de Nemours & Co.*, 353 U.S. 586 (1957). https://supreme.justia.com/cases/federal/us/353/586/case.html.

[16] D. Mattioli and D. Cimilluca, "Stock Surge Fuels Deal Boom," *Wall Street Journal*, November 18, 2014, A-1, B-1; D. Gelles, "Mega-Mergers Popular Again on Wall Street," *New York Times*, November 18, 2014, A-1, B-4.

hence, its market power. The same holds true when the courts are ruling on mergers: The market share of the new, combined company will have a strong bearing on the court's decision as to the legality of the merger.

The primary criterion the courts use in determining the relevant product market is, again, substitutability or cross-elasticity of demand for a product. Other factors used are (1) public recognition of the product market, (2) distinct customer prices, (3) the product's sensitivity to price changes, (4) whether unique facilities are necessary for production, and (5) peculiar product characteristics.

When identifying the geographic market, what the courts are interested in is where the merging companies compete. The courts may decide that this geographic market includes all cities with a population of more than 10,000, or they may judge this market to be regional, national, or international.

Probable Impact on Competition. The courts have traditionally gauged a merger's impact on competition by examining factors such as:

1. Market foreclosure, resulting from the merger of a customer and its supplier, so that competing customers may be foreclosed from the market if the supplier's goods are in demand and that demand exceeds supply.
2. Potential elimination of competition from a market if two competing firms merge.
3. Entrenchment of a smaller firm in a market if a large firm with "deep pockets" acquires it and supplies the capital the small firm needs to eliminate competitors.
4. Trends in the market revealing a high rate of concentration, as measured by percentage of the market that the leading four to six competitors in an industry have.
5. Postmerger evidence revealing anticompetitive effects on a market.

Types of Mergers. The courts have distinguished between horizontal, vertical, and conglomerate mergers because each type has a potentially different impact on competition. **Horizontal mergers** involve the acquisition of one firm by another that is at the same competitive level in the distribution system. This type of merger usually leads to the elimination of a competitor. For example, in 1984, Chevron's purchase of the Gulf Oil Corporation eliminated one oil company at Chevron's level in the industry.

Vertical mergers involve the acquisition of one firm by another that is at a different level in the distribution system. For example, if a shoe manufacturer acquires a company that has many retail shoe outlets, the merger is termed vertical because one company is at the manufacturing level and the other is at the retailing level of the distribution system.

Conglomerate mergers involve the acquisition by one firm of another that produces products or services that are not directly related to those of the acquiring firm. For example, the acquisition by GM (an automobile company) of EDS (a technology company) merged two companies that did not produce directly related products and services.

Horizontal Mergers. In the 1960s and early 1970s, whenever a merger would result in what was labeled undue concentration in a particular market, there was a presumption of illegality. In a landmark case,[17] the Supreme Court termed a postacquisition market share of 30 percent or more prima facie illegal. In another equally important case involving the merger of two retail grocery store chains,[18] the Court, perceiving a trend toward fewer competitors in the retail-store market,

horizontal merger A merger between two or more companies producing the same or a similar product and competing for sales in the same geographic market.

vertical merger A merger that integrates two firms that have a supplier–customer relationship.

conglomerate merger A merger in which the businesses of the acquiring and the acquired firm are totally unrelated.

---
[17] *United States v. Philadelphia National Bank*, 374 U.S. 321 (1963).
[18] *United States v. Vons Grocery*, 384 U.S. 270 (1966).

held a postacquisition share of 8.9 percent presumptively illegal. In both cases, the Court's initial determination of the relevant product and geographic markets and the percentage of market shares the merged company would have become determinative of the result.

Vertical Mergers. Vertical mergers are termed *backward* when a retailer attempts to acquire a supplier and *forward* when a supplier attempts to acquire a retailer. In vertical merger cases, unlike horizontal merger cases, the courts have tended not to put great emphasis on market share percentage. Instead, they have generally examined the potential for foreclosing competition in the relevant market. For example, if a retailer acquires a supplier of widgets, will other widget suppliers be foreclosed from selling to the retailer? What impact will that foreclosure have on the widget market? The courts also look at the trend in the supplier's market toward concentration, barriers to entry, and the financial health of the acquired firm.

Conglomerate Mergers. As with horizontal and vertical mergers, the courts, using a case-by-case approach, have developed criteria they use in conglomerate merger situations to determine whether Section 7 of the Clayton Act has been violated. Because conglomerate mergers result in the combining of firms in different fields that are not competing with each other, the courts have found for the plaintiffs when it can be shown that the acquiring firm was already planning to move into the field and did not move into it only because it had "acquired" its way in; in effect, the conglomerate merger had prevented a company that was a potential entrant from entering and increasing the number of competitors.

APPLYING THE LAW TO THE FACTS...

TK Electrics was a powerful battery company with locations in almost every state in America. It heard of a smaller battery company that had a couple locations and no website. TK Electrics decided to take over the smaller company. What type of merger is this? Why?

Defenses to Section 7 Complaints. In cases brought under Section 7 of the Clayton Act, defendants have met complaints by private plaintiffs, as well as those filed by the Justice Department and the FTC, by asserting the following defenses:

1. *The merger does not have a substantial effect on interstate commerce.* For the Clayton Act to be applicable, the merger must be shown to have a substantial effect on interstate commerce, for the federal government may act—and a federal statute may be applied—only if interstate activity, as opposed to intrastate activity, is involved. As noted in Chapter 5, however, activities involving interstate commerce have been broadly interpreted under the Commerce Clause of the Constitution by the federal courts.

2. *The merger does not have the probability of substantially lessening competition or tending to create a monopoly.* Since the 1980s, firms have argued that mergers are procompetitive and beneficial to the economy and the nation because they improve economic efficiency and enable U.S.-based companies to compete with state-subsidized and state-owned foreign multinationals.

3. *One of the companies to the merger is failing.* This defense must meet three criteria: (a) The failing company had little hope of survival without the merger; (b) the acquiring company is the only one interested in purchasing the failing company, or if there are several interested purchasers, it is the least threat to competition in the relevant market; and (c) all possible methods of saving the failing company have been tried and have been unsuccessful.

4. *The merger is solely for investment purposes.* Section 7 does not apply to a corporation's purchase of stock in another company "solely for investment purposes," so long as the acquiring corporation does not use its stock purchase for "voting or otherwise to bring about, or attempting to bring about, the substantial lessening of competition." The courts look on this defense skeptically, especially when purchases of a company's stock by another company exceed 5 percent of the shares outstanding.

The Stichting Defense

(*Teva v. Mylan*)[a]

"Stichting" is a hostile takeover defense under Dutch law that is applied by firms seeking to ward off a merger. It allows a firm to place its shares in a trust and gives trustees the right to act if there is a need to block a takeover. In the case of Teva Pharmaceuticals' attempt to take over Mylan, Teva needed to own 4.6 percent of Mylan's shares to bring suit under Dutch law in the Enterprise Chamber of the Amsterdam Court of Appeals. This court has handled this type of litigation for many years.

---

[a]See Steven Solomon, "Maneuvers and Dutch Defenses That May Complicate Mylan-Teva Takeover War," *New York Times*, June 23, 2015, 135.

Enforcement. The Justice Department, the FTC, and private individuals and corporations can all enforce Section 7. The Justice Department divides authority with the FTC on the basis of areas of historical interest as well as according to the expertise of the staff of each agency.

When the Clayton Act was enacted, it provided no criminal punishment for violators, but merely allowed the Justice Department to obtain injunctions to prevent further violations. Recall that Sections 1 and 2 of the Sherman Act do establish criminal sanctions and that Section 4 of the Clayton Act (see section titled "Private Enforcement") allows individuals to sue on their own behalf and to obtain triple damages, court costs, and attorney's fees if they can show injury based on violations of either the Sherman Act or the Clayton Act. The Clayton Act also allows individuals to obtain injunctions. Furthermore, if a business is found guilty of violating the Sherman Act in a suit brought by the Justice Department, this finding is prima facie evidence of a violation when a private party sues for treble damages under the Clayton Act. That is, the private party need not prove a violation of the antitrust statutes all over again, but merely introduces into evidence a copy of the court order that found the defendant guilty of a Sherman Act violation.

Premerger Notification. The Hart–Scott–Rodino Act of 1976, which amended Section 7 of the Clayton Act, introduced a **premerger notification requirement** into the area of mergers. If the acquiring company has sales of $100 million or more, if the acquired firm has sales of $10 million or more, and if either affects interstate commerce, both firms must file notice of the pending merger with the Justice Department and the FTC 30 days before the merger is finalized. This notice enables the department and the FTC to assess the probable competitive impact of the merger before it takes place.

premerger notification requirement The legislatively mandated requirement that certain types of firms notify the FTC and the Justice Department 30 days before finalizing a merger so that these agencies can investigate and challenge any mergers they find anticompetitive.

Remedies. When parties decide to go ahead with a merger despite being advised that an enforcement action will be brought, the Justice Department and the FTC have three basic civil remedies available: civil injunctions, cease-and-desist orders, and divestiture. The Antitrust Division of the Justice Department, however, tries to avoid using these remedies. Instead, it seeks compromise. Thus, at times, it has succeeded in getting the acquiring firm to agree to a divestiture of some subsidiaries of the postmerger firm. At other times, it has prevailed on the acquiring firm to agree that the postmerger firm will refrain from some form of business conduct—for example, that it will not compete in certain geographic areas for a period of years.

Individuals and corporations may also bring private civil actions for triple damages against a firm that violates Section 7 of the Clayton Act. These private actions, which far outnumber government antitrust cases, are important for preserving a competitive business environment.

SECTION 8: INTERLOCKING DIRECTORATES

Section 8 prohibits an individual from becoming a director in two or more corporations if any of them has capital, surplus, and individual profits aggregating more than $21,327,000 or competitive sales of $2,132,000 (in 2005; the amount is to be adjusted each year by the FTC) when engaged in interstate commerce, if any of them were or are competitors, or where agreements to eliminate competition between such corporations would be a violation of the antitrust law.

With the growing number of conglomerates and the rise of the "professional" director who sits on many companies' boards for a fee, this long-dormant section of the Clayton Act has been the basis of some private civil litigation in recent years. The trend toward diversification by many large firms has resulted in overlapping areas of competition in many corporations, so there are potential violations of Section 8 for outside directors of these firms.

It should be noted that Section 8 excludes from its coverage banks, banking associations, and trust companies. Directors of corporations in these industries, therefore, do not have to be concerned about a potential Section 8 violation.

Other Antitrust Statutes

FEDERAL TRADE COMMISSION ACT OF 1914

The outburst of reform that produced the Clayton Act also produced the FTCA, which prohibits "unfair methods of competition." This broad, sweeping language and the courts' interpretation of it allow the FTC to bring antitrust enforcement actions against business conduct prohibited by the Sherman and Clayton Acts. When prosecution may be difficult because of the level of proof required under those acts, the FTC may bring a civil action under the FTCA. Also, business conduct that may not quite reach the level of prohibition under either the Sherman or the Clayton Act may be actionable under the "unfair" competition language of the FTCA.

The following case illustrates the reach of this statute.

CASE 24-8

California Dental Association v. Federal Trade Commission
United States Supreme Court
526 U.S. 756 (1999)

The California Dental Association (CDA) (defendant) is a nonprofit association of local dentists' organizations to which about 75 percent of California's dentists belong. The CDA provides insurance arrangements and other benefits for its members and engages in lobbying, litigation, marketing, and public relations on its members' behalf.

The CDA's members agree to abide by the association's Code of Ethics, which, among other things, prohibits false or misleading advertising. The CDA has issued interpretive advisory opinions and guidelines relating to advertising. These guidelines included restrictions on two types of truthful, nondeceptive advertising: price advertising, particularly discounted fees, and advertising relating to the quality of dental services.

The FTC filed a complaint with an ALJ, alleging that the CDA violated Section 5 of the FTCA in applying its guidelines to restrict price and quality advertising. The ALJ found that the CDA violated Section 5. On appeal, the FTC upheld this finding, as did the U.S. Court of Appeals for the 9th Circuit.

Justice Souter

Even on the view that bars [prohibitions] on truthful and verifiable price and quality advertising are prima facie anticompetitive and place the burden of procompetitive justification on those who agree to adopt them, the very issue at the threshold of this case is whether professional price and quality advertising is sufficiently verifiable in theory and in fact to fall within such a general rule. . . . [I]t seems to us that the CDA's advertising restrictions might plausibly be thought to have a net procompetitive effect, or possibly no effect at all on competition. The restrictions on . . . advertising are, at least on their face, designed to avoid false or deceptive advertising in a market characterized by striking disparities between the information available to the professional and the patient. In a market for professional services, in which advertising is relatively rare and the comparability of service packages not easily established, the difficulty for customers or potential competitors to get and verify information about the price and availability of services magnifies the dangers to competition associated with misleading advertising. What is more, the quality of professional services tends to resist either calibration or monitoring by individual patients or clients, partly because of the specialized knowledge required to evaluate the services, and partly because of the difficulty in determining whether, and the degree to which, an outcome is attributable to the quality of services (like a poor job of tooth-filling) or to something else (like a very tough walnut). Patients' attachments to particular professionals, the rationality of which is difficult to assess, complicate the picture even further. The existence of such significant challenges to informed decision making by the customer for professional services immediately suggests that advertising restrictions arguably protecting patients from misleading or irrelevant advertising call for more than cursory treatment as obviously comparable to classic horizontal agreements to limit output or price competition.*

Judgment for the Defendants. *Vacated* and *remanded* to Court of Appeals.

CRITICAL THINKING ABOUT THE LAW

One of the central ideas of critical thinking is that the quality of anyone's conclusion, including that of Justice Souter, is closely related to the quality of the evidence used to support that particular conclusion. Although Justice Souter is not making the final decision in Case 24-8, what he did certainly supports the CDA's efforts to continue its current restrictions on its members. Evaluating evidence is often made easier by asking what kinds of evidence would have been ideal.

1. To form his argument, Justice Souter makes a number of assertions. Name two claims he made in Case 24-8 for which additional evidence would enhance your confidence in their accuracy.

    Clue: Find places where Justice Souter makes factual claims but offers no evidence beyond his statement to support their validity.

2. To demonstrate to yourself that evidence is necessary for you to believe someone, locate two assertions made by Justice Souter in Case 24-8 that have no supporting evidence. Write two statements that say the opposite of what Justice Souter said. Do your statements assist the plaintiff in this case?

    Clue: If your statements were true, would the CDA's restrictions on advertising be more likely to harm consumers than if Justice Souter's statements were true?

BANK MERGER ACT

The Bank Merger Act of 1966 requires that all bank mergers be approved in advance by the banking agency having jurisdiction—that is, the Federal Reserve Board, the Federal Deposit Insurance Corporation (FDIC), or the Comptroller of

---

*California Dental Association v. Federal Trade Commission, United States Supreme Court, 526 U.S. 756 (1999).

the Currency. Before making a decision, the agency with jurisdiction must obtain a report "on the competitive factors involved" from the U.S. attorney general and from the other two agencies.

Even if the agency approves the merger, the Justice Department may bring a suit within 30 days. This action automatically stays the merger, and a federal district court must then review all issues concerning the merger de novo (newly or from the beginning). If not challenged by the U.S. attorney general within 30 days, a bank merger is still subject to liability under Section 2 of the Sherman Act if it is shown to have resulted in a monopoly. Acquisitions by bank holding companies are subject to the same antitrust standards that are applied to other industries.

The Bank Merger Act has become more significant in light of a 1985 decision of the U.S. Supreme Court approving regional banking and acquisitions by banks across state lines, when state legislatures have given prior approval. Furthermore, major bank, insurance, and brokerage companies (e.g., Travelers and Citicorp) have merged, and large banks have merged with each other (e.g., Bank One and National Bank of Chicago).

Global Dimensions of Antitrust Statutes

TRANSNATIONAL REACH OF U.S. ANTITRUST LEGISLATION

Sections 1 and 2 of the Sherman Act explicitly apply to "trade or commerce . . . with foreign nations," so they obviously have transnational reach. In contrast, Sections 2 and 3 of the Clayton Act, because they apply to price discrimination, tying arrangements, and exclusive-dealing contracts for commodities sold for "use, consumption, or resale within the United States," have no transnational reach. Section 5 of the FTCA, however, is given express transnational reach by Section 4 of the Export Trading Company Act (Webb–Pomerene Export Act), which extends the meaning of "unfair methods of competition" to practices in export trade against other competitors engaged in such trade even though acts constituting unfair methods of competition "are done without the territorial jurisdiction of the United States." The courts, using a case-by-case approach, have interpreted the language of these statutes so as to establish principles of law that guide companies and their management in determining whether certain activities are illegal because of their transnational impact.

In 1994, Congress passed the Antitrust Enforcement Assistance Act of 1994, which gave the Department of Justice authority to negotiate "mutual assistance" agreements with foreign antitrust enforcers. Also in 1994, the Justice Department and the FTC issued guidelines, based on current statutory and case law, that tell foreign and U.S. companies when either of the agencies is likely to act against alleged anticompetitive action in international trade. The following kinds of behavior may be investigated under the guidelines:

1. A merger of foreign companies that has significant sales in the United States
2. Conduct by foreign companies that has a direct, substantial, and reasonably foreseeable effect on commerce within the United States or on U.S. companies' export business
3. Anticompetitive schemes by importers that have a significant impact on the United States
4. Anticompetitive actions by foreign firms selling to the U.S. government

CASE 24-9

Carrier Corp. v. Outokumpu Oyj
United States Court of Appeals, Sixth Circuit
673 F.3d 430 (2012)

Carrier Corporation is a U.S. firm that manufactures air-conditioning and refrigeration (ACR) equipment. To make these products, Carrier used ACR copper tubing bought from Outokumpu Oyj, a Finnish company. Carrier is one of the world's largest purchasers of ACR copper tubing. The Commission of the European Committees (EC) found that Outokumpu had conspired with other companies to fix ACR tubing prices in the United States by agreeing that only Outokumpu would sell ACR tubing in the U.S. market. The District Court dismissed Carrier's claim for lack of jurisdiction. Carrier appealed.

Judge Nelson

Carrier's complaint describes, in some detail, an elaborate worldwide conspiracy in which the U.S market for ACR copper tubing was assigned to Outokumpu. Furthermore, Carrier alleges that this conspiracy caused the price of goods purchased within the United States to increase, which in turn caused a direct antitrust injury. In support of these allegations, the complaint references numerous specific dates during which the cartel met and the various agreements its members entered into. Assuming that these allegations are true, as we must, we conclude that Carrier has met any applicable requirement that it alleges a [substantial] effect on U.S. commerce.

Outokumpu, which attached the full EC decision to its motion to dismiss, counters that many of the details contained in the complaint are drawn from [an] EC decision that found no evidence that the cartel's focus extended beyond Europe. As a consequence, Outokumpu argues that any details regarding specific meetings and agreements occurring during the [cartel] meetings are of no assistance to Carrier because they related only to a European conspiracy.

We are [not] persuaded by this argument. The EC decision clearly states that "insofar as the activities of the cartel relate to sales in countries that are not members of the Community they lie outside the scope of this Decision." Thus, any silence on the part of the EC decision as to the U.S. markets may simply reflect the limited scope of the decision.

Furthermore, Carrier offers additional circumstantial allegations that corroborate its claim that the market-allocation scheme extended to the United States. Although Carrier's complaint provides numerous circumstantial allegations, of particular interest is its claim that [Outokumpu's competitors] initially refrained from aggressively competing for Carrier's U.S. business until 2003, and they suddenly began doing so at that time. It is true that the mere fact that competitors do not intrude upon one another's markets does not necessarily mean that an illegal market-allocation scheme is taking place. When two companies refrain from entering a market and then suddenly do so after a cartel dissolves, however, there are good grounds for suspicion.

Defendants question why it is that KME and Wieland waited until two years after the Cuproclima conspiracy dissolved to enter the United States if all that was stopping them was a supposed allocation conspiracy. It is certainly plausible, however, that these two companies would require time to reconfigure their operations so as to enable themselves to enter a new market. This analysis demonstrates that, contrary to Outokumpu's argument—and the district court's conclusion—Carrier's complaint does not fall within the rare exception which permits district courts to dismiss complaints that are "wholly insubstantial and frivolous."*

Decision of the District Court is *reversed* and *remanded*.

GLOBAL DIMENSIONS OF U.S. ANTITRUST LAWS

The general principle guiding the courts in the application of the Sherman Act (and other U.S. antitrust laws) has been that if the U.S. or foreign private companies enter into an agreement forbidden by Section 1 and that agreement affects the foreign commerce of the United States, then the U.S. courts have jurisdiction. The question then becomes: How much commerce must be affected before U.S. courts will assume jurisdiction? The Department of Justice's guidelines on

---

*Carrier Corp. v. Outokumpu OYJ, United States Court of Appeals Sixth Circuit, 673 F.3d 430 (2012).

its foreign antitrust enforcement policy announced a jurisdictional standard that requires business practices to have a "substantial and foreseeable effect on the foreign commerce of the United States." An example is an agreement by U.S. corporations selling roller bearings to divide up markets in Latin America. The department has stated that this country's antitrust laws will not be applied to certain business practices if they have no "direct or intended effect," and most commentators agree that trivial restraints affecting the foreign commerce of the United States are not likely to be prosecuted. (The "Assignment on the Internet" section at the end of this chapter encourages readers to deal with some of the issues raised here.)

U.S. appellate courts have held that U.S. courts do have jurisdiction over business conduct by a foreign corporation that is based on a decision by the corporation's government to replace a competitive economic model with a state-regulated model. State-regulated models encourage price-fixing and collaboration among competitors, especially when a government actually prohibits competition between firms. The courts of the United States do not evaluate the lawfulness of acts of foreign sovereigns performed within their own territories, even if the foreign commerce of the United States is affected by those acts, because the act-of-state doctrine forbids them to do so. Under this doctrine (discussed in Chapter 8), when the illegal conduct is that of a foreign government (as opposed to that of foreign individuals), the courts are not permitted to examine and decide the merits of any alleged claim. This approach applies to any case in which a statute, a decree, an order, or a resolution of a foreign government or governments is alleged to be unlawful under U.S. law. For instance, when the International Association of Machinists brought a suit claiming that an agreement by member states of the Organization of Petroleum Exporting Countries (OPEC) to increase the price of crude oil through taxes and price-setting was in violation of Sections 1 and 2 of the Sherman Act, the federal district court dismissed the case for lack of jurisdiction based on the act-of-state doctrine.

ENFORCEMENT

A court decision condemning certain business practices prohibited by U.S. antitrust laws—such as price-fixing, allocation of markets, or boycotts—may give a plaintiff satisfaction but no equitable relief. For example, if a U.S. corporation enters into a price-fixing agreement with a foreign corporation to determine the price of uranium worldwide, the foreign corporation may be made a defendant in a U.S. court and the plaintiff may win the case. Can the foreign defendant, however, be forced to pay triple damages? Usually not, unless it has assets in the United States that can be seized or there is a treaty of friendship and commerce between the United States and the foreign corporation's home country providing for the implementation of judicial decrees of U.S. courts in that country's courts. The second possibility is limited by the fact that very few treaties contain such terms. The first possibility is more promising because many foreign corporations have assets, such as bank accounts, in the United States. The plaintiff can get a decree freezing those assets until the corporation pays the court-ordered damages, because Section 6 of the Sherman Act provides for forfeiture of property to enforce an antitrust decree.

In the OPEC case referred to earlier, a group of foreign governments colluded to fix prices, but the crude oil was extracted, transported, and sold by U.S. oil corporations. Would it have been appropriate for the plaintiffs in that case to obtain a court decree ordering the seizure of the assets of the oil companies on the ground that they were coconspirators in the price-fixings? Would OPEC have cared about U.S. oil companies' assets? If it did care, would it have ceased selling oil to the United States? The answer to that question would probably depend on the supply and demand of oil on the world market.

COMPARATIVE LAW CORNER

Antitrust Laws in the United States and the European Union

In this chapter, we reviewed the major antitrust laws in the United States (Table 24-2). The European Union is governed by Articles 8 and 82 of the Treaty of Amsterdam and the Merger Control Regulation.

If you are an employee or an officer of a U.S. company doing business in any of the 28 EU countries, you may want to look at the comparative differences of antitrust laws. For example, you know that Section 1 of the Sherman Act prohibits "concerted" or conspiratorial practices (two or more) such as price-fixing, output restrictions, tying arrangements, and the like, as outlined in this chapter. In contrast, the EU through its Competition Directorate has issued a number of "blocks" or group exceptions that exempt whole categories of agreements. Even if there is not a block exemption, individual exemptions can be granted for specific agreements. Exemptions are binding on all national authorities and the courts of member nations. Case law and legislation have created exemptions to U.S. antitrust laws (Table 24-4), but they are more difficult to obtain.

Although civil penalties in the form of fines are permitted under EU regulations, both criminal and civil actions may be brought in the United States; no private enforcement of EU laws is possible. Thus, while doing business in the EU, it behooves the businessperson to know a great deal about EU law, most particularly block and individual exemptions of certain practices (e.g., price-fixing) that may be punishable in the United States civilly and/or criminally, but not in the EU.

Also, the standards by which a practice may be judged differ considerably. For example, judging of monopoly positions under Article 8-2 of the EU treaty may be stricter, because "the abuse" of the company's dominant position is broader than the concept of "monopolization" and "attempted monopolization" under Section 2 of the Sherman Act. It would be well to seek the advice of counsel if you were a Microsoft or Google company officer, for example, or an employee selling Microsoft or Google products in EU countries. In September 2007, the European Court of First Instance upheld a $600 million fine against Microsoft, levied initially by the European Commission, because of "abuse" of the company's "dominant position" in Intel-compatible PC operating systems markets. In the period from 2010 to 2012, the European Union led the way in dealing with firms such as Google, Apple, Intel, MasterCard, and others that allegedly violated the EU Treaty. Most of the cases settled with fines. In April 2013, Google agreed to a settlement with the EU. For the first time, a binding legal commitment to make minor changes to the look of its Web search engine were made to allay concerns that competitors might be hurt.[19] In April 2015, the EU's Bureau of Competition filed a set of formal charges (statement of objections) against Google. The company has 10 weeks to respond to the charges that it has abused its search engine's large market share (its "dominance") by favoring its own products over those of its rivals. The U.S. Federal Trade Commission had closed its inquiry into Google in 2013 without finding any wrongdoing. Also, in April 2015, EU regulators filed charges against Gazprom, the Russian energy giant, alleging that it is abusing its dominant position in the natural gas market of the EU. EU members are heavily reliant on Russia for its energy needs. Russia could interpret this action by EU regulators as part of sanctions imposed by Western powers following the Ukraine crisis. Also, the Kremlin earns large sums from the sale of gas supplied by Gazprom.[20]

In May 2015 the EU opened a separate investigation into whether large companies were impeding competition in online shopping. The investigation focused on how electronics, clothing, shoes, and online content are bought and sold online. The question of whether commercial companies have created artificial barriers that stopped citizens of EU countries (28) from buying goods across other parts of the continent. This seems to be directed at companies such as Amazon. It was the leader in European sales in 2015 with $15.4 billion, the latest figure available according to Internet Relations.

---

[19] Amir Efrati, "Google Makes EU Concessions," *Wall Street Journal*, April 15, 2013, B-3.
[20] Jessica Guynn, "EU Enforcer Means Business," *USA Today*, April 20, 2015, 5B; A. Kramer, "Weakened Gazprom Is Target," *New York Times*, Business Day, April 23, 2015, B-12.

SUMMARY

This chapter includes a history and summary of sections of the Sherman Act, Clayton Act, Federal Trade Commission Act, and Bank Merger Act. We have sought to examine the policy implications of these acts for business managers and consumers. We also examined the international dimensions of U.S. antitrust statutes in light of multinationals doing business in the United States, as well as the impact of the U.S. antitrust statutes on those doing business in other countries. With this transnational reach of antitrust legislation, enforcement by the U.S. federal courts in both types of circumstances has become more frequent.

REVIEW QUESTIONS

24-1 Who is responsible for enforcement of the Sherman Antitrust Act? Explain.

24-2 What industries are exempt from U.S. antitrust law? Explain.

24-3 Explain the difference between horizontal and vertical restraints under Section 1 of the Sherman Act.

24-4 Define a conglomerate merger. Explain.

24-5 Do other countries have antitrust laws? Explain.

REVIEW PROBLEMS

24-6 Gerber, Heinz, and Beech-Nut, manufacturers of baby food, account for essentially the entire market in the United States. A number of retail stores filed an antitrust class action against the companies, alleging that they had violated Section 1 of the Sherman Act by engaging in an unlawful conspiracy to fix prices. The class presented evidence that sales representatives employed by the three companies exchanged price information about their products, including, on occasion, sending each other advance notice of price increases. The class of store owners also introduced evidence of emails indicating that the companies were aware of the anticipated price increase before they were announced in the market. One memo stated that Heinz would not try to secure a majority base of distribution in a sales area because it had agreed to a "truce" with Gerber. Did the companies violate the antitrust laws? Would you need any additional evidence to make this determination? Explain your answer.

24-7 Dayton Superior Corp. sells its products in interstate commerce to several companies, including Spa Steel Products, Inc. The purchasers often compete directly with each other for customers. From 2005 to 2007, one of Spa Steel's customers purchased Dayton Superior's products from two of Spa Steel's competitors. According to the customer, Spa Steel's prices are always 10 to 15 percent higher for the same products. As a result, Spa Steel loses sales to at least that customer and perhaps others. Spa Steel wants to sue Dayton Superior for price discrimination. Which requirements for such a claim under Section 2 of the Clayton Act does Spa Steel satisfy? What additional facts will it need to prove? *Dayton Superior Corp. v. Spa Steel Products, Inc.*, 2012 WL 113663 (N.D.N.Y. 2012).

24-8 Falstaff Brewing Company was the fourth-largest brewer in the United States, with 5.9 percent of the national market. Falstaff acquired a local New England brewery, Narragansett, to penetrate the New England market. Narragansett had 20 percent of that market. The Justice Department filed suit against Falstaff under Section 7 of the Clayton Act. What was the result? Explain.

24-9 Dentsply International, Inc., is one of a dozen manufacturers of artificial teeth for dentures and other restorative devices. Dentsply sells its teeth to 23 dealers of dental products. The dealers supply the teeth to dental laboratories, which fabricate dentures for sale to dentists. Hundreds of other dealers compete with each other on the basis of price and service. Some manufacturers sell directly to the laboratories. There are also thousands of laboratories that compete with each other on the basis of price and service. Because of advances in dental medicine, however, artificial tooth manufacturing is marked by low growth potential, and Dentsply prohibits its dealers from marketing competitors' teeth unless they were selling teeth before 1993. The federal government filed a suit in a federal district court against Dentsply, alleging, among other things, a violation of Section 2 of the Sherman Act. What must the government show to succeed in its suit? Are those elements present in the case? What should the court rule?

24-10 Ford Motor acquired Autolite, an independent manufacturer of spark plugs that accounted for 15 percent of national sales of spark plugs. GM, through its AC brand, accounted for another 30 percent of national sales. After Ford's acquisition of Autolite, Champion was the only independent manufacturer of spark plugs remaining in the market. Its market share declined from 50 percent to 33 percent after the acquisition. The Justice Department sued Ford for violation of Section 7, and the federal district court ruled in favor of the United States. The court ordered Ford to do the following: (1) divest itself of Autolite, (2) stop manufacturing spark plugs for 10 years, and (3) purchase 50 percent of its requirements from Autolite for 5 years. Ford appealed. Who won and why?

CASE PROBLEMS

24-11 Gene Washington, Diron Talbert, and Sean Lumpkin, former professional football players, instituted a suit on behalf of themselves and a putative class of similarly situated players alleging that the National Football League (NFL) had monopolized the market for their likeness in violation of the Sherman Act. The players argued that because they were not allowed the rights to game footage and images from the games in which they had played, the NFL had committed an antitrust violation. The players alleged that they received little compensation from game footage in which they appeared, while the NFL reaped large sums. The game footage at issue was owned either by the NFL alone or by the NFL and a particular team. In other words, the individual teams did not exclusively own the footage of their own players. Thus, the teams had to "cooperate to produce and sell" the footage because "no one entity can do it alone." The district court held that the NFL and its teams could "conspire to market each [team's] individually owned property, but not property the teams and the NFL can only collectively own." The court therefore found no illegal concert action under the Sherman Act. Should the retired players be permitted to sell their likeness in the footage? Why or why not? *Washington v. National Football League*, 880 F. Supp. 2d 1004 (D. Minn. 2012).

24-12 When Deer Valley Resort Co. (DVRC) was developing its ski resort in the Wasatch Mountains near Park City, Utah, it sold parcels of land in the resort village to third parties. Each sales contract reserved the right of approval over the conduct of certain businesses on the property, including ski rentals. For 15 years, DVRC permitted Christy Sports, LLC, to rent skis in competition with DVRC's ski rental outlet. When DVRC opened a new mid-mountain ski rental outlet, it revoked Christy's permission to rent skis. This meant that most skiers who flew into Salt Lake City and shuttled to Deer Valley had few choices: They could carry their ski equipment with them on their flight, take a shuttle into Park City and look for cheaper ski rentals there, or rent from DVRC. Christy filed a suit in a federal district court against DVRC. Was DVRC's action an attempt to monopolize in violation of Section 2 of the Sherman Act? Why or Why not? *Christy Sports, LLC v. Deer Valley Resort Co.*, 555 F.3d 1188 (10th Cir. 2009).

24-13 Together, EMI, Sony BMG Music Entertainment, Universal Music Group Recordings, Inc., and Warner Music Group Corp. produced, licensed, and distributed 80 percent of the digital music sold in the United States. The companies formed MusicNet to sell music to online services that sold the songs to consumers. MusicNet required all of the same restrictions. Digitization of music became cheaper, but MusicNet did not change its prices. Did MusicNet violate the antitrust laws? Explain. *Starr v. Song BMG Music Entertainment*, 592 F.3d 314 (2d Cir. 2010).

24-14 MacDonald Group, Ltd., owned and operated the Fresno Fashion Fair Mall and leased space to Edmond's, a California retail jeweler. The lease contained a covenant that limited MacDonald to one additional jewelry store as a tenant in the mall. The lease was entered into in 1969. In 1978, MacDonald was involved in the construction of an expansion to the mall and began negotiations to include a retail jeweler in the new space. Edmond's objected and brought suit. The covenant provides that only two jewelry stores would be tenants in the Fresno Fashion Fair Mall. The expansion would still be part of Fresno Fashion Fair Mall and would not have a separate name. Would the covenant apply to the new addition to the mall? Explain. *Edmond's of Fresno v. MacDonald Group, Ltd.*, 217 Cal. Rptr. 375 (1985).

24-15 John Sheridan owned a Marathon gas station franchise. He sued Marathon Petroleum Co. under Section 1 of the Sherman Act and Section 3 of the Clayton Act, charging it with illegally tying the processing of credit card sales to the gas station. As a condition of obtaining a Marathon dealership, dealers had to agree to let the franchisor process credit cards. They could not shop around to see if credit card processing could be obtained at a lower price from another source. The district court dismissed the case for failure to state a claim. Sheridan appealed. Is there a tying arrangement? If so, does it violate the law? Explain. *Sheridan v. Marathon Petroleum Co.*, 530 F.3d 590 (7th Cir. 2009).

THINKING CRITICALLY ABOUT RELEVANT LEGAL ISSUES

Our antitrust laws are far too weak. If you take a look at the business world today, it is blatantly obvious that companies are gaining too much power.

For example, Google is being sued by the European Union because it no longer is concerned with competition in 26 countries. Other companies are powerful, too. Consider Nike, a company that can pay its workers only pennies per hour and then charge consumers hundreds of dollars for its products. Another example is General Electric. Besides being powerful in the electronics industry, General Electric has branched out into other industries as well. It owns NBC, a network that many Americans depend on for news. In the past, Americans could count on the news to give them an objective report of what's going on in the nation. Now, they hear only the stories that the corporate bigwigs want them to hear.

These corporations are so powerful and so impersonal that they'll just get up and leave if they think they can get more money elsewhere. New York City lost over a million jobs in the manufacturing industry between 1970 and 1984 because companies would rather put their factories in Third World nations than pay Americans a decent wage.

If society is to return to its status as the greatest in the world, Congress and the courts need to broaden the scope of antitrust laws. We need to make sure that any corporation powerful enough to control public opinion or to ignore its obligations to society is broken up.

1. What ethical values seem to support this author's conclusion? Explain.
2. What words or phrases are ambiguous in this essay? Explain.
3. Construct an essay giving an opinion that is opposite that of the author of this essay. (Chapter 1 might be of assistance in answering these questions.)

ASSIGNMENT ON THE INTERNET

This chapter introduces you to antitrust law in the United States and the global dimensions of these laws. A Supreme Court decision, *Hoffman-LaRoche Ltd. v. Empagran S.A.*, clarified the reach of the Foreign Trade Antitrust Improvement Act. Do an online search to find the case and then read it. What specific issue did the Court address? Do you agree with the Court's reasoning? Why or why not? (See the section in this chapter on "Global Dimensions of Antitrust Statutes.")

Finally, use the Internet to search for commentaries on the case. What do other legal scholars have to say about the Court's decision?

ON THE INTERNET

www.stolaf.edu/people/becker/antitrust The Antitrust Case Browser located at this address provides a collection of U.S. Supreme Court case summaries dealing with violations of antitrust statutes. This is a site where you can begin your search for legal resources related to antitrust law and policy.
www.law.cornell.edu/topics/antitrust.html The Legal Information Institute provides an overview to antitrust law as well as links to recent antitrust court decisions.
www.usdoj.gov/atr This is the Web address of the Department of Justice Antitrust Division.
www.antitrustinstitute.org The American Antitrust Institute home page contains news and information about antitrust enforcement.
www.ftc.gov/bc/index.shtml This is the Federal Trade Commission's antitrust website, which contains numerous links to current antitrust issues.

FOR FUTURE READING

Gellhorn, Ernest, William E. Kovacic, and Stephen Calkins. *Antitrust Law and Economics in a Nutshell.* St. Paul, MN: West, 2004.

Miller, Sandra K., Penelope Sue Greenberg, and Ralph H. Greenberg. "An Empirical Glimpse into Limited Liability Companies: Assessing the Need to Protect Minority Investors." *American Business Law Journal* 43 (2006): 609.

Sagers, Christopher. *Examples & Explanations: Antitrust.* New York: Wolters Kluwer, 2011.

CHAPTER TWENTY FIVE

Laws of Debtor–Creditor Relations and Consumer Protection

- DEBTOR–CREDITOR RELATIONS
- THE FEDERAL BANKRUPTCY CODE AND THE INCORPORATION OF THE BANKRUPTCY ABUSE PREVENTION AND CONSUMER PROTECTION ACT OF 2005
- THE EVOLUTION OF CONSUMER LAW
- FEDERAL REGULATION OF BUSINESS TRADE PRACTICES AND CONSUMER–BUSINESS RELATIONSHIPS
- FEDERAL LAWS REGULATING CONSUMER CREDIT AND BUSINESS DEBT-COLLECTION PRACTICES
- DODD-FRANK ACT AND CONSUMER PROTECTION
- STATE CONSUMER LEGISLATION
- GLOBAL DIMENSIONS OF CONSUMER PROTECTION LAWS

Chapters 9 and 10 covering contract law offered a view of private law and how it governs the relationship between two individuals or corporations that buy and sell items. When you went to the bookstore (or on the Internet) to buy this textbook, you entered into a legally binding contract. The bookstore made an offer, and you, as the buyer, accepted that offer. Your acceptance was signified by your picking out the book and paying for it at the cash register. Assuming that there was consideration, mutual assent, competent parties, and a legal object (and we are convinced that this text is a legal object), you entered into an enforceable contract. Before taking this course, you probably never thought of the act of buying a textbook as a legally binding transaction. Most consumers do not. They generally see it as an exchange of money for something they want or are required to buy.

Because consumers do not think of buying a product as a formal legal transaction, they are usually unaware of the legal implications of an exchange of money for a product or service until they have problems. The posttransaction business–consumer relationship then becomes the basis for angry exchanges, hurt feelings, and sometimes litigation. If both business managers and consumers had some knowledge of the requirements of contract law, as well as federal, state, and local statutes governing consumer transactions, there would be less friction between these two important parties in our economy (and fewer disputes over the types of product and service liabilities discussed in Chapter 12).

In this chapter, we describe debtor–creditor relationships; the Bankruptcy Act of 2005,[1] which amends the federal Bankruptcy Code; and the evolution of consumer law through legislation and case law. We then examine major federal legislation governing such trade practices as advertising, labeling, and the issuance of warranties on products. Federal laws pertaining to the credit arrangements entered into by consumers and the debt-collection practices of businesses are discussed, as are state laws governing consumer transactions. The chapter ends with an examination of the global dimensions of consumer protection laws.

Debtor–Creditor Relations

In this section, we briefly describe the rights of and remedies for creditors and debtors. Both the case law of federal and state courts and federal and state statutory law play a significant role. The U.S. economy has more and more become a credit-based economy. One can use a Visa or MasterCard to purchase everything from a home and automobile to clothes and a computer. When these transactions take place, a creditor and a debtor are created. We will define the **creditor** as the lender in the transaction (e.g., the bank that issues the credit card) and the **debtor** as the borrower (e.g., the business or individual that uses the credit card).

creditor The lender in a transaction.

debtor The borrower in a transaction.

CRITICAL THINKING ABOUT THE LAW

When a legal entity is unable to pay its debts, bankruptcy law provides various options for the entity or individual to resolve those debts. Bankruptcy remedies are available to individuals, partnerships, and corporations (both private and public). In 2009, more than 1.5 million bankruptcies were filed; more than 30,000 business firms filed for bankruptcy. The legal actions that result involve basic ethical norms, particularly those of fairness and compassion. Debtors, whether they are large firms or your neighbors, are in trouble. Should we help them?

Before we can answer that question, we should consider several issues. What are the costs of that help? Should people or firms be able to consume resources without paying for them? What kinds of incentive effects do we create if we allow debtors to escape their obligations?

Consider the following case example and, specifically, the effect of ethical norms in shaping your reaction to the situation.

Woodcock graduated from law school and finished his MBA in 1983. His student loans came due nine months later. Because he was a part-time student until 1990, he requested that payment be deferred. Because he was not in a degree program, payment should not have been deferred under the terms of the loan, but the lender incorrectly approved the deferral. Woodcock filed for bankruptcy in 1992, more than seven years after the loans first became due. Hence, that debt was discharged unless there was an "applicable suspension of the repayment period." The Dodd-Frank financial overhaul statute of 2010 created the Consumer Financial Protection Bureau, which regulates lenders that provide private student loans.[a] In 2015, the Obama administration was working on amendments to the bankruptcy act that would forgive students' loans under certain conditions.

1. What are the relevant ethical norms that affect your reaction to this case?

    Clue: Go back to Chapter 1 where the concept of ethical norms first appears. Which of the norms discussed there apply to the proper legal reaction to Woodcock's debt?

2. Our legal system is in many regards governed by principles of personal responsibility. When a person signs a contract or accepts credit, we ordinarily expect that person to fulfill the terms of what we see as that person's choice. Which ethical norms attach most closely to this theme of personal responsibility?

    Clue: For each ethical norm, ask yourself: Does this norm strengthen or weaken personal responsibility?

[a]*Woodcock v. Chemical Bank*, 144 F.3d 1340 (10th Cir. 1998).

[1] 11 U.S.C. § 101 (Pub. L. No. 109-8).

RIGHTS OF AND REMEDIES FOR CREDITORS

The following rights and remedies are most commonly used by creditors to enforce their rights. They include liens, garnishments, creditor's composition agreements, mortgage foreclosures, and debtor's assignment of assets for the credits.

Liens. A **lien** is a claim on the debtor's property that must be satisfied before any creditor can make a claim. There are both statutory liens (mechanic's liens) and common-law liens (artisan's and innkeeper's liens).

A **mechanic's lien** is placed on the real property of a debtor when the latter does not pay for the work done by the creditor. In effect, a debtor–creditor relationship is created, in which the real property becomes the security interest for the debt owed. For example, when a contractor adds a room onto the house of the debtor and payment is not made, the contractor becomes a lienholder on the property after a period of time (usually 60 to 120 days), and foreclosure may take place. Notice of foreclosure must be given to the debtor in advance.

An **artisan's lien** is created by common law that enables a creditor to recover payment from a debtor for labor and services provided on the latter's personal property. For example, Andrew leaves a lawn mower at Jake's repair shop. Jake repairs the lawn mower, but Andrew never picks up his personal property. After a period of time, Jake can attach the lawn mower.

Once a debt is due and the creditor brings legal action, the debtor's property may be seized by virtue of a judicial lien. Types of judicial liens include attachment, writ of execution, and garnishment.

Attachment involves a court-ordered judgment allowing a local officer of the court (e.g., a sheriff) to seize property of a debtor. On the motion of the creditor, this order of seizure may take place after all procedures have been followed according to state law. This is usually, but not always, a prejudgment remedy. If at trial the creditor prevails, the court will order the seized property to be sold to satisfy the judgment rendered.

If the debtor refuses to pay the creditor or cannot pay, usually a clerk of the court will direct the sheriff, in a **writ of execution**, to seize any of the debtor's (nonexempt) real or personal property within the court's jurisdiction. Any excess after the sale will be returned to the debtor.

A creditor may ask for a **garnishment** order of the court, usually directed at wages owed by an employer or a bank where the debtor has an account. This can be either a postjudgment or a prejudgment remedy, although the latter requires a hearing. Both federal and state laws limit the amount of money for which a debtor's take-home pay may be garnished.[2]

Mortgage Foreclosure. Creditors called *mortgage holders* (*mortgagees*) have a right to foreclose on real property when a debtor (*mortgagor*) defaults. There are statutes in each of the 50 states calling for the process of foreclosure. In general, a court-ordered sale of the property takes place when the debtor receives notice and cannot pay. After the costs of foreclosure and the mortgage debt have been satisfied, the mortgagor (debtor) may receive the surplus. It should be noted that mortgage foreclosures are an important source of income for those advising debtors, including lawyers and finance firms.

Suretyship and Guaranty Contracts. A contract of **suretyship** allows a third person to pay the debt of another (debtor) that is owed to a creditor in the event the debtor does not pay. The suretyship creates an express contract with the creditor, under which the surety is primarily liable. In Chapter 9, we discussed this matter with regard to third-party beneficiary contracts.

lien A claim on a debtor's property that must be satisfied before any creditor can make a claim.

mechanic's lien A lien placed on the real property of a debtor when the latter does not pay for the work done by the creditor.

artisan's lien A lien that enables a creditor to recover payment from a debtor for labor and services provided on the debtor's personal property (e.g., fixing a lawn mower).

attachment A court-ordered judgment allowing a local officer of the court to seize property of a debtor.

writ of execution An order by a clerk of the court directing the sheriff to seize any of the nonexempt real or personal property of a debtor who refuses to or cannot pay a creditor.

garnishment An order of the court granted to a creditor to seize wages or bank accounts of a debtor.

suretyship A contract between a third party and a creditor that allows the third party to pay the debt of the debtor; the surety is primarily liable.

---

[2]Consumer Credit Protection Act of 1968, 15 U.S.C. § 60 *et seq.*, allows a debtor to retain 75 percent of the debtor's disposable weekly income per week, or the sum equivalent to 30 hours paid at federal minimum wage, whichever is greater.

744 PART THREE • Public Law and the Legal Environment of Business

APPLYING THE LAW TO THE FACTS...

Lisa has a debt to Paula, and she cannot afford the debt. Lisa's mother contacts Paula and agrees to take over her daughter's debt and begins making payments. Due to unforeseen circumstances before she has finished paying off the debt, Lisa's mother can no longer afford the payments and Paula comes to collect the debt. Who is liable for the debt, Lisa or her mother?

guaranty Similar to a suretyship except that the third person is secondarily liable (i.e., required to pay only after the debtor has defaulted).

A **guaranty** contract is similar to a suretyship arrangement except that the third person is secondarily liable to the creditor. The guarantor is required to pay the debtor's obligation only after the debtor has defaulted and usually only after the creditor has made an attempt to collect. In Chapter 9, we discussed primary and secondary liability under the statute of frauds, which requires that a guaranty contract be in writing. The case of primary liability is the main exception to that requirement.

RIGHTS AND REMEDIES FOR DEBTORS

Debtors as well as creditors are protected by the law. For example, property is exempt from creditors' actions. Federal and state consumer protection statutes are discussed at length in this chapter. We also discuss bankruptcy laws as they apply to debtors and creditors.

Exemptions to Attachments. We have indicated that creditors can attach, or *levy* on, real and personal property. To protect debtors, however, certain exemptions are made. The best-known exemption for debtors is the *homestead exemption*. Historically, people have been allowed to retain a home up to a specified dollar amount or in the entirety. The purpose is to prevent a person from losing his or her home if forced into bankruptcy or faced with a claim from an unsecured creditor. Texas and California offer by far the most generous homestead exemption in bankruptcy proceedings. Many people move to one of these states when faced with bankruptcy.

Some personal properties are exempt, depending on state statutory law. Some examples include household furniture, a vehicle used to get to work, equipment used in a trade, and animals used on a farm.

Digging a Deeper Hole for Debtors

Between 2008 and 2010, in the midst of a deep recession, television advertisements were run (sometimes with the White House pictured in the background!), claiming that they would solve debtors' credit card and other financial problems—although there were fine-print disclaimers below the ad. The debt settlement industry is made up of companies that charge fees ranging from 15 to 20 percent of an individual's credit card balances. Such fees are usually collected up front. Worried about credit scores and possible bankruptcy, many of those seeing these ads are willing to pay whatever is necessary. Using figures from the debt settlement industry, only approximately one-third of applicants either completed a debt settlement program or were still involved with a company to save money in order to pay off their debts.

The Dodd-Frank Financial Regulatory Reform Act of 2010 (Dodd-Frank), previously discussed in Chapter 23, constrains the debt settlement industry and offers some exemptions, as follows:

- It directs the Federal Reserve Board (the Fed) to cap debit card fees at a level that is "reasonable and proportional to the cost of processing transactions." Fee caps apply only to debit cards issued by banks with more than

$10 billion in assets. This covers about 12 of the largest U.S. banks, which control about two-thirds of debit transactions.
- It directs the Fed to consider the cost of fraud involving debit cards.
- It excludes debit cards issued by banks on behalf of state and federal agencies to recipients of beneficiaries of programs such as unemployment insurance and child support.
- It exempts prepaid debit cards generally used by people who do not have bank accounts.

Even after the passage of Dodd-Frank, some questions remain: (1) What does "reasonable and proportional" to the cost of processing a transaction mean? Who will determine this? Explain. (2) What are the arguments in favor of debt settlement companies? Explain. (3) What are the arguments of those that would regulate debt settlement companies? Explain. (4) What sources would you consult to answer these questions? Explain.

The Federal Bankruptcy Code and the Incorporation of the Bankruptcy Abuse Prevention and Consumer Protection Act of 2005

HISTORY AND BACKGROUND

The Constitution of the United States provides another debtor right, namely, the right to petition for bankruptcy. Congress has the authority to establish "uniform laws" on the subject of bankruptcy throughout the United States (Article 1, Section 8). The U.S. Bankruptcy Code (Code) has several major goals: (1) to bring about the equitable distribution of the debtor's property among the debtor's creditors; (2) to discharge the debtor from its debts, enabling the debtor to rehabilitate itself and start fresh; (3) to preserve ongoing business relations; and (4) to stabilize commercial usage.

The Code was based in large part on a series of amendments such as the Bankruptcy Reform Act of 1978. In 2005, the most significant changes ever to be made to the Code took place with the enactment of the Bankruptcy Abuse Prevention and Consumer Protection Act of 2005 (the 2005 Act). This act was intended to overhaul, in large part, certain provisions of the Bankruptcy Code. The 2005 Act was an attempt to meet the complaints of the business community with regard to the increase in the number of filings for personal bankruptcy. From 1980 to 2000, these filings increased markedly (by 300,000 to 1.5 million per year), allegedly for debtors to evade the payment of debts.

Provisions. The U.S. Bankruptcy Code (Code) is composed of nine chapters. Chapters 1, 3, and 5 apply to the management and administration of the Code, particularly the substantive law set out in Chapters 7, 9, 11, 12, and 13 that will be examined here. *Straight* bankruptcy (Chapter 7) provides for the liquidation of the debtor's property. Other proceedings generally apply to the *reorganization* and *adjustment* of the debtor's debts (Chapters 11, 12, and 13). The 2005 Act added Chapter 15 to the Code for cross-border solvency cases. It incorporates the Model Law on Cross-Border Insolvency promulgated by the UN Commission on International Trade Law. Chapter 1 and some sections of Chapters 2 and 5 apply to the definition of terms and administration of the new Chapter 15, which seeks to make cross-border filings easier to accomplish. It also encourages cooperation between the United States and foreign countries with regard to transnational insolvency cases.

BANKRUPTCY MANAGEMENT AND PROCEEDINGS

Courts. The Bankruptcy Code, inclusive of all amendments, including those of the 2005 Act, grants to the U.S. district courts original and exclusive jurisdiction over all bankruptcy cases. Bankruptcy courts are attached to each of the 96 federal district courts across the nation to prevent the district courts from being overwhelmed by bankruptcy cases that are both numerous and complex. *Bankruptcy judges* are specialists who hear only bankruptcy proceedings. They are appointed for a period of 14 years by the U.S. court of appeals for the circuit in which the bankruptcy court is located. Among other things, the 2005 Act increased the number of bankruptcy judges to handle the burgeoning workload. The federal district courts may hear appeals from the bankruptcy courts. A Bankruptcy Appellate Panel service is available if all parties consent to such a hearing.

Federal law provides for a federal government official called a *United States Trustee* who supervises and handles many of the administrative details associated with these complex cases. For example, the trustee may appoint an interim trustee to take control of a debtor's estate before the appointment of a permanent trustee who is elected by the creditors as set forth in cases under Chapters 7, 11, 12, and 13.

Bankruptcy Petition. The filing of a bankruptcy petition initiates a case. Petitions usually are set forth in one of two forms:

- *Voluntary Petition.* Usually filed by a debtor under the following chapters: Chapter 7 (liquidation), Chapter 11 (reorganization), Chapter 12 (family farmer or fisherman), and Chapter 13 (adjustment-of-debts cases). The petition must clearly state the debts of the debtor.
- *Involuntary Petition.* A creditor(s) places the debtor in bankruptcy. An involuntary petition must set forth a statement that the debtor is not paying its debts as they become due. If the debtor has 12 or more creditors, the petition must be signed by a minimum of 3. If there are fewer than 12 creditors, any number can sign the petition. Those who sign the involuntary petition must have unsecured claims of at least $12,300 in the aggregate (this number and others will most likely increase over time).

Bankruptcy Schedules. An individual debtor must submit the schedules (lists and statements) noted here after it files a voluntary petition:

- A list of creditors with addresses
- A list of all property the debtor owns
- A statement of the debtor's financial affairs
- A statement of the debtor's monthly income
- A current income and expense statement
- A copy of the debtor's federal income tax returns for the most recent years before the filing of the petition

Usually, attorneys who specialize in bankruptcy law assist in preparing these schedules. The 2005 Act requires the attorneys involved, under penalty of perjury, to certify the information contained in the petition and the schedules. It is now necessary for attorneys to make a thorough investigation of the financial position of any debtor the attorney represents. This adds to the cost of filing for bankruptcy.

The bankruptcy court will file an *order for relief* unless the debtor challenges an involuntary petition. In that instance, a trial will be held to determine whether the order for relief should be granted.

Creditors' Meeting. Within a reasonable time after an order of relief is granted (10 to 30 days), the court will call a meeting of the creditors (*first meeting of creditors*). Without a judge present, the debtor must appear for questioning by the creditors. The debtor may be accompanied by his or her attorney. A *proof of claim* must be filed by each of the creditors in a timely manner.

Bankruptcy Trustee. Trustees are appointed in Chapter 7, 12, and 13 cases. In a Chapter 13 proceeding, if there is a showing of fraud, dishonesty, or incompetence, a trustee is appointed. He or she becomes a legal representative of the debtor's estate. Trustees generally are lawyers, accountants, or business professionals. They have wide powers that include but are not limited to the following:

- Taking immediate possession of the debtor's property
- Under the 2005 Act, protecting domestic support creditors (e.g., children of the debtor)
- Separating secured and unsecured debts
- Under the 2005 Act, filing a statement as to whether the case is presumed to be an abuse under the means test (Chapter 7 proceedings only)
- Setting aside exempt property
- Investigating the debtor's financial affairs
- Employing a professional to assist in administration of the case
- Distributing the proceeds of the estate

Automatic Stay. When a debtor petitions for bankruptcy, certain activities of the creditors are immediately suspended (*stayed*). A stay applies to collection activities of both secured and unsecured creditors. Many debtors file for bankruptcy before foreclosure to prevent the loss of crucial assets or to stave off litigation from creditors. Some *creditors'* actions are automatically stayed:

- Legal action to collect debts before petitioning for bankruptcy
- Enforcing judgments obtained against a debtor
- Enforcing liens against property of the debtor
- Nonjudicial collection efforts by a creditor (e.g., repossession of an automobile)

There are exceptions to stays:

- Creditors can recover prior domestic support (e.g., alimony and child support)
- Criminal actions against debtors are not stayed
- Certain securities and financial transactions are not stayed
- Action taken pursuant to certain government or police regulatory/powers are not stayed

Relief from a stay may be granted by a court on petition by a *secured* creditor when certain assets are depreciating and are not protected during the bankruptcy procedure.

Discharge. In bankruptcy cases filed under Chapters 7, 11, 12, and 13, if all the requirements are met, the court relieves the debtor of responsibility to pay its debts, and it grants the debtor a *discharge* of all or some of its debts.

Certain debts, however, are nondischargeable:

- Certain taxes and customs duties and debt incurred to pay such taxes or custom duties

- Legal liabilities resulting from obtaining money, property, or services by false pretenses, false representations, or actual fraud
- Legal liability for willful and malicious injuries to the person or property of another
- Domestic support obligations and property settlements arising from divorce or separation proceedings
- Student loans unless the debt would impose undue hardship
- Debts that were or could have been listed in a previous bankruptcy in which the debtor waived or was denied a discharge
- Consumer debts for luxury goods or services in excess of $500 per creditor, if incurred by an individual debtor on or within 90 days before the order for relief (these are presumed to be nondischargeable)
- Cash advances aggregating more than $750 obtained by an individual debtor under an open-ended credit plan within 70 days before the order for relief (these are presumed to be nondischargeable)
- Fines, penalties, or forfeitures owed to a governmental entity

Bankruptcy Estate. When a bankruptcy case is commenced, a separate legal entity is created, which is often referred to in the proceedings as an *estate*. This estate consists of all legal and equitable interests of the debtor in nonexempt property. The estate includes property of the debtor that the debtor acquires, within 180 days after the filing of the petition, by inheritance, by a property settlement, by divorce decree, or as a beneficiary of a life insurance policy. In addition, the estate includes proceeds, rents, and profits from the property. The 2005 Act excludes from the bankruptcy estate savings for postsecondary education through education individual retirement accounts (education IRAs) and 529 plans.

Estate Exemptions. The federal Bankruptcy Code establishes a list of estate property and assets that the debtor can claim as *exempt property*. Such exemptions are *adjusted* every three years to reflect the *consumer price index*. The following exemptions and dollar limits were set by the 2005 Act:

- Interest up to $22,975 in equity in property used as a residence and burial plots (called the *homestead exemption*)
- Interest up to $2,950 in value in one motor vehicle. New York State increased the value up to $4,000 or $10,000 for a disabled debtor in December 2010
- Interest up to $475 per item in household goods and furnishings, wearing apparel, appliances, books, animals, crops, or musical instruments, up to an aggregate value of $9,850 for all items
- Interest in jewelry up to $1,225
- Interest in any property the debtor chooses (including cash) up to $975, plus up to $9,250 of any unused portion of the homestead exemption. New York State has increased the value of property that people can retain when they declare bankruptcy or when creditors win judgments against them. The homestead exemption has been increased from $50,000 to $75,000, $125,000, or $250,000 depending on county of residence
- Interest up to $1,850 in value in implements, tools, or professional books used in the debtor's trade
- Professionally prescribed health aids
- Many government benefits regardless of value, including Social Security benefits, welfare benefits, unemployment compensation, veteran's benefits, disability benefits, and public assistance benefits

- Certain rights to receive income, including domestic support payments (e.g., alimony and child support), certain pension benefits, profit-sharing payments, and annuity payments
- Interests in wrongful death benefits and life insurance proceeds to the extent necessary to support the debtor or his or her dependents
- Personal injury awards up to $18,450

Retirement funds that are in a fund or an account that is exempt from taxation under the Internal Revenue Code shall not exceed $1 million for an individual unless the interests of justice require this amount to be increased. There is an exemption for IRAs.

State Exemption. The Code also provides for certain exemptions by states; that is, the states are permitted to enact their own legislation allowing exemptions. If they do so, they may (1) give debtors the option of choosing between federal and state exemptions or (2) require debtors to follow state law.

Fraudulent Transfers. The 2005 Act gave the bankruptcy courts the power to void certain fraudulent transfers of the debtor's property and obligations incurred by the debtor within two years of the filing of the petition for bankruptcy:

- Debtor's transfer of property for less than a reasonable equivalent consideration when he or she is insolvent
- Debtor's transfer of property with the intent to delay or defraud the creditor
- Debtor's transfer of assets to a living will; these can be voided if (1) the transfer was made within 10 years before the date of the filing of the petition for bankruptcy, (2) the transfer was made to a self-settled trust, and (3) the debtor is the beneficiary of the trust

CHAPTER 7

Under Chapter 7 (straight bankruptcy), a debtor turns over all assets to a trustee. The trustee sells the nonexempt assets and distributes the proceeds to creditors. The remaining debts are discharged. The nonexempt estate assets for a debtor were discussed previously. Before the 2005 Act, Chapter 7 bankruptcy proceedings became part of many individuals' financial planning, as debts were easily discharged and the debtor was freed to start over. The 2005 Act restricts debtors' ability to obtain a Chapter 7 bankruptcy. Now, a dollar-based *means test*, as well as a *medium income test* based on the debtor's state of residence, must be met before a debtor may discharge its debts. If these tests are not met, the 2005 Act provides that the debtor's Chapter 7 proceeding may, with the debtor's consent, be dismissed or converted to a Chapter 13 or Chapter 11 bankruptcy. The 2005 Act in effect pushed many debtors out of Chapter 7 and into Chapter 13 debt-adjustment bankruptcy. Debtors were forced to pay some of their future income over a period of five years to pay off debts owed before petitioning for bankruptcy.

The goals of the 2005 Act are to (1) deny Chapter 7 discharge to debtors who have the means to pay some of their unsecured debts from earnings following the filing of a petition for bankruptcy and (2) to steer most Chapter 7 bankruptcies to Chapter 13 instead.

Under Section 707(b) of the 2005 Act, the bankruptcy court may dismiss a Chapter 7 liquidation filing by an individual debtor whose debts are primarily consumer in nature if the court finds that granting relief to the debtor would be an abuse of Chapter 7 (the means and median income tests will be determinative). *Consumer debts* are defined as debts incurred for personal, family, or household use. Within 10 days after the first meeting of creditors, as previously discussed, the trustee must file a statement with the court as to whether the debtor filing is abusive.

Discharge of Debts. Discharge enables the debtor to say that it is no longer legally responsible for paying certain claims by creditors on unsecured debts. In a Chapter 7 proceeding, such a discharge of debts is important because it is granted soon after the petition is filed. Under the 2005 Act, the debtor can be granted Chapter 7 relief only after eight years following Chapter 7 or Chapter 11 bankruptcy and only after six years following Chapters 12 or 13 relief.

Discharge is not available to partnerships, limited liability companies, and corporations under the federal Bankruptcy Code. All of these must liquidate under state law before or upon completion of a Chapter 7 proceeding.

Acts Barring Discharge. A debtor will be denied discharge of his or her debts based on certain acts:

- False representation of his or her financial position when credit is extended
- Falsifying, destroying, or concealing financial records
- Failing to account for assets
- Failing to submit to questioning of creditors
- Failing to complete an instructional course on financial management as required by the 2005 Act when filing under Chapter 7

Student loans cannot be discharged under Chapter 7. *Student loans* are defined by the Code as those made, or guaranteed, by governmental units. The 2005 Act added student loans made by nongovernmental commercial institutions such as banks, as well as funds for scholarships, benefits, and stipends granted by educational institutions. The Code states that student loans can be discharged in bankruptcy only if a denial of discharge would cause "undue hardship" (defined as "severe mental or physical disability of debtor or inability to pay for necessities of him, her, or dependent"). In the following case excerpt, the court seeks to define *undue hardship*.

CASE 25-1

In re Savage v. United States Bankruptcy
Appellate Panel (First Circuit)
311 Bankr. 835 (2004)

Brenda Savage attended college in the mid-1980s—taking out five student loans—but she did not graduate. In 2003, at the age of 41, single, and in good health, she lived with her 15-year-old son in an apartment in Boston, Massachusetts. Her son attended Boston Trinity Academy, a private school. Savage worked 37.5 hours per week for Blue Cross/Blue Shield of Massachusetts. Her monthly gross wages were $3,079.79. Her employment provided health insurance, dental insurance, life insurance, a retirement savings plan, and paid vacations and personal days. She also received monthly child-support income of $180.60. After deductions, her total net monthly income was $2,030.72. Her monthly expenses included, among other things, $607 for rent, $221 for utilities, $76 for phone, $23.99 for an Internet connection, $430 for food, $75 for clothing, $12.50 for laundry and dry cleaning, $23 for medical expenses, $95.50 for transportation, $193.50 for charitable contributions, $43 for entertainment, $277.50 for her son's tuition, and $50 for his books. In February, Savage filed a petition for bankruptcy, seeking to discharge her student loan obligations to Educational Credit Management Corporation (ECMC). At the time, she owed $32,248.45. The court ordered a discharge of all but $3,120. ECMC appealed to the U.S. Bankruptcy Appellate Panel for the First Circuit.

Judge Haines

Under 11 U.S.C. Section 523(a)(8), debtors are not permitted to discharge educational loans unless excepting the loans from discharge will impose an undue hardship on the debtor and the debtor's dependents.

Under "totality of the circumstances" analysis, a debtor seeking discharge of student loans must prove by a preponderance of [the] evidence that (1) her past, present, and reasonably reliable future financial resources; (2) her and her dependents' reasonably necessary living expenses; and (3) other relevant facts or circumstances unique to the case prevent her from paying the student loans in question while still maintaining a minimal standard of living, even when aided by a discharge of other pre-petition debts.

The debtor must show not only that her current income is insufficient to pay her student loans, but also that her prospects for increasing her income in the future are too limited to afford her sufficient resources to repay the student loans and provide herself and her dependents with a minimal (but fair) standard of living.

Ms. Savage has not demonstrated that her current level of income and future prospects warrant discharge of her loans. Her present income may be insufficient to pay her student loans and still maintain precisely the standard of living she now has. But it would enable her to repay the loans without undue hardship. Moreover, the record plainly establishes that her prospects for a steady increase in income over time are promising. She has been steadily employed at the same job and regularly receives annual raises. Nothing indicates change is in the wind. Moreover, Ms. Savage currently works 37½ hours a week, leaving time for some part-time work (or longer hours at her present job).

To prove undue hardship for purposes of Section 523(a)(8), a debtor must show that her necessary and reasonable expenses leave her with too little to afford repayment.

Private school tuition is not *generally* considered a reasonably necessary expense in bankruptcy cases. Although compelling circumstances may distinguish a given case, the [courts] uniformly hold that a debtor's mere preference for private schooling is insufficient to qualify the attendant expense as necessary and reasonable.

Ms. Savage did not demonstrate a satisfactory reason why her son needs to attend private school at a monthly cost of $277.50 (plus $50 for books). When asked to explain why she did so, she testified:

> *There were a lot of fights, a lot of swearing, a lot of other things going on. I mean he would wake up every morning crying because he didn't want to go to school. So I had to find a school to put him in—where he was going to—I mean, he didn't do well that whole year. I had to keep going down to the school several times. He was just a mess the whole school year. So I had to find another school.*

Although we understand why Ms. Savage prefers that her son attend private school, she has not demonstrated that the public school system cannot adequately meet her son's educational needs. Her preference appears sincere, but that alone is not sufficient to sustain the bankruptcy court's implicit conclusion that forgoing this expense would constitute undue hardship.

Given the fact that at least $322.50 (private school tuition and books) in expense can be eliminated from Ms. Savage's budget without creating undue hardship, her student loans cannot be discharged under Section 523(a)(8). It is worth noting, as well, that Ms. Savage's son will reach majority in just a few years, a consequence that will reduce her required expenses considerably.*

Reversed the order of the bankruptcy court and *remanded* the case for judgment in ECMC's favor.

CRITICAL THINKING ABOUT THE LAW

The legal rule in a case is expressed in words, but in this case, as in any legal reasoning, words are slippery characters. They need a great deal of attention if we are to avoid being misled by them. It would have been quite possible, for instance, for Savage to have known the rule of law in this situation and to have honestly believed that she deserved to have the debt discharged.

1. What is the legal rule in this case?

    Clue: When it was time for the court to make its decision, what standard did it use to reach its conclusion?

2. What is the key ambiguous phrase in this decision? What alternative meanings does that phrase have?

    Clue: Can you imagine any reasonable person believing that private school in Savage's situation is a necessary expenditure for her family?

A bankruptcy court may revoke a discharge within one year after it is granted if it was obtained through fraud of the debtor, concealment, or destruction of property or through a proceeding in which the debtor may be guilty of a felony.

---

*In re Savage v. United States Bankruptcy Appellate Panel (First Circuit), 311 Bankr. 835 (2004).

DEBTORS–CREDITORS AND THE LEGAL ENVIRONMENT

Chapter 9: Bankrupt Cities

In 1934, Congress first created Chapter 9 (Pub. Law 25, 48 Stat. 798), declared unconstitutional by the U.S. Supreme Court in 1936 (298 U.S. 519) and revised in 1941 (Pub. Law 37). The largest filing ($18 billion to $20 billion) for a municipality took place with the city of Detroit, Michigan, in July 2013 (M. Davey and M. Walsh, *New York Times*, "Billions in Debt: Detroit Tumbles into Insolvency," pp. 1, 3, July 19, 2013).

A federal judge allowed Detroit to restructure its finances under bankruptcy protection, but indicated that the city would have to cut payments to its pension fund. This would set a precedent for other cities in the United States. Detroit could become one of the first municipalities to use bankruptcy protection to force bondholders to take less than the principal they are owed under Chapter 9. Other cities include Stockton, California; Jefferson County, Alabama; and Orange County, California. A total of 11 cities have historically filed for bankruptcy under Chapter 9. The Detroit case is being watched closely to see whether a city in debt to its pension fund is immune from cuts imposed by a bankruptcy judge. If so, more cities might file for bankruptcy. The effect this and other cases could have on the municipal bond market may be serious. Many individuals, pension funds, and cities may have to seek other ways to finance their debt.

In 2014, after several years of negotiation, the Detroit bankruptcy ended with companies among the city, its employees, retirees, and its creditors (100 in number), with the help of the bankruptcy judge Gerald Rosen and what became known as the "grand bargain," agreeing that the distinguished Detroit Institute of Arts (DIA) would be rescued from bankruptcy property by the country's largest foundation (Ford Foundation) and the State of Michigan through the pledge of $816 million. The funds raised helped to pay public workers' pensions over a 10-year period. The ownership of the museum has been transferred from the city of Detroit to an independent charitable trust.

Many people wondered whether this Detroit bankruptcy set a bad precedent for any poorly managed city. Others saw this as the only way out for a city whose population was 80 percent black and 38 percent below the poverty line. Only 12 states allowed a full city bankruptcy. Chapter 9 of Title II of the U.S. Bankruptcy Code is entitled "Adjustment of Debts of a Municipality." *Municipality* was defined in this case as a "political subdivision or public agency or instrumentality of the States." Approximately 18 large and small cities have filed for bankruptcy under Chapter 9. The possibility of lowering its bond ratings to "junk" due to action by Standard & Poor's usually follows. Chapter 9, Title II, Section 109(c) of the U.S. Bankruptcy Code.

Statutory Distribution of Property. If a debtor qualifies for Chapter 7 bankruptcy, the nonexempt property of the bankruptcy estate (previously examined) must be distributed to secured and unsecured creditors. Under the Code, priorities are established. The trustee in bankruptcy collects and distributes property in most voluntary bankruptcies.

CHAPTER 13

Under Chapter 13 of the Code (the wage earner's plan), a portion of the consumer-debtor's earnings is paid into the court for distribution to creditors over three years or, with court approval, over five years. Both wage earners and individuals engaged in business whose unsecured debts are not in excess of $307,675 and who have secured debts not in excess of $922,975 may qualify under Chapter 13. Only voluntary petitions for bankruptcy may be filed under this chapter. Creditors cannot force petitioners into Chapter 13 bankruptcy. Often, creditors agree to a *composition plan*, whereby each creditor receives a percentage of what the debtor owes in exchange for releasing the debtor from the debt.

The plan of payment set out under the 2005 Act may be up to three years or five years based on the following:

- If the debtor's or debtor's and spouse's monthly income multiplied by 12 is less than the state's median income for the year for the same size family up to four members plus $525 per month for each member in excess of four members, the plan period may not exceed three years, unless the court approves a period up to five years for cause.
- If the debtor's or debtor's and spouse's monthly income multiplied by 12 is equal to or more than the state's median income for the year for the same size family up to four members plus $525 per month for each member in excess of four members, the plan period may not be longer than five years.

The bankruptcy court can confirm a Chapter 13 plan if it (1) was proposed in good faith, (2) passes a feasibility test, and (3) is in the best interest of the creditors.

Some Advantages and Disadvantages of Bankruptcy

	Advantages	*Disadvantages*
Debtors	*Automatic Stay* • Instantly suspends most litigation and collection activities against the debtor, its property, or the bankruptcy estate. *Control* • Debtor retains possession of the bankruptcy estate (unless a trustee is appointed). • Chapter 11 permits the debtor to operate in the ordinary course of business.	*Administrative Costs* • Legal and accounting expenses. • Official creditors' committee fees. *Reduction in Autonomy* • Creditor oversight. • Management's ability to make and implement decisions rapidly and autonomously is curtailed. *Stigma of Bankruptcy* • Morale or confidence problems among staff, vendors, or customers. • Customer anxiety regarding future warranty claims or product support.
Creditors	*Enhanced Value and Participation* • Preserves going-concern value of an insolvent business. • Debtor-in-possession is more accountable due to bankruptcy reporting and notice requirements. *Equitable Distribution* • When inequitable conduct by any creditor (typically an insider) has prejudiced others, the bankruptcy court has the authority to subordinate all or part of the transgressor's claim to payment of other creditors. *Involuntary Petitions* • Creditors may file an involuntary petition for relief under Chapter 7 or (more rarely) Chapter 11 and force the debtor into bankruptcy.	*Suspension of Individual Remedies* • Automatic stay stalls foreclosure. • Nondebtor parties to executor contracts and unexpired leases are left in limbo. *Reduced Distribution* • Only a small fraction of Chapter 11 cases filed result in a successful reorganization. Continued operation results in less funds to distribute at liquidation.

CHAPTER 11

Chapter 11 of the Bankruptcy Code is generally aimed at financially troubled businesses, but individuals (with the exception of stockbrokers) are also eligible.

Its purpose is to allow a business to reorganize and continue to function while it is arranging for the discharge of its debts. Note the contrast to Chapter 7, which discharges debts by selling off all assets; you can see why Chapter 11 is more advantageous for individuals who qualify. Reorganization under Chapter 11 may be voluntary or involuntary. The court, after receiving the debtor's petition and ordering relief, appoints committees representing stockholders in the business as well as creditors. If these groups can agree to a fair and reasonable plan that satisfies their constituencies, the court will order its implementation. If some creditors or stockholders disagree on the plan, the court will still order it implemented if the judge finds it fair and reasonable under the circumstances.

In most cases, the debtor continues to operate the business during the reorganization process (as a *debtor-in-possession*, or DIP). He or she can enter into contracts, purchase supplies, incur debts, and carry on other activities of the business. The bankruptcy court may appoint a trustee at any time if there is a showing of fraud, dishonesty, or gross mismanagement by the debtor.

The debtor has the exclusive right to file a *plan of reorganization* with the bankruptcy court within 120 days after the order for relief; under the 2005 Act, this period is extended to 18 months from the date of the order of relief. The debtor has the right to obtain creditors' approval of a plan within 180 days from the date of the order of relief. If the debtor fails to file a plan, any *interested party* (e.g., a trustee, a creditor, or an equity holder) may do so within 20 months of the order. A plan of reorganization may be *confirmed* by the bankruptcy court. The conditions for confirmation usually include the following: (1) The plan is in the best interest of the creditors—that is, they receive at least what they would have received under a Chapter 7 liquidation proceeding; (2) the plan is feasible; and (3) each class of creditors accepts the plan. If a class of creditors does not accept the plan, a bankruptcy court, under the Code, may confirm the plan under a *cram-down provision*. At least one class of creditors must have voted to accept the plan if the court is to use this provision, and no creditors may be discriminated against.

Individuals Filing under Chapter 11 Reorganization. Under the 2005 Act, special rules are established when individuals, as opposed to firms, apply for reorganization. The 2005 Act states that the plan for payment to creditors must provide for the payment of a portion of the debtor's earnings from personal services that are earned after commencement of the case. These payments must be made by the debtor under the plan and be completed before the court will grant a discharge of any unpaid debts to the debtor.

CASE 25-2

RadLAX Gateway Hotel, LLC v. Amalgamated Bank
Supreme Court of the United States
132 S. Ct. 2065 (2012)

In 2007, RadLAX Gateway Hotel, LLC and RadLAX Gateway Deck, LLC (debtors) purchased the Radisson Hotel at Los Angeles International Airport, together with an adjacent lot on which the debtors planned to build a parking structure. To finance the purchase, the renovation of the hotel, and construction of the parking structure, the debtors obtained a $142 million loan from Longview Ultra Construction Loan Investment Fund, for which Amalgamated Bank (creditor or Bank) served as trustee. The lenders obtained a blanket lien on all of the debtors' assets to secure the loan.

Within two years, the debtors had run out of funds and were forced to stop construction. By August 2009, they owed more than $120 million on the loan, with over $1 million in interest accruing every month and no prospect for obtaining additional funds to complete the project. Both debtors filed voluntary petitions under Chapter 11 of the Bankruptcy Code.

Pursuant to Section 1129(b)(2)(A) of the Bankruptcy Code, the debtors sought to confirm a "cramdown" bankruptcy plan over the Bank's objection. That plan proposed selling substantially all of the debtors' property at an auction and using the sale proceeds to repay the Bank. Under the debtors' proposed auction procedures, however, the Bank would not be permitted to bid for the property using the debt it was owed to offset the purchase price, a practice known as "credit-bidding." Instead, the Bank would be forced to bid cash. The Bankruptcy Court denied the debtors' request, concluding that the auction procedures did not comply with the Bankruptcy Code's requirements for cramdown plans. The 7th Circuit affirmed, holding that Section 1129(b)(2)(A) does not permit debtors to sell an encumbered asset free and clear of a lien without permitting the lienholder to credit-bid.

Justice Scalia

Chapter 11 bankruptcy is implemented according to a "plan" typically proposed by the debtor, which divides claims against the debtor into separate "classes" and specifies the treatment each class will receive. Generally, a bankruptcy court may confirm a Chapter 11 plan only if each class of creditors affected by the plan consents. Section 1129(b) creates an exception to that general rule, permitting confirmation of nonconsensual plans—commonly known as "cramdown" plans—if "the plan does not discriminate unfairly, and is fair and equitable, with respect to each class of claims or interests that is impaired under, and has not accepted, the plan." Section 1129(b)(2)(A) establishes criteria for determining whether a cramdown plan is "fair and equitable" with respect to secured claims like the Bank's.

A Chapter 11 plan confirmed over the objection of a "class of secured claims" must meet one of three requirements in order to be deemed "fair and equitable" with respect to the nonconsenting creditor's claim. The plan must provide:

(i)(I) that the holders of such claims retain the liens securing such claims, whether the property subject to such liens is retained by the debtor or transferred to another entity, to the extent of the allowed amount of such claims; and (II) that such claim deferred cash payments totaling at least the allowed amount of such claim, of a value, as of the effective date of the plan, of at least the value of such holder's interest in the estate's interest in such property;

(ii) for the sale, subject to section 363(k) of the title, of any property that is subject to the liens securing such claims, free and clear of such liens, with such liens to attach to the proceeds of such sale, and the treatment of such liens on proceeds under clause (i) or (iii) of this subparagraph; or

(iii) for the realization by such holders of the indubitable equivalent of such claims. 11 U.S.C. § 1129(b)(2)(A).

Under clause (i), the secured creditor retains its lien on the property and receives deferred cash payments. Under clause (ii), the property is sold free and clear of the lien, "subject to section 363(k)," and the creditor receives a lien on the proceeds of the sale. Section 363(k), in turn, provides that "unless the court for cause orders otherwise the holder of such claim may bid at such sale, and, if the holder of such claim purchases such property, such holder may offset such claim purchases such property, such holder may offset such a claim against the purchase price of such property"—*i.e.*, the creditor may credit-bid at the sale, up to the amount of its claim. Finally, under clause (iii), the plan provides the secured creditor with the "indubitable equivalent" of its claim.

The debtors in this case have proposed to sell their property free and clear of the Bank's liens, and to repay the Bank using the sale proceeds—precisely, it would seem, the disposition contemplated by clause (ii). Recognizing this problem, the debtors instead seek plan confirmation pursuant to clause (iii), which—unlike clause (ii)—does not expressly foreclose the possibility of a sale without credit-bidding. According to the debtors, their plan can satisfy clause (iii) by ultimately providing the Bank with the "indubitable equivalent" of its secured claim, in the form of cash generated by the auction.

We find the debtors' reading of § 1129(b)(2)(A)—under which clause (iii) permits precisely what clause (ii) proscribes—to be hyperliteral and contrary to common sense. A well established canon of statutory interpretation succinctly captures the problem: "[I]t is a commonplace of statutory construction that the specific governs the general." That is particularly true where as in § 1129(b)(2)(A), "Congress has enacted a comprehensive scheme and has deliberately targeted specific problems with solutions."

Here, clause (ii) is a detailed provision that spells out the requirements for selling collateral free of liens, while clause (iii) is broadly worded provision that says nothing about such a sale. The general/specific canon explains that the "general language" of clause (iii), "although broad enough to include it, will not be held to apply to a matter specifically dealt with" in clause (ii).

The structure here suggests that (i) is the rule for plans under which the creditor's lien remains on the property, (ii) is the rule for plans under which the property is sold free and clear of the creditor's lien, and (iii) is a residual provision covering dispositions under all other plans—for example, one under which the creditor receives the property itself, the "indubitable equivalent" of its secured claim. Thus, debtors may not sell their property free of liens under § 1129(b)(2)(A) without allowing lienholders to credit-bid, as required by clause (ii).*

Judgment of Court of Appeals *affirmed*.

---

*RadlaxGatewat Hotel, LLC v. Amalgamated Bank Supreme Court of The United States 132 S. Ct. 2065, (2012).

Small Business Bankruptcy. A "test track" Chapter 11 exists for small businesses if their liabilities do not exceed $2 million and they do not own or manage real property. This allows a bankruptcy proceeding without the appointment of a creditors' committee, saving time and costs. A small business debtor has 180 days from the order of relief to file a reorganization plan. If such a debtor fails to do so, creditor(s) or other interested parties may do so within 300 days after the order of relief. The bankruptcy court must confirm the small business debtor's plan within 45 days after the plan is filed if it meets the requirements of Chapter 11.

CHAPTER 12

Under Chapter 12, family farmers and fishermen have a right to file for bankruptcy reorganization under the Code, as amended by the 2005 Act (adjustment of debts of a family farmer or fisherman with regular income). In the 2005 Act, "family farmers" and "fisherman" are defined as follows:

- *Family farmers* are individuals or spouses with a debt of less than $3,273,000 and at least 50 percent related to farming and whose gross income for the preceding petitioned year or second and third year was at least 50 percent earned from farming operations. Corporations or partnerships owned by a family or relatives with more than 80 percent of its assets related to farming operation and a total business debt not exceeding $3,273,000 (2005 figures that are adjusted yearly) are defined as *family partners*.
- *Family fisherman* has about the same definition of family and spouse as for farmers. The total debt must not exceed $1,500,000, and 80 percent of it must be related to a commercial fishing operation. The same percentages are used for a corporation or partnership as for farmers. The total debt must not exceed $1,500,000.

A debtor can convert a Chapter 12 reorganization to a Chapter 7 at any time. A family farmer or fisherman debtor, as defined here, must file a plan of reorganization. Usually, the plan should provide for payments to creditors over no longer than a three-year period. The plan must be confirmed by the bankruptcy court after a hearing at which creditors can appear and object to the plan. Priority is given to unsecured and secured creditors under Chapter 12 for distribution. As stated previously in this chapter, Section 5 of the Bankruptcy Code defines certain unsecured claims as priority claims (Table 25-1).

TABLE 25-1 TYPES OF BANKRUPTCY PROCEEDINGS

	Chapter 7	*Chapter 11*	*Chapter 12*	*Chapter 13*
Purpose	Liquidation	Reorganization	Adjustment	Adjustment
Eligible Debtors	Most debtors	Most debtors, including railroads	Family farmer who meets certain debt limitations	Individual with regular income who meets certain debt limitations
Type of Petition	Voluntary or involuntary	Voluntary or involuntary	Voluntary	Voluntary

DEBTORS AND CREDITORS (2014–2015) AND THE LEGAL ENVIRONMENT

Who Won with the Enactment of the Bankruptcy Abuse Prevention and Consumer Protection Act of 2005?

The Bankruptcy Reform Act of 1978 was the last major piece of legislation Congress enacted that amended the Federal Bankruptcy Code. Many claimed that it made bankruptcy too easy. To support such claims, they pointed to the increase in bankruptcy filings between 1978 and 2003, which peaked at 1,619,097. Business groups, credit card companies, and firms providing automobile loans claimed that the bankruptcy process was being *abused* (note the title of the 2005 Act).

In contrast, consumer groups opposed the proposed reforms. They claimed that "special interests" wrote the legislation as it moved through congressional committees, most particularly in the House–Senate Conference Committee. They also argued that it was the loose credit card policy of banks and credit card companies that drove people into bankruptcy. They further argued that medical costs were a large reason for bankruptcy filings. One study found that 46.4 percent of those filing for personal bankruptcy did so for medical reasons.

Under the 2005 Act, whenever a debtor has an annual income in excess of the mean income of his or her state of residence, the debtor may be forced into a Chapter 13 reorganization plan and thus make periodic payments over a period of five years. Previously, under Chapter 7, most debtors had few durable assets; thus, few Chapter 13 proceedings (20 percent or less) were filed as personal bankruptcies. Most creditors saw debtors walk away from their debts.

Critics of the 2005 Act complain that it is too costly to declare personal bankruptcy under a Chapter 13 repayment plan, which mandates that they pay a portion of their debts over a five-year period. Also, the paperwork increases the costs to all people who work on bankruptcies. For example (as noted previously), attorneys have to certify the accuracy of petitions and schedules or be subject to sanctions by the court. Private investigators may have to be employed. These costs are passed on to the debtor. Finally, there is a serious problem as to whether Chapter 13 reorganization plans are completed or whether debtors default on those plans.

Those who favored the 2005 Act argued that debtors would no longer look at bankruptcy as a planning tool. Critics argued that the act would prevent debtors from obtaining a financial "fresh start," which is a basic purpose of the bankruptcy laws.

Who wins? It might depend on whether you are a debtor or a creditor.

Chapter 12 Discharge. After all payments are made by a farmer or fisherman debtor (usually over a period of three years), the court will grant the debtor(s) discharge of all debts covered by the plan. For example, if the plan calls for 35 percent of the outstanding unsecured debt to be repaid to unsecured creditors and if it is in fact paid, the court will grant discharge of the unpaid 65 percent barring statutory or creditor's objections.

THE NEW BANKRUPTCY LAW—2011

The New Bankruptcy Law—2011 is technically an amendment rather that an all-out reform as the 2005 Bankruptcy Abuse Prevention and Consumer Protection Act (2005 Act) was. The president signed H.R. 6198: Bankruptcy Technical Corrections Act of 2010 (BTCA) to correct technical errors in the 2005 Act described by scholars and practitioners. The BTCA made numerous corrections ranging from incorrect referrals from one part of the code to another to inserting missing language and words and removing misplaced words.

The technical corrections often affected (1) the power of the court, (2) public access to papers, (3) debtor reporting requirements, (4) case administration, and (5) priorities of creditors' claims.

> **Proposed New Bankruptcy Plan**
>
> The Financial Institution Bankruptcy Act was introduced into Congress in 2014. It is a bankruptcy plan for big banks. Cosponsored by Republicans and Democrats, it created a new sector of the bankruptcy code. The 2010 Dodd-Frank law allows regulations to rescue such institutions. This approach will still be allowed despite Dodd-Frank.
>
> Under the proposed new law, a failing giant firm or the Federal Reserve Boards of Governors may trigger the bankruptcy filing if it's "necessary to prevent serious adverse effects on the financial stability of the United States." The affected company can contest the Federal Reserve's decision in a federal court of law.

The Evolution of Consumer Law

As Adam Smith's laissez-faire philosophy, with its revolt against government intervention in the economy, gained popularity in the eighteenth and nineteenth centuries, the **freedom-to-contract doctrine** evolved through case law out of our state court systems. The courts said that assuming that parties were legally competent, they should be allowed to enter into whatever contracts they wished. Neither the courts nor any other public authority should intervene except in cases of fraud, undue influence, duress, or some other illegality. Governed by this doctrine, the U.S. courts throughout the nineteenth century generally refused to interfere in contractual relations merely because one party was more economically powerful or better able to drive a hard bargain. In effect, they upheld the principle of *caveat emptor* (let the buyer beware).

Since the 1930s, state courts (and some federal courts) have curtailed the traditional freedom to contract by establishing rules of public policy and doctrines of unconscionability and fundamental breach that allow the courts to interfere in contractual relationships, especially when the seller is in the stronger economic position and a consumer has no other source to buy from. The doctrine of freedom to contract has also been limited by the implied warranty doctrine, as well as by the courts' relaxation of strict privity relationships between manufacturers and consumers. (These aspects of contract and product and service liability law were discussed in Chapters 9, 10, and 12.)

freedom-to-contract doctrine Parties who are legally competent are allowed to enter into whatever contracts they wish.

ECONOMICS

In previous chapters, we studied the role of business organizations (Chapters 16 and 17); here, we study the interaction between consumers and government (federal, state, and local). The role of government in this area has been that of an actor and a referee between consumers and the business community.

As an actor, the government at all levels consumes about 38 percent of the nation's total output and employs approximately 22 million individuals. State and local governments employ more than 19 million people and spend $3.3 trillion a year.[3]

In its role as referee, we have seen that rules (laws) that have been set out over a period of time may interfere in classical economic theory (marketplace economics). The laws of supply and demand may not function when competition is affected by consumer legislation such as the Fair Credit Reporting Act (FCRA), the Fair Credit Billing Act (FCBA), and the Equal Credit Opportunity Act

---

[3] Accessed October 10, 2013 at www.usgovernmentspending.com and www.governing.com/gov-data/public-workforce-salaries/monthly-government-employment-changes-totals.html.

(ECOA). These rules or laws may be based on political rather than economic foundations and decisions. Before these forms of legislation in the 1960s and 1970s, consumers were told that caveat emptor (consumer beware) was the golden rule. Today, though, the government's role as referee has become prominent. Nevertheless, debate continues among economists, political scientists, and legal scholars as to what role, if any, the government should play in the relationship between consumers and business organizations.[4]

Federal Regulation of Business Trade Practices and Consumer–Business Relationships

THE FEDERAL TRADE COMMISSION: FUNCTIONS, STRUCTURE, AND ENFORCEMENT POWERS

Functions. The Federal Trade Commission (FTC) has been discussed throughout this book. It was created by the Federal Trade Commission Act (FTCA) expressly to enforce Section 5 of that act, which forbids "unfair methods of competition." Section 5 was originally intended to be used to regulate anticompetitive business practices not reached by the Sherman Act. In 1938, it was amended by the Wheeler–Lea Act to prohibit "unfair or deceptive acts or practices." From then on, even if a business practice did not violate the Sherman or Clayton antitrust statutes, the FTC could use the broad "unfair or deceptive" language of Section 5 to protect consumers against misleading advertising and labeling of goods, as well as against other anticompetitive conduct by business. Thus, the FTC became the leading federal consumer protection agency.

Since 1938, Congress has passed several statutes delegating further administrative and enforcement authority to the FTC. Among them are the Fair Packaging and Labeling Act; the Lanham Act; the Magnuson-Moss Warranty–Federal Trade Improvement Act; the Telemarketing and Consumer Fraud and Abuse Prevention Act; important sections of the Consumer Credit Protection Act; the Bankruptcy Reform Act; the Hobby Protection Act; the Wool Products Labeling Act; the Hart–Scott–Rodino Antitrust Improvement Act; and the Food, Drug, and Cosmetics Act.

Structure and Enforcement Powers. In Chapter 18, we used the FTC in our example of the adjudicative process for federal administrative agencies, and we presented a diagram of the agency's structure in Exhibit 18-2. Before you read any further, you may want to turn back to that exhibit to get a quick picture of how the agency is organized.

Selected Consumer Protection Laws

Agency, Department, or Commission	Regulation of Consumer Credit Assets	Regulation of Fraud Issues	Regulation of Consumer Health and Safety Issues
Federal Trade Commission Established: 1914	Credit advertising; Fair Credit Reporting Act; Fair Debt Collection Practices Act; Credit Card Accountability, Responsibility and Disclosure Act of 2009 (effective February 2010); Dodd-Frank Wall Street Reform and Consumer Protection Act of 2010	Advertising; sales practices	

(Continued)

---
[4]R. Posner, *Economic Analysis of Law* (7th ed.) (New York: Aspen/LittleBrown, 2007), ch. 2.

Agency, Department, or Commission	Regulation of Consumer Credit Assets	Regulation of Fraud Issues	Regulation of Consumer Health and Safety Issues
Food and Drug Administration (U.S. Department of Health and Human Services) Established: 1930			Labeling of food (except meat, poultry, and eggs), drugs, and cosmetics; adulterated food and cosmetics; approval of drugs and medical devices
U.S. Department of Agriculture Established: 1862			Labeling of meat, poultry, and eggs; inspection of meat-, poultry-, and egg-processing facilities
Consumer Product Safety Commission Established: 1972			Consumer Product Safety Act (CPSA)
Federal Communications Commission Established: 1934		Telemarketing	Broadcast standards
U.S. Postal Service Established: 1775		Sales practices	
Securities and Exchange Commission Established: 1934		Securities fraud	
Federal Reserve Board Established: 1913	Truth in Lending Act (Regulation Z); Consumer Leasing Act (Regulation M); ECOA (Regulation B); Electronic Fund Transfer Act (Regulation E)		
U.S. Department of Labor Established: 1913			
Bankruptcy Courts	Chapter 13 consumer bankruptcy		

The commission is composed of a chairperson and four commissioners, who are nominated by the president and confirmed by the Senate. No more than three commissioners may come from the same political party.

The FTC's Bureau of Competition is responsible for the investigation of complaints of "unfair or deceptive practices" under Section 5 of the FTCA. If the bureau finds merit in the complaint and cannot get the offending party to voluntarily stop the deceptive or unfair practice or to enter into a consent order, it issues a formal complaint, which leads to a hearing before an administrative law judge.

The FTC also has other enforcement weapons at its disposal. It may assess fines, obtain injunctive orders, order corrective advertising, order rescissions of contracts and refunds to consumers, and obtain court orders forcing sellers to pay damages to consumers.

DECEPTIVE AND UNFAIR ADVERTISING

Deceptive Advertising. In passing the 1938 Wheeler–Lea Act amending Section 5 of the FTCA, Congress made it clear that it wished to give the

FTC the power "to cover every form of advertising deception over which it would be humanly practicable to exercise government control." Through their interpretation of the "unfair or deceptive" language of Section 5 over the years, the commission and the courts have evolved a three-part standard whereby the FTC staff must show that:

1. there is a misrepresentation or omission in the advertising likely to mislead consumers;

2. consumers are acting reasonably under the circumstances; and

3. the misrepresentation or omission is material. Neither intent to deceive nor reliance on the advertising need be shown.

CASE 25-3

Federal Trade Commission v. Verity International, Ltd.
United States Court of Appeals, Second Circuit
443 F.3d 48 (2006)

The incessant demand for pornography, some have said, is an engine of technological development. The telephonic system at dispute in this appeal is an example of that phenomenon—it was designed and implemented to ensure that consumers paid charges for accessing pornography and other adult entertainment. The system identified the user of an online adult-entertainment service by the telephone line used to access that service and then billed the telephone-line subscriber for the cost of that service as if it were a charge for an international phone call to Madagascar. This system had the advantage that the user's credit card never had to be processed, but it had a problem as well: It was possible for someone to access an adult-entertainment service over a telephone line without authorization from the telephone-line subscriber who understands himself or herself to be contractually bound to pay all telephone charges, including those that disguised fees for the adult entertainment.

The Federal Trade Commission brought suit to shut down such a telephone business as a deceptive and unfair trade practice within the meaning of Section 5(a)(1) of the Federal Trade Commission Act. The FTC sued Verity International, Ltd. (Verity) and Automatic Communications, Ltd. (ACL), corporations that operated this billing system, as well as Robert Green and Marilyn Shein, who controlled these corporations.

The court entered an order against the defendants-appellants for a total of $17.9 million. The defendants-appellants appealed to the U.S. Court of Appeals.

Justice Walker

[To prove a deceptive act or practice under Section 5(a)(1) of the FTC Act, the FTC must show that an act or practice was false or misleading, that consumers acted reasonably under the circumstances, and that the practice or representation complained of was material.]

The FTC contends that the first element is satisfied by proof that the defendants-appellants caused telephone-line subscribers to receive explicit and implicit representations that they could not successfully avoid paying charges for adult entertainment that had been accessed over their phone lines—what we call a "representation of incontestability." The defendants-appellants caused charges for adult entertainment to appear on phone bills as telephone calls, thereby capitalizing on the common and well-founded perception held by consumers that they must pay their telephone bills, irrespective of whether they made or authorized the calls. [This] conveyed a representation of incontestability.

[As for the second element, the] FTC contends that the representation of incontestability was false and therefore likely to mislead consumers who did not use or authorize others to use the adult entertainment in question; the defendants-appellants contend that the representation was rendered true by agency principles.

Under common law agency principles, a person is liable to pay for services that she does not herself contract for if another person has authority to consent on her behalf to pay for the services.

[But a] computer is not primarily understood as a payment mechanism, and in the ordinary habits of human behavior, one does not reasonably infer that because a person is authorized to use a computer, the subscriber to the telephone line connected to that computer has authorized the computer user to purchase online content.

The FTC proved the second element of its claim.

Finally [with respect to the third element,] telephone-line subscribers found the representation material to their decision whether to pay the billed charges because of the worry of telephone-line disconnection, the perception of the futility of challenging the charges, the desire to avoid credit-score injury, or some combination of these factors.

The district court measured the appropriate amount of [the judgment] as "the full amount lost by consumers." This was error. The appropriate measure is the benefit unjustly received by the defendants.

[Phone service providers] received some fraction of the money before any payments were made to the defendants-appellants.

[Also,] some fraction of consumers actually used or authorized others to use the services at issue.*

We *affirm* all components of the District Court's order of relief except for the monetary judgment. The case is *remanded* to the District Court consistent with this opinion.

puffery An exaggerated recommendation made in a sales talk to promote the product.

We look at three types of deceptive advertising in this section: (1) that involving prices; (2) that involving product quality and quantity; and (3) testimonials by well-known sports, entertainment, and business figures. *False price comparisons* are one form of deceptive price advertising. Another is offers of a "free" item to a customer who buys one at a "regular" price, when, in fact, the "regular" price covers the cost of the "free" good. The classic deceptive price advertising is the "bait-and-switch" tactic, which consists of advertising one product at a low price to entice customers into the store and then switching their attention to a higher-priced product.

The second type of advertising about *product quality* and *quantity* is often found to be deceptive under Section 5 of the FTCA. For example, when a car sales representative tells a customer, "This car is the best-running car that has ever been sold," is this mere puffery or is it deception? This kind of hype is usually considered an acceptable form of **puffery**. However, had the sales representative said, "This car will run at least 50,000 miles without a change of oil," the claim would cross the line into the territory of deception.

The FTC staff does not have to show that a claim is expressly deceptive; it is sufficient to show that deception is implicit. Also, the FTC must present evidence that there is no basis for the claim made by the advertiser. When an advertiser claims a certain quality in a product on the basis of what "studies show," it must possess reasonable substantiation of its claim. The commission will consider the cost of substantiation, the consequences of a false claim, the nature of the product, and what experts believe constitutes reasonable substantiation before deciding whether the advertising was deceptive.

Years ago, American Home Products, manufacturers of Anacin, advertised that its drug had a unique painkilling formula that was superior to the formulas of all other drugs containing analgesics. The FTC staff charged that this claim was deceptive because there was insufficient substantiation to show that Anacin was either unique or superior to other nonprescription drugs containing the same ingredients (aspirin and caffeine).[5] The full commission and the court of appeals agreed after examining the evidence presented by American Home Products. In a similar case in 1984, the FTC filed a complaint against General Nutrition (GN), charging it with deceptive advertising for claiming, purportedly on the basis of a National Academy of Science report, that consumption of its dietary supplement Healthy Greens was related to reduced rates of cancer in humans. GN entered into a consent order in which it agreed to cease such advertising.[6]

---

[5]*American Home Products v. FTC*, 695 F.2d 681 (3rd Cir. 1982).

[6]"General Nutrition Inc. Prohibited Trade Practices," 54: 9198 *Federal Register* (Mar. 6, 1989).

Federal Trade Commission v. Verity International, Ltd., United States Court of Appeals, Second Circuit, 443 F.3d 48 (2006).

CASE 25-4

Federal Trade Commission v. QT, Inc.
United States Court of Appeals, Seventh Circuit
512 F.3d 858 (2008)

QT, Inc., and assorted related companies, heavily promoted the Q-Ray Ionized Bracelet on television infomercials as well as on its website. In its promotions, the company made many claims about the pain-relief powers of these bracelets, including that the bracelet offered immediate, significant, or complete pain relief and could cure chronic pain. At trial in the U.S. District Court for the Northern District of Illinois, the presiding judge labeled all such claims as fraudulent; forbade further promotional claims; and ordered the company to pay $16 million, plus interest, into a fund to be distributed to all customers. QT, Inc., appealed.

Justice Easterbrook

According to the district court's findings, almost everything that defendants have said about the bracelet is false. Here are some highlights:

- Defendants promoted the bracelet as a miraculous cure for chronic pain, but it has no therapeutic effect.
- Defendants told consumers that claims of "immediate, significant or complete pain relief" had been "test proven," they hadn't.
- Defendants represented that the therapeutic effect wears off in a year or two, despite knowing that the bracelet's properties do not change. This assertion is designed to lead customers to buy new bracelets. Likewise the false statement that the bracelet has a "memory cycle specific to each individual wearer" so that only the bracelet's original wearer can experience pain relief is designed to increase sales by eliminating the second hand market and "explaining" the otherwise embarrassing fact that the buyer's friends and neighbors can't perceive any effect.

The magistrate judge did not commit a clear error, or abuse his discretion, in concluding that the defendants set out to bilk unsophisticated persons who found themselves in pain from arthritis and other chronic conditions.

Defendants maintain that the magistrate judge subjected their statements to an excessively rigorous standard of proof.

The Federal Trade Commission Act forbids false and misleading statements, and a statement that is plausible but has not been tested in the most reliable way cannot be condemned out of hand.

For the Q-Ray Ionized Bracelet, all statements about how the product works—Q-Ray, ionization, enhancing the flow of bio energy, and the like—are blather. Defendants might as well have said: "Beneficent creatures from the 17th Dimension use this bracelet as a beacon to locate people who need pain relief, and whisk them off to their homeworld every night to provide help in ways unknown to our science."

Proof is what separates an effect new to science from a swindle. Defendants themselves told customers that the bracelet's efficacy had been "test-proven," but defendants have no proof of the Q-Ray Ionized Bracelet's efficacy. The "tests" on which they relied were bunk. What remain are testimonials, which are not a form of proof. That's why the "testimonial" of someone who keeps elephants off the streets of a large city by snapping his fingers is the basis of a joke rather than proof of cause and effect.

Physicians know how to treat pain. Why pay $200 for a Q-Ray Ionized Bracelet when you can get relief from an aspirin tablet that costs 1¢?*

Affirmed, in favor of Plaintiff.

The U.S. Court of Appeals for the 7th Circuit affirmed the district court's decision. QT, Inc., was required to stop its deceptive advertising and to pay the $16 million, plus interest, so that its customers could be reimbursed.

CRITICAL THINKING ABOUT THE LAW

Based on the facts in this case, is there any reasonable basis on which a firm could sell this Ionized Bracelet and believe that it was not engaged in deceptive advertising? Reasons are tools, and in most situations, one can provide a reason for what seems like even the most bizarre behavior. This case provides you with an opportunity to think about the need to be very careful to examine the existence of reasons. The existence of a reason is only the first step in propounding an argument on which we should rely.

1. What logic might QT use to justify its claim that the district court used an overly rigorous standard of proof in its finding against QT?

---

*Federal Trade Commission v. QT, Inc., United States Court of Appeals, Seventh Circuit, 512 F.3d 858 (2008).

> *Clue:* Does deception require more than the possibility of potential deception before it rises to the level of illegal advertising? Might the claims QT was making be so outrageous that no reasonable consumer would have believed them?
>
> 2. What is the relevance of Judge Easterbrook's elephants-in-the-streets analogy to an assessment of QT's reasoning?
>
> *Clue:* Why do we even require claims to have reliable proof?

testimonial A statement by a public figure professing the merits of some product or service.

The third form of deceptive advertising discussed here is **testimonials** by public figures (e.g., athletes and entertainers) endorsing a product. The FTC guidelines require that such figures actually use the product they are touting and prefer it to competitive products. The FTC's Bureau of Consumer Protection also monitors claims by public figures that they have superior knowledge of a product. For example, singer Pat Boone represented Acne Stain as a cure for acne when there was no scientific basis for the claim; he also failed to disclose a financial interest he had in Acne Stain. After the FTC filed a complaint against him, Boone agreed to enter into a consent order.[7]

Unfair Advertising. Section 5 of the FTCA also forbids "unfair" advertising. The FTC guidelines consider advertising to be unfair if consumers cannot reasonably avoid injury, the injury is harmful in its net effect, or it causes substantiated harm to a consumer. In 1975, the commission promulgated a rule for public comment that forbade all television advertising related to children. This action came about after complaints that advertising by cereal and toy companies on Saturday morning television programs was addressed to a select age group that could not weigh the advertising rationally; thus, the advertising was "unfair." In 1980, Congress terminated the FTC rule-making proceedings dealing with children's advertising for political reasons and, for good measure, forbade the FTC to initiate rule-making proceedings of any type based on the concept of "unfairness." The commission may still challenge individual acts or practices as unfair under Section 5 of the FTCA in an adjudicatory context. In 2000, the FTC issued new guidelines to help online businesses comply with existing laws.[8]

Private-Party Suits and Deceptive Advertising. A company may sue a competitor under the Lanham Act of 1947 (see the discussion in Chapter 14 about trademarks), which forbids "false description or representation." Parties bringing actions based on violations of this act may request an injunction or corrective advertising. For example, McDonald's and Wendy's, in separate cases, accused Burger King of falsely portraying its hamburgers as superior to rivals' burgers on the basis of an alleged taste test. In their suits, McDonald's and Wendy's questioned the scientific basis of Burger King's survey and its analysis of the results. Both cases were settled out of court.[9]

The FTC and Deceptive Labeling and Packaging. Under the Fair Packaging and Labeling Act (FPLA), the Department of Health and Human Services (DHHS) promulgates rules governing the labeling and packaging of products. The DHHS has issued rules governing the packaging of foods, drugs, and cosmetics—all of which the FTC has enforcement jurisdiction over. The FPLA and the enacted rules require that such product packaging contain the name and address of

---

[7] *In re Cooga Mooga, Inc., & Charles E. Boorea*, 92 FTC 310 (1978).

[8] *Advertising and Marketing on the Internet: Rules of the Road* (Sept. 2000).

[9] *McDonald's v. Burger King*, 82/2005 (S.D. Fla. 19982); *Wendy's International Inc. v. Burger King*, C-2-82-1179 (S.D. Ohio 1982).

the manufacturer or distributor; the net quantity, which must be placed in a conspicuous location on the package front; and an accurate description of all contents. The purpose of the FPLA is to allow consumers to compare prices on the basis of some uniform measure of content. The Nutrition Labeling and Education Act of 1990 requires standard nutrition facts on all labels and regulates the use of such terms. The FTC enforces terms with Food and Drug Administration (FDA) approval.

In the following case the buyer of a new car complained that the car had failed to achieve the advertised fuel economy in the automaker brochure and listed on the car's label.

CASE 25-5

Paduano v. American Honda Motor Co.
Court of Appeals, Fourth District
88 Col. Rptr. 3d 90 (2009)

In 2004, Gaetano Paduano bought a new Honda Civic Hybrid in California. The information label on the car states that the fuel economy estimates from the Environmental Protection Agency (EPA) were 47 miles per gallon (mpg) for city driving and 48 mpg for highway driving. Honda's sales brochure added, "Just drive the Hybrid like you would a conventional car and save on fuel bills." Paduano soon became frustrated with the car's fuel economy, which was less than half the EPA's estimate. When American Honda Motor Company refused to repurchase the vehicle, Paduano filed a suit in a California state court against the automaker, alleging deceptive advertising in the federal Energy Policy and Conservation Act (EPCA), which prescribed the EPA's fuel economy estimate, and preempted Paduano's claims (see Chapter 5). The Court issued a summary judgment in Honda's favor. Paduano appealed to a state intermediate appellate court.

Judge Aaron

The basic rules of preemption are not in dispute: *Under the supremacy clause of the United States Constitution, Congress has the power to preempt state law concerning matters that lie within the authority of Congress. In determining whether federal law preempts state law, a court's task is to discern congressional intent.* Congress's express intent in this regard will be found when Congress explicitly states that it is preempting state authority.

Honda argues that [the EPCA] prevents Paduano from pursuing his claims. That provision states in pertinent part,

When a requirement under [the EPCA] is in effect, a State or a political subdivision of a State may adopt or enforce a law or regulation on disclosure of fuel economy or fuel operating costs for an automobile covered by [the EPCA] only if the law or regulation is identical to that requirement.

Honda goes on to assert that "Paduano's deceptive advertising and misrepresentation claims would impose *non* identical disclosure requirements."

Contrary to Honda's characterization, Paduano's claims are based on statements Honda made in its advertising brochure to the effect that one may drive a Civic Hybrid in the same manner as one would a conventional car, and need not do anything "special," in order to achieve the beneficial fuel economy of the EPA estimates. Paduano is challenging Honda's commentary in which it alludes to those estimates in a manner that may give consumers the misimpression that they will be able to achieve mileage close to the EPA estimates while driving a Honda Hybrid in the same manner as they would a conventional vehicle. Paduano does not seek to require Honda to provide "additional alleged facts" regarding the Civic Hybrid's fuel economy, as Honda suggests, but rather, seeks to prevent Honda from making misleading claims about how easy it is to achieve better fuel economy. Contrary to Honda's assertions, if Paduano were to prevail on his claims, Honda would not have to do anything differently with regard to its disclosure of the EPA mileage estimates.*

Summary judgment in favor of Honda is *reversed* and case *remanded*.

CONSUMER LEGISLATION

Franchising Relationships. Misrepresentation by franchisors of the potential profits to be made by franchisees is a violation of Section 5 of the FTCA, which prohibits "unfair" methods of competition as well as "unfair and deceptive

---

*Paduano v. American Honda Motor Co., Court of Appeals, 4th District 88. Col. Rptr.3d 90 (2009).

trade practices." To combat blatant fraud involved in franchising (discussed in Chapter 16), the FTC in 1979 promulgated rules governing franchise systems.

In its 1979 Franchising Rule, the commission defines a *franchise* as a commercial operation in which the franchisee pays a minimum fee of $500 to use the trademark of, or to sell goods and services supplied by, a franchisor that exercises significant control over, or promises significant aid to, the franchisee's business operation. The fee must be paid within six months after the business is begun. For example, McDonald's franchisees receive the trademark (the "golden arches") in return for an initial fee and a percentage of revenues that go to the McDonald's Corporation (the franchisor). McDonald's, in its franchising agreement, specifies the products that must be bought from McDonald's, the quality of food to be served, the store hours, cleanliness standards, and grounds for termination of the franchising agreement.

The FTC rule governing franchising requires each franchisor to provide a disclosure document to prospective franchisees that sets out such pertinent information as the names and addresses of the officers of the franchisor, any felony convictions of those officers, their involvement in any bankruptcy proceedings, all restrictions on a franchisee's territories or the customers it may sell to, and any training or financing the franchisor makes available to franchisees. If a franchisor suggests a potential level of sales, income, or profits, all materials that form the basis of those predictions must be made available to prospective franchisees and the FTC.

Violations of the FTC Franchising Rule may lead to a fine of up to $10,000. The FTC can bring a civil action for damages on behalf of franchisees in federal district court as well as administrative enforcement actions before an administrative law judge.

Consumer Warranties. The FTC is also in charge of enforcing the Magnuson-Moss Warranty Act–Federal Trade Improvement Act of 1975, which applies to manufacturers and sellers of consumer products that make an express written warranty. You will recall from the discussion of product and service liability law in Chapter 12 that an *express warranty* is a guarantee or promise by the seller or manufacturer that goods (products) meet certain standards of performance. Note that the act does not cover oral warranties, whether express or implied. *Consumer products*, as defined by the act, are goods that are normally purchased for personal, family, or household use. The courts have interpreted this definition liberally.

The purpose of the Magnuson-Moss Warranty Act is to prevent sellers and manufacturers from passing on confusing and misleading information to consumers. To that end, the act requires that all conditions of a warranty be clearly and conspicuously disclosed for any product sold in interstate commerce that costs more than $10. Furthermore, consumers must be told what to do if a product is defective.

Before this act was passed, a consumer purchasing a video recorder, for instance, could not be sure whether the "limited warranty" covered all labor and parts or only some labor and parts. Thus, after sending the video recorder back to the manufacturer or authorized dealer for repairs, the consumer might discover that the warranty covered a $50 part but not $150 in labor costs—a rather nasty surprise.

full warranty Under the Magnuson-Moss Warranty Act, a written protection for buyers that guarantees free repair of a defective product. If the product cannot be fixed, the consumer must be given a choice of a refund or a replacement free of charge.

Full or Limited Warranties. All written warranties of consumer products that cost more than $10 must be designated as "full" or "limited." A **full warranty** means that the manufacturer or dealer must fix a defective product; if it does not, the warranty is breached and the consumer has grounds for a breach-of-warranty suit. If efforts to fix the product fail, the consumer must be given a choice of refund or replacement free of charge. A manufacturer or supplier that gives only a limited warranty on its product can restrict the duration of implied warranties if the limit is designated conspicuously on the product. In effect, a **limited warranty** is any warranty that does not meet the conditions of a full warranty.

Second buyers and bailees, as well as bystanders, are covered by the Magnuson-Moss Warranty Act and pertinent FTC regulations. Also, manufacturers or sellers cannot limit the time period within which implied warranties of the product are effective. Finally, damages to the consumer cannot be limited unless the limits are expressly stated on the face of the product.

Remedies. The Magnuson-Moss Warranty Act gives an individual consumer or a class of consumers the right to bring a private action for a breach of a written (although not an oral) warranty. Consumers who bring such actions can recover the costs of the suit, including attorney's fees, if they win. Before they file suit, however, they must give the manufacturer or seller a reasonable opportunity to "cure" the breach of warranty by replacing or fixing the product.

Telemarketing Legislation. The Telephone Consumer Protection Act of 1991 restricts the activities of telemarketers and bans certain interstate telephone sales practices altogether. Some of the prohibited practices include:

- Calling a person's residence at any time other than between 8 a.m. and 8 p.m.
- Claiming an affiliation with a governmental agency at any level when such an affiliation does not exist
- Claiming an ability to improve a consumer's credit records or to obtain loans for a person regardless of that person's credit history
- Not telling the receiver of the call that it is a sales call
- Claiming an ability to recover goods or money lost by a consumer

Each violation of these regulations is punishable by a fine of up to $10,000; exempted are insurers, franchisers, online services, stocks and bonds salespeople regulated by the SEC, and not-for-profit organizations.

The FTC's regulations are an outgrowth of a federal statute, the Telemarketing and Consumer Fraud and Abuse Prevention Act of 1994, and the Telemarketing Sales Rule of 1995. Consumers can recover actual monetary loss, or $500 for each violation of the act. Treble damages are allowed. Congress passed this statute after holding hearings that revealed that some telemarketing firms ("budget shops") used imaginary sweepstakes and other schemes to bilk $40 billion a year from consumers (particularly the elderly) and small businesses. The Telemarketing Sales Rule fleshed out details of the 1994 legislation.

The statute and the FTC regulations were made enforceable in the federal courts by the 50 state attorneys general as well as by the FTC. This broad scope was intended to put an end to the situation in which telemarketers who engaged in fraud moved quickly from one state to another without being caught because state attorneys general did not have the authority to pursue them under federal law. Moreover, state law often did not provide for suitable punishment when such telemarketers were caught.

The "Do Not Call" Registry, which allows consumers to sign up with the FTC to eliminate some calls coming from "pitch men" or telemarketers, has had some 50 million registrants since it became effective in October 2003. Most states now have "baby" Do Not Call laws, which cover intrastate telemarketing calls.

Many exceptions are allowed under the Telephone Consumer Protection Act (e.g., calls from people conducting surveys or raising funds for charities or politicians). Calls are allowed from people with whom consumers have "established relationships" (e.g., firms to which consumers pay bills or obtain a delivery service). It is possible that some businesses will attempt to exploit this exemption by using questionnaires, raffle entries, or coupons to establish a "relationship."

limited warranty Under the Magnuson-Moss Warranty Act, any written warranty that does not meet the conditions of a full warranty.

Just as consumer dislike of junk telephone calls eventually led to congressional action, the advent of spam or junk email has elicited frustration and disgusted reactions from consumers who use the Internet. Yahoo! and other Internet service providers are also opposed to the overwhelming amount of spam they are forced by the nature of their business to carry, albeit inadvertently.

The FTC has brought a small number of enforcement actions under the Controlling the Assault of Non-Solicited Pornography and Marketing Act (CAN-SPAM Act) of 2003, which was intended to reduce the quantity of unsolicited emails. The law bars senders of commercial email from using fictitious identities and requires them to provide ways for recipients to remove themselves from mailing lists. The difficulty in these cases is based largely on gaining jurisdiction over elusive defendants.

According to some industry estimates, spam now accounts for 85 percent of all email traffic. The cost to U.S. business may be as high as $2,000 per employee in wasted productivity, according to one analysis by Nucleus Research, a technology research firm. Compliance with the CAN-SPAM Act has dwindled to 1 percent based on a study of a quarter million email messages by MX Logic, Inc., a software company that produces spam-blocking programs.[10]

At the state level, several states have enacted antispam statutes. See, for example, *State v. Heckel*,[11] in which the State of Washington's Supreme Court upheld an antispam statute that prohibits false and misleading email messages. As of October 2004, the FTC had joined with 19 agencies from 15 countries to combat unsolicited email or spam in an Action Plan for Enforcement.

TECHNOLOGY AND THE LEGAL ENVIRONMENT

E-Commerce: Junk Telephone Calls, Junk Faxes, "Do Not Call," and Antispam Legislation

The Telephone Consumer Protection Act of 1991 restricts automated devices that make thousands of calls an hour to play recorded advertising messages without prior written consent of the called parties. The act exempts emergency calls, calls made to businesses, and calls made to or by nonprofit organizations. Personal calls are not covered by the act. The Federal Communications Commission enforces the act. In a court of appeals decision, the law was upheld as a valid restriction on commercial speech.[a]

Another provision of the 1991 act bans unsolicited faxes that contain advertisements. Junk faxes were outlawed to prevent owners of fax machines from having to pay for paper as a receiver. Private and political faxes are exempt. The act has a damage maximum of $500 in damages for each violation. A class action suit against Hooters of America in a federal district court in Georgia, in which a local businessman, Sam Nicholson (and 1,320 other people), sued, resulted in a jury verdict of $12 million. In the *Hooters* case, 42 reams of paper with Hooters' advertisements in the form of coupons were introduced into evidence.

Public response to unwanted (unsolicited) advertising has led to legislation and an increasing number of cases. Some have argued that we need to learn to push the "delete" button instead of requesting more legislation. Advertisers blame the trial lawyers and fax or telephone marketers.

---

[a]*Moser v. Federal Communications Commission*, 46 F.3d 970 (9th Cir. 1995).

---

[10]H. Witt, "The Spam King," *Toledo Blade* (Knight-Ryder), D-1 (Aug. 7, 2004); D. Nasaw, "Federal Law Fails to Lessen Flow of Junk E-Mail," *Wall Street Journal*, D-2 (Aug. 10, 2004). See *Hypertouch, Inc. v. Value Click, Inc.*, 123 Cal.3d. 8 (2011) or more on the Federal CAN-SPAM Act.

[11]Pub. L. 90-321 (1969), 80. U.S.C. 146 *et seq.*, 15 U.S.C. 1601.

Federal Laws Regulating Consumer Credit and Business Debt-Collection Practices

Consumer credit means buyer power, which translates into demand for products, which, in turn, increases the supply of products produced by manufacturers. In short, this nation's economy runs on credit. With Americans owing more than $1.5 trillion and businesses and banks mailing out thousands of credit card applications almost daily, it is imperative that both business managers and consumers understand the rules governing credit arrangements.

Until 1969, when Congress passed the Consumer Credit Protection Act, most consumer protection was left to the states. There were many abuses in the issuance and reporting of credit terms. Often, consumer-debtors were ignorant of the annual percentage rates they were being charged, which made it impossible for them to shop around and compare rates. The Consumer Credit Protection Act (CCPA) of 1969 was designed to give consumers a fair shake in all areas of credit. We examine here several important sections of this comprehensive act under their popular titles: the Truth-in-Lending Act, the Electronic Fund Transfer Act (EFTA), the FCRA, the ECOA, the Fair Credit Billing Act, and the Fair Debt Collection Practices Act (FDCPA) (see Exhibit 25-1).

TRUTH-IN-LENDING ACT

Goals. The Truth-in-Lending Act of 1969 (TILA), as amended in 1982, seeks to make creditors disclose all terms of a credit arrangement before they enter into an agreement with a consumer-debtor. By mandating uniform terms and standards, it also seeks to give consumers a basis for comparative shopping: Consumers' ability to shop around for the lowest interest rates or finance charges promotes competition in the consumer credit market.

Scope. The TILA applies to creditors that regularly extend credit for less than $25,000 to natural persons for personal and family purposes. Corporations and persons applying for more than $25,000 in credit (except for buying a home) are not covered by the act. To be covered by the TILA, creditors must regularly extend credit (e.g., banks, finance companies, retail stores, credit card issuers, and savings and loans institutions), demand payment in more than four installments, or assess a finance charge.

Provisions. The TILA is a complex, detailed, and costly act for both management (creditors) and consumers (debtors). Six important provisions of the TILA deal with general disclosure, finance charges, the annual percentage rate, the right to cancel a contract, open- and closed-end credit transactions, and credit advertising.

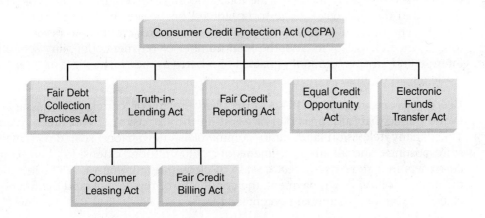

EXHIBIT 25-1

SIGNIFICANT SECTIONS OF THE CONSUMER CREDIT PROTECTION ACT

Regulation Z A group of rules set forth by the Federal Reserve Board to implement some provisions of the Truth-in-Lending Act, requiring lenders to disclose certain information to borrowers.

annual percentage rate (APR) The effective annual rate of interest being charged a consumer by a creditor that depends on the compounding period the creditor is using.

open-end credit Credit extended on an account for an indefinite time period so that the debtor can keep charging on the account, up to a certain amount, while paying the outstanding balance either in full or in installment payments.

closed-end credit A credit arrangement in which credit is extended for a specific period of time and the exact number of payments and the total amount due have been agreed upon between the borrower and the creditor.

General Disclosure. The general disclosure provisions of the TILA require all qualified creditors to disclose all terms clearly and conspicuously in meaningful sequence and to furnish the consumer with a copy of the disclosure requirements. Additional information may be incorporated into the disclosure statement, so long as it is not confusing.

Finance Charges. The finance charge provisions of the TILA—as well as the Federal Reserve Board's **Regulation Z**, which implements some provisions of the act—require a system of charge disclosure so that consumers can compare credit costs using uniform standards. A *finance charge* includes any dollar charges that make up the cost of credit to the consumer. These may be interest rates; service, carrying, or transaction charges; charges for mandatory credit life insurance on an installment loan in the event of the death of the debtor; loan fees; "points" when buying a home; and appraisal fees.

Annual Percentage Rate. The **annual percentage rate (APR)**—the effective annual rate of interest being charged by the creditor—must be disclosed in a meaningful and sequential way to all consumers of credit. Annual percentage rates differ according to the compounding period being used by the creditor, so this disclosure requirement makes it easy for consumers to see exactly what interest rate they are paying and to do some comparison shopping. The Federal Reserve Board publishes Regulation Z annual percentage tables. If creditors use these tables and follow their instructions, there is a legal presumption of correctness.

Right to Cancel. The right to cancel a contract applies only to home loans. A consumer has a right to cancel such a loan three days after entering the contract. The three-day period begins after the proper truth-in-lending disclosures have been made, usually at the closing by a bank employee. Because so many documents are being signed at this time, the disclosure statement is sometimes overlooked by both the bank employee and the borrower. The TILA was amended in 1995 to prevent borrowers from rescinding loans for minor clerical errors in closing documents.

Open- and Closed-End Credit Transactions. Regulation Z distinguishes between open-end credit and closed-end credit; each has separate disclosure requirements. An **open-end credit** transaction (e.g., MasterCard and a revolving charge account in a retail store such as Sears) extends credit for an indefinite time period and gives the debtor the option to pay in full or in installments. The initial statement to the debtor with such an account must include such items as the elements of any finance charge and the conditions under which it will be imposed, the APR as estimated at the time of credit extension, if and when overcharges will be imposed, and the minimum payment for each periodic statement.

A **closed-end credit** transaction (e.g., a personal consumer loan, a student loan, or a car loan) is extended by the creditor for a limited time period. All conditions of the loan, including the total amount financed, the number of payments, and the due dates, have to be agreed on before the loan is extended. For this type of credit transaction, Regulation Z requires creditors to disclose the total finance charges, the APR, the total number of payments due, any security interest, and any prepayment penalties or rebates to the consumer if payment is made ahead of time.

Credit Advertising. The TILA governs all advertising or "commercial messages" to the public that "aid, promote, or assist" in the extension of consumer credit. Thus, many television and radio commercials, newspaper ads, direct mail, store postings, and all announcements of a "blue ribbon" extension of credit to customers in a store must meet certain disclosure requirements. These include the amount of the down payment, the conditions of repayment, and the finance charges expressed in annual percentage terms.

Remedies. The FTC and seven other federal agencies are responsible for enforcement of the TILA. All these agencies have at their disposal uniform corrective actions that can be brought on behalf of a consumer-debtor.

First, a consumer can be reimbursed for a creditor's overcharging and for billing errors with regard to finance charges. Second, if the creditor has exhibited a pattern of negligence, misleading statements, or an intentional failure to disclose, the case may be referred to the Justice Department for criminal action. Conviction in such cases can bring a fine of up to $5,000, imprisonment up to 1 year, or both.

Private parties are the major enforcers of the TILA. A private party who brings suit must first show that the transaction affected interstate commerce; it must then show that the creditor failed to comply with the TILA or Regulation Z. Private parties (usually consumer-debtors) do not have to show that they were injured by the failure to disclose. Moreover, the creditor's noncompliance need only be slight. Damages recovered are usually actual damages plus a penalty of twice the finance charges imposed in connection with the transaction. "Reasonable attorney fees," as determined by the court, can also be collected.

Class actions brought on behalf of consumer-debtors are also permissible under the TILA. Usually, these are brought by legal aid societies or other non-profit groups against creditors with a history of blatantly unscrupulous dealings with consumers. See Case 25-6 for the U.S. Supreme Court's view of TILA material.

CASE 25-6

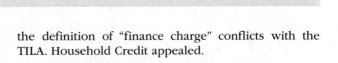

Household Credit Services, Inc. v. Pfenning
United States Supreme Court
541 U.S. 232 (2004)

Sharon Pfenning holds a credit card initially issued by Household Credit Services, Inc., but in which MBNA American Bank, N.A., now holds an interest through the acquisition of Household's credit card operation. Although the terms of Pfenning's credit card agreement set her credit limit at $2,000, Pfenning was able to make charges exceeding that limit, subject to a $29 "over-limit fee" for each month in which her balance exceeded $2,000.

On August 24, 1999, Pfenning filed a complaint in the U.S. District Court for the Southern District of Ohio on behalf of a purported nationwide class of all consumers who were charged over-limit fees by Household or MBNA (the defendants). Pfenning alleged that the defendants allowed her and the other members of the class to exceed their credit limits, thereby subjecting them to over-limit fees. Pfenning claims that the defendants violated the TILA by failing to classify the over-limit fees as "finance charges" and thereby "misrepresented the true cost of credit." The defendants moved to dismiss the complaint on the ground that Regulation Z specifically excludes over-limit fees from the definition of "finance charge." The District Court granted the petitioners' motion to dismiss. On appeal, Pfenning argued, and the court of appeals agreed, that Regulation Z's explicit exclusion of over-limit fees from the definition of "finance charge" conflicts with the TILA. Household Credit appealed.

Justice Thomas

TILA itself does not explicitly address whether over-limit fees are included within the definition of "finance charge." Congress defined "finance charge" as "all charges, payable directly or indirectly by the person to whom the credit is extended, and imposed directly or indirectly by the creditor as an incident to the extension of credit." § 1605(a). Because petitioners would not have imposed the over-limit fee had they not "granted [respondent's] request for additional credit, which resulted in her exceeding her credit limit," the Court of Appeals held that the over-limit fee in this case fell squarely within § 1605(a)'s definition of "finance charge."

The Court of Appeals' characterization of the transaction in this case, however, is not supported even by the facts as set forth in respondent's complaint. Respondent alleged in her complaint that the over-limit fee is imposed for each month in which her balance exceeds the original credit limit. If this were true, however, the over-limit fee would be imposed not as a direct result of an extension of credit for a purchase that caused respondent to exceed her $2,000 limit, but rather as a result of the fact that her charges exceeded her $2,000 limit at the time respondent's monthly charges were officially calculated. Because over-limit fees, regardless of a creditor's particular billing practices, are imposed only when a consumer exceeds his credit limit, it is perfectly reasonable to characterize an over-limit fee not as a charge imposed for obtaining an extension of credit over a consumer's credit limit, but rather as a penalty for violating the credit agreement.

Moreover, an examination of TILA's related provisions, as well as the full text of § 1605 itself, casts doubt on the Court of Appeals' interpretation of the statute. A consumer holding an open-end credit plan may incur two types of charges—finance charges and "other charges which may be imposed as part of the plan." . . . TILA does not make clear which charges fall into each category. But TILA's recognition of at least two categories of charges does make clear that Congress did not contemplate that all charges made in connection with an open-end credit plan would be considered "finance charges." And where TILA does explicitly address over-limit fees, it defines them as fees imposed "in connection with an extension of credit," rather than "incident to the extension of credit," § 1605(a).

Regulation Z's exclusion of over-limit fees from the term "finance charge" is in no way manifestly contrary to § 1605. Regulation Z defines the term "finance charge" as "the cost of consumer credit." . . .

Because over-limit fees, which are imposed only when a consumer breaches the terms of his credit agreement, can reasonably be characterized as a penalty for defaulting on the credit agreement, the Board's decision to exclude them from the term "finance charge" is surely reasonable.*

The Supreme Court *reversed* the Court of Appeals and ruled in *favor* of the Defendant, Household.

CREDIT CARD ACCOUNTABILITY, RESPONSIBILITY AND DISCLOSURE ACT OF 2009

The Credit Card Accountability, Responsibility and Disclosure Act of 2009, which became effective in February 2010, instituted some new provisions regarding rates, fees, disclosures and notices, and billing practices. It also left untouched more than a few issuer practices about which consumers had complained.

Rates. The 2009 Act did not limit the rates an issuer can charge new customers, nor did it limit how much issuers can raise rates on future purchases. However, credit card issuers (banks and others) can no longer raise interest rates on existing balances. Issuers also cannot raise rates on new accounts for 12 months.

Fees. Credit card issuers are no longer allowed to charge a fee when the debtor exceeds the credit limit unless the debtor agrees to this arrangement and signs up for this service. In essence, issuers cannot charge extra because of the way a person pays his or her bills. However, credit card companies can still charge annual fees, inactivity fees, and other fees.

Disclosure and Notice. Credit card issuers must give 45 days' notice before raising interest rates, before charging certain fees such as annual fees or cash advance fees, and before making other significant account changes. Card issuers can close your account or lower your credit limit for any reason without giving advance notice, though. In the months before this law became effective, credit card companies sharply lowered limits for thousands of cardholders.

Billing Practices. Many customers (borrowers) have several lines of credit with different interest rates on the same credit card. For example, you may get one rate for a cash advance and another for purchases on the same card. Issuers are now required to apply any amount paid by the customer above the minimum to the balance with the highest rate. The new law made no change to another common situation, though: When cardholders have credit lines with different interest rates, card issuers are still allowed to apply the minimum payment to the lowest-rate debt.

Credit card companies must now mail or deliver the debtor's bill at least 21 days before payment is due, and the due date must be the same every month. Banks can no longer use a customer's average daily balance over two months to calculate interest.

---

*Household Credit Services, Inc. v. Pfenning United States Supreme Court 541 U.S. 232 (2004).

THE ELECTRONIC FUND TRANSFER ACT[12]

Goals. In 1978, Congress amended the TILA and enacted the Electronic Fund Transfer Act (EFTA) to regulate financial institutions that offer electronic fund transfers involving an account held by a customer. The Federal Reserve Board was empowered to enforce the provisions of the EFTA and adopted Regulation E to further interpret the act. Types of consumer electronic funds include automated teller machines (ATMs), point-of-sale terminals (e.g., debit cards), pay-by-phone systems, and direct deposit and withdrawals.

Provisions. Consumer rights established by the EFTA apply in the following areas:

1. *Unsolicited cards.* Banks can send an unsolicited EFTA card to a consumer only if the card is not valid for use when received.
2. *Errors in billing.* Customers have 60 days from the receipt of a bank statement to notify the bank of its error. The bank has 10 days to investigate. The bank can recredit the customer account to gain 45 more days to investigate.
3. *Lost or stolen debit cards.* If the bank is notified within 2 days from the time a card is lost, the customer can be liable for only $50 in unauthorized use. Liability increases to $300 up to 60 days and more than $500 after 60 days when no notice is given.
4. *Transactions.* A bank has to provide written evidence of a transaction made through a computer terminal.
5. *Statements.* Banks must provide a monthly statement to an EFTA customer at the end of a month in which a transaction is made. If no transactions are made, a quarterly statement must be provided.

The EFTA covers only transactions involving accounts held by natural persons for personal, family, or household purposes. Many fund transfers fall outside the EFTA. An attempt to create a uniform body of law in the 50 states took place in 1989 with the creation of Article 4A of the UCC, but Article 4A and the EFTA are mutually exclusive. Article 4A does not apply to or regulate any part of an electronic fund transfer that is subject to the EFTA.

Remedies. Under the EFTA, a financial institution is liable to any customer for all damages caused by its failure to make an electronic transfer in a timely manner and when instructed to do so by the customer. The institution is liable for damages caused by its failure to credit a deposit of funds and by its failure to make a preauthorized transfer from a customer account.

If an act of God or other circumstance beyond its control or a technical malfunction takes place, the financial institution is not liable.

A PLASTIC SOCIETY

With credit cards, consumers can spend more money more quickly than when using cash. As of 2004, consumer debt was at an estimated $838 billion. Card issuers obtain anywhere from 1 to 5 percent commission on each transaction.[13] The facts that credit cards are omnipresent and susceptible to theft made the EFTA necessary for consumer protection.

---

[12]15 U.S.C. § 1693g(a), *et seq.*, Regulation E, 12 C.F.R. § 205.6 *et seq.* Available at www.fdic.gov/regulations/laws/rules/6500-3100.html.

[13]See J. Sapsford, "As Cash Wanes, America Becomes a Plastic Nation," *Wall Street Journal*, A-1, A-6 (Jul. 23, 2004); 15 U.S.C. 1681 *et seq.*

Technological advances may limit the use of cards in the future. With an increase in online shopping, the credit industry is looking forward to using advanced identification methods such as silicon wafers embedded in the computer keyboard, which would read a fingerprint and match it online with a bank copy so that the bank can then authorize (or be alerted to) the sale. Could the "plastic society" be changing?

THE FAIR CREDIT REPORTING ACT[14]

Goals. The Fair Credit Reporting Act (FCRA) was enacted by Congress in 1970 as an amendment to the CCPA to ensure that credit information obtained by credit agencies would remain confidential. Individual privacy is a major concern in an era in which three major credit bureaus (TRW, TransUnion, and Equifax) hold files on a majority of U.S. citizens. At the same time, Congress wanted to force the agencies to adopt reasonable procedures to allow lenders such as banks and finance corporations to have access to the information they need to make decisions on whether to lend money.

The FCRA sought especially to correct three common abuses by credit agencies: the failure to set uniform standards for keeping information confidential, the retention of irrelevant and sometimes inaccurate information in their files, and the failure to respond to consumer requests for information.

Provisions. The following are the most important provisions of the FCRA:

1. The creditor or lender must give notice to a consumer whenever that consumer has been unfavorably affected by an adverse credit report from a consumer reporting agency. A *consumer reporting agency* is defined as any entity that "regularly engages in the practice of assembling or evaluating consumer credit or other information on consumers for the purpose of furnishing consumer reports to third parties." If a company infrequently furnishes information to a third party or collects it for internal use only, it does not fall within the FCRA.

2. The consumer may go to the credit agency that issued the adverse report and be "informed of the nature and substance" of the information on file. A written request for information by a consumer must also be honored.

3. Credit reporting agencies are required to keep files up to date and delete inaccuracies. If these inaccuracies have been passed on to lenders, the agencies are also required to notify them of the errors. Agencies cannot retain stale information and must follow reasonable procedures to update information dealing with bankruptcies, tax liens, criminal records, and bad debts. If a credit transaction involves $50,000 or more, these agency actions are not necessary before issuing a credit report.

4. If consumers do not agree with what is in their files or with what has been reported by the credit agency to a lender, they can file a written report (of 100 words or less) giving their side of the dispute.

5. A consumer credit agency may issue credit information reports to (a) a court in response to a court order; (b) the consumer to whom the report relates (upon written request); (c) a person or entity whom the agency has reason to believe will use the information in connection with making a credit transaction, obtaining employment, licensing, or obtaining personal or family insurance; and (d) anyone having a legitimate business need for the information to carry on a business transaction with a consumer.

6. In 1997, the act was amended to restrict the use of credit reports by employers. An employer must notify a job applicant or current employee that a

---

[14]15 U.S.C. 1681 *et seq.*

report may be used and must obtain the applicant's consent before requesting an individual credit report from a credit bureau. Additionally, before refusing to hire or terminating or denying a promotion, the employer must provide an individual with a "pre-adverse action disclosure," which must contain the individual's credit report and a copy of the FTC's "A Summary of Your Rights under the Fair Credit Reporting Act."

The following case sets out the standard for civil liability under the FCRA.

CASE 25-7

Safeco Insurance Co. v. Burr
United States Supreme Court
127 S. Ct. 2201 (2007)

Safeco Insurance Company and GEICO General Insurance Company issued automobile insurance policies to three applicants without telling them that the companies had obtained credit reports on them. One applicant filed a lawsuit against Safeco and two applicants sued GEICO under the Fair Credit Reporting Act.

Justice Souter

The Fair Credit Reporting Act requires notice to any consumer subjected to "adverse action based in whole or in part on any information contained in a consumer credit report." Anyone who "willfully fails" to provide notice is civilly liable to the consumer. The questions in these consolidated cases are whether willful failure covers a violation committed in reckless disregard of the notice violation, and, if so, whether petitioners Safeco and GEICO committed reckless violations. We hold that reckless action is covered, that GEICO did not violate the statute, and that while Safeco might have, it did not act recklessly.

Congress enacted the Act in 1970 to ensure fair and accurate credit reporting, promote efficiency in the banking system, and protect consumer privacy. The Act requires, among other things, that "any person who takes any adverse action with respect to any consumer that is based in whole or in part on any information contained in a consumer report" must notify the affected consumer. The notice must point out the adverse action, explain how to reach the agency that reported on the consumer's credit, and tell the consumer that he can get a free copy of the report and dispute its accuracy with the agency. As it applies to an insurance company, "adverse action" is "a denial or cancellation of, an increase in any charge for, or a reduction or other adverse or unfavorable change in the terms of coverage or amount of any insurance, existing or applied for."

In GEICO's case, the initial rate offered to Edo [one of the applicants] was the one he would have received if his credit score had not been taken into account, and GEICO owed him no adverse action notice under the Act.

Safeco did not give Burr and Massey (the other applicants) any notice because it thought the Act did not apply to an initial application, a mistake that left the company in violation of the statute if Burr and Massey received higher rates "based in whole or in part" on their credit reports; if they did, Safeco would be liable to them on a showing of reckless conduct (or worse). The first issue we can forget, however, for although the record does not reliably indicate what rights they would have obtained if their credit reports had not been considered, it is clear enough that if Safeco did violate the statute, the company was not reckless in falling down in its duty.

There being no indication that Congress had something different in mind, we have no reason to deviate from the common law understanding in applying the statute. Thus, a company subject to the Act does not act in reckless disregard of it unless the action is not only a violation under a reasonable reading of the statute's terms, but shows that the company ran a risk of violating the law substantially greater than the risk associated with a reading that was merely careless. Here, there is no need to pinpoint the negligence/recklessness line, for Safeco's reading of the statute, albeit erroneous, was not objectively unreasonable.*

Reversed, in favor of Defendant and *remanded*.

The Court of Appeals correctly held that reckless disregard of a requirement of the Act would qualify as a willful violation within the meaning of the Act. But there was no need for that court to remand the cases for factual development. GEICO's decision to issue no adverse action notice to Edo was not a violation of the Act, and Safeco's misreading of the statute was not reckless. The judgments of the Court of Appeals are therefore reversed in both cases, which are remanded for further proceedings consistent with this opinion.†

---

*Safeco Insurance Co. v. Burr, United States Supreme Court 127 S. Ct. 2201 (2007).
†Jerman v. Carlisle, McNellie, Rini, Kramer & Ulrich, LPA, Supreme Court of US, 130 S. Ct. 160.

CRITICAL THINKING ABOUT THE LAW

Very early in our lives, most of us are told that "ignorance of the law is no excuse." Are Safeco and GEICO being permitted to violate the law on the ground that they were unaware of the meaning of the law?

1. Express the legal rule in this case in a manner that addresses the Court's apparent understanding of reckless disregard of a statute.

    Clue: Does Justice Souter believe that reckless disregard constitutes violation of a statute? When would disregard become reckless?

2. What facts would have made this case go in favor of the plaintiffs?

    Clue: Construct a set of facts under which each insurance company would have violated the statute.

Remedies. The FTC may bring actions in the federal courts to obtain cease-and-desist orders against credit agencies and users of information, or it may seek to obtain administrative enforcement orders from an administrative law judge. Violations of the FCRA are considered "unfair or deceptive practices" under Section 5 of the FTCA. An agency that fails to comply with FCRA is liable for actual damages plus additional damages not to exceed $1,000, plus attorney's fees. Six other government entities can also enforce the FCRA: the Federal Reserve Board, the Comptroller of the Currency, the Federal Home Loan Bank Board, the National Credit Union, the Interstate Commerce Commission, and the Secretary of Agriculture.

IDENTITY THEFT AND CREDIT RATINGS

Identity theft may be defined as the use of another's name to obtain some illegal gain through financial instruments held or owned by another. Often, this includes, for example, stolen credit cards, Social Security numbers, computer codes, and telephone numbers.

The Fair and Accurate Credit Transaction Act of 2003[15] (FACTA) (which amends the FCRA) seeks to combat identity theft by allowing consumers to place fraud alerts in their credit files and to block information from being placed in a file if caused by identity theft or fraud. In addition, FACTA provides the FTC and other federal regulatory agencies with rule-making authority to combat identity theft, which the agencies previously lacked. The FTC may receive reports and studies on identity theft by nongovernmental national credit agencies and other private-sector firms, and consumers may receive the same.

FACTA provides that the FTC and other agencies may study the effect of credit scores collected by credit agencies and their effect on the affordability of financial products such as mortgages and insurance. Consumers may now have access to their individual credit reports once a year free of charge from one of the three national nongovernmental credit agencies (Equifax, Experian, and TransUnion). In applying for mortgages, insurance, loans, and other financial tools, credit reports will be helpful.

Criminal liability is incurred by a person who "knowingly and willfully obtains information on a consumer from a consumer reporting agency under false pretenses." Anyone convicted of this charge is subject to a fine up to $5,000 and possible imprisonment for a maximum of 1 year.

---

[15] 15 U.S.C. § 168; Pub. L. No. 108-159 (2003).

Civil liability is incurred by a credit agency for any user of credit agency information (e.g., a bank) if a consumer, in a private action, can show that the agency willfully violated the FCRA through repetitive errors. For such violations, the court may assess punitive and actual damages, court costs, and attorney's fees. A consumer who is able to show negligence on the part of the credit agency or user will obtain actual damages, court costs, and reasonable attorney's fees.

EQUAL CREDIT OPPORTUNITY ACT[16]

Goals. When Congress enacted the ECOA in 1974, it was trying to eliminate all forms of discrimination in granting credit, including those based on race, sex, color, religion, national origin, marital status, receipt of public assistance—and exercise of one's rights under the act. The Federal Reserve Board, which is charged with implementing ECOA's regulations, may exempt any "classes of transactions not primarily for household or family purposes." Thus, most commercial transactions are exempt from the ECOA.

Provisions. The ECOA and Regulation B, promulgated by the Federal Reserve Board, provide that:

1. A creditor may not request information from a credit applicant about a spouse or a former spouse, the applicant's marital status, any alimony and child support received, gender, childbearing, race, color, religion, or national origin.

2. A creditor must notify the applicant of what action has been taken on the application within 30 days of receiving it. The notification must contain (a) a statement of the action taken and, if the application is denied, either a statement of the reasons for the denial or a disclosure of the applicant's right to receive a statement of such reasons; (b) a statement of the basic provisions of the ECOA; and (c) the name and address of the relevant administrative agency that deals with compliance by creditors.

A two-year statute of limitations applies to suits brought under the ECOA. This, however, sometimes does not apply when the act is used as a defense. For example, there is a conflict among state and federal court decisions in cases in which the wife is illegally required under ECOA to cosign a loan guarantee for her husband's business. When the bank later brings suit to collect on the promissory note on which the husband has defaulted, the wife raises the defense of a violation of the act even though the two-year statute of limitations has run.

Other ECOA violations, besides requiring a spouse to cosign, include (a) asking for information about an applicant's spouse or former spouse when it is not relevant; (b) taking race, sex, or national origin into account when making a credit decision; (c) requiring certain types of life insurance before issuing a loan; (d) basing a credit decision on the area in which the applicant lives; and (e) asking about an applicant's intent to have children.

Remedies. The FTC and other agencies can bring administrative actions on behalf of consumers before an administrative law judge, as well as civil injunctive actions in the federal district courts. In addition, if a creditor violates the ECOA, individuals who believe they have been injured under the act can bring an action in federal district court for actual and punitive damages. Punitive damages may not exceed $10,000 for an individual successful plaintiff, but they may go as high as $500,000 for successful class action plaintiffs. Plaintiffs may also ask for injunctive relief to prohibit future discriminatory actions by the creditor.

---

[16] Reg. B, Sec. 202.1-15.

THE FAIR CREDIT BILLING ACT[17]

Goals. Congress enacted the FCBA as an amendment to the TILA in 1974 to eliminate inaccurate and unfair billing practices as well as to limit the liability of consumer debtors for the unauthorized use of their credit cards.

Provisions. The FCBA provisions cover (1) issuers of credit cards, (2) creditors who extend credit over more than four monthly installments, and (3) creditors who assess finance charges. The following provisions of the act establish a procedure for correcting billing errors:

1. All creditors must notify consumer-debtors of their rights and duties when an account is opened and every six months thereafter. They must notify debtors on the billing statement where they may inquire (provide contact information) when they notice a billing error.

2. Consumer-debtors who believe their billing statement contains an error must notify the creditor in writing within 60 days, identifying themselves, their account number, the item, and the amount in dispute. Within 30 days, the creditor must notify the debtor that it has received notice of the alleged billing error. Within 90 days, or two billing cycles, the creditor must notify the consumer-debtor of the outcome of its investigation. During this period, the creditor cannot take any action to collect the debt in dispute. It may continue to send billing statements listing the disputed item, but these statements must give notice to the consumer that the item in dispute does not have to be paid.

If after investigation the creditor finds that there was a billing error, it must correct the error and notify the consumer. If it finds no error, it must notify the consumer-debtor and substantiate its reason. A bad credit report cannot be filed until 10 days after the substantiation has been sent out.

If a faulty product is purchased on credit, the consumer can withhold payment until the dispute is settled, provided he or she notifies the creditor immediately after finding the fault in the product. The creditor is then obligated to attempt to negotiate the dispute between the seller of the product and the consumer-debtor.

Remedies. Individual consumer-debtors as well as government agencies can bring actions against creditors who violate the FCBA. The only penalty set forth in the act is that creditors forfeit their right to collect up to $50 on each item in dispute on each periodic statement. This amount includes interest and finance charges on the amount in dispute.

THE FAIR DEBT COLLECTION PRACTICES ACT[18]

Goals. The purpose of the FDCPA is to prevent creditors or debt collectors from harassing consumer-debtors at their places of work or at home. The act defines *debt collectors* as those who are in the business of collecting debts from others. In 1987, attorneys who regularly perform debt activities were brought under the provisions of the FDCPA.[19]

Approximately 5,000 debt-collection agencies seek about $5 billion in debts from some 8 million consumers annually. Many use sophisticated WATS telephone lines and computers. They are paid a 20 to 50 percent commission on what they collect, so they often are quite aggressive in their collection methods,

---
[17]15 U.S.C. 1601 *et seq.*
[18]Pub. L. 95; 91 Stat. 874; 15 U.S.C. Section 1692; September 20, 1978.
[19]See *Heinz v. Jenkins*, 513 U.S. 1109 (1995).

as well as successful—so successful that in the 1980s, the federal government turned over many outstanding federal loans (including student loans) to these private collection agencies.

States are exempt from FDCPA enforcement within their boundaries if they have laws meeting the FDCPA requirements. Actually, state laws are often more vigorously enforced than the federal law.

Provisions. The FDCPA prohibits the following activities by debt collectors who are covered by the act:

1. They may not contact a third party (other than the debtor's family and lawyer) except to find out where the debtor is. The idea behind this provision is that the debtor's name (reputation) among friends, acquaintances, or employers should not be ruined.

2. They may not contact a debtor during "inconvenient" hours. This provision seeks to prevent creditors from harassing a debtor in the middle of the night. "Inconvenient hours" are considered to be from 9 p.m. to 8 a.m. for a debtor whose workday is normally 8 a.m. to 5 p.m. If the credit collection agency knows that the debtor is represented by a lawyer, it may not contact the debtor at all.

3. They cannot contact a debtor in an abusive, deceptive, or unfair way. For example, posing as a lawyer or police officer is forbidden.

The act requires a debt collector, within five days of the initial communication with a consumer, to provide the consumer with a written notice that includes (1) the amount of the debt, (2) the name of the current creditor, and (3) a statement informing the consumer that he or she can request verification of the alleged debt. The consumer can recover damages from the collection agency for violations of the act.

The following case illustrates the importance of language in statutory interpretation—in this instance, of the FDCPA.

CASE 25-8

Jerman v. Carlisle, McNellie, Rini, Kramer & Ulrich, LPA
Supreme Court of the United States
559 U.S. 573 (2010)

Respondents in this case are a law firm, Carlisle, McNellie, Rini, Kramer & Ulrich, LPA [Legal Professional Association], and one of its attorneys, Adrienne S. Foster (collectively Carlisle). In April 2006, Carlisle filed a complaint in Ohio state court on behalf of a client, Countrywide Home Loans, Inc. Carlisle sought foreclosure of a mortgage held by Countrywide in real property owned by petitioner Karen L. Jerman. The complaint included a "Notice," later served on Jerman, stating that the mortgage debt would be assumed to be valid unless Jerman disputed it in writing. Jerman's lawyer sent a letter disputing the debt, and Carlisle sought verification from Countrywide. When Countrywide acknowledged that Jerman had, in fact, already paid the debt in full, Carlisle withdrew the foreclosure lawsuit.

Jerman then filed her own lawsuit seeking damage under the FDCPA [Fair Debt Collection Practices Act], contending that Carlisle violated [the act] by stating that her debt would be assumed valid unless she disputed it in writing. While acknowledging a division of authority on the question, the District Court held that Carlisle had violated [the act] by requiring Jerman to dispute the debt in writing. The court ultimately granted summary judgment to Carlisle, however, concluding that Section 1692k(c) [of the FDCPA] shielded it from liability because the violation was not international, resulted from a bona fide error, and occurred despite the maintenance of procedures reasonably adapted to avoid any such error. The Court of appeals for the Sixth Circuit affirmed. Acknowledging that the Courts of Appeals are divided regarding the scope of the bona fide error defense and that the "majority view is that the defense is available for clerical and factual errors only," the Sixth Circuit nonetheless held that Section 1692(c) extends to "mistakes of law."

Justice Sotomayor

The parties disagree about whether a "violation" resulting from a debt collector's misinterpretation of the legal requirements of the FDCPA can ever be "not intentional" under 1692k(c). Jerman contends that when a debt collector intentionally commits the act giving rise to the violation (here, sending a notice that included the "in writing" language), a misunderstanding about what the Act requires cannot render the violation "not intentional," given the general rule that mistake or ignorance of law is no defense. Carlisle and the dissent, in contrast, argue that nothing in the statutory text excludes legal errors from the category of "bona fide error[s]" covered by 1692k(c) and note that the Act refers not to an unintentional "act" but rather an unintentional "violation." The latter term, they contend, evinces [makes clear] Congress's intent to impose liability only when a party knows its conduct is unlawful. Carlisle urges us, therefore, to read 1692k(c) to encompass "all types of error," including mistakes of law.

We decline to adopt the expansive reading of Section 1692k(c) that Carlisle proposes. *We have long recognized the "common maxim, familiar to all minds, that ignorance of the law will not excuse any person, either civilly or criminally." Our law is therefore no stranger to the possibility that an act may be "intentional" for purposes of civil liability, even if the actor lacked actual knowledge that her conduct violated the law.* [Emphasis added.]

When Congress has intended to provide a mistake-of-law defense to civil liability, it has often done so more explicitly than here. In particular, the FTC [Federal Trade Commission] Act's administrative penalty provisions—which Congress expressly incorporated into the FDCPA—apply only when a debt collector acts with "actual knowledge or knowledge fairly implied on the basis of objective circumstances" that its action was "prohibited by [the FDCPA]." Given the absence of similar language in Section 1692k(c) it is fair inference that Congress chose to permit injured consumers to recover actual damages, costs, fees, and modest statutory damages for "intentional" conduct, including violations resulting from mistaken interpretation of the FDCPA, while reserving the more onerous penalties of the FTC Act for debt collectors whose intentional actions also reflected "knowledge fairly implies on the basis of objective circumstances" that the conduct was prohibited.

We draw additional support for the conclusion that bona fide errors in Section 1692k(c) do not include mistaken interpretations of the FDCPA from the requirement that a debt collector maintain "procedures reasonably adapted to avoid any such error." The dictionary defines "procedures" as "a series of steps followed in a regular orderly definite way." In that light, the statutory phrase is more naturally read to apply to processes that have mechanical or other such "regular orderly" steps to avoid mistakes—for instance, the kind of internal controls a debt collector might adopt to ensure its employees do not communicate with consumers at the wrong time of day or make false representations as to the amount of a debt. We do not dispute that some entities may maintain procedures to avoid legal errors. But legal reasoning is not a process. For this reason, we find that the broad statutory requirement of the procedures reasonably designed to avoid "any" bona fide error indicates that the relevant procedures are ones that help to avoid errors like clerical or factual mistakes. Such procedures are more likely to avoid error than those applicable to legal reasoning particularly in the context of a comprehensive and complex federal statute such as the FDCPA that imposes open-ended prohibitions on, *inter alia* [among other things], "false, deceptive," or "unfair" practices.

For the reasons discussed the judgment of the United States Court of Appeals for the Sixth Circuit is *reversed*, and the case is *remanded* for further proceedings consistent with this opinion.

Remedies. A violation of the FDCPA is considered a violation of Section 5 of the FTCA. The FTC and individual debtors may both bring actions. The FTC may issue cease-and-desist orders and levy fines after an internal administrative agency proceeding.

Individual debtors may bring civil actions to recover actual damages, including those for embarrassment and mental distress. An additional $1,000 may be assessed for each violation for malicious damages. Attorney's fees are recoverable by debtors who win their suits and in the event that the creditor brings an action against the debtor that is found to be "harassing."

E-Commerce and Consumer Protection

The FTC has indicated that more than 160,000 Internet-related fraud complaints were received in 2003, with estimated losses of nearly $200 million.[a] About 50 percent of the complaints involved online auctions. Despite these numbers,

---

[a] N. Wingfield, "Problem for Cops on eBay Beat," *Wall Street Journal*, A-1, A-8 (Aug. 8, 2007).

very few cases are actually filed in courts or with administrative agencies because fraud on the Internet is often practiced by people in foreign countries. Jurisdiction over these cases has become important. In addition, the staffs of local, state, and federal governments combined have too few enforcers. Also, there is a fear of disclosing some suspects because it may assist other alleged criminals in evading law enforcement authorities.

Government regulators at all levels have sought to keep up with scammers who use "phishing" techniques. "Phishers" send emails to Internet users, asking for passwords and other information and disguising the messages as official communications from companies such as eBay. In eBay's case, this method is used to swipe the identities of users, in particular eBay sellers who have built a reputation for honesty.[b] This identity scam allows those seeking to defraud others to obtain thousands of potential targets.

The FTC (and other federal and state agencies) has become a referee for business and consumer transactions on the Internet. Everything from airline tickets to books to pianos may be purchased on the Web. As this activity increases, so does the number of techniques for committing fraud. As discussed in this chapter, deceptive advertising never ceases; it seems to be an entrenched part of transactions involving business and consumer. "Junk" faxes, telephone calls, and spam lead to cases of fraud. The best weapon for consumer protection may be an educated consumer working with business and government agencies.

At the state level, consumer protection statutes have been amended to cover Internet transactions. Often, the most effective consumer protection is initiated at local levels. The Internet complaint process has become an important tool.

[b]*Id.*

Dodd-Frank Act and Consumer Protection

CREDIT AND DEBIT CARDS

The act includes a provision at reducing "interchange fees." These are fees that banks charge retailers when consumers pay with debit cards. Under the Dodd-Frank Act, the Federal Reserve Board (Fed) will determine what constitutes a "reasonable and proportional" fee for debit card transactions, which usually run about 1 percent of the transaction, and are passed on to the consumer. If they are lowered, prices to consumers may be lowered. Retailers are allowed to require a minimum purchase before they accept a debit or credit card. Banks have indicated that they may have to eliminate debit card reward programs and increase other fees to make up for lost revenue.

CONSUMER LOANS

The new Consumer Financial Protection Bureau will regulate mortgage credit cards, some payday lenders and check cashing companies, and lenders that provide private student loans. Auto dealers' financing and insurance arms are exempt from the bureau's jurisdiction, after extensive lobbying.

CREDIT SCORES

Consumers who are turned down for a loan are entitled to receive a copy of the credit score the lender used to make the decision. Consumers are entitled to a free score if they were offered a loan at a rate higher than the one provided to borrowers with excellent credit. Consumers can obtain a free credit score if the score results in an "adverse action" such as loss of a job applied for or a higher insurance rate or other matters that depend on such scores.

RESIDENTIAL MORTGAGES

Lenders are no longer allowed to pay mortgage brokers a commission based on the interest rate for a home loan.

The Dodd-Frank Act would prevent borrowers from paying a portion of the closing cost up front and rolling the rest into the loan in the form of a higher interest rate:

- Prepayment penalties would be limited or prohibited depending on the type of loan.
- Lenders would be required to determine whether the borrower could afford the monthly mortgage/payments, combined with insurance and assessments.

State Consumer Legislation

It has often been argued that state, city, county, and private agencies (e.g., the Better Business Bureau) are closer geographically to the problems the average consumer encounters and thus are more effective at resolving them than are federal agencies, particularly when relatively small amounts of money are involved. In this section, we examine some consumer-oriented legislation originating with the states. Although state law is often overlooked in treatments of consumer protection law, it is important because it touches the lives of many Americans daily.

UNIFORM CONSUMER CREDIT CODE

The Uniform Consumer Credit Code (UCCC) was drafted by the National Conference of Commissioners on State Laws in 1968 and revised in 1974 and 1982. The commissioners' aim was to replace the patchwork of differing state consumer laws with a uniform state law in the area of consumer credit.

The UCCC takes a disclosure approach to consumer credit similar to that of the federal legislation discussed in this chapter. It regulates interest and finance rates, sets out creditors' remedies, and prohibits fine-print clauses. (It incorporates the TILA by reference.) Like the FTC regulations, the UCCC gives the consumer three days to cancel a sale when it is made as a result of home solicitations. So far, only 10 states have adopted the UCCC, and each of them has somewhat altered the original uniform statute.

UNFAIR AND DECEPTIVE PRACTICES STATUTES

All states and the District of Columbia have statutes forbidding deceptive acts and practices in a way similar to Section 5 of the FTCA. So closely are they modeled on that act, in fact, that they are often called the *baby FTC laws*.

State attorney general offices typically have consumer fraud divisions that investigate consumer fraud and false advertising and that seek injunctions, fines, or restitution in state courts. Often, notice of investigation by a state attorney general's office is sufficient to discourage a practice such as false advertising. Furthermore, private consumer actions as well as class actions are permitted under most state statutes. Usually, consumers may obtain actual and punitive damages as well as court costs and attorney's fees.

Attempts to combat consumer fraud at the state level range from mandatory disclosure statutes, requiring merchants to set out all terms and conditions in a financing agreement, to laws requiring cooling-off periods that allow consumers a set number of days to cancel a purchase sold by a door-to-door salesperson. One class of state consumer laws, the lemon laws, gives consumers warranty and refund rights on used cars when a material defect can be shown. Mandatory seat-belt-use laws and license-suspension statutes are also consumer-oriented in that they protect buyers and drivers of automobiles.

ARBITRATION OF DISPUTES

State attorneys general have been encouraging private groups such as the Better Business Bureau to play a role in exposing fraudulent sales tactics and in arbitrating disputes. One excellent example in this area is a formal agreement between General Motors (GM) and the Better Business Bureau that allows consumers to bring their complaints about car engines to the bureau. GM has agreed to be bound by the bureau's decisions, although consumers have the right to go to court if they disagree with a decision.

Global Dimensions of Consumer Protection Laws

As companies have become multinational, they have had to look at varying national consumer protection laws and how they differ. For example, a company such as Coca-Cola seeks to standardize its advertising for purposes of reducing costs and improving quality and to appeal to internationally mobile consumers. One of the factors that prevents complete standardization of advertising is legality. Differing national views on consumer protection, protection of competition, and standards of morality and nationalism may prevent a multinational company from delivering the same advertising message in each nation it sells goods.

In the area of consumer protection, countries differ on the amount of deception permitted in advertising. For example, the United Kingdom and the United States allow competing companies to advertise in a comparative way (e.g., Burger King and McDonald's). In contrast, the Philippines prohibits this form of advertising. The United States is concerned with sexism in advertising as well as advertising of tobacco. Most countries in Europe, Asia, and Latin America have few, if any, prohibitions in those areas.

In 1984, the European Union's Commission adopted what was termed a "Misleading Advertising Directive." Similar to Section 5 of the FTCA, the directive called on member states of the European Union to prohibit misleading advertising by statute and to create means to enforce such laws. Similar to the U.S. laws, the directive requires that courts and agencies within member states be given the power to require companies to substantiate claims made in advertisements. Member nations have gradually enacted legislation that fits the cultural mores to which it is to be applied.

In Mexico, the Federal Consumer Protection Act of 1975 (FCPA) was modeled in large part after several U.S. statutes. Some provisions dealing with advertising include the "principle of truthfulness" between customers and merchants. Labeling instructions must be clear as to content. There must be warnings on all advertised products, as well as truthfulness in advertising on radio and television. In addition to advertising, the FCPA covers areas such as warranties, consumer credit disclosure, and unconscionable clauses in contracts. Both private parties and the federal attorney general for consumer affairs may bring actions in courts of law. With the advent of NAFTA, the FCPA has become more important as the United States, Canada, and Mexico seek to bring some uniformity to their consumer protection laws.

SUMMARY

This chapter examined debtor–creditor relations, particularly as related to the Bankruptcy Reform Act of 2005. Consumer law originated in the 1930s in case law and evolved quickly during the consumer rights movement of the 1960s and 1970s. Federal regulation of business and trade practices is highly dependent on the Federal Trade Commission, the watchdog agency charged with enforcing

Section 5 of the FTCA, which forbids unfair or deceptive business practices and unfair methods of competition. Prohibited trade practices include deceptive and unfair advertising, misrepresentation by franchisors, deceptive or confusing warranties, and deceptive telemarketing practices.

Federal laws regulating consumer credit all come under the comprehensive umbrella of the Consumer Credit Protection Act. Important parts of this umbrella act are the TILA, which forces creditors to disclose all terms of a credit arrangement to consumer-debtors before they sign the agreement; the FCRA, which seeks to ensure that credit agencies keep accurate, confidential records; the ECOA; the Fair Credit Billing Act; the FDCPA; and the Consumer Leasing Act. The Dodd-Frank Act of 2010 deals with credit and debit cards, consumer loans, and residential mortgages.

State consumer legislation is important because state and local agencies are often closer to consumers' problems than are federal agencies and, therefore, are more effective.

As international business grows, companies need to be aware of their rights and duties under consumer protection laws around the world.

REVIEW QUESTIONS

25-1 What is meant by the term *garnishment* when it is used in the context of debtor–creditor?

25-2 How does the Bankruptcy Act of 2005 change the federal Bankruptcy Code?

25-3 Explain the general types of deceptive advertising with which the FTC is most concerned.

25-4 Explain the suretyship agreement.

25-5 What must be disclosed by a consumer credit reporting agency under the FCRA?

25-6 What enforcement weapon does the FTC use against parties that violate the ECOA?

REVIEW PROBLEMS

25-7 Brian Cleary and Rita Burke filed a suit against cigarette maker Philip Morris USA, Inc., seeking class action status for a claim of deceptive advertising. Cleary and Burke claimed that "light" cigarettes such as Marboro Lights were advertised as safer than regular cigarettes even though the health effects are the same. They contended that the tobacco companies concealed the true nature of light cigarettes. Philip Morris correctly claimed that it was authorized by the government to advertise cigarettes, including light cigarettes. Assuming that is true, should the plaintiffs still be able to bring a deceptive advertising claim against the tobacco company? Why or why not? *Cleary v. Phillip Morris USA, Inc.*, 683 F.Supp.2d 730 (N.D.Ill. 2010).

25-8 Montoro petitioned himself into voluntary bankruptcy. There were three major claims against his estate. One was made by Carlton, a friend who held Montoro's negotiable promissory note for $2,500; one was made by Elmer, an employee who was owed three months' back wages of $4,500; and one was made by the United Bank of the Rockies on an unsecured loan of $5,000. In addition, Dietrich, an accountant retained by the trustee, was owed $500, and property taxes of $1,000 were owed to Rock County. Montoro's nonexempt property was liquidated, with proceeds of $5,000. Discuss fully what amount each party will receive and why.

25-9 Onondaga Bureau of Medical Economics (OBME), a collection agency for physicians, sent the plaintiff, Seabrook, a letter demanding payment for a $198 physician's bill. In addition to demanding payment, the letter stated that OBME's client could commence against Seabrook a legal action that could result in a garnishment of his wages. Does OBME's letter violate the FDCPA in that it (a) does not give Seabrook the required notice or (b) threatened legal action against him? Explain your answer.

25-10 Sears formulated a plan to increase sales of its top-of-the-line Lady Kenmore brand dishwasher. Sear's plan sought to change the Lady Kenmore's image without reengineering or making any mechanical improvements to the dishwasher itself. To accomplish this, Sears undertook a 4-year $8 million advertising campaign that claimed that the Lady Kenmore completely eliminated the need to prerinse and prescrape dishes. As a result of this campaign, sales rose by more than 300 percent. The "no scraping, no prerinsing" claim was not true, however, and Sears had no reasonable basis for asserting the claim. In addition, the owner's manual that customers received after they purchased the dishwasher contracted the claim.

After a thorough investigation, the FTC filed a complaint against Sears, alleging that the advertisements were false and misleading. The final FTC order required Sears to stop making the "no scraping, no prerinsing" claim. The order also prevented Sears from (1) making any "performance claims" for "major home appliances" without first possessing a reasonable basis consisting of substantiating tests or other evidence; (2) misrepresenting any test, survey, or demonstration regarding "major home appliances"; and (3) making any advertising statements not consistent with statements in postpurchase materials supplied to purchasers of "major home appliances." Sears contends that the order is too broad because it covers appliances other than dishwashers and includes "performance claims." Explain whether Sears is correct.

25-11 Millstone applied for a new automobile insurance policy after he moved from Washington, DC, to St. Louis. He was told that a background investigation would be conducted in connection with the application. One week later, he was notified that the policy would not be granted because of a report that the insurance company had received from Investigative Reports, a credit bureau. After repeated efforts to obtain his file, Millstone was informed by Investigative Reports that his former neighbors in Washington considered him a "hippie," a drug user, and a possible political dissident. Investigative Reports refused to discuss the matter further. Has Investigative Reports fulfilled its obligations to Millstone? Explain.

CASE PROBLEMS

25-12 55th Management Corp. in New York City owns residential property that it leases to various tenants. In June 2000, claiming that one of the tenants, Leslie Goldman, owed more than $13,000 in back rent, 55th retained Jeffery Cohen, an attorney, to initiate nonpayment proceedings. Cohen filed a petition in New York state court against Goldman, seeking recovery of the unpaid rent and at least $3,000 in attorneys' fees. After receiving notice of the petition, Goldman filed a suit in a federal district court against Cohen. Goldman contended that the notice of the petition constituted an initial contact that, under the FDCPA, required a validation notice at the time, or within five days, of the notice of the petition. Therefore, Goldman argued that Cohen was in violation of the FDCPA. Should the filing of a suit in state court be considered "communication," requiring a debt collector to provide a validation notice under the FDCPA? Why or why not? *Goldman v. Cohen*, 445 F.3d 152 (2nd Cir. 2006).

25-13 Barry Sussman graduated from law school but also served time in prison for attempting to collect debts by posing as an FBI agent. He theorized that if a debt-collection business collected only debts that it owned as a result of buying checks written on accounts with insufficient funds (NSF checks), it would not be subject to the FDCPA. Sussman formed Check Investors, Inc., to act on his theory. Check Investors bought more that 2.2 million NSF checks, with an estimated face value of about $348 million, for pennies on the dollar. Check Investors, Inc., added a fee of $125 or $130 (more than the legal limit in most states) to the face amount of each check and aggressively pursued its drawer of being a criminal and threatened the person with arrest and prosecution. The threats were false. Check Investors never took steps to initiate a prosecution. The employee contacted the drawer's family members and used "saturation phoning"—phoning drawer numerous times in a short period. Employees used abusive language, referring to drawers as "deadbeats," "retards," "thieves," and "idiots." Check Investors netted more than $10.2 million from its efforts. *Federal Trade Commission v. Check Investors, Inc.*, 502 F.3d 159 (3d Cir. 2007).

25-14 The Nutrition Labeling and Education Act (NLEA) requires packaged food to have a "Nutrition Fact" panel that sets out "nutritional information," including "the total number of calories" per serving. Before the 2010 health care reforms enacted provisions on menu labeling, restaurants were exempt from this requirement. The NLEA also regulated nutritional content claims such as "low sodium" that a purveyor might choose to add to a label. The NLEA permitted a state or city to require restaurants to disclose nutrition information about the food they serve, but expressly preempted state or local attempts to regulate nutritional content claims. New York City Health Code Section 81.50 requires 10 percent of the restaurants in the city, including McDonald's, Burger King, and KFC, to post calorie content information on their menus. The New York State Restaurant Association (NYSRA) filed a suit in a federal district court, contending that the NLEA preempts laws that conflict with federal laws. Was the NYSRA correct? Explain. *New York State Restaurant Association v. New York City Board of Health*, 556 F.3d 114 (2d Cir. 2009).

25-15 CrossCheck, Inc., provides check authorization services to retail merchants. When a customer presents a check, the merchant contacts CrossCheck, which estimates the probability that the check will clear the bank. If the check is within an acceptable statistical range, CrossCheck notifies the merchant. If the check is dishonored, the merchant sends it to CrossCheck, which pays it. CrossCheck then attempts to redeposit the check. If this fails, CrossCheck takes further steps to collect the amount. William Winterstein took his truck to C&P Auto Service Center, Inc., for a tune-up and paid for the service with a check. C&P contacted CrossCheck and, on its recommendation, accepted the check. When the check was dishonored, C&P mailed it to CrossCheck, which reimbursed C&P and sent a letter to Winterstein, requesting payment. Winterstein filed a suit in a federal district court against CrossCheck, asserting that the letter violated the FDCPA. CrossCheck filed a motion for summary judgment. Who won? Explain. *Winterstein v. CrossCheck, Inc.*, 149 F. Supp. 2d 466 (N.D. Ill. 2001).

25-16 Source One Associates, Inc., is based in Poughquag, New York. Peter Easton, Source One's president, is responsible for its daily operations. Between 1995 and 1997, Source One received requests from persons in Massachusetts seeking financial information about individuals and businesses. To obtain this information, Easton first obtained the targeted individuals' credit reports through Equifax Information Services by claiming that the reports would be used only in connection with credit transactions involving the consumers. From the reports, Easton identified financial institutions at which individuals held accounts and then called the institutions to learn the account balances by impersonating as officers of the institutions or as the account holders. The information was then provided to Source One's customers for a fee. Easton did not know why the customers wanted the information. The Commonwealth of Massachusetts filed a suit in a Massachusetts state court against Source One and Easton, alleging violations of the FCRA. Did the defendants violate the FCRA? Explain. *Commonwealth v. Source One Associates, Inc.*, 436 Mass. 118, 763 N.E.2d 42 (2002).

THINKING CRITICALLY ABOUT RELEVANT LEGAL ISSUES

The FTC needs to set tighter standards with regard to advertising aimed at children. The FTCA is supposed to protect consumers from deceptive advertising, and it is about time the FTC does its job and enforces the law.

Children are less sophisticated than adults, and they're unable to separate reality from fiction. Therefore, they are more susceptible to the cunning ploys of marketing and advertising wizards. These people show no shame, endlessly manipulating small children just to make money.

"How are our children being manipulated?" you ask. It's obvious. Every time they turn on the TV, they're subjected to a plethora of commercial advertisements. Many of the TV shows that kids watch are nothing more than half-hour advertisements for a particular toy. In addition, the ads mislead children. In the ads, toy companies show kids looking as happy and satisfied as possible while they play with the toys. The children who see these images are convinced that if they only had the toy, they would be just as happy. When they actually receive the toy, however, they find that it's fun to play with for a few hours, but not much longer. They never experience the continuing climax of joy that the advertisers lead them to think they will. Such disappointments are likely to harm the children psychologically, making them become cynical at a young age.

For these reasons, the FTC must step in to protect our children from these money-hungry marketers. To fail to do so is to jeopardize America's future: its children.

See Chapter 1 of this text for assistance in answering these questions.

1. What primary ethical norm is downplayed by this argument?

2. In this argument, what is the relevant rule of law to which the author refers?

3. What reasons does the author give for tighter control of advertising aimed at children?

4. State opposing arguments to those set out by the author in this essay.

ASSIGNMENT ON THE INTERNET

As e-commerce continues to grow in the United States and abroad, new consumer protection laws are needed. Now that you know something about current consumer protection laws in the United States, use the Internet to research recent developments in consumer protection for transactions through cyberspace. Make a list of recommendations for new regulations or rules you would like to see enacted. What ethical norms are implicit in your recommendations?

ON THE INTERNET

www.lectlaw.com//tcos.html Here is a law library site that is a good place to begin your research about consumer protection issues.

www.law.cornell.edu/wex/debtor_and_creditor The Legal Information Institute provides an overview of debtor–creditor law, as well as links to recent debtor–creditor law decisions.

www.ct.gov/dcp/site/default.asp The Connecticut State Department of Consumer Protection website provides citizens of that state with consumer information. Many states similarly offer some form of consumer protection information online.

www.consumer.ftc.gov The FTC maintains a website with consumer information.

www.nclc.org This is the site of the National Consumer Law Center, which provides a wealth of information on the topic of debtor–creditor relations and consumer protection.

FOR FUTURE READING

Carlson, David Gray. "Cars and Homes in Chapter 13 after the 2005 Amendments to the Bankruptcy Code." *American Bankruptcy Institute Law Review* 14 (2006): 301.

Nickles, Steve H. "Behavioral Effect of New Bankruptcy Law on Management and Lawyers: Collage of Recent Statutes and Cases Discouraging Chapter 11 Bankruptcy." *Arkansas Law Review* 59 (2006): 329.

APPENDIX A
The Constitution of the United States

PREAMBLE

We the People of the United States, in Order to form a more perfect Union, establish Justice, insure domestic Tranquility, provide for the common defence, promote the general Welfare, and secure the Blessings of Liberty to ourselves and our Posterity, do ordain and establish this Constitution for the United States of America.

ARTICLE I

Section 1. All legislative Powers herein granted shall be vested in a Congress of the United States, which shall consist of a Senate and a House of Representatives.

Section 2. [1] The House of Representatives shall be composed of Members chosen every second Year by the People of the several States, and the Electors in each State shall have the Qualifications requisite for Electors of the most numerous Branch of the State Legislature.

[2] No Person shall be a Representative who shall not have attained to the Age of twenty five Years, and been seven Years a Citizen of the United States, and who shall not, when elected, be an Inhabitant of that State in which he shall be chosen.

[3] Representatives and direct Taxes shall be apportioned among the several States which may be included within this Union, according to their respective Numbers, which shall be determined by adding to the whole Number of free Persons, including those bound to Service for a Term of Years, and excluding Indians not taxed, three fifths of all other Persons. The actual Enumeration shall be made within three Years after the first Meeting of the Congress of the United States, and within every subsequent Term of ten Years, in such Manner as they shall by Law direct. The Number of Representatives shall not exceed one for every thirty Thousand, but each State shall have at Least one Representative; and until such enumeration shall be made, the State of New Hampshire shall be entitled to choose three, Massachoosetts eight, Rhode Island and Providence Plantations one, Connecticut five, New York six, New Jersey four, Pennsylvania eight, Delaware one, Maryland six, Virginia ten, North Carolina five, South Carolina five, and Georgia three.

[4] When vacancies happen in the Representation from any State, the Executive Authority thereof shall issue Writs of Election to fill such Vacancies.

[5] The House of Representatives shall choose their Speaker and other Officers and shall have the sole Power of Impeachment.

Section 3. [1] The Senate of the United States shall be composed of two Senators from each State, chosen by the Legislature thereof, for six Years; and each Senator shall have one vote.

[2] Immediately after they shall be assembled in Consequence of the first Election, they shall be divided as equally as may be into three Classes. The Seats of the Senators of the first Class shall be vacated at the Expiration of the Second Year, of the second Class at the Expiration of the fourth Year, and of the third Class at the Expiration of the sixth Year, so that one third may be chosen every second Year, and if Vacancies happen by Resignation, or otherwise, during the Recess of the Legislature of any State, the Executive thereof may make temporary Appointments until the next Meeting of the Legislature, which shall then fill such Vacancies.

[3] No Person shall be a Senator who shall not have attained to the Age of thirty Years, and been nine Years a Citizen of the United States, and who shall not, when elected, be an Inhabitant of that State for which he shall be chosen.

[4] The Vice President of the United States shall be President of the Senate, but shall have no Vote, unless they be equally divided.

[5] The Senate shall choose their other Officers, and also a President pro tempore, in the Absence of the Vice President, or when he shall exercise the Office of President of the United States.

[6] The Senate shall have the sole Power to try all Impeachments. When sitting for that Purpose, they shall be on Oath or Affirmation. When the President of the United States is tried, the Chief Justice shall preside: And no Person shall be convicted without the Concurrence of two thirds of the Members present.

[7] Judgment in Cases of Impeachment shall not extend further than to removal from Office, and disqualification to hold and enjoy any Office of honor, Trust, or Profit under the United States: but the Party convicted shall nevertheless be liable and subject to Indictment, Trial, Judgment, and Punishment, according to Law.

Section 4. [1] The Times, Places and Manner of holding Elections for Senators and Representatives, shall be prescribed in each State by the Legislature thereof; but the Congress may at any time by Law make or alter such Regulations, except as to the Places of choosing Senators.

[2] The Congress shall assemble at least once in every Year, and such Meeting shall be on the first Monday in December, unless they shall by Law appoint a different Day.

Section 5. [1] Each House shall be the Judge of the Elections, Returns, and Qualifications of its own Members, and a Majority of each shall constitute a Quorum to do Business, but a smaller Number may adjourn from day to day, and may be authorized to compel the Attendance of absent Members, in such Manner, and under such Penalties as each House may provide.

[2] Each House may determine the Rules of its Proceedings, punish its Members for Disorderly Behavior, and, with the Concurrence of two thirds, expel a Member.

[3] Each House shall keep a Journal of its Proceedings, and from time to time publish the same, excepting such Parts as may in their Judgment require Secrecy; and the Yeas and Nays of the Members of either House on any question shall, at the Desire of one fifth of those Present, be entered on the Journal.

[4] Neither House, during the Session of Congress, shall, without the Consent of the other, adjourn for more than three days, nor to any other Place than that in which the two Houses shall be sitting.

Section 6. [1] The Senators and Representatives shall receive a Compensation for their Services, to be ascertained by Law, and paid out of the Treasury of the United States. They shall in all Cases, except Treason, Felony and Breach of the Peace, be privileged from Arrest during their Attendance at the Session of their respective Houses, and in going to and returning from the same; and for any speech or Debate in either House, they shall not be questioned in any other Place.

[2] No Senator or Representative shall, during the Time for which he was elected, be appointed to any civil Office under the Authority of the United States, which shall have been created, or the Emoluments whereof shall have been increased during such time and no Person holding any Office under the United States, shall be a Member of either House during his Continuance in Office.

Section 7. [1] All Bills for raising Revenue shall originate in the House of Representatives; but the Senate may propose or concur with Amendments as on other Bills.

[2] Every Bill which shall have passed the House of Representatives and the Senate, shall, before it become a Law, be presented to the President of the United States; If he approve he shall sign it, but if not he shall return it, with his Objections to the House in which it shall have originated, who shall enter the Objections at large on their Journal, and proceed to reconsider it. If after such Reconsideration two thirds of that House shall agree to pass the Bill, it shall be sent together with the Objections, to the other House, by which it shall likewise be reconsidered, and if approved

by two thirds of that House, it shall become a Law. But in all such Cases the Votes of both Houses shall be determined by Yeas and Nays, and the Names of the Persons voting for and against the Bill shall be entered on the Journal of each House respectively. If any Bill shall not be returned by the President within ten Days (Sundays excepted) after it shall have been presented to him, the Same shall be a Law, in like Manner as if he had signed it, unless the Congress by their Adjournment prevent its Return in which Case it shall not be a Law.

[3] Every Order, Resolution, or Vote, to Which the Concurrence of the Senate and House of Representatives may be necessary (except on a question of Adjournment) shall be presented to the President of the United States; and before the Same shall take Effect, shall be approved by him, or being disapproved by him, shall be repassed by two thirds of the Senate and House of Representatives, according to the Rules and Limitations prescribed in the Case of a Bill.

Section 8. [1] The Congress shall have Power To lay and collect Taxes, Duties, Imposts and Excises, to pay the Debts and provide for the common Defence and general Welfare of the United States; but all Duties, Imposts and Excises shall be uniform throughout the United States;

[2] To borrow money on the credit of the United States;

[3] To regulate Commerce with foreign Nations, and among the several States, and with the Indian Tribes;

[4] To establish an uniform Rule of Naturalization, and uniform Laws on the subject of Bankruptcies throughout the United States;

[5] To coin Money, regulate the Value thereof, and of foreign Coin, and fix the Standard of Weights and Measures;

[6] To provide for the Punishment of counterfeiting the Securities and current Coin of the United States;

[7] To Establish Post Offices and Post Roads;

[8] To promote the Progress of Science and useful Arts, by securing for limited Times to Authors and Inventors the exclusive Right to their respective Writings and Discoveries;

[9] To constitute Tribunals inferior to the Supreme Court;

[10] To define and punish Piracies and Felonies committed on the high Seas, and Offenses against the Law of Nations;

[11] To declare War, grant Letters of Marque and Reprisal, and make Rules concerning Captures on Land and Water;

[12] To raise and support Armies, but no Appropriation of Money to that Use shall be for a longer Term than two Years;

[13] To provide and maintain a Navy;

[14] To make Rules for the Government and Regulation of the land and naval Forces;

[15] To provide for calling forth the Militia to execute the Laws of the Union, suppress Insurrections and repel Invasions;

[16] To provide for organizing, arming, and disciplining, the Militia, and for governing such Part of them as may be employed in the Service of the United States, reserving to the States respectively, the Appointment of the Officers, and the Authority of training the Militia according to the discipline prescribed by Congress;

[17] To exercise exclusive Legislation in all Cases whatsoever, over such District (not exceeding ten Miles square) as may, by Cession of particular States, and the Acceptance of Congress, become the Seat of the Government of the United States, and to exercise like Authority over all Places purchased by the consent of the Legislature of the State in which the Same shall be, for the Erection of Forts, Magazines, Arsenals, dock-Yards, and other needful Buildings;—And

[18] To make all Laws which shall be necessary and proper for carrying into Execution the foregoing Powers, and all other Powers vested by this Constitution in the Government of the United States, or in any department or Officer thereof.

Section 9. [1] The Migration or Importation of Such Persons as any of the States now existing shall think proper to admit, shall not be prohibited by the Congress prior to the Year one thousand eight hundred and eight, but a Tax or duty may be imposed on such Importation, not exceeding ten dollars for each Person.

[2] The privilege of the Writ of Habeas Corpus shall not be suspended, unless when in Cases of Rebellion or Invasion the public Safety may require it.

[3] No Bill of Attainder or ex post facto law shall be passed.

[4] No Capitation, or other direct, Tax shall be laid, unless in Proportion to the Census or Enumeration herein before directed to be taken.

[5] No Tax or Duty shall be laid on articles exported from any State.

[6] No Preference shall be given by any Regulation of Commerce or Revenue to the Ports of one State over those of another: nor shall Vessels bound to, or from, one State be obliged to enter, clear, or pay Duties in another.

[7] No money shall be drawn from the Treasury, but in Consequence of Appropriations made by Law; and a regular Statement and Account of the Receipts and Expenditures of all public Money shall be published from time to time.

[8] No Title of Nobility shall be granted by the United States: And no Person holding any Office of Profit or Trust under them, shall, without the Consent of the Congress, accept of any present, Emolument, Office, or Title, of any kind whatever, from any King, Prince, or foreign State.

Section 10. [1] No State shall enter into any Treaty, Alliance, or Confederation; grant Letters of Marque and Reprisal; coin Money; emit Bills of Credit; make any Thing but gold and silver Coin a Tender in Payment of Debts; pass any Bill of Attainder, ex post facto Law, or Law impairing the Obligation of Contracts, or grant any Title of Nobility.

[2] No State shall, without the Consent of the Congress, lay any Imposts or Duties on Imports or Exports, except what may be absolutely necessary for executing its inspection Laws: and the net Produce of all Duties and Imposts, laid by any States on Imports or Exports, shall be for the Use of the Treasury of the United States; and all such Laws shall be subject to the Revision and Control of the Congress.

[3] No State shall, without the Consent of Congress, lay any Duty of Tonnage, keep Troops, or Ships of War in time of Peace, enter into any Agreement or Compact with another State, or with a foreign Power, or engage in War, unless actually invaded, or in such imminent Danger as will not admit of delay.

ARTICLE II

Section 1. [1] The executive Power shall be vested in a President of the United States of America. He shall hold his Office during the Term of four Years, and, together with the Vice President, chosen for the same Term, be elected, as follows:

[2] Each State shall appoint, in such Manner as the Legislature thereof may direct, a Number of Electors, equal to the whole Number of Senators and Representative to which the State may be entitled in the Congress; but no Senator or Representative, or Person holding an Office of Trust or Profit under the United States, shall be appointed as Elector.

[3] The Electors shall meet in their respective States, and vote by Ballot for two Persons, of whom one at least shall not be an Inhabitant of the same State with themselves. And they shall make a List of all the Persons voted for, and of the Number of Votes for each; which List they shall sign and certify, and transmit sealed to the Seat of the Government of the United States, directed to the President of the Senate. The President of the Senate shall, in the Presence of the Senate and House of Representatives, open all the Certificates, and the Votes shall then be counted. The Person having the greatest Number of Votes shall be the President, if such Number be a Majority of the whole Number of Electors appointed; and if there be more than one who have such Majority, and have an equal Number of Votes, then the House of Representatives shall immediately choose by Ballot one of them for President; and if no Person have a Majority, then from the five highest on the List the said House shall in like Manner choose the President. But in choosing the President, the Votes shall be taken by States the Representation from each State having one Vote; A quorum for this Purpose shall consist of a Member or Members from two thirds of the States, and a Majority of all the States shall be necessary to a Choice. In every Case, after the Choice of the President, the Person having the greater Number of Votes of the Electors shall be the Vice President. But if there should remain two or more who have equal Votes, the Senate shall choose from them by Ballot the Vice President.

[4] The Congress may determine the Time of choosing the Electors, and the Day on which they shall give their Votes; which Day shall be the same throughout the United States.

[5] No person except a natural born Citizen, or a Citizen of the United States, at the time of the Adoption of this constitution, shall be eligible to the Office of President; neither shall any Person be eligible to that Office who shall not have attained to the Age of thirty five Years, and been fourteen Years a Resident within the United States.

[6] In case of the removal of the President from Office, or of his Death, Resignation or Inability to discharge the Powers and Duties of the said Office, the Same shall devolve on the Vice President, and the Congress may by Law provide for the Case of Removal, Death, Resignation or Inability, both of the President and Vice President, declaring what Officer shall then act as President, and such Officer shall act accordingly, until the disability be removed, or a President shall be elected.

[7] The President shall, at stated Times, receive for his Services, a Compensation, which shall neither be increased nor diminished during the Period for which he shall have been elected, and he shall not receive within that Period any other Emolument from the United States, or any of them.

[8] Before he enter on the Execution of his Office, he shall take the following Oath or Affirmation: "I do solemnly swear (or affirm) that I will faithfully execute the Office of President of the United States, and will to the best of my Ability, preserve, protect and defend the Constitution of the United States."

Section 2. [1] The President shall be Commander in Chief of the Army and Navy of the United States, and of the militia of the several States, when called into the actual Service of the United States; he may require the Opinion, in writing, of the principal Officer in each of the Executive Departments, upon any Subject relating to the Duties of their respective Offices, and he shall have Power to grant Reprieves and Pardons for Offenses against the United States, except in Cases of Impeachment.

[2] He shall have Power, by and with the Advice and Consent of the Senate to make Treaties, provided two thirds of the Senators present concur, and he shall nominate, and by and with the Advice and Consent of the Senate, shall appoint Ambassadors, other public Ministers and Consuls, Judges of the supreme Court, and all other Officers of the United States, whose Appointments are not herein otherwise provided for, and which shall be established by Law; but the Congress may by Law vest the Appointment of such inferior Officers, as they think proper, in the President alone, in the Courts of Law, or in the Heads of Departments.

[3] The President shall have Power to fill up all Vacancies that may happen during the Recess of the Senate, by granting Commissions which shall expire at the End of their next Session.

Section 3. He shall from time to time give to the Congress Information of the State of the Union, and recommend to their Consideration such Measures as he shall judge necessary and expedient; he may, on extraordinary Occasions, convene both Houses, or either of them, and in Case of Disagreement between them, with Respect to the Time of Adjournment, he may adjourn them to such Time as he shall think proper; he shall receive Ambassadors and other public Ministers; he shall take Care that the Laws be faithfully executed, and shall Commission all the Officers of the United States.

Section 4. The President, Vice President and all civil Officers of the United States shall be removed from Office on Impeachment for, and Conviction of, Treason, Bribery, or other high Crimes and Misdemeanors.

ARTICLE III

Section 1. The judicial Power of the United States, shall be vested in one supreme Court, and in such inferior Courts as the Congress may from time to time ordain and establish. The Judges, both of the supreme and inferior Courts, shall hold their Offices during good Behaviour, and shall, at stated Times, receive for their Services a Compensation, which shall not be diminished during their Continuance in Office.

Section 2. [1] The judicial Power shall extend to all Cases, in Law and Equity, arising under this Constitution, the Laws of the United States, and Treaties made, or which shall be made, under their Authority;—to all Cases affecting Ambassadors, other public Ministers and Consuls;—to all Cases of admiralty and maritime Jurisdiction;—to Controversies to which the United States shall be a Party;—to Controversies between two or more States;—between a State and Citizens of another State;—between Citizens of different States;—between Citizens of the same State claiming Lands under the Grants of different States, and between a State, or the Citizens thereof, and foreign States, Citizens or Subjects.

[2] In all Cases affecting Ambassadors, other public Ministers and Consuls, and those in which a State shall be a Party, the supreme Court shall have original Jurisdiction. In all the other Cases before mentioned, the supreme Court shall have appellate Jurisdiction, both as to Law and Fact, with such Exceptions, and under such Regulations as the Congress shall make.

[3] The trial of all Crimes, except in Cases of Impeachment, shall be by Jury; and such Trial shall be held in the State where the said Crimes shall have been committed; but when not committed within any State, the Trial shall be at such Place or Places as the Congress may by Law have directed.

Section 3. [1] Treason against the United States, shall consist only in levying War against them, or, in adhering to their Enemies, giving them Aid and Comfort. No Person shall be convicted of Treason unless on the Testimony of two Witnesses to the same overt Act, or on Confession in open Court.

[2] The Congress shall have Power to declare the Punishment of Treason, but no Attainder of Treason shall work Corruption of Blood, or Forfeiture except during the Life of the Person attainted.

ARTICLE IV

Section 1. Full Faith and Credit shall be given in each State to the public Acts, Records, and judicial Proceedings of every other State. And the Congress may by general Laws prescribe the Manner in which such Acts, Records and Proceedings shall be proved, and the Effect thereof.

Section 2. [1] The Citizens of each State shall be entitled to all Privileges and Immunities of Citizens in the Several States.

[2] A Person charged in any State with Treason, Felony, or other Crime, who shall flee from Justice, and be found in another State, shall on demand of the executive Authority of the State from which he fled, be delivered up, to be removed to the State having Jurisdiction of the Crime.

[3] No Person held to Service or Labour in one State, under the Laws thereof, escaping into another, shall, in Consequence of any Law or Regulation therein, be discharged from such Service or Labour, but shall be delivered up on Claim of the Party to whom such Service or Labour may be due.

Section 3. [1] New States may be admitted by the Congress into this Union; but no new State shall be formed or erected within the Jurisdiction of any other State; nor any State be formed by the Junction of two or more States, or Parts of States, without the Consent of the Legislatures of the States concerned as well as of the Congress.

[2] The Congress shall have Power to dispose of and make all needful Rules and Regulations respecting the Territory or other Property belonging to the United States; and nothing in this Constitution shall be so construed as to Prejudice any Claims of the United States, or of any particular State.

Section 4. The United States shall guarantee to every State in this Union a Republican Form of Government, and shall protect each of them against Invasion; and on Application of the Legislature, or of the Executive (when the Legislature cannot be convened) against domestic Violence.

ARTICLE V

The Congress, whenever two thirds of both Houses shall deem it necessary, shall propose Amendments to this Constitution, or, on the Application of the Legislatures of two thirds of the several States, shall call a Convention for proposing Amendments, which, in either Case, shall be valid to all Intents and Purposes, as part of this Constitution, when ratified by the Legislatures of three fourths of the several States, or by Conventions in three fourths thereof, as the one or the other Mode of Ratification may be proposed by the Congress; Provided that no Amendment which may be made prior to the Year One thousand eight hundred and eight shall in any Manner affect the first and fourth Clauses in the Ninth Section of the first Article; and that no State, without its Consent, shall be deprived of its equal Suffrage in the Senate.

ARTICLE VI

[1] All Debts contracted and Engagements entered into, before the Adoption of this Constitution shall be as valid against the United States under this Constitution, as under the Confederation.

[2] This Constitution, and the Laws of the United States which shall be made in Pursuance thereof; and all Treaties made, or which shall be made, under the Authority of the United States, shall be the supreme Law of the Land; and the Judges in every State shall be bound thereby, any Thing in the Constitution or Laws of any State to the Contrary notwithstanding.

[3] The Senators and Representatives before mentioned, and the Members of the several State Legislatures, and all executive and judicial Officers, both of the United States and of the several States, shall be bound by Oath or Affirmation, to support this Constitution; but no religious Test shall ever be required as a Qualification to any Office or public Trust under the United States.

ARTICLE VII

The Ratification of the conventions of nine States shall be sufficient for the Establishment of this Constitution between the States so ratifying the Same.

Articles in addition to, and amendment of, the constitution of the united states of america, proposed by congress, and ratified by the legislatures of the several states pursuant to the fifth article of the original constitution.

AMENDMENT I [1791]

Congress shall make no law respecting an establishment of religion, or prohibiting the free exercise thereof; or abridging the freedom of speech, or of the press; or the right of the people peaceably to assemble, and to petition the Government for a redress of grievances.

AMENDMENT II [1791]

A well regulated Militia, being necessary to the security of a free State, the right of the people to keep and bear Arms, shall not be infringed.

AMENDMENT III [1791]

No Soldier shall, in time of peace be quartered in any house, without the consent of the Owner, nor in time of war, but in a manner to be prescribed by law.

AMENDMENT IV [1791]

The right of the people to be secure in their persons, houses, papers, and effects, against unreasonable searches and seizures, shall not be violated, and no Warrants shall issue, but upon probable cause, supported by Oath or affirmation, and particularly describing the place to be searched, and the persons or things to be seized.

AMENDMENT V [1791]

No person shall be held to answer for a capital, or otherwise infamous crime, unless on a presentment or indictment of a Grand Jury, except in cases arising in the land or naval forces, or in the Militia, when in actual service in time of War or public danger, nor shall any person be subject for the same offence to be twice put in jeopardy of life or limb; nor shall be compelled in any criminal case to be a witness against himself, nor be deprived of life, liberty, or property, without due process of law; nor shall private property be taken for public use, without just compensation.

AMENDMENT VI [1791]

In all criminal prosecutions, the accused shall enjoy the right to a speedy and public trial, by an impartial jury of the State and district wherein the crime shall have been committed, which district shall have been previously ascertained by law, and to be informed of the nature and cause of the accusation; to be confronted with the witnesses against him; to have compulsory process for obtaining witnesses in his favor, and to have the Assistance of Counsel for his defence.

AMENDMENT VII [1791]

In Suits at common law, where the value in controversy shall exceed twenty dollars, the right of trial by jury shall be preserved, and no fact tried by jury, shall be otherwise re-examined in any Court of the United States, than according to the rules of the common law.

AMENDMENT VIII [1791]

Excessive bail shall not be required, nor excessive fines imposed, nor cruel and unusual punishments inflicted.

AMENDMENT IX [1791]

The enumeration in the Constitution, of certain rights, shall not be construed to deny or disparage others retained by the people.

AMENDMENT X [1791]

The powers not delegated to the United States by the Constitution, nor prohibited by it to the States, are reserved to the States respectively, or to the people.

AMENDMENT XI [1798]

The Judicial power of the United States shall not be construed to extend to any suit in law or equity, commenced or prosecuted against one of the United States by Citizens of another State, or by Citizens or Subjects of any Foreign State.

AMENDMENT XII [1804]

The Electors shall meet in their respective states and vote by ballot for President and Vice-President, one of whom, at least, shall not be an inhabitant of the same state with themselves; they shall name in their ballots the person voted for as President, and in distinct ballots the person voted for as Vice-President, and they shall make distinct lists of all persons voted for as President, and of all persons voted for as Vice-President, and of the number of votes for each, which lists they shall sign and certify, and transmit sealed to the seat of the government of the United States, directed to the President of the Senate;—The President of the Senate shall, in the presence of the Senate and House of Representatives, open all the certificates and the votes shall then be counted;—The person having the greatest number of votes for President, shall be the President, if such number be a majority of the whole number of Electors appointed; and if no person have such majority, then from the persons having the highest numbers not exceeding three on the list of those voted for as President, the House of Representatives shall choose immediately, by ballot, the President. But in choosing the President, the votes shall be taken by states, the representation from each state having one vote; a quorum for this purpose shall consist of a member or members from two-thirds of the states, and a majority of all the states shall be necessary to a choice. And if the House of Representatives shall not choose a President whenever the right of choice shall devolve upon them before the fourth day of March next following, then the Vice-President shall act as President, as in the case of the death or other constitutional disability of the President.—The person having the greatest number of votes as Vice-President, shall be the Vice-President, if such number be a majority of the whole number of Electors appointed, and if no person have a majority, then from the two highest numbers on the list, the Senate shall choose the Vice-President; a quorum for the purpose shall consist of two-thirds of the whole number of Senators, and a majority of the whole number shall be necessary to a choice. But no person constitutionally ineligible to the office of President shall be eligible to that of Vice-President of the United States.

AMENDMENT XIII [1865]

Section 1. Neither slavery nor involuntary servitude, except as a punishment for crime whereof the party shall have been duly convicted, shall exist within the United States, or any place subject to their jurisdiction.

Section 2. Congress shall have power to enforce this article by appropriate legislation.

AMENDMENT XIV [1868]

Section 1. All persons born or naturalized in the United States, and subject to the jurisdiction thereof, are citizens of the United States and of the State wherein they reside. No State shall make or enforce any law which shall abridge the privileges or immunities of citizens of the United States; nor shall any State deprive any person of life, liberty, or property, without due process of law; nor deny to any person within its jurisdiction the equal protection of the laws.

Section 2. Representatives shall be apportioned among the several States according to their respective numbers, counting the whole number of persons in each State excluding Indians not taxed. But when the right to vote at any election for the choice of electors for President and Vice President of the United States, Representatives in Congress, the Executive and Judicial officers of a State, or the members of the Legislature thereof, is denied to any of the male inhabitants of such State, being twenty-one years of age, and citizens of the United States, or in any way abridged, except for participation in rebellion, or other crime, the basis of representation therein shall be reduced in the proportion which the number of such male citizens shall bear to the whole number of male citizens twenty-one years of age in such State.

Section 3. No person shall be a Senator or Representative in Congress, or elector of President and Vice President, or hold any office, civil or military, under the United States, as a member of any State, who having previously taken an oath, as a member of Congress, or as an officer of the United States, or as a member of any State legislature, or as an executive or judicial officer of any State, to support the Constitution of the United States, shall have engaged in insurrection or rebellion against the same, or given aid or comfort to the enemies thereof. But Congress may by a vote of two-thirds of each House, remove such disability.

Section 4. The validity of the public debt of the United States, authorized by law, including debts incurred for payment of pensions and bounties for services in suppressing insurrection or rebellion, shall not be questioned. But neither the United States nor any State shall assume or pay any debt or obligation incurred in aid of insurrection or rebellion against the United States, or any claim for the loss or emancipation of any slave; but all such debts, obligations and claims shall be held illegal and void.

Section 5. The Congress shall have power to enforce, by appropriate legislation, the provisions of this article.

AMENDMENT XV [1870]

Section 1. The right of citizens of the United States to vote shall not be denied or abridged by the United States or by any State on account of race, color, or previous condition of servitude.

Section 2. The Congress shall have power to enforce this article by appropriate legislation.

AMENDMENT XVI [1913]

The Congress shall have power to lay and collect taxes on incomes, from whatever source derived, without apportionment among the several States, and without regard to any census or enumeration.

AMENDMENT XVII [1913]

[1] The Senate of the United States shall be composed of two Senators from each State, elected by the people thereof, for six years and each Senator shall have one vote. The electors in each State shall have the qualifications requisite for electors of the most numerous branch of the State legislatures.

[2] When vacancies happen in the representation of any State in the Senate, the executive authority of such State shall issue writs of election to fill such vacancies: **Provided,** That the legislature of any State may empower the executive thereof to make temporary appointments until the people fill the vacancies by election as the legislature may direct.

[3] This amendment shall not be so construed as to affect the election or term of any Senator chosen before it becomes valid as part of the Constitution.

AMENDMENT XVIII [1919]

Section 1. After one year from the ratification of this article the manufacture, sale, or transportation of intoxicating liquors within, the importation thereof into, or the exportation thereof from the United States and all territory subject to the jurisdiction thereof for beverage purposes is hereby prohibited.

Section 2. The Congress and the several States shall have concurrent power to enforce this article by appropriate legislation.

Section 3. This article shall be inoperative unless it shall have been ratified as an amendment to the Constitution by the legislatures of the several States, as provided in the Constitution, within seven years from the date of the submission hereof to the States by the Congress.

AMENDMENT XIX [1920]

[1] The right of citizens of the United States to vote shall not be denied or abridged by the United States or by any State on account of sex.

[2] Congress shall have power to enforce this article by appropriate legislation.

AMENDMENT XX [1933]

Section 1. The terms of the President and Vice President shall end at noon on the 20th day of January, and the terms of Senators and Representatives at noon on the 3d day of January, of the years in which such terms would have ended if this article had not been ratified; and the terms of their successors shall then begin.

Section 2. The Congress shall assemble at least once in every year, and such meeting shall begin at noon on the 3d day of January, unless they shall by law appoint a different day.

Section 3. If, at the time fixed for the beginning of the term of the President, the President elect shall have died, the Vice President elect shall become President. If the President shall not have been chosen before the time fixed for the beginning of his term, or if the President elect shall have failed to qualify, then the Vice President elect shall act as President until a President shall have qualified; and the Congress may by law provide for the case wherein neither a President elect nor a Vice President elect shall have qualified, declaring who shall then act as President, or the manner in which one who is to act shall be selected, and such person shall act accordingly until a President or Vice President shall have qualified.

Section 4. The Congress may by law provide for the case of the death of any of the persons from whom the House of Representatives may choose a President whenever the right of choice shall have devolved upon them, and for the case of the death of any of the persons from whom the Senate may choose a Vice President whenever the right of choice shall have devolved upon them.

Section 5. Sections 1 and 2 shall take effect on the 15th day of October following the ratification of this article.

Section 6. This article shall be inoperative unless it shall have been ratified as an amendment to the Constitution by the legislatures of three-fourths of the several States within seven years from the date of its submission.

AMENDMENT XXI [1933]

Section 1. The eighteenth article of amendment to the Constitution of the United States is hereby repealed.

Section 2. The transportation or importation into any State, Territory, or possession of the United States for delivery or use therein of intoxicating liquors, in violation of the laws thereof, is hereby prohibited.

Section 3. This article shall be inoperative unless it shall have been ratified as an amendment to the Constitution by conventions in the several States, as provided in the Constitution, within seven years from the date of the submission hereof to the States by the Congress.

AMENDMENT XXII [1951]

Section 1. No person shall be elected to the office of the President more than twice, and no person who has held the office of President, or acted as President, for more than two years of a term to which some other person was elected President shall be elected to the office of President more than once. But this Article shall not apply to any person holding the office of President when this Article was proposed by the Congress, and shall not prevent any person who may be holding the office of President, or acting as President, during the term within which this Article becomes operative from holding the office of President or acting as President during the remainder of such term.

Section 2. This article shall be inoperative unless it shall have been ratified as an amendment to the Constitution by the legislatures of three-fourths of the several States within seven years from the date of its submission to the States by the Congress.

AMENDMENT XXIII [1961]

Section 1. The District constituting the seat of Government of the United States shall appoint in such manner as the Congress may direct:

A number of electors of President and Vice President equal to the whole number of Senators and Representatives in Congress to which the District would be entitled if it were a State, but in no event more than the least populous state; they shall be in addition to those appointed by the states, but they shall be considered, for the purposes of the election of President and Vice President, to be electors appointed by a state; and they shall meet in the District and perform such duties as provided by the twelfth article of amendment.

Section 2. The Congress shall have power to enforce this article by appropriate legislation.

AMENDMENT XXIV [1964]

Section 1. The right of citizens of the United States to vote in any primary or other election for President or Vice President, for electors for President or Vice President, or for Senator or Representative in Congress, shall not be denied or abridged by the United States, or any State by reason of failure to pay any poll tax or other tax.

Section 2. The Congress shall have power to enforce this article by appropriate legislation.

AMENDMENT XXV [1967]

Section 1. In case of the removal of the President from office or of his death or resignation, the Vice President shall become President.

Section 2. Whenever there is a vacancy in the office of the Vice President, the President shall nominate a Vice President who shall take office upon confirmation by a majority vote of both Houses of Congress.

Section 3. Whenever the President transmits to the President pro tempore of the Senate and the Speaker of the House of Representatives his written declaration that he is unable to discharge the powers and duties of his office, and until he transmits to them a written declaration to the contrary, such powers and duties shall be discharged by the Vice President as Acting President.

Section 4. Whenever the Vice President and a majority of either the principal officers of the executive departments or of such other body as Congress may by law provide, transmit to the President pro tempore of the Senate and the Speaker of the House of Representatives their written declaration that the President is unable to discharge the powers and duties of his office, the Vice President shall immediately assume the powers and duties of the office as Acting President.

Thereafter, when the President transmits to the President pro tempore of the Senate and the Speaker of the House of Representatives his written declaration that no inability exists, he shall resume the powers and duties of his office unless the Vice President and a majority of either the principal officers of the executive department or of such other body as Congress may by law provide, transmit within four days to the President pro tempore of the Senate and the Speaker of the House of Representatives their written declaration and the President is unable to discharge the powers and duties of his office. Thereupon Congress shall decide the issue, assembling within forty-eight hours for that purpose if not in session. If the Congress, within twenty-one days after receipt of the latter written declaration, or, if Congress is not in session, within twenty-one days after Congress is required to assemble, determines by two-thirds vote of both Houses that the President is unable to discharge the power and duties of his office, the Vice President shall continue to discharge the same as Acting President; otherwise, the President shall resume the powers and duties of his office.

AMENDMENT XXVI [1971]

Section 1. The right of citizens of the United States, who are eighteen years of age or older, to vote shall not be denied or abridged by the United States or by any State on account of age.

Section 2. The Congress shall have power to enforce this article by appropriate legislation.

AMENDMENT XXVII [1996]

No law, varying the compensation for the services of the Senators and Representatives, shall take effect, until an election of Representatives shall have intervened.

Glossary

24-hour rule Prohibits both union representatives and employers from making speeches to "captive audiences" of employees within 24 hours of a representative election.

absolute privilege The right to make any statement, true or false, about someone and not be held liable for defamation.

acid rain Precipitation with a high acidic content (pH level of less than 5) caused by atmospheric pollutants.

act-of-state doctrine A state that each nation is bound to respect the independence of another and the courts of one nation will not sit in judgment on the acts of the courts of another nation.

actual authority Includes expressed authority as well as implied authority, or that authority customarily given to an agent in an industry, a trade, or a profession.

administrative agency Any body that is created by the legislative branch to carry out specific duties.

administrative law judge (ALJ) A judge, selected on the basis of a merit exam, who is assigned to a specific administrative agency.

administrative law Any rule (statute or regulation) that directly or indirectly affects an administrative agency.

Administrative Procedure Act (APA) Law that establishes the standards and procedures that federal administrative agencies must follow in their rulemaking and adjudicative functions.

adversarial system System of litigation in which the judge hears evidence and arguments presented by both sides in a case and then makes an objective decision based on the facts and the law as presented by each side.

adverse possession Acquiring ownership of realty by openly treating it as one's own, with neither protest nor permission from the real owner, for a statutorily established period of time.

affirm Term for an appellate court's decision to uphold the decision of a lower court in a case that has been appealed.

affirmative action plans Programs adopted by employers to increase the representation of women and minorities in their workforces.

Age Discrimination in Employment Act of 1967 (ADEA) Statute that prohibits employers from refusing to hire, discharging, or discriminating against people in terms or conditions of employment on the basis of age.

agency A fiduciary relationship between two persons in which one (the agent) acts on behalf of, and is subject to the control of, the other (the principal).

agency by estoppel (apparent authority) An agency relationship in which the principal is estopped from denying that someone is the principal's agent after leading a third party to believe that the person is an agent.

agency by implied authority Agency relationship in which customs and circumstances, rather than a detailed formal agreement, determine the agent's authority.

agency by ratification Agency relationship in which an unauthorized agent commits the principal to an agreement and the principal later accepts the unauthorized agreement, thus ratifying the agency relationship.

alternative dispute resolution (ADR) Resolving legal disputes through methods other than litigation, such as negotiation and settlement, arbitration, mediation, private trials, minitrials, summary jury trials, and early neutral case evaluation.

ambiguous Possessing two or more possible interpretations.

Americans with Disabilities Act of 1991 (ADA) Statute requiring that employers make reasonable accommodations to the known disabilities of an otherwise qualified job applicant or employee with a disability, unless the necessary accommodation would impose an undue burden on the employer's business.

analogy A comparison based on the assumption that if two things are alike in some respect, they must be alike in other respects.

annual percentage rate (APR) The effective annual rate of interest being charged a consumer by a creditor that depends on the compounding period the creditor is using.

appellate jurisdiction The power to review a decision previously made by a trial court.

appropriate bargaining unit May be an entire plant, a single department, or all employees of a single employer, as long as there is a mutuality of interest among the proposed members of the unit.

appropriation A privacy tort that consists of using a person's name or likeness for commercial gain without the person's permission. The decision of the arbitrator is binding on both parties.

arbitration A dispute resolution method whereby the disputing parties submit their disagreement to a mutually agreed-upon neutral decision maker or one provided for by statute. The decision of the arbitrator is binding on both parties.

arraignment Formal appearance of the defendant in court to answer the indictment by entering a plea of guilty or not guilty.

arrest To seize and hold under the authority of the law.

artisan's lien A lien that enables a creditor to recover payment from a debtor for labor and services provided on the debtor's personal property (e.g., fixing a lawn mower).

assault Intentional placing of a person in fear or apprehension of an immediate, offensive bodily contact.

assignment The present transfer of an existing right.

assumption of the risk A defense to negligence based on showing that the plaintiff voluntarily and unreasonably encountered a known risk and that the harm the plaintiff suffered was the harm that was risked.

attachment A court-ordered judgment allowing a local officer of the court to seize property of a debtor.

attorney–client privilege Provides that information furnished by a client to an attorney in confidence, in conjunction with a legal matter, may not be revealed by the attorney without the client's permission.

award The arbitrator's decision.

bail An amount of money the defendant pays to the court upon release from custody as security that he or she will return for trial.

bailment A relationship in which one person (the bailor) transfers possession of personal property to another (the bailee) to be used in an agreed-on manner for an agreed-on period of time.

battery Intentional, unwanted, and offensive bodily contact.

bilateral contract The exchange of one promise for another promise.

bilateral investment treaty (BIT) Treaty between two parties to outline conditions for investment in either country.

binding arbitration clause A provision in a contract mandating that all disputes arising under the contract be settled by arbitration.

bonds Long-term loans secured by a lien or mortgage on corporate assets.

boycott A refusal to deal with, purchase goods from, or work for a business.

breach of contract Failure of one of the parties to perform their obligations under the contract at the time performance is due.

bribery The offering, giving, soliciting, or receiving of money or any object of value for the purpose of influencing the judgment or conduct of a person in a position of trust, especially a government official.

broker A person engaged in the business of buying and selling securities for others' accounts.

business ethics The study of what makes up good and bad conduct as related to business activities and values.

business judgment rule A rule that says that corporate officers and directors are not liable for honest mistakes of business judgment.

callable bonds A bond which the issuer has the right to redeem prior to its maturity date, under certain conditions.

capital structure The percentage of each type of capital—debt, preferred stock, and common equity—used by a corporation.

case law Law resulting from judicial interpretations of constitutions and statutes.

Chicago School An approach to antitrust policy that is based solely on the goal of economic efficiency or the maximization of consumer welfare.

civil law Law governing litigation between two private parties.

Civil Rights Act of 1866 Statute guaranteeing that all persons in the United States have the same right to make and enforce contracts and have the full and equal benefit of the law.

Civil Rights Act of 1871 Statute that prohibits discrimination by state and local governments.

class action suit A lawsuit brought by a member of a group of persons on behalf of all members of the group.

Clayton Act Prohibits price discrimination, tying, exclusive-dealing arrangements, and corporate mergers that substantially lessen competition or tend to create a monopoly in interstate commerce.

closed-end credit A credit arrangement in which credit is extended for a specific period of time and the exact number of payments and the total amount due have been agreed upon between the borrower and the creditor.

closely held corporation A corporation whose stock is not traded on the national securities exchanges but is privately held by a small group of people.

collective bargaining Negotiations between an employer and a union primarily over wages, hours, and terms and conditions of employment.

collusion Concerted action by two or more individuals or business entities in violation of the Sherman Act.

Commerce Clause Empowers Congress to regulate commerce with foreign nations, with Indian tribes, and among the states; found in Article I.

commercial impracticability Situation that makes performance of a contract unreasonably expensive, injurious, or costly to a party.

common stock A class of stock that entitles its owner to vote for the corporation's board of directors, receive dividends, and participate in the net assets upon liquidation of the corporation.

comparative negligence A defense that allocates recovery based on percentage of fault allocated to plaintiff and defendant; available in either pure or modified form.

compensatory damages Monetary damages awarded for a breach of contract that results in higher costs or lost profits for the injured party.

competency A person's ability to understand the nature of the transaction and the consequences of entering into it at the time the contract was entered into.

complaint The initial pleading in a case that states the names of the parties to the action, the basis for the court's subject matter jurisdiction, the facts on which the party's claim is based, and the relief that the party is seeking.

complete performance Completion of all the terms of the contract.

conclusion A position or stance on an issue; the goal toward which reasoning pushes us.

concurrent jurisdiction Applies to cases that may be heard in either the federal or the state court system.

condemnation The process whereby the government acquires the ownership of private property for a public use over the protest of the owner.

condition precedent A particular event that must take place to give rise to a duty of performance of a contract.

condition subsequent A particular event that, when it follows the execution of a contract, terminates the contract.

conditional estate The right to own and possess the land, subject to a condition whose happening (or nonhappening) will terminate the estate.

conditional privilege The right to make a false statement about someone and not be held liable for defamation provided the statement was made without malice.

conflict of interest A conflict that occurs when a corporate officer or director enters into a transaction with the corporation in which he or she has a personal interest.

conglomerate merger A merger in which the businesses of the acquiring and the acquired firm are totally unrelated.

conscious parallelism Identical actions (usually price increases) that are taken independently but nearly simultaneously by two or more leading companies in an industry.

consent order An agreement by a business to stop an activity that an administrative agency alleges to be unlawful and to accept the remedy the agency imposes; no admission of guilt is necessary.

consideration A bargained-for exchange of promises in which a legal detriment is suffered by the promisee.

contingency fee Agent's compensation that consists of a percentage of the amount the agent secured for the principal in a business transaction.

contract A legally enforceable exchange of promises or an exchange of a promise for an act.

contributory negligence A defense to negligence that consists of proving that the plaintiff did not exercise the ordinary degree of care to protect against an unreasonable risk of harm and that this failure contributed to causing the plaintiff's harm.

conversion Intentional permanent removal of property from the rightful owner's possession and control.

convertible bonds Bonds that can be exchanged at a fixed rate for another type of security.

co-ownership Ownership of land by multiple persons or business organizations; all tenants have an equal right to occupy all of the property.

copyright The exclusive legal right to reproduce, publish, and sell the fixed form of an expression of an original creative idea.

corporate opportunity doctrine A doctrine, established by case law, that says that corporate officers, directors, and agents cannot take personal advantage of an opportunity that in all fairness should have belonged to the corporation.

corporation An entity formed and authorized by state law to act as a single legal person and to raise capital by issuing stock to investors who are the owners of the corporation.

counterclaim Defendant's statement of facts showing cause for action against the plaintiff and a request for appropriate relief.

creditor The lender in a transaction.

creditor–beneficiary contract A contract in which the promisee obtains a promise from the promisor to fulfill a legal obligation of the promisee to a third party.

criminal fraud Intentional use of some sort of misrepresentation to gain an advantage over another party.

criminal law Composed of federal and state statutes prohibiting wrongful conduct ranging from murder to fraud.

critical thinking skills The ability to understand the structure of an argument and apply a set of evaluative criteria to assess its merits.

cross-elasticity of demand or substitutability If an increase in the price of one product leads consumers to purchase another product, the two products are substitutable and there is said to be cross-elasticity of demand.

cross-licensing An illegal practice in which two patent holders license each other to use their patented objects only on condition that neither will license anyone else to use those patented objects without the other's consent.

culture Learned norms of society based on values and beliefs.

dealer A person engaged in the business of buying and selling securities for his or her own account.

debentures Unsecured long-term corporate loans.

debtor The borrower in a transaction.

deed Instrument of conveyance of property.

defamation Intentional publication (communication to a third party) of a false statement that is harmful to the plaintiff's reputation.

defendant Party against whom an action is being brought.

deposition Pretrial testimony by witnesses who are examined under oath.

disclaimer Disavowal of liability for breach of warranty by the manufacturer or seller of a good in advance of the sale of the good.

disclosed principal One whose identity is known by the third party when the latter enters into an agreement negotiated by the agent.

discovery The pretrial gathering of information from each other by the parties.

disparagement Intentionally defaming a business product or service.

disparate impact Occurs when the employer's facially neutral policy or practice has a discriminatory effect on employees who belong to a protected class.

disparate treatment Occurs when the employer treats one employee less favorably than another because of that employee's color, race, religion, sex, or national origin.

donative intent Intent to transfer ownership to another at the time the donor makes actual or constructive delivery of the gift to the donee.

donee–beneficiary contract A contract in which the promisee obtains a promise from the promisor to make a gift to a third party.

due diligence defense An affirmative defense raised in lawsuits charging misrepresentation in a registration statement. It is based on the defendant's claim to have had reasonable grounds to believe that all statements in the registration statement were true and no omission of material fact had been made. This defense is not available to the issuer of the security.

Due Process Clause Provides that no one can be deprived of life, liberty, or property without "due process of law"; found in the Fifth Amendment.

duress Any wrongful act or threat that prevents a party from exercising free will when executing a contract.

duress defense An affirmative defense claiming that the defendant was forced to commit the wrongful act by threat of immediate bodily harm or loss of life.

early neutral case evaluation When parties explain their respective positions to a neutral third party who then evaluates the strengths and weaknesses of the cases. This evaluation then guides the parties in reaching a settlement.

easement An irrevocable right to use some portion of another's land for a specific purpose.

economic strike A nonviolent work stoppage for the purpose of obtaining better terms and conditions of employment under a collective bargaining agreement.

effluent limitations Maximum allowable amounts of pollutants that can be discharged from a point source within a given time period.

embezzlement The wrongful conversion of the property of another by one who is lawfully in possession of that property.

eminent domain The constitutional right of the government to take privately owned real property for a public purpose in exchange for just compensation to the owner.

employer–employee relationship Relationship in which an agent (employee) who works for pay and is subject to the control of the principal (employer) may enter into contractual relationships on the latter's behalf.

employer–independent contractor relationship Relationship in which the agent (independent contractor) is hired by the principal (employer) to do a specific job but is not controlled with respect to physical conduct or details of work performance.

employment-at-will doctrine A contract of employment for an indeterminate term is terminable at will by either the employer or the employee; the traditional American rule governing employer–employee relations.

enabling legislation Legislation that grants lawful power to an administrative agency to issue rules, investigate potential violations of rules or statutes, and adjudicate disputes.

entrapment An affirmative defense claiming that the idea for the crime did not originate with the defendant but was put into the defendant's mind by a police officer or other government official.

Environmental Impact Statement (EIS) A statement that must be prepared for every major federal activity that would significantly affect the quality of the human environment.

Environmental Protection Agency (EPA) The federal agency charged with the responsibility of conducting an integrated, coordinated attack on all forms of pollution of the environment.

Equal Pay Act of 1963 Statute that prohibits wage discrimination based on sex.

equitable remedies Nonmonetary damages awarded for breach of contract when monetary damages would be inadequate or impracticable.

estoppel A legal bar to either alleging or denying a fact because of one's previous words or actions to the contrary.

ethical norms Standards of conduct that we consider good or virtuous.

ethics The study of what makes up good and bad conduct, inclusive of related actions and values.

exchange market A securities market that provides a physical facility for the buying and selling of stocks and prescribes the number and qualifications of its broker-members. These brokers buy and sell stocks through the exchange's registered specialists, who are dealers on the floor of the exchange.

exclusive-dealing contract Agreement in which one party requires another party to sell and promote only the brand of goods supplied by the first party.

exclusive federal jurisdiction Applies to cases that may be heard only in the federal court system.

executed contract A contract of which all the terms have been performed.

executive administrative agency An agency located within a department of the executive branch of government; heads and appointed members serve at the pleasure of the president.

executive exemption Exemption to the ADEA that allows mandatory retirement of executives at age 65.

executive power The power delegated by Congress to an administrative agency to investigate whether the rules enacted by the agency have been properly followed by businesses and individuals.

express contract An exchange of oral or written promises between parties, which are enforceable in a court of law.

express warranty A warranty that is clearly stated by the seller or manufacturer.

expressed agency (agency by agreement) Agency relationship formed through oral or written agreement.

expressed authority Authority that arises from specific statements made by the principal (employer) to the agent (employee).

expropriation The taking of private property by a host-country government for political or economic reasons.

fair use doctrine A legal doctrine providing that a portion of a copyrighted work may be reproduced for purposes of "criticism, comment, news reporting, teaching (including multiple copies for classroom use), scholarships, and research."

false light A privacy tort that consists of intentionally taking actions that would lead observers to make false assumptions about the person.

Family and Medical Leave Act (FMLA) A law designed to guarantee that workers facing a medical catastrophe or certain specified family responsibilities will be able to take needed time off from work without pay but without losing medical benefits or their jobs.

federal preemption Constitutional doctrine stating that in an area in which federal regulation is pervasive, state legislation cannot stand.

federal supremacy Principle declaring that any state or local law that directly conflicts with the federal Constitution, laws, or treaties is void.

federalism A system of government in which power is divided between a central authority and constituent political units.

fee simple absolute The right to own and possess the land against all others, without conditions.

felony A serious crime that is punishable by death or imprisonment in a penitentiary.

Fifth Amendment Protects individuals against self-incrimination and double jeopardy and guarantees them the right to trial by jury; protects both individuals and businesses through the Due Process Clause and the Takings Clause.

First Amendment Guarantees freedom of speech, press, and religion and the right to peacefully assemble and to petition the government for redress of grievances.

first appearance Appearance of the defendant before a magistrate, who determines whether there was probable cause for the arrest.

fixture An item that is initially a piece of personal property but is later attached permanently to the realty and is treated as part of the realty.

foreign subsidiary A company that is wholly or partially owned and controlled in a company based in another country.

Fourteenth Amendment Applies the entire Bill of Rights, excepting parts of the Fifth Amendment, to the states.

Fourth Amendment Protects the right of individuals to be secure in their persons, homes, and personal property by prohibiting the government from conducting unreasonable searches of individuals and seizing their property.

franchising A commercial agreement between a party that owns a trade name or trademark (the franchisor) and a party that sells or distributes goods or services using that trade name or trademark (the franchisee).

fraud Misrepresentation of a material fact made with intent to deceive the other party to a contract, who reasonably relied on the misrepresentation and was injured as a result. *See also* criminal fraud.

freedom-to-contract doctrine Parties who are legally competent are allowed to enter into whatever contracts they wish.

full warranty Under the Magnuson-Moss Warranty Act, a written protection for buyers that guarantees free repair of a defective product. If the product cannot be fixed, the consumer must be given a choice of a refund or a replacement free of charge.

future interest The present right to possess and own the land in the future.

garnishment An order of the court granted to a creditor to seize wages or bank accounts of a debtor.

general partnership A partnership in which management responsibilities and profits are divided (usually equally) among the partners and all partners have unlimited personal liability for the partnership's debts.

general warranty deed A deed that promises that the grantor owns the land and has the right to convey it and that the land has no encumbrances other than those stated in the deed.

genuine assent Assent to a contract that is free of fraud, duress, undue influence, and mutual mistake.

good-faith bargaining Following procedural standards laid out in Section 8 of the NLRA; failure to bargain in good faith, by either the employer or the union, is an unfair labor practice.

grand jury A group of 12 to 23 citizens convened in private to decide whether enough evidence exists to try the defendant for a felony.

guaranty Similar to a suretyship except that the third person is secondarily liable (i.e., required to pay only after the debtor has defaulted).

Harvard School An approach to antitrust policy that is based on the desirability of preserving competition to prevent the accumulation of economic and political power, the dislocation of labor, and market inefficiency.

hazardous waste Any waste material that is ignitable, corrosive, reactive, or toxic when ingested or absorbed.

horizontal boycott A concerted refusal by a trade association to deal with members that do not follow the association's regulations.

horizontal division of markets Collusion between two or more competitors to divide markets, customers, or product lines among themselves.

horizontal merger A merger between two or more companies producing the same or a similar product and competing for sales in the same geographic market.

horizontal price-fixing Collusion between two or more competitors to set prices for a product or service, directly or indirectly.

horizontal restraint of trade Restraint of trade that occurs between competitors at the same level of the marketing structure.

hostile bid A tender offer that is opposed by the management of the target company.

implied (implied-in-fact) contract A contract that is established by the conduct of a party rather than by the party's written or spoken words.

implied covenant of good faith and fair dealing An exception to the employment-at-will doctrine, based on the theory that every employment contract, even an unwritten one, contains the implicit understanding that the parties will deal fairly with each other.

implied warranty A warranty that automatically arises out of a transaction.

implied warranty of fitness for a particular purpose A warranty that arises when the seller tells the consumer a good is fit for a specific use.

implied warranty of merchantability A warranty that a good is reasonably fit for ordinary use.

impossibility of performance Situation in which the party cannot legally or physically perform the contract.

in personam jurisdiction Jurisdiction over the person; the power of a court to render a decision that affects the legal rights of a specific person.

in rem jurisdiction The power of a court to render a decision that affects property directly rather than the owner of the property.

income bond A bond whose payment of interest is dependant on sufficient earnings.

indemnity Obligation of the principal to reimburse the agent for any losses the agent incurs while acting on the principal's behalf.

independent administrative agency An agency whose appointed heads and members serve for fixed terms and cannot be removed by the president except for reasons defined by Congress.

indictment A formal written accusation in a felony case.

information A formal written accusation in a misdemeanor case.

informational picketing Picketing designed to truthfully inform the public of a labor dispute.

injunction Temporary or permanent court order preventing a party to a contract from doing something.

insanity defense An affirmative defense claiming that the defendant's mental condition precluded him or her from understanding the wrongful nature of the act committed or from distinguishing wrong from right in general.

insider trading The use of material, nonpublic information received from a corporate source by someone who has a fiduciary obligation to shareholders and potential investors and who benefits from trading on such information.

intangible property Personal property that does not have a physical form and is usually evidenced in writings (e.g., an insurance policy).

intentional infliction of emotional distress Intentionally engaging in outrageous conduct that is likely to cause extreme emotional pain to the person toward whom the conduct is directed.

intentional interference with a contract Knowingly and successfully taking action for the purpose of enticing a third party to breach a valid contract with the plaintiff.

intentional tort A civil wrong that involves taking some purposeful action that the defendant knew, or should have known, would harm the person, property, or economic interests of the plaintiff.

international franchising Contractual agreement whereby a company (licensor) permits another company (licensee) to market its trademarked goods or services in a particular nation.

international licensing Contractual agreements by which a company (licensor) makes its intellectual property available to a foreign individual or company (licensee) for payment.

international trade The export of goods and services from a country, and the import of goods and services into a country.

invasion of privacy A privacy tort that consists of encroaching on the solitude, seclusion, or personal affairs of someone who has the right to expect privacy.

joint and several liability The legal principle that makes two or more people liable for a judgment, either as individuals or in any proportional combination. Under this principle, a person who is partially responsible for a tort can end up being completely liable for damages.

joint stock company A partnership agreement in which members of the company own shares that are transferable but all goods are held in the name of the members, who assume partnership liability.

joint tenancy Form of co-ownership of real property in which all owners have equal shares in the property, may sell their shares without the consent of the other owners, and may have their interest attached by creditors.

joint venture Relationship between two or more persons or corporations or an association between a foreign multinational and an agency of the host-country government or a host-country national set up for a business undertaking for a limited time period.

judicial activism A judicial philosophy that says the courts should take an active role in encouraging political, economic, and social change.

judicial power The power delegated by Congress to an administrative agency to adjudicate cases through an administrative proceeding; includes the power to issue a complaint, have a hearing held by an administrative law judge, and issue either an initial decision or a recommended decision to the head(s) of an agency.

judicial restraint A judicial philosophy that says courts should refrain from determining the constitutionality of a legislative act unless absolutely necessary and that social, political, and economic change should be products of the political process.

jurisdiction The power of a court to hear a case and render a binding decision.

jurisdictional picketing Picketing by one union to protest an assignment of jobs to another union's members.

jurisprudence The science or philosophy of law; law in its most generalized form.

labor–management committee A forum in which workers communicate directly with upper management. May be illegal under the NLRA if the committee has an impact on working conditions, unless all workers in a bargaining unit or a plant participate or the employees on the committee are carrying out a traditional management function.

Landrum–Griffin Act Governs the internal operation of labor unions.

larceny The secretive and wrongful taking and carrying away of the personal property of another with the intent to permanently deprive the rightful owner of its use or possession.

lease The contract that transfers possessory interest in a property from the owner (lessor) to the tenant (lessee).

leasehold The right to possess property for an agreed-upon period of time stated in a lease.

legal acceptance An acceptance that shows objective intent to enter into the contract, that is communicated by proper means to the offeror, and that mirrors the terms of the offer.

legal object Contract subject matter that is lawful under statutory and case law.

legal offer An offer that shows objective intent to enter into the contract, is definite, and is communicated to the offeree.

legislative power The power delegated by Congress to an administrative agency to make rules that must be adhered to by individuals and businesses regulated by the agency; these rules have the force of law.

letter of comment or deficiency Informal letter issued by the SEC indicating what corrections must be made in a registration statement for it to become effective.

libel Publication of a defamatory statement in permanent form.

license A temporary, revocable right to be on someone else's property.

lien A claim on a debtor's property that must be satisfied before any creditor can make a claim.

life estate The right to own and possess the land until one dies.

limited liability company (LLC) A hybrid corporation-partnership similar to the Subchapter S corporation but with far fewer restrictions.

limited partnership A partnership that has one general partner who is responsible for managing the business and one or more limited partners who invest in the partnership but do not participate in its management and whose liability is limited to the amount of capital they contribute.

limited warranty Under the Magnuson-Moss Warranty Act, any written warranty that does not meet the conditions of a full warranty.

liquidated damages Monetary damages for nonperformance that are stipulated in a clause in the contract.

litigation A dispute resolution process going through the judicial system; a lawsuit.

long-arm statute A statute authorizing a state court to obtain jurisdiction over an out-of-state defendant when that party has sufficient minimum contacts with the state.

malpractice suits Service liability suits brought against professionals, usually based on a theory of negligence, breach of contract, or fraud.

mandatory subjects of collective bargaining Subjects over which the parties must bargain, including rates of pay, wages, hours of employment, and other terms and conditions of employment.

manifest program A program that attempts to see that hazardous wastes are properly transported to disposal facilities licensed by the EPA so that the agency will have an accurate record (manifest) of the location and amount of all hazardous wastes.

market share theory A theory of recovery in liability cases according to which damages are apportioned among all the manufacturers of a product based on their market share at the time the plaintiff's cause of action arose.

mechanic's lien A lien placed on the real property of a debtor when the latter does not pay for the work done by the creditor.

mediation An alternative dispute resolution method in which the disputant parties select a neutral party to help them reconcile their differences by facilitating communication and suggesting ways to solve their problems.

merger One company's acquisition of another company's assets or stock in such a way that the second company is absorbed by the first.

mineral rights The legal ability to dig or mine the minerals from the earth below the surface of one's land.

minitrial An alternative dispute resolution method in which lawyers for each side present the case for their side at a proceeding refereed by a neutral adviser, but settlement authority usually resides with senior executives of the disputing corporations.

Miranda rights Certain legal rights—such as the right to remain silent to avoid self-incrimination and the right to an attorney—that a suspect must be immediately informed of upon arrest.

misappropriation Use of an unsolicited idea for a product, service, or marketing method without compensating the originator of the idea.

misdemeanor A crime that is less serious than a felony and is punishable by fine or imprisonment in a local jail.

mistake Error as to material fact. A bilateral mistake is one made by both parties; a unilateral mistake is one made by only one party to the contract.

mistake-of-fact defense An affirmative defense claiming that a mistake made by the defendant vitiates criminal intent.

mock jury Group of individuals, demographically matched to the actual jurors in a case, in front of whom lawyers practice their arguments before presenting the case to the actual jury.

modify Term for an appellate court's decision that, although the lower court's decision was correct, it granted an inappropriate remedy that should be changed.

monetary damages Dollar sums awarded for a breach of contract; "legal" remedies.

monopoly An economic market situation in which a single business has the power to fix the price of goods or services.

motion to dismiss Defendant's application to the court to put the case out of judicial consideration because even if the plaintiff's factual allegations are true, the plaintiff is not entitled to relief.

multinational (transnational) corporation A corporation whose production, distribution, ownership, and management span several nations.

National Ambient Air Quality Standards (NAAQS) A two-tiered set of standards developed for the chief conventional air pollutants: primary standards designed to protect public health and secondary standards designed to protect public welfare.

National Institute for Occupational Safety and Health (NIOSH) A research facility established by the OSH Act to identify occupational health and safety problems; develop controls to prevent occupational accidents and diseases; and disseminate its findings, particularly to OSHA.

National Labor Relations Board (NLRB) The administrative agency set up to interpret and enforce the Wagner Act (NLRA).

negligence Failure to live up to the standard of care that a reasonable person would meet to protect others from an unreasonable risk of harm.

negligence per se Legal doctrine that says when a statute has been enacted to prevent a certain type of harm and the defendant violates that statute, causing that type of harm to befall the plaintiff, the plaintiff may use proof of the violation as proof of negligence.

negligent tort A civil wrong that involves a failure to meet the standard of care a reasonable person would meet and, because of that failure, harm to another results.

negotiation and settlement An alternative dispute resolution method in which the disputant parties come together informally to try to resolve their differences.

nolo contendere A plea of no contest that subjects the defendant to punishment but is not an admission of guilt.

nominal damages Monetary damages of a very small amount (e.g., $1) awarded to a party that is injured by a breach of contract but cannot show real damages.

noninvestment company A company whose primary business is not in investing or trading in securities.

nonprice vertical restraint Restraint used by a manufacturer to limit the territory in which a retailer may sell the manufacturer's products and the number of stores the retailer can operate, as well as the customers the retailer can serve in a location.

norm An expected standard of conduct.

notes Short-term loans.

nuisance An unreasonable interference with someone else's use and enjoyment of his or her land.

Occupational Safety and Health Act (OSH Act) A regulatory act designed to provide a workplace free from recognized hazards that are likely to cause death or serious harm to employees.

Occupational Safety and Health Administration (OSHA) The agency responsible, under the OSH Act, for setting and enforcing standards for occupational health and safety.

Occupational Safety and Health Review Commission (OSHRC) An independent body that reviews OSHA citations, penalties, and abatement periods when they are contested by an employer.

open-end credit Credit extended on an account for an indefinite time period so that the debtor can keep charging on the account, up to a certain amount, while paying the outstanding balance either in full or in installment payments.

organizational (recognitional) picketing Picketing designed to force the employer to recognize and bargain with an uncertified union.

original jurisdiction The power to initially hear and decide (try) a case.

over-the-counter (OTC) market A securities market that has no physical facility and no membership qualifications and whose broker-dealers are market makers who buy and sell stocks directly from the public.

par value The nominal or face value of a stock or bond.

parens patriae suit A lawsuit brought by a state attorney general on behalf of the citizenry of that state.

parol (oral) evidence rule When parties have executed a written agreement that is complete on its face, oral agreements made before, or at the same time as, the written agreement that vary, alter, or contradict the written agreement are invalid.

partially disclosed principal One whose identity is not known to the third party at the time of the agreement, although the third party does know that the agent represents a principal.

participating bond A bond that pays both interest at a fixed rate and to dividends.

partnership A voluntary association of two or more persons formed to carry on a business as co-owners for profit.

patent Grants the holder the exclusive right to produce, sell, and use a product, process, invention, machine, or asexually reproduced plant for 20 years.

per se standard A legal standard applicable to restraints of trade that are inherently anticompetitive. Because such restraints are automatically in violation of Section 1 of the Sherman Act, courts do not balance procompetitive and anticompetitive effects in such cases.

permissive subjects of collective bargaining Subjects that are not primarily about conditions of employment and, therefore, need not be bargained over.

personal property All property that is not real property; may be tangible or intangible.

pesticide Any substance designed to prevent, destroy, repel, or mitigate any pest or to be used as a plant regulator or defoliant.

petit jury A jury of 12 citizens impaneled to decide on the facts at issue in a criminal case and to pronounce the defendant guilty or not guilty.

petty crime A minor crime punishable under federal statutes, by fine or incarceration of no more than six months.

picketing The stationing of individuals outside an employer's place of business to inform passersby of the fact of a labor dispute.

plaintiff Party on whose behalf the complaint is filed.

plea bargaining The negotiation of an agreement between the defendant's attorney and the prosecutor, whereby the defendant pleads guilty to a certain charge or charges in exchange for the prosecution reducing the charges.

pleadings Papers filed by a party in court and then served on the opponent in a civil lawsuit.

point sources Distinct places from which pollutants are discharged into water, such as paper mills, electric utility plants, sewage treatment facilities, and factories.

police power The states' retained authority to pass laws to protect the health, safety, and welfare of the community.

power of attorney An agency agreement used to give an agent authority to sign legal documents on behalf of the principal.

predatory pricing Pricing below the average variable cost to drive out competition.

preferred stock A class of stock that entitles its owner to special preferences relating to either dividends or the distribution of assets.

premerger notification requirement The legislatively mandated requirement that certain types of firms notify the FTC and the Justice Department 30 days before finalizing a merger so that these agencies can investigate and challenge any mergers they find anticompetitive.

preponderance of the evidence A legal standard whereby a bare majority (51 percent) of the evidence is sufficient to justify a ruling.

price discrimination A price differential that is below the average variable cost for the seller; considered predatory, and therefore illegal, under the Clayton Act.

primary ethical norms The four norms that provide the major ethical direction for the laws governing business behavior: freedom, stability, justice, and efficiency.

primary-line injury A form of price discrimination in which a seller attempts to put a local competitive firm out of business by lowering its prices only in the region where the local firm sells its products.

principal–agent relationship Relationship in which the principal gives the agent expressed or actual authority to act on the former's behalf.

private international law Law that governs the relationships between private parties involved in transactions across borders.

private law Law dealing with the enforcement of private duties.

private trial An alternative dispute resolution method in which cases are tried, usually in private, by a referee who is selected by the disputants and empowered by statute to enter a binding judgment.

probable cause The reasonable inference from the available facts and circumstances that the suspect committed the crime.

procedural due process Procedural steps to which individuals are entitled before losing their life, liberty, or property.

procedural rule A rule that governs the internal processes of an administrative agency.

professional corporation A corporation that is organized by doctors, dentists, lawyers, accountants, or other professionals and is specified in state statutes.

property A bundle of rights, in relation to others, to possess, use, and dispose of a tangible or intangible object.

prospectus The first part of the registration statement the SEC requires from issuers of new securities. It contains material information about the business and its management, the offering itself, the use to be made of the funds obtained, and certain financial statements.

proxy A document by which shareholders or a publicly held company can transfer its rights to vote at a shareholders' meeting to a second party.

public disclosure of private facts A privacy tort that consists of unwarranted disclosure of a private fact about a person.

public international law Law that governs the relationships between nations.

public law Law dealing with the relationship of government to individual citizens.

public policy exception An exception to the employment-at-will doctrine that makes it unlawful to dismiss an employee for taking certain actions in the public interest.

publicly held corporation A corporation whose stock is traded on at least one national securities exchange.

puffery An exaggerated recommendation made in a sales talk to promote the product.

punitive damages Monetary damages awarded in excess of compensatory damages for the sole purpose of deferring similar conduct in the future.

quasi-contract A court-imposed agreement to prevent the unjust enrichment of one party when the parties had not really agreed to an enforceable contract.

quitclaim deed A deed that simply transfers to the grantee the interest that the grantor owns in the property.

Racketeer Influenced and Corrupt Organizations Act (RICO) Federal statute that prohibits persons employed by or associated with an enterprise from engaging in a pattern of racketeering activity, which is broadly defined to include almost all white-collar crimes as well as acts of violence.

real property Land and everything permanently attached to it.

reason An explanation or justification provided as support for a conclusion.

rebuttal A brief additional argument by the plaintiff to address any important matters brought out in the defendant's closing argument.

red herring A preliminary prospectus that contains most of the information that will appear in the final prospectus, except for the price of the securities. The "red herring" prospectus may be distributed to potential buyers during the waiting period, but no sales may be finalized during this period.

redirect examination Questioning by the directing attorney following cross-examination. The scope of the questions during redirect is limited to questions asked in the cross-examination.

reformation Correction of terms in an agreement so that they reflect the true understanding of the parties.

Regulation Z A group of rules set forth by the Federal Reserve Board to implement some provisions of the Truth-in-Lending Act, requiring lenders to disclose certain information to borrowers.

Rehabilitation Act of 1973 Prohibits discrimination in employment against otherwise qualified persons who have a handicap. Applies only to the federal government, employers that have contracts with the federal government, and parties who administer programs that receive federal financial assistance.

remand Term for an appellate court's decision that an error was committed that may have affected the outcome of the case and that the case should therefore be returned to the lower court.

res ipsa loquitur Legal doctrine that allows a judge or a jury to infer negligence on the basis of the fact that accidents of the type that happened to the plaintiff generally do not occur in the absence of negligence on the part of someone in the defendant's position.

rescission Cancellation of a contract.

respondeat superior Legal doctrine imposing liability on a principal for torts committed by an agent who is employed by the principal and subject to the principal's control.

restraint of trade Action that interferes with the economic law of supply and demand.

restrictive covenants Promises by the owner, generally included in the deed, to use or not to use the land in particular ways.

reverse Term for an appellate court's decision that the lower court's decision was incorrect and cannot be allowed to stand.

reverse discrimination Discrimination in favor of members of groups that have been previously discriminated against; claim usually is raised by white males.

rule-of-reason standard A legal standard holding that only unreasonable restraints of trade violate Section 1 of the Sherman Act. If the court determines that an action's anticompetitive effects outweigh its procompetitive effects, the restraint is unreasonable.

rules of civil procedure The rules governing proceedings in a civil case; federal rules of procedure apply in all federal courts, and state rules apply in state courts.

scienter Knowledge that a representation is false.

secondary boycott A boycott against one employer to force it to cease doing business with another employer with which the union has a dispute.

secondary-line injury A form of price discrimination in which a seller offers a discriminatory price to one buyer but not to another buyer.

secured bond Type of bond secured by the issuer's pledge of a specific asset.

Securities and Exchange Commission (SEC) The federal administrative agency charged with overall responsibility for the regulation of securities, including ensuring that investors receive "full and fair" disclosure of all material facts with regard to any public offering of securities. It has wide enforcement powers to protect investors against price manipulation, insider trading, and other dishonest dealings.

security A stock, a bond, or any other instrument of interest that represents an investment in a common enterprise with reasonable expectations of profits that are derived solely from the efforts of those other than the investor.

separation of powers Constitutional doctrine whereby the legislative branch enacts laws and appropriates funds, the executive branch sees that the laws are faithfully executed, and the judicial branch interprets the laws.

service Providing the defendant with a summons and a copy of the complaint.

sexual harassment Unwelcome sexual advances, requests for sexual favors, and other verbal or physical conduct of a sexual nature that explicitly or implicitly makes submission a term or condition of employment or creates an intimidating, hostile, or offensive environment.

shadow jury Group of individuals, demographically matched to the actual jurors in a case, that sits in the courtroom during a trial and then "deliberates" at the end of each day so that lawyers have continuous feedback as to how their case is going.

shelf registration Procedure whereby a large corporation can file a registration statement for securities it wishes to sell over a period of time rather than immediately.

Sherman Act Federal statutes that make illegal every combination, contract, or conspiracy that is an unreasonable restraint of trade when this concerted action involves interstate commerce.

short-swing profits Profits made by directors, officers, or owners of 10 percent or more of the securities of a corporation as a result of buying and selling the securities within a six-month period.

slander Spoken defamatory statement.

social responsibility Concern of business entities about profit-seeking and non-profit-seeking activities and their unintended impact on others directly or indirectly involved.

sole proprietorship A business owned by one person who has sole control over management and profits.

sovereign immunity doctrine States that a foreign-owned private property that has been expropriated is immune from the jurisdiction of courts in the owner's country.

specific performance A court order compelling a party to perform in such a way as to meet the terms of the contract.

state court jurisdiction Applies to cases that may be heard only in the state court system.

state implementation plan (SIP) A plan required of every state that explains how the state will meet federal air pollution standards.

state-of-the-art defense A product liability defense based on adherence to existing technologically feasible standards at the time the product was manufactured.

statute of limitations A statute that bars actions arising more than a specified number of years after the cause of the action arises.

statute of repose A statute that bars actions arising more than a specified number of years after the product was purchased.

statutory law Law made by the legislative branch of government.

stock The capital that a corporation raises through the sale of shares that entitle the shareholders to certain rights of ownership.

stock option A stock warrant issued to employees; cannot be traded.

stock warrant A document authorizing its holder to purchase a stated number of shares of stock at a stated price, usually for a stated period of time; may be freely traded.

strict liability offense An offense for which no state of mind or intent is required.

strict liability tort A civil wrong that involves taking action that is so inherently dangerous under the circumstances of its performance that no amount of due care can make it safe.

strike A temporary, concerted withdrawal of labor.

Subchapter C corporation A corporation that organizes and operates as a domestic business but for tax purposes is treated like a partnership.

Subchapter S corporation A business that is organized like a corporation but, under IRC Subchapter S, is treated like a partnership for tax purposes so long as it abides by certain restrictions pertaining to stock, shareholders, and affiliations.

subject matter jurisdiction The power of a court to render a decision in a particular type of case.

submission agreement Separate agreement providing that a specific dispute be resolved through arbitration.

substantial performance Completion of nearly all the terms of the contract plus an honest effort to complete the rest of the terms, coupled with no willful departure from any of the terms.

substantive due process Requirement that laws depriving individuals of liberty or property be fair.

substantive rule A rule that creates, defines, or regulates the legal rights of administrative agencies and the parties they regulate.

summary jury trial An alternative dispute resolution method that consists of an abbreviated trial, a nonbinding jury verdict, and a settlement conference.

summons Order by a court to appear before it at a certain time and place.

Superfund A fund authorized by CERCLA to cover the costs of cleaning up hazardous waste disposal sites whose owners cannot be found or are unwilling or unable to pay for the cleanup.

Supremacy Clause Provides that the U.S. Constitution and all laws and treaties of the United States constitute the supreme law of the land; found in Article VI.

suretyship A contract between a third party and a creditor that allows the third party to pay the debt of the debtor; the surety is primarily liable.

syndicate An investment group that privately agrees to come together for the purpose of financing a large commercial project that none of the syndicate members could finance alone.

Taft–Hartley Act Bars unions from engaging in specified unfair labor practices, makes collective bargaining agreements enforceable in U.S. district courts, and provides a civil damages remedy for parties injured by certain prohibited union activities.

Takings Clause Provides that if the government takes private property for public use, it must pay the owner just compensation; found in the Fifth Amendment.

tangible property Personal property that is material and movable (e.g., furniture).

technology-driven standards Standards that take account of existing levels of technology and require the best control system possible given the limits of that technology.

technology-forcing standards Standards of pollution control set primarily on the basis of health considerations, with the assumption that once regulators have set the standards, industry will be forced to develop the technology needed to meet them.

tenancy by the entirety Form of co-ownership of real property, allowed only to married couples, in which one owner cannot sell without the consent of the other, and the creditors of only one owner cannot attach the property.

tenancy in common Form of co-ownership of real property in which owners may have equal or unequal shares of the property, may sell their shares without the consent of the other owners, and may have their interest attached by creditors.

tender offer A public offer by an individual or a corporation made directly to the shareholders of another corporation in an effort to acquire the targeted corporation at a specific price.

tertiary-line injury A form of price discrimination in which a discriminatory price is passed along from a secondary-line party to a favored party at the next level of distribution.

testimonial A statement by a public figure professing the merits of some product or service.

title Ownership of property.

Title VII Statute that prohibits discrimination in hiring, firing, or other terms and conditions of employment on the basis of race, color, religion, sex, or national origin.

tort An injury to another's person or property; a civil wrong.

toxic substance Any chemical or mixture whose manufacture, processing, distribution, use, or disposal presents an unreasonable risk of harm to human health or the environment.

trade dress The overall appearance and image of a product that has acquired secondary meaning.

trade secret A process, product, method of operation, or compilation of information used in a business that is not known to the public and that may bestow a competitive advantage on the business.

trademark A distinctive mark, word, design, picture, or arrangement used by the producer of a product that tends to cause consumers to identify the product with the producer.

trespass to personalty Intentionally exercising dominion and control over another's personal property.

trespass to realty (trespass to real property) Intentionally entering the land of another or causing an object to be placed on the land of another without the landowner's permission.

trust A business arrangement in which owners of stocks in several companies place their securities with trustees, who jointly manage the companies and pay out a specific share of their earnings to the securities holders.

tying arrangement A restraint of trade wherein the seller permits a buyer to purchase one product or service only if the buyer agrees to purchase a second product or service. For example, a patent holder issues a license to use a patented object on condition that the licensee also agrees to buy nonpatented products from the patent holder.

Ultramares Doctrine Rule making accountants liable only to those in a privity-of-contract relationship with the accountant.

underwriter Investment banking firm that agrees to purchase a securities issue from the issuer, usually on a fixed date at a fixed price, with a view to eventually selling the securities to brokers, who, in turn, sell them to the public.

undisclosed principal One whose identity and existence are unknown to the third party.

undue influence Mental coercion exerted by one party over the other party to the contract.

unfair competition Entering into business for the sole purpose of causing a loss of business to another firm.

unfair labor practice strike A nonviolent work stoppage for the purpose of protesting an employer's commission of an unfair labor practice.

unilateral contract An exchange of a promise for an act.

valid contract A contract that meets all legal requirements for a fully enforceable contract.

variance Permission given to a landowner to use a piece of his or her land in a manner prohibited by the zoning laws; generally granted to prevent undue hardship.

venue Where a case is brought (usually the county of the trial court); prescribed by state statute.

vertical merger A merger that integrates two firms that have a supplier–customer relationship.

vertical price-fixing Stipulation by a manufacturer to a retailer to which it sells products as to what price the retailer must charge for those products.

vertical restraint of trade Restraint that occurs between individuals or corporations at different levels of the marketing structure.

virus A computer program that destroys, damages, rearranges, or replaces computer data.

void contract A contract that at its formation has an illegal object or serious defects.

voidable contract A contract that gives one of the parties the option of withdrawing.

voir dire Process whereby the judge and/or the attorneys question potential jurors to determine whether they will be able to render an unbiased opinion in the case.

Wagner Act Guarantees the rights of workers to organize and bargain collectively and forbids employers from engaging in specified unfair labor practices. Also called the National Labor Relations Act (NLRA).

warranty A guarantee or binding promise that goods (products) meet certain standards of performance.

water rights The legal ability to use water flowing across or underneath one's property.

white-collar crime A crime committed in a commercial context by a member of the professional–managerial class.

winding-up The process of completing all unfinished transactions, paying off outstanding debts, distributing assets, and dividing remaining profits after a partnership has been terminated or dissolved.

workers' compensation laws State laws that provide financial compensation to covered employees or their dependents when employees are injured on the job.

work-product doctrine Provides that formal and informal documents prepared by an attorney in conjunction with a client's case are privileged and may not be revealed by the attorney without the client's permission.

worm A program that travels from one computer to another but does not attach itself to the operating system of the computer it "infects." It differs from a "virus," which is also a migrating program but one that attaches itself to the operating system of any computer it enters and can infect any other computer that uses files from the infected computer.

writ of execution An order by a clerk of the court directing the sheriff to seize any of the nonexempt real or personal property of a debtor who refuses to or cannot pay a creditor.

zoning Government restrictions placed on the use of private property to ensure the orderly growth and development of a community and to protect the health, safety, and welfare of citizens.

Index

A

AAA. *See* American Arbitration Association
Aaliyah (singer), 320–321
ABA-Approved Attorney Referral System, 506
Abandoned property, 378
In re Abbott Laboratories Derivative Shareholders Litigation, 465–466
Absolute advantage, 223
Absolute deontology, 194
Absolute privilege, 313
ACA. *See* Affordable Care Act
Academy of Professional Neutrals, 79
Acceptance, legal, 260–263
Accident Compensation Act, 355
Accord and satisfaction, 282
Accountability
  corporate, 660
  duty of, 420
Accountants' liability, 353, 354
Accounting, 173, 199–200
  duty of, 418
  financial, 468
  land and, 273
  new regulations of, 660
  triple-bottom-line, 639
Accredited investor, 667
Accutane, 337
Acid rain, 629–630
Act-of-state doctrine, 231–233, 736
Actual authority, 409
Actual cause, 320
Act utilitarianism, 193–194
ADA. *See* Americans with Disabilities Act
ADAAA. *See* Americans with Disabilities Act Amendments Act
Adaptive skills area, 603
Adarand Constructors, Inc. v. Pena, 606, 607
ADEA. *See* Age Discrimination in Employment Act
Adelphia, 464
Adjudication, 489–493
Administrative activities, 494
Administrative agencies, 482–483
  creation of, 485–486
  functions of, 486–494
  global dimensions of, 498
  introduction to, 483–485
  local, 496–497
  power limitations and, 494–496
  reasons for growth, 485
  as source of law, 27
  state, 496–497
  types of, 484
Administrative law, 29, 483–484
Administrative law judge (ALJ), 489, 491–492, 547, 548, 672, 706
Administrative Procedure Act (APA), 483, 486, 494
ADR. *See* Alternative dispute resolution
Adversarial process, 52–53
Adverse possession, 370
Advertising, 307
  children and, 786–787
  credit, 770
  deceptive, 760–765

false, 198
  negligent, 334
Affirm, 65
Affirmative action, 606–610
Affirmative action plans, 606, 611
Affirmative defenses, 58
Affordable Care Act (ACA), 120
AFSCME. *See* American Federation of State, County, and Municipal Employees
AFT. *See* American Federation of Teachers
After-acquired evidence of employee misconduct, 600
Age Discrimination in Employment Act (ADEA), 86, 572, 596–601
Agency, 409
  by agreement, 413
  by estoppel, 415–416
  by implied authority, 413–414
  by ratification, 415
Agency law, 408
  creation of agency relationship and, 412–416
  definition and types of agency relationships, 409–412
  duties of agents and principals, 416–419
  finance and, 411
  global dimensions of, 426–427
  liability to third parties and, 420–424
  summary of, 428
  termination of principal–agent relationship, 424–426
Agreement
  agency by, 413
  CBA, 87
  consent, 166
  in general partnership, 437, 441–442
  hot cargo, 563
  mandatory arbitration, 86, 93
  mutual assistance, 734
  submission, 83–84
  termination by, 424–425
  terms of agreement, 446
Agricultural activities, 707
Agricultural Adjustment Act, 112, 113
AICPA. *See* American Institute of Certified Public Accountants
Aid, 642
Air quality regulations, 627–631
Alien corporations, 454
Alien Tort Claims Act, 231
ALJ. *See* Administrative law judge
Allenwood (prison camp), 159
Alternative dispute resolution (ADR), 76–77
  arbitration and, 80–93, 102–103
  court-annexed, 95–97
  early neutral case evaluation and, 94
  future of, 98
  global dimension of, 98–100
  mediation and, 78–80
  minitrials and, 93–94
  negotiation and settlement and, 78
  private trials and, 95
  summary jury trials and, 95
Alternative Dispute Resolution Act, 96
Altomare, Richard, 665
Ambiguity, 8–9, 12–13

Amendments. *See specific amendments*
America Invents Act, 395
American Arbitration Association (AAA), 91, 93, 98
American Bar Association, 50, 80, 176, 200
American Broadcasting Company, Inc. et. al. v. Aereo, Inc., 399–400
American Express Co. vs. Italian Colors Restaurant, 89–90
American Federation of State, County, and Municipal Employees (AFSCME), 549
American Federation of Teachers (AFT), 549
American Institute of Certified Public Accountants (AICPA), 199
American Law Institute, 26, 252, 292, 350, 673
American legal system
  actors in, 49–52
  adversarial process and, 52–53
  civil litigation and, 53–69
  court system structure and, 46–49
  global dimensions of, 69–70
  jurisdiction and, 35–44
  summary of, 71
  venue and, 44–45
American Medical Association, 202
American Needle, Inc. v. National Football League, 702–703
American Psychological Association, 587
American realist school, 22
American Recovery and Reinvestment Act, 636
American Society of Chartered Life Underwriters (ASCLU), 200
Americans with Disabilities Act (ADA), 87, 528, 572, 601, 602–606
Americans with Disabilities Act Amendments Act (ADAAA), 603
American Telephone and Telegraph (AT&T), 707
American Trucking Association, Inc. v. Michigan Public Service Commission, 115
Amerling, Kristin, 178
Analogies, 10, 13–14
Andean Common Market (ANCOM), 234
Anderson, Pamela Lee, 260–262
Anderson, Warren, 188, 190
Annual percentage rate (APR), 770
Antitakeover regulations, 682
Anti-Terrorism Act (ATA), 231
Antitrust Enforcement Assistance Act, 734
Antitrust laws, 89, 222, 699–700. *See also* Clayton Act; Sherman Act
  Bank Merger Act and, 733–734
  enforcement of, 705–707, 736
  in EU, 737
  exemptions from, 707
  FTCA and, 732–733
  global dimensions of, 735–736
  introduction to, 700–704
  summary of, 738
  transnational reach of, 734–735
APA. *See* Administrative Procedure Act
APEC. *See* Asia-Pacific Economic Cooperation
Apparent authority, 415–416
Appellate judges, 50

808

Appellate jurisdiction, 35
Appellate procedure, 64–66
Apple Computer Company, 196, 310, 395, 401
Appropriate bargaining unit, 556–559
Appropriation, 315
APR. *See* Annual percentage rate
Aquinas, Thomas, 20
Arab Bank, 231–232
Arbitral panels, 241
Arbitration, 80–93
 common uses of, 92
 court-annexed, 97
 for global dispute resolution, 243
 malpractice and, 102–103
 methods of securing, 83–91
 problems with, 92–93
 selection of arbitrator and, 91–92
Arbitration briefs, 81
Arbitrator's Code of Ethics, 91
Architectural Systems, Inc. v. Gilbane Building Co., 283
Army Corps of Engineers, 94, 115, 179
Arraignment, 152
Arrest, 149
Arrowhead School District No. 75, Park County, Montana v. James A. Klyap, Jr., 288–289
Artisan's lien, 743
ASCLU. *See* American Society of Chartered Life Underwriters
ASEAN. *See* Association of Southeast Asian Nations
Ashton, Catherine, 237
Asian Development Bank, 241
Asia-Pacific Economic Cooperation (APEC), 17, 234
Aspen Skiing Company v. Aspen Highlands Skiing Corp., 722
Assault, 307–309
Assignment, of rights, 274
Association of Southeast Asian Nations (ASEAN), 235
Assumption of the risk, 323
ATA. *See* Anti-Terrorism Act
In re Atlanta Pipe Corp, 97
AT&T. *See* American Telephone and Telegraph
Attachment, 743, 744
AT&T Mobility LLC v. Concepcion et ux, 88, 90
Attorney–client privilege, 49–50
Attorneys, 49–50
Audito v. City of Providence, 255–256
Auer v. Paliath, 423
Authentication, 293
Authorized noncitizen workers, 530–532
Automatic stay, 747
Award, 81

B

Baby FTC laws, 782
Bacanovic, Peter, 5–6, 7
Back pay, 595
Backward vertical mergers, 730
BACT. *See* Best available control technology
Baer v. Chase, 258–259
Bail, 151
Bailments, 379
Banana bills, 318
Bank bailouts, 188, 239–240

Bank Merger Act, 699, 733–734
Bankrupt cities, 752
Bankruptcy. *See* U.S. Bankruptcy Code
Bankruptcy Abuse Prevention and Consumer Protection Act, 649, 651, 745, 757
Bankruptcy Act, 742
Bankruptcy judges, 746
Bankruptcy Reform Act, 745, 759
Bankruptcy Technical Corrections Act (BTCA), 757
Barbara Schreiber v. Burlington Northern, Inc., 680–681
Bartlett, Karen, 340–341
Base fine, 162
BATS. *See* Better Alternative Trading System
Batson v. Kentucky, 60, 61–62
Battery, 308, 309
BBB. *See* Better Business Bureau
Beachem v. Unemployment Compensation Board of Review, 508
Beam v. Stewart, 467–468
Bear Stearns, Inc., 654
Behavioral approach, to management, 556
Bench trial, 152
Benefit corporations, 454
Bennett Amendment, to Equal Pay Act, 574
Berman v. Parker, 373
Bernard L. Madoff Investment Securities LLC, 146
Berne Convention, 403
Best available control technology (BACT), 626
Best practicable control technology (BPCT), 626
Better Alternative Trading System (BATS), 676
Better Business Bureau (BBB), 78, 92, 783
BFOQ. *See* Bona fide occupational qualification
Biculturalism, 596
Bielby, William, 68
Bilateral contracts, 255
Bilateral investment treaties (BITs), 229
Bilateral mistake, 267
Bill of Rights, 121, 139
Bilski v. Kappos, 393–394
Binding arbitration clause, 83–84, 87, 102–103
Bin Laden, Osama, 37
BITs. *See* Bilateral investment treaties
Black Star Farms v. Oliver, 119
Blakely v. Washington, 162–163
Blight, 373
Blue-sky laws, 155, 692
Blurring, 391
BMW v. Gore, 301, 303, 304
Board of directors, 464–465
Board of Patent Appeals and Interferences, 393
Board rules, 548–549
Boesky, Ivan, 686
Bona fide occupational qualification (BFOQ), 586–587, 599–600
Bonds, 458
Bonkowski, Jeffrey, 514–516
Booking, 150–151
Boomer et al. v. Atlantic Cement Co., 618–619
Boston Scientific Corp., 338
Boycotts, 563, 712–713
Boyer-Liberto, Reya C., 584–586
BPCT. *See* Best practicable control technology

The Bramble Bush (Llewellyn), 22
Brandeis, Louis D., 708
Bras, Antonio, 165–166
Braswell v. United States, 138–139
Breach of contract, 56, 281–282, 286–290
Breach of sales contract, 290
Breach of warranty, 341–346
Brehm v. Eisner, 471
Brennan's Inc. v. Colbert, 457–458
Brian Pietrylo and Doreen Marino, Plaintiffs v. Hillstone Restaurant Group dba *Houston's, Defendant*, 524–526
Bribery, 30, 165–166, 194, 216–218, 693–694
Briefs, 65, 81
Brodsky v. United States Nuclear Regulatory Commission, 624
Brokers, 673–675
Brown & Brown, Inc. v. Johnson, 270
Browner, Carol, 622
Brown & Williamson, 202
Bruney, Brian, 92
Brzonkala v. Morrison, 112, 114
BTCA. *See* Bankruptcy Technical Corrections Act
Bubba's Bar-B-Q Oven v. Holland Co., 389
Buchanan Energy, 450
Bundesverfassungsgericht (Federal Constitutional Court), 70
Burden of persuasion, 152
Burden of production of evidence, 152
Burden of proof, 152–153, 313, 579
Bureau of Competition, 490
Bureau of Consumer Protection, 490, 764
Burger King, 764, 783
Bush, George W., 69, 121, 373, 493, 506, 596, 659
Business and Government in the Global Marketplace (Wiedenbaum), 244
Business associations, 250. *See also* Corporations
 choice of organization forms, 434
 common organization forms of, 434–444
 comparison of organization forms, 432–433
 global dimensions of, 447–449
 LLCs, 433, 473–475
 specialized forms of, 444–447
 summary of, 449, 476–477
Business ethics, 189, 190–191, 193
Business judgment rule, 469
Business trade practices
 consumer legislation and, 765–768
 deceptive advertising and, 760–765
 FTC and, 759–760
 unfair advertising and, 764
Bystanders, product liability to, 351

C

California Dental Association v. Federal Trade Commission, 732–733
Callable bonds, 459
Camel cigarettes, 202–203
Canada, 17, 181, 219, 240, 631, 640, 694
 corporate speech in, 127
 cybercrime in, 174
Cancun Agreement, 631
Canons of Professional Responsibility, 200
CAN-SPAM Act. *See* Controlling the Assault of Non-Solicited Pornography and Marketing Act

810 INDEX

Capital structure, 460
Carefirst Pregnancy Center (CPC), 44–45
Carley Capital Group, 283
Carrier Corp. v. Outokumpu Oyj, 735
Carter, Jimmy, 27
CASA. *See* Community Alternative Systems Agency
Case law, 25–26, 27, 251–252, 269–270
Cash-for-grades scheme, 172
Cassandra Jenkins v. American Express Financial Corp., 509–510
Castes, 221
Categorical imperative, 194–195
Caveat emptor, 758
CBA. *See* Collective bargaining agreement
CCBFOIBT. *See* Convention on Combating Bribery of Foreign Officials in International Business Transactions
CCPA. *See* Consumer Credit Protection Act
CC&Rs. *See* Declaration of Covenants, Conditions, and Restrictions
CDA. *See* Communications Decency Act
Cendant Corporation, 147
Central Hudson Gas & Electric Corp. v. Public Service Commission of New York, 123–125, 126
Central Intelligence Agency (CIA), 179, 196
Central planning, 215
CEOs. *See* Chief executive officers
CERCLA. *See* Comprehensive Environmental Response, Compensation, and Liability Act
CERT/CC. *See* Computer Emergency Response Team Coordination Center
Certification marks, 386
CFAA. *See* Computer Fraud and Abuse Act
CFCs. *See* Chlorofluorocarbons
CFOs. *See* Chief financial officers
CFSL. *See* Codification of the Federal Securities Law
CFTC. *See* Commodity Futures Trading Commission
Challenger (space shuttle), 196
Chapter 7 bankruptcy, 745, 749–752
Chapter 11 bankruptcy, 753–756
Chapter 12 bankruptcy, 756–757
Chapter 13 bankruptcy, 752–753
Charge, 592–593, 594
Chase Bank USA v. Fortunato, 37
Chase Manhattan Mortgage Corporation, 516
Checks and balances, 107, 110
Cheerleading, 674
Chevron U.S.A. Inc. v. National Resources Defense Council, Inc., 488–489
Chicago Board of Trade, 460
Chicago School, 700–701, 704
Chief executive officers (CEOs), 659
Chief financial officers (CFOs), 659
Child Internet Protection Act (CIPA), 122
Child Online Protection Act (COPA), 122
Children, 111, 786–787
China, 357–358, 379, 445, 449, 640, 645, 694
   corruption in, 207
   franchising in, 225–226
   import controls in, 232–233
   international environment of business and, 214–215
Chlorofluorocarbons (CFCs), 642
CIA. *See* Central Intelligence Agency
Cinerama Inc. v. Technicolor Inc., 471
CIPA. *See* Child Internet Protection Act

CISG. *See* Convention on the International Sale of Goods
Citizens United v. Federal Election Commission, 127
City of Arlington v. Federal Communications Commission, 488–498
Civil law, 28, 219, 220
Civil litigation
   appellate procedure, 64–66
   class actions, 66–69
   pretrial stage of, 53–59
   trial, 59–64
Civil Rights Act (1866), 572–573
Civil Rights Act (1871), 572–573
Civil Rights Act (1964), 29, 572, 575–596. *See also* Title VII
Civil Rights Act (1991), 572, 575–596, 610
Civil rights laws, 572
Claims examiner, 512
Class Action Fairness Act, 69, 306
Class actions, 66–69, 707
Classifications of law, 27–29
   civil, 28
   criminal, 28
   cyberlaw, 29
   private, 28
   procedural, 29
   public, 28
   substantive, 29
Clayton Act, 89, 494, 699, 704–706, 713, 717, 759
   exclusive-dealing contracts and, 727
   interlocking directorates and, 732
   mergers and acquisitions and, 727–732
   price discrimination and, 725–727
   tying arrangements and, 727
Clean Air Act, 341
Clean Air Act Amendments, 628, 629
Clean Water Act, 115
Clear Channel, 84
Climate change, 630–631, 645
Clinton, Bill, 107–109, 400, 581
Closed-circuit cameras, 526
Closed-end credit, 770
Closely held corporations, 454
Close-Up, 225
Closing arguments, 64
Coalition Against Insurance Fraud, 145–146
COBRA. *See* Consolidated Omnibus Budget Reconciliation Act
Coca-Cola, 483, 783
Code of Professional Ethics for Certified Public Accountants, 199
Codes of ethics, 195–205
Codification of the Federal Securities Law (CFSL), 673
COE. *See* Council of Europe
Coker v. Pershad, 411–412
Colbert, Edward, 457
Collective bargaining, 540, 559–560
Collective bargaining agreement (CBA), 87
Collective marks, 386
Collusion, 708
Commerce Clause, 110–119
Commercial activity, 230
Commercial impracticability, 284
Commission on Wartime Contracting in Iraq and Afghanistan, 179
Commodities, 726
Commodity Futures Trading Commission (CFTC), 484, 498, 658

Common enterprise, 661
Common law, 70, 219–220, 251, 275, 294, 699
Common Sense Legal Reform Act, 306
Common stock, 459
Communications Decency Act (CDA), 122, 310–311
Communications Workers of America, 549
Communist Party, 220
Community Alternative Systems Agency (CASA), 604
Community service, 163
Comparative advantage, 223
Comparative negligence, 322
Compensation
   disclosure of, 675–676
   duty of, 417, 420
   executive, 656–657
   unemployment, 503, 506–510
   workers' compensation laws, 503, 510–513
Compensatory damages, 286–287, 300–301, 595
Competency, 267–268
Competition Directorate, 737
Competitive injury, 726
Complaint, 36, 55
Complete performance, 280
Composition plan, 752
Comprehensive Drug Abuse Prevention and Control Act, 112
Comprehensive Environmental Response, Compensation, and Liability Act (CERCLA), 635–636
Comptroller of the Currency, 498, 733–734
Comptroller of the Treasury of Maryland v. Wynne et ux, 118
Computer crimes, 170–172
Computer Emergency Response Team Coordination Center (CERT/CC), 174
Computer Fraud and Abuse Act (CFAA), 171
Computer information transaction, 293
Conciliation, 593
Conclusions, 4, 7–8, 11, 12
Concurrent federal jurisdiction, 41
Concurring opinion, 65
Condemnation, 371
Conditional estate, 362–363
Conditional privilege, 313
Conditions precedent, 282
Conditions subsequent, 283
Condominiums, 365–367
Conferences, 641
Conflict of interest, 469, 674, 675
Conforming goods, 281
Conglomerate mergers, 729, 730
Congress, 23, 25
   Commerce Clause and, 110–112
   positivist school and, 21
   separation of powers and, 107
   spending powers and, 119–120
Congressional Record, 26
Congressional Review Act, 495
Conrad, Constance, 177
Conrad, James, 177
Conroy v. Owens-Corning Fiberglass, 305
Conscious parallelism, 708
Consent agreement, 166
Consent order, 491
Consequential theories, 193–194
Consideration, 263–264
Consolidated Omnibus Budget Reconciliation Act (COBRA), 503, 510

Constitution, 23, 105, 222, 788–791. *See also* specific amendments
   Due Process Clause of, 38, 132–133, 135–136, 139, 302
   Equal Protection Clause of, 60–61, 139
   judicial review and, 51
   Speech and Debate Clause of, 313
   Supremacy Clause of, 27, 106, 107, 222, 497
Constitutional law, 28–29
Constitutional principles
   amendments and, 121–139
   Commerce Clause and, 110–119
   Constitution and, 105
   federalism and, 105–107
   separation of powers and, 107–110
   summary of, 140
   taxes and spending powers of federal government, 119–121
Construct validity, 587
Consumer credit
   Credit Card Accountability, Responsibility and Disclosure Act and, 772
   ECOA and, 777
   EFTA and, 773
   FCBA and, 778
   FCRA and, 774–776
   FDCPA and, 778–780
   identity theft and, 775–777
   TILA and, 769–772
Consumer Credit Protection Act (CCPA), 759, 769
Consumer debts, 749
Consumer Financial Protection Bureau, 658, 781
Consumer law. *See also* Dodd-Frank Wall Street Reform and Consumer Protection Act; Federal Trade Commission
   consumer warranties and, 766
   evolution of, 758–759
   franchising relationships and, 765–766
   remedies and, 767
   telemarketing legislation and, 767–768
Consumer loans, 781
Consumer price index, 748
Consumer Product Safety Commission (CPSC), 482, 484, 760
Consumer protection, 658
   credit and debit cards and, 781
   credit scores and, 781
   e-commerce and, 780–781
   global dimensions of, 783
   loans and, 781
   residential mortgages and, 782
Consumer reporting agency, 774
Consumer Sales Practices Act, 88
Consumer warranties, 766
Consumer welfare, 704
Content validity, 587
Continental TV, Inc. v. GTE Sylvania, 716–717
Contingency fee, 417
Contract law, 29, 250, 279
   breach of contract remedies, 286–291
   case law and, 251–252
   classifications of contracts, 253–257
   discharge methods and, 280–286
   e-contracts, 262–263, 291–293
   elements of, 257–270
   global dimensions of, 293–294
   parol (oral) evidence rule and, 273
   sources of, 251–253

   summary of, 275–276, 294–295
   third-party beneficiary contracts, 274
   UCC and, 251, 252
   written contracts, 271–273
Contractual liability, 420–421
Contributory negligence, 322, 512
Controlled Substances Act (CSA), 112
Controlling the Assault of Non-Solicited Pornography and Marketing Act (CAN-SPAM Act), 315, 768
Convention on Combating Bribery of Foreign Officials in International Business Transactions (CCBFOIBT), 216, 648, 694
Convention on Cybercrime, 181
Convention on the International Sale of Goods (CISG), 282, 294
Conventions, 641
Conversion, 308, 317–318
Convertible bonds, 459
Convertible preferred stock, 460
Conveyance, 368
Cook, Randolph, 310
Coomer v. Kansas City Royals, 323–324
Cooperation
   duty of, 417, 420
   international, 639
   substantial, 162
Cooperatives, 367–368
Cooper Industries v. Leatherman Tool Group, Inc., 205
Co-ownership, 364–365
COPA. *See* Child Online Protection Act
Copenhagen Agreement, 631
Copyright Act, 225, 398, 400
Copyrights, 222, 397–402
Corporate accountability, 660
Corporate attorneys, 50
Corporate codes of ethics, 197–198
Corporate commercial speech, 123–126
Corporate culture, 67, 68, 197
Corporate ethics, 197–198
Corporate executives, liability on, 157–160
Corporate fraud, 145
Corporate investment decision-making, 236
Corporate liability, 155–157
Corporate mismanagement, 685, 689
Corporate opportunity doctrine, 466
Corporate personnel, 156
Corporate political speech, 126–127, 143
Corporate powers, 457
Corporations, 435, 453–454
   board of directors and, 464–465
   classification of, 454–456
   creation of, 456–458
   as criminal, 154–155
   fiduciary duties and, 466–472
   financing of, 458–461
   global dimensions of, 476
   hierarchy in, 462
   managers and, 466
   officers and, 466
   operation of, 462–472
   shareholders and, 459, 462–463
Corruption Perception Index (CPI), 218–219
Cost-benefit analysis, 70
Council of Europe (COE), 181
Council of Ministers, 237
Counterclaim, 56, 57
Counterfeit Access Device and Computer Fraud and Abuse Act, 171
County courts, 47

County of Wayne v. Hathcock et al., 372
Court-annexed ADR, 95–97
Courts of appeal, 46, 47, 49
Courts of common pleas, 47
Court system structure, 46–49
Cousins v. Realty Ventures, Inc., 417–418
CPC. *See* Carefirst Pregnancy Center
CPCU. *See* Society of Chartered Property and Casualty Underwriters
CPI. *See* Corruption Perception Index
CPSC. *See* Consumer Product Safety Commission
Cram-down provision, 754
Credit advertising, 770
Credit Card Accountability, Responsibility and Disclosure Act, 772
Credit card fraud, 181
Credit cards, 781
Creditor, 742, 743–747
Creditor–beneficiary contract, 274
Creditors' meeting, 747
Credit rating agencies, 657
Credit Rating Agency Reform Act, 650, 651
Credit scores, 781
Crime, 148–149. *See also* White collar crime
   computer, 170–172
   cybercrime, 174, 181
   financial, 172
   gender-motivated, 115
   intrabusiness, 164
Criminal fraud, 167–169
Criminal law, 28, 148–149
Criminal liability, to third parties, 424
Criminal procedure, 150–154
Criteria pollutants, 627
Criterion-related validity, 587
Critical legal studies school, 22
Critical thinking, 1
   applying, 15–16
   importance of, 2–4
   legal environment of business and, 18
   legal norms and, 43
   legal reasoning and, 11–15
   legal system and, 36
   model of, 4–7
   steps of, 7–11
Critical thinking questions, 6
Critical thinking skills, 2, 18
Crosby v. National Foreign Trade Council, 222–223
Cross-elasticity of demand or substitutability, 720, 729
Cross-examination, 63
Cross-licensing, 395
Cross-State Air Pollution Transport Rule, 628
CSA. *See* Controlled Substances Act
Culpability score, 162
Culturally homogeneous, 215
Culture, 215–216
   of caring, 23
   corporate, 67, 68, 197
Cumulative preferred stock, 459
Customs, 221
Cybercrime, 174, 181
Cyberlaw, 29
Czarist legal system, 220

D

Daniels, Claudia, 439
Daniels, William, 439

812 INDEX

Data, deconstruction and unlawful appropriation of, 172
Datakey Electronics, Inc., 438–439
Davis, Saundra, 583
Davis–Bacon Act, 504
DEA. *See* Drug Enforcement Administration
Dealers, 673–675
Debarment, of officers and directors, 210–211
Debentures, 458
Debit cards, 781
Debt
　consumer, 749
　discharge of, 750
　liquidated and unliquidated, 264
Debt collectors, 778
Debt financing, 458–459
Debtor, 742, 744–745
Debtor–creditor relations, 742–745
Debtor-in-possession (DIP), 754
Deceptive advertising, 760–765
Decertification, 551
Declaration of Covenants, Conditions, and Restrictions (CC&Rs), 365, 374
Deed, 363, 368, 369–370
Defalcation, 167
Defamation, 308, 309–314
Defendant, 36, 56, 57, 64
Defense Intelligence Agency (DIA), 179, 196
Defense of another, 309
Defense of Marriage Act (DOMA), 133, 134–136
Defenses, 153
　affirmative, 58
　to battery, 309
　to breach-of-warranty action, 345–346
　to defamation, 313–314
　Equal Pay Act and, 574–575
　mergers and, 730–731
　to negligence, 322–326
　to negligence-based product liability action, 339–341
　under Securities Act, 669–673
　Securities Exchange Act and, 679–682
　to strict product liability action, 351
　Title VII and, 586–587
Democracy, 382–383, 501
"Demonstration" workplaces, 521
Deontological theories, 193, 194–195
Department of Agriculture, 630, 760
Department of Commerce, 232
Department of Defense, 232
Department of Defense Authorization Law, 517
Department of Health and Human Services (DHHS), 764
Department of Homeland Security (DHS), 484
Department of Justice (DOJ), 155, 178, 460–461, 622, 705–706, 731, 734–735
Department of Labor (DOL), 505, 506, 517, 760
Department of Motor Vehicles (DMV), 115
Department of Public Works (DPW), 165
Department of State, 232
Department of Transportation, 484
Deponit and Hyoscyamine Sulfate Extended Release (Hyoscyamine Sulfate ER), 177
Deposition, 58
Derivatives, 657–658
DES. *See* Diethylstilbestrol
Design for environment (DFE), 640
Detroit, Michigan, 752
DFE. *See* Design for environment

Dharmasastra (Hindu law), 219, 220, 221
DHHS. *See* Department of Health and Human Services
DHS. *See* Department of Homeland Security
DIA. *See* Defense Intelligence Agency
Diablo Valley College, 172
Diamond Woodworks, Inc. v. Argonaut Insurance Co., 305
Diethylstilbestrol (DES), 352
Digital Millennium Copyright Act (DMCA), 400–402
DIP. *See* Debtor-in-possession
DiPascali, Frank, 146
Direct examination, 63
Director generals, 238
Direct regulation, 621
Disability, 602. *See also* Americans with Disabilities Act
Disbarment, of lawyers, 210–211
Discharge, 626, 747–748
　contract law and, 280–286
　of debts, 750
　marketable discharge permits, 620–621
Disclaimers, 345
Disclosed principal, 422
Discovery, 58–59
Discrimination. *See* Employment discrimination
Disney, Roy, 472
Disparagement, 308, 318
Disparate impact, 576–579
Disparate treatment, 576, 579
Disposable income, 215
Dissenting opinions, 65
Dissolution, 441
Diversification, 728
Diversity of citizenship, 41, 69
DMCA. *See* Digital Millennium Copyright Act
DMV. *See* Department of Motor Vehicles
DNA evidence, 131
Dodd-Frank Wall Street Reform and Consumer Protection Act (Dodd-Frank), 482, 648, 651, 655–659, 744–745, 781–782
DOJ. *See* Department of Justice
DOL. *See* Department of Labor
Dolan v. City of Tigard, 137, 377–378
DOMA. *See* Defense of Marriage Act
Domain names, 390
Domestic corporations, 454
Donative intent, 378
Donee–beneficiary contract, 274
"Do Not Call" Registry, 767, 768
Dow Jones Industrial Average, 188, 649
DPPA. *See* Driver Privacy Protection Act
DPW. *See* Department of Public Works
Draft Regulatory Program, 496
Driver Privacy Protection Act (DPPA), 115
Dr. Miles Medical Co. v. John D. Park & Sons Co., 714
Drogin, Richard, 68
Drug Enforcement Administration (DEA), 112
Drug-Free Workplace Act, 527
Drug testing, 527–528, 535–536
Dual court system, 35, 40
Due diligence defense, 670
Due process
　economic substantive, 133
　procedural, 132
　substantive, 133
Due Process Clause, 38, 132–133, 135–136, 139, 302

DuPont, 639
Duress, 153, 265
Duty, 319, 420
　of agents and principals, 416–419
　extra, 574
　fiduciary, 466–472
　preexisting, 263
　separation of, 109
Dynergy, 146

E

EA. *See* Environmental assessment
Eagle v. Fred Martin Motor Co., 88
Early neutral case evaluation, 94
Earnings before interest and taxes (EBIT), 468
Easement, 271, 364
EB-1 visa, 531–532
eBay, 781
Ebbers, Bernie, 147
EBIT. *See* Earnings before interest and taxes
ECB. *See* European Central Bank
ECOA. *See* Equal Credit Opportunity Act
E-commerce, 421, 692, 768, 780–781
Economic efficiency, 700
Economic Espionage Act, 397
Economics, 758–759
　microeconomics, 688
　neoclassical theory, 191
Economic strikes, 561
Economic substantive due process, 133
Economies of scale, 728
E-contracts, 262–263, 291–293
ECPA. *See* Electronic Communications Privacy Act
EDGAR. *See* Electronic data gathering, analysis, and retrieval
EDS. *See* Electronic Data Systems
Edwards, Charles, 662
EEO-1 Form, 577–578
EEOC. *See* Equal Employment Opportunity Commission
EEOC v. Luce, Forward, Hamilton & Scripps, 87
EEOC v. R.G. & G.R. Harris Funeral Homes, Inc., 591
EEOC v. Waffle House, Inc., 87
Effective vindication, 89
Efficient markets, 688
Effluent limitations, 625
Effort, 574
EFTA. *See* Electronic Fund Transfer Act; European Free Trade Association
E.I. DuPont de Nemours and Co. v. Kolon Industries, 719–720
Eighteenth Amendment, 792
Eighth Amendment, 29, 791
EIS. *See* Environmental Impact Statement
Eisner, Michael, 472
Elders, Jocelyn, 203
Electronic Communications Privacy Act (ECPA), 523–524
Electronic data gathering, analysis, and retrieval (EDGAR), 655
Electronic Data Systems (EDS), 728
Electronic Fund Transfer Act (EFTA), 773
Electronic media, 655
Electronic monitoring and communication, 523–527
Electronic service of process, 37

INDEX

Electronic Signatures in Global and National Commerce Act (E-SIGN), 292
Eleemosynary institutions, 456
Eleventh Amendment, 791
Elgin (prison camp), 159
Ellerth v. Burlington, 581
Ellsberg, Daniel, 210
Elmore v. Owens Illinois, Inc., 351
Email policy, 527
Embezzlement, 170
Emerson, Ralph Waldo, 12
Emine Bayram v. City of Binghamton and City of Binghamton Zoning Board of Appeals, 375–376
Eminent domain, 371, 375, 382–383
Emission charges, 620
Employee absenteeism, 517
Employee committees, 552
Employee Health Act, 14
Employee Polygraph Protection, 529
Employee privacy rights, 523–529, 535–536
Employee wages and benefits, 503
Employer–employee relationship, 409
Employer–independent contractor relationship, 410
Employment-at-will doctrine, 570–571
Employment discrimination, 569–570
    ADA and, 602–606
    ADEA and, 596–601
    affirmative action and, 606–610, 611
    Civil Rights Acts (1866 and 1871) and, 572
    Civil Rights Acts (1964 and 1991) and, 575–596
    constitutional provisions and, 572
    employment-at-will doctrine and, 570–571
    Equal Pay Act and, 572, 573–575
    federal statutes prohibiting, 572
    global dimensions of, 610–611
    Rehabilitation Act and, 601–602
    summary of, 611–612
Employment relationship, 503
    COBRA and, 510
    employee privacy rights and, 523–529, 535–536
    FMLA and, 513–518
    global dimensions of, 532
    immigration law and, 529–532
    OSH Act and, 518–523
    summary of, 533
    unemployment compensation and, 503, 506–510
    wage and hour laws and, 504–506
    workers' compensation laws and, 503, 510–513
Enabling legislation, 485
Endispute, 79, 95, 98
Enea v. Superior Court of Monterey County, 439–440
Energiewende, 632
Energy and Commerce Committee, 496
Enron, 143, 145, 147, 177, 197, 464, 659
Entrapment, 153
Environmental assessment (EA), 624
Environmental impacts, 624
Environmental Impact Statement (EIS), 623–625
Environmental law, 617–618
    air quality regulations and, 627–631
    alternative approaches to, 618–621
    EPA and, 621–622
    global dimensions of, 639–642

hazardous waste and toxic substance regulations and, 631–638
    NEPA and, 623–625
    Pollution Prevention Act and, 638–639
    water quality regulations and, 625–627
Environmental Protection Agency (EPA), 484–485, 490, 617, 621–622, 628
Environmental sustainability, 640
EPA. *See* Environmental Protection Agency
EPA v. EME Home City Generation, L.P., 628
Equal Credit Opportunity Act (ECOA), 758–759, 777
Equal Employment Opportunity Commission (EEOC), 27, 86, 484, 575, 579, 588–589, 593
Equal Pay Act, 572, 573–575
Equal Protection Clause, 60–61, 139
Equal Rights Amendment, 712
Equal work, 574
Equitable remedies, 289–290
Equity financing, 459–461
Escola v. Coca-Cola, 321
E-SIGN. *See* Electronic Signatures in Global and National Commerce Act
E-signatures, 292
Estate exemptions, 748–749
Ethical norms, 9–10, 13
Ethics
    business, 189, 190–191, 193
    codes of ethics, 195–201
    corporate, 197–198
    global dimensions of, 207
    summary of, 207–208
    theories of, 193–195
EU. *See* European Union
Euro-Arab Chamber of Commerce, 99
European Central Bank (ECB), 239, 476
European Court of Justice, 221, 239
European Free Trade Association (EFTA), 234
European Monetary Union, 239
European Union (EU), 17, 226, 233–234, 236–240, 427, 476, 737
Excelsior list, 549
Exchange Act (1934). *See* Securities Exchange Act
Exchange markets, 676
Exclusive agency contract, 413
Exclusive-dealing contracts, 718, 727
Exclusive federal jurisdiction, 41
Executed contracts, 256
Execution, of voluntary transfer of real property, 368–370
Executive administrative agencies, 484
Executive branch, 27, 494–495
Executive compensation, 656–657
Executive exemption, 600
Executive orders, 27, 496
Executive power, 485
Executory contracts, 256
Exemplary damages, 286–289
Exempted rulemaking, 487–488
Exempt property, 748
Exigency leave, 517
Exit visas, 222
Ex parte Kia Motors America, Inc., 45
Expatriates, 224
Export activities, 707
Export Administration Act and the Arms Export Control Act, 232
Export controls, 232

Export Trading Act, 224
Export Trading Company Act, 707
Express contracts, 253
Expressed agency, 413
Expressed authority, 409
Express warranty, 341, 342, 345, 766
Expropriation, 229
External Action Service, 237
Externalities, 202, 205
Extortion, 165
Extra duties, 574
In re Exxon Valdez, 191–192
Exxon Valdez oil spill, 191–192, 304

F

FAA. *See* Federal Arbitration Act
Facilitating payments, 216
FACTA. *See* Fair and Accurate Credit Transaction Act
Facto v. Pantagis, 285–286
Facts, 7, 11–12
Fair and Accurate Credit Transaction Act (FACTA), 776
Fair Credit Billing Act (FCBA), 758, 778
Fair Credit Reporting Act (FCRA), 758, 774–776
Fair Debt Collection Practices Act (FDCPA), 778–780
Fair Labor Standards Act (FLSA), 504–506, 507, 573
Fair Packaging and Labeling Act (FPLA), 759, 764–765
Fair use doctrine, 398
False advertising, 198
False Claims Act, 177–178
False entry, 167
False imprisonment, 308, 316
False light, 315
False price comparisons, 762
False token, 167
Family and Medical Leave Act (FMLA), 136, 503, 513–518
Family farmers, 756
Family fisherman, 756
Family Winemakers of California, et al. v. Jenkins, 119
Fannie Mae, 659
Faragher v. City of Boca Raton, 584
FASB. *See* Financial Accounting Standards Board
Fastow, Andrew, 177
FBI. *See* Federal Bureau of Investigation
FCBA. *See* Fair Credit Billing Act
FCC. *See* Federal Communications Commission
FCPA. *See* Federal Consumer Protection Act; Foreign Corrupt Practices Act
FCRA. *See* Fair Credit Reporting Act
FD. *See* Regulation Fair Disclosure
FDA. *See* Food and Drug Administration
FDCPA. *See* Fair Debt Collection Practices Act
FDIC. *See* Federal Deposit Insurance Corporation
Federal Alcohol Administration, 125
Federal Arbitration Act (FAA), 82–83, 89–90
Federal Bureau of Investigation (FBI), 179
Federal Communications Act, 29
Federal Communications Commission (FCC), 482, 484, 488–489, 493, 496, 711, 760

Federal Constitutional Court (Bundesverfassungsgericht), 70
Federal Consumer Protection Act (FCPA), 783
Federal court judges, 51
Federal court system, 46, 95–97
Federal Deposit Insurance Corporation (FDIC), 484, 649, 733
Federal Food, Drug, and Cosmetic Act (FFDCA), 338, 638, 759
Federal Insecticide, Fungicide, and Rodenticide Act (FIFRA), 637–638
Federalism, 105–107, 238
Federal Judicial Center, 306
Federal laws
  franchising and, 445
  securities and, 649–651
  against white collar crime, 175–180
Federal Mediation and Conciliation Service (FMCS), 78, 79, 91
Federal Mine Safety and Health Act, 111, 132
Federal Motor Vehicle Safety Standard Act, 341
Federal preemption, 106–107
Federal Privacy Act (FPA), 494, 495
Federal Register, 27, 486, 487, 488, 501
Federal Register Act, 494, 495
Federal regulations, 166, 496
Federal Reserve Board, 498, 658, 733, 760, 777
Federal Rules of Civil Procedure, 26
Federal Rules of Criminal Procedure, 8, 10
Federal supremacy, 106
Federal Tort Claims Act (FTCA), 494, 495
Federal Trade Commission (FTC), 27, 169, 445, 482, 484, 490, 759–760
Federal Trade Commission Act (FTCA), 418, 699, 704, 705, 732–733, 759
Federal Trade Commission v. QT, Inc., 763
Federal Trade Commission v. Verity International, Ltd., 761–762
Federal Trademark Dilution Act (FTDA), 391
Federal Unemployment Tax Act, 508
Federal Water Pollution Control Act (FWPCA), 625–626
Federal Whistleblower Enhancement Act, 180
Fee simple absolute, 362
Felony, 28, 149, 151
Feminist school, 22–23
Fen-Phen, 336
FFDCA. *See* Federal Food, Drug, and Cosmetic Act
Fiduciary duties, 466–472
Field examiners, 547
Fielding, Lewis J., 210
FIFRA. *See* Federal Insecticide, Fungicide, and Rodenticide Act
Fifteenth Amendment, 792
Fifth Amendment, 52, 132–139, 572, 672, 791
File-sharing networks and technologies, 401
Filing suit, 593
Final Policy on Penalty Reductions (EPA), 622
Finance, 109, 200, 411
Finance charge, 770
Financial accounting, 468
Financial Accounting Standards Board (FASB), 461
Financial crimes, 172
Financial Industry Regulatory Authority (FINRA), 677
Financial Institution Bankruptcy Act, 758
Financial sanctions, 155
Financial Services Authority (FSA), 498
Financial Stability Oversight Council, 656

Financing of corporations, 458–461
Finding of no significant impact (FONSI), 624
FINRA. *See* Financial Industry Regulatory Authority
First Amendment, 23, 121–127, 143, 398, 791
First appearance, 150–151
First Florida Bank v. Max Mitchell & Co., 353
First meeting of creditors, 747
First National Bank of Boston v. Bellotti, 126, 143
Fitl v. Strek, 290–291
Fixtures, 362
Flammable Fabrics Act, 338
Florida Bar v. Went for It and John T. Blakely, 125
Florida Transportation Services, Inc. v. Miami-Dade County, 119
Florida v. Harris, 129
Florida v. Jardines, 129–130, 131
FLSA. *See* Fair Labor Standards Act
FMCS. *See* Federal Mediation and Conciliation Service
FMLA. *See* Family and Medical Leave Act
FOIA. *See* Freedom of Information Act
FONSI. *See* Finding of no significant impact
Food and Drug Administration (FDA), 5, 157, 195, 338, 358, 465, 760, 765
Food disparagement, 318
Forbes, Walter, 147
Ford, Gerald, 109
Foreign corporations, 454
Foreign Corrupt Practices Act (FCPA), 30, 194–195, 207, 216–217, 427, 648–651
Foreign direct investment, 226–229
Foreign-natural test, 343–344
Foreign Sovereign Immunities Act (FSIA), 230
Foreign Tire Sales, Inc. (FTS), 358
Forgery, 167
Formal rulemaking, 486, 501
Form I-9, 529
Forum non conveniens, 44–45, 227
Forward vertical mergers, 730
Foster, Gene, 146
Fourteenth Amendment, 38, 121, 123–124, 125, 139, 572, 792
Fourth Amendment, 128–132, 210, 488, 672, 791
Fox Television Stations, Inc. v. Federal Communications Commission, 493
FPA. *See* Federal Privacy Act
FPLA. *See* Fair Packaging and Labeling Act
Franchising, 444
  fees, 446
  international, 224–226, 448
  on Internet, 447
  relationships, 765–766
  terms of agreement, 446
Frankel, Martin, 685–686
Franklin American Corporation, 685
Frankson v. Browne & Williamson, 305
Fraud
  corporate, 145
  credit card, 181
  criminal, 167–169
  genuine assent and, 265
  health care, 177
  insurance, 146
  securities, 682–689, 697–698
Fraud-on-the-market theory, 682, 689
Fraudulent concealment, 167
Fraudulent records, 172
Fraudulent transfers, 749

Freddie Mac, 659
Fred Martin Motors, 88
Freedom of Information Act (FOIA), 494, 495, 694
Freedom-to-contract doctrine, 758
Free speech
  corporate commercial speech, 123–126
  corporate political speech, 126–127, 143
  hate speech, 123
  student speech, 121–122
Friehling, David, 146
Frobose, Hallie, 675
FSA. *See* Financial Services Authority
FSIA. *See* Foreign Sovereign Immunities Act
FTC. *See* Federal Trade Commission
FTCA. *See* Federal Tort Claims Act; Federal Trade Commission Act
FTDA. *See* Federal Trademark Dilution Act
FTS. *See* Foreign Tire Sales, Inc.
Full warranty, 766–767
Future interest, 363
FWPCA. *See* Federal Water Pollution Control Act

G

GAAP. *See* Generally Accepted Accounting Principles
GAAS. *See* Generally Accepted Auditing Standards
Gaddy v. Douglass, 425–426
Gaetano & Associates, Inc. v. National Labor Relations Board, 553–554
GAF Corporation, 188, 204
Galanter, Marc, 53
Gambling, 242, 269
GAO. *See* General Accounting Office; Government Accountability Office
Garnishment, 743
Garvin, David A., 2
Gator.com Corp. v. L.L. Bean, Inc., 45
GATT. *See* General Agreement on Tariffs and Trade
Gatz Properties, LLC v. Auriga Capital Corporation, 474
Gazprom, 737
GCC. *See* Gulf Cooperation Council
GDP. *See* Gross domestic product
Gender
  discrimination, 590–591
  peremptory challenges, 60–62
Gender-motivated crimes, 115
General Accounting Office (GAO), 93
General Agreement on Tariffs and Trade (GATT), 233, 235–236
General Dynamics Land Systems, Inc. v. Dennis Cline et al., 596–597
General jurisdiction, 46
Generally Accepted Accounting Principles (GAAP), 353
Generally Accepted Auditing Standards (GAAS), 353
General Motors (GM), 229, 371, 463, 703, 783
General partnership, 432, 435, 436–443
  creating, 437–439
  items included in agreements, 437
  model agreement form, 441–442
  pros and cons of, 443
  relationships between partners and, 439–440, 442
  terminating, 440–441

General public director (GPD), 173
General Services Administration (GSA), 484
General warranty deed, 369–370
Genuine assent, 264–267
Geragos v. Borer, 305
Germany, 70, 216, 240, 241, 419, 476, 694
   eminent domain in, 375
   insider trading in, 691
   solar energy in, 632
Gifts, 378
Gilmer v. Interstate/Johnson Lane Corp., 86, 90, 103
GlaxoSmithKline, 505
Global business, 224
Global climate change, 630–631, 645
Global commons, 641
Global dimensions
   of administrative agencies, 498
   of ADR, 98–100
   of agency law, 426–427
   of American legal system, 69–70
   of antitrust laws, 735–736
   of business associations, 447–449
   of consumer protection, 783
   of contract law, 293–294
   of corporations, 476
   of employment discrimination, 610–611
   of employment relationship, 532
   of environmental law, 639–642
   of ethics, 207
   of intellectual property, 402–403
   of labor–management relations, 564
   of legal environment of business, 30
   of product liability, 353–355
   of property, 379–380
   of securities, 693–695
   of torts, 326–327
   of white collar crime, 181
Global dispute resolution, 242–244
Globalization, 17, 224, 243–244
GM. *See* General Motors
Godfrey, Karen, 347–349
Goldman Sachs, 663
Gone with the Wind (novel), 398
Gonzales v. Raich, 112–113, 114, 115
Good-faith bargaining, 559
Goods, 110–111, 252, 279, 281
Google, 214, 395, 737, 740
Gossels v. Fleet National Bank, 419
Government
   contracts, 284–286
   spending powers of, 119–121
   subsidies, 620
Government Accountability Office (GAO), 179–180
Government in Sunshine Act (Sunshine Act), 494
Government of India (GOI), 227
Government Printing Office, 26
GPD. *See* General public director
Grand jury, 52, 151
Grand larceny, 170
Granholm v. Heald, 115
Grasso, Richard, 464
Gray, Kevin, 167–168
Grease payments, 216
Great Depression, 482, 485, 538, 660
Greece, 239–240, 476
Green cards, 530–531
Greenhouse, "Bunny," 179
Greenman v. Yuba Power Products Co., 346

Green marketing, 640
Griggs v. Duke Power Co., 579
Groh v. Ramirez, 128
Gross domestic product (GDP), 215
Groupthink, 161, 196–197
Grutter v. Bollinger, 606, 608
GSA. *See* General Services Administration
GTO Investments, Inc., 450
Guarantees, 343
Guaranty contracts, 744
Guillen, Ignacio, 114
Guilty mind (*mens rea*), 148, 157, 186
Gulf Cooperation Council (GCC), 235
Gun Control Act, 132
Gun-Free School Zone Act, 111

H

H-1B visa, 531–532
Hackers, 170
Hague Convention on the Service of Judicial and Extrajudicial Documents in Civil and Commercial Matters, 354
Hallmark Cards, Inc. v. Murley, 287
Hall Street Associates, L.L.C. v. Mattel, Inc., 82–83
Handicapped individual, 601, 602
Hansen, Joel, 509
Harassment, 579–582
   same-sex, 582, 590
   sexual, 12, 107–109, 579, 587, 591
Harris, Teresa, 580
Hart–Scott–Rodino Antitrust Improvement Act, 731, 759
Harvard School, 701, 704
Harvard University, 609
Hate speech, 123
Haughton, Aaliyah Dana, 320–321
Hazard Elimination Program, 114, 115
Hazardous and Solid Waste Amendments, 633
Hazardous Substances Labeling Act, 338
Hazardous waste and toxic substance regulations, 631–638
Hazelwood, Joseph, 191–192
HEALTH. *See* Help Efficient, Accessible, Low-cost, Timely Healthcare Act
Health and Safety Executive, 180
Health care fraud, 177
Helewicz, Joseph, 202
Help Efficient, Accessible, Low-cost, Timely Healthcare Act (HEALTH), 306
Hertz Corporation v. Friend, 41–43
Hierarchy of needs, 523
Hill, Anita, 591
Hindu law (*Dharmasastra*), 219, 220, 221
Hobbs Act, 164, 165
Hobby Protection Act, 759
Hodgins, Leo, 417–418
Hodgson v. Greyhound Lines, Inc., 600
Holiday Act, 532
Holiday Inn Franchising, Inc. v. Hotel Associates, Inc., 446–447
Holmes, Oliver Wendell, 19
Home confinement, 164
Homeland Security Act, 177
Homestead exemption, 744, 748
Homosexuality, 615
Honda Motor Co. v. Oberg, 301–302
Hopwood v. State of Texas, 607
Horizontal boycotts, 712–713
Horizontal division of markets, 711–712

Horizontal mergers, 729–730
Horizontal price-fixing, 709–711
Horizontal restraints of trade, 709–718
Horvath, Karl, 450
Host-country nationals, 224
Hostile bid, 679
Hostile environment, 582
Hot cargo agreement, 563
House arrest, 164
House Committee on Energy and Commerce, 25
Household Credit Services, Inc. v. Pfenning, 771–772
HSI. *See* Office of Homeland Security Investigations
Humanist theories, 193, 195
Human relations movement, 556
Human rights, 207
Hutton, E. F., 161
Hybrid rulemaking, 487
Hyoscyamine Sulfate ER. *See* Deponit and Hyoscyamine Sulfate Extended Release

I

Iacono v. Lyons, 272
IBM. *See* International Business Machines
IBT. *See* International Brotherhood of Teamsters
ICA. *See* Investment Company Act
ICANN. *See* Internet Corporation for Assigned Names and Numbers
ICC. *See* Interstate Commerce Commission
ICE. *See* Immigration and Customs Enforcement
ICSID. *See* International Center for the Settlement of Investment Disputes
IDCA. *See* International Development Corporation Agency
Identity theft, 169, 775–777
Identity Theft and Assumption Deterrence, 169
Ignazio v. Clear Channel Broadcasting, Inc. et al., 84–85
Illegal consideration, 264
Illusory promises, 264
ILO. *See* International Labor Organization
ImClone, 5, 7, 197
IMF. *See* International Monetary Fund
Immigration and Customs Enforcement (ICE), 530–532
Immigration law, 529–532
Immigration Reform and Control Act (IRCS), 529–530
Implied (implied-in-fact) contracts, 253
Implied covenant of good faith and fair dealing exception, 571
Implied warranty, 341
   of fitness for a particular purpose, 344–345
   of merchantability, 342–344
Import controls, 232–233
Impossibility of performance, 283–284
Imposter terminal, 170
Incidental beneficiary contracts, 274
Income bonds, 458
Incorporation, 456–457
Indemnity, 417, 420
Independent administrative agencies, 484
Independent judgment, 543
Indictment, 52, 151–152
Individual codes of ethics, 195–197
Individual with a disability, 602

In Dubious Battle (Steinbeck), 548
Industrial Revolution, 538
Informal negotiations, 54–55
Informal rulemaking, 486–487, 501
Information. *See* Indictment
Informational picketing, 563
In-house counsel, 49
Initial public offering (IPO), 670
Injunctions, 290
In personam jurisdiction, 36, 44–45
In rem jurisdiction, 40
Insanity, 153, 268
Insider Trader Sanctions Act, 678, 679, 690
Insider trading, 685, 686, 689, 691
Insider Trading and Securities Enforcement Act, 690
Institute of Internal Auditors, 199
Institutional school, of social responsibility, 201, 204–205
Insurance companies, 301
Insurance fraud, 146
Intangible property, 378
Intellectual disability, 603
Intellectual property, 224
   copyrights and, 222, 397–402
   global dimensions of, 402–403
   introduction to, 385
   patents and, 222, 393–396
   trademarks and, 385–393
   trade secrets and, 396–397
Intended beneficiary contracts, 274
Intentional infliction of emotional distress, 308, 317
Intentional interference with a contract, 308, 318–319
Intentional torts
   against economic interest, 308, 318–319
   against persons, 307–317
   against property, 308, 317–318
Interchange fees, 781
Interest in land, contract for sale of, 271
Interest of justice, 8
Interlocking directorates, 732
Intermediate courts of appeal, 46, 49
Intermediate scrutiny, 139
Internal controls, 109
Internal Revenue Code (IRC), 455
Internal Revenue Service (IRS), 154, 169, 200, 412, 437
International Association of Machinists, 736
International Brotherhood of Teamsters (IBT), 549
International Business Machines (IBM), 703, 704
International Center for the Settlement of Investment Disputes (ICSID), 243
International Chamber of Commerce, 99, 243
International cooperation, 639
International Court of Justice, 221
International Covenant on Economic, Social, and Cultural Rights, 532
International Development Corporation Agency (IDCA), 484
International franchising, 224–226, 448
International Labor Organization (ILO), 564
International law, 221–223
International legal environment of business, 213
   act-of-state doctrine and, 231–233
   dimensions of, 214–223
   global dispute resolution and, 242–244

   legal and economic integration and, 233–242
   methods of engaging, 223–229
   risks of engaging in, 229–231
International licensing, 224–226
International Monetary Fund (IMF), 239–240, 476
International Securities Enforcement Cooperation Act (ISECA), 650, 694
International Space Station, 196
International trade, 223–224
International Trade Administration (ITA), 233
International Trade Commission (ITC), 233
Internet
   case law precedents and, 26
   computer crimes and, 170–172
   consumer protection and, 780–781
   copyrights and, 398–402
   cybercrime and, 174, 181
   cyberlaw and, 29
   defamation and, 310, 314
   dispute resolution on, 81
   domain names and, 390
   e-commerce and, 421, 692, 768, 780–781
   e-contracts, 262–263, 291–293
   franchising on, 447
   in personam jurisdiction and, 44–45
   legal acceptance on, 262–263
   mediation on, 79
   online gambling and, 242
   as public accommodation, 609
   SEC and, 655
   spam and, 316
   taxation of, 120–121
   white collar crime on, 145
Internet Corporation for Assigned Names and Numbers (ICANN), 390
Internet protocol (IP), 497
Internet Tax Freedom Act, 120
Internet Tax Nondiscrimination Act, 121
Interstate commerce, 693, 726
Interstate Commerce Commission (ICC), 25, 482, 711
Interstate Oil Compact, 707
Intoxication, 268
Intrabusiness crime, 164
Intrastate activities, 707
Invasion of privacy, 308, 315
Investigators, 53
Investment Advisers Act, 649, 652
Investment Company Act (ICA), 649, 651
Investment contract, 662
Investor's Overseas Services (IOS), 181
Involuntary bankruptcy petition, 746
Involuntary transfer
   of personal property, 378–379
   of real property, 370–374
IOS. *See* Investor's Overseas Services
IP. *See* Internet protocol
IPO. *See* Initial public offering
Iraq reconstruction, 178, 179
IRC. *See* Internal Revenue Code
IRCS. *See* Immigration Reform and Control Act
IRS. *See* Internal Revenue Service
ISECA. *See* International Securities Enforcement Cooperation Act
Islam, 215, 588–589
Islamic law, 219–220
Issue, 7, 12
ITA. *See* International Trade Administration
ITC. *See* International Trade Commission

J

Jackson v. Rich's, 316
JAMA. *See Journal of the American Medical Association*
James, Lesa, 416
JAMS Resolution Services, 79, 95, 98
Janis, Irving, 161
Japan, 70, 181, 241, 449, 631, 694
   agency law in, 427
   cybercrime in, 174
   environmental law in, 624
   insider trading in, 691
   punitive damages in, 306
Javelin Strategy & Research, 169
J.E.B. v. Alabama, ex rel. T.B., 60–62
Jeffrey Bonkowski v. Oberg Industries, Inc., 514–516
Jehovah J. God, Jesus J. Christ, the Jehovah Witness Foundation Inc., 58
Jenkins, Cassandra, 509–510
Jennifer Johnson v. Board of Regents of the University of Georgia, 608
Jerman v. Carlisle, McNellie, Rini, Kramer & Ulrich, LPA, 779–780
Joe Camel, 202–203
Johannesburg Earth Summit, 641
Johnson, Lyndon B., 27
Johnson Construction Co. v. Shaffer, 206–207
Johnson & Johnson, 204
Johnson v. Santa Clara County Transportation Agency, 607
Jointly and several liability, 422
Joint ownership, 365
Joint stock company, 444–445
Joint tenancy, 365
Joint ventures, 229, 444, 445
Jones, Robert A., 258
Jones v. Clinton, 107–109, 581
Jones v. National American University, 598–599
Joseph Oncale v. Sundowner Offshore Services, 582, 590
Journal of the American Medical Association (JAMA), 202
JPMorgan Chase & Company, 654
Judges, 50–51
   bankruptcy, 746
   election of, 73–74
Judicial activism, 52
Judicial branch, 25–26, 496
Judicial power, 485
Judicial restraint, 51–52
Judicial review, 51, 488, 491, 492
Junk bonds, 459
Junk email, 768
Junk faxes, 768
Junk telephone calls, 768
Jurisdiction, 35
   general, 46
   in personam, 36, 44–45
   NLRB and, 543–544
   over persons and property, 36–40
   subject matter, 40–44
Jurisdictional picketing, 563
Jurisdictional strikes, 562
Jurisprudence, 19–23
Jury, 43–44
   grand, 52, 151
   instructions of, 64
   selection of, 59–63
   summary jury trials, 95

INDEX

K

Kant, Immanuel, 194–195
Kay, David, 216–218
Keller v. Central Bank of Nigeria, 230
Kenyon & Kenyon LLP, 457–458
In re KeyTronics, 438–439
Kidder Peabody, 663
Kimel v. Florida Board of Regents, 597
King, Martin Luther, Jr., 20–21
Klein, Calvin, 309
Koczur, Thomas, 258
Kohel v. Bergen Auto Enterprises, L.L.C., 280–281
Koontz v. St. John's Rangement District, 377
Korean Conflict, 27
Kozlowski, Dennis, 147
Krogh, Egil, Jr., 210
Krupny Plan, 225
Kyllo, Danny, 128
Kyoto Protocol, 631, 645

L

Laborers' International Union of North America, Local 872, AFL–CIO, and Stephanie Shelby, 554–555
Labor–management committees, 551–552
Labor–management relations, 538–539
  boycotts, 563
  global dimensions of, 564
  legislation and, 540–548
  organizing, 548–559
  picketing, 563
  strikes, 561–562, 567
Labor organizing
  appropriate bargaining unit and, 556–559
  board rules and, 548–549
  domination of, 551–552
  support of, 551–552
  unfair labor practices by employers and, 549–554
Land, 273
  interest in, 271
  ownership, 362
  use restrictions, 374–378
Landrum–Griffin Act, 539, 542
Language, 215
Lanham Act, 386, 391, 759
Larceny, 169–170
Larry King Live, 309
Last-clear-chance doctrine, 322
Latex allergies, 342
Law and economics school, 23
Lawful strikes, 561–562
Laws. *See also* Antitrust laws; Federal laws; State laws
  baby FTC, 782
  blue-sky, 155, 692
  civil rights, 572
  classification of, 27–29
  cyberlaw, 29
  lemon, 782
  sources of, 23–27
  unjust, 20–21
  wage and hour laws and, 504–506
  workers' compensation laws, 503, 510–513
Law School Admissions Test (LSAT), 609
Lay, Kenneth, 147, 177
Lead, 627
Lease, 364
Leasehold estates, 364

Leatherman Tool Group, Inc., 205
Ledbetter v. Goodyear Tire & Rubber Co., 595
Leegin Creative Leather Products, Inc. v. PSKS, Inc., dba Kay's Kloset, Kay's Shoes, 714–715
Legal acceptance, 260–263
Legal action, initiation of, 55–56
Legal environment of business, 17. *See also* International legal environment of business
  classifications of law and, 27–29
  definition of, 18–19
  global dimensions of, 30
  jurisprudence and law and, 19–23
  reasons for studying, 19
  sources of law and, 23–27
Legal literacy, 18
Legal norms, 43
Legal object, 269–270
Legal offer, 257–260
Legal reasoning, 5, 11–15
Legal remedies, 286–289
Legislative branch, 496
Legislative power, 485
Legitimate businesses, 176
Lemon laws, 782
Lenin, Vladimir, 220
Lessee, 364
Lessor, 364
Letters of comment or deficiency, 664
Levy, 744
Liability. *See also* Product liability; Third parties, liability to
  accountants', 353, 354
  contractual, 420–421
  corporate, 155–157
  on corporate executives, 157–160
  jointly and several, 422
  LLCs, 433, 473–475
  LLLP, 443
  on lower-level corporate criminals, 160
  market share, 351–352
  under Securities Act, 669–673
  under Securities Exchange Act, 689–690
  service, 332, 352–353
  strict, 154, 341–346
  strict liability torts, 307, 308, 326, 346–351
  tort, 422, 423
Libel, 309, 313
Library of Congress, 397, 401
License, 364
Licensing
  cross-licensing, 395
  international, 224–226
Liebeck v. McDonald's, 305
Liens, 743
Life estate, 363
Lilly Ledbetter Fair Pay Act, 595–596
Limited liability companies (LLCs), 433, 473–475
Limited liability limited partnership (LLLP), 443
Limited partnership (limited liability partnership), 433, 443–444
Limited warranty, 767
Linde v. Arab Bank, PLC, 231–232
Liquidated damages, 287–289
Liquidated debts, 264
Liquidation preferred stock, 460
Lisbon Treaty, 237
Litigation, 243. *See also* Civil litigation

Litigation Sciences, 62
Litigation Section Task Force on Alternative Dispute Resolution, 80
Litwin v. Blackstone Group, LP, 670
LLCs. *See* Limited liability companies
Llewellyn, Karl, 22
LLLP. *See* Limited liability limited partnership
Loans
  consumer, 781
  student, 169, 750
Loan-sharking, 111
Local administrative agencies, 496–497
Local Government Antitrust Act, 707
Lockhard v. Pizza Hut Inc., 582
Lodsys, 395
Lok Adalat (people's court), 98
London Commercial Court, 221
London Court of International Arbitration, 99
Long-arm statutes, 37
Lore, Nicholas, 516–517
Lorillard Tobacco Co. et al. v. Thomas F. Reilly, 126
Los Angeles County Flood Control District v. Natural Resources Defense Council, Inc. et al., 626
Los Angeles Times, 667
Lost property, 378
Love bug virus, 174
Lower-level corporate criminals, liability on, 160
Loyalty, duty of, 417, 420
LSAT. *See* Law School Admissions Test
Lucas v. South Carolina Coastal Commission, 136–137, 138
Lynch, Marshawn, 386

M

Maastricht Summit, 236
Maastricht Treaty, 237–239
MacPherson v. Buick Motor Co., 334
Macy v. Dept. of Justice, 590
Madoff, Bernard, 146
Magistrate, 149
Magnuson-Moss Warranty–Federal Trade Improvement Act, 759, 766–767
Mail fraud, 167
Major federal regulation, 496
Majority opinion, 65
Major life activities, 602–603
Malice, 313
Malpractice, 102–103, 353
Malpractice suits, 352
Management, 70, 94. *See also* Labor–management relations
  affirmative action plans and, 611
  behavioral approach to, 556
  biculturalism and, 596
  consumers and, 250
  environmental sustainability and, 640
  hierarchy of needs and, 523
  production and, 284
  responsibility and, 175
Managerial employee, 543
Managerial school, of social responsibility, 201, 204
Managers, 466
Mandatory arbitration agreements, 86, 93
Mandatory subjects of collective bargaining, 560
Manifest program, 633–634
Manufacturing defect, 350

Marbury v. Madison, 26, 51
Marijuana, medical, 112, 115
Marino, Doreen, 524–526
Marketable discharge permits, 620–621
Market economy, 215
Marketing, 307
   environmental sustainability and, 640
   oil, 707
   telemarketing, 767–768
Marketplace economics, 758
Market Reform Act, 650, 651
Market share liability, 351–352
MARS. *See* Mediations Arbitration Services
Martha Stewart Living, Inc., 197, 467, 659
Marx, Karl, 220
Marxism, 214
Maryland v. King, 131
Maslow, Abraham, 523
Massachusetts v. Environmental Protection Agency, 628
Massed picketing, 563
Master–servant relationship, 409
Material, 669
Material breach, 280
Matsushita Electric Industrial Co. v. Zenith Radio Corp., 725
Matthew Shepard and James Byrd, Jr. Hate Crime Prevention Statute, 123
Maximum contaminant level goals (MCLGs), 627
Maximum contaminant levels (MCLs), 627
McAuliffe, Christa, 196
McBride v. Taxman Corp., 421
McConnell International LLC, 174
McDonald's, 214–215, 305, 335, 448, 764, 766, 783
McKennon v. Nashville Banner, 600
MCLGs. *See* Maximum contaminant level goals
MCLs. *See* Maximum contaminant levels
McMillan v. City of New York, 604–605
McNeil Consumer Products Company, 335
McNulty, Paul, 155
MDOC. *See* Minnesota Department of Commerce
Means test, 749
Mechanic's lien, 743
Mediation, 78–80
   court-annexed, 97
Mediations Arbitration Services (MARS), 81
Medical marijuana, 112, 115
Medical testing, 529
Medium income test, 749
Meeting-the-competition defense, 727
Members, 473
Mens rea (guilty mind), 148, 157, 186
Mercado Commun del Ser Mercosul (Mercosul), 234
Merchantability, implied warranty of, 342–344
Mercosul. *See* Mercado Commun del Ser Mercosul
Merger Control Regulation, 737
Mergers, 727–732
   defenses to complaints, 730–731
   enforcement of, 731
   legality of, 728
   premerger notification requirement and, 731
   probable impact on competition and, 729
   reasons for increase in, 728
   relevant product and geographic markets and, 728–729
   remedies and, 731–732
   types of, 729–730

Merit, 587
Merit-based wage systems, 575
Meritor, 580, 581
"Merit" workplaces, 521
Merkur Steel, 450
Merrill Lynch, 5, 674
Methylisocyanate (MIC), 188, 227
Methyl tertiary butyl ether (MTBE), 341
MIC. *See* Methylisocyanate
Microeconomics, 688
Microsoft Corporation, 412, 703, 706, 723–724
Military caregiver's leave, 517
Milken, Michael, 686
Miller, Steven, 589
Miller v. AT&T Corp., 514
Mineral rights, 362
Minimum wage, 504, 508, 561
Ministerial Conference, 235
Minitrials, 93–94
Minneapolis & St. Louis Railroad Co. v. Beckwith, 143
Minnesota Department of Commerce (MDOC), 497
Minority groups, arbitration and, 93
Minors, 267–268, 409
Miranda rights, 149–150
Miranda v. Arizona, 149, 150
Misappropriation, 308, 319
Misdemeanors, 28, 149, 151
Mislaid property, 379
Misleading Advertising Directive, 783
Missing information, 10–11, 15
Missouri v. Seibert, 150
Misstatement of corporation, 685, 687
Mistake, 267
Mistake-of-fact defense, 153
Mitsubishi Motors Corp. v. Soler Chrysler-Plymouth, 99–100
Mixed motives, 587
Mock jury, 63
Model Code of Professional Responsibility, 199
Model Law on Cross-Border, 745
Modify, 65
Monetary damages, 286–289
Monetary union, 236, 238
Money laundering, 693–694
Monopolies, 718–725
   attempt to monopolize and, 724–725
   intent to monopolize and, 723
   overwhelming power in market and, 722–723
   relevant product and geographic markets and, 720–721
Monopolization, 719
Monson, Diane, 112
Montreal Protocol, 641–642
Moody's Global Long Term Rating Scale, 459
Moore, William E., 58
Moral law, 20
Morale, 517
Moral obligation, 264
Morgan, Barbara, 196
Morgantown (prison camp), 159
Morse v. Frederick, 122
Mortgage foreclosure, 743
Motions
   to dismiss, 56
   posttrial, 64
   pretrial, 57–58
Motorsport Marketing, Inc. v. Wiedmaier, Inc., 416

MTBE. *See* Methyl tertiary butyl ether
Multinational corporations, 454, 455
Municipality, 752
Murley, Janet, 287
Murphy, Douglas, 216–218
Mutual assistance agreements, 734
Mutual consent, 261
Mutual mistake, 267
Mutual Pharmaceutical Company, Inc. v. Bartlett, 340–341
MySpace, 37, 524

N

NAAQS. *See* National Ambient Air Quality Standards
Nacchio, Joseph, 146
NAD. *See* National Advertising Division
Nader, Ralph, 198
NAFTA. *See* North American Free Trade Agreement
Napster, 401
NASA. *See* National Aeronautics and Space Administration
NASD. *See* National Association of Securities Dealers
NASDAQ. *See* National Association of Securities Dealers Authorized Quotations
National Advertising Division (NAD), 198
National Aeronautics and Space Administration (NASA), 484
National Ambient Air Quality Standards (NAAQS), 628
National American University (NAU), 598–599
National Association of Broadcasters, 198
National Association of Optometrists and Opticians, 116
National Association of Securities Dealers (NASD), 87, 93, 498, 676
National Association of Securities Dealers Authorized Quotations (NASDAQ), 695
National Conference of Commissioners on Uniform State Laws, 292, 445, 473
National Consumer Arbitration Program, 92
National Cooperation Research Act, 701
National Defense Authorization Act, 179
National Education Association (NEA), 549
National Environmental Policy Act (NEPA), 41, 623–625, 639
National Federation of the Blind (NFB), 609
National Guard, 517
National Highway Safety Transportation Board (NHSTB), 482
National Highway Transportation and Safety Administration, 628
National Institute for Occupational Safety and Health (NIOSH), 518, 521–523
National Labor Relations Act (NLRA), 4, 79, 539, 540–541, 548, 570
National Labor Relations Board (NLRB), 482, 484, 552, 556, 558
   jurisdiction and, 543–544
   procedures in representation cases, 544–546
   procedures in unfair labor practice cases, 546–548
   structure of, 542–543
National Organization of Women (NOW), 712
National origin, 588
National Pollutant Discharge Elimination System (NPDES), 625

INDEX 819

National Sanitation Foundation (NSF), 712
National Science Foundation (NSF), 484
National Security Agency (NSA), 179, 196
National sovereignty, 207
National Transportation Safety Board (NTSB), 484
Nat'l Ass'n of Optometrists & Opticians v. Brown, 116–117
Natural law school, 20–21, 33, 195
NAU. *See* National American University
NEA. *See* National Education Association
Needs, hierarchy of, 523
Negligence, 319, 512
   defenses for, 322–326, 339–341
   environmental law and, 620
   privity limitation and, 333–334
   product liability and, 333–341
   tort liability and, 423
Negligence per se, 321–322, 338
Negligent advertising, 334
Negligent design, 334, 338
Negligent failure to warn, 334–338
Negligent hiring, 423
Negligent manufacture, 334
Negligent provision of an inadequate warning, 334
Negligent testing or failure to test, 334
Negligent torts, 307, 308, 319–322
Negotiation and settlement, 78
Nellis (prison camp), 159
Nelson, Melissa, 570
Nemet Chevrolet, Ltd. v. Consumeraffairs.com, Inc., 310–312
Neoclassical economic theory, 191
Neo-Keynesians, 23
NEPA. *See* National Environmental Policy Act
Nerve center approach, 42–43
NET. *See* No Electronic Theft Act
Network Solutions, Inc. (NSI), 390
New Bankruptcy Law, 757–758
Newcal Industries, Inc. v. Ikon Office Solutions, 721–722
New Groupthink Process, 196–197
New Source Review (NSR), 629
New York Stock Exchange (NYSE), 86, 93, 188, 649, 676, 712
New York Times, 679
NFB. *See* National Federation of the Blind
NHSTB. *See* National Highway Safety Transportation Board
Nicholson, Sam, 768
Nike, 391, 740
9/11. *See* September 11, 2001 terrorist attacks
Nineteenth Amendment, 60, 792
Ninth Amendment, 791
NIOSH. *See* National Institute for Occupational Safety and Health
Nixon, Richard, 210, 693
NLRA. *See* National Labor Relations Act
NLRB. *See* National Labor Relations Board
NLRB v. Cabot Carbon Co., 552
NLRB v. Jones & Laughlin Steel Corp, 111
No Electronic Theft Act (NET), 400
Nolo contendere, 152
Nominal damages, 287, 301
Nonattainment areas, 628
Nonbusiness contracts, 273
Nonemployee organizers, 551
Non-Federal Employee Whistleblower Protection Act, 179
Noninvestment company, 668

Nonprice vertical restraints, 715
Nonprofit corporations, 454, 456
Nonreporting issuer, 663
Normative ethical theorists, 30
Norms
   ethical, 9–10, 13
   legal, 43
North American Free Trade Agreement (NAFTA), 17, 222, 233–234, 240–242, 426, 642
No-solicitation rules, 550
Notes, 458
Notice of Hearing, 545
Notice of Petition for Election, 545
Notification, duty of, 420
Novation, 282
Novello, Antonia, 202
NOW. *See* National Organization of Women
NPDES. *See* National Pollutant Discharge Elimination System
NRC. *See* Nuclear Regulatory Commission
NSA. *See* National Security Agency
NSF. *See* National Sanitation Foundation; National Science Foundation
NSI. *See* Network Solutions, Inc.
NSR. *See* New Source Review
NTSB. *See* National Transportation Safety Board
Nuclear Regulatory Commission (NRC), 484, 624
Nuisance, 618
Nur, Bilan, 589
Nutrition Labeling and Education Act, 765
NYSE. *See* New York Stock Exchange

O

Obama, Barack, 180, 240–241, 396, 448, 595–596
Obedience, duty of, 418, 420
Obergefell et al. v. Hodges et al, 136, 139
Objectively impossible, 283
Occupational disqualification, 163–164
Occupational injury and illness, incidence rates of, 518
Occupational Safety and Health Act (OSH Act), 154, 166, 417, 484, 518–523
Occupational Safety and Health Administration (OSHA), 27, 484, 518–521
   enforcement by, 519
   inspection procedure and, 519
   penalties and, 520–521
   public education by, 521
   standard setting by, 519
   state plans and, 521
Occupational Safety and Health Review Commission (OSHRC), 518, 521
O'Connor v. Consolidated Caterers Corp., 597
OECD. *See* Organization for Economic Cooperation and Development
Offer, legal, 257–260
Office of Homeland Security Investigations (HSI), 530
Office of Management and Budget (OMB), 27, 486, 496
Office of Personnel Management (OPM), 27, 484
Office of Regulatory Enforcement, 622
Office of the General Counsel, 491
Officers, 466
Offshore transactions, 695

Oil marketing, 707
Old Joe Camel, 202–203
Olins, Robert, 254
Olis, Jamie, 146
OMB. *See* Office of Management and Budget
Omnibus Crime Control and Safe Streets Act, 523
One-shotters (OSs), 53
Online gambling, 242, 269
Online legal acceptance, 262–263
OPEC. *See* Organization of Petroleum Exporting Countries
Open-end credit, 770
Opening statements, 63
OPIC. *See* Overseas Private Investment Corporation
OPM. *See* Office of Personnel Management
Option, 259
Oral (parol) evidence rule, 273
Order for relief, 746
Organizational (recognitional) picketing, 563
Organization for Economic and Cultural Development, 207
Organization for Economic Cooperation and Development (OECD), 216, 234, 694
Organization of Petroleum Exporting Countries (OPEC), 736
Organized Crime Control Act, 175
Original jurisdiction, 35
Ortho Evra, 337
OSHA. *See* Occupational Safety and Health Administration
OSH Act. *See* Occupational Safety and Health Act
Oshana, George, 450
OSHRC. *See* Occupational Safety and Health Review Commission
OSs. *See* One-shotters
Ossai, Arthur, 230
OTC. *See* Over-the-counter (OTC) markets
Outsourcing, 448–449
Overlapping responsibility, 175
Overseas Private Investment Corporation (OPIC), 229
Over-the-counter (OTC) markets, 676
Overtime, 505
Oxley, Michael, 659

P

Pacific Gas and Electric Company (PG&E), 76–77
PACs. *See* Political action committees
Paduano v. American Honda Motor Co., 765
Palazzolo v. Rhode Island, 137
Pan Handle Realty, LLC v. Olins, 254
Pantagis Renaissance, 285–286
Paramount Contracting Co. v. DPS Industries, Inc., 252–253
Parens patriae suits, 707
Parents Involved in Community Schools v. Seattle School District No. 1, 510, 608
Parents Television Council, 493
Paris Convention, 403
Parker v. Brown, 707
Parliament, 239
Parol (oral) evidence rule, 273
Partial performance, 271
Partial strikes, 562
Participating bonds, 458
Participating preferred stock, 459

Partnership
  general, 432, 435, 436–443
  limited, 433, 443–444
Par value, 461
Patel, Marilyn Hall, 402
Patent and Trademark Office, 396
Patent Cooperation Treaty, 403
Patent Office, 386, 395
Patents, 222, 393–396
Patent trolls, 395–396
Patti, Frank, 167–168
Paxil, 336
PCAOB. *See* Public Company Accounting Oversight Board
PDA. *See* Pregnancy Discrimination Act
Peer-to-peer networking, 401
Pelman v. McDonald's, 335
Penn Central Railroad, 668
Penthouse International v. Barnes, 414
PeopleClaim, 81
People's court (Lok Adalat), 98
Peremptory challenges, 59–62
Perez v. United States, 111
Perfect tender rule, 281
Performance
  complete, 280
  duty of, 418–419, 420
  impossibility of, 283–284
  partial, 271
  specific, 289–290
  substantial, 280
Perjury, 7
Permissive subjects of collective bargaining, 560
Per se standard, 709
Personal injury cases, 299
Personal property, 252, 378–379
Pesticides, 637
Petit juries, 52
Petty crimes, 149
Pew Research Center, 538
Pfenning, Sharon, 771
PG&E. *See* Pacific Gas and Electric Company
Phishing, 781
Picketing, 563
PIDA. *See* Public Interest Disclosure Act of 1998
Pierce County v. Guillen, 114, 115
Piercing the corporate veil, 457
Pietrylo, Brian, 524–526
Piggybacking, 170
Pike v. Bruce Church, Inc., 118
Pinnacle Project, 372
Piracy, 226
Plaintiff, 36, 63–64
Plan of reorganization, 754
Plea bargaining, 152
Pleadings, 55
PLIVA, Inc. v. Mensing, 340
Plum Creek C.A. v. Oleg Borman, 365–368
PMN. *See* Premanufacturing notice
Point sources, 625
Poletown Neighborhood Council v. City of Detroit and the Detroit Economic Development Corporation, 371–372
Police power, 116, 374
Political action committees (PACs), 198, 703
Political intelligence, 689
Political union, 236, 238
Pollution prevention, 640
Pollution Prevention Act, 638–639
Polygraph testing, 529
Positivist school, 21
Postal Service, 760

Postindustrial society, 191
Posttrial motions, 64
Potentially responsible parties, 635
Power of attorney, 413
Precedents, 12, 14, 21, 26
Predatory conduct, 725
Predatory pricing, 725
Preemployment testing, 528
Preexisting duty rule, 263
Preferred stock, 459–460
Pregnancy Discrimination Act (PDA), 592
Prejudicial error, 65
Premanufacturing notice (PMN), 637
Premerger notification requirement, 731
Premises rule, 511
Prempro, 336
Preponderance of the evidence, 492
Pre-rulemaking action, 496
Prescription drugs, 335, 336–337
Presidents, 495–496
Pretext plus, 598
Pretrial diversion (PTD), 152
Pretrial stage, 53–59
Prevailing wage, 504
Price discrimination, 725–727
Price-fixing, 424, 709–711, 713
Primary ethical norms, 13
Primary-line injury, 726
Prince William Sound, 191–192
Principal, 409
  disclosed, 422
  duties of, 416–419
  partially disclosed, 422
  undisclosed, 422
Principal–agent relationship, 409, 410, 424–426
Principal place of business, 42
Principle of truthfulness, 783
Principles of Federal Prosecution, 151
Privacy
  employee rights to, 523–529, 535–536
  invasion of, 308, 315
  torts, 314–316
Private development, 373–374
Private international law, 221
Private law, 29
The Private Movie Company, Inc. v. Pamela Lee Anderson et al., 260–262
Private Securities Litigation Reform Act (Reform Act), 650, 651
Private trials, 95
Privity, 342
Privity limitation, 333–334
Probable cause, 149
Procedural due process, 132
Procedural justice, 94
Procedural law, 29
Procedural rule, 483
Production
  burden of production of evidence, 152
  management and, 284
Productivity-based wage systems, 575
Product liability, 306, 332
  breach of warranty and, 341–346
  to bystanders, 351
  global dimensions of, 353–355
  market share liability and, 351–352
  negligence and, 333–341
  service liability and, 352–353
  strict liability in tort, 346–351
  theories of recovery and, 333–351
Product quality and quantity, 762
Product stewardship, 640

Product trademarks, 386
Professional codes of ethics, 198–201
Professional corporations, 454, 455–456
Professional obligation school, of social responsibility, 201, 205–206
Profit-oriented school, of social responsibility, 201–204
Profits derived solely from the efforts of others, 661
Promises enforceable without consideration, 263–264
Proof of claim, 747
Property, 360. *See also* Intellectual property; Real property
  as bundle of rights, 361
  exempt, 748
  global dimensions of, 379–380
  personal, 252, 378–379
Property Firsters, 137
Prospectus, 663
Protected classes, 575
  national origin, 588
  PDA and, 592
  race and color, 588
  religion, 588–589
  sex, 590–591
Protestant work ethic, 216
Proximate cause, 320
Proxy, 462–463
  solicitations, 677–679
PTD. *See* Pretrial diversion
Publication, 318
Public Company Accounting Oversight Board (PCAOB), 178–179, 200, 659
Public corporation, 433
Public disclosure of private facts, 315
Public figure for a limited purpose, 314
Public figure for all purposes, 314
Public figures, 313–314
Public Interest Disclosure Act of 1998 (PIDA), 180
Public international law, 221
Public law, 28–29, 481
Publicly held corporations, 454, 455
Public policy exception, 571
Public safety, 13
Public use, 373
Public Utility Holding Company Act, 649, 651
Public water supply system, 627
Puffery, 762
Punitive damages, 287, 301–307, 330, 595
Pure Food, Drug, and Cosmetics Act, 157, 158, 164, 166

Q

Quality, 284
Quality Car & Truck Leasing, Inc. v. Sark, 435–436
Quasi-contracts, 256–257
Questions of fact, 59
Quid pro quo sexual harassment, 579
Quitclaim deed, 370
Quotas, 233

R

Race, as protected class, 588
Race-based peremptory challenges, 60–62
Racketeer Influenced and Corrupt Organizations (RICO), 175–176, 230
RadLAX Gateway Hotel, LLC v. Amalgamated Bank, 754–755

Raich, Angel, 112, 114
Railroad Revitalization Act, 668
Ramadan, 589
Random drug testing, 528, 535–536
Rapanos v. United States, 115
Rational basis test, 139
RCRA. *See* Resource Conservation and Recovery Act
Reagan, Ronald, 27, 496
RealNetworks, Inc. v. DVD Control Copy Association, Inc. et al., 402
Real property, 252, 361–362
   interests in, 362–368
   involuntary transfer of, 370–374
   land use restrictions and, 374–378
   trespass to realty, 308, 317
   voluntary transfer of, 368–370
Reasonable cause, 593
Reasonable expectations of profit, 661
Reasonable expectation test, 344
Reasonable likelihood, 8–9
Reasonable-person standard, 319
Reasons, 7–8, 12
Rebuttal, 64
Recognitional picketing, 563
Recording, 370
Recovery theories
   liability to bystanders, 351
   negligence, 333–341
   strict liability in breach of warranty, 341–346
   strict liability in tort, 346–351
Red herring, 664
Redirect examination, 63
Reeves v. Sanderson Plumbing Products, Inc., 598
Reform Act, 650, 651
Reformation, 289
Regents of the University of California v. Bakke, 607
Registrars, 390
Registration statements, 669–671
Regulation Fair Disclosure (FD), 675
Regulation school, of social responsibility, 201, 206–207
Regulation Z, 770
Regulatory crimes, 148
Regulatory takings, 136
Rehabilitation Act, 572, 601–602
Reimbursement, duty of, 417, 420
Relationships. *See also* Employment relationship
   agency relationships, 409–416
   employer–employee, 409
   employer–independent contractor relationship, 410
   franchising, 765–766
   master–servant, 409
   between partners, 439–440
   principal–agent, 409, 410, 424–426
Religion, 215, 588–589, 615
Remand, 65
Remedies
   breach of contract, 286–291
   consumer law and, 767
   equitable, 289–290
   legal, 286–289
   mergers and, 731–732
   under Securities Act, 669–673
   Securities Exchange Act and, 679–682, 689–690
   Title VII and, 595
Remedies Act, 650, 651, 672

Reno v. Condon, 115
Rent-a-judge, 95
Repeat players (RPs), 53
Reply, 56
Representative democracy, 501
Resale price maintenance, 713
Rescission, 289
Research, 641
Residential mortgages, 782
Residential service, 36
Res ipsa loquitur, 321, 347, 348
Resource Conservation and Recovery Act (RCRA), 633–635
Respondeat superior, 409, 422
Responsibility, 175, 574
Restatement (Second) of Torts, 346, 349, 353
Restatement (Third) of Torts, 350–351
Restatements, 26
Restraint of trade, 708
   horizontal, 709–718
   vertical, 713–718
Restrictive covenants, 374
Retaliation, 583–586
Reverse discrimination, 606, 607–608
Revised Model Business Corporations Act (RMBCA), 458, 459, 461, 464, 469, 681
Revised Uniform Limited Partnership Act (RULPA), 443
Revised Uniform Partnership Act (RUPA), 436, 437, 439
Revocation, of legal offer, 259–260
Reya C. Boyer-Liberto v. Fontainbleu Corporation, 584–586
Reyes, Gregory, 461
Reynolds, R. J. (RJR), 202–203, 710–711
RICO. *See* Racketeer Influenced and Corrupt Organizations
Right of removal, 43
Riley v. California, 131
Ringling Brothers–Barnum & Bailey, 391
Rio Earth Summit, 641
Risk-utility test, 350
RJR. *See* Reynolds, R. J.
RMBCA. *See* Revised Model Business Corporations Act
Robinson–Patman Act, 725, 726
ROBS. *See* Rollover business corporation
Rodman, Dennis, 317
Role conflict, 596
Role overload, 596
Rollover business corporation (ROBS), 455
Romano-Germanic civil law, 219
Rome Treaty, 236
Rosen, Gerald, 752
Rousso v. State, 119
Royalties, 395
RPs. *See* Repeat players
Rubin v. Coors Brewing Co., 125
Rule 10b-5, of Securities Exchange Act, 683, 684, 687, 689
Rule 33, 5, 7, 8, 10
Rule 144, of Securities Act, 669
Rule 144(a), of Securities Act, 669, 695
Rule 147, of Securities Act, 667, 668
Rulemaking, 486–489, 501
Rule-of-reason standard, 708–709, 713
Rules of civil procedure, 26, 54
Rules of law, 8, 12
Rule utilitarianism, 30, 193–194
RULPA. *See* Revised Uniform Limited Partnership Act
Rumsey, Phillip R., 376

RUPA. *See* Revised Uniform Partnership Act
Rüsen, Hein, 450
Russian Entertainment Wholesale, Inc. v. Close-Up International, Inc., 225

S

Safeco Insurance Co. v. Burr, 775
Safe Drinking Water Act (SDWA), 627
Safe residue level, 638
Safe working conditions, duty to provide, 417, 420
Sagan, Carl, 310
Saint Francis College et al. v. Majid Ghaidan Al-Khazraji, 573
Salami slicing, 170
Salary arbitration, 92
Sale of Goods Act, 343
Sales, 252, 726. *See also* Contract law
Saliva drug test, 535–536
Same-sex harassment, 582, 590
Same-sex marriage, 133, 134–136
SARA. *See* Superfund Amendment and Reauthorization Act
Sarbanes, Paul, 659
Sarbanes-Oxley Act, 147, 177–179, 197–200, 648–651, 659–660, 692
Sark, Michael, 435–436
Sastras, 221
In re Savage v. United States Bankruptcy, 750–751
SBA. *See* Small Business Administration
SBREFA. *See* Small Business Regulatory Enforcement Fairness Act
SCA. *See* Stored Communications Act
Scheidler v. National Organization for Women, Inc., 165
School District No. 403 v. Fraser, 122
Schreiber, Barbara, 680
Schuette v. Coalition to Defend Affirmative Action, 609
Scienter, 687
Scrushy, Richard, 147
SCS. *See* Soil Conservation Service
SDWA. *See* Safe Drinking Water Act
SEA. *See* Single European Act
Seasoned issuer, 663
SEC. *See* Securities and Exchange Commission
Second Amendment, 791
Secondary boycotts, 563
Secondary-line injury, 727
Secondary strike, 562
Secretariat, 241
Secured bonds, 458
Securities, 30, 647–648
   Dodd-Frank Act and, 655–659
   e-commerce and, 692
   federal legislation and, 649–651
   global dimensions of, 693–695
   introduction to regulation of, 648–655
   marketplace of, 692
   Sarbanes-Oxley Act and, 659–660
   SEC and, 652–653
   state laws, 692
Securities Act (1933), 485, 648, 649, 651, 660–673, 694
   communications and, 664–665
   definition of securities and, 661
   disclosure and, 663
   exemptions from registration under, 666–668
   intrastate offering exemptions, 666, 667

Securities Act (1933) (*continued*)
  liability, remedies, and defenses under, 669–673
  posteffective period and, 663, 664
  prefiling period and, 663–664
  private placement exemptions, 666
  purpose and goals of, 662
  registration of securities under, 662–665
  registration statement and process, 662–663
  resale restrictions and, 668–669
  shelf registration and, 665
  small business exemptions, 666, 667–668
  waiting period and, 663, 664
Securities and Exchange Commission (SEC), 27, 198–199, 458, 482, 496, 652–653, 760
  General Accounting Office and, 93
  Internet and, 655
  proxy and, 462
Securities and Exchange Commission v. Edwards, 662
Securities and Exchange Commission v. Texas Gulf Sulphur Co., 687–688
Securities Enforcement Remedies and Penny Stock Reform Act (Remedies Act), 650, 651, 672
Securities Exchange Act (1934), 485, 648, 649, 651, 673–692
  compensation disclosure and, 675–676
  liability and remedies under, 689–690
  proxy solicitations and, 677–679
  registration of securities issuers, brokers, and dealers, 673–675
  remedies and defensive strategies, 679–682
  securities fraud and, 682–689
  securities markets and, 676–677
  short-swing profits and, 691–692
  takeover bids and, 678–679
  tender offers and, 678–679
Securities fraud, 682–689, 697–698
Securities Investor Protection Act (SIPA), 649, 651, 674
Securities Investor Protection Corporation (SIPC), 649
Securities Litigation Uniform Standards Act, 650, 651
Securities markets, 676–677
SEC v. Howey, 661, 662
Seisin, 369
SEIU. *See* Service Employees International Union
Sekhar v. United States, 164–165
Self-auditing, 622
Self-defense, 309
Self-regulating organizations (SROs), 676–677
Senate–House Conference Committee, 25
Seniority-based wage systems, 575
Sentencing, of white-collar criminals, 162–164
Sentencing Guidelines (1991), 162–163
Separation of duties, 109
Separation of powers, 107–110
September 11, 2001 terrorist attacks, 37, 196, 588, 589
Seroquel, 337
Service Employees International Union (SEIU), 549
Service liability, 332, 352–353
Service marks, 386
Service of process, 36, 37, 56
Seventeenth Amendment, 792
Seventh Amendment, 791
Sex, as protected class, 590–591

Sex-based peremptory challenges, 61–62
Sexual harassment, 12, 107–109, 579, 587, 591
Sexual orientation, 615
Shadow jury, 63
Shaffer, Bubba, 206–207
Shareholders, 459, 462–463, 677–678
Shari'a law, 220
Sharkey, Helen, 146
Shelby, Stephanie, 554–555
Shelf registration, 665
Sherman Act, 12, 424, 699, 704, 735, 759
  combinations and restraints of trade and, 708–718
  monopolies and, 718–725
Shopkeepers' tort, 316
Short-swing profits, 691–692
Signal picketing, 563
Similar working conditions, 574
Sindell v. Abbott Laboratories, 352
Single European Act (SEA), 236, 239
SIP. *See* State implementation plan
SIPA. *See* Securities Investor Protection Act
SIPC. *See* Securities Investor Protection Corporation
Siracusano v. Matrixx Initiatives, Inc., 683–684
Sit-down strikes, 562
Sixteenth Amendment, 792
Sixth Amendment, 29, 61, 163, 672, 791
Skill, 574
Skilling, Jeffrey, 145, 177
Skills testing, 529
Slander, 309, 313
Small Business Administration (SBA), 484
Small business bankruptcy, 756
Small Business Regulatory Enforcement Fairness Act (SBREFA), 495
Small public offerings, 667
Smith v. Van Gorkom, 469–470
Social framework analysis, 68
Socialist law, 219, 220
Social networking, 37
Social responsibility, 189, 191–192, 201–207
Society of Chartered Property and Casualty Underwriters (CPCU), 200
Sociocultural values, 207
Sociological school, 21–22
Soft variables, 609
Soil Conservation Service (SCS), 642
Sole proprietorship, 432, 434–436
Solid Waste Agency of Northern Cook County v. United States Army Corps of Engineers, 115
Sources of law, 23–27
South American Common Market, 233, 234
Southern Prestige Industries, Inc. v. Independence Plating Co., 39
South Fla. Water Management Dist. v. Miccosukee Tribe, 626
Sovereign acts doctrine, 284–286
Sovereign immunity doctrine, 230–231
Spam, 315, 768
SPDs. *See* Special public directors
Special-hazards exception, 511
Special public directors (SPDs), 173
Specialty Healthcare and Rehabilitation Center of Mobile and United Steelworkers, District 9, Petitioner, 557–558
Specific performance, 289–290
Speech and Debate Clause, 313
Spending powers, of federal government, 119–121
Spirito, Tony, 258

SROs. *See* Self-regulating organizations
Stambovsky v. Ackley and Ellis Realty, 265–266
Stanford, Allen, 146
"Star" workplaces, 520
State action doctrine, 707
State administrative agencies, 496–497
State consumer legislation, 782
State court judges, 50–51
State court jurisdiction, 40–41
State court system, 46–49, 95–97
Stated capital, 461
State district courts, 47
State Farm v. Campbell, 302, 304
State implementation plan (SIP), 628
State laws
  franchising and, 445
  securities, 692
  against white collar crime, 180
Statement of Position, 545
State of incorporation, 42
State-of-the-art defense, 340, 351
State securities regulations, 155
State v. Heckel, 768
Statutes, 23, 25
  against employment discrimination, 572
  of limitation, 339, 346
  long-arm, 37
  of repose, 339
Statutory law, 21, 23–25, 27, 252, 269
Steinbeck, John, 548
Stewart, Lawrence F., 5, 7, 8–9, 10
Stewart, Martha, 5–6, 7, 467
Stichting Defense, 731
Sting operations, 153
Stock, 453, 459–460
Stock options, 460–461
Stockton Steel, 589
Stock warrant, 460
Stone, Christopher, 173
Stored Communications Act (SCA), 132, 524
Straight bankruptcy, 745, 749–752
Strek, Mark, 290–291
Strict liability
  in breach of warranty, 341–346
  offenses, 154
  torts, 307, 308, 326, 346–351
Strict scrutiny, 139
Strikes, 561–562, 567
Student loans, 169, 750
Student speech, 121–122
Subchapter S corporation, 433, 454, 455
Subcommittee on Consumer Finance and Telecommunications, 496
Subject matter jurisdiction, 40–44
Submission agreement, 83–84
Substantial cooperation, 162
Substantial performance, 280
Substantive due process, 133
Substantive law, 29
Substantive rule, 483
Summary jury trials, 95
Summons, 36
Sunshine Act. *See* Government in Sunshine Act
Superfund, 635, 636
Superfund Amendment and Reauthorization Act (SARA), 635
Supervisor, 543
Supremacy Clause, of Constitution, 27, 106, 107, 222, 497
Supreme Court, 46, 51. *See also specific cases*
Suretyship, 743

INDEX 823

Surface Mining and Reclamation Act, 137
Susette Kelo et al., Petitioners v. City of New London, Connecticut et al., 372–373
Sustainability vision, 640
Sustainable development, 639
Sutherland, Edwin, 154
Swartz, Mark, 147
Sydney, Laura, 254
Syndicate, 444, 445

T

Taft–Hartley Act, 541
Tailpipe Rule, 628
Takeover bids, 678–679
Takings Clause, 136–137
Taliban, 37
Tangible property, 378
Tarnishment, 391
TARP. *See* Troubled Asset Relief Program
Tax credits, 728
Taxes, 119–121, 222, 474
Taxman v. Board of Education, 606, 607
Taxpayers Against Fraud, 178
Technology. *See* Internet
Technology-driven standards, 621
Technology-forcing standards, 621
Telecommunications Act, 488
Telemarketing and Consumer Fraud and Abuse Prevention Act, 759, 767
Telemarketing legislation, 767–768
Telephone Consumer Protection Act, 767, 768
Temps, 412
Tenancy by the entirety, 365
Tenancy in common, 365
Tender offers, 678–679
Tenth Amendment, 791
Teresa Harris v. Forklift Systems, Inc., 580
Termination by agreement, 424–425
Termination by operation of law, 425–426
Tertiary-line injury, 727
Testimonials, 764
Thermal imaging, 129
Third Amendment, 791
Third-country nationals, 224
Third parties, liability to
  contractual liability, 420–421
  criminal liability, 424
  of disclosed, partially disclosed, and undisclosed principals, 422
  e-commerce, 421
  tort liability, 422
  tort liability and negligence, 423
Third-party beneficiary contracts, 274
Thirteenth Amendment, 572, 792
Thomas, Carl, 450
Thomas, Clarence, 591
Thompson, David, 465
Thota, Venkateswarlu, 324–326
Thunor Trust, 685
TILA. *See* Truth-in-Lending Act
Tinker v. Des Moines Independent School District, 121
Tips, 504–505
Title, 378
Title VII, of Civil Rights Act (1964), 12, 610
  anatomy of case, 593
  applicability of, 575–576
  enforcement procedures for, 592–594
  Lilly Ledbetter Fair Pay Act and, 595–596
  proof and, 576–583
  protected classes and, 588–592

remedies and, 595
retaliation and, 583–586
statutory defenses and, 586–587
Tobacco industry, 202, 783
"Too big to fail," 657
Tortfeasor, 300
Tort liability, 422, 423
Torts, 410
  classification of, 307
  damages available and, 300–307
  environmental law and, 618–620
  global dimension of, 326–327
  goals of, 299–300
  intentional, 307–319
  negligent, 307, 308, 319–322
  strict liability, 307, 308, 326, 346–351
Toxic assets, 656
Toxic substances and products, 357–358, 637. *See also* Hazardous waste and toxic substance regulations
Toxic Substances Control Act (TSCA), 636–637
Toys "R" Us, Inc. v. Canarsie Kiddie Shop, Inc., 386–388
TPP. *See* Trans-Pacific Partnership
Trade Commission, 241
Trade dress, 389
Trade fixtures, 362
Trademarks, 222, 225, 385–393
Trade Related Aspects of International Property Rights (TRIPS), 402–403, 406
Trade secrets, 396–397
Transgender discrimination, 590–591
Transnational corporations, 207, 455
Trans-Pacific Partnership (TPP), 448
Transparency International, 218
Transportation infrastructure, 215
Transportation v. Moriel, 304–305
Treason, 149
Treaties, 27, 30, 221, 641–642
Treaty of Amsterdam, 737
Trespass to personalty, 308, 317
Trespass to realty (trespass to real property), 308, 317
Trial court judges, 50
Trials
  bench, 152
  jury selection and, 59–63
  minitrials, 93–94
  opening statements and, 63
  plaintiff's case and, 63–64
  private, 95
  summary jury, 95
Triple-bottom-line accounting, 639
TRIPS. *See* Trade Related Aspects of International Property Rights
Trojan horse, 170
Troubled Asset Relief Program (TARP), 657
Truman, Harry S., 27
Trustees, 747
Trust Indenture Act, 649, 651
Trusts, 700
Truth-in-Lending Act (TILA), 769–772
TSCA. *See* Toxic Substances Control Act
Twelfth Amendment, 791–792
Twentieth Amendment, 792–793
Twenty-first Amendment, 793
Twenty-second Amendment, 793
Twenty-third Amendment, 793
Twenty-fourth Amendment, 793
Twenty-fifth Amendment, 793
Twenty-sixth Amendment, 793
Twenty-seventh Amendment, 793

24-hour rule, 549
Two Pesos v. Taco Cabana, 389
Tyco International, 143, 147, 197, 659
Tying arrangements, 395, 717–718, 727
Tylenol, 204, 335
Typing tests, 528–529

U

UAW. *See* United Auto Workers
UCC. *See* Uniform Commercial Code; Union Carbide India Ltd.
UCCC. *See* Uniform Consumer Credit Code
UCITA. *See* Uniform Computer Information Transaction Act
UETA. *See* Uniform Electronic Transcription Act
UFCW. *See* United Food and Commercial Workers International Union
UFOC. *See* Uniform Franchise Offering Circular
UGESP. *See Uniform Guidelines on Employee Selection Procedures*
ULLCA. *See* Uniform Limited Liability Company Act
Ultramares Doctrine, 353
Unconscionability, 342
Under 2 MOU, 631
Undervalued assets, 728
Underwriters, 663
Undisclosed principal, 422
Undue hardship, 750
Undue influence, 267
Unemployment compensation, 503, 506–510
Unfair advertising, 764
Unfair competition, 308, 319
Unfair labor practices, 541, 549–559
Unfair labor practice strike, 562
Uniform Commercial Code (UCC), 251–252, 262, 275, 279, 292, 294, 341–342
Uniform Computer Information Transaction Act (UCITA), 292–293, 294
Uniform Consumer Credit Code (UCCC), 782
Uniform Domain Name Dispute Resolution Policy, 390
Uniform Electronic Transcription Act (UETA), 292, 421
Uniform Franchise Law, 445
Uniform Franchise Offering Circular (UFOC), 445
Uniform Guidelines on Employee Selection Procedures (UGESP), 587
Uniform Limited Liability Company Act (ULLCA), 473
Uniform Partnership Act (UPA), 436
Uniform Securities Act, 692
Unilateral contracts, 255
Unilateral mistake, 267
In re Union Carbide Corp. Gas Plant Disaster v. Union Carbide Corp., 227–228
Union Carbide India Ltd. (UCC), 188, 201, 204, 226–229
Unions. *See* Labor–management relations
United Auto Workers (UAW), 548
United Food and Commercial Workers International Union (UFCW), 549
United Haulers Association, Inc. v. Oneida-Herkimer Solid Waste Management Authority, 115
United Nations Climate Change Conference, 631
United Nations Commission of International Trade Law, 99

United Nations Commission on Human Right, 532
United Nations Conference on the Human Environment, 641
United Nations Convention on the Recognition and Enforcement of Foreign Arbitral Awards, 98
United Nations Convention on the Recognition of Foreign Arbitral Awards, 243
United Nations Multinational Code, 30
United States Department of Agriculture v. United Foods, Inc., 127
United States of America v. Martha Stewart and Peter Bacanovic, 5–6, 7
United States Trustee, 746
United States v. American Library Association, 122
United States v. Booker, 163
United States v. Doe, 138
United States v. Enmons, 165
United States v. Fanfan, 163
United States v. Gray, 167–168
United States v. Kay, 216–218
United States v. Lake, 111
United States v. Lopez, 114, 116
United States v. Microsoft Corporation, 723–724
United States v. Morrison, 115
United States v. Park, 157–158, 159
United States v. Patane, 150
United States v. Wallach, 10
United States v. Warshak, 132
United States v. Windsor, 134–136
United Steelworkers, 549
United Steelworkers v. Weber, 607
Universal Copyright Convention, 403
University of Arizona, 58
University of California v. Bakke, 608–610
University of Texas Southwestern Medical Center v. Nassar, 584
Unjust laws, 20–21
Unlawful appropriation of data or services, 172
Unlawful strikes, 562
Unliquidated debts, 264
Unreasonable search and seizure, 128
Unseasoned issuer, 663
Unsecured bonds, 458
UPA. *See* Uniform Partnership Act
Uruguay Round Accord, 235, 235n10, 402
USA PATRIOT Act, 171
U.S. Bankruptcy Code, 742
    Chapter 7 (straight bankruptcy) of, 745, 749–752
    Chapter 11 of, 753–756
    Chapter 12 of, 756–757
    Chapter 13 (wage earner's plan) of, 752–753
    history and background of, 745
    management and proceedings, 746–749
    New Bankruptcy Law and, 757–758
    types of proceedings, 756
USCIS. *See* U.S. Citizenship and Immigration Services
U.S. Citizenship and Immigration Services (USCIS), 529
U.S. Code Congressional News and Administrative Reports, 26
U.S. v. Tappen, 171

V

VA. *See* Veterans Administration
Valid contracts, 256
Vance v. Ball State University, 583
Van Rompuy, Herman, 237
Variance, 374
Veasey, E. Norman, 471
Venkateswarlu Thota and North Texas Cardiology Center v. Margaret Young, 324–326
Venue, 44–45
Vertical mergers, 729, 730
Vertical price-fixing, 713
Vertical restraints of trade, 713–718
Vertical territorial and customer restraints, 715–716
Veterans Administration (VA), 46, 484
Veterans Court of Appeals, 46
Victor Moseley and Cathy Moseley et al., dba Victor's Little Secret v. V Secret Catalogue, Inc. et al., 391–393
Violence Against Women Act, 112
Violent picketing, 563
Virus, 172, 174
Voice over Internet Protocol (VoIP), 497
Void and voidable contracts, 256
VoIP. *See* Voice over Internet Protocol
Voir dire, 59, 62
Voluntary bankruptcy petition, 746
Voluntary compliance, 490
Voluntary participation program (VPP), 520
Voluntary pollution prevention, 639
Voluntary transfer
    of personal property, 378
    of real property, 368–370
Vonage Holdings Corp. v. Minnesota Public Utilities Commission, 497
VPP. *See* Voluntary participation program

W

Wage and Hour Division (WHD), 506
Wage and hour laws, 503, 504–506
Wage earner's plan, 752–753
Wagner Act (National Labor Relations Act), 539, 540–541, 548
Waksal, Samuel, 5
Wall Street Journal, 16, 465, 679, 686
Wal-Mart, 505, 526
Wal-Mart Stores Inc. v. Dukes, 67
Walt Disney Company, 472
Warrantless searches, 131
Warranty
    express, 341, 342, 345, 766
    full, 766–767
    general warranty deed, 369–370
    implied, 341–345
Washington Convention, 243
Water quality regulations, 625–627
Water rights, 362
Watkins, Sherron, 177
Watson, James W., 203
Webb-Pomerene Trade Act, 707
Welge v. Planters Lifesavers Co., 347–349
Well-known seasoned issuer, 663
Wendy's, 448, 764
The Wharf (Holdings) Limited v. United International Holdings, Inc., 690
WHD. *See* Wage and Hour Division
Wheeler–Lea Act, 759, 760
Whistleblower Protection Act (WPA), 177, 179–180
White, Miles, 465
White collar crime, 145–148
    blame for, 186
    common crimes, 164–172
    crime and criminal procedure, 148–154
    distinguishing features of, 154–164
    federal laws against, 175–180
    global dimensions of, 181
    prevention of, 173–175
    state laws against, 180
    summary of, 182
White-Collar Crime Penalty Act, 167
Whitney Benefits, Inc. v. United States, 137
Wickard v. Filburn, 111, 112–113
Wiedenbaum, Murray, 244
Wildcat strikes, 562
William Jefferson Clinton v. Paula Corbin Jones, 107–109, 110, 581
Williams Act, 677, 679
Williamson Oil Co. v. Philip Morris, USA, 710–711
Williams v. Braum Ice Cream Stores, Inc., 343–344
Wilson, Woodrow, 725
The Wind Done Gone (novel), 398
Winding-up, 441
Winterbottom v. Wright, 334
Wire fraud, 167
Wool Products Labeling Act, 759
Workers' compensation laws, 503, 510–513
Work permits, 222
Work-product doctrine, 49
World Bank, 241, 243
WorldCom, 143, 147, 197, 464, 675
World Trade Organization (WTO), 17, 222, 226, 233–236, 402–403, 426
World-Wide Volkswagen Corp. v. Woodson, District Judge of Cook County, 37–39
Worms, 172
Wozniak, Steve, 196
WPA. *See* Whistleblower Protection Act
Wright v. Universal Maritime Service, 87
Writ of certiorari, 66
Writ of execution, 743
Written contracts, 271–273
WTO. *See* World Trade Organization
Wynn v. Wynn Resorts, 472

Y

Yarborough v. Alvarado, 150
Yaz, 337
Yould, Rachel, 169
Young, Margaret, 324
Young, William "Ronnie," 324–326
Young v. Becker & Poliakoff, 303–304
Young v. UPS, 592

Z

Zabner v. Howard Johnson's Inc., 344
Zicam, 337
Zoning, 374–377
Zyprexa, 337